Encyclopedia of Gerontology and Population Aging

Danan Gu • Matthew E. Dupre
Editors

Encyclopedia of Gerontology and Population Aging

Volume 3

E–G

With 284 Figures and 166 Tables

Editors
Danan Gu
Population Division
Department of Economics and
Social Affairs
United Nations
New York, NY, USA

Matthew E. Dupre
Department of Population Health
Sciences, Department of Sociology
Duke University
Durham, NC, USA

ISBN 978-3-030-22008-2   ISBN 978-3-030-22009-9 (eBook)
ISBN 978-3-030-22010-5 (print and electronic bundle)
https://doi.org/10.1007/978-3-030-22009-9

© Springer Nature Switzerland AG 2021
This work is subject to copyright. All rights are reserved by the Publisher, whether the whole or part of the material is concerned, specifically the rights of translation, reprinting, reuse of illustrations, recitation, broadcasting, reproduction on microfilms or in any other physical way, and transmission or information storage and retrieval, electronic adaptation, computer software, or by similar or dissimilar methodology now known or hereafter developed.
The use of general descriptive names, registered names, trademarks, service marks, etc. in this publication does not imply, even in the absence of a specific statement, that such names are exempt from the relevant protective laws and regulations and therefore free for general use.
The publisher, the authors, and the editors are safe to assume that the advice and information in this book are believed to be true and accurate at the date of publication. Neither the publisher nor the authors or the editors give a warranty, expressed or implied, with respect to the material contained herein or for any errors or omissions that may have been made. The publisher remains neutral with regard to jurisdictional claims in published maps and institutional affiliations.

This Springer imprint is published by the registered company Springer Nature Switzerland AG.
The registered company address is: Gewerbestrasse 11, 6330 Cham, Switzerland

Preface

Gerontology and population aging encompass a wide range of disciplines, often as much distinct as they are interconnected, that are dedicated to advancing our understanding of human aging and its population dynamics. Many areas of gerontology and population aging continue to develop and refine as new theories are tested, new methods are developed, and new age-related issues emerge over time. In many ways, it is the multidisciplinary breadth of this field that has kept it at the forefront of scientific discovery and provided new knowledge in almost all aspects of aging. The tremendous toll of the COVID-19 pandemic on the health and survival of older adults is one such profound example of how human biology, social conditions, public health, and many other factors can intersect at the societal and global levels. Although there's still much to be learned about this outbreak, as well as many other ongoing and pressing issues, we have no doubt that research in gerontology and population aging will continue to ask the right questions, leverage insights from diverse perspectives and approaches, and ultimately provide us with answers to the most important challenges facing older adults.

Our primary motivation for this encyclopedia was to develop a comprehensive resource that consolidates and highlights the major theories, methods, datasets, and research findings across fields that collectively define gerontology and population aging. With the somewhat unique multidisciplinary foundation of this area, it has been inspiring to be a part of this important encyclopedia and having the immeasurable support of contributions from leading experts in their respective fields.

The encyclopedia brings together a comprehensive collection of work highlighting established research and emerging science in all relevant disciplines in gerontology and population aging. This reference book covers the breadth of the field, gives readers access to all major sub-fields, and illustrates their interconnectedness with other disciplines. With more than 970 entries, organized under more than 60 thematic sections, and written by more than 1,300 experts from around the world – including anthropologists, biologists, economists, psychiatrists, public policy experts, sociologists, and others – this encyclopedia delves deep into key areas of gerontology and population aging, both new and classic, such as biogerontology, geriatrics, behavioral gerontology, healthcare, geropsychology, gerontechnology, ageism in social media and the visual arts, policies on population aging, the roles of major NGOs, cultural dimensions of aging, and many more areas.

We believe this encyclopedia is a timely and valuable resource for scholars, researchers, and practitioners who are interested in gerontology and population aging. Whether it be learning about a new concept or delving deep into issues of ageism, possible mechanisms of longevity, or the disablement process, this resource can be beneficial to a wide audience at all levels, including researchers, teachers and students, policy makers, nongovernmental agencies, public health practitioners, business marketing planners, futurists and strategists, and many other individuals and organizations.

In preparing this encyclopedia, we are especially grateful to our family members whose love, understanding, and support were essential to navigate the successful completion of this work. We are also particularly grateful to the more than 1,300 accomplished colleagues worldwide who contributed more than 970 chapters that make up this reference work. Special thanks also go out to the more than 60 renowned experts in their fields who served as section editors – dedicating their time, expertise, and diligence in coordinating the large breadth of entries. Our appreciation also extends to our 15 advisory board members who provided guidance in the process of developing and preparing the encyclopedia.

Lastly, the editors and authors of this encyclopedia are collectively grateful to the many members of the editorial team at Springer Nature who have been nothing short of amazing. In particular, Evelien Bakker, Raasika Dhandapani, Monika Garg, Michael Hermann, Salmanul Faris Nedum Palli, and Audrey Wong have provided invaluable support and guidance throughout the process of developing and publishing this work.

NY, USA   Danan Gu
NC, USA   Matthew E. Dupre
October 2021

List of Topics

Ageism

Section Editor: *Sheri R. Levy*

Ableism and Ageism
Age Discrimination in the Workplace
Age Segregation
Age Stereotypes
Age-Based Stereotype Threat
Ageism Around the World
Ageism in Healthcare
Ageism in the Family
Antiaging Movement in the Mass Media
Anxiety About Aging
Career Choices and Ageism
Children's Attitudes Toward Aging and Older Adulthood
Heterosexism and Ageism
History of Ageism
Intergenerational Programs on Anti-Ageism
Intergenerational Resource Tensions
Leadership, Politics, and Ageism
Older Adults Abuse and Neglect
Racism and Ageism
Reducing Ageism
Self-Reported Ageism
Sexism and Ageism
Stereotype Embodiment Theory
SuperAgers and Ageism

Aging and Diseases

Section Editor: *Mari Armstrong-Hough*

Aging and Health Disparities
Aging and Tuberculosis
Cataract
Chronic Obstructive Pulmonary Disease (COPD)
Diabetes Management in Older Adults
Parkinson's Disease
Pneumonia
Rheumatic Diseases Among Older Adults

Aging and Financial Well-Being

Section Editor: *Andrzej Klimczuk*

Age Management and Labor Market Policies
Autonomy and Aging
Bridge Jobs
Corporate Social Responsibility and Creating Shared Value
Developmental Stake Hypothesis
Digital Divide and Robotics Divide
Economics of Aging
Employment and Outplacement Services
Encore Jobs
Financial Assets
Financial Exploitation by Family Members
Financial Frauds and Scams
Financial Gerontology
Financial Literacy
Financial Social Work
Financial Socialization
Financial Therapy
Human Wealth Span
Joint Tenancy and Tenancy in Common
Life Cycle Theories of Savings and Consumption
Moral Economy Theory
Objectively Measured Financial Resources
Senior Entrepreneurship

Socioeconomic Status
Subjectively Measured Financial Resources
Wage Gap

Aging and Housing

Section Editor: *Debbie Faulkner*

Accommodation Downsizing
Affordable Housing
Aged Care Homes
Cohousing
Congregate Housing
Home Equity Conversion
Home Modification
Home Ownership
Homelessness
Homeshare
Housing
Housing Assistance
Intergenerational Housing
Mobile and Manufactured Homes
Retirement Villages
Senior Centers
Smart Homes
Social Housing
Supportive Housing

Aging and Nutrition

Section Editor: *Virginia Boccardi*

Aging and Cholesterol Metabolism
Anorexia of Aging
Anthropometric Indices and Nutritional Parameters in Centenarians
Artificial Nutrition at Old Age
Basal Metabolic Rate
Body Mass Index, Overweight, and Insulin Resistance
Diet and Calorie Restriction
Healthy Diet
Improving Nutrition in Older Adults
Malnutrition
Mediterranean Diet
Nut Consumption and Health
Nutrition and Aging: Nutrition Balance and Dietary Protein Needs
Nutrition and Aging: Surgical Issues
Obesity Paradox
Slow-Ageing Diets

Aging and Public Policy

Section Editors: *Magdalena Klimczuk-Kochańska* and *Andrzej Klimczuk*

Access to Public Transportation for Older Adults
Active Aging and Active Aging Index
Advocacy Organizations for Older Adults
AGE Platform Europe
Aging Network
Aging Policy Analysis and Evaluation
Aging Policy Cycle and Governance
Aging Policy Ideas
Aging Policy Transfer, Adoption, and Change
Alzheimer Europe
American Aging Association
Area Agencies on Aging
Caregiver Credits
Entitlement Program
Eurocarers
European Year of Active Ageing and Solidarity between Generations
Family and Medical Leave Act
Global AgeWatch Index and Insights
Human Rights Convention for Older Persons
Independent Aging
Inheritance Laws
International Year of Older Persons
John A. Hartford Index of Societal Aging
Laws for Older Adults
Lifelong Learning Platform
Madrid International Plan of Action on Ageing
Older Americans Act
Political Activism
Political Gerontology
Politics of Aging and Interest Groups
Productive Aging
Social Entrepreneurship and Social Innovation in Aging
Sustainable Development Goals and Population Aging
Tax Policies and Older Adults
Townsend Movement
Vienna International Plan of Action on Ageing
World Assembly on Ageing

List of Topics

Aging in a Global Context

Section Editor: *Danan Gu*

Older Adults in Conflicts and Crises

Aging in the Arts, Media, and Culture-1

Section Editor: *Raquel Medina*

Ageing and Detective Fiction
Aging and the Road Movie
Aging in Children's Literature
Aging in the Short Story
Care Home Stories
Feminism and Aging in Literature
Poetry and Age
Reifungsroman/Vollendungsroman/Bildungsroman
Television Series on Aging: Aging and Serial Narration

Aging in the Arts, Media, and Culture-2

Section Editor: *Sarah Falcus*

Ageing and Dance
Aging and Photography
Aging and Popular Music
Aging, Stardom, and "The Economy of Celebrity"
Cougars and Silver Foxes in Film and TV
Decline and Progress Narrative
Dementia as Cultural Metaphor
Dementia Narratives
Gendered Aging and Sexuality in Audiovisual Culture
Intersections of Race/Ethnicity and Age in Film and Literature
Late-Life Creativity

Aging in the Arts, Media, and Culture-3

Section Editor: *Aagje Swinnen*

History of Longevity Discourses
Perpetual Adolescence in Literature and Film
Queering Gerontology
Silvering Screen
Staging Age
The Mirror Stage of Old Age
The Performativity of Age
The Youthful Structure of the Look

Aging Medicine and Geriatrics

Section Editor: *M. Cristina Polidori*

Comprehensive Geriatric Assessment
Disorders of Water and Electrolytes in Older Adults
Drug Reactions, Resistance, and Polypharmacy
Geriatric Rehabilitation, Instability, and Falls
Nutrition and Lifestyle
Pain in Older Persons
Personalized Medicine and Decision-Making
Physiology of Aging as a Basis of Complexity in Aging Medicine and Geriatrics
Prevention of Age-Related Cognitive Impairment, Alzheimer's Disease, and Dementia
Sarcopenia
Teaching Aging Medicine
Technology and Telemedicine
Vascular Diseases of Ageing

Aging, Migration, and Geographical Gerontology

Section Editor: *Danan Gu*

Aging Migrants
Aging Refugees
Asian Aging Refugees and Immigrants in the United States
Geographical Gerontology
Health Disparities Among Aging Migrants
Migration Theories
Retirement Migration
Rural Ageing
Urbanization and Aging

Behavioral Gerontology

Section Editor: *Rainbow Tin Hung Ho*

Affect Regulation
Behavioral Gerontology

Benign Prostate Hyperplasia
Benzodiazepines
Biobehavioral Effects
Disruptive Behaviors
Expressive Arts Therapy
Health Literacy and Health Behavior
Home Health Therapies
Prosocial Behavior
Qigong Intervention
Risk Avoidance
Self-Regulation

Biogerontology-General-1

Section Editor: *Giacinto Libertini*

Aging and Cancer
Aging as Phenoptotic Phenomenon
Aging Definition
Aging for Perennial Cells
Aging Mechanisms
Aging Pathology
Aging Theories
Animal Models of Aging
Antagonistic Pleiotropy Aging Theory
Antiaging Strategies
Biogerontology
Cell Senescence
Cessation of Somatic Growth Aging Theory
Disposable Soma Aging Theory
Dykeratosis Congenita
Effects of Telomerase Activation
Evolvability Theory of Aging
Gradual Cell Senescence
Human Aging, Mitochondrial and Metabolic Defects (The Novel Protective Role of Glutathione)
Hutchinson-Gilford Progeria Syndrome
Kin Selection Aging Theory
Mitochondrial Reactive Oxygen Species Aging Theory
Mutation Accumulation Aging Theory
Non-evolutionary and Evolutionary Aging Theories
Non-programmed (Nonadaptive) Aging Theories
Oxidation Damage Accumulation Aging Theory (The Novel Role of Glutathione)
Phenoptosis and Supra-individual Selection

Programmed (Adaptive) Aging Theories
Senolytic Drugs
Subtelomere
Telomerase
Telomere-Subtelomere-Telomerase System
Telomeres
Timeline of Aging Research
Werner Syndrome

Biogerontology-General-2

Section Editor: *Michael J. Rae*

Cell Replacement
Chronobiology and Aging
Elastin, Aging-Related Changes in
Geroscience
Human Factors and Ergonomics for Aging
Multimorbidity in Aging
Nicotinamide Adenine Dinucleotide (NAD+) in Aging
Sirtuins in Aging

Business of Aging

Section Editor: *Danan Gu*

Gray Is the New Green: Opportunities of Population Aging
Longevity Industry

Cardiovascular Diseases

Section Editor: *Xiao-Li Tian*

Atherosclerosis
Atrial Natriuretic Peptide
Brachial Artery
Cardiovascular Response
Cardiovascular System
Cerebral Palsy
Cholesterol Levels
Circadian Amplitude
Circulatory System
Electrocardiogram Abnormalities
Electroconvulsive Therapy
Electroencephalography
End-Diastolic Volume Index

Heart Failure
Hypertension
Hypertensive Cardiovascular Diseases
Ischemic Attack
Ischemic Heart Disease
Middle Cerebral Artery Strokes
Myocardial Infarction
Peripheral Artery Disease
Stroke
Thrombus
Ventricular–Vascular Interaction

Caregiving in Changing Family and Social Contexts

Section Editor: *Xue Bai*

Care Needs
Caregiver Stress
Critical Care Nursing
Custodial Care
Fee-for-Service
Formal and Informal Care
Kin Availability
Nursing Home Policy and Regulations in the United States
Personal Care
Post-acute Care
Primary Caregivers
Quality of Care
Respite Care
Skilled Care
Special Care Units

Cellular Aging

Section Editor: *Wenhua Zheng*

Autophagy in Aging
Cell Damage and Transformation in Aging
Cell Morphology in Aging
Cellular Aging/Senescence
Cellular Proteostasis in Aging
Cellular Repair Processes
Programmed Cell Death
Redox Signaling
Reproductive Cell-Cycle Theory of Aging
Stem Cells Aging

Climate Change and Older People

Section Editor: *Gary Haq*

Climate, Vulnerability, and Older People
Climate Gerontology
Climate Resilience and Older People
Flooding and Older People
Gray Consumption
Heatwaves and Older People
Older Adults and Environmental Voluntarism
Tropical Cyclones and Older People
Wildfires and Older People

Cognitive Aging

Section Editor: *Adam J. Woods*

Apolipoprotein Epsilon 4
Cognitive Behavioral Therapy (II)
Cognitive Compensatory Mechanisms
Cognitive Processes
Cognitive Training
Communication Technologies and Older Adults
Geriatric Mental Health
Global Deterioration Scale
Information-Processing Theory
Intellectual Disability (Cognitive Disability)
Intelligence (Crystallized/Fluid)
Language and Communication Disorders
Memory Disorders: Drug Treatments
Mental Disorders
Mental Health Services
Metamemory
Mild Cognitive Impairment
Mini-Mental State Examination (MMSE)
Montreal Cognitive Assessment
NIH Toolbox Cognition Battery
Positivity Effect
Prospective Memory in Older Adults
Reality Orientation
Short-Term/Long-Term Memory
Speech Capability
Speed of Processing
Validation Therapy
Vascular Dementia
Working Memory

Culture and Longevity

Section Editor: *Danan Gu*

Ayurveda, Longevity, and Aging
Bushen and Longevity in Chinese Culture
Confucian Culture and Filial Piety
Confucian Stages of Life
Feng Shui and Aging in Place
Geriatric Management in Persian Medicine
Longevity in Korean Culture
Siddha Practice and Management of Geriatrics
Traditional Chinese Medicine
Unani Medicine and Healthy Living
Yangsheng and Longevity in Chinese Culture
Yoga Practices and Health Among Older Adults

Demographic Transition

Section Editor: *Yuying Tong*

Baby Boom/Baby Bust
Birth Control Policy and Population Aging
Causes of Population Aging
Demographic Transition Theories
Graying of Hair
Lowest-Low Fertility
One-Child Policy and Population Aging in China
Population Decline
Population Explosion and Implosion
Population Momentum
Population Pyramid
Probability of Dying
Second Demographic Transition

Disability and Functional Limitation

Section Editor: *Qiushi Feng*

Bedsore
Biosocial Model of Disability
Chewing Ability
Disability Measurement
Disability Types
Disability-Affirmative Therapy
Falls
Functional Limitation
Handicap
Hip Fracture
Impairment
OARS Multidimensional Functional Assessment Questionnaire (OMFAQ)
Sensory Disability
Social Disability

Disability and Pension

Section Editor: *Na Yin*

Medicaid for People with Disabilities
Medicare for People with Disabilities
Social Security Disability Insurance
Supplemental Security Income Program
Workers' Compensation, "Veterans" Disability Compensation, and Long-Term Disability Insurance

Employment and Retirement

Section Editor: *Li-Hsuan Huang*

Delaying Retirement
Layoffs and Subjective Well-Being
Mandatory and Statutory Retirement Ages
Pre-retirees' Preparation for Retirement
Re-employment of (Early) Retirees
Replacement Rate
Retirement Contracts
Retirement Patterns
Retirement Transition
Self-Employment Among Older Adults

End-of-Life Issues

Section Editor: *Amy Yin Man Chow*

Advance Care Planning
Advance Care Planning: Advance Directives
Advance Care Planning: Medical Orders at the End of Life (MOLST, POLST)
Anticipatory Grief
Assisted Dying
Bereavement Care
Caregiving at the End of Life
Death Anxiety
Death Denial

Death Trajectory
Death with Dignity
Dying in Place
End-of-Life Care
End-of-Life Decision-Making in Acute Care Settings
Euthanasia and Senicide
Funerals and Memorial Practices
Good Death
Professional Grief and Burnout
Rumination in Bereavement
Social Support in Bereavement
Terminal Change
Thanatology
Theory of Harmonious Death

Environmental Health

Section Editor: *Danan Gu*

Age-Friendly Cities and Communities: New Directions for Research and Policy
Environmental Gerontology
Longevity Areas and Mass Longevity
Neighborhood Social Environment and Health

Family Demography

Section Editor: *Naomi J. Spence*

Aging Households
Beanpole Family Structure
Clan Families
Co-residence
Cohabitation
Crowded Nest
Empty Nest
Family Demography
Family Diversity
Family Formation and Dissolution
Family Poverty Among Older Adults
Family Tree
Living Arrangements in Later Life
Sandwich Generation
Stem Family
Wealth Flows and Population Aging
Widowhood

Frailty

Section Editor: *Olga Theou*

Biology of Frailty
Cognition and Frailty
Epidemiology of Frailty
Frailty and Social Vulnerability
Frailty in Clinical Care
Measurement of Frailty
Mobility and Frailty
Nutrition and Frailty
Physical Activity, Sedentary Behaviors, and Frailty
Polypharmacy and Frailty

Gender Issues

Section Editor: *Susan Pickard*

Feminist Theories and Later Life
Gender and Caring in Later Life
Gender and Employment in Later Life
Gender Disparities in Health in Later Life
Gender Equality in Later Life
Gender Inequity
Gender Issues in Age Studies
Menopause
Midlife and Gender
Poverty and Gender Issues in Later Life in the UK
Psychology, Aging Women, and the Life Course: Women's Negotiation of Menopause
Representations of Older Women and White Hegemony
Women Empowerment

Genetics of Aging-1

Section Editor: *Lok Ting Lau*

Alleles
Chromosome
DNA Chip
DNA Damage Theory
Genetics: Ethnicity
Genetics: Gender
Genetics: Gene Expression
Genetics: Parental Influence

Genome Instability
Genome-Wide Association Study
Genotype
Human Genome Project
Mitochondrial DNA Mutations
Mutation
Nucleotides
Phenotype
Somatic Cell Nuclear Transfer
Somatic Mutations and Genome
Transgenic Mice
Whole Genome Sequencing

Genetics of Aging-2

Section Editor: *Diddahally R. Govindaraju*

Familial Dyslipidemias
Fragile X Syndrome and Premutation Aging Disorders
Genes that Delay Aging
Genetic Control of Aging
Genetic Theories of Aging
Genomics of Aging and Longevity
Molecular and Epigenetic Clocks of Aging
Mutation Load and Aging
Progeria: Humans
Progeria: Model Organisms

Geriatric Psychiatry

Section Editor: *Natalia Ojeda Del Pozo*

Delirium
Dementia
Psychoneuroimmunology
Schizophrenia in Older Adults
Sundown Syndrome

Gerontechnology

Section Editor: *Angela Y. M. Leung*

Cybercounseling
Digital Health
Gerontechnology
Home Health Technology
Telerehabilitation (Remote Therapy)

Gerontology-General

Section Editors: *Danan Gu* **and** *Leilani Feliciano*

Age Heaping
Ageism in Media and Visual Arts
Aging and Plastic Surgery
Andersen Model
Behavioral Interventions in Dementia
Centenarian Rate, Life Expectancy, and Autoimmune Diseases
COVID-19 Pandemic and Healthy Aging
Critical Gerontology
Dehydration in Older Adults
Demographic Dividend
Educational Gerontology
Gateway to Global Aging Data
Gerontocracy
Healthy Aging
Hearing Loss
Human Mortality Database
Lifetime Primary Occupation and Health/Longevity in Old Age
National Transfer Accounts Project
Place of Death
Podiatric Medicine
Regional Inequality in Longevity and Lifestyle in Europe
Saint Louis University Mental Status Examination and Older Adults
Senior Learning
Skin Aging
Socioeconomic Differentials in Health: Divergence, Convergence, and Persistent Inequality Theories
Subjective Cognitive Decline
Ultrasound Morphology of Longevous Persons
Universities of the Third Age
(World) Supercentenarian Database
Wound Repair and Healing in Older Adults

Health and Retirement Studies Series

Section Editors: *David Weir* **and** *Mary Beth Ofstedal*

Brazilian Longitudinal Study of Aging (ELSI-Brazil)
China Health and Retirement Longitudinal Study (CHARLS)
Costa Rican Longevity and Healthy Aging Study
English Longitudinal Study of Ageing
Health and Aging in Africa: A Longitudinal Study of an INDEPTH Community in South Africa
Health and Retirement Studies Series
Healthy Ageing in Scotland (HAGIS)
Indonesia Family Life Survey
Korean Longitudinal Study of Ageing
Longitudinal Aging Study in India
Malaysia Ageing and Retirement Survey
Mexican Health and Aging Study
The Health and Retirement Study
The Irish Longitudinal Study on Ageing
The Northern Ireland Cohort for the Longitudinal Study of Ageing (NICOLA)
The Survey of Health, Ageing and Retirement in Europe
WHO's Study on Global AGEing and Adult Health (SAGE)

Health Informatics

Section Editor: *Ping Yu*

Computerized Provider Order Entry
Consumer Health Informatics
Electronic Health Record
Electronic Nursing Documentation
Health 2.0
Health 3.0
Health Informatics
Health Information Exchange
Health Information Standard
Managing Long-Term Conditions: Wearable Sensors and IoT-Based Monitoring Applications
MedlinePlus
mHealth
Patient Opinion Leaders
Patient Portal
Telehealth
Telemedicine
Telenursing
Wearable Technology

Healthcare and Healthcare Financing

Section Editors: *Karen L. Fortuna* **and** *Nazmi Sari*

Access to Care
Acute Care
Adverse Selection and Long-Term Care
Community Health for Older Adults
Doctor-Patient Relations
Emergency Care
Family Medicine
Grossman Model
Health Insurance
Health-Care Financing
Healthcare Utilization
Managed Care and Aging
Medicaid
Medicare: Coverage, Evolution, and Challenges
Medigap
Risk Adjustment: Applications in Healthcare Markets
User Fees (Coinsurance, Copayment, and Deductibles)

Historical Gerontology

Section Editor: *Ilia Stambler*

Aging Research in the Late Nineteenth to Early Twentieth Century
Endocrine Rejuvenation in a Historical Perspective
Gerocomia in the History of Aging Care
Homeostasis in the History of Aging Research
Life-Extensionism in a Historical Perspective
Medical Alchemy in History
Status of Older People: Ancient and Biblical
Strategies for Engineered Negligible Senescence

Immune System and Aging

Section Editor: *Graham Pawelec*

Age-Related Changes in the Murine Immune System
Autoimmune Hemolytic Anemia
Autoimmune Thyroiditis
Consequences of Pneumonia in Older Adults
Cytomegalovirus and Human Immune System Aging
Follicular Dendritic Cells
Human Immune System in Aging
Immune Aging, Autoimmunity, and Autoinflammation
Immunological Theory of Aging
Influenza Vaccination in Older Adults

Informal/Family Caregiving

Section Editor: *Neena L. Chappell*

Abuse and Caregiving
Benefits of Caregiving
Caregiver Identity
Caregivers' Outcomes
Caregiving Across Neurodegenerative Diseases
Caregiving Among the LGBT Community
Caregiving and Ethnicity
Caregiving and Social Support: Similarities and Differences
Double-Duty and Triple-Duty Caregivers
Effectiveness of Respite Care for Caregivers of Older Adults
Employment and Caregiving
Fictive Kin
Filial Piety and Responsibilities Among the Chinese
Former Carers
Hospice and Caregiving
Intergenerational Family Caregiving in Welfare Policy Context
Interventions for Caregivers of Older Adults
Precarity and Aged Care
Preparation for Future Care: The Role of Family Caregivers
Rural-Urban Comparisons in Caregiving for Older Adults
Self, Informal, and Formal Long-Term Care: The Interface
Social Exchange Theory and Its Implications for Caregiving

Intergenerational Relations

Section Editor: *Christine A. Mair*

Domestic Violence in Older Adults
Intergenerational Exchange and Support
Intergenerational Family Dynamics and Relationships
Intergenerational Family Structures
Intergenerational Migration and Relations in the International Context
Intergenerational Programs
Intergenerational Relationships
Intergenerational Solidarity
Intergenerational Stake
Kinship Networks
Multigenerational Families
Multigenerational Workforce
Older Adults and Information and Communication Technologies in the Global North
Parenting and Grandparenting

Lifestyle and Aging

Section Editor: *Wei Zhang*

Aerobic Exercise Training and Healthy Aging
Alcohol Consumption and Health
Exercise Adherence
Exercise and Healthy Cardiovascular Aging
Insomnia, Sleep Disorders, and Healthy Aging
Laboring Work and Healthy Aging
Leisure Activities and Healthy Aging
Physical Activities
Sleep and Aging
Smoking, Aging, and Longevity
Sport and Healthy Aging
Tourism and Well-Being

Longevity

Section Editor: *Kirill Andreev*

Best Performance Countries in Longevity
Decomposing a Difference in Life Expectancies
Maximum Lifespan
Modal Age at Death
Race Crossover in Longevity
Rectangularization of Survival Curve
Reproduction and Longevity in Humans
The Male-Female Health-Mortality Paradox

Long-Term Care

Section Editor: *Bei Wu*

Gerontological Nursing
Length of Stay in Long-Term Care Settings
Long-Term Care Financing
Person-Centered Care for Older Adults
Technology-Enabled Long-Term Care Services and Supports (T-eLTCSS) in Home Settings

Methods in Gerontology

Section Editors: *Anthony R. Bardo* and *Kenneth Carl Land*

Age-Period-Cohort Models
Big Data
Cross-Sectional and Longitudinal Studies
Dyad/Triad Studies
Ethnography
Experimental Studies and Observational Studies
False Negative/False Positive
Hierarchical Models
Item Response Theory and Modeling
Latent Class Analysis
Life Course Perspective
Likert Scale
Missing Data Concepts
Mobile Data Collection with Smartphones
Narrative Analysis
Qualitative Research/Quantitative Research
Recruitment and Retention in Aging Research
Repeated Cross-Sectional Design
Selective Bias in Longitudinal Studies
Semiparametric Methods
Structural Equation Models
Survival Analysis

Mortality

Section Editor: *Patrick Gerland*

Force of Mortality
Life Tables
Mortality at Older Ages and Mean Age at Death
Mortality Leveling
Mortality Modeling
Mortality Projection

Neurosciences of Aging

Section Editor: *Adam J. Woods*

Alzheimer's Disease
Autonomic Nervous System
Brain Atrophy
Central Nervous System
Cerebral Metabolism
Electrophysiology
Leukoencephalopathy
Neurochemistry
Neuroendocrine Theory of Aging
Neuroimaging
Neuroinflammation
Neuromuscular System
Neuroplasticity
Neuroscience of Aging
Neurotransmitter in the Aging Brain
Neurotrophic Factors
Neurotrophic Factors Link to Alzheimer's Disease
Non-invasive Brain Stimulation
White Matter Hyper-intensities

NGOs and Civil Society

Section Editor: *Danan Gu*

AARP
Aging Analytics Agency
Alliance for Aging Research

American Federation for Aging Research
European Network in Aging Studies
Global Coalition on Aging
HelpAge International
HelpAge USA
International Council on Social
  Welfare (ICSW)
International Federation on Ageing
International Network for the Prevention of Elder
  Abuse (INPEA)
International Network on Health Expectancies
  and the Disablement Process (REVES)
International Society on Aging and
  Diseases (ISOAD)
NGO Committee on Ageing
North American Network in Aging Studies
SENS Research Foundation
The International Association of Gerontology and
  Geriatrics
Vetek (Seniority): The Movement for Longevity
  and Quality of Life

Oncology and Treatment

Section Editors: *Paul Hofman* and *Rabia Boulahssass*

Aging and Cancer: Concepts and Prospects
Cancer Diagnosis
Cancer Screening
Chemotherapy Toxicity
Clinical Trial Design for Older Cancer Patients
Cognitive Disorders in Older Patients with
  Cancer
COVID-19 Pandemic and Geriatric Oncology
Epidemiology, Aging, and Cancer
Frailty Screening
Geriatric Assessment for Older Adults with
  Cancer
Geriatric Interventions
Immunotherapy
Mini Datasets for Research in Geriatric
  Oncology
Monitoring and Clinical Research in
  Oncogeriatrics
OncoAge
Prediction of Outcomes Among Cancer
  Patients
Prehabilitation in Older Adults with Cancer
Quality of in Older Patients with Cancer
Radiation Therapy
Rehabilitation in Older Adults
  with Cancer
Targeted Therapies
Tumors: Brain
Tumors: Breast
Tumors: Colorectal
Tumors: Gastrointestinal Cancers
Tumors: Gynecology
Tumors: Lymphomas
Tumors: Melanoma
Tumors: Non-small Cell Lung Cancer
Tumors: Oto-Rhino-Laryngology
Tumors: Urologic Cancer

Psychogerontology-General-1

Section Editors: *Susanne Wurm* and *Anna E. Kornadt*

Age Group Dissociation
Awareness of Age-Related Change
Cross-Cultural Psychogerontology
Dual Process Theory of Assimilation and
  Accommodation
Generativity and Adult Development
Geropsychology
Goal Setting
Life-Span Development
Life-Span Theory of Control
Locus of Control
Multidimensional Views of Aging and
  Old Age
Personality in Later Life
Plasticity of Aging
Possible Selves Theory
Psychological Theories of Health and Aging
Selective Engagement of Resources
Self-Continuity
Self-Perceptions of Aging
Subjective Age
Terror Management Theory and Its Implications
  for Older Adults
The Model of Selection, Optimization, and
  Compensation
Time Perspective Across Adulthood

Psychogerontology-General-2

Section Editor: *Sofia von Humboldt*

Acceptance and Commitment Therapy
Adjustment to Aging
Aging in Place
Altruism
Cyberpsychology and Older Adults
Learned Helplessness
LGBT in Old Age
Motivation: Theory/Human Model
Neuropsychology
Obesity, Perceived Weight Discrimination, and Well-Being
Obsessive–Compulsive Personality Disorder
Performance Anxiety
Pet-Raising and Psychological Well-Being
Psychogeriatrics
Psychological Assessment
Psychological Fatigue
Psychometrics
Psychopathology
Psychotherapy
Ryff's Psychological Well-Being Scale
Self-Actualization
Sexuality and Sexual Well-Being
Substance Abuse and Addiction

Psychological Well-Being-1

Section Editors: *Sharpley Hsieh* and *Lei Feng*

ATUS Well-Being Module
Buffering Hypothesis
Cognitive Behavioral Therapy (I)
Depression and Antidepressants
Diagnostic and Statistical Manual of Mental Disorders-5 (DSM-5)
Distress
Geriatric Anxiety Inventory
Geriatric Anxiety Scale
Geriatric Depression Scale
Interpersonal Psychotherapy
Loneliness
Psychiatric Diseases in Relation to Physical Illness
Psychosocial Behavioral Intervention
Severity of Psychosocial Stressors Scale
SF-12 and SF-36 Health Survey
Social Isolation
The Center for Epidemiologic Studies Depression (CES-D) Scale

Psychological Well-Being-2

Section Editor: *Rongjun Yu*

Behavioral and Psychological Symptoms of Dementia
Emotion
Hamilton Anxiety Rating Scale
Hamilton Rating Scale for Depression
Inhibitory Deficit Hypothesis
Mood Disorders
Motivation
Perceived Control
Personal Perception of Aging
Positive Affect
Posttraumatic Stress Disorder
Social Cognition
Stress and Coping Among Older Adults
Stress Theory of Aging
Subjective Well-Being

Quality of Life

Section Editor: *Danan Gu*

Aging in Place and Quality of Life
Aging in Place: Maintaining Quality of Life for Older Persons at Home
Aging in the Right Place
Indigenous Cultural Generativity: Teaching Future Generations to Improve Our Quality of Life
Quality of Life
Residential Happiness and Quality of Life
Subjective Well-Being of Eldercare Recipients and Providers

Religion, Spirituality, and Aging

Section Editor: *Rhonda Shaw*

Aging: An Islamic Perspective
Atheism

Beliefs of Life After Death and Social Partner Choice
Biblical Perspectives on Aging
Daoism and Longevity
Hinduism Teachings and Aging
Meditation and Mindfulness: Resources for Aged Care
Quran on Aging
Religion and Spirituality in End-of-Life Care
Religion and Spirituality: Older Inmates
Religiosity
Religious Involvement, Health, and Longevity

Resilience

Section Editor: *Heather Farmer*

Personality and Healthy Aging
Purpose in Life Among Older Adults
Resilience: Measures and Models
Self-Efficacy
Self-Esteem

Science, Technology, and Society

Section Editor: *Ilia Stambler*

Clinical Translation Acceleration
Crowdsourcing and Crowdfunding in Aging Research
Ethics of Lifespan Extension
Health Technology Assessment of Aging Therapies
Longevity Activism
Longevity Advocacy
Longevity Dividend
Medical Ethics
Regulation of Geroprotective Medications
Targeting Aging with Metformin (TAME)

Social Capital and Social Relations

Section Editor: *Joonmo Son*

Bereavement and Loss
Buffering Function
Grandparenting
Intergenerational Exchanges
Marginalization
Marital Relationships
Marriage and Remarriage
Mutual Support Groups
Neighborhood Coherence
Social Network Index
Social Stress
Social Support
Socioemotional Selectivity Theory

Social Participation

Section Editor: *Danan Gu*

Digital Participation
Disengagement Theory
Social Integration
Social Participation
Volunteering and Health Outcomes Among Older Adults

Social Security and Pension Systems

Section Editor: *Bernardo Lanza Queiroz*

Annuities
Defined Benefit and Defined Contribution
Fiscal Welfare
Individual Retirement Arrangements
Joint and Survivor Annuities
Pension Systems
Reverse Mortgages
Social Security Around the World
Social Security in Latin America
Social Security in the United States
Social Security System in China
Social Security: History and Operations
Social Security: Long-Term Financing and Reform
Welfare States

Social Services

Section Editors: *Xiaoling Xiang* and *Emily J. Nicklett*

Adult Day Services
Adult Foster Homes

List of Topics

Adult Protective Services
Ambulatory and Outpatient Care
Animal-Assisted Therapy
Care Coordination
Care Management
Case Management
Chronic Disease Self-Management
Comprehensive Digital Self-Care Support System (CDSSS)
Congregate and Home-Delivered Meals
Elastic Demand/Inelastic Demand in Social Services
Faith-Based Social Services
Familism
Geriatric Social Workers
Guardianship
Health Maintenance Organizations
Home- and Community-Based Services (HCBS)
Legal Services
Meals on Wheels (MOW)
Participation and Co-production in the Care and Support of Older People
Personal Assistant Services
Preventive Care
Program of All-Inclusive Care for the Elderly (PACE)
Registered Nurse
Social Services Utilization
Social Work
Veterans Care

Sociogerontology

Section Editors: *Melanie Sereny Brasher* and *Karen Siu Lan Cheung*

Activity Theory
Continuity Theory
Modernization Theory
Racial and Ethnic Disparities in Health
Social Clock
Social Gerontology
Social Roles
The European Social Model
Theory of Cumulative Disadvantage/Advantage
Urban-Rural Health Disparities at Older Ages

Special Subpopulations

Section Editor: *Danan Gu*

Centenarians
Childless Older Adults
Older Prisoners
Oldest-Old Adults
Only-Child Older Parents in China
Supercentenarians
"Three-No" and "Five-Guarantee" Older Adults

Successful Aging

Section Editor: *Qiushi Feng*

Aging Well
Health Expectancy
Layperson-Defined Successful Aging
Objectively Rated Successful Aging
Optimal Aging
Subjectively Rated Successful Aging
Successful Aging 2.0
Successful Aging Life Expectancy

Surveys for Aging Studies

Section Editor: *Matthew E. Dupre*

Advanced Cognitive Training for Independent and Vital Elderly (ACTIVE)
Berlin Aging Study II (BASE-II)
Cambodia Elderly Survey
China Health and Nutrition Survey, 1989–2019
Chinese Longitudinal Healthy Longevity Survey (CLHLS)
Cohort Study of Centenarians in Hainan, China (CHCCS)
Consortium on Health and Aging: Network of Cohorts in Europe and the United States (CHANCES)
Consortium on Interplay of Genes and Environment Across Multiple Studies
Danish Centenarians Studies
German Ageing Survey (DEAS)
Hispanic EPESE (Established Population for the Epidemiological Study of the Elderly)
Interdisciplinary Longitudinal Study on Adult Development and Aging
Iowa Centenarian Study

IPUMS
Italian Centenarians and Semi-supercentenarians Surveys
Myanmar Aging Survey
National Long Term Care Survey
National Social Life, Health, and Aging Project (NSHAP)
National Survey of Older Persons in Thailand
New Zealand Health, Work, and Retirement Longitudinal Study
Panel Survey and Study on Health, Aging, and Retirement in Thailand
The 2007 Philippine Study on Aging
The Australian Longitudinal Study of Ageing
The Demographic and Health Surveys (DHS)
The Georgia Centenarian Study
The Long Life Family Study (LLFS)
The Longitudinal Aging Study Amsterdam (LASA): An Overview
The National Health Interview Survey
The Older Persons and Informal Caregivers Survey Minimum Dataset (TOPICS-MDS)
The Social Environment and Biomarkers of Aging Study (SEBAS)
The Tokyo Centenarian Study
Women's Health and Aging Studies

About the Editors

Danan Gu works at the United Nations Population Division. He has an extensive publication record, with research in the areas of health and longevity, population aging, family demography and its applications, urbanization, estimates, and projections. He serves as a senior editorial board member of *BMC Geriatrics*, and he is an editor for the Springer book series Advances in Studies of Aging and Health. He also serves as an editorial board member of the *Journal of Gerontology: Social Sciences* and *Journal of Aging and Health*.

Matthew E. Dupre is an associate professor in the Department of Population Health Sciences and the Department of Sociology at Duke University. He is also a senior fellow at the Center for the Study of Aging and Human Development and faculty at the Duke Clinical Research Institute. Dr. Dupre is a nationally recognized medical sociologist who studies how social factors impact the development and progression of chronic disease in older adults. His work has appeared in the *American Journal of Epidemiology, American Journal of Public Health, Demography, JAMA, Journals of Gerontology, Social Forces,* and other leading journals in aging, medicine, and the social sciences.

About the Section Editors

Kirill Andreev
Department of Economic and Social Affairs
United Nations
New York, NY, USA

Mari Armstrong-Hough
Department of Epidemiology
Department of Social and Behavioral Sciences
NYU School of Global Public Health
New York, NY, USA

Xue Bai
Department of Applied Social Sciences
The Hong Kong Polytechnic University
Hung Homm, Hong Kong

Anthony R. Bardo
Department of Sociology
University of Kentucky
Lexington, KY, USA

Virginia Boccardi
Section of Geriatrics Santa Maria della Misericordia Hospital
University of Perugia
Perugia, Italy

Rabia Boulahssass
Geriatric Coordination Unit for Geriatric Oncology (UCOG) PACA Est Nice University Hospital Centre Hospitalier Universitaire de Nice, Université Côte d'Azur, FHU OncoAge, Nice, France

Melanie Sereny Brasher
Department of Sociology and Anthropology
University of Rhode Island
Kingston, RI, USA

About the Section Editors

Neena L. Chappell
Department of Sociology
Institute on Aging and Lifelong Health
University of Victoria
Victoria, BC, Canada

Amy Yin-man Chow
Department of Social Work and
Social Administration
The University of Hong Kong
Hong Kong, SAR, China

Natalia Ojeda Del Pozo
Department of Psychology
Faculty of Health Science
University of Deusto
Spain

Sarah Falcus
Reader in Contemporary Literature
School of Music, Humanities and Media
University of Huddersfield
Huddersfield, UK

Heather Farmer
Department of Human Development
and Family Sciences
University of Delaware
Newark, DE, USA

Debbie Faulkner
University of South Australia
Adelaide, Australia

Leilani Feliciano
Department of Psychology
University of Colorado Colorado Springs
Colorado Springs, CO, USA

Qiushi Feng
Department of Sociology and Centre for
Family and Population Research (CFPR)
National University of Singapore
Singapore, Singapore

About the Section Editors

Karen L. Fortuna
Geisel School of Medicine
Dartmouth College
Hanover, NH, USA

Patrick Gerland
Population Division
Department of Economic and Social Affairs
United Nations
New York, NY, USA

Diddahally R. Govindaraju
Institute for Aging Research
Albert Einstein College of Medicine
Bronx, NY, USA

Museum of Comparative Zoology
Harvard University
Cambridge, MA, USA

Gary Haq
Department of Environment and Geography
Stockholm Environment Institute
University of York
York, UK

Rainbow Tin Hung Ho
The University of Hong Kong
Centre on Behavioral Health and
Department of Social Work and
Social Administration
Hong Kong, SAR, China

Paul Hofman
Department of Pathology
FHU OncoAge, University Côte d'Azur
Nice, France

Sharpley Hsieh
Department of Psychology
The University of Queensland
Brisbane, Australia

Li-Hsuan Huang
Department of Economics
National Central University
Chungli, Taoyuan, Taiwan

About the Section Editors

Sofia von Humboldt
ISPA – Instituto Universitário
William James Center for Research
Lisboa, Portugal

Cheung Siu-Lan Karen
Sau Po Centre on Ageing
The University of Hong Kong
Hong Kong, SAR, China

Andrzej Klimczuk
Department of Public Policy
SGH Warsaw School of Economics
Warsaw, Poland

Magdalena Klimczuk-Kochańska
Faculty of Management
University of Warsaw
Warsaw, Poland

Anna E. Kornadt
Department of Behavioural and
Cognitive Sciences
University of Luxembourg
Esch-sur-Alzette, Luxembourg

Kenneth Carl Land
Social Science Research Institute
Duke University
Durham, NC, USA

Lok Ting Lau
Innovation and Technology Development Office
The Hong Kong Polytechnic University
Hung Hom, Hong Kong SAR, China

Angela Y. M. Leung
School of Nursing
The Hong Kong Polytechnic University
Hong Kong, SAR, China

Sheri R. Levy
Department of Psychology
Stony Brook University
Stony Brook, NY, USA

Giacinto Libertini
Italian National Health Service
ASL NA2 Nord
Frattamaggiore, Italy

Department of Translational Medical Sciences
Federico II University
Naples, Italy

Christine A. Mair
Department of Sociology
Anthropology, and Public Health
University of Maryland Baltimore County
Baltimore, MD, USA

Raquel Medina
Visiting Research Fellow
Aston University
Birmingham, UK

Emily J. Nicklett
Department of Social Work, College for Health
Community and Policy
The University of Texas at San Antonio
San Antonio, TX, USA

Mary Beth Ofstedal
Institute for Social Research
University of Michigan
Ann Arbor, MI, USA

Graham Pawelec
Department of Immunology
University of Tübingen
Tübingen, Germany

Health Sciences North Research Institute
Sudbury, Ontario, Canada

Susan Pickard
Department of Sociology
Social Policy and Criminology
University of Liverpool
Liverpool, UK

About the Section Editors

M. Cristina Polidori
Ageing Clinical Research, Department II of Internal Medicine and Center for Molecular Medicine Cologne
University of Cologne, Faculty of Medicine and University Hospital Cologne
Cologne, Germany

CECAD
University of Cologne, Faculty of Medicine and University Hospital Cologne
Cologne, Germany

Bernardo Lanza Queiroz
Demography
Universidade Federal de Minas Gerais
Belo Horizonte, Brazil

Michael J. Rae
SENS Research Foundation
Mountain View, CA, USA

Nazmi Sari
Department of Economics
University of Saskatchewan, Arts 815
Saskatoon, SK, Canada

Rhonda Shaw
School of Psychology
Charles Sturt University
Port Macquarie, NSW, Australia

Joonmo Son
Department of Sociology
National University of Singapore
Singapore, Singapore

Naomi J. Spence
Department of Sociology
Lehman College, City University of New York
Bronx, NY, USA

Ilia Stambler
Department: Science
Technology and Society
Bar Ilan University
Ramat Gan, Israel

About the Section Editors

Aagje Swinnen
Department of Literature and Art
Maastricht University
The Netherlands

Olga Theou
Dalhousie University
Physiotherapy and Geriatric Medicine
Halifax, Nova Scotia, Canada

Xiao-Li Tian
Jiangxi Key Laboratory of Human Aging
Nanchang University
Nanchang, Jiangxi Province, China

Yuying Tong
Department of Sociology
The Chinese University of Hong Kong
Sha Tin, Hong Kong

David Weir
Survey Research Center
Institute for Social Research
University of Michigan
Ann Arbor, MI, USA

Adam J. Woods
Department of Clinical and Health Psychology
McKnight Brain Institute University of Florida
Gainesville, FL, USA

Bei Wu
Rory Meyers College of Nursing
New York University
New York, NY, USA

Susanne Wurm
Department for Prevention Research and
Social Medicine
Institute for Community Medicine
University Medicine Greifswald
Greifswald, Germany

About the Section Editors

Xiaoling Xiang
University of Michigan
School of Social Work
Ann Arbor, MI, USA

Na Yin
Austin W. Marxe School of Public
and International Affairs
Baruch College, City University of New York and
CUNY Institute for Demographic Research
New York City, NY, USA

Ping Yu
School of Computing and Information
Technology
Faculty of Engineering and Information Sciences
University of Wollongong
Wollongong, NSW, Australia

Rongjun Yu
Department of Management
Hong Kong Baptist University
Hong Kong, SAR, China

Wei Zhang
Sociology
University of Hawaii at Manoa
Honolulu, HI, USA

Wenhua Zheng
Faculty of Health Sciences
University of Macau
Taipa, Macau SAR, China

Advisory Board

John M. "Johnny" Adams Carl I. Bourhenne Medical Research Foundation (Aging Intervention Foundation), Newport Beach, CA, USA

Gerontology Research Group, Newport Beach, CA, USA

John Beard Centre of Excellence in Population Ageing Research (CEPAR), University of New South Wales, Sydney, Australia

Kaare Christensen Danish Aging Research Center, Department of Public Health, University of Southern Denmark, Odense, Denmark

Department of Clinical Genetics, Odense University Hospital, Odense, Denmark

Department of Clinical Biochemistry and Pharmacology, Odense University Hospital, Odense, Denmark

Deborah Carr Department of Sociology, Boston University, Boston, MA, USA

Eileen M. Crimmins Davis School of Gerontology, University of Southern California, Los Angeles, CA, USA

Lesley H Curtis Department of Population Health Sciences, Duke University, Durhjam, NC, USA

Linda K. George Arts and Sciences Professor of Sociology Emerita, Duke University, Durham, NC, USA

Mark D. Hayward Department of Sociology and Population Research Center, University of Texas at Austin, Austin, TX, USA

Peter A. Lichtenberg Institute of Gerontology and Merrill Palmer Skillman Institute, Wayne State University, Detroit, MI, USA

Marc Luy Vienna Institute of Demography/Austrian Academy of Sciences, Wittgenstein Centre (IIASA, VID/ÖAW, WU), Vienna, Austria

Kyriakos Markides University of Texas Medical Branchx, Galveston, TX, USA

Jean-Marie Robine Institut National de la Santé et de la Recherche Médicale (INSERM) and Ecole Pratique des Hautes Etudes (EPHE), Paris and Montpellier, France

Redford B. Williams Department of Psychiatry and Behavioral Sciences, Division of Behavioral Medicine, Duke University, Durham, NC, USA

Yi Zeng Center for the Study of Aging and Human Development and Geriatrics Division, School of Medicine, Duke University, Durham, NC, USA

Center for Healthy Aging and Development Studies, National School of Development, Peking University, Beijing, China

Zachary Zimmer Mount Saint Vincent University, Halifax, NS, Canada

Contributors

Pamela Abbott Centre for Global Development and School of Education, University of Aberdeen, Aberdeen, Scotland, UK

Noorman Abdullah Department of Sociology, National University of Singapore, Singapore, Singapore

Dominic Abrams Centre for the Study of Group Processes, School of Psychology, Keynes College, University of Kent, Canterbury, Kent, UK

Jenna Abrams Silver School of Social Work, New York University, New York, NY, USA

Giulia Accardi Laboratory of Immunopathology and Immunosenescence, Department of Biomedicine, Neurosciences and Advanced Diagnostics, University of Palermo, Palermo, Italy

Wilco P. Achterberg Department of Public Health and Primary Care, Leiden University Medical Center, Leiden, The Netherlands

Shweta Majumdar Adur Department of Sociology, California State University, Los Angeles, Los Angeles, CA, USA

Deepali Advani Economics, The Graduate Center, CUNY, New York, NY, USA

Shereen Afifi School of Engineering, Computer and Mathematical Sciences, Auckland University of Technology, Auckland, New Zealand

Sumra Afzal Medical Investigation of Neurodevelopmental Disorders (MIND) Institute, University of California Davis, Sacramento, CA, USA

Department of Pediatrics, University of California Davis School of Medicine, Sacramento, CA, USA

Arunika Agarwal Department of Global Health and Population, Harvard T. H. Chan School of Public Health, Boston, MA, USA

Anna Aiello Laboratory of Immunopathology and Immunosenescence, Department of Biomedicine, Neurosciences and Advanced Diagnostics, University of Palermo, Palermo, Italy

Sumio Akifusa School of Oral Health Sciences, Faculty of Dentistry, Kyushu Dental University, Kitakyushu, Japan

I. Akushevich Biodemography of Aging Research Unit, Social Science Research Institute, Duke University, Durham, NC, USA

David T. Albee AARP, Washington, DC, USA

Isabelle Albert University of Luxembourg, Esch-sur-Alzette, Luxembourg

Alejandro Albizu Center for Cognitive Aging and Memory, Department of Clinical and Health Psychology, McKnight Brain Institute, University of Florida, Gainesville, FL, USA

Mariana Asmar Alencar Departamento de Fisioterapia, Escola de Educação Física, Fisioterapia e Terapia Ocupacional, Universidade Federal de Minas Gerais, Belo Horizonte, MG, Brazil

Joanne Allen Massey University, Palmerston North, New Zealand

Gordon Alley-Young Department of Communications and Performing Arts, Kingsborough Community College – City University of New York, Brooklyn, NY, USA

Jennifer Alonso-García Faculty of Economics and Business, Department Economics, Econometrics and Finance, University of Groningen, Groningen, The Netherlands

UNSW Business School, ARC Centre of Excellence in Population Ageing Research (CEPAR), UNSW Sydney, Kensington, NSW, Australia

Fiona M. Alpass Massey University, Palmerston North, New Zealand

Bader Alshoumr Centre for IT-enabled Transformation, School of Computing and Information Technology, University of Wollongong, Wollongong, NSW, Australia

College of Public Health and Health Informatics, University of Hail, Hail, Saudi Arabia

Alper Altinanahtar Department of Economics, Yeditepe University, Istanbul, Turkey

Nicole M. Amada Brooklyn College and The Graduate Center of the City University of New York, New York, NY, USA

Howard Amital Department of Medicine 'B', Sheba Medical Center, Tel-Hashomer and Sackler Faculty of Medicine, Tel Aviv University, Tel Aviv, Israel

Gabriel Amitsis University of West Attica, Athens, Greece

Ruopeng An Brown School, Washington University, St. Louis, MO, USA

Department of Kinesiology and Community Health, College of Applied Health Sciences, University of Illinois, Champaign, IL, USA

Dararatt Anantanasuwong Center for Aging Society Research, Research Center, National Institute of Development Administration, Bangkok, Thailand

Karen Andersen-Ranberg Danish Aging Research Center, Department of Public Health, University of Southern Denmark, Odense, Denmark

Department of Geriatrics, Odense University Hospital, Odense, Denmark

Department of Clinical Research, University of Southern Denmark, Odense, Denmark

Keith A. Anderson School of Social Work, University of Montana, Missoula, MT, USA

Melissa K. Andrew Division of Geriatric Medicine, Dalhousie University and Nova Scotia Health Authority, Halifax, NS, Canada

Georgia J. Anetzberger Case Western Reserve University, Cleveland, OH, USA

Anne Annink Leyden Academy on Vitality and Ageing, Leiden, The Netherlands

Alessandro Antonelli Department of Clinical and Experimental Medicine, University of Pisa, Pisa, Italy

Thomas Aparicio Gastroenterology and Digestive Oncology Department, Saint Louis Hospital, APHP, Paris, France

MaryBeth Apriceno Department of Psychology, Stony Brook University, Stony Brook, NY, USA

Yasumichi Arai Centre for Supercentenarian Medical Research, Keio University School of Medicine, Tokyo, Japan

Walter Arancio Department of Biological, Chemical and Pharmaceutical Sciences and Technologies (STEBICEF), University of Palermo, Palermo, Italy

Ane Arbillaga-Etxarri Physical Activity and Sports Sciences, Faculty of Psychology and Education, University of Deusto, Donostia-San Sebastián, Spain

Eleni Aretouli Cognitive Neuroscience Lab, School of Psychology, Aristotle University of Thessaloniki, Thessaloniki, Greece

Elizabeth Arias Mortality Statistics Branch, Division of Vital Statistics, National Center for Health Statistics, Hyattsville, MD, USA

Elva Dolores Arias-Merino Departamento de Salud Pública, Centro Universitario de Ciencias de la Salud, Universidad de Guadalajara, Guadalajara, Jalisco, Mexico

Hideki Ariizumi Department of Economics, Wilfrid Laurier University, Waterloo, ON, Canada

Beatrice Arosio Geriatric Unit, Fondazione IRCCS Ca' Granda, Ospedale Maggiore Policlinico, Milan, Italy

Department of Clinical Sciences and Community Health, University of Milan, Milan, Italy

Felix Arrieta Social Work and Sociology Department, Faculty of Social and Human Sciences, University of Deusto, Donostia-San Sebastian, Spain

Abdolrahim Asadollahi Department of Aging Health, Faculty of Health, Shiraz University of Medical Sciences, Shiraz, Iran

Australian Centre for Quality of Life, Deakin University, Melbourne, VIC, Australia

Kunnathil Muhammed Aslam Maulana Azad National Urdu University, Hyderabad, India

Christian Aspalter United International College, Beijing Normal University-Hong Kong Baptist University, Zhuhai, China

Sandrine Aspeslagh Department of Medical Oncology, UZ Brussel, Vrije Universiteit Brussel, Brussel, Belgium

Gil Atzmon Faculty of Natural Science, University of Haifa, Haifa, Israel

A. Auperin Service de biostatistique et d'épidémiologie, Gustave Roussy, Oncostat 1018, INSERM, labélisé Ligue contre le Cancer, Université Paris-Saclay, Villejuif, France

Siobhan Austen School of Economics, Finance and Property, Curtin Business School, Perth, WA, Australia

Giulia Avancini Sociology, Catholic University of Milan, Milan, Italy

Robert Averbuch Department of Psychiatry, University of Florida, Gainesville, FL, USA

Halimah Awang Social Wellbeing Research Centre (SWRC), University of Malaya, Kuala Lumpur, Malaysia

Katarina Babnik Faculty of Arts, Department of Psychology, University of Ljubljana, Ljubljana, Slovenia

Marijana Bađun Institute of Public Finance, Zagreb, Croatia

Maarouf Baghdadi Max Planck Institute for Biology of Ageing, Cologne, Germany

Gülistan Bahat Department of Internal Medicine, Division of Geriatrics, Istanbul University, Istanbul Medical School, Istanbul, Turkey

Chen Bai School of Labor and Human Resources, Renmin University of China, Beijing, China

Xue Bai Department of Applied Social Sciences, The Hong Kong Polytechnic University, Hung Homm, Hong Kong

Mirza Mansoor Baig School of Engineering, Computer and Mathematical Sciences, Auckland University of Technology, Auckland, New Zealand

Chitra Balasubramanian Siddha Regional Research Institute (Under Central Council for Research in Siddha, Ministry of AYUSH, Govt. of India), Puducherry, India

Maria Adela Balderas-Cejudo Oxford Institute of Population Ageing, University of Oxford, Oxford, United Kingdom

Esic Business and Marketing School, Madrid, Spain

C. Baldini Département d'Innovation Thérapeutique et d'Essais Précoces (DITEP), Gustave Roussy, Université Paris-Saclay, Villejuif, France

Lodovico Balducci Moffitt Cancer Center, Tampa, FL, USA

Péter Balogh Department of Immunology and Biotechnology, University of Pécs Clinical Center, Pécs, Hungary

Karen Bandeen-Roche Johns Hopkins Bloomberg School of Public Health, Baltimore, MD, USA

Luoman Bao Department of Sociology, California State University, Los Angeles, Los Angeles, CA, USA

Mario Barangea Faculty of Philosophy, University of Bucharest, Bucharest, Romania

Elisabetta Barbi Department of Statistical Sciences, Sapienza University of Rome, Rome, Italy

Magali Barbieri Department of Demography, University of California, Berkeley, CA, USA

French Institute for Demographic Studies (INED), Paris, France

Anthony R. Bardo Department of Sociology, University of Kentucky, Lexington, KY, USA

Richard Barker New Medicine Partners, London, UK

Paul James Barr The Dartmouth Institute for Health Policy and Clinical Practice, Geisel School of Medicine at Dartmouth College, Lebanon, NH, USA

The Center for Technology and Behavioral Health, Geisel School of Medicine at Dartmouth College, Lebanon, NH, USA

J. M. Barratt IFA, Toronto, ON, Canada

Richard E. Barrett University of Illinois at Chicago, Chicago, IL, USA

Elizabeth Barry University of Warwick, Coventry, UK

Constanza Bartolotti-Herrera Unidad de Geriatría Aguda, Hospital Clínico de la Fuerza Aérea de Chile, Universidad Mayor Santiago, Santiago, Chile

Servicio de Salud Araucanía Sur, Centro de Apoyo para Personas con Demencia, Kimünche, Temuco, Chile

Servicio de Medicina Interna, Hospital Hernán Henríquez Aravena, Temuco, Chile

Duygu Basaran Sahin Department of Sociology, The Graduate Center, City University of New York, New York, NY, USA

Ugofilippo Basellini Laboratory of Digital and Computational Demography, Max-Planck-Institute for Demographic Research, Rostock, Germany

Institut national d'études démographiques (INED), Aubervilliers, France

Marco Battaglini Italian National Institute of Statistics, Rome, Italy

Kate Batz Aging Analytics Agency, San Francisco, CA, USA

Shawn Bauldry Department of Sociology, Purdue University, West Lafayette, IN, USA

Piers Bayl-Smith Department of Psychology, Macquarie University, Sydney, NSW, Australia

Mst Marzina Begum Department of Public Administration, University of Rajshahi, Rajshahi, Bangladesh

David Bell Stirling Management School, University of Stirling, Stirling, UK

Janna Belser-Ehrlich University of Florida, Gainesville, FL, USA

Hiram Beltrán-Sánchez Department of Community Health Sciences, California Center for Population Research, Fielding School of Public Health, University of California, Los Angeles (UCLA), Los Angeles, CA, USA

Susanny J. Beltran School of Social Work, University of Central Florida, Orlando, FL, USA

Svetlana Di Benedetto Center for Lifespan Psychology, Max Planck Institute for Human Development, Berlin, Germany

Leila Bengrine Department of Medical Oncology, Centre Georges-Francois Leclerc, Dijon, France

Ivy Benjenk University of Maryland School of Public Health, College Park, MD, USA

Ester Benko Nursing Care Department, Faculty of Health Sciences, University of Primorska, Izola, Slovenia

Thomas Benzing Department II of Internal Medicine and Center for Molecular Medicine Cologne, Excellence Cluster CECAD, University of Cologne, Faculty of Medicine and University Hospital Cologne, Cologne, Germany

Éva Berde Department of Microeconomics, Corvinus University of Budapest, Budapest, Hungary

Demography and Economics Research Centre, School of Economics, Corvinus University of Budapest, Budapest, Hungary

Maria Berghs Leicester School of Allied Health Sciences, De Montfort University, Leicester, UK

Yoav S. Bergman School of Social Work, Ariel University, Ariel, Israel

Veronica N. Z. Bergstrom Department of Psychology, University of Toronto, Toronto, ON, Canada

M. Berg-Weger School of Social Work, Gateway Geriatric Education Center, Saint Louis University, St. Louis, MO, USA

Lisa Berkman Harvard Center for Population and Development Studies, Harvard University, Cambridge, MA, USA

School of Public Health, Faculty of Health Sciences, University of the Witwatersrand, Johannesburg, South Africa

Lars Bertram Lübeck Interdisciplinary Platform for Genome Analytics, Institutes of Neurogenetics and Integrative and Experimental Genomics, University of Lübeck, Lübeck, Germany

Department of Psychology, University of Oslo, Oslo, Norway

Nicole Beutell Department of Psychology, University of Colorado at Colorado Springs, Colorado Springs, CO, USA

Tirth R. Bhatta Department of Sociology, University of Nevada, Las Vegas, NV, USA

Arundhati Bhattacharyya Department of Political Science, University of Burdwan, Burdwan, West Bengal, India

Hazel Maridith Barlahan Biag Medical Investigation of Neurodevelopmental Disorders (MIND) Institute, University of California Davis, Sacramento, CA, USA

Department of Pediatrics, University of California Davis School of Medicine, Sacramento, CA, USA

Nicholas Bienko Department of Psychiatry and Behavioral Sciences, Johns Hopkins School of Medicine, Baltimore, MD, USA

Holly Birkett Birmingham Business School, University of Birmingham, Birmingham, UK

Elise S. Bisset Department of Pharmacology, Dalhousie University, Halifax, NS, Canada

R. A. Bitenc Albert-Ludwigs-Universität, Freiburg, Germany

Uwe Bittlingmayer Institute of Sociology, University of Education, Freiburg, Germany

Anita Björklund Carlstedt School of Health and Welfare, Jönköping University, Jönköping, Sweden

Cecilia Bjursell School of Education and Communication, Jönköping University, Jönköping, Sweden

Anne Blawert Institute of Psychogerontology, Friedrich-Alexander-Universität Erlangen-Nürnberg (FAU), Nürnberg, Germany

Andy Bleaden ECHAlliance, Stockport, UK

Kenneth A. Blocker Department of Educational Psychology, College of Education, University of Illinois at Urbana-Champaign, Champaign, IL, USA

David E. Bloom Department of Global Health and Population, Harvard T.H. Chan School of Public Health, Boston, MA, USA

Melissa J. Bloomer School of Nursing and Midwifery, Centre for Quality and Patient Safety Research, Deakin University, Geelong, VIC, Australia

Adrian Blundell Geriatric Medicine, Nottingham University Hospitals NHS Trust, Queens Medical Centre, Nottingham, UK

Patricia A. Bomba Geriatrics, Excellus BlueCross BlueShield, Rochester, NY, USA

Axel Börsch-Supan Munich Center for the Economics of Aging, Max-Planck-Institute for Social Law and Social Policy, Munich, Germany

Economics of Ageing, Technical University Munich, Munich, Germany

Rabia Boulahssass Geriatric Coordination Unit for Geriatric Oncology (UCOG) PACA Est Nice University Hospital, Centre Hospitalier Universitaire de Nice, Université Côte d'Azur, FHU OncoAge, Nice, France

Emanuel M. Boutzoukas Center for Cognitive Aging and Memory, Department of Clinical and Health Psychology, McKnight Brain Institute, University of Florida, Gainesville, FL, USA

Dawn M. E. Bowdish Department of Pathology and Molecular Medicine, McMaster University, Hamilton, ON, Canada

Dawn Bowers Department of Clinical and Health Psychology, College of Public Health and Health Professions, University of Florida, Gainesville, FL, USA

Nicola Luigi Bragazzi Postgraduate School of Public Health, Department of Health Sciences (DISSAL), University of Genoa, Genoa, Italy

Swgwmkhang Brahma Peace and Conflict Studies and Management, Sikkim University, Gangtok, Sikkim, India

Nady Braidy Centre for Healthy Brain Ageing, School of Psychiatry, University of New South Wales, Sydney, NSW, Australia

NPI, Euroa Centre, UNSW School of Psychiatry, Prince of Wales Hospital, Randwick, Sydney, NSW, Australia

Etienne Brain Department of Medical Oncology, Institut Curie, Paris/Saint-Cloud, France

Martina Brandt Institute of Sociology, TU Dortmund University, Dortmund, Germany

Jochen Brandtstädter Institut für Psychologie, Universität Trier, Trier, Germany

Melanie Sereny Brasher University of Rhode Island, Kingston, RI, USA

Elizabeth Breaux Department of Sociology, Appalachian State University, Boone, NC, USA

Sophia Brebou Hellenic Health Foundation, Athens, Greece

Gilbert Brenes Centro Centroamericano de Población, Universidad de Costa Rica, San Jose, Costa Rica

Catherine Bridge Faculty of the Built Environment, UNSW Sydney, Sydney, NSW, Australia

Nicole Brooke The Salvation Army, Aged Care, Redfern/Sydney, NSW, Australia

J. Scott Brown Department of Sociology and Gerontology, Miami University, Oxford, OH, USA

Patricia Brownell Graduate School of Social Service, Fordham University, Cleveland, OH, USA

Jonathan Browning Department of Psychiatry, University of Florida, Gainesville, FL, USA

Xiana Bueno Centre d'Estudis Demogràfics, Barcelona, Spain

Tine Buffel Manchester Institute for Collaborative Research on Ageing (MICRA), The University of Manchester, Manchester, UK

David Bunce College of Humanities, Arts and Social Sciences, Flinders University, Adelaide, SA, Australia

M. Haroon Burhanullah Department of Psychiatry and Behavioral Sciences, Johns Hopkins School of Medicine, Baltimore, MD, USA

James R. Burke Department of Neurology, Duke University, Durham, NC, USA

Erin Y. Burk-Leaver Columbia University, New York, NY, USA

A Burns IFA, Toronto, ON, Canada

Frances Burns Centre for Public Health, Queen's University Belfast, Institute of Clinical Sciences, Belfast, UK

Holger Busch Department of Developmental Psychology, Trier University, Trier, Germany

Sandra S. Butler School of Social Work, University of Maine, Orono, ME, USA

Kerryn Butler-Henderson Digital Health, College of Health and Medicine, University of Tasmania, Launceston, TAS, Australia

Angela Byrnes Department of Nutrition and Dietetics, Royal Brisbane and Women's Hospital, Herston, QLD, Australia

Emilia Cabras CES Don Bosco – Centro de Enseñanza Superior en Humanidades y Ciencias de la Educación, Madrid, Spain

Cheri Cabrera College of Natural and Health Sciences, University of Northern Colorado, Greeley, CO, USA

Susan Cadell School of Social Work, Renison University College, University of Waterloo, Waterloo, ON, Canada

H. Le Caer Service de pneumologie, Centre Hospitalier de Saint-Brieuc, Saint-Brieuc, France

Philippe Caillet Service de Gériatrie, Unité de Coordination en Onco-Gériatrie, Hôpitaux Universitaire Paris Ouest – Hôpital Européen Georges Pompidou, Paris, France

Darwin Caldwell Istituto Italiano di Tecnologia, Genoa, Italy

Kristin Calfee Department of Clinical and Health Psychology, Center for Cognitive Aging and Memory, McKnight Brain Institute, University of Florida, Gainesville, FL, USA

Riccardo Calvani Institute of Internal Medicine and Geriatrics, Università Cattolica del Sacro Cuore, Rome, Italy

Fondazione Policlinico Universitario "Agostino Gemelli" IRCCS, Rome, Italy

Emmanuelle Cambois CERMES3, UMR CNRS 8211 - Unité Inserm 988 – EHESS, Université Paris Descartes, Paris, France

Andrew Cameron School of Theology, Charles Sturt University, Canberra, ACT, Australia

Valentina Cannella University of Barcelona, Barcelona, Spain

Phillip Cantu University of Texas Medical Branch, Galveston, TX, USA

Vladimir Canudas-Romo School of Demography, RSSS, CASS The Australian National University, Canberra, ACT, Australia

Giorgia Capacci Italian National Institute of Statistics, Rome, Italy

Paula Carder OHSU-PSU School of Public Health, Portland, OR, USA

Elwood Carlson Center for Demography and Population Health, Florida State University, Tallahassee, FL, USA

Requena Carmen Department of Psychology, Faculty of Education, University of Leon, Leon, Spain

Fiona Carmichael Birmingham Business School, University of Birmingham, Birmingham, UK

Elisabeth Carola Department of Medical Oncology, Groupe Hospitalier Public du Sud de l'Oise, Creil, France

María Teresa Carreras Rodríguez Memory Unit, Neurology Department, Hospital de La Princesa, Madrid, Spain

Calogero Caruso Laboratory of Immunopathology and Immunosenescence, Department of Biomedicine, Neurosciences and Advanced Diagnostics, University of Palermo, Palermo, Italy

Maria Irene Carvalho Universidade de Lisboa, Institute of Social and Political Sciences, Center for Public Administration and Public Policies, Lisbon, Portugal

Célia Casaca Soares School of Health, Centre for Interdisciplinary Applied Research in Health (CIIAS/IPS), Polytechnic Institute of Setubal, Setubal, Portugal

Núria Casado-Gual Grup Dedal-Lit, University of Lleida, Lleida, Spain

Martina Casati Geriatric Unit, Fondazione IRCCS Ca' Granda, Ospedale Maggiore Policlinico, Milan, Italy

Graziella Caselli Department of Statistical Sciences, Sapienza University of Rome, Rome, Italy

Kelsie Cassell Department of Epidemiology of Microbial Diseases, Yale School of Public Health, New Haven, CT, USA

Paula Maria Machado Arantes de Castro Departamento de Fisioterapia, Escola de Educação Física, Fisioterapia e Terapia Ocupacional, Universidade Federal de Minas Gerais, Belo Horizonte, MG, Brazil

Montserrat Celdrán University of Barcelona, Barcelona, Spain

Gokce Cerev Department of Labor Economics and Industrial Relations, Fırat University, Elazığ, Turkey

Jingwen Chai Department of Psychology, National University of Singapore, Singapore, Singapore

Kalliopi Chainoglou Department of International and European Studies, University of Macedonia, Thessaloniki, Greece

Chi Kwok Chan S. H. Ho Urology Centre, Department of Surgery, Prince of Wales Hospital, The Chinese University of Hong Kong, Hong Kong, China

Ka Ki Chan Department of Social Work, Hong Kong Baptist University, Hong Kong, China

Kitty Chan Centre for Gerontological Nursing, School of Nursing, The Hong Kong Polytechnic University, Hong Kong, China

Siu Han Chan United International College, Beijing Normal University-Hong Kong Baptist University, Zhuhai, China

Wai Chi Chan Department of Psychiatry, The University of Hong Kong, Hong Kong, China

Wallace Chi Ho Chan Department of Social Work, The Chinese University of Hong Kong, Hong Kong, China

Cynthia K. Chandler Department of Counseling and Higher Education, University of North Texas, Denton, TX, USA

Yuanqing Chang School of Psychological and Cognitive Sciences, Peking University, Beijing, China

Neena L. Chappell Institute on Aging and Lifelong Health and Department of Sociology, University of Victoria, Victoria, BC, Canada

Sarah Chard Department of Sociology, Anthropology, and Health Administration and Policy, University of Maryland Baltimore County, Baltimore, MD, USA

Susan T. Charles Department of Psychological Science, University of California, Irvine, CA, USA

Karen Charlton School of Medicine, Faculty of Science, Medicine and Health, University of Wollongong, Wollongong, NSW, Australia

Alison L. Chasteen Department of Psychology, University of Toronto, Toronto, ON, Canada

Somnath Chatterji Division of Data, Analytics and Delivery for Impact, World Health Organization, Geneva, Switzerland

Kaushik Chattopadhyay Division of Epidemiology and Public Health, School of Medicine, University of Nottingham, Nottingham, UK

Hing-Wah Chau Victoria University, Melbourne, VIC, Australia

Habib Chaudhury Department of Gerontology, Simon Fraser University, Vancouver, BC, Canada

Swapnali Chavan Department of Sociology, IUPUI, Indianapolis, IN, USA

Yanin Chavarri-Guerra Department of Hemato-Oncology, Instituto Nacional de Ciencias Médicas y Nutrición Salvador Zubirán, Mexico City, Mexico

Tji Tjian Chee Department of Psychological Medicine, National University Hospital, Singapore, Singapore

Department of Psychological Medicine, Yong Loo Lin School of Medicine, National University of Singapore, Singapore, Singapore

Chen Chen Department of Global Health, School of Health Sciences, Wuhan University, Wuhan, China

Dan Chen The Chinese University of Hong Kong, Shatin, Hong Kong, China

Huashuai Chen Xiangtang University, Xiangtan, China

Center for the Study of Aging and Human Development and Geriatrics Division, School of Medicine, Duke University, Durham, NC, USA

Jie Chen School of Public Health, University of Maryland School of Public Health, College Park, MD, USA

Julian J.-L. Chen School of Molecular Sciences, Arizona State University, Tempe, AZ, USA

Li-Li Chen The Second Affiliated Hospital of Nanchang University, Nanchang City, China

Luxi Chen Centre for Family and Population Research, National University of Singapore, Singapore, Singapore

Qi Chen Department of Cardiovascular Medicine, Second Affiliated Hospital of Nanchang University, Nanchang, China

Xinxin Chen Institute of Social Science Survey, Peking University, Beijing, China

Jun Cheng National Center for Tuberculosis Control and Prevention, Chinese Center for Disease Control and Prevention, Beijing, China

Pak Wing Cheng Department of Psychiatry, The University of Hong Kong, Hong Kong, China

S.-T. Cheng Department of Health and Physical Education, The Education University of Hong Kong, Tai Po, Hong Kong, China

Katie E. Cherry Department of Psychology, Louisiana State University, Baton Rouge, LA, USA

Ka To Cheung Sau Po Centre on Ageing, Department of Social Worker and Social Administration, The University of Hong Kong, Pokfulam, Hong Kong, China

Iris Chi Suzanne Dowark-Peck School of Social Work, University of Southern California, Los Angeles, CA, USA

Vico Chiang School of Nursing, The Hong Kong Polytechnic University, Hong Kong, China

Elisa M. Childs School of Social Work, University of Georgia, Athens, Georgia

Vladimir A. Chistyakov Ivanovsky Academy of Biology and Biotechnology, Institute of Biology, Southern Federal University, Rostov-on-Don, Russia

Teresa T. W. Chiu Centre on Behavioral Health, The University of Hong Kong, Hong Kong, China

Sally Chivers Trent Centre for Aging and Society, Trent University, Peterborough, ON, Canada

Jinmyoung Cho Center for Applied Health Research, Baylor Scott & White Health, Dallas, TX, USA

Sarah E. Choate School of Social Work, Louisiana State University, Baton Rouge, LA, USA

Seung-won Choi Department of Sociology, Michigan State University, East Lansing, MI, USA

Sherry O. K. Chong Psychology Laboratories – Department of Social and Behavioural Sciences, City University of Hong Kong, Hong Kong, China

Wayne F. W. Chong Nanyang Technological University, Singapore, Singapore

GeroPsych Consultants Pte Ltd, Singapore, Singapore

Jill M. Chonody School of Social Work, Boise State University, Boise, IN, USA

Amy Yin Man Chow Department of Social Work and Social Administration, The University of Hong Kong, Pokfulam, Hong Kong, China

Joan C. Chrisler Department of Psychology, Connecticut College, New London, CT, USA

Kaare Christensen Danish Aging Research Center, Department of Public Health, University of Southern Denmark, Odense, Denmark

Department of Clinical Genetics, Odense University Hospital, Odense, Denmark

Department of Clinical Biochemistry and Pharmacology, Odense University Hospital, Odense, Denmark

Charlene H. Chu Lawrence S. Bloomberg Faculty of Nursing, University of Toronto, Toronto, ON, Canada

KITE-Toronto Rehabilitation Institute, University Health Network, Toronto, Canada

Sunny L. H. Chu Emerging Viral Diagnostics (HK) Limited, Hong Kong, China

Yongqiang Chu School of Public Administration, Guangdong University of Finance and Economics, Guangzhou, China

Hwei-Lin Chuang Department of Economics, National Tsing Hua University, Hsinchu, Taiwan

Christie Chung Psychology Department, Mills College, Oakland, CA, USA

Teresa Chung Department of Applied Biology and Chemical Technology, The Hong Kong Polytechnic University, Hung Hom, Kowloon, Hong Kong

Chen Chuqian Department of Social Work and Social Administration, The University of Hong Kong, Hong Kong, People's Republic of China

Christine Cigolle VA Ann Arbor Healthcare System, Geriatric Research, Education and Clinical Center (GRECC), Ann Arbor, MI, USA

Department of Family Medicine, University of Michigan, Ann Arbor, MI, USA

Department of Internal Medicine, Division of Geriatric and Palliative Medicine, University of Michigan, Ann Arbor, MI, USA

Kenis Cindy Department of General Medical Oncology and Geriatric Medicine, University Hospitals Leuven, Leuven, Belgium

Roberto Cingolani Istituto Italiano di Tecnologia, Genoa, Italy

Caitlin Clarke Department of Kinesiology and Community Health, College of Applied Health Sciences, University of Illinois, Champaign, IL, USA

Lindsay Clarke Alliance for Aging Research, Washington, DC, USA

Lindy Clemson Faculty of Health Sciences, The University of Sydney, Sydney, NSW, Australia

Alan A. Cohen PRIMUS Research Group, Department of Family Medicine and Emergency Medicine, University of Sherbrooke, Sherbrooke, QC, Canada

Ronald A. Cohen Center for Cognitive Aging and Memory, Department of Clinical and Health Psychology, University of Florida, Gainesville, FL, USA

Steven A. Cohen Department of Health Studies, University of Rhode Island, Kingston, RI, USA

J. Michael Collins School of Human Ecology and La Follette School of Public Affairs, University of Wisconsin-Madison, Madison, WI, USA

Sarah Collins Department of Sociology, University of Memphis, Memphis, TN, USA

Keith Comito Life Extension Advocacy Foundation/Lifespan.io, Seaford, NY, USA

Valeria Conti Department of Medicine, Surgery and Dentistry "Scuola Medica Salernitana", University of Salerno, Salerno, Italy

Nico A. Contreras Department of Immunobiology and the University of Arizona Center on Aging, University of Arizona College of Medicine-Tucson, Tucson, AZ, USA

Christopher P. Coplen Department of Immunobiology and the University of Arizona Center on Aging, University of Arizona College of Medicine-Tucson, Tucson, AZ, USA

Graziamaria Corbi Department of Medicine and Health Sciences, University of Molise, Campobasso, Italy

Victoria Cornell Research and Evaluation, Business Development, ECH, Adelaide, Australia

Housing and the Built Environment Special Interest Group, Australian Association of Gerontology, Melbourne, Australia

Jennifer C. Cornman Jennifer C. Cornman Consulting, Granville, OH, USA

Francesca Corradini Sociology, Catholic University of Milan, Milan, Italy

Romain Corre Pneumology unit, Centre Hospitalier de Cornouaille, Quimper, France

Jorge Correia Jesuino Center for Philosophy of Sciences of the University of Lisbon (CFCUL), Lisbon, Portugal

Franco Cortese Aging Analytics Agency, Toronto, ON, Canada

Theodore D. Cosco Gerontology Research Centre, Department of Gerontology, Simon Fraser University, Vancouver, BC, Canada

Oxford Institute of Population Aging, University of Oxford, Oxford, UK

Andrea Costa ISPA – Instituto Universitário, Lisbon, Portugal

Alexandra Crampton Department of Social and Cultural Sciences, Marquette University, Milwaukee, WI, USA

Beth R. Crisp School of Health and Social Development, Deakin University, Geelong, VIC, Australia

Anca Cristofovici Université de Caen-Normandie, Caen, France

Valorie A. Crooks Department of Geography, Simon Fraser University, Burnaby, BC, Canada

Sharon M. Cruise Centre for Public Health, Queen's University Belfast, Institute of Clinical Sciences, Belfast, UK

Patrick J. Cruitt Department of Psychological and Brain Sciences, Washington University in St. Louis, St. Louis, MO, USA

Gibran Cruz-Martinez Institute of Public Goods and Policies, Consejo Superior de Investigaciones Científicas (CSIC), Madrid, Spain

Tristan Cudennec Hôpitaux Antoine-Béclère, Geriatric Unit of Pr Teillet AP-HP, Université Paris Saclay, Saint-Aubin, France

Seyran Gursoy Cuhadar Kocaeli University/Industrial Relations and Labour Economics, Kocaeli, Turkey

Qi Cui School of Demography, RSSS, CASS The Australian National University, Canberra, ACT, Australia

Margaret Currie Social Economic and Geographical Sciences, The James Hutton Institute, Aberdeen, UK

Cassie Curryer School of Humanities and Social Sciences, University of Newcastle (UoN), Callaghan, NSW, Australia

Hippolyte d'Albis Paris School of Economics, CNRS, Paris, France

Rubens A. da Silva Département des Sciences de la Santé, Programme de physiothérapie de l'université McGill offert en extension à l'Université du Québec à Chicoutimi (UQAC), Centre intersectoriel en santé durable, Lab BioNR – UQAC, Saguenay, QC, Canada

Aimin Dang Department of Special Care Center, National Clinical Research Center for Cardiovascular Diseases, Fuwai Hospital, National Center for Cardiovascular Diseases, Chinese Academy of Medical Sciences and Peking Union Medical College, Beijing, People's Republic of China

Carina Dantas Innovation Department, Cáritas Diocesana de Coimbra, Coimbra, Portugal

Gabrielle D. Daoust Department of International Relations, The University of Sussex, Sussex, UK

Mitra Das University of Massachusetts Lowell, Lowell, MA, USA

Amaury Daste Department of Medical Oncology, Hôpital Saint-André, Bordeaux University Hospital-CHU Bordeaux, Bordeaux, France

Fabíola Bof de Andrade Departamento de Epidemiologia, Fundação Oswaldo Cruz, Belo Horizonte, Minas Gerais, Brazil

Laure de Decker Société Francophone d'Oncogériatrie, CHU Nantes, Geriatric Unit, Université de Nantes, Saint Herblain, France

Aubrey D. N. J. de Grey SENS Research Foundation, Mountain View, CA, USA

Lorenzo De Michieli Istituto Italiano di Tecnologia, Genoa, Italy

Cesar de Oliveira Department of Epidemiology and Public Health, University College London, London, UK

José Manuel Sousa de São José Faculty of Economics, University of Algarve, Faro, Portugal and Interdisciplinary Centre of Social Sciences, New University of Lisbon, Lisbon, Portugal

Wouter De Tavernier Center for Social and Cultural Psychology, KU Leuven, Leuven, Belgium

Lore Decoster Department of Medical Oncology, UZ Brussel, Vrije Universiteit Brussel, Brussel, Belgium

Joris Deelen Max Planck Institute for Biology of Ageing, Cologne, Germany

Christian Deindl Institute of Medical Sociology, Medical Faculty, Heinrich-Heine-University Düsseldorf, Düsseldorf, Germany

Steven T. DeKosky Department of Neurology and McKnight Brain Institute, University of Florida, Gainesville, FL, USA

Ilja Demuth Department of Endocrinology and Metabolic Diseases, Charité – Universitätsmedizin Berlin, BCRT – Berlin Institute of Health Center for Regenerative Therapies, Berlin, Germany

Chao Deng Illawarra Health and Medical Research Institute, University of Wollongong, Wollongong, NSW, Australia

Medical School, University of Wollongong, Wollongong, Australia

Yuanyuan Deng ARC Centre of Excellence in Population Ageing Research, UNSW Business School, The University of New South Wales, Sydney, NSW, Australia

Yuri V. Denisenko Ivanovsky Academy of Biology and Biotechnology, Institute of Biology, Southern Federal University, Rostov-on-Don, Russia

Nicole DePasquale Division of General Internal Medicine, Duke University School of Medicine, Durham, NC, USA

Vanessa di Lego Vienna Institute of Demography/Austrian Academy of Sciences, Wittgenstein Centre (IIASA, VID/ÖAW, WU), Vienna, Austria

Pilar Díaz Department of Social Psychology, University of Granada, Granada, Spain

Irma Fabiola Díaz-García Social Sciences Department, Universidad de Guadalajara CUCS, Guadalajara, Jalisco, Mexico

Manfred Diehl Department of Human Development and Family Studies, Colorado State University, Fort Collins, CO, USA

Congcong Ding Department of Cardiovascular Medicine, The Second Affiliated Hospital of Nanchang University, Nanchang, Jiangxi, China

Ellen Dingemans Netherlands Interdisciplinary Demographic Institute (NIDI-KNAW), The Hague, The Netherlands

Het PON, Tilburg, The Netherlands

Vinicius do Rosario School of Medicine, Faculty of Science, Medicine and Health, University of Wollongong, Wollongong, NSW, Australia

Stefanie Doebler Department of Sociology, Social Policy and Criminology, University of Liverpool, Liverpool, UK

Josephine Dolan Research Centre for Women, Ageing and Media, University of Gloucestershire, Cheltenam, UK

Vanessa Dominguez Center for Cognitive Aging and Memory, Department of Clinical and Health Psychology, McKnight Brain Institute, University of Florida, Gainesville, FL, USA

Emma Domínguez-Rué Department of English, University of Lleida, Lleida, Catalonia, Spain

Gretchen Donehower Center for the Economics and Demography of Aging, Department of Demography, University of California, Berkeley, CA, USA

Yu-Gang Dong Department of Cardiology, Heart Center, The First Affiliated Hospital of Sun Yat-sen University, Guangzhou, People's Republic of China

Timothy A. Donlon Honolulu Heart Program (HHP)/Honolulu-Asia Aging Study (HAAS), Department of Research, Kuakini Medical Center, Honolulu, HI, USA

Departments of Cell and Molecular Biology and Pathology, John A. Burns School of Medicine, University of Hawaii, Honolulu, HI, USA

Orna A. Donoghue The Irish Longitudinal Study on Ageing (TILDA), Trinity College Dublin, Dublin, Ireland

Elaine Douglas Stirling Management School, University of Stirling, Stirling, UK

Delali Adjoa Dovie Department of Sociology, University of Ghana, Accra, Ghana

William H. Dow School of Public Health, University of California, Berkeley, Berkeley, CA, USA

Johanna Drewelies Humboldt University Berlin, Berlin, Germany

Shufa Du Department of Nutrition and Carolina Population Center, The University of North Carolina at Chapel Hill, Chapel Hill, NC, USA

Yan Du School of Nursing, University of Texas Health Science Center at San Antonio, San Antonio, TX, USA

Joanne Duberley Birmingham Business School, University of Birmingham, Birmingham, UK

Roman Dubianski Department of Breast Cancer and Reconstructive Surgery, Maria Sklodowska-Curie National Research Institute of Oncology, Warsaw, Poland

Nicole Dubus San Jose State University, San Jose, CA, USA

Jerome Dugan Department of Health Services in School of Public Health, University of Washington, Seattle, WA, USA

Matthew E. Dupre Department of Population Health Sciences, Department of Sociology, Duke University, Durham, NC, USA

Sandra Düzel Center for Lifespan Psychology, Max Planck Institute for Human Development, Berlin, Germany

Pearl A. Dykstra Department of Public Administration and Sociology, Erasmus University, Rotterdam, The Netherlands

William W. Eaton Bloomberg School of Public Health, Department of Mental Health, Johns Hopkins University, Baltimore, MD, USA

Alexander Eckersley Division of Cell Matrix Biology and Regenerative Medicine, School of Biological Sciences, Faculty of Biology, Medicine and Health, The University of Manchester, Manchester, UK

Dalkhat M. Ediev North-Caucasian State Academy, Cherkessk, Russia

International Institute for Applied Systems Analysis, Laxenburg, Austria

Department of Demography (HSMSS), Lomonosov Moscow State University, Moscow, Russia

Jerri D. Edwards Department of Psychiatry and Behavioral Neurosciences, University of South Florida, Tampa, FL, USA

Ryan D. Edwards University of California, Berkeley, CA, USA

Viviana Egidi Department of Statistical Sciences, Sapienza University of Rome, Rome, Italy

Hans-Jörg Ehni Institute for the Ethics and History of Medicine, University of Tübingen, Tübingen, Germany

Stefanie El Madawi University of Huddersfield, Huddersfield, UK

Richard Elliott Newcastle University, Newcastle upon Tyne, UK

Catherine Eng UCSF School of Medicine, Division of Geriatrics, On Lok Senior Health Services, San Francisco, CA, USA

Tuğba Erdoğan Department of Internal Medicine, Division of Geriatrics, Istanbul University, Istanbul Medical School, Istanbul, Turkey

Mary Ann Erickson Ithaca College Gerontology Institute, Ithaca, NY, USA

Marla J. Erwin Department of Psychology, Louisiana State University, Baton Rouge, LA, USA

María Claudia Espinel-Bermudez Departamento de Salud Pública, Centro Universitario de Ciencias de la Salud, Universidad de Guadalajara e Instituto Mexicano del Seguro Social, Guadalajara, Jalisco, Mexico

Nicole D. Evangelista Center for Cognitive Aging and Memory, McKnight Brain Institute, Department of Clinical and Health Psychology, University of Florida-Gainesville, Gainesville, FL, USA

Ludovic Evesque Digestive Oncology Unit, Department of Medical Oncology, Antoine-Lacassagne Cancer Center, Nice, France

Bonnie Ewald Social Work and Community Health, Rush University Medical Center, Chicago, IL, USA

Martine Extermann Moffitt Cancer Center, University of South Florida, Tampa, FL, USA

Elisa Fabbri Translational Gerontology Branch, National Institute on Aging, NIH, Baltimore, MD, USA

J. Kaci Fairchild Department of Psychiatry and Behavioral Sciences, Sierra Pacific Mental Illness Research Education and Clinical Center VA Palo Alto, Stanford University School of Medicine, Palo Alto, CA, USA

Ghadeer Falah Faculty of Natural Science, University of Haifa, Haifa, Israel

Claire Falandry Geriatric Unit, Lyon Sud University Hospital, Hospices Civils de Lyon and Lyon University, Pierre-Bénite, France

CarMEN Laboratory of Lyon University, INSERM U.1060/Université Lyon1/INRA U. 1397/INSA Lyon/Hospices Civils Lyon, Pierre-Bénite, France

Sarah Falcus University of Huddersfield, Huddersfield, UK

Alexander T. Falk Department of Radiation Oncology, Centre Antoine Lacassagne, Nice, France

Fédération Claude Lalanne, Université Nice Cote d'Azur, Nice, France

Poupak Fallahi Department of Translational Research of New Technologies in Medicine and Surgery, University of Pisa, Pisa, Italy

Gilbert Fan Division of Supportive and Palliative Care, National Cancer Centre Singapore, Singapore, Singapore

Gracy B. Y. Fang Department of Applied Social Sciences, The Hong Kong Polytechnic University, Hong Kong, Hong Kong

Hui Fang The First Affiliated Hospital of Zhengzhou University, Zhengzhou, China

Heather Farmer Department of Population Health Sciences, Duke University, Durham, NC, USA

Center for the Study of Aging and Human Development, Duke University, Durham, NC, USA

Debbie Faulkner Centre for Housing, Urban and Regional Planning, School of Social Sciences, The University of Adelaide, Adelaide, SA, Australia

Leilani Feliciano Department of Psychology, University of Colorado at Colorado Springs, Colorado Springs, CO, USA

Jorge Felix University of Sao Paulo, Sao Paulo, Brazil

Center for Longevity Economy Studies, Sao Paulo, Brazil

Qiushi Feng Department of Sociology, Centre for Family and Population Research, National University of Singapore, Singapore, Singapore

Zhixin Feng Primary Care and Population Sciences, Faculty of Medicine, University of Southampton, Southampton, Hampshire, UK

Marisa Fernández Sánchez Memory Unit, Neurology Department, Hospital de La Princesa, Madrid, Spain

Beatriz Fernández-Varas Instituto de Investigaciones Biomedicas, CSIC/UAM. CIBER de Enfermedades Raras (CIBERER), Madrid, Spain

Joel Fernandes Department of Psychiatry, University of Florida, Gainesville, FL, USA

Malcom Randall VA Medical Center, Gainesville, FL, USA

Heshan J. Fernando Department of Clinical and Health Psychology, College of Public Health and Health Professions, University of Florida, Gainesville, FL, USA

Sam Fernes Manchester, UK

Nicola Ferrara Department of Translational Medical Sciences, Federico II University, Naples, Italy

Silvia Martina Ferrari Department of Clinical and Experimental Medicine, University of Pisa, Pisa, Italy

Fabiane Ribeiro Ferreira Departamento de Fisioterapia, Escola de Educação Física, Fisioterapia e Terapia Ocupacional, Universidade Federal de Minas Gerais, Belo Horizonte, MG, Brazil

Evelyn Ferri Geriatric Unit, Fondazione IRCCS Ca' Granda, Ospedale Maggiore Policlinico, Milan, Italy

Luigi Ferrucci Translational Gerontology Branch, National Institute on Aging, NIH, Baltimore, MD, USA

Amelia Filippelli Department of Medicine, Surgery and Dentistry "Scuola Medica Salernitana", University of Salerno, Salerno, Italy

Matthew G. Fillingim Center for Cognitive Aging and Memory, Department of Clinical and Health Psychology, University of Florida, Gainesville, FL, USA

Brian Karl Finch Center for Economic and Social Research, University of Southern California, Los Angeles, CA, USA

Department of Sociology and Spatial Sciences, Univerity of Southern California, Los Angeles, CA, USA

Michael Fine Department of Sociology, Macquarie University, Sydney, NSW, Australia

Koren L. Fisher Center for Successful Aging, Department of Kinesiology, College of Health and Human Development, California State University, Fullerton, CA, USA

Malcolm John Fisk Centre for Computing and Social Responsibility, De Montfort University (DMU), Leicester, UK

James Fleet Department of Ageing and Health, Guy's and St Thomas' NHS Foundation Trust, London, UK

Jerome L. Fleg Division of Cardiovascular Sciences, National Heart, Lung, and Blood Institute, National Institutes of Health, Bethesda, MD, USA

Aisling M. Fleury Logan Hospital, Metro South Hospital and Health Service, Meadowbrook, QLD, Australia

Centre for Health Services Research (Ageing and Geriatric Medicine Research Program), Faculty of Medicine, University of Queensland, St Lucia, QLD, Australia

María Elena Flores-Villavicencio Social Sciences Department, Universidad de Guadalajara CUCS, Guadalajara, Jalisco, Mexico

Fabio Folgheraiter Sociology, Catholic University of Milan, Milan, Italy

Joelle H. Fong Lee Kuan Yew School of Public Policy, National University of Singapore, Singapore, Singapore

Kenneth N. K. Fong Department of Rehabilitation Sciences, The Hong Kong Polytechnic University, Hung Hom, HongKong SAR

Marvin Formosa Department of Gerontology and Dementia Studies, Faculty for Social Wellbeing, University of Malta, Msida, Malta

Brent M. Foster Department of Clinical and Health Psychology, College of Public Health and Health Professions, Center for Cognitive Aging and Memory, McKnight Brain Institute, University of Florida, Gainesville, FL, USA

Claudio Franceschi IRCCS, Institute of Neurological Sciences of Bologna, Bologna, Italy

Linda Smith Francis Human Development, Warner School of Education, University of Rochester, Rochester, NY, USA

Sarah L. Francis Department of Food Science and Human Nutrition, Iowa State University, Ames, IA, USA

Eric Francois Digestive Oncology Unit, Department of Medical Oncology, Antoine-Lacassagne Cancer Center, Nice, France

Annette Franke Protestant University of Applied Sciences Ludwigsburg, Ludwigsburg, Germany

Daniela Frasca Department of Microbiology and Immunology, University of Miami Miller School of Medicine, Miami, FL, USA

Alexandra M. Freund Department of Psychology, Developmental Psychology: Adulthood, University of Zurich, Zurich, Switzerland

Linda P. Fried Columbia University Mailman School of Public Health, New York, NY, USA

Esther M. Friedman RAND Corporation Santa Monica, Santa Monica, CA, USA

Eric Frost Department of Microbiology and Infectious diseases, University of Sherbrooke, Sherbrooke, QC, Canada

Tsung-Hsi Fu Department of Social Work, National Taiwan University, Taipei, Taiwan

Yuanyuan Fu Beijing Normal University, Beijing, China

Maja Fuglsang Palmer Centre for Research on Ageing and ESRC Centre for Population Change, Faculty of Social Sciences, University of Southampton, Southampton, UK

Tamas Fulop Research Center on Aging, Graduate Program in Immunology, Faculty of Medicine and Health Sciences, University of Sherbrooke, Sherbrooke, QC, Canada

A. W. T. Fung Department of Applied Social Sciences, The Hong Kong Polytechnic University, Kowloon, Hong Kong, China

Laura M. Funk Department of Sociology and Criminology, University of Manitoba, Winnipeg, MB, Canada

Miyoko Fuse Portland Community College, Portland, OR, USA

Simon Galas IBMM, University of Montpellier, CNRS, ENSCM, Montpellier, France

Thais P. Galletti Department of Statistics, Universidade Federal de Minas Gerais, Belo Horizonte, Brazil

Javier Ganzarain AFEdemy, Ltd., Barcelona, Spain

Yuan Gao Neurology Department, the First Affiliated Hospital of Zhengzhou University, Zhengzhou, Henan, China

Paolo Garagnani Department of Experimental, Diagnostic and Specialty Medicine (DIMES), University of Bologna, Bologna, Italy

Irma Fabiola Díaz García Public Health Department, Universidad de Guadalajara CUCS, Guadalajara, Jalisco, Mexico

Manuel García-Goñi Department of Applied and Structural Economics and History, Faculty of Economics and Business, Universidad Complutense de Madrid, Madrid, Spain

Isabella Gattas-Vernaglia Department of Geriatric Oncology, Instituto do Câncer do Estado de São Paulo (ICESP), São Paulo, Brazil

Margaret Gatz Center for Economic and Social Research, University of Southern California, Los Angeles, CA, USA

Department of Psychology, University of Southern California, Los Angeles, CA, USA

Uma Gaur Faculty of Health Sciences, University of Macau, Taipa, Macau, China

Noralie H. Geessink Department of Geriatric Medicine, Radboud University Medical Center, Nijmegen, The Netherlands

Romain Geiss Service de Gériatrie, Unité de Coordination en Onco-Gériatrie, Hôpitaux Universitaire Paris Ouest – Hôpital Européen Georges Pompidou, Paris, France

Tracey Gendron Department of Gerontology, College of Health Professions, Virginia Commonwealth University, Richmond, VA, USA

Linda K. George Sociology Emerita, Duke University, Durham, NC, USA

Fabiola H. Gerpott Management Group, WHU - Otto Beisheim School of Management, Duesseldorf, Deutschland

ARC Centre of Excellence in Population Ageing Research (CEPAR), Perth, Australia

Denis Gerstorf Humboldt University Berlin, Berlin, Germany

T. R. Gettinger School of Social Work, Saint Louis University, St. Louis, MO, USA

Hamid GholamHosseini School of Engineering, Computer and Mathematical Sciences, Auckland University of Technology, Auckland, New Zealand

Anna Giardini Psychology Unit of Montescano, Istituti Clinici Scientifici Maugeri IRCCS, Montescano (PV), Italy

Roseann Giarrusso Department of Sociology, California State University-Los Angeles, Los Angeles, CA, USA

Thomas Gilbert Geriatrics Unit, Centre Hospitalier Lyon Sud, Hospices Civils de Lyon, Pierre-Bénite, France

Geriatrics Unit, Centre Hospitalier de Vienne, Vienne, France

Megan Gilligan Human Development and Family Studies, Iowa State University, Ames, IA, USA

Nikesha Gilmore Department of Surgery, Division of Cancer Control, University of Rochester Medical Center, Rochester, NY, USA

Mathilde Gisselbrecht Service de Gériatrie, Unité de Coordination en Onco-Gériatrie, Hôpitaux Universitaire Paris Ouest – Hôpital Européen Georges Pompidou, Paris, France

Cristina Giuliani Laboratory of Molecular Anthropology and Centre for Genome Biology, Department of Biological, Geological and Environmental Sciences (BiGeA), University of Bologna, Bologna, Italy

School of Anthropology and Museum Ethnography, University of Oxford, Oxford, UK

Dana A. Glei Center for Population and Health, Graduate School of Arts and Sciences, Georgetown University, Washington, DC, USA

Judith Godin Division of Geriatric Medicine, Dalhousie University and Nova Scotia Health Authority, Halifax, NS, Canada

Jorming Goh NUSMed Healthy Longevity Translational Research Programme, Yong Loo Lin School of Medicine, National University of Singapore, Singapore, Singapore

Centre for Healthy Longevity, National University Health System, Singapore, Singapore

Noreen Goldman Office of Population Research, Princeton University, Princeton, NJ, USA

Theodore C. Goldsmith Azinet LLC, Annapolis, MD, USA

Yasuyuki Gondo Graduate School of Human Sciences, Osaka University, Suita, Osaka, Japan

Ernest Gonzales Silver School of Social Work, New York University, New York, NY, USA

Michael Gordon Department of Clinical and Health Psychology, College of Public Health and Health Professions, Center for Cognitive Aging and Memory, McKnight Brain Institute, University of Florida, Gainesville, FL, USA

Ana Fontoura Gouveia Banco de Portugal, Nova School of Business and Economics and Economics for Policy, Lisbon, Portugal

Diddahally Govindaraju The Institute for Aging Research, Albert Einstein College of Medicine, Bronx, NY, USA

Museum of Comparative Zoology, Harvard University, Cambridge, MA, USA

Helen L. Graham Helen and Arthur E. Johnson College of Nursing and Health Sciences, University of Colorado Colorado Springs, Colorado Springs, CO, USA

Heather Grain Global eHealth Collaborative, Melbourne, Australia

Health and Biomedical Informatics Centre, Melbourne University, Melbourne, Australia

Katie L. Granier Department of Psychology, University of Colorado at Colorado Springs, Colorado Springs, CO, USA

Pamela Gravagne Arts-In-Medicine Program, University of New Mexico, Albuquerque, NM, USA

Mary L. Greaney Department of Health Studies, University of Rhode Island, Kingston, RI, USA

Tjaša Grebenšek Anton Trstenjak Institute of Gerontology and Intergenerational Relations, Ljubljana, Slovenia

Jeff Greenberg Department of Psychology, The University of Arizona, Tucson, AZ, USA

Emily Grundy Institute for Social and Economic Research, University of Essex, Colchester, UK

Danan Gu Population Division, Department of Economics and Social Affairs, United Nations, New York, NY, USA

Joana Guedes Instituto Superior de Serviço Social do Porto (ISSSP), Centro Lusíada de Investigação em Serviço Social e em Intervenção Social (CLISSIS), Porto, Portugal

Olivier Guerin Geriatric Coordination Unit for Geriatric Oncology (UCOG) PACA Est CHU de Nice, Nice, France

FHU OncoAge, Nice, France

University Côte d'Azur, Nice, France

Joël Guigay Medical Oncology Department, Centre Antoine Lacassagne, FHU OncoAge, Université Côte d'Azur, Nice, France

Joseph M. Gullett Center for Cognitive Aging and Memory, Department of Clinical and Health Psychology, University of Florida, Gainesville, FL, USA

Margaret Morganroth Gullette Women's Studies Research Center, Brandeis University, Waltham, MA, USA

Caixia Guo CAS Key Laboratory of Genomics and Precision Medicine, Beijing Institute of Genomics, University of Chinese Academy of Sciences, Chinese Academy of Sciences, Beijing, China

Yu Guo Renmin University of China, Beijing, China

Alisha Gupta Division of General Internal Medicine and Geriatrics, Northwestern University Feinberg School of Medicine, Chicago, IL, USA

Jack M. Guralnik University of Maryland School of Medicine, Baltimore, MD, USA

Ashita S. Gurnani Psychiatry and Cognitive Neurology (BIDMC), Beth Israel Deaconess Medical Center/Harvard Medical School and Massachusetts Mental Health Center, Boston, MA, USA

Danielle Gutman Faculty of Natural Science, University of Haifa, Haifa, Israel

Gloria Gutman Department of Gerontology/Gerontology Research Centre, Simon Fraser University, Vancouver, BC, Canada

Trish Hafford-Letchfield School of Social Work and Social Policy, Glasgow, Scotland

Randi J. Hagerman Medical Investigation of Neurodevelopmental Disorders (MIND) Institute, University of California Davis, Sacramento, CA, USA

Department of Pediatrics, University of California Davis School of Medicine, Sacramento, CA, USA

David Hailey Centre for IT-Enabled Transformation, School of Computing and Information Technology, University of Wollongong, Wollongong, NSW, Australia

Umaima Zahra Halim Centre for IT-enabled Transformation, School of Computing and Information Technology, University of Wollongong, Wollongong, Australia

Cal J. Halvorsen Center on Aging and Work, School of Social Work, Boston College, Chestnut Hill, MA, USA

M. E. Hamaker Department of Geriatric Medicine, Diakonessenhuis Utrecht, Utrecht/Zeist/Doorn, The Netherlands

Benjamin M. Hampstead Department of Psychiatry, University of Michigan, Ann Arbor, MI, USA

Mental Health Service, VA Ann Arbor Healthcare System, Ann Arbor, MI, USA

David J. Hancock Memorial University of Newfoundland, St. John's, NL, Canada

Jean-Michel Hannoun-Lévi Department of Radiation Oncology, Centre Antoine Lacassagne, Nice, France

Fédération Claude Lalanne, Université Nice Cote d'Azur, Nice, France

Gary Haq Stockholm Environment Institute, Department of Environment and Geography, University of York, Heslington, York, UK

Gabriella M. Harari Department of Communication, Stanford University, Stanford, CA, USA

Cheshire Hardcastle Department of Clinical and Health Psychology, University of Florida, Gainesville, FL, USA

Department of Public Health and Health Professions, University of Florida, Gainesville, FL, USA

Molly-Gloria Harper Western University, London, ON, Canada

Andrea L. Harris Geriatric Clinics, University of Utah, Salt Lake City, UT, USA

Peter Hartley Department of Public Health and Primary Care, University of Cambridge, Cambridge, UK

Department of Physiotherapy, Cambridge University Hospitals NHS Foundation Trust, Cambridge, UK

Heike Hartung Department of English and American Studies, University of Potsdam, Potsdam, Germany

Philip D. Harvey Department of Psychiatry and Behavioral Sciences, University of Miami Miller School of Medicine, Miami, FL, USA

Bruce W. Carter VA Medical Center, Miami, FL, USA

Jaroslava Hasmanová Marhánková Department of Sociology, University of West Bohemia, Pilsen, Czech Republic

Department of Gender Studies, Charles University in Prague, Prague, Czech Republic

Department of Sociology, Charles University, Prague, Czech Republic

Hanna Hausman Department of Clinical and Health Psychology, Gainesville, FL, USA

Leah Haverhals VA Eastern Colorado Health Care System, Aurora, CO, USA

Alexandra J. Hawkey Translational Health Research Institute, School of Medicine, Western Sydney University, Sydney, NSW, Australia

Louise C. Hawkley Academic Research Centers, NORC at the University of Chicago, Chicago, IL, USA

Wan He Washington, DC, USA

Yao He Institute of Geriatrics, the 2nd Medical Center, Beijing Key Laboratory of Aging and Geriatrics, National Clinical Research Center for Geriatrics Diseases, Chinese PLA General Hospital, Beijing, China

Jean M. Hébert Departments of Neuroscience and Genetics, Albert Einstein College of Medicine, New York, NY, USA

Arthur E. Helfand Department of Community Health and Aging, Temple University School of Podiatric Medicine, Philadelphia, PA, USA

Peter J. Helm Department of Psychology, The University of Arizona, Tucson, AZ, USA

Marie Hennecke Department of Psychology, University of Siegen, Siegen, Germany

Utz Herbig Department of Microbiology, Biochemistry, and Molecular Genetics and Center for Cell Signaling, Rutgers Biomedical and Health Sciences, New Jersey Medical School, Rutgers University, Newark, NJ, USA

Moritz Hess SOCIUM Research Center on Inequality and Social Policy, University of Bremen, Bremen, Germany

Thomas M. Hess Psychology, North Carolina State University, Raleigh, NC, USA

Mira Hidajat Population Health Sciences, Bristol Medical School, University of Bristol, Bristol, UK

Christine Hildesheim Institute of Psychology, Heidelberg University, Heidelberg, Germany

Patrick L. Hill Department of Psychological and Brain Sciences, Washington University in St. Louis, St. Louis, MO, USA

Sarah N. Hilmer NHMRC Cognitive Decline Partnership Centre, Kolling Institute of Medical Research, Northern Clinical School, Faculty of Medicine and Health, University of Sydney, Sydney, NSW, Australia

Departments of Clinical Pharmacology and Aged Care, Royal North Shore Hospital, St Leonards, NSW, Australia

Sharron Hinchliff University of Sheffield, Sheffield, UK

Helena M. Hinterding Max Planck Institute for Biology of Ageing, Cologne, Germany

Yosuke Hirayama Graduate School of Human Development and Environment, Kobe University, Kobe, Japan

Nobuyoshi Hirose Centre for Supercentenarian Medical Research, Keio University School of Medicine, Tokyo, Japan

Andy Hau Yan Ho Psychology Programme, School of Social Sciences, Nanyang Technological University, Singapore, Singapore

Centre for Population Health Sciences (CePHaS), Lee Kong Chian School of Medicine, Nanyang Technological University, Singapore, Singapore

Palliative Care Centre for Excellence in Research and Education (PalC), Singapore, Singapore

Rainbow Tin Hung Ho Department of Social Work and Social Administration, The University of Hong Kong, Hong Kong, China

Centre on Behavioral Health, The University of Hong Kong, Hong Kong, China

Sau Po Centre on Ageing, The University of Hong Kong, Hong Kong, China

Roger Chun Man Ho Department of Psychological Medicine, National University of Singapore, Singapore, Singapore

Jan Hofer Department of Developmental Psychology, Trier University, Trier, Germany

Jaco Hoffman Ageing and Generational Dynamics in Africa (AGenDA), Optentia Research Focus Area, North-West University, Vanderbijlpark, South Africa

Oxford Institute of Population Ageing, University of Oxford, Oxford, UK

Paul Hofman FHU OncoAge, Nice, France

University Côte d'Azur, Nice, France

Hospital-Related Biobank (BB-0033-00025), Nice, France

CNRS UMR7284, Inserm U1081, Institut de Recherche sur le Cancer et le Vieillissement (IRCAN), Laboratoire de Pathologie Clinique et Expérimentale et Biobanque, Hôpital Pasteur, Nice, France

Jasmon W. T. Hoh Department of Sociology, National University of Singapore, Singapore, Singapore

Charles J. Holahan Department of Psychology, University of Texas at Austin, Austin, TX, USA

Richard Holbert Department of Psychiatry, University of Florida, Gainesville, FL, USA

Daniel Holman Department of Sociological Studies, University of Sheffield, Sheffield, UK

Sarah Holmes Doctoral Program in Gerontology, University of Maryland Baltimore, Baltimore, MD, USA

Julianne Holt-Lunstad Department of Psychology, Brigham Young University, Provo, UT, USA

Kui Hong Department of Cardiovascular Medicine, Second Affiliated Hospital of Nanchang University, Nanchang, China

Michael Hong Department of Epidemiology and Biostatistics, Western University, London, ON, Canada

Emiel O. Hoogendijk Department of Epidemiology and Biostatistics, Amsterdam UMC, VU University Medical Center, Amsterdam Public Health Research Institute, Amsterdam, The Netherlands

Karen Hooker School of Social and Behavioral Health Sciences, College of Public Health and Human Sciences, Oregon State University, Corvallis, OR, USA

Shiro Horiuchi CUNY Institute for Demographic Research, New York, NY, USA

Haiman Hou The first affiliated hospital of Zhengzhou University, Zhengzhou, China

Susan E. Howlett Department of Pharmacology, Dalhousie University, Halifax, NS, Canada

Department of Medicine (Geriatric Medicine), Dalhousie University, Halifax, NS, Canada

William G. Hoy Medical Humanities Program, Baylor University, Waco, TX, USA

Sharpley Hsieh School of Psychology, The University of Queensland, St. Lucia, QLD, Australia

Department of Psychology, Royal Brisbane and Women's Hospital, Herston, QLD, Australia

David Hsiou Psychology and Neuroscience Department, Baylor University, Waco, TX, USA

Bo Hu Care Policy and Evaluation Centre (CPEC), Department of Health Policy, London School of Economics and Political Science (LSE), London, UK

Jinzhu Hu Department of Cardiovascular Medicine, Second Affiliated Hospital of Nanchang University, Nanchang, China

Cassandra L. Hua Department of Sociology and Gerontology, Miami University, Oxford, OH, USA

Kui Huang Department of Geriatrics, Tongji Hospital, Tongji Medical College, Huazhong University of Science and Technology, Wuhan, Hubei, China

Lingli Huang Nanyang Technological University, Singapore, Singapore

Li-Hsuan Huang Department of Economics, National Central University, Chungli, Taoyuan, Taiwan

Xiao Huang Department of Cardiovascular Medicine, The Second Affiliated Hospital of Nanchang University, Nanchang, Jiangxi, China

Ying Huang Department of Demography, University of Texas at San Antonio, San Antonio, TX, USA

Ruth E. Hubbard Centre for Health Services Research (Ageing and Geriatric Medicine Research Program), Faculty of Medicine, University of Queensland, St Lucia, QLD, Australia

Princess Alexandra Hospital, Metro South Hospital and Health Service, Woolloongabba, QLD, Australia

Luke Huber Human Development and Family Studies, Iowa State University, Ames, IA, USA

Tjaša Hudobivnik Anton Trstenjak Institute of Gerontology and Intergenerational Relations, Ljubljana, Slovenia

Mark Hughes Southern Cross University, Gold Coast, QLD, Australia

Ka Yi Hui Biochemistry, Department of Biology, University of Fribourg, Fribourg, Switzerland

Martijn Huisman Department of Epidemiology and Biostatistics, Amsterdam UMC, VU University Medical Center, Amsterdam Public Health Research Institute, Amsterdam, The Netherlands

Department of Sociology, Faculty of Social Sciences, Vrije Universiteit Amsterdam, Amsterdam, The Netherlands

Andrea N. Hunt Department of Sociology and Family Studies, University of North Alabama, Florence, AL, USA

Meredith Rucker Hunter AARP, Washington, DC, USA

Bettina S. Husebo Department of Global Public Health and Primary Care, Centre for Elderly and Nursing Home Medicine, University of Bergen, Municipality of Bergen, Norway

Teresa Iannaccone Department of Medicine, Surgery and Dentistry "Scuola Medica Salernitana", University of Salerno, Salerno, Italy

Laura Iarriccio Instituto de Investigaciones Biomedicas, CSIC/UAM. CIBER de Enfermedades Raras (CIBERER), Madrid, Spain

Steve Iliffe Department of Primary Care and Population Health, University College London, London, UK

Namrah Ilyas Centre for Clinical Psychology, University of the Punjab, Lahore, Punjab, Pakistan

Zhen Jie Im Faculty of Social Sciences, University of Helsinki, Helsinki, Finland

Aprinda Indahlastari Department of Clinical and Health Psychology, College of Public Health and Health Professions, Center for Cognitive Aging and Memory, McKnight Brain Institute, University of Florida, Gainesville, FL, USA

Rebecca E. Ingram Department of Psychology, University of Colorado Colorado Springs, Colorado Springs, CO, USA

Ian Inkster Aging Analytics Agency, Biogerontology Research Foundation, London, UK

Hideki Innan Graduate University for Advanced Studies, Hayama, Kanagawa, Japan

Anthea Innes Salford Institute for Dementia, University of Salford, Salford, Manchester, UK

Yoshiko Lily Ishioka Graduate School of Science and Technology, Keio University, Yokohama, Kanagawa, Japan

Malik Itrat Department of Tahaffuzi wa Samaji Tib, National Institute of Unani Medicine, Bengaluru, Karnataka, India

Maya Izumi School of Oral Health Sciences, Faculty of Dentistry, Kyushu Dental University, Kitakyushu, Japan

Håkan Jönson School of Social Work, Lund University, Lund, Sweden

Sarah E. Jackson Department of Behavioural Science and Health, University College London, London, UK

Jacquelyn B. James Sloan Research Network on Aging and Work, School of Social Work, Boston College, Chestnut Hill, MA, USA

Elzbieta A. Jarmuzewska Department of Internal Medicine, Polyclinic IRCCS, Ospedale Maggiore, University of Milan, Milan, Italy

Domantas Jasilionis Max Planck Institute for Demographic Research, Rostock, Germany

Demographic Research Centre, Vytautas Magnus University, Kaunas, Lithuania

S. Michal Jazwinski Tulane University Health Sciences Center, New Orleans, LA, USA

Dmitri Jdanov Max Planck Institute for Demographic Research, Rostock, Germany

Higher School of Economics, National Research University, Moscow, Russia

Ilija Jeftic Department of Immunobiology and the University of Arizona Center on Aging, University of Arizona College of Medicine-Tucson, Tucson, AZ, USA

Department of Pathophysiology, Faculty of Medical Sciences, University of Kragujevac, Kragujevac, Serbia

Ana Jegundo Innovation Department, Cáritas Diocesana de Coimbra, Coimbra, Portugal

Hannah Jensen-Fielding School of Psychology, The University of Queensland, Brisbane, QLD, Australia

Bernard Jeune Danish Aging Research Center, Department of Public Health, University of Southern Denmark, Odense, Denmark

Li Li Ji Laboratory of Physiological Hygiene and Exercise Science, School of Kinesiology, University of Minnesota Twin Cities, Minneapolis, MN, USA

Yan Ji The first affiliated hospital of Zhengzhou University, Zhengzhou, China

Yinze Ji Fuwai Hospital, Chinese Academy of Medical Sciences and Peking Union Medical College, Beijing, People's Republic of China

Yu-Peng Jian Division of Cardiac Surgery, Heart center, The First Affiliated Hospital, Sun Yat-sen University, Guangzhou, China

Lin Jiang School of Social Work, University of Texas Rio Grande Valley, Edinburg, TX, USA

Quanbao Jiang Institute for Population and Development Studies, School of Public Policy and Administration, Xi'an Jiaotong University, Xi'an, China

Yizhou Jiang Faculty of Health Sciences, University of Macau, Taipa, Macau, China

Kunlin Jin Department of Pharmacology and Neuroscience, University of North Texas Health Science Center, Fort Worth, TX, USA

International Society on Aging and Disease, Fort Worth, TX, USA

Xin Jin Sau Po Centre on Ageing, Department of Social Worker and Social Administration, The University of Hong Kong, Pokfulam, Hong Kong, China

Mary Ann Johnson University of Nebraska, Lincoln, NE, USA

Florence Joly Clinical Research Department, Centre François Baclesse, Caen, France

Normandie University, UNICAEN, INSERM, ANTICIPE, Caen, France

Cancer and Cognition Platform, Ligue Nationale Contre le Cancer, Caen, France

University Hospital of Caen, Caen, France

Vanessa Joosen Department of English Literature, University of Antwerp, Antwerp, Belgium

B. H. Judd City Futures Research Centre, University of New South Wales, Sydney, NSW, Australia

Łukasz Jurek Wrocław University of Economics, Wrocław, Poland

Sindhuja Kadambi Wilmot Cancer Institute, University of Rochester Medical Center, Rochester, NY, USA

Laura Kadowaki Department of Gerontology, Simon Fraser University, Vancouver, BC, Canada

Kathleen Kahn MRC/Wits rural Public Health and Health Transitions Research Unit, School of Public Health, Faculty of Health Sciences, University of Witwatersrand, Johannesburg, South Africa

Robert M. Kaiser Department of Medicine, Division of Geriatrics and Palliative Medicine, The George Washington University School of Medicine, Washington, DC, USA

Geriatrics and Extended Care, Veterans Affairs Medical Center, Washington, DC, USA

Mary Kalfoss Faculty of Health Sciences, VID Specialized University, Oslo, Norway

Alice E. Kane Department of Genetics, Harvard Medical School, Boston, MA, USA

Charles Perkins Centre, University of Sydney, Sydney, NSW, Australia

Soon-Hock Kang School of Humanities and Behavioural Sciences, Singapore University of Social Sciences, Singapore, Singapore

Sung-wan Kang School of Social Work, Missouri State University, Springfield, MO, USA

Daniel B. Kaplan School of Social Work, Adelphi University, Garden City, NY, USA

Matthew S. Kaplan Department of Agricultural Economics, Sociology, and Education, The Pennsylvania State University, University Park, PA, USA

David Karasik Azrieli Faculty of Medicine, Bar Ilan University, Safed, Israel

Marcus Institute for Aging Research, Hebrew SeniorLife, Boston, MA, USA

Eswarappa Kasi Department of Tribal Studies, Indira Gandhi National Tribal University (IGNTU), Amarkantak, Madhya Pradesh, India

Ruth Katz Departments of Human Services, Yezreel Academic College, Yezreel Valley, Israel

The University of Haifa, Haifa, Israel

Dustin Scott Kehler Department of Medicine, Dalhousie University, Halifax, NS, Canada

Vincent W. Keng Department of Applied Biology and Chemical Technology, The Hong Kong Polytechnic University, Kowloon, Hong Kong

Brian K. Kennedy NUSMed Healthy Longevity Translational Research Programme, Yong Loo Lin School of Medicine, National University of Singapore, Singapore, Singapore

Centre for Healthy Longevity, National University Health System, Singapore, Singapore

Singapore Institute of Clinical Sciences, A*STAR, Singapore, Singapore

Rose Anne Kenny The Irish Longitudinal Study on Ageing (TILDA), Trinity College Dublin, Dublin, Ireland

Mercer's Institute for Successful Ageing (MISA), St James's Hospital, Dublin, Ireland

Abdul Haseeb Khan Human Aging Research Institute (HARI), School of Life Science, Nanchang University, Nanchang, Jiangxi, China

Jessica Khan Department of Psychiatry, University of Florida, Gainesville, FL, USA

Tariq Nadeem Khan Department of Kulliyate Tib, National Institute of Unani Medicine, Bengaluru, Karnataka, India

Y. M. Khoo School of Social Work, Saint Louis University, St. Louis, MO, USA

Eva Kiesswetter Institute for Biomedicine of Aging, Friedrich-Alexander-Universität Erlangen-Nürnberg, Nuremberg, Germany

Seoyoun Kim Department of Sociology, Texas State University, San Marcos, TX, USA

Yijung K. Kim Department of Gerontology, John W. McCormack Graduate School of Policy and Global Studies, University of Massachusetts Boston, Boston, MA, USA

Yun Jin Kim Xiamen University Malaysia, Kuala Lumpur, Malaysia

Andrew King Department of Sociology, University of Surrey, Guildford, UK

Stephanie Kirkland The University of Alabama at Birmingham, Birmingham, AL, USA

Thomas B. L. Kirkwood Institute of Cell and Molecular Biosciences, Newcastle University, Newcastle upon Tyne, UK

Institute for Ageing, Campus for Ageing and Vitality, Newcastle University, Newcastle upon Tyne, UK

Sunita Kishor The Demographic and Health Surveys (DHS) Program, ICF, Rockville, MD, USA

Miia Kivipelto Division of Clinical Geriatrics, Center for Alzheimer Research, Karolinska Institutet, Stockholm, Sweden

Stockholms Sjukhem, Research and Development Unit, Stockholm, Sweden

Institute of Public Health and Clinical Nutrition, University of Eastern Finland, Kuopio, Finland

The Ageing Epidemiology Research Unit, School of Public Health, Imperial College London, London, UK

Daniela Klaus German Centre of Gerontology (DZA), Berlin, Germany

Andrzej Klimczuk SGH Warsaw School of Economics, Warsaw, Poland

Martin Knapp Care Policy and Evaluation Centre (CPEC), Department of Health Policy, London School of Economics and Political Science (LSE), London, UK

John Knodel University of Michigan, Ann Arbor, MI, USA

Kim Knudson Department of Psychiatry, University of Florida, Gainesville, FL, USA

Pei-Chun Ko Singapore University of Social Sciences, Singapore, Singapore

Livia Kohn Department of Religion, Boston University, Boston, MA, USA

Gotaro Kojima Department of Primary Care and Population Health, University College London, London, UK

Klara Komici Department of Medicine and Health Sciences, University of Molise, Campobasso, Italy

Kathrin Komp-Leukkunen Faculty of Social Sciences, University of Helsinki, Helsinki, Finland

Jooyoung Kong School of Social Work, University of Wisconsin-Madison, Madison, WI, USA

Dorien T. A. M. Kooij Human Resource Studies, Tilburg University, Tilburg, The Netherlands

Anna E. Kornadt Bielefeld University, Bielefeld, Germany

Paul Kowal Division of Data, Analytics and Delivery for Impact, World Health Organization, Geneva, Switzerland

Axel Kowald Institute for Ageing, Campus for Ageing and Vitality, Newcastle University, Newcastle upon Tyne, UK

Łukasz Kozar University of Lodz, Lodz, Poland

Jessica N. Kraft Center for Cognitive Aging and Memory, University of Florida, Gainesville, FL, USA

J. Kravchenko Department of Surgery, Duke University School of Medicine, Durham, NC, USA

Ulla Kriebernegg Center for Inter-American Studies, University of Graz, Graz, Austria

Wui Sing Kua School of Humanities and Behavioural Sciences, Singapore University of Social Sciences, Singapore, Singapore

Fumie Kumagai Kyorin University-Inokashira Campus, Mitaka, Tokyo, Japan

Ute Kunzmann Life-Span Developmental Psychology Laboratory, University of Leipzig, Leipzig, Germany

Hoi Shan Kwan School of Life Sciences, The Chinese University of Hong Kong, Shatin, NT, Hong Kong

Rick Yiu Cho Kwan Centre for Gerontological Nursing, School of Nursing, The Hong Kong Polytechnic University, Hong Kong, China

Corinna E. Löckenhoff Department of Human Development, Cornell University/Weill Cornell Medicine, Ithaca, NY, USA

Margie E. Lachman Department of Psychology, Brandeis University, Waltham, MA, USA

Madeline R. Lag University of Colorado, Colorado Springs, Colorado Springs, CO, USA

Jenny A. Lagervall Department of Psychology, University of Colorado Colorado Springs, Colorado Springs, CO, USA

Daniel W. L. Lai Department of Applied Social Sciences, The Hong Kong Polytechnic University, Hong Kong, Hong Kong

F. H. Y. Lai Department of Rehabilitation Sciences, The Hong Kong Polytechnic University, Kowloon, Hong Kong, China

Francisca Yuen-ki Lai Center for General Education, National Tsing Hua University, Hsinchu, Taiwan

Mun Sim Lai Population Division, Department of Economic and Social Affairs, United Nations, New York, NY, USA

Wing-Fu Lai School of Life and Health Sciences, The Chinese University of Hong Kong (Shenzhen), Shenzhen, Guangdong, China

Department of Applied Biology and Chemical Technology, Hong Kong Polytechnic University, Hong Kong, China

Julia M. Laing Department of Psychiatry, University of Michigan, Ann Arbor, MI, USA

Yan Yan Nelly Lam Innovation and Technology Development Office, The Hong Kong Polytechnic University, Hung Hom, Kowloon, Hong Kong

Jack Lam Institute for Social Science Research, University of Queensland, Indooroopilly, QLD, Australia

Damon G. Lamb Center for Cognitive Aging and Memory, College of Medicine, University of Florida, Gainesville, FL, USA

Brain Rehabilitation Research Center, Malcom Randall Veterans Affairs Medical Center, Gainesville, FL, USA

Center for OCD and Anxiety Related Disorders, Department of Psychiatry, College of Medicine, University of Florida, Gainesville, FL, USA

Sarah Lamb Department of Anthropology, Brandeis University, Waltham, MA, USA

Ruth A. Lamont Psychology: Centre for Research in Ageing and Cognitive Health (REACH), University of Exeter Medical School, St. Luke's Campus, Exeter, UK

Kenneth Carl Land Department of Sociology and Social Science Research Institute, Duke University, Durham, NC, USA

Frieder R. Lang Institute of Psychogerontology, Friedrich-Alexander-Universität Erlangen-Nürnberg, Nuremberg, Germany

Marie Lange Clinical Research Department, Centre François Baclesse, Caen, France

Normandie University, UNICAEN, INSERM, ANTICIPE, Caen, France

Cancer and Cognition Platform, Ligue Nationale Contre le Cancer, Caen, France

Audrey Laporte Institute of Health Policy Management and Evaluation, Dalla Lana School of Public Health, University of Toronto, Toronto, ON, Canada

Anis Larbi Singapore Immunology Network (SIgN), Aging and Immunity Program, Agency for Science Technology and Research (A∗STAR), Singapore, Singapore

Joseph T. Lariscy Department of Sociology, University of Memphis, Memphis, TN, USA

Mary Larkin Faculty of Wellbeing, Education and Language Studies, The Open University, Milton Keynes, UK

Kenzie Latham-Mintus Department of Sociology, IUPUI, Indianapolis, IN, USA

Bobo Hi Po Lau Department of Counselling and Psychology, Hong Kong Shue Yan University, Hong Kong, Hong Kong

Johnson Y. N. Lau Avalon Genomics (Hong Kong) Limited, Shatin, Hong Kong

Department of Applied Biology and Chemical Technology, The Hong Kong Polytechnic University, Hung Hom, Hong Kong

Lok Ting Lau Department of Applied Biology and Chemical Technology, The Hong Kong Polytechnic University, Hung Hom, Kowloon, Hong Kong

Innovation and Technology Development Office, The Hong Kong Polytechnic University, Hung Hom, Kowloon, Hong Kong

Pui Yan Flora Lau Department of Sociology, Hong Kong Shue Yan University, Hong Kong, Hong Kong

Daniela Laudisio Dipartimento di Medicina Clinica e Chirurgia, Unit of Endocrinology, Federico II University Medical School of Naples, Naples, Italy

David R. Lawrence Newcastle University School of Law, Newcastle-upon-Tyne, UK

John C. Layke Beverly Hills Plastic Surgery Group, Beverly Hills, CA, USA

Patrick Lazarevič Vienna Institute of Demography/Austrian Academy of Sciences, Wittgenstein Centre (IIASA, VID/ÖAW, WU), Vienna, Austria

Philip Lazarovici School of Pharmacy, Institute for Drug Research, Faculty of Medicine, The Hebrew University of Jerusalem, Jerusalem, Israel

Joel M. Le Forestier Department of Psychology, University of Toronto, Toronto, ON, Canada

Emilie Le Rhun Department of Neurology and Brain Tumor Center, University Hospital and University of Zurich, Zurich, Switzerland

University of Lille, Lille, France

Neuro-oncology, General and Stereotaxic Neurosurgery Service, University Hospital of Lille, Lille, France

Breast Cancer Department, Oscar Lambret Center, Lille, France

Isabel Leal William James Center for Research, ISPA – Instituto Universitário, Lisbon, Portugal

Clémence Lecardonnel Geriatrics Unit, Centre Hospitalier Lyon Sud, Hospices Civils de Lyon, Pierre-Bénite, France

Geriatrics Unit, Centre Hospitalier de Vienne, Vienne, France

Stephanie Lederman American Federation for Aging Research (AFAR), New York, NY, USA

Eunji Lee School of Social Work, University of Wisconsin-Madison, Madison, WI, USA

Geok Ling Lee Department of Social Work, Faculty of Arts and Social Sciences, National University of Singapore, Singapore, Singapore

Gina Lee Iowa State University, Ames, IA, USA

Jaewon Lee School of Social Work, Michigan State University, East Lansing, MI, USA

Janet Lok Chun Lee Department of Social Work and Social Administration, The University of Hong Kong, Hong Kong, China

Jinkook Lee Center for Economic and Social Research, University of Southern California, Los Angeles, CA, USA

Department of Economics, University of Southern California, Los Angeles, CA, USA

RAND Corporation, Santa Monica, CA, USA

Kathy Lee School of Social Work, University of Texas at Arlington, Arlington, TX, USA

Sang E. Lee School of Social Work, San Jose State University, San Jose, CA, USA

Tatia M. C. Lee The State Key Laboratory of Brain and Cognitive Sciences, The University of Hong Kong, Hong Kong, Hong Kong

Laboratory of Neuropsychology, The University of Hong Kong, Hong Kong, Hong Kong

Institute of Clinical Neuropsychology, The University of Hong Kong, Hong Kong, Hong Kong

Terence Kin Wah Lee Department of Applied Biology and Chemical Technology, The Hong Kong Polytechnic University, Kowloon, Hong Kong

Wonik Lee Department of Social Welfare, Pusan National University, Busan, South Korea

Yeonjung Lee Faculty of Social Work, University of Calgary, Calgary, Alberta, Canada

George W. Leeson Oxford Institute of Population Ageing, University of Oxford, Oxford, United Kingdom

Mckay Lefler Department of Psychology, Brigham Young University, Provo, UT, USA

Wei Lei Psychiatry Department, The Affiliated Hospital, Southwest Medical University, Luzhou, China

Martina M. L. LEI The Hong Kong Polytechnic University, Hong Kong, China

Angela Y. M. Leung Centre for Gerontological Nursing, School of Nursing, The Hong Kong Polytechnic University, Hong Kong, SAR, China

Lilia S. Lens-Pechakova Life Extension Beyond Borders Association, Nice, France

Nimkit Lepcha Department of Peace and Conflict Studies and Management, Sikkim University, Gangtok, Sikkim, India

Magdalena Leszko Institute of Psychology, Szczecin, Poland

Doris Y. P. Leung The Hong Kong Polytechnic University, Hong Kong, China

Diane Levin-Zamir Department of Health Education and Promotion, Clalit Health Services, Tel Aviv, Israel

School of Public Health, University of Haifa, Haifa, Israel

Becca R. Levy Department of Social and Behavioral Sciences, Yale School of Public Health, New Haven, CT, USA

Cari Levy VA Eastern Colorado Health Care System, Aurora, CO, USA

Nicole A. Levy Columbia University, New York, NY, USA

Sheri R. Levy Department of Psychology, Stony Brook University, Stony Brook, NY, USA

Judith Lewis Centre for Housing, Urban and Regional Planning, Adelaide University, Adelaide, SA, Australia

Jordan P. Lewis WWAMI School of Medical Education, University of Alaska Anchorage, Anchorage, AK, USA

National Resource Center for Alaska Native Elders, School of Social Work, University of Alaska Anchorage, Anchorage, AK, USA

Hua-Ming Li Division of Cardiac Surgery, Heart Center, The First Affiliated Hospital, Sun Yat-sen University, Guangzhou, China

Jenny X. Li The Chinese University of Hong Kong, Hong Kong, China

Jia Li Department of Applied Social Sciences, The Hong Kong Polytechnic University, Hong Kong, China

Jie Li Department of Psychology, Renmin University of China, Beijing, China

Juxiang Li Department of Cardiovascular Medicine, Second Affiliated Hospital of Nanchang University, Nanchang, China

Ke Li School of Social Work, University of Pittsburgh, Pittsburgh, PA, USA

Libby C. W. Li Emerging viral diagnostics (HK) Limited, Hong Kong, China

Shuo Li The first affiliated hospital of Zhengzhou University, Zhengzhou, China

Ting Li Center for Population and Development Studies, Renmin University of China, Beijing, China

Yapeng Li Department of Neurology, The First Affiliated Hospital of Zhengzhou University, Zhengzhou, Henan, China

Yue Li China Population and Development Research Center, Beijing, China

Zuo-Zhi Li Department of Cardiology, Fuwai Hospital, Chinese Academy of Medical Sciences and Peking Union Medical College, Beijing, People's Republic of China

Giacinto Libertini Italian National Health Service, ASL NA2 Nord, Frattamaggiore, Italy

Department of Translational Medical Sciences, Federico II University, Naples, Italy

Stuart M. Lichtman Memorial Sloan Kettering Cancer Center, Commack, NY, USA

Christina M. Lill Genetic and Molecular Epidemiology Group, Lübeck Interdisciplinary Platform for Genome Analytics (LIGA), Institutes of Neurogenetics and Cardiogenetics, University of Lübeck, Lübeck, Germany

Ageing Epidemiology Research Unit, School of Public Health, Imperial College, London, UK

M. Fernanda Lima-Costa Departamento de Epidemiologia, Fundação Oswaldo Cruz, Belo Horizonte, Minas Gerais, Brazil

Programa de Pós Graduação em Saúde Pública, Universidade Federal de Minas Gerais, Belo Horizonte, Minas Gerais, Brazil

Hongmei Lin School of Psychological and Cognitive Sciences, Peking University, Beijing, China

Li Lin Division of Cardiology, Department of Internal Medicine, Tongji Medical College, Tongji Hospital, Huazhong University of Science and Technology, Wuhan, Hubei, China

Ziyong Lin Center for Lifespan Psychology, Max Planck Institute for Human Development, Berlin, Germany

Ulman Lindenberger Center for Lifespan Psychology, Max Planck Institute for Human Development, Berlin, Germany

Max Planck UCL Centre for Computational Psychiatry and Ageing Research, Berlin, Germany

Max Planck UCL Centre for Computational Psychiatry and Ageing Research, London, UK

Jennifer L. Lindsey Vanderbilt University Medical Center, Nashville, TN, USA

Valerie Barnes Lipscomb University of South Florida, Sarasota, FL, USA

Howard Litwin Paul Baerwald School of Social Work and Social Welfare, The Hebrew University, Jerusalem, Israel

Chang Liu Department of Applied Social Sciences, The Hong Kong Polytechnic University, Hong Kong, China

Chen Liu Department of Cardiology, Heart Center, The First Affiliated Hospital of Sun Yat-sen University, Guangzhou, People's Republic of China

Cuizhen Liu National University of Singapore, Singapore, Singapore

Edgar Liu City Futures Research Centre, Faculty of Built Environment, UNSW, Sydney, NSW, Australia

Junhao Liu Department of Risk and Insurance, Wisconsin School of Business, University of Wisconsin-Madison, Madison, WI, USA

Kai Liu Neurology Department, the First Affiliated Hospital of Zhengzhou University, Zhengzhou, Henan, China

Lei Liu Department of Psychology, College of Teacher Education, Ningbo University, Ningbo, China

Pi-Ju Liu School of Nursing, Purdue University, West Lafayette, IN, USA

Xinjing Liu Neurology Department, the First Affiliated Hospital of Zhengzhou University, Zhengzhou, Henan, China

Zhen Liu Department of Sociology, Zhejiang University, Hangzhou, China

Daniel Lloret Department of Health Psychology, UMH (Universidad Miguel Hernández), Alicante, Spain

Lilian H. Lo Department of Applied Biology and Chemical Technology, The Hong Kong Polytechnic University, Kowloon, Hong Kong

Temmy Lee Ting Lo Department of Social Work and Social Administration, The University of Hong Kong, Hong Kong, China

Dhenugen Logeswaran School of Molecular Sciences, Arizona State University, Tempe, AZ, USA

Elke Loichinger Federal Institute for Population Research, Wiesbaden, Germany

Jeanne F. Loring CSO, Aspen Neuroscience, La Jolla, CA, USA

Allura F. Lothary Psychology, North Carolina State University, Raleigh, NC, USA

Ann Therese Lotherington Centre for Women's and Gender Research and Faculty of Humanities, Social Sciences and Education, UiT The Arctic University of Norway, Tromsø, Norway

Vivian Weiqun Lou Sau Po Centre on Ageing, Department of Social Worker and Social Administration, The University of Hong Kong, Pokfulam, Hong Kong, China

Peter Louras Palo Alto University, Palo Alto, CA, USA

Gail Low Faculty of Nursing, Edmonton Clinical Health Academy, University of Alberta, Edmonton, AB, Canada

Kelvin E. Y. Low Department of Sociology, National University of Singapore, Singapore, Singapore

Timothy Low Singapore Management University, Singapore, Singapore

Ariela Lowenstein Center for Research and Study of Aging, Graduate Department of Gerontology, Faculty of Health and Welfare, The University of Haifa, Haifa, Israel

Siyao Lu Department of Sociology, National University of Singapore, Singapore, Singapore

Shuyu Lu School of Library and Information Studies, The University of Oklahoma, Norman, OK, USA

Yi Lu School of Psychological and Cognitive Sciences, Peking University, Beijing, China

J. Lubben Boston College, Chestnut Hill, MA, USA

Albert Lukas Competence Center of Geriatric Medicine, Helios Medical Center Bonn/Rhein-Sieg, Academic Teaching Hospital, University Bonn, Bonn, Germany

Linfei Luo Department of Cardiovascular Medicine, The Second Affiliated Hospital of Nanchang University, Nanchang, Jiangxi, China

Ye Luo Department of Sociology, Anthropology and Criminal Justice, Clemson University, Clemson, SC, USA

G. Luscombe The AGEncy Project, Sydney, NSW, Australia

Alexandre A. Lussier Department of Molecular Biology and Genetics, Cornell University, Ithaca, NY, USA

Department of Computational Biology, Cornell University, Ithaca, NY, USA

Mary A. Luszcz College of Education, Psychology and Social Work, Flinders University, Adelaide, SA, Australia

Emily Lux School of Social Work, University of Illinois at Urbana-Champaign, Urbana, IL, USA

Marc Luy Vienna Institute of Demography/Austrian Academy of Sciences, Wittgenstein Centre (IIASA, VID/ÖAW, WU), Vienna, Austria

Faqin Lv Department of Ultrasonography, Hainan Hospital of Chinese People's Liberation Army General Hospital, Sanya, China

K. G. Lyamzaev A.N. Belozersky Institute of Physico-Chemical Biology and Faculty of Bioengineering and Bioinformatics, Institute of Mitoengineering, M.V. Lomonosov Moscow State University, Moscow, Russia

Scott M. Lynch Department of Sociology, Duke University, Durham, NC, USA

Heidi A. Lyons Department of Sociology, Anthropology, Social Work and Criminal Justice, Oakland University, Rochester, MI, USA

Ashley Lytle College of Arts and Letters, Stevens Institute of Technology, Hoboken, NJ, USA

Flora Ma Palo Alto University, Palo Alto, CA, USA

Ning Ma Department of Urology, The First Hospital, Jilin University, Changchun, China

Xiaolu Ma College of Biomedical Engineering, Taiyuan University of Technology, Taiyuan, China

Yue-Dong Ma Department of Cardiology, Heart Center, The First Affiliated Hospital of Sun Yat-sen University, Guangzhou, People's Republic of China

Jamie L. Macdonald Department of Psychology, Stony Brook University, Stony Brook, NY, USA

Elizabeth MacKinlay School of Theology, Charles Sturt University, Canberra, ACT, Australia

Samaneh Madanian School of Engineering, Computer and Mathematical Sciences, Auckland University of Technology, Auckland, New Zealand

Marina Maffoni Department of Brain and Behavioral Sciences, University of Pavia, Pavia, Italy

Stefania Maggi National Research Council, Neuroscience Institute, Padova, Italy

Raul Magni-Berton Sciences Po Grenoble, Université Grenoble-Alpes, PACTE, Grenoble, France

Allison Magnuson Wilmot Cancer Institute, University of Rochester Medical Center, Rochester, NY, USA

Atiya Mahmood Department of Gerontology, Simon Fraser University, Vancouver, BC, Canada

Roberta Maierhofer Center for Inter-American Studies, University of Graz, Graz, Austria

Christine A. Mair Department of Sociology, Anthropology, and Public Health, University of Maryland, Baltimore County, Baltimore, MD, USA

Aaron Maitland Division of Health Interview Statistics, National Center for Health Statistics, Hyattsville, MD, USA

Mui Hing June Mak Department of Cultural and Religious Studies, The Chinese University of Hong Kong, Hong Kong, China

Patrick Mallea Ingénieur et doctorat Mines Paris Tech/centre de recherche sur les risques et les crises, Groupe NEHS, Paris, France

Francesca Mangialasche Division of Clinical Geriatrics and Aging Research Center, Center for Alzheimer Research, Karolinska Institutet, Stockholm, Sweden

Arduino A. Mangoni Discipline of Clinical Pharmacology, College of Medicine and Public Health, Flinders University and Flinders Medical Centre, Adelaide, Australia

Cristina Manguan-García Experimental Models for Human Diseases, Instituto de Investigaciones Biomedicas CSIC/UAM, CIBER de Enfermedades raras (CIBERER), Madrid, Spain

Norma Mansor Faculty of Economics and Administration, University of Malaya, Kuala Lumpur, Malaysia

Social Wellbeing Research Centre (SWRC), University of Malaya, Kuala Lumpur, Malaysia

Valentina Manzo Department of Medicine, Surgery and Dentistry "Scuola Medica Salernitana", University of Salerno, Salerno, Italy

Luigi Marano Department of Medicine, Surgery and Neurosciences, Unit of General Surgery and Surgical Oncology, University of Siena, Siena, Italy

Christopher Steven Marcum National Institutes of Health, Bethesda, MD, USA

Kyriakos Markides University of Texas Medical Branchx, Galveston, TX, USA

Michael Marsiske Department of Clinical and Health Psychology, University of Florida, Gainesville, FL, USA

Ricardo Iván Martínez-Zamudio Department of Microbiology, Biochemistry, and Molecular Genetics and Center for Cell Signaling, Rutgers Biomedical and Health Sciences, New Jersey Medical School, Rutgers University, Newark, NJ, USA

Finbarr C. Martin Population Health Sciences, King's College London, London, UK

Kelsey Martin Alliance for Aging Research, Washington, DC, USA

Peter Martin Iowa State University, Ames, IA, USA

Susanne Martin École Polytechnique Fédérale de Lausanne, Lausanne, Switzerland

Puts Martine Canada Research Chair in the Care for Frail Older persons, Lawrence S. Bloomberg Faculty of Nursing, University of Toronto, Toronto, Canada

Emanuele Marzetti Institute of Internal Medicine and Geriatrics, Università Cattolica del Sacro Cuore, Rome, Italy

Fondazione Policlinico Universitario "Agostino Gemelli" IRCCS, Rome, Italy

Anne R. Mason Centre for Health Economics, University of York, York, UK

Tahir Masud Geriatric Medicine, Nottingham University Hospitals NHS Trust, Queens Medical Centre, Nottingham, UK

Rikyia Matsukura College of Economics, Nihon University, Tokyo, Japan

Leonardo De Mattos Istituto Italiano di Tecnologia, Genoa, Italy

C. Matz-Costa Boston College, Chestnut Hill, MA, USA

Justin T. Max Capital One Financial Corporation, Richmond, VA, USA

Andrea Mayo Department of Medicine, Dalhousie University, Halifax, NS, Canada

Faculty of Health, Dalhousie University, Halifax, NS, Canada

Mark T. Mc Auley Faculty of Science and Engineering, Department of Chemical Engineering, University of Chester, Chester, UK

Janet S. McCord Department of Thanatology, Marian University, Fond du Lac, WI, USA

Lynn McDonald University of Toronto, Toronto, ON, Canada

Janet E. McElhaney Health Sciences North Research Institute, Sudbury, ON, Canada

Christine A. McGarrigle The Irish Longitudinal Study on Ageing (TILDA), Trinity College Dublin, Dublin, Ireland

Dennis McGonagle Section of Musculoskeletal Disease, Leeds Institute of Molecular Medicine, University of Leeds, NIHR Leeds Musculoskeletal Biomedical Research Unit, Chapel Allerton Hospital, Leeds, UK

June M. McKoy Division of General Internal Medicine and Geriatrics, Northwestern University Feinberg School of Medicine, Chicago, IL, USA

Logan McLeod Department of Economics, Wilfrid Laurier University, Waterloo, ON, Canada

Patrizia Mecocci Institute of Gerontology and Geriatrics, Department of Medicine, University of Perugia and Clinical Unit of Geriatrics, S. Maria della Misericordia Hospital, Perugia, Italy

Kate de Medeiros Department of Sociology and Gerontology, Miami University, Oxford, OH, USA

Anthony Medford Interdisciplinary Centre on Population Dynamics, University of Southern Denmark, Odense, Denmark

Raquel Medina Languages and Social Sciences, Aston University, Birmingham, UK

Magdalena Medrano-Ramos Departamento de Salud Pública, Centro Universitario de Ciencias de la Salud, Universidad de Guadalajara, Guadalajara, Jalisco, Mexico

R. J. F. Melis Department of Geriatrics/Radboudumc Alzheimer Center, Radboud University Medical Center, Radboud Institute for Health Sciences, Nijmegen, The Netherlands

Sara Melo Instituto Superior de Serviço Social do Porto (ISSSP), Instituto de Sociologia da Universidade do Porto (ISUP), Porto, Portugal

Sandra Mendes Universidade de Trás-os-Montes e Alto Douro (UTAD), Centro de Investigação e Estudos de Sociologia (CIES-ISCTE-IUL), Vila Real, Portugal

Neyda Ma. Mendoza-Ruvalcaba Health Sciences Division, Universidad de Guadalajara CUTONALA, Guadalajara, Jalisco, Mexico

C. Mertens Service d'Oncogériatrie, Institut Bergonié, Bordeaux, France

Giorgio Metta Istituto Italiano di Tecnologia, Genoa, Italy

S. F. Metzelthin Health Services Research, CAPHRI – Ageing and Long-Term Care, Academische Werkplaats Ouderenzorg, Maastricht University, Maastricht, The Netherlands

Tanguy Meunier Department of Geriatrics, Geriatric Oncology Unit, APHP, Paris Cancer Institute CARPEM, Hôpital Europeen Georges Pompidou, Paris, France

Université de Paris, Paris, France

Alejandra Michaels Obregon Sealy Center on Aging, University of Texas Medical Branch, Galveston, TX, USA

Sylwia Sulimiera Michalak Faculty of Medicine and Health Sciences, Department of Pharmacology and Toxicology, University of Zielona Góra, Zielona Góra, Poland

C. Michel Medical Oncology Department, Centre Antoine Lacassagne, FHU OncoAge, Université Côte d'Azur, Nice, France

Luís Midão UCIBIO, REQUIMTE and Faculty of Pharmacy, University of Porto, Porto, Portugal

Roberto J. Millar Gerontology Doctoral Program, University of Maryland Baltimore, Baltimore, MD, USA

Lois Miller University of Wisconsin – Madison, Madison, WI, USA

Tim Miller Population and Development, United Nations, New York, NY, USA

Vivian J. Miller Department of Social Work, College of Human Services, Bowling Green State University, Bowling Green, OH, USA

Elena Milova Life Extension Advocacy Foundation, New York City, NY, USA

Beyon Miloyan Department of Psychology, School of Health and Life Sciences, Federation University, Ballarat, VIC, Australia

Chivon A. Mingo Gerontology Institute, Georgia State University, Atlanta, Georgia

Mario G. Mirisola Medical Oncology Unit, DiChirOnS, Medical School, Università di Palermo, Palermo, Italy

Farhaan Mirza School of Engineering, Computer and Mathematical Sciences, Auckland University of Technology, Auckland, New Zealand

Barbara A. Mitchell Department of Sociology/Anthropology and Department of Gerontology, Simon Fraser University, Burnaby, BC, Canada

Leander K. Mitchell School of Psychology, University of Queensland, Brisbane, Australia

Melissa Mitchell Global Coalition on Aging, New York, NY, USA

Zhi-Wei Mo Division of Cardiac Surgery, Heart Center, The First Affiliated Hospital, Sun Yat-sen University, Guangzhou, People's Republic of China

Sanjay Kumar Mohanty Department of Fertility Studies, International Institute for Population Sciences, Mumbai, MH, India

Supriya G. Mohile Wilmot Cancer Institute, University of Rochester Medical Center, Rochester, NY, USA

Md. Awal Hossain Mollah Department of Public Administration, University of Rajshahi, Rajshahi, Bangladesh

Md Nurul Momen Department of Public Administration, University of Rajshahi, Rajshahi, Bangladesh

Caitlin Monahan Department of Psychology, Stony Brook University, Stony Brook, NY, USA

Livia Montana Harvard Center for Population and Development Studies, Harvard University, Cambridge, MA, USA

Henri Montaudié Department of Dermatology, Université Nice Côte d'Azur, Centre Hospitalier Universitaire, Nice, France

INSERM, U1065, The Mediterranean Center for Molecular Medicine Team 12, Nice, France

Rudolf H. Moos Department of Psychiatry and Behavioral Sciences, Stanford University, Stanford, CA, USA

Amílcar Moreira Institute of Social Sciences, University of Lisbon, Lisbon, Portugal

Masanori Mori Palliative Care Team, Seirei Mikatahara General Hospital, Hamamatsu, Shizuoka, Japan

Juan A. Moriano Department of Social and Organizational Psychology, UNED (Universidad Nacional de Educación a Distancia), Madrid, Madrid, Spain

Alan Morris Institute for Public Policy and Governance, University of Technology Sydney, Sydney, NSW, Australia

Brian J. Morris Basic and Clinical Genomics Laboratory, School of Medical Sciences and Bosch Institute, Sydney Medical School, University of Sydney, Sydney, NSW, Australia

Honolulu Heart Program (HHP)/Honolulu-Asia Aging Study (HAAS), Department of Research, Kuakini Medical Center, Honolulu, HI, USA

Department of Geriatric Medicine, John A. Burns School of Medicine, University of Hawaii, Honolulu, HI, USA

Rebecca S. Morse Department of Thanatology, Marian University, Fond du Lac, WI, USA

Paul Moss Institute of Immunology and Immunotherapy, University of Birmingham, Birmingham, West Midlands, UK

Loic Mourey Department of Medical Oncology, Institut Claudius Regaud, Institut Universitaire du Cancer Toulouse Oncopole, Toulouse, France

Laurence D. Mueller Department of Ecology and Evolutionary Biology, University of California, Irvine, CA, USA

Anita Mukherjee Department of Risk and Insurance, Wisconsin School of Business, University of Wisconsin-Madison, Madison, WI, USA

Ludmila Müller Center for Lifespan Psychology, Max Planck Institute for Human Development, Berlin, Germany

Cynthia A. Munro Department of Psychiatry and Behavioral Sciences, Department of Neurology, Johns Hopkins School of Medicine, Baltimore, MD, USA

Anjana Muralidharan Mental Illness Research, Education, and Clinical Center (MIRECC), Veterans Affairs Capitol Health Care Network (VISN 5), Baltimore, MD, USA

Department of Psychiatry, University of Maryland School of Medicine, Baltimore, MD, USA

Aidan J. Murphy Center for Cognitive Aging and Memory, College of Medicine, University of Florida, Gainesville, FL, USA

Kaitlin Murray Department of Gerontology, Simon Fraser University, Vancouver, BC, Canada

Giovanna Muscogiuri Dipartimento di Medicina Clinica e Chirurgia, Unit of Endocrinology, Federico II University Medical School of Naples, Naples, Italy

Laura Naegele Department of Ageing and Work, Institute of Gerontology, University of Vechta, Vechta, Germany

Nirmala Naidoo Division of Data, Analytics and Delivery for Impact, World Health Organization, Geneva, Switzerland

Sigal Pearl Naim Departments of Human Services, Yezreel Academic College, Yezreel Valley, Israel

Department of Gerontology, Ben-Gurion University of the Negev, Beer Sheva, Israel

Priya Nambisan Department of Health Informatics and Administration, University of Wisconsin – Milwaukee, Milwaukee, WI, USA

Joshua Kin-man Nan Hong Kong Baptist University, Hong Kong, China

Christopher M. Napolitano Department of Educational Psychology, University of Illinois at Urbana-Champaign, Champaign, IL, USA

Department of Psychology, Developmental Psychology: Adulthood, University of Zurich, Zurich, Switzerland

Josefina N. Natividad Population Institute – College of Social Sciences and Philosophy, University of the Philippines Diliman, Quezon City, Philippines

Ana Patricia Navarrete-Reyes Department of Geriatrics, Instituto Nacional de Ciencias Médicas y Nutrición Salvador Zubirán, Mexico City, Mexico

Humaira Naz Centre for Clinical Psychology, University of the Punjab, Lahore, Punjab, Pakistan

Vir Singh Negi Department of Clinical Immunology, Jawaharlal Institute of Post-graduate Medical Education and Research, Pondicherry, India

Yan Yan Nelly Lam Innovation and Technology Development Office, The Hong Kong Polytechnic University, Hung Hom, Kowloon, Hong Kong

Peter Nelson University of Kentucky, Lexington, KY, USA

Marília R. Nepomuceno Research Group on Lifespan Inequalities, Max-Planck-Institute for Demographic Research, Rostock, Germany

Charlotte E. Neville Centre for Public Health, Queen's University Belfast, Institute of Clinical Sciences, Belfast, UK

Ka Lun Ng Geriatric Medicine, Nottingham University Hospitals NHS Trust, Queens Medical Centre, Nottingham, UK

Tiia Ngandu Public Health Promotion Unit, Finnish Institute for Health and Welfare, Helsinki, Finland

Hoa Hong Nguyen School of Engineering, Computer and Mathematical Sciences, Auckland University of Technology, Auckland, New Zealand

Uyen-Sa D. T. Nguyen Department of Biostatistics and Epidemiology, School of Public Health, University of North Texas Health Science Center, Fort Worth, TX, USA

Georgeta Niculescu Romanian Association for Person Centred Psychotherapy (ARPCP), Bucharest, Romania

Tuomo Nieminen Päijät-Häme Central Hospital, Lahti, Finland

Janko Nikolich-Zugich Department of Immunobiology and the University of Arizona Center on Aging, University of Arizona College of Medicine-Tucson, Tucson, AZ, USA

Nicole R. Nissim Department of Neuroscience, McKnight Brain Institute, University of Florida, Gainesville, FL, USA

Department of Clinical and Health Psychology, College of Public Health and Health Professions, Center for Cognitive Aging and Memory, McKnight Brain Institute, University of Florida, Gainesville, FL, USA

Olivia Noel Department of Psychology, University of Colorado at Colorado Springs, Colorado Springs, CO, USA

Michael S. North Stern School of Business, New York University, New York, NY, USA

Mark Novak California State University, Bakersfield, Bakersfield, CA, USA

Andrew M. O'Shea Department of Clinical and Health Psychology, College of Public Health and Health Professions, Center for Cognitive Aging and Memory, McKnight Brain Institute, University of Florida, Gainesville, FL, USA

Vanessa Obetz University of Colorado Colorado Springs, Department of Psychology, Colorado Springs, CO, USA

Burcu Özdemir Ocaklı Faculty of Health Sciences, Department of Social Work, Ankara University, Ankara, Turkey

Mary Beth Ofstedal Health and Retirement Study, Survey Research Center, Institute for Social Research, University of Michigan, Ann Arbor, MI, USA

Natalia Ojeda Del Pozo Department of Methods and Experimental Psychology, University of Deusto, Bilbao, Spain

Michele L. Okun BioFrontiers Center, University of Colorado Colorado Springs, Colorado Springs, CO, USA

Marcel G. M. Olde Rikkert Department of Geriatric Medicine, Radboud University Medical Center, Nijmegen, The Netherlands

Rhoda Olkin Clinical Psychology Program, California School of Professional Psychology at Alliant International University, Emeryville, CA, USA

Alexey M. Olovnikov Institute of Biochemical Physics, Russian Academy of Sciences, Moscow, Russia

National Medical Research Center for Obstetrics, Gynecology and Perinatology named after Academician V. I. Kulakov of the Ministry of Health of the Russian Federation, Moscow, Russia

S. Jay Olshansky School of Public Health, University of Illinois at Chicago, Chicago, IL, USA

Anthony D. Ong Department of Human Development, Cornell University, Ithaca, NY, USA

Carlos Orihuela Department of Microbiology, School of Medicine, The University of Alabama at Birmingham, Birmingham, AL, USA

Karina Orozco-Rocha Faculty of Economics, Universidad de Colima, Colima, COL, Mexico

Maricel Oró-Piqueras Faculty of Arts, Universitat de Lleida, Lleida, Spain

Sofia Ortet Innovation Department, Cáritas Diocesana de Coimbra, Coimbra, Portugal

Ulrich Orth University of Bern, Bern, Switzerland

C. Ortholan Service d'oncologie – radiothérapie, Centre Hospitalier Princesse-Grace, Monaco, France

Jing-Song Ou Division of Cardiac Surgery, Heart center, The First Affiliated Hospital, Sun Yat-sen University, Guangzhou, China

Xiaoting Ou Department of Applied Social Sciences, The Hong Kong Polytechnic University, Hong Kong, Hong Kong

Zhi-Jun Ou Division of Hypertension and Vascular Diseases, Heart center, The First Affiliated Hospital, Sun Yat-sen University, Guangzhou, China

Nadine Ouellette Department of Demography, Université de Montréal, Montreal, QC, Canada

Matiss Ozols Division of Cell Matrix Biology and Regenerative Medicine, School of Biological Sciences, Faculty of Biology, Medicine and Health, The University of Manchester, Manchester, UK

Nancy A. Pachana School of Psychology, The University of Queensland, Brisbane, QLD, Australia

Milind M. Padki Vetek - the Movement for Longevity and Quality of Life (US Branch), Teaneck, NJ, USA

Adaixa Padron Department of Clinical and Health Psychology, University of Florida, Gainesville, FL, USA

Katia Padvalkava Oxford Internet Institute, University of Oxford, Oxford, UK

Elena Paillaud Department of Geriatrics, Geriatric Oncology Unit, APHP, Paris Cancer Institute CARPEM, Hôpital Europeen Georges Pompidou, Paris, France

Clinical Epidemiology and Ageing Unit, EA 7376, Université Paris-Est, Créteil, France

Karen Pak Human Resource Studies, Tilburg University, Tilburg, The Netherlands

Human Resource Management, HAN University of Applied Sciences, Arnhem/Nijmegen, The Netherlands

Ximena Palacios-Espinosa Department of Psychology, Universidad del Rosario, Bogotá, Colombia

Xi Pan Texas State University, San Marcos, TX, USA

Sophie Panel Sciences Po Bordeaux/Centre Emile Durkheim, PACTE, Bordeaux, France

Laurence Lloyd Parial Centre for Gerontological Nursing, School of Nursing, The Hong Kong Polytechnic University, Hong Kong, SAR, China

Marius D. Pascariu Biometric Risk Modelling Chapter, SCOR Global Life SE, Paris, France

University of Southern Denmark, Odense, Denmark

Ian Patterson The University of Queensland, School of Tourism, Queensland, Australia

William R. Patterson Alexandria, VA, USA

Salvin Paul Department of Peace and Conflict Studies and Management, Sikkim University, Gangtok, Sikkim, India

Graham Pawelec Department of Immunology, University of Tübingen, Tübingen, Germany

Health Sciences North Research Institute, Sudbury, ON, Canada

Nancy L. Pedersen Department of Medical Epidemiology and Biostatistics, Karolinska Institute, Stockholm, Sweden

Sara Pedro da Silva ISPA – Instituto Universitário, Lisbon, Portugal

Lu-lu Pei The First Affiliated Hospital of Zhengzhou University, Zhengzhou, China

Rong Peng National Economics Research Center and School of Economics, Guangdong University of Finance and Economics, Guangzhou, Guangdong, China

Charles Pennell Global Coalition on Aging, New York, NY, USA

Margaret J. Penning Department of Sociology, University of Victoria, Victoria, BC, Canada

Deidre B. Pereira Department of Clinical and Health Psychology, University of Florida, Gainesville, FL, USA

Jolanta Perek-Białas Institute of Sociology, Jagiellonian University, Cracow, Poland

Warsaw School of Economics, Warsaw, Poland

Arokiasamy Perianayagam Department of Development Studies, International Institute for Population Sciences, Mumbai, MH, India

Fatima Perkins Western Reserve Area Agency on Aging, Cleveland, OH, USA

Rosario Perona Experimental Models for Human Diseases, Instituto de Investigaciones Biomedicas CSIC/UAM, CIBER de Enfermedades raras (CIBERER), Madrid, Spain

Pasqualina Perrig-Chiello Department of Psychology, University of Bern, Bern, Switzerland

Daniel Perry Alliance for Aging Research, Washington, DC, USA

Sean Perryman Center for Cognitive Aging and Memory, Department of Clinical and Health Psychology, McKnight Brain Institute, University of Florida, Gainesville, FL, USA

Melita Peršolja University of Primorska, Department of Nursing, Faculty of Health Sciences, Nova Gorica, Slovenia

Janette Perz Translational Health Research Institute, School of Medicine, Western Sydney University, Sydney, NSW, Australia

Susan Peschin Alliance for Aging Research, Washington, DC, USA

Gabrielle N. Pfund Department of Psychological and Brain Sciences, Washington University in St. Louis, St. Louis, MO, USA

Lorna Philip Department of Geography and Environment, School of Geosciences, University of Aberdeen, Aberdeen, UK

David R. Phillips Department of Sociology and Social Policy, Lingnan University, Hong Kong, China

Drystan Phillips Center for Economic and Social Research, University of Southern California, Los Angeles, CA, USA

Kristin Phillips VA Ann Arbor Healthcare System, Geriatric Research, Education and Clinical Center (GRECC), Ann Arbor, MI, USA

Chris Phillipson Manchester Institute for Collaborative Research on Ageing (MICRA), The University of Manchester, Manchester, UK

Anna Picca Institute of Internal Medicine and Geriatrics, Università Cattolica del Sacro Cuore, Rome, Italy

Fondazione Policlinico Universitario "Agostino Gemelli" IRCCS, Rome, Italy

Susan Pickard Department of Sociology, Social Policy and Criminology, School of Law and Social Justice, University of Liverpool, Liverpool, UK

John A. J. Pickering Department of Geography, Simon Fraser University, Burnaby, BC, Canada

Marissa A. Pifer Department of Psychology, University of Colorado at Colorado Springs, Colorado Springs, CO, USA

Alberto Pilotto Department of Geriatric Care, Orthogeriatrics and Rehabilitation, Galliera Hospital, Genoa, Italy

Department of Interdisciplinary Medicine, University of Bari, Bari, Italy

Martin Pinquart Psychology, Philipps University, Marburg, Germany

Martin P. Piotrowski Department of Sociology, The University of Oklahoma, Norman, OK, USA

Christopher I. Platt Division of Cell Matrix Biology and Regenerative Medicine, School of Biological Sciences, Faculty of Biology, Medicine and Health, The University of Manchester, Manchester, UK

Johanne Poisson Department of Geriatrics, Geriatric Oncology Unit, APHP, Paris Cancer Institute CARPEM, Hôpital Europeen Georges Pompidou, Paris, France

Université de Paris, Paris, France

Courtney A. Polenick Department of Psychiatry, University of Michigan Medical School, Ann Arbor, MI, USA

M. Cristina Polidori Ageing Clinical Research, Department II of Internal Medicine and Center for Molecular Medicine Cologne, University of Cologne, Faculty of Medicine and University Hospital Cologne, Cologne, Germany

CECAD, University of Cologne, Faculty of Medicine and University Hospital Cologne, Cologne, Germany

Anne Poljak Mark Wainwright Analytical Centre, University of New South Wales, Sydney, NSW, Australia

School of Medical Sciences, University of New South Wales, Sydney, NSW, Australia

Leonard W. Poon University of Georgia, Athens, GA, USA

Eric C. Porges Department of Clinical and Health Psychology, College of Public Health and Health Professions, Center for Cognitive Aging and Memory, McKnight Brain Institute, University of Florida, Gainesville, FL, USA

Eric S. Porges Department of Clinical and Health Psychology, College of Public Health and Health Professions, Center for Cognitive Aging and Memory, McKnight Brain Institute, University of Florida, Gainesville, FL, USA

Conceição Portela ByCaring.PT, Parede, Portugal

Wiraporn Pothisiri College of Population Studies, Chulalongkorn University, Bangkok, Thailand

Lucia Pozzi Department of Economics and Business, University of Sassari, Sassari, Italy

Catherine Price Department of Clinical and Health Psychology, College of Public Health and Health Professions, University of Florida, Gainesville, FL, USA

Victoria Prieto-Rosas Universidad de la República, Montevideo, Uruguay

Michael A. Province Division of Statistical Genomics, Department of Genetics, Washington University School of Medicine, St. Louis, MO, USA

Alexia Prskawetz Wittgenstein Centre of Demography and Global Human Capital (IIASA, VID/ÖAW, WU), Wien, Austria

Institute of Statistics and Mathematical Methods in Economics, Research Unit Economics, TU Wien, Vienna, Austria

Gabriella Pugliese Dipartimento di Medicina Clinica e Chirurgia, Unit of Endocrinology, Federico II University Medical School of Naples, Naples, Italy

Tom Pullum The Demographic and Health Surveys (DHS) Program, ICF, Rockville, MD, USA

Gregory L. Purser School of Social Work, Louisiana State University, Baton Rouge, LA, USA

Norella M. Putney Department of Sociology, California State University-Los Angeles, Los Angeles, CA, USA

Siyu Qian Centre for IT-enabled Transformation, School of Computing and Information Technology, Faculty of Engineering and Information Sciences, University of Wollongong, Wollongong, NSW, Australia

Li Qiu New York, NY, USA

Lien Quach Boston VA Research Institute, Inc., Boston, MA, USA

Department of Gerontology, University of Massachusetts Boston, Boston, MA, USA

Anabel Quan-Haase Western University, London, ON, Canada

Bernardo Lanza Queiroz Department of Demography, Universidade Federal de Minas Gerais, Belo Horizonte, Brazil

Michael J. Rae SENS Research Foundation, Mountain View, CA, USA

Md. Mostafijur Rahman Department of Law, Prime University, Dhaka, Bangladesh

Annalise M. Rahman-Filipiak Department of Psychiatry, University of Michigan, Ann Arbor, MI, USA

Maheema Rai Gangtok, Sikkim, India

Srilakshmi M. Raj Department of Molecular Biology and Genetics, Cornell University, Ithaca, NY, USA

Tata Cornell Institute, Cornell University, Ithaca, NY, USA

Luisa Ramírez Department of Psychology, Universidad del Rosario, Bogotá, Colombia

Ana Ramovš Anton Trstenjak Institute of Gerontology and Intergenerational Relations, Ljubljana, Slovenia

Jože Ramovš Anton Trstenjak Institute of Gerontology and Intergenerational Relations, Ljubljana, Slovenia

Ksenija Ramovš Anton Trstenjak Institute of Gerontology and Intergenerational Relations, Ljubljana, Slovenia

Marta Ramovš Anton Trstenjak Institute of Gerontology and Intergenerational Relations, Ljubljana, Slovenia

Rammohan V. Rao California College of Ayurveda, Nevada City, CA, USA

Nur Fakhrina Ab Rashid Social Wellbeing Research Centre (SWRC), University of Malaya, Kuala Lumpur, Malaysia

Barbara Ratzenboeck Center for Inter-American Studies, University of Graz, Graz, Austria

Roland Rau University of Rostock, Rostock, Germany

Max Planck Institute for Demographic Research, Rostock, Germany

Rachel S. Rauvola Department of Psychology, Saint Louis University, Saint Louis, MO, USA

Tenko Raykov Measurement and Quantitative Methods, Michigan State University, East Lansing, MI, USA

George W. Rebok Department of Mental Health and Center on Aging and Health, Johns Hopkins University, Baltimore, MD, USA

Emily Reeve College of Medicine, University of Saskatchewan, Saskatoon, SK, Canada

NHMRC Cognitive Decline Partnership Centre, Kolling Institute of Medical Research, Northern Clinical School, Faculty of Medicine and Health, University of Sydney, Sydney, NSW, Australia

Geriatric Medicine Research and College of Pharmacy, Dalhousie University and Nova Scotia Health Authority, Halifax, NS, Canada

Colin D. Rehm Office of Community and Population Health, Montefiore Medical Center, Bronx, NY, USA

Department of Epidemiology and Population Health, Albert Einstein College of Medicine, Bronx, NY, USA

M. Carrington Reid Division of Geriatrics, Weill Cornell Medical College, New York City, NY, USA

Eric N. Reither Department of Sociology, Social Work and Anthropology, Utah State University, Logan, UT, USA

Samuèle Rémillard-Boilard Manchester Institute for Collaborative Research on Ageing (MICRA), The University of Manchester, Manchester, UK

Adriana M. Reyes Policy Analysis and Management, Cornell University, Ithaca, NY, USA

Contributors

Carla Ribeirinho Universidade de Lisboa, Institute of Social and Political Sciences, Lisbon, Portugal

Oscar Ribeiro Center for Health Technology and Services Research (CINTESIS), Porto/Aveiro, Portugal

Department of Education and Psychology, University of Aveiro, Aveiro, Portugal

Anita Richert-Kaźmierska Faculty of Management and Economics, Gdansk University of Technology, Gdansk, Poland

Ronald Richman Johannesburg, South Africa

Lauren Richmond Department of Psychology, Stony Brook University, Stony Brook, NY, USA

Gwenola Ricordeau California State University, Chico, CA, USA

Tim Riffe Laboratory of Population Health, Max-Planck-Institute for Demographic Research, Rostock, Germany

Jürgen A. Ripperger Biochemistry, Department of Biology, University of Fribourg, Fribourg, Switzerland

Justin Roberts The Cambridge Centre for Sport and Exercise Sciences, Anglia Ruskin University, Cambridge, UK

Jean-Marie Robine Institut National de la Santé et de la Recherche Médicale (INSERM) and Ecole Pratique des Hautes Etudes (EPHE), Paris and Montpellier, France

Stephanie A. Robinson Center for Healthcare Organization and Implementation Research, Department of Veterans Affairs, Edith Nourse Rogers Memorial Veterans Hospital, Bedford, MA, USA

K. Rockwood Division of Geriatric Medicine, Dalhousie University, Halifax, NS, Canada

Rafael Rofman The World Bank, Buenos Aires, Argentina

Wendy A. Rogers Department of Kinesiology and Community Health, College of Applied Health Sciences, University of Illinois at Urbana-Champaign, Champaign, IL, USA

Jason Roh Corrigan Minehan Heart Center, Division of Cardiology, Department of Internal Medicine, Massachusetts General Hospital, Boston, USA

Harvard Medical School, Boston, MA, USA

Kaitlyn P. Roland Institute on Aging and Lifelong Health, University of Victoria, Victoria, BC, Canada

Darryl B. Rolfson Geriatric Medicine, University of Alberta, Edmonton, AB, Canada

Joana Rolo ISPA – Instituto Universitário, Lisbon, Portugal

Roman Romero-Ortuno Trinity College Dublin, Discipline of Medical Gerontology, Mercer's Institute for Successful Ageing, St James's Hospital, Dublin, Ireland

Global Brain Health Institute, Lloyd Building, Trinity College Dublin, Dublin, Ireland

Walter Rosenberg Social Work and Community Health, Rush University Medical Center, Chicago, IL, USA

Anthony Rosenzweig Corrigan Minehan Heart Center, Division of Cardiology, Department of Internal Medicine, Massachusetts General Hospital, Boston, USA

Harvard Medical School, Boston, MA, USA

Luis Rosero-Bixby Centro Centroamericano de Población, Universidad de Costa Rica, San Jose, Costa Rica

Jörg Rössel Institute of Sociology, University of Zurich, Zurich, Switzerland

Klaus Rothermund Institut für Psychologie, Friedrich-Schiller-Universität Jena, Jena, Germany

John W. Rowe Columbia University, New York, NY, USA

Theresa A. Rowe Division of General Internal Medicine and Geriatrics, Northwestern University Feinberg School of Medicine, Chicago, IL, USA

Raquel Royo Faculty of Social and Human Sciences, University of Deusto, Bilbao, Spain

Lei Ruan Department of Geriatrics, Tongji Hospital, Tongji Medical College, Huazhong University of Science and Technology, Wuhan, Hubei, China

Maksim Rudnev Instituto Universitário de Lisboa (ISCTE-IUL), CIS-IUL, Lisbon, Portugal

National Research University Higher School of Economics, Moscow, Russia

Cort W. Rudolph Department of Psychology, Saint Louis University, Saint Louis, MO, USA

Carmelinda Ruggiero Department of Medicine, Medical Faculty, Geriatric Institute, S. Maria della Misericordia Hospital, University of Perugia, Perugia, Italy

Steven Ruggles Institute for Social Research and Data Innovation, University of Minnesota, Minneapolis, MN, USA

Fiona S. Rupprecht Institute of Psychogerontology, Friedrich-Alexander-Universität Erlangen-Nürnberg, Nuremberg, Germany

Jane E. Ruseski Department of Economics, West Virginia University, Morgantown, WV, USA

David Russell Department of Sociology, Appalachian State University, Boone, NC, USA

Kimberley Ruxton Department of Orthopaedic and Trauma Surgery and Discipline of Clinical Pharmacology, College of Medicine and Public Health, Flinders University and Flinders Medical Centre, Adelaide, Australia

Amanda M. Rymal California State University, San Bernardino, CA, USA

Albert Sabater Faculty of Business and Economic Sciences, University of Girona, Girona, Spain

Luis Saboga-Nunes Institute of Sociology, University of Education, Freiburg, Germany

Institute for Evidence Based Medicine, Faculdade de Medicina, Universidade de Lisboa, Lisbon, Portugal

Public Health Research Centre, Universidade NOVA de Lisboa, Lisbon, Portugal

Coimbra Polytechnic Institute, Coimbra, Portugal

Perminder Sachdev Centre for Healthy Brain Ageing, School of Psychiatry, University of New South Wales, Sydney, NSW, Australia

Neuropsychiatric Institute, Euroa Centre, Prince of Wales Hospital, Sydney, NSW, Australia

Muhammad Saeed School of Life Science, Human Aging Research Institute (HARI), Nanchang University, Nanchang, Jiangxi, China

Atrayee Saha Department of Sociology, Muralidhar Girls' College, Kolkata, West Bengal, India

Diana Sahrai Pädagogische Hochschule Fachhochschule Nordwestschweiz, Institut Spezielle Pädagogik und Psychologie, Basel, Switzerland

K. Sako Faculty of Economics, Keio University, Tokyo, Japan

Maria Jimena Salcedo-Arellano Medical Investigation of Neurodevelopmental Disorders (MIND) Institute, University of California Davis, Sacramento, CA, USA

Department of Pediatrics, University of California Davis School of Medicine, Sacramento, CA, USA

Carlos Alberto Aguilar Salinas Unidad de Investigación en Enfermedades Metabólicas, Instituto Nacional de Ciencias Médicas y Nutrición Salvador Zubirán, Mexico City, Mexico

Departamento de Endocrinología y Metabolismo, Instituto Nacional de Ciencias Médicas y Nutrición Salvador Zubirán, Mexico City, Mexico

Escuela de Medicina y Ciencias de la Salud, Tecnologico de Monterrey, Monterrey, Nuevo Leon, Mexico

Mavis Salt Aged Care Plus, The Salvation Army, Redfern, NSW, Australia

Paul Salvin Department of Peace and Conflict Studies and Management, Sikkim University, Gangtok, Sikkim, India

Joze Sambt School of Economics and Business, University of Ljubljana, Ljubljana, Slovenia

Mariano Sánchez Department of Sociology, University of Granada, Granada, Spain

Miguel Sánchez-Romero Wittgenstein Centre of Demography and Global Human Capital (IIASA, VID/ÖAW, WU), Wien, Austria

Institute of Statistics and Mathematical Methods in Economics, Research Unit Economics, TU Wien, Vienna, Austria

Linn J. Sandberg School of Culture and Education, Södertörn University, Stockholm, Sweden

Karen Sands Ageless Way Academy and Sands and Associates, LLC, Roxbury, CT, USA

Roger Sapsford Aberdeenshire, Scotland, UK

Nazmi Sari Department of Economics, University of Saskatchewan, Saskatoon, SK, Canada

Doğa Başar Sariipek Labour Economics and Industrial Relations Department, Kocaeli University, Kocaeli, Turkey

Sisira Sarma Department of Epidemiology and Biostatistics, Western University, London, ON, Canada

Institute for Clinical Evaluative Sciences, Toronto, ON, Canada

Lawson Health Research, London, ON, Canada

Leandro Sastre Experimental Models for Human Diseases, Instituto de Investigaciones Biomedicas CSIC/UAM, CIBER de Enfermedades raras (CIBERER), Madrid, Spain

Jessica M. Sautter University of the Sciences, Philadelphia, PA, USA

Benjamin Schüz Institute of Public Health and Nursing Research, University of Bremen, Bremen, Germany

Annette Scherpenzeel Economics of Ageing, Technical University Munich, Munich, Germany

Jordana E. Schiralli Department of Psychology, University of Toronto, Toronto, ON, Canada

Joergen Schlundt School of Chemical and Biomedical Engineering, Nanyang Technological University (NTU), Singapore, Singapore

Nanyang Technological University Food Technology Centre (NAFTEC), Nanyang Technological University (NTU), Singapore, Singapore

Josef Schmid Institute of Political Science, Tübingen University, Tübingen, Germany

Wyatt Schmitz Department of Sociology, The University of Oklahoma, Norman, OK, USA

Eva-Luisa Schnabel Network Aging Research, Heidelberg University, Heidelberg, Germany

Lauren M. Schneider Clinical Psychology, University of Colorado at Colorado Springs, Colorado Springs, CO, USA

Yvonne Schoon Department of Geriatric Medicine, Radboud University Medical Center, Nijmegen, The Netherlands

Josje D. Schoufour Faculty of Sports and Nutrition and Faculty of Health, Center of Expertise Urban Vitality, Amsterdam University of Applied Sciences, Amsterdam, The Netherlands

Johannes Schröder Section of Geriatric Psychiatry, Heidelberg University, Heidelberg, Germany

Anastacia Schulhoff Department of Sociology, Appalachian State University, Boone, NC, USA

John G. Schumacher University of Maryland, Baltimore County (UMBC), Baltimore, MD, USA

Ella Schwartz Israel Gerontological Data Center, Paul Baerwald School of Social Work and Social Welfare, The Hebrew University of Jerusalem, Jerusalem, Israel

School of Social Work, Bar-Ilan University, Ramat Gan, Israel

Susan Wile Schwarz Global Coalition on Aging, New York, NY, USA

Dietrich Schwela Stockholm Environment Institute, Dept. of Environment and Geography, University of York, York, UK

Stacey Scott Department of Psychology, Stony Brook University, Stony Brook, NY, USA

Bonnie M. Scott Department of Clinical and Health Psychology, College of Public Health and Health Professions, University of Florida, Gainesville, FL, USA

Julia E. T. Scott College of Education, Psychology and Social Work, Flinders University, Adelaide, SA, Australia

Kiersten Scott Psychology and Neuroscience Department, Baylor University, Waco, TX, USA

F. Scotté Department of Medical Oncology and Supportive Care, Foch Hospital, Suresnes, France

Michael Scullin Psychology and Neuroscience Department, Baylor University, Waco, TX, USA

Kishore Seetharaman Department of Gerontology, Simon Fraser University, Vancouver, BC, Canada

Daniel L. Segal Department of Psychology, University of Colorado at Colorado Springs, Colorado Springs, CO, USA

Alexander Seifert Institute of Sociology, University of Zurich, Zurich, Switzerland

University Research Priority Program (URPP) Dynamics of Healthy Aging, University of Zurich, Zurich, Switzerland

Rajagopal V. Sekhar Section of Endocrinology, Diabetes and Metabolism, Baylor College of Medicine, Houston, TX, USA

T. V. Sekher Department of Population Policies and Programmes, International Institute for Population Sciences, Mumbai, MH, India

Erik Selecky Center for Continuing Education, Technical University, Zvolen, Slovakia

Maha Sellami Padeh and Ziv Hospitals, Azrieli Faculty of Medicine, Bar-Ilan University, Ramat Gan, Israel

Rodrigo Serrat University of Barcelona, Barcelona, Spain

Sanket Shah Department of Clinical Immunology, Jawaharlal Institute of Post-graduate Medical Education and Research, Pondicherry, India

Jacob Shane Brooklyn College and The Graduate Center of the City University of New York, New York, NY, USA

Rhonda Shaw School of Psychology, Charles Sturt University, Port Macquarie, NSW, Australia

Deirdre M. Shea Center for Cognitive aging and Memory, Department of Aging and Geriatric Research, McKnight Brain Institute, University of Florida, Gainesville, FL, USA

Department of Clinical and Health Psychology, University of Florida, Gainesville, FL, USA

Yat-Fung Shea Department of Medicine, Li Ka Shing Faculty of Medicine, Queen Mary Hospital, University of Hong Kong, Hong Kong, Hong Kong

Jie Shen College of Life Information Science and Instrument Engineering, Hangzhou Dianzi University, Hangzhou, China

Ke Shen School of Social Development and Public Policy, Fudan University, Shanghai, China

Michael J. Sherratt Division of Cell Matrix Biology and Regenerative Medicine, School of Biological Sciences, Faculty of Biology, Medicine and Health, The University of Manchester, Manchester, UK

Mao-Mao Shi Division of Cardiac Surgery, Heart Center, The First Affiliated Hospital, Sun Yat-sen University, Guangzhou, China

Ida Shiang HelpAge USA, Washington, DC, USA

Chonggak Shin Korea Employment Information Service, Seoul, Republic of Korea

Vladimir Shkolnikov Max Planck Institute for Demographic Research, Rostock, Germany

Higher School of Economics, National Research University, Moscow, Russia

Ismail Shogo Centre for Family and Population Research, National University of Singapore, Singapore, Singapore

Destin D. Shortell Department of Clinical and Health Psychology, College of Public Health and Health Professions, Center for Cognitive Aging and Memory, McKnight Brain Institute, University of Florida, Gainesville, FL, USA

Binbin Shu The Chinese University of Hong Kong, Hong Kong, China

Olga S. Shubernetskaya National Medical Research Center for Obstetrics, Gynecology and Perinatology named after Academician V. I. Kulakov of the Ministry of Health of the Russian Federation, Moscow, Russia

Clayton J. Shuman University of Michigan School of Nursing, Ann Arbor, MI, USA

Olga Yurievna Shvarova New Medicine Partners, London, UK

Cornel C. Sieber Institute for Biomedicine of Aging, Friedrich-Alexander-Universität Erlangen-Nürnberg, Nuremberg, Germany

Jelena S. Siebert Institute of Psychology, Heidelberg University, Heidelberg, Germany

Janet Sigal Fairleigh Dickinson University, Teaneck, NJ, USA

International Council of Psychologists Tantra Park Circle, Boulder, CO, USA

Jussiane G. Silva Department of Statistics, Universidade Federal de Minas Gerais, Belo Horizonte, Brazil

Marta Silva Faculty of Health Sciences, Centre of Reproduction, Development and Aging, Institute of Translational Medicine, University of Macau, Macau SAR, China

Sara Silva ISPA- Instituto Universitário, Lisbon, Portugal

Michelle Pannor Silver University of Toronto, Toronto, ON, Canada

María Silvestre Faculty of Social and Human Sciences, University of Deusto, Bilbao, Spain

Peter Simonsen University of Southern Denmark, Odense, Denmark

Julia Simonson German Centre of Gerontology (DZA), Berlin, Germany

Pachitjanut Siripanich School of Applied Statistics, National Institute of Development Administration, Bangkok, Thailand

V. P. Skulachev A.N. Belozersky Institute of Physico-Chemical Biology and Faculty of Bioengineering and Bioinformatics, Institute of Mitoengineering, M.V. Lomonosov Moscow State University, Moscow, Russia

Lee Smith The Cambridge Centre for Sport and Exercise Sciences, Anglia Ruskin University, Cambridge, UK

Matthew Lee Smith Center for Population Health and Aging, Texas A&M University, College Station, TX, USA

Department of Environmental and Occupational Health, School of Public Health, Texas A&M University, College Station, TX, USA

Department of Health Promotion and Public Health, College of Public Health, The University of Georgia, Athens, GA, USA

Sherri L. Smith Department of Head and Neck Surgery and Communication Sciences, Duke University, Durham, NC, USA

Department of Population Health Sciences, Duke University, Durham, NC, USA

Center for the Study of Aging and Human Development, Duke University, Durham, NC, USA

Michael A. Smyer Bucknell University, Lewisburg, PA, USA

Anne Snick SAPIRR—Systems Approach to Public Innovation and Responsible Research, Oud-Heverlee, Belgium

Matthew Sobek Institute for Social Research and Data Innovation, University of Minnesota, Minneapolis, MN, USA

Anna S. Solovieva National Medical Research Center for Obstetrics, Gynecology and Perinatology named after Academician V. I. Kulakov of the Ministry of Health of the Russian Federation, Moscow, Russia

Susan B. Somers International Network for the Prevention of Elder Abuse, Nassau, NY, USA

Joonmo Son Department of Sociology, National University of Singapore, Singapore, Singapore

Samir Soneji Department of Health Behavior, University of North Carolina, Chapel Hill, NC, USA

Bo Song The First Affiliated Hospital of Zhengzhou University, Zhengzhou, China

Ting Song Centre for Digital Transformation, School of Computing and Information Technology, Faculty of Engineering and Information Sciences, University of Wollongong, Wollongong, NSW, Australia

Illawarra Health and Medical Research Institute, University of Wollongong, Wollongong, NSW, Australia

Wen-Xia Song Department of Pathology, The First Hospital of Jilin University, Changchun, Jilin, China

Yajun Song Department of Social Work, East China University of Science and Technology, Shanghai, China

Zhi-Ming Song Department of Sports Medicine, The First Hospital of Jilin University, Changchun, Jilin, China

Amanda Sonnega Health and Retirement Study, Survey Research Center, Institute for Social Research, University of Michigan, Ann Arbor, MI, USA

Silvia Sörensen Department of Counseling and Human Development, Warner School for Education and Human Development, University of Rochester, Rochester, NY, USA

Enrique Soto-Perez-de-Celis Department of Geriatrics, Instituto Nacional de Ciencias Médicas y Nutrición Salvador Zubirán, Mexico City, Mexico

Pierre Soubeyran Department of Medical Oncology, Institut Bergonié, Université de Bordeaux, Bordeaux Cedex, France

Pierre-Louis Soubeyran Department of Hematology, Institut Bergonié, Inserm U1218, SIRIC BRIO, Université de Bordeaux, Bordeaux, France

Kathrin Speh Department of Psychology, Universität Konstanz, Konstanz, Germany

Naomi J. Spence Department of Sociology, Lehman College, City University of New York, Bronx, NY, USA

Dominik Spira Department of Endocrinology and Metabolic Diseases, Charité – Universitätsmedizin Berlin, Berlin, Germany

Eric Stallard Biodemography of Aging Research Unit at the Social Science Research Institute, Duke University, Durham, NC, USA

Ilia Stambler Science, Technology and Society, Bar Ilan University, Ramat Gan, Israel

Ursula M. Staudinger Robert N. Butler Columbia Aging Center, Columbia University, New York, NY, USA

Department of Sociomedical Sciences, Mailman School of Public Health, Columbia University, New York, NY, USA

Laetitia Stefani CarMEN Laboratory of Lyon University, INSERM U.1060/ Université Lyon1/INRA U. 1397/INSA Lyon/Hospices Civils Lyon, Pierre-Bénite, France

Sara E. Stemen Department of Sociology and Gerontology, Miami University, Oxford, OH, USA

Yannick Stephan University of Montpellier, Montpellier, France

Christine V. Stephens Massey University, Palmerston North, New Zealand

Sanchez Stéphane Responsable de l'unité de recherche Clinique, Centre Hospitalier de Troyes, Troyes, France

Andrew Steptoe Department of Epidemiology and Public Health, University College London, London, UK

Department of Behavioral Science and Health, University College London, London, UK

Bruce A. Stevens Theology, Charles Sturt University, Barton, ACT, Australia

Martina Štípková Department of Sociology, University of West Bohemia, Pilsen, Czech Republic

Jeffrey E. Stokes Department of Gerontology, John W. McCormack Graduate School of Policy and Global Studies, University of Massachusetts Boston, Boston, MA, USA

Lisa E. Stone Department of Psychology, University of Colorado at Colorado Springs, Colorado Springs, CO, USA

Timo E. Strandberg University of Helsinki, Helsinki University Hospital, Helsinki, Finland

Center for Life Course Health Research, University of Oulu, Oulu, Finland

Barbara Strasser Medical School, Sigmund Freud Private University, Vienna, Austria

John Strauss Department of Economics, University of Southern California, Los Angeles, CA, USA

John Strauss University of Southern California, Los Angeles, CA, USA

Justyna Stypinska Free University Berlin, Berlin, Germany

Fei Sun School of Social Work, Michigan State University, East Lansing, MI, USA

Sai Sun Division of Biology and Biological Engineering/Computation and Neural Systems, California Institute of Technology, Pasadena, CA, USA

School of Psychology, Center for Studies of Psychological Application and Key Laboratory of Mental Health and Cognitive Science of Guangdong Province, South China Normal University, Guangzhou, China

Zhijun Sun Department of Cardiovascular Medicine, The First Medical Center of Chinese PLA General Hospital, Beijing, China

Pildoo Sung Centre for Ageing Research and Education, Duke-NUS Medical School, Singapore, Singapore

Zainab Suntai Silver School of Social Work, New York University, New York, NY, USA

Uma Suryadevara Department of Psychiatry, University of Florida, Gainesville, FL, USA

Malcom Randall VA Medical Center, Gainesville, FL, USA

Ajda Svetelšek Anton Trstenjak Institute of Gerontology and Intergenerational Relations, Ljubljana, Slovenia

Emma Swain HelpAge International, London, UK

Hannah J. Swift Centre for the Study of Group Processes, School of Psychology, Keynes College, University of Kent, Canterbury, Kent, UK

Aagje Swinnen Maastricht University and University of Humanistic Studies, Utrecht, The Netherlands

Kathy Sykes USEPA, Washington, DC, USA

Zofia Szweda-Lewandowska Collegium of Socio-Economics, Warsaw School of Economics, Warsaw, Poland

Gunhild Tøndel NTNU Social Research/Unit of Diversity and Inclusion, Trondheim, Norway

Annika Taghizadeh Larsson Department of Social and Welfare Studies, Linköping University, Norrköping, Sweden

Deborah M. S. Tai Emerging viral diagnostics (HK) Limited, Hong Kong, China

Xiujun Tai School of Economics and Management, Shanxi Normal University, Linfen City, China

Tiina Tambaum Estonian Institute for Population Studies, Tallinn University, Tallinn, Estonia

Fengyan Tang School of Social Work, University of Pittsburgh, Pittsburgh, PA, USA

Jennifer Yee Man Tang Sau Po Centre on Ageing, The University of Hong Kong, Hong Kong, China

Suqin Tang Department of Sociology, Law School, Shenzhen University, Shenzhen, China

Geraldine Tan-Ho Psychology Programme, School of Social Sciences, Nanyang Technological University, Singapore, Singapore

Samia Tasmim Department of Sociology, University of Memphis, Memphis, TN, USA

Brad Taylor Department of Clinical and Health Psychology, University of Florida, Gainesville, FL, USA

Department of Public Health and Health Professions, University of Florida, Gainesville, FL, USA

Bussarawan Teerawichitchainan Department of Sociology and the Centre for Family and Population Research, National University of Singapore, Singapore, Singapore

Soraia Teles Center for Health Technology and Services Research (CINTESIS), Porto/Aveiro, Portugal

Department of Behavioral Sciences, Institute of Biomedical Sciences Abel Salazar, University of Porto (ICBAS-UP), Porto, Portugal

Fei Teng School of Psychology, South China Normal University, Guangzhou, China

Jeremy Yuen Chun Teoh S. H. Ho Urology Centre, Department of Surgery, Prince of Wales Hospital, The Chinese University of Hong Kong, Hong Kong, China

Clemens Tesch-Römer German Centre of Gerontology (DZA), Berlin, Germany

Furtuna Tewolde Department of Clinical and Health Psychology, Center for Cognitive Aging and Memory, McKnight Brain Institute, University of Florida-Gainesville, Gainesville, FL, USA

Leng Leng Thang Department of Japanese Studies, National University of Singapore, Singapore, Singapore

Duanpen Theerawanviwat School of Applied Statistics, National Institute of Development Administration, Bangkok, Thailand

Olga Theou Department of Medicine, Dalhousie University, Halifax, NS, Canada

Hein Thet Ssoe HelpAge International Myanmar, Nay Pyi Taw, Myanmar

Kelsey R. Thomas Department of Psychiatry, University of California, San Diego School of Medicine, La Jolla, CA, USA

VA San Diego Healthcare System, San Diego, CA, USA

Sable Thompson Department of Clinical and Health Psychology, College of Public Health and Health Professions, University of Florida, Gainesville, FL, USA

Xiao-Li Tian School of Life Science, Human Aging Research Institute (HARI), Nanchang University, Nanchang, Jiangxi, China

Erik J. Timmermans Department of Epidemiology and Biostatistics, Amsterdam UMC, VU University Medical Center, Amsterdam Public Health Research Institute, Amsterdam, The Netherlands

Emily Kate Timms School of English, University of Leeds, Leeds, UK

Stephen Tollman MRC/Wits rural Public Health and Health Transitions Research Unit, School of Public Health, Faculty of Health Sciences, University of Witwatersrand, Johannesburg, South Africa

Cecilia Tomassini Department of Economics, University of Molise, Campobasso, Italy

Yuying Tong Department of Sociology, The Chinese University of Hong Kong, Sha Tin, Hong Kong

Catalina Torres Interdisciplinary Centre on Population Dynamics (CPOP), University of Southern Denmark, Odense, Denmark

Lois C. Towart Faculty of Design Architecture and Building, University of Technology Sydney, Ultimo, NSW, Australia

Laura Tradii Department of Social Anthropology, University of Cambridge, Cambridge, UK

Kelsey R. Traeger Center for Cognitive Aging and Memory, Department of Clinical and Health Psychology, McKnight Brain Institute, University of Florida, Gainesville, FL, USA

Thanh V. Tran Boston College School of Social Work, Chestnut Hill, MA, USA

Antonia Trichopoulou Hellenic Health Foundation, Athens, Greece

Erin Trifilio Department of Clinical and Health Psychology, University of Florida, Gainesville, FL, USA

Parker Y. L. Tsang Emerging Viral Diagnostics (HK) Limited, Hong Kong, China

Nancy B. Y. Tsui Avalon Genomics (Hong Kong) Limited, Shatin, Hong Kong

Shelbie Turner School of Social and Behavioral Health Sciences, College of Public Health and Human Sciences, Oregon State University, Corvallis, OR, USA

Laura Upenieks Department of Sociology, University of Texas at San Antonio, San Antonio, TX, USA

Bogusława Urbaniak University of Lodz, Lodz, Poland

Jane M. Ussher Translational Health Research Institute, School of Medicine, Western Sydney University, Sydney, NSW, Australia

Carlos Ernesto Vázquez-Arias Universidad de Guadalajara, Guadalajara, Jalisco, Mexico

Matthew Vail Social Work and Community Health, Rush University Medical Center, Chicago, IL, USA

Marie Valéro Geriatrics Unit, Centre Hospitalier Lyon Sud, Hospices Civils de Lyon, Pierre-Bénite, France

H. P. J. van Hout Amsterdam UMC, Amsterdam Public Health, EMGO+, Amsterdam, The Netherlands

Willeke van Staalduinen AFEdemy, Ltd., Gouda, The Netherlands

Quentin Vanhaelen Insilico Medicine Inc., Rockville, MD, USA

Maria Varlamova Marie Skłodowska-Curie Actions Innovative Training Network EuroAgeism, Institute of Sociology, Jagiellonian University, Cracow, Poland

Christin-Melanie Vauclair Instituto Universitário de Lisboa (ISCTE-IUL), CIS-IUL, Lisbon, Portugal

Haydeé Cristina Verduzco-Aguirre Department of Hemato-Oncology, Instituto Nacional de Ciencias Médicas y Nutrición Salvador Zubirán, Mexico City, Mexico

Nicola Veronese National Research Council, Neuroscience Institute, Aging Branch, Padova, Italy

Mary Jo Vetter NYU Rory Meyers College of Nursing, New York, NY, USA

Lucie Vidovićová Research Institute for Labour and Social Affairs, Brno, Czech Republic

Edgar Ramos Vieira Department of Physical Therapy, Florida International University, Miami, FL, USA

Alba Vieira Campos Memory Unit, Neurology Department, Hospital de La Princesa, Madrid, Spain

Kriti Vikram Department of Sociology, National University of Singapore, Singapore, Singapore

Abhijit Visaria Centre for Ageing Research and Education, Duke-NUS Medical School, Singapore, Singapore

Athina Vlachantoni Centre for Research on Ageing and ESRC Centre for Population Change, Faculty of Social Sciences, University of Southampton, Southampton, UK

Claudia Vogel German Centre of Gerontology (DZA), Berlin, Germany

Dorothee Volkert Institut for Biomedicine of Aging, University Erlangen-Nürnberg, Nürnberg, Germany

Sofia von Humboldt William James Center for Research, ISPA – Instituto Universitário, Lisbon, Portugal

Allyssa A. Wadsworth Department of Economics, West Virginia University, Morgantown, WV, USA

Gert G. Wagner Center for Lifespan Psychology, Max Planck Institute for Human Development, Berlin, Germany

German Socio-Economic Panel Study (SOEP), Berlin, Germany

Hans-Werner Wahl Institute of Psychology, Heidelberg University, Heidelberg, Germany

Network Aging Research, Heidelberg University, Heidelberg, Germany

Linda J. Waite Department of Sociology, University of Chicago, Chicago, IL, USA

Allison Walden University of Colorado at Colorado Springs, Colorado Springs, CO, USA

Richard F. Walker ProSoma Biosciences LLC, Indian Rocks Beach, FL, USA

Wendell C. Wallace Centre for Criminology and Criminal Justice, Department of Behavioural Sciences, The University of the West Indies, St. Augustine, Trinidad and Tobago

C. L. Wallace School of Social Work, Saint Louis University, St. Louis, MO, USA

L. M. K. Wallace Division of Geriatric Medicine, Dalhousie University, Halifax, NS, Canada

Karen Walton School of Medicine, Faculty of Science, Medicine and Health, University of Wollongong, Wollongong, NSW, Australia

Adrian H. Y. Wan Department of Social Work and Social Adminsitration, The University of Hong Kong, Hong Kong, China
Centre on Behavioral Health, The University of Hong Kong, Hong Kong, China

Haitao Wang Department of Neuropharmacology and Drug Discovery, School of Pharmaceutical Sciences, Southern Medical University, Guangzhou, China

Huijun Wang National Institute for Nutrition and Health, Chinese Center for Disease Control and Prevention, Beijing, China

Jianyun Wang Faculty of Economics and Management, East China Normal University, Shanghai, China

Jing Wang Fudan University School of Nursing, Shanghai, China

Lina Wang School of Medicine, Huzhou University, Huzhou Central Hospital, Huzhou, Zhejiang, China

Lina Wang School of Public Policy and Administration, Xi'an Jiaotong University, Xi'an, China

Ning Wang School of Nursing, Southern Medical University, Guangzhou, People's Republic of China

Ning Wang School of Social and Public Administration, East China University of Science and Technology, Shanghai, China

Q. R. Wang Faculty of Social Sciences, The University of Hong Kong, Hong Kong, China

Wenjuan Wang The Demographic and Health Surveys (DHS) Program, ICF, Rockville, MD, USA

Yafeng Wang Institute of Social Science Survey, Peking University, Beijing, China

Zhiyi Wang School of Business, University of Michigan at Ann Arbor, Ann Arbor, MI, USA

David F. Warner Department of Sociology, University of Alabama at Birmingham, Birmingham, AL, USA

Karla T. Washington Department of Family and Community Medicine, University of Missouri, Columbia, MO, USA

Tiffany R. Washington School of Social Work, University of Georgia, Athens, Georgia

Abdulla Watad Department of Medicine 'B', Sheba Medical Center, Tel-Hashomer and Sackler Faculty of Medicine, Tel Aviv University, Tel Aviv, Israel

Section of Musculoskeletal Disease, Leeds Institute of Molecular Medicine, University of Leeds, NIHR Leeds Musculoskeletal Biomedical Research Unit, Chapel Allerton Hospital, Leeds, UK

Mary Miu Yee Waye The Nethersole School of Nursing, The Croucher Laboratory for Human Genomics, The Chinese University of Hong Kong, Hong Kong, China

Helen Wear London Deanery Postgraduate Training Programme, London, UK

Julie D. Weeks Division of Analysis and Epidemiology, National Center for Health Statistics, Hyattsville, MD, USA

Heminxuan Wei Renmin University of China, Beijing, China

Wei Wei Department of Urology, The First Hospital, Jilin University, Changchun, China

Joyce Weil College of Natural and Health Sciences, University of Northern Colorado, Greeley, CO, USA

Maxine Weinstein Center for Population and Health, Graduate School of Arts and Sciences, Georgetown University, Washington, DC, USA

David Weir Health and Retirement Study, Survey Research Center, Institute for Social Research, University of Michigan, Ann Arbor, MI, USA

David Weiss Institute of Psychology, Leipzig University, Leipzig, Germany

Michael Weller Department of Neurology and Brain Tumor Center, University Hospital and University of Zurich, Zurich, Switzerland

Barry Wellman NetLab Network, Toronto, ON, Canada

Jessica S. West Department of Sociology, Duke University, Durham, NC, USA

Center for the Study of Aging and Human Development, Duke University, Durham, NC, USA

Sally West Age UK, London, UK

Markus Wettstein German Centre of Gerontology (DZA), Berlin, Germany

Keith Wheaton Department of Biochemistry, Immunology and Microbiology, Faculty of Medicine, University of Ottawa, Ottawa, ON, Canada

Emily E. Wiemers Department of Public Administration and International Affairs and Aging Studies Institute, Syracuse University, Syracuse, NY, USA

Margaret E. Wiggins Department of Clinical and Health Psychology, College of Public Health and Health Professions, University of Florida, Gainesville, FL, USA

Hans Wildiers University Hospitals Leuven and Katholieke Universiteit Leuven (KULeuven), Leuven, Belgium

Jenny Wilkens Center for Economic and Social Research, University of Southern California, Los Angeles, CA, USA

Scott E. Wilks School of Social Work, Louisiana State University, Baton Rouge, LA, USA

Bradley J. Willcox Honolulu Heart Program (HHP)/Honolulu-Asia Aging Study (HAAS), Department of Research, Kuakini Medical Center, Honolulu, HI, USA

Department of Geriatric Medicine, John A. Burns School of Medicine, University of Hawaii, Honolulu, HI, USA

John B. Williamson Center for Cognitive Aging and Memory, College of Medicine, University of Florida, Gainesville, FL, USA

Brain Rehabilitation Research Center, Malcom Randall Veterans Affairs Medical Center, Gainesville, FL, USA

Center for OCD and Anxiety Related Disorders, Department of Psychiatry, College of Medicine, University of Florida, Gainesville, FL, USA

Sherry L. Willis Department of Psychiatry and Behavioral Sciences; Department of Psychology, University of Washington, Seattle, WA, USA

Donna M. Wilson Faculty of Nursing, University of Alberta, Edmonton, AB, Canada

Leslie A. Winters Human Development and Family Studies, Iowa State University, Ames, IA, USA

Rainer Wirth Department for Geriatric Medicine, Ruhr-University Bochum, Herne, Germany

Jacek M. Witkowski Department of Pathophysiology, Medical University of Gdańsk, Gdańsk, Poland

Firman Witoelar Australian National University, Canberra, ACT, Australia

Raphael Wittenberg Care Policy and Evaluation Centre (CPEC), Department of Health Policy, London School of Economics and Political Science (LSE), London, UK

Anita Wohlmann Department of English and Linguistics, Obama Institute for Transnational American Studies, Johannes Gutenberg University, Mainz, Germany

Mary K. Wojczynski Division of Statistical Genomics, Department of Genetics, Washington University School of Medicine, St. Louis, MO, USA

Joseph D. Wolfe The University of Alabama at Birmingham, Birmingham, AL, USA

Eliza M. L. Wong Centre for Gerontological Nursing, School of Nursing, The Hong Kong Polytechnic University, Hong Kong, SAR, China

Esther Wong NUSMed Healthy Longevity Translational Research Programme, Yong Loo Lin School of Medicine, National University of Singapore, Singapore, Singapore

Centre for Healthy Longevity, National University Health System, Singapore, Singapore

Gloria Wong Department of Social Work and Social Administration, The University of Hong Kong, Hong Kong, China

Rebeca Wong Sealy Center on Aging, University of Texas Medical Branch, Galveston, TX, USA

University of Texas Medical Branch, University Boulevard, Galveston, Texas, USA

Wai-Tat Wong Department of Anaesthesia and Intensive Care, The Chinese University of Hong Kong, Hong Kong, China

John L. Woodard Wayne State University, Detroit, MI, USA

Adam J. Woods Department of Clinical and Health Psychology, College of Public Health and Health Professions, Center for Cognitive Aging and Memory, McKnight Brain Institute, University of Florida, Gainesville, FL, USA

Department of Neuroscience, University of Florida, Gainesville, FL, USA

Christine Woods School of Special Education, School Psychology and Early Childhood Studies, University of Florida, Gainesville, FL, USA

Dorian R. Woods Nijmegen School of Management, Radboud University, Nijmegen, Gelderland, The Netherlands

David Kenneth Wright School of Nursing, Faculty of Health Sciences, University of Ottawa, Ottawa, Canada

Matthew R. Wright Department of Criminology, Sociology, and Geography, Arkansas State University, Jonesboro, AR, USA

Anise M. S. Wu Department of Psychology, Faculty of Social Sciences (FSS), University of Macau, Macau, China

Bei Wu Rory Meyers College of Nursing and NYU Aging Incubator, New York University, New York, NY, USA

Liyun Wu School of Social Work, Norfolk State University, Norfolk, VA, USA

Qiao Wu Leonard Davis School of Gerontology, University of Southern California, Los Angeles, CA, United States

Shuyi Wu Cambridge Center for Ageing and Neuroscience, University of Cambridge, Cambridge, UK

Susanne Wurm Institute of Psychogerontology, Friedrich-Alexander-Universität Erlangen-Nürnberg (FAU), Nürnberg, Germany

Benlu Xin Business School, Jilin University, Changchun, Jilin, China

Hanzhang Xu Department of Family Medicine and Community Health, Duke University, Durham, NC, USA

School of Nursing, Duke University, Durham, NC, USA

Hongwei Xu Department of Sociology, CUNY-Queens College, Queens, NY, USA

Qingwen Xu School of Social Work, New York University, New York, NY, USA

Shuping Xu Department of Social Work, School of Social Development, East China University of Political Science and Law, Shanghai, China

Xin Yi Xu Centre for Gerontological Nursing, School of Nursing, The Hong Kong Polytechnic University, Hong Kong, SAR, China

Yuming Xu Neurology Department, The First Affiliated Hospital of Zhengzhou University, Zhengzhou, Henan, China

Zhenyan Xu Department of Cardiovascular Medicine, Second Affiliated Hospital of Nanchang University, Nanchang, China

Takashi Yamashita Department of Sociology, Anthropology, and Health Administration and Policy, University of Maryland Baltimore County, Baltimore, MD, USA

Elsie Yan Department of Applied Social Sciences, The Hong Kong Polytechnic University, Hong Kong, Hong Kong

Jinhua Yan Department of Geriatrics, Tongji Hospital, Tongji Medical College, Huazhong University of Science and Technology, Wuhan, Hubei, China

Fang Yang Department of Social Work, School of Sociology and Political Sciences, Shanghai University, Shanghai, China

Department of Sociology, Shanghai University, Shanghai, China

Jing Yang Department of Neurology, The First Affiliated Hospital of Zhengzhou University, Zhengzhou, Henan, China

Lin Yang Department of Epidemiology, Centre for Public Health, Medical University of Vienna, Vienna, Austria

Shanshan Yang Institute of Geriatrics, the 2nd Medical Center, Beijing Key Laboratory of Aging and Geriatrics, National Clinical Research Center for Geriatrics Diseases, Chinese PLA General Hospital, Beijing, China

Zhaoyang Yang Fujian University of Traditional Chinese Medicine, Fuzhou, China

Yao Yao Center for Healthy Aging and Development Studies, National School of Development, Peking University, Beijing, China

A. Yashkin Biodemography of Aging Research Unit, Social Science Research Institute, Duke University, Durham, NC, USA

Joshua C. Y. Yau Centre on Behavioral Health, The University of Hong Kong, Hong Kong, China

Bora Yenihan Labour Economics and Industrial Relations Department, Kocaeli University, Kocaeli, Turkey

Dongwook Yeo Laboratory of Physiological Hygiene and Exercise Science, School of Kinesiology, University of Minnesota Twin Cities, Minneapolis, MN, USA

Dannii Y. Yeung Psychology Laboratories – Department of Social and Behavioural Sciences, City University of Hong Kong, Hong Kong, China

Stecy Yghemonos Eurocarers, Brussels, Belgium

Yi Huang Department of Psychology, University of Pennsylvania, Singapore, Singapore

Na Yin Marxe School of Public and International Affairs, Baruch College, City University of New York, CUNY Institute for Demographic Research, New York, NY, USA

Yin Yin Department of Sociology, Central University of Finance and Economics, Beijing, China

Virpi Ylänne Centre for Language and Communication Research, School of English, Communication and Philosophy, Cardiff University, Cardiff, UK

Adrienne M. Young Department of Nutrition and Dietetics, Royal Brisbane and Women's Hospital, Herston, QLD, Australia

School of Exercise and Nutrition Sciences, Queensland University of Technology, Kelvin Grove, QLD, Australia

Robert D. Young Supercentenarian Research and Database Division, Gerontology Research Group, Sandy Springs, GA, USA

Andy Yu Corrigan Minehan Heart Center, Division of Cardiology, Department of Internal Medicine, Massachusetts General Hospital, Boston, USA

Harvard Medical School, Boston, MA, USA

Jianhua Yu Department of Cardiovascular Medicine, Second Affiliated Hospital of Nanchang University, Nanchang, China

Junhong Yu The State Key Laboratory of Brain and Cognitive Sciences, The University of Hong Kong, Hong Kong, Hong Kong

Laboratory of Neuropsychology, The University of Hong Kong, Hong Kong, Hong Kong

Ping Yu Faculty of Engineering and Information Sciences, School of Computing and Information Technology, Centre for Digital Transformation, University of Wollongong, Wollongong, NSW, Australia

Illawarra Health and Medical Research Institute, University of Wollongong, Wollongong, NSW, Australia

Smart Infrastructure Facility, Faculty of Engineering and Information Sciences, University of Wollongong, Wollongong, NSW, Australia

Rongjun Yu Department of Management, Hong Kong Baptist University, Hong Kong, SAR, China

Shu M. Yu Department of Psychology, Faculty of Social Sciences (FSS), University of Macau, Macau, China

Hao-Xiang Yuan Division of Cardiac Surgery, Heart center, The First Affiliated Hospital, Sun Yat-sen University, Guangzhou, China

Bo Yuan Department of Psychology, Ningbo University, Ningbo, China

Asghar Zaidi Government College University, Lahore, Pakistan
Oxford Institute of Population Ageing, University of Oxford, Oxford, UK

Frances Zainoeddin International Federation on Ageing Bridgepoint Drive, Toronto, ON, Canada

Matteo Zallio Center for Design Research, Mechanical Engineering, Stanford University, Stanford, CA, USA

Emma Zang Department of Sociology, Yale University, New Haven, CT, USA

Paola Zaninotto Department of Epidemiology and Public Health, University College London, London, UK

Arman Zargaran Department of Traditional Pharmacy, School of Persian Medicine, Tehran University of Medical Sciences, Tehran, Iran

Steven H. Zarit Department of Human Development and Family Studies, Penn State University, University Park, PA, USA

Mohammad M. Zarshenas Department of Phytopharmaceuticals (Traditional Pharmacy), School of Pharmacy, Shiraz University of Medical Sciences, Shiraz, Iran
Research Office for the History of Persian Medicine, Shiraz University of Medical Sciences, Shiraz, Iran

Barbara Zecchi University of Massachusetts Amherst, Amherst, MA, USA

Hannah Zeilig London College of Fashion, University of the Arts, London (UAL), London, UK
University of East Anglia, Norwich, UK

Sergei Zelenev International Council on Social Welfare, Bronx, NY, USA

Wen Zeng Macau Sun Yat-Sen Association of Medical Sciences, Macau, China

Yi Zeng Center for the Study of Aging and Human Development and Geriatrics Division, School of Medicine, Duke University, Durham, NC, USA
Center for Healthy Aging and Development Studies, National School of Development, Peking University, Beijing, China

Andreas Zenthöfer Department of Prosthodontics, University Hospital Heidelberg, Heidelberg, Germany

Bing Zhang National Institute for Nutrition and Health, Chinese Center for Disease Control and Prevention, Beijing, China

Chun-Quan Zhang The Second Affiliated Hospital of Nanchang University, Nanchang City, China

Cuntai Zhang Department of Geriatrics, Tongji Hospital, Tongji Medical College, Huazhong University of Science and Technology, Wuhan, Hubei, China

Division of Cardiology, Department of Internal Medicine, Tongji Medical College, Tongji Hospital, Huazhong University of Science and Technology, Wuhan, Hubei, China

Han Zhang Department of Global Health and Population, Harvard University, Cambridge, MA, USA

Li Zhang Rehabilitation Department, Chinese PLA General Hospital, Beijing, China

Liqing (Alice) Zhang Department of Psychiatry, University of Florida, Gainesville, FL, USA

Junan Zhang Department of Humanities and Law, Shanghai Business School, Shanghai, China

Keqing Zhang Department of Sociology, University of Hawaii at Manoa, Honolulu, HI, USA

Le Zhang Department of Geriatrics, Tongji Hospital, Tongji Medical College, Huazhong University of Science and Technology, Wuhan, Hubei, China

Meng Xuan Zhang Department of Psychology, Faculty of Social Sciences (FSS), University of Macau, Macau, China

Mengxi Zhang School of Public Health, Ball State University, Muncie, IN, USA

Wei Zhang Department of Sociology, University of Hawaii at Manoa, Honolulu, HI, USA

Xin Zhang School of Psychological and Cognitive Sciences, Peking University, Beijing, China

Zhenmei Zhang Department of Sociology, Michigan State University, East Lansing, MI, USA

Zhenyu Zhang Faculty of Engineering and Information Sciences, School of Computing and Information Technology, University of Wollongong, Wollongong, NSW, Australia

Cheng-Wu Zhao Department of Sports Medicine, The First Hospital of Jilin University, Changchun, Jilin, China

Lu Zhao Department of Neurology, The First Affiliated Hospital of Zhengzhou University, Zhengzhou, Henan, China

Yali Zhao Central Laboratory, Hainan Hospital of Chinese PLA General Hospital, Sanya, China

Yaohui Zhao National School of Development, Peking University, Beijing, China

Shanshan Zhen Department of Psychology, National University of Singapore, Singapore, Singapore

Zhihong Zhen School of Sociology and Political Science, Shanghai Univerisity, Shanghai, China

Wenhua Zheng Faculty of Health Sciences, Centre of Reproduction, Development and Aging, Institute of Translational Medicine, University of Macau, Macau SAR, China

Renyao Zhong Faculty of Economics and Management, East China Normal University, Shanghai, China

Xuefeng Zhong Health Education Institute of Anhui Provincial Center for Disease Control and Prevention, Hefei, China

Haiyan Zhu Department of Sociology, Virginia Polytechnic Institute and State University, Virginia Tech, Blacksburg, VA, USA

Yunshu Zhu Centre for IT-enabled Transformation, School of Computing and Information Technology, University of Wollongong, Wollongong, Australia

Zachary Zimmer Mount Saint Vincent University, Halifax, NS, Canada

Mohd Zulkifle Department of Kulliyate Tib, National Institute of Unani Medicine, Bengaluru, Karnataka, India

E

Ear, Nose, Throat (ENT) Cancer

▶ Tumors: Oto-Rhino-Laryngology

Early Detection of Cancer

▶ Cancer Diagnosis

Early Diagnosis of Cancer

▶ Cancer Diagnosis

Early Mortality Contraction

▶ Rectangularization of Survival Curve

Early Patient Access to Innovations

▶ Clinical Translation Acceleration

Early Satiety

▶ Anorexia of Aging

Earnings Disparity

▶ Wage Gap

Ecogerontology

▶ Environmental Gerontology

Ecological Momentary Assessment

▶ Mobile Data Collection with Smartphones

Ecology of Aging

▶ Environmental Gerontology

Economic Abuse of Older Adults by Family Members

▶ Financial Exploitation by Family Members

© Springer Nature Switzerland AG 2021
D. Gu, M. E. Dupre (eds.), *Encyclopedia of Gerontology and Population Aging*,
https://doi.org/10.1007/978-3-030-22009-9

Economic Gerontology

▶ Economics of Aging

Economics of Aging

Kathrin Komp-Leukkunen
Faculty of Social Sciences, University of Helsinki, Helsinki, Finland

Synonyms

Economic gerontology; Economics of old age; Economics of population aging

Definition

The economics of aging consider the phenomenon of aging from an economic perspective. They understand aging in two ways: as the aging of individuals and as the aging of populations. They usually study the aging of individuals at the micro-level, exploring how aging individuals navigate the economic sphere as a consumer, worker, or pensioner, among other things. Moreover, they study the aging of individuals at the meso-level, showing especially how workplaces can adapt to the increasing age of their employees. In contrast, they study the aging of populations at the macro-level, determining how governments can handle the growing number of older people with limited budgets, for example.

Introduction

Population aging is intrinsically linked to economics. On the one hand, it partly results from the shift from an agricultural to an industrial economy. According to the Demographic Transition Theory, this shift created the preconditions for longer lives (Kirk 1996). On the other hand, population aging affects the economy, leading to an aging workforce, a higher number of pensioners, and the need for new products and services.

Consequently, economic considerations of old age and population aging are common, and they deliver essential insight into the causes and consequences of population aging. The following paragraphs will outline discussions in the economics of aging. They will start with the micro-level, which focuses on the individual older person, continue with the meso-level, which considers the workplace, and then move on to the macro-level, which looks at aging populations. The article concludes with an outline of future directions of research.

The Micro-Level: An Economic Perspective on Aging Individuals

The micro-level concerns itself with the aging individual. It may look at individuals who already reached old age, which aligns it with the study of aging (gerontology). Alternatively, it may look at individuals as they age, which aligns it with the study of how individuals develop over time (life-course research). Within economics, it falls into the area of microeconomics.

Studying aging individuals is particularly interesting from the economic perspective because old age is defined according to economic status. In modern Western society, old age is understood as the time after people retired from paid work. People usually retired around age 65, making the age of 65 a commonly accepted marker for the onset of old age (Kohli 2007). As a result, studies on old age often are studies on retirement, exploring, for example, pension income, poverty, or economic activity in retirement.

Studies on the process of aging, in contrast, often consider influences on the retirement transition and on how lives change when people retire. The timing of the retirement transition depends, among other things, on pension regulations, a person's work biography, and the previously accumulated wealth. When people experience the retirement transition, their lives change in manifold ways. A substantial change happens in the income streams when income from paid work stops and pension schemes become a new income source. Another change that occurs simultaneously is the one in consumption patterns, where work-related expenses subside and health- and leisure-related expenses increase (Perek-Bialas and Schippers 2013).

The Meso-Level: An Economic Perspective on Aging Workplaces

The meso-level draws attention to workplaces. It focuses on these institutions as a social context for aging individuals. Additionally, it considers how these institutions are affected by the aging of their members. Within economics, companies and workplaces are the subjects of microeconomic considerations.

Workplaces act as a social context for aging individuals. They structure part of the aging individuals' daily lives, influencing what these individuals do when and where. Beyond this, they may also provide goals, a sense of meaning and orientation to older workers. In a best-case scenario, the aging workers identify with their jobs, and their colleagues are an important part of their social networks. As a result, workplace attachment forms and possibly prevents individuals from retiring early (Carr et al. 2016). In a worst-case scenario, the aging workers feel disconnected from their jobs, which can lead to early retirement. A job disconnect can, for example, form when employers practice ageism, excluding older workers from social staff activities or workplace qualification measures (Abuladze and Perek-Bialas 2018).

Workplaces change when their members age. During the aging process, workers may start to work more effectively because they accumulated experiential knowledge, meaning knowledge gained from the experiences made at work (Starks 2013). At the same time, their formal education may become outdated, and they may need to gain new knowledge especially in the area of information technologies (Behaghel et al. 2014). Moreover, emerging health problem may force them to organize their work differently, reducing the physical strain caused by their work. Because of these shifts, the skill sets, training needs, and most productive organizational structures in a workplace change when the workforce ages. As a result, employers may need to engage in age management practices, retraining older workers, restructuring workplaces, and ensuring a harmonious collaboration between the different generations in a workplace (Truxillo et al. 2015).

The Macro-Level: An Economic Perspective on Aging Populations

The macro-level considers populations in their entirety. It explores what happens when the share of older individuals in these populations increases, meaning when a higher percentage of individuals is aged 65 years or older. The effects most often considered are those on pension schemes, on the workforce, and on markets for goods and services. Within economics, these questions are covered by macroeconomic considerations.

Population aging has substantial effects on pension schemes. Where pension schemes are pay-as-you-go-financed, they focus on redistribution between individuals. They use the pension contributions of today's workers to finance the pension benefits of today's pensioners. As a result, the sustainability of these schemes depends on the ratio of workers to pensions at any point of time: if many workers support few pensioners, the redistribution works; if few workers support many pensioners, the redistribution fails. Population aging increases the share of pensions in a population, thereby threatening the functioning of pay-as-you-go-financed schemes (Godinez-Olivares et al. 2016).

Where pension schemes are capital-stock-based, they focus on redistribution over time. These schemes use the pension contributions that an individual accumulated during their working years to finance this individual's pension benefits. This redistribution functions as long as the time spent in retirement is stable or decreasing. However, if the time spent in retirement increases, the redistribution fails. Population aging entails that people spend more time in retirement, which threatens the sustainability of capital-stock-based pensions (Esping-Andersen et al. 2002). Because of the demographic pressure on both kinds of pension schemes, policymakers and researchers actively search for promising pension reform strategies, debating especially lowering the amount of pension benefits distributed, increasing the pension contributions collected, delaying retirement, and strengthening the private responsibility for old age income (Organisation for Economic Co-Operation and Development 2017).

Population aging likewise affects the workforce as the share of retirees in a population increases, the share of individuals of working-age decreases. Moreover, the average worker becomes older. These shifts trigger worries about the size of the workforce, questioning whether there will be enough individuals of working age in a country to fill all the job openings. Prominent strategies for solving this dilemma are to increase workforce participation rates especially among women and to encourage immigration of skilled workers (Bloom et al. 2010; Esping-Andersen et al. 2002).

Finally, population aging affects the markets for goods and services. When populations age, there are more older consumers acting on the market. Older people often have savings, which makes them interesting as a target group of customers. Therefore, consumers increasingly target older individuals, creating what is called a "silver economy" or "longevity economy" that caters to them (Klimczuk 2016). Traditionally, these economic systems provided especially health-related goods and services to older people. However, over the last decades, the number of healthy retirees increased. This shift changed the demands that the market has to meet. In response, the culture, leisure, and travel offers for older people increased (Perek-Bialas and Schippers 2013). How exactly the silver economy develops differs across countries (Klimczuk 2016). In European countries, it largely focuses on technology. In addition to age-friendly universities and silver tourism, it mainly consists of computer and technology-driven solutions, such as: robots, self-driving cars, wearable technology, electronic health and care services, and smart homes. Many more of these products are assumed to enter the market within the next 5–10 years (European Commission 2018).

Future Directions of Research

Research on the economics of aging will continue to flourish in the future. This research will highlight how the image of older people and the transition into old age change on the heels of pension reforms. Moreover, it will have to determine how best to adapt workplaces to aging populations. These adaptations are necessary to maintain high productivity of older workers and to reduce incentives for early retirement. Moreover, it will look into options to allow older people to become entrepreneurs, setting up their own businesses and creating their own work environments when their established work places do not fit their needs anymore (The Gerontological Society of America 2018).

Additionally, it has to determine how best to increase women's workforce participation rates, which includes measures to strike a work-family-balance. Also, research has to outline how to best reform pension schemes to guarantee their financial sustainability and to prevent old age poverty. Finally, research on the economics of aging will follow the emerging silver economy and explore its functioning. An interesting question is how older individuals – who may have struggled with the increase of technology in workplaces – will react to the increasing offer of technological solutions in the silver economy.

Conclusion

The economics of aging play an essential role in aging populations, helping us to understand the consequences of population aging and highlighting possible reactions to it. At the micro-level, they explore the situation of retirees and the transition into retirement. At the meso-level, they consider how workplaces can best be matched to older workers. At the macro-level, they reflect on the implications of population aging on pension schemes, the workforce, and markets for goods and services. Taken together, these considerations underline the diverse effects of population aging. Moreover, they underline that population aging requires reactions at a different level and that these levels should best be coordinated for increased reform impact. Because it combines considerations from economics, aging research, labor market research, sociology, and social policy, the economics of aging can make a reflected contribution to these discussions.

Cross-References

- Age Management and Labor Market Policies
- Bridge Jobs
- Delaying Retirement
- Encore Jobs
- Financial Gerontology
- Life Cycle Theories of Savings and Consumption
- Re-employment of (Early) Retirees
- Retirement Contracts
- Retirement Patterns
- Retirement Transition
- Senior Entrepreneurship
- Tax Policies and Older Adults

References

Abuladze L, Perek-Bialas J (2018) Measures of ageism in the labour market in international social studies. In: Ayalon L, Tesch-Römer C (eds) Contemporary perspectives on ageism. International perspectives on ageing, vol 19. Springer, Cham, pp 461–491. https://doi.org/10.1007/978-3-319-73820-8_28

Behaghel L, Caroli E, Roger M (2014) Age-bias technological technical and organizational change, training and employment prospects of older workers. Economica 81(322):368–389. https://doi.org/10.1111/ecca.12078

Bloom D, Canning D, Fink G (2010) Implications of population ageing for economic growth. Oxf Rev Econ Policy 26(4):583–612. https://doi.org/10.1093/oxrep/grq038

Carr E, Hagger-Johnson G, Head J, Shelton N, Stafford M, Stansfeld S, Zaninotto P (2016) Working conditions as predictors of retirement intentions and exit from paid employment: a 10-year follow-up of the English Longitudinal Study of Ageing. Eur J Ag 13(1):39–48. https://doi.org/10.1007/s10433-015-0357-9

Esping-Andersen G, Gallie D, Hemerijck A, Myles J (eds) (2002) Why we need a new welfare state. Oxford University Press, Oxford. https://doi.org/10.1093/0199256438.001.0001

European Commission (2018) The silver economy – final report. Publications Office of the European Union, Luxembourg. https://doi.org/10.2759/685036

Godinez-Olivares H, Boado-Penas M, Haberman S (2016) Optimal strategies for pay-as-you-go pension finance: a sustainability framework. Insur Math Econ 69:117–126. https://doi.org/10.1016/j.insmatheco.2016.05.001

Kirk D (1996) Demographic transition theory. Pop Stud-J Demog 50(3):361–387. https://doi.org/10.1080/0032472031000149536

Klimczuk A (2016) Comparative analysis of national and regional models of the silver economy in the European Union. Int J Ageing Later Life 10(2):31–59. https://doi.org/10.3384/ijal.1652-8670.15286

Kohli M (2007) The institutionalization of the life course: looking back to look ahead. Res Hum Dev 4(3-4):253–271. https://doi.org/10.1080/15427600701663122

Organisation for Economic Co-Operation and Development (2017) Pensions at a glance 2017: OECD and G20 indicators. OECD Publishing, Paris. https://doi.org/10.1787/pension_glance-2017-en

Perek-Bialas J, Schippers JJ (2013) Economic gerontology: older people as consumers and workers. In: Komp K, Aartsen M (eds) Old age in Europe. A textbook of gerontology. Springer, Dordrecht, pp 79–96. https://doi.org/10.1007/978-94-007-6134-6_6

Starks A (2013) The forthcoming generational workforce transition and rethinking organizational knowledge transfer. J Intergener Relatsh 11(3):223–237. https://doi.org/10.1080/15350770.2013.810494

The Gerontological Society of America (2018) Longevity economics: leveraging the advantages of an aging society. The Gerontological Society of America, Washington, DC

Truxillo D, Cadiz D, Hammer L (2015) Supporting the aging workforce: a review and recommendations for work intervention research. Annu Rev Organ Psych Organ Behav 2:351–381. https://doi.org/10.1146/annurev-orgpsych-032414-111435

Economics of Old Age

- Economics of Aging

Economics of Population Aging

- Economics of Aging

E-Counseling

- Cybercounseling

Educational Gerontology

Marvin Formosa
Department of Gerontology and Dementia Studies, Faculty for Social Wellbeing, University of Malta, Msida, Malta

Definition

The term "educational gerontology" was first used in a 1970 doctoral program at the University of Michigan to denote those "activities and study that occur at the interface of education and gerontology" (Peterson 1980: 68). It achieved academic prominence some years later with the publication of the first issue of the international journal *Educational Gerontology*, where "educational gerontology" was defined as the:

> ... study and practice of instructional endeavors for and about aged and aging individuals. It can be viewed as having three distinct, although interrelated, aspects: (1) educational endeavors designed for persons who are middle aged and older; (2) educational endeavors for a general or specific public about aging and older people; and (3) educational preparation of persons who are working or intend to be employed in serving older people in professional or paraprofessional capacities. (Peterson 1976: 62)

Peterson (1978, 1980) elaborated upon his original definition in his essays in *Introduction to Educational Gerontology* and *Educational Gerontology* where he embedded the subfield of educational gerontology in a "3 × 2 matrix" that postulated the major elements and activities of this area of study and practice. While across the top of the matrix, Peterson (1980: 68) situated three instructional audiences – namely, "(1) instruction of older people, (2) instruction of general or specific audiences about aging, and (3) instruction of persons who work with or in behalf of older people," and on its side he included two categories of functions – namely, "study" which included "research on and teaching about the needs, theory, philosophy, and environment in which older people function and the educational implications of this knowledge" and "practice" which comprised "the design, implementation, administration, and evaluation of instructional programs for older people." For Peterson (1980), this matrix led to six key components for the practice of educational gerontology:

Instructional gerontology. Research and theory building activities to clarify the elements involved in instructing older people, directed toward the conceptualization process of the circumstances under which older people can learn most effectively and efficiently.

Senior adult education. It includes the planning and conduct of educational endeavors for older people with the purpose of increasing their knowledge and skills so that they might enjoy life more and become more competent to meet the challenges of contemporary life.

Social gerontology. This includes research and theory building designed to understand the condition of older people and to explore the methods for communicating this information to families of older people, decision makers, or agency personnel who could assist senior citizens.

Advocacy gerontology. The dissemination of what is known about the processes of aging to both the general public and specific elements of that public who are in the unique position to be of assistance to older people through direct service, policy formation, or resource allocation.

Gerontological education. The study of instruction of professionals preparing for employment in the field of aging such as personnel who provide direct services in social service agencies, nursing homes, or recreation centers or who work in community and voluntary agencies.

Professional gerontology. Training of gerontologists is designed to lead to the development of skilled practitioners who can design, implement, and carry out the services needed in educating older people, meeting their social service and health needs, and changing society's attitudes.

Overview

Although Peterson's contributions remain influential to academic debates in "educational gerontology," his operational definition quickly fell out of favor as researchers held his model as being too ambitious. Indeed, the infrastructure of adult education and gerontology departments in most academic institutions was, and continues to be, too limited to meet all of Peterson's listed competencies (Glendenning 1985). In reaction, Glendenning (1983) drew a distinction between "educational gerontology" as focusing on the processes of older adult education and "gerontological education" as the preparation of students and professionals for a specialized career in aging studies. Moreover, he reserved the area of "public education about aging" to the realm of nongovernmental advocacy. On similar lines, Jarvis (1990) contended that Peterson's third element (instruction of professional and paraprofessionals) is actually included within the second (instruction about older people), since the latter refers to "the general dissemination of knowledge about the processes of human aging and the facilitation of empathy towards those who are old" (Peterson 1985: 13). Jarvis (1990: 402) argued that "there are only two major divisions within the body of knowledge... educational gerontology and gerontological education." He reasoned that while gerontology is a specialist field of practice and study rather, and a unique combination of knowledge drawn from a variety of disciplines, older adult education is a specialty within education and a subspecialty within the education of adults. Jarvis anticipated correctly that in the future one would expect educational gerontology to separate itself from gerontological education "given the fact that they have profoundly different knowledge bases in the first place" (Jarvis 1990: 408).

At the turn of the 1990s, a consensus was reached in both adult education and gerontological academia that Peterson's components of educational gerontology are best broken up into two areas: educational gerontology (older adult learning) and gerontological education (teaching gerontology). Since then, while researchers seeking to progress our knowledge of the optimal way to prepare students to work in aging settings grouped their work under the term of "gerontological education," others strictly focusing on some aspect of learning in the later years confined the term "educational gerontology" to refer to the teaching and learning of older adults (Findsen and Formosa 2011). Accordingly, for the past three decades, the boundaries of educational gerontology were taken to stretch around the following four areas of interest:

Instructional gerontology: How older people function; environmental context; educational motivation; early school leavers 40–50 years ago and the learning situation; the psychology of learning; memory and intelligence; learning aptitude; program models; teaching method; good practice, theory, and research.

Senior adult education: Enabling older adults to extend their range of knowledge and skills; assessment of student needs; training of tutors; curriculum development; marketing and delivery; evaluation.

Self-help instructional gerontology: Learning and helping others to learn; how to teach and how to learn in a self-help mode; establishing a curriculum; quality control; establishing standards; access to educational institutions; encountering distrust of formalism; need for independence; consumer sensitivity; developmental potential; relationship of teacher and taught; good practice, theory, and research.

Self-help senior adult education: Learning groups; coping skills...helping the homebound, the institutionalized, the frail elderly; reminiscence; administration; assistance; problem of travel.

(Glendenning 1989: 125 – italics in original)

The subsequent parts of this entry focus on the key theoretical underpinnings, research findings, and innovative learning practices that permeate the field of educational gerontology.

Theoretical Underpinnings

Early rationales advocating an improved educational provision for older persons were located within the functionalist paradigm, as researchers highlighted how late-life education can enable persons to later life. In a seminal article, Groombridge (1982) outlined five reasons why education is beneficial to an aging population: to promote self-reliance and independence, to enable older people to cope more effectively, to boost their contribution to society, to encourage older persons to impart their experiences to each other and to other generations, and to enhance self-actualization. A theoretical rupture occurred during the late 1980s as critical educational gerontologists – while not disputing the possibility that older adult education may offset a range of benefits for participants – decried that such a standpoint harbored a degree of "instrumental rationality" and, as a result, is too preoccupied with means rather than ends (Allman 1984). The first wave of critical educational gerontology was grounded upon four principles:

> (i) a focus on the linkage between the relationship of capitalism and ageing on one hand, and education in later life on the other, (ii) challenging that education for older people is a neutral enterprise, (iii) a sensitivity towards concepts such as empowerment, transformation, and consciousness-raising, and (iv), a praxeological approach based on dialogue between tutors and learners. (Glendenning and Battersby 1990: *passim*)

In retrospect, critical educational gerontology was a welcome counterpoint to conventional philosophies of late-life education whose *raison d'être* had always been closely utilitarian, rather than normative, in character (Formosa 2011). While Cusack (1999) advanced a community program of research and teaching that enables older learners to become aware of stereotypical assumptions about what it means to be old, Formosa (2000, 2005, 2007) conducted fieldwork at the University of the Third Age in Malta and highlighted how it is possible for older adult educational practice to arise as yet another euphemism for glorified and oppressive practice on class, gender, and ethnic grounds. Yet, critical educational gerontology was not immune to criticism, as educationists championing a humanist paradigm in educational gerontology perceived its principles as too "dubious," "comprehensive," and "wide-ranging" (Percy 1990) and pushed forward for yet another theoretical break in educational gerontology. Learning in later life, for educational gerontologists backing a humanist paradigm, constituted:

> ...a matter of personal quest. Learners begin from where they are; they follow the thrust of their own curiosities in order to make what is around them more meaningful; ideally they should be free of external constraints so that they can learn until they are satisfied, until they have achieved the potential that is within them. (Percy 1990: 23)

To this effect, Withnall's (2006: 30) empirical study on the experiences of older adult learners concluded that "the drive towards emancipation and empowerment implicit within [critical educational gerontology] is inappropriate in that it assumed an unjustifiable homogeneity among older people and appears to be imposing a new kind of ideological constraint." Indeed, educators such as Nye (1998) provided strong documentation highlighting the difficulties in enabling older learners to reach satisfactory levels of emancipation. Taking stock of this critical-humanist debate is not straightforward as both sides possess valid arguments in favor of their standpoint. While persons may be inherently "good," they are ultimately situated in a "turbo-capitalist" social reality characterized by an irreversible destruction of nature and cultures, so that everyday experiences may be anything but "humanizable." As such, the search for an alternative formulation that conceptualizes educational gerontology in terms of a really continuous "lifelong" learning that would straddle economic, democratic, personal, and other concerns across the life course in an inclusive way is currently still elusive.

Key Research Findings

Despite the burgeoning number of research articles in the field of educational gerontology, research on the participation rates of older persons in learning activities remains sparse, as studies tend to be based on relatively small samples gathered in particular geographical areas. In the United States, the 2005 National Household Education survey found that 23% of adults aged between 65-plus participated in a nonaccredited learning activity organized by community or business institutions in the previous year (O'Donnell 2006). Recent data from Eurostat reported that in

2017, across the EU-28 Member States, 4.9% of the population aged 55–74 participated in formal and nonformal educational activities, although this figure reaches 5.8% if one takes in consideration only the EU-15 Member States (Eurostat 2018). However, if one researches the participation of older adults in solely nonformal learning activities, one elicits higher results as in the case of the survey conducted by the Spanish National Institute of Statistics which found that 22.8% of the sample aged 55–74 participated in a learning activity in the previous 12 months (Villar and Celdrán 2013). Research found that contrary to stereotypical notions on third age learning, older adults do not only want to learn about a large range of topics resembling academic disciplines but also want to learn practical life and vocational skills (Boulton-Lewis 2010). As Talmage et al. (2015) pointed out, many learners want to learn about art and learn through artistic experiences such as drawing, painting, theatre, and music but also life skills regarding their changing bodies and lifestyles as they age, as well as computer skills and subjects in the natural and physical sciences such as biology, marine habitats, geology, and astronomy.

Initial research suggested that older adults are motivated to learn so as to meet five types of needs – namely, coping, expression, contribution, influence, and transcendence (McClusky 1974). It is noteworthy that while the first four types may be found in adult education studies, the need for transcendence is unique for older learners in that contemplative needs or needs for life review are unique to older adults. Another influential direction in the study of motivations for engagement in late-life learning was O'Connor's (1987) distinction between expressive needs (personal development and social relations) and instrumental needs (work, career, and skills requirements). Some motivations, however, defy a simply expressive-instrumental dichotomy as many older adults were also found to engage in learning pursuits to expand their social support networks following divorce, widowhood, and "empty nest" transitions. Hodkinson et al.' (2008) notion of "acquisition versus becoming" – namely, acquiring knowledge and skills and undergoing a process of personal reconstruction, respectively – is yet another useful typology when deciphering the motivations underlying older adult learning. Although one finds some evidence in favor of instrumental motivations for learning in later life – to sustain employment or gaining job-specific skills or qualifications (Phillipson and Ogg 2010) – research revealed a major shift in favor of expressive motivations. While Tam (2016) found that "learning is the broadening of my horizons" and that "learning allows me to continue employment or to rejoin the workforce after retirement" were the two statements most and least strongly agreed among survey respondents, Kim and Kim (2015) reported that self-actualization (sense of enjoyment, satisfaction, and achievement) was found to be the most influential motive among older adults learning English as a foreign language. Such a line of research has been, however, criticized for adhering to the "misery perspective," which leads to a research direction focused on age-related problems such as diseases and overall health decline (Talmage et al. 2015). As a result, researchers like Boulton-Lewis (2010) and Villar and Celdrán (2013) have begun to focus on third agers' potential for new knowledge acquisition and fulfilment of learning needs, rather than focusing solely on how learning can meet physiological, psychological, and social needs.

Innovative Learning Practices

Older adult learning holds a rich tradition in the United States (Findsen and Formosa 2011). There is a now a general consensus that the inaugural lifelong learning institute targeting older persons was the Institute for Retired Professionals, which was established in 1962, by "a group of 152 retired New York City schoolteachers...in Greenwich village," and later renamed as "Institutes for Learning in Retirement" (Manheimer et al. 1995). Such institutes have operated uninterruptedly for the past half-century, and although there is no single model of operation, they all share a similar feature in that they are hosted by a college or university with a similar culture and sense of mission. Throughout the 1970s and 1980s, other organizations replicated or adapted the Institute for Retired Professionals model. While 1972 witnessed the launch of a lifelong learning program for older persons in the faith-based, volunteer-run, Shepherds Centers, in 1976, the Fromm Institute for Lifelong Learning was established to provide retirees with daytime, noncredit, college-level courses in a variety of academic subjects (Manheimer et al. 1995). Elderhostel (Road Scholar since 2011) was founded in 1975 to organize weeklong courses of instruction and discussion in colleges and universities, and by 2006, it was successfully organizing some 8,000 programs throughout the world to about 160,000 members (Jarvis 2012). Another key player in delivering learning programs for older persons, the Bernard Osher Foundation, was founded in 1977. Its mission remains to enhance the quality of life in later life through a lifelong learning network under the aegis of the Osher Lifelong Learning Institutes, a consortium of some 130 institutes who, despite their different modes of organization, provide noncredit educational programs to adults aged 50 years or older (Shinagel 2012). Harvard University also recognized the lifelong learning movement and in 1977 founded the Harvard Institute for Learning in Retirement, whose size – since the early 2010s – was maintained at a steady annual number of 500 to 550 members (Manheimer et al. 1995).

In Europe, the practice of educational gerontology is centered around the Universities of the Third Age (U3As) and University Programs for Older People (UPOPs). The U3A, which was founded in 1973, can be defined as sociocultural centers where older persons acquire new knowledge of significant issues or validate the knowledge which they already possess, in an agreeable milieu and in accordance with easy and acceptable methods (Formosa 2014). UPOPs refer to the tendency of European universities, but mostly in Spain and Germany, to open their degree programs to older persons (Fernández-Ballesteros et al. 2013). A similar movement to the UPOPs is found in Japan, and in 1993, China began establishing Universities for Seniors in line with the government's 7-year development plan which decrees that all universities should open their doors to older persons (Jarvis 2012). Another popular learning movement targets older men (Formosa et al. 2014) and chiefly typified by the Men's Sheds organization which originated in Australia and defined as:

> ...any community-based, non-profit, non-commercial organisation that is accessible to all men, and whose primary activity is the provision of a safe and friendly environment where men are able to work on projects at their own pace in their own time in the company of other men. (Australian Men's Sheds Association, as cited in Golding 2015: 10)

Another innovative organization is the Elder Academy, founded in Hong Kong in 2007, whereby local schools and welfare organizations team up to run "elder academies" in which older persons have the opportunity to interact and work with younger learners, thus promoting intergenerational learning (Tam 2016). Objectives include "to maintain healthy physical and mental well-being, to realize the objective of fostering a sense of worthiness between the elders and the young, to optimize existing resources, to promote harmony between the elders and the young, [and] to strengthen civic education" (Chui 2012: 152).

Future Directions of Research

Much water has passed under the bridge since Peterson's definition of educational gerontology saw the light of the day. Educational gerontology has passed from being an emergent field of study to a recognized domain in both adult education and gerontology faculties, distinguished by a thriving sum of publications that expound its theoretical, empirical, and policy boundaries. A running theme is that educational gerontology is "empowering and transformational by meeting the diverse and sometimes different personal, social and well-being needs of older adults" (Findsen et al. 2017: 509). However, this is not the same as saying that the field does not include research lacunae and new directions in future research are especially warranted. First, educational gerontology requires more attention to participation studies that, rather than simply uncovering the characteristics and motivations of distinctive learners, also understand the causes as why working class older adults, older men, older persons living in rural communities, and ethnic minorities are reluctant to participate in older adult learning. Second, more research is required on the valuing and recognition of late-life learning that takes place outside formal and nonformal contexts and on how older adults engage in self-directed learning, sometimes in isolation, and at other times with family members and friends, through various institutions that range from religious centers to libraries to the social media. Third, the educational system that spends some 18 years, and substantial financial capital, to prepare citizens for adulthood is clearly ageist. Research is required on pre-retirement learning models that would include subjects as diverse as the formalities surrounding pensions, the drawing of wills, and strategies active and successful aging. Fourth, the challenges that older persons face in their attempt to enroll in formal learning have been largely overlooked, and thus far, older learners remain largely outside higher education. Research is required to identify how higher education may enable older adults to play a leading role in creating a new type of aging for the twenty-first century (built around extended economic, family, and citizenship roles), unlocking mental capital and promoting well-being in later life, and supporting a range of professional and voluntary groups working on behalf of older people. Finally, the field of older adult learning tends to be hijacked by the "successful aging" paradigm, thus rendering the presence of physical and cognitive frailty as a persona non grata. This is unjustifiable since many older people experience mobility and cognitive challenges to the extent of becoming housebound or having to take up residence in care homes. Thus, there warrants a strong research drive as how older adult learning, both in theory and practice, can bridge third and fourth age avenues so that older persons with physical disabilities and dementia have an equal opportunity for inclusion in late-life learning.

Summary

In the 1970s, educational gerontology referred to all those activities and study that occur at the interface of education and gerontology. Yet, such an operational definition was found to be overly ambitious by gerontologists and educationists alike, and in the 1980s, a distinction was made between "educational gerontology" as focusing on the processes of older adult education and "gerontological education" as the preparation of students and professionals for a specialized career in aging studies. Rationales in favor of educational gerontology were in consensus on the beneficial impacts of older adult learning, although a rift emerged between conservatives and radicals who stressed adjustment and empowerment, respectively. The 1970s were an extremely fertile ground for the establishment of institutions providing nonformal learning opportunities to older adults – most notably, Institutes for Learning in Retirement, Elderhostel, and Universities of the Third Age. Future research directions should focus on that interface between educational gerontology on one hand and participation studies, informal learning, pre-retirement learning, higher education, and fourth age learning.

Cross-References

► European Civil Society Platform on Lifelong Learning
► Senior learning

References

Allman P (1984) Self-help learning and its relevance for learning and development in later life. In: Midwinter E (ed) Mutual aid universities. Croom Helm, Beckenham, pp 72–90

Boulton-Lewis GM (2010) Education and learning for the elderly: why, how, what. Educ Gerontol 36(3):213–228. https://doi.org/10.1080/03601270903182877

Chui E (2012) Elderly learning in Chinese communities: China, Hong Kong, Taiwan, Singapore. In: Boulton-Lewis G, Tam M (eds) Active ageing, active learning: issues and challenges. Springer, Dordrecht, pp 141–161. https://doi.org/10.1007/978-94-007-2111-1_9

Cusack SA (1999) Critical educational gerontology and the imperative to empower. Educ Ageing 14(1):21–37

Eurostat (2018) Participation rate in education and training (last 4 weeks) by type, sex and age[trng_lfs_09]. Last update: 20-04-2018. http://appsso.eurostat.ec.europa.eu/nui/submitViewTableAction.do Accessed 24 September 2018

Fernández-Ballesteros R, Caprara M et al (2013) Effects of university programs for older adults: changes in cultural and group stereotype, self-perception of aging, and emotional balance. Educ Gerontol 39(2):119–131. https://doi.org/10.1080/03601277.2012.699817

Findsen B, Formosa M (2011) Lifelong learning in later life: a handbook on older adult learning. Sense Publishers, Rotterdam. https://doi.org/10.1007/978-94-6091-651-9

Findsen B, Golding B, Jekenc Krasovec S, Schmidt-Hertha B (2017) Ma te ora ka mohio/'Through life there is learning'. AJAL 57(3):509–526

Formosa M (2000) Older adult education in a Maltese University of the Third Age: a critical perspective. Educ Ageing 15(3):315–339

Formosa M (2005) Feminism and critical educational gerontology: An agenda for good practice. Ageing Int 30 (4):396–411. https://doi.org/10.1007/s12126-005-1023-x

Formosa M (2007) A Bourdieusian interpretation of the University of the Third Age in Malta. J Malt Educ Res 4 (2):1–16

Formosa M (2011) Critical Educational gerontology: a third statement of principles. IJEA 2(1):317–332

Formosa M (2014) Four decades of Universities of the Third Age: past, present, and future. Aging Soc 34 (1):42–66. https://doi.org/10.1017/S0144686X12000797

Formosa M, Chetcuti Galea R, Farrugia Bonello M (2014) Older men learning through religious and political affiliations: case studies from Malta. Androgogic Perspect 20(3):57–69. https://doi.org/10.4312/as.20.3.57-69

Glendenning F (1983) Educational gerontology: a review of American and British developments. Int J Lifelong Educ 2(1):63–82. https://doi.org/10.1080/0260137830020106

Glendenning F (1985) Education for older adults in Britain: a developing movement. In: Glendenning (ed) Educational gerontology: international perspectives. Croom Helm, Beckenham, pp 100–141

Glendenning F (1989) Educational gerontology in Britain as an emerging field of study and practice. Educ Gerontol 15(2):121–131. https://doi.org/10.1080/0380127890150202

Glendenning F, Battersby D (1990) Why we need educational gerontology and education for older adults: a statement of first principles. In: Glendenning, Percy K (eds) Ageing, education and society: Readings in educational gerontology. Association for Educational Gerontology, Keele, pp 219–231

Golding B (2015) The Men's Shed movement: the company of men. Common Ground Publishing, Champaign

Groombridge B (1982) Learning, education and later life. Adult Educ 54(4):314–325

Hodkinson P, Ford G, Hodkinson H, Hawthorn R (2008) Learning as a retirement process. Educ Gerontol 34 (3):167–184. https://doi.org/10.1080/03601270701835825

Jarvis P (1990) Trends in education and gerontology. Educ Gerontol 16(4):401–409. https://doi.org/10.1080/0380127900160409

Jarvis P (2012) The age of learning: seniors learning. In: Rubenson K (ed) Adult learning and education. Academic, Amsterdam, pp 163–170

Kim T-Y, Kim Y-K (2015) Elderly Korean learners' participation in English learning through lifelong education: focusing on motivation and demotivation. Educ Gerontol 41(2):120–135. https://doi.org/10.1080/03601277.2014.929345

Manheimer RJ, Snodgrass DD, Moskow-McKenzie D (1995) Older adult education: a guide to research, programs, and policies. Greenwood, Westport

McClusky HY (1974) Education for ageing: the scope of the field and perspectives for the future. In: Grabowski S, Mason WD (eds) Learning for ageing. Adult Education Association of the USA, Washington, DC, pp 324–355

Nye EF (1998) A Freirean approach to working with elders or: conscientização at the Jewish community centre. J Aging Stud 12(2):107–116

O'Connor DM (1987) Elders and higher education: instrumental or expressive goals? Educ Gerontol 13(6):511–519. https://doi.org/10.1080/0360127870130607

O'Donnell K (2006) Adult Education Participation in 2004–2005 (NCES 2006-077). U.S. Department of Education. National Center for Education Statistics, Washington, DC

Percy K (1990) The future of educational gerontology: a second statement of first principles. In: Glendenning F, Percy K (eds) Ageing, education and society: readings in educational gerontology. Association for Educational Gerontology, Keele, pp 232–239

Peterson DA (1976) Educational gerontology: the state of the art. Educ Gerontol 1(1):61–73. https://doi.org/10.1080/03601277.1976.12049517

Peterson DA (1978) Toward a definition of educational gerontology. In: Sherron RH, Lumsden DB (eds) Introduction to educational gerontology. Hemisphere Publishing Corporation, Washington, DC, pp 1–29

Peterson DA (1980) Who are the educational gerontologists? Educ Gerontol 5(1):65–77. https://doi.org/10.1080/0360hyp800050105

Peterson DA (1985) Towards a definition of older learners. In: Sherron RH, Lumsden B (eds) Introduction to educational gerontology, 2nd edn. Hemisphere Publishing Corporation, Washington, DC, pp 185–204

Phillipson C, Ogg J (2010) Active ageing and universities: engaging older learners. Universities UK, London

Shinagel M (2012) Demographics and Lifelong Learning Institutes in the 21st Century. Contin High Educ Rev 76:20–29

Talmage CA, Lacher RG, Pstross M et al (2015) Captivating lifelong learners in the third age: Lessons learned from a university based institute. Adult Educ Q 65

(3):232–249. https://doi.org/10.1177/0741713615577109

Tam M (2016) Later life learning experiences: listening to the voices of Chinese elders in Hong Kong. Int J Lifelong Educ 35(5):569–585. https://doi.org/10.1080/02601370.2016.1224042

Villar F, Celdrán M (2013) Learning in later life: participation in formal, non-formal and informal activities in a nationally representative Spanish sample. Eur J Ageing 10:135–144. https://doi.org/10.1007/s10433-012-0257-1

Withnall A (2006) Exploring influences on later life learning. Int J Lifelong Educ 25(1):29–49. https://doi.org/10.1080/02601370500309477

EEG

▶ Electrophysiology

Effectiveness of Care

▶ Quality of Care

Effectiveness of Respite Care for Caregivers of Older Adults

Anne R. Mason
Centre for Health Economics, University of York, York, UK

Synonyms

Adult day care; Short breaks; Social support

Definition

Respite care is an umbrella term for a range of services provided intermittently in the home, community, or institution to provide temporary relief to the principal informal caregiver. Services include sitting services, day care, host family care, and overnight care. Some services also provide activities or interventions for the care recipient and/or the caregiver.

Overview

The overarching aim of respite care is to promote the well-being of the caregiver by providing substitution for the normal caring duties of the unpaid caregiver (Shaw et al. 2009). Studies of respite care are enormously heterogeneous (Mason et al. 2007a; Shaw et al. 2009; Thomas et al. 2017). They vary in terms of their study design, the intervention investigated (setting, staffing, duration, access costs, flexibility), the type of caregiver and type of care recipient who participate, the benefits, costs and harms measured, and whether or not effects on care recipients are assessed. As there is unlikely to be a "one-size-fits-all" model for respite care, its effectiveness will necessarily be context-specific (Thomas et al. 2017).

The evidence base is disappointingly weak. Empirical studies are generally small, of low methodological quality, and with short-term follow-up. There remains a dearth of evidence on the economics of respite care (Knapp et al. 2013).

Qualitative studies have documented the barriers to uptake of respite care, including caregiver attitudes, awareness of services and their perceived quality, acceptability, and flexibility. This evidence is useful for informing intervention design and policy decisions.

Key Research Findings

For all types of respite, the effects upon caregivers are small or not statistically significant. Higher quality studies identify modest benefits only for certain subgroups (Mason et al. 2007b).

There is no reliable evidence that respite care delays entry to long-term residential care, and some evidence that the likelihood of institutionalization is higher in respite users (Shaw et al. 2009). For care recipients with dementia, there is no evidence that respite affects the risk of institutionalization one way or the other (Maayan et al. 2014).

Many studies report high levels of satisfaction with respite care by caregivers, although satisfaction appears to be lower if care recipients have challenging behaviors or if the quality of respite care is suboptimal (Victor 2009). Qualitative studies suggest that many caregivers value respite services highly with positive improvements in self-reported measures of health and well-being (Victor 2009), but studies employing objective measures of mental or physical health offer little support for this finding (Thomas et al. 2017). Evidence for beneficial health effects is limited to certain subgroups – such as care recipients without a partner, or caregivers with significant support needs – and is based on small numbers of studies (Mason et al. 2007a; Shaw et al. 2009). Similarly, evidence on the impact on caregiver burden is mixed, and meta-analyses find no significant effect (Mason et al. 2007b; Shaw et al. 2009).

In terms of other effects, services provided during the working day appear most effective in supporting caregiver employment; short breaks alone are not effective (Brimblecombe et al. 2018) (▶ "Employment and caregiving"). Respite can improve caregivers' sense of social inclusion, and caregivers may benefit indirectly through information, reassurance, and support provided either informally by respite staff (Victor 2009) or through the use of decision aids (Stirling et al. 2012).

Evidence on the adverse effects of respite care on care recipients is sparse. There is no evidence of an increased risk of mortality (Mason et al. 2007a). Hospital-based respite may disrupt sleep patterns in older people (Lee et al. 2007) while the impact on their physical health is mixed (Shaw et al. 2009; Victor 2009). In terms of psychological impacts, care recipients can become distressed, confused, and may resist or refuse respite care (Shaw et al. 2009). The negative unintended consequences of respite on caregivers include guilt and anxiety, and a sense of "failure" (Shaw et al. 2009; Victor 2009).

Future Directions of Research

There remains an urgent need for well-designed primary studies to test the impact of respite on caregivers and care recipients and to identify appropriate, affordable support services. As the decision to access respite depends on caregivers' perceptions of quality, acceptability, and flexibility, it is important that interventions are co-produced with caregivers. The study duration should allow sufficient follow-up to capture "steady-state" effects. Studies need to assess outcomes that matter to caregivers and care recipients and to assess costs and any unintended consequences.

Cross-References

▶ Adult Day Services
▶ Caregiver Interventions
▶ Caregivers' Outcomes
▶ Caregiver Stress
▶ Respite Care

References

Brimblecombe N, Fernandez JL, Knapp M, Rehill A, Wittenberg R (2018) Review of the international evidence on support for unpaid carers. J Long-Term Care, September, 25–40. https://doi.org/10.21953/lse.ffq4txr2nftf

Knapp M, Iemmi V, Romeo R (2013) Dementia care costs and outcomes: a systematic review. Int J Geriatr Psychiatry 28:551–561. https://doi.org/10.1002/gps.3864

Lee D, Morgan K, Lindesay J (2007) Effect of institutional respite care on the sleep of people with dementia and their primary caregivers. J Am Geriatr Soc 55:252–258

Maayan N, Soares-Weiser K, Lee H (2014) Respite care for people with dementia and their carers. Cochrane Database Syst Rev: CD004396. https://doi.org/10.1002/14651858.CD004396.pub3

Mason A et al (2007a) A systematic review of the effectiveness and cost-effectiveness of different models of community-based respite care for frail older people and their carers. Health Technol Assess 11:iii-88

Mason A, Weatherly H, Spilsbury K, Golder S, Arksey H, Adamson J, Drummond M (2007b) The effectiveness and cost-effectiveness of respite for caregivers of frail older people. J Am Geriatr Soc 55:290–299

Shaw C et al (2009) Systematic review of respite care in the frail elderly. Health Technol Assess 13:1–224., iii. https://doi.org/10.3310/hta13200

Stirling C, Leggett S, Lloyd B, Scott J, Blizzard L, Quinn S, Robinson A (2012) Decision aids for respite service choices by carers of people with dementia: development and pilot RCT. BMC Med Inform Decis 12:21. https://doi.org/10.1186/1472-6947-12-21

Thomas S, Dalton J, Harden M, Eastwood A, Parker G (2017) Updated meta-review of evidence on support for carers. Health Serv Deliv Res 5:1–162. https://doi.org/10.1177/1355819618766559

Victor E (2009) A systematic review of interventions for carers in the UK: outcomes and explanatory evidence. Princess Royal Trust for Carers, Woodford Green

Effects of Telomerase Activation

Dhenugen Logeswaran and Julian J.-L. Chen
School of Molecular Sciences, Arizona State University, Tempe, AZ, USA

Synonyms

Telomerase-dependent cellular immortality

Definition

Telomerase counteracts telomere shortening and prevents cell senescence. Activation of telomerase confers unlimited cell growth in cancer cells and is viewed as a potential means for antiaging therapeutics.

Overview and Key Findings

Telomerase is the key enzyme that confers cellular immortality in eukaryotes. In humans, the highly proliferative germline and stem cells maintain replicative capacity by synthesizing telomerase enzyme to counteract telomere shortening that results from the end-replication problem associated with linear chromosomes. In contrast, differentiated somatic cells lack telomerase and have limited replicative capacity. Most cancers acquire cellular immortality by reactivating synthesis of the telomerase catalytic protein component. Understanding the regulatory mechanisms of the human telomerase enzyme would offer plausible means to delay human aging at cellular and organismal levels.

Cellular Aging and Telomere Shortening

Normal human somatic cells are mortal with limited replicative capacity. Leonard Hayflick discovered that cell cultures derived from human tissues are capable of only a finite number of cell divisions before undergoing growth arrest or senescence, which is termed the Hayflick limit (Hayflick 1965). The replicative capacity of mortal cells correlates to the length of telomeres at chromosome ends (Harley et al. 1990). Human telomeres consist of telomeric DNA with simple TTAGGG repeats and the shelterin complexes of six proteins that bind specifically to the telomeric DNA, which protects chromosome ends from fusion or being recognized as DNA breaks (ref to ▶ "Telomeres" entry). During each cell division cycle, chromosomal DNA molecules are replicated by cellular DNA polymerase, which does not fully replicate the telomeric DNA at the ends of linear chromosomes, known as the "end-replication problem" (Olovnikov 1973; Watson 1972). Incomplete replication of chromosome ends leads to gradual shortening of telomeric DNA (Kelleher et al. 2002). Telomere length is viewed as a "mitotic clock" counting down as the cells divide. Upon reaching a critical length, short telomeres lose protective function and trigger cellular senescence, limiting the life span of cells and contributing to the cellular aging (Harley et al. 1992).

Critically short telomeres induce cell senescence by activating DNA damage responses (ref to ▶ "Cell Senescence" entry). The protective function of telomeres relies on the proper complex formation between telomeric DNA and the shelterin proteins which cooperatively forms a unique lariat structure called the telomere-loop (t-loop) wherein the terminal single-stranded telomeric DNA folds back to invade the double-stranded telomeric region. Extremely short telomeric DNA loses binding of the shelterin proteins and fails to form the protective t-loop structure. The unprotected telomeres are then recognized as DNA breaks, which induce DNA damage responses (DDR) and activate double-strand break (DSB) DNA repair pathways. Three major DDR enzymes ataxia telangiectasia mutated (ATM), ataxia telangiectasia and Rad3-related (ATR) kinases, and poly(ADP-ribose) polymerase 1 (PARP1) can be activated upon sensing dysfunctional telomeres (Fagagna et al. 2003). Activated ATM and ATR kinases are part of a phosphorylation cascade which phosphorylates CHK1 and CHK2 that in turn activate p53 by phosphorylation. Phosphorylated p53 then activates p21 at the transcriptional level which inhibits cyclin-dependent kinases and arrests cell cycle, resulting in cellular senescence (de Lange 2018).

In addition to soliciting DDR pathways, the unprotected telomeric DNAs can be repaired as DSBs by either non-homologous end joining (NHEJ) or homology-directed repair (HDR) pathways. Through the NHEJ pathway, unprotected chromosome ends are ligated, resulting in chromosome end-to-end fusions and genomic instability. On the other hand, homologous sister telomeres can undergo HDR resulting in telomere sister chromatid exchanges. Due to the repetitive nature of telomeric DNA, an offset in breakpoints during recombination within sister telomeres causes unequal exchange of sister chromatids and a length disparity between sister telomeres. The inheritance of the shorter telomeres thus limits the proliferative potential of the offspring cells.

Telomerase and Cellular Immortality

Telomerase counteracts telomere shortening by adding telomeric DNA repeats to chromosome ends, which is the most common solution in eukaryotes to the end-replication problem of linear chromosomes. Telomerase functions as a ribonucleoprotein enzyme composed of two catalytically essential core subunits: the telomerase reverse transcriptase (TERT) protein component and the telomerase RNA (TR) component (ref to ▶ "Telomerase" entry). The TERT component constitutes the active site that catalyzes DNA polymerization reaction, while the TR component provides a short RNA template for each DNA repeat synthesis and facilitates the processive addition of repeats. In addition to the two core subunits, telomerase holoenzyme contains a number of accessory proteins that are essential for biogenesis and regulation of telomerase in vivo. These telomerase accessory proteins are divergent among different groups of eukaryotic species and are often shared with other noncoding RNAs (Podlevsky and Chen 2016). For example, vertebrate TR and box H/ACA snoRNA share common protein components for a similar biogenesis pathway. Both core components, TERT and TR, and telomerase accessory proteins are necessary for the proper function of telomerase holoenzyme in the cells for maintaining sufficient telomere length.

Telomerase maintains telomere length and sustains replicative capacity, which is crucial for cells that are highly proliferative or immortal. In humans, telomerase activity can be readily detected in highly proliferative cells, such as germline and stem cells (Meyerson et al. 1997), that can maintain telomere length even after an extremely large number of cell divisions (Fig. 1). However, telomerase activity is undetectable in

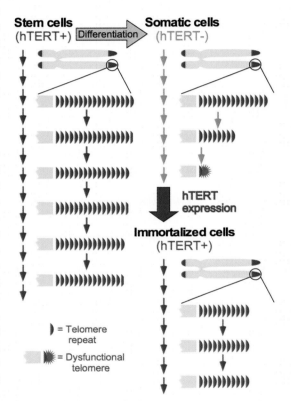

Effects of Telomerase Activation, Fig. 1 Telomerase maintains telomere length in stem cells and immortalized somatic cells. Stem cells are telomerase-positive and have a significantly low rate of telomere shortening. In contrast, the differentiated somatic cells are telomerase-negative with the hTERT expression repressed and have a high rate of telomere shortening which eventually results in dysfunctional telomeres. Ectopic expression of the hTERT protein gene or reactivation of the repressed hTERT gene in somatic cells restores a sufficient level of telomerase activity and maintains telomere length during successive cell divisions human somatic cells which experience significant telomere shortening after a limited number of cell divisions (Fig. 1). The lack of telomerase in somatic cells is due to transcriptional repression of the human TERT (hTERT) protein gene during stem cell differentiation, which primarily involves epigenetic modifications and chromatin remodeling in the promoter region (Lewis and Tollefsbol 2016). In contrast, human TR (hTR) and telomerase accessory proteins are constitutively expressed in both telomerase-positive and -negative cells, indicating that the hTERT protein is the limiting component for producing active telomerase in human cells. Indeed, the expression of only the hTERT protein gene in telomerase-negative somatic cells is sufficient to restore telomerase activity, prevent telomere shortening, and confer cellular immortality (Bodnar et al. 1998).

Human cancer cells are immortal and capable of maintaining stable telomere length. More than 85% of cancer types in humans reactivate telomerase by activating transcription of hTERT protein gene to prevent telomere shortening (Kim et al. 1994). In some cancer cells, the upregulation of hTERT expression is mediated by the transcription factor c-Myc which is central to cell growth and metabolism (Wang et al. 1998). Many cancer cells harbor mutations in the hTERT promoter region that activate hTERT transcription (Horn et al. 2013; Huang et al. 2013). The remaining cancer types employ a telomerase-independent, alternative lengthening telomere (ALT) mechanism mediated by homologous recombination (O'Sullivan and Almouzni 2014). ALT cancer cells lack telomerase activity and characteristically have long and heterogeneous telomeric DNA. Although most cancers reactivate the hTERT gene for cellular immortality, telomerase genes are not considered oncogenes as the presence of telomerase enzyme alone does not trigger tumorigenesis (Harley 2002).

In addition to hTERT transcription regulation, alternative splicing of hTERT pre-mRNA produces isoforms of hTERT protein and alters telomerase activity in cells (Ulaner et al. 2000). Moreover, posttranslational modifications and subcellular localization of telomerase enzyme have been reported as additional means to regulate telomerase function in vivo (Liu et al. 2001). In addition to hTERT, biogenesis of hTR may be regulated to alter stability, posttranscriptional processing, and subcellular localization, which provide additional layers of telomerase activity regulation (Schmidt and Cech 2015). Thus, the complex interplay between genetic and epigenetic factors regulates telomerase activity among various cell types in the human body.

Telomerase Mutations and Short Telomere Syndrome

The level and activity of telomerase in human germline and stem cells are crucial for maintaining telomere length. Mutations in genes of telomerase components or telomere-binding proteins lead to defects or deficiency of telomerase holoenzyme which result in a spectrum of disorders collectively called "short telomere syndromes." Patients with these disorders show significantly reduced telomere lengths and accelerated aging phenotypes in stem cells. Multiple organs and organ systems suffer consequences when highly proliferative tissue become severely affected due to short telomeres and increased cell senescence (Mangaonkar and Patnaik 2018).

Dyskeratosis congenita (DC), the first reported short telomere syndrome, is genetically linked to mutations in genes encoding for several telomerase components including hTERT, hTR, and dyskerin complex proteins (Heiss et al. 1998; Mitchell et al. 1999; Vulliamy et al. 2001). The DC patients typically have the onset of symptoms, including aberrant skin pigmentation, nail dystrophy, and leukoplakia, later in life when their telomeres have shortened to a critical length. Furthermore, patients with DC often develop a plethora of additional abnormalities in highly proliferative tissues such as hematopoietic stem cells in bone marrow, which eventually lead to bone marrow failure and death.

In addition to DC, patients with familial idiopathic pulmonary fibrosis (IPF) and aplastic anemia are often found with short telomeres and mutations in telomerase genes. IPF causes scar tissue formation in the lungs which affects gas exchange. Prevalence of IPF increases in older individuals over 50 and strongly associates with smoking. Aplastic anemia is a blood disorder characterized by the insufficient production of new blood cells by the bone marrow. Clinical manifestation of these diseases is a result of haploinsufficiency where a single functional copy of the TERT gene is insufficient to prevent telomeric DNA shortening in older age after a large number of cell divisions. The onset of disease symptoms correlates to the remaining level and activity of the affected telomerase enzyme and the shortening rate of telomeres. Patients with mutations that severely reduce telomerase activity could have early onset of the disease at a younger age. Therapeutics for patients with short telomere diseases would require means to increase telomerase function or level in order to stop or reverse telomere shortening in stem cells.

Telomerase and Organismal Aging

The role of telomerase in organismal aging has been studied mostly using the mouse model system. Mice with deletion of both copies of the mouse TR (mTR) gene exhibit progressive telomere shortening over 4–5 generations without morphological abnormalities (Blasco et al. 1997). Intriguingly, the late-generation mice show critically short telomeres and increased chromosomal abnormalities including end-to-end fusions and other chromosomal rearrangements. While the late-generation telomerase-negative mice did not exhibit classic aging symptoms, they show shortened life span and various types of organ degenerative phenotypes such as deficient wound healing, intestinal atrophy, and infertility (Hemann et al. 2001; Herrera et al. 1999; Lee et al. 1998; Rudolph et al. 1999).

Mice with mTERT deleted show phenotypes similar to mTR-deficient mice with no apparent aging characteristics in early generations, but the later-generation mice show impaired functions in testes, bone marrow, and intestinal tissue (Chiang et al. 2010; Meznikova et al. 2009). These studies support the pivotal role of telomerase in maintaining telomere function in highly proliferative cell populations at the organismal level. Interestingly, the degenerative abnormalities observed in the late-generation telomerase-negative mice are reversible upon reactivation of telomerase. The restoration of telomerase activity in the aging telomerase-deficient mice shows increased telomere length, improved fertility, better median survival, and reduced apoptosis (Jaskelioff et al. 2011). Furthermore, telomerase-reactivated mice show a restored brain size with improved olfactory response and enhanced neural fitness.

Telomerase also plays a critical role in tissue regeneration. Zebrafish have the ability to completely repair various organs including the heart following lesions. Fish that lack active TERT protein fail to regenerate the damaged heart tissue (Bednarek et al. 2015), suggesting that functional telomerase is required for cardiac rejuvenation in zebrafish. In a recent study, a subset of mouse hepatocytes was found to have high levels of telomerase activity. It was proposed that these hepatocytes, following liver injury, can repopulate an injured liver by accelerated clonal proliferation (Lin et al. 2018). These studies suggest the importance of telomerase in tissue regeneration.

While the mouse model has offered unparalleled insights into the effects of telomerase activation at the organismal level, it is essential to be aware of some fundamental differences between mice and humans in telomere biology, telomerase regulation, and cell senescence when applying knowledge learned from mice to humans. In contrast to the 10–15 kb telomere length in humans, laboratory mice have significantly longer telomeres ranging from 40 to 150 kb (or 10–60 kb measured by fluorescence in situ hybridization) but much shorter life spans of only 18–24 months. Thus, it remains unclear whether telomere attrition is causative to organismal aging. Additionally, mouse somatic cells have constitutive telomerase expression, and murine cells in culture do not undergo telomere-induced replicative senescence. Certain wild-derived mouse strains have telomere lengths similar to humans and can be used as a better model system for the study of human short telomere syndromes (Strong et al. 2011). In the future, it is crucial to validate the mouse findings in humans or other primate model organisms.

Applications: Telomerase-Mediated Therapeutics

As supported by a large body of research studies, telomerase activation has a significant effect on reversing cellular aging by replenishing telomere repeats. Small molecules that activate hTERT expression and restore functional telomerase in human cells have been identified by screening natural products or synthetic compound libraries and proposed to have a positive effect on preventing telomere shortening and delaying cellular aging (ref to ▶ "Antiaging Strategies"). For example, a bioactive compound isolated from the plant *Astragalus membranaceus* was shown to activate telomerase with noticeable increases in telomere length in mice and humans (Bernardes de Jesus et al. 2011; Fauce et al. 2008). Furthermore, sex hormone therapy was found to activate telomerase, mitigate symptoms of aplastic anemia, and increase survival in an aplastic anemia mouse model (Bär et al. 2015). Following administration of a synthetic male sex hormone in patients with short telomeres, telomere elongation was observed in their leukocytes (Townsley et al. 2016). The specificity and the exact activation mechanism of these compounds need to be carefully elucidated before any safe therapeutic use. Moreover, telomerase activation also has a potential risk of increased cancer incidence. Therefore, careful control of telomerase activity in human cells and tissues is crucial and desired.

An interesting study tests the possibility of telomerase activation gene therapy in adult mice utilizing adeno-associated viral vectors (AAV9). Mice treated with AAV9-mTERT showed telomerase reactivation in various tissues conferring enhanced epithelial tissue blockade, improved cognitive function, insulin metabolism, and bone density (Bernardes de Jesus et al. 2012). Moreover, gene therapy-treated mice showed a significant increase in median life spans following therapeutic intervention. Mice introduced with catalytically inactive mTERT, however, did not have regenerative capabilities, suggesting telomere length maintenance function of telomerase being responsible for the fitness improvement. The AAV-mTERT-mediated telomerase activation system does not seem to increase cancer incidence, possibly because it targets cells that have undergone mitosis rather than highly proliferative tissues, and the non-integrative AAV vectors being lost upon successive cell divisions (Bernardes de Jesus et al. 2012; Muñoz-Lorente et al. 2018).

Adult stem cells are telomerase-positive and thus capable of self-renewal over a vast number of cell divisions. However, adult stem cells from older adults show considerably shorter telomere length than younger adults, suggesting that adult stem cells are not truly immortal and experience notable shortening of telomeres over the life span of an individual (Vaziri et al. 1994). Therefore, compounds that enhance the already existing telomerase activity in adult stem cells, but not activating telomerase synthesis in somatic cells, would be an attractive and safer strategy for delaying aging at the cellular and/or organismal levels, as overexpression of telomerase may increase cancer incidence.

Telomerase inhibitors, on the other hand, are considered potential anticancer drugs, as telomerase is activated in most cancer cells (85–90%). Although a plethora of anticancer approaches targeting telomerase has previously been evaluated, only a handful of compounds have reached clinical trials. These include a hTR complementary modified oligonucleotide inhibitor (Asai et al. 2003) and anticancer vaccines based on hTERT peptides (Jafri et al. 2016). A major challenge of telomerase inhibitory approaches in tumor reduction is the time required for telomeres to be shortened to a critical length and induce cell death from the time of treatment. This time lag may either allow tumors to increase cell mass or activate additional pathways to maintain telomere length (Shay 2016). Combinatorial therapy integrating telomerase-mediated and conventional cancer treatments may overcome such challenge.

Future Directions

Telomerase activity and levels are delicately regulated in human stem cells. Any defects or deficiencies of telomerase in stem cells lead to short telomere syndromes, while overexpression of telomerase may lead to increased cancer incidence. Understanding the fundamental mechanisms of telomerase regulation would pave the ways to developing means for precise control or fine-tuning of telomerase activity in specific cell types, which could lead to effective antiaging therapeutics or treatments for patients with short telomere syndromes.

Cross-References

▶ Antiaging Strategies
▶ Cell Senescence
▶ Telomerase
▶ Telomeres

References

Asai A, Oshima Y, Yamamoto Y et al (2003) A novel telomerase template antagonist (GRN163) as a potential anticancer agent. Cancer Res 63:3931–3939

Bär C, Huber N, Beier F et al (2015) Therapeutic effect of androgen therapy in a mouse model of aplastic anemia produced by short telomeres. Haematologica 100:1267–1274. https://doi.org/10.3324/haematol.2015.129239

Bednarek D, González-Rosa Juan M, Guzmán-Martínez G et al (2015) Telomerase is essential for zebrafish heart regeneration. Cell Rep 12:1691–1703. https://doi.org/10.1016/j.celrep.2015.07.064

Bernardes de Jesus B, Schneeberger K, Vera E et al (2011) The telomerase activator TA-65 elongates short telomeres and increases health span of adult/old mice without increasing cancer incidence. Aging Cell 10:604–621. https://doi.org/10.1111/j.1474-9726.2011.00700.x

Bernardes de Jesus B, Vera E, Schneeberger K et al (2012) Telomerase gene therapy in adult and old mice delays aging and increases longevity without increasing cancer. EMBO Mol Med 4:691–704. https://doi.org/10.1002/emmm.201200245

Blasco MA, Lee HW, Hande MP et al (1997) Telomere shortening and tumor formation by mouse cells lacking telomerase RNA. Cell 91:25–34. https://doi.org/10.1016/S0092-8674(01)80006-4

Bodnar AG, Ouellette M, Frolkis M et al (1998) Extension of life-span by introduction of telomerase into normal human cells. Science 279:349–352. https://doi.org/10.1126/science.279.5349.349

Chiang YJ, Calado RT, Hathcock KS et al (2010) Telomere length is inherited with resetting of the telomere set-point. Proc Natl Acad Sci U S A 107:10148. https://doi.org/10.1073/pnas.0913125107

de Lange T (2018) Shelterin-mediated telomere protection. Annu Rev Genet 52:223–247. https://doi.org/10.1146/annurev-genet-032918-021921

Fagagna dA, Reaper PM, Clay-Farrace L et al (2003) A DNA damage checkpoint response in telomere-initiated senescence. Nature 426:194. https://doi.org/10.1038/nature02118

Fauce SR, Jamieson BD, Chin AC et al (2008) Telomerase-based pharmacologic enhancement of antiviral function of human CD8+ T lymphocytes. J Immunol 181:7400–7406. https://doi.org/10.4049/jimmunol.181.10.7400

Harley CB (2002) Telomerase is not an oncogene. Oncogene 21:494. https://doi.org/10.1038/sj.onc.1205076

Harley CB, Futcher AB, Greider CW (1990) Telomeres shorten during ageing of human fibroblasts. Nature 345:458–460. https://doi.org/10.1038/345458a0

Harley CB, Vaziri H, Counter CM et al (1992) The telomere hypothesis of cellular aging. Exp Gerontol 27:375–382. https://doi.org/10.1016/0531-5565(92)90068-B

Hayflick L (1965) The limited in vitro lifetime of human diploid cell strains. Exp Cell Res 37:614–636. https://doi.org/10.1016/0014-4827(65)90211-9

Heiss NS, Knight SW, Vulliamy TJ et al (1998) X-linked dyskeratosis congenita is caused by mutations in a highly conserved gene with putative nucleolar functions. Nat Genet 19:32. https://doi.org/10.1038/ng0598-32

Hemann MT, Rudolph KL, Strong MA et al (2001) Telomere dysfunction triggers developmentally regulated germ cell apoptosis. Mol Biol Cell 12:2023–2030. https://doi.org/10.1091/mbc.12.7.2023

Herrera E, Samper E, Martin-Caballero J et al (1999) Disease states associated with telomerase deficiency appear earlier in mice with short telomeres. EMBO J 18:2950–2960. https://doi.org/10.1093/emboj/18.11.2950

Horn S, Figl A, Rachakonda PS et al (2013) TERT promoter mutations in familial and sporadic melanoma. Science 339:959–961. https://doi.org/10.1126/science.1230062

Huang FW, Hodis E, Xu MJ et al (2013) Highly recurrent TERT promoter mutations in human melanoma. Science 339:957–959. https://doi.org/10.1126/science.1229259

Jafri MA, Ansari SA, Alqahtani MH et al (2016) Roles of telomeres and telomerase in cancer, and advances in telomerase-targeted therapies. Genome Med 8:69. https://doi.org/10.1186/s13073-016-0324-x

Jaskelioff M, Muller FL, Paik JH et al (2011) Telomerase reactivation reverses tissue degeneration in aged telomerase-deficient mice. Nature 469:102–107. https://doi.org/10.1038/nature09603

Kelleher C, Teixeira MT, Forstemann K et al (2002) Telomerase: biochemical considerations for enzyme and substrate. Trends Biochem Sci 27:572–579. https://doi.org/10.1016/S0968-0004(02)02206-5

Kim N, Piatyszek M, Prowse K et al (1994) Specific association of human telomerase activity with immortal cells and cancer. Science 266:2011–2015. https://doi.org/10.1126/science.7605428

Lee HW, Blasco MA, Gottlieb GJ et al (1998) Essential role of mouse telomerase in highly proliferative organs. Nature 392:569–574. https://doi.org/10.1038/33345

Lewis KA, Tollefsbol TO (2016) Regulation of the telomerase reverse transcriptase subunit through epigenetic mechanisms. Front Genet 7:1–12. https://doi.org/10.3389/fgene.2016.00083

Lin S, Nascimento EM, Gajera CR et al (2018) Distributed hepatocytes expressing telomerase repopulate the liver in homeostasis and injury. Nature 556:244–248. https://doi.org/10.1038/s41586-018-0004-7

Liu K, Hodes RJ, N-p W (2001) Cutting edge: telomerase activation in human T lymphocytes does not require increase in telomerase reverse transcriptase (hTERT) protein but is associated with hTERT phosphorylation and nuclear translocation. J Immunol 166:4826–4830. https://doi.org/10.4049/jimmunol.166.8.4826

Mangaonkar AA, Patnaik MM (2018) Short telomere syndromes in clinical practice: bridging bench and bedside. Mayo Clin Proc 93:904–916. https://doi.org/10.1016/j.mayocp.2018.03.020

Meyerson M, Counter CM, Eaton EN et al (1997) hEST2, the putative human telomerase catalytic subunit gene, is up-regulated in tumor cells and during immortalization. Cell 90:785–795. https://doi.org/10.1016/S0092-8674(00)80538-3

Meznikova M, Erdmann N, Allsopp R et al (2009) Telomerase reverse transcriptase-dependent telomere equilibration mitigates tissue dysfunction in mTert heterozygotes. Dis Model Mech 2:620–626. https://doi.org/10.1242/dmm.004069

Mitchell JR, Wood E, Collins K (1999) A telomerase component is defective in the human disease dyskeratosis congenita. Nature 402:551. https://doi.org/10.1038/990141

Muñoz-Lorente MA, Martínez P, Tejera Á et al (2018) AAV9-mediated telomerase activation does not accelerate tumorigenesis in the context of oncogenic K-Ras-induced lung cancer. PLoS Genet 14:e1007562. https://doi.org/10.1371/journal.pgen.1007562

Olovnikov AM (1973) A theory of marginotomy. The incomplete copying of template margin in enzymic synthesis of polynucleotides and biological significance of the phenomenon. J Theor Biol 41:181–190. https://doi.org/10.1016/0022-5193(73)90198-7

O'Sullivan RJ, Almouzni G (2014) Assembly of telomeric chromatin to create ALTernative endings. Trends Cell Biol 24:675–685. https://doi.org/10.1016/j.tcb.2014.07.007

Podlevsky JD, Chen JJ-L (2016) Evolutionary perspectives of telomerase RNA structure and function. RNA Biol 13:720–732. https://doi.org/10.1080/15476286.2016.1205768

Rudolph KL, Chang S, Lee H-w et al (1999) Longevity, stress response, and cancer in aging telomerase-deficient mice. Cell 96:701–712. https://doi.org/10.1016/S0092-8674(00)80580-2

Schmidt JC, Cech TR (2015) Human telomerase: biogenesis, trafficking, recruitment, and activation. Genes Dev 29:1095–1105. https://doi.org/10.1101/gad.263863.115

Shay JW (2016) Role of telomeres and telomerase in aging and cancer. Cancer Discov 6:584–593. https://doi.org/10.1158/2159-8290.CD-16-0062

Strong MA, Vidal-Cardenas SL, Karim B et al (2011) Phenotypes in mTERT$^{+/-}$ and mTERT$^{-/-}$ mice are due to short telomeres, not telomere-independent functions of telomerase reverse transcriptase. Mol Cell Biol 31:2369–2379. https://doi.org/10.1128/MCB.05312-11

Townsley DM, Dumitriu B, Liu D et al (2016) Danazol treatment for telomere diseases. N Engl J Med 374:1922–1931. https://doi.org/10.1056/NEJMoa1515319

Ulaner GA, Hu JF, Vu TH et al (2000) Regulation of telomerase by alternate splicing of human telomerase reverse transcriptase (hTERT) in normal and neoplastic ovary, endometrium and myometrium. Int J Cancer 85:330–335. https://doi.org/10.1002/(SICI)1097-0215(20000201)85:3<330::AID-IJC6>3.0.CO;2-U

Vaziri H, Dragowska W, Allsopp RC et al (1994) Evidence for a mitotic clock in human hematopoietic stem cells: loss of telomeric DNA with age. Proc Natl Acad Sci U S A 91:9857–9860. https://doi.org/10.1073/pnas.91.21.9857

Vulliamy T, Marrone A, Goldman F et al (2001) The RNA component of telomerase is mutated in autosomal dominant dyskeratosis congenita. Nature 413:432. https://doi.org/10.1038/35096585

Wang J, Xie LY, Allan S et al (1998) Myc activates telomerase. Genes Dev 12:1769–1774. https://doi.org/10.1101/gad.12.12.1769

Watson JD (1972) Origin of concatemeric T7 DNA. Nat New Biol 239:197–201. https://doi.org/10.1038/newbio239197a0

Efficacy Belief

▶ Perceived Control

Efficient Performance vs Extended Longevity

▶ Cessation of Somatic Growth Aging Theory

eHealth

▶ Digital Health
▶ Health Informatics

Elastic Demand/Inelastic Demand in Social Services

Liyun Wu
School of Social Work, Norfolk State University, Norfolk, VA, USA

Synonyms

Elasticity of demand of social services; Price elasticity of demand of social services

Definition

The concept of elasticity of demand plays a crucial role in many economic discussions and studies. The price elasticity of demand is a ratio which is defined by the percentage change in the quantity demanded over the percentage change in the price of a good or service (Dean et al. 2016). Based on different levels of responsiveness to changes in price, elasticities can be divided into three broad categories: elastic, inelastic, and unitary. The demand of a good or service is said to be elastic when the quantity demanded changes significantly with a change in price, and the elasticity is greater than one. The demand is inelastic when the quantity demanded does not change significantly with a change in price, and the elasticity is less than one. The demand is unitary when the percentage change in the quantity demanded equals the percentage change in price, and the elasticity equals to exactly one.

Overview

An understanding of price elasticities is of growing importance for social services sectors as governments increasingly consider new types of taxes and subsidies toward service provision. For instance, school vouchers programs allow education tax dollars to be diverted from public schools to private schools. The housing choice vouchers (HCV) programs provide affordable housing to low-income individuals, seniors, and people with disabilities. An analysis of those demand responsiveness can provide a greater understanding of consumer needs, client behaviors, spending trends, and cost-sharing strategies in order to improve the well-being of individuals, families, communities, and the larger society.

The modern social welfare system is built and strengthened in order to fight poverty, illiteracy, and other social ills which are caused by massive immigration, rapid urbanization, and large-scale industrialization (Karger and Stoesz 2018). As a result of the collective efforts of individuals, religious groups, nonprofit organizations, government entities, and other segments of benevolent society, comprehensive programs and services have been established to address poverty, education, medical care, public housing, homeless, social security, mental health, and substance abuse services to name a few. With the advancement of sciences and technologies in the twenty-first century, there is a growing trend of privatizing social services and building long-lasting public-private partnerships toward more effective social service delivery (Jomo et al. 2016). As the social welfare system has evolved over time, the influx of social service programs makes it inevitable to review the effects of price changes on the demand for service provision.

Key Research Findings

Population aging has become a worldwide phenomenon as people are living longer. According to the most recent statistics from World Health Organization (WHO 2018), the world's population aged 60 years and older was 900 million in 2015, and the number grows at an unprecedented rate and is predicted to become two billion by 2050. Therefore, the demand for health and social services among older population has significantly increased. By taking into consideration the variations in individual choices, the demand or use of particular services can vary substantially. The literature has demonstrated the following factors to have an influential effect on service use: age, chronic illnesses, functional ability, health status, living arrangement, marital status, and other socio-economic variables (Wilkinson and Marmot 2006).

Researchers have applied innovative methodological techniques to understand demand responses for various types of social services. Some studies have utilized natural experiments to estimate elasticities of demand and impacts of policy interventions. Brot-Goldberg et al. (2017) studied a large self-insured firm which shifted the primary health insurance option from the zero cost-sharing into high-deductible plan (HDHP) at the end of 2012, and this policy change affected 160,000 employees and their dependents. There was a considerable demand response to the price change. Sick consumers substantially reduced spending when under deductible, even when expected price was low. By using individual level administrative data from the Chilean private health insurance market, Duarte (2012) estimated price elasticity of demand across a variety of health care services and identified that the demand for acute service (appendectomy) was completely inelastic (0), and the demand for psychological visit was very elastic (-2.08). Moreover, the results revealed two factors which contributed to the size of elasticity: the number of visits and intensity of each visit. Last, elasticities varied among different segments of populations, in that low-income and older individuals were less price-sensitive than high-income and young individuals, respectively.

Some studies used an instrumental variable strategy to get inferential statistics on the price elasticity. Ellis et al. (2017) controlled the endogenous price by instrumenting individual monthly cost shares by employer-year-plan-month average cost shares. Their study analyzed a dataset with 171 million person-months spanning 73 employers from 2008 to 2014, and they found that price elasticities varied a lot across types of service, ranging from high elasticities for pharmaceuticals (-0.44), specialists visits (-0.32), MRIs (-0.29), and mental health and substance abuse (-0.26), to lower elasticities for preventive visits (-0.02) and emergency room visits (-0.04). Kowalski (2016) also utilized instrumental variable strategy to control the endogenous medical care cost. By using a family member's injury to induce price variation, the researcher found that price elasticities varied from -0.76 to -1.49, larger than previous estimates reported by RAND Health Insurance Experiment.

In addition to discussions of methodological challenges in price elasticities, studies also presented empirical evidences on private provision of public insurance. Einav et al. (2018) examined how coinsurance varied across drugs in privately-provided health insurance plans in the context of Medicare Part D. The analytical sample was obtained by using two administrative datasets: one from Medicare Part D and the other from the Centers for Medicare and Medicaid Services. After estimating price elasticities of demand across more than 150 drugs and more than 100 therapeutic classes, the researchers found that private insurers participating in the Medicare Part D program created unique cost-sharing strategies by charging higher consumer coinsurance for drugs with higher-price elasticities.

Prospects

Because of the complexity of human nature, the demand for social services is somewhat more complicated than the demand for typical commercial products or services, and case management practices have to initiate individual service plans (ISP) before finalizing the service delivery. Therefore, it is less straightforward to estimate its elasticity of demand. When developing care policies and practices for older adults, stakeholders need to know why the older adults are demanding social services and what factors contribute to their decision-making process. In addition, researchers need to take caution for endogenous choice of social services and price variables and carefully select the identification strategy in their empirical studies. Research findings have to be viewed with caution because different studies may differ with respect to model assumptions, time periods, data sources, data structure, analytical strategies, and so on. Moreover, future research needs to understand the welfare consequences of large-price responsiveness. Future research also needs to further advance knowledge about private provision of social insurance by taking into consideration moral hazard and risk protection. For instance, what are social benefits for private insurers to set higher price for drugs with higher elasticities?

Cross-References

▶ Care Needs
▶ Case Management
▶ Financial Social Work
▶ Home- and Community-Based Services
▶ Social Services Utilization
▶ Social Work

References

Brot-Goldberg Z, Chandra A, Handel BR, Kolstad J (2017) What does a deductible do? The impact of cost-sharing on health care prices, quantities, and spending dynamics. Q J Econ. https://doi.org/10.1093/qje/qjx013

Dean E, Elardo J, Green M, Wilson B, Berger S (2016) Principles of microeconomics: scarcity and social provisioning. Creative commons attribution. Adapted from Openstax Principles of Economics. Retrieved from https://openoregon.pressbooks.pub/socialprovisioning/

Duarte F (2012) Price elasticity of expenditure across health care services. J Health Econ 31(6):824–841. https://doi.org/10.1016/j.jhealeco.2012.07.002

Einav L, Finkelstein A, Polyakova M (2018) Private provision of social insurance: drug-specific price elasticities and cost sharing in Medicare Part D. Am Econ J Econ Pol 10(3):122–153. https://doi.org/10.1257/pol.20160355

Ellis RP, Martins B, Zhu W (2017) Health care demand elasticities by type of service. J Health Econ 55:232–243. https://doi.org/10.1016/j.jhealeco.2017.07.007

Jomo KS, Chowdhury A, Sharma K, Platz D (2016) Public–private partnerships and the 2030 agenda for sustainable development: fit for purpose. United Nations Department of Economics & Social Affairs (DESA) working paper no. 148. https://www.un.org/esa/desa/papers/2016/wp148_2016.pdf. Accessed 3 Aug 2020

Karger HJ, Stoesz D (2018) American social welfare policy: a pluralist approach, 8th edn. Pearson Publishing, London

Kowalski A (2016) Censored quantile instrumental variable estimates of the price elasticity of expenditure on medical care. J Bus Econ Stat 34(1):107–117. https://doi.org/10.1080/07350015.2015.1004072

Wilkinson RG, Marmot MG (2006) In: Wilkinson RG, Marmot MG (eds) Social determinants of health, 2nd edn. Oxford University Press, Oxford

World Health Organization (2018) Ageing and health. Retrieved from https://www.who.int/news-room/factsheets/detail/ageing-and-health. Accessed 4 Oct 2019

Elasticity of Demand of Social Services

▶ Elastic Demand/Inelastic Demand in Social Services

Elastin, Aging-Related Changes in

Christopher I. Platt, Alexander Eckersley, Matiss Ozols and Michael J. Sherratt
Division of Cell Matrix Biology and Regenerative Medicine, School of Biological Sciences, Faculty of Biology, Medicine and Health, The University of Manchester, Manchester, UK

Synonyms

Elastin, Aging-Related modification; Elastin, Aging-Related degradation

Definition

Age-related changes in elastin refer to modifications such as nonenzymatic cross-linking and/or proteolytic degradation, which adversely affect the mechanical and biochemical properties of elastin-rich tissues and organs. Elastin (in combination with other extracellular matrix components) forms elastic fibers. These structures are primarily responsible for driving passive recoil in organs such as arteries, lungs, and skin.

Overview

Elastin is an abundant extracellular matrix component which is found in many dynamic tissues including skin, arteries, the lungs and ligaments (Yanagisawa and Wagenseil 2019). Within these tissues, elastin is deposited on a scaffold of fibrillin microfibrils to form elastic fibers. These fibers, in conjunction with other extracellular matrix components, mediate passive recoil (sometimes referred to as elasticity) (Sherratt 2009). Elastin is synthesized predominantly in fetal and neonatal development and, in common with many other structural extracellular matrix proteins, persists in human tissues for many decades (Shapiro et al. 1991). As a consequence of their longevity, elastic fibers are subject to the accumulation of damage (due to processes such as mechanical wear and tear, enzymatic activity, oxidation, ultraviolet radiation [UVR], and aberrant cross-linking) (Sherratt 2009). In lungs and arteries, this damage impacts on tissue stiffness and recoil and hence on pulmonary vital capacity and vascular stiffness, respectively (Lai-Fook and Hyatt 2000) (Mitchell 2018). Damage to elastin may also induce systemic and tissue-specific effects via the action of elastin breakdown products which can enter the circulation and induce tissue remodelling in disparate organs (Le Page et al. 2019). In skin, loss of elastic fiber architecture, principally in sun-exposed areas, is a key histological marker of photoaging which in turn is associated with loss of resilience, increased stiffness and fragility, and the development of external signs of aging such as wrinkles (Naylor et al. 2011; Bonte et al. 2019). A key element of the elastic fiber system in skin, the fibrillin-rich oxytalan fibers in the papillary dermis, can be partially restored via the action of topical treatments, such as retinoic acid and extracellular matrix-derived peptides (Watson et al. 2009). In other organs, therapeutic interventions to restore elastic fiber structure and function are limited, although recent advances have been made in the synthesis of elastin-rich biomaterials to replace damaged tissues. Given that decline in tissue elasticity plays a fundamental role in age-related morbidity and mortality, new strategies are urgently required to prevent and/or restore elastin and elastic fiber architecture and function.

Key Research Findings

Elastin and Elastic Fiber Synthesis and Function

In vertebrates, elastin is the key molecular mediator of passive recoil (elasticity) restoring organs such as the skin, lungs, and arteries to their resting position (Yanagisawa and Wagenseil 2019; Subramaniam et al. 2017; O'Brien et al. 2019). Unlike polymers such as rubber, elastin requires water to function as an elastic molecule (Wang et al. 2018). Mature elastin, which is synthesized as the soluble precursor tropoelastin, is a highly insoluble extracellular matrix protein composed of alternating hydrophobic and cross-linking domains (Keeley et al. 2002). In tissues, elastin functions in combination with other proteins in the form of elastic fibers (Sherratt 2009). The lungs and skin contain fibers which are discrete structures connected into a larger network, while in arteries elastin and the associated components form concentric lamellae in the medial layer (Kielty et al. 2007). The second component of the elastic fiber, the fibrillin microfibril, is an elastic macromolecular biopolymer composed predominately of fibrillin-1 but also of other associated proteins (Eckersley et al. 2018). During early development, fibrillin microfibrils, form scaffolds onto which elastin is deposited. The elastin becomes cross-linked by the enzyme lysyl oxidase to eventually form mature elastic fibers in which the elastin core is surrounded by a mantle of fibrillin microfibrils (Godwin et al. 2019). Elastic fibers can interact with cells and other extracellular components via multiple binding sites on fibrillin microfibrils (Thomson et al. 2019). Within the fiber, mature elastin is highly insoluble and relatively resistant to the action of tissue proteases, UVR, and potentially oxidative damage (Hibbert et al. 2019; Schmelzer et al. 2012). Also, in contrast to intracellular proteins, which commonly have half-lives measured in hours or days, measurements of aspartic acid racemization (the rate at which L-aspartic acid spontaneously converts to the

D-form) indicate that once synthesized elastin persists for the lifetime of the individual (Shapiro et al. 1991; Sherratt 2009). As synthesis of new elastin is inhibited in humans postnatally by the action of microRNAs (Ott et al. 2011) elastin and elastic fiber-associated proteins, are prone to accumulating damage with age. Although this entry concentrates on elastin aging in three organs (lungs, arteries and skin), elastin remodelling is also important in other organs and tissues.

Pulmonary Aging

In the aging lung, loss of elastic fibers is associated with reduced elasticity (both compliance and recoil) which stimulates emphysema and compromises forced vital capacity and expiratory volume (Lai-Fook and Hyatt 2000). This loss of mechanical function impacts on quality of life for many older individuals and can be a major contributing factor to mortality as a consequence of pulmonary infections and chronic disorders in the older population (Brandenberger and Muhlfeld 2017). Cellular senescence (see entry for "Cell senescence"), oxidative stress (see entry for "Oxidation damage accumulation aging theory"), fibrosis (primarily deposition of collagen), and a pro-inflammatory environment have all been associated with this age-related loss of function (Parikh et al. 2019; Hecker 2018).

Arterial Aging

In the cardiovascular system, aging of elastin and elastic fibers is associated with arterial stiffening (arteriosclerosis) and raised blood pressure (hypertension) which in turn are linked to heart and renal failure, stroke and aortic aneurysm (see entries for "Circulatory system" and "Vascular diseases of aging") (Mitchell 2018; Sherratt 2009). While the link between increased arterial stiffness, reduced recoil and raised blood pressure is well established, it is still unclear whether structural and mechanical remodelling of the arterial wall causes hypertension (Wilkinson et al. 2009). It is clear, however, that arterial elastin undergoes age-related fragmentation and that, increasingly with age, the microstructure of the wall undergoes profound remodelling. The age-related changes in the mechanical properties of the artery are mediated by remodelling of both elastic lamellae and collagen fibers (Mitchell 2018; Lopez-Guimet et al. 2018). The cellular and biochemical mechanisms underpinning aging of the vasculature are likely to also include nonenzymatic glycation and cross-linking (via the addition of sugar residues), medial layer calcification, mechanical fatigue and the action of elastases (Lee et al. 2019; Azpiazu et al. 2019; Hodis and Zamir 2011; Brankovic et al. 2019). These molecular changes may bring about further remodelling due to altered cell signalling, as is seen in the inherited connective tissue disorder Marfan syndrome in which fibrillin-1 mutations affect sequestration of cell signalling molecules (Wohl et al. 2016).

Skin Aging

In skin, the elastic fiber system consists of elastin-rich elastic fibers in the deep reticular dermis which give way to fibrillin-enriched fibers (elaunin) and ultimately bundles of fibrillin microfibrils (oxytalan) in the papillary dermis (Naylor et al. 2011). This ordered architecture undergoes gradual remodelling in aging photo-protected skin (termed intrinsic aging) and more rapid and extensive remodelling in UVR-exposed skin (known as extrinsic or photoaging) (see entry for "Skin aging") (Bonte et al. 2019). Loss of oxytalan fibers from the papillary dermis is a key histological marker of early photoaging (Watson et al. 1999). Severe photoaging which is characterized by loss of elasticity and the ridged appearance of the dermal-epidermal junction is additionally recognized histologically by the presence of large deposits of disorganized material known as solar elastosis (Han et al. 2014). This elastotic material may be due to breakdown of existing elastic fibers or the aberrant deposition of new material (Knott et al. 2009). While elastin itself is relatively resistant to protease activity and to UVR, fibrillin microfibrils are particularly susceptible to damage via both direct UVR exposure and the indirect photodynamic production of reactive oxygen species (Hibbert et al. 2015; Hibbert et al. 2019). Other factors which may affect dermal elastic fibers include adipocyte hypertrophy in individuals with raised BMI (see entry for "Body mass index, overweight, and insulin resistance"), smoking, pollution and cellular senescence (Ezure and Amano 2015; Morita 2007; Blume-Peytavi et al. 2016; Herbig et al. 2006) (Fig. 1).

Examples of Application

Regardless of the causative mechanisms which induce elastic fiber remodelling, it has been established in skin that some elements of the histological architecture of the elastic fiber system (the oxytalan fibers of the papillary dermis) along with the mechanical function and external appearance of the organ can be at least partially restored by the application of exogenous treatments including retinoic acid and matrix-derived peptides (Watson et al. 2008; Watson et al. 2009; Bradley et al. 2015). However, long-term use of topical retinoic acid can also induce skin thinning and inflammation. The stimulus of inflammatory pathways in treatments such as microneedling may be beneficial in promoting repair of dermal collagen and elastic fibers (El-Domyati et al. 2018)

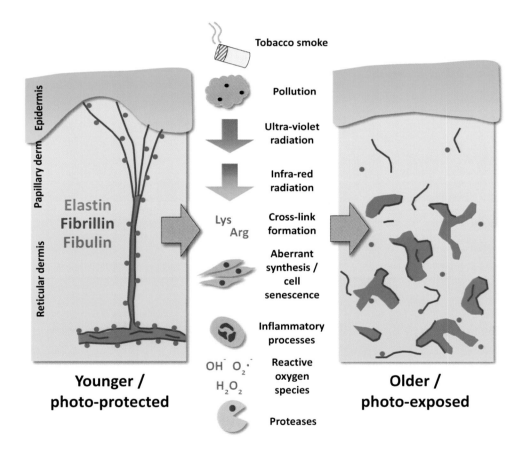

Elastin, Aging-Related Changes in, Fig. 1 Elastin and elastic fiber composition and architecture in young and old skin. In young, photo-protected skin, elastin combines with other components such as fibrillin microfibrils and fibulins to form large diameter fibers in the reticular dermis. In the papillary dermis, bundles of fibrillin microfibrils (oxytalan fibers) intercalate in the dermal-epidermal junction. Numerous factors including tobacco smoke, pollution and radiation may bring about remodelling of the elastic fiber system via the action of inflammatory processes including proteases and oxidative stress (reactive oxygen species) in addition to the gradual formation of insoluble cross-links with sugar molecules and the impact of cellular senescence. In older, photo-exposed skin, the characteristic architecture of the elastic fiber system is replaced with a disorganized material known as solar elastosis which is primarily confined to the reticular dermis and, in an in vitro system, plant extracts can stabilize collagen and elastin subjected to enzymatic degradation (Ku and Mun 2008). In internal organs such as arteries, calcification of medial elastic fibers and protease activity have been inhibited by aluminum ions (Bailey et al. 2004). More recently, this group has shown that intravenously administered nanoparticles, incorporating a calcium-chelating agent and polyphenol (pentagalloyl glucose), can reduce medial elastin calcification, macrophage recruitment and protease activity while avoiding the side effects associated with systemic chelation therapy (Lei et al. 2014; Nosoudi et al. 2016).

Future Directions of Research

Despite over 100 years of research (Richards and Gies 1902), key questions remain with regard to elastic fiber composition, function and regeneration. The precise mechanisms via which elastic fibers assemble in vivo and hence the methods required to synthesize elastic fibers in vitro are not yet fully characterized. Furthermore, the extent to which elastic fiber composition and function (i) are tailored to individual tissues, (ii) vary with developmental stage, (iii) are compromised in aging and disease, and (iv) can be mimicked in vitro to produce functional tissue constructs also remains poorly understood (Eckersley et al. 2018). Finally, the potential to manipulate cell-matrix signalling via elastin and elastic fiber-derived peptides (matrikines) is an area ripe for exploration (Duca et al. 2004; Le Page et al. 2019).

Summary

Elastin is an insoluble extracellular matrix polymer which, in the form of multicomponent elastic fibers, mediates passive recoil (elasticity) in many vertebrate tissues. In humans, the protein is synthesized in early development, while its translation is inhibited in adults. As a consequence, elastin and elastic fibers accumulate damage with age profoundly compromising both mechanical and biochemical tissue function in lungs, arteries and skin. In skin, enhanced synthesis of elastic fiber components can be induced by exogenous treatments, although the precise mechanisms of elastic fiber assembly (both in vitro and in vivo) remain to be elucidated. The authors would like to acknowledge the support of a Walgreens Boots Alliance program grant.

Cross-References

▶ Body Mass Index, Overweight, and Insulin Resistance
▶ Cell Senescence
▶ Circulatory System
▶ Oxidation Damage Accumulation Aging Theory (The Novel Role of Glutathione)
▶ Skin Aging
▶ Vascular Diseases of Ageing

References

Azpiazu D, Gonzalo S, Villa-Bellosta R (2019) Tissue non-specific alkaline phosphatase and vascular calcification: a potential therapeutic target. Curr Cardiol Rev 15:91–95. https://doi.org/10.2174/1573403x14666181031141226

Bailey M, Pillarisetti S, Jones P, Xiao H, Simionescu D, Vyavahare N (2004) Involvement of matrix metalloproteinases and tenascin-C in elastin calcification. Cardiovasc Pathol 13:146–155. https://doi.org/10.1016/s1054-8807(04)00009-2

Blume-Peytavi U, Kottner J, Sterry W, Hodin MW, Griffiths TW, Watson REB, Hay RJ, Griffiths CEM (2016) Age-associated skin conditions and diseases: current perspectives and future options. Gerontologist 56:S230–S242. https://doi.org/10.1093/geront/gnw003

Bonte F, Girard D, Archambault J-C, Desmouliere A (2019) Skin changes during ageing. Subcell Biochem 91:249–280. https://doi.org/10.1007/978-981-13-3681-2_10

Bradley EJ, Griffiths CEM, Sherratt MJ, Bell M, Watson REB (2015) Over-the-counter anti-ageing topical agents and their ability to protect and repair photoaged skin. Maturitas 80:265–272. https://doi.org/10.1016/j.maturitas.2014.12.019

Brandenberger C, Muhlfeld C (2017) Mechanisms of lung aging. Cell Tissue Res 367:469–480. https://doi.org/10.1007/s00441-016-2511-x

Brankovic SA, Hawthorne EA, Yu XJ, Zhang YH, Assoian RK (2019) MMP12 deletion preferentially attenuates axial stiffening of aging arteries. J Biomech Eng Trans Asme 141:9. https://doi.org/10.1115/1.4043322

Duca L, Floquet N, Alix AJP, Haye B, Debelle L (2004) Elastin as a matrikine. Crit Rev Oncol Hematol 49:235–244. https://doi.org/10.1016/j.critrevonc.2003.09.007

Eckersley A, Mellody KT, Pilkington S, Griffiths CEM, Watson REB, O'cualain R, Baldock C, Knight D, Sherratt MJ (2018) Structural and compositional diversity of fibrillin microfibrils in human tissues. J Biol Chem 293:5117–5133. https://doi.org/10.1074/jbc.RA117.001483

El-Domyati M, Abdel-Wahab H, Hossam A (2018) Combining microneedling with other minimally invasive procedures for facial rejuvenation: a split-face comparative study. Int J Dermatol 57:1324–1334. https://doi.org/10.1111/ijd.14172

Ezure T, Amano S (2015) Decrease of skin elasticity with increment of subcutaneous adipose tissue is associated with decrease of elastic fibers in the dermal layer. Exp Dermatol 24:924–929

Godwin ARF, Singh M, Lockhart-Cairns MP, Alanazi YF, Cain SA, Baldock C (2019) The role of fibrillin and microfibril binding proteins in elastin and elastic fibre assembly. Matrix Biol J Int Soc Matrix Biol. https://doi.org/10.1016/j.matbio.2019.06.006

Han A, Chien AL, Kang S (2014) Photoaging. Dermatol Clin 32:291. https://doi.org/10.1016/j.det.2014.03.015

Hecker L (2018) Mechanisms and consequences of oxidative stress in lung disease: therapeutic implications for an aging populace. Am J Phys Lung Cell Mol Phys 314:1642–1653. https://doi.org/10.1152/ajplung.00275.2017

Herbig U, Ferreira M, Condel L, Carey D, Sedivy JM (2006) Cellular senescence in aging primates. Science 311:1257–1257. https://doi.org/10.1126/science.1122446

Hibbert SA, Watson RE, Gibbs NK, Costello P, Baldock C, Weiss AS, Griffiths CE, Sherratt MJ (2015) A potential role for endogenous proteins as sacrificial sunscreens and antioxidants in human tissues. Redox Biol 5:101–113. https://doi.org/10.1016/j.redox.2015.04.003

Hibbert SA, Watson REB, Griffiths CEM, Gibbs NK, Sherratt MJ (2019) Selective proteolysis by matrix metalloproteinases of photo-oxidised dermal extracellular matrix proteins. Cell Signal 54:191–199. https://doi.org/10.1016/j.cellsig.2018.11.024

Hodis S, Zamir M (2011) Mechanical events within the arterial wall under the forces of pulsatile flow: a review. J Mech Behav Biomed Mater 4:1595–1602. https://doi.org/10.1016/j.jmbbm.2011.01.005

Keeley FW, Bellingham CM, Woodhouse KA (2002) Elastin as a self-organizing biomaterial: use of recombinantly expressed human elastin polypeptides as a model for investigations of structure and self-assembly of elastin. Philos Trans R Soc Lond Ser B Biol Sci 357:185–189

Kielty CM, Stephan S, Sherratt MJ, Williamson M, Shuttleworth CA (2007) Applying elastic fibre biology in vascular tissue engineering. Philos Trans R Soc B Biol Sci 362:1293–1312

Knott A, Reuschlein K, Lucius R, Stab F, Wenck H, Gallinat S (2009) Deregulation of versican and elastin binding protein in solar elastosis. Biogerontology 10:181–190. https://doi.org/10.1007/s10522-008-9165-3

Ku CS, Mun SP (2008) Stabilization of extracellular matrix by Pinus radiata bark extracts with different molecular weight distribution against enzymatic degradation and radicals. Wood Sci Technol 42:427–436. https://doi.org/10.1007/s00226-008-0182-9

Lai-Fook SJ, Hyatt RE (2000) Effects of age on elastic moduli of human lungs. J Appl Physiol 89:163–168

Le Page A, Khalil A, Vermette P, Frost EH, Larbi A, Witkowski JM, Fulop T (2019) The role of elastin-derived peptides in human physiology and diseases. Matrix Biol. https://doi.org/10.1016/j.matbio.2019.07.004

Lee T-W, Kao Y-H, Chen Y-J, Chao T-F, Lee T-I (2019) Therapeutic potential of vitamin D in AGE/RAGE-related cardiovascular diseases. Cell Mol Life Sci. https://doi.org/10.1007/s00018-019-03204-3

Lei Y, Nosoudi N, Vyavahare N (2014) Targeted chelation therapy with EDTA-loaded albumin nanoparticles regresses arterial calcification without causing systemic side effects. J Control Release 196:79–86. https://doi.org/10.1016/j.jconrel.2014.09.029

Lopez-Guimet J, Pena-Perez L, Bradley RS, Garcia-Canadilla P, Disney C, Geng H, Bodey AJ, Withers PJ, Bijnens B, Sherratt MJ, Egea G (2018) Micro CT imaging reveals differential 3D micro-scale remodelling of the murine aorta in ageing and Marfan syndrome. Theranostics 8:6038–6052. https://doi.org/10.7150/thno.26598

Mitchell GF (2018) Aortic stiffness, pressure and flow pulsatility, and target organ damage. J Appl Physiol 125:1871–1880. https://doi.org/10.1152/japplphysiol.00108.2018

Morita A (2007) Tobacco smoke causes premature skin aging. J Dermatol Sci 48:169–175. https://doi.org/10.1016/j.jdermsci.2007.06.015

Naylor EC, Watson RE, Sherratt MJ (2011) Molecular aspects of skin ageing. Maturitas 69:249–256

Nosoudi N, Chowdhury A, Siclari S, Karamched S, Parasaram V, Parrish J, Gerard P, Vyavahare N (2016) Reversal of vascular calcification and aneurysms in a rat model using dual targeted therapy with EDTA- and PGG-loaded nanoparticles. Theranostics 6:1975–1987. https://doi.org/10.7150/thno.16547

O'Brien ME, Chandra D, Wilson RC, Karoleski CM, Fuhrman CR, Leader JK, Pu JT, Zhang YZ, Morris A, Nouraie S, Bon J, Urban Z, Sciurba FC (2019) Loss of skin elasticity is associated with pulmonary emphysema, biomarkers of inflammation, and matrix metalloproteinase activity in smokers. Respir Res 20:11. https://doi.org/10.1186/s12931-019-1098-7

Ott CE, Grunhagen J, Jager M, Horbelt D, Schwill S, Kallenbach K, Guo G, Manke T, Knaus P, Mundlos S, Robinson PN (2011) Micro RNAs differentially expressed in postnatal aortic development down-regulate elastin via 3′ UTR and coding-sequence binding sites. PLoS One 6:12. https://doi.org/10.1371/journal.pone.0016250

Parikh P, Wicher S, Khandalavala K, Pabelick CM, Britt RD, Prakash YS (2019) Cellular senescence in the lung across the age spectrum. Am J Phys Lung Cell Mol Phys 316: L826–L842. https://doi.org/10.1152/ajplung.00424.2018

Richards AN, Gies WJ (1902) Chemical studies of elastin, mucoid, and other proteids in elastic tissue, with some notes on ligament extractives. Am J Phys 7:93–134

Schmelzer CEH, Jung MC, Wohlrab J, Neubert RHH, Heinz A (2012) Does human leukocyte elastase degrade intact skin elastin? FEBS J 279:4191–4200. https://doi.org/10.1111/febs.12012

Shapiro SD, Endicott SK, Province MA, Pierce JA, Campbell EJ (1991) Marked longevity of human lung parenchymal elastic fibers deduced from prevalence of D-aspartate and nuclear weapons-related radiocarbon. J Clin Invest 87:1828–1834

Sherratt MJ (2009) Tissue elasticity and the ageing elastic fibre. Age 31:305–325. https://doi.org/10.1007/s11357-009-9103-6

Subramaniam K, Kumar H, Tawhai MH (2017) Evidence for age-dependent air-space enlargement contributing to loss of lung tissue elastic recoil pressure and increased shear modulus in older age. J Appl Physiol 123:79–87. https://doi.org/10.1152/japplphysiol.00208.2016

Thomson J, Singh M, Eckersley A, Cain SA, Sherratt MJ, Baldock C (2019) Fibrillin microfibrils and elastic fibre proteins: functional interactions and extracellular regulation of growth factors. Semin Cell Dev Biol 89:109–117. https://doi.org/10.1016/j.semcdb.2018.07.016

Wang YJ, Hahn J, Zhang YH (2018) Mechanical properties of arterial elastin with water loss. J Biomech Eng Trans Asme 140:8. https://doi.org/10.1115/1.4038887

Watson RE, Griffiths CE, Craven NM, Shuttleworth CA, Kielty CM (1999) Fibrillin-rich microfibrils are reduced in photoaged skin. Distribution at the dermal-epidermal junction. J Invest Dermatol 112:782–787

Watson REB, Long SP, Bowden JJ, Bastrilles JY, Barton SP, Griffiths CEM (2008) Repair of photoaged dermal matrix by topical application of a cosmetic 'antiageing' product. Br J Dermatol 158:472–477. https://doi.org/10.1111/j.1365-2133.2007.08364.x

Watson REB, Ogden S, Cotterell LF, Bowden JJ, Bastrilles JY, Long SP, Griffiths CEM (2009) A cosmetic 'anti-ageing' product improves photoaged skin: a double-blind, randomized controlled trial. Br J Dermatol 161:419–426. https://doi.org/10.1111/j.1365-2133.2009.09216.x

Wilkinson IB, Mceniery CM, Cockcroft JR (2009) Arteriosclerosis and atherosclerosis guilty by association. Hypertension 54:1213–1215. https://doi.org/10.1161/hypertensionaha.109.142612

Wohl AP, Troilo H, Collins RF, Baldock C, Sengle G (2016) Extracellular regulation of bone morphogenetic protein activity by the microfibril component Fibrillin-1. J Biol Chem 291:12732–12746. https://doi.org/10.1074/jbc.M115.704734

Yanagisawa H, Wagenseil J (2019) Elastic fibers and biomechanics of the aorta: insights from mouse studies. Matrix Biol J Int Soc Matrix Biol. https://doi.org/10.1016/j.matbio.2019.03.001

Elastin, Aging-Related Degradation

▶ Elastin, Aging-Related Changes in

Elastin, Aging-Related Modification

▶ Elastin, Aging-Related Changes in

Elder Abuse

▶ Abuse and Caregiving
▶ Older Adults Abuse and Neglect

Elder Abuse by Spouse/Partner

▶ Domestic Violence in Older Adults

Elder Family Financial Exploitation

▶ Financial Exploitation by Family Members

Elder Financial Abuse in the Family

▶ Financial Exploitation by Family Members

Elder Intimate Violence

▶ Domestic Violence in Older Adults

Elder/Dependent Financial Abuse by Family Members

▶ Financial Exploitation by Family Members

Elder/Vulnerable Financial Abuse by Family Members

▶ Financial Exploitation by Family Members

Elderly

▶ Tumors: Brain

Electrocardiogram Abnormalities

Zhenyan Xu, Jinzhu Hu, Juxiang Li, Jianhua Yu, Qi Chen and Kui Hong
Department of Cardiovascular Medicine, Second Affiliated Hospital of Nanchang University, Nanchang, China

Synonyms

Abnormal electrocardiogram; Arrhythmias

Definition

Age is an independent risk factor for the development of many cardiovascular diseases. With the progression of aging, sinus node pacemaker cells are progressively reduced, and myocardial cells and interstitial fibrosis are significantly increased. The expression and distribution of sodium potassium-calcium channel and gap junction protein become abnormal. Cardiac weight increases, and a series of cardiac structures and electrophysiological changes, such as the incidence and mortality of arrhythmia also increased significantly.

Overview

As societies age and urbanization processes accelerate, older adult are at high cardiovascular disease risks with various abnormal electrocardiogram (ECG). Aging is an independent risk factor for the development of many cardiovascular diseases (Donato et al. 2018). With aging, the incidence and mortality due to abnormal electrocardiogram diseases such as sick sinus syndrome (SSS), atrial fibrillation (AF), and ventricular arrhythmia increase significantly.

At present, the molecular mechanisms of age-related changes in senile arrhythmia are still unclear. The sinus node pacemaker cells progressively reduces with age, while myocardial cells and interstitial fibrosis significantly increases; as a result, sodium potassium-calcium channel expression becomes abnormal, and gap junction protein expression and distribution become abnormal as well (Stöckigt et al. 2017). A series of cardiac structures and electrophysiological changes controlled by microRNA, facilitate the formation of myocardial reentry, directly affecting the electrical triggering of the heart and electrical conduction may be the basic treatment for the development of senile arrhythmia (Cannon et al. 2017; Donato et al. 2018). For arrhythmias in older adults, timely and accurate diagnoses and reasonable and effective treatments should be done to prevent the disease from worsening, reduce the incidence of adverse events such as syncope and malignant arrhythmia, and improve the quality of life in older people.

Key Research Findings

Myocardial Fibrosis and Aging

Cardiomyocytes, fibroblasts, and vascular smooth muscle cells in the heart are connected by a complex matrix composed primarily of collagen to maintain structural integrity and plasticity (Gyöngyösi et al. 2017). Myocardial fibrosis, a common pathological feature of many cardiovascular disorders, is characterized by dysregulated collagen turnover, that collagen synthesis increased predominates and excessive diffuse collagen accumulation in the interstitial and perivascular spaces (Li et al. 2014). The relationship between aging and the development of myocardial fibrosis is of great importance, and aging myocardium is associated with progressive and significant collagen deposition, which is called age-related fibrosis.

The mechanisms of age-related myocardial fibrosis have not been fully characterized yet; nevertheless, the hypothesis is the imbalance of collagen synthesis and degradation. In addition, the established age-related myocardial fibrosis might be reversed when disrupting the collagen homeostasis in animal models (Rosin et al. 2015). Studies have confirmed that atrial fibrosis is closely related to age, and left atrial fibrosis shows a more significant correlation with age (Luo et al. 2013). Myocardial fibrosis disrupts the myocardial architecture and is associated with many cardiac diseases, such as hypertension, heart failure (HF), coronary heart disease (CHD), and several arrhythmias, including atrial fibrillation (AF) and sick sinus syndrome (SSS). Usually, patients with these diseases above can improve clinical symptoms after standard treatment but do not reverse myocardial fibrosis (Gyöngyösi et al. 2017).

Older adults are prone to conduction disorders, which may be related to the infiltration of fibrous tissue and the reduction of muscle components in the conduction tissues after aging. Myocardial fibrosis leads to a decrease in the number of pacemaker cells in the sinus node, a decrease in local myocardial electrical conduction velocity, an increase in conduction heterogeneity, and promotion of micro-reentry, providing a necessary matrix for the occurrence and maintenance of arrhythmia (Luo et al. 2013).

Sick Sinus Syndrome in Older People

The sinoatrial node (SN) is located in the right atrium at the junction of the superior vena cava without a definitive border from the atrial tissue and is the natural pacemaker of the human heart (Liang et al. 2017). A special set of ion channels on the cell membrane is responsible for generating myocardial action potentials (AP) that cause electrical impulses (See ▶ "Smart Homes") (Evans and Shaw 1977; Monfredi et al. 2010). If the SN function is impaired, the heartbeat will be affected subsequently.

Sick sinus syndrome (SSS) is a spectrum of arrhythmias and different electrocardiographic symptoms caused by SN impulse formation dysfunction or sinus node to atrial impulse conduction disorder due to primordial ganglion or surrounding tissue (De Ponti et al. 2018). It has various presentations, such as bradycardia, sinus arrest, sinoatrial block, and alternant episodes of bradycardia and tachycardia (Walsh-Irwin and Hannibal 2015). It may lead to major cardiovascular events, syncope, thromboembolism, and inadequate heart rate response to exercise or stress. The incidence rate of SSS in the United States is expected to increase from 78,000 in 2012 to 172,000 in 2060 (Jensen et al. 2014). This disease usually occurs in older people, and its prevalence is about 1:600 in patients over 65 years old; 78–80% of patients with SSS implanted pacemakers are older than 65 years (Mozaffarian et al. 2016; Rodriguez and Schocken 1990).

Myocardial degenerative fibrosis is one of the main reasons for aging-related SSS (De Ponti et al. 2018). Increased fibrous connective tissue in myocardial tissue, decreased myocardial cells, and continuous progression can lead to organ structural damage and dysfunction, and even failure. As early as 1954, it was first noted that aging is associated with fibrosis of SN (Lev 1954). However, fibrosis does not explain all SSS cases. Although morphological studies in the 1970s showed that most patients with SSS were associated with SN fibrosis, the same study showed that other patients with the same clinical manifestations had normal SN histology (Evans and Shaw 1977). In experimental studies, extensive electrical remodeling and reduced expression of related genes are considered to be possible causes of SSS (Verkerk and Wilders 2015). Therefore, aging is associated with both structural and molecular remodeling in older people.

Due to atrial remodeling, aging causes both intrinsic heart rate palliation and SN conduction prolongation (Dobrzynski et al. 2007). Prolonged atrial conduction is associated with aging and is consistent with the age-related development of interstitial fibrosis. Age is negatively correlated with left atrial wavelength, which may explain age-related changes in atrial matrix and increased prevalence of atrial fibrillation (AF) (Kistler et al. 2004). Structural and anatomic abnormalities are observed. There are both global and regional reductions in atrial voltage with an increase in the heterogeneity of voltage with aging. P-wave duration prolongs and a significant increase in atrial ERP with increasing age occurs (Kistler et al. 2004). Therefore, the correlation between SSS observed during aging supports the hypothesis that both diseases have the same pathophysiological characteristics, and myocardial degenerative fibrosis is a possible link.

Patients with SSS usually complain of fatigue, dizziness, and poor work tolerance. The initial performance of SSS was mainly severe and symptomatic sinus bradycardia. Although normal people might have a slow heartbeat due to increased parasympathetic tone at night, bradycardia caused by SSS is usually observed during the day and has little or no change during physical activity (De Ponti et al. 2018). In some cases, sudden onset of SN arrest and long-term cardiac arrest in SSS can cause syncope if accompanied by significant organic heart disease and without self-protective ventricular escape rhythm. The pathophysiologic mechanism determining SSS also modifies the atrial myocardium to generate arrhythmogenic substrate leading to development of atrial arrhythmias, including AF. It is also called bradycardia-tachycardia syndrome (Tse et al. 2017).

If there is a clear correlation between an abnormal ECG and the above symptoms, SSS can be directly diagnosed. When SSS occurs intermittently, Holter monitoring can be used to assess the association between ECG abnormalities and symptoms. It may also help detect other clinical arrhythmias, such as bradycardia-tachycardia syndrome. When dynamic electrocardiographic evidence is insufficient and highly suspected based on clinical performance, consider using an implantable circular recorder to monitor selected cases, especially in patients with syncope (Brignole et al. 2018). An exercise stress test might help to identify the inherent form of SSS and bradycardia caused by hyperkinetic vagal tone, which may be similar to the electrocardiographic findings caused by SSS.

Symptoms related to SSS can be prevented by pacemaker implantation, but there are still more challenges in therapy. Patients with sinus bradycardia and heart rate greater than 50 beats/min and one-degree atrioventricular block do not require treatment. For patients with long-term sinus bradycardia that heart rate slows down to less than 40 beats/min, and often accompanied by obvious symptoms, permanent pacemaker implantation is the only effective method (Brignole et al. 2018). The prognosis of patients with SSS mostly depends on whether they have basic heart diseases, and there is no significant correlation with whether or not to implant a pacemaker. Older patients with SSS often have multiple heart diseases, which have brought new challenges to the diagnosis and treatment of SSS. Administration of antiarrhythmic drugs to control atrial arrhythmias often worsens SSS, so pacing methods and drug selection often require comprehensive consideration.

Atrial Fibrillation in Older Adults

Atrial fibrillation (AF) is the most common type of arrhythmias in older adults, occurring in 1–2% of the general population. In China, the prevalence rate of AF in adults is 0.77% (Zhou et al. 2004). In all populations studied, AF prevalence and incidence increases with age and aging populations (Proietti et al. 2016). The incidence of AF is more than 9% in people older than 80 years in the United States (Epstein et al. 2013). The average age of male patients with AF is 66.8 years, and the average age of female patients is 74.6 years. According to the American Heart Association 2020 statistical update, the prevalence of AF worldwide is estimated to be 5 million/year and is expected to rise to 12.1–15.9 million by 2050 in the United States (Chung et al. 2020). It is expected that in the European Union, the number of adults age ≥55 years with AF will become more than double between 2010 and 2060, from 8.8 to 17.9 million (Krijthe et al. 2013). The risk of developing AF doubles with each progressive decade of age and exceeds 20% by age of 80 years (Magnani et al. 2016). Older adults with AF are more likely to have comorbid medical conditions, which raises the cost of hospitalizations.

The cause of AF in older adults may be different from that in younger adults. Older patients are more prone to persistent or permanent AF than younger patients. Their comorbidities are also high, including coronary artery disease, chronic heart failure, chronic kidney disease, chronic obstructive pulmonary disease, valvular heart disease, and hypertension. History of bleeding events and transient ischemic attack (TIA) in the past is more common in older patients (Fumagalli et al. 2015). In addition, they have fewer symptoms of palpitations, but main clinical manifestation is dyspnea.

AF can induce heart failure, lead to tachycardia, increase the risk of thromboembolism, especially stroke. AF is associated with fivefold stroke risk and can cause up to 25% of strokes in older adults. The incidence of stroke in patients with AF is significantly higher than that in non-AF. Fifteen percent of patients with stroke were caused by AF, and AF was the first independent cause of ischemic stroke in people over 75 years of age. The risk of death in patients with AF is 1.5 times that of patients with non-AF (Mairesse et al. 2017). Asymptomatic or clinically silent AF is common and patients may not report any symptom commonly attributable to arrhythmia or may experience both symptomatic and asymptomatic episodes of AF (Mairesse et al. 2017). In addition, older-adult patients with AF are prone to hypertension, diabetes, heart failure, and other diseases, and their risk of stroke increases. Older age plays a pivotal role and is a major risk factor for AF leading to ischemic stroke. Therefore, it is included in the $CHADS_2$ and CHA_2DS_2-VASc score systems for predicting stroke risk in patients with AF.

Age as a risk factor for AF, aging involves the degradation of most biochemical and physiological functions in the body. How the pathophysiologic mechanisms of aging increase the likelihood for AF development remain poorly understood. A longer time period during which the atrial myocardium is exposed to external stressors likely plays a role in the association of age and AF as well. Electrical remodeling and structural remodeling of atrial muscles with age may play an important role in the initiation and persistence of atrial fibrillation. Fibrosis is ubiquitous in the atrium of the aging heart and is also a hallmark of the matrix of structural atrial fibrillation. Increased cardiac fibrosis and atrial conduction slowing in patients with atrial fibrillation is observed (Nattel et al. 2014). It has been demonstrated in animal models that ion current changes are associated with age-related changes in electrical activity. Arterial stiffness increases with age, and recent studies have found it to be associated with increased left atrial diameter and increased risk of AF (Fumagalli et al. 2016). Most older patients present with one or more comorbidities. So it appears very difficult if not impossible to distinguish the impact of these comorbidities from true "age"-related factors (Wasmer et al. 2017).

Because complications of AF are serious, it is clinically important for early screening and prevention of stroke. Active screening for AF in high-risk populations is critical to develop a correct stroke prevention strategy, especially in older persons. For patients over the age of 65 years, the presence of atrial fibrillation should be observed at the same time as routine physical examination of pulse or ECG. A 72-h ambulatory electrocardiogram should be performed in patients with unexplained hemorrhagic stroke or transient ischemic attack to screen for paroxysmal atrial fibrillation. If not found, a longer-term ECG recording is recommended. For patients over the age of 75 or at high risk of stroke, an electrocardiogram is also recommended to screen for atrial fibrillation (Mairesse et al. 2017).

Heart rate control seems to be superior to rhythm control, which is also the treatment of optimal choice for older patients. AF is associated with SSS, which complicates the rate control agents treatment, and antiarrhythmic drugs are used to control the ventricular response in AF and facilitate maintenance of normal sinus rhythm, respectively. Antiarrhythmic drugs have been classified according to the Vaughan-Williams classification scheme by their mechanism of action: sodium channel blockers (class I), β-blockers (class II), potassium channel blockers (class III), and calcium channel blockers (class IV). For the rate control treatment of AF, the application of class I and III antiarrhythmic drugs has become a consensus, but due to the obvious systemic adverse reactions of amiodarone, older-adult patients should closely monitor thyroid, liver, and lung function (January et al. 2019).

According to the CHA_2DS_2-VASc scoring system, patients aged >75 years have an increased 2 score and their risk of stroke increases. Advanced age plays a pivotal role in the risk of stroke in AF. Therefore, the anticoagulant therapy of older patients should be given more attention. For patients with AF and an elevated CHA_2DS_2-VASc score of 2 or greater in men, or 3 or greater in women, oral anticoagulants are recommended. Optional drugs include warfarin, dabigatran, rivaroxaban, apixaban, and edoxaban, the later four drugs are currently recommended over warfarin (January et al. 2019). Although AF-related ischemic stroke is a greater risk in the older population, anticoagulant therapy presents some challenges in this population. Older age is associated with a higher risk of bleeding complications, including intracranial hemorrhage, gastrointestinal bleeding, and traumatic hemorrhage secondary to falls, all of which may exacerbate anticoagulants. For the majority of patients with AF who are less than 75 years old, the daily monitoring international standardization ratio (INR) is controlled at 2.0–3.0, and for those older than 75 years, the INR is 1.6–2.5 (Edholm et al. 2015). Elderly patients who undergo coronary artery revascularization are at higher risk of bleeding, especially those over 75 years old. In this setting, if possible, shorter antithrombotic regimens are recommended, and anticoagulation monotherapy is endorsed 6–12 months after an acute coronary syndrome (Diez-Villanueva and Alfonso 2019). In patients with contraindications for long-term anticoagulant therapy, especially those with high risk of bleeding or previous fatal events under this therapy, occlusion or exclusion of the left atrial appendage may be recommended (Yerasi et al. 2018).

Ventricular Arrhythmia in Older People

The incidence of arrhythmia in the elderly population ranges from 16% to 36%, with more than half of them having ventricular arrhythmias. Approximately 3/4 of the population older than 70 years can face a single ventricular premature beat. The prevalence of ventricular premature beats is 8.6% in a healthy male population, which is significantly higher than that of men aged 20–40 years (the incidence was only 0.5%). Persistent ventricular tachycardia and ventricular fibrillation is common cause of sudden cardiac death (SCD). In the United States, there are 18.4–46.2 million older people who die of sudden cardiac death, which is the primary reason for death among older persons (Goldberger et al. 2008).

For older patients with asymptomatic and ventricular premature contractions, no special treatment is required. When the number of pre-contractions is greater than 10,000 times, appropriate antiarrhythmic drugs (such as propafenone) may be considered for treatment (Al-Khatib et al. 2018). In particular, it is necessary to pay attention to whether the patient is accompanied by hemodynamic instability such as syncope. For elderly people without contraindications, beta-blockers can significantly reduce the risk of sudden cardiac death and reduce the occurrence of ventricular arrhythmias. For patients with frequent ventricular premature contraction, persistent monomorphic ventricular tachycardia, idiopathic polymorphic ventricular tachycardia, and idiopathic ventricular fibrillation, transcatheter radiofrequency ablation may be considered. Non-sustained and persistent ventricular tachycardia in elderly patients is associated with structural heart disease (Al-Khatib et al. 2018). On the basis of active treatment of the cause, β-blockers and amiodarone can be given; termination of persistent ventricular tachycardia in non-acute myocardial infarction can be given to lidocaine; patients without cardiac dysfunction can be given sotalol, especially persistent ventricular tachycardia in cardiac disease should be treated with intracardiac defibrillator. For older patients, due to a series of complications such as heart fai*lure and coronary heart disease (especially myocardial infarction), it is more important to be alert to malignant ventricular arrhythmia and electrical storms (Al-Khatib et al. 2018).

Future Directions of Research

Overall, with degenerative changes in the structure and function of the heart caused by aging of the human body, the incidence of arrhythmia is increasing, the pathogenesis tends to be complicated, and the clinical manifestations and treatments are unique. Thus a comprehensive understanding of older adults is needed. Pathogenesis, clinical manifestations, and treatment characteristics can effectively reverse the occurrence of malignant arrhythmia and improve the survival rate of older adults. Human aging is accompanied by increased sympathetic nervous system (SNS) activity, which can lead to a range of adverse cardiovascular consequences including arrhythmia (Balasubramanian et al. 2019). Studying the relationship between aging and sympathetic nerve activity and cardiovascular disease is helpful for the prevention and treatment of cardiovascular diseases related to age and sympathetic nerves. Aging involves a progressive decline of body function, and the level of mitochondrial DNA (MtDNA) deletion is high, resulting in mitochondrial function defects (Baris et al. 2019). Whether this mosaic mitochondrial defect can explain the typical arrhythmia predisposition in the older-adult heart is worthy of further study.

In patients with atrial tachycardia, advanced age is a predictor of multi-source atrial tachycardia. There are no large trials to improve information on complications and success rates in older adults with atrial tachycardia during ablation. Among older-adult patients, weakness, falls, dementia, and increased prevalence of cerebral microbleeds and cerebral amyloid angiopathy can increase the risk of bleeding events. Therefore, anticoagulation therapy for older patients with AF requires a more optimized approach. In addition, the decline in liver and kidney function, weakness, limited life expectancy, and the presence of competitive comorbidities in older adults further complicate the treatment of arrhythmias in patients (Curtis et al. 2018). Although there is a large literature to guide the treatment of patients with arrhythmias, the proportion of older patients in clinical trials is often insufficient, which requires more attention in the future.

There are critical knowledge gaps in the management of arrhythmias in the very elderly population. Current evidence is mostly based on observational studies, registries, or post hoc analyses of randomized trials and is prone to inherent selection bias. Therefore, future randomized studies should include the elderly population and look at outcomes, such as quality of life, functional status, and healthcare costs, in addition to clinical endpoints and mortality.

Summary

In summary, the older adult population is a high-risk group for arrhythmias, and its occurrence may be related to age-related myocardial fibrosis and cardiac structural remodeling and electrical remodeling. The most common types of senile arrhythmia are atrial fibrillation and sick sinus node symptoms. Clinical management should be screened as soon as possible, actively treating organ diseases, standardizing reasonable diagnosis and treatment, preventing complications, and delaying progression of disease.

References

Al-Khatib SM, Stevenson W, Ackerman M, Bryant W, Callans D, Curtis A, Page R (2018) 2017 AHA/ACC/HRS guideline for management of patients with ventricular arrhythmias and the prevention of sudden cardiac death: a report of the American College of Cardiology/American Heart Association Task Force on Clinical Practice Guidelines and the Heart Rhythm Society. J Am Coll Cardiol 72(14):e91–e220. https://doi.org/10.1016/j.jacc.2017.10.054

Balasubramanian P, Hall D, Subramanian M (2019) Sympathetic nervous system as a target for aging and obesity-related cardiovascular diseases. Geroscience 41(1):13–24. https://doi.org/10.1007/s11357-018-0048-5

Baris O, Ederer S, Neuhaus J, von Kleist-Retzow J, Wunderlich C, Pal M, Wunderlich F, Peeva V, Zsurka G, Kunz W (2019) Mosaic deficiency in mitochondrial oxidative metabolism promotes cardiac arrhythmia during aging. Cell Metab 21(5):667–677. https://doi.org/10.1016/j.cmet.2015.04.005

Brignole M, Moya A, de Lange FJ, Deharo J-C, Elliott PM, Fanciulli A, . . . Martín A (2018) 2018 ESC guidelines for the diagnosis and management of syncope. Eur Heart J 39(21):1883–1948. https://doi.org/10.1093/eurheartj/ehy037

Cannon L, Zambon AC, Cammarato A, Zhang Z, Vogler G, Munoz M, . . . Bodmer R (2017) Expression patterns of cardiac aging in *Drosophila*. Aging Cell 16(1):82–92. https://doi.org/10.1111/acel.12559

Chung MK et al (2020) Lifestyle and risk factor modification for reduction of atrial fibrillation: a scientific statement from the American Heart Association. Circulation 141:e750–e772. https://doi.org/10.1161/CIR.0000000000000748

Curtis A, Karki R, Hattoum A, Sharma U (2018) Arrhythmias in patients ≥80 years of age: pathophysiology, management, and outcomes. J Am Coll Cardiol 71(18):2041–2057. https://doi.org/10.1016/j.jacc.2018.03.019

De Ponti R, Marazzato J, Bagliani G, Leonelli FM, Padeletti L (2018) Sick sinus syndrome. Card Electrophysiol Clin 10:183–195. https://doi.org/10.1016/j.ccep.2018.02.002

Diez-Villanueva P, Alfonso F (2019) Atrial fibrillation in the elderly. J Geriatr Cardiol 16:49–53. https://doi.org/10.11909/j.issn.1671-5411.2019.01.005

Dobrzynski H, Boyett MR, Anderson RH (2007) New insights into pacemaker activity: promoting understanding of sick sinus syndrome. Circulation 115:1921–1932. https://doi.org/10.1161/CIRCULATIONAHA.106.616011

Donato A, Machin D, Lesniewski L (2018) Mechanisms of dysfunction in the aging vasculature and role in age-related disease. Circ Res 123(7):825–848. https://doi.org/10.1161/CIRCRESAHA.118.312563

Edholm K, Ragle N, Rondina MT (2015) Antithrombotic management of atrial fibrillation in the elderly. Med Clin N Am 99:417–430. https://doi.org/10.1016/j.mcna.2014.11.012

Epstein AE et al (2013) 2012 ACCF/AHA/HRS focused update incorporated into the ACCF/AHA/HRS 2008 guidelines for device-based therapy of cardiac rhythm abnormalities: a report of the American College of Cardiology Foundation/American Heart Association Task Force on Practice Guidelines and the Heart Rhythm Society. Am Coll Cardiol 61:e6–e75. https://doi.org/10.1016/j.jacc.2012.11.007

Evans R, Shaw DB (1977) Pathological studies in sino-atrial disorder (sick sinus syndrome). Br Heart J 39:778–786

Fumagalli S et al (2015) Age-related differences in presentation, treatment, and outcome of patients with atrial fibrillation in Europe: the EORP-AF General Pilot Registry (EURObservational Research Programme-Atrial Fibrillation). JACC Clin Electrophysiol 1:326–334. https://doi.org/10.1016/j.jacep.2015.02.019

Fumagalli S et al (2016) Atrial fibrillation after electrical cardioversion in elderly patients: a role for arterial stiffness? Results from a preliminary study. Aging Clin Exp Res 28:1273–1277. https://doi.org/10.1007/s40520-016-0620-8

Goldberger JJ et al (2008) American Heart Association/American College of Cardiology Foundation/Heart Rhythm Society scientific statement on noninvasive risk stratification techniques for identifying patients at risk for sudden cardiac death. A scientific statement from the American Heart Association Council on Clinical Cardiology Committee on Electrocardiography and Arrhythmias and Council on Epidemiology and Prevention. J Am Coll Cardiol 52:1179–1199. https://doi.org/10.1016/j.jacc.2008.05.003

Gyöngyösi M et al (2017) Myocardial fibrosis: biomedical research from bench to bedside. Eur J Heart Fail 19:177–191. https://doi.org/10.1002/ejhf.696

January CT et al (2019) 2019 AHA/ACC/HRS focused update of the 2014 AHA/ACC/HRS guideline for the management of patients with atrial fibrillation: a report of the American College of Cardiology/American Heart Association Task Force on Clinical Practice Guidelines and the Heart Rhythm Society Heart Rhythm. Circulation. https://doi.org/10.1016/j.hrthm.2019.01.024

Jensen PN, Gronroos NN, Chen LY, Folsom AR, DeFilippi C, Heckbert SR, Alonso A (2014) Incidence of and risk factors for sick sinus syndrome in the general population. J Am Coll Cardiol 64:531–538. https://doi.org/10.1016/j.jacc.2014.03.056

Kistler PM et al (2004) Electrophysiologic and electro-anatomic changes in the human atrium associated with age. J Am Coll Cardiol 44:109–116. https://doi.org/10.1016/j.jacc.2004.03.044

Krijthe BP et al (2013) Projections on the number of individuals with atrial fibrillation in the European Union, from 2000 to 2060. Eur Heart J 34:2746–2751. https://doi.org/10.1093/eurheartj/eht280

Lev M (1954) Aging changes in the human sinoatrial node. J Gerontol 9:1–9

Li AH, Liu PP, Villarreal FJ, Garcia RA (2014) Dynamic changes in myocardial matrix and relevance to disease: translational perspectives. Circ Res 114:916–927. https://doi.org/10.1161/CIRCRESAHA.114.302819

Liang X, Evans SM, Sun Y (2017) Development of the cardiac pacemaker. Cell Mol Life Sci 74:1247–1259. https://doi.org/10.1007/s00018-016-2400-1

Luo T, Chang CX, Zhou X, Gu SK, Jiang TM, Li YM (2013) Characterization of atrial histopathological and electrophysiological changes in a mouse model of aging. Int J Mol Med 31:138–146. https://doi.org/10.3892/ijmm.2012.1174

Magnani JW et al (2016) Atrial fibrillation and declining physical performance in older adults: the health, aging, and body composition study. Circ Arrhythm Electrophysiol 9:e3525. https://doi.org/10.1161/CIRCEP.115.003525

Mairesse GH et al (2017) Screening for atrial fibrillation: a European Heart Rhythm Association (EHRA) consensus document endorsed by the Heart Rhythm Society (HRS), Asia Pacific Heart Rhythm Society (APHRS), and Sociedad Latinoamericana de Estimulacion Cardiaca y Electrofisiologia (SOLAECE). Europace 19:1589–1623. https://doi.org/10.1093/europace/eux177

Monfredi O, Dobrzynski H, Mondal T, Boyett MR, Morris GM (2010) The anatomy and physiology of the sinoatrial node – a contemporary review. Pacing Clin Electrophysiol 33:1392–1406. https://doi.org/10.1111/j.1540-8159.2010.02838.x

Mozaffarian D et al (2016) Heart disease and stroke statistics – 2016 update: a report from the American Heart Association. Circulation 133:e38–e360. https://doi.org/10.1161/CIR.0000000000000350

Nattel S et al (2014) Early management of atrial fibrillation to prevent cardiovascular complications. Eur Heart J 35:1448–1456. https://doi.org/10.1093/eurheartj/ehu028

Proietti M et al (2016) A population screening programme for atrial fibrillation: a report from the Belgian Heart Rhythm Week screening programme. Europace 18:1779–1786. https://doi.org/10.1093/europace/euw069

Rodriguez RD, Schocken DD (1990) Update on sick sinus syndrome, a cardiac disorder of aging. Geriatrics 45:26–30, 33–36

Rosin NL, Sopel MJ, Falkenham A, Lee TDG, Légaré J (2015) Disruption of collagen homeostasis can reverse established age-related myocardial fibrosis. Am J Pathol 185:631–642. https://doi.org/10.1016/j.ajpath.2014.11.009

Stöckigt F, Beiert T, Knappe V, Baris OR, Wiesner RJ, Clemen CS, ... Schrickel JW (2017) Aging-related mitochondrial dysfunction facilitates the occurrence of serious arrhythmia after myocardial infarction. Biochem Biophys Res Commun 493:604–610. https://doi.org/10.1016/j.bbrc.2017.08.145

Tse G et al (2017) Tachycardia–bradycardia syndrome: electrophysiological mechanisms and future therapeutic approaches (review). Int J Mol Med 39:519–526. https://doi.org/10.3892/ijmm.2017.2877

Verkerk AO, Wilders R (2015) Pacemaker activity of the human sinoatrial node: an update on the effects of mutations in HCN4 on the hyperpolarization-activated current. Int J Mol Sci 16:3071–3094. https://doi.org/10.3390/ijms16023071

Walsh-Irwin C, Hannibal GB (2015) Sick sinus syndrome. AACN Adv Crit Care 26:376–380. https://doi.org/10.1097/NCI.0000000000000099

Wasmer K, Eckardt L, Breithardt G (2017) Predisposing factors for atrial fibrillation in the elderly. J Geriatr Cardiol 14:179–184. https://doi.org/10.11909/j.issn.1671-5411.2017.03.010

Yerasi C et al (2018) An updated systematic review and meta-analysis of early outcomes after left atrial appendage occlusion. J Interv Cardiol 31:197–206. https://doi.org/10.1111/joic.12502

Zhou Z, Hu D, Chen J, Zhang R, Li K, Zhao X (2004) An epidemiological survey of atrial fibrillation in China. Zhonghua Nei Ke Za Zhi 43(7):491–494. (in Chinese)

Electroconvulsive Therapy

Yan Ji[1], Haiman Hou[1], Shuo Li[1] and Yuming Xu[2]
[1]The first affiliated hospital of Zhengzhou University, Zhengzhou, China
[2]Neurology Department, The First Affiliated Hospital of Zhengzhou University, Zhengzhou, Henan, China

Synonyms

Psychiatric therapy; Psychiatric treatment

Definition

Electroconvulsive therapy (ECT) is a psychiatric treatment that provides relief in patients with mental disorders by electrically inducing seizures (Rudorfer et al. 2003). It is a safe and effective interventional treatment for major depression, mania, and catatonia (Rudorfer et al. 2003). ECT has short-term and long-term efficiencies. In the short term, ECT works as an anticonvulsant effect in the frontal lobes, while it works as a neurotrophic effect in the medial temporal lobe in the long run (Abbott et al. 2014).

ECT should be performed with informed consent. It is administered with anesthesia and a muscle relaxant (Ghafur et al. 2012). A round of ECT is usually performed two or three times per week for successive 2–4 weeks or until obvious relief is observed (Read and Bentall 2010). For patients with major depression who do not respond well to drug therapy, ECT provides symptom relief in about half of these patients within 1 year (Dierckx et al. 2012). The most common adverse effect of ECT includes transient confusion and transient memory loss after treatment, especially in those receiving bilateral electrode placement compared with unilateral settings (Gisselmann 2001).

Overview

The ECT, first applied in the 1930s, showed promising improvement in patients with mental disorders. From the 1930s to 1950s, the use of ECT was conflicting. This treatment showed essential improvement in mental disorders, while its overuse brought a negative reputation. The modifications of ECT included introduction of general anesthesia, muscle relaxing agents, oxygenation, and refined electrode placement. These modifications improved therapeutic efficiency of ECT. However, the use of ECT was dramatically reduced by influence of both the psychopharmacology revolution and distorted media presentations in the meantime. During the 1960s and 1970s, the utility of ECT had been reexamined by professional organizations. These replications showed good efficiency and safety of ECT.

Moreover, drug medications and other treatments had shown limitations in clinical practice. Thus, ECT came back into the mainstream of treatments for mental illness (Goodwin 1994).

The efficacy of ECT has been demonstrated in many studies. However, the exact mechanism of ECT remains undetermined. A recent review summarized neuroimaging findings in patients who responded well with ECT. It indicated that immediate relief after ECT was associated with reduced blood flow and metabolism in frontal lobes, while long-term efficiency resulted from increased perfusion and metabolism in medial temporal lobe and hippocampus (Abbott et al. 2014). Previous observational studies demonstrated good safety even in patients who received more than 100 treatments of ECT without obvious damages (Devanand et al. 1991). The electrical stimulus levels of ECT were set above patients' seizure threshold. For bilateral ECT, it was set at approximately 1.5 folds of seizure threshold, while it could be as high as 12 times for unilateral ECT (Rudorfer et al. 2003).

Key Research Findings

Equipment About ECT

ECT treatment and recovery areas should include the standard monitor by anesthesiologists. During an ETC procedure, prepared equipment includes a stethoscope, blood pressure monitor, electrocardiography (EKG) monitor, pulse oximeter, suction apparatus, and an oxygen delivery system. Ventilator and resuscitation equipment are also available along with anesthetic induction supplies and medication. It is necessary for a nasal cannula or face mask to provide supplementary oxygen and a nerve stimulator to assess neuromuscular blockade. Electromyography (EMG), electroencephalography (EEG) leads, and multiple blood pressure cuffs are also needed during the entire procedure.

Preparation

Consent

In some countries, e.g., the United States and Canada, written informed consent is a necessary part of ECT. For the consent to be deemed valid, it should include information about the benefits and risks of the treatment, information about alternatives to ECT, and an assessment of the patient's decision-making capacity (Mankad et al. 2010).

Preparations Before ECT

Patients should be appropriately nil per os (NPO) for the procedure, which includes no light meals for 6 h, no full-fat meals for 8 h, and no clear liquids for 2 h before anesthesia.

Technique

ETC Technique

ECT can be performed both on an inpatient basis and an outpatient basis. Usually, a dedicated suite, a postanesthesia care unit, or an ambulatory surgery site are most frequently used on an outpatient basis, while patients with severe debilitation including substantial medical or psychiatric illness may start on an inpatient basis and transition to an outpatient basis as needed. The ECT stimulus is commonly a brief pulse (0.5–2.0 ms) or ultra-brief pulse (less than 0.5 ms) waveform. While the brief pulse is considered standard, the ultra-brief is considered more tolerable (Sienaert et al. 2018). The electricity dose should be met with certain standards, due to affecting efficacy, the speed of response, and adverse cognitive effects. Seizure threshold should be established via trials, and higher doses of current are needed to find error during the primary treatment session (Petrides and Fink 1996; Petrides et al. 2009). Following initial dose calculation, the dose at subsequent ECT sessions for bilateral ECT is 1.5–2 times of seizure threshold and for right unilateral is 6 times the seizure threshold. During a course of ECT treatment, the seizure threshold commonly increases as the patient develops tolerance.

Surveillance of Life Indicators

Vital signs, including blood oxygen saturation, ECG, and EEG activity, are recorded continuously during ECT procedure. EMG is recorded on the right foot to measure the motor component of seizure activity. During the entire procedure, a nerve stimulator is utilized to monitor succinylcholine and a depolarizing muscle relaxant used to reduce tonic-clonic contractions. As an alternative to EMG, a blood pressure cuff is inflated on the patient's ankle to prevent succinylcholine from entering the foot, allowing a visual monitor of seizure activity with measurement of tonic-clonic contractions. Following intravenous induction, a bite block is placed to protect the patient's tongue and teeth. The initiation and termination of a cerebral seizure being monitored via EEG are recorded from right and left frontal and mastoid positions. Seizure induction is via two electrodes placed bitemporally or a right unilateral electrode; both allow the electrical current to pass into the scalp (Bjølseth et al. 2015).

During ECT, the goal of the anesthesiologist is to facilitate a safe and pain-free experience for the patient (Bryson et al. 2017). Oxygenation pretreatment in patient via nasal cannula or face mask is followed by anesthetic induction and paralysis. Administration of an anticholinergic medication before ECT may prevent arrhythmias, such as bradycardia, asystole, and excessive oral secretions. To induce cerebral vasoconstriction via hypocarbia, the patient is often hyperventilated via bag valve mask before delivery of the electrical stimulus to increase seizure intensity (Bergsholm et al. 1984). The gold standard for induction of anesthesia is methohexital, given at 0.75–1 mg/kg (Bryson et al. 2017). Skeletal muscle relaxation is integral to minimize a motor seizure and avoid musculoskeletal injury. The depolarizing neuromuscular relaxant succinylcholine is used at 0.75–1 mg/kg with an elimination half-life of 41 s (Bryson et al. 2018). In cases where succinylcholine is contraindicated including neuromuscular disease or injury, burn injury, pseudocholinesterase deficiency, or hyperkalemia, nondepolarizing neuromuscular relaxants are the preferred option (American Psychiatric Association 1990).

Standard of ECT Seizures

Although most of the therapeutic ECT seizures last from 15 to 70 s, EEG recording lasts about 25% longer than the motor seizure (Mayur et al. 1999). Seizures lasting less than 15 s may not be clinically effective, while prolonged seizures may cause cognitive impairment. A missed or short seizure should have a follow-up with a short period of hyperventilation and restimulation with a higher electrical current. If a patient is experiencing a prolonged seizure persisting more than 2 min, an induction agent such as propofol or methohexital is either given at a half dose or a benzodiazepine, to suppress seizure activity and avoid neurologic injury. In a patient with numerous missed seizures (Datto et al. 2002), anesthetic induction agents, such as etomidate or ketamine, may be useful as they exhibit less anticonvulsant effects compared to methohexital. While caffeine had been previously administered to prolong seizures, it is no longer the recommendation due to its uncertain safety profile for this purpose.

ETC for Pregnant Patients

In pregnant patients receiving ECT (Rabheru 2001), hyperventilation should be avoided as it can reduce placental blood flow and cause fetal hypoxia. Minimization of NPO time and adequate intravenous fluid hydration is essential to avoid dehydration and premature uterine contractions. In a female with a gestational age greater than 20 weeks, left lateral uterine displacement is essential to optimize maternal venous return and maintain optimal uterine blood flow. In pregnancies greater than 24 weeks gestation considered viable, fetal heart rate and uterine activity should undergo continuous monitoring 30 min before and after each treatment by an obstetrician that can manage obstetric and neonatal emergencies. Because pregnant patients are at higher risk for aspiration pneumonitis due to their full stomach status, anticholinergic drugs that reduce lower esophageal sphincter tone should be avoided, as they can increase reflux (Anderson and Reti 2009). In this case, a no particulate antacid such as sodium citrate is safer for aspiration prophylaxis.

Contraindication

There are few absolute contraindications for ECT. These contraindications include pheochromocytoma and elevated intracranial pressure with mass effect. Relative contraindications include elevated intracranial pressure without mass effect, cardiovascular conduction defects, high-risk pregnancies, aortic, and cerebral aneurysms. Many researches confirmed the efficiency and safety of ETC treatment (Pompili et al. 2014). ECT induced small tonic-clonic seizures, which can cause transient increases in blood pressure, myocardial oxygen consumption, heart rate, and intracranial pressure. Therefore, intensive care is necessary for patients with cardiovascular, pulmonary, or central nervous system compromise. ECT also should be performed with caution under the following circumstances: patients had poorly controlled epilepsy, untreated aneurysms, recent stroke, and solid brain tumor. Some literature summarized the results of a meta-analysis about the utility of ECT in patients with cranial metallic objects (cMO). They concluded that cMO might not represent an absolute contraindication for the performance of ECT. However, the indication for ECT should be put in place thoroughly in patients with cMO (Gahr et al. 2014).

Risks

For patients who plan to receive ECT, a complete history and physical examination need to be collected. The history of cardiac ischemia or arrhythmia, heart failure, and intracranial pathology are significant risk factors (Tess and Smetana 2009). A review of laboratory data can be tailored to the patient's medical history and medications, and electrocardiograms are advisable in patients aged 50 years and older (Devanand et al. 1988). Serum glucose levels require checking both preoperatively and in the recovery room, as ECT treatments can raise blood glucose levels. The equipment for external defibrillation should be available at the patient's bedside, and detection mode should be turned off during the procedure of ETC. Noninvasive fetal monitoring is recommended for pregnant patients after 14–16 weeks, and a no stress test with a tocometer is also recommended after 24 weeks.

The usage of medications should be noted. *Ginkgo biloba*, ginseng, St. John's wort, valerian, and kava may interfere with ECT. Theophylline is a risk in patients with status epilepticus (Devanand et al. 1988). Other drugs, such as short-acting intravenous beta blockers, which may reduce ECT-related hypertension and tachycardia, may also shorten seizure duration and reduce therapeutic efficacy. But these cardiac medications including aspirin, statins, antihypertensive agents, antianginal medications, and clopidogrel can be continued the day of the procedure.

ECT utilizes general anesthesia. There are many kinds of anesthetic induction medications, including barbiturates (such as thiopental and methohexital) and nonbarbiturate agents (such as propofol and etomidate). It is well-known that seizure induced by ECT should last longer than 30 s. Many anesthetic induction medications have an effect on seizure duration. Among them, methohexital is the most commonly used induction agent due to its quick onset, effectiveness, low cost, and minimal effect on seizure duration. Propofol and thiopental have been shown to reduce seizure duration. Etomidate has correlations with myoclonus and increased seizure duration.

Side Effects

Headache and nausea are the most common immediate side effects of ETC, which can be treated with analgesic and antiemetic drugs. The usage of hypnotics and muscle relaxants has aroused serious complications such as muscle tears or bone fractures. General disorientation in the immediate postictal phase is a common but self-limiting occurrence (Ingram et al. 2008). Some patients may experience difficulties in acquiring and retaining new information. This anterograde memory impairment will recover to the baseline level at 1-month follow-up (Ingram et al. 2008). It is easy to arouse the occurrence of retrograde amnesia especially after bilateral ECT (Ingram et al. 2008). Objective measures found it to be usually less than 6 months posttreatment (Fraser et al. 2008). Systematic examinations are recommended to find the nature and extent of retrograde memory impairment (Rose et al. 2003). Efforts are made to further minimize cognitive side effects, by using different electrodepositions and waveforms. However, the three commonly used electrode placements are highly effective, and their cognitive profile does not seem to differ dramatically. Previous studies showed that ultra-brief pulse width (0.3 ms) produced no deterioration in a wide range of cognitive measures (Sienaert et al. 2010). However, a concern of lower AD efficacy in ultra-brief ECT and additional treatment sessions to achieve results comparable with those achieved with standard pulse ECT existed (Kellner 2009).

Attention After Successful ECT

ECT should not be terminated early after an evident remission is achieved. This is analogous to stopping psychotropic drugs early, which will carry a high risk of relapse. Studies reported a relapse rate as high as 64–84%, especially in patients within the first 6 months after a successful treatment course (Prudic et al. 2004). However, over a longer year follow-up, patients receiving ECT had fewer re-hospitalizations, decreased emergency room visits, and lower overall healthcare costs than those who only received treatment with medications. A study observed randomly receiving continuation treatment with placebo, nortriptyline, or a combination of nortriptyline and lithium in patients with MDD remitted with ECT and found patients treated with nortriptyline relapsed or the combination of nortriptyline and lithium had a decreased relapse rate compared to the patients with only placebo (Sackeim et al. 2001b). Another study observed a 37% relapse rate in patients with unipolar depression, who received the lithium-nortriptyline combination and C-ECT (continuation ECT) treatment after remitted with a course of bilateral ECT. Most clinicians are convinced that relapse rates can be further reduced by adapting treatment schedules to symptom emergence in each patient (Kellner and Lisanby 2008), and individuality of treatment, administering the minimum number of treatments, is recommended to achieve sustained remission. Adequate post-ECT treatment, either pharmacotherapy or C-ECT, is indispensable. There are some recommendations about the length of C-ECT. Just like the usage of maintenance pharmacotherapy being advised to be lifelong in patients who are severely ill and medication refractory, C-ECT should also be open-ended (Sienaert and Peuskens 2006), with reevaluations of the treatment plan at least every 6 months. There is neither a reason to set a lifetime maximum number of treatments nor evidence that tolerance develops to ECT. Relapses during C-ECT usually respond rapidly to further ECT and to increases in treatment frequency (Fox 2001).

ECT in Psychological Disease

Major Depression

ECT can be referred to patients with major depression who failed at least three trials of antidepressant medications. The typical ECT stimulus settings are seizure threshold titration or fixed-dose method. During ECT, EEG should be used to monitor seizure onset, regularity, duration, and suppression (Perera et al. 2004). Approximately 50–60% of these patients would benefit from ECT with obvious symptom remission (Kho et al. 2003). A meta-analysis indicated that ECT was equally effective in patients with unipolar and bipolar depression. Symptoms were remised in approximately 51.5% of patients with unipolar depression and 50.9% in those with bipolar depression within 1 year after ECT (Dierckx et al. 2012). Antidepressants after ECT, such as tricyclics, provided longer remission compared to ECT alone (Jelovac et al. 2013). A meta-analysis compared efficiency between ECT and transcranial magnetic stimulation in treatment-resistant major depression. It indicated that ECT provided almost twofold relief compared to transcranial magnetic stimulation (Micallef-Trigona 2014).

Seizure potentiation should be considered if a patient did not respond well or even deteriorated after three to six rounds of treatments. More evidence is needed to establish standardized treatment to induce seizures to guide common clinical practice. Increasing pulse width and/or switch from unilateral to bitemporal stimulus may be effective in inducing seizures when patients failed current treatments. During this procedure, the seizure threshold should be set via stimulus dose-titration method (Weiner and Reti 2017).

Catatonia, Mania, and Schizophrenia

ECT is among the first line of treatments in patients with severe or even life-threatening catatonia (Sienaert et al. 2014). ECT showed temporary and sustained benefits for catatonic patients with autism spectrum disorders (DeJong et al. 2014). However, more evidence is needed to validate the efficiency of ECT in catatonia. The evidence is the same as the treatment of mania. ECT was identified as an effective tool in severe or life-threatening situations with mania, especially in those failed other treatments (Kanba et al. 2013). In addition, ECT was also performed in patients with schizophrenia under circumstances where rapid improvement was needed or other antipsychotic treatments failed (Tharyan and Adams 2005).

Example of Application/Future Directions

ECT is used in the treatment of severe major depression, catatonia, prolonged or severe mania, and acute episodes of schizophrenia. It is also recommended as a treatment for suicidality, severe psychosis, and food refusal secondary to depression. It is highly effective in individuals with treatment resistance when a rapid response is required and for patients that have exhibited a favorable response to ECT previously. ECT for the treatment of psychiatric disorders involves the application of electricity to the scalp in order to induce seizure activity. The efficacy of ECT is highly dependent on technique, including electrode placement and dosage, with remission rates ranging from 20% to more than 80% (Sackeim et al. 2000). ECT can be performed on an inpatient or an outpatient basis. It is administered under anesthesia with a muscle relaxant in order to prevent skeletal muscle contraction and possible injury during tonic-clonic activity. The electroencephalogram is monitored during ECT to confirm seizure activity and to document seizure duration. A usual course of ECT involves multiple administrations, typically given two or three times per week until the patient is no longer suffering from symptoms. ECT is a relatively safe and low-risk procedure that requires interprofessional care coordination among anesthesiologists, psychiatrists, and nurses. In a controlled setting, ECT administered causes a very low mortality rate, but it continues to cause mild memory loss in the long term. ECT can be used safely in elderly patients, in pregnancy, in debilitated patients, in breastfeeding patients, and in persons with cardiac pacemakers or implantable cardioverter defibrillators (Lisanby 2007).

Future Directions

Perhaps the most pressing issue in the field of ECT is to determine an appropriate threshold of electrical stimulus intensity, which can make a balance between better therapeutic response and less cognitive side effects. Ongoing research is needed to determine the individualized optimum therapeutic electrical stimulus and the refinement of administration to reduce undesirable side effects. ECT has been proven to be effective in treating major depression; however, it is still not the first-line treatment (Samalin et al. 2020). More trials are needed to confirm its efficiency in the future, in order to guide clinical practice. Meanwhile, electroencephalogram (EEG) monitoring during a seizure is becoming a promising method to improve the safety and effectiveness of ECT. This would serve as another useful means to administer the lowest effective electric stimulus intensity. Work to date with magnetic seizure therapy has been encouraging, but it is still in the research phase.

Another important problem with ECT is the optimal approach to maintenance with ECT, in order to prevent relapse, which is often seen in treated patients. Sackeim et al. has reported that patients had a relapse rate of 84%, 6 months after-ECT when they did not receive any form of maintenance therapy (Sackeim et al. 2001). Till now, very little is known about its mode of action and underlying mechanism. A recent study showed that the beneficial effects from ECT differed between groups stratified by sex and age (Gueney et al. 2020). These problems are still under research, and forthcoming results will be useful for better application of ECT.

Summary

ECT is now used widely with informed consent as a safe and effective intervention for major depressive disorder and other psychiatric disorders. The efficacy has been well proven by many studies. Before undergoing ECT, patients need a full medical evaluation, as well as an assessment for the risk of general anesthesia. However, follow-up treatment, especially for the maintenance with ECT, is still poorly understood. Meanwhile, cognitive side effects are further important problems that need to be solved or avoided. Additional well-designed studies are needed to further refine administration and monitoring of ECT.

Cross-References

▶ Cognitive Behavioral Therapy (I)
▶ Electroencephalography
▶ Mental Disorders
▶ Psychosocial Behavioral Intervention

References

Abbott CC et al (2014) A review of longitudinal electroconvulsive therapy: neuroimaging investigations. J Geriatr Psychiatry Neurol 27:33–46. https://doi.org/10.1177/0891988713516542

American Psychiatric Association (1990) Task Force on Electroconvulsive Therapy. The practice of ECT: recommendations for treatment, training and priviliging. Convuls Ther 6(2):85–120

Anderson EL, Reti IM (2009) ECT in pregnancy: a review of the literature from 1941 to 2007. Psychosom Med 71(2):235–242. https://doi.org/10.1097/PSY.0b013e318190d7ca

Bergsholm P, Gran L, Bleie H (1984) Seizure duration in unilateral electroconvulsive therapy. The effect of hypocapnia induced by hyperventilation and the effect of ventilation with oxygen. Acta Psychiatr Scand 69(2):121–128

Bjølseth TM et al (2015) Clinical efficacy of formula-based bifrontal versus right unilateral electroconvulsive therapy (ECT) in the treatment of major depression among elderly patients: a pragmatic, randomized, assessor-blinded, controlled trial. J Affect Disord 175:8–17. https://doi.org/10.1016/j.jad.2014.12.054

Bryson EO et al (2017) Individualized anesthetic management for patients undergoing electroconvulsive therapy: a review of current practice. Anesth Analg 124(6):1943–1956. https://doi.org/10.1213/ANE.0000000000001873

Bryson EO et al (2018) Extreme variability in succinylcholine dose for muscle relaxation in electroconvulsive therapy. Australas Psychiatry 26(4):391–393. https://doi.org/10.1177/1039856218761301

Datto C et al (2002) Augmentation of seizure induction in electroconvulsive therapy: a clinical reappraisal. J ECT 18(3):118–125

DeJong H, Bunton P, Hare DJ (2014) A systematic review of interventions used to treat catatonic symptoms in people with autistic spectrum disorders. J Autism Dev Disord 44(9):2127–2136. https://doi.org/10.1007/s10803-014-2085-y

Devanand DP et al (1988) Status epilepticus following ECT in a patient receiving theophylline. J Clin Psychopharmacol 8(2):153

Devanand DP et al (1991) Absence of cognitive impairment after more than 100 lifetime ECT treatments. Am J Psychiatry 148:929–932. https://doi.org/10.1176/ajp.148.7.929

Dierckx B et al (2012) Efficacy of electroconvulsive therapy in bipolar versus unipolar major depression: a meta-analysis. Bipolar Disord 14(2):146–150. https://doi.org/10.1111/j.1399-5618.2012.00997.x

Fox HA (2001) Extended continuation and maintenance ECT for long-lasting episodes of major depression. J ECT 17:60–64

Fraser LM, O'Carroll RE, Ebmeier KP (2008) The effect of electroconvulsive therapy on autobiographical memory: a systematic review. J ECT 24:10–17. https://doi.org/10.1097/YCT.0b013e3181616c26

Gahr M et al (2014) Safety of electroconvulsive therapy in the presence of cranial metallic objects. J ECT 30(1):62–68. https://doi.org/10.1097/YCT.0b013e318295e30f

Ghafur MS et al (2012) Comparison between the effect of liothyronine and piracetam on personal information, orientation and mental control in patients under treatment with ECT. Indian J Psychiatry 54:154–158. https://doi.org/10.4103/0019-5545.99536

Gisselmann A (2001) Practice in electroconvulsive therapy. Ann Fr Anesth Reanim 20:180–186

Goodwin FK (1994) New directions for ECT research. Introduction. Psychopharmacol Bull 30:265–268

Gueney P et al (2020) Electroconvulsive therapy in depression: improvement in quality of life depending on age and sex. J ECT. https://doi.org/10.1097/YCT.0000000000000671

Ingram A, Saling MM, Schweitzer I (2008) Cognitive side effects of brief pulse electroconvulsive therapy: a review. J ECT 24:3–9. https://doi.org/10.1097/YCT.0b013e31815ef24a

Jelovac A, Kolshus E, McLoughlin DM (2013) Relapse following successful electroconvulsive therapy for major depression: a meta-analysis. Neuropsychopharmacology 38:2467–2474. https://doi.org/10.1038/npp.2013.149

Kanba S, Kato T, Terao T, Yamada K (2013) Guideline for treatment of bipolar disorder by the Japanese

Society of Mood Disorders, 2012. Psychiatry Clin Neurosci 67:285–300. https://doi.org/10.1111/pcn.12060

Kellner CH (2009) Ultrabrief pulse right unilateral ECT: a new standard of care? Psychiatr Times 26:1–4

Kellner C, Lisanby SH (2008) Flexible dosing schedules for continuation electroconvulsive therapy. J ECT 24:177–178. https://doi.org/10.1097/YCT.0b013e318185d2f5

Kho KH et al (2003) A meta-analysis of electroconvulsive therapy efficacy in depression. J ECT 19:139–147

Lisanby SH (2007) Electroconvulsive therapy for depression. N Engl J Med 357(19):1939–1945. https://doi.org/10.1056/NEJMct075234

Mankad MV et al (2010) Clinical manual of electroconvulsive therapy. American Psychiatric Publishing, Washington, DC

Mayur PM et al (1999) Motor seizure monitoring during electroconvulsive therapy. Br J Psychiatry 174:270–272

Micallef-Trigona B (2014) Comparing the effects of repetitive transcranial magnetic stimulation and electroconvulsive therapy in the treatment of depression: a systematic review and meta-analysis. Depress Res Treat 2014(7):135049. https://doi.org/10.1155/2014/135049

Perera TD et al (2004) Seizure expression during electroconvulsive therapy: relationships with clinical outcome and cognitive side effects. Neuropsychopharmacology 29:813–825. https://doi.org/10.1038/sj.npp.1300377

Petrides G, Fink M (1996) The "half-age" stimulation strategy for ECT dosing. Convuls Ther 12(3):138–146

Petrides G et al (2009) Seizure threshold in a large sample: implications for stimulus dosing strategies in bilateral electroconvulsive therapy: a report from CORE. J ECT 25(4):232–237

Pompili M et al (2014) Electroconvulsive treatment during pregnancy: a systematic review. Expert Rev Neurother 14(12):1377–1390. https://doi.org/10.1586/14737175.2014.97237

Prudic J et al (2004) Effectiveness of electroconvulsive therapy in community settings. Biol Psychiatry 55:301–312

Rabheru K (2001) The use of electroconvulsive therapy in special patient populations. Can J Psychiatry 46(8):710–719. https://doi.org/10.1177/070674370104600803

Read J, Bentall R (2010) The effectiveness of electroconvulsive therapy: a literature review. Epidemiol Psichiatr Soc 19:333–347

Rose D et al (2003) Patients' perspectives on electroconvulsive therapy: systematic review. BMJ 326(7403):1363–1368. https://doi.org/10.1136/bmj.326.7403.1363

Rudorfer M, Henry ME, Sackeim HA (2003) Electroconvulsive therapy. Psychiatry 2003:6

Sackeim HA et al (2000) A prospective, randomized, double blind comparison of bilateral and right unilateral electroconvulsive therapy at different stimulus intensities. Arch Gen Psychiatry 57:425–434

Sackeim H et al (2001a) Ultra-brief pulse ECT and the affective and cognitive consequence of ECT. J ECT 17:77

Sackeim HA et al (2001b) Continuation pharmacotherapy in the prevention of relapse following electroconvulsive therapy: a randomized controlled trial. JAMA 285:1299–1307

Samalin L et al (2020) Adherence to treatment guidelines in clinical practice for using electroconvulsive therapy in major depressive episode. J Affect Disord 264:318–323

Sienaert P, Peuskens J (2006) Electroconvulsive therapy: an effective therapy of medication-resistant bipolar disorder. Bipolar Disord 8:304–306. https://doi.org/10.1111/j.1399-5618.2006.00317.x

Sienaert P et al (2010) Randomized comparison of ultra-brief bifrontal and unilateral electroconvulsive therapy for major depression: cognitive side-effects. J Affect Disord 122:60–67. https://doi.org/10.1016/j.jad.2009.06.011

Sienaert P et al (2014) A clinical review of the treatment of catatonia. Front Psychiatry 5:181. https://doi.org/10.3389/fpsyt.2014.00181

Sienaert P et al (2018) Pulse width in electroconvulsive therapy: how brief is brief? J ECT 34(2):73–74

Tess AV, Smetana GW (2009) Medical evaluation of patients undergoing electroconvulsive therapy. N Engl J Med 360(14):1437–1444. https://doi.org/10.1056/NEJMra0707755

Tharyan P, Adams CE (2005) Electroconvulsive therapy for schizophrenia. Cochrane Database Syst Rev CD000076. https://doi.org/10.1002/14651858.CD000076.pub2

Weiner RD, Reti IM (2017) Key updates in the clinical application of electroconvulsive therapy. Int Rev Psychiatry 29:54–62. https://doi.org/10.1080/09540261.2017.1309362

Electroencephalogram

▶ Electrophysiology

Electroencephalography

Yuming Xu

Neurology Department, The First Affiliated Hospital of Zhengzhou University, Zhengzhou, Henan, China

Synonyms

Encephalogram

Definition

Electroencephalography(EEG) is a graph obtained by magnifying and recording the spontaneous and rhythmic electrical activities of the brain from the scalp through the electrodes of precision electronic equipment (Torres 1983).

Overview

In the earliest, Richard Caton (Caton 1875) presented his findings about electrical phenomenon of the exposed cerebral hemispheres in rabbits and monkeys and thought it may be related to brain function. Since then, several researchers have devoted themselves to studying brain activity in animals and human (Pravdich-Neminsky 1913). With studies of the EEG in humans by Hans Berger until 1929, EEG was used for functional analysis of the brain (Berger 1929). First, he named the waves appeared in quiet, closed eyes alpha waves, and the waves observed in the blink of an eye beta waves. Finally, he called these electrical activities EEG which is affected by age, sensory stimuli, disease, and physical physiology.

EEG includes routine EEG, audio EEG, video EEG (short- and long-term video EEG) (Bettini et al. 2014), and pharmaco-EEG. Routine clinical EEG is a useful tool in diagnosing and distinguishing epilepsy, sleep disorders, depth of anesthesia, coma, syncope (Nalbantoglu and Tan 2019), cerebrovascular diseases, encephalitides (Freund and Ritzl 2019), craniocerebral trauma, dementia (Radic et al. 2019), and brain death. It is also used to evaluate prognosis for coma survivors resuscitated after cardiac arrest (Westhall 2017). Video-EEG has a remarkably good yield in diagnosing psychogenic nonepileptic seizures and is routinely used to investigate paroxysmal events (Bettini et al. 2014; Zanzmera et al. 2019). Video-EEG has been used to characterize genetic generalized epilepsies (De Marchi et al. 2017) and refractory epilepsy (Hupalo et al. 2017), while video ambulatory EEG (V-AEEG) is a new technique which could add increased capacity for long-term EEG monitoring to overstretched inpatient video telemetry (IPVT) services and offers a convenient, economical alternative to IPVT. Audio EEG is an emerging technology that helps to understand the complex structure of EEG (Vialatte et al. 2012). It provides a meaningful triage tool for fast evaluation of patients with suspected subclinical seizures (Parvizi et al. 2018), although it cannot replace the traditional EEG. Pharmaco-EEG is another new noninvasive method used to assess the effects of pharmacological compounds on the central nervous system by processing the EEG signals which directly reveal the spontaneous synchronized postsynaptic neuronal activity of the cortex with high temporal resolution (Jobert and Wilson 2015).

Key Research Findings

Electroencephalography records electrical activity generated by the cerebral cortex. This activity reflects the electrical currents that flow in the extracellular spaces of the brain that are the summated effects of innumerable excitatory and inhibitory synaptic potentials upon cortical neurons. Although CT and MRI have developed rapidly, EEG is still an important and irreplaceable technology for localizing brain lesions.

Normal Waves

Patients who are examined should keep their eyes closed and be relaxed in a comfortable position. During the procedure, the patient is asked to breathe deeply 20 times a minute for 3 min. Then, a strong light is placed at about 15 inches before the patient's eyes and flashed at frequencies of 1–20 times per second with the patient blinking. EEG waveforms were characterized by their location, amplitude, frequency continuity, morphology, synchrony, symmetry, and reactivity. Based on the frequency, there are four basic types of normal waves, including alpha (8–12 Hz), beta (above 13–30 Hz), theta (4–7 Hz), and delta (less than 4 Hz) waves. Alpha waves are the feature of the normal background rhythm of the adult EEG recording. They present in normal awake EEG and are usually the dominant rhythm seen in the posterior region of the brain in relaxed older children and adults (Aird and Gastaut 1959). Beta waves are most frequently seen in healthy adults and children, particularly for infants and young children. They are dominant in the frontal and central head regions. When the patient in drowsiness and N1 sleep, the amplitude will increase (Frost Jr. et al. 1973). Theta waves are prominent in the frontocentral head regions and are normally found during drowsiness and sleep. They are normal in wakefulness in children. Delta waves are the most prominent in the frontocentral head regions. It is one important feature of the sleeping EEG (Cordeau 1959). Spikes and sharp waves which are common in the EEG of normal newborns, if detected in adults, generally indicate epilepsy (Constant and Sabourdin 2012).

EEG in Epilepsy

Elderly people are one of the fastest-growing populations in the world, and the incidence of epilepsy in older people is much higher than in other population subgroups (Kirmani et al. 2014). EEG is the most valuable method for the diagnosis of epilepsy and helps to determine its clinical management. In most cases of generalized epilepsies, an MRI scan is normal. Therefore, clinical history and EEG are essential for an accurate diagnosis (King et al. 1998). Different seizure patterns are associated with different clinical syndromes. Generalized spike wave complex with a frontal maximum and the frequency range of 2–5 Hz is the feature EEG of patients with generalized epilepsies (Seneviratne et al. 2012). The slow spike-wave complex usually on a slow background EEG is usually seen in Lennox-Gastaut syndrome, symptomatic generalized epilepsy. During tonic seizures of Lennox-Gastaut syndrome, a low-amplitude fast activity is typical (Armstrong 1993; Markand 2003). In patients with childhood absence epilepsy, the typical interictal EEG characterized by bilateral, synchronous, symmetrical 3 Hz spike-wave complex on a normal background, which appears abruptly in all leads of the EEG simultaneously and disappears almost as suddenly at the end of the seizure could be detected (Seneviratne et al. 2012). Besides, EEG could also be used in therapeutic monitoring in patients with generalized epilepsies (Andersson et al. 1997).

EEG in Cerebral Ischemic Stroke

EEG was not often used for the diagnosis of stroke. It was considered that EEG was only helpful for distinguishing a transient ischemic attack from seizure. Transient ischemic attack is a common disease in the elderly. However, in recent years, scientists found that, in patients with stroke, EEG changes including generalized and focal slowing waves could be detected. The EEG abnormalities are associated with the disease severity, which could be slow activity in the lesion area, faster activity loss in surrounding area, and slow activity in contralateral hemisphere (Jordan 2004). Higher values of the power of delta activity relative to faster activity detected by quantitative EEG analysis are related to the cerebral ischemic lesions and penumbra (Sheorajpanday et al. 2011; Finnigan et al. 2016). For patients with acute ischemic stroke, the most common EEG wave was focal slowing. Abnormal EEG including generalized and focal slowing was significantly associated with clinical deterioration including higher NIHSS score on admission and discharge as well as with hemorrhagic transformation of the ischemic lesion (Wolf et al. 2017).

EEG in Encephalitis

In patients with herpes simplex encephalitis, EEG is of considerable value in the diagnosis. Autoimmune encephalitis is becoming an increasing risk factor of subsequent epilepsy in older people. The characteristic waves are periodic high-voltage sharp waves and slow-wave complexes at intervals of 1–3 per second in the temporal regions (Smith et al. 1975). For many patients with anti-NMDAR encephalitis, beta: delta power ratio (BDPR) is significantly higher, and extreme delta brush is considered to be a novel EEG finding, which is associated with a more prolonged illness (Foff et al. 2017) (Schmitt et al. 2012). Low-voltage EEG activity, which refers to less than 20 mV of background activity without the appearance of a hump or spindle during the acute phase of encephalitis, is one of the most important risk factors for postencephalitic epilepsy (Hosoya et al. 2002).

In Creutzfeldt–Jakob disease (CJD), the typical EEG patterns of periodic sharp complexes were of high diagnostic value (Steinhoff et al. 2004). The other infectious encephalitis is often associated with sharp or spike activity, especially for patients who got seizures. The diagnosis would be more accurate when integrating with CT and MRI.

EEG in Impaired Consciousness

The utility of EEG in the evaluation of prognosis in coma is underestimated. The electrographic seizure activity findings could suggest diagnostic possibility which may be overlooked. Based on the etiology of disease, EEG tends to be slower as consciousness is depressed. In patients with unexplained altered level of consciousness, the frequency of epileptiform activity was high (Ricardo et al. 2012). The EEG response to external stimulation could guide diagnosis and prognosis. A lighter level of coma has electrocerebral responsiveness. Neocortical death may present electrocerebral silence in a technically adequate record implies, in the absence of hypothermia or drug overdose (Hirsch 2011). For patients with locked-in syndrome, consciousness is, in fact, preserved, although the patients appear to be comatose. The patients have quadriplegia and supranuclear paralysis of the facial and bulbar muscles, but the EEG is usually normal.

EEG in Depression

EEG is being widely used in depression and prediction of the effect of antidepressive drugs. The global connectivity among cerebral regions of patients with depression is disturbed. Patients with lower EEG frequency bands during sleep exhibit relative loss of small-word network characteristics and lower synchronization likelihood. EEG derived biomarkers, including EEG frequency bands activity, theta cordance, hemispheric alpha asymmetry, and evoked potentials could predict the response to antidepressive drugs (Baskaran et al. 2012). Pretreatment EEG of an increased propensity toward global processing appears to facilitate response to antidepressant drugs (Jaworska et al. 2018).

EEG in Analgesics

Pharmaco-EEG provides reliable quantitative measures of the psychophysical and neurophysiological response to certain stimuli, such as pain. Both spontaneous EEG and evoked potentials (Eps) are sensitive to changes induced by analgesics. Pentazocine changed the EEG in an opiate-like manner, while flupirtine increased relative power in the theta and beta range (Malver et al. 2014). Pharmaco-EEG may emerge as a biomarker for prediction of treatment response to analgesics in the individual patient.

Prospects

The EEG has been used for many purposes besides the conventional uses of clinical diagnosis and conventional cognitive neuroscience. Long-term EEG recordings in epilepsy patients are still used today for seizure prediction (Keiper 2006). Long-term EEG monitoring compared with normal EEG, sustained monitoring, clinical manifestations of patients with seizures can be observed from video. The patient's clinical seizures and EEG effectively linked to the detection rate of epileptiform abnormal discharge is improved, the detection rate of epilepsy is increased, which helps the diagnosis and clinical classification of epilepsy (Z 2018). Some studies to evaluate EEG differences among syndromes in genetic generalized epilepsy based on quantified data, which suggest that the density and duration of epileptiform discharges can help differentiate among genetic generalized epilepsy syndromes (Seneviratne et al. 2017). Neurofeedback remains an important extension, and in its most advanced form is also attempted as the basis of brain–computer interfaces (Wan et al. 2016). The EEG is also used quite extensively in the field of neuromarketing. The EEG is altered by drugs that affect brain functions, the chemicals that are the basis for psychopharmacology. Berger's early experiments recorded the effects of drugs on EEG. The science of pharmaco-electroencephalography has developed methods to identify substances that systematically alter brain functions for therapeutic and recreational use. EEG abnormality risk varied widely among specific antipsychotics. The risk was particularly high with clozapine and olanzapine, moderate with risperidone and typical neuroleptics, and low with quetiapine (Centorrino et al. 2002). EEGs have been used as evidence in criminal trials in the Indian state of Maharashtra. Brain Electrical Oscillation Signature Profiling (BEOS), an EEG technique, was used in the trial of State of Maharashtra to show that Sharma remembered using arsenic to poison her ex-fiancé, although the reliability and scientific basis of BEOS is disputed (https:// encyclopedia. thefreedictionary.com/Electroencephalography).

A sizeable amount of research is currently being carried out to make EEG devices smaller, more portable, and easier to use. So-called "wearable EEG" is based upon creating low power wireless collection electronics and "dry" electrodes which do not require a conductive gel to be used (A. Casson et al. 2010). Wearable EEG aims to provide small EEG devices that are present only on the head and which can record EEG for days, weeks, or months at a time, as ear-EEG (A. J. Casson et al. 2008).

Such prolonged and easy-to-use monitoring could make a step-change in the diagnosis of chronic conditions such as epilepsy and greatly improve the end-user acceptance of brain control interface systems. Research is also being carried out to identify specific solutions to increase the battery lifetime of wearable EEG devices through the use of the data reduction approach. For example, in the context of epilepsy diagnosis, data reduction has been used to extend the battery lifetime of wearable EEG devices by intelligently selecting, and only transmitting, i.e., electrodes with built-in readout circuit, increasingly being implemented in wearable healthcare and lifestyle applications due to active electrodes' robustness to environmental interference.

Developing a wearable EEG system, with medical-grade signal quality on noise, electrode offset tolerance, common-mode rejection ratio, input impedance, and power dissipation, remains a challenging task (Xu et al. 2017). The emergence of conformal "tattoo" type EEG electrodes is highlighted as a key next step for giving very small and socially discrete units. In addition, new recommendations for the performance validation of novel electrode technologies are given, with standards in this area seen as the current main bottleneck to the wider take up of wearable EEG (Zanzmera et al. 2019).

Summary

EEG is a commonly used examination method in clinical practice. Traditional regular EEG, video EEG, and audio EEG are mainly used for the diagnosis and differential diagnosis of epilepsy, sleep disorders, depth of anesthesia, coma, syncope, cerebrovascular disease, encephalitides, craniocerebral trauma, dementia, and brain death in the older people. As an emerging examination method, pharmaco-EEG is still in the research stage and was expected to be used in clinical to provide a support for diagnosis, treatment, and prognosis evaluation of disease in future.

Cross-Reference

▶ Atherosclerosis
▶ Cerebral Metabolism
▶ Cerebral Palsy
▶ Middle Cerebral Artery Strokes
▶ Stroke

References

Aird RB, Gastaut Y (1959) Occipital and posterior electroencephalographic rhythms. Electroencephalogr Clin Neurophysiol 11:637–656. https://doi.org/10.1016/0013-4694(59)90104-x

Andersson T, Braathen G, Persson A, Theorell K (1997) A comparison between one and three years of treatment in uncomplicated childhood epilepsy: a prospective study. Ii. The eeg as predictor of outcome after withdrawal of treatment. Epilepsia 38:225–232. https://doi.org/10.1111/j.1528-1157.1997.tb01101.x

Armstrong DL (1993) Epileptic syndromes in infancy, childhood and adolescence. J Neuropathol Exp Neurol 52:431–431. https://doi.org/10.1097/00005072-199307000-00012

Baskaran A, Milev R, McIntyre RS (2012) The neurobiology of the eeg biomarker as a predictor of treatment response in depression. Neuropharmacology 63:507–513. https://doi.org/10.1016/j.neuropharm.2012.04.021

Berger H (1929) Uber das elektrenkephalogramm des menschen. Arch Psychiatr Nervenkr 87:527–570. https://doi.org/10.1007/BF01814645

Bettini L, Croquelois A, Maeder-Ingvar M, Rossetti AO (2014) Diagnostic yield of short-term video-eeg monitoring for epilepsy and pness: a european assessment. Epilepsy & behavior: E&B 39:55–58. https://doi.org/10.1016/j.yebeh.2014.08.009

Casson AJ, Smith S, Duncan JS, Rodriguez-Villegas E (2008) Wearable eeg: what is it, why is it needed and what does it entail? Conference proceedings: ... annual international conference of the IEEE engineering in medicine and biology society. IEEE Engineering in Medicine and Biology Society Annual Conference 2008:5867–5870. https://doi.org/10.1109/iembs.2008.4650549

Casson A, Yates D, Smith S, Duncan J, Rodriguez-Villegas E (2010) Wearable electroencephalography. What is it, why is it needed, and what does it entail? IEEE engineering in medicine and biology magazine: the quarterly magazine of the Engineering in Medicine & Biology Society 29:44–56. https://doi.org/10.1109/memb.2010.936545

Caton R (1875) Electrical currents of the brain. J Nerv Ment Dis 2:610

Centorrino F, Price BH, Tuttle M, Bahk WM, Hennen J, Albert MJ, Baldessarini RJ (2002) Eeg abnormalities during treatment with typical and atypical antipsychotics. Am J Psychiatry 159:109–115. https://doi.org/10.1176/appi.ajp.159.1.109

Constant I, Sabourdin N (2012) The eeg signal: a window on the cortical brain activity. Paediatr Anaesth 22:539–552. https://doi.org/10.1111/j.1460-9592.2012.03883.x

Cordeau JP (1959) Monorhythmic frontal delta activity in the human electroencephalogram: a study of 100 cases. Electroencephalogr Clin Neurophysiol 11:733–746. https://doi.org/10.1016/0013-4694(59)90113-0

De Marchi LR, Corso JT, Zetehaku AC, Uchida CGP, Guaranha MSB, Yacubian EMT (2017) Efficacy and safety of a video-eeg protocol for genetic generalized epilepsies. Epilepsy & behavior: E&B 70:187–192. https://doi.org/10.1016/j.yebeh.2017.03.029

Finnigan S, Wong A, Read S (2016) Defining abnormal slow eeg activity in acute ischaemic stroke: delta/alpha ratio as an optimal qeeg index. Clin Neurophysiol 127:1452–1459. https://doi.org/10.1016/j.clinph.2015.07.014

Foff EP, Taplinger D, Suski J, Lopes MB, Quigg M (2017) Eeg findings may serve as a potential biomarker for anti-nmda receptor encephalitis. Clin EEG Neurosci 48:48–53. https://doi.org/10.1177/1550059416642660

Freund B, Ritzl EK (2019) A review of eeg in anti-nmda receptor encephalitis. J Neuroimmunol 332:64–68. https://doi.org/10.1016/j.jneuroim.2019.03.010

Frost JD Jr, Carrie JR, Borda RP, Kellaway P (1973) The effects of dalmane (flurazepam hydrochloride) on human eeg characteristics. Electroencephalogr Clin Neurophysiol 34:171–175. https://doi.org/10.1016/0013-4694(73)90044-8

Hirsch LJ (2011) Classification of eeg patterns in patients with impaired consciousness. Epilepsia 52(Suppl 8):21–24. https://doi.org/10.1111/j.1528-1167.2011.03228.x

Hosoya M, Ushiku H, Arakawa H, Morikawa A (2002) Low-voltage activity in eeg during acute phase of encephalitis predicts unfavorable neurological outcome. Brain and Development 24:161–165. https://doi.org/10.1016/s0387-7604(02)00011-6

Hupalo M, Wojcik R, Jaskolski DJ (2017) Intracranial video-eeg monitoring in presurgical evaluation of patients with refractory epilepsy. Neurol Neurochir Pol 51:201–207. https://doi.org/10.1016/j.pjnns.2017.02.002

Jaworska N, Wang H, Smith DM, Blier P, Knott V, Protzner AB (2018) Pre-treatment eeg signal variability is associated with treatment success in depression. Neuro Image Clinical 17:368–377. https://doi.org/10.1016/j.nicl.2017.10.035

Jobert M, Wilson FJ (2015) Advanced analysis of pharmaco-eeg data in humans. Neuropsychobiology 72:165–177. https://doi.org/10.1159/000431096

Jordan KG (2004) Emergency eeg and continuous eeg monitoring in acute ischemic stroke. Journal of clinical neurophysiology: official publication of the American Electroencephalographic Society 21:341–352

Keiper A (2006) The age of neuroelectronics. New Atlantis (Washington, DC) 11:4–41

King MA, Newton MR, Jackson GD, Fitt GJ, Mitchell LA, Silvapulle MJ, Berkovic SF (1998) Epileptology of the first-seizure presentation: a clinical, electroencephalographic, and magnetic resonance imaging study of 300 consecutive patients. Lancet (London, England) 352:1007–1011. https://doi.org/10.1016/s0140-6736(98)03543-0

Kirmani BF, Robinson DM, Kikam A, Fonkem E, Cruz D (2014) Selection of antiepileptic drugs in older people. Curr Treat Options Neurol 16:295. https://doi.org/10.1007/s11940-014-0295-4

Malver LP, Brokjaer A, Staahl C, Graversen C, Andresen T, Drewes AM (2014) Electroencephalography and analgesics. Br J Clin Pharmacol 77:72–95. https://doi.org/10.1111/bcp.12137

Markand ON (2003) Lennox-gastaut syndrome (childhood epileptic encephalopathy). Journal of clinical neurophysiology: official publication of the American Electroencephalographic Society 20:426–441. https://doi.org/10.1097/00004691-200311000-00005

Nalbantoglu M, Tan OO (2019) The yield of electroencephalography in syncope. Ideggyogyaszati szemle 72:111–114. https://doi.org/10.18071/isz.72.0111

Parvizi J, Gururangan K, Razavi B, Chafe C (2018) Detecting silent seizures by their sound. Epilepsia 59:877–884. https://doi.org/10.1111/epi.14043

Pravdich-Neminsky VV (1913) Ein versuch der registrierung der elektrischen gehirnerscheinungen. Zentralblatt für Physiologie 27:951–960

Radic B, Petrovic R, Golubic A, Bilic E, Borovecki F (2019) Eeg analysis and spect imaging in alzheimer's disease, vascular dementia and mild cognitive impairment. Psychiatr Danub 31:111–115. https://doi.org/10.24869/psyd.2019.111

Ricardo JA, Franca MC Jr, Lima FO, Yassuda CL, Cendes F (2012) The impact of eeg in the diagnosis and management of patients with acute impairment of consciousness. Arq Neuropsiquiatr 70:34–39. https://doi.org/10.1590/s0004-282x2012000100008

Schmitt SE, Pargeon K, Frechette ES, Hirsch LJ, Dalmau J, Friedman D (2012) Extreme delta brush: a unique eeg pattern in adults with anti-nmda receptor encephalitis. Neurology 79:1094–1100. https://doi.org/10.1212/WNL.0b013e3182698cd8

Seneviratne U, Cook M, D'Souza W (2012) The electroencephalogram of idiopathic generalized epilepsy. Epilepsia 53:234–248. https://doi.org/10.1111/j.1528-1167.2011.03344.x

Seneviratne U, Hepworth G, Cook M, D'Souza W (2017) Can eeg differentiate among syndromes in genetic generalized epilepsy? Journal of clinical neurophysiology: official publication of the American Electroencephalographic Society 34:213–221. https://doi.org/10.1097/wnp.0000000000000358

Sheorajpanday RV, Nagels G, Weeren AJ, van Putten MJ, De Deyn PP (2011) Quantitative eeg in ischemic stroke: correlation with functional status after 6 months. Clin Neurophysiol 122:874–883. https://doi.org/10.1016/j.clinph.2010.07.028

Smith JB, Westmoreland BF, Reagan TJ, Sandok BA (1975) A distinctive clinical eeg profile in herpes simplex encephalitis. Mayo Clin Proc 50:469–474

Steinhoff BJ, Zerr I, Glatting M, Schulz-Schaeffer W, Poser S, Kretzschmar HA (2004) Diagnostic value of periodic complexes in creutzfeldt-jakob disease. Ann Neurol 56:702–708. https://doi.org/10.1002/ana.20261

Torres F (1983) Electroencephalography: basic principles, clinical applications and related fields. Arch Neurol 40:191–192. https://doi.org/10.1001/archneur.1983.04050030085025

Vialatte FB, Dauwels J, Musha T, Cichocki A (2012) Audio representations of multi-channel eeg: a new tool for diagnosis of brain disorders. Am J Neurodegener Dis 1:292–304

Wan F, da Cruz JN, Nan W, Wong CM, Vai MI, Rosa A (2016) Alpha neurofeedback training improves ssvep-based bci performance. J Neural Eng 13:036019. https://doi.org/10.1088/1741-2560/13/3/036019

Westhall E (2017) Electroencephalography as a prognostic tool after cardiac arrest. Semin Neurol 37:48–59. https://doi.org/10.1055/s-0036-1595815

Wolf ME, Ebert AD, Chatzikonstantinou A (2017) The use of routine eeg in acute ischemic stroke patients without seizures: generalized but not focal eeg pathology is associated with clinical deterioration. Int J Neurosci 127:421–426. https://doi.org/10.1080/00207454.2016.1189913

Xu J, Mitra S, Van Hoof C, Yazicioglu RF, Makinwa KAA (2017) Active electrodes for wearable eeg acquisition: review and electronics design methodology. IEEE Rev Biomed Eng 10:187–198. https://doi.org/10.1109/rbme.2017.2656388

Z l (2018) Application effect of long range video eeg monitoring in the diagnosis of epilepsy. China Foreign Med Treatment 37:47–49. https://doi.org/10.16662/j.cnki.1674-0742.2018.04.047

Zanzmera P, Sharma A, Bhatt K, Patel T, Luhar M, Modi A, Jani V (2019) Can short-term video-eeg substitute long-term video-eeg monitoring in psychogenic nonepileptic seizures? A prospective observational study. Epilepsy & behavior: E&B 94:258–263. https://doi.org/10.1016/j.yebeh.2019.03.034

Electronic Health Record

Ping Yu
Faculty of Engineering and Information Sciences, School of Computing and Information Technology, Centre for Digital Transformation, University of Wollongong, Wollongong, NSW, Australia
Illawarra Health and Medical Research Institute, University of Wollongong, Wollongong, NSW, Australia
Smart Infrastructure Facility, Faculty of Engineering and Information Sciences, University of Wollongong, Wollongong, NSW, Australia

Synonyms

Electronic medical record

Definition

According to International Organization for Standardization (ISO) (ISO/TR20514 2005), electronic health record (EHR) is repository of information regarding the health status of a subject of care, in computer-processable form.

The information can be retrospective, concurrent, and prospective. It can be transmitted securely and accessible by multiple authorized users. The primary purpose of EHR is to support continuing, efficient, and quality integrated healthcare.

Healthcare Information and Management Systems Society (HIMSS), the largest international health information management association, defines EHR as a longitudinal electronic record of patient health information (https://www.himss.org/library/ehr). The information is generated by one or more encounters in any care delivery setting. The content of EHR includes patient demographics, progress notes, problems, medications, vital signs, past medical history, immunizations, laboratory data, and radiology reports. Electronic health record can be a complete record of a clinical patient encounter. It can also support other care-related activities, i.e., evidence-based decision support, quality management, and outcome reporting, directly or indirectly via interface.

Overview

With the ability to capture, store, process, and exchange health information electronically, the benefits of EHR are enormous and well recognized. These include helping healthcare providers to provide higher-quality and safer care for patients while creating tangible enhancement for their organization (Technology 2019). EHRs can also help providers better provide and manage care for patients by providing accurate, up-to-date, and complete information about patients at the point of care and enabling more coordinated, efficient care. EHR is one of the core clinical tools for healthcare providers (Braunstein 2019). In the 1960s and 1970s, the development of new computer technology laid the foundation for the inception of EHR (Evans 2016). By 2019, EHR has been widely used by healthcare providers to manage patient health information in most developed countries in Europe, North America, Asia, and Oceania. The definition, structure, content, use, benefits, and risks of using an EHR system have been extensively studied and reviewed in the literature (Entzeridou et al. 2018; Evans 2016; Gardner 2016; Häyrinen et al. 2008; ISO/TR20514 2005).

Key Research Findings

The Initiation of EHR

In the early days, EHRs were implemented in primary healthcare in the UK (Detmer 2000) and were in-house developed and used at large, tertiary hospitals in the USA (Ornstein et al. 1993; Stead et al. 1983; Weed 1971). Due to its complexity, some information was stored in EHR and some on paper. Up to 2015, the goal of replacing the entire paper charts with EHR has not been fully achieved in most EHRs (Evans 2016). By 2015, most primary healthcare providers in the developed countries used commercial EHR. The pioneer hospitals in the USA replaced their expensive, in-house built, first-generation EHR systems with the relatively lower maintenance cost, vendor EHR products that promised to meeting the legislative requirements for the meaningful use in the USA (Gardner 2016).

The Building Blocks of EHR

The essential building blocks of EHR include software functionality, standards, implementation, and safety.

New Functionality of EHR

In addition to the basic functionalities as described in the HIMSS definition, new EHR functionalities are continuously emerging in response to the healthcare providers' increasing realization of the opportunities that digitized data and interoperable EHR systems can offer to diagnosis, treatment, holistic patient management, and health administration. Some examples are the images from the bedside which are captured by high-resolution cameras in mobile devices and inserted into the EHR (Abramson et al. 2012). To improve medication safety, medication ordering is coupled with clinical decision support functions and includes e-prescribing interfaces to local pharmacies (Abramson et al. 2012). Mental and behavioral health data have been included in the

EHR to better tailor medical services to personal needs and increase patient usage of the services (Druss et al. 2014). Electronic health records have also been introduced into Australian nursing homes to improving nursing documentation and care for the older people (Munyisia et al. 2014; Zhang et al. 2011) and have been used in home health or hospice care in the USA (Bercovitz et al. 2010). Other emerging trends include allowing patients to access their EHR data and enter family health histories and personal stories, both medical and nonmedical, into EHR, integrating human genome data with clinical data (Evans 2016), and linking biobanks to personal and family health information in EHRs to accurately identify subjects with specific diseases (Pathak et al. 2012). Public health has also harvested increased value in EHRs, i.e., biosurveillance systems are linked to EHRs to detect events of public health significance (Gundlapalli et al. 2007).

Health Information Standards that Underpin EHR Interoperability

Patient care is undertaken by various healthcare providers, each uses a certain EHR system that supports own clinical practice. Therefore, a functional digital health ecosystem is a network of health information systems including EHR that can securely transfer the relevant information in response to healthcare provider and patient needs. To enable the establishment of such an interoperable ecosystem, in 1987, Health Level Seven International (HL7), a not-for-profit, ANSI-accredited organization that is dedicated to the development of system interoperability standard, was founded. In 1991, its standard HL7 was adopted as the main interface standard to connect different EHR systems, i.e., homegrown EHRs and vendor systems (Hammond 1991). In 2018 HL7 International has members in more than 55 countries. The need for standardized terminology and dictionary codes (semantics) has seen the development of the common interface terminology SNOMED Clinical Terms, Logical Observation Identifiers Name and Codes (LOINC), and Unified Medical Language System (UMLS). DICOM (Digital Imaging and Communications in Medicine) is the standard for medical image interoperability.

Electronic Health Record Implementation

Financial incentives are needed to stimulate EHR adoption. Due to the high initial and continuous maintenance costs, and the uncertainty associated with the change processes in EHR introduction, implementing EHR is a significant organizational undertaken that many healthcare organizations may initially hesitate to embark on. Financial incentives are thus needed to stimulate adoption and use of EHR. A remarkably successfully story is the US government's allocation of nearly $30 billion through the Health Information Technology for Economic and Clinical Health (HITECH) Act in 2009 to stimulate the implementation of EHR in clinicians' offices and hospitals. Associated with the HITECH is the meaningful use, a financial strategy to reward performance and penalize non-compliance (Gardner 2016). In 2014, the country has witnessed the positive effects of this stimulus package on the nationwide uptake of interoperable EHRs in 75% of hospitals and 82% of primary care physician offices in the USA (Gardner 2016). The new EHR platform has opened the door for further digitization of healthcare, as seen above the emergence of various new functionalities.

Appropriate management is required in implementation to realize EHR value. Perceived value and use of EHR are largely dependent on alignment between system functionality, workflow, and user's information needs (Evans 2016; Fragidis and Chatzoglou 2018). Therefore, implementing useful and operational EHR in healthcare services requires continuous attention to identify and monitor the emerging hurdles for change (Gardner 2016). The hurdles for the successful implementation of EHR in a healthcare management group include decreasing provider communication, steeping learning curve for users, incomplete data migration, intentional decrease in volume of work, redistribution of staff or work, resistance to learning or using a new EHR, system configuration, workarounds, change in care pathways, health insurance changes, etc. (Colicchio et al. 2019). For the nationwide EHR, commitment and involvement of all stakeholders are key success factors; conversely, the lack of support and the negative reaction to any change from the medical, nursing, and administrative community is the most critical failure factor (Fragidis and Chatzoglou 2018).

Safety of Electronic Health Record

Due to its critical role, patients, providers, and healthcare facilities continue to demand assurance about access to EHR data, patient privacy, data ownership, liability, informed consent for secondary use of patient data, etc. (Evans 2016). Concerns about EHR safety involve both unsafe technology and unsafe use. Poor EHR system design and improper use can adversely affect patient safety by causing errors such as incorrect patient identification, incorrect calculation of medication dosages, and poor availability of patient data (Virginio and Ricarte 2015). These can jeopardize the integrity of the information in the EHR, leading to errors that endanger patient safety or decrease the quality of care. The associated legal implications may include increasing fraud and abuse and litigations. Therefore, this will require EHR implementations to include a robust mechanism to monitor and learn from adverse events, errors, and potential adverse events and errors (Meeks et al. 2014).

EHR Use and Evolution by 2015

By 2015, primary healthcare and hospitals in many developed countries have introduced EHR into routine healthcare (Evans 2016). The Canada Health Infoway initiative developed interoperable EHRs nationwide. Australia launched a consumer self-controlled MyHealth Record. Estonia implemented a nationwide EHR system that gives full access to its citizens. Hong Kong and Taiwan all have successful EHR. Most EHRs are web-/client-server-based and use mouselike scrolling and pointer devices to access data and entry screen and relational databases (Evans 2016). The Veterans Administrative in the US Health Exchange has advanced health information exchange (HIE) interoperability standards and patient consent policies (Byrne et al. 2014). Increasing amount of information in EHRs motivates the development of clinical decision support systems, a new, significant domain of medical informatics (Evans 2016). Simple clinical decision support systems included reminder systems, drug-allergy, drug-drug interaction, and abnormal laboratory test results. Knowledge bases began to be designed as separate databases and used within EHRs to feed clinical decision support functions.

Challenges for the Implementation of EHR

Up to 2015, many of the early expectations for EHRs have not been realized (Evans 2016; Gardner 2016). For example, although many healthcare providers appreciated that EHR supports for administrative functions and allows printouts, they do not believe that EHR have saved them time (Kainz et al. 1992). Although the large amount of data collected in EHR are valuable assets for epidemiological research, the poor quality of these data is problematic for reuse because misinformation can cause potential patient harm (Walsh et al. 2018). The negative impact of EHR includes administrative burdens on physicians (Jamoom et al. 2016); suboptimal workarounds for handling exceptions to normal workflow unintendedly imposed by EHRs, which may jeopardize patient safety, effectiveness, and efficiency of care (Blijleven et al. 2017); and team communication issues, which may pose threat to patient safety and quality of care (Assis-Hassid et al. 2019). It is particularly dangerous for older people with multiple complexities and chronic diseases. In order to make the most use of EHR for improving quality and safety of patient care, there is a mounting request from the healthcare providers and researchers to decrease or remove the need for a patient's consent to access their medical information. Conversely mental health information is sensitive and potentially damaging if privacy is breached and can result in patients being reluctant to seek treatment if they are not assured of confidentiality (Clemens 2012).

Future Directions of Research

Despite the concerted effort by HL7 and ISO, the international standards for interoperable application still need to be developed (Evans 2016). Future development is needed to make use of health, social, economic, behavioral, and environmental data to communicate, interpret, and act intelligently upon complex healthcare information to foster precision medicine and a learning health system. Research is also required on the prevalence of EHR risks and their impact on quality and safety of patient care and strategies for reducing the risks (Bowman 2013). Innovation is needed to drive major policy changes in healthcare and EHR advancement to improve patient safety, i.e., use EHR with decision support capabilities for error detection and prevention, for chronic disease management and develop a safety culture in the healthcare organizations. Robust evaluation is also required to quantify the value of EHR for patient care. Overall, with increasing use of EHR, the major effort in R&D in EHR has shifted to the procedural, professional, social, political, and especially ethical issues and the need for compliance with standards and information security, rather than the technical issues (Evans 2016).

Summary

As one of the core clinical tools with many foreseeable benefits for healthcare, EHR has been introduced into primary healthcare and hospitals in the major developed countries in the world. To achieve interoperability, EHR development in many countries has been underpinned by robust international standards. The implementation of EHR in healthcare organizations is facilitated by incentive programs, and it requires continuous identification, monitoring, and managing barriers for change. The establishment of EHR system has led to many new opportunities for further social technical innovation to improve safety, quality, and efficiency of healthcare and for improving healthcare for older people. Future development needs to make use of health, social, economic, behavioral, and environmental data to communicate, interpret, and act intelligently upon complex healthcare information to foster precision medicine and a learning health system.

Cross-References

▶ Patient Portal

References

Abramson EL, Patel V, Malhotra S et al (2012) Physician experiences transitioning between an older versus newer electronic health record for electronic prescribing. Int J Med Inform 81(8):539–548. https://doi.org/10.1016/j.ijmedinf.2012.02.010

Assis-Hassid S, Grosz BJ, Zimlichman E et al (2019) Assessing EHR use during hospital morning rounds: a multi-faceted study. PLoS One 14(2):e0212816. https://doi.org/10.1371/journal.pone.0212816

Bercovitz A, Sengupta M, Jamison P (2010) Electronic medical record adoption and use in home health and hospice. NCHS Data Brief 45(45):1–8

Blijleven V, Koelemeijer K, Wetzels M et al (2017) Workarounds emerging from electronic health record system usage: consequences for patient safety, effectiveness of care, and efficiency of care. JMIR Hum Factors 4(4):e27. https://doi.org/10.2196/humanfactors.7978

Bowman S (2013) Impact of electronic health record systems on information integrity: quality and safety implications. Perspect Health Inf Manag 10:1c

Braunstein ML (2019) Health care in the age of interoperability part 5: the personal health record. IEEE Pulse 10(3):19–23. https://doi.org/10.1109/mpuls.2019.2911804

Byrne CM, Mercincavage LM, Bouhaddou O et al (2014) The department of veterans affairs' (VA) implementation of the virtual lifetime electronic record (VLER): findings and lessons learned from health information exchange at 12 sites. Int J Med Inform 83(8):537–547. https://doi.org/10.1016/j.ijmedinf.2014.04.005

Clemens NA (2012) Privacy, consent, and the electronic mental health record: the person vs. the system. J Psychiatr Pract 18(1):46–50. https://doi.org/10.1097/01.pra.0000410987.38723.47

Colicchio TK, Borbolla D, Colicchio VD et al (2019) Looking behind the curtain: identifying factors contributing to changes on care outcomes during a large commercial EHR implementation. EGEMS (Wash DC) 7(1):21. https://doi.org/10.5334/egems.269

Detmer DE (2000) Information technology for quality health care: a summary of United Kingdom and United States experiences. Qual Health Care 9(3):181–189. https://doi.org/10.1136/qhc.9.3.181

Druss BG, Ji X, Glick G et al (2014) Randomized trial of an electronic personal health record for patients with serious mental illnesses. Am J Psychiatry 171(3): 360–368. https://doi.org/10.1176/appi.ajp.2013.13070913

Entzeridou E, Markopoulou E, Mollaki V (2018) Public and physician's expectations and ethical concerns about electronic health record: benefits outweigh risks except for information security. Int J Med Inform 110:98–107. https://doi.org/10.1016/j.ijmedinf.2017.12.004

Evans RS (2016) Electronic health records: then, now, and in the future. Yearb Med Inform (Suppl 1):S48–S61. https://doi.org/10.15265/IYS-2016-s006

Fragidis LL, Chatzoglou PD (2018) Implementation of a nationwide electronic health record (EHR). Int J Health Care Qual Assur 31(2):116–130. https://doi.org/10.1108/ijhcqa-09-2016-0136

Gardner RM (2016) Clinical information systems – from yesterday to tomorrow. Yearb Med Inform 25(Suppl 1):S62–S75. https://doi.org/10.15265/IYS-2016-s010

Gundlapalli AV, Olson J, Smith SP et al (2007) Hospital electronic medical record-based public health surveillance system deployed during the 2002 Winter Olympic Games. Am J Infect Control 35(3):163–171. https://doi.org/10.1016/j.ajic.2006.08.003

Hammond WE (1991) Health level 7: an application standard for electronic medical data exchange. Top Health Rec Manage 11(4):59–66

Häyrinen K, Saranto K, Nykänen P (2008) Definition, structure, content, use and impacts of electronic health records: a review of the research literature. Int J Med Inform 77(5):291–304. https://doi.org/10.1016/j.ijmedinf.2007.09.001

ISO/TR20514 (2005) Electronic health record. https://www.iso.org/obp/ui/#iso:std:iso:tr:20514:en. Accessed 9 June 2019

Jamoom EW, Heisey-Grove D, Yang N et al (2016) Physician opinions about EHR use by EHR experience and by whether the practice had optimized its ehr use. J Health Med Inform 7(4):1000240. https://doi.org/10.4172/2157-7420.1000240

Kainz C, Lassmann R, Schaffer H et al (1992) Survey of computerized obstetric information systems in Austria. Arch Gynecol Obstet 252(2):87–91

Meeks DW, Smith MW, Taylor L et al (2014) An analysis of electronic health record-related patient safety concerns. J Am Med Inform Assoc 21(6):1053–1059. https://doi.org/10.1136/amiajnl-2013-002578

Munyisia E, Yu P, Hailey D (2014) The effect of an electronic health record system on nursing staff time in a nursing home: a longitudinal cohort study. Australas Med J 7(7):285–293. https://doi.org/10.4066/AMJ.2014.2072

Ornstein SM, Garr DR, Jenkins RG (1993) A comprehensive microcomputer-based medical records system with sophisticated preventive services features for the family physician. J Am Board Fam Pract 6(1):55–60

Pathak J, Kiefer RC, Bielinski SJ et al (2012) Applying semantic web technologies for phenome-wide scan using an electronic health record linked biobank. J Biomed Semantics 3(1):10. https://doi.org/10.1186/2041-1480-3-10

Stead WW, Hammond WE, Straube MJ (1983) A chartless record – is it adequate? J Med Syst 7(2):103–109

Technology TOotNCfHI (2019) What are the advantages of electronic health records? https://www.healthit.gov/faq/what-are-advantages-electronic-health-records. Accessed 9 June 2019

Virginio LA Jr, Ricarte IL (2015) Identification of patient safety risks associated with electronic health records: a software quality perspective. Stud Health Technol Inform 216:55–59. https://doi.org/10.3233/978-1-61499-564-7-55

Walsh KE, Marsolo KA, Davis C et al (2018) Accuracy of the medication list in the electronic health record-implications for care, research, and improvement. J Am Med Inform Assoc 25(7):909–912. https://doi.org/10.1093/jamia/ocy027

Weed LL (1971) The problem oriented record as a basic tool in medical education, patient care and clinical research. Ann Clin Res 3(3):131–134

Zhang Y, Yu P, Shen J (2011) The benefits of introducing electronic health records in residential aged care facilities: a multiple case study. Int J Med Inform 81(10):690–704. https://doi.org/10.1016/j.ijmedinf.2012.05.013

Electronic Healthcare Record

▶ Electronic Nursing Documentation

Electronic Medical Record

▶ Electronic Health Record

Electronic Nursing Documentation

Ning Wang
School of Nursing, Southern Medical University, Guangzhou, People's Republic of China

Synonyms

Electronic healthcare record

Definition

Electronic nursing documentation is a digital format of nursing documentation which records nursing care planned and delivered to individual clients by qualified nurses or other caregivers under the direction of a qualified nurse (Urquhart et al. 2009). Electronic nursing documentation constitutes a significant part of a healthcare client's electronic health records (EHRs)) (see ▶ "Electronic Health Record").

Overview

Electronic nursing documentation has been commonly applied in aged care organizations to replace the traditional paper-based documentation (Kelley et al. 2011; Zhang et al. 2012). In aged care setting, nursing documentation is the principal clinical record collecting and storing older clients' information about their health and life conditions, care services provided to them, and their responses to the care (Wang 2012). It functions as a communication tool to facilitate continuity and individuality of care and safety of the clients. Nursing documentation can also serve other purposes such as quality assurance, legal purposes, health planning, allocation of resources, nursing development, and research (Wang 2012; Rajkovic et al. 2006; American Nurses Association 2014). Because of its capacity to store, process, retrieve, display, analyze, and report information (see ▶ "Health Informatics"), a meaningful electronic nursing documentation system that can effectively meet all the above functions can contribute to time-saving and efficiency gain in data management (Monsen et al. 2010).

Key Research Findings

Models and Components of Electronic Nursing Documentation in Aged Care

In nursing there are various documentation models such as continual sequential documentation, process-oriented documentation, problem-oriented documentation, and focus documentation (Thoroddsen et al. 2009; New Zealand Nurses Organization 2017). Continual sequential documentation is featured by its ongoing and narrative nature, recording free-text data in chronological order over a specific time period (New Zealand Nurses Organization 2017). Progress notes are this type of nursing documentation. Process-oriented documentation is usually based on the nursing process model which involves stages of assessment, problem/diagnosis, outcome planning, implementation, and evaluation (American Nurses Association 2019; Karkkainen and Eriksson 2003). The nursing process model has been widely applied as the theoretical basis of nursing documentation in aged and other healthcare settings across the world (Thoroddsen et al. 2009; Gjevjon and Hellesø 2010; Semachew 2018). Based on this model, electronic nursing documentation consists of nursing history, assessment forms, and the nursing care plan (Wang 2012; Akhu-Zaheya et al. 2018). Problem-oriented documentation refers to chartings relating to a particular clinical issue such as flow sheets and checklists about fall, pain, and wound (New Zealand Nurses Organization 2017). Focused documentation is similar to problem-oriented documentation but records data about a specific care event. Examples may include nursing transfer report and nursing discharge note. Research has shown that electronic nursing documentation systems in aged care could contain all types of these nursing documentation models (Wang 2012).

Quality of Electronic Nursing Documentation in Aged Care

It has been well recognized that electronic documentation systems can improve health professionals' access to more complete, accurate, legible, and up-to-date patient data (Oroviogoicoechea et al. 2008). Wang (2012) has assessed the quality of different components of Australian aged care nursing documentation in respect to three quality attributes derived from international literature: format and structure, documentation process, and documentation content (Wang et al. 2011). The quality of documentation format and structure represents the physical presentations of nursing data such as legibility, completeness, and redundancy, while the quality of documentation process concerns procedural issues of data collection such as timeliness, accuracy, and signature. The quality of documentation content refers to the meaning of data about care process. Based on these, the results of evaluation showed that data contained in electronic nursing documentation had better structure and process quality in comparison to the paper-based counterpart. This reflected on more complete and comprehensive clients' electronic admission and assessment forms and more legible data with complete signature and dating in all types of the electronic records (Wang et al. 2013a, b). Concerning the content quality, the electronic care plan recorded more signs and symptoms of client problems and evaluation of care (Wang et al. 2015). Other research findings have also revealed quality characteristics of electronic nursing documentation in the context of different study environments such as having more comprehensive data about the nursing process, using standardized language, and recording specific items about particular client issues and better relevance of the message (Gunningberg et al. 2009; Müller-Staub et al. 2007, 2008). Rykkje (2009) has reported limited use of abbreviations and symbols when the documentation was electronic. These results provide reference for the development and application of electronic nursing documentation system in aged care environment.

Standardized Nursing Languages and Electronic Nursing Documentation

With increased adoption of electronic nursing documentation in healthcare settings, nurses need to communicate the care they provided to the clients with their nursing colleagues and the other healthcare professionals through the electronic nursing documentation systems (see ▶ "Health Informatics"). While older clients more frequently use multiple healthcare services, to ensure the accuracy and completeness of their information transferred between different healthcare providers, the documented data from each provider's electronic record system need to be standardized, with the term providing the same meaning for all the providers involved in the care (see ▶ "Health Information Standard"). Standardized terminologies, including standardized nursing languages (SNLs), provide the terminology standards. Therefore, they allow the collection of sharable and comparable data elements across disciplines and care settings for better interdisciplinary communication, accurate measurement of care quality, use of data for various analysis, and demonstration of nursing contributions (Welton and Harper 2016; Rutherford 2008).

A range of SNLs has been developed to document nursing care. The American Nurses Association recognizes 12 languages for nursing which are being applied in electronic nursing documentation systems globally (Monsen et al. 2010; Tastan et al. 2014). These include ABC Codes, Clinical Care Classification (CCC), International Classification of Nursing Practice (ICNP), Logical Observation Identifiers Names and Codes (LOINC), NANDA International, Nursing Interventions Classification (NIC), Nursing Minimum Data Set (NMDS), Nursing Management Minimum Data Set (NMMDS), Nursing Outcomes Classification (NOC), Systematic Nomenclature of Medicine Clinical Terms (SNOMED CT), Omaha System, and Perioperative Nursing Data Set (PNDS). However, research supporting the terminology use in EHR in aged care has been limited. Tastan et al.'s (2014) review has reported only 12 studies published in English on SNL application in long-term care facilities between 1985 and 2011. There was lack of strong evidence supporting the terminology application in this setting. An American study (Huard and Monsen 2017) has investigated the use of SNLs in EHR across different care settings involving aged care in Minnesota healthcare system. The results showed that majority of electronic documentation entries collected did not identify a SNL. Wang et al.'s study (2015) on the quality of nursing documentation in Australian aged care has identified a trend of abandoning nursing diagnoses in the nursing care plan. Especially in the electronic format of nursing care plan, the term Nursing Problem/Diagnose was replaced with the term of Observation, resulting in more vague description of older clients' issues for care.

Summary

As an integral component of a healthcare client's EHR, electronic nursing documentation records nursing care that are planned and delivered to individual clients. To facilitate continuity and individuality of care and safety of clients, nursing documentation needs to be of high quality and to use SNLs that are interoperable with the records of the other healthcare professionals who serve the same client.

Future research needs to continuously pay attention on mapping existing SNLs with reference terminologies to produce interoperability in EHRs for better data utilization (see ▶ "Health Information Exchange"). Methods and approaches for effectively engaging healthcare clients in accessing and documenting their health data are also a promising research direction. In addition, existing research lacks evidence to support the influence of SNL use on patient outcomes or other important healthcare-related outcomes. This should be a focus of future research. Furthermore, researchers need to strengthen study designs to promote continuous improvement of the SNLs and to further improve the scientific base that supports the nursing practice-focused nursing language systems.

Cross-References

▶ Consumer Health Informatics
▶ Digital Health
▶ Electronic Health Record
▶ Health Informatics
▶ Health Information Exchange
▶ Health Information Standard
▶ Mobile Health
▶ Telehealth
▶ Telemedicine
▶ Telenursing

References

Akhu-Zaheya L, Al-Maaitah R, Hani SB (2018) Quality of nursing documentation: Paper-based health records versus electronic-based health records. J Clin Nurs 27:e578–e589. https://doi.org/10.1111/jocn.14097
American Nurses Association (2014) Defending yourself through documentation. Am Nurse Today 9(2). https://www.americannursetoday.com/defending-yourself-through-documentation/. Accessed 23 Jan 2019
American Nurses Association (2019) The nursing process. https://www.nursingworld.org/practice-policy/workforce/what-is-nursing/the-nursing-process/. Accessed 12 Jan 2019
Gjevjon ER, Hellesø R (2010) The quality of home care nurses' documentation in new electronic patient records. J Clin Nurs 19(1–2):100–108. https://doi.org/10.1111/j.1365-2702.2009.02953
Gunningberg L, Fogelberg-Dahm M, Ehrenberg A (2009) Improved quality and comprehensiveness in nursing documentation of pressure ulcers after implementing an electronic health record in hospital care. J Clin Nurs 18(11):1557–1564. https://doi.org/10.1111/j.1365-2702.2008.02647
Huard RJC, Monsen KA (2017) Standardized nursing terminology use in electronic health records in Minnesota. Mod Clin Med Res 1(1):13–19. https://doi.org/10.22606/mcmr.2017.11003
Karkkainen O, Eriksson K (2003) Evaluation of patient records as part of developing a nursing care classification. J Clin Nurs 12(2):198–205
Kelley TF, Brandon DH, Docherty SL (2011) Electronic nursing documentation as a strategy to improve quality of patient care. J Nurs Scholarsh 43(2):154–162. https://doi.org/10.1111/j.1547-5069.2011.01397
Monsen K, Honey M, Wilson S (2010) Meaningful use of a standardized terminology to support the electronic health record in New Zealand. Appl Clin Inform 1(4):368–376. https://doi.org/10.4338/ACI-2010-06-CR-0035
Müller-Staub M, Needham I, Odenbreit M et al (2007) Improved quality of nursing documentation: Results of a nursing diagnoses, interventions, and outcomes implementation study. Int J Nurs Terminol Classif 18(1):5–17
Müller-Staub M, Lunney M, Lavin MA et al (2008) Testing the Q-DIO as an instrument to measure the documented quality of nursing diagnoses, interventions, and outcomes. Int J Nurs Terminol Classif 19(1):20–27. https://doi.org/10.1111/j.1744-618X.2007.00075
New Zealand Nurses Organization (2017) Guideline: documentation. https://www.nzno.org.nz/LinkClick.aspx?fileticket=GH84aNBNd64%3D&portalid=0. Accessed 23 Jan 2019
Oroviogoicoechea C, Elliott B, Watson S (2008) Review: evaluating information systems in nursing. J Clin Nurs 17(5):567–575
Rajkovic V, Sustersic O, Rajkovic U (2006) E-nursing documentation as a tool for quality assurance. Stud Health Technol Inform 122:298–303
Rutherford M (2008) Standardized nursing language: what does it mean for nursing practice? Online J Issues Nurs 13(1):1–12
Rykkje L (2009) Implementing electronic patient record and VIPS in medical hospital wards: evaluating change in quantity and quality of nursing documentation by using the audit instrument Cat-Ch-Ing. Nord J Nurs Res 29(2):9–13
Semachew A (2018) Implementation of nursing process in clinical settings: the case of three governmental hospitals in Ethiopia, 2017. BMC Res Notes 11(1):173. https://doi.org/10.1186/s13104-018-3275-z

Tastan S, Linch GC, Keenan GM et al (2014) Evidence for the existing American Nurses Association-recognized standardized nursing languages: a systematic review. Int J Nurs Stud 51(8):1160–1170. https://doi.org/10.1016/j.ijnurstu.2013.12.004

Thoroddsen A, Saranto K, Ehrenberg A et al (2009) Models, standards and structures of nursing documentation in European countries. Stud Health Technol Inform 146:327–331

Urquhart C, Currell R, Grant MJ et al (2009) Nursing record systems: effects on nursing practice and healthcare outcomes. Cochrane Database Syst Rev (1):1–66. https://doi.org/10.1002/14651858

Wang N (2012) Evaluation of quality of paper-based versus electronic nursing documentation in Australian residential aged care homes. University of Wollongong Thesis Collection, Wollongong

Wang N, Hailey D, Yu P (2011) Quality of nursing documentation and approaches to its evaluation: a mixed-method systematic review. J Adv Nurs 67(9):1858–1875. https://doi.org/10.1111/j.1365-2648.2011.05634

Wang N, Hailey D, Yu P (2013a) Description and comparison of quality of electronic versus paper-based resident admission forms in Australian aged care facilities. Int J Med Inform 82(5):313–324. https://doi.org/10.1016/j.ijmedinf.2012.11.011

Wang N, Hailey D, Yu P (2013b) Description and comparison of documentation of nursing assessment between paper-based and electronic systems in Australian aged care homes. Int J Med Inform 82(9):789–797. https://doi.org/10.1016/j.ijmedinf.2013.05.002

Wang N, Hailey D, Yu P (2015) The quality of paper-based versus electronic nursing care plan in Australian aged care homes: a documentation audit study. Int J Med Inform 84(8):561–569. https://doi.org/10.1016/j.ijmedinf.2015.04.004

Welton J, Harper E (2016) Measuring nursing care value. Nurs Econ 34(1):7–14

Zhang Y, Yu P, Shen J (2012) The benefits of introducing electronic health records in residential aged care facilities: a multiple case study. Int J Med Inform 81:690–704. https://doi.org/10.1016/j.ijmedinf.2012.05.013

Electronic Prescribing (e-Prescribing)

▶ Computerized Provider Order Entry

Electronic Therapy

▶ Cybercounseling

Electrophysiology

Vanessa Dominguez[1] and Adam J. Woods[2,3]
[1]Center for Cognitive Aging and Memory, Department of Clinical and Health Psychology, McKnight Brain Institute, University of Florida, Gainesville, FL, USA
[2]Department of Clinical and Health Psychology, College of Public Health and Health Professions, Center for Cognitive Aging and Memory, McKnight Brain Institute, University of Florida, Gainesville, FL, USA
[3]Department of Neuroscience, University of Florida, Gainesville, FL, USA

Synonyms

EEG; ERP; Electroencephalogram; Event-related potentials

Definition

Electrophysiology is the study of electrical currents in living cells and tissues. Through electrophysiological measuring techniques, it is possible to measure the electrical activity of biological cells within an organism (in vivo) and in environments outside living organisms (in vitro) (Verkhratsky and Parpura 2014; Graziane and Dong 2016). Electrophysiology recordings and measurement is referred to as electrography. Studies can be done from the level of whole organs down to single ion channel proteins in the membrane of nerve cells.

Overview

In the late eighteenth century, Luigi Galvani introduced the idea of bioelectricity using animal models (Piccolino 1997) by using a metal wire to contract the muscles in a pair of exposed frog leg nerves (Verkhratsky et al. 2006). Since Galvani's experiments with frog preparations, the field of electrophysiology has evolved to more sophisticated methods of measuring electrical excitability. Electrophysiological recording techniques measure voltage change or electrical current flow of ions through intracellular and extracellular recording methods across the cell membrane (Graziane and Dong 2016; Verkhratsky et al. 2006). Extracellular recording methods measure the electrical activity from a group of cells or adjacent cells. There are two types of extracellular recordings: single-unit and multi-unit recordings (Graziane and Dong 2016). Single-unit recordings measure electrical activity by placing the microelectrode close to the cell membrane (Humphrey and Schmidt 1990). On the other hand, multi-unit recordings can measure activity in a group of nearby neurons (Graziane and Dong 2016). Intracellular recording methods measure the electrical activity from a single cell. Intracellular recordings are acquired by inserting the tip of an electrode into a cell. The three most widely used intracellular recording methods are voltage-clamping, current-clamping, and patch-clamping. Voltage-clamp techniques measure membrane current by manipulating the membrane voltage (Verkhratsky and Parpura 2014). Current-clamp techniques measure membrane potentials by applying a time-varying or constant current (Verkhratsky and Parpura 2014). The patch-clamp technique is one of the most used electrophysiology methods. Unlike traditional intracellular recordings, the patch-clamp technique involves placing a micropipette next to a cell and a small amount of the membrane being sucked inside the tip of the microelectrode to create a seal on the cell membrane (Hamill et al. 1981; Sakmann and Neher 1984; Verkhratsky and Parpura 2014; Rubaiy 2017).

In neuroscience, electrophysiological techniques measure the electrical and chemical signals in neurons at a cellular and tissue level delivered to the brain and from the brain. Electrical signals traveling toward the central nervous system are called afferent signals, and electrical signals traveling across neurons away from the central nervous system are called efferent signals. Electrophysiology is a crucial tool to neuroscience research in aging populations to gain knowledge on cortical activity as well as cardiovascular, neuromuscular, and other systems (Rubaiy 2017).

Common electrophysiology human recording approaches used in neuroscience research in aging populations are electroencephalography (EEG), electrocardiography (ECG/EKG), and electromyography (EMG). Electroencephalography is a noninvasive method of recording brain activity measured by placing electrodes on the scalp in a multielectrode array, with newer electrode systems ranging from 16-channels to 256-channel systems (Pizzagalli 2006; Tivadar et al. 2019). EEG is commonly used to look at Evoked Potentials (EPs) which are responses in the brain time-locked to a stimulus (Oken and Phillips 2009) and Event-Related Potentials (ERPs) which are brain responses to specific events or stimuli. In older adults, EEG signal analysis is often used to monitor neuron activity during aging (Al-Qazzaz et al. 2014).

Electrocardiography (ECG/EKG) is a recording method used to monitor the electrical activity of the heart by placing electrodes on the skin. Typically, two electrodes are placed on the surface of the skin in an area free of fatty tissue and muscle, for example, on the arms, chest, or legs. However, for the purposes of diagnosing any abnormalities in the heart rhythm, there is a standard system (Berntson et al. 2007) of electrode placement with 12 leads placed in ten electrodes (one on each arm and leg and six on the chest). In aging populations, ECGs are commonly used to differentiate between normal aging and changes from other more serious conditions that often happens comorbidly with cardiac disease.

Electromyography (EMG) is a recording method of the electrical activity of muscle tissue. EMG measures voluntary and involuntary muscle contraction. There are two types of EMG recordings techniques: surface and intramuscular. Surface EMG electrodes are noninvasive and are applied to the skin on top of the muscle, as opposed to electrodes in intramuscular EMG which are invasive, and a needle electrode is inserted through the skin and into the muscle (Chowdhury et al. 2013). EMG recordings have a multitude of applications and are commonly used to diagnose neuromuscular diseases. EMG studies in older adults examine the neuromuscular system in order to detect changes in motor function as the aging process occurs.

Key Research Findings

The growing body of neuroscience research using electrophysiological recording techniques has led to much of the current knowledge on aging. One of the most significant findings in electrophysiology research was done by Hubel and Wiesel in 1959 by using single cell recordings in cats to learn about visual cortex (Hubel and Wiesel 1959). Hubel and Wiesel's early research in cats eventually led to research in the cortical neuron organization in monkeys due to their similarity in visual system as humans. Inspired by Hubel and Wiesel's work, Michael Merzenich's work in cortical reorganization (Merzenich et al. 1984) after central nervous system (CNS) damage was one of the pioneering works in neuroplasticity. This work was pivotal in learning how the CNS is able to respond to stimuli by reorganizing its connections, structure, and function, even in older adults. Research suggests plastic changes can increase neural connections, but also the weakening of neural connections in such diseases like dementia and Alzheimer's disease using EEG to study neuronal degeneration (Al-Qazzaz et al. 2014; Cassani et al. 2018). Using EEG to study cognition in older populations has elucidated what comprises normal cognitive changes and what cognitive changes are pathologic and related to disease.

Research using ECG in older adults has largely revolved around diagnosing and monitoring heart conditions. Due to the changes that occur in the heart and conduction, prevalence of heart rhythm disorders increases along with age according to Goyal and Rich (2016). Globally, cardiovascular disease is the leading cause of death according to the World Health Organization (WHO 2018) and because of its high prevalence, research in cardiovascular disease is important. Abnormalities in the electrical impulses being conducted through the heart can occur comorbidly with a large number of diseases (Strait and Lakatta 2012), posing a problem in diagnosing cardiac problems due to overlap in symptomology. Fortunately, research in cardiac abnormalities using ECG and EKGs is widely used and abnormalities in these measures have been shown to be an indicator of coronary heart disease (Auer et al. 2012). ECG is used to test and treat heart-related abnormalities called arrhythmias. There are four main types of arrhythmias: ventricular arrhythmias, bradyarrhythmias, supraventricular tachycardias, and extra beats. In treating arrhythmias, automated heart monitoring and the delivery of electrical impulse to treat abnormal rhythms is attributed to pacemakers. Pacemakers are implanted into a person and it uses electrode technology to monitor the electrical activity of the heart, sends information out, and then adjusts the pace of the heart rhythm (Graetzer 2018).

Findings using EMG in aging populations has proven to be useful in understanding the body's behaviors and motor functioning. The degeneration of neural and motor mechanisms in older adults effects mobility and the ability to perform certain motor tasks (Coscrato Cardozo et al. 2013). Research studies have found that impaired neuromuscular activation with aging is due to reduced motor unit discharge rates (Clark and Fielding 2011; Connelly et al. 1999). These findings have helped build on the knowledge on neurological conditions such as Parkinson's disease. Parkinson's disease is often associated with older adults, and EMG is used to monitor motor reflexes in gait patterns and falling (Jenkins et al. 2009).

The combination of different electrophysiology recording techniques in research of older adults has been crucial in gaining understanding of the human aging process from different systems.

Future Directions

Advancements in electrophysiology recording techniques over the past several decades have greatly progressed understanding of ion channels in living cells. However, there is still a need for improving recording technology systems. Advancements in recording techniques would allow for increased sensitivity, specificity, and improved spatial resolution. Future uses for electrophysiology should emphasize combining EEG and neuroimaging. A combination of both approaches to research can atone for some of the pitfalls of EEG only approaches with poor spatial resolution and noise (Pizzagalli 2006). Currently, research in electrophysiology stops at the ion channel proteins, but with future advancements in technology, electrophysiology recording techniques may be able to increase spatial resolution, attain better mapping performance, and achieve recordings of micromolecules. With the emergence of targeted cell therapy to treat certain cancers, future developments in electrophysiology can allow for target therapy at a molecular level (Rubaiy 2017). Future directions in research should examine using electrophysiology techniques to identify biomarkers in disease (Cassani et al. 2018). EEG can add significant knowledge to multivariate measures of functioning and cognition adding to how disease is diagnosed, monitored, and how outcomes are predicted.

Finally, as technology and research evolve, efforts should be made to increase the accessibility and portability of electrophysiology measuring techniques. Improvements in technology and the emergence of telemedicine could potentially lead to remote health monitoring and immediate feedback in older adults (Evans et al. 2016). Home monitoring systems could increase access for older adults with mobility issues or living in rural areas, minimizing a gap in care. Research using smartphone or wearable technology measuring heart and associated cardiovascular disease and electrodermal activity are already in use, but future research monitoring brain activity (EEG) could help with seizure detection and monitoring age-related cognitive decline (Khanna et al. 2015).

Summary

Starting with Luigi Galvani's experiments in the eighteenth century to the development of modern electrophysiology recordings, the field of electrophysiology has led to the understanding of ion channels in vital cell processes of biological systems (Rubaiy 2017). Electrophysiology is important for monitoring and observation of changes in cells that occur during the aging process, such as age-related decline in cognitive processes. The application of electrophysiological techniques in research has advanced knowledge of varying disorders and diseases affecting older adults such as neurological, cardiovascular, and motor disorders. Future research and advances in technology to further the insights of the brain framework will lead the way to electrophysiology recording approaches increasing the degrees of sensitivity and specificity looking at ion channels to a molecular level.

Cross-References

▶ Alzheimer's Disease
▶ Cardiovascular system
▶ Central Nervous System
▶ Electroencephalography
▶ Neuroimaging
▶ Neuromuscular System
▶ Neuroplasticity
▶ Neuroscience of Aging
▶ Parkinson's Disease
▶ Peripheral Nervous System
▶ Prevention of Age-Related Cognitive Impairment, Alzheimer's Disease, and Dementia
▶ Vascular Diseases of Ageing

References

Al-Qazzaz NK, Ali SH, Ahmad SA, Chellappan K, Islam MS, Escudero J (2014) Role of EEG as biomarker in the early detection and classification of dementia. Sci World J 2014:906038. https://doi.org/10.1155/2014/906038

Auer R, Bauer DC, Marques-Vidal P, Butler J, Min LJ, Cornuz J, Satterfield S, Newman AB, Vittinghoff E, Rodondi N, Study HABC (2012) Association of major and minor ECG abnormalities with coronary heart disease events. JAMA 307(14):1497–1505. https://doi.org/10.1001/jama.2012.434

Berntson GG, Quigley KS, Lozano D (2007) Handbook of psychophysiology: cardiovascular psychophysiology. https://doi.org/10.1017/CBO9780511546396.008

Cassani R, Estarellas M, San-Martin R, Fraga FJ, Falk TH (2018) Systematic review on resting-state EEG for Alzheimer's disease diagnosis and progression assessment. Dis Markers 2018:5174815. https://doi.org/10.1155/2018/5174815

Chowdhury RH et al (2013) Surface electromyography signal processing and classification techniques. Sensors (Basel) 13(9):12431–12466. https://doi.org/10.3390/s130912431

Clark DJ, Fielding RA (2011) Neuromuscular contributions to age-related weakness. J Gerontol A Biol Sci Med Sci 67(1):41–47. https://doi.org/10.1093/gerona/glr041

Connelly D, Rice C, Roos M, Vandervoort A (1999) Motor unit firing rates and contractile properties in tibialis anterior of young and old men. J Appl Physiol 87(2):843–852. https://doi.org/10.1152/jappl.1999.87.2.843

Coscrato Cardozo A, Gonçalves M, Zamfolini Hallal C, Ribeiro Marques N (2013) Age-related neuromuscular adjustments assessed by EMG. Electrodiagnosis in new Frontiers of Clinical Research. InTech. https://doi.org/10.5772/55053

Evans J, Papadopoulos A, Silvers CT et al (2016) Remote health monitoring for older adults and those with heart failure: adherence and system usability. Telemed J E Health 22(6):480–488. https://doi.org/10.1089/tmj.2015.0140

Goyal P, Rich M (2016) Electrophysiology and heart rhythm disorders in older adults. J Geriatr Cardiol 13(1):645–651. https://doi.org/10.11909/j.issn.1671-5411.2016.08.001

Graetzer HG (2018) Pacemaker implantation. Magill's Medical Guide (Online Edition). Retrieved from http://lp.hscl.ufl.edu/login?url=http://search.ebscohost.com/login.aspx?direct=true&AuthType=ip,uid&db=ers&AN=87690599&site=eds-live

Graziane N, Dong Y (2016) Extracellular and intracellular recordings. In: Electrophysiological analysis of synaptic transmission. Neuromethods, vol 112. Humana Press, New York

Hamill O, Marty A, Neher E, Sakmann B, Sigworth F (1981) Improved patch-clamp techniques for high-resolution current recording from cells and cell-free membrane patches. Pflugers Arch – Eur J Physiol 391(2):85–100. https://doi.org/10.1007/BF00656997

Hubel DH, Wiesel TN (1959) Receptive fields of single neurones in the cat's striate cortex. J Physiol 148(3):574–591

Humphrey DR, Schmidt EM (1990) Extracellular single-unit recording methods. In: Boulton AA, Baker GB, Vanderwolf CH (eds) Neurophysiological techniques. Neuromethods, vol 15. Humana Press, Totowa

Jenkins ME, Almeida QJ, Spaulding SJ, van Ooostveen RB, Holmes JD, Johnson AM, Perry SD (2009) Plantar cutaneous sensory stimulation improves single-limb support time, and EMG activation patterns among individuals with Parkinson's disease. Parkinsonism Relat Disord 15(9):697–702. https://doi.org/10.1016/j.parkreldis.2009.04.004

Khanna A, Pascual-Leone A, Michel C, Farzan F (2015) Microstates in resting-state EEG: current status and future directions. J Neu Biobehav Rev 49(1):105–113. https://doi.org/10.1016/J.NEUBIOREV.2014.12.010

Merzenich M, Nelson R, Stryker M, Cynader M, Schoppmann A et al (1984) Somatosensory cortical map changes following digit amputation in adult monkeys. J Comp Neuro 224(4):591–605. https://doi.org/10.1002/cne.902240408

Oken B, Phillips T (2009) Evoked potentials: clinical. Encyclopedia of neuroscience, pp 19–28. https://doi.org/10.1016/B978-008045046-9.00587-8

Piccolino M (1997) Luigi Galvani and animal electricity: two centuries after the foundation of electrophysiology. Trends Neurosci 20(10):443–448. https://doi.org/10.1016/S0166-2236(97)01101-6

Pizzagalli DA (2006) Electroencephalography and high-density electrophysiological source localization. In: Cacioppo JT, Tassinary LG, Berntson G (eds) Foundations of Psychophysiology, Cambridge University Press. pp 56–84

Rubaiy HN (2017) A short guide to electrophysiology and ion channels. J Pharm Pharm Sci 20:48–67. https://doi.org/10.18433/J32P6R

Sakmann B, Neher E (1984) Patch clamp techniques for studying ionic channels in excitable membranes. Annu Rev Physiol 46(1):455–472

Strait JB, Lakatta EG (2012) Aging-associated cardiovascular changes and their relationship to heart failure. Heart Fail Clin 8(1):143–164

Tivadar R, Murray M, Helmholtz V, Du Bois-Reymond E, Huxley A et al (2019) A primer on electroencephalography and event-related potentials for organizational neuroscience. Organ Res Methods 22(1):69–94. https://doi.org/10.1177/1094428118804657

Verkhratsky A, Parpura V (2014) History of electrophysiology and the patch clamp. Humana Press, New York, pp 1–19

Verkhratsky A, Krishtal O, Petersen O (2006) From Galvani to patch clamp: the development of electrophysiology. Pflügers Arch Eur J Physiol 453(3):233–247. https://doi.org/10.1007/s00424-006-0169-z

WHO (2018) The top 10 causes of death. (2018, May 24). Retrieved from https://www.who.int/en/news-room/fact-sheets/detail/the-top-10-causes-of-death

E-Mail Counseling

▶ Cybercounseling

Embolism

▶ Thrombus

Emergency Care

John G. Schumacher
University of Maryland, Baltimore County (UMBC), Baltimore, MD, USA

Synonyms

Critical care; Emergency room; Geriatric emergency medicine; Senior emergency center; Urgent care

Definition

Emergency care in the form of emergency medicine is the medical subspecialty that addresses the acute, immediate, and unscheduled illness or injury needs of a population (ACEP 2015). Typically, emergency medicine is provided in emergency departments (EDs) open 24 h a day for emergencies that could reasonably be expected to result in loss of life or limb. By law, EDs are required to medically evaluate any presenting patient regardless of circumstance. Specialized EDs called geriatric emergency departments (GEDs) are units that provide geriatric-trained ED staff and services tailored to the unique emergency care needs of older adults. In 2017, the American College of Emergency Physicians created an accreditation credential for GEDs (ACEP 2017; Huff 2018).

Overview

Older adults in the USA make more than 21 million emergency care visits to EDs annually (Rui and Kang 2015), and EDs are frequently a site for hospital admission decisions (Schuur and Venkatesh 2012). Older adults represent a challenging ED population due to factors including their functional heterogeneity, atypical presentations, comorbidities, risks for polypharmacy, and relatively longer medical histories. Hwang and Morrison's (2007) seminal article highlighted an emerging focus on geriatric emergency medicine (GEM) by articulating the unique elements for a GED including geriatric ED staff training, tailored triage procedures, cognitive status screening, falls screening, modifications to the ED physical environment, and enhanced medication reconciliation. For GEDs, an interdisciplinary approach based on principles of geriatric medicine is advocated in order to optimize the emergency care of older adults. This approach involves emergency physicians, emergency nurses, pharmacists, social workers, hospital administrators, and extends to all hospital nodes touched by the ED including the medical and surgical floors, patient transport, imaging, dietary, housekeeping, medical records, and volunteer services. The first dedicated GED opened in 2007, and nationwide more than 80 GED were operating by 2018 (Schumacher et al. 2018).

Key Research Findings

Specific GEM research demonstrates that compared to younger patients, older adults are hospitalized more frequently (Greenwald et al. 2016) and for different reasons (Lo et al. 2017), yet, their ED use is generally appropriate (Pines et al. 2013). Research on GEDs shows the efficacy of a range of model including the use of nurse liaisons (Aldeen et al. 2014), transitional care nurses (Hwang et al. 2018), social workers (Hamilton et al. 2015), pharmacists (Thompson 2015), and consultation services (Yuen et al. 2013). Key research themes focus on triage screening (Southerland et al. 2017), falls risk (Tirrell et al. 2015), delirium (Han and Suyama 2018), palliative care (Rosenberg et al. 2013), and frailty (Brousseau et al. 2018). Research documenting the outcomes of GED care is sparse but includes the reduction in admission to ICUs (Grudzen et al. 2015), reduced ED recidivism, hospital admission, and length of stay (Keyes et al. 2014) and compliance with post-ED care recommendations (Biese et al. 2014).

Examples of Application

Nationally, the majority of GEDs are promoted as senior emergency centers or senior-friendly emergency rooms without using the word geriatric in their title. The educational resources for GEM are limited but expanding. Textbooks on GEM (Sanders 1996; Kahn and Magauran 2014; Mattu et al. 2016) focus narrowly on clinical issues like cardiovascular dysfunction, altered mental status, abdominal pain, syncope, pharmacology interactions, and falls. Descriptive research suggests older adult patients frequently present to the ED with nonspecific complaints and an unclear etiology for their wide-ranging symptoms (Wachelder et al. 2017). For example, chest pain, abdominal pain, dizziness, and shortness of breath are older adults' most frequently reported symptoms in the ED making a rapid, clear differential diagnosis challenging. A number of informative GEM resources now exist to guide emergency medicine providers including the Geriatric Emergency Medicine Competencies (Hogan et al. 2010), the national Geriatric Emergency Department Guidelines (Carpenter et al. 2014), and resources for EDs seeking Geriatric Emergency Department Accreditation (GEDA) by the American College of Emergency Physicians (2017). Internationally, European emergency providers adopted a GEM curriculum (Bellou et al. 2016) and the International Federation of Geriatric Emergency Medicine has promoted minimum standards for the care of older ED patients (Ellis et al. 2018).

Future Directions for Research

Demographics suggest that EDs will treat increasing numbers of older adults (US Census 2017). Researchers have described an agenda that includes a focus on evidenced-based medicine, clinical outcomes, transitions of care, and reducing ED recidivism (Wilber et al. 2006). The rapid increase of older adults with dementia suggests the need to systematically examine the dementia-friendliness of emergency medicine settings and how this patient population can be best served. Researchers also note the alarming omission of GEM from a recent national report on the future of the field of emergency medicine that suggests a broader lack of recognition of the emergency care needs of the older adult population (Carpenter et al. 2011). Future research on GED settings can identify the key correlates of changes and innovations in GEM services (McCusker et al. 2017). It can look at examining new care services such as in urgent care centers as well as in potentially expanding the scope of practice for community paramedics to deliver more community health and preventive care in the community (Shah et al. 2018).

Summary

Emergency care provided in emergency departments delivers fundamental care for older adults with acute health care needs. The development of GEDs represents a key innovation for more comprehensively addressing the unique care needs of older adults. The GED model of care will continue to evolve to optimize care for the heterogeneous older adult population and their caregivers. Advancements in ED interdisciplinary teams, models of care, geriatric screening tools, care processes, outcome measures, and related factors will contribute to the continuous improvement of emergency care for older adults.

Cross-References

▶ Ageism in Healthcare
▶ End-of-Life Care
▶ Teaching Aging Medicine

References

Aldeen AZ, Courtney DM, Lindquist LA, Dresden SM, Gravenor SJ (2014) Geriatric emergency department innovations: preliminary data for the geriatric nurse liaison model. J Am Geriatr Soc 62(9):1781–1785
American College of Emergency Physicians (ACEP) (2015) Definition of emergency medicine. https://www.acep.org/patient-care/policy-statements/definition-of-emergency-medicine. Accessed 19 Mar 2019

American College of Emergency Physicians (ACEP) (2017) Geriatric emergency department accreditation program. https://www.acep.org/geda/#sm.001t3x9s916zfdaazrc1akbrhrndh. Accessed 19 Mar 2019

Bellou A, Conroy SP, Graham CA (2016) The European curriculum for geriatric emergency medicine. Eur J Emerg Med 23(4):239

Biese K, Lamantia M, Shofer F, McCall B, Roberts E, Stearns SC et al (2014) A randomized trial exploring the effect of a telephone call follow-up on care plan compliance among older adults discharged home from the emergency department. Acad Emerg Med 21(2):188–195

Brousseau AA, Dent E, Hubbard R, Melady D, Emond M, Mercier E et al (2018) Identification of older adults with frailty in the Emergency Department using a frailty index: results from a multinational study. Age Ageing 47(2):242–248

Carpenter CR, Shah MN, Hustey FM, Heard K, Gerson LW, Miller DK (2011) High yield research opportunities in geriatric emergency medicine: prehospital care, delirium, adverse drug events, and falls. J Gerontol A Biol Sci Med Sci 66:775–783

Carpenter CR, Bromley M, Caterino JM, Chun A, Gerson LW, Greenspan J, Wilber S (2014) Optimal older adult emergency care: introducing multidisciplinary geriatric emergency department guidelines from the American College of Emergency Physicians, American Geriatrics Society, Emergency Nurses Association, and Society for Academic Emergency Medicine. J Am Geriatr Soc 62:1360–1363

Ellis B, Carpenter C, Lowthian J, Mooijaart S, Nickel C, Melady D (2018) Statement on minimum standards for the care of older people in emergency departments by the geriatric emergency medicine special interest group of the international federation for emergency medicine. Can J Emerg Med 20(3):368–369

Greenwald PW, Estevez RM, Clark S, Stern ME, Rosen T, Flomenbaum N (2016) The ED as the primary source of hospital admission for older (but not younger) adults. Am J Emerg Med 34(6):943–947

Grudzen C, Richardson LD, Baumlin KM, Winkel G, Davila C, Ng K et al (2015) Redesigned geriatric emergency care may have helped reduce admissions of older adults to intensive care units. Health Aff 34(5):788–795

Hamilton C, Ronda L, Hwang U, Abraham G, Baumlin K, Morano B et al (2015) The evolving role of geriatric emergency department social work in the era of health care reform. Soc Work Health Care 54(9):849–868

Han JH, Suyama J (2018) Delirium and dementia. Clin Geriatr Med 34:327–354

Hogan TM, Losman ED, Carpenter CR, Sauvigne K, Irmiter C, Emanuel L, Leipzig RM (2010) Development of geriatric competencies for emergency medicine residents using an expert consensus process. Acad Emerg Med 17:316–324

Huff C (2018) ACEP accrediting geriatric emergency departments move to standardize special needs elder care. Ann Emerg Med 71(5):21A–24A

Hwang U, Morrison RS (2007) The geriatric emergency department. J Am Geriatr Soc 55:1873–1876

Hwang U, Dresden SM, Rosenberg MS, Garrido MM, Loo G, Sze J et al (2018) Geriatric emergency department innovations: transitional care nurses and hospital use. J Am Geriatr Soc 66(3):459–466

Kahn JH, Magauran B (2014) Geriatric emergency medicine: principles and practice. Wiley-Blackwell, New York

Keyes DC, Singal B, Kropf CW, Fisk A (2014) Impact of a new senior emergency department on emergency department recidivism, rate of hospital admission, and hospital length of stay. Ann Emerg Med 63(5):517–524

Lo AX, Flood KL, Biese K, Platts-Mills TF, Donnelly JP, Carpenter CR (2017) Factors associated with hospital admission for older adults receiving care in us emergency departments. J Gerontol A Biol Sci Med Sci 72(8):1105–1109

Mattu A, Grossman S, Rosen P (2016) Geriatric emergencies: a discussion-based review. Wiley, Hoboken

McCusker J, Vadeboncoeur A, Cossette S, Veillette N, Ducharme F, Minh Vu TT et al (2017) Changes in emergency department geriatric services in quebec and correlates of these changes. J Am Geriatr Soc 65(7):1448–1454

Pines JM, Mullins PM, Cooper JK, Feng LB, Roth KE (2013) National trends in emergency department use, care patterns, and quality of care of older adults in the United States. J Am Geriatr Soc 61(1):12–17

Rosenberg M, Lamba S, Misra S (2013) Palliative medicine and geriatric emergency care: challenges, opportunities, and basic principles. Clin Geriatr Med 29(1):1–29

Rui P, Kang K (2015) National Hospital Ambulatory Medical Care Survey: 2015 emergency department summary tables. http://www.cdc.gov/nchs/data/ahcd/nhamcs_emergency/2015_ed_web_tables.pdf. Accessed 19 Mar 2019

Sanders AB (1996) Emergency care of the elder person. Beverly-Cracom Publications, St. Louis

Schumacher JG, Hirshon JM, Magidson P, Chrisman M, Hogan T (2018) Tracking the rise of geriatric emergency departments in the us. J Appl Gerontol:1–18. https://doi.org/10.1177/2F0733464818813030

Schuur JD, Venkatesh AK (2012) The growing role of emergency departments in hospital admissions. N Engl J Med 367:391–393

Shah MN, Hollander MM, Jones CM, Caprio TV, Conwell Y, Cushman JT, Coleman EA (2018) Improving the ED-to-home transition: the community paramedic-delivered care transitions intervention-preliminary findings. J Am Geriatr Soc. https://doi.org/10.1111/jgs.15475. Advance online publication

Southerland LT, Slattery L, Rosenthal JA, Kegelmeyer D, Kloos A (2017) Are triage questions sufficient to assign fall risk precautions in the ED? Am J Emerg Med 35(2):329–332

Thompson CA (2015) Pharmacists integrate into geriatric emergency department. Am J Health Syst Pharm 72(2):92

Tirrell G, Sri-on J, Lipsitz LA, Camargo CA, Kabrhel C, Liu SW (2015) Evaluation of older adult patients with falls in the emergency department: discordance with national guidelines. Acad Emerg Med 22(4):461–467

US Census Bureau (2017) The nation's older adult population is still growing, June 22. https://www.census.gov/newsroom/press-releeases/2017/cb17-100.html. Accessed 19 Mar 2019

Wachelder JJ, Stassen PM, Hubens LP, Brouns SH, Lambooij SL, Dieleman JP et al (2017) Elderly emergency patients presenting with non-specific complaints: characteristics and outcomes. PLoS One 12(11):e0188954

Wilber S, Gerson LW, Terrell KM, Carpenter CR, Shah MN, Heard K, Hwang U (2006) Geriatric emergency medicine and the 2006 institute of medicine reports from the committee on the future of emergency care in the us health system. Acad Emerg Med 13:1345–1351

Yuen T, Lee L, Or I, Yeung K, Chan J, Chui C et al (2013) Geriatric consultation service in emergency department: how does it work? Emerg Med J 30(3):180–185

Emergency Care Nursing

▶ Critical Care Nursing

Emergency Room

▶ Emergency Care

Emerging Adults

▶ Perpetual Adolescence in Literature and Film

Emotion

Lei Liu
Department of Psychology, College of Teacher Education, Ningbo University, Ningbo, China

Synonyms

Affection; Feeling; Mood

Definition

Emotion is a mental state identified from four major aspects – cognitions/appraisals, feelings, physiological responses, and behaviors (Shiota and Kalat 2011). Emotion is often intertwined with these terms such as motivation, disposition, personality, and temperament.

Overview

The research literature on emotion and aging that has accumulated over the past 70 years covers a wide range of topics, such as emotional experience and aging, emotion regulation and aging, emotion perception and aging, and affective disorders and aging. Generally, researchers measure older adults' emotions by relying mainly on self-reports, physiological measurements, and behavioral observations.

Key Research Findings

Emotions can be represented by a few continuous dimensions, mainly including valence and arousal dimensions. Most studies about emotion experience and aging suggest that as people grow older, they report less negative affective experience and more positive affective experience (Carstensen and Charles 1998; Mroczek 2004; Shiota and Kalat 2011). However, emotions can also be divided into basic emotions and complex emotions. Some studies show that older and younger adults experience specific emotions at different frequencies. Compared with younger adults, older adults report less anger, less regret, and more contentment and experience similar or increasing sadness (Mather and Ponzio 2016; Pasupathi et al. 1998).

Increasing studies find that there is an age-related positive effect in older age. Compared with younger adults, older adults are more likely to favor positive over negative information at attention and memory and to choose emotionally meaningful social contacts (Mather and Carstensen 2005). The positive nature of everyday emotional experience in later age is that older adults get better at regulating their emotions (Urry and Gross 2010). Recent research show more spontaneous emotional regulation in older age. For example, compared with younger adults, older adults had lower responsiveness and better recovery to a negative social situation and maintained longer positive pre-task feelings (Luong and Charles 2014; Voelkle et al. 2013). However, some studies suggest that there are no consistent age advantages for either older adults or younger adults (Isaacowitz et al. 2017; Mather 2012). Furthermore, there are age differences in regulation strategies people tend to use. Older adults are more likely to use suppression and avoid emotional situations and less likely to use reappraisal, rumination, and active coping than do younger adults (Mather and Ponzio 2016; Scheibe et al. 2015).

Most studies about emotion perception show negative effects of age. Compared with younger adults, older adults are typically worse at recognizing sadness, fear, and angry expressions (Isaacowitz and Stanley 2011). Furthermore, recent studies have examined how aspects of stimuli characteristics and perceiver characteristics influenced older adults' emotion perception performance. These findings suggested that older adults still performed worse than younger adults (Isaacowitz et al. 2017). Because older adults feel a greater diversity of emotions compared with younger adults, they may misidentify emotional expressions (Kellough and Knight 2012; Seungyoun et al. 2015).

Older adults are obviously at greater risk for ill health than younger adults. However, research show that older adults have lower incidence and prevalence rates of depression and anxiety than younger adults (Blazer 2003; Singleton et al. 2003). Instead, older adults report more depressive symptoms than younger adults (Fiske et al. 2009). That is, older adults do experience negative feelings, but they do not become depressed.

Summary

Increasing evidences showed that age has different effects on emotion across different emotional domains. In general, emotion is a domain in which older adults perform well especially taking into account the declines in cognitive and physical domains. Far more research is needed in the area of emotion and aging.

Cross-References

▶ Affect Regulation
▶ Emotion Regulation
▶ Mood Disorders
▶ Positive Affect
▶ Positivity Effect

References

Blazer DG (2003) Depression in late life: review and commentary. J Gerontol Ser A Biol Med Sci 58: M249–M265

Carstensen LL, Charles ST (1998) Emotion in the second half of life. Curr Dir Psychol Sci 7:144–149

Fiske A, Wetherell JL, Gatz M (2009) Depression in older adults. Annu Rev Clin Psychol 5:363–389

Isaacowitz DM, Stanley JT (2011) Bringing an ecological perspective to the study of aging and recognition of emotional facial expressions: past, current, and future methods. J Nonverbal Behav 35:261–278

Isaacowitz DM, Livingstone KM, Castro VL (2017) Aging and emotions: experience, regulation, and perception. Curr Opin Psychol 17:79–83

Kellough JL, Knight BG (2012) Positivity effects in older adult's perception of facial emotion: the role of future time perspective. J Gerontol Ser B Psychol Sci Soc Sci 67:150–158

Luong G, Charles ST (2014) Age differences in affective and cardiovascular responses to a negative social interaction: the role of goals, appraisals, and emotion regulation. Dev Psychol 50:1919–1930

Mather M (2012) The emotion paradox in the aging brain. Ann NY Acad Sci 1251:33–49

Mather M, Carstensen LL (2005) Aging and motivated cognition: the positivity effect in attention and memory. Trends Cogn Sci 9:496–502

Mather M, Ponzio A (2016) Emotion and aging. In: Barrett LF, Lewis M, Haviland-Jones JM (eds) Handbook of emotions, 4th edn. Guilford Publications, New York

Mroczek DK (2004) Positive and negative affect at midlife. In: Brim OG, Ryff CD, Kessler RC (eds) How healthy are we? A national study of well-being at midlife. University of Chicago Press, Chicago

Pasupathi M, Carstensen LL, Turk-Charles S, Tsai J (1998) Emotion and aging. In: Encyclopedia of mental health, vol 2. Academic, San Diego, pp 91–101

Scheibe S, Sheppes G, Staudinger UM (2015) Distract or reappraise? Age-related differences in emotion-regulation choice. Emotion 15:677–681

Seungyoun K, Geren JL, Knight BG (2015) Age differences in the complexity of emotion perception. Exp Aging Res 41:556–571

Shiota MN, Kalat JW (2011) Emotion, 2nd edn. Wadsworth, Belmont

Singleton N, Bumpstead R, O'Brien M, Lee A, Meltzer H (2003) Psychiatric morbidity among adults living in private households, 2000. Int Rev Psychiatry 15:65–73

Urry HL, Gross JJ (2010) Emotion regulation in older age. Curr Dir Psychol Sci 19:352–357

Voelkle MC, Ebner NC, Linderberger U, Riediger M (2013) Here we go again: anticipatory and reactive mood responses to recurring unpleasant situations throughout adulthood. Emotion 13:424–433

Emotion Regulation

▶ Affect Regulation

Emotion Self-Regulation

▶ Affect Regulation

Emotional Loneliness

▶ Loneliness

Employment

▶ Employment and Caregiving

Employment and Caregiving

Yeonjung Lee
Faculty of Social Work, University of Calgary, Calgary, Alberta, Canada

Synonyms

Employment; Family caregiving; Informal caregiving; Labor force participation; Work

Overview

Employed family caregivers face difficult choices as they try to balance work and caregiving commitments. A large number of family caregivers to older adults are employed or working part time in addition to their caregiving responsibilities. In Canada, approximately 72% of women and 83% of men caregivers are employed (Lilly 2011). Role responsibilities divided between caregiving and working frequently compete and conflict; employed caregivers are described as feeling "sandwiched" (Neal and Hammer 2007) as they struggle with balancing and combining both responsibilities (Wang et al. 2018). Consequently, some working caregivers are likely to make adjustments in their work and caregiving activities (Scharlach et al. 2007; Lee et al. 2015a; Lee and Tang 2015). It is estimated that about 25% of working caregivers in Canada made work-related adjustments in order to take on caregiving responsibilities (Duxbury et al. 2009; Fast 2015). However, working can also provide the advantage of taking a respite from caregiving for some caregivers (Hansen and Slagsvold 2015).

Key Research Findings

In research on caregiving and employment, many studies have examined the relationship independently, assuming a unidirectional relationship (Pavalko and Henderson 2006; Young and Grundy 2008; Lee and Tang 2015). That is, they distinguish caregiving interference with work from work interference with caregiving, and address each model separately, with more attention directed to the impact of caregiving on work.

Several studies (Pavalko and Woodbury 2000; Dentinger and Clarkberg 2002) have shown that caregivers are likely to work. As family caregiving often causes a financial burden, caregivers may want to remain employed due to financial considerations or health insurance (Carmichael and Charles 2003; Dentinger and Clarkberg 2002; Eales et al. 2015). In addition, some caregivers consider their workplace a respite from the demands of caregiving (Carmichael and Charles

2003; Hansen and Slagsvold 2015). From this perspective, employed caregivers consider their employment as a means of buffering the strain and stress of caregiving demands, (Pavalko and Woodbury 2000). However, a large body of literature has also documented that caregiving is negatively related to employment, showing that many caregivers left the labor force or retired or shifted from full-time to part-time employment as a result of providing care to parents (Pavalko and Henderson 2006; Spiess and Schneider 2003; Longacre et al. 2016).

Compared to the literature on the effects of caregiving on work, relatively fewer studies have examined how caregivers' employment is associated with the decision to take on caregiving responsibilities. Some research has suggested that employment limits the likelihood and amount of time that family members provide care (Scharlach et al. 2007; Young and Grundy 2008; Feinberg et al. 2011), though women are likely to become caregivers regardless of their employment status unlike men. This suggests that employment was not significantly associated with stopping caregiving for women (Moen et al. 1994). However, the cross-sectional designs of these studies limit the ability to infer causality. Mentzakis et al. (2009) examine the determinants of informal care using a longitudinal data and show that participation in the labor force negatively affects the decision to be a caregiver for men and women.

A few studies have examined the reciprocal relationship between employment and caregiving (Berecki-Gisolf et al. 2008; Pavalko and Artis 1997; Lee et al. 2015a). Boaz and Muller (1992) are among the first researchers to suggest the potential for the simultaneity of both relationships based on cross-sectional design. By distinguishing full-time and part-time employment, their findings show that full-time employment of caregivers reduces the hours of caregiving and women caregivers are much less likely than their male counterparts to have full-time employment. However, there is no significant effect for part-time employment. More recently, Lee et al. (2015a) test the association using longitudinal panel data and show no reciprocal relationship between caregiving and labor force participation. Instead, the findings show that there is negative effect of caregiving on employment for women, whereas employment reduces the possibility of being a caregiver for men.

Areas for future research include more diverse approaches to address employment and caregiving. Rather than focusing on employment status itself and distinguishing those who are employed and those who are not, the experiences and perceptions of employment need to be examined. Caregiving should be measured the same way. Instead of simply indicating whether a person is providing care or not, how caregivers perceive their care experiences might help an understanding of not only the negative outcomes but also the positive perspectives of caregiving. These perspectives have been raised previously by pointing out the lack of insight into the relationship between caregiving and employment and applied mostly to examine caregiver outcomes such as caregiver burden, well-being, and self-esteem (Reid et al. 2010). For example, Reid and colleagues (2010) examine the effects of employment status and work interferences on caregiver well-being, respectively, and suggest that subjective assessment of work interferences may play a more important role than does employment status.

Implications for Policy and Practice

These empirical findings regarding the association between caregiving and employment have implications for policy and practice. In the long term, the substantial earnings loss for caregivers raises questions about their retirement income because caregivers accumulate fewer future pension benefits (Lee et al. 2015b). Therefore, efforts to address poverty issues among caregivers, such as providing family caregiver credits, may improve their pension entitlements and add value to their caregiving work. Also, providing direct cash transfers or offering tax credits for purchasing long-term care insurance may compensate caregivers for their financial loss (Mellor 2000).

The workplace is a primary arena for supporting working caregivers to manage care and work responsibilities (Neal and Hammer 2007; Ireson et al. 2018). Workplace supports including flexible work schedules, paid leave, or supportive supervisors, and co-workers positively influence family caregivers' employment outcomes by reducing stress and role strain and helping them to meet their caregiving responsibilities (Kossek et al. 2001; Feinberg 2018). On the other hand, limited job flexibility and fewer workplace supports are likely to increase lateness for work and absenteeism, thus decreasing the productivity of employed caregivers (Scharlach 1994; Dembe et al. 2011) as well as their mental health outcomes (Li and Lee under review).

Paid family leave and supportive social services are important to assist family caregivers in the labor force and improve caregiving outcomes. Increased availability of publicly supported home care systems and caregiving leave can both increase flexibility in the workplace and increase the possibility for caregivers to remain in the labor force (Pavalko and Henderson 2006).

Summary

Given that many older adults are likely to live in the community, the number of family members and friends who provide care to these individual is and will be increasing. Research is needed to identify who are most vulnerable in the labor force and how the intersectionality of socioeconomic status affects employment as well as caregiving outcomes among caregivers.

References

Berecki-Gisolf J, Lucke J, Hockey R et al (2008) Transitions into informal caregiving and out of paid employment of women in their 50s. Soc Sci Med 67:122–127

Boaz RF, Muller CF (1992) Paid work and unpaid help by caregivers of the disabled and frail elders. Med Care 30:149–158

Carmichael F, Charles S (2003) The opportunity costs of informal care: does gender matter? J Health Econ 22:781–803

Dembe AE, Partridge JS, Dugan E et al (2011) Employees' satisfaction with employer-sponsored elder-care programs. Int J Work Health Manag 4:216–227

Dentinger E, Clarkberg M (2002) Informal caregiving and retirement timing among men and women: gender and caregiving relationships in late midlife. J Fam Issues 23:857–879

Duxbury LE, Schroeder B, Higgins CA (2009) Balancing paid work and caregiving responsibilities: a closer look at family caregivers in Canada. Can Policy Res Netw Ott, Ontario

Eales J, Keating N, Donalds S et al (2015) Assessing the needs of employed caregivers and employers. University of Alberta, Research on Aging, Policies and Practice, Edmonton

Fast J (2015) Caregiving for older adults with disabilities. Available via Institute for Research on Public Policy

Feinberg LF (2018) Breaking new ground: supporting employed family caregivers with workplace leave policies. AARP Public Policy Inst 136:1–28

Feinberg LF, Reinhard SC, Houser A et al (2011) Valuing the invaluable: 2011-the growing contributions and costs of family caregiving. http://assets.aarp.org/rgcenter/ppi/ltc/i51-caregiving.pdf. Accessed 27 Aug 2011

Hansen T, Slagsvold B (2015) Feeling the squeeze? The effects of combining work and informal caregiving on psychological well-being. Eur J Ageing 12:51–60

Ireson R, Sethi B, Williams A (2018) Availability of caregiver-friendly workplace policies (CFWP s): an international scoping review. Health Soc Care Community 26:e1–e14

Kossek EE, Colquitt JA, Noe RA (2001) Caregiving decisions, well-being and performance: the effects of place and provider type as a function of dependent type and work-family climates. Acad Manag J 44:29–44

Lee Y, Tang F (2015) More caregiving, less working: caregivers roles and gender differences. J Appl Gerontol 34:465–483

Lee Y, Tang F, Kim KH et al (2015a) Exploring gender differences in the relationships between eldercare and labour force participation. Can J Aging/La Rev Can Vieillissement 34:14–25

Lee Y, Tang F, Kim KH et al (2015b) The vicious cycle of parental caregiving and financial well-being: a longitudinal study of women. J Gerontol Ser B Psychol Sci Soc Sci 70:425–431

Li L, Lee Y (under review) Employment adjustment and mental health of employed family caregivers in Canada

Lilly MB (2011) The hard work of balancing employment and caregiving: what can Canadian employers do to help? Healthc Policy 7:23

Longacre ML, Valdmanis VG, Handorf EA et al (2016) Work impact and emotional stress among informal caregivers for older adults. J Gerontol B Psychol Sci Soc Sci 72:522–531

Mellor JM (2000) Filling in the gaps in long term care insurance. In: Meyer MH (ed) Care work: gender, labor and the welfare state. Routledge, New York

Mentzakis E, McNamee P, Ryan M (2009) Who cares and how much: exploring the determinants of co-residential informal care. Rev Econ Househ 7:283–303

Moen P, Robison J, Felds V (1994) Women's work and caregiving role: a life course approach. J Gerontol: Soc Sci 49:S176–S186

Neal MB, Hammer LB (2007) The sandwiched generation: introduction. In: Neal MB, Hammer LB (eds) Working couples caring for children and aging parents. Lawrence Erlbaum Associates, Mahwah

Pavalko EK, Artis JE (1997) Women's caregiving and paid work: causal relationships in late midlife. J Gerontol: Soc Sci 52B:S170–S179

Pavalko EK, Henderson KA (2006) Combining care work and paid work: do workplace policies make a difference? Res Aging 28:359–374

Pavalko EK, Woodbury S (2000) Social roles as process: caregiving careers and women's health. J Health Soc Behav 41:91–105

Reid CR, Stajduhar KI, Chappell NL (2010) The impact of work interferences on family caregiver outcomes. J Appl Gerontol 29:267–289

Scharlach AE (1994) Caregiving and employment: competing or complementary roles? Gerontologist 34:378–385

Scharlach AE, Gustavson K, Dal Santo TS (2007) Assistance received by employed caregivers and their care recipients: who helps care recipients when caregivers work full time? Gerontologist 47:752–762

Spiess CK, Schneider AU (2003) Interactions between care-giving and paid work hours among European midlife women, 1994 to 1996. Ageing Soc 23:41–68

Wang YN, Hsu WC, Yang PS et al (2018) Caregiving demands, job demands, and health outcomes for employed family caregivers of older adults with dementia: structural equation modeling. Geriatr Nurs 39:676–682

Young H, Grundy E (2008) Longitudinal perspectives on caregiving, employment history and marital status in midlife in England and Wales. Health Soc Care Community 16:388–399

Employment and Health/Longevity

▶ Lifetime Primary Occupation and Health/Longevity in Old Age

Employment and Outplacement Services

Bogusława Urbaniak
University of Lodz, Lodz, Poland

Synonyms

Employment offices and older adults; Job centers and older adults; Work agencies and older adults

Definition

The providers of employment and outplacement services (EOS) are private and public organizations which act as an intermediary between job seekers, including unemployed people aged 50 and above, and employers with vacancies (See ▶ "Age Management and Labor Market Policies"). The re-employment initiatives for people in the age group 50+ (See ▶ "Re-employment of (Early) Retirees") are part of employment services provided by public employment service and specialized private firms. In industrialized countries, public authorities made first interventions in the labor market and set about embedding them in the institutional framework already in the years 1880–1910 (Thuy et al. 2001). The public employment service (PES) is traditionally focused on supporting unemployed persons with intermediate and low skills through a range of counseling, employment, and vocational services and on helping employers to fill vacancies from the PES register. The PES is also a vehicle for delivering active labor market programs (ALMPs) designed to create jobs or improve the relative labor market position of various target groups or both. The PES is especially committed to difficult-to-place clients such as unemployed aged 50+ who have the status of a vulnerable group of unemployed because of above-average problems they face in the labor market.

Overview

As a remedy for rapidly rising unemployment levels in Europe hit by the oil crisis of the 1970s, measures suppressing the supply of labor such as early retirement, readily available disability pensions, and social welfare schemes were offered. To create job opportunities for young unemployed in 1978, Denmark introduced early retirement (Foden and Jepsen 2002), as well as other alternative exit routes for older workers to encourage them to leave the labor market. However, the austerity policies introduced in Europe in the late 1980s and early 1990s, as well as other factors, made it necessary for public employment services to reach for ALMP instruments for bringing older workers back in the workforce. The governments of many European countries decided to raise the mandatory retirement age gradually but consistently (See ▶ "Retirement Patterns"). The extension of the nominal period of economic activity increased the importance of ALMP's instruments dedicated to older workers. The early customers of the PES were low-skilled older unemployed with clearly obsolete skills. It was only in the years following the most recent crisis (2007–2014) that an increase in the share of high-qualified customers of the PES (including those aged 50+) was noted (McGrath 2016).

Theories Related to Employment and Outplacement Services

Human capital theory: a large part of workers' firm-specific capital that they acquire from human capital investments made by their employers and by developing company-specific skills over long years of work for the same employer may not be easily transferred to other organizations (Farber 2017). The capital is certainly an asset as long as a one does his or her old job, but its usability for an older worker who loses a job is frequently problematic. What they certainly need more in such circumstances is the knowledge about where to find a new job and the skill of successfully applying for it but having spent a long time in the same organization they have neither. They are also frequently troubled by multiple health problems because health, an essential component of human capital, deteriorates with age (See ▶ "Gender Disparities in Health in Later Life"). The combination of the factors causes that older worker to face mounting problems in the labor market and slide into long-term unemployment, and, as their health continues to deteriorate, their odds of leaving unemployment behind grow smaller. The lower value of their human capital decreases the probability that they will reenter the labor force unassisted by employment and outplacement services.

New institutional economy: as a labor market institution helping unemployed people of all ages find jobs, the PES can be counted among the institutions of the new institutional economics, a theoretical concept which has been developing since the 1970s. The division of the labor market into different segments in the years 1950s–1960s paved the way for the creation of the dual labor market concept and for understanding the distinctive character of older workers as a segment of the labor market. Public employment services form an institutional system that influences employers' decisions through labor market instruments and labor market programs for older unemployed workers. Similarly, to other institutions created for this group of customers, such as legal protection of their employment and the tax wedge, they adjust the scale of qualitative and quantitative changes in the labor market. This activity must naturally lead to a discussion about the optimal level of state intervention in economic processes (Kwiatkowski 2017).

Main Types of Employment and Outplacement Services Dedicated to Older Job Seekers

The PES' assistance for older unemployed is based on two main approaches (Maksim et al. 2018). The purpose of the target group approach is to ensure equal employment opportunities (See ▶ "Reducing Ageism") for some selected problem group, such as unemployed people aged 50+, through the assistance of counselors specially trained for work with older unemployed. The counselors offer them a standard package of services and access to general ALMP instruments, for example, selected omitting of the needs of individual customers. The preferential treatment for beneficiaries aged 50+ consists of a greater range and intensity of support activities (See ▶ "Productive Aging"). The weakness of the approach is that it assumes that all unemployed in this age group have the same problems in the labor market and a resulting lack of a more detailed analysis of individual needs, stigmatization, the perpetuation of stereotypes about people aged 50 and above (See ▶ "Sexism and Ageism"), and professionals with a lot of experience likely to perceive the support or advice they are offered as trivial and inadequate for their situation. Also, the rates of beneficiaries returning to the labor force are low.

The individualized approach consists of a counselor and an unemployed person aged 50+, developing together a customized activation strategy. The cooperation between them goes on throughout the activation period. Early on, the customer's needs and problems are identified using profiling methods, and then appropriate activation measures are selected. This approach does not involve the creation of individual ALMP instruments for some target groups of unemployed people. Its advantages include the equality of opportunities, a nondiscriminatory approach, and higher and more sustainable rates of beneficiaries returning to employment. A survey conducted by the European Commission showed that the individualized approach was applied in countries such as Austria, Germany, France, the United Kingdom (UK), the Czech Republic, and Hungary, but whether it was used toward all unemployed or only in the case of risk groups such as people aged 50+ was not reported (Hake 2011).

Government Programs for Job Seekers Aged 50+

Older job seekers can participate in unrestricted-access programs meant for the general population of displaced workers and in special programs for the older unemployed (See ▶ "Encore Jobs"), which are funded from local, national, and federal sources, and in the EU-28 also from the European Social Fund (ESF). The source of funding for such programs is federal or national budgets, and in the European Union (EU)-28 countries also the European Social Fund (ESF). The focus of the programs is on enabling unemployed persons aged 50 and older to reenter the labor force and on reducing periods out of employment to the possible extent. The programs are enhanced by standard measures accompanying the national labor market policy. Some of them, such as job offers for low-income pensioners, aim to secure beneficiaries' social welfare. A good example of this approach is the South Korean Senior Employment Program, which supports older individuals experiencing financial hardships due to Korea's inadequate public and private retirement income systems. The United States Senior Community Service Employment Program offers subsidized minimum-wage jobs in community service positions in public and not-for-profit organizations (Choi 2016).

For labor market programs to be effective, they need to be holistic, long-term, tailored to the characteristics and needs of the displaced workers, and capable of recognizing upfront their strengths and transferable skills. Their design must allow for a combination of training, retraining, and other activities, as many older workers have obsolete skills (See ▶ "Senior Learning"). The process of assisting an unemployed person aged 50+ to return to the labor market should start with advice, career planning, and screening for training, followed by the planning and delivery of training, employment assistance, and ongoing monitoring. A range of auxiliary services is also necessary, such as online job search assistance, help with résumé writing and specific job cover letter preparation, coaching in job interview techniques, and active self-marketing (PES 2018).

Private Employment Agencies and Outplacement Services

Article 1 of Convention No. 181 of the International Labour Organization of 1997 defines a private employment agency as any natural or legal person, independent of the public authorities, which provides one or more of the following labor market services aimed at (a) matching offers of and applications for employment, (b) employing workers with a view to making them available to a third party ("user enterprise"), and (c) providing any other form of support for job seekers (ILO 1997). It is interesting to note that some private employment agencies specialize in the delivery of employment services for older workers.

Outplacement services for workers older than 50 years of age are a form of customized assistance delivered by private agencies on behalf of older persons who were laid off or lost a job. In most cases, its legal basis is an agreement concluded between the agency and the former employer to assist employees who have left the employer, either voluntarily or involuntarily, in the transition to a new job (See ▶ "Recruitment and Retention in Aging Research"). The costs of the service are usually covered by the employer, although in some cases they may be paid by the final beneficiaries, for example, former employees. Having employers pay part or all of the costs of outplacement services can be understood as a penalty for laying off older workers (Chéron et al. 2011).

An outplacement service can also be included in a severance package for regular employees (and almost always for senior-level employees) as an expression of the humane lay-off approach assuming that employees who leave the organization need both financial assistance and outplacement assistance (See ▶ "Bridge Jobs"). An outplacement package offers individual or group career counseling, coaching, advising, job interview training, and helping candidates improve their resumes and profiles on social networking sites facilitating professional and business contacts. According to Eurofound, Bulgaria, Croatia, Cyprus, Greece, and Hungary do not use outplacement services as an instrument supporting the labor market integration of older workers. In other countries, such as Austria, they are either available to workers of all ages (fit2work) or are dedicated to older workers (Early Intervention 50+) (Mandl et al. 2018). A large number of institutions and organizations supporting people aged 50+ in the labor market point to the existence of an assistance network consisting of (a) public and private employment services; (b) temporary work agencies, which are a sort of private employment agencies; (c) nongovernmental organizations; and (d) education and training providers.

There are at least three models of outplacement services characterized by Koral (2009) and Klimczuk-Kochańska and Klimczuk (2015): classic outplacement, adapted outplacement, and environmental outplacement. The classic model focuses on the diagnosis of potential and needs of the laid-off workers, psychological support, career counseling, training, and job placement. An employer, who has decided to implement outplacement, establishes the rules of layoffs and a list of employees for dismissal. It then instructs the consulting firm, employment agencies, or nongovernmental organizations (NGOs), which prepare the team to provide services. This model is considered as the most effective for the laid-off workers 50+. The adapted outplacement is characterized by greater involvement in the process of supporting laid-off workers by their social environment, including local governments, business environment, and NGOs. Because of the use of instruments such as public works, intervention works, internships, temporary employment arrangements, exemptions from local taxes, the creation of business incubators, and the organization of trade fairs, outplacement has also come into the focus of interest of several labor market institutions. The last model – environmental outplacement – is designed to prevent the effects of long-term unemployment in the case of the collapse of the local labor market and needs to establish the activation center in a region. This model includes the use of instruments of classic outplacement and elements of group work and social animation.

Key Research Findings

Although older workers who have long service with the same employer are at a lower risk of being terminated than younger workers, they are also less likely to find a new job after a lay-off (EC 2012), not only because of their obsolete skills and limited adaptability to new working environments, but also due to negative stereotypes and discrimination against older people in the labor market (See ▶ "Reducing Ageism"). Faced with so many barriers, they gradually lose motivation to find work over the period of unemployment. For unemployed people aged 50+ to have another chance, a series of actions is necessary, from profiling of their individual needs to the selection of appropriate, customized services.

The Anglo-Saxon countries design their ALMP measures around a principle called "welfare-to-work" (workfare or W2W), resembling the German "fördern und fordern." It assumes that the unemployed welfare beneficiaries can be encouraged, or even pressed, to find employment.

In the early 1990s, the UK government adopted a public-private partnership approach under which the support of long-time unemployed started to be contracted out to private sector welfare-to-work (W2W) providers. In the case of workers aged 50+ who have been without work for 6 months or longer and draw unemployment benefits, the main source of support continues to be state-run Jobcentre Plus shops whose effectiveness is regarded as moderate. The weakness of the solution based on private W2W providers is that their revenues depend on job outcomes (successful placement of customers in jobs), so they tend to "shelve" unemployed customers over 50 because of relatively poor prospects of putting them in jobs compared with younger persons ("creaming"). The New Deal 50 plus program, one of the UK government's W2W initiatives, offers job seekers the assistance of personal counselors and a tax-free wage top-up (employment credit) for those who found paid work, for up to 1 year, and a training grant. In 2017, the Work Programme and Work Choice Schemes were replaced by the Health and Work Programme created for people with disabilities aged 50+ and those who have long been jobless.

A survey of the Work Programme's beneficiaries conducted between April 2013 and July 2014 showed that unemployed people aged 50+ receiving JSA (Jobseeker's Allowance for persons capable of working) had a low predicted probability of job start in the first 6 months of the program (female 15%, male 18%), which decreased to 5.8% and 6.8% for females and males, respectively, in an average 3-month period in the second year of the program. Predicted probability of job starts in the first 3 months of the program for people aged 50+ receiving the Employment and Support Allowance (ESA) for people with disability was estimated at 2.4% (Brown et al. 2018).

The Hartz reform of the German market policy and institutions turned labor offices into labor agencies and changed the system of unemployment benefits, also those dedicated to older unemployed workers (until the end of 2007, unemployed aged over 58 were eligible for unemployment benefits until retirement without having to seek employment actively). In the years 2005–2015, the federal government ran a program for unemployed people aged 50 and over (Perspective 50 Plus – Employment pacts for older workers in the regions), which proved effective in reducing unemployment in this age group. The most popular of its instruments was the assistance of labor offices' counselors in organizing contacts with employers and their involvement in recruitment and selection processes. Another frequently used instrument was an analysis of clients' competence capital aimed to identify competencies they were not aware of having and which could help them to find an alternative career when the chances of getting a job similar to the one they lost were slim. The third of the most popular instruments were individual coaching meetings which significantly helped the program beneficiaries overcome their low spirits, disbelief in their strengths, and a sense of inferiority associated with long-term unemployment that was the common experience of most of them. A measure of the program's effectiveness is the fact that every third beneficiary found a new job (according to the 2005–2015 data). This outcome was better than the average for other assistance programs supporting unemployed people in the same age group (Maksim et al. 2018).

Future Directions of Research

The general improvement in the labor market and declining unemployment rates seem to have little influence on the situation of unemployed people aged 50+, among whom the share of long-term unemployed remains relatively unchanged. It is thought that their labor market situation could be improved by the PES, offering them the assistance of an individual coach-counselor, especially during the first 6 months after a job loss and cooperating with local employers. Exposure to jobs through work trials and the introduction of emeritus apprenticeships are advised as a source of new opportunities for unemployed people aged 50+ (See ▶ "Retirement Transition").

Given the character of demographic changes in the labor market, it can be expected that the PES will undertake in the near future activities to reach economically inactive people (inactive older workers, working-age women, ethnic minorities, and migrants) who are central to current policy discussions on increasing labor force participation and social inclusion (EC 2019, p. 12). The deficits faced by national pension systems (See ▶ "Economics of Aging") will increase the pressure on employees extending the duration of their economic activity (See ▶ "Retirement Patterns"), the introduction of flexible labor market policies, and broader use of employment and outplacement services.

Summary

The probability of older job seekers finding jobs after being laid off is exceptionally low. This leaves no doubt that they are among unemployed people who are less advantaged in the labor market than younger individuals and that they need institutional assistance to avoid long-term unemployment or economic inactivity. The mission of the public employment service (PES) and private organizations that, respectively, provide employment and outplacement services (EOS) includes support for older job seekers. The range of services they are offered includes guidance from, among others, counselors, career counseling, training, public works, and internships. In order to make the services more effective, counselors customize them to individual needs. Older job seekers are also offered participation in general programs designed for displaced workers as well as in special programs meant for particular groups of older unemployed people.

Cross-References

▶ Age Management and Labor Market Policies
▶ Bridge Jobs
▶ Delaying Retirement
▶ Employment and Outplacement Services
▶ Encore Jobs
▶ Gender and Employment in Later Life
▶ Self-Employment Among Older Adults
▶ Social Entrepreneurship and Social Innovation in Aging

References

Brown J, Katikireddi SV, Leyland AH, McQuaid RW, Frank J, Macdonald EB (2018) Age, health and other factors associated with return to work for those engaging with a welfare-to-work initiative: a cohort study of administrative data from the UK's Work Programme. BMJ Open 8:e024938. https://doi.org/10.1136/bmjopen-2018-024938

Chéron A, Hairault JO, Langot F (2011) Age-dependent employment protection. Econ J 121(December):1477–1504. https://doi.org/10.1111/j.1468-0297.2011.02453.x

Choi E (2016) Older workers and federal work programs: the Korean Senior Employment Program (KSEP). J Aging Soc Policy. https://doi.org/10.1080/08959420.2016.1153993

European Commission (2012) PES and older workers. Toolkit for public employment services. DG Employment, Social Affairs and Inclusion. Available via Budapest Institute. http://www.budapestinstitute.eu/uploads/P2P_older_workers_toolkit_20120809.pdf. Accessed 20 Feb 2019

European Commission (2019) Annual report. European Network of Public Employment Services (PES), Luxembourg

Farber HS (2017) Employment, hours, and earnings consequences of job loss: US evidence from the displaced workers survey. J Labor Econ 35:S235–S272. https://doi.org/10.1086/692353

Foden D, Jepsen M (2002) Active strategies for older workers in the European Union. A comparative analysis of recent experiences. In: Jepsen D, Foden M, Hutsebaut M (eds) Active strategies for older workers in the European Union. European Trade Union Institute, Brussels, pp 437–460

Hake BJ (2011) The role of public employment services in extending working lives: sustainable employability for older workers. European Commission, Brussels. Available via PES Network. https://ec.europa.eu/social/BlobServlet?docId=9690&langId=en. Accessed 10 Jan 2019

International Labor Organization (1997) Private employment agencies convention C 181. https://www.ilo.org/dyn/normlex/en/f?p=NORMLEXPUB:12100:0::NO::P12100_INSTRUMENT_ID:312326

Klimczuk-Kochańska M, Klimczuk A (2015) Outplacement: the Polish experience and plans for development in the labour market. In: Romano S, Punziano G (eds) The European social model adrift: Europe, social cohesion and the economic crisis. Ashgate, Burlington, pp 89–106

Koral J (2009) Outplacement – sposób na bezrobocie (Outplacement: a way to fight unemployment). Fundacja Inicjatyw Społeczno-Ekonomicznych, Warsaw

Kwiatkowski E (ed) (2017) Instytucje rynku pracy w krajach OECD. Istota, tendencje i znaczenie ekonomiczne (Labor market institutions in OECD countries: essence, tendencies, and economic significance). Wydawnictwo Uniwersytetu Łódzkiego, Łódź

Maksim M, Wiśniewski Z, Wojdyło M (2018) Strategie aktywizacji zawodowej bezrobotnych w wieku 50+ dla publicznych służb zatrudnienia. Teoria i praktyka (Strategies of vocational activation of the unemployed aged 50+ for public employment services: theory and practice). Wydawnictwo Uniwersytetu Mikołaja Kopernika, Toruń

Mandl I, Patrini V, Jalava J, Lantto E, Muraille M (2018) State initiatives supporting the labour market integration of older workers. Working paper. Available via Eurofound. https://www.eurofound.europa.eu/sites/default/files/wpef18003.pdf. Accessed 27 Dec 2018

McGrath J (2016) EU labour market dynamics and PES activity. Report no. 1. Years of crisis 2007–2014. Luxembourg. Available via European Commission. https://ec.europa.eu/social/BlobServlet?docId=16334&langId=en. Accessed 27 Dec 2018

PES (2018) EU Network of Public Employment Services. Strategy to 2020 and beyond. Available via European Commission. https://ec.europa.eu/social/main.jsp?catId=1100&langId=en. Accessed 10 Feb 2019

Thuy P, Hansen E, Price D (2001) The Public Employment Service in a changing labour market. International Labour Organisation, Geneva

Employment Offices and Older Adults

▶ Employment and Outplacement Services

Empty Nest

Barbara A. Mitchell
Department of Sociology/Anthropology and Department of Gerontology, Simon Fraser University, Burnaby, BC, Canada

Definition

Empty nest refers to a family life course transition and postparental phase that occurs when children have moved out and left the parental home.

Overview

The term empty nest is commonly used to refer to a family transition whereby children have grown up and have left the parental home. Historically, the empty nest phase of the family life cycle was relatively rare or of a short duration. With rising longevity and lower fertility rates, parents began to experience an extended postparental phase without active childrearing and dependent children in the household. This shift in household living arrangements typically occurs when parents are middle-to-older-aged and children are young adults and includes a reorganization and readjustment of parental roles and family relationships. Diversity among families is also found in the specific timing, nature, and circumstances of this transition. Key research findings, examples of selected studies, and future research directions on this topic will be reviewed.

Key Research Findings and Applications

Research on the empty nest has focused on a number of interrelated substantive topics: (1) socio-demographic and economic changes in parent-child living arrangements; (2) the impact of the empty nest on marital/coupled relationships; (3) parental reactions to the empty nest, including gender and ethnic differences; and (4) intergenerational relations during/after empty nest transitions.

Socio-Demographic and Economic Changes

The empty nest is a relatively new post twentieth-century phenomenon. From a historical perspective, this life cycle transition used to be of a relatively short duration and occurred when a couple was older-aged. For instance, it was fairly common for elderly people who lived and owned farm property households in nineteenth-century America to reside with at least one child for reciprocal support until the older generation died (Chudacoff and Hareven 1979; Ruggles 2011). Since the mid-twentieth century, the length of the empty nest period began to expand as life expectancy began to rise and fertility patterns changed, including a compression of the childbearing years. Fertility and family size also began to shrink due to changes in birth control technologies and fertility values that promoted smaller families (Bouchard 2014).

The transition to the empty nest also started to emerge during middle-age as life expectancy rose. For example, the average couple who married in 1960 and had children during the first five years of their marriage would have launched their last child in 1985 and both parents were likely in their 40s (Barber 1989; Glick 1977). Moreover, once children left home they were generally "gone for good." Thus, parents could generally anticipate their transition to retirement and later years without children living in the home (See ▶ "Aging Households").

Over the past several decades the "complete" transition to the empty nest has become delayed, less linear, and permanent (Mitchell et al. 2019) (See ▶ "Co-residence"). Increasing numbers of young adult children return home after an initial departure to refill the empty nest as "fledgling" or "incompletely launched adults" (Clemens and Axelson 1985; Schnaiberg and Goldenberg 1989) (See ▶ "Crowded Nest"). Since the Great Recession of 2008, the proportion of young adult children remaining in and/or returning home rose to its highest share in more than forty years (Davis et al. 2018; U.S. Census Bureau 2014). This counter pattern of children returning home created the phenomenon dubbed 'boomerang kids' (e.g., see Mitchell 2007). This trend of refilled nests is largely attributed to economic and other sociodemographic shifts, such as the increasing need for postsecondary education, precarious labor markets, the high cost of housing, and later and/or more unstable partnership union and family formation (See ▶ "Living Arrangements").

Consequently, static life cycle depictions of family events or transitions are less applicable to contemporary family life. These "newer" life course models are also more likely to emphasize variability and the possibility of life course transition reversals. These transformations in empty nest living arrangements have significant implications for aging families and relationship dynamics.

Impact of the Empty Nest on Marital/Coupled Relationships

Previous scholarship on martial adjustment and marital happiness over the life course has often found a U-shaped association. In a review of research on marital quality during the 1980s, Glenn (1990, p. 823) observed that, "a curvilinear relationship between family stage and some aspects of marital quality is about as close to being certain anything ever is in the social sciences." This conceptual depiction is characterized by high marital happiness in the early (or pre-parent) years of marriage, a decline in marital happiness during the middle (or parental) years, and a rise in the marital happiness in the later (or postparental) years.

While the adoption of this curvilinear model continues to be widespread, it has been criticized for methodological limitations. Cross-sectional data using nonprobability samples have often been used to make inferences about the population of married parents. Reaching conclusions about changes over time, therefore, is problematic. For example, this may artificially increase the mean happiness among couples in long-term marriages since many unhappy couples may have divorced by the time children leave home (e.g., see Mitchell 2016; VanLaningham et al. 2001).

Despite the weaknesses of the U-shaped model, it is well established that when children leave home parents will experience some positive changes in roles and time allocation demands that can affect the quality of their relationships (Bouchard and McNair 2016; White and Edwards 1990). Typically, the day-to-day physical and emotional burden of household responsibilities is alleviated, providing partners more freedom to spend together as a couple and to pursue other social roles that can enhance life satisfaction. Less work-family conflict is also reported (Erikson et al. 2010) and greater equity in the relationship (especially among wives) can be experienced (e.g., see Kulik et al. 2010) (See ▶ "Midlife and Gender").

While many positive benefits for couples have been reported, there is also some evidence to show that the transition to the empty nest can increase the risk of marital dissolution, especially for those who reach this phase relatively early in their marriage. Some parents may have stayed together for "the sake of the children" and women are more likely than men to initiate a separation or divorce at this stage of life (Bouchard 2014; Heidemann et al. 1998).

Parental Reactions to the Empty Nest

Folklore and early social scientific research left the strong impression that children's leaving home is an unhappy or traumatic event (Glenn 1975). The term empty nest syndrome technically refers to a clinical or depressive condition. It became popularized in the 1970s based on some studies that showed that woman (many of whom were middle-class full time homemakers) became depressed after their children left home. It was assumed that mothers, in particular, were negatively affected because they suffered a major role loss and void in life which could not easily be replaced. However, this body of research has been criticized for being based on small non-representative clinical samples that often included women who were hospitalized for depression.

Overall, the stereotype of the empty nest syndrome has largely been debunked as a cultural myth by researchers, although popular media and literature continue to use this term to indicate any unhappy or negative feeling of distress that parents have when their children leave home. A classic study conducted by American sociologist Lillian Rubin (1979) found that stay-at-home-mothers were generally relieved when children left home. More recent studies (e.g., Mitchell and Lovegreen 2009; Mitchell and Wister 2015) also confirm that the majority of parents do not experience severe or long-lasting negative emotional consequences. These researchers also observe that fathers are similar to mothers in their reactions to children's leaving home, and that ethno-cultural background may shape individual parental responses.

Parents from traditional collectivist cultural groups (e.g., Asian, Southern-European countries) are found to have different normative expectations and reactions to the departure of their children relative to those from individualistic ethnic backgrounds (e.g., North American and British). For instance, parents from collectivist cultures may view their children's leaving home for marriage or education (rather than to seek independence) as a signifier of successful parenting. Moreover, in some ethnic groups (e.g., traditional South Asian), there may be the expectation that the nest never empties and that at least one child (i.e., the eldest son and his wife) remains in the home to form a multigenerational or extended household. Violations of these cultural norms and expectations, therefore, may create conflict, stress, and anxiety for parents (Mitchell and Wister 2015).

Yet, despite relatively consistent findings that most parents – and regardless of cultural background – do not experience traumatic adverse effects, psychologists recognize that it is fairly normal for parents to experience a temporary sense of loss when their nests empty. There is also growing agreement that empty nests can bring many personal opportunities to parents' lives, as evidenced by the large self-help industry of books targeted to help "empty nester" parents successfully adjust to, and cope with their new roles.

Intergenerational Relations Before/After Empty Nest Transitions

Research on intergenerational relations before/after empty nest transitions tends to focus on how it is a process of separation rather than a singular event and how parent-child relations affect, or are affected by, this phase. Indeed, the process of leaving the parental home is a real-life separation that might be emotionally and psychologically challenging at the entire family level with emerging adults (e.g., see Kins et al. 2013). Furthermore, the decision to leave home is often framed in terms of a joint decision that is negotiated by the child and the parent(s) (Mitchell et al. 1989; Billari and Liefbroer 2007). The family environment and quality of intergenerational relationships can influence this decision-making process and subsequent empty nest family relations of support and long-term trajectories (See ▶ "Intergenerational Exchange and Support" and ▶ "Intergenerational Relations").

Notably, adverse family environments, including those with high levels of conflict, stress, and/

or poverty, are found to promote earlier homeleaving (e.g., Mitchell et al. 1989). Leaving home prematurely and emptying the nest under unfavorable conditions is generally found to reflect poorer quality relationships prior to leaving. This homeleaving situation can affect post-departure contact patterns, support exchanges, and family well-being, including feelings of ambivalence. For instance, children may have less desire to maintain contact with their parents and parents may worry about their children's safety and security. Yet, as suggested by the literature on intergenerational ambivalence, parents can also feel obligated to maintain frequent interactions with their children even when the quality of the relationships is poor or low (Lüscher 2011).

However, for most parents, the quality of parent-child relations tends to remain unchanged (Ward and Spitze 2007) after homeleaving, although intergenerational relations can improve once children empty the nest. Parents can now interact with their children on a more peer-like level due to their children's greater maturity and adoption of self-sufficient adult roles and statuses (Mitchell 2007). Both early and more recent studies have also reported that adult children increasingly become resources for their aging parents, and especially mothers (e.g., see Barber 1989; Leopold 2012).

Finally, with regard to other patterns and exchanges of support after the transition to the empty nest, it is no longer assumed that once children leave home that parental support to children discontinues. Once children leave home, parents may continue to provide considerable financial and other types of social/emotional support, especially when children are attending college or they trying to establish themselves as independent adults in separate households.

Future Directions of Research

There are several promising avenues for future work in this area. Current research findings suggest that theory and research should widen to include more diversity and unpredictability in empty nest transitions. Studies tend to focus on the experiences of two parent, married, heterosexual couples, and have not sufficiently included other types of family structures (e.g., single parent or stepfamily households) and alternative partnerships (e.g., cohabiting unions, LGBTQ). Gender- and class-based analyses have also been limited.

There is also a need for more cross-cultural research and cross-national studies. There has been a growing body of recent studies on this topic in China (e.g., Zhang et al. 2017) and Europe (e.g., see Tanis et al. 2017; Tosi 2017), but research from other countries is virtually nonexistent. Future comparative work could shed further light on the cultural, socio-demographic, and economic context of empty nest transitions and related global family change in both Western and non-Western societies.

Moreover, longitudinal studies that can trace changes in family situations, age-related processes (e.g., maturational, biological), and other macro-level developments over time (advances in medical and social media technologies) would be useful. Finally, another significant area of future inquiry is to more carefully connect empty nest experiences to other intersecting family transitions (e.g., retirement, eldercare, grandparenthood), and against the backdrop of rapid population aging (See ▸ "Delaying Retirement" and ▸ "Retirement Transition").

Summary

The concept of an empty nest is a relatively new phenomenon and represents a significant life course transition and process rather than a singular event. Moreover, the occurrence of an empty nest in contemporary family life may not be permanent since a rising number of young adults return home after an initial departure to refill the nest. Generally, parental marital relationships are not negatively affected by the departure of children and parents tend to report the same or better marital quality. Similarly, research has dispelled the myth of the empty nest syndrome as a pervasive and predominantly negative experience for parents (especially mothers). However, parents will experience some degree of role reorganization and changes in family relationships (e.g., in terms of parent-child patterns of support and contact patterns) following the departure of their children. Research on this topic also documents diversity in experiences, and particularly in terms of the specific timing and nature of this transition, in addition to the impact on parental lives and their intergenerational relationships.

Cross-References

- Aging Households
- Co-residence
- Crowded Nest
- Intergenerational Exchange and Support
- Intergenerational Relations
- Living Arrangements
- Midlife and Gender
- Retirement Transition

References

Barber CE (1989) Transition to the empty nest. In: Bahr SJ, Peterson ET (eds) Aging and the family. Lexington Books, Lexington, pp 15–32

Billari FC, Liefbroer AC (2007) Should I stay or should I go? The impact of age norms on leaving home. Demography 44(1):181–198

Bouchard G (2014) How do parents react when their children leave home? An integrative review. J Adult Dev 21:69–79

Bouchard G, McNair JL (2016) Dyadic examination of the influence of family relationships on life satisfaction at the empty-nest stage. J Adult Dev 23:174–182

Chudacoff HP, Hareven TK (1979) From the empty nest to family dissolution: life course transitions into old age. J Fam Hist 4:69–83

Clemens A, Axelson L (1985) The not-so-empty nest: the return of the fledgling adult. Fam Relat 34:259–264

Davis EM, Kim K, Fingerman KL (2018) Is an empty nest best? Coresidence with adult children and parental marital quality before and after the great recession. J Gerontol: Psychol Sci 73(3):372–381

Erikson, JJ, Martinengo G, Hill EJ (2010) Putting work and family experiences in context: differences by family life stage. Hum Relat 63:955–979

Glenn ND (1975) Psychological well-being in the postparental stage: some evidence from national surveys. J Marriage Fam 37:105–110

Glenn ND (1990) Quantiative research on marital quality in the 1980s: a critical review. J Marriage Fam 52(4):818–831

Glick PC (1977) Updating the life cycle of the family. J Marriage Fam 39:5–13

Heidemann B, Suhomlinova O, O'Rand AM (1998) Economic independence, economic status, and empty nest in midlife marital disruption. J Marriage Fam 60:219–231

Kins E, Soenens B, Beyers W (2013) Separation anxiety in families with emerging adults. J Fam Psychol 27(3):495–505

Kulik, L, Shilo-Levin, S, Liberman, G (2010) Work family role conflict and well-being among women and men. J Career Assessment 24(4):651–668

Leopold T (2012) The legacy of leaving home: the long-term effects of coresidence on parent-child relationships. J Marriage Fam 74:399–412

Lüscher (2011) Ambivalence: a 'sensitizing construct' for the study and practice of intergenerational relationships. J Intergener Relatsh 9(2):191–206

Mitchell BA (2007) The boomerang age: transitions to adulthood in families. Aldine Transaction, New Brunswick

Mitchell BA (2016) Happily ever after? Marital satisfaction during the middle adulthood years. In: Bookwala J (ed) Couple relationships in the middle and later years: their nature, complexity, and role in health and illness. American Psychological Association, Washington, DC, pp 17–36

Mitchell BA, Lovegreen L (2009) The empty nest syndrome in midlife families: a multi-method exportation of parental gender differences and cultural dynamics. J Fam Issues 30:1654–1670

Mitchell BA, Wister AV (2015) Midlife challenge or welcome departure? Cultural and family-related expectations of empty nest transitions. Int J Aging Hum Dev 81(4):260–280

Mitchell BA, Wister AV, Burch TK (1989) The family environment and leaving the parental home. J Marriage Fam 51(3):605–613

Mitchell BA, Wister AV, Zdaniuk B (2019) Are the parents all right? parental stress, ethnic culture and intergenerational relations in aging families. J Comp Fam Stud 50(1): 51–74

Rubin L (1979) Women of a certain age: the midlife search for self. Harper & Row, New York

Ruggles S (2011) Intergenerational coresidence and family transitions in the United States, 1850–1880. J Marriage Fam 73:136–148

Schnaiberg A, Goldenberg S (1989) From empty nest to crowded nest: the dynamics of incompletely-launched young adults. Soc Probl 36:251–269

Tanis M, van der Louw M, Buijzen M (2017) From empty nest to social networking site: what happens in cyberspace when children are launched from the parental home? Comput Hum Behav 68:56–63

Tosi M (2017) Leaving-home transition and later parent-child relationships: proximity and contact in Italy. Eur Soc 19(1):69–90

U.S. Census Bureau (2014) Current population survey, annual social and economic supplement, 1967–2014. Retrieved from https://www.census.gov/hhes/families/data/adults.html

VanLaningham J, Johnson DR, Amato P (2001) Marital happiness, marital duration, and the u-shaped curve: evidence from a five-wave panel study. Soc Forces 78(4):1313–1341

Ward RA, Spitze GD (2007) Nestleaving and coresidence by young adult children: the role of family relations. Res Aging 29:257–277

White L, Edwards J (1990) Emptying the nest and parental well-being: evidence from national panel data. Am Sociol Rev 55:235–242

Zhang J, Zhang J-P, Cheng Q-M, Huang F-F, Li S-W, Wang A-N, Su P (2017) The resilience status of empty-nest elderly in a community: a latent class analysis. Arch Gerontol Geriatr 68:161–167

ENAS

▶ European Network in Aging Studies

Encephalogram

▶ Electroencephalography

Encore Careers

▶ Encore Jobs

Encore Jobs

Burcu Özdemir Ocaklı
Faculty of Health Sciences,
Department of Social Work,
Ankara University, Ankara, Turkey

Synonyms

Encore careers; Re-careering; Second careers

Definitions

The term "encore jobs" (or "encore careers") has many definitions, all of which reflect the essence of the term as the process of engaging in paid and unpaid (voluntary) work activity during the stage between conventional adulthood and old age. Freedman (2006) described encore jobs as opportunities for individuals between the end of midlife and the arrival of true old age to engage in work that matters personally and socially. The conventional retirement model prescribes a transition from full-time paid work to full-time leisure activities, whereas encore careers allow for a smooth transition where paid work, leisure, and volunteer work can all be incorporated into retirement years. Bridge jobs and unretirement are among the most popular ways for paid work during the encore years. Encore careers also include "second careers," meaning engaging in a substantially different occupation and/or industry than a worker's prior career after the age of 40 (Helppie-McFall and Sonnega 2017). There may even be third or more careers, and this process of "re-careering" might continue throughout the life course – embarking on a new career in a new industry later in life, after leaving a long-term primary career (Rice 2015). Since most of the bridge jobs are part-time or short-term, they do not qualify as second careers (Helppie-McFall and Sonnega 2017). Unretirement, on the other hand, encompasses second careers, bridge jobs, and encore careers as an umbrella concept (Maestas 2010).

Overview

Due to the demographic wave created by the boomers' generation, there has been an increased interest in researching the older generation. The initial interest in the older population was fixing the functional problems experienced by older people. Due to their adherence to the betterment of society, boomers have profoundly changed the conventional understanding of old age. The initial life course approach (childhood, adolescence, education, employment, retirement, and old age) was disrupted by the boomers' population which is aging with extended healthy life expectancy (Moen 2016). In order to secure the sustainable economic growth in the presence of frequent global crises and an aging society, states have diverted their attention on the inclusion of the previously inactive population into work in the last few decades.

Nevertheless, economic inclusion of the population who are more likely to be inactive has been placed at the core of employment policies (Kleinman 2002; van Berkel and Moller 2002). With the extended healthy life expectancy, improved medical care, and technological advances, life course (school-work-retirement) approach to employment has also been redefined, which, in turn, brings about new approaches to tackling the challenges risen from reoccurring economic crises and aging population (Moen 2016). Moen argues that encore jobs are a product of "time-shifting" in the new life course debate (Moen 2016) where biographical pacing, timing, and sequencing of roles over the life course are no longer the case (Moen and Flood 2013). Such as postponing parenting and going back to school, encore careers are part of this new phenomenon of time-shifting. Even though "baby boomers" have majorly preferred to follow a conventional life course, starting with "generation X," time-shifting has spread out, and encore jobs and careers will be on the rise as they enter into retirement age (Moen 2016).

Encore career movement that emerged from these endeavors provides governments a basis, on which they devise policies to create active older citizens (Alboher 2013). Staying in paid work at older ages has been increasingly endorsed as a movement which brings about value not only to the older person at the individual level but also to the society and economy in general (Bambra 2012). The onset of encore careers may differentiate according to personal circumstances and demographics (e.g., marital status, gender, intergenerational support, health condition) and may last up to 20 years depending on the personal financial conditions (Alboher 2013). While providing a continued income, encore jobs also offer a medium to socialize and learn new skill set (Moen 2016). Briefly, encore jobs merge personal meaning with social ends.

Discussion on encore jobs has become more relevant with the ever-increasing debate on "future of work." Technological advancements and the extensive use of artificial intelligence have started to reduce the need for human labor. Moreover, the change in the labor force composition and demand for a new set of skills are forcing people to adapt to new forms of work. Some jobs will disappear, while others will flourish. Older people, as one of the fast-growing populations, will play a significant role in shaping these future jobs. Hence, encore jobs lie at the heart of discussions on the "future of work."

Supporting Organizations

There are a number of nongovernmental organizations (NGOs) that promote encore careers which provide comprehensive guidance and information for the encore job seekers and taking on lobbying and advocacy roles for influencing policy. Encore.org is one of the prominent examples that place its primary focus primarily on creating engaging older citizens. Through their Gen2Gen, Encore Fellowship, Encore Network, Encore Prize, Encore Leader Convening, and Encore Public Voices Fellowship Programmes, Encore.org not only design but also shape the future of the society by connecting all the related stakeholders (Encore 2018). Encore Fellowship Program exemplifies one of the best practices in the field. It is a yearlong fellowship that is designed to merge the older persons' expertise with the NGO sector. During this year, a fellow could earn 25,000 USD for 1000 h of work. In this innovative initiative, people who have a prominent level of expertise on specific topics are connected to the NGOs that need such expertise for capacity building. The Encore.org actively pairs both counterparts and sets up an interview. Upon a successful interview, the fellow starts their position. The salary of the fellow is usually paid by the nonprofit partner. However, in some cases, the salary is covered by organizations. In 2012, San Francisco (USA)-based

Encore.org has created 200 fellowship positions in 20 different cities in the USA. The successful application by San Francisco-based Encore.org has later inspired the creation of federal fellowships for those who are 55 and older in the USA (Quinton 2013).

Key Research Findings

Current research shows that there has been an increase in the number of encore jobs that people have started to take up. In a study carried out in the USA, more than half of the population aged between 65 and 74 reported that they are holding encore jobs (Gerson 2010). Another study demonstrated that between one third and one half of older workers have switched to bridge jobs before completely withdrawing from labour force (Cahill et al. 2006). Similar findings were also reported in Slaughter's book entitled as *Unfinished Business* (2015). With the allure that boomers have created, neither turning 65 nor the being retired no longer suggest an inactive period of life. They were rather found to be actively seeking encore jobs (Moen 2016).

Gender has been an essential focus of inquiry for the encore career research, and the gender gap in primary careers seems to persist also in encore careers. A study conducted by Moen and Flood (2013) shows that men exit from full-time employment following a conventional model of retirement. As the age range increases, the number of retirees among men also increases. Between the ages of 70 and 74, only 8% of men are employed full-time, and this percentage drops to 5% between the ages 75 and 79. However, a significant amount of men continue part-time work, become self-employed, or do voluntary work. 6–11% of retirees pursue part-time work in their 60s and 70s, whereas 12–14% of men in their 50s and 60s choose self-employment. Self-employment drops to 10% among men in their 70s. Ten percent of men in their 70s report formal volunteering, and 9% report informal volunteering during their retirement time. Even though women's full-time employment is lower than men of all ages, the gap becomes more visible during the encore years for paid work. However, women are more likely to engage in voluntary work and provide informal help to others compared to men. This gap between men and women is evidence for the continuation of the gendered division of labor also during the encore years.

Educational attainment is also a determinant in influencing the type of encore career that older generations pursue. Moen and Flood's study (2013) shows that having a college degree is found to decrease the odds of engaging in paid work during encore years for both men and women. However, college education doubles the odds for engaging in voluntary work after retirement. In other studies, baby boomers, particularly those who have college degrees, were also reported to engage in voluntary activities (Powell and DiMaggio 1991; Moen 2016). Ethnicity is also a significant component for the prediction of engaging in paid work during the encore years. Black men were found to have lower odds of engaging in paid work compared to white men, whereas Hispanic men were more likely to engage in paid work compared to Caucasian men (Moen and Flood 2013). There is also research that shows that compared to other races/ethnicities, white people are more likely to formally volunteer (Brown and Warner 2008).

Stress levels and enjoyment of the job have also been a focused subject for encore jobs research. Johnson's et al. research (2009) suggests that moving into encore careers puts less stress on older workers compared to their previous career jobs. When the enjoyment of the job was in question that most of the respondents (81.8%) reported enjoying their primary careers, however, this percentage has increased to 91.4% implying a greater enjoyment of second careers (Johnson et al. 2009). The same study also shows that encore career jobs are less prestigious compared to their primary careers (Johnson et al. 2009). With decreased prestige, the earnings also decrease with encore jobs compared to primary career jobs (Johnson and Kawachi 2007; Sonnega et al. 2016). Similarly, Kohli (2007) drew attention to the fact that older workers are highly populated in low-paying and less advantaged occupations.

Flexibility is also a key component of encore jobs. In the Johnson et al. study (2009), the respondents also reported flexibility and flexible working hours are significant characteristics of encore career jobs. Rau and Adams' study (2005) also confirms this with the finding that older workers are more likely to apply to a job if the post indicated part-time and flexible work opportunities.

Future Directions of Research

Even though some of the research findings on encore jobs are presented here, there is still a great deal of unknown about the future of encore careers. With changing modes of production and technological advancements, there has been a great interest in pointing out the most popular and income-generating jobs for young people, who are about to start their career. However, future estimations for prospective jobs should also be made for encore jobs since demand will be on the rise. For this reason, it is crucial to assess the current skill profile of the older generation transitioning to retirement. The average age of retirement, educational attainment level, soft skills, and manual skills are among the crucial factors in determining the types of encore jobs available for the retired. In addition to researching skill profile for encore jobs, it is also significant to assess to what extent retired people are willing to carry on with an encore career after retirement. Individual attitudes, along with ethnic, racial, and cultural varieties, should also be considered in future research. Moreover, family relations and health status should also be a focus of inquiry in order to assess the supply in the encore job market.

Summary

Encore jobs, also referred to as encore careers and second careers, define a period of paid or unpaid work that retirees engage in after their primary careers. Increased healthy life expectancy along with early retirement plans has pushed older people to continue economic activity/work-life, especially in the light of recent economic crises. Starting with baby boomers, following a second career has become a widespread phenomenon in the fast-aging countries. In addition to the economic benefit that it provides to older people and the community, encore jobs help older people lead an active, productive, and meaningful life. NGOs such as Encore.org are playing an active role in the creation of encore jobs and engaging older individuals with these jobs. Research shows that encore jobs tend to be more enjoyable compared to primary careers and have more flexible working hours while they are less prestigious and pay less. Being female, being Caucasian, and having higher education decrease the odds of taking up an encore job but increase the odds for doing voluntary work later in life. However, there is an excellent gap in the literature regarding the factors that affect taking on encore jobs. Given the fact that demographic aging is on the rise, demand for such jobs is likely to increase in the future. Moreover, the debate on the "future of work" will add to the discussions about the availability and nature of these encore jobs.

Cross-References

▶ Age Management and Labor Market Policies
▶ Bridge Jobs
▶ Delaying Retirement
▶ Employment and Outplacement Services
▶ Gender and Employment in Later Life
▶ Re-employment of (Early) Retirees
▶ Senior Entrepreneurship

References

Alboher M (2013) The encore career handbook: how to make a living and a difference in the second half of life. Workman Publishing Company, New York
Bambra C (2012) Work, worklessness, and the political economy of health inequalities. Oxford University Press, London. https://doi.org/10.1136/jech.2009.102103
Brown TH, Warner DF (2008) Divergent pathways? Racial/ethnic differences in older women's labor force

withdrawal. J Gerontol 63B(3):S122–SS34. https://doi.org/10.1093/geronb/63.3.S122
Cahill KE, Giandrea MD, Quinn JF (2006) Retirement patterns from career employment. The Gerontologist 46(4):514–523. https://doi.org/10.1093/geront/46.4.514
Encore (2018) Programs. https://encore.org/. Accessed 01 Dec 2018
Freedman M (2006) The social-purpose encore career: baby boomers, civic engagement, and the next stage of work. Generations 30(4):43–46
Gerson K (2010) The unfinished revolution: how a new generation is reshaping family, work, and gender in America. Oxford University Press, New York. https://doi.org/10.1111/j.1741-3737.2010.00807.x
Helppie-McFall B, Sonnega A (2017) Characteristics of second-career occupations: a review and synthesis. University of Michigan Retirement Research Center (MRRC) Working Paper, WP 2017-375, Ann Arbor. http://mrrc.isr.umich.edu/wp375/. Accessed 01 Dec 2018
Johnson RW, Kawachi J (2007) Job changes at older ages: effects on wages, benefits, and other job attributes. Center for Retirement Research WP2007-4. http://webarchive.urban.org/publications/311435.html. Accessed 01 Dec 2018
Johnson RW, Kawachi J, Lewis E (2009) Older workers on the move: recareering in later life. https://assets.aarp.org/rgcenter/econ/2009_08_recareering.pdf. Accessed 01 Dec 2018
Kleinman MP (2002) A European welfare state? European Union social policy in context. Palgrave, Basingstoke/New York
Kohli M (2007) The institutionalization of the life course: looking back to look ahead. Res Hum Dev 4(3–4):253–271. https://doi.org/10.1080/1542760070 1663122
Maestas N (2010) Back to work: expectations and realizations of work after retirement. J Hum Resour 45(3):718–748. https://doi.org/10.1353/jhr.2010.001
Moen P (2016) Encore adulthood: boomers on the edge of risk, renewal, & purpose. Oxford University Press, New York. https://doi.org/10.1093/acprof:oso/978019 9357277.001.0001
Moen P, Flood S (2013) Limited engagements? Women's and men's work/volunteer time in the encore life course stage. Soc Probl 60(2):206–233. https://doi.org/10.15 25/sp.2013.60.2.206
Powell WW, DiMaggio PJ (1991) The new institutionalism in organizational analysis. University of Chicago Press, Chicago
Quinton S (2013) Work, forever: why interning at 60 is the new retirement plan. The Atlantic https://www.theatlantic.com/business/archive/2013/09/work-forever-why-interning-at-60-is-the-new-retirement-plan/279381/. Accessed 01 Dec 2018
Rau BL, Adams GA (2005) Attracting retirees to apply: desired organizational characteristics of bridge employment. J Organ Behav 26(6):649–660. https://doi.org/10.1002/job.330
Rice CK (2015) The phenomenon of later-life recareering by well-educated baby boomers. J Psychol Issues Organ Cult 6(2):7–38. https://doi.org/10.1002/jpo c. 21179
Slaughter AM (2015) Unfinished business: women, men, work, family. Random House, New York
Sonnega A, Helppie-McFall B, Willis RJ (2016) Occupational transitions at older ages: what moves are people making? http://www.mrrc.isr.umich.edu/publications/papers/pdf/wp352.pdf. Accessed 01 Dec 2018
van Berkel R, Moller IH (2002) Active social policies in the EU: inclusion through participation? Policy Press, Bristol

End of Life

▶ Beliefs of Life After Death and Social Partner Choice

End of Natural Reproductive Life

▶ Menopause

End-Diastolic Volume Index

Chun-Quan Zhang and Li-Li Chen
The Second Affiliated Hospital of Nanchang University, Nanchang City, China

Synonyms

Global end diastolic volume index; Left ventricular end-diastolic volume index; Right ventricular end-diastolic volume index

Definition

The end-diastolic volume index (EDVI) is the ratio of the end-diastolic volume of the heart chamber to the body surface area. The ratio of left ventricular end-diastolic volume to body surface area is called left ventricular end-diastolic volume index (LVEDVI); similarly, the ratio of right ventricular end-diastolic volume to body surface area is called right ventricular end-diastolic volume index (RVEDVI). The global end diastolic volume index (GEVDI) is calculated by dividing the global end diastolic volume (the sum of left atria end-diastolic volume, right atria end diastolic volume, left ventricular end diastolic volume, and right ventricular end diastolic volume) by the body surface area (Kapoor et al. 2016). LVEDVI, RVEDVI, and GEVDI are indices of cardiac preload. The accuracy of GEDVI is higher than that of RVEDVI and LVEDVI because it refers to the end-diastolic volume of each chamber of the heart, so it is often used for clinical monitoring. The recommended normal reference value range is 680–800 ml/m^2 (Grindheim et al. 2016). GEDVI measurement is mainly monitored by the pulse indicator continuous cardiac output (PICCO) monitor using the transpulmonary thermodilution (TPTD) technique. PICCO reflects the whole heart condition by volume monitoring and can obtain extravascular lung water (EVLW) index, which makes continuous real-time cardiac output monitoring possible. Compared to global end-diastolic volume (GEDV), GEDVI can accurately compare patients of different body types, as GEDVI takes the body surface area parameters into account.

Overview

With the aging of the world's population and the extension of life expectancy, the number of older adults seeking emergency medical treatment and the number of emergency deaths are increasing. The number of critically ill patients at older ages is also increasing proportionally. In addition, the most common cause of death in older adults is organ failure, especially respiratory failure and heart failure. Most of the causes of respiratory failure and heart failure are infection. Sepsis and septic shock are the most frequent diseases. At the same time, due to the aging of organs in older adults, the body compensatory function decreases, infection is serious and difficult to control, and treatment is difficult. Therefore, it is important and necessary to accurately monitor the cardiac volume of critically ill patients at older ages to seek the cause and predict fluid responsiveness. Some studies have demonstrated that GEDVI compared with the central venous pressure (CVP) more correctly predicted volume responsiveness in patients with septic shock (Michard et al. 2003). The GEDVI can accurately reflect the blood volume, which is the state of cardiac preload, and it is not affected by external factors such as intra-abdominal hypertension, ventilator, and other factors. The traditional CVP monitoring is the pressure in cavity vein entering the right atria, which indirectly reflects cardiac preload and guides clinical treatment, but it cannot accurately depict blood volume (Sasai et al. 2014). The measurement result is influenced by many factors, such as the use of vascular active drugs, depth of cannula, changes in patient position, and mechanical ventilation. Comparing with CVP and pulmonary artery occlusion pressure (PAOP), GEDVI shows more obvious advantages in monitoring cardiac preload. It was also noted that GEDVI target-directed fluid resuscitation showed a better clinical effect than CVP in patients with septic shock (Yu et al. 2017). GEDVI can accurately reflect the blood volume. In one pediatric animal model, GEDV derived from TPTD was a reliable indicator of cardiac preload. Moreover, it is also proved that GEDVI rather than pulmonary capillary wedge pressure (PAWP) and CVP accurately reflected fluid responsiveness (Renner et al. 2007). PAOP is equally controversial in predicting fluid responsiveness. Studies have shown that GEDVI was preferable to PAOP in predicting fluid responsiveness (Breukers et al. 2009). However, some scholars have found that PAOP and GEDVI had their own advantages in predicting fluid response. In patients with preserved left ventricular function, GEDVI is a better predictor of fluid responsiveness than filling pressure, whereas in patients with impaired left ventricular function, filling pressure is more important in identifying nonresponders to a fluid challenge (Trof et al. 2011).

Key Research Findings

Hemodynamic monitoring is crucial for early identification and management of key changes in hemodynamic parameters to optimize tissue oxygen delivery. The three main factors that affect cardiac ejection are preload, myocardial contractility, and afterload. According to the Frank-Starling mechanism, cardiac preload is a determinant of cardiac function. Therefore, a reliable estimate of cardiac preload is helpful in the treatment of severe circulatory dysfunction. Parameters such as CVP, right atrial pressure (RAP), right ventricular pressure (RVP), pulmonary artery pressure (PAP), and pulmonary artery occlusion pressure (PAOP) monitored by pulmonary catheters (PAC) have been the main contents of cardiac preload hemodynamic monitoring. However, studies have shown that PAOP and CVP cannot predict ventricular filling volume, cardiac performance, or response to volume infusion in normal subjects (Kumar et al. 2004). In addition, with the development of TPTD technology and echocardiography, it is no longer difficult to measure GEDV, which has been proved to be a reliable predictive marker of preload compared with the pressure preload parameters derived from traditional PACs.

Echocardiography can predict preload by assessing end-diastolic volume and area. In addition, with the development and application of real-time three-dimensional echocardiography, the ventricular volume can be accurately estimated, even EDVI of each ventricle including the right ventricle with irregular shape and complicated anatomical structure can be reliably measured three-dimensional echocardiographic measurement of right ventricular volume and function with cardiac magnetic resonance verification studies has shown that it is a good tool for accurate and reproducible assessment of right ventricular volume and right ventricular function (Park et al. 2016). The research found that the cardiac output measured by transcatheter three-dimensional echocardiography and pulse indicator continuous cardiac output (PICCO) technology using the principle TPTD has a good correlation (Hammoudi et al. 2017). Echocardiography is convenient, noninvasive, and highly accurate. However, echocardiography can only reflect the function of a single time slot and cannot provide continuous hemodynamic monitoring. Although there have been reports of a new single-use mini transesophageal ultrasound probe that can be placed in a critical patient for 72 h, which is well tolerated, the mini probe can be used for continuous hemodynamic evaluation in patients receiving mechanical ventilation, but due to the lack of professional knowledge in the echocardiography among intensive care personnel, its use is limited.

TPTD Technology

TPTD is a continuously developing technique in the field of hemodynamic monitoring, which does not require invasive pulmonary catheter implantation posing less trauma and can be placed for a longer time duration than PACs. The single cardiac output (CO) is measured by TPTD, and the continuous CO is obtained by analyzing the area under the arterial pressure wave curve; meanwhile, intrathoracic blood volume (ITBV), EVLW, and GEDVI can be calculated. ITBV and GEDVI have been proved by many studies to be a reproducible, sensitive, and more accurate indicator of cardiac preload than PAOP, right ventricular end-diastolic pressure (RVEDVP), and CVP. The working principle of PICCO is to insert a central venous catheter and a femoral artery catheter through the TPTD technique, and the femoral artery catheter is equipped with a temperature detector. After the central vein is injected with a temperature indicator, such as iced saline, it is distributed in each chamber of the heart. According to the temperature attenuation curve measured by the temperature indicator of the femoral artery, the distribution of the volume of the heart chamber can be obtained. A certain amount of iced saline is injected from the central vein, and the volume and temperature are quickly dispersed into the heart and lungs. When the thermal signal is detected by the temperature detector, the temperature difference can be identified and merged into a TPTD curve. The computer analyzes the curve by itself. A basic parameter, combined with the femoral artery pressure waveform and a series of important clinical parameters with special significance such as GEDV, cardiac output (CO), cardiac index (CI), arterial pressure, and EVLW, is obtained.

GEDV and the Principle of Measuring GEDV with TPTD Technology

GEDV is a hypothetical volume, assuming that the four chambers are simultaneously in diastole. GEDV also contains the volume of the central vein and aorta from the injection site to the measurement site, in addition to the EDV of the four heart chambers. GEDV is the difference between chest heat volume (intrathoracic thermal volume (ITTV)) and lung heat volume (pulmonary thermal volume (PTV)). It is a reliable indicator of cardiac preload in severely ill patients. The PICCO system is used to perform hemodynamic monitoring using TPTD technology. The PICCO system implementation analysis is based on CO, mean transit time (Mtt), and TPTD curve of downslope time (Dst). The calculation relationship between them is as follows (Kapoor et al. 2016):

$$PTV = CO \cdot Dst$$
$$ITTV = CO \cdot Mtt$$
$$GEDV = ITTV - PTV$$
$$GEDV\ (PICCO) = CO\ (Mtt - Dst).$$

GEDV has been proved to be a reliable predictive marker of preload. TPTD method can provide more reliable preload predictors of cardiac capacity, such as GEDV and ITBV, better than some versions of GEDV estimates derived from PACs.

Application of GEDVI

GEDVI and EVLWI can be used to compare cardiac preload of different body types (Wiesenack et al. 2005). GEDVI is shown to be a more reliable indicator to reflect incremental cardiac preload than CVP, PCWP, or LVEDA. However, GEDVI failed to become a variable of fluid reactivity. The reference range of GEDVI suggested by the expert opinion is 680~800 ml/m^2. A study found that GEDV and ITBV depend on age and gender. The age and gender dependence of GEDV and ITBV persisted even after switching to GEDVI and ITBVI considering body surface area (Wolf et al. 2009). Some researchers believe that the reason for this difference is not because of the difference in sex and age, but the influence of the difference in aortic volume. Their results suggest that the aortic volume is a substantial confounding variable for GEDV measurements performed with TPTD (Akohov et al. 2020). Therefore, it would not be appropriate to use a fixed range of GEDVI or ITBVI for recovery, regardless of age or gender. In addition, young children in the process of growing differences in weight gain and organ weight growth, lead to measure GEDVI value on the high side and EVLW value on the low side, so GEDVI value and EVLW value should be corrected in the childhood period (Lemson et al. 2011). A study has shown that GEDVI-based algorithm guided therapy can shorten and reduce the need for vasopressor, catecholamine, mechanical ventilation, and ICU stay in patients undergoing cardiac surgery (Goepfert et al. 2007). Research has shown that changes in BNP concentration are associated with changes in GEDVI, and continuous measurement of BNP concentration may be a useful tool for monitoring cardiac chamber volume status (Zhang et al. 2013). In a multicenter, prospective observational subgroup, there was no consistent relationship between GEDVI and fluid reserve reactivity in the early stages of severe sepsis in mechanically ventilated patients, regardless of the presence or absence of sepsis-induced myocardial dysfunction (SIMD) (Endo et al. 2013). Therefore, GEDVI, as a parameter of early cardiac preload in severe sepsis, has limitations. In addition, volume parameter GEDVI plays an important role in hemodynamic monitoring of certain diseases, such as pancreatitis, burns, seizure syndrome, and subarachnoid hemorrhage. It was found that the accurate use of variables measured by the thermal dilution method of the lung, such as GEDV and GEDVI, can predict the occurrence of subarachnoid hemorrhage, pulmonary edema, and DIC (Watanabe et al. 2012). A large number of studies have found that liquid resuscitation under the guidance of GEDVI in patients with septic shock can reduce the fluid load of patients, achieve better oxygenation, reduce heart failure, and shorten mechanical ventilation time and ICU residence time. GEDVI is also of high value in guiding fluid resuscitation in patients with COPD complicated with septic shock.

Future Directions of Research

GEDVI is a predictor of cardiac preload, which is better than CVP and PAOP in predicting fluid responsiveness in patients with septic shock. GEDVI-guided fluid management in critically ill patients, especially in critically ill patients at older ages, can reduce complications and ICU stay time after cardiac surgery. However, the GEDVI index also has a couple of defects. There is no uniform reference value of GEDVI, and it is affected by age and sex, so it cannot apply in respect of all patients with a single reference value of GEDVI. A meta-analysis found that patients with sepsis had significantly higher average GEDVI than patients undergoing major surgery and concluded that treatment targets were required to be adjusted for different groups of patients (Huber et al. 2017). In general, GEDVI is a satisfactory index for predicting cardiac preload, but its application needs to be specified, and specific GEDVI should be referred to for patients with different diseases and different ages. Therefore, academic circles should unify the reference standard of GEDVI to resolve the current problems. The measurement of GEDVI by the TPTD technique is influenced by some confounding factors, such as the volume of ascending aorta. TPTD technology needs to be improved further to reduce the impact of confounding factors on GEDVI so that it can be better employed in the clinic.

Summary

GEDVI is an effective indicator of cardiac preload, and it can reflect the changes of cardiac preload more reliably than traditional parameters such as CVP and PAWP (pulmonary arterial wedge pressure). GEDVI obtained by TPTD can effectively monitor the preload of the heart, guide clinical treatments, and benefit patients.

Cross-References

▶ Atherosclerosis
▶ Cardiovascular Response
▶ Cardiovascular System
▶ Electrocardiogram Abnormalities
▶ Heart Failure
▶ Hypertension
▶ Hypertensive Cardiovascular Diseases
▶ Ischemic Attack
▶ Ischemic Heart Disease
▶ Stroke

References

Akohov A, Barner C, Grimmer S, Francis RC, Wolf S (2020) Aortic volume determines global end-diastolic volume measured by transpulmonary thermodilution. Intensive Care Med Exp 8. https://doi.org/10.1186/s40635-019-0284-8

Breukers RBGE, de Wilde RBP, van den Berg PCM, Jansen JRC, Faes TJC, Twisk JWR, Groeneveld ABJ (2009) Assessing fluid responses after coronary surgery: role of mathematical coupling of global end-diastolic volume to cardiac output measured by transpulmonary thermodilution. Eur J Anaesthesiol 26:954–960. https://doi.org/10.1097/EJA.0b013e32833098c6

Endo T, Kushimoto S, Yamanouchi S, Sakamoto T, Ishikura H, Kitazawa Y, Taira Y, Okuchi K, Tagami T, Watanabe A, Yamaguchi J, Yoshikawa K, Sugita M, Kase Y, Kanemura T, Takahashi H, Kuroki Y, Izumino H, Rinka H, Seo R, Takatori M, Kaneko T, Nakamura T, Irahara T, Saito N, Picco PESG (2013) Limitations of global end-diastolic volume index as a parameter of cardiac preload in the early phase of severe sepsis: a subgroup analysis of a multicenter, prospective observational study. J Intensive Care 1:11. https://doi.org/10.1186/2052-0492-1-11

Goepfert MSG, Reuter DA, Akyol D, Lamm P, Kilger E, Goetz AE (2007) Goal-directed fluid management reduces vasopressor and catecholamine use in cardiac surgery patients. Intensive Care Med 33:96–103. https://doi.org/10.1007/s00134-006-0404-2

Grindheim G, Eidet J, Bentsen G, Ramamoorthy C (2016) Transpulmonary thermodilution (PiCCO) measurements in children without cardiopulmonary dysfunction: large interindividual variation and conflicting reference values. Pediatr Anesth 26:418–424. https://doi.org/10.1111/pan.12859

Hammoudi N, Hékimian G, Laveau F, Achkar M, Isnard R, Combes A (2017) Three-dimensional transoesophageal echocardiography for cardiac output in critically ill patients: a pilot study of ultrasound versus the thermodilution method. Arch Cardiovasc Dis 110:7–13. https://doi.org/10.1016/j.acvd.2016.04.009

Huber W, Mair S, Götz SQ, Tschirdewahn J, Frank J, Höllthaler J, Phillip V, Schmid RM, Saugel B (2017) A systematic database-derived approach to improve indexation of transpulmonary thermodilution-derived global end-diastolic volume. J Clin Monit Comput 31:143–151. https://doi.org/10.1007/s10877-016-9833-9

Kapoor P, Bhardwaj V, Sharma A, Kiran U (2016) Global end-diastolic volume an emerging preload marker vis-a-vis other markers – have we reached our goal? Ann Card Anaesth 19:699–704. https://doi.org/10.4103/0971-9784.191554

Kumar A, Anel R, Bunnell E, Habet K, Zanotti S, Marshall S, Neumann A, Ali A, Cheang M, Kavinsky C, Parrillo JE (2004) Pulmonary artery occlusion pressure and central venous pressure fail to predict ventricular filling volume, cardiac performance, or the response to volume infusion in normal subjects. Crit Care Med 32:691–699. https://doi.org/10.1097/01.ccm.0000114996.68110.c9

Lemson J, Merkus P, van der Hoeven JG (2011) Extravascular lung water index and global end-diastolic volume index should be corrected in children. J Crit Care 26:432–437. https://doi.org/10.1016/j.jcrc.2010.10.014

Michard F, Alaya S, Zarka V, Bahloul M, Richard C, Teboul J (2003) Global end-diastolic volume as an indicator of cardiac preload in patients with septic shock. Chest 124:1900–1908. https://doi.org/10.1378/chest.124.5.1900

Park J, Lee S, Lee J, Yoon YE, Park E, Kim H, Lee W, Kim Y, Cho G, Sohn D (2016) Quantification of right ventricular volume and function using single-beat three-dimensional echocardiography: a validation study with cardiac magnetic resonance. J Am Soc Echocardiogr 29:392–401. https://doi.org/10.1016/j.echo.2016.01.010

Renner J, Meybohm P, Gruenewald M, Steinfath M, Scholz J, Boening A, Bein B (2007) Global end-diastolic volume during different loading conditions in a pediatric animal model. Anesth Analg 105:1243–1249. https://doi.org/10.1016/j.echo.2016.01.010

Sasai T, Tokioka H, Fukushima T, Mikane T, Oku S, Iwasaki E, Ishii M, Mieda H, Ishikawa T, Minami E (2014) Reliability of central venous pressure to assess left ventricular preload for fluid resuscitation in patients with septic shock. J Intensive Care 2:58. https://doi.org/10.1186/s40560-014-0058-z

Trof RJ, Danad I, Reilingh MWL, Breukers RBGE, Groeneveld ABJ (2011) Cardiac filling volumes versus pressures for predicting fluid responsiveness after cardiovascular surgery: the role of systolic cardiac function. Crit Care 15:R73. https://doi.org/10.1186/cc10062

Watanabe A, Tagami T, Yokobori S, Matsumoto G, Igarashi Y, Suzuki G, Onda H, Fuse A, Yokota H (2012) Global end-diastolic volume is associated with the occurrence of delayed cerebral ischemia and pulmonary edema after subarachnoid hemorrhage. Shock 38:480–485. https://doi.org/10.1097/SHK.0b013e31826a3813

Wiesenack C, Fiegl C, Keyser A, Laule S, Prasser C, Keyl C (2005) Continuously assessed right ventricular end-diastolic volume as a marker of cardiac preload and fluid responsiveness in mechanically ventilated cardiac surgical patients. Crit Care 9:R226–R233. https://doi.org/10.1186/cc3503

Wolf S, Riess A, Landscheidt JF, Lumenta CB, Friederich P, Schürer L (2009) Global end-diastolic volume acquired by transpulmonary thermodilution depends on age and gender in awake and spontaneously breathing patients. Crit Care 13:R202. https://doi.org/10.1186/cc8209

Yu J, Zheng R, Lin H, Chen Q, Shao J, Wang D (2017) Global end-diastolic volume index vs CVP goal-directed fluid resuscitation for COPD patients with septic shock: a randomized controlled trial. Am J Emerg Med 35:101–105. https://doi.org/10.1016/j.ajem.2016.10.015

Zhang Z, Ni H, Lu B, Xu X (2013) Changes in brain natriuretic peptide are correlated with changes in global end-diastolic volume index. J Thorac Dis 5:156–160. https://doi.org/10.3978/j.issn.2072-1439.2012.11.05

Endocrine Rejuvenation in a Historical Perspective

Ilia Stambler
Science, Technology and Society,
Bar Ilan University, Ramat Gan, Israel

Synonyms

Hormone replacement therapy; Opotherapy; Organotherapy; Rejuvenation

Definition

The historical practice of using endocrine gland extracts, transplants, and stimulants, as well as synthetic hormone replacements to ameliorate degenerative aging processes and extend healthy lifespan.

Overview

Extracts from various endocrine glands have been utilized for rejuvenation purposes since ancient times (Gruman 1966). The origins of this practice can be traced back to the ancient art of "organotherapy" or "opotherapy" (from the Greek *opos* "juice") or using animal parts and extracts for therapy, including animal sex organs and extracts from animal horns. The use of rejuvenative opotherapy can be observed as early as the ancient Chinese Mawangdui medical manuscripts (c. 200 BCE), the Ayurvedic Sushruta Samhita (c. 300 BCE), and the Talmud (c. 200–500 CE) (Unschuld 2003; Bhishagratna 1911; Rosner 1995). The use of sex gland extracts and other animal tissues for rejuvenation purposes spread to Europe in the Middle Ages, as animal materials alongside mineral and plant materials were common ingredients in the preparation of elixirs (Lloyd 1883; Walsh 1911). Since the early modern period, in the eighteenth and nineteenth century, there were emerging attempts to graft or implant gland tissues into animals. One of the earliest recorded examples of such attempts includes John Hunter's experiments with grafting the testicles of a cock into a hen (1770). These experiments were repeated by Arnold Adolf Berthold (1849), Jean Marie Philippeaux (1858), and Paolo Mantegazza (1864) (Voronoff 1925; Sengoopta 2006). In the late nineteenth century and the beginning of the twentieth century, the experiments and therapies started to be translated to people.

At the end of the nineteenth and the beginning of the twentieth century, there emerged the hopes for immediate and rather far-reaching rejuvenation and life extension thanks to medical science. Since that time and well until WWII, most of such hopes were associated with supplementation, replacement, and stimulation of endocrine glands, mainly the sex glands. This entire medical movement received the name "endocrine rejuvenation." Though there were much earlier precedents, in both Western and Eastern medical traditions, the most recognizable founding figure of this stream was the French biologist Charles-Édouard Brown-Séquard, famous for his rejuvenation experiments with dogs' and guinea pigs' sex gland extracts (Brown-Séquard 1889). Yet he had a cohort of prominent followers. One of the most famous (even notorious) was the Parisian physician Serge Voronoff (1866–1951) – one of the pioneers of xeno-transplantation, conducted for rejuvenation purposes. His signature method was the transplantation ("grafting") into humans of animal (mainly male ape) sex gland tissues, as he believed that the deterioration of aging is mainly due to a lack of balance or failure of endocrine function which could be reestablished by transplanting the sex gland tissue (Voronoff 1925). Beside Voronoff, a cohort of French pioneers of endocrine rejuvenation (mainly practicing sex gland tissue transplantations and various other means of endocrine stimulation) included Placide Mauclaire, Lois Dartigues, Raymond Petit, Léopold Lévi, Henri de Rothschild, and others, forming a "school" (Stambler 2014).

Similarly, in the German-speaking world, from the beginning of the twentieth century and through the 1920s, the "endocrine rejuvenation" movement flourished. The practitioners utilized a wide variety of organotherapeutic (endocrine gland) supplements, gland transplants, or gland stimulants (predominantly those of the sex glands). In this area, Austria was prominent, on a par with France, leading the way with the pioneering works of the Viennese physician Eugen Steinach (1861–1944) on sexual rejuvenation, most famously the "Steinach procedure." The procedure involved spermatic cord vasoligation, as the suppression of sperm-producing activity was supposed to stimulate sex-hormone-producing activity, allegedly leading to increasing the blood flow and other reinvigorating effects (Steinach and Loebel 1940). Steinach was followed by a host of other Austrian rejuvenators – Karl Doppler, Robert Lichtenstern, Paul Kammerer, Erwin Last, August Bier, Emerich Ullmann, Otto Kauders, and others – performing the "Steinach Procedure" by the hundreds and thousands, as well as other modified versions of sexual stimulation. Sigmund Freud (1856–1939) had Steinach's rejuvenating operation performed on himself in 1923 by the surgeon Victor Blum. In Germany too, in the first quarter of the twentieth century, the endocrine rejuvenation movement boomed, though perhaps to a lesser extent than in Austria, exemplified by the works of the physicians Jürgen Harms, Peter Schmidt, Richard Mühsam, and Ludwig Levy-Lenz. Indeed, in the 1920s, Steinach's procedure was widely applied all across the world. In Germany and Austria, however, the application of Steinach's operation was among the widest. Beside operative interventions, by the late 1920s, hormonal preparations for rejuvenation also became an object of keen interest in Germany and Austria. Sex gland extracts and stimulants for men and women enjoyed by far the greatest vogue in the rejuvenation business (Romeis 1931).

There was also strong interest in this area across the Atlantic. Americans were among the first to conduct operations on sex glands – vasectomy and sex gland transplantations – for rejuvenative and other therapeutic purposes. Among the prominent users of "rejuvenative" vasectomy and vasoligation techniques were Harry Clay Sharp and Harry Benjamin. Rejuvenation by transplantation of sex gland tissues (male and female) were also practiced in the USA. Some of the first practitioners included: Victor Darwin Lespinasse, Levi Jay Hammond and Howard Anderson Sutton, George Frank Lydston, Max Thorek, and Leo Leonidas Stanley. Endocrine extracts were also popular in the USA, even before Charles-Édouard Brown-Séquard's experiment (as attested by Brown-Séquard himself). Endocrine rejuvenation surgery and supplementation were also practiced in other countries. For example, in Russia, the rejuvenative extract "spermin" was produced by Alexander Poehl of St. Petersburg (1850–1908). Thus "endocrine rejuvenation" formed a global therapeutic movement.

In later assessments, however, reductionist rejuvenation techniques did not appear to live up to their promise. For the endocrine "grafting" methods, the problem of graft rejection by the host organism appeared almost insurmountable (Hamilton 1986). On the other hand, replacing, supplementing, or stimulating a single endocrine gland did not appear to durably forestall the deterioration of the entire organism, with a particular lack of evidence for efficacy in elderly patients, and no conclusive evidence for extending either the life span or health span by such means were established (Romeis 1931). Consequently, a recoil from immediate rejuvenation attempts occurred, accompanied by a recoil from their underlying mechanism and reductionism (Lumière 1932; Stambler 2017). Hence, the "endocrine rejuvenation" movement was largely abandoned and even discredited. Still, it represented a significant chapter in the history of aging research and therapy.

Mainly due to the therapeutic failures and methodological shortcomings of endocrine rejuvenation, toward WWII the interest in such rejuvenation attempts strongly subsided, and virtually disappeared during the war. The aftermath of WWII (roughly since the 1950s) may mark a transition with reference to rejuvenation attempts. By the 1950s, the hopes for immediate, far reaching rejuvenation, mainly through endocrine interventions, so strong in the early twentieth century, were virtually abandoned. Instead, great expectations for future amelioration of aging and extension of healthy life were pinned on continuous fundamental aging research, informed by the explosive scientific and technological advances of the mid-century, especially replacement medicine and molecular biology. Thus, a trend can be observed toward and after the 1950s: a marked decrease of interest in immediate rejuvenation and increase of interest in biological theories of aging that endeavored first to fully elucidate its mechanisms and thereafter indicate targets for intervention. In the 1950s, the means suggested for potential amelioration of aging and healthy longevity shifted their focus. The geographic focus also notably shifted, demonstrating the increasing, even dominant, presence of the USA. In therapeutic approaches, the central emphasis was gradually placed on antioxidants, following Denham Harman's seminal work of 1956, delineating the "free radical theory of aging" (Harman 1956).

Yet "endocrine rejuvenation" by no means died but reemerged in new forms and is still present. The rejuvenative opotherapy or organotherapy of the 1900s–1920s, using animal organ extracts or derivatives as the chief means of supplementation, gave birth to synthetic hormone replacements developed during the 1930s–1950s. The road to such synthetic therapies was paved by the success of the German biochemist Adolf Butenandt in synthesizing female sex hormones (estrone in 1929 and progesterone in 1935), as well as by the Croatian/Swiss Leopold (Lavoslav) Ruzicka's and the Polish/Dutch Ernst Laqueur's successful synthesis of male sex hormones ("androsterone" and "testosterone" from 1931 to 1935). In the late 1950s–early 1960s, hormonal therapy appears to resume its prominence in aging therapy, when synthetic hormones cheapened and became more widely available. Thus, the declining levels of 17-ketosteroids, and their particular forms, such as dehydroepiandrosterone (DHEA), androsterone, and estrone, were said to play a crucial role in aging by the American inventor of oral contraceptives, Gregory Pincus, whose group sought to develop steroid supplementation regimens designed to restore their patterns to youthful levels (Rubin et al. 1955). In the 1960s, in the USA, thyroxine (a thyroid hormone) was popularized for anti-aging by Charles Brusch and Murray Israel, estrogen and progesterone by Robert Wilson, and thymosin (a thymus hormone) by Allan Goldstein. Hormone levels of the hypothalamus (such as dopamine and growth hormone-releasing hormone) were assigned a crucial role in the onset of aging by the American researchers William Donner Denckla and Caleb Finch (c. 1975), who referred to the hypothalamus as a hormonal "aging clock" (Finch 1975). This echoed the work of the Russian-Soviet researcher of aging, Vladimir Dilman (1925–1994), on the sensitivity threshold elevation theory of aging and impairment of homeostasis as the main cause of aging and the role of the hypothalamus as "the large biological clock" (Dilman 1958). Presently, the role of hormones in aging and the potential uses of hormone therapy or hormone modulation to ameliorate aging-related ill-health are far from being fully elucidated.

Summary and Future Directions of Research

Endocrine rejuvenation represented some of the earliest forms of attempted therapeutic interventions into degenerative aging processes for the extension of healthy life. Though such therapies have often fallen short of their original promise, their modified forms retain their significance in the therapeutic aging arsenal to the present. In the scientific domain, thanks to extensive research, it is hoped to minimize the potential drawbacks and side effects, while maximizing the potential health benefits by regulating, fine-tuning, optimizing, and personalizing endocrine rejuvenative therapies. In the domain of historical research, it will be worthwhile to draw additional parallels of the hopefulness boom and bust cycle associated with early endocrine rejuvenation attempts and later attempts of other forms of rejuvenative therapies. Such a comparative historical research could reveal both the vicissitudes of scientific validation and of the human psychology.

Cross-References

▶ Homeostasis in the History of Aging Research
▶ Life-Extensionism in a Historical Perspective
▶ Medical Alchemy in History

References

Bhishagratna KKL (1911) An English translation of the Sushruta samhita, based on original Sanskrit text, vol 2, Chikitsasthanam (Therapeutics). Calcutta

Brown-Séquard CE (1889) Des effets produits chez l'homme par des injections sous-cutanées d'un liquide retiré des testicules frais de cobaye et de chien (Effects in man of subcutaneous injections of freshly prepared liquid from guinea pig and dog testes). Comptes Rendus des Séances de la Société de Biologie, Série 9(1):415–419. Reprinted and translated in: Emerson GM (ed) (1977) Benchmark papers in human physiology, vol 11, Aging. Dowden, Hutchinson and Ross, Stroucodsburg, pp 68–76

Dilman VM (1958) O vozrastnom povyshenii deiatelnosti nekotorikh hypotalamicheskikh centrov. Trudy Instituta Physiologii imeni I.P. Pavlova 7:326–336 (On senescent elevation of the activity of some

hypothalamic centers. In: Works of I.P. Pavlov's Institute of Physiology)
Finch C (1975) Neuroendocrinology of aging: a view of an emerging area. Bioscience 25:645–650
Gruman GJ (1966) A history of ideas about the prolongation of life. The evolution of prolongevity hypotheses to 1800. Trans Am Philos Soc 56:1–102
Hamilton D (1986) The monkey gland affair. Chatto and Windus, London
Harman D (1956) Aging: a theory based on free radical and radiation chemistry. J Gerontol 11:298–300
Lloyd JU (1883) Pharmaceutical preparations. Elixirs, their history, formulae, and methods of preparation. Including practical processes for making the popular elixirs of the present day, and those which have been official in the old pharmacopoeias. Together with a résumé of unofficial elixirs from the days of Paracelsus, 2nd edn. Robert Clarke and Company, Cincinnati
Lumière A (1932) Sénilité et Rajeunissement (Aging and Rejuvenation). Librairie J.-B. Baillière et Fils, Paris
Romeis B (1931) Altern und Verjüngung. Eine Kritische Darstellung der Endokrinen "Verjüngungsmethoden", Ihrer Theoretischen Grundlagen und der Bisher Erzielten Erfolge. Verlag von Curt Kabitzsch, Leipzig (Aging and rejuvenation. A critical presentation of endocrine "rejuvenation methods," Their theoretical foundations and up-to-date successes)
Rosner F (1995) Medicine in the Bible and the Talmud. Ktav, Hoboken
Rubin BL, Dorfman RI, Pincus G (1955) 17-Ketosteroid excretion in aging subjects. In: Wolstenholme GEW, Cameron MP (eds) CIBA foundation colloquia on ageing, vol 1, General aspects. J. & A. Churchill Ltd, London, pp 126–137
Sengoopta C (2006) The most secret quintessence of life. Sex, glands, and hormones, 1850–1950. The University of Chicago Press, Chicago
Stambler I (2014) A history of life-extensionism in the twentieth century. Longevity History, Rishon Lezion. http://www.longevityhistory.com/. Accessed 15 Feb 2019
Stambler I (2017) Reductionism and holism in the history of aging and longevity research: Does the whole have parts? In: Longevity promotion: multidisciplinary perspectives. Longevity History, Rishon Lezion. http://www.longevityhistory.com/. Accessed 15 Feb 2019
Steinach E, Loebel J (1940) Sex and life. Forty years of biological and medical experiments. Faber, London
Unschuld PU (2003) Huang Di Nei Jing Su Wen: nature, knowledge and imagery in an ancient Chinese medical text. University of California Press, Berkeley
Voronoff S (1925) Rejuvenation by grafting (trans: Imianitoff FF). George Allen and Unwin Ltd, London
Walsh JJ (1911) Old-time makers of medicine. The story of the students and teachers of the sciences related to medicine during the Middle Ages. Fordham University Press, New York

End-of-Life Care

Melissa J. Bloomer
School of Nursing and Midwifery, Centre for Quality and Patient Safety Research, Deakin University, Geelong, VIC, Australia

Synonyms

Dying care; Hospice care; Palliation; Palliative care; Terminal care

Definition

End-of-life care	It is more commonly used to describe the last days of hours of life (Lunney et al. 2003). While the provision of palliative care is not possible or necessary for every person who dies, the demand for end-of-life care is increasing worldwide, particularly in inpatient and hospital settings (Burbeck et al. 2014; Calanzani et al. 2013).
Palliative care	It is defined as an approach that improves the quality of life of patients with life-threatening illness, through the prevention and relief of suffering through early identification, impeccable assessment, and treatment of pain and other physical, psychosocial, and spiritual symptoms (World Health Organization [WHO] 2018).

Overview

End-of-life care and Palliative care are terms that are often used interchangeably. Yet the definitions provided above clearly articulate the differences. While it would be ideal for every person to receive palliative care from the point of diagnosis with a life-limiting illness until death, the reality is that many people die without forewarning or a formal diagnosis, and not everyone will accept specialist palliative care services. This may be because historically, specialist palliative care services have been associated with treating the needs of cancer patients who are known to have high symptom burden (Worldwide Palliative Care Alliance 2017). None-the-less, the goal should be to ensure all people have access to end-of-life care that meets their individual needs.

Worldwide, the demand for end-of-life care is increasing. There are multiple factors contributing to the increasing demand. The population is rapidly ageing worldwide (WHO 2020). While ageing may be seen as a positive, equally it may be viewed as a threat to society when individuals are experiencing negative health effects associated with ageing (Chang et al. 2019). Ageing is characterized by a progressive loss of physiological integrity, leading to impaired function and increased vulnerability to death (López-Otín et al. 2013, p. 1194). Ageing is also associated with an increased risk of a person suffering from multi-morbidity (Beard et al. 2016) and frailty (see "► Epidemiology of Frailty" in this volume). Frailty is a progressive age-related deterioration in physiological systems resulting in increased vulnerability to stressors and higher risk of adverse outcomes including falls and hospitalization (Clegg et al. 2013; Fried et al. 2001) and has been shown to be significantly associated with higher mortality (Kojima et al. 2018). Older age is also associated with high risk of cognitive impairment and dementia (Lavrencic et al. 2017), functional impairments including disability (Perrels et al. 2014; Zhao et al. 2010), with a predictable and steady decline towards death (Bravell et al. 2010). Suffering is also expected to increase with age towards the end of life, due to widespread under-acknowledgement, under-assessment, and under-treatment of the older person's needs (Davies and Higginson 2004; WHO 2017b). Decreased independence (Bravell et al. 2010), loss of spouse (Bloomer et al. 2019b), limited social support networks and social isolation (Bravell et al. 2010; Ng et al. 2015), and a lack of emotional and spiritual support (Hunt et al. 2014) further compound the needs of older people towards the end of life.

Population ageing and health decline including frailty and disability impact all aspects of society, including healthcare (WHO 2017a). Changes in the causes of mortality have also resulted in changes to the nature of how older people die. Older people are less likely to have a record of their care and end-of-life preferences leading to unwanted interventions (Hunt et al. 2014) and limiting the opportunity for appropriate, timely, and high quality end-of-life care (Porock et al. 2009). Increased attention to the provision of supports and services to meet individual's needs at the end of life, in terms of policy and planning and end-of-life care provision, is essential given the dying phase is now described as a longer period of time, characterized by difficult symptom management, psychological problems, spiritual distress, and complicated treatment decisions (Voumard et al. 2018).

Key Research Findings

Prognostication

Variations in illness trajectories and complex multi-morbidity make prognostication of dying difficult. Sudden death is now less common in Western or developed countries, with more people dying from a progressive illness with a predictable pattern of decline (Murray et al. 2005). Yet recognizing when a person may be dying remains difficult particularly given that advanced disease, multiple comorbidities, general physical decline, increasing dependence and need for support, increasing frailty, and decreasing activity are all indicators that a person may be approaching the end of life (Thomas et al. 2016). In 2005, Murray et al. identified three typical illness trajectories resulting in death. The first typifies the scenario for a person with a short and predictable period of decline over a few weeks, months of years, most likely associated with a cancer diagnosis. Several prognostication tools exist to aid clinicians to determine when a person may be dying (Murray et al. 2005). The second trajectory represents more long-term conditions such as heart failure or other chronic disease where the person may be ill for many months or years with occasional acute severe exacerbations; each resulting in poorer overall health and decline towards death (Murray et al. 2005). The third illness trajectory, describes those with prolonged dwindling, associated with progressive disability from an already low baseline of cognitive or physical functioning, such as an older person with dementia. With prolonged dwindling, a person's functioning is already low with the timeframe to death quite variable (Murray et al. 2005). These trajectories can aide clinicians to consider when a patient may be approaching the end of life, which is essential to good end-of-life care planning (Morgan et al. 2019).

The Gold Standards Framework Prognostic Indicator (Thomas et al. 2016) is similarly used around the world as a tool to assist clinicians in determining when a person may be dying. The use of a surprise question, "Would you be surprised if the patient died within the next year?" as a trigger to identify patients for who may benefit from end-of-life planning and care, has been shown to be very helpful in prognostication (O'Callaghan et al. 2014). In the acute hospital setting, screening using the Gold Standards Framework Prognostic Indicator was found to be sensitive and specific, suggesting it was useful in identifying those approaching death and enabling an organizational approach to care interventions towards the end of life (O'Callaghan et al. 2014). Yet a subsequent systematic review and meta-analysis of the accuracy of the surprise question demonstrated wide degree of accuracy, from poor to reasonable, with the risk of at least as many "false positives" as "true positives" (White et al. 2017). Therefore, the surprise question is best used as part of a wider prognostic assessment that includes both general measures of performance status and disease burden, as well as disease-specific measures relevant to each individual person's diagnosis (White et al. 2017), with clinician interpretation an essential component.

Even when it is identified that a person may be approaching the end of life, workplace cultural factors have been shown to impede overt declarations that a patient may be dying, with clinicians preferring to wait for formal notification from the head of the treating team that the older person is to receive end-of-life care (Bloomer et al. 2013). Delayed recognition of dying inhibits the conveying of information among the greater treating team, can result in delayed referrals to specialist palliative care and support personnel, and limits opportunities for discussions and shared decision-making about end-of-life care and preferences (Bloomer et al. 2013; Swagerty 2017).

End-of-Life Care Settings

Historically, death occurred in the home; prior to this, hospitals were largely only involved in end-of-life care for travelers, the poor, and the orphaned (Howarth 2007). In the past, the ability to provide end-of-life care at home was possible, mostly because of the presence of informal carers in the home who managed all aspects of care including intimate care, symptom control, and social, spiritual, and emotional support (Howarth 2007). Dying and death were viewed as a normal and inevitable part of life, not a condition necessarily needing intervention or prevention (Rosenberg 2011). Changes in modern society impacting family structure and function, and creation of a view that mortality is a phenomenon that can be controlled by medical science, have meant that who provides care for those who are dying, and where, has changed (Howarth 2007).

Many Western countries continue to favor enabling older people to die in their own homes (Lloyd et al. 2011), with death at home regarded as a marker of a "good death" and an indicator of success of the healthcare system (Gerber et al. 2019). But there is clear evidence to demonstrate that death at home is not universally as common as desired. While there is a trend towards death in community and home settings, recent data demonstrates nearly 30% of deaths in the USA (National Center for Health Statistics 2017), 54.8% of UK deaths (UK Office for National Statistics 2018), and 56.8% of Canadian deaths (Burge et al. 2015) occur in inpatient settings. While it is not possible to definitively say that avoiding admission to hospital results in a better death (Perrels et al. 2014), without considering multiple other factors, death in an inpatient hospital setting is still common, with older people much less likely to die at home (Higginson and

Sen-Gupta 2000). UK data shows that the majority of older people die in settings other than their usual place of residence (Perrels et al. 2014). Across Europe, the situation is much the same, with most older people dying in institutional care settings rather than at home (Houttekier et al. 2010; Ní Chróinín et al. 2011). In Australia acute hospitals now provide the majority of end-of-life care (Australian Commission on Safety and Quality in Health Care 2015). While Australia ranks second to the United Kingdom in terms of Quality of Death, an index that measures the quality of palliative care available to adults in 80 countries (Economist Intelligence Unit 2015), acute hospital settings can fall short in the provision of end-of-life care for older people. Hence, optimizing hospital provision of end-of-life care needs is a particularly pressing priority (Gott et al. 2019).

For older people, who are less likely to seek help than any other age group (Zhao et al. 2010), ensuring end-of-life care aligns with their needs is even more important. The "Ten Priorities Towards a Decade of Healthy Ageing" (WHO 2017a) describes the need for health systems to align to the needs of older people, so that older people get the health care they need when and where they need it. Yet in comparison to other age groups, it is only when increased unmet needs including but not limited to symptom relief and emotional support become too much to manage independently (Bravell et al. 2010), that older people seek help that results in an admission to hospital.

Acute health systems have a strong bias towards curative care (Worldwide Palliative Care Alliance 2017), and are designed to deal with acute and mostly irreversible conditions, rather than more complex and chronic health conditions associated with ageing (WHO 2017a) or death. Therefore the care needs of older people as they approach death are not always met (WHO 2016). The traditional medical model, upon which most acute hospital facilities operate, provide siloed care where the focus is on treatment of a single disease (Carter et al. 2020). In contrast, the uncertain prognosis of chronic but terminal conditions that typify the end of life in older age, mean it may be difficult to identify who is in the last year of life or approaching death (Zhao et al. 2010) and ensure the right care. Cognitive impairment also complicates matters further, with a greater likelihood of transfer to a hospital for care management prior to death (Perrels et al. 2014), possibly related to the inability of the person to express their preferences for care. Hence there is a clear need to incorporate and accommodate cognition care into end-of-life care provision (Brayne et al. 2006).

No matter the location of death, many have attempted to define what constitutes a good death. Some may argue that the notion of a "good death" to be an oxymoron, because death cannot be good. A good death has been described as one where the person dies quickly, with minimal suffering (Kastbom et al. 2017). A recent review to explore patients' perspectives identified a good death as one in which there is control of pain and symptoms, preparation for death, clear decision-making, a sense of closure, and respect for personhood (Krikorian et al. 2020). Open acknowledgement of the imminence of death provides advance warning to allow affairs to be settled, the person's preferences to be sought and prioritized, and for the death to occur in the person's place of preference, where possible (Gott et al. 2008).

Dementia and End-of-Life Care

Dementia is the leading cause of disability and death in older people (Sousa et al. 2010). Concerns exist about the quality of care provided to people with dementia at the end of life, when compared to those who are dying from other diagnoses; perhaps because there is an assumption that dementia is a natural part of ageing, rather than a terminal illness. Cognitive impairment, however, also impacts a person's ability to contribute to decisions about the care they receive. Older people want an opportunity to discuss their preferences, including for their end-of-life care, but few are afforded the opportunity to do so while they are able (Weathers et al. 2016). Hence, timely discussions and formal recording of the care preferences for older people, particular before dementia impacts their ability to express preferences, is essential (Bamford et al. 2018; Bloomer et al. 2018).

For those with dementia, the uncertainty of dying, typified by the "prolonged dwindling" trajectory (Murray et al. 2005) described earlier in this chapter, makes recognition of dying problematic (Bamford et al. 2018). Yet recognition is essential to facilitating a change in care to focus on the provision of high-quality end-of-life care (Bamford et al. 2018). Consideration for the setting for end-of-life care is equally essential and best guided by knowing the person's wishes. Even when a dementia diagnosis does exist, transfers from the usual place of residence to an inpatient setting still occur (Bamford et al. 2018). Due consideration should be given to whether the older person with dementia is likely to benefit from a transfer from their usual place of residence, and to avoid unnecessary hospital admissions (Bamford et al. 2018).

Advance Care Planning

Advance care planning, found in many countries worldwide, is a process of communication between individuals, their family/significant others, and clinicians to understand, discuss, and plan an individual's future health care decisions in the event they are no longer able to contribute (Detering et al. 2010). Considered a marker of good quality care for older people (Bamford et al. 2018), advance care planning has been shown to improve care from the perspective of the person and their family, diminishing stress, anxiety, and depression in surviving relatives (Detering et al. 2010). Yet, there are multiple barriers to advance care planning.

The extent to which people want to engage in advance care planning is variable (Weathers et al. 2016), likely indicative of peoples' comfort with thinking and talking about mortality. For clinicians, there is a need for training and education to ensure they are willing and comfortable to initiate conversations about death and advance care planning (Weathers et al. 2016). Even when people are open to completing an advance care plan, the time it takes and with multiple conversations between the individual, their family or significant others and members of their treating team can impede completion (Detering et al. 2010). Advance care planning requires a proactive approach, considered well in advance of an acute illness, founded on good communication between all parties (Threapleton et al. 2017). Ideally, an advance care plan would be initiated by a person's primary care provider, while the person is well (Bloomer et al. 2010) as a part of routine care (Detering et al. 2010), and stored in the patient's medical record so that it is readily available when needed. In Australia, even with national promotion campaigns targeting older people, and Government endorsement of advance care planning as a key component of quality end-of-life care (Australian Commission on Safety and Quality in Health Care 2013), the evidence suggests that advance care planning had little to impact inpatient end-of-life care (Bloomer et al. 2019b, c).

What is needed is a coordinated, systematic model of advance care planning (Detering et al. 2010), that is founded on open and honest conversations with patients about their future care (Merlane and Armstrong 2020), as a matter of routine care, or at the point of diagnosis with a life-limiting illness. When used, advance care plans have benefits not only for the person, but also their family and significant others and clinicians, reducing transfer and unnecessary hospitalizations, and instead prioritizing comfort at the end of life (Weathers et al. 2016).

Cultural Considerations at the End of Life

What constitutes a good death is not a universally explained concept (see "▶ Good Death" in this volume). Rather, the perception of a good death is dependent on each person's culture, which can include ethnicity and ancestry, language spoken, religious affiliation, ideas and belief systems, customs, and social behaviors (Australian Bureau of Statistics 2017). A person's culture explains who an individual is, how they connect with others, the sense of identity and belonging, and the meaning a person places on death (Lloyd et al. 2011).

Common to Western cultures, is an assumption of openness to talking about a person's diagnosis and dying status (Lloyd et al. 2011). To assume this for other ethnic or cultural groups is problematic. Even within Western settings, it should not be assumed that an individual and their family or significant others are accepting of full disclosure of health information such as prognosis (Bellamy and Gott 2013). A New Zealand study reported that patients of Chinese descent were not in favor of receiving news of a life-limiting diagnosis, expressing a preference instead for this information to be relayed to a family member, an ideal that may run counter to the assumed Western thinking of New Zealanders (Bellamy and Gott 2013). Conversely, Maori beliefs that death was central to life meant talking about death and dying was less difficult. In essence, heterogeneity of views in relation to talking about dying and death across cultural groups should be anticipated.

To similarly assume that care needs at the end of life are the same for every dying older person is short-sighted and problematic for end-of-life care provision. A "one-size-fits-all" approach to end-of-life care is therefore almost always likely to fail to consider the older person's individual needs (Bloomer et al. 2019a). The Australian context is a good example. Australia is one of the most culturally and linguistically diverse (CALD) countries in the world, with more than 29% of Australians born overseas (Australian Bureau of Statistics 2019), one quarter speaking a language other than English and approximately 130 religions followed (Victorian State Government 2016). Despite Australia's diversity, a recent study of end-of-life care in hospital settings found that clinicians lacked understanding of the potential breadth of cultural rituals and practices to be considered as part of the provision of end-of-life care (Bloomer et al. 2019a). Patients' cultural needs before and after death were not always accommodated, and negatively impacted by organizational constraints and inadequate communication (Bloomer et al. 2019a).

What is needed is a global commitment to ensuring care for older people is culturally responsive and respectful (Broom et al. 2013). To provide such services, significant changes to health care environments are required (Betancourt et al. 2016). The first step is for healthcare providers to take early steps to understand a person's culture and how it may influence their care needs and preferences towards the end of life (Betancourt et al. 2016). In the context of large and complex health systems, this may not be as simple as it sounds. One suggestion is for clinicians to act as cultural brokers, who can provide links and support to minimize cultural boundaries between older people from culturally diverse backgrounds and the health care system (Crawford et al. 2017) is a positive step forward. But the suggestion assumes that clinicians are culturally aware and have the know-how for providing culturally sensitive care (Bloomer et al. 2019a).

Clinical Priorities for End-of-Life Care

The obvious first step in discussing the clinical priorities for end-of-life care for older people is the nexus between geriatric medicine and palliative care. Geriatric medicine as a specialty developed as a response to the multi-morbidity in older people and is primarily concerned with the health care of older people (Voumard et al. 2018). Given the complexity of need for older people with multi-morbidity, the goals of geriatric medicine are to maintain and restore functional capabilities and improve quality of life for older people (Voumard et al. 2018). While geriatric medicine typically focuses on those aged 65 years or over, it is those aged 80 and over, whose needs are more complex, who benefit the most form a geriatric medicine approach to care (Voumard et al. 2018).

Specialist Palliative Care, as a discipline, focuses on symptom management and control and quality of life (WHO 2018). However, it does not automatically imply that actions and interventions focused on life-saving cannot also occur. Rather, specialist palliative care can be offered concurrent with other interventions aimed at life-saving (Brown and Ashcraft 2019). This approach should also not be associated with clinicians giving up, because at its best, a dual approach provides excellent symptom management and exceptional patient, family, and clinician-team communication regarding the diagnosis, prognosis, hopes, goals, and treatment expectations over time (Swagerty 2017).

By combining aspects of geriatric medicine with specialist palliative care, the focus of end-of-life care is centered on the older person, their unique situation, care needs and priorities

(Voumard et al. 2018), and realistic of expected outcomes. The ideal approach is one that is founded on a multidisciplinary team who provide an integrated approach to care to ensure continuity between existing service/care providers in the person's choice of care setting (Voumard et al. 2018) that is inclusive of the person and their family/significant others as members of the team (Swagerty 2017).

Ongoing assessment should focus on the assessment of symptoms, with the goal of relieving suffering and distress in order to provide a good death for the older person (Swagerty 2017). Overly aggressive, burdensome, and futile treatments should be avoided, instead focusing on treatments that provide comfort, and psychosocial and spiritual support should be offered (van der Steen et al. 2013).

Simultaneous assessment of family needs and coping is essential to family coping, particularly given that dying is not only what the older person experiences but also what the family experiences (Donnelly and Dickson 2013). The evidence is clear that families want sensitive and timely information, regular updates about changes in the health status of the person, and opportunities to discuss preferences for end-of-life care (Moore et al. 2017). The opportunity to spend time in close proximity to the person before and after death allows for a process of farewell-taking (Mossin and Landmark 2011) and facilitates cultural rites and family traditions (Slatyer et al. 2015). Even though family presence may be thought to affect the provision of care for the dying person (Donnelly and Dickson 2013), supporting family involvement in a manner that is appropriate for them is imperative.

Future Directions of Research

Given the population is ageing globally (World Health Organization 2020), future research must focus on identifying innovative interventions that improve the quality of end-of-life care provided to those who need it. Further work is needed to clearly articulate the gap between current end-of-life care provision and the unique and often diverse needs of individuals (Bravell et al. 2010). How end-of-life care is communicated among multidisciplinary teams, across practice settings and into the community also needs further investigation. What is also underrepresented in the research literature is research to examine how psychosocial and spiritual needs of the dying and their families/significant others can be better supported, ideally, with interventions developed to complement clinical care.

Further research is needed to improve the quality of end-of-life care, irrespective of place of death. While the notion of a good death is often idealized, how this is enacted is not as clear. Acknowledging that a person's views of a good death are very individual and unique (Krikorian et al. 2020), it is imperative to also acknowledge a person's views are likely dynamic in nature and may even change over time and as the person approaches death (Vig et al. 2002). This suggests that further work is necessary to examine health professionals' understanding of the most common contributors to a good death, and whether their own notions and expectations may influence end-of-life care decisions (Krikorian et al. 2020). Only then will we see the good death, and perceived by the person, actualized.

Consideration for end-of-life care provision in developing countries is also essential. While the factors that contribute to or impede quality end-of-life care across the globe are not fully understood, what is known is that income levels are a strong indicator of the availability and quality of end-of-life care provision (Economist Intelligence Unit 2015). Many developing countries are still unable to provide basic pain management, and in others, palliative care services are not embedded as part of health care systems (Economist Intelligence Unit 2015), and/or legal frameworks are needed to support and facilitate end-of-life decision-making practices (Gendeh et al. 2016). Research is needed to identify and evaluate sustainable strategies that address service delivery models, workforce needs, leadership and governance for improving access to palliative and end-of-life care across low- and middle-income countries (Donkor et al. 2018).

Summary

Death is a certainty for all. Advancing technologies may allow us to delay death, but we cannot prevent it. Therefore, the focus of end-of-life care should be on providing the best possible death, which is one in which the impending death is acknowledged; clinical care prioritizes comfort and symptom control; care aligns with each person's wishes, preferences, and cultural beliefs; and in the person's preferred location. To achieve this, requires a shift in thinking away from life-saving, to one that values the person, the life lived, prioritizes respect, and honors their person's wishes till death.

Cross-References

- ▶ Advance Care Planning
- ▶ Advance Care Planning: Advance Directives
- ▶ Advance Care Planning: Medical Orders at the End of Life (MOLST, POLST)
- ▶ Bereavement Care
- ▶ Caregiving at the End of Life
- ▶ Death Anxiety
- ▶ Death Denial
- ▶ Death Trajectory
- ▶ Death with Dignity
- ▶ Dying in Place
- ▶ End-of-Life Decision-Making in Acute Care Settings
- ▶ Euthanasia and Senicide
- ▶ Good Death
- ▶ Hospice and Caregiving
- ▶ Professional Grief and Burnout
- ▶ Religion and Spirituality in End-of-Life Care
- ▶ Social Support in Bereavement
- ▶ Terminal Change
- ▶ Thanatology
- ▶ Theory of Harmonious Death

References

Australian Bureau of Statistics (2017) Cultural diversity in Australia [Press release]. Retrieved from http://www.abs.gov.au/ausstats/abs@.nsf/Lookup/by%20Subject/2071.0~2016~Main%20Features~Cultural%20Diversity%20Data%20Summary~30

Australian Bureau of Statistics (2019) 3412.0 – migration, Australia, 2017–18. ABS, Canberra. Retrieved from https://www.abs.gov.au/ausstats/abs@.nsf/Latestproducts/3412.0Main%20Features22017-18?opendocument&tabname=Summary&prodno=3412.0&issue=2017-18&num=&view=

Australian Commission on Safety and Quality in Health Care (2013) Safety and quality of end-of-life care in acute hospitals. A background paper. ACSQHC, Sydney. Retrieved from https://www.safetyandquality.gov.au/publications-and-resources/resource-library/safety-and-quality-end-life-care-acute-hospitals-background-paper

Australian Commission on Safety and Quality in Health Care (2015) National consensus statement: essential elements for safe and high quality end-of-life care. Retrieved from https://www.safetyandquality.gov.au/publications-and-resources/resource-library/national-consensus-statement-essential-elements-safe-and-high-quality-paediatric-end-life-care

Bamford C, Lee R, McLellan E, Poole M, Harrison-Dening K, Hughes J, ... Exley C (2018) What enables good end of life care for people with dementia? A multi-method qualitative study with key stakeholders. BMC Geriatr 18(1):302. https://doi.org/10.1186/s12877-018-0983-0

Beard JR, Officer A, de Carvalho IA, Sadana R, Pot AM, Michel JP, ... Mahanani WR (2016) The world report on ageing and health: a policy framework for healthy ageing. Lancet 387(10033):2145–2154. https://doi.org/10.1016/S0140-6736(15)00516-4

Bellamy G, Gott M (2013) What are the priorities for developing culturally appropriate palliative and end-of-life care for older people? The views of healthcare staff working in New Zealand. Health Soc Care Community 21(1):26–34. https://doi.org/10.1111/j.1365-2524.2012.01083.x

Betancourt J, Green A, Carrillo J, Ananeh-Firempong O (2016) Defining cultural competence: a practical framework for addressing racial/ethnic disparities in health and health care. Public Health Rep 118(4):293–302. https://doi.org/10.1093/phr/118.4.293

Bloomer M, Tan H, Lee S (2010) End of life care – the importance of advance care planning. Aust Fam Physician 39(10):734–737

Bloomer MJ, Endacott R, O'Connor M, Cross W (2013) The 'dis-ease' of dying: challenges in nursing care of the dying in the acute hospital setting. A qualitative observational study. Palliat Med 27(8):757–764. https://doi.org/10.1177/0269216313477176

Bloomer M, Botti M, Runacres F, Poon P, Barnfield J, Hutchinson A (2018) Communicating end-of-life care goals and decision-making among a multi-disciplinary geriatric inpatient rehabilitation team: a qualitative descriptive study. Palliat Med 32(10):1615–1623. https://doi.org/10.1177/0269216318790353

Bloomer M, Botti M, Runacres F, Poon P, Barnfield J, Hutchinson A (2019a) Cultural considerations at end of life in a geriatric inpatient rehabilitation setting.

Collegian 26:165–170. https://doi.org/10.1016/j.colegn.2018.07.004

Bloomer M, Botti M, Runacres F, Poon P, Barnfield J, Hutchinson A (2019b) End-of-life care for older people in subacute care: a retrospective clinical audit. Collegian 26:22–27. https://doi.org/10.1016/j.colegn.2018.02.005

Bloomer M, Hutchinson A, Botti M (2019c) End-of-life care in hospital: an audit of care against Australian national guidelines. Aust Health Rev 43(4):578–584. https://doi.org/10.1071/AH18215

Bravell ME, Malmberg B, Berg S (2010) End-of-life care in the oldest old. Palliat Support Care 8(3):335–344. https://doi.org/10.1017/S1478951510000131

Brayne C, Gao L, Dewey M, Matthews FE (2006) Dementia before death in ageing societies – the promise of prevention and the reality. PLoS Med 3(10):e397. https://doi.org/10.1371/journal.pmed.0030397

Broom A, Good P, Kirby E, Lwin Z (2013) Negotiating palliative care in the context of culturally and linguistically diverse patients. Intern Med J 43(9):1043–1046. https://doi.org/10.1111/imj.12244

Brown KL, Ashcraft AS (2019) Comfort or care: why do we have to choose? Implementing a geriatric trauma palliative care program. J Trauma Nurs 26(1):2–9. https://doi.org/10.1097/JTN.0000000000000410

Burbeck R, Candy B, Low J, Rees R (2014) Understanding the role of the volunteer in specialist palliative care: a systematic review and thematic synthesis of qualitative studies. BMC Palliat Care 13(1):3. https://doi.org/10.1186/1472-684X-13-3

Burge F, Lawson B, Johnston G, Asada Y, McIntyre PF, Flowerdew G (2015) Preferred and actual location of death: what factors enable a preferred home death? J Palliat Med 18(12):1054–1059. https://doi.org/10.1089/jpm.2015.0177

Calanzani N, Higginson IJ, Gomes B (2013) Current and future needs for hospice care: an evidence-based report. Commission into the Future of Hospice Care, London. Retrieved from https://www.hospiceuk.org/

Carter AJE, Earle R, Grégoire M-C, MacConnell G, MacDonald T, Frager G (2020) Breaking down silos: consensus-based recommendations for improved content, structure, and accessibility of advance directives in emergency and out-of-hospital settings. J Palliat Med 23(3):379–388. https://doi.org/10.1089/jpm.2019.0087

Chang AY, Skirbekk VF, Tyrovolas S, Kassebaum NJ, Dieleman JL (2019) Measuring population ageing: an analysis of the global burden of disease study 2017. Lancet Public Health 4(3):e159–e167. https://doi.org/10.1016/S2468-2667(19)30019-2

Clegg A, Young J, Iliffe S, Rikkert MO, Rockwood K (2013) Frailty in elderly people. Lancet 381(9868):752–762. https://doi.org/10.1016/S0140-6736(12)62167-9

Crawford R, Stein-Parbury J, Dignam D (2017) Culture shapes nursing practice: findings from a New Zealand study. Patient Educ Couns 100(11):2047–2053. https://doi.org/10.1016/j.pec.2017.06.017

Davies E, Higginson IJ (2004) Better palliative care for older people. WHO, Copenhagen. Retrieved from https://apps.who.int/iris/handle/10665/107563

Detering KM, Hancock AD, Reade MC, Silvester W (2010) The impact of advance care planning on end of life care in elderly patients: randomised controlled trial. BMJ:340. https://doi.org/10.1136/bmj.c1345

Donkor A, Luckett T, Aranda S, Phillips J (2018) Barriers and facilitators to implementation of cancer treatment and palliative care strategies in low- and middle-income countries: systematic review. Int J Public Health 63(9):1047–1057. https://doi.org/10.1007/s00038-018-1142-2

Donnelly S, Dickson M (2013) Relatives' matched with staff's experience of the moment of death in a tertiary referral hospital. QJM 106(8):731–736. https://doi.org/10.1093/qjmed/hct095

Economist Intelligence Unit (2015) The 2015 quality of death index: ranking palliative care across the world. A report from the economist intelligence unit. Retrieved from http://www.lienfoundation.org/sites/default/files/2015%20Quality%20of%20Death%20Report.pdf

Fried LP, Tangen CM, Walston J, Newman AB, Hirsch C, Gottdiener J et al (2001) Frailty in older adults: evidence for a phenotype. J Gerontol Ser A 56(3):M146–M157

Gendeh HS, Bhar AS, Gendeh MK, Yaakup H, Gendeh BS, Kosai NR, bin Abdul Rahman R (2016) Caring for the dying in a developing country, how prepared are we. Med J Malaysia 71(5):259

Gerber K, Hayes B, Bryant C (2019) 'It all depends!': a qualitative study of preferences for place of care and place of death in terminally ill patients and their family caregivers. Palliat Med 33(7):802–811. https://doi.org/10.1177/0269216319845794

Gott M, Small N, Barnes S, Payne S, Seamark D (2008) Older people's views of a good death in heart failure: implications for palliative care provision. Soc Sci Med 67(7):1113–1121. https://doi.org/10.1016/j.socscimed.2008.05.024

Gott M, Robinson J, Moeke-Maxwell T, Black S, Williams L, Wharemate R, Wiles J (2019) 'It was peaceful, it was beautiful': a qualitative study of family understandings of good end-of-life care in hospital for people dying in advanced age. Palliat Med 33(7):793–801. https://doi.org/10.1177/0269216319843026

Higginson IJ, Sen-Gupta G (2000) Place of care in advanced cancer: a qualitative systematic literature review of patient preferences. J Palliat Med 3(3):287–300. https://doi.org/10.1089/jpm.2000.3.287

Houttekier D, Cohen J, Bilsen J, Addington-Hall J, Onwuteaka-Philipsen BD, Deliens L (2010) Place of death of older persons with dementia. A study in five European countries. J Am Geriatr Soc 58(4):751–756. https://doi.org/10.1111/j.1532-5415.2010.02771.x

Howarth G (2007) The social context of death in old age. Work Older People 11(3):17–20. https://doi.org/10.1108/13663666200700045

Hunt KJ, Shlomo N, Addington-Hall J (2014) End-of-life care and preferences for place of death among the oldest old: results of a population-based survey using VOICES-Short Form. J Palliat Med 17(2):176–182. https://doi.org/10.1089/jpm.2013.0385

Kastbom L, Milberg A, Karlsson M (2017) A good death from the perspective of palliative cancer patients. Support Care Cancer 25(3):933–939. https://doi.org/10.1007/s00520-016-3483-9

Kojima G, Iliffe S, Walters K (2018) Frailty index as a predictor of mortality: a systematic review and meta-analysis. Age Ageing 47(2):193–200. https://doi.org/10.1093/ageing/afx162

Krikorian A, Maldonado C, Pastrana T (2020) Patient's perspectives on the notion of a good death: a systematic review of the literature. J Pain Symptom Manag 59(1):152–164. https://doi.org/10.1016/j.jpainsymman.2019.07.033

Lavrencic LM, Richardson C, Harrison SL, Muniz-Terrera G, Keage HA, Brittain K, … Stephan BC (2017) Is there a link between cognitive reserve and cognitive function in the oldest-old? J Gerontol Ser A 73(4):499–505. https://doi.org/10.1093/gerona/glx140

Lloyd L, White K, Sutton E (2011) Researching the end-of-life in old age: cultural, ethical and methodological issues. Ageing Soc 31(3):386–407. https://doi.org/10.1017/S0144686X10000966

López-Otín C, Blasco MA, Partridge L, Serrano M, Kroemer G (2013) The hallmarks of aging. Cell 153(6):1194–1217. https://doi.org/10.1016/j.cell.2013.05.039

Lunney JR, Lynn J, Foley DJ, Lipson S, Guralnik JM (2003) Patterns of functional decline at the end of life. JAMA 289(18):2387–2392. https://doi.org/10.1001/jama.289.18.2387

Merlane H, Armstrong L (2020) Advance care planning. Br J Nurs 29(2):96–97. https://doi.org/10.12968/bjon.2020.29.2.96

Moore KJ, Davis S, Gola A, Harrington J, Kupeli N, Vickerstaff V et al (2017) Experiences of end of life amongst family carers of people with advanced dementia: longitudinal cohort study with mixed methods. BMC Geriatr 17(1):135. https://doi.org/10.1186/s12877-017-0523-3

Morgan DD, Tieman JJ, Allingham SF, Ekström MP, Connolly A, Currow David C (2019) The trajectory of functional decline over the last 4 months of life in a palliative care population: a prospective, consecutive cohort study. Palliat Med 33(6):693–703. https://doi.org/10.1177/0269216319839024

Mossin H, Landmark B (2011) Being present in hospital when the patient is dying – a grounded theory study of spouses experiences. Eur J Oncol Nurs 5:382. https://doi.org/10.1016/j.ejon.2010.11.005

Murray SA, Kendall M, Boyd K, Sheikh A (2005) Illness trajectories and palliative care. BMJ (Clin Res) 330(7498):1007–1011. https://doi.org/10.1136/bmj.330.7498.1007

National Center for Health Statistics (2017) Health, United States, 2017: with special feature on mortality. Retrieved from https://www.cdc.gov/nchs/data/hus/hus17.pdf

Ng TP, Jin A, Feng L, Nyunt MSZ, Chow KY, Feng L, Fong NP (2015) Mortality of older persons living alone: Singapore longitudinal ageing studies. BMC Geriatr 15:126–126. https://doi.org/10.1186/s12877-015-0128-7

Ní Chróinín D, Haslam R, Blake C, Ryan K, Kyne L, Power D (2011) Death in long-term care facilities: attitudes and reactions of patients and staff. A qualitative study. Eur Geriatr Med 2(1):56–59. https://doi.org/10.1016/j.eurger.2010.12.002

O'Callaghan A, Laking G, Frey R, Robinson J, Gott M (2014) Can we predict which hospitalised patients are in their last year of life? A prospective cross-sectional study of the Gold Standards Framework Prognostic Indicator Guidance as a screening tool in the acute hospital setting. Palliat Med 28(8):1046–1052. https://doi.org/10.1177/0269216314536089

Perrels AJ, Fleming J, Zhao J, Barclay S, Farquhar M, Buiting HM, Brayne C (2014) Place of death and end-of-life transitions experienced by very old people with differing cognitive status: retrospective analysis of a prospective population-based cohort aged 85 and over. Palliat Med 28(3):220–233. https://doi.org/10.1177/0269216313510341

Porock D, Pollock K, Jurgens F (2009) Dying in public: the nature of dying in an acute hospital setting. J Hous Elder 23(1-2):10. https://doi.org/10.1080/02763890802664521

Rosenberg JP (2011) Whose business is dying? Death, the home and palliative care. Cult Stud Rev 17(1):15–30. https://doi.org/10.5130/csr.v17i1.1971

Slatyer S, Pienaar C, Williams AM, Proctor K, Hewitt L (2015) Finding privacy from a public death: a qualitative exploration of how a dedicated space for end-of-life care in an acute hospital impacts on dying patients and their families. J Clin Nurs 24(15/16):2164–2174. https://doi.org/10.1111/jocn.12845

Sousa RM, Ferri CP, Acosta D, Guerra M, Huang Y, Jacob K, … Pichardo GR (2010) The contribution of chronic diseases to the prevalence of dependence among older people in Latin America, China and India: a 10/66 Dementia Research Group population-based survey. BMC Geriatr 10(1):53. https://doi.org/10.1186/1471-2318-10-53

Swagerty D (2017) Integrating quality palliative and end-of-life care into the geriatric assessment: opportunities and challenges. Clin Geriatr Med 33(3):415–429. https://doi.org/10.1016/j.cger.2017.03.005

Thomas K, Armstrong Wilson J, the GSF Team (2016) The gold standards framework Proactive Indicator Guidance (PIG). Retrieved from http://www.goldstandardsframework.org.uk/cd-content/uploads/files/PIG/NEW%20PIG%20-%20%20%202020.1.17%20KT%20vs17.pdf

Threapleton D., Chung R., Wong SY, Wong E, Kiang N, Chau P, … Yeoh E (2017) Care toward the end of life in older populations and its implementation facilitators and barriers: a scoping review. J Am Med Dir Assoc 18(12):1000–1009.e1004. https://doi.org/10.1016/j.jamda.2017.04.010

UK Office for National Statistics (2018) Deaths by place of occurrence, 2016. Retrieved from https://www.ons.gov.uk/peoplepopulationandcommunity/birthsdeathsa

ndmarriages/deaths/adhocs/007620deathsbyplaceofoccurrence2016

van der Steen JT, Radbruch L., Hertogh CMPM, de Boer ME, Hughes JC, Larkin P, ... Volicer L (2013) White paper defining optimal palliative care in older people with dementia: a Delphi study and recommendations from the European Association for Palliative Care. Palliat Med 28(3):197–209. https://doi.org/10.1177/0269216313493685

Victorian State Government (2016) 2016 census: a snapshot of our diversity. Office of Multicultural Affairs and Citizenship, Melbourne. Retrieved from https://www.multicultural.vic.gov.au/population-and-migration/victorias-diversity/2016-census-a-snapshot-of-our-diversity

Vig EK, Davenport NA, Pearlman RA (2002) Good deaths, bad deaths, and preferences for the end of life: a qualitative study of geriatric outpatients. J Am Geriatr Soc 50(9):1541–1548

Voumard R, Truchard ER, Benaroyo L, Borasio G, Büla C, Jox R (2018) Geriatric palliative care: a view of its concept, challenges and strategies. BMC Geriatr 18(1):220. https://doi.org/10.1186/s12877-018-0914-0

Weathers E, O'Caoimh R, Cornally N, Fitzgerald C, Kearns T, Coffey A, ... Molloy DW (2016) Advance care planning: a systematic review of randomised controlled trials conducted with older adults. Maturitas 91:101–109. https://doi.org/10.1016/j.maturitas.2016.06.016

White N, Kupeli N, Vickerstaff V, Stone P (2017) How accurate is the 'Surprise Question' at identifying patients at the end of life? A systematic review and meta-analysis. BMC Med 15(1):139–139. https://doi.org/10.1186/s12916-017-0907-4

World Health Organization (2016) Global strategy and action plan on ageing and health (2016–2020): a framework for coordinated global action by the world health organization, member states, and partners across the sustainable development goals. Retrieved from http://www.who.int/ageing/GSAP-Summary-EN.pdf?ua=1

World Health Organization (2017a) 10 priorities towards a decade of healthy ageing. Retrieved from https://www.who.int/ageing/10-priorities/en/

World Health Organization (2017b) Health topics. Ageing. Retrieved from WHO https://www.who.int/ageing/en/

World Health Organization (2018) WHO definition of palliative care. Retrieved from http://www.who.int/cancer/palliative/definition/en/

World Health Organization (2020) Ageing and life course. Retrieved from https://www.who.int/ageing/en/

Worldwide Palliative Care Alliance (2017) WHO global atlas of palliative care at the end of life. WHO, Geneva. Retrieved from http://www.thewhpca.org/resources/global-atlas-on-end-of-life-care

Zhao J, Barclay S, Farquhar M, Kinmonth AL, Brayne C, Fleming J (2010) The oldest old in the last year of life: population-based findings from Cambridge City over-75s cohort study participants aged 85 and older at death. J Am Geriatr Soc 58(1):1–11. https://doi.org/10.1111/j.1532-5415.2009.02622.x

End-of-Life Care Decision-Making

▶ Advance Care Planning

End-of-Life Care Planning

▶ Advance Care Planning

End-of-Life Care Preferences

▶ Advance Care Planning

End-of-Life Decision-Making in Acute Care Settings

Wai-Tat Wong
Department of Anaesthesia and Intensive Care, The Chinese University of Hong Kong, Hong Kong, China

Synonyms

Advance care planning; Advance directives

Definition

Acute care settings refer to hospital facilities with patients being directly admitted from the community to the casualty medicine department. Those patients usually present with acute illnesses including medical diseases (e.g., sepsis, myocardial infarction), surgical diseases (e.g., intestinal obstruction, appendicitis, cholecystitis), and traumas (e.g., accidental fall, road traffic accident) (Hirshon et al. 2013). These acute illnesses are life-threatening and may cause organs failure, which requires patients to be supported by life-sustaining treatments to save their lives. The life-threatening nature of the organ failure from acute illnesses is usually unexpected by the patients and their families.

Life-sustaining treatments (LSTs) include mechanical ventilation, vasopressor infusion, renal replacement therapy, and cardiopulmonary resuscitation for patients suffering from respiratory, cardiovascular, renal system failure, and cardiac arrest eventually. LSTs are usually conducted in intensive care units (ICUs), but sometimes they can be provided in some intermediate care units, depending on the resources in different health care systems. Withholding (WH) LSTs mean not to initiate or escalate LSTs while withdrawing (WD) LSTs mean to actively discontinue the ongoing LSTs, when the chance of meaningful recovery is judged to be dismal (Wong and Joynt 2016).

End-of-life (EOL) decision is the decision on whether to initiate or continue the use of LSTs during acute life-threatening illnesses when the chance of meaningful recovery is dismal. Before and after the initiation of LSTs, patients may deteriorate very rapidly because of the acute nature of the diseases. Therefore, the available time for EOL decision is usually very limited in acute care settings.

Overview

Older adults are particularly vulnerable to acute life-threatening illnesses because of frailty. The decision on WH/WD LST needs to be made at a certain stage of acute illnesses for frail older adults, to avoid unnecessary prolonged suffering during hospital stay and after discharge from the hospital if they survive. Older adults often are incapable of communicating with physicians because of the acute illness or the unavoidable use of sedative drugs for LSTs. Patients' autonomy should be respected by gathering substituted opinions from their close families. Eventually, a shared decision, contributed by physicians and patients' family, on appropriate EOL care plan regarding LSTs should be finalized through a well-structured communication process.

Key Research Findings

Frailty

Frailty is the multisystem decline in health status resulting in cumulative impairment in physical and cognitive reserve (Montgomery and Bagshaw 2017). Clinical Frailty Scale (CFS), a scale of nine scores with five as a cut-off point to define frailty, is a commonly used scoring system to evaluate the severity of frailty (Juma et al. 2016). Frailty in older adults increases their risk of hospital mortality, and more importantly, even if they survived, jeopardizes the quality of their future life. Frailty evaluation is known to be a more comprehensive tool, compared to chronological age, in assessing prognosis and potential of recovery in critically ill older adults. Frailty was found in 43.1% of older adults ($\geq$ 80 years) requiring ICU admissions in a multicenter study conducted in Europe. The 30-day mortality of frail older adults was 59%, which is significantly higher than nonfrail older adults (Flaatten et al. 2017). In the United States, frailty was found in up to one-third of patients admitted to ICU and incremental worsening of survival with increasing CFS was noticed in the 1-year survival analysis (Brummel et al. 2017). Besides mortality, frail older adults, if survived, had poor outcome, including greater disability, impaired quality of life, and likely requirement of institutional care (Montgomery and Bagshaw 2017).

Beneficence and Nonmaleficence

WH/WD LSTs is a common practice in ICU and other acute care facilities around the world based on the ethical principles of beneficence and nonmaleficence (Buckley et al. 2004; Phua et al. 2015; Sprung et al. 2003); despite this, the decision usually end up in mortality. Physicians should be the one to initiate the decision-making process (Joynt et al. 2014). However, time is usually limited in the process due to the acute nature of the deterioration. The incharge physicians in the acute care facilities should seriously consider the burden associated with the LSTs and the possible impact during and after hospitalization (Reignier et al. 2019).

Autonomy and Substituted Decision-Making

An advance directive (AD) is a legal binding document signed by the patient, when he or she is mentally competent, in the presence of a doctor and a witness. The AD should indicate the form of treatment he or she would or would not accept when he or she is terminally ill and no longer competent. AD is legalized in the United States, United Kingdom, Germany, Australia, Singapore, and some other western countries. It is not yet legalized in Hong Kong, China, and other developing countries in Asia (Brown 2003; Chu et al. 2011; Lush 1993).

Without a legalized AD, patients' autonomy should be respected by gathering substituted opinions from patients' close families, when direct communication with patients is impractical. The prerequisite for a valid substituted opinion is that patient's thoughts and feelings with regards to serious illnesses, acceptable quality of life, dying, and death are known or can be presumed by their families (Wendler 2017). Families ought to be reminded to express opinions based upon the patients' values and preferences, but not their own.

Shared Decision-Making

Shared decision-making (SDM) on EOL care is a dynamic process where different stakeholders, including physicians and families, share responsibilities for decisions on WH/WD LSTs (Fig. 1) (Grignoli et al. 2018). Medical specialists, who have been taking care of the patients' current acute illnesses or chronic diseases, should contribute their opinions on treatment options, prognostications of mortality, and expected outcomes in case of survival. Medical consensus on the appropriate use of LSTs, or WH/WD LSTs, should be reached after thorough discussions by different specialists. Once the medical consensus is achieved, SDM should be applied to ensure adequate and appropriate involvement by patients' close family acting as surrogates.

Communication

The SDM process should be accomplished by early and repeated goal of care discussions with the critically ill patients and/or their families (Joynt et al. 2014). While EOL care discussion may not be appropriate in the early phase of acute life-threatening illnesses, the early goal of care discussion within 24 h of acute deterioration could pave the way for subsequent communications regarding WH/WD LSTs for those frail older adults subsequently deemed to have minimal chance of meaningful recovery.

A well-structured communication with families during critical illness should include providing accurate medical information, clarifying substituted and shared decision-making concept, addressing families' concern, and managing their emotions. This form of communication

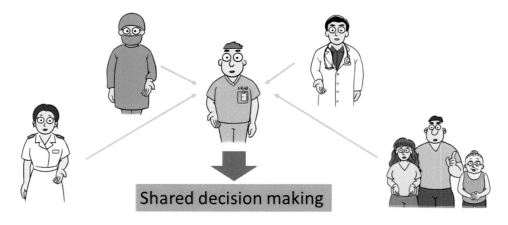

End-of-Life Decision-Making in Acute Care Settings, Fig. 1 Shared decision-making model

intervention has been shown to be effective in shortening the dying process of 1.26 days in ICU (Lee et al. 2019). All parties should maintain mutual understanding and respect to conclude a beneficial shared decision on WH/WD LSTs for the patients.

Effect of video-based educational intervention on cardiopulmonary resuscitation (CPR) and intubation has been evaluated on patients admitted to the acute hospital. Older adults who have viewed the video were more likely to choose limitation of support, compared to patients who received the same information by verbal description only (Merino et al. 2017). Another study, evaluating the effect of video displaying the use of LSTs including intubation, CPR, and vasopressor drug infusion showed that patients suffering from terminal illnesses were less likely to opt for LSTs after viewing the video (El-Jawahri et al. 2015). Audio-visual assistance can be sought to facilitate the communication process in end-of-life care discussion.

Illustration by a Clinical Case

Pneumonia is the most common cause of death in the hospitals around the world (Guidet et al. 2020). Invasive mechanical ventilation can be used as the LST to sustain life to allow recovery of the diseased lung from pneumonia. However, a significant proportion of patients, particularly frail older people, cannot survive from the disease even after prolonged ventilator support and hospital stay. While waiting for the recovery, they may require a temporary tracheostomy to facilitate the weaning of the ventilator and clearance of sputum. Even after weaning off the ventilator support, they may need a prolonged rehabilitation in the hospital before going home. Frail older patients, particularly those suffering from chronic diseases like dementia, heart failure, chronic lung disease, renal failure, and malignancy, may not be able to regain their premorbid functional capacity even after a prolonged hospital stay (Fig. 2).

A temporary tracheostomy, a prolonged hospitalization, and subsequent worsening of functional state may not be acceptable to some older adults. Therefore, when the patient is admitted to the acute care facility for management of pneumonia, an early goal of care discussion should be conducted by the physician to explain the disease, treatment options, prognosis, and expected outcome. In case of deterioration that necessitates the consideration of invasive mechanical ventilation, the physician should specifically clarify the possible outcomes of temporary tracheostomy, prolonged hospitalization, worsening of patients' functional state, and mortality if treatment fails. Substituted opinion should then be sought from families regarding the use of LSTs. Family should be guided and supported by physicians in the discussion process. They should be reminded that their opinion should be based on

End-of-Life Decision-Making in Acute Care Settings, Fig. 2 Worsening functional state after recovery from acute illnesses

patient's previous expressed view on similar encounters of acute life-threatening illnesses. For example, patient may have expressed their feeling toward treatment in the hospitals and care in the institutions when they visit other sick relatives in the hospitals or elderly homes. If such previous expressed view is not present, families need to make their judgment based on their understanding of patient's value and preference from their daily living. Video-based educational tools can be utilized to help the family to envisage the situation of organ support during hospitalization and impaired functional state after discharge from the hospital. Initiating or withholding invasive mechanical ventilation should be finalized by reaching consensus among physicians and patients' families.

Similar discussion and decision-making process can be repeated after initiation of mechanical ventilation or other LSTs, as changes in mind by families are not uncommon after witnessing the real situation of LSTs in ICUs. Changes in decision, which can be initiated by either physicians or patients' families, can be made according to the patient's clinical progress.

Future Direction of Research

With the ageing population, the prevalence of frailty is as high as 1 in 6 of the population of older adults in the community (Ofori-Asenso et al. 2019). However, they are more likely to be suffering from multiple comorbidities, making them prone to acute life-threatening illness. Most of the existing research findings relevant to end-of-life decision were focusing on patients suffering from chronic progressive organ failure (e.g., chronic obstructive pulmonary disease (COPD), end-stated heart failure, renal failure, and advanced dementia), or metastatic malignancy rather than frail older adults in the community. Regarding acute life-threatening illnesses, investigations were frequently conducted in ICUs looking into the communication and decision-making after LSTs being initiated. A timely end-of-life decision before initiation of LSTs is the best option to avoid the suffering that is unacceptable to older people. Lack of interdisciplinary approach was found to be an important barrier in embedding EOL care in acute hospitals (Noble et al. 2018). Data regarding decision on WH LSTs as part of the EOL care in the acute care facilities should be the research direction in the future. The acceptable decision-making model and appropriate component in the communication process before LSTs initiated in ICU should be evaluated.

While a well-structured communication and video-based education intervention are proven to be beneficial to patients and families in the hospital settings (Merino et al. 2017; White et al. 2018), the acceptance and effectiveness of video-based public educations regarding WH/WD LSTs in the community should be evaluated. Substituted and shared decision-making process is important, but difficult for the public to understand. The present public's understanding of these two concepts and the appropriate way to improve their comprehension is another important research topic in the future.

Summary

EOL decision is difficult in acute care setting because of the time constraint from potential rapid deterioration from acute illnesses and frequently impossible direct communication with patients. The decision-making should be based on the shared decision-making model with medical input from different specialists taking care of the patients and substituted patients' opinion from families. A well-structured communication is the key to success in the decision-making process. Future research should focus on potential benefits of earlier decision-making before referring to ICU and public education regarding EOL decision in acute care setting.

Cross-References

▶ Advance Care Planning
▶ Advance Care Planning: Advance Directives
▶ Advance Care Planning: Medical Orders at the End of Life (MOLST, POLST)
▶ Assisted Dying

- Caregiving at the End of Life
- End-of-Life Care
- Frailty in Clinical Care
- Good Death
- Pneumonia
- Religion and Spirituality in End-of-Life Care
- Terminal Change
- Thanatology

References

Brown BA (2003) The history of advance directives. A literature review. J Gerontol Nurs 29(9):4–14

Brummel NE, Bell SP, Girard TD, Pandharipande PP, Jackson JC, Morandi A, ... Ely EW (2017) Frailty and subsequent disability and mortality among patients with critical illness. Am J Respir Crit Care Med 196(1):64–72. https://doi.org/10.1164/rccm.201605-0939OC

Buckley TA, Joynt GM, Tan PY, Cheng CA, Yap FH (2004) Limitation of life support: frequency and practice in a Hong Kong intensive care unit. Crit Care Med 32(2):415–420. https://doi.org/10.1097/01.CCM.0000110675.34569.A9

Chu LW, Luk JK, Hui E, Chiu PK, Chan CS, Kwan F, ... Woo J (2011) Advance directive and end-of-life care preferences among Chinese nursing home residents in Hong Kong. J Am Med Dir Assoc 12(2):143–152. https://doi.org/10.1016/j.jamda.2010.08.015

El-Jawahri A, Mitchell SL, Paasche-Orlow MK, Temel JS, Jackson VA, Rutledge RR, ... Volandes AE (2015) A randomized controlled trial of a CPR and intubation video decision support tool for hospitalized patients. J Gen Intern Med 30(8):1071–1080. https://doi.org/10.1007/s11606-015-3200-2

Flaatten H, De Lange DW, Morandi A, Andersen FH, Artigas A, Bertolini G, ... Guidet B (2017) The impact of frailty on ICU and 30-day mortality and the level of care in very elderly patients (>/= 80 years). Intensive Care Med 43(12):1820–1828. https://doi.org/10.1007/s00134-017-4940-8

Grignoli N, Di Bernardo V, Malacrida R (2018) New perspectives on substituted relational autonomy for shared decision-making in critical care. Crit Care 22(1):260. https://doi.org/10.1186/s13054-018-2187-6

Guidet B, de Lange DW, Boumendil A, Leaver S, Watson X, Boulanger C, ... Flaatten H (2020) The contribution of frailty, cognition, activity of daily life and comorbidities on outcome in acutely admitted patients over 80 years in European ICUs: the VIP2 study. Intensive Care Med 46(1):57–69. https://doi.org/10.1007/s00134-019-05853-1

Hirshon JM, Risko N, Calvello EJ, Stewart de Ramirez S, Narayan M, Theodosis C, ... Acute Care Research Collaborative at the University of Maryland Global Health Initiative (2013) Health systems and services: the role of acute care. Bull World Health Organ 91(5):386–388. https://doi.org/10.2471/BLT.12.112664

Joynt GM, Lipman J, Hartog C, Guidet B, Paruk F, Feldman C, ... Sprung CL (2014) The Durban world congress ethics round table IV: health care professional end-of-life decision making. J Crit Care. https://doi.org/10.1016/j.jcrc.2014.10.011

Juma S, Taabazuing MM, Montero-Odasso M (2016) Clinical frailty scale in an acute medicine unit: a simple tool that predicts length of stay. Can Geriatr J 19(2):34–39. https://doi.org/10.5770/cgj.19.196

Lee HW, Park Y, Jang EJ, Lee YJ (2019) Intensive care unit length of stay is reduced by protocolized family support intervention: a systematic review and meta-analysis. Intensive Care Med. https://doi.org/10.1007/s00134-019-05681-3

Lush D (1993) Advance directives and living wills. J R Coll Physicians Lond 27(3):274–277. Retrieved from https://www.ncbi.nlm.nih.gov/pmc/articles/PMC5396760/pdf/jrcollphyslond90361-0066.pdf

Merino AM, Greiner R, Hartwig K (2017) A randomized controlled trial of a CPR decision support video for patients admitted to the general medicine service. J Hosp Med 12(9):700–704. https://doi.org/10.12788/jhm.2791

Montgomery C, Bagshaw SM (2017) Frailty in the age of VIPs (very old intensive care patients). Intensive Care Med 43(12):1887–1888. https://doi.org/10.1007/s00134-017-4974-y

Noble C, Grealish L, Teodorczuk A, Shanahan B, Hiremagular B, Morris J, Yardley S (2018) How can end of life care excellence be normalized in hospitals? Lessons from a qualitative framework study. BMC Palliat Care 17(1):100. https://doi.org/10.1186/s12904-018-0353-x

Ofori-Asenso R, Chin KL, Mazidi M, Zomer E, Ilomaki J, Zullo AR, ... Liew D (2019) Global incidence of frailty and prefrailty among community-dwelling older adults: a systematic review and meta-analysis. JAMA Netw Open 2(8):e198398. https://doi.org/10.1001/jamanetworkopen.2019.8398

Phua J, Joynt GM, Nishimura M, Deng Y, Myatra SN, Chan YH, ... the Asian Critical Care Clinical Trials Group (2015) Withholding and withdrawal of life-sustaining treatments in intensive care units in Asia. JAMA Intern Med. https://doi.org/10.1001/jamainternmed.2014.7386

Reignier J, Feral-Pierssens AL, Boulain T, Carpentier F, Le Borgne P, Del Nista D, ... French Intensive Care Society (2019) Withholding and withdrawing life-support in adults in emergency care: joint position paper from the French Intensive Care Society and French Society of Emergency Medicine. Ann Intensive Care 9(1):105. https://doi.org/10.1186/s13613-019-0579-7

Sprung CL, Cohen SL, Sjokvist P, Baras M, Bulow HH, Hovilehto S, ... Ethicus Study Group (2003) End-of-life practices in European intensive care units: the Ethicus study. JAMA 290(6):790–797. https://doi.org/10.1001/jama.290.6.790

Wendler D (2017) The theory and practice of surrogate decision-making. Hast Cent Rep 47(1):29–31. https://doi.org/10.1002/hast.671

White DB, Angus DC, Shields AM, Buddadhumaruk P, Pidro C, Paner C, ... PARTNER Investigators (2018) A randomized trial of a family-support intervention in intensive care units. N Engl J Med 378(25):2365–2375. https://doi.org/10.1056/NEJMoa1802637

Wong WT, Joynt GM (2016) End-of-life care. In: Guidet B, Valentin A, Flaatten H (eds) Quality management in intensive care. Cambridge University Press, Cambridge, UK/New York, pp 85–91

End-of-Life Issues

▶ Religion and Spirituality in End-of-Life Care

End-of-Life Trajectory

▶ Death Trajectory

Endometrial Cancer

▶ Tumors: Gynecology

Endophenotypes

▶ Genetic Control of Aging

Endpoints of Cancer Treatments

▶ Prediction of Outcomes Among Cancer Patients

Endurance Training

▶ Aerobic Exercise Training and Healthy Aging

Energy Metabolism

▶ Basal Metabolic Rate
▶ Cerebral Metabolism

Energy Use, Greenhouse Gas Emissions

▶ Gray Consumption

Enforced Retirement Age

▶ Mandatory and Statutory Retirement Ages

Engineering Approach

▶ Strategies for Engineered Negligible Senescence

Engineering Psychology

▶ Human Factors and Ergonomics for Aging

English Longitudinal Study of Ageing

Paola Zaninotto[1] and Andrew Steptoe[1,2]
[1]Department of Epidemiology and Public Health, University College London, London, UK
[2]Department of Behavioral Science and Health, University College London, London, UK

Overview

The English Longitudinal Study of Ageing (ELSA) was established to address complex, policy-relevant questions within a rigorous scientific framework. ELSA tracks the multiple characteristics of participants as they move through middle to older ages, resulting in an invaluable source of high quality data. The broad research areas and questions motivating the design of the study include health trajectories, disability, and healthy life expectancy; the nature and determinants of economic position and resources in old age; the timing of retirement and postretirement labor market activity; the nature of social networks, social support and social participation at older ages; household and family structure and the transfer of resources; and most importantly, the linked dynamic relationships between all these domains. ELSA has been designed to maximize comparability with other international longitudinal studies on aging, such as the Health and Retirement Study (HRS) in the USA (See ▶ "Health and Retirement Study"), and the Survey of Health, Ageing and Retirement in Europe (SHARE) (See ▶ "The Survey of Health, Ageing and Retirement in Europe"). ELSA is supported by UK government departments and the United States National Institute on Aging.

Study Design

The ELSA sample is selected to be representative of people aged 50 and older, living in private households in England. The study sample is periodically refreshed with new participants at younger ages enrolled to ensure that it remains representative of the full age spectrum. Each sample is drawn from households that had previously responded to the Health Survey for England (HSE), an annual cross-sectional survey that is designed to monitor the health of the general population. HSE households who had agreed to be recontacted at some time in the future containing adults 50 years or older provide the basis for the ELSA samples. Within households, two types of individual are eligible to take part in the study: "core" sample members and their cohabiting partners. Core sample members are individuals aged 50 and over living in private residential addresses in England at the time of the interview who took part in the relevant HSE assessment. The cohabiting partners include people who were not assessed in the HSE, individuals who were aged less than 50 years, and others. Cohabiting partners are offered a full interview to make it possible to carry out analyses of a representative sample of couples in which at least one spouse is 50 or older.

Sample

An overview of the sample sizes achieved for the complete study (core members and partners) is given in Fig. 1. The main ELSA sample is formed of several cohorts who joined the study at different waves of data collection.

Data Collection Schedule

The first wave of ELSA took place in 2002/2003, and individuals are reinterviewed every 2 years. On every occasion, the survey comprises a computer-assisted personal interview (CAPI) and a self-completion questionnaire. On alternate waves, a nurse visit is carried out to collect biomarkers and more detailed measures of function. There is provision for collection of a reduced set of information by proxy where a sample member is unable to take part through poor health, or through physical or cognitive disability. Additional one-off assessments have also been made, and are discussed in Ad-Hoc Modules. To date, eight waves of data have been collected and are available for analysis. The ninth wave of the study is being collected in 2018/2019.

Response Rates

The analysis of response rates and attrition from ELSA is complicated because of variations in responses to different elements of the study, deaths, and differences between core and refreshment cohorts. Some participants may fail to respond to one wave but take part in later waves, and there is the additional issue of whether nonresponse to the original HSE survey from which the sample was drawn should be taken into account (Banks et al. 2018). In Table 1, the study response rates at each wave of the study are reported for each cohort. The number of deaths is also reported for each wave. It must be noted

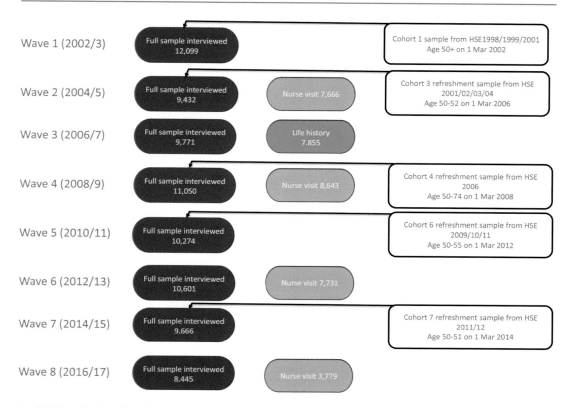

English Longitudinal Study of Ageing, Fig. 1 Wavs of Data Collection in the English Longitudinal Study of Ageing

that information on deaths is periodically updated and numbers might differ from previously published data. To maintain the representativeness of the sample when carrying out analyses, cross-sectional and longitudinal weights are computed and made available with the main data set.

Content

The content of ELSA balances repeat measures of core issues to ensure a sustained time series across waves with new material introduced in each wave. Details of the exact measures can be found at www.elsa-project.ac.uk, but can broadly be divided into the content of the CAPI, self-completion questionnaire, and nurse visit. Most of the questions are asked at every wave, whereas some others are rotated to be asked every 4 or 6 years. Supplementary modules have been included at various points during the history of ELSA to address additional topics.

Core Face-to-Face Interview Topics

The CAPI questionnaire is divided into modules. In the household demographics module, information is collected (or updated) about everyone living in the household, including sex, age, and relationship, and about children living outside the household. In the individual demographics module, details are collected about respondents' legal marital status, parents' age and cause of death, and number of living children, number of grandchildren, great-grandchildren, and siblings, and the respondent's circumstances in childhood. In the health module, information is collected about self-reported general health, long-standing illness or disability, eyesight, hearing, dental health, specific diagnoses and symptoms, pain, difficulties with daily activities, smoking, physical activity, mental health, urinary incontinence, sleep, balance, falls and fractures, and quality of care for cardiovascular disease, depression, diabetes, falls, and osteoarthritis. Information has been

English Longitudinal Study of Ageing, Table 1 ELSA achieved interview sample counts for core members and study response rates

Wave	1 2002–03	2 2004–05	3 2006–07	4 2008–09	5 2010–11	6 2012–13	7 2014–15	8 2016–17
Cohort 1 sample	11,391	8780	7535	6623	6242	5659	4894	4219
Response rates	67%	82%	73%	74%	69%	66%	61%	55%
Cohort 3 refreshment sample	–	–	1275	972	936	888	787	723
Response rates	–	–	61%	63%	75%	72%	65%	60%
Cohort 4 refreshment sample	–	–	–	2291	1912	1796	1606	1470
Response rates	–	–	–	–	85%	82%	75%	70%
Cohort 6 refreshment sample	–	–	–	–	-	826	661	582
Response rates	–	–	–	–	–	55%	82%	72%
Cohort 7 refreshment sample	–	–	–	–	–	–	301	229
Response rates	–	–	–	–	–	–	61%	77%
Number of deaths	–	281	748	1234	2216	2790	3382	3827

The "study response rate" at a given wave is the proportion of the remaining eligible cohort interviewed at that wave
The totals for productive interviews include those in care homes/institutions. Those in care homes/institutions are excluded from the response rate calculations

collected on participation in national screening programs for breast and colorectal cancer. Topics covered in the social care module include the nature of care received, who it was received from, the amount received, and payments made for care. Additional questions about expectations around the funding of social care and questions on short stays in residential/nursing homes were also recently added. The use of public transports is covered in the social participation module.

A substantial portion of the CAPI focuses on economic and financial issues. Information is collected about current work activities, current and past pensions, reasons for job change, health-related job limitations, and working beyond the State Pension Age and state pension deferral. If the respondent is retired and receiving a pension, details are collected about pensions and amount received. As the state pension age is changing in the UK, questions have been added about respondents' knowledge of these issues. The income and assets module assesses the income that respondents received from a variety of sources over the last 12 months. It also collects details of financial and nonfinancial assets held, any income from these assets, regular transfers from nonhousehold members, and one-off payments in the last year. Questions about lifetime receipt of gifts and inheritances have been included in recent waves. In the housing module, information is collected about the current housing situation (including size and quality), housing-related expenses, adaptations to accommodation for those with physical impairments, ownership of durable goods and cars, and consumption.

Cognitive function is assessed in ELSA using a series of objective neurocognitive tests evaluating memory, speed, executive function, and mental flexibility. Numeracy and health literacy are also assessed. The expectations module collects information about expectations for the future in several dimensions, financial decision making, and relative deprivation. The effort and reward module gathers information about care provided to others and volunteering, and the reasons for these activities. In more recent waves, new questions about care provided to grandchildren have been added.

During the face-to-face interview, objective walking speed is also assessed for those aged 60 and over.

An End-of-Life module was introduced at wave 2 to collect information about a respondent who has died since their last interview. A close friend or relative is interviewed and asked the respondent's health, social, and economic circumstances in the last 2 years of their life, as well as their assets after they died. Details about the date, place, and circumstances of death are also collected.

Self-completion Questionnaire

Topics covered in the self-completion questionnaire vary between waves, but include measures of quality of life, social and cultural participation and social networks, loneliness, internet use, altruism, control at work, life satisfaction, subjective well-being, social networks, age discrimination, religion, and health behaviors such as consumption of fruit and vegetables and alcohol. Measures of time use have been added in recent waves. At wave 3, two supplementary self-completion questionnaires containing anchoring vignettes about disability and work were introduced. At waves 6 and 8, a detailed self-completion questionnaire was introduced to cover sexual experience, attitudes, and desire (Lee et al. 2016).

Nurse Visit

At alternate waves, the respondents are offered a nurse visit in their homes. The nurse visit involves measurement of physical functioning (balance and grip strength), anthropometric measurements, information about prescribed medications, and collection of blood samples for extraction of biomarkers and DNA. Detailed information of the content of the nurse visit is reported in Table 2. In wave 8, half of the respondents were offered the nurse interview, with the other half of the nurse interviews being carried out at wave 9.

Ad-Hoc Modules

In wave 3 of ELSA, a life history interview was administered to eligible ELSA sample members. The aim was to collect retrospective information in a number of areas including: children (natural and adopted); fertility; cohabiting and important

English Longitudinal Study of Ageing, Table 2 Information collected at each nurse visit

Wave 2 (2004/5)	Wave 4 (2008/9)	Wave 6 (2012/13)	Wave 8 (2016/17)
Weight, height, waist	Weight, height, waist	Weight, height, waist	Weight (in main interview) Waist (nurse visit)
Grip, balance, chair rise, tandem stand, leg raise, timed walk	Grip, balance, chair rise, tandem stand, leg raise, timed walk	Grip, balance, chair rise, tandem stand, leg raise, timed walk	Grip Timed walk
Blood pressure, lung function	Blood pressure, lung function	Blood pressure, lung function	Blood pressure
Lipids, triglycerides, HbA1c, glucose	Lipids, triglycerides, HbA1c, glucose	Lipids, triglycerides, HbA1c, glucose	Lipids, triglycerides HbA1c, glucose
C-reactive protein, fibrinogen	C-reactive protein, fibrinogen, white cell count	C-reactive protein, fibrinogen, white cell count	C-reactive protein, fibrinogen, white cell count
Hemoglobin, ferritin	Hemoglobin, ferritin	Hemoglobin, ferritin	Hemoglobin, ferritin
DNA	(DNA)	(DNA)	(DNA)
	IGF-1, DHEAS	IGF-1, Vitamin D	Vitamin D
Apolipoprotein E		Cortisol, DHEA, cortisone, testosterone, progesterone (hair)	PAXgene tubes
$N = 7666$ (88%) Bloods = 6231	$N = 8218$ (86%) Bloods = 6438	$N = 7699$ (88%) Bloods = 6180; Hair = 2602	$N = 3525$ Bloods = 2762

noncohabiting relationships; housing and geographical mobility; living situation when they were 10 years old; jobs and earnings; health including injuries, childhood health, smoking and gynecological issues; relationship with parents when a child; and other important and difficult events in their lives.

An experimental module on "risk and time preferences" was administered to a subsample in wave 5 of ELSA. This behavioral economic module was designed to measure attitudes toward financial risk-taking and objective preferences for certainty versus risk and self-control.

In wave 9 of ELSA, a nutrition assessment was administered using the Oxford WebQ (Liu et al. 2011) dietary questionnaire. This is an internet-based self-administered tool specifically designed for use in large-scale prospective studies in the UK. The questionnaire collects information on the types and quantities of foods and beverages consumed in the previous 24 h from each of the 21 food groups; from this information, the nutrient intakes are automatically calculated. Participants are asked to complete this assessment for two separate 24 h periods.

The Harmonized Cognitive Assessment Protocol (HCAP)

The Harmonized Cognitive Assessment Protocol (HCAP) is a substudy carried out within ELSA to investigate dementia risk using a cognitive assessment battery standardized across aging studies. The HCAP has been designed to leverage the HRS international network of studies to produce comparable data on cognition and to provide research diagnoses of dementia and mild cognitive impairment. Interviews were carried out with 1274 men and women aged 60 and over in 2018, recruited using a stratified sampling procedure based on current cognitive ability. Release of these data to the UK Data Service is planned for 2019.

Auxiliary Data

Written consent has been obtained by ELSA respondents to link their data to official records. To date, the ELSA data has been successfully linked to Hospital Episode Statistics (HES) data, mortality and cancer registration records, and financial records including National Insurance contributions and state benefits.

Key Findings

Numerous analyses of ELSA data have been carried out by investigators across the world from several disciplines, and the information is also used by UK Government Departments for policy analyses. More than 300 peer-reviewed articles have been published since 2014, as detailed at www.elsa-project.ac.uk. As more waves of data are collected and available for analyses, the findings produced are increasingly based on complex modeling of longitudinal data to provide better evidence of the health trajectories of older people. Examples include studies on cognitive function and dementia. One study examined age-trajectories of cognitive function over an 8-year period and factors related to them. The results suggested cognitive function declined significantly over time, and the decline was steeper at older ages. Factors related to a steeper decrease in cognitive function included older age, dementia, low education, poor physical functioning, depression, and modifiable risk factors such as alcohol consumption, smoking, and physical inactivity (Zaninotto et al. 2018). Another study estimated age-specific dementia incidence in ELSA from 2002 to 2013, confirming the decline in incidence over recent years observed in other high income countries, while projecting substantial increases in prevalence over the next 20 years (Ahmadi-Abhari et al. 2017). Evidence on the factors related to dementia suggested that lower wealth in late life, but not education, is associated with increased risk of dementia (Cadar et al. 2018) and factors such as loneliness and hearing impairment are precursors of dementia (Rafnsson et al. 2017; Davies et al. 2017).

Flourishing in later life has been the focus of a series of ELSA-based studies. Eudemonic well-being and sustained enjoyment of life have been shown to be associated with increased survival (Steptoe et al. 2015; Zaninotto et al. 2016). Importantly, life skills (such as conscientiousness, emotional stability, determination, control, and optimism) and the sense of life being worthwhile in later life are related to economic success, higher well-being, and favorable health outcomes (Steptoe and Wardle 2017; Steptoe and Fancourt 2019).

In other studies, factors associated with extended working lives have been in focus, showing that good working conditions, such as higher levels of job control and autonomy and strong attachment to the labor market during adulthood encourage extended working (Carr et al. 2016; Wahrendorf et al. 2018). Furthermore, current and lifetime health influences the decisions and the ability to work beyond state pension age (Di Gessa et al. 2017).

Future Plans and Innovations

Plans to continue data collection of ELSA beyond 2019 are currently in place. The study is continuously evolving to cover current and future issues in aging. Innovations are being introduced both in terms of methods, such as mixed-mode data collection, and content. There are numerous policy and research priorities in the aging field, and ELSA endeavors to remain sufficiently flexible to respond to these challenges.

Data Access

The data set is available to all researchers for free upon registration through the UK Data Service at https://www.ukdataservice.ac.uk. The data are deposited as soon as possible after collection.

Cross-References

▶ Aging and Health Disparities
▶ China Health and Retirement Longitudinal Study (CHARLS)
▶ Health and Retirement Study
▶ Health and Retirement Studies Series
▶ Healthy Ageing in Scotland (HAGIS)
▶ Longitudinal Aging Study in India
▶ Mexican Health and Aging Study
▶ The Irish Longitudinal Study on Ageing
▶ The Northern Ireland Cohort for the Longitudinal Study of Ageing (NICOLA)
▶ The Survey of Health, Ageing and Retirement in Europe

References

Ahmadi-Abhari S, Guzman-Castillo M, Bandosz P et al (2017) Temporal trend in dementia incidence since 2002 and projections for prevalence in England and Wales to 2040: modelling study. BMJ 358:j2856

Banks J, Batty GD, Nazroo J, Oksala A, Steptoe A (eds) (2018) The dynamics of aging. Evidence from the English Longitudinal Study of Ageing 2002–16. Institute for Fiscal Studies, London

Cadar D, Lassale C, Davies H et al (2018) Individual and area-based socioeconomic factors associated with dementia incidence in England: evidence from a 12-year follow-up in the English Longitudinal Study of Ageing. JAMA Psychiat 75:723–732

Carr E, Hagger-Johnson G, Head J et al (2016) Working conditions as predictors of retirement intentions and exit from paid employment: a 10-year follow-up of the English Longitudinal Study of Ageing. Eur J Ageing 13:39–48

Davies HR, Cadar D, Herbert A et al (2017) Hearing impairment and incident dementia: findings from the English Longitudinal Study of Ageing. J Am Geriatr Soc 65:2074–2081

Di Gessa G, Corna LM, Platts LG et al (2017) Is being in paid work beyond state pension age beneficial for health? Evidence from England using a life-course approach. J Epidemiol Community Health 71:431–438

Lee DM, Nazroo J, O'Connor DB et al (2016) Sexual health and well-being among older men and women in england: findings from the English Longitudinal Study of Ageing. Arch Sex Behav 45:133–144

Liu B, Young H, Crowe FL et al (2011) Development and evaluation of the Oxford WebQ, a low-cost, web-based method for assessment of previous 24 h dietary intakes in large-scale prospective studies. Public Health Nutr 14:1998–2005

Rafnsson SB, Orrell M, d'Orsi E et al (2017) Loneliness, social integration, and incident dementia over 6 years: prospective findings from the English Longitudinal Study of Ageing. J Gerontol B Psychol Sci Soc Sci. https://doi.org/10.1093/geronb/gbx087

Steptoe A, Fancourt D (2019) Leading a meaningful life at older ages and its relationship with social engagement, prosperity, health, biology, and time use. PNAS 116:1207–1212

Steptoe A, Wardle J (2017) Life skills, wealth, health, and wellbeing in later life. PNAS 114:4354–4359

Steptoe A, Deaton A, Stone AA (2015) Subjective wellbeing, health, and ageing. Lancet 385:640–648

Wahrendorf M, Zaninotto P, Hoven H et al (2018) Late life employment histories and their association with work and family formation during adulthood: a sequence analysis based on ELSA. J Gerontol B Psychol Sci Soc Sci 73:1263–1277

Zaninotto P, Wardle J, Steptoe A (2016) Sustained enjoyment of life and mortality at older ages: analysis of the English Longitudinal Study of Ageing. BMJ 355:i6267

Zaninotto P, Batty GD, Allerhand M et al (2018) Cognitive function trajectories and their determinants in older people: 8 years of follow-up in the English Longitudinal Study of Ageing. J Epidemiol Community Health 72:685–694

Enteral Nutrition (EN)

▶ Artificial Nutrition at Old Age

Entitlement Program

Eswarappa Kasi[1] and Atrayee Saha[2]
[1]Department of Tribal Studies, Indira Gandhi National Tribal University (IGNTU), Amarkantak, Madhya Pradesh, India
[2]Department of Sociology, Muralidhar Girls' College, Kolkata, West Bengal, India

Synonyms

Exemption program; Grant program

Definition

Entitlement programs, according to Yglesias (2015), refer to those programs which are endowed to spend money "automatically" unless there is any regulation to change or reform the programs by the US Congress. In other words, such programs which have already been listed mainly through the Older Americans Act (OAA) 1965 to provide economic and social benefit for better ways of livelihood to the aged persons residing in the United States are considered to be "entitlement programs" and are realized through the state funds to the older citizens of the country. Yglesias (2015) emphasizes that such forms of funding take place in two ways. One refers to the "discretionary" spending. According to this, the US Congress passes laws to spend certain amount of money on certain sets of programs for a certain period, typically 1 year, and then ends this spending when there is a newer program for which a newer fund appropriation process is necessary. The second refers to the entitlement programs, in which Congress establishes an "enduring formula-driven program." Congress accordingly defines the eligibility criteria for these programs; for example, only senior citizens are eligible for social security programs. Accordingly, Congress also forms a formula to calculate the amount which these citizens are eligible for payment under the "entitlement program." Once the procedure is done, the money flows accordingly, until there is a newer law passed by the government to stop this funding. This form of funding, according to Yglesias (2015), is called as "mandatory funding" or "entitlements."

The OAA (1965: 1–2) emphasizes on the provision of equal opportunity to all the older American citizens who need economic and social security conforming to the "traditional concept" of "inherent dignity" of every individual residing in American democratic society. The OAA declares the provision of such security measures as the responsibility of the several states, the political subdivisions, and the Indian tribes to their older people. According to the OAA (1965), such measures will ensure (1) an adequate income in retirement in accordance with the American standard of living; (2) the best possible physical and mental healthcare; (3) making provision for affordable housing based on the unique needs of the older citizens; (4) provision of community services for long-term care, low-cost transportation, improving health and happiness, and prevention against "abuse," "neglect," and "exploitation"; (5) opportunity for employment; and (6) retirement in "health," "honor," and "dignity" and many other programs and services which will enable the American older citizens to live independently and happily in their community.

Overview

In 2002, the United Nations General Assembly (UNGA) endorsed the *Political Declaration and the Madrid International Plan of Action on Ageing* (UNGA 2013). This plan of action pointed out several such measures which should be taken up for the development of resources in order to provide for a suitable environment for the older population. Some of the essential measures included in the plan are promoting health and well-being throughout life; ensuring universal and equal access to healthcare services; providing appropriate services for older persons with HIV or AIDS; training care providers and health professionals; meeting the mental health needs of older persons; providing appropriate services for older persons with disability (addressed in the health priority); providing care and support for caregivers; and preventing neglect and abuse of, and violence against, older people (addressed in the environments priority) (WHO 2015: 5).

The OAA (1965: 3–10) has notified some of the following services which constitute the "entitlement program" in the United States for the American older citizens. In brief these include (1) services dealing with "comprehensive assessment" or assessment on a regular basis of the "psychological, physical, and social needs of the individuals"; (2) periodic reassessment of the needs of the older individual or the family and also make provisions for family caregiver; (3) developing collective action for addressing "public concern or an unmet human, educational, healthcare, environmental, or public safety need"; (4) engaging in "disease prevention and health-promotion services," such as regular health checkup for glaucoma, hypertension, cancer, diabetes, vision, and hearing; (5) promoting "evidence-based health-promotion programs and programs regarding physical fitness, music therapy, improved nutrition, substance abuse reduction, and other health-related malpractices"; (6) making provisions for education on different health-related problems; (7) prevention of the older persons from any form of elder abuse, neglect, and exploitation; (8) services for prevention of family violence and saving the older persons from being affected as a consequence of such violence; (9) services of "homemakers," "home health aides," "visiting and telephone reassurance," "core maintenance," "personal care services," and similar such services; (10) "integrated long-term care" provided under the Medicaid program established under title XIX of Social Security Act, including nursing facilities, home- and community-based services, personal care services, and case management services; (11) providing legal assistance in case of seeking help for compensation and others and also for gaining knowledge on the various legal provisions of the healthcare; and many others.

The above list provides only a few of the services notified by the OAA as important services to be provided under the scheme of the entitlement programs in the United States. However, Medicaid and Medicare programs introduced under the social security act are considered to be the most well-known and commonly used entitlement programs. Few other programs include payments for defense services and also reduction in taxes and pension facilities to the older Americans. The OAA also declares that such services are to be provided to the American citizens, which include the American Indians (AI), Alaskan Natives (AN), and several tribal organizations. The OAA has also declared the formation of the "Aging Networks" in the form of the State Units on Aging (SUAs), Area Agencies on Aging (AAAs), and various regional organizations which look after the effective implementation of the schemes and regular assessment of these as well.

The WHO report on aging and health, *World Report on Ageing and Health* (2015), emphasized that, given the current social and economic condition along with the availability of the medical technologies, most people in the history can expect to live in their 60s and beyond. The demographic transition toward an increasing number of the older population throughout the world is both a positive sign and also a matter of concern for the different states and the governmental and non-governmental agencies. This changing demographic condition imposes on the state that the government should consider the several ways in which it can help to promote "healthy aging." That the older population can give back to the society and that this function of the old population can only be ensured through various normative measures which will take into account the provision of medical-care facilities, caregiving arrangements, training to the health practitioners and the caregivers, and improved financial and health schemes for the old population are some of the essential components which are to be looked at to provide entitlement to the older population worldwide.

Along with the improvement in the basic social and economic needs of the older population, a key role is to be played by the government to ensure the equal contribution of the older person in the growing economy; while it is difficult to expect that the older population aged above the age group of 80 or 90 years to work productively, it is also not exceedingly difficult to think that the population in the age group of 60–70 years can work equally well and contribute to the growth of the economy. For instance, the World Health Organization (WHO) notes that in the United States during 2011–2012, 23% of the entrepreneurs were aged between 55 and 64 and twice so many successful entrepreneurs belonged to the age group above 50 years of age than those who are under the age of 25 (WHO 2015: 10). However, the story is not uniform for all countries across the world. The low-income countries and also countries affected by poor aging and retirement policies and lack of infrastructure to deal with the problem of increasing older population do not put forth a colorful picture of the social, political, and economic environment of the older population.

The international legal and policy frameworks which encode the international human rights law emphasize on the provision of "universal freedoms" and "entitlements of individuals" which include the "civil" and "political rights" entailing the "right to life," "social," "economic," and "cultural rights" including the "right to health, social security, and housing" to all those who are to be protected by the law. The WHO emphasizes that these rights cannot be sacrificed for some sections of the population because of their age. In other words, the WHO points out the necessity of social security to the aged and their protection through laws and rights as a process of "entitlement" to the aged population (WHO 2015: 5). For instance, the United States Department of Agriculture (USDA) has committed to the provision of healthy food to the underserved and economically dependent Americans who cannot access the market to buy food and also cannot afford for the same. The Supplemental Nutrition Assistance Program (SANP) of the USDA serves four million seniors. Almost 42% of the "eligible elderly" population are covered under this program (USDA 2015).

Key Research Findings

In common parlance, "entitlement" refers to the different rights of people which are ensured by the various normative measures in different countries. In terms of entitlement program for the older population or to address the problem of aging, this entails in the different forms of financial and medical aid given to the older population for a better social and environmental condition for the older population. As mentioned before, this form of entitlement program can also be in the form of provision of better food and housing facility to the older population.

According to Gist (2007), "entitlements" refer to the category of benefits defined in the Congressional Budget Reform and Impoundment Control Act of 1974 which are conferred on any person or groups of persons who may belong to such criteria as to be considered under this regulation. In the United States, Gist (2007: 14–15) pointed out that there are two types of entitlements provided by the government. The author categorized them to as "means-tested" and "non-means-tested" entitlements. The "non-means-tested" entitlements refer to those entitlements or social insurance programs which provide insurance based on the "work histories" and not on the need of the person or any group of persons, including the social security, Medicare, unemployment insurance, and civilian and military retirement programs. On the other hand, the "means-tested" entitlements are those insurance programs which have "asset test" or a "limiting" eligibility for the benefits. Programs in this category include Supplemental Security Income (SSI), food stamps, and Medicaid (Gist 2007: 15; Yglesias 2015). The "non-means-tested" entitlement programs have greater extension, almost four to five times more than the "means-tested" entitlement programs in the United States.

Yglesias (2015) pointed out that the entitlement programs in the United States which are focused on providing financial benefits in the form of unemployment insurance, retirement benefits for the civilians and the military personnel, tax credits, veteran benefits, and farm subsidies use a sizable portion of the exchequer of the government. Out of these expanded funding of $2.2 trillion in 2014 on entitlements, $1.5 trillion was spent on social security and Medicare, and about half of the rest was spent on the healthcare provisions in the country. This emphasizes that the United States has made an extensive effort to take up the issue of the entitlement needed in terms of healthcare for the majority of its population. In 1970 the entitlement program spent about 5% of the national income of the United States. Currently, the expenditure has risen to 15% of the total national income and is projected to steadily rise in the future as well. To a considerable extent as Yglesias (2015) pointed out, this form of spending has increased recently with an increase in the number of the older population and the need for the provision of health and social benefits to this population in the United States.

In 2017, the population of 60 years and above was estimated to be 962 million constituting 16% of the global population. This aged population is growing at a rate of 3% every year. Currently, Europe has an estimated 25% of its population in the age group of 60 or above. The number of older persons is projected to be 1.4 billion in 2030 and 2.1 billion in 2050 and is estimated to rise to 3.1 billion in 2100. The number of persons aged 80 and above is projected to triple in 2050, and by 2100, it is estimated to rise to 909 million, almost 7 times the population in 2017 (UN 2017). Bloom et al. (2011) emphasized on the importance of the implementation of policies which would help in providing social protection to the older persons through an analysis of the retirement policies, pension, and health schemes in countries such as Africa, India, Latin America, Caribbean cities, and other countries with an increasing population of the older persons.

Bloom et al. (2011: 3) point out that the problem faced by older persons is not only due to poverty. The older persons are "vulnerable" because of the lack of "companionship" and "physical care" and "assistance" from the missing younger population. In traditional societies such as those of Japan, he points out that the proportion of the older Japanese people living with their children is as low as 42% which is much lower than the 87% mark in the 1960s. Similarly, in Central American countries, only 10–23% of older people lives on their own compared to more than 50% in Argentina and Uruguay. Eighty-three percent of the labor force in the Organisation for Economic Co-operation and Development (OECD) countries is covered under the mandatory pension schemes. The coverage is only 21% in China and less than 10% in India. The pension system, according to Bloom et al. (2011: 6–7), functions in two ways across the world. In one system, the benefits to the retired individuals are financed from the contributions made by the current employers and workers through their savings and savings accumulated from past contributions as well.

A typical example of this is the social security scheme of the United States. The other is a "fully funded" system, which functions based on the "defined contributions" made by the workers to individual accounts. For lower-income countries, both the systems are difficult to implement. Some examples of implementation of social security schemes for the older adults taken up in various parts of the world and their impact on the separate economies are discussed in the following section.

Examples of Application

Latin American countries have several pension systems. In countries such as Brazil, publicly operated plans have covered 52% of the employed individuals and 14% in Paraguay. In India, there are both "defined-benefit" and "defined-contribution" pension systems. However, according to Bloom et al. (2011), the reach of the pension system is limited to the workers in the formal sector. The National Pension Scheme (NPS) was introduced in India (i) to provide old-age income, (ii) to ensure reasonable market-based returns over the long term, and (iii) to extend old-age security coverage to all citizens (PFRDA 2014, 2019). Under this scheme, there are two accounts: one is a pension account, also called as the Tier I account, which is mandatory for the government employees, and the second is the savings account, also called as the Tier II account, which is a withdrawal account, providing the facility of withdrawing money according to the requirement of the individual employers. Another important aspect of this system is that the amount saved in this account is invested by the employer through his or her choice. However, if there is no choice opted by the employer, then the investment is made through an "Auto Choice" option.

Health insurance also plays a crucial role in providing social security to the older population. Medications for older persons have been made affordable in countries such as Australia and the United States. In countries such as Germany, Japan, and South Korea, long-term care insurance is available universally. In Scandinavian countries, there are tax-funded strategies to help in the form of social protection to older persons (Bloom et al. 2011: 8). In European countries, public pensions are the source of 60% of the income of older persons. The OECD countries including Austria, Belgium, New Zealand, Japan, and many others account that 59% of the household income of persons aged above 65 years come from the public pension transfers, 24% comes from employment and self-employment, and 17% comes from the private pension schemes. A similar amount of spending is also seen for such entitlements in countries such as the Republic of Korea, Mexico, and Chile (ILO 2014: 2). The report has also emphasized that the share of public transfers in older persons' income is maximum in the Republic of Korea followed by Mexico, Chile, Australia, the United States, Japan, Turkey, Canada, the United Kingdom, Sweden, Denmark, Norway, Iceland, the Netherlands, and New Zealand. In countries such as Austria, France, Belgium, Czech Republic, and many others, the share of public transfers to older persons' income is much less (ILO 2014: 3–4).

China introduced pension schemes in 2009 and 2011, which aimed to cover both rural and the urban aged population. It introduced two pensions systems: social pension scheme, which was to be funded by the government, and individual savings account, which was to be paid by the insured persons, collective entities (if any), and the government. By 2013, China could bring nearly 75% of the population aged 15 and above under the pension scheme (ILO 2014: 14). Such means of social protection in different countries which provide care beyond the family-based care is possible only through government financing. In lower-income countries, often, these availabilities function in the form of various impediments to the implementation of the schemes and the provision of social protection to the older persons.

The problem faced by the older persons in sub-Saharan Africa (SSA) is much more complicated than these countries where there is lack of infrastructure to provide facilities for social protection to the older population and are also vulnerable due to health conditions, retirement in the rural areas, 67% illiteracy in Africa, gender-based discrimination against the women of the African society, migration, HIV/AIDS, and so on and so forth (Parmar et al. 2014). The HelpAge India report on social security to the older persons of SSA points out that there is a necessity to provide security through "social" pension. This is necessary for those residing in the countries of South Africa, Botswana, and Tanzania, where 50% and 60% of children orphaned by epidemics live with their grandparents (HelpAge International 2008).

Parmar et al. (2014) emphasized on the importance of the Ghanaian National Health Insurance Scheme (NHIS) and Senegal's Plan Sesame in playing a key role for providing social security to the older population in SSA. The NHIS, launched in 2003, provides national health financing system with "decentralized operations." Every district is equipped with insurance fund which is financed by the central-level funds and premiums. The premiums are collected from the formal sector employees and cover almost 95% of the disease burden in Ghana. These aided medical services include outpatient, inpatient, and emergency care, deliveries, dental care, essential drugs, and others. Children under age 18, pregnant women, indigents (including the poor and destitute), and all people over 70 years old are exempted from paying the premiums. In 2011, 8.2 million people (33% of the population) had registered and renewed their registration in the scheme, and 4.9% of the active members were over 70s (Parmar et al. 2014: 36–37).

Senegal's Plan Sesame provides free access to healthcare services to all citizens above the age of 60, which is an estimated 5.9% of the total population. Costs of consultations, diagnostics, essential drugs, and hospitalization are covered under the scheme. However, the analysis done by Parmar et al. (2014) emphasized that the availability of these entitlement programs does not ensure equal access to the older population in SSA. According to the critical analysis of the schemes, the older people belonging to richer households are more likely to enroll for both the NHIS and the Plan Sesame. Parmar et al. (2014) in their study on the survey of households which have had benefitted from the two schemes have shown that there are "persistent inequities" in enrollment for older persons in NHIS and Plan Sesame due to a combination of political, economic, and sociocultural dimensions. The study has shown that older persons from rented houses and with no income are least likely to enroll in NHIS and Plan Sesame, respectively. Similarly, older persons with greater participation in political activities and political networks are more likely to enroll in such plans than those who are devoid of political influences. Parmar et al. (2014) also emphasize the influence of sociocultural networks on older persons to enroll in such schemes.

Schemes such as the pension schemes in India; NHIS and Plan Sesame in Ghana; "social" pension in sub-Saharan African regions; health insurance policy schemes in the United States, Australia, India, and other developing economies; social security, Medicare, unemployment insurance, and civilian and military retirement programs; Medicaid in the United States; and the establishment of Aging Network agencies, Area Agencies on Aging, and the State Units on Aging in the United States are some of the important social security and support service schemes which have been implemented for safe and secure livelihood of the older citizens residing in different parts of the world. In the research based on the evaluation of the various schemes which have been functioning in various parts of the world, it is seen that emphasis has been made on the provision of social security as an important requirement for the well-being of the older citizens. Not all the schemes and plans are remarkably successful. However, these schemes and plans for social security are paving a way forward for a secure environment for older residents to leave peacefully and independently in their communities. For example, the Indian government has emphasized on sensitizing the youth for increasing care and responsibility toward the older parents and family support to the older citizens living in their communities (Berkman et al. 2012). For example, in India the Ministry of Health and Government of New Zealand has made provisions for support services for older people, care plus schemes for the older persons who need extreme help and rest home audit facilities provided in the form of services to the older citizens (MOH 2018a, b, c, d). The support services are provided by the district health boards (DHBs) which provide personal care, support to the career or the person who stays with the older persons, household services such as cleaning and cooking, and also equipment for safety maintenance at home. Care plus is used for services to the chronically ill older persons. In this scheme, the older citizen must be enrolled with one of the public health organizations (PHOs). The older persons enrolled in this scheme receive regular checkup and assessment of their health condition. The rest home audit certification helped in assessing the quality of the rest homes to which an older citizen wishes to move to. The government brings out a list of the rest homes which are suitable for the residence of the older persons in New Zealand.

Future Directions of Research

In order to make provision for such social security measures for the older persons, the government has to pay off a sizable portion of its finances for such entitlements. In the United States, the three entitlements for older persons, Social Security, Medicare, and Medicaid, use up almost 75% of the total outlay for entitlements. The expenditure was 8.7% of the American GDP in 2007 and is estimated to rise to about 19% in 2050 (Gist 2007). According to Gist (2007), Social Security has helped in bringing down the poverty rate among older persons in the United States and has also helped in making them independent.

On the one hand, there is an increasing concern on the availability of healthcare measure for older persons. On the other hand, there is an increasing burden faced by the government in terms of finances to support such social security measures which help the older persons to stay healthy. India has taken a step forward in this regard by increasing the age of retirement from 60 years to 65 years, which ideally increases the workforce available for a longer period and also ensures the economic gain in terms of financial help if the older persons remain at work for a more extended period.

Labor force participation among the older men was highest in Africa (52.2%) followed by Latin America and the Caribbean (38.1%), Asia (34.8%), Northern America (23.5%), and Europe (10.2%) (UN 2015). The implementation of the various schemes is a crucial step ahead for equitable development and growth of the older population. This further imposes to look into the question as Gist (2007) emphasizes on the introduction of an affordable healthcare system, extending coverage to those without insurance, adjusting benefits, and improving the work provisions. Other key areas of research include the policies and plans for improving the chances of equity to suit the domestic and the international demands toward the aging problem. Increasing tax revenues, longer work lives, and increased national and personal savings are some of the important sources of social protection in the form of "entitlements" which can help to provide a healthy life for the older persons across the world.

Summary

Choudhary (2019) emphasized that with the increasing older population, this is a challenge to the government of every country to make provision for healthy aging of older persons. The only way to cater to this problem is to look for provision of different "entitlements" or "entitlement programs" which will provide social protection, economic independence, political rights, legal awareness, and other opportunities in the form of food aid and housing aid, through pension schemes or increasing the age of retirement, and others.

While some countries such as the United States have worked toward ensuring such "entitlements" through public transfers to cope up with the problem of increasing population in the age group of 60 years and above, there are many other countries where public transfers are not very high to take the burden of providing services to older persons. On the one hand, there is the issue of equitable distribution of the entitlements or the facilities to the older citizens. On the other hand, there is an increasing pressure on the governments to bear the cost of such "entitlements." Thus, in the issue of aging and entitlements, though have been taken up in most of the countries, there are many such countries which lag behind, and there are many other countries which have the provision for entitlements but may not have the provision for equitable entitlements to the aged cohort from different socioeconomic, political, and ethnic backgrounds.

Cross-References

▶ Fiscal Welfare
▶ Medicare for People with Disabilities
▶ Pension Systems

▶ Program of All-Inclusive Care for the Elderly (PACE)
▶ Social Security Around the World
▶ Welfare States

References

Berkman L, Sekhar TV, Capistrant B, Zheng Y (2012) Social networks, family, and care giving among older adults in India. In: Smith JP, Majumdar M (eds) Aging in Asia: findings from new and emerging data initiatives. The National Academies Press, Washington, DC. https://www.ncbi.nlm.nih.gov/books/NBK109207/. Accessed 09 June 2019

Bloom DE, Jiminez E, Rosenberg L (2011) Social protection of older people. PGDA working paper no. 83: Program on the Global Demography of Aging. https://cdn1.sph.harvard.edu/wpcontent/uploads/sites/1288/2013/10/PGDA_WP_83.pdf/. Accessed 10 Mar 2019

Choudhary I (2019) Do we care enough for the elderly? A survey of old age homes in Delhi. Econ Polit Wkly 54(11):14–16. https://www.epw.in/journal/2019/11/commentary/do-we-care-enough-elderly.html. Accessed 10 Mar 2019

Gist, RJ (2007) Population Aging, Entitlement Growth, and the Economy. AARP Public Policy Institute. http://www.aarp.org/ppi

HelpAge International (2008) Protecting the rights of older people in Africa. https://www.helpage.org/silo/files/protecting-the-rights-of-older-people-in-africa.pdf/. Accessed 12 Mar 2019

ILO (International Labour Office) (2014) Social protection for older persons: key policy trends and statistics. International Labour Office, Geneva

MOH (Ministry of Health), New Zealand (2018a) Services for older people. https://www.health.govt.nz/your-health/services-and-support/health-care-services/services-older-people. Accessed 24 June 2019

MOH (Ministry of Health), New Zealand (2018b) Support services for older people. https://www.health.govt.nz/your-health/services-and-support/health-care-services/services-older-people/support-services-older-people. Accessed 24 June 2019

MOH (Ministry of Health), New Zealand (2018c) Care Plus. https://www.health.govt.nz/your-health/services-and-support/health-care-services/care-plus. Accessed 24 June 2019

MOH (Ministry of Health), New Zealand (2018d) Rest home certification and audits. https://www.health.govt.nz/your-health/services-and-support/health-care-services/services-older-people/rest-home-certification-and-audits. Accessed 24 June 2019

OAA (1965/2018) Older Americans Act of 1965. As Amended Through P.L. 114–144, Enacted April 19, 2016. https://legcounsel.house.gov/Comps/Older%20Americans%20Act%20Of%201965.pdf. Accessed 08 June 2019

Parmar, D, G. W. F. Dkhimi, A. Ndiaye, F.A. Asante, D.K. Arhinful and P. Mladovski (2014) 'Enrolment of older people in social health protection programs in West Africa – Does social exclusion play a part?', Social Science & Medicine, 119: 36–44. https://doi.org/10.1016/j.socscimed.2014.08.011

PFRDA (Pension Fund Regulatory and Development Authority) (2014) Understanding National Pension Scheme (NPS). https://www.pfrda.org.in/index1.cshtml?lsid=86. Accessed 09 June 2019

PFRDA (Pension Fund Regulatory and Development Authority) (2019) NPS for Non-Resident Indians (NRIs). https://www.pfrda.org.in/writereaddata/links/nps%20for%20nris%20-presentation17fc71e-ccda-40a8-abf9-982a78fbfbe3.pdf. Accessed 08 June 2019

UN (United Nations) (2015) World population ageing. http://www.un.org/en/development/desa/population/publications/pdf/ageing/WPA2015_Report.pdf/. Accessed 12 Mar 2019

UN (United Nations) (2017) World population prospects: key findings and advance tables. United Nations Department of Economic and Social Affairs, New York. https://esa.un.org/unpd/wpp/Publications/Files/WPP2017_KeyFindings.pdf/. Accessed 08 June 2019

UNGA (United Nations General Assembly) (2013) Resolution Adopted by the General Assembly on 20 December 2012. https://undocs.org/A/RES/67/139/. Accessed 12 Mar 2019

USDA (United States Department of Agriculture) (2015) Fact Sheet: USDA Support for Older Americans, Press release no: 0202.15 https://www.usda.gov/media/press-releases/2015/07/13/fact-sheet-usda-support-older-americans. Accessed 13 Mar 2019

WHO (World Health Organization) (2015) World report on ageing and health. World Health Organization, Geneva

Yglesias M (2015) Entitlement reform. VOX. https://www.vox.com/2014/5/12/18076886/entitlement-reform. Accessed 12 Mar 2019

Entrepreneurship in Aging Population

▶ Social Entrepreneurship and Social Innovation in Aging

Environmental Context of Aging

▶ Environmental Gerontology

Environmental Gerontology

Jasmon W. T. Hoh[1], Siyao Lu[1], Yin Yin[2], Qiushi Feng[3], Matthew E. Dupre[4] and Danan Gu[5]

[1]Department of Sociology, National University of Singapore, Singapore, Singapore
[2]Department of Sociology, Central University of Finance and Economics, Beijing, China
[3]Department of Sociology, Centre for Family and Population Research, National University of Singapore, Singapore, Singapore
[4]Department of Population Health Sciences, Department of Sociology, Duke University, Durham, NC, USA
[5]Population Division, Department of Economics and Social Affairs, United Nations, New York, NY, USA

Synonyms

Aging and environments; Aging in context; Climate gerontology; Ecogerontology; Ecology of aging; Environmental context of aging; Environmental impacts on aging and longevity; Environments and longevity

Definition

Environmental gerontology is a branch of gerontology that studies the relationships and interactions between older adults and their socio-physical environments, often aiming to modify and optimize the person-environment relationship to improve quality of life and healthy longevity (Wahl and Weisman 2003). In scope, it can cover many subdomains of gerontology, including physiology, geropsychology (or psychogeorontology), behavioral gerontology, biogernontology, geroscienes, sociogerontology, ecogerontololgy, geogerontology, financial gerontology, and geriatrics.

Overview

Origin

The pursuit of knowledge about the interactions between humans and nature can be traced back to ancient times. In traditional Chinese culture, people stood in awe of nature, believing that every aspect of daily living was significantly associated with the natural world. Accordingly, health and disease were deemed to be strongly related to natural elements (Li 2011). Ancient Chinese philosophers, in particular *Laozi* (sixth century – fourth century BC; dates of birth/death unknown) and *Confucius* (551–479 BC), established a mature view of ecological ethics (Li et al. 2016). *Confucius* advocated for the "harmony of man and nature," insisting that human beings are part of nature. *Laozi* claimed "Tao follows nature," and stressed the superiority of nature in which human activities should follow and respect (Jiang and Xu 2010). Deeply influenced by these ideas, through history, Chinese have always been mindful of the protection of the natural environment, and its sustainable development, though not in the modern sense (Li 2011).

The environmental ethics of Europeans has been largely influenced by ancient Greeks, Hebrews, and Judeo-Christians (e.g. Chemhuru 2017; Kay 1989; Silecchia 2004). Different notions of nature – theological views, humanistic views, and scientific views – were often coexistent throughout European history (Chemhuru 2017; Hoffmann 2004; Hou 1997; Zhilin 1989). Ancient Greeks believed that nature had its own "mind," and developed a scientific view of nature, which regarded nature as an independent system with its own inherent logic (Chemhuru 2017). The theological view of nature came from the Bible: "And God said: Let us make man in our image, after our likeness: and let them have dominion over the fish of the sea, and over the fowl of the air, and over the cattle, and over all the earth, and over every creeping thing that creepeth upon the earth" (Genesis 1:26). It claimed that nature was created to meet the needs of human beings. A humanistic view of nature emerged in the Renaissance period. It centered on human beings, regarding the human experience as the starting point for the understanding of himself, of God, and of nature (Hoffmann 2004; Hou 1997; Zhilin 1989). All of these views of nature were similar in their disposition of describing humans and the environment as counterbalancing entities, with an essence of anthropocentrism. During the Industrial Revolution, in particular, the centrality of "humans" in relation to nature became especially strengthened (Roka 2020).

With the expansion of industrialization worldwide, the relationship between humans and nature was increasingly described by separation and opposition. With the increasing demand for natural resources, human beings started to exploit nature in an unprecedented scale, and various environmental issues soon arose. Most notably, forest resources were severely damaged and there were serious issues of air pollution in many industrialized cities (Masters and Ela 2014). This worldwide environmental crisis in the late nineteenth and early twentieth century challenged people (and societies) to reflect on the relationship between humans and nature. Accordingly, anthropocentrism became a target of criticism; and the environmental ethics against anthropocentrism has developed since then (Roka 2020).

Modern Environmental Gerontology

Modern scientific studies on environmental ethics were not present until the early twentieth century. As an (in)direct consequence of the Industrial Revolution, the population explosion and environmental crises that occurred since the early twentieth century called for a "basic change of values" in our connection with the environment (Bai 2018). In 1923, Albert Schweitzer first coined the concept of "environmental ethics" and later Aldo Leopold adopted this idea and proposed that human beings should be treated as equals with nature (Keulartz and Korthals 2019). These views had a significant and far-reaching impact on the development of environmental ethics in the West. A new field emerged – environmental ethics – and quickly developed as a science in three countries: The United States, Australia, and Norway (Brennan and Lo 2015). In the 1970s, contemporary environmental ethics became widely recognized as an academic discipline.

Modern environmental gerontology first originated from studies on the relationship between social factors, behavioral factors, and health. In 1951, Kurt Lewin proposed the seminal field theory, stating that behavior is a function of the person in their environment (Lewin 1951). In the 1970s and 1980s, research further underscored the importance of the physical and social environment in understanding the aging population and improving the quality of life in older adults. A series of related theoretical frameworks were subsequently proposed in gerontology, including the age-loss continuum theory (Pastalan 1970), environmental congruence theory (Kahana 1982), and the hypothesis of changing emphasis (Rowles 1978; Rowles and Bernard 2013). Perhaps the most impactful was Lawton and Nahemow's general ecological model of aging. This model emphasizes how human behaviors are affected by interactions at personal, interpersonal, and environmental levels (Lawton and Nahemow 1973). Contemporaneously, with the development of geographical gerontology, gerontologists began highlighting the role of spatial experiences and attachment to place for understanding issues in later life (Golant 1989; Warnes 1999).

During the 1990s and the first decade of the twenty-first century, environmental gerontology developed into a transdisciplinary field. Researchers from diverse disciplines – such as biomedicine, psychology, and architecture – launched a new generation of scientific inquiry. Likewise, social and behavioral scientists in this field have also benefited from interdisciplinary collaborations with engineers, computer scientists, ergonomists, designers, and other experts (Rowles and Bernard 2013). In recent decades, sociogerontology, psychogerontology, biogerontology, and ecogerontology have developed as related fields to the discipline.

Major Related Fields

Sociogerontology

Sociogerontology (See ▶ "Social Gerontology") examines the social, economic, and demographic aspects of older adults, and how social structures affect them (Philips et al. 2013). As human aging is inextricably embedded within the context of social environments, Sociogerontology provides a basis to understand how social forces such as healthcare, retirement, and social support can impact the aging process. For example, long-term care systems for older adults have evolved to complement the traditional healthcare delivery system and assist older adults to receive adequate care outside of hospitals, either within their own homes or their communities. With increases in average life expectancies, retirement-planning (i.e., age at retirement) is a related issue that will have significant implications on the maintenance of pension systems and national healthcare insurance systems (OECD 2011). Likewise, social support networks play an important role in ensuring the well-being of older adults. Informal social support networks, such as those given by family, friends, or members of the community, provide valuable emotional support and companionship to older adults. Formal support includes services provided by paid professionals or formal organizations, such as nursing facilities and assistance with daily living. Taken together, these social factors form an integral part of the social-physical environment of older adults and are a core theme of environmental gerontology.

Geropsychology/Psychogerontology

Geropsychology (See ▶ "Geropsychology") is a subfield within psychology that applies psychological methods and research to the study of older adults and aging (American Psychological Association 2016). It focuses on a broad range of issues, such as mental health, personality, well-being, and the cognitive functioning of older adults – perhaps the most researched topic in this area. Longitudinal studies show that cognitive function remains relatively stable before the age of 70, and individual differences can be attributed in part to an accumulation of many socio-structural factors (Schaie 2013). Personality development has also been studied in this field from a life course perspective (Specht et al. 2011). Environmental gerontology is an important part of Geropsychologic research as older adults often spend a large portion of their older years within limited social and/or physical environments – such as in the home or assisted living facilities (Oswald and Wahl 2004). Research in this area includes theories such as place identity, control beliefs, and place attachment, which are widely applicable to both Geropsychology and Environmental Gerontology (Oswald et al. 2007).

Biogerontology/Geriatrics

Biogerontology (See ▶ "Biogerontology") refers to the study of aging and age-related diseases from a biological point of view (Rattan 2018). Through research on the biological processes contributing to aging, biogerontology aims to better understand the process of aging and delay the onset of aging-related diseases, thus increasing the years lived in good health for older adults (Farrelly 2012). With technological advances, the field has made tremendous inroads in understanding the potential pathways of aging that may be possible to manipulate (Rattan 2018). Such breakthroughs in our understanding of these mechanisms could facilitate the development of much-needed interventions. Addressing the biological mechanisms of Alzheimer's disease, for example, may enable researchers to target specific genes and/or introduce new treatments to slow (or prevent) disease pathology (Alvarez-Erviti et al. 2011; Fenech 2017; Nilsson et al. 2010). With advances in biotechnology, some areas of biogerontological research focus on the interactions between genes and the environment (GxE) and genome-environment-wide interactions. Genome-environment-wide interaction studies (GEWIS) are now a central field of biogerontology that considers the joint effects of genetic and environmental risk factors in understanding the mechanisms of disease and longevity (Assary et al. 2018; Simonds et al. 2016). As such, the rapid development of GEWIS in the literature has broadened its scope and become directly relevant to the study of environmental gerontology.

In the field of geriatrics, with regard to environmental gerontology, studies have shown that enriched environments can have benefits for cognitive function and protect against the degenerative effects of aging (Hertzog et al. 2009; Leal-Galicia et al. 2008; Mora et al. 2007). Such studies have also highlighted the critical role of environments in Alzheimer's disease (Jankowsky et al. 2005; Lores-Arnaiz et al. 2006). However, research that integrates data and perspectives from Biogerontology and Environmental gerontology remains relatively underdeveloped and warrants more attention in the literature (Wahl et al. 2012).

Ecogerontology/Climate Gerontology

Climate gerontology (see ▶ "Climate Gerontology") is an interdisciplinary field that examines the challenges that older adults encounter in the face of climate change and extreme weather events (Haq and Gutman 2014). A recent report by the Intergovernmental Panel on Climate Change (IPCC) has substantiated the harmful impacts of global warming on people around the world, including the melting of ice caps, rising sea levels, higher frequency of extreme weather events such as heatwaves, floods, droughts, and wildfires (IPCC 2018). Older adults are disproportionately at risk of suffering harm from these events due to a higher likelihood of pre-existing medical conditions (McCann and Ames 2011), vulnerability to abuse or neglect (Gutman and Yon 2014), lower mobility (Fernandez et al. 2002), and higher susceptibility to diseases (Wu et al. 2015), and consequently face risks of mortality during such natural disasters (see ▶ "Climate Change, Vulnerability, and Older People", ▶ "Older Adults in Conflicts and Crises"). Social and economic factors may further exacerbate the disadvantages of older adults and reduce their ability to respond to stresses stemming from climate-related events (See ▶ "Climate Change, Vulnerability, and Older People"). Studies in this field have sought to identify factors that can aid older adults in building resilience to climate change and extreme weather events. Several factors (See ▶ "Climate Resilience and Older People") such as quality neighborhood amenities, being in good health and economic standing, and having adequate social networks have been identified as particularly important in improving the coping capacity of older adults facing environmental crises. With the increasing incidence of climate-related events, efforts to modify and/or adapt to the physical-social environment may be highly relevant in building resilience and reducing vulnerability among older adults (Sánchez-González and Chávez-Alvarado 2016).

Geographic Gerontology

Along with advances in human geography and social gerontology, geographic gerontology has emerged as a transdisciplinary field studying the relationships between older adults and the spaces (and places) in which and through which age and aging occur (see ▶ "Aging in Place", ▶ "Aging in the Right Place", ▶ "Geographical Gerontology"). Geographic gerontology focuses on the spatial perspectives of population aging, the distribution and mobility of older adults, and the relationships with their living environments and services at different geographic scales. It also examines why and how space and place matter for older adults in their aging processes. For example, it would be ineffective for service planners/providers to implement social services in an area for older adults without knowing the spatial distribution/location of this population — in particular those who need such services. To develop effective services, service planners/providers not only need to know who requires services, who needs what services, and what changes there will be in the next few years, but they also need to know where they live, where they go, and whether and where they need to set up more service sites, etc. Furthermore, research has shown that factors such as climate, weather, trace elements in the soil, and blue zones (areas with high concentrations of long-lived adults) are associated with aging and longevity (see ▶ "Longevity Areas and Mass Longevity"). All of these factors underscore how geographic environments can play a crucial role in older adult's health and well-being with advancing age.

Cultural Gerontology

Cultural gerontology focuses on how social norms, values, morals, and practices influence older adults and their aging processes. Such cultural elements, and related environmental/contextual factors, shape the social roles and images of older people in society, and can influence stereotypes of old age (Edmondson 2013; see ▶ "History of Ageism"). Due to rapid epistemological and historico-social transformations over the last two decades in many countries, culture has gained "an unprecedented role in the constitution of social identities and realities" (Twigg and Martin 2015a, 353). For example, ageism, an important concept in cultural gerontology, refers to people's negative views and attitudes toward people because of their age – including negative stereotypes, prejudice, and discrimination (Twigg and Martin 2015b; see ▶ "History of Ageism"). Research has shown that older adults who experience ageism tend to have poorer psychological well-being, more impaired memory, delayed physical recovery, reduced hearing function, and a shorter life span (see ▶ "Ageism Around the World", ▶ "Critical Gerontology"). Such negative views toward older adults are largely shaped by social norms, and vary by culture, which underscore the importance of social and cultural environments in the aging process.

Future Directions

After several decades of development, environmental gerontology has become an established multidisciplinary science. Numerous topics are encompassed within environmental gerontology, and many theories have been developed, which have collectively improved our understanding of how environments influence age and the aging process. Moving forward, the following areas may warrant additional attention in the literature.

First, as the proportion of older adults continues to increase worldwide (United Nations 2019), research on the development of livable communities for older adults is becoming an increasingly vital issue. Creating livable and age-friendly communities, free of physical and/or social barriers, is important for promoting healthy aging and increasing active life expectancy (See ▶ "Age-Friendly Cities and Communities: New Directions for Research and Policy"). In 2007, the World Health Organization (WHO) released an official report on age-friendly cities that has been widely used to evaluate the quality and sustainability of cities to promote healthy aging (Plouffe et al. 2016; WHO 2007; see ▶ "Age-Friendly Cities and Communities: New Directions for Research and Policy"). More than a decade later, WHO (2018) published another report to chart the development of this initiative and to identify challenges that have arisen. Although there has been substantial progress in building age-friendly cities and communities around the world, there remain critical gaps in implementing projects that minimize inequities around the world; as well as geographic gaps in promoting their coverage (WHO 2018). With regard to scientific research, more studies are needed on the mechanisms through which enhanced urban environments improve the quality of life among older adults and to better quantify the contributions of these projects.

Second, with the increasing availability of various high-resolution geographic information system (GIS) datasets on both the natural and social environments, more studies using cross- or multi-scale analytic strategies, as well as comparative studies, are needed to investigate how environmental contextual factors influence older adult's access to and use of health services, health and well-being, and overall longevity. The need for such research is heightened with the increased frequency of environmental issues such as air pollution, extreme climate, ecosystem degradation, and epidemics/pandemics that are challenging the contemporary world (see ▶ "Climate Change, Vulnerability, and Older People").

Third, with ageism being common in many cultures around the world (see ▶ "Ageism Around the World"), and its apparent amplification during the COVID-19 pandemic (Ng et al. 2021; Xi et al. 2020), it is important to better understand (and promote) anti-ageism to help reduce and eventually eliminate such negative social environments toward older adults.

Fourth, research on GxE and genome-environment-wide interactions has greatly improved our understanding of how biological and environmental factors interact to influence human health and longevity. However, an understudied area in GxE and genome-environment-wide interactions is how to incorporate life course perspectives to recognize not only the role of environmental factors in later life, but also the individual and contextual factors that shaped early life.

Finally, with advances in technology and digitalization, the utilization of smart digital devices for daily life is becoming more accepted and prevalent among older adults. However, the overall rate of utilization among older adults continues to be much lower than younger adults – primarily because of the limited knowledge in using these devices (see ▶ "Gerontechnology"). Therefore, it remains a priority to remove such IT barriers to promote the understanding and use of smart devices among older adults. More notably, environmental gerontology should contribute to research on the environmental/contextual factors that influence utilization among older adults and how this usage may subsequently impact health and well-being.

Summary

Environmental gerontology is a modern multidisciplinary branch of gerontology that studies the relationships and interactions between older adults and their socio-physical environments. In many ways, this field encompasses nearly every branch of gerontology and broadly focuses on understanding and optimizing person-environment relationships to improve the health and well-being of older adults. With rapid increases in population aging, and the mounting challenges of climate change, it is expected that environmental gerontology will continue to play an important role in gerontological research.

Cross-References

▶ Aging in Place
▶ Aging in the Right Place
▶ Biogerontology
▶ Climate Gerontology
▶ Critical Gerontology
▶ Geographical Gerontology
▶ Gerontechnology

Disclaimer: The views expressed in this chapter are solely those of the authors and do not reflect those of the National University of Singapore, the Central University of Finance and Economics, Duke University, or the United Nations.

References

Alvarez-Erviti L, Seow Y, Yin H et al (2011) Delivery of siRNA to the mouse brain by systemic injection of targeted exosomes. Nat Biotechnol 29(4):341–345. https://doi.org/10.1038/nbt.1807

American Psychological Association (2016) Geropsychology: it's your future. https://www.apa.org/pi/aging/resources/geropsychology. Accessed 24 Aug 2020

Assary E, Vincent JP, Keers R et al (2018) Gene-environment interaction and psychiatric disorders: review and future directions. Semin Cell Dev Biol 77: 133–143. https://doi.org/10.1016/j.semcdb.2017.10.016

Bai Q (2018) Philosophical thoughts and value embodiment in western ecological literature. Lan Zhou Xue Kan 10:92–101. [Chinese]

Brennan A, Lo YS (2015) Environmental ethics. The Stanford encyclopedia of philosophy (Summer 2020 Ed). https://plato.stanford.edu/archives/sum2020/entries/ethics-environmental

Chemhuru M (2017) Elements of environmental ethics in ancient Greek philosophy. Phronimon 18(1):15–30. https://doi.org/10.17159/2413-3086/2017/1954

Edmondson R (2013) Cultural gerontology: valuing older people. In: Komp K, Aartsen M (eds) Old age in Europe: a textbook of gerontology. Springer, Dordrecht, pp 113–130

Farrelly C (2012) 'Positive biology' as a new paradigm for the medical sciences: focusing on people who live long, happy, healthy lives might hold the key to improving human well-being. EMBO Rep 13(3):186–188. https://doi.org/10.1038/embor.2011.256

Fenech M (2017) Vitamins associated with brain aging, mild cognitive impairment, and Alzheimer disease: biomarkers, epidemiological and experimental evidence, plausible mechanisms, and knowledge gaps. Adv Nutr 8(6):958–970. https://doi.org/10.3945/an.117.015610

Fernandez LS, Byard D, Lin CC et al (2002) Frail elderly as disaster victims: emergency management strategies. Prehosp Disaster Med 17(2):67–74. https://doi.org/10.1017/S1049023X00000200

Golant SM (1989) The residential moves, housing locations, and travel behavior of older people: inquiries by geographers. Urban Geogr 10(1):100–108. https://doi.org/10.2747/0272-3638.10.1.100

Gutman GM, Yon Y (2014) Elder abuse and neglect in disasters: types, prevalence and research gaps. Int J Disaster Risk Reduct 10(Pt. A):38–47. https://doi.org/10.1016/j.ijdrr.2014.06.002

Haq G, Gutman G (2014) Climate gerontology. Z Gerontol Geriatr 47(6):462–467. https://doi.org/10.1007/s00391-014-0677-y

Hertzog C, Kramer AF, Wilson RS et al (2009) Enrichment effects on adult cognitive development: can the functional capacity of older adults be preserved and enhanced? Psychol Sci Public Interest 9(1):1–65. https://doi.org/10.1111/j.1539-6053.2009.01034.x

Hoffmann RC (2004) An environmental history of medieval Europe. Cambridge University Pres, Cambridge

Hou W (1997) Reflections on Chinese traditional ideas of nature. Environmental History 2(4):482–493. https://doi.org/10.2307/3985610

IPCC (2018) Global warming of 1.5°C: an IPCC special report on the impacts of global warming of 1.5°C above pre-industrial levels and related global greenhouse gas emission pathways, in the context of strengthening the global response to the threat of climate change, sustainable development, and efforts to eradicate poverty. Summary for policymakers. https://www.ipcc.ch/site/assets/uploads/sites/2/2018/07/SR15_SPM_High_Res.pdf

Jankowsky JL, Melnikova T, Fadale DJ et al (2005) Environmental enrichment mitigates cognitive deficits in a mouse model of Alzheimer's disease. J Neurosci 25(21):5217–5224. https://doi.org/10.1523/JNEUROSCI.5080-04.2005

Jiang L, Xu G (2010) Research on environmental protection thought and its contemporary value in Chinese traditional culture. Environ Econ. 80:38–41. [Chinese]

Kahana E (1982) A congruence model of person-environment interaction. In: Lawton MP, Windley PG, Byerts TO (eds) Aging and the environment: theoretical approaches. Springer Publishing Company, New York, pp 97–121

Kay J (1989) Human dominion over nature in the Hebrew Bible. Annals of the Association of American Geographers 79(2):214–232

Keulartz J, Korthals M (2019). Environmental ethics. In Kaplan DM (ed). Ecncyclopedia of Food and Agricultural Ethics (2 edn). Springer Nature, Dordrecht, The Netherlands, pp 713–722

Lawton MP, Nahemow L (1973) Ecology and the aging process. In: Eisdorfer C, Lawton MP (eds) Psychology of adult development and aging. American Psychological Association, pp 619–674. https://doi.org/10.1037/10044-020

Leal-Galicia P, Castañeda-Bueno M, Quiroz-Baez R et al (2008) Long-term exposure to environmental enrichment since youth prevents recognition memory decline and increases synaptic plasticity markers in aging. Neurobiol Learn Mem 90:511–518. https://doi.org/10.1016/j.nlm.2008.07.005

Lewin K (1951) Field theory in social science: selected theoretical papers (Dorwin Cartwright, ed.). Harpers, Oxford

Li W (2011) The ideological basis of environmental protection in ancient China – based on the analysis of the pre-Qin and Han Dynasties. J Xi'an Jiaotong Univ (Social Sciences) 31(1):12–18. [Chinese]

Li Y, Cheng H, Beeton RJS et al (2016) Sustainability from a Chinese cultural perspective: the implications of harmonious development in environmental management. Environ Dev Sustain 18:679–696. https://doi.org/10.1007/s10668-015-9671-9

Lores-Arnaiz S, Bustamante J, Arismendi M et al (2006) Extensive enriched environments protect old rats from the aging dependent impairment of spatial cognition, synaptic plasticity and nitric oxide production. Behav Brain Res 169:294–302. https://doi.org/10.1016/j.bbr.2006.01.016

McCann JC, Ames BN (2011) Adaptive dysfunction of selenoproteins from the perspective of the triage theory: why modest selenium deficiency may increase risk of diseases of aging. FASEB J 25(6):1793–1814. https://doi.org/10.1096/fj.11-180885

Mora F, Segovia G, del Arco A (2007) Aging, plasticity and environmental enrichment: structural changes and neurotransmitter dynamics in several areas of the brain. Brain Res Rev 55:78–88. https://doi.org/10.1016/j.brainresrev.2007.03.011

Ng R, Chow YTJ, Yang W (2021) Culture linked to increasing ageism during Covid-19: evidence from a 10-billion-word corpus across 20 countries. J Gerontol Soc Sci . Online first. https://doi.org/10.1093/geronb/gbab057

Nilsson P, Iwata N, Muramatsu S et al (2010) Gene therapy in Alzheimer's disease – potential for disease modification. J Cell Mol Med 14(4):741–757. https://doi.org/10.1111/j.1582-4934.2010.01038.x

Organisation for Economic Co-operation and Development (2011) Trends in retirement and in working at older ages. In: Pensions at a Glance 2011: retirement-income systems in OECD and G20 countries. OECD Publishing, Paris

Oswald F, Wahl H-W (2004) Housing and health in later life. Rev Environ Health 19:223–252

Oswald F, Wahl H-W, Schilling O et al (2007) Relationships between housing and healthy aging in very old age. Gerontologist 47:96–107. https://doi.org/10.1093/geront/47.1.96

Pastalan LA (1970) Privacy as an expression of human territoriality. In: Pastalan LA, Carson DH (eds) Spatial behavior of older people. University of Michigan-Wayne State University Institute of Gerontology, Ann Arbor, pp 88–101

Philips J, Walford N, Hockey A et al (2013) Older people and outdoor environment: Pedestrian anxieties and barriers in the use of familiar and unfamiliar spaces. Geoforum 47:113–124. https://doi.org/10.1016/j.geoforum.2013.04.002

Plouffe L, Kalache A, Voelcker I (2016) A critical review of the WHO age-friendly cities methodology and its implementation. In: Moulaert T, Garon S (eds) Age-friendly cities and communities in international comparison. International perspectives on aging. Springer, Cham. https://doi.org/10.1007/978-3-319-24031-2_2

Rattan SIS (2018) Biogerontology: research status, challenges and opportunities. Acta Biomed 89(2):291–301. https://doi.org/10.23750/abm.v89i2.7403

Roka K (2020) Anthropocene and climate change. In: Filho WL, Azul AM, Brandli L, et al (eds) Climate Action. Springer Nature, Switzerland AG, pp 20–32

Rowles GD (1978) Prisoners of space? Exploring the geographic experience of older people. Westview Press, Boulder, p 1978

Rowles GD, Bernard M (2013) Environmental gerontology: making meaningful places in old age. Springer Publishing Company, New York, 336 pp

Sánchez-González D, Chávez-Alvarado R (2016) Adjustments to physical-social environment of the elderly to climate change: proposals from environmental gerontology. In: Sánchez-González D, Rodríguez-Rodríguez V (eds) Environmental gerontology in Europe and Latin America. International perspectives on aging, vol 13. Springer, Cham. https://doi.org/10.1007/978-3-319-21419-1_6

Schaie KW (2013) Developmental influences on adult intelligence: the Seattle Longitudinal Study, 2nd edn. Oxford University Press, New York

Silecchia LA (2004) Environmental ethics from the perspectives of NEPA and Catholic social teaching: Ecological guidance for the 21st Century, Wm. & Mary Envtl. L. & Policy Rev 28(3):659–797. https://scholarship.law.wm.edu/wmelpr/vol28/iss3/3

Simonds NI, Ghazarian AA, Pimentel CB et al (2016) Review of the gene-environment interaction literature in cancer: what do we know? Genet Epidemiol 40(5): 356–365. https://doi.org/10.1002/gepi.21967

Specht J, Egloff B, Schmukle SC (2011) Stability and change of personality across the life course: the impact of age and major life events on mean-level and rank-order stability of the Big Five. J Pers Soc Psychol 101: 862–882. https://doi.org/10.1037/a0024950

Twigg J, Martin W (2015a) The challenge of cultural gerontology. The Gerontologist 55(3):353–359. https://doi.org/10.1093/geront/gnu061

Twigg J, Martin W (2015b) Routledge handbook of cultural gerontology. The Routledge, New York

United Nations (2019) World population prospects: the 2019 revision. UN Department of Economic and Social Affairs Population Division, New York

Wahl H-W, Wiesman GD (2003) Environmental gerontology at the beginning of the new millennium: reflections on its historical, empirical, and theoretical development. The Gerontologist 43(5):616–627

Wahl H-W, Iwarsson S, Oswald F (2012) Aging well and the environment: toward an integrative model and research agenda for the future. Gerontologist. https://doi.org/10.1093/geront/gnr154

Warnes AM (1999) UK and western European late-age mortality: trends in cause-specific death rates, 1960–1990. Health Place 5:111–118

World Health Organization (2007) Global age-friendly cities: a guide. WHO, Geneva

World Health Organization (2018) The global network for age-friendly cities and communities. WHO, Geneva

Wu J, Xiao J, Li T et al (2015) A cross-sectional survey on the health status and the health-related quality of life of the elderly after flood disaster in Bazhong City, Sichuan, China. BMC Public Health 15. https://doi.org/10.1186/s12889-015-1402-5

Xi W, Xu W, Zhang X et al (2020) A thematic analysis of Weibo topics (Chinese Twitter hashtags) regarding older adults during the COVID-19 outbreak. J Gerontol Soc Sci, Online first. https://doi.org/10.1093/geronb/gbaa148

Zhilin L (1989) The differences between Chinese and Western concepts of nature and the trend toward their convergence. Chinese Studies in Philosophy 21(1):20–49. https://doi.org/10.2753/csp1097-1467210120

Environmental Impacts on Aging and Longevity

▶ Environmental Gerontology

Environments and Longevity

▶ Environmental Gerontology

Epidemiology of Frailty

Gotaro Kojima and Steve Iliffe
Department of Primary Care and Population Health, University College London, London, UK

Synonyms

Impairment; Progressive Dwindling

Definition

Frailty is "a state of vulnerability to poor resolution of homeostasis after a stressor event and a consequence of cumulative decline in many physiological systems during a lifetime"(Clegg et al. 2013, p. 752).

Overview

During the past century, life expectancy has considerably improved over the world, mainly due to improvements in public health (World Health Organization 2012). This transformation of global populations has resulted in an increasing number of older adults in most developed and developing countries. According to the forecast by the United Nations in World Population Prospects: The 2017 Revision, the number of people aged 60 or older is expected to more than double by 2050 and to more than triple by 2100, from 962 million in 2017 to 2.1 billion in 2050 and 3.1 billion in 2100, corresponding to 11.2%, 21.4%, and 27.7% of the world population, respectively (United Nations 2017). As we age we generally develop health problems and become more frail. Increasing life expectancy brings chronic diseases and decline of physical and cognitive functions, eventually leading to disability or dependency. Ongoing population aging on a worldwide scale is having considerable demographic impacts on current healthcare systems. Therefore, transformation of healthcare systems and long-term care services is urgently needed to effectively and sustainably meet the needs of older adults with complex and interacting medical problems (Kojima et al. 2019a).

Older adults are a highly heterogeneous population with different biological, environmental, experiential and genetic backgrounds, and their health status and its trajectories may vary substantially (Kojima et al. 2019b). The course of health status may also be affected by physical, psychological, and social factors. As a result, older adults with the same chronological age can have a different biological age, and it is challenging to properly describe the health status of older adults (Mitnitski et al. 2002). In recent years frailty has been gaining scientific attention as one of the potential concepts that captures the heterogeneity and overall health diversity of older adults (Clegg et al. 2013).

Two distinct conceptualizations of frailty were advocated in 2001, the frailty phenotype and the Frailty Index. The frailty phenotype was described by Fried and colleagues using the Cardiovascular Health Study cohort (Fried et al. 2001). This approach argues that frailty can be captured as a clinical syndrome characterized by five specific physical components, which are unintentional weight loss, self-reported exhaustion, weakness, slow walking speed, and low physical activity (Fried et al. 2001). A combination of three or more features is defined as frail and having 0 and 1–2 features are defined as robust and prefrail, respectively (Fried et al. 2001). In contrast, the other approach tries to define frailty as a state of age-related deficit accumulation, rather than as a specific clinical syndrome, and to quantitatively describe it as a "Frailty Index" (Mitnitski et al. 2001). The Frailty Index can be computed as a ratio of the number of deficits present to the number of deficits assessed for a given individual (Searle et al. 2008). The deficits can be symptoms, signs, diseases, disabilities, laboratory, radiographic or electrocardiographic abnormalities, and social characteristics (Searle et al. 2008).

Before these conceptualizations, frailty was often used as synonymous with aging, disability, or comorbidities (Fried et al. 2004). These entities are closely related to each other, become more common with aging, and often coexist and overlap (Fried et al. 2004). People of advanced age are not necessarily frail and vulnerable to negative health outcomes, while people in any age group can be frail, suffering from chronic medical problems and disabilities and declining functionally and mentally and eventually dying early (Schuurmans et al. 2004). Disability is generally defined as difficulty or dependency in performing activities of daily living (ADL) (Katz et al. 1970) and instrumental ADL (IADL) (Lawton and Brody 1969). Comorbidity is often described as having two or more medically diagnosed diseases. People with comorbidities and/or disabilities are not always frail and can be robust, with a good quality of life. In contrast, frailty is now conceptualized as a state of decreased physiological reserve and compromised capacity to maintain homeostasis as a consequence of age-related multiple accumulated deficits (Clegg et al. 2013; Fried et al. 2001; Mitnitski et al. 2001; Morley et al. 2013). This conceptualization of frailty does make it distinguishable from aging, disability. or comorbidities (Fried et al. 2004). However there is an ongoing argument about how best to define frailty precisely, and there is as yet no internationally accepted gold standard definition of frailty.

Since the two operationalizations of frailty, the phenotypic and cumulative deficit approaches, were developed, various operationalized definitions of frailty have been proposed (Dent et al. 2016). However, the phenotype and the Index remain the most widely used tools in research (Buta et al. 2016). Measurement of frailty is out of the scope of this chapter and will not be discussed in detail. The operationalization of these two frailty definitions has enabled researchers to capture frailty, quantify its magnitude objectively, and examine its consequences for older adults. As a result, an increasing number of frailty studies have been undertaken in aging populations, in a wide range of settings (Rockwood and Howlett 2018). These studies have accumulated mounting evidence on and have contributed to our fundamental understanding of frailty. Common clinical symptoms of frailty are fatigue, weight loss, sarcopenia, loss of resilience, and inability to recover from acute stressors (Clegg et al. 2013). Frail individuals are vulnerable to adverse health outcomes when they are exposed to internal and external stressors, including falls (Kojima 2015a), fractures (Kojima 2016a), emergency department visits (Kojima 2019), hospitalization (Kojima 2016b), nursing home placement (Kojima 2018), disabilities (Kojima 2017a), and mortality (Kojima et al. 2018) all of which may lead to poor quality of life (Kojima et al. 2016a).

Key Research Findings

In light of the negative health outcomes associated with frailty and its huge burden on patients, their families, society, and healthcare systems, it is important to document the prevalence of frailty to inform allocation of healthcare resource, healthcare policy making, and the provision of appropriate interventions for individuals.

A systematic review paper published by Collard and colleagues in 2012 showed that prevalence of frailty based on 21 studies involving a total of 61,500 community-dwelling older adults aged 65 years or older varied widely, ranging from 4.0% (the USA, the frailty phenotype) to 59.1% (the Netherlands, Sherbrooke Postal Questionnaire) (Collard et al. 2012). The overall weighted prevalence of frailty was 10.7% (95% confidence interval = 10.5–10.9%) (Collard et al. 2012). It should be noted that various frailty tools were included in the review: the most common one being the frailty phenotype and others including the Frailty Index, the Study of Osteoporotic Fracture index, the FRAIL scale, Groningen Frailty Indicator, Tilburg Frailty Indicator, and the Sherbrooke Postal Questionnaire (Collard et al. 2012). The prevalence of frailty based on physical phenotype ranged from 4.0% (the USA) to 17.0% (10 European countries and the USA) with the overall weighted prevalence of 9.9% (95%CI = 9.6–10.2%, 15 studies, 44,894 participants), while the prevalence of frailty based on broad definitions or measurement instruments ranged from 4.2% to 59.1% with overall weighted prevalence of 13.6% (95%CI = 13.2–14.0%, 8 studies, 24,072 participants) (Collard et al. 2012).

Another systematic review by Choi and colleagues in 2015 searched for studies examining the prevalence of frailty (based on the frailty phenotype) in nationally representative cohorts of community-dwelling older adults aged 65 or older (Choi et al. 2015). The review included six studies from the USA, ten European countries, the UK, Ireland, Taiwan, and Korea. The frailty prevalence ranged from 4.9% (Taiwan) to 27.3% (Spain) (Choi et al. 2015). Pooled estimates of frailty prevalence were not provided by this review.

Japan is one of the countries with the fastest aging population in the world (Kojima et al. 2016b). One review study from Japan demonstrated that a pooled frailty prevalence among a total of 11,940 Japanese community-dwelling older adults aged 65 years or more was 7.4% (Kojima et al. 2016b). All of five included studies used the frailty phenotype to define frailty, and the prevalence of frailty from each study was in a relatively narrow range between 4.6% and 9.5% (Kojima et al. 2016b).

These systematic reviews focused mainly on developed countries, such as the USA and European countries. Evidence of frailty in other areas and developing countries has been scarce and identified much higher prevalence of frailty. One systematic review focused on community-dwelling older adults aged 60 or older in countries in Latin America and the Caribbean and showed a frailty prevalence ranging from 7.7% to 42.6%, with a pooled prevalence of 19.6% (29 studies, 43,083 participants, 95%CI = 15.4–24.3%) (Da Mata et al. 2016). Of the 29 studies included, 26 studies (90.0%) used the frailty phenotype or modified versions of it (Da Mata et al. 2016). Another review examined frailty among community-dwelling older adults aged 60 or greater living in low- and middle-income countries, based on the World Bank country classification (Siriwardhana et al. 2018). A little more than half of 47 included studies were from Latin America or the Caribbean areas (24/47, 51.1%). The frailty prevalence was available in 67 cohorts from the 47 studies, corresponding to a total of 75,133 participants, ranging from 3.9% (China) to 26.0% (India), and the pooled frailty prevalence was 17.4% (95%CI = 14.4–20.7%) (Siriwardhana et al. 2018). Similar to the other systematic review, the majority of the studies (37/47, 78.7%) used the frailty phenotype or modified versions of it (Siriwardhana et al. 2018).

Age

Frailty tends to progress, and the prevalence of frailty increases exponentially with age (Clegg et al. 2013). Some of the reviews conducted age-stratified meta-analysis and consistently showed higher prevalence of frailty with more advanced age. The review by Collard showed that prevalence of frailty was approximately 3%, 7%, and 10% in those 65–69, 70–74, and 75–79 years of age, while it increased up to 16% in those aged 80–84 and 36% in those aged 85 or older (Collard et al. 2012). Age-stratified analysis by the review from Japan showed prevalence of frailty among those aged 65–69, 70–74, 75–79, 80–84, and 85 or older was 1.9% (95%CI = 0.9–3.3%), 3.8% (95% CI = 2.3–5.7%), 10.0% (95%CI = 6.6–14.2%), 20.4% (95%CI = 18.2–22.6%), and 35.1% (95%CI = 30.6–39.8%), respectively (Kojima et al. 2016b). The review of low- and middle-income countries showed prevalence of frailty was 6.2% (95%CI = 4.0–8.8), 8.2% (95% CI = 6.3–10.3%), 10.3% (95%CI = 8.2–12.6%), 15.4% (95%CI = 12.6–18.4%), 22.6% (95%CI = 18.5–26.9%), and 29.8% (95%CI = 25.6–34.2%) for those aged 60–64, 65–69, 70–74, 75–79, 80–84, and 85 or older (Siriwardhana et al. 2018). In contrast, the mean age of the study cohorts included in the analysis did not explain the between-study heterogeneity in a review paper of Latin America and the Caribbean (Da Mata et al. 2016).

Gender

In general women tend to be frailer than men, while men are more likely to die at a younger age than women (Clegg et al. 2013). Four reviews performed meta-analyses stratifying by gender, and all of them showed higher prevalence of frailty among women than men. In the review by Collard, the prevalence of frailty in women was 9.6% (11 studies, 17,746 female participants, 95%CI = 9.2–10.0%), which was significantly higher than that of men (5.2%, 95%CI = 4.9–5.5%, 11 studies, 22,596 male participants) (Collard et al. 2012). The Japanese systematic review showed the prevalence of frailty among women and men to be 8.1% (95%CI = 6.1–10.3%) and 7.6% (95%CI = 6.9–8.3%), respectively (Kojima et al. 2016b). The systematic review of countries of Latin America and the Caribbean showed a greater gender disparity: women 23.4% (95%CI = 16.6–30.9%, 19 studies) among 17,669 older women and 15.0% (95%CI = 11.1–19.4%, 19 studies) among 12,282 older men (Da Mata et al. 2016). The review which included low- to middle-income countries showed higher prevalence of frailty among women (15.2%, 95%CI = 12.5–18.1%) than men (11.1%, 95%CI = 8.9–13.4%) based on 24 studies using the frailty phenotype with objective tests (Siriwardhana et al. 2018).

Selected Populations

The prevalence of frailty has been studied in selected populations. Compared with general community-dwelling populations, a much higher prevalence was observed in samples of patients with specific diseases, such as Alzheimer's disease (pooled prevalence 31.9%) (Kojima et al. 2017), cancer (median prevalence 42%) (Handforth et al. 2015), heart failure (overall estimated prevalence 44.5%) (Denfeld et al. 2017), or end-stage renal disease (pooled prevalence 36.8% by objectively measured frailty phenotype and 67.0% by self-reported frailty phenotype) (Kojima 2017b). One systematic review identified nine studies of nursing home patients and found that frailty prevalence ranged from 19.0% to 75.6% and that more than 90% were either frail (pooled prevalence = 52.3%, 95%CI = 37.9–66.5%) or prefrail (pooled prevalence = 40.2%, 95%CI = 28.9–52.1%) (Kojima 2015b).

There is considerable variability in the prevalence of frailty observed across the studies of community-dwelling older adults. This may be due to differences in sociodemographic characteristics of the study cohorts or measurements used to capture frailty. According to the currently available data from the systematic reviews described above, age and gender have been extensively investigated, and advanced age and female gender were consistently shown to be related to higher prevalence of frailty (Collard et al. 2012; Da Mata et al. 2016; Kojima et al. 2016b; Siriwardhana et al. 2018). The prevalence of frailty also varied depending on what measurement tools were used (Collard et al. 2012; Da Mata et al. 2016; Kojima et al. 2016; Siriwardhana et al. 2018). Even among studies using the frailty phenotype showed a wide diversity in the frailty prevalence, ranging from 4.9% to 27.3% (Choi et al. 2015). This may be due to the modifications of the frailty phenotype criteria (Theou et al. 2015).

Future Directions for Research

Frailty research should be focused on how frailty research can be translated into clinical settings and utilized to improve health of older adults. One of the current challenges hampering the translation of frailty research is the lack of uniform conceptual and operational definition of frailty. Although most of the frailty tools can identify at-risk individuals, they identify different groups of people (Theou et al. 2013). The variety of frailty definitions used in previous studies may affect interpretation of the evidence, comparison with other studies, generalizability of the findings, and implementation in healthcare policy. Another challenge will be identification of frailty, which will lead to timely interventions and potentially better health outcomes. Population-based screening for frailty could be costly and resource-intensive, and there is currently no supporting evidence to justify it (British Geriatrics Society 2014). Nevertheless, some expert panels and societies recommended frailty screening for people with specific conditions, for example, for people aged 70 or older (Morley et al. 2013; Rodríguez-Laso et al. 2018) or people aged 75 or older with disability in IADL (Rolland et al. 2011).

From a public health perspective, evidence on the prevalence of frailty among community-dwelling older adults is of great importance and may be useful information for policy making or population-based case finding. For example, if frail older adults with high risk can effectively be searched for and identified, they could benefit from a multidimensional interdisciplinary review of medical, functional, and psychosocial needs, such as comprehensive geriatric assessment, and followed by individually tailored interventions in a timely manner. Intervention studies of potential methods to halt, delay, or even reverse frailty are also needed.

Summary

Current conceptualizations of frailty give us a way to understand the heterogeneity of health status in older adults, independent of chronological age. They have contributed to advances in frailty research and to a better understanding of frailty. The prevalence of frailty varies widely across studies which may be due to age, gender, frailty tools used, and other factors. Translation of frailty research into clinical practice would be facilitated by agreement about a standard frailty definition which would allow effective identification of frail older adults at risk, potentially leading to better care for older adults.

Cross-References

- Biology of Frailty
- Frailty in Clinical Care
- Measurement of Frailty

References

British Geriatrics Society (2014) British Geriatrics Society, Fit for Frailty Part 1. https://www.bgs.org.uk/sites/default/files/content/resources/files/2018-05-23/fff_full.pdf. Access data: 13 September 2019

Buta BJ, Walston JD, Godino JG et al (2016) Frailty assessment instruments: systematic characterization of the uses and contexts of highly-cited instruments. Ageing Res Rev 26:53–61. https://doi.org/10.1016/j.arr.2015.12.003

Choi J, Ahn A, Kim S et al (2015) Global prevalence of physical frailty by Fried's criteria in community-dwelling elderly with National Population-Based Surveys. J Am Med Dir Assoc 16(7):548–550. https://doi.org/10.1016/j.jamda.2015.02.004

Clegg A, Young J, Iliffe S et al (2013) Frailty in elderly people. Lancet 381(9868):752–762. https://doi.org/10.1016/s0140-6736(12)62167-9

Collard RM, Boter H, Schoevers RA et al (2012) Prevalence of frailty in community-dwelling older persons: a systematic review. J Am Geriatr Soc 60(8):1487–1492. https://doi.org/10.1111/j.1532-541 5.2012.04054.x

Da Mata FA, Pereira PP, Andrade KR et al (2016) Prevalence of frailty in Latin America and the Caribbean: a systematic review and meta-analysis. PLoS One 11(8):e0160019. https://doi.org/10.1371/journal.pone.0160019

Denfeld QE, Winters-Stone K, Mudd JO et al (2017) The prevalence of frailty in heart failure: a systematic review and meta-analysis. Int J Cardiol 236:283–289. https://doi.org/10.1016/j.ijcard.2017.01.153

Dent E, Kowal P, Hoogendijk EO (2016) Eur J Intern Med 31:3–10. https://doi.org/10.1016/j.ejim.2016.03.007

Fried LP, Tangen CM, Walston J et al (2001) Frailty in older adults: evidence for a phenotype. J Gerontol A Biol Sci Med Sci 56(3):M146–M156

Fried LP, Ferrucci L, Darer J et al (2004) Untangling the concepts of disability, frailty, and comorbidity: implications for improved targeting and care. J Gerontol A Biol Sci Med Sci 59(3):255–263

Handforth C, Clegg A, Young C et al (2015) The prevalence and outcomes of frailty in older cancer patients: a systematic review. Ann Oncol 26(6):1091–1101. https://doi.org/10.1093/annonc/mdu540

Katz S, Downs TD, Cash HR et al (1970) Progress in development of the index of ADL. Gerontologist 10(1):20–30

Kojima G (2015a) Frailty as a predictor of future falls among community-dwelling older people: a systematic review and meta-analysis. J Am Med Dir Assoc 16(12):1027–1033. https://doi.org/10.1016/j.jamda.2015.06.018

Kojima G (2015b) Prevalence of frailty in nursing homes: a systematic review and meta-analysis. J Am Med Dir Assoc 16(11):940–945. https://doi.org/10.1016/j.jamda.2015.06.025

Kojima G (2016a) Frailty as a predictor of fractures among community-dwelling older people: a systematic review and meta-analysis. Bone 90:116–122

Kojima G (2016b) Frailty as a predictor of hospitalisation among community-dwelling older people: a systematic review and meta-analysis. J Epidemiol Community Health 70(7):722–729

Kojima G (2017a) Frailty as a predictor of disabilities among community-dwelling older people: a systematic review and meta-analysis. Disabil Rehabil 39(19):1897–1908

Kojima G (2017b) Prevalence of frailty in end-stage renal disease: a systematic review and meta-analysis. Int Urol Nephrol 49(11):1989–1997. https://doi.org/10.1007/s11255-017-1547-5

Kojima G (2018) Frailty as a predictor of nursing home placement among community-dwelling older adults: a systematic review and meta-analysis. J Geriatr Phys Ther 41(1):42–48

Kojima G (2019) Frailty as a predictor of emergency department utilization among community-dwelling older people: a systematic review and meta-analysis. J Am Med Dir Assoc 20(1):103–105. https://doi.org/10.1016/j.jamda.2018.10.004

Kojima G, Iliffe S, Jivraj S et al (2016a) Association between frailty and quality of life among community-dwelling older people: a systematic review and meta-analysis. J Epidemiol Community Health 70(7):716–721. https://doi.org/10.1136/jech-2015-206 717

Kojima G, Iliffe S, Taniguchi Y et al (2016b) Prevalence of frailty in Japan: a systematic review and meta-analysis. J Epidemiol 27(8):347–353

Kojima G, Liljas A, Iliffe S et al (2017) Prevalence of frailty in mild to moderate Alzheimer's disease: a systematic review and meta-analysis. Curr Alzheimer Res. https://doi.org/10.2174/1567205014666170417104236

Kojima G, Iliffe S, Walters K (2018) Frailty index as a predictor of mortality: a systematic review and meta-analysis. Age Ageing 47(2):193–200. https://doi.org/10.1093/ageing/afx162

Kojima G, Liljas A, Iliffe S (2019a) Frailty syndrome: implications and challenges for healthcare policy. Risk Manag Healthc Policy 12:23–30. https://doi.org/10.2147/rmhp.S168750

Kojima G, Taniguchi Y, Iliffe S et al (2019b) Transitions between frailty states among community-dwelling older people: a systematic review and meta-analysis. Ageing Res Rev 50:81–88. https://doi.org/10.1016/j.arr.2019.01.010

Lawton MP, Brody EM (1969) Assessment of older people: self-maintaining and instrumental activities of daily living. Gerontologist 9(3):179–186

Mitnitski AB, Mogilner AJ, Rockwood K (2001) Accumulation of deficits as a proxy measure of aging. ScientificWorldJournal 1:323–336. https://doi.org/10.1100/tsw.2001.58

Mitnitski AB, Graham JE, Mogilner AJ et al (2002) Frailty, fitness and late-life mortality in relation to chronological and biological age. BMC Geriatr 2:1

Morley JE, Vellas B, van Kan GA et al (2013) Frailty consensus: a call to action. J Am Med Dir Assoc 14(6):392–397. https://doi.org/10.1016/j.jamda.201 3.03.022

Rockwood K, Howlett SE (2018) Fifteen years of progress in understanding frailty and health in aging. BMC Med 16(1):220. https://doi.org/10.1186/s12916-018-1223-3

Rodríguez-Laso Á, Mora M Á C, Sánchez I G et al (2018) State of the art report on the prevention and management of frailty

Rolland Y, Benetos A, Gentric A et al (2011) Frailty in older population: a brief position paper from the French society of geriatrics and gerontology. Geriatr Psychol Neuropsychiatr Vieil 9(4):387–390. https://doi.org/10.1684/pnv.2011.0311

Schuurmans H, Steverink N, Lindenberg S et al (2004) Old or frail: what tells us more? J Gerontol A Biol Sci Med Sci 59(9):M962–M965

Searle SD, Mitnitski A, Gahbauer EA et al (2008) A standard procedure for creating a frailty index. BMC Geriatr 8:24. https://doi.org/10.1186/1471-2318-8-24

Siriwardhana DD, Hardoon S, Rait G et al (2018) Prevalence of frailty and prefrailty among community-dwelling older adults in low-income and middle-income countries: a systematic review and meta-analysis. BMJ Open 8(3):e018195. https://doi.org/10.1136/bmjopen-2017-018195

Theou O, Brothers TD, Mitnitski A et al (2013) Operationalization of frailty using eight commonly used scales and comparison of their ability to predict all-cause mortality. J Am Geriatr Soc 61(9):1537–1551. https://doi.org/10.1111/jgs.12420

Theou O, Cann L, Blodgett J et al (2015) Modifications to the frailty phenotype criteria: systematic review of the current literature and investigation of 262 frailty phenotypes in the survey of health, ageing, and retirement in Europe. Ageing Res Rev 21:78–94. https://doi.org/10.1016/j.arr.2015.04.001

United Nations (2017) World Population Prospecets: The 2017 Revision. https://www.un.org/development/desa/publications/world-population-prospects-the-2017-revision.html. Access date: 13 September 2019

World Health Organization (2012) Are you ready? What you need to know about ageing. https://www.who.int/world-health-day/2012/toolkit/background/en/. Access data: 13 September 2019

Epidemiology, Aging, and Cancer

Constanza Bartolotti-Herrera
Unidad de Geriatría Aguda, Hospital Clínico de la Fuerza Aérea de Chile, Universidad Mayor Santiago, Santiago, Chile
Servicio de Salud Araucanía Sur, Centro de Apoyo para Personas con Demencia, Kimünche, Temuco, Chile
Servicio de Medicina Interna, Hospital Hernán Henríquez Aravena, Temuco, Chile

Overview

Aging can be defined as a progressive functional decline, or a gradual deterioration of physiological functions, an intrinsic age-related process of loss of viability and increase in vulnerability (López-Otín et al. 2013). Though universal, these changes occur earlier in some people and later in others. Aging is a multidimensional, highly individualized process, and physiological decline cannot be predicted purely on the basis of patient chronological age. Even within the same person, various functions and domains are affected at different rates.

This heterogeneity has to be considered when determining, for example, optimal treatment approaches for older subjects. At most, chronologic age represents a landmark beyond which the majority of older individuals are found. This landmark is found around age 70. While the majority of older people are over 70, it would be a major mistake to consider all individuals aged over 70 years as old.

The construct of successful aging is related to the concept of physiologic age. For some it represents a delay in comorbidity and functional decline to the last few weeks or days of life; for others it represents adequate adaptation to the medical, functional, and social changes of aging. In the absence of laboratory tests, the estimate of one's physiologic age is currently best performed applying a geriatric assessment (Balducci et al. 2010). Attempts to define the causes of aging have been impeded by the complexity of the phenotype coupled with the costs and duration of longevity studies. Recently, progress has accelerated, bringing geroscience to the forefront (UN Department of Economic and Social Affairs 2017).

Historically most people died young and, of course, not from age-related diseases but rather from starvation and epidemics (cholera, smallpox, tuberculosis, and other infections) as well as from physical violence. Just three centuries ago, life expectancy was less than 16 years of age and 75% of people born in London in 1662 died before they reached the age of 26 (Graunt's life table) (Morobia 2013). The progress of civilization eliminated many causes of death that killed young people in the past. This dramatically increased the average lifespan. Nowadays death from aging is technically death from age-related diseases, which are manifestations of advanced aging (Blagosklonny 2010).

A critical issue is to increase one's active life expectancy. The Healthy Life Years (HLY) indicator corresponds to the number of remaining years that a person of a certain age is expected to live without disability. Loss of independence is a major cause of quality of life deterioration for older individuals. There is currently no consistent evidence that the lengthening of life expectancy goes with a reduction in the total lifetime days of disability, the so-called compression of morbidity (Chatterji et al. 2015).

The comprehensive geriatric assessment is a multidimensional assessment tool that examines different age-related domains including comorbidity, function, physical performance, cognition, nutrition, emotional status, polypharmacy, social support, and living environment. The CGA provides a more accurate estimation of functional reserve than a standard clinical evaluation. In addition, the CGA unearths unsuspected conditions that may compromise treatment outcome if not addressed. These include comorbidity, malnutrition, and absence of reliable social support.

An Aging World

In many underdeveloped and developing countries, such as most Latin American countries, like Chile, the "older adults" benefits granted by the government start at 60 years of age. Global population aged 60 or over is growing faster than all younger age groups. As fertility declines and life expectancy rises, the proportion of the population above a certain age rises as well. This phenomenon, known as population aging, is occurring throughout the world. In 2017, there was an estimated 962 million people aged over 60 in the world, comprising 13 percent of the global population. The population is growing at a rate of about 3% per year. Currently, Europe has the greatest percentage of population aged 60 or over, accounting for 25% of Europe's population. By 2050 all regions of the world except Africa will have nearly a quarter or more of their population at age 60 and above. The number of older persons in the world is projected to be 1.4 billion in 2030, and 2.1 billion in 2050, and could rise up to 3.1 billion in 2100. Over the next few decades, a further increase in the population of older persons is almost inevitable, given the size of the cohorts born in recent decades (UN Department of Economic and Social Affairs 2017).

The number of people aged 80 or over is projected to triple by 2050 and by 2100 to increase to nearly seven times its value in 2017. It's projected to increase from 137 million in 2017 to 425 million in 2050 and further to 909 million in 2100 worldwide. In 2017, 27% of people aged 80 or over resided in Europe, but that share is expected to decline to 17% in 2050 and to 10% in 2100 as the populations of other regions continue to increase in size and to grow older themselves (UN Department of Economic and Social Affairs 2017).

The 2017 Revision of the World Population Prospects confirms that significant gains in life expectancy have been achieved in recent years. Globally, life expectancy at birth rose by 3.6 years between 2000–2005 and 2010–2015, or from 67.2 to 70.8 years. All regions shared the rise of life expectancy over this period, but the greatest gains were in Africa, where life expectancy rose by 6.6 years between these two periods after rising by less than 2 years over the previous decade. Life expectancy in Africa in 2010–2015 stood at 60.2 years, compared to 71.8 in Asia, 74.6 in Latin America and the Caribbean, 77.2 in Europe, 77.9 in Oceania, and 79.2 in Northern America (Fig. 1) (UN Department of Economic and Social Affairs 2017).

Key Research Findings

Facing the Challenges of Aging

Consequently, human aging and age-associated diseases are emerging as among the greatest challenges and financial burdens faced by developed and developing countries (Christensen et al. 2009). Although average life expectancy has increased dramatically in the last 100 years, this has not been accompanied by an equivalent increase in healthy life expectancy, which has been termed as *healthspan*. While life expectancy continues to rise, *healthspan* is not keeping pace because current disease treatment often decreases mortality without preventing or reversing the decline in overall health. Elders are sick longer, often juggling with multiple chronic diseases simultaneously. Thus, there is an urgent need to extend *healthspan* (Kennedy et al. 2014). The Healthy Life Years (HLY) indicator corresponds to the number of remaining years that a person of a certain age is expected to live without disability (Bogaert et al. 2018).

Targeting diseases individually for an aging population is very complicated because most older adults have multiple morbidities that

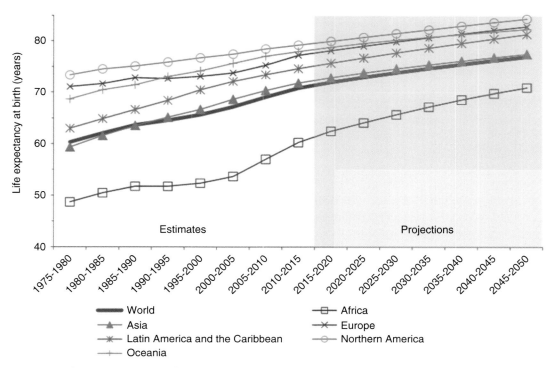

Epidemiology, Aging, and Cancer, Fig. 1 Life expectancy at birth (years) by region: estimates 1975–2015 and projection 2015–2050

interact, thus confounding therapeutic strategies. By understanding how aging enables pathology, new therapeutics will arise for multiple chronic diseases, providing an opportunity to extend human *healthspan* by targeting aging directly.

This remarkable accomplishment also poses a formidable challenge to public health. Aging is characterized by a progressive loss of physiological integrity, leading to impaired function and increased vulnerability to death. This deterioration is the primary risk factor for major human pathologies, including cardiovascular disease, cancer, diabetes, arthritis, osteoporosis, blindness, and neurodegenerative diseases like Alzheimer's disease among others with profound consequences and enormous medical costs. For many, aging also leads to the incapacity to respond to stress and return to homeostasis. The age-related loss of resiliency and increased vulnerability is now widely considered a geriatric health condition, termed *frailty* (Kennedy et al. 2014; Stuck et al. 1993).

Association of Aging and Disease

Though aging promotes disease, new findings suggest the reciprocal may also be true: diseases and/or their treatments may also accelerate aging. For instance, long-term cytomegalovirus (CMV) infection can induce chronic inflammation and exhaust the adaptive immune response, accelerating unrelated age-associated pathologies (Pawelec et al. 2012). This is also observed in human immunodeficiency virus (HIV) patients, attributable either to virus-induced inflammation or, ironically, treatment modalities (Pathai et al. 2014). Similarly, children subjected to chemotherapies often present with accelerated aging features decades later (Robison and Hudson 2014). These findings highlight relationships between disease and aging.

Increasingly, inflammation is being linked to aging and chronic disease. Acute inflammatory responses to insults such as injury and infection are critical for organismal health and recovery. However, the basal inflammatory response rises with age, leading to low-level chronic inflammation that is likely maladaptive, promoting aging; this concept is dubbed *inflammaging* (Calcinotto et al. 2019).

There is a need to identify the pathways by which adaptive and chronic inflammation are induced and to identify the outcomes of this *inflammaging*. One of the pathways is the accumulation of senescent cells in multiple tissues during aging. These senescent cells have a unique senescence-associated secretory profile (SASP) that includes many proinflammatory cytokines. Cell senescence, whereby cells irreversibly cease proliferation in response to stress, was long suspected of driving organismal aging. However, the number of senescent cells in most aged tissues is limited. The SASP potentially explains how a few senescent cells can have broad, adverse effects by secreting proinflammatory factors with autocrine, paracrine, and endocrine activities. It is essential to understand the in vivo consequences of senescent cells and identify interventional strategies that may mitigate their effects. Recent genetic strategies to ablate senescent cells in mice set the stage for determining to what extent they drive aspects of normal aging (Kennedy et al. 2014).

Metabolic dysregulation also accompanies and induces aging and is exacerbated by chronic diseases such as type 2 diabetes. Among the critical questions to be addressed is how overnutrition and obesity affect the aging metabolome, whether prolongevity interventions suppress age-related metabolic dysfunction, and how these interventions act in individuals with type 2 diabetes. Understanding links between diet and aging is also quite challenging. For example, the role of the microbiome in response to diet and aging also requires elaboration. Preliminary data suggest that the gut microbiome changes dramatically with age (Heintz and Mair 2014), although causes and effects remain undetermined. These issues are relevant to inflammation as well, as adipose tissue is a major source of inflammatory cytokines. Determining how the proinflammatory response in adipose tissue is initiated and propagated and the systemic effects of this response on aging should be a high priority. Links between altered metabolism and inflammation may underlie connections between aging pathways previously thought to be independent.

Cancer and Aging

The current situation of aging research exhibits many parallels with that of cancer research in previous decades. The cancer field gained a major momentum in 2000, with the publication of a landmark paper that enumerated the hallmarks of cancer. Nine candidate hallmarks are proposed by López-Otín et al. 2013 that are generally considered to contribute to the aging process and together determine the aging phenotype, namely, genomic instability, telomere attrition, epigenetic alterations, loss of proteostasis, deregulated nutrient sensing, mitochondrial dysfunction, cellular senescence, stem cell exhaustion and altered intercellular communication. The interaction of these hallmarks result in DNA and genomic damage accumulation. This categorization has helped to conceptualize the essence of cancer and its underlying mechanisms. Aging research has experienced an unprecedented advance over recent years, particularly with the discovery that the rate of aging is controlled, at least to some extent, by these genetic pathways and biochemical processes conserved in evolution.

At first sight, cancer and aging may seem to be opposite processes: cancer is the consequence of an aberrant gain of cellular fitness, whereas aging is characterized by a loss of fitness. At a deeper level, however, cancer and aging share common origins. The time-dependent accumulation of cellular, genomic, and DNA damage is widely considered to be the general cause of aging (López-Otín et al. 2013; UN Department of Economic and Social Affairs 2017). Therefore cancer and aging can be regarded as two different manifestations of the same underlying process, specifically the accumulation of cellular damage also involving uncontrolled cellular overgrowth or hyperactivity (Blagosklonny 2010).

Due to these demographic and epidemiological transitions, the global burden of cancer is rapidly increasing. In the Global Cancer Incidence in Older Adults Trial, the cancer burden was examined according to cancer sites and world regions. In 2012, an estimated 6.7 million new cases of cancer were diagnosed in adults aged 65 years and older, representing 47.5% of the total number of new cancer cases worldwide, while 8% of the world population was aged 65 years and older (Pilleron et al. 2018).

Most of the new cancer cases occurred in Europe (1.9 million, 29% of the world total) and Asia (2.8 million, 42%) including 1.5 million cases diagnosed in China (22% of the world total), whereas people aged 65 years and older in these respective regions represented 22%, 39%, and 31% of the world older population, respectively. Nearly 48% of all new cancer cases among older adults occurred in less developed regions where more than 60% of older adults live (Pilleron et al. 2018). The geographical pattern of age-standardized rate (ASR) is shown for both males and females in Fig. 2.

Overall, the number of new cancer cases (all cancer sites combined) is expected to double by 2035 (14 million) among older adults; from 3.9 to 8.5 million (a 118% increase) among older males, and from 2.8 to 5.7 million among older females (a 104% increase). This rise is observed in all world regions, with the greatest relative increase expected in Middle East and North Africa (157%) and the smallest relative increase in Europe (47%). The increase is expected to be much larger in less developed regions with an estimated increase of 144% by 2035, as compared to 54% in the more developed regions.

In a recent study of cancer incidence, mortality, and prevalence of more than 20 different, major cancers in the US population in the period 1998–2002 and for ages 0–114 years (Harding et al. 2012), the cancer incidence rates for most cancer types increased up to a maximum at ages 75–90 years and dropped quite abruptly afterward. Mortality rates for the same cancer types showed similar patterns and, with some exceptions, peaked within 5 years of the age of cancer incidence peak. A trend of both incidence and mortality rates toward zero among centenarians led the authors to suggest that centenarians may be asymptomatic or insusceptible to cancer. Also a study from the Netherlands (de Rijke et al. 2000) during the years 1989–1995 for ten specific cancers in the older population with age categories up to ages 95 and over found that most incidence rates peaked at ages 75–84 or 85–94 years and declined afterward, while the corresponding mortality rates mostly increased until ages 95+ years. The patterns of cancer incidence peak and decline

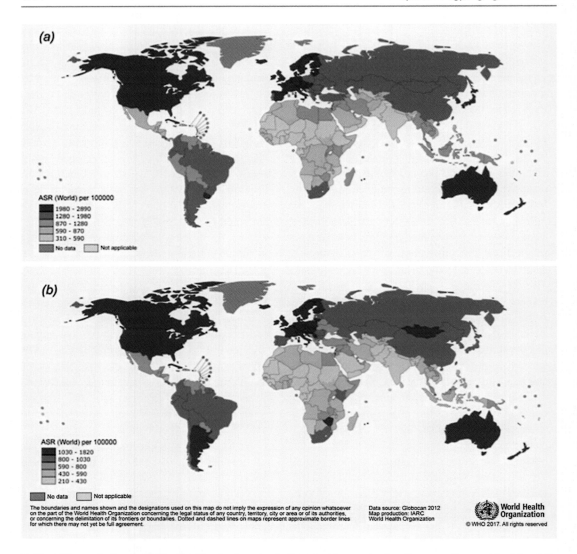

Epidemiology, Aging, and Cancer, Fig. 2 Age-standardized incidence rates for all cancer sites combined in adults aged 65 years and older (per 100,000 inhabitants) in 184 countries in 2012 among (**a**) males and (**b**) females

at similar ages are also seen in a study using Japanese, US, former German Democratic Republic, and Canadian data from the International Agency for Research on Cancer (IARC) (Arbeev et al. 2005).

Major Cancer Sites in Older People

The major cancer site among older males is prostate cancer in all regions of the world, except in Asia where lung cancer exceeded all other cancer sites (Fig. 3). Lung and colorectal cancer were also common cancers in most regions. Liver cancer was frequent among older males in Asia, Middle East, and North Africa and sub-Saharan Africa, and stomach cancer was the second most diagnosed cancer in Asia. These five cancers represented over two-thirds of the total burden of cancer among those aged 65 years and older in all regions (Pilleron et al. 2018).

Among older females, breast cancer was the leading cancer worldwide (Fig. 3) and in most world regions, except in Asia including China

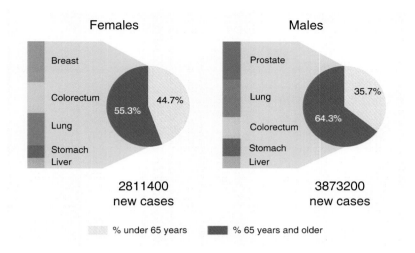

Epidemiology, Aging, and Cancer, Fig. 3 Proportion of the five leading cancer types among females and males aged 65 years and older in 2012 worldwide (Maione 2005)

and sub-Saharan Africa where colorectal, lung, and cervical cancers were more common, respectively. Similar to males, liver was a leading cancer site in Asia, the Middle East, and North Africa and sub-Saharan Africa, while colorectal cancer was one of the five main cancer sites in all regions. The proportion of the total burden of cancers in this age group explained by the top five cancers was less than among males but still represented between 46% in the Middle East and North Africa and 66% in China.

These projections for all cancer sites combined at global and regional levels are likely to be underestimated. Worldwide GLOBOCAN observed increases in the incidence of breast, colorectal, and prostate cancer, especially in countries in transition (Ferlay et al. 2012). Persons over 65 account for 70% of all cancer deaths and over 75 years for 50%. The age-adjusted cancer mortality rate for those over 65 is 1068/100,000 compared to 67/100,000 for those under 65. Mortality of older subjects with cancer is complex to characterize as comorbidity competes with cancer as a cause of death (Assouan, ASCO 2019). Moreover, several reasons may explain over-mortality of older patients: suboptimal treatment even if the prescription of chemotherapy increased substantially in the last 20 years, delay for diagnosis, restricted access to innovation, and/or increased toxicities compared to younger years (Doat et al. 2014; Lund et al. 2018).

Future Direction of Research

Cancer will predominantly develop in the older population; nevertheless, patients aged 75 and over remain dramatically low subrepresented in clinical oncology trials (Xie 2018). This situation results in an absence in the clarity of the most adequate treatment, with a narrowing therapeutic proposal for older patients with cancer and with worse outcomes, including an increased risk in toxicity and early mortality. With the rise of older patients with cancer and a considerable heterogeneity from one patient to the other, the need for more precise and effective evaluations is growing and becoming more and more relevant for oncologists (Maione 2005).

Emerging studies draw attention to the fact that several factors beyond age can affect an older person's ability to tolerate cancer therapy. Broadening clinical oncology trials to capture not only chronological age but also domains within the geriatric assessment would allow clinicians to better identify which characteristics and subsets of older adults who may benefit from treatment versus others who are most vulnerable to morbidity and/or mortality. Existing gaps create an urgent need for future research studies to include a robust and detailed characterization of how geriatric assessment domains including cognition, comorbidity, functional status, geriatric syndromes, nutrition, polypharmacy, and socioeconomic can influence the pharmacology of cancer therapies (Nightingale et al. 2019).

Summary

Older people are well-known for the heterogeneity in the way they age, which means important diversity in their general health status and functional performance. So when cancer is diagnosed and cancer therapy proposed, the decision of the best therapy will commonly be influenced by the coexistence of other diseases or conditions (comorbidity). The prevalence of comorbid conditions among older cancer patients is high, for example, in the United States one-third to one-half of older cancer patients were found to have comorbid conditions. Comorbidities are associated with poorer survival, lower treatment receipt, poorer quality of life, higher healthcare costs, and longer (and, therefore, more expensive) hospital stay, especially for cancers that generally have better prognosis (Pilleron et al. 2018). Other main prognosis factors, independent from oncological factors, are altered nutritional, mobility, and functional status (Caillet et al. 2014). Evidence regarding depressive mood and cognition are more heterogeneous and need further studies. Due to this heterogeneity in the health status and prognosis of older cancer patients, chronological age is not a meaningful guide to treatment and long-term follow-up (Stuck et al. 1993; Maione 2005).

Cross-References

▶ Active Aging and Active Aging Index
▶ Aging and Cancer: Concepts and Prospects
▶ Aging as Phenoptotic Phenomenon
▶ Aging Definition
▶ Aging Mechanisms
▶ Aging Theories
▶ Cancer Diagnosis
▶ Cancer Screening
▶ Healthy Aging

References

Arbeev KG, Ukraintseva SV, Arbeeva LS, Yashin AI (2005) Mathematical models for human cancer incidence rates. Demogr Res 12:237–272

Assouan D, Paillaud E, Caillet P, Kempf E, Vincent H, Brain E, de Lempdes GR, Touboul C, Martinez-Tapia C, Allain M, Bastuji-Garin S, Hanon O, Laurent M, Canoui-Poitrine F, (2019) Cancer-specific mortality and competing causes of death in older adults: A prospective, multicenter cohort study (ELCAPA-19). J Clin Oncol 37 (15_suppl):11547–11547

Balducci L, Colloca G, Cesari M, Gambassi G (2010) Assessment and treatment of elderly patients with cancer. Surg Oncol 19:117–123

Blagosklonny M (2010) Why human lifespan is rapidly increasing: solving "longevity riddle" with "revealed-slow-aging" hypothesis. Aging 2:177–182

Bogaert P, Van Oyen H, Beluche I, Cambois E, Robine J (2018) The use of the global activity limitation Indicator and healthy life years by member states and the European Commission. Arch Pub Health 76(1):30–37. Springer Link

Caillet P, Laurent M, Bastuji-Garin S, Liuu E, Culine S, Lagrange J et al (2014) Optimal management of elderly cancer patients: usefulness of the Comprehensive Geriatric Assessment. Clin Interv Aging 29(9):1645–1660. https://doi.org/10.2147/CIA.S57849

Calcinotto A, Kohli J, Zagato E, Pellegrini L, Demaria M, Alimonti A (2019) Cellular senescence: aging, cancer, and injury. Physiol Rev 99:1047–1078

Chatterji S, Byles J, Cutler D, Seeman T, Verdes E (2015) Health, functioning, and disability in older adults – present status and future implications. Lancet 385:563–575

Christensen K, Doblhammer G, Rau R, Vaupel J (2009) Ageing populations: the challenges ahead. Lancet 374:1196–1208

de Rijke JM, Schouten LJ, Hillen HF, Kiemeney LA, Coebergh J-WW, van den Brandt PA (2000) Cancer in the very elderly Dutch population. Cancer 89:1121–1133

Doat S, Thiébaut A, Samson S, Ricordeau P, Guillemot D, Mitry E (2014) Elderly patients with colorectal cancer: Treatment modalities and survival in France. National data from the ThInDiT cohort study. Eur J Cancer 50(7):1276–1283. https://doi.org/10.1016/j.ejca.2013.12.026

Ferlay J, Soerjomataram I, Dikshit R, Eser S, Mathers C, Rebelo M et al (2012) Cancer incidence and mortality worldwide: sources, methods and major patterns in GLOBOCAN. Int J Cancer 136:E359–E386

Harding C, Pompei F, Wilson R (2012) Peak and decline in cancer incidence, mortality, and prevalence at old ages. Cancer 118:1371–1386

Heintz C, Mair W (2014) You are what you host: microbiome modulation of the aging process. Cell 156:408–411

Kennedy B, Berger S, Brunet A, Campisi J, Cuervo A, Epel E et al (2014) Geroscience: linking aging to chronic disease. Cell 159:709–713

López-Otín C, Blasco M, Partridge L, Serrano M, Kroemer G (2013) The hallmarks of aging. Cell 153:1194–1217

Lund JL, Sanoff HK, Peacock Hinton S, Muss H, Pate V, Stürmer T (2018) Potential Medication-Related Problems in Older Breast, Colon, and Lung Cancer Patients in the United States. Cancer Epidem Biomar 27 (1):41–49 https://doi.org/10.1158/1055-9965.EPI-17-0523

Maione P (2005) Pretreatment quality of life and functional status assessment significantly predict survival of elderly patients with advanced non-small-cell lung cancer receiving chemotherapy: a prognostic analysis of the multicenter Italian lung cancer in the elderly study. J Clin Oncol 23:6865–6872

Morobia A (2013) Epidemiology's 350th anniversary: 1662–2012. Epidimiology 24:179–183

Nightingale G, Schwartz R, Kachur E, Dixon B, Cote C, Barlow A et al (2019) Clinical pharmacology of oncology agents in older adults: a comprehensive review of how chronologic and functional age can influence treatment-related effects. J Geriatr Oncol 10:4–30

Pathai S, Bajillan H, Landay A, KP H (2014) Is HIV a model of accelerated or accentuated aging? J Gerontol A Biol Sci Med Sci 69:833–842

Pawelec G, McElhaney J, Aiello A, Derhovanessian E (2012) The impact of CMV infection on survival in older humans. Curr Opin Immunol 24:507–511

Pilleron S, Sarfati D, Janssen-Heijnen M, Vignat J, Ferlay J, Bray F et al (2018) Global cancer incidence in older adults, 2012 and 2035: a population-based study. Int J Cancer 144:49–58

Robison L, Hudson M (2014) Survivors of childhood and adolescent cancer: life-long risk and responsibilities. Nat Rev Cancer 14:61–70

Stuck A, Siu A, Wieland G, Adams J, Rubenstein L (1993) Comprehensive geriatric assessment: a meta-analysis of controlled trials. Lancet 342:1032–1036

UN Department of Economic and Social Affairs (15 de January de 2017) The 2017 Revision. Obtenido de United Nations: https://www.un.org/development/desa/publications/world-population-prospects-the-2017-revision.html

Xie H (2018) Industry collaboration when developing novel agents in oncology. In: Kelly W, Halabi S (eds) Oncology clinical trials. DemosMedical, Durham, pp 27–28

E-psychology

▶ Cyberpsychology and Older Adults

E-Psychotherapy

▶ Cybercounseling

Equity Credit Line

▶ Home Equity Conversion

ERP

▶ Electrophysiology

Eternal Life

▶ Beliefs of Life After Death and Social Partner Choice

E-Therapy

▶ Cybercounseling

Ethics of Lifespan Extension

David R. Lawrence
Newcastle University School of Law, Newcastle-upon-Tyne, UK

Synonyms

Ethics of longevity; Moral debate over life extension

Definition

Debate around the appropriate moral principles that should govern the use and application of antiaging science where it may be used to radically extend *Homo sapiens* lifespan.

Introduction to the Debate

Radical life extension is an example of a human enhancement technology, one designed to grant a new, or augment an existing, capacity to *Homo sapiens*. Life extension is subject to fierce debate within certain sections of global bioethics (broadly conceived), and it is a topic that will likely never be agreed upon. A vast proportion of literature concerning human enhancement of any type is to some extent morally partisan, arguing from solidified pro- or anti-enhancement positions. Well-reasoned discourse in the field is not rare; however, while the debate broadly concentrates on the virtues of pursuing the technologies, it is likely to remain at an academic stalemate. Those in favor promote the potentialities with regard to increased enjoyment, reduced suffering, and personal liberty, whereas those skeptical highlight the risks of devaluing life and experience, of overpopulation, and of subverting the "natural order," among other issues.

The essential case for the positive view of life extension is presented by Christine Overall (2003):

> [O]ther things being equal, a longer life is a better one, provided that one is in a minimally good state of health. The case for longer life ... is founded on a genuine appreciation of human potential, of what people want in their lives and are capable of doing and experiencing when given more opportunities. An increased lifespan gives human beings the chance for activities and experiences that they would not otherwise have enjoyed. Collectively, extending average life expectancy provides for the society in which it occurs the value of increased experience, know-how, labor, loving relationships, and so on – that is, whatever healthy old(er) people can contribute.

The essential case for the negative view of life extension can be seen in Kass (2017): "The pursuit of perfect bodies and further life extension will deflect us from realizing more fully the aspirations to which our lives naturally point: for living well, rather than merely staying alive"; and in Fukuyama's fears of a "nursing home world."

Overview

History

A desire for some version of biological immortality or the radical extension of life has been a preeminent human fixation throughout history, with proponents and critics strongly present in historic cultural and scientific narratives including ancient Egyptian medical texts (Breasted 1991), mythologies such as the Epic of Gilgamesh, historic treatises on health including Luigi Cornaro's *The Sure and Certain Method of Attaining a Long and Healthful Life* and Hufeland's *Macrobiotics: The Art of Prolonging Life* (Freeman 1979), enlightenment texts such as Francis Bacon's *New Atlantis* (Bacon 1627), and the works of early modern scientists from Robert Boyle to Alexis Carrel (Hughes 2011). This preoccupation has continued into the present, particularly among futurists, with the first book on life extension to capture popular interest and spark modern debate coming in 1976 (Kurtzman and Gordon 1977) and subsequently sparking a strong academic discourse as outlined below (See ▶ "Life-Extensionism in a Historical Perspective").

Key Issues

Aging as Illness

The leading bioethicist Arthur Caplan suggests that "ageing is in no way an intrinsic part of human nature... there is no reason why it is intrinsically wrong to try to reverse or cure aging" (Caplan 2004). This demonstrates a frequently stated view of aging as an illness in and of itself amongst advocates of life extension. Eric Jeungst more specifically places it in that arena: "As long as anti-aging interventions serve to forestall the morbidities associated with the aging process, they have a legitimate place in the armamentarium of preventive medicine." (Jeungst 2004) (See ▶ "Bioethics"). This attitude follows from the view that scientific research into radical life extension stems from work in the arena of understanding the aging process and related diseases including the prevalence of certain cancers, degenerative illnesses such as Alzheimer's disease, heart disease, and osteoporosis; and the desire to find treatments and cures for these.

Moral Imperatives

Related to this view is the idea espoused by Harris, among others, that there is a moral duty to treat illness and save lives where possible (Harris 2007). While there may be debate over the desirability of immortality or deliberate life extension, there is no such debate around efforts to cure and prevent disease. This therefore leads to an imperative to develop and provide treatments, which is likely to lead to extended lifespan by default, rather than it being a deliberate goal of biogerontological research (Rantanen 2013). The scientific breakthroughs necessary in therapeutic medicine that are inherent to successful life extension may provide their own moral imperative to pursue it, since doing so.

There may be a further moral imperative to pursue the project of life extension directly, in that success in doing so will inherently require the development of treatments and cures for the various associated causes of mortality. Whether or not life extension is ultimately deemed acceptable, to not undertake research towards, it may also prevent the alleviation of other harms. It is not clear that such a decision should be made on the behalf of future generations (Intelligence Squared 2016).

Therapy/Enhancement Divide

Where aging is recognized as illness or disease, treatments are likely to fall within the ongoing and extensive debate around the so-called "therapy-enhancement divide" within bioethical literature. This divide is related to the foundational definitions of "enhancement" and can be reduced at a basic level to whether or not there is some baseline or "normal" (Daniels 2000) level of function for the human body. The view that there is a divide rests on the idea that any intervention that restores this "species-typical" function is a therapy, and any that goes beyond it would be an enhancement. The alternative view denies the notion of a useful "normal" as variance within *Homo sapiens* is so wide and claims that therapies are themselves enhancement as they relate to the patient's state before the intervention (Harris 2007). This debate holds important implications for any regulatory development around life extension technologies – for instance, it is likely that medical interventions and enhancement technologies are likely to come under different regimes, and the classification of life extension may determine its availability within healthcare provision.

Gerontologiphobia

A potential issue with the treatment of aging as illness is the tendency of the public to consider aging research (as a collective of late-life diseases) as:

> ...a public menace bound to produce a world filled with nonproductive, chronically disabled, unhappy senior citizens consuming more resources than they produce. ... Pointing out that such an argument would inveigh, with equally fallacious force, against research on heart attacks, diabetes, and cancer (whose goals, like those of gerontology, are to allow people to live longer and healthier lives) does little good in practice to dispel this fixed belief. (Miller 2004)

This "gerontologiphobic" (Miller 2004) view is encapsulated by Kass (1985):

> simply to covet a prolonged life span for ourselves is both a sign and a cause of our failure to open ourselves to procreation and to any higher purpose... [The] desire to prolong youthfulness is not only a childish desire to eat one's life and keep it; it is also an expression of a childish and narcissistic wish incompatible with devotion to posterity.

It is not clear how this attitude could or should be overcome, though it lies in direct conflict with the Harrisian moral imperatives that might reasonably take precedence.

Overpopulation

As indicated by Miller, perhaps the primary fear expressed by bioconservative commentators around life extension is one of the risk of overpopulation, in particular its possible effects on society and the environmental issues it may engender. The problems of overcrowding and resource exhaustion that we presently face may be greatly exacerbated if aging no longer reduces populations at the current rate (Kass 2003). However, there are many counterpoints to this popular argument. There is evidence to suggest a decline in the growth of worldwide population growth rate and predictions of eventual stabilization and ultimate decline, which would suggest overpopulation beyond any current issues may not transpire (More 2004). Extended lifespan would, in this scenario, only slow or mitigate that decline. James Hughes (2004) argues that any risks of social and resourcing issues should and could be addressed by political and technological means including universal basic income and the use of alternative energy technologies; and further that these problems are not unique to the scenario of overpopulation through life extension but rather are versions of issues, we have the means to address today. Further, it has been suggested that an increased lifespan would give people a greater stake in the future and therefore lead to the implementation of more sustainable environmental policies (Bostrom 2005b).

Equity, Equality, Liberty, and Justice
One position common to both sides of the debate around life extension is the idea that access to any life extension technologies must be equitable and not available only to the wealthy (Sutherland 2006). Critics of life extension argue that, as with all enhancement technologies, this is unlikely to be the reality and so their use will create a two-tiered society of "haves" and "have-nots" (Fukuyama 2002, Hauskeller 2013), which would serve to further entrench existing inequalities between rich and poor (Harris 2002). Socio-economic injustice is no new phenomenon and is therefore likely, it is argued, to restrict access to the best available options for life extension in the same manner as it now restricts access to particular goods or services. For instance, the rich may live longer than the poor as they have better access to health care, and this injustice is made worse if the rich are able to live radically longer lives than those who do not have access to life extension technology. Relatedly, there is an enforced disadvantage to one party where another gains an advantage over them, which suggests that any system wherein equal access is not guaranteed cannot ever be just (Lin, Allhoff and Steinberg 2011, Agar 2014).

To this end, the option of "equality by denial" is sometimes proposed, in which inequality is prevented by banning the technology outright (Glannon 2002). This could be interpreted as being in disregard of the concept of liberty. Per John Stuart Mill, liberty – the freedom to act as one chooses – ought be the primary concern in deciding what is good for a given individual (Mill 1863). However, this freedom is only limited by the "harm principle"; that there is a duty not to act injuriously to the interests of another person. It is possible to envisage scenarios in which those with extended lifespan may cause harm, including through unequal access.

However, proponents offer counterpoints to these fears. A common rejoinder to the idea of unequal access is the suggestion that technological access is generally subject to a "trickle-down" effect, wherein early adopters pay a premium but prices decrease as the technology is refined over time. Hughes has argued that these inequalities can also be mitigated by public policy, such as the introduction of a comprehensive universal healthcare which includes access to enhancement technologies such as life extension. Bans, it is claimed, are more likely to cause harm by encouraging black-market trade and usage (Hughes 2004). Furthermore, it is argued, the overall goal of human life is to survive; and so it is compatible with human nature to desire to undertake life extension and other enhancement procedures (Berry 2010).

Immortality, Meaning, and Identity
The prospect of (biological) immortality through life extension interventions is the root of much of the historical interest outlined above, which frequently saw it in positive aspect. On the other hand, literary sources such as Mary Wollstonecraft Shelley's *The Mortal Immortal* have long explored the idea of an immortal life as torturous wherein one must see loved one after loved one die. It has been argued that the inevitable problem of immortality would be abject boredom unless a person changed significantly, in which case they may become a different person entirely (Williams 1973) or at least be subject to such altered self-conception that they no longer retain their

narrative identity (DeGrazia 2005). This change may be such that we would not care what happened to this different person, with our present selves having ceased to exist (Kagan 2012). This would also likely affect our relationships with others (Hauskeller 2013). This appears to be in conflict with the apparent desire of life extensionists to retain their ability to grow and experience existence, to enjoy new hobbies, jobs, and sights (Häyry 2011). However, it is not clear that these pitfalls are reason enough to not pursue the life extension project; save where personal preference suggests otherwise (Stambler 2017).

A further criticism of immortality relates to a range of questions around the meaning of life, the meaning of death, and the meaning of human experience. It is seen as important to value the life we have today, and to uphold that value as one of "human heritage." Characteristic of this view is an attitude that death or the prospect thereof gives life meaning, which implies that to have no prospect of death removes or decreases this meaning, perhaps reducing the value of our experiences and choices – as any consequence can be overcome over time (Kass 2003). The concept of meaning is deeply experiential, personal, and subjective, however, and it is not necessarily possible to say what grants meaning to the life of a given individual. It is also possible that extended lifespans (and the holders of these lifespans) may themselves give rise to new meanings and enjoyments that cannot be understood by those who have not experienced them (Häyry 2011).

Accompanying concerns around meaning, life extension as it is frequently presented is susceptible to criticisms that it risks dehumanization. Popular works commonly frame aging as an engineering problem and the body as a machine to be repaired – making life extension a question of maintenance (de Grey and Rae 2008). Within literature on the "posthuman" there is an idea of "the body as instrument" (Hayles 1999), which when combined with the appeals to nature frequently found in bioconservative commentary suggest a "contempt for the flesh" which devalues the meaning associated with the natural body (Midgely 2002). These arguments are also often associated with the transhumanist movement, which has been defended on the basis that the transcendence of the limits of the human body, including the desire to regain and maintain youth, is an attempt to achieve our oldest hopes, evident throughout history (Bostrom 2005a).

Long-Term Consequence

Some concerns have been voiced around the potential indirect social and anthropological issues that might result from radical life extension. As evidenced throughout history, younger generations tend to enter political discourse with newer and more progressive ideas. Older generations have, generally, been more conservative than their offspring, and there is a tendency to lean towards further conservatism as one ages. It can be argued that there is some progressive benefit to allowing older generations disappear in order to allow for this generational shift. Wolpe highlights the values of pre-American Civil War generations and the relative unacceptability today of many elements of society at that time (Intelligence Squared 2016).

This maintenance of older generations may have other affects beyond the political sphere. Where older generations remain reproductively viable and the overall population growth rates stabilize or regress (as above), it is possible that the human evolutionary potential may be reduced. It is unlikely that this would lead to atrophy or direct harm given the large population number, but it could introduce susceptibility to existential risks and gradually affect well-being over time (Gyngell 2015).

Summary and Future Directions of Research

The above presents only a brief overview of the key areas of bioethical and broader philosophical attention within the radical life extension debate. The issues engendered align closely with the larger ongoing conversation around human enhancement, including questions of equality, liberty to act in one's own benefit, distributive justice, the question of therapy versus enhancement and the potential ramifications that may have on access, and the spectre of contempt for the flesh, but life extension raises a number of unique issues. The most commonly voiced are around the risks of overpopulation on social and environmental matters in a world which already struggles to support our resource needs, but more abstract questions include whether aging can be considered illness and therefore subject to a moral imperative to save lives, and whether biological immortality implies a less meaningful existence or even removes the moral value of an individual life or its achievements. Furthermore, there are concerns around the potential stagnation of our species and society that radical life extension might bring with it.

Given the nature of ethical discourse, it is difficult to envision the future major topics that may emerge, but it is certain that the debate will enter a more popular and public sphere as the life extension movement and the science behind it develops and makes breakthroughs. Public bioethical debate tends to wax only with when a given technology is a present reality that might directly affect the lives of that public, and so it will be a fascinating area of discussion to track as it reacts to the science. It may be the case that attitudes towards life extension technologies soften as it becomes a reality, as it will be just one of many enhancement technologies that increasingly pervade our daily lives.

Cross-References

▶ Bioethics
▶ Life Extension
▶ Life-Extensionism in a Historical Perspective
▶ Life-Span Extension
▶ Longevity Advocacy
▶ Longevity Areas and Mass Longevity
▶ Longevity Dividend
▶ Medical Ethics
▶ Regulation of Geroprotective Medications

References

Agar N (2014) The moral case against radical life extension. In: Truly human enhancement: a philosophical defence of limits by Nicholas Agar.: MIT Press, Cambridge, MA
Bacon F (1627) New Atlantis; reproduced in: Bacon, F (2008). Vickers, Brian (ed.). The Major Works. Oxford University Press
Berry R (2010) A polemic for human enhancement. Metascience 19(2):263–266. https://doi.org/10.1007/s11016-010-9361-z
Bostrom N (2005a) A history of transhumanist thought. J Evol Technol 14(1):1–25
Bostrom N (2005b) Recent developments in the ethics science and politics of life-extension. Ageing Horiz 3:28–33
Breasted JH (1991) The Edwin smith surgical papyrus: published in facsimile and hieroglyphic transliteration with translation and commentary in two volumes. University of Chicago Press, Chicago
Caplan A (2004). An unnatural process: why it is not inherently wrong to seek a cure for aging) In: Post S, Binstock R (eds) The fountain of youth: cultural scientific and ethical perspectives on a biomedical goal 283. Oxford University Press, Oxford
Daniels N (2000) Normal functioning and the treatment-enhancement distinction. Camb Q Healthc Ethics 9(3):309–322. https://doi.org/10.1017/S0963180100903037
de Grey A, Rae M (2008) Ending aging: the rejuvenation breakthroughs that could reverse human aging in our lifetime. St. Martin's Griffin, London
DeGrazia D (2005) Enhancement technologies and human identity. J Med Phil 30(3):261–283. https://doi.org/10.1080/03605310590960166
Freeman JT (1979) Ageing: its history and literature. Human Sciences Press, New York
Fukuyama F (2002) Our posthuman future: consequences of the biotechnology revolution. Farrar Straus & Giroux, New York
Glannon W (2002) Reply to Harris. Bioethics 16(3):292–297. https://doi.org/10.1111/1467-8519.00287
Gyngell C (2015) The ethics of human life extension: the second argument from evolution. J Med Phil 40(6):696–713. https://doi.org/10.1093/jmp/jhv027
Harris J (2002) Intimations of immortality – the ethics and justice of life expanding therapies. Curr Legal Probs 55(1):65–95. https://doi.org/10.1093/clp/55.1.65
Harris J (2007) Enhancing evolution: the ethical case for making better people. Princeton University Press, Princeton
Hauskeller M (2013) Better humans?: understanding the enhancement project. Acumen, Durham
Hayles K (1999) How we became posthuman: virtual bodies in cybernetics literature and informatics. University of Chicago Press, Chicago
Häyry M (2011) Considerable life extension and three views on the meaning of life. Camb Q Healthc Ethics 20(1):21–29
Hughes J (2004) Citizen cyborg: why democratic societies must respond to the redesigned human of the future. Westview Press, Colorado

Hughes J (2011) Transhumanism. In: Bainbridge W (ed) Leadership in science and technology: a reference handbook. Sage, Thousand Oaks
Jeungst E (2004) Anti-aging research and the limits of medicine. In Post S, Binstock R The fountain of youth: cultural scientific and ethical perspectives on a biomedical goal Oxford University Press, Oxford at 336
Kagan S (2012) Death. Yale University Press, New Haven
Kass L (1985) Toward a more natural science: biology and human affairs. Free Press, New York
Kass L (2003) Chapter 9: L'Chaim and its limits: why not immortality? In: Kass L (ed) Life liberty and the defense of dignity: the challenge for bioethics. Encounter books, New York
Kass L (2017) Leading a worthy life: finding meaning in modern times. Encounter Books, New York
Kurtzman J, Gordon P (1977) No more dying: the conquest of aging and the extension of human life. Dell, New York
Lin P, Allhoff F, Steinberg J (2011) Ethics of human enhancement: an executive summary. Sci Eng Ethics 17(2): 201–212. https://doi.org/10.1007/s11948-009-9191-9
Midgely M (2002) Science as salvation. Routledge, London
Mill JS (1863) On liberty. Ticknor and Fields, London
Miller R (2004) Extending life: scientific prospects and political obstacles. In: Post S, Binstock R (eds) The fountain of youth: cultural scientific and ethical perspectives on a biomedical goal. Oxford University Press, Oxford. At 243
More M (2004) Superlongevity without overpopulation. In: Klein B (ed) The scientific conquest of death: essays on infinite lifespans. Immortality Institute, Buenos Aires. at 169
Overall C (2003) Aging death and human longevity: a philosophical enquiry. University of California Press, Berkeley
Rantanen R (2013) Issues in the debate on the ethics of considerable life extension. Res Cogn 1:34–51
Intelligence Squared (2016) Debate: lifespans are long enough. February 3.. https://intelligencesquaredus.org/debates/lifespans-are-long-enough. Accessed April 9 2019
Stambler I (2017) Longevity promotion: multidisciplinary perspectives. Lulu Press, Rison Lezion
Sutherland J (2006) The ideas interview: Nick Bostrom. The Guardian May 9. https://www.theguardian.com/science/2006/may/09/academicexperts.genetics. Accessed 9 Apr 2019
Williams B (1973) The Makropulos case: reflections on the tedium of immortality. In: Williams B (ed) Problems of the self: philosophical papers 1956–1972. Cambridge University Press, Cambridge, pp 82–100

Ethics of Longevity

▶ Ethics of Lifespan Extension

Ethnic Caregivers

▶ Caregiving and Ethnicity

Ethnic Caregiving Roles

▶ Caregiving and Ethnicity

Ethnic Family Caregivers

▶ Caregiving and Ethnicity

Ethnic Family Caregiving Network

▶ Caregiving and Ethnicity

Ethnography

Sarah Chard
Department of Sociology, Anthropology, and Health Administration and Policy, University of Maryland Baltimore County, Baltimore, MD, USA

Definition

Ethnography is the study of a culture or cultural group from an emic or insider's perspective. The term also can refer to the resulting text.

Overview

Ethnography traces its roots to early twentieth-century British social anthropologists (e.g., Malinowski 1922; Macdonald 2001) and the "Chicago School" sociologists who promoted studying culture through fieldwork (Deegan 2001). While early ethnographies often presented static, functionalistic descriptions of artificially bounded social groups (see Clifford and Marcus 1986), contemporary ethnographers more often employ grounded theory in their projects (Strauss and Corbin 1990; Charmaz 2006), as well as critical (Madison 2012), interpretive (Geertz 1973), and reflexive lenses (Davies 1999).

Methodologically, ethnography relies heavily on participant observation, which involves immersion in a group's activities in order to uncover explicit and tacit symbols, beliefs, and practices (Agar 1996; DeWalt and DeWalt 2011). Participant observation also provides unique opportunities for informal discussions that reveal variations in meanings, actions, or expectations (Agar 1996). Field notes of participant observation must be systematically recorded as they serve as the primary dataset (Emerson et al. 2007).

Ethnographic research frequently includes in-depth interviews, which let the researcher explore key topics among defined sets of persons (Spradley 1979; Kaufman 1994). Utilizing interviews with participant observation allows for member checking and data triangulation, two common mechanisms for increasing validity (Fetterman 1998). Ethnographic approaches additionally may be utilized alongside quantitative methods in large projects focused on cultural context, with the ethnography playing a central role in identifying emic perspectives (e.g., Newman 2003).

As ethnographic studies often result in sizeable text-based datasets, qualitative software can facilitate data management. Programs vary in their features and cost, making it important to investigate the fit between the software and the analytical plan. Ethnographers employ a range of analytical methods, including thematic (Luborsky 1994; Wutich et al. 2015) and narrative approaches (Holstein and Gubrium 2012; see also Saldana 2016).

Examples of Application

Ethnographic research has provided numerous insights into the cultural dimensions of aging. Cross-cultural studies have illuminated how the meaning of age and aging praxis are situated within cultural understandings of the life course, social roles, and institutions (Lynch and Danely 2013). Researchers also have critically identified aging experiences as heavily informed by the intersections of economics, ethnicity/race (Newman 2003), and gender (Lamb 2000; Cliggett 2005). Such studies have offered fine-grained insights into the nature of suffering, as one example, within and across gender and ethnic groups (Black 2006; Black and Rubinstein 2009). Ethnographies of institutions, e.g., long-term care and assisted living facilities (AL), have identified the tensions surrounding the definition and enactment of "quality" care (Gubrium 1975; Eckert et al. 2009; Morgan et al. 2011). By employing an ethnographic approach, these researchers have revealed AL residents often perceive quality as a process; thus, quality requires institutions to balance formal rules and residents' desires (Morgan et al. 2011). Finally, ethnographic research has revealed changing care practices, particularly with increasing inequality (Buch 2018) and transnational migration (Lamb 2009; Coe 2017). Global economic and social flows are placing limits on informal care networks (Sokolovsky 2009) and creating new models of care and social relationships (Lamb 2009; Coe 2018; Pauli and Bedorf 2018).

Summary

Ethnography provides important insights on the cultural dimensions of aging. Ethnographic methods can be employed to investigate cultural groups, institutions, or social problems.

Cross-References

▶ Qualitative Research/Quantitative Research

References

Agar M (1996) The professional stranger: an informal introduction to ethnography. Academic, New York

Black H (2006) Soul pain: the meaning of suffering in later life. Baywood Publishing Company, Amityville

Black H, Rubinstein R (2009) The effect of suffering on generativity: accounts of elderly African American men. J Gerontol Soc Sci 64B:296–303
Buch E (2018) Inequalities of aging: paradoxes of independence in American home care. New York University Press, New York
Charmaz K (2006) Constructing grounded theory: a practical guide through qualitative analysis. Sage, London
Clifford J, Marcus G (1986) Writing culture: the poetics and politics of ethnography. University of California Press, Berkeley
Cliggett L (2005) Grains from grass: aging, gender and famine in rural Africa. Cornell University Press, Ithaca
Coe C (2017) Transnational migration and the commodification of eldercare in urban Ghana. Identities 24:542–556
Coe C (2018) Imagining institutional care, practicing domestic care: inscriptions around aging in southern Ghana. Anthropol Aging 39:18–32
Davies CA (1999) Reflexive ethnography: a guide to researching selves and others. Routledge, London
Deegan MJ (2001) The Chicago school of ethnography. In: Atkinson P, Coffey A, Delamont S et al (eds) Handbook of ethnography. Sage, Thousand Oaks, pp 11–25
DeWalt K, DeWalt B (2011) Participant observation. A guide for fieldworkers. AltaMira Press, Lanham
Eckert JK, Morgan LA, Carder PC et al (2009) Inside assisted living: the search for home. Johns Hopkins University Press, Baltimore
Emerson RM, Fretz RI, Shaw LL (2007) Writing ethnographic fieldnotes. University of Chicago Press, Chicago
Fetterman DM (1998) Ethnography step by step. Sage, Newbury Park
Geertz C (1973) The interpretation of cultures; selected essays. Basic Books, New York
Gubrium JF (1975) Living and dying at Murray Manor. St. Martin's Press, New York
Holstein J, Gubrium J (eds) (2012) Varieties of narrative analysis. Sage, Thousand Oaks
Kaufman S (1994) In-depth interviewing. In: Qualitative methods in aging research. Sage, Thousand Oaks, pp 123–136
Lamb S (2000) White saris and sweet mangoes. University of California Press, Berkeley
Lamb S (2009) Aging and the Indian diaspora. Indiana University Press, Bloomington
Luborsky M (1994) The identification and analysis of themes and patterns. In: Qualitative methods in aging research. Sage, Thousand Oaks, pp 189–210
Lynch C, Danely J (eds) (2013) Transitions & transformations: cultural perspectives on aging and the life course. Berghahn Books, New York
Macdonald S (2001) British social anthropology. In: Atkinson P, Coffey A, Delamont S et al (eds) Handbook of ethnography. Sage, Thousand Oaks, pp 60–79
Madison S (2012) Critical ethnography: methods, ethics, and performance. Sage, Thousand Oaks
Malinowski B (1922) Argonauts of the western Pacific. Waveland Press, Prospect Heights
Morgan LA, Frankowski AC, Roth E et al (2011) Quality assisted living: informing practice through research. Springer, New York
Newman KS (2003) A different shade of gray: midlife and beyond in the inner city. The New Press, New York
Pauli J, Bedorf F (2018) Retiring home? House construction, age inscriptions, and the building of belonging among Mexican migrants and their families in Chicago and rural Mexico. Anthropol Aging 39:48–65
Saldana J (2016) The coding manual for qualitative researchers. Sage, Los Angeles
Sokolovsky J (2009) Ethnic elders and the limits of family support in a globalizing world. In: The cultural context of aging. Praeger, Westport, pp 289–301
Spradley JP (1979) The ethnographic interview. Holt, Rinehart and Winston, New York
Strauss A, Corbin J (1990) The basics of qualitative research: grounded theory procedures and techniques. Sage, Newbury Park
Wutich A, Ryan G, Bernard HR (2015) Text analysis. In: Bernard HR, Gravlee C (eds) Handbook of methods in cultural anthropology. Rowman & Littlefield, Lanham, pp 533–559

Eudaimonic Well-Being Scale

▶ Ryff's Psychological Well-Being Scale

Eurocarers

Ana Ramovš[1], Marta Ramovš[1] and Stecy Yghemonos[2]
[1]Anton Trstenjak Institute of Gerontology and Intergenerational Relations, Ljubljana, Slovenia
[2]Eurocarers, Brussels, Belgium

Synonyms

European Association Working for Carers

Definition

Eurocarers (European Association Working for Carers) is the European Union's (EU) nonprofit umbrella organization working with and for informal carers across Europe. The Eurocarers network brings together more than 65 carers' organizations, research institutes, and universities from 25 countries. It pursues charitable, scientific, educational, and advocacy purposes and is aimed at promoting and defending the rights relating to the representation and social inclusion of carers at both national and EU levels. Eurocarers' secretariat is based in Brussels.

Overview

The Eurocarers network was officially established in Luxembourg in December 2006. Its origin lies in two European networks: *Carmen*, a network on integrated care, and *Eurofamcare*, a research network on carers of older persons. In the *Carmen* project, researchers, practitioners, and policy-makers, among them, representatives of the carers' movement, found each other and concluded that it was time for carers to be heard at the European level. The Eurofamcare network – which mapped the situation of carers of older persons and the policy measures available to support them in six countries – also diagnosed a strong need for carers to make themselves heard in order to ensure that EU policy developments take account of the specific challenges and issues faced by carers.

The Eurocarers secretariat was opened in 2014, following the signature of a partnership agreement with the European Commission's Directorate-General for Employment, Social Affairs and Inclusion. This resulted from the Commission's recognition that informal carers play an increasingly significant role in the provision of long-term care across Europe and this, very often to the detriment of carers' social inclusion, employment prospects and health. Engaging in a dialogue on the best ways to address their needs with an organization conveying their voice is therefore seen as necessary for the achievement of the EU's objectives.

Mission and Vision

The Association aims to enhance the recognition of informal care as a crucial element in European long-term care systems and to represent and act on behalf of carers and ex-carers (as well as their organizations), whatever may be their age and the specific needs of the persons receiving care (Eurocarers 2014). According to the Association, an "informal carer" means a person who provides unpaid care for someone with chronic illness, disability, or other long-lasting health or care need, outside of a professional or formal framework.

Eurocarers' work and activities are set against the backdrop of demographic aging in Europe, which leads to a growing incidence of age-related conditions and a growing demand for care. In Europe, 80% of long-term care (LTC) is provided by informal carers (Hoffmann and Rodrigues 2010), and estimates of the economic value of this unpaid care – as a percentage of the overall cost of formal long-term care provision in the EU – range from 50% to 90% (Triantafillou et al. 2010). The projected budgetary impact of a gradual shift from informal to formal care by 2070 would imply an increase of the share of GDP devoted to long-term care by 130% on average for the EU (European Commission 2018). The value of informal care in Europe is not only a matter of finances. Informal care and solidarity also have an intrinsic value from a moral standpoint, for example, standing and caring for vulnerable groups (e.g., people who are chronically ill, persons with disabilities, and frail older people) not because of any personal interest but because they need this support.

At the same time, the combination of various demographic and socioeconomic developments – such as lower birth rates, the trend toward smaller families, increasing mobility (leading to greater physical distances between relatives), the rising number of women entering the labor market, and a prolonged working life due to delayed retirement (partly following explicit policies aiming at increasing labor force participation of women and older workers) – is rapidly leading to increased strain on carers. In many cases – and even more so when no adequate support is available – informal care can be a determinant of adverse health outcomes, a barrier to education and employment, and a driver of discrimination, social exclusion, and poverty.

The work of the Eurocarers network builds on the idea that carers should receive the recognition they deserve for the significant role they play in community care, and this should be reflected in all policies affecting carers. Without the vital contribution of carers, European care systems would not be sustained. Informal care should, therefore, be "normalized" by ensuring that carers are seen as equal partners in the planning and delivery of care and have access to the financial, practical, and emotional support they require. Initiatives targeting carers should focus on identifying, assessing, and supporting them in a personalized and outcome-focused way as well as a consistent and comprehensive manner (Eurocarers 2018a).

Furthermore, people should have an option to choose freely if they want to be (and remain) a carer and to what extent they want to be involved in caring; at the same time, people needing care should have the right to choose whom they wish to be their carers. People who do not want – or are not able – to be involved in the provision of care for their relatives or who only want to play a limited role in these activities should be able to rely on affordable and professional care options of good quality, in particular home-care and community-based services. Investments into formal care, therefore, remain the central element of universal and carer-friendly care systems (Eurocarers 2018a).

Carers who want and are able to care for their dependent relatives/friends should have easy access to the information, guidance, advocacy, advice, and training they need – fitting to the stage of their carer's career. They should have the opportunity of taking time off. Therefore, adequate relief, for example, respite care arrangements, acceptable to both the carer and the cared-for person, must be readily available and tailored to carers' needs. Carers should also have the possibility to combine caring with paid employment. This presupposes labor market policies that allow for care activities as well as formal care available during working hours. Carers' needs of social life and healthcare should be recognized, receiving adequate support and protection. They should have financial security, being covered by social security schemes such as income replacement benefits, accident insurance, and old-age pensions, in order to avoid impoverishment as a consequence of caring (Eurocarers 2018a).

Work and Activities

Advocacy and Policy Development

Eurocarers works on national and European level. They are involved in policy debates by promoting evidence-based recommendations that genuinely reflect the opportunities and challenges of informal carers across Europe. Their advocacy work encompasses a portfolio of EU policy developments that have the potential to improve (or deteriorate) the situation of carers; they are active in debates on care and employment; care, health, and well-being; young carers; and poverty, social exclusion, and care (Eurocarers 2018b).

Since 2009, Eurocarers acts as the secretariat of the European Parliament Interest Group on Informal Carers, which they created. The group brings together more than 30 MEPs from 11 member states. It provides Eurocarers access to a broad and powerful coalition which can work as an internal pressure group within the European Parliament. Also, when an action for the Interest Group as a whole is not appropriate, Eurocarers can call and rely on its members to support policy initiatives (such as propose amendments, ask Parliamentary Questions, and host and chair meetings).

Since 2014, Eurocarers works in close partnership with the European Commission's Directorate-General for Employment, Social Affairs and Inclusion (DG EMPL) in the framework of the Employment and Social Innovation (EaSI) program. The partnership allows the network to play a leading consultative role in the framework of the Open Social Method of Coordination and the implementation of the EU 2020 strategy on issues related to the development of adequate, sustainable, and equitable long-term care services across the EU. In order to shape policy environment that is more favorable to carers, Eurocarers also collaborates with other interest and advocacy groups at national and EU levels – including organizations representing disabled people, women's organizations, and organizations campaigning against social exclusion and poverty.

Research and Development

In recent years, the Eurocarers network has been involved in many research and implementation projects aiming to generate evidence-based recommendations and action for the recognition of carers and carers' needs. An example of this is INNOVAGE project, which is dedicated to developing, implementing, evaluating four potentially cost-effective social innovations, and creating a new web-based platform consisting of social innovations from anywhere in the world. The platform provides public information on care and support services, legal and financial information, information about the most common impairments of older people, strategies and information about coping with caregiving, and suggestions on how to reconcile care with family and work.

Another project with involvement of the Eurocarers was Me-We (Psychosocial Support for Promoting Mental Health and Well-being among Adolescent Young Carers in Europe; 2018–2021) that aims to mitigate the risk factor of being an adolescent young carer by empowering the young with improved resilience and enhanced social support (from family, schools, peers, and services). The specific objectives of the project are to systematize knowledge on young adolescent carers, co-design, develop, and test – together with young adolescent carers – a framework of effective and multicomponent psychosocial interventions for primary prevention and to carry out wide knowledge translation actions for dissemination, awareness promotion, and advocacy.

Eurocarers was involved into EdyCare that aimed to empower teachers and other school staff in upper secondary education to recognize young adolescent carers (16–19 years old) in classes and maximize their learning opportunities while ensuring their social inclusion. The project partners will develop an assessment tool to help teachers and school staff to identify young carers; develop a package of educational strategies, didactical approaches, and organizational adjustments to help schools to facilitate young carers and produce a handbook providing guidelines and recommendations on how teachers and school staff can work at best with young carers; and develop a massive open online course (MOOC) for training teachers and school staff on the young carers phenomenon and their needs and preferences.

Another project with Eurocarers was the CARE4DEM project, which focuses on dementia carers and aims to develop a new and innovative model of mutual aid groups which promotes carers' involvement by introducing web-based tools and integrating it with another type of interventions, in order to enhance caregivers' satisfaction with care and reduce burnout. Mutual aid groups are commonly associated with several benefits, including mental and social well-being, by increasing self-confidence, resilience, and knowledge. They are a powerful opportunity for non-formal learning, as they help carers cope with difficulties, by sharing experiences, interacting with peers, and building connection and affectivity. The project also intends to expand the professional development of mutual aid groups' facilitators by creating a network for mutual and peer learning across Europe.

The ENhANCE (EuropeaN curriculum for fAmily aNd Community nursE) project with the involvement of Eurocarers seeks to target the existing mismatch between the skills currently offered by nurses working in primary healthcare (PHC) and those actually demanded by both public healthcare institutions and private service providers when applying innovative healthcare models centered on PHC. Currently, no standardized EU-wide professional profile has been defined for family and community nurses taking into account the World Health Organization (WHO) and EU recommendations. The competency-based professional profile will be the baseline for the definition of a European, innovative, learning outcome-oriented modular vocational education and training (VET) curriculum for nurses.

Through these projects and their follow-up, strong links have been established with the EU institutions as well as with a vast variety of decision-makers and stakeholders from the public and private sector, including the WHO Europe office in Copenhagen and the WHO Primary Health Care Advisory Group, in the fields such as research, health, employment and social affairs, education, human rights, and information and communications technology-based solutions for care and caring. In 2015, Eurocarers Research Working Group was established, bringing together Eurocarers' research members and carer organization members actively involved in the research. The group has at least once a year face-to-face meeting.

Empowerment of Carers' Organizations

Eurocarers supports carers and their organizations through the collection, exchange, and dissemination of information, experience, expertise, and good practice, as well as innovations, through seminars, working groups, events, newsletter, webpage, etc. It interprets relevant EU policy developments for member organizations working at national and regional levels and involves these organizations in EU policy developments. Eurocarers stimulates and supports the development of representative and sustainable carers' organizations in all EU states, especially in countries and regions where these do not exist.

Organization and Financing

The Association has three categories of membership: full members, associate members, and observer members. The bodies of the Association are the General Assembly (the highest instance of the Association; it must be held at least once a year) and the Board of Directors. They are assisted in their activities by the Secretariat of the Association. The Association's income includes the membership fees, donations, grants, subsidies, and legacies awarded to support either the general aims or a specific goal, which does not run counter to the Association's aims and tasks (Eurocarers 2016).

Future Directions of Research

Taking into account the EU political and demographic context, and based on the current knowledge gaps and scientific challenges in the area of care and caring, Eurocarers has identified the following eight research priorities: (1) cultural change in health and social care systems; (2) legislative and policy aspects of carers' situation; (3) the value and costs of caring; (4) health and well-being; (5) coordination of care systems; (6) new technologies; (7) the role of friends, neighbors, and volunteers; and (8) "what's worked well and why?", documenting and exchanging national experiences and young carers (Eurocarers 2015).

Summary

Eurocarers is designed as a platform for collaboration, support, exchange, awareness raising, and policy change. It brings together carers' organizations as well as relevant universities and research institutes interested and involved in the care and caring. Eurocarers' collective efforts seek to ensure that the existing and growing care needs of an aging European population are addressed in a universal and equitable way and that the essential contribution of unpaid/informal carers in the provision of care is valued, recognized as central to the sustainability of health and long-term care systems, and supported to prevent the negative impact of care on carers themselves. The work of Eurocarers builds on the idea that carers' know-how and needs are worth listening to and that people should have the right to choose freely whether they want to be a carer and to what extent they want to be involved in caring.

The mission of the Eurocarers network is, therefore, to act as a voice for informal carers, irrespective of their age or the particular health need of the person they are caring for, by (1) documenting and raising awareness about the significant contribution made by carers to health and social care systems and the economy as a whole, the impact of caring on carers' ability to maximize their life prospects and maintain an active and productive life, as well as the need to address the daily challenges facing carers across Europe; (2) contributing to evidence-based policy development at national as well as European level that takes account of carers, i.e., promote their social inclusion and the development of support services and enable them to remain active in paid employment and maintain a social life; and (3) promote mutual learning and an exchange of good practice and innovation throughout the EU.

Cross-References

▶ Abuse and Caregiving
▶ Benefits of Caregiving
▶ Caregiver Identity
▶ Caregiver Interventions
▶ Caregivers' Outcomes
▶ Caregiving and Ethnicity
▶ Double-Duty and Triple-Duty Caregivers
▶ Effectiveness of Respite Care for Caregivers of Older Adults
▶ Employment and Caregiving
▶ Formal and Informal Care
▶ Former Caregivers
▶ Self, Informal, and Formal Long-Term Care: The Interface

References

Eurocarers (2014) Eurocarers statutes. http://www.eurocarers.org/userfiles/files/Annex%203%20-%20Approved%20amendments%20to%20statutes%20-%20AGM%202016%20%2B%20integration%20from%20minutes.pdf. Accessed 7 Jan 2019
Eurocarers (2015) Eurocarers research priorities. http://www.eurocarers.org/userfiles/files/research/Eurocarers%20Research%20Priorities%20-%20August%202015.pdf. Accessed 7 Jan 2019
Eurocarers (2016) Internal rules. http://www.eurocarers.org/userfiles/files/Approved%20amendments%20to%20Eurocarers%20Internal%20Rules%20(EN)%20-%20AGM%202016.pdf. Accessed 7 Jan 2019
Eurocarers (2018a) About us. http://www.eurocarers.org/about_principles?lang=. Accessed 7 Jan 2019
Eurocarers (2018b) Our work. http://www.eurocarers.org/Policy-Work?lang=. Accessed 7 Jan 2019
European Commission Directorate-General for Economic and Financial Affairs (2018) The 2018 ageing report, economic and budgetary projections for the EU Member States (2016–2070). https://ec.europa.eu/info/sites/info/files/economy-finance/ip079_en.pdf. Accessed 17 Dec 2018
Hoffmann F, Rodrigues R (2010) Informal carers: who takes care of them? Policy Brief. European Centre for Social Welfare Policy and Research. https://www.euro.centre.org/downloads/detail/1256. Accessed 17 Dec 2018
Triantafillou J, Naiditch M, Repkova K, Stiehr K, Carretero S, Emilsson T, Di Santo P, Bednarik R, Brichtova L, Ceruzzi F, Cordero L, Mastroyiannakis T, Ferrando M, Mingot K, Ritter J, Vlantoni D (2010) Informal care in the long-term care system, European overview paper. https://www.euro.centre.org/downloads/detail/768. Accessed 17 Dec 2018

European Association Working for Carers

▶ Eurocarers

European Civil Society Platform on Lifelong Learning

▶ Lifelong Learning Platform

European Network in Aging Studies

Ulla Kriebernegg and Barbara Ratzenboeck
Center for Inter-American Studies, University of Graz, Graz, Austria

Synonyms

ENAS

Definition

The European Network in Aging Studies (ENAS) is an association of researchers concerned with the study of cultural aging. It facilitates international and interdisciplinary collaboration among Aging

Studies scholars at all career levels, most notably in the humanities and social sciences. The network promotes academic research on cultural dimensions of human aging. Among others this includes research on cultural narratives and popular (self-)images of aging and old age and their histories, literary and other artistic representations of aging and old age, as well as collective memory, and media and policy discourses on aging and old age. Furthermore, it encompasses research on generational identities; intersections of age with other social categories such as gender, sexuality, ethnicity, religion or belief, (dis)ability, or class; the social meaning and organization of life course transitions such as retirement or relocation; and people's aging experiences in relation to (their) bodies and material objects. ENAS contributes to fighting ageism through research on cultural aspects of aging and old age.

Overview

ENAS was first established in 2010 within the framework of the project "Live to be a Hundred: The Cultural Fascination with Longevity," funded by the Netherlands Organization for Scientific Research in the program Internationalization of the Humanities. Its inaugural conference "Theorizing Age: Challenging the Disciplines" took place in Maastricht in 2011 and resulted in a special issue of the *International Journal of Aging and Later Life* in 2012 (7(2)). In 2013, ENAS was launched as an official association (ZVR number: 298202943, Central Register of Associations in Austria) with its seat in Graz, Austria. Institutional founding members of the network are the Center for Gender and Diversity (Maastricht University, the Netherlands); the Center for Inter-American Studies (University of Graz, Austria); the Women, Ageing and Media Research Group (University of Gloucestershire, UK); the Group Dedal-Lit (University of Lleida, Spain); NISAL (Linköping University, Sweden); and the German Aging Studies Group (Germany). Associated founding partners from the USA include the Aging and Ageism Caucus of the National Women's Studies Association (NWSA) and the Age Studies Discussion Group of the Modern Language Association (MLA).

Parallel to the development of the European Network in Aging Studies, North American partners established the North-American Network in Aging Studies (NANAS) in 2013. ENAS and NANAS have since been sister organizations. Since 2017, the collaboration of the two networks has included a biannual joint international conference. With more than 300 participants, the first joint conference "AgingGraz2017," held at the University of Graz and the Medical University of Graz, Austria, in 2017, was one of the biggest academic events in the field of *Cultural Aging Studies* based in the humanities and social sciences since its beginning in the 1990s.

In addition to academic events, ENAS also facilitates knowledge dissemination through the Aging Studies book series that focuses on interdisciplinary research in cultural aging and is published by Transcript. In terms of research, ENAS provides members with a platform for projects and has been acting as project partner in a number of research endeavors. Another key objective of the network is the promotion of up-and-coming researchers. This is realized through summer schools and seminars organized by member organizations of ENAS (e.g. Women, Ageing and Media International Summer School at the University of Gloucestershire, UK, and the Aging Studies Module at the Graz International Summer School Seggau at the University of Graz, Austria) and biannual awards granted for best MA and PhD theses in the field of Aging Studies.

In 2018, ENAS had more than 120 members from more than 15 different countries and continues to grow. Researchers and students at all career levels interested in the cultural study of human aging and old age are invited to apply for membership of the association. This can easily be done by visiting the website of the association, www.agingstudies.eu (ENAS 2019), and following the instructions provided in the "Membership" section.

Future activities of ENAS will include regional workshops of European members and partner institutions as well as joint conferences with its North American sister organization NANAS.

Through academic events and other activities, ENAS will continue to support the institutionalization of Aging Studies research in Europe and beyond by facilitating interdisciplinary collaboration between scholars of all academic levels working in the field.

Cross-References

▶ North American Network in Aging Studies

Acknowledgment The authors thank Aagje Swinnen (University of Maastricht/University of Humanistic Studies, Utrecht, the Netherlands) for her assistance in preparing this entry.

References

ENAS (2019) Website of the European Network in Aging Studies. www.agingstudies.eu. Accessed 19 Mar 2019

Swinnen A, Port C (eds) Age, narrative, performance: Essays from the humanities. Special issue. International Journal of Aging and Later Life (2012) 7(2). https://doi.org/10.3384/ijal.1652-8670.1272

European Year of Active Ageing and Solidarity Between Generations

Gabriel Amitsis
University of West Attica, Athens, Greece

Definition

The European Year for Active Ageing and Solidarity between Generations (EY2012) is a nonbinding policy instrument adopted in 2011 by the lawmaking bodies of the European Union (the Council and the European Parliament) to address the challenges of an aging Europe through active aging in the areas of employment, participation, and independent living, looking both at the needs and rights of older people as well as their potentials and their contribution to the economy and society. It coincided with the tenth anniversary of the Madrid International Plan of Action on Ageing (United Nations 2002), which stressed that *"persons, as they age, should enjoy equal access to active participation in the economic, social, cultural and political life of their societies as the main development objective."*

Overview

The active aging discourse (Walker 2002; Zaidi and Howse 2017) is linked to a rather challenging context in the global aging social research agenda (Estes et al. 2003). Phillipson and Baars (2007) provide a useful distinction in terms of critical stages in the evolution of approaches to aging as: (a) the 1940s–1960s, when aging was viewed as the social problems for an individual arising from disengagement and dependency in old age; (b) the 1970s–1980s, when aging emerged as an economic and employment issue; and (c) the 1990s onward, when aging was viewed as a global phenomenon requiring concerted efforts of learnings from varying practices around the world. Moreover, it should be added the recent emphasis of active and healthy aging emerging from the European Union (EU) non-binding policy initiatives, such as the European Year for Active Ageing and Solidarity between Generations and the European Innovation Partnership–Active and Healthy Ageing (European Commission 2014a), and also from the World Health Organization Global Strategy on Ageing and Health (WHO 2015).

Active aging is, according to the European Commission, the process of creating better opportunities so that older women and men can play their part in the labor market, combating poverty, particularly that of women, and social exclusion, fostering volunteering and active participation in family life and society, and encouraging healthy aging with dignity (European Commission 2011). This involves, among other things, adapting working conditions, combating negative age stereotypes and age discrimination, improving health and safety at work, adapting lifelong learning systems to the needs of an aging workforce, and ensuring that social protection systems, particularly pension schemes, are adequate and provide the right incentives (Foster and Walker 2015).

Active aging is, according to the World Health Organization, the process of optimizing opportunities for health, participation, and security in order to enhance the quality of life as people age (WHO 2002). Active aging allows people to realize their potential for physical, social, and mental well-being throughout the life course and to participate in society while providing them with adequate protection, security, and care when they need it. Accordingly, the promotion of active aging requires a multidimensional approach and ownership by and lasting support among all generations (Walker 2009).

Taking into account the non-exclusive competences of the EU in the social policy field (as defined in article 5 of the Treaty on the functioning of the European Union – TFEU), this entry focuses on the most important flagship initiative of the EU active aging discourse, the European Year for Active Ageing and Solidarity between Generations (EY2012). It discusses its policy objectives and principles, administration, funding, and content, while it considers its implementation and evaluation process in the light of future directions of research.

The Adoption of the 2012 European Year for Active Ageing and Solidarity between Generations

The first calls for a European Year for Active Ageing and Solidarity between Generations (EY2012) within the EU social policy agenda were made during the Slovenian Presidency of the European Council in the first semester of 2008. In June 2009, the European Commission launched a public consultation with the aim of collecting ideas and suggestions from the key stakeholders and experts on how to achieve the greatest possible impact with a European Year for Active Ageing and Intergenerational Solidarity.

Based on relevant results, the European Commission had proposed, in September 2010, to designate the year 2012 as the European Year for Active Ageing (European Commission 2010a). It stressed that the preservation of solidarity between generations would, in particular, depend on ensuring that the baby boom cohorts stay longer in the labor market and remain healthy, active, and autonomous as long as possible. The Council and the European Parliament adopted the decision on this European Year in September 2011 (European Parliament and the Council 2011) adding "solidarity between generations" to its title and defined a budget envelope of five million euro for the period between January 2011 and December 2012.

The Context of the EY2012

The EY2012 sought to address the challenges of an aging Europe through active aging in the areas of employment, participation, and independent living, looking both at the needs and rights of older people as well as their potentials and their contribution to the economy and society. The overall objective of the Year was to mobilize relevant actors in the promotion of active aging and intergenerational solidarity. These actors include the Member States, their regional and local authorities, social partners, civil society, and the business community, including small- and medium-sized enterprises.

The specific objectives were set in Article 2 of the Decision No 940/2011/EU and can be summarized as follows: (1) to raise general awareness of the value of active aging and ensure that it is accorded a prominent position in the policy agenda; (2) to stimulate debate, exchange information, and develop mutual learning to promote active aging policies; (3) to create a framework for commitment and concrete action by Union and Member States with the involvement of stakeholders; and (4) to promote activities which will help to combat age discrimination, to overcome age-related stereotypes, and to remove barriers.

The Implementation of the EY2012 at the EU Level

Most of the activities at the EU level were designed to support stakeholders in the

Member States and facilitate their participation in the European Year. The EY2012 was thus implemented in close cooperation with national coordinators for the European Year and the Stakeholder Coalition. The main activity at European level was communication and promotional campaign implemented by a contractor through a media relations network. The central hub of this campaign was the EY website (http://europa.eu/ey2012), which presented information in 23 languages, although much of the content on specific initiatives could only be made available in the national language of the country concerned as well as in English. Five publications were released to support stakeholders: (a) the "Eurobarometer Special Survey on Active Ageing" (European Commission 2012a); (b) the Eurostat statistical portrait on active aging and solidarity between generations (European Commission 2012b); (c) the brochure "How to promote active ageing in Europe" (Age Platform Europe, European Commission and Committee of the Regions 2011); (d) the Social Europe Guide "Demography, active ageing and pensions" (European Commission 2012c); and (e) the brochure "The EU contribution to Active Ageing and Solidarity between generations" (European Commission 2012d).

The European Commission also organized several events and conferences throughout the year. Entitled "Stay Active – what does it take," the opening event took place in Copenhagen on 18–19 January 2012, hosted by the Danish EU Presidency. An awards ceremony of the European Year for Active Ageing and Solidarity between Generations took place on 13 November 2012 in Brussels. It highlighted inspiring people and initiatives that made a significant contribution to active aging and solidarity between generations. The European Commissioner for Employment, Social Affairs, and Inclusion, Mr. László Andor, announced the winners for the seven award categories.

The annual EU Access City Award, handed over by Vice-President Viviane Reding on 3 December 2012, paid particular attention to the accessibility needs of older persons. Representatives of older persons' organizations were included in the EU jury. The EY2012 closing event "From visions to actions" was held in Nicosia on 10 December 2012. It started with a statement from the President of the Republic of Cyprus, followed by a speech by Commissioner László Andor. The program covered a range of issues, including the current situation of aging in Europe and the still untapped potentials of older people.

The Implementation of the EY2012 at the National Level

All participating countries (27 EU Member States, Iceland, Liechtenstein, and Norway) appointed each a national coordinator (NC). Their work started well in advance of the European Year, with a first meeting held on 30 November of 2010. The NC prepared public activities, promoting local and regional activities and trying to involve all relevant stakeholders in the European Year. They presented National Work Programmes (NWPs) to the European Commission outlining the national activities planned throughout 2012.

The NWPs reflected different policy contexts, notably regarding active aging needs, policy priorities and distribution of competences, and the availability of resources. The Commission proposed, however, some common events and activities and supported in particular the organization national opening and closing events as well as the *Generations at School* and *Seniorforce Day* campaigns and the European award scheme. The opening events involved in whole 4,500 participants with a prominent level of political support demonstrated notably by the participation of high-level politicians. Fifteen countries appointed 115 national ambassadors of EY2012 in order to reach a wider public. They came from varied backgrounds: academics, nongovernmental organization (NGO) representatives, journalists, actors, and businesspeople. The number of ambassadors appointed varied from 1 (in Estonia, Hungary, and Poland) to 38 in Austria.

The EY2012 triggered a large number of initiatives for promoting active aging in the

Member States. A total of 748 national and transnational initiatives were implemented during the Year, counting only those featuring on the EU website. Civil society organizations were the promoter for 291 of these initiatives. Germany, Spain, and Italy were the countries with most initiatives on the EU website, followed by France, Austria, and Poland. The focus of these activities was mostly on exchanging knowledge and experience and on awareness-raising.

The Evaluation of the EY2012

In line with Article 11 of the Decision No 940/2011/EU, the Commission submitted in September 2014 its report containing an overall assessment of the initiatives provided for in this Decision with details of implementation and results to serve as a basis for future EU policies, measures, and actions in this field (European Commission 2014a), based on the conclusions from an external evaluation process (Ecorys 2012). According to this external evaluation, the Year had a positive impact on the conditions for active aging, both at European and national levels. The objectives and the activities of the Year were relevant, and the approach was successful in reaching the objectives (awareness-raising, stimulating debate and mutual learning, creating a framework for commitment and concrete action, combating age discrimination).

The objective of raising awareness has been primarily achieved through EU level and national initiatives and events, which were more often targeted to relevant organizations and civil society groups than individuals. The Generations@school initiative mobilized around 480 schools and an estimated 27.000 participants, while the European award scheme triggered 1.300 submissions (including most of the Generations@school events). The European "Seniorforce Day" involved over 11.000 participants. These initiatives attracted much media attention to the EY2012, drawing attention to success stories and positive examples highlighted in the award scheme. Altogether, the EY2012 generated significant press coverage with 6.162 printed articles, 3.432 online articles, and 329 broadcasts.

The EY2012 succeeded in stimulating public debate and fostering mutual learning around the themes of active aging and solidarity between generations. There was also a strong focus on good practice dissemination. The EU database was a powerful instrument to share information on success stories and positive examples. A number of national and transnational projects promoted mutual learning throughout the Year.

The need to combat age discrimination was highlighted in many public events and throughout various projects. In this context, the importance of improving the opportunities for active aging in employment was particularly emphasized.

The EY2012 offered a framework indeed to make commitments and to take concrete action. The mobilization of civil society organizations around the theme of the EY can be regarded as a significant achievement. The involvement of regional and local governments was also significant but did not happen in all countries. The EY2012 encouraged some countries to either strengthen an already existing national policy agenda on active aging and intergenerational solidarity or to develop new strategies. Austria, for example, launched the Federal Plan for Senior Citizens in January 2012. In Spain, a white paper was adopted. It analyzed the main aspects that determine the quality of life of older people.

A number of countries launched policy initiatives on the occasion of the EY: Belgium established in November 2012 a new Federal Advisory Council for older people; Ireland decided that every local authority area in the country should have its own age-friendly county program by the end of 2013; Poland adopted in August 2012 a government program for senior citizens' social activity for the years 2012–2013. Also, the external evaluation highlighted the strong complementarity between the EY2012 and other EU policy activities, such as the *White Paper on Pensions*, the *Demographic Forum*, the 2nd *Health Programme*, the *Accessibility Act*, and the *European Innovation Partnership on Active and Healthy Ageing*.

Follow-Up Initiatives of the EY2012

The EY2012 has managed to mobilize numerous governmental and nongovernmental actors. It has helped convey a more positive image of population aging by highlighting the potentials of older people and promoting their active participation in society and the economy (Tymowski 2015). The many EU Member States and civil society organizations have used the European Year as an opportunity to develop new initiatives or strengthen their existing ones. The activities triggered by the EY2012 demonstrated the growing interest in active aging.

Several tools have been or are being developed for this purpose and in particular (1) the *Guiding Principles for Active Ageing and Solidarity between the Generations* and (2) the *Active Ageing Index (AAI)*. The 19 *Guiding Principles for Active Ageing and Solidarity between the Generations* were endorsed by the EU's Council of Social Affairs Ministers on 6 December 2012 (European Council 2012). They are structured under the headings of the Year (employment, participation in society, and independent living), addressed to the EU Member States and other relevant levels of government and organizations which have a role to play in further improving conditions for active aging. The application of these guiding principles would also contribute to the attainment of the employment and poverty reduction targets of the Europe 2020 Strategy (European Commission 2010b), notably as a result of more people being able to work longer and earning better pension entitlements (European Commission 2015).

The *Active Ageing Index*, presented in December 2012 at the EY2012 Closing Conference in Cyprus, is a product of a joint project undertaken in 2012 by the European Commission Directorate General for Employment, Social Affairs, and Inclusion together with the Population Unit of the United Nations Economic Commission for Europe (UNECE) and the European Centre for Social Welfare Policy and Research in Vienna. It consists of the overall AAI, as well as its gender and the domain-specific breakdown and the constituting individual indicators. AAI measures the level to which older people live independent lives and participate in paid employment and social activities as well as their capacity to actively age. It has 22 indicators grouped into 4 domains: employment, social participation, independent living, and capacity for active aging. The first three domains measure achievements, while the fourth is a measure of preparedness for achieving positive results. All indicators and their aggregation into composite measures are available separately for men and women (Walker and Zaidi 2016).

The latest available results presented in the 2015 AAI report (Zaidi and Stanton 2015) give a clear indication that a healthy and active life during old age is no longer considered just an ideal in European countries; rather it is a reality for many and a genuine possibility for many more. The fact that the countries at the top of the AAI score have done consistently well across all domains is an indication that active aging is a coherent policy area where a balanced and well-founded approach can lead to achievements that leave nobody behind. At the same time, no country scores consistently at the very top in each domain, indicating that there is progress to be made for everyone but in different dimensions. Robust progress of about 2 points in the AAI on average happened over the 4 years since 2008 in the EU countries. This improvement is observed even though many countries experienced the 2008 financial and economic crisis and introduced severe fiscal austerity measures during this period (e.g., Amitsis 2013).

Future Directions of Research

According to the current division of powers between the EU and its Member States, national policymakers are still free to make key decisions about active aging, while the EU bodies continue to support the development of a healthy and active aging agenda as part of the European social model legacy in challenging times (Vaughan-Whitehead 2015). Taking as a starting point the outcomes of the EY2012, new research agendas should pay attention to the context and the impact of the most promising EU initiatives in the field. These have taken so far the form of (1) policy guidance, notably in the context of the European Semester where recommendations on longer employment careers have been addressed to most Member States (Azzopardi-Muscat and Brand 2014); (2) country-specific recommendations on long-term care (European Commission 2018a); (3) funding opportunities through the European Social Fund in the 2014–2020 Programming Period for Cohesion Policy (European Commission 2014b); and (4) the framework for the development of a Silver Economy Strategy (European Commission 2018b).

Summary

Over the past two decades, active aging has emerged in Europe as a fundamental policy response to the challenges of population aging. Although the EU has not succeeded yet to change dramatically the current status quo in the formulation of national active aging policies (due to complex political and legal constraints), the implementation of the EY2012 has marked a shift from a narrow economic or productivist perspective to a broader social paradigm, based on two pillars: active aging and intergenerational solidarity. It is expected that the new EU leadership (after the European Parliament elections in May 2019) will highlight the future directions of this shift during challenging policy processes between the EU bodies and national stakeholders (the Member States, their regional and local authorities, social partners, civil society, interest groups, and the business community, including small- and medium-sized enterprises).

Cross-References

▶ Active Aging and Active Aging Index
▶ Healthy Aging
▶ International Year of Older Persons
▶ The European Social Model

References

Age Platform Europe, European Commission and Committee of the Regions (2011) How to promote active ageing in Europe. Publications Office of the European Union, Luxembourg. https://doi.org/10.2767/36193

Amitsis G (2013) Challenging statutory pensions reforms in an aging Europe – adequacy versus sustainability. In: Phellas C (ed) Aging in European societies. Springer, New York, pp 9–32. https://doi.org/10.1007/978-1-4419-8345-9_2

Azzopardi-Muscat N, Brand H (2014) The 'European semester' – a growing force shaping health systems policy and reform in the European Union. Eur J Pub Health 24(Suppl 2). https://doi.org/10.1093/eurpub/cku164.089

Ecorys (2012) Evaluation of the European year for Active Ageing and Solidarity between Generations 2012. Report for the European Commission. Ecorys, Brussels

Estes C, Biggs S, Phillipson C (2003) Social theory, social policy and ageing. Open University Press, Maidenhead

European Commission (2010a) Proposal for a decision on the European Year for Active Ageing and solidarity between generations, Brussels, COM(2010) 462/6.9.2010

European Commission (2010b) Europe 2020: a strategy for smart, sustainable and inclusive growth. COM(2010) 2020. Communication from the Commission, Brussels. http://eur-lex.europa.eu/LexUriServ/LexUriServ.do?uri=COM:2010:2020:FIN:EN:PDF Accessed 1 Feb 2019

European Commission (2012a) Eurobarometer special survey on active ageing. http://ec.europa.eu/public_opinion/archives/eb_special_379_360_en.htm. Accessed 1 Feb 2019

European Commission (2012b) Active ageing and solidarity between generations. http://www.epp.eurostat.ec.europa.eu/portal/page/portal/product_details/publication?p_product_code=KS-EP-11-001. Accessed 1 Feb 2019

European Commission (2012c) Demography, active ageing and pensions. Publications Office of the European Union, Luxembourg. https://doi.org/10.2767/65869

European Commission (2012d) The EU contribution to active ageing and solidarity between generations. Publications Office of the European Union, Luxembourg. https://doi.org/10.2767/67267

European Commission (2014a) Report on the implementation, results and overall assessment of the 2012 European Year for Active Ageing and Solidarity between Generations. Brussels, COM/2014/0562/15.9.2014

European Commission (2014b) Investing in people: EU funding for employment and social inclusion. Social Europe guide, vol 7. Publications Office of the European Union, Luxembourg. https://doi.org/10.2767/23781

European Commission (2015) The 2015 ageing report – economic and budgetary projections for the 28 EU Member States (2013–2060), European Economy 3/2015.

http://ec.europa.eu/economy_finance/publications/european_economy/2015/pdf/ee3_en.pdf. Accessed 1 Feb 2019

European Commission (2018a) Communication 2018 European semester – Country-specific recommendations. Brussels, COM(2018) 400/23.5.2018

European Commission (2018b) The silver economy – final report. Publications Office of the European Union, Luxembourg. https://doi.org/10.2759/685036

European Council (2012) Declaration on the European Year for Active Ageing and Solidarity between Generations – the way forward. Brussels, 7 December 2012

European Parliament and the Council (2011) Decision no 940/2011/EU of 14 September 2011 on the European Year for Active Ageing and Solidarity between Generations. http://data.europa.eu/eli/dec/2011/940/oj. Accessed 1 Feb 2019

Foster R, Walker A (2015) Active and successful aging: a European policy perspective. The Gerontologist 55(1):83–90. https://doi.org/10.1093/geront/gnu028

Phillipson C, Baars J (2007) Social theory and social ageing. In: Westerhoff G, Bond J, Peace S, Dittman-Kohli F, Westerhof GJ (eds) Ageing in society: European perspectives on gerontology. Sage, London, pp 68–84. https://doi.org/10.4135/9781446278918.n4

Tymowski J (2015) European Year for Active Ageing and Solidarity between Generations: European implementation assessment. European Parliament, Brussels. https://doi.org/10.2861/230093

United Nations (2002) Report of the second world assembly on ageing. Madrid, 8–12 April 2002. A/CONF.179/9

Vaughan-Whitehead D (ed) (2015) The European social model in crisis – is Europe losing its soul? Edward Elgar Publishing, Cheltenham. https://doi.org/10.4347/9781783476565

Walker A (2002) A strategy for active ageing. Int Soc Secur Rev 55(1):121–140. https://doi.org/10.1111/1468-246X.00118

Walker A (2009) The emergence of active ageing in Europe. J Ageing Soc Policy 21(1):75–93. https://doi.org/10.1080/08959420802529986

Walker A, Zaidi A (2016) New evidence on active ageing in Europe. Intereconomics 51(3):139–144. https://doi.org/10.1007/s10272-016-0592-0

World Health Organization (2002) Active aging: a policy framework. Geneva. WHO reference number: WHO/NMH/NPH/02.8

World Health Organization (2015) World report on ageing and health. Geneva. https://www.who.int/ageing/.../world-report-2015.../en/

Zaid A, Stanton D (2015) Active Ageing Index 2014: analytical report. Centre for Research on Ageing, University of Southampton. http://www.southampton.ac.uk/assets/sharepoint/groupsite/Administration/SitePublisher-document-store/Documents/aai_report.pdf. Accessed 1 Feb 2019

Zaidi A, Howse K (2017) The policy discourse of active ageing: some reflections. J Popul Ageing 10(1):1–10. https://doi.org/10.1007/s12062-017-9174-6

Euthanasia and Senicide

Andy Hau Yan Ho[1,2,3] and Geraldine Tan-Ho[1]
[1]Psychology Programme, School of Social Sciences, Nanyang Technological University, Singapore, Singapore
[2]Centre for Population Health Sciences (CePHaS), Lee Kong Chian School of Medicine, Nanyang Technological University, Singapore, Singapore
[3]Palliative Care Centre for Excellence in Research and Education (PalC), Singapore, Singapore

Synonyms

Assisted death; Assisted dying; Assisted suicide; Mercy killing; Physician-assisted suicide

Definition

The word euthanasia was conceived from the Greek words "eu," meaning "good," and "thanatos," meaning "death" to refer to an easy and painless death. In modern society, euthanasia is understood as an employment or omission of procedures with the aim of accelerating or bringing about death in patients with incurable illnesses in order to release them from unbearable suffering. There are two forms of euthanasia: active euthanasia is when death is caused by a direct and deliberate act such as a doctor giving a patient a dose of lethal medication, while passive euthanasia is when the process of natural death is allowed to take place by either withdrawing or withholding life-sustaining treatment; for example, switching off a machine to allow a person to die of their illness. Euthanasia can further be categorized into nonvoluntary and voluntary euthanasia. Nonvoluntary euthanasia takes place without determining the patient's will; such cases would involve patients in a persistent vegetative state or without adequate cognitive abilities. Voluntary euthanasia is administered in fulfilment of the wishes expressed by the patient. The process of euthanasia is directly carried out by a physician, usually intravenously with a lethal substance; as such, it differs from assisted dying or assisted suicide, in which the physician provides oral lethal drugs to patients, who would then carry out the procedure themselves (Castro et al. 2016). Nonetheless, euthanasia and assisted dying are intricately linked in contemporary research and ethical debates. Senicide is the abandonment to death, suicide, or killing of older adults. It is a customary practice once found but no longer exist among certain indigenous cultures in Germany, Greenland, Canada, and Japan. The only country where senicide still exists in the contemporary era is India; despite being deemed unconstitutional under the Indian Penal Code (Chatterjee 2017), senicide, or better known as "Thalaikoothal," is still being practiced silently in the region of Tamil Nadu (Mathew 2016). While euthanasia and senicide both result in death, the intention of each practice differs greatly. Euthanasia is usually voluntary and carried out with the intentions of alleviating the patient's suffering. Senicide is usually involuntary and carried out because the family cannot afford to care for the sick older adults or perceives that older adults have lost his or her economic and practical value due to illness (Post 1990).

Overview

The oldest known practice of euthanasia comes from Ancient Greece and Rome, in which lethal substances such as hemlock were used to hasten one's death. This practice was supported by philosophers such as Socrates, Plato, and Seneca the Elder (Mystakidou et al. 2005). Euthanasia continued to be fairly accepted and common in the early eighteenth century, until the beginning of the contemporary euthanasia debate in the late 1800s (Kemp 2002). By the 1970s, a right-to-die movement emerged in the United States, which advocated for law reforms that allow individuals to decide on the manner of their own deaths. In 1990, the Patient Self-Determination Act (i.e., PSDA) was passed by the United States Congress, which assures individuals would be given an opportunity to participate in and direct their own healthcare decision. More importantly, the PSDA empowers patients with the right to make their own decisions on the type and intensity of medical care that they want and the conditions and timing of their deaths (Humphry and Clement 2000). The fundamental value that drives this movement is centered on the principle of respect of autonomy, which encompasses the values of individual choice, liberty rights, privacy, and freedom of will (Beauchamp and Childress 2001). In the years that follow, numerous organizations in the United States continue to push forth the right-to-die movement, while increasing number of court cases had also emerged to underscore the need for a clear legislation to support patients' right to die. By 1997, the state of Oregon enacted with Death with Dignity Act the legalization of physician-assisted suicide which allows terminally ill patients "to end their lives through the voluntary self-administration of lethal medication, expressive prescribed by a physician for that purpose" (Oregon Health Authority 2019). During the same period, similar movements on the right to die for pushing forth legislation for euthanasia and assisted suicide were also observed in Europe.

As of 2019, euthanasia and assisted suicide are legal in four Western European countries including the Netherlands, Belgium, Luxembourg, and Switzerland; in two North American countries including Canada and the United States, specifically in California, Colorado, District of Columbia, Hawaii, Oregon, Vermont, and Washington; as well as in South America, namely, Colombia. In both the Netherlands and Belgium, child euthanasia is also legal despite vigorous debates on its ethics, philosophy, and morality. While each country and state have their differing criteria and laws governing the administration of euthanasia and assisted suicide, the common requirement is that the patient must be suffering from a terminal disease that causes intense physical and psychological pain that cannot be managed successfully and consciously make a request for euthanasia or assisted suicide. This request must be supervised and approved by a doctor, a lawyer, and a psychiatrist or psychologist. With current and anticipated population aging which inevitably translate into greater prevalence of chronic life-threatening illnesses among older adults, a renewed emphasis on promoting quality of life, quality of death, and quality of palliative end-of-life care for terminally ill patients has prompted public discourses on assisted dying procedures in many countries around the world. While controversial, discussions and debates on the topic have become increasingly widespread and frequent (Castro et al. 2016).

Key Research Findings

Statistics and Trends

Steck and colleagues (2016) conducted a first-of-its-kind systematic review that examines the statistics, characteristics, and trends for assisted dying in Belgium, Luxembourg, the Netherlands, Switzerland, and the United States since the inception of relevant legislations till 2012. According to this report, the percentage of euthanasia and assisted deaths among all deaths ranged from 0.1% to 0.2% in the United States and Luxembourg to 1.8–2.9% in Belgium and the Netherlands, with the percentage of reported cases and the percentage of assisted deaths compared to all deaths increasing over time in most countries. The researchers further reported that the typical person dying from an assisted death was a well-educated male cancer patient between the ages of 60–85 years. Alongside this report, a handful of systematic review that examined the attitudes and perspectives toward euthanasia and assisted deaths among patients, family carers, public, as well as professional caregivers had also been published.

Attitudes and Perspectives of Patients, Carers, and the Public

An extensive amount of research using both quantitative and qualitative methods have been conducted in Western European societies to examine the attitudes and perspectives of patients, family carers, and the general public about euthanasia and assisted dying. Through a qualitative synthesis of 16 quantitative studies together with a systematic review of 94 surveys, Hendry et al. (2013) reported 4 major themes related to reasons for and against assisted deaths. First, concerns about poor quality of life as a result of serious illness, causing unbearable suffering, dependency, sense of burden, hopelessness, and loss of self, were identified as important reasons for considering assisted death among all participants regardless of illness experiences. While some believed they could potentially cope with the initial losses of functional capacity by redirecting their energy and interests on other more manageable activities, the ultimate loss of one's ability to live a meaningful and purposeful life leading to the decay of self-identity and personhood would render life not worth living (Kuuppelomaki 2000; Mak and Elwyn 2005). While some patients and carers acknowledged that intolerable physical pain and related symptoms of illness may influence the desire of assisted death as they are detrimental to quality of life, many argued that pain can be controlled and managed with good palliative care (Chapple et al. 2006; Eliott and Olver 2008) and therefore should not be the reason for a hastened death (Johansen et al. 2005; Winland-Brown 2001). However, incontinence and major cognitive impairments due to dementia were deemed particularly difficult to endure and were seen as reasons to cease living among patients and carers (Wilson et al. 2000).

Second, the desire for a good quality of death with one's capacity for choice, autonomy, and control over the manner of living and dying was deemed important reason that influences the decision for assisted death among all participants (Pearlman et al. 2005). It was believed that euthanasia and assisted dying can potentially contribute to individuals' quality of death through preventing prolonged and unnecessary suffering, respecting personal wishes, and upholding dignity. Third, there were concerns among study participants that the legislation of euthanasia and assisted dying could lead to a "slippery slope" in which vulnerable individuals such as those with major disabilities may be abused and coerced into a hastened death by medical professionals (Fadem et al. 2003; Dees et al. 2011). Other concerns include financial pressure on the cost of care which could lead patients and carers to request assisted death (Drum et al. 2010), as well as discriminated and marginalized groups may be offered limited care options for life-sustaining treatment under a legal system that supports euthanasia (Baeke et al. 2011). As such, participants believed there needs to be mechanisms and legislations to safeguard the rights to quality care for all patients. Finally, Hendry et al. (2013) reported that moral and religious views, as well as personal experience with mortality, have profound impact on one's view toward assisted dying. Particularly, participants who hold a strong belief system toward faith and ethics believed that euthanasia is morally wrong, while others who had witnessed the pain and suffering of a loved one's end of life may be more motivated to support assisted dying. Overall, around two-thirds of participants regardless of health status felt that euthanasia and assisted dying is acceptable, about half support its legislation, and around one-third would considered it themselves under specific circumstances.

Attitudes and Perspectives of Physicians, Nurses, and Social Workers

A vast body of research has examined the attitudes toward euthanasia and physician-assisted deaths among healthcare providers. Dickinson et al. (2005) systematically reviewed 39 high-quality studies that investigated US physicians' views toward these two practices and found that many more doctors agreed with physician-assisted death and supported its registration rather than active voluntary euthanasia. The reason for this may be attributed to the Hippocratic Oath; to help a patient take his or her own life may be more acceptable than taking the life directly. Dickinson et al. (2005) further reported that while doctors supported physician-assisted death in theory, they were not prepared to perform the task as they did not believe they have adequate knowledge to do so. Another systematic review conducted by Vézina-Im and colleagues (2014) examined 27 quantitative studies that investigated the variables associated with motivations of physicians and nurses to practice voluntary euthanasia. It was found that the single most significant variable associated with the willingness to perform euthanasia was past behavior – health professionals who were familiar with administering or being involved in euthanasia practices were more motivated to perform a euthanasia procedure. Another variable that influenced motivation for physicians and nurses was the type of medical specialty or work setting of the healthcare professional. The researchers reported that those working in specialties and settings that were exposed to patients with chronic suffering would be more motivated to carry out euthanasia than those who did not. The last significant motivation was the psychological state of the patient – physicians and nurses were more inclined to perform euthanasia for patients who were not assessed to be depressed and who had a short life expectancy. Religion, despite being the single variable that was most frequently assessed, was not found to be significantly related to motivation in more than half of the studies. Finally, McCormick (2011) reviewed 54 articles that examined self-determination and the right to die from a social work perspective and reported the majority of social workers in the United States support the principle of patient self-determination, and self-determination is highly correlated with the support of both euthanasia and physician-assisted deaths. Specifically, 50% of social workers in Texas and 65–70% of social worker in Washington supported both practices. The researcher concluded that social workers are generally supportive of a right to die through the mean of passive euthanasia; he further urged the social work profession to undertake a more public stance in their support for self-determination so as to protect patients' rights particularly in the context of palliative care and end-of-life care.

Future Directions of Research

So far little is known about the beliefs, attitudes, and perspectives toward such practices in greater Asia. The reason for this gap of knowledge may well be due to a long-standing emphasis on respecting and cultivating life within the Eastern collectivist cosmology. Nonetheless, with rapid population aging across all major Asian countries including China, Japan, Korea, and Singapore, the demand for greater self-determination, control, and autonomy among the up-and-coming generations of older adults will inevitably put euthanasia and assisted death into the spotlight of public discourse. Hence, the need to examine public and professional views on such practices is warranted for informing relevant healthcare policies.

Summary

Euthanasia and assisted death will continue to be a topic of debate under the rubric of population aging, self-determination, and human right ethics. While there may never be an absolute consensus regarding the morality of such practices, one thing that all people can agree upon is the necessity to improve and enhance holistic palliative care and end-of-life care provisions around the world. As the rights to die may be contested under different circumstances, the imperative need to protect and uphold the dignity, quality of life, and quality of death of every terminally ill patient is incontestable as these acts of kindness and compassion define the very fabric of our existence and our common humanity.

Cross-References

▶ Critical Care Nursing
▶ Death with Dignity
▶ End-of-Life Care
▶ End-of-Life Decision-Making in Acute Care Settings
▶ Personalized Medicine and Decision-Making

References

Baeke G, Wils JP, Broeckaert B (2011) 'We are (not) the master of our body': elderly Jewish women's attitudes towards euthanasia and assisted suicide. Ethn Health 16(3):259–278
Beauchamp TL, Childress JF (2001) Principles of biomedical ethics. Oxford University Press, New York
Castro MR, Antunes GC, Marcon LMP, Andrade LS, Ruckl S, Andrade VLA (2016) Euthanasia and assisted suicide in western countries: a systematic review. Rev Bioetic 24(2):355–367
Chapple A, Ziebland S, McPherson A, Herxheimer A (2006) What people close to death say about euthanasia and assisted suicide: a qualitative study. J Medial Ethics 32(12):706–710
Chatterjee P (2017) The customary practice of Senicide, with special reference to India. Gin Verlag, Munich
Dees M, Vernooij-Dassen M, Dekkers W, Vissers KC, van Well C (2011) 'Unbearable suffering': a qualitative study on the perspectives of patients who request assistance in dying. J Med Ethics 37(12):727–734
Dickinson GE, Clark D, Winslow M, Marples R (2005) US physicians' attitudes concerning euthanasia and physician-assisted death: a systematic literature review. Morality 10(1):43–52
Drum CE, White G, Taitano G, Horner-Johnson W (2010) The Oregon death with dignity act: results of a literature review and naturalistic inquiry. Disabil Health J 3(1):3–15
Eliott JA, Olver IN (2008) Dying cancer patients talk about euthanasia. Soc Sci Med 67(4):647–656
Fadem P, Minkler M, Perry M, Blum K, Moore LF, Rogers J, Williams L (2003) Attitudes of people with disabilities toward physician-assisted suicide legislation: broadening the dialogue. J Health Polit Policy Law 28(6):977–1001
Hendry M, Pasterfield D, Lewis R, Carter B, Hodgson D, Wilkinson C (2013) Why do we want the right to die? A systematic review of the international literature on the views of patients, carers and the public on assisted dying. Palliat Med 27(1):13–26
Humphry D, Clement M (2000) Freedom to die: people, politics, and the right-to-die movement. St. Martin's Griffin, New York
Johansen S, Holen JC, Kaasa S, Loge HJ, Materstvedt LJ (2005) Attitudes towards, and wishes for, euthanasia in advanced cancer patients at a palliative medicine unit. Palliat Med 19(6):454–460
Kemp N (2002) Merciful release: a history of the British euthanasia movement. Manchester University Press, Manchester
Kuuppelomaki M (2000) Attitudes of cancer patients, their family members and health professionals toward active euthanasia. Eur J Cancer Care 9(1):16–21
Mak YYW, Elwyn G (2005) Voices of the terminally ill: uncovering the meaning of desire for euthanasia. Palliat Med 19(4):343–350
Mathew S (2016) Thalaikoothal: killing of the already withering. Retrieved from: https://soumyamathew94.wordpress.com/2016/04/16/thalaikoothal-killing-of-the-already-withering/
McCormick AJ (2011) Self-determination, the right to die, and culture: a literature review. Soc Work 56(2):119–128
Mystakidou K, Parpa E, Tsilika E, Katsouda E, Vlahos L (2005) The evolution of euthanasia and its perceptions in Greek culture and civilization. Perspect Biol Med 48(1):95–104
Oregon Health Authority (2019) Death with Dignity Act. Retrieved from: https://www.oregon.gov/oha/PH/

PROVIDERPARTNERRESOURCES/EVALUATION
RESEARCH/DEATHWITHDIGNITYACT/Pages/
index.aspx

Pearlman RA, Hsu C, Starks H, Back AL, Gordon JR, Bharucha AJ, Koenig BA, Battin MP (2005) Motivations for physician- assisted suicide: patient and family voices. J Gen Intern Med 20(3):234–239

Post S (1990) Severely demented elderly people: a case against senicide. J Am Geriatr Soc 38(6):1532–5415

Steck N, Egger M, Maessen M, Reisch T, Zwahlen M (2016) Euthanasia and assisted suicide in selected European countries and US states: systematic literature review. Med Care 51(10):938–944

Vezina-Im LA, Lavoie M, Krol P, Olivier-D'Avignon M (2014) Motivations of physicians and nurses to practice voluntary euthanasia: a systematic review. BMC Palliat Care 13(1):20

Wilson KG, Scott JF, Graham ID, Kozak JF, Chater S, Viola RA, de Faye BJ, Wwaver LA, Curran D (2000) Attitudes of terminally ill patients toward euthanasia and physician-assisted suicide. JAMA Intern Med 160 (16):2454–2460

Winland-Brown JE (2001) John, and Mary Q. Public's perceptions of a good death and assisted suicide. Issues Interdiscip Care 3(2):137–144

Event History Analysis

▶ Survival Analysis

Event-Related Potentials

▶ Electrophysiology

Evolutionary Aging Theories

▶ Non-evolutionary and Evolutionary Aging Theories

Evolutionary Connections Between Aging and Cancer

▶ Aging and Cancer

Evolutionary Theories of Aging

▶ Mutation Accumulation Aging Theory

Evolvability Theory of Aging

Theodore C. Goldsmith
Azinet LLC, Annapolis, MD, USA

Synonym

Theory to the effect that senescence aids evolutionary adaptation

Definition

Evolvability theory of aging: Theory proposing that senescence increases a population's ability to evolve resulting in the acquisition of biological mechanisms that cause senescence. This is one of a family of theories to the effect that aging is genetically programmed because senescence produces an evolutionary advantage for a possessing population.

Overview

Efforts to develop an aging theory based on Darwin's evolutionary mechanics theory (1859) failed to explain observed mammal aging despite nearly a century of effort. This led to the development of modern "evolutionary" non-programmed aging theories based on modifications to Darwin's mechanics (Medawar 1952; Williams 1957) proposing that senescence has little negative impact on a species *population* despite being eventually catastrophic as seen from an *individual's* viewpoint.

Modern programmed aging theories are additionally based on any of a family of post-1962 population-oriented evolutionary mechanics theories that propose that an evolved organism design characteristic (trait) that benefits the survival and growth of a species population can evolve and be retained despite creating some disadvantage for individual members. Evolutionary mechanics theories in this category include group selection (Wynne-Edwards 1962), kin selection (Hamilton 1964), and small-group selection (Travis 2004) in addition to evolvability (Wagner and Altenberg 1996).

Evolvability-based aging theories propose that senescence benefits the ability of a population to adapt and therefore survive and grow and that this benefit caused the evolution and retention of programmed aging. Evolvability concepts also provide explanations for other observations that appear to conflict with Darwin's mechanics such as animal altruism, sexual reproduction, and reproductive behaviors that produce an individual disadvantage.

Darwin's evolutionary mechanics concept (1859) assumes that the ability to evolve (we can use the term *evolvability*) is an inherent property of all living organisms that did not vary between species and therefore could be considered a constant. His concept assumed that all organisms possessed the ability to transfer information about their designs to descendants, along with the ability to copy the information in order to produce multiple descendants. Darwin further assumed all organisms were susceptible to mutations that would change that information and therefore their subsequently inherited designs and that all evolving organisms were living and competing for survival under wild conditions in which their populations were limited by external circumstances. Darwin's idea was very individual-oriented: The evolution process causes the selection of traits that cause *possessing individuals* to produce more adult descendants.

Subsequent genetics discoveries showed that some of these assumptions are incorrect and that, in complex, diploid, sexually reproducing species, most and possibly essentially all of an organism's ability to evolve is itself the result of evolved traits and other characteristics of the population involved (Goldsmith 2014). We can more definitively define evolvability as encompassing the precision and the rate at which the evolution process could adapt an organism to changes in its external world. Darwin's "tiny steps" incremental evolution concept, still widely accepted, requires that the evolution process be capable of high precision or be capable of distinguishing between very small differences in designs: Are slightly longer claws better? Consequently, the precision with which the evolution process operates would be important. Similarly the adaptation rate or time required to accomplish a particular increment of adaptation would be important. A population that could adapt more rapidly would logically have an evolutionary advantage.

Adult Death Rate Affects Evolvability

Whether a single individual lives longer and breeds more than another individual having a slightly different design is a matter of luck or chance. What we can say is that individuals having a particular design possess a greater probability of producing descendants than those having some other slightly different design. We can therefore consider the life of an individual to represent a trial in the probability sense of the design possessed by that individual and that the rate and precision with which evolution would proceed would depend on the rate at which trials were executed. In other words evolvability would depend on death rate and the consequent statistics associated with the evolution process in a particular population.

Because adult traits are not fully expressed in juveniles, deaths in juveniles generally do not affect the evolution of adult traits. This is because latent traits do not affect the performance of the organism in surviving and reproducing. Therefore we can modify this idea to state that evolvability is a function of *adult death rate*, which in turn is proportional to the size of the population and inversely proportional to the length of an average adult life (Goldsmith 2008). Figure 1 describes the idea that an optimum trial (organism life) would entail living long enough to become reproductively and physically mature followed by a period in which the evolution process would evaluate the merits of the organism's design relative to surviving and reproducing (the trial period), followed by the end of the trial enforced by the organism's

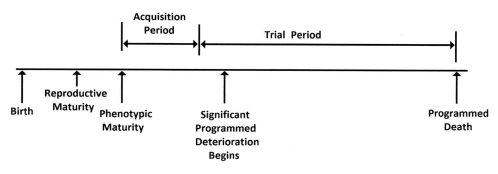

Evolvability Theory of Aging, Fig. 1 Life of an organism seen as a trial of the organism's inherited design

internal design. According to this concept, the internal ability to live longer than a species-specific age creates an evolutionary *disadvantage*.

This analysis suggests that because larger and more complex organisms necessarily require more time for growth (and consequently longer lives) and necessarily have smaller populations, that evolvability nominally declines with increasing organism size and complexity. However, complex organisms possess many evolved traits that plausibly greatly increase evolvability, compensating for this problem. In particular, traits that act to increase the evolutionary effect of each individual life would help compensate for smaller populations and longer life spans as further discussed below.

The Evolution of Acquisition Traits

Acquisition traits in animals complicate the evolvability issue (Goldsmith 2017). We can define an acquisition trait as one in which the evolutionary value of the trait depends on the acquisition of something that accumulates during the lifetime of the organism. *Intelligence* could be defined as the ability to process acquired information about the external world in such a way as to increase an animal's probability of surviving and reproducing. We can then define *wisdom* as essentially the product of intelligence and experience where experience is the totality of information acquired during the animal's life. Experience accumulates during an animal's life, and older animals therefore have more experience. This is the basis of the intelligence quotient or IQ concept. The problem here is that the selectable property is wisdom and wisdom nominally increases with age. Therefore, an older less intelligent but more experienced individual could likely be more fit than a younger more intelligent but less experienced individual. In a non-senescent population, this would interfere with the evolution of intelligence. Therefore, senescence, by gradually compensating for the age advantage, aids the evolution of intelligence.

Immunity is another acquisition trait in which an animal progressively acquires immunity during its life because of incrementally adding exposures to pathogens. Consequently an older, non-senescing animal is more likely to avoid infection than an otherwise identical younger animal. Here again, the evolution of the complex mechanisms that allow immunity to be acquired would be aided by senescence.

The Evolution of Variation

Darwin proposed that *natural variation* in inheritable characteristics was essential to the evolution process. Without variation there would be nothing for the natural selection process to select. We could add to this discussion the definition *local variation* or variation between individuals that could plausibly interact with each other in a competitive natural selection context. Darwin could also reasonably assume that the variation in designs seen between close relatives such as siblings was caused by an inherent property of biological inheritance similar to the inevitable presence of noise in an analog communication.

However, as famously described by Watson and Crick (1953), biological inheritance is a *digital data communications process*. Information concerning an organism's design is conveyed by the sequence in which four different base pairs appear in DNA molecules. Variation is *not* a "natural" property of a digital data communications scheme, which inherently produces exact copies of data or nonfunctional corrupted data (see Goldsmith 2014). Genetics discoveries showed that the variation we see between individuals in a population of a complex species is actually the result of extremely complex and obviously evolved mechanisms that handle the digital data such as sexual reproduction, diploid genomic structure, recombination, unequal crossover, and transposons (Krebs et al. 2017). "Natural" variation is actually the result of the operation of many evolved traits. For example, a species population possessing a reproductive behavior that caused individuals to seek mates that were remotely located or phenotypically different from themselves would tend to have more variation than a population having a reproductive behavior that caused it to select mates locally. As another example, recombination, occurring in virtually every instance of sexual reproduction, can produce variation between even close relatives such as siblings, obviously aiding the existence of local variation. From this viewpoint, identical twins represent a malfunction in the variation-producing mechanisms.

Evolvability and Reproduction

Relative to asexual reproduction, sexual reproduction in animals certainly appears to involve a major Darwinian disadvantage. Because of the relative reproductive uselessness of males, sexual reproduction produces a nominal factor-of-two reduction in an individual's ability to produce descendants. In animals that tend to their young like birds and mammals, a male could provide nurturing and protective functions that partly offset the disadvantage. This would not be true for species in which this was not the case such as reptiles. The discussion above suggests that sexual reproduction provides a major evolvability advantage that compensated for its reproductive disadvantage.

As another example, diploid sexual reproduction creates a situation in which a single individual can contain substantial genetic diversity in the form of differences between its two sets of genetic data. This allows a population that has been decimated by some event to recover into one that immediately possesses genetic diversity and local variation.

In a semelparous species, an individual either does or does not reproduce a somewhat binary result although the number of surviving descendants could vary. The life of a multiparous individual is more nuanced because each subsequent reproduction represents another instance of survival and reproduction. Therefore a multiparous life would have an evolvability advantage by contributing more information per life lived and therefore increasing evolvability relative to a semelparous population having the same adult death rate. Similarly, some believe (Goldsmith 2014; Skulachev 1997) that gradual multi-symptom senescence such as seen in mammals and other multiparous species provides an evolvability advantage over acute single-symptom suicide. Aging could have a challenge effect that in more complex animals would increase evolvability.

The bighorn sheep (*Ovis canadensis*) have a mating behavior that causes them to have head-butting contests to determine mating rights. Winners are the larger and stronger and consequently tend to be older and more developed.

This behavior is individually adverse and reduces the probability that an individual will reproduce. While individuals are sexually mature at 2 years of age, observers (Valdez and Krausman 1999) suggest a typical wild sheep will not mate until substantially older.

This behavior acts as a population-density-sensitive limitation on reproduction and would represent a smaller restriction in sparsely populated areas. This would act to benefit a population by acting to prevent it from overpopulating its food supply.

The behavior directly acts to encourage the subsequent evolution of particular phenotypic properties such as strong necks, cushioned brains, and large horns that would presumably not have evolved in the absence of the mating ritual. In effect, the sheep evolved their own branch of evolutionary mechanics theory.

In a non-senescent context, this behavior would tend to ruin genetic diversity and variation by allowing a few individuals to dominate reproduction. The sheep illustrate how an individually adverse trait can benefit a population and further that evolutionary mechanics is a very complex subject. The preceding discussion suggests that evolvability and design-limited life span would be more significant in complex organisms such as mammals. This may help explain extremely long life spans seen in some simple organisms such as trees and clams.

Reproduction is clearly highly associated with senescence (Libertini 2017). For example, an increase in reproductive maturity would affect the life span needed by an organism.

Evolvability-Based Programmed Aging Theories

Weismann suggested (1882) that limiting individual life span by means of an evolved programmed death mechanism would aid the evolution process by freeing resources for younger members of a population. Because younger members were presumably minutely more evolved, doing this would aid the evolution process to operate more rapidly. Until recently, Weismann's idea was widely dismissed because at the time there did not exist any rationale for rejecting Darwin's individual-oriented evolutionary mechanics concept and Weismann eventually recanted (1892).

A more recent evolvability-based programmed aging theory (Goldsmith 2017) suggests that programmed aging would benefit a population in multiple ways:

```
Increases adult death rate and thereby
evolvability
Enhances the evolution of acquisition
traits
Enhances the evolution of animals with
social structure
Increases local variation and genetic
diversity
Provides a challenge effect and enhances
per-life evolutionary benefit
```

Objections to Evolvability-Based Programmed Aging Theories

Proponents of modern non-programmed aging theories such as the antagonistic pleiotropy theory (Williams 1957) and disposable soma theory (Kirkwood and Holliday 1979) have long argued that all of the post-1962 population-oriented evolutionary mechanics theories are invalid because a long-term benefit (e.g., survival of a population) cannot offset a short-term disadvantage (e.g., reduced probability that an individual will produce adult descendants). However, genetics discoveries have exposed many aspects of biological inheritance that act toward validating all of the population-oriented theories (see Goldsmith 2014). Note that the non-programmed theories mentioned above are themselves based on the population-oriented concepts of Medawar and Williams (see the entry "▶ Timeline of Aging Research"). Proponents of programmed aging have extensively criticized the non-programmed theories (e.g., Goldsmith 2013; Skulachev 2011). In addition, evolvability-based aging theories have counter-arguments of their own as described below.

A common objection to programmed aging theories (e.g., de Grey 2007) is the idea that only a small surviving portion of a population would be affected by the internal ability (or nonability) to live longer than a species-specific age and therefore there would be little evolutionary force toward living longer *or* toward limiting life. Consequently there would be limited evolutionary motivation to evolve and retain a suicide mechanism. However, if internally limiting life span increases evolvability, we would expect that such increase would generally improve the ability of an organism to evolve and therefore affect its ability to evolve any trait, even those that benefit young individuals. The negative impact of long-surviving individuals could exceed a simple numerical assessment in ways such as described above.

Another objection (de Grey 2015) is that in a wild population, *average* life span is clearly limited by external conditions and so it is not clear why an internal program or suicide mechanism would be necessary. However, in the absence of internal life span limitation, some individuals could be expected to live very long lives and produce a very large number of descendants. Assuming the population size was limited by external conditions, many other individuals would necessarily die without producing adult descendants. This would reduce adult death rate, variation, genetic diversity, and therefore evolvability. This effect would be worse in animals having a social organization or "pecking order," which is, at least to some extent, an acquisition trait. The "king" is much less likely to die from combat, predators, environmental conditions, or starvation than a "foot soldier" and thus, if internally immortal, could live a very long time and produce a very large number of descendants.

Future Directions of Research

Evolvability-based aging theories and other programmed aging theories suggest that senescence is an evolved biological function and is therefore likely to be scheduled and implemented in a manner similar to other life-cycle functions such as growth and reproduction. Such functions typically involve nervous or chemical signaling and can involve detection of external conditions in the coordination of the function's effect on diverse tissues and systems. Research toward exploring signaling pathways and other aspects of an aging program has begun. See the entry "▶ Timeline of Aging Research" and (Longo et al. 2015; Mitteldorf 2018).

Summary

Programmed aging theories based on evolvability suggest that biological mechanisms that cause senescence have evolved because senescence increases the ability of a possessing population to survive and grow by increasing its ability to adapt. Programmed aging theories suggest that aging is itself a treatable condition and suggest potential intervention targets.

Cross-references

▶ Non-programmed Aging Theories
▶ Programmed Aging Theories
▶ Timeline of Aging Research

References

Darwin CR (1859) On the origin of species by means of natural selection, or the preservation of the favoured races in the struggle for life. John Murray, London
de Grey A (2007) Calorie restriction, post-reproductive life span, and programmed aging: a plea for rigor. Ann N Y Acad Sci 1119:296–305. https://doi.org/10.1196/annals.1404.029
de Grey A (2015) Do we have genes that exist to hasten aging? New data, new arguments, but the answer is still no. Curr Aging Sci 8(1):24–33
Goldsmith TC (2008) Aging, evolvability, and the individual benefit requirement; medical implications of aging theory controversies. J Theor Biol 252:764–768. https://doi.org/10.1016/j.jtbi.2008.02.035
Goldsmith T (2013) Arguments against non-programmed aging theories. Biochem Mosc 78:971–978. https://doi.org/10.1134/S0006297913090022
Goldsmith TC (2014) The evolution of aging, 3rd ed. Azinet Press, Annapolis. ISBN 13: 978-0-9788709-0-4
Goldsmith TC (2017) Evolvability, population benefit, and the evolution of programmed aging in mammals. Biochem Mosc 82(12):1423–1429. https://doi.org/10.1134/S0006297917120021
Hamilton WD (1964) The genetical evolution of social behaviour, I, II. J Theor Biol 7:1–52
Kirkwood TB, Holliday R (1979) The evolution of ageing and longevity. Proc R Soc Lond B Biol Sci 205:531–546. https://doi.org/10.1098/rspb.1979.0083
Krebs J, ed. et al. (2017) Lewin's genes XII. Jones & Bartlett Learning, Burlington. ISBN 1284104494
Libertini G (2017) Sex and aging: a comparison between two phenoptotic phenomena. Biochemistry (Mosc) 82(12):1435–1455. https://doi.org/10.1134/S0006297917120045
Longo VD, Antebi A, Bartke A, Barzilai N, Brown-Borg HM, Caruso C et al (2015) Interventions to slow aging in humans: are we ready? Aging Cell 14:497–510. https://doi.org/10.1111/acel.12338
Medawar PB (1952) An unsolved problem in biology. H. K. Lewis, London. Reprinted in: Medawar PB (1957) The uniqueness of the individual. Methuen, London
Mitteldorf J (2018) An incipient revolution in the testing of anti-aging strategies. Biochem Mosc 83(12):2018. https://doi.org/10.1134/S000629791812009X
Skulachev V (1997) Aging is a specific biological function rather than the result of a disorder in complex living systems: biochemical evidence in support of Weismann's hypothesis. Biochem Mosc 62(11):1191–1195
Skulachev V (2011) Aging as a particular case of phenoptosis, the programmed death of an organism. (A response to Kirkwood-Melov "on the programmed/non-programmed nature of aging within the life history"). Aging (Albany NY) 3(11):1120–1123. https://doi.org/10.18632/aging.100403
Travis J (2004) The evolution of programmed death in a spatially structured population. J Gerontol A Biol Sci Med Sci 59A(4):301–305

Valdez R, Krausman PR (1999) Mountain sheep of North America. The University of Arizona Press, Tucson

Wagner G, Altenberg L (1996) Perspective: complex adaptations and the evolution of evolvability. Evolution 50:967–976. https://doi.org/10.1111/j.1558-5646.1996.tb02339.x

Watson JD, Crick FH (1953) Molecular structure of nucleic acids: a structure for deoxyribose nucleic acid. Nature 171:737–738

Weismann A (1882) Uber die Dauer des Lebens. Fischer, Jena

Weismann A (1892) Essays upon heredity and kindred biological problems, vol II. Clarendon Press, Oxford

Williams GC (1957) Pleiotropy, natural selection and the evolution of senescence. Evolution 11:398–411

Wynne-Edwards VC (1962) Animal dispersion in relation to social behaviour. Oliver & Boyd Ltd, Edinburgh

E-Writing Therapy

▶ Cybercounseling

Ex-caregivers

▶ Former Carers

Ex-carers

▶ Former Carers

Exclusion

▶ Age Segregation

Executive Function

▶ Working Memory

Exemption Program

▶ Entitlement Program

Exercise

▶ Exercise Adherence
▶ Physical Activities

Exercise Adherence

Helen L. Graham
Helen and Arthur E. Johnson College of Nursing and Health Sciences, University of Colorado Colorado Springs, Colorado Springs, CO, USA

Synonyms

Adherence; Compliance; Exercise; Exercise maintenance; Geriatric medicine; Nonadherence; Physical activity; Rehabilitation

Definition

Exercise is a structured and repetitive physical conditioning program and a subcategory of physical activity (PA) (Rivera-Torres et al. 2019). Major purposes of exercise are conditioning, maintaining, and improving individual health (Room et al. 2017). Adherence, once referred to as compliance (Rosner 2006), has not been consistently defined. The World Health Organization defines adherence by the degree of how well a person's behavior corresponds to the instructions received from medical or other health-care providers (Sabate 2003). Adherence to exercise can be intentional or nonintentional (Horne et al. 2013). Exercise adherence is a modifiable health risk factor over which individuals have a degree of control. Older adults' willingness and ability to commit to a routine exercise plan affect health outcomes.

Overview

Scientific evidence supports the benefits of physical activity and regular exercise for adults with and without chronic health conditions (Hoffmann et al. 2016; Eckel et al. 2014). The American Heart Association recommends physician referral to a cardiac rehabilitation exercise program for adults of all ages following a heart attack and for other cardiac conditions (Labarthe et al. 2016). For older adults, the World Health Organization (Sabate 2003) recommends 150 min of weekly aerobic exercise including 2–3 days of resistance training (Franco et al. 2015). A physician assessment ideally precedes the initiation of an exercise program, and recommendations should be made for the appropriate level of exercise intensity (low, moderate, or high) for older adults. Rarely however do older adults engage in moderate to vigorous levels of exercise (Sparling et al. 2015; Garber et al. 2011; Hill, et al. 2011). Health-care providers prescribe exercise regimes for patients with the expectation recommendations will be followed. According to Picorelli et al. (2014), variability in adherence measures makes comparisons between studies difficult. Although there is not an agreed-upon optimal exercise adherence measure, it has been measured as (1) average number of home, gym, or program exercise sessions attended per week, (2) number of weeks exercising, (3) portion of days undertaken, or (4) number of minutes exercising/weekly (Rivera-Torres et al. 2019).

Meeting physical activity recommendations is a challenge for adults, especially for older adults (Franco et al. 2015). Exercise adherence is multifactorial and affected largely by physical ability, personality, psychosocial characteristics, and demographic factors (Picorelli et al. 2014). For older adults, there is a fear of falling, of strains, or of getting hurt (Hill et al. 2011). According to Franco et al. (2015), social influence, physical limitations, competing priorities, access difficulties, personal benefits, motivation, and beliefs influence exercise adherence. Long-term adherence to health-promoting behaviors is more complex and difficult to sustain than short-term adherence (D'Angelo et al. 2014). Problems associated with poor exercise adherence are not solely influenced by individual factors but also by factors associated with provider and health-care systems (Hawley-Hague et al. 2014). Attitudes of the therapist as well as the patient regarding exercise should be considered when setting exercise goals (Franco et al. 2015).

Individual commitment and adherence to routine exercise are largely behavioral in nature. Behavioral theories are the basis for measuring exercise adherence and testing physical activity interventions. Frequently cited theories include Social Cognitive Theory (Bandura 1997) and the Theory of Planned Behavior (Madden et al. 1992). Both provide a comprehensive framework taking into account individual and social factors effecting adherence and nonadherence behaviors. Bandura's Social Cognitive Theory, the most often cited health behavioral change model, stresses, regardless of an individual's behavior, that adherence will be dependent on the individual's outcome expectations and his or her belief in their ability to engage in a particular behavior (Bandura 1997; Brosse et al. 2002). Keller et al. (1999) found Social Cognitive Theory resourceful in predicting initiation and maintaining physical activity behaviors among adults. Attitudinal components affecting exercise adherence in older adult are cognitive, affective, and behavioral; each influences intention to exercise and to maintain exercise behaviors. Ajzen and Madden (1986) studied behavior, attitudes, and subjective norms when testing interventions designed to increase physical activity and found intention to act correlated more with behavior than with attitudes and subjective norms. Perceived behavioral control, a concept of the Theory of Reasoned Action, includes both personal and external factors of time, money, willpower, and opportunity (Shumaker et al. 1998) which influence health behavior decisions. Exercise programs based on health behavior theoretical constructs are effective in promoting and increasing physical activity (Garber et al. 2011).

Key Research Findings

Nonadherence to a healthy lifestyle including physical activity and exercise is a significant health concern nationally and globally (Franco et al. 2015). Despite strong evidence of physical and mental benefits from regular exercise (Chodzko-Zajko et al. 2009; Garber et al. 2011), fewer than 80% of individuals in the United States meet the PA recommended guidelines set forth by the US Department of Health and Human Services (2002). In fact, fewer than 50% of the population in developed countries adhere to long-term health therapies, and rates for adherence are even lower in developing countries (Sabate 2003). Researchers report that 50% of older adults have trouble staying with exercise (Rosner 2006). The barriers related to exercise adherence are known, but the motivations of individuals who maintain exercise adherence are not as easily understood.

Empirical research for predicting adherence guides interventional physical activity studies. Supervised exercise programs demonstrate better adherence outcomes than unsupervised programs. Effective communication and positive relationships between providers, therapists, and patients are other factors contributing to increased exercise program attendance (Picorelli et al. 2014). Additional adherence factors include individual health condition and psychological (Brosse et al. 2002) as well as physical status (Liu and Miyawaki 2019). A Cochrane study reviewed interventions designed to improve exercise adherence in adults 65 years and older and found that motivation classes, information messages, and other interventions were unsuccessful as compared to interventions offering participant monitoring, feedback, and booster sessions (Room et al. 2017). The perspectives of older people to physical activity are found in comments from systematic qualitative studies. Predominant themes include personal benefits of physical activity, physical limitations, social influences, and competing priorities (Franco et al. 2015).

Health-care providers and policy makers recognize the need to increase exercise adherence among adults, especially older adults. Randomized control trials (RCT), the gold standard of studies, overwhelmingly demonstrate an association between supervised exercise participation and increased longevity, including reduced hospitalizations in older adults with cardiac conditions (Heran et al. 2011). In a review of 16 meta-analyses, investigators observed exercise, compared to certain medications, decreased mortality. This finding however was limited to *specific* disease conditions (Naci and Ioannidis 2013).

Social support and encouragement to exercise coming from external sources have been shown to be exceptionally beneficial. Family members, neighbors, community members, social network, and media all influence individual exercise behavior (Rivera-Torres et al. 2019). Social support is a benefit of attending community center-based adult exercise programs (Picorelli et al. 2014), a benefit not acquired from home-based programs. In one community center, physical strength and conditioning program researchers found long-term training promising in individuals 75 years and older (Aartolahti et al. 2015). Overall, studies find group-based exercise programs appear to have better adherence rates compared with home-based programs (Kohn et al. 2016).

Future Directions for Research

The complexity of issues surrounding exercise adherence, especially for older adults, is significant. Researching interventions, which optimally support older adults exercising, should continue in developed and developing countries. Interventional research will be instrumental in developing methods to increase exercise adherence and for tackling nonadherence issues. Identifying what exercise is enjoyable and meets individuals' values and needs could lessen nonadherence. An interdisciplinary systematic theory-based approach to studying individual and population-based exercise adherence is strongly encouraged. Up to now, studies have focused on exercise adherence associated with specific disease conditions. There is also a need for more general older adult non-disease-specific studies (Room et al. 2017) and ones that address gender, ethnic, and disability issues. The World Health Organization (Sabate 2003) recommends a patient-tailored prescribed exercise plan recognizing that the "one-size-fits-all" approach to exercise adherence has the potential to overlook significant individual distinctions and result in negative outcomes. Exploring self-determined motivation through more qualitative studies may provide needed insight into what exercise programs are valuable and appeal to older adults. Additionally, increased use of Internet-based technologies to include activity monitoring, telecommunication, and systems providing prompt individual feedback is another option for improving exercise adherence outcomes (Room et al. 2017). More RCT trials are also needed comparing medication treatment with exercise prescriptions for specific diseases.

Finally, patient education cannot be overlooked. Compared with younger adults, older adults have more difficulty understanding recommendations (Dimatteo et al. 1992). Exercise recommendations from providers and trainers include type, duration, and intensity, and for some older adults, these details can be overwhelming. Developing simple measures to assess individual knowledge and capacity for learning and maintaining an exercise plan prior to developing a plan would be a significant step forward. Disseminating research findings to practitioners and trainers related to assessing exercise adherence and program development should be a constant goal. Exercise adherence research in years to come will require ingenuity, innovative strategies, and substantial financial funding.

References

Aartolahti E, Tolppanen AM, Lonnroos E, Hartikainen S, Hakkinen A (2015) Health condition and physical function as predictors of adherence in long-term strength and balance training among community-dwelling older adults. Arch Gerontol Geriatr 61:452–457. https://doi.org/10.1016/j.archger2015.06.016

Ajzen I, Madden TJ (1986) Prediction of older adults directed behavior, attitudes, intentions, and perceived behavioural control. J Exp Soc Psychol 22:453–474

Bandura A (1997) Self-efficacy: the exercise of control. Freeman, New York

Brosse AL, Sheets ES, Lett HS, Blumenthal JA (2002) Exercise and the treatment of clinical depression in adults recent findings and future directions. Sports Med 32(12):741–760

Chodzko-Zajko W, Proctor DN, Singh MA, Minson CT, Nigg CR, Salem GJ, Skinne JS (2009) Exercise and physical activity for older adults. Med Sci Sports Exerc:1510–1530. https://doi.org/10.1249/MSS.0b013e33181a0c95c

D'Angelo ME, Pelletier LG, Reid RD, Huta V (2014) The roles of self-efficacy and motivation in the prediction of short-and long-term adherence to exercise among patients with coronary heart disease. Health Pyschol 33(11):1344–1353

Dimatteo MR, Hays RD, Sherbourne CD (1992) Adherence to cancer regimens: implications for treating the older patient. Oncology 6(2 Supplement):50–57

Eckel RH, Jakicic JM, Ard JD, DeJesus JM, Miller NH, Hubbard VS et al (2014) AHA/ACC guideline on lifestyle management to reduce cardiovascular risk: a report of the American College of Cardiology/American Heart Association task force on practice guidelines. Circulation 129(suppl 2):S76–S99

Franco MR, Tong A, Howard K, Sherrington C, Ferreira PH, Pinto RZ, Ferreira ML (2015) Older people's perspectives on participation in physical activity: a systematic review and thematic synthesis of qualitative literature. Br J Sports Med 49:1268–1276. https://doi.org/10.1136/bjsports-2014-094015

Garber CE, Blissmer B, Deschenes MR, Franklin B, Lamonte MJ, Lee IM, Nieman DC, Swain DP (2011) Quantity and quality of exercise for developing and maintaining cardiorespiratory, musculoskeletal, and neuromotor fitness in apparently healthy adults: guidance for prescribing exercise. Med Sci Sports Exerc 43:1334–1359

Hawley-Hague H, Horne M, Campbell M, Demack S, Skelton D, Todd C (2014) Multiple levels of influence on older adults' attendance and adherence to community exercise classes. The Gerontologist 54(4): 599–610. https://doi.org/10.1093/geront/gnt075

Heran BS, Chen JM, Ebrahim S, Moxham T, Oldridge N, Rees K, Thompson DR, Taylor RS (2011) Exercise-based cardiac rehabilitation for coronary heart disease. Cochrane Database Syst Rev Cochrane Collab 7:10–11

Hill AM, Hoffmann T, McPhail S, Beer C, Hill KD, Brauer GG, Hainers TP (2011) Factors associated with older patients' engagement in exercise after hospital discharge. Arch Phys Med Rehabil 92:1395–1405. https://doi.org/10.1016/j.apmr.2011.04.009

Hoffmann TC, Maher CG, Riffa T et al (2016) Prescribing exercise interventions for patients with chronic conditions. CMAJ 188(7):510–518

Horne R, Chapman SC, Parham R, Freemantle N, Forbes A, Cooper V (2013) Understanding patients' adherence-related beliefs about medicines prescribed for long-term conditions: a metanalytic review of the necessity-concerns framework. PLoS One 8(12). https://doi.org/10.1371/journal.pone.0080633

Keller C, Fleury J, Gregor-Holt N, Thompson T (1999) Predictive ability of social cognitive theory in exercise research: an integrated literature review. Worldviews on Evidence-based Nursing presents the archives of Online Journal of Knowledge Synthesis for Nursing, 6(1):19–31.

Kohn M, Belza B, Petrescu-Prahova M, Miyawaki CE (2016) Beyond strength: participants perceptions on the benefits of an older adult exercise program. Health Educ Behav 43(3):305–312

Labarthe DR, Goldstein LB, Antman EM, Arnett D, Fonarow G, Albets MJ, Whitsel L et al (2016) Evidence-based policy making: assessment of the American Heart Association 's strategic policy portfolio: a policy statement from the American Heart Association. Circulation 133(18):e615–e653

Liu M, Miyawaki CE (2019) What types of physical function predict program adherence in older adults? Rehabil Nurs J Off J Assoc Nurs:1–8. https://doi.org/10.1097/rnj.0000000000000209

Madden TJ, Ellen PS, Ajzen I (1992) A comparison of the theory of planned behavior and the theory of reasoned action. Pers Soc Psychol Bull 18(1):3–9

Naci H, Ioannidis JP (2013) Comparative effectiveness of exercise and drug interventions on mortality outcomes: metaepidemilogical study. BMJ 347:5577. https://doi.org/10.1136/bmj.f5577

Picorelli AM, Pereira LS, Pereira DS, Felicio D, Sherrington C (2014) Adherence to exercise programs for older people is influenced by program characteristics and personal factors: a systematic review. J Physiother 60:151–156

Rivera-Torres S, Fahey DT, Rivera MA (2019) Adherence to exercise programs in older adults: informative report. Gerontol Geriatr Med 5:I–10

Room J, Hannink E, Dawes H, Barker K (2017) What interventions are used to improve exercise adherence in older people and what behavioral techniques are they based on? A systematic review. BMJ Open 7:e019221. https://doi.org/10.1136/bmjopen-2017-019221

Rosner F (2006) Patient noncompliance: causes and solutions. Mt Sinai J Med 73(2):553–559

Sabate E (2003) Adherence to long-term therapies: evidence for action. World Health Organization, Geneva

Shumaker SA, Achron EB, Ockene JK, McBee WL (1998) The handbook of health behavior change, 2nd edn. Springer Publishing Company, New York

Sparling PB, Howard BJ, Dunstan DW, Owen N (2015) Recommendations for physical activity in older adults. BMJ (6):350. https://doi.org/10.1136/bmj.h100

U.S. Department of Health and Human Services (2002) Physical activity fundamental to preventing disease. Washington, DC. Retrieved from https://aspe.hhs.gov/basic-report/physical-activity-fundamental-preventing-disease

U.S. Department of Health and Human Services (2008) Physical activity guidelines for Americans. Washington, DC. Retrieved from https://health.gov/paguidelines/pdf/paguide.pdf

Exercise and Healthy Cardiovascular Aging

Jason Roh, Andy Yu and Anthony Rosenzweig
Corrigan Minehan Heart Center, Division of Cardiology, Department of Internal Medicine, Massachusetts General Hospital, Boston, USA
Harvard Medical School, Boston, MA, USA

Synonyms

Physical activity

Definition

Exercise: A subset of physical activity that is planned, structured, repetitive, and purposeful in the sense that improvement or maintenance of physical fitness is the objective (Thompson et al. 2003).

Overview

Cardiovascular disease (CVD) is a leading cause of morbidity and mortality in older adults and is increasing in prevalence with the aging of populations worldwide. Global efforts to mitigate the growing burden of CVD have placed substantial emphasis on physical activity and exercise in maintaining cardiovascular health and preventing CVD in older adults. Over the past 50 years, advancements in exercise science, molecular biology, and imaging technology have begun to unravel the physiological and molecular mechanisms by which exercise attenuates (and potentially reverses) cardiovascular aging phenotypes associated with CVD. Here, we present a brief review on the role of cardiovascular aging in CVD, the role of exercise in optimizing cardiovascular health in older adults, and lastly, provide some insights into how exercise's ability to modulate cardiovascular aging phenotypes may lead to novel therapeutic strategies for older adults with CVD.

Key Research Findings

Impact of Population Aging on Cardiovascular Disease Epidemiology

Advanced age is the dominant risk factor for nearly every subtype of CVD. The incidence of coronary artery disease, heart failure, arrhythmias, hypertension, peripheral vascular disease, and stroke all increase with age (Yazdanyar and Newman 2009). Not unexpectedly, as the global population has continued to age, so has the prevalence of CVD. Recent estimates indicate that there are ~422.7 million cases of CVD worldwide, which account for over one-third of the global mortality (Roth et al. 2017). Older adults represent the vast majority of individuals with CVD and account for >80% of all CVD-related deaths in many parts of the world (Yazdanyar and Newman 2009; Roth et al. 2017). Thus, understanding the role of cardiovascular aging in CVD and identifying strategies to effectively intervene in this process are becoming critical to maximizing human healthspan and maintaining viable healthcare infrastructures.

Cardiovascular Aging

Similar to all organ systems in the body, the cardiovascular system undergoes a biological aging process that leads to structural and functional changes in the heart and vasculature that can predispose older adults to disease. In the simplest sense, this process can be divided into cardiac and vascular aging, although the processes are fundamentally intertwined.

Cardiac Aging

Cross-sectional studies in healthy adults have demonstrated that the structure of the heart changes with age. Left ventricular (LV) wall thickness and left atrial (LA) dimensions increase, LV geometry becomes more spherical, and progressive degeneration occurs in the underlying valvular and conduction systems (Lakatta and Levy 2003a). At the cellular level, there is a significant loss of cardiomyocytes that occurs with aging and results in a compensatory hypertrophic response with increased interstitial and perivascular fibrosis (Lakatta and Levy 2003a; Olivetti et al. 1991). With aging, the remaining cardiomyocytes progressively lose their capacities for renewal and repair (Bergmann et al. 2009) and develop impairments in Ca^{2+} handling, contractility, protein homeostasis, fatty acid metabolism, and mitochondrial function (Bodyak et al. 2002). Collectively, these structural, cellular, and molecular changes result in a number of functional impairments in the aged heart that increase the risk of heart failure, a leading cause of morbidity and mortality in older adults (Lakatta and Levy 2003a; McMurray and Steward 2000; Kane et al. 2011). Age-associated deficits in Ca^{2+} handling and ß-adrenergic desensitization, along with increased myocardial hypertrophy and fibrosis, result in impaired relaxation and stiffness of the myocardium, a hallmark feature of cardiac aging (Lakatta and Levy 2003a). The resting contractile function of the heart is generally preserved in older adults. However, more sensitive imaging methods have recently demonstrated that even the "healthy" aged heart displays some level of subclinical systolic dysfunction (Hung et al. 2017), which becomes even more apparent in response to exercise, or physiological stress. While recruitable stroke volume is preserved, chronotropic and contractile reserves are significantly impaired in older adults and contribute to the progressive decline in exercise capacity that occurs with chronological aging (Stratton et al. 1994; Roh et al. 2016).

Vascular Aging

Similar to the heart, the vasculature undergoes numerous age-related changes, which contribute to increased CVD risk in older adults. Most notable are an increase in central arterial stiffness and generalized endothelial dysfunction (Mitchell et al. 2004; Celemajer et al. 1994; Lakatta and Levy 2003b). The increase in arterial stiffness is predominantly due to a pathophysiologic vascular remodeling process that occurs with normal aging and results in increased elastin degradation, calcification, and collagen accumulation (Lakatta and Levy 2003b). This phenomenon is responsible for the isolated systolic hypertension commonly seen in older adults, which can subsequently lead to pathologic cardiac hypertrophy, microvascular remodeling, and heart failure (Lakatta and Levy 2003b). At the molecular level, age-associated increases in oxidative stress and low-grade inflammation lead to a generalized state of endothelial dysfunction and reductions in nitric oxide (NO) availability, a hallmark feature of vascular aging (Lakatta and Levy 2003b; Taddei et al. 2001; Seals et al. 2012). Decreased NO impairs the vasodilatory, antithrombotic, and anti-inflammatory properties of the vasculature, increasing the risk for atherosclerotic-related CVD, including myocardial infarction, stroke, and peripheral vascular disease (Lakatta and Levy 2003b; Taddei et al. 2001; Seals et al. 2012). Furthermore, defective repair processes, along with reduced angiogenic and regenerative potential, in the aged vasculature impair the older adult's protective mechanisms to ischemic injury, resulting in increased disease burden after a vascular insult (Torella et al. 2004; Lahteenvuo and Rosenzweig 2012).

Beneficial Effects of Physical Activity and Exercise on Cardiovascular Disease Prevention in Older Adults

Although cardiovascular aging is not synonymous with CVD, the progression of these age-related changes is generally believed to contribute to the development of CVD in older adults. Thus, identifying interventions that can potentially reverse or attenuate these aging phenotypes has become a central theme in the prevention and treatment of CVD in older adults. In this regard, exercise has emerged as perhaps the single most effective lifestyle intervention in maintaining physical fitness, independence, and preventing CVD in older adults (Stratton et al. 1994; Roh et al. 2016; Shiroma and Lee 2010; Blair et al. 1995; Manson et al. 2002; Chomistek et al. 2013; O'Connor et al. 2009; Fei et al. 2013; Sui et al. 2007; Manini et al. 2006; Bauman et al. 2016). Indeed, the World Health Organization, along with numerous other medical professional societies, recommends that older adults perform at least 150 min of moderate-intensity aerobic exercise or > 75 min of high-intensity aerobic activity per week (Thompson et al. 2003; Fei et al. 2013). The rationale for these guidelines is largely based on a well-established inverse relationship between physical activity and all-cause and cardiovascular-related mortality (Shiroma and Lee 2010; Blair et al. 1995; Manson et al. 2002; Sui et al. 2007; Manini et al. 2006). Although the cardiovascular benefits associated with exercise were initially identified in predominantly middle-aged adults, they have been consistently reproduced in older adults, in whom there is not only a clear dose-dependent inverse relationship between physical activity and function, but also relative risk reductions in diabetes, hypertension, myocardial infarction, stroke, claudication, heart failure hospitalizations, and overall mortality (Thompson et al. 2003; Stratton et al. 1994; Roh et al. 2016; Fei et al. 2013; Sui et al. 2007; Manini et al. 2006; Bauman et al. 2016).

Physiological, Cellular, and Molecular Effects of Exercise on Cardiovascular Aging

Advancements in molecular biology and imaging technology over the past few decades have provided enormous insights into the physiological and molecular mechanisms by which exercise mediates its beneficial effects on cardiovascular aging phenotypes associated with disease pathophysiology. Cross-sectional studies comparing sedentary and athletic older adults have suggested that lifelong exercise attenuates age-related changes in the heart and vasculature, particularly impairments in the relaxation or compliance properties of the heart and central arteries (Roh et al. 2016; Arbab-Zadeh et al. 2004; Vaitkevicius et al. 1993; Seals et al. 2008). Moreover, a growing body of prospective exercise training studies in both healthy older adults and those with established CVD are now beginning to elucidate how exercise exerts these physiological benefits in the context of aging. Aerobic exercise interventions, which include walking, running, cycling, and swimming, not only consistently increase peak oxygen consumption, a key metric of functional capacity, in older adults, but have also been shown to potentially improve resting systolic and diastolic function, chronotropic and contractile reserves, peripheral oxygen extraction, arterial stiffness, microvascular density, and vasodilatory properties of the vasculature (Stratton et al. 1994;

Roh et al. 2016; Howden et al. 2018; Edelmann et al. 2011; Haykowsky et al. 2011; Tanaka et al. 2000; Maeda et al. 2005; DeSouza et al. 2000; Taddei et al. 2000).

The underlying molecular mechanisms by which exercise mediates these changes on cardiovascular aging phenotypes have mostly been studied in experimental animal models. Unlike the pathologic cardiac hypertrophic remodeling that occurs with aging and disease, exercise induces a distinct physiological hypertrophic process in the heart that is not only cardioprotective (Bostrom et al. 2010; Liu et al. 2015) but also seems to reverse many of the deleterious molecular changes associated with cardiac aging, including impaired Ca^{2+} handling, mitochondrial dysfunction, fibrosis, and angiogenesis (Stratton et al. 1994; Roh et al. 2016; Iemitsu et al. 2006; Kwak et al. 2011). Interestingly, recent work in rodent models has also shown that exercise can induce cardiomyocyte regeneration in the adult heart (Vujic et al. 2018), raising the exciting potential that exercise might be able to attenuate or reverse the substantial decline in cardiomyocyte numbers and replicative capacity associated with aging (Olivetti et al. 1991; Bergmann et al. 2009). Aerobic exercise training seems to have similar benefits in terms of reversing the age-associated deficits in vascular compliance and regeneration. By enhancing NO synthesis and bioavailability, exercise not only increases vasodilatory properties while reducing oxidative stress and inflammation in the vasculature, but also increases the mobilization and recruitment of endothelial progenitor cells, which are a key element in the vascular repair process (Lesniewski et al. 2011; Schuler et al. 2013).

Future Directions for Research

Extensive research in humans and experimental animal models have made it abundantly clear that the health benefits of exercise extend into older age, and that regular exercise is an essential component to maintaining cardiovascular fitness and health in older adults. However, despite the clear association between exercise and healthy aging, recent data indicate that the percentage of individuals not meeting the minimum recommended physical activity guidelines markedly increases with age, to the point where nearly half of those over 80 do not meet the minimal threshold for health (Bauman et al. 2016). With CVD being the leading cause of mortality in older adults and the prevalence only expected to increase with population aging, it is imperative that we develop effective strategies to successfully implementing exercise into the medical care of older adults. As such, we propose that further research is needed in the following areas:

1. Development of effective strategies to increase compliance with physical activity and exercise guidelines in older adults
2. Determine optimal exercise type and intensity for older adults, particularly as it relates to specific age, sex, and race groups, as well as CVD subtypes.
3. Further investigation into the underlying molecular mechanisms driving exercise-mediated cardiovascular benefits in older adults in the hope of identifying and validating novel targets for therapeutic intervention in this population

Summary

In summary, increasing human longevity and population aging are leading to an increasing global burden of chronic noncommunicable diseases, most notably CVD. Exercise is one of the most effective strategies to maintaining function and preventing CVD in older adults and needs to be further integrated into the daily lifestyle and medical care of all older adults. Since the ability to exercise can diminish with frailty and advanced age, elucidating the underlying molecular mechanisms by which exercise modulates cardiovascular aging phenotypes associated with disease may provide a novel approach to developing effective new therapeutics for CVD in older adults.

Cross-References

- Aerobic Exercise Training and Healthy Aging
- Cardiovascular Response
- Cardiovascular System
- Congestive Heart Failure
- Exercise Adherence
- Healthy Aging
- Hypertension
- Hypertensive Cardiovascular Diseases
- Ischemic Heart Disease
- Leisure Activities and Healthy Aging
- Myocardial Infarction
- Peripheral Artery Disease
- Physical Activities
- Physical Activity, Sedentary Behaviors, and Frailty
- Sport and Healthy Aging
- Stroke
- Vascular Diseases of Ageing
- Yoga Practices and Health Among Older Adults

References

Arbab-Zadeh A, Dijk E, Prasad A et al (2004) Effect of aging and physical activity on left ventricular compliance. Circulation 110(13):1799–1805

Bauman A, Merom D, Bull FC et al (2016) Updating the evidence for physical activity: summative reviews of the epidemiological evidence, prevalence, and interventions to promote "active aging". Gerontologist 56:S268–S280

Bergmann O, Bhardway RD, Bernard S et al (2009) Evidence for cardiomyocyte renewal in humans. Science 324(5923):98–102

Blair SN, Kohl HW III, Barlow CE et al (1995) Changes in physical fitness and all-cause mortality. A prospective study of healthy and unhealthy men. JAMA 273(14):1093–1098

Bodyak N, Kang PM, Hiromura M et al (2002) Gene expression profiling of the aging mouse cardiac myocytes. Nucleic Acids Res 30(17):3788–3794

Bostrom P, Mann N, Wu J et al (2010) C/EBPß controls exercise-induced cardiac growth and protects again pathological cardiac remodeling. Cell 143(7):1072–1083

Celemajer DS, Sorensen KE, Spiegelhalter DJ et al (1994) Aging is associated with endothelial dysfunction in healthy men years before the age-related decline in women. J Am Coll Cardiol 24(2):471–476

Chomistek AK, Manson JE, Stafnick ML et al (2013) The relationship of sedentary behavior and physical activity to incident cardiovascular disease: results from the Women's health initiative. J Am Coll Cardiol 61:2346–2354

DeSouza CA, Shapiro LF, Clevenger CM et al (2000) Regular aerobic exercise prevents and restores age-related declines in endothelium-dependent vasodilation in health mean. Circulation 102(12):1351–1357

Edelmann F, Gelbrich G, Dungen HD et al (2011) Exercise training improves exercise capacity and diastolic function in patients with heart failure with preserved ejection fraction: results of the ex-DHF (exercise training in diastolic heart failure) pilot study. J Am Coll Cardiol 17:1780–1791

Fei S, Norman IJ, While AE (2013) Physical activity in older people: a systematic review. BMC Public Health 13:1–17

Haykowsky MJ, Brubaker PH, John JM et al (2011) Determinants of exercise intolerance in elderly heart failure patients with preserved ejection fraction. J Am Coll Cardiol 3:265–174

Howden EJ, Sarma S, Lawley JS et al (2018) Reversing the cardiac effects of sedentary aging in middle age- a randomized controlled trial. Circulation 137:1549–1560

Hung CL, Goncalves A, Shah AM et al (2017) Age- and sex-related influences on left ventricular mechanics in elderly individuals free of prevalent heart failure: the ARIC study (atherosclerosis risk in communities). Circ Cardiovasc Imaging 10(1):1–9

Iemitsu M, Maeda S, Jesmin S et al (2006) Exercise training improves aging-induced downregulation of VEGF angiogenic signaling cascade in hearts. Am J Physiol Heart Circ Physiol 291(3):H1290–H1308

Kane GC, Karon BL, Mahoney DW et al (2011) Progression of left ventricular diastolic dysfunction and risk of heart failure. JAMA 306:856–863

Kwak HB, Kim J, Hoshi K et al (2011) Exercise training reduces fibrosis and matrix metalloproteinase dysregulation in the aging rat heart. FASEB J 25(3):1106–1117

Lahteenvuo J, Rosenzweig A (2012) The role of angiogenesis in cardiovascular aging. Circ Res 110(9):1252–1264

Lakatta EG, Levy D (2003a) Arterial and cardiac aging: major shareholders in cardiovascular disease enterprises. Circulation 107:346–354

Lakatta EG, Levy D (2003b) Arterial and cardiac aging: major shareholders in cardiovascular disease enterprises: part I: aging arteries: a "set up" for vascular disease. Circulation 107(1):139–146

Lesniewski LA, Durrant JR, Connell ML et al (2011) Aerobic exercise reverses arterial inflammation with aging in mice. Am J Physiol Heart Circ Physiol 301(3):H1025–H1032

Liu X, Xiao J, Zhu H et al (2015) miR-222 is necessary for exercise-induced cardiac growth and protects against pathological cardiac remodeling. Cell Metab 21(4):584–595

Maeda S, Iemitsu M, Miyauchi T et al (2005) Aortic stiffness and aerobic exercise: mechanistic insight from microarray analyses. Med Sci Sports Exerc 37(10):1710–1716

Manini TM, Everhart JE, Patel KV et al (2006) Daily activity energy expenditure and mortality among older adults. JAMA 296(2):171–179

Manson JE, Greenland P, LaCroix AZ et al (2002) Walking compared with vigorous exercise for the prevention of cardiovascular events in women. N Engl J Med 347:716–725

McMurray JJ, Steward S (2000) Epidemiology, aetiology, and prognosis of heart failure. Heart 83:596–602

Mitchell GF, Parise H, Benjamin EJ et al (2004) Changes in arterial stiffness and wave reflection with advancing age in healthy men and women: the Framingham heart study. Hypertension 43(6):1239–1245

O'Connor CM, Whellan DJ, Lee KL et al (2009) Efficacy and safety of exercise training in patients with chronic heart failure: HF-ACTION randomized controlled trial. JAMA 301(14):1439–1450

Olivetti G, Melissari M, Capasso JM et al (1991) Cardiomyopathy of the aging human heart: myocyte loss and reactive cellular hypertrophy. Circ Res 68:1560–1568

Roh JD, Rhee J, Chaudhari V et al (2016) The role of exercise in cardiac aging: from physiology to molecular mechanisms. Circ Res 118:279–295

Roth GA, Johnson C, Abajobir A et al (2017) Global, regional, and National Burden of cardiovascular diseases for 10 causes, 1990 to 2015. J Am Coll Cardiol 70(1):1–25

Schuler G, Adams V, Goto Y (2013) Role of exercise in the prevention of cardiovascular disease: results, mechanisms, and new perspective. Eur Heart J 34:1790–1799

Seals DR, Desouza CA, Donato AJ et al (2008) Habitual exercise and arterial aging. J Appl Phys 105(4):1323–1332

Seals DR, Jablonski KL, Donato AJ (2012) Aging and vascular endothelial function in humans. Clin Sci (Lond) 120(9):357–375

Shiroma EJ, Lee IM (2010) Physical activity and cardiovascular health: lessons learned from epidemiological studies across age, gender, and race/ethnicity. Circulation 122(7):743–752

Stratton JR, Levy WC, Cerqueira MD et al (1994) Effects of aging and exercise training in healthy men. Circulation 89:1648–1655

Sui X, LaMonte MJ, Laditka JN et al (2007) Cardiorespiratory fitness and adiposity as mortality predictors in older adults. JAMA 298(21):2507–2516

Taddei S, Galetta F, Virdis A et al (2000) Physical activity prevents age-related impairment in nitric oxide availability in elderly athletes. Circulation 101(25):2896–2901

Taddei S, Virdis A, Ghiadoni L et al (2001) Age-related reduction in NO availability and oxidative stress in humans. Hypertension 38:274–279

Tanaka H, Dinenno FA, Monahan KD et al (2000) Aging, habitual exercise, and dynamic arterial compliance. Circulation 102:1270–1275

Thompson PD, Buchner D, Pina IL et al (2003) Exercise and physical activity in the prevention and treatment of atherosclerotic cardiovascular disease: a statement from the council on clinical cardiology (subcommittee on exercise, rehabilitation, and prevention) and the council on nutrition, physical activity, and metabolism (subcommittee on physical activity). Circulation 107:3109–3116

Torella D, Leosco D, Indolfi C et al (2004) Aging exacerbates negative remodeling and impairs endothelial regeneration after balloon injury. Am J Physiol Heart Circ Physiol 287(6):H2850–H2860

Vaitkevicius PV, Fleg JL, Engel JH et al (1993) Effects of aging and aerobic capacity on arterial stiffness in healthy adults. Circulation 88:1456–1462

Vujic A, Lerchenmuller C, Wu TD et al (2018) Exercise induces new cardiomyocyte regeneration in the adult mammalian heart. Nat Commun 9(1):1–7

Yazdanyar A, Newman AB (2009) The burden of cardiovascular disease in the elderly: morbidity, mortality, and costs. Clin Geriatr Med 25(4):563–577

Exercise Maintenance

▶ Exercise Adherence

Exhaustion

▶ Psychological Fatigue

Experience Sampling Methods

▶ Mobile Data Collection with Smartphones

Experimental Studies and Observational Studies

Martin Pinquart
Psychology, Philipps University, Marburg, Germany

Synonyms

Experimental studies: Experiments, Randomized controlled trials (RCTs); Observational studies: Non-experimental studies, Non-manipulation studies, Naturalistic studies

Definitions

The experimental study is a powerful methodology for testing causal relations between one or more explanatory variables (i.e., independent variables) and one or more outcome variables (i.e., dependent variable). In order to accomplish this goal, experiments have to meet three basic criteria: (a) experimental manipulation (variation) of the independent variable(s), (b) randomization – the participants are randomly assigned to one of the experimental conditions, and (c) experimental control for the effect of third variables by eliminating them or keeping them constant.

In observational studies, investigators observe or assess individuals without manipulation or intervention. Observational studies are used for assessing the mean levels, the natural variation, and the structure of variables, as well as for determining the strength of the association between variables and – in the case of longitudinal studies – the temporal direction of the association between variables.

Overview

Research in the field of gerontology and population aging relies on a variety of research designs that provide distinct and potentially complementary answers to different questions (Bergeman and Boker 2016; Freund and Isaacowitz 2013; Piccinin et al. 2011; Schaie and Caskle 2005). Different study designs are needed for the description of aging and for the analysis of the explanatory mechanisms that cause age-associated change. Scientists should use the most appropriate design to study their research questions. At a general level, observational (non-experimental) studies and experimental studies can be distinguished (Fig. 1). The fundamental difference between both groups of studies rests in the use of a manipulation (experimental variation) of the variable(s) that is/are assumed to cause an effect.

Observational Studies

In observational (non-experimental) studies, investigators observe individuals without experimental manipulation or intervention. There is an inadequacy about the term "observational study" because the outcome variable of an experiment could also be observed.

Observational studies can be further categorized into descriptive and correlational studies. The goal of descriptive studies is to describe a phenomenon, such as the prevalence of physical or mental illnesses in old age, or older adults' attitudes and behaviors. Surveys are the most common type of descriptive studies. If researchers strive to collect data from every individual of a defined population, the descriptive study is called

Experimental Studies and Observational Studies, Fig. 1 Research methods in the field of gerontology and population aging

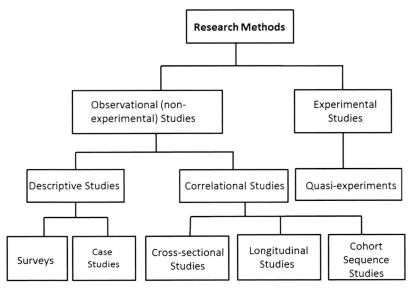

a census. Census studies provide data that are representative of the defined population and therefore have high external validity. On the other extreme, case studies involve only one or few participants who are described in great detail. This approach may be useful when researchers are interested in a rare phenomenon (e.g., an uncommon disease), and it would be difficult to find a large number of study participants. While most surveys collect quantitative data, case studies often collect qualitative data or a mix of quantitative and qualitative data (see ▶ "Qualitative Research/Quantitative Research"). A main limitation of case studies is the limited generalizability of their results to larger populations.

Because most observational studies assess more than one variable, their data can also be used for correlational study designs. Correlational studies analyze the relationship between two or more quantitative variables. They provide important information about the strength of the relationship between variables (which is often reflected in the size of the correlational coefficient). As most researchers in the field of gerontology are interested in age differences or age-associated change, it is important to distinguish between two kinds of correlational studies – cross-sectional studies that collect data on individuals of different ages at one point of measurement and longitudinal studies that assess the same individuals over two or more points of measurement (see ▶ "Cross-Sectional and Longitudinal Studies"). Longitudinal studies can be used for determining the temporal direction of the association between variables, for example, whether an active lifestyle predicts positive changes in intelligence in old age and/or whether more intelligent older adults are more likely to maintain an active lifestyle than their peers.

Because studies on aging often address a large age range, observed age differences may be confounded with cohort differences (the effects of different historical times on aging of distinct birth cohorts), and longitudinal changes in a cohort may not be the same as the changes observed among earlier or later cohorts (see ▶ "Age-Period-Cohort Models"). Sequential designs combine cross-sectional and longitudinal data and allow for investigating intraindividual change, interindividual differences in change parameters, as well as differences between cohorts (Schaie and Caskle 2005).

Benefits and Limitations. As a main benefit, observational studies can be done in a naturalistic setting which promotes the external validity (generalizability) of their results. In addition, a large number of variables can be assessed. Longitudinal and sequential studies provide an important insight on average, as well as differential patterns of change across large time intervals (Bergeman and Boker 2016).

As a main limitation, observed correlation does not imply causality. To conclude that one variable causes or influences another variable, three criteria must be met: covariation (which is tested in all correlational studies), directionality – the presumed cause precedes the presumed effect in time (which can be tested in longitudinal studies) – and nonspuriousness, the elimination of alternate explanations (Menard 2002). To show nonspuriousness, investigators must ideally demonstrate that the relationship is maintained when all extraneous variables that might explain the association between two variables are kept constant. This is very hard to do and often impossible in correlational research, although statistically controlling for known, relevant third variables increases the internal validity of the results. Longitudinal research in the field of aging has some unique challenges because of changes in sample composition due to mortality and the emergence of severe illnesses (e.g., dementia) that preclude further participation.

Experiments and Quasi-experiments

Experimental studies introduce particular levels of the independent variable and measure whether this manipulation has an effect on the outcome variable(s). As a second key element, individuals are randomly assigned to an experimental condition, such as a treatment (e.g., cognitive training) or no-treatment (i.e., the control condition). While most experiments compare members of different experimental groups (by use of a between-subject design), others use a within-subject design in which all participants carry out all conditions in a randomized order. The third critical element is the experimental control. The effect of confounding variables is eliminated by keeping all factors other than the independent variable(s) constant, by or allowing them to vary non-systematically between the experimental conditions. Carrying out the study in the controlled setting of a laboratory makes the conditions as similar as possible for all participants.

Experiments provide a more rigorous form of evidence than observational, non-experimental studies. They are used for drawing causal inferences because alternative explanations can be ruled out (Patten and Newhart 2018). In the field of evidence-based medicine, the experimental RCTs are considered the "gold standard" due to their ability to eliminate rival hypotheses or explanations (Moher 1998).

Although experiments are the best design for testing causal hypotheses, an experimental manipulation of a large number of variables is not possible (e.g., age cannot be randomly assigned), unethical (e.g., confronting participants with serious negative life events), or not practical (e.g., experimentally testing whether participation in a cognitive stimulation program in early childhood would have an effect on cognitive performance in old age). In these cases, quasi-experiments can be used. Campbell and Stanley (1966) initially introduced the term quasi-experiment to distinguish the "randomized experiment" (in which participants are randomly assigned to experimental conditions) from experiments in which researchers do not apply random assignment (quasi-experiments), but all other criteria for an experiment are met. Here comparisons are made between people in groups that already exist or that are built based on existing characteristics. As with "randomized experiments," the goal of quasi-experiments is to draw causal inferences from the study. In aging research, many quasi-experimental designs are used when testing the reactions of two or more age groups toward experimental demands while holding potential confounding variables constant. But, by definition, these quasi-experiments lack random assignment. For example, participants cannot be assigned to being young or old. Although there are procedures for reducing the possibility of alternative explanations of quasi-experimental results (such as propensity score matching for making the compared groups in third variables as similar as possible), causation still cannot be fully established because the experimenter does not have total control over third variables (Shadish et al. 2002).

Natural experiments are a subtype of quasi-experiments. The term natural experiment describes a naturally occurring contrast between two conditions. For example, when studying effects of pension eligibility age on the age at retirement, researchers could identify a country that changed eligibility age and compare age at retirement before and after this social change occurred (Atalay and Barrett 2015). For interpreting reasons for between-group differences in age at retirement, researchers would have to exclude the possibility that other cohort differences (e.g., amount of money in savings, living costs) and other aspects of social change that occurred in parallel and might have caused the observed change in the age at retirement.

Benefits and limitations. The main benefits of experiments are the test of causal hypotheses and the high internal validity that their results achieve. Randomization guarantees that there is no systematic bias in the selection of the experimental conditions. Nonetheless, chance factors may lead to between-group differences, particularly if few persons participate in each experimental condition. Therefore, the first step of analyzing experimental data is to check whether the randomization worked (i.e., the experimental groups did not differ at pretest).

To a large extent, internal validity is achieved through experimental control, but the artificial experimental conditions in the laboratory reduce the generalizability of the results to daily life – the external validity of the results. The conflict between internal and external validity has been called the experimenter's dilemma (Jung 1971). When confronted with this dilemma, experimenters tend to opt in favor of internal validity because results could not be generalized anyway if the internal validity is weak, and one could not draw confident conclusions about the effects of the independent variable(s). As another limitation, experiments that manipulate one or two conditions (independent variables) cannot detect causality if only a combination of a larger number of variables causes an outcome (Rutter 2007). In addition, experimental studies tend to assess short-term effects of the experimental manipulation, while longitudinal, observational studies assess associations between variables over longer time intervals, in some cases over decades.

As already reported, the fact that some variables cannot be experimentally manipulated constitutes a further limitation that results in the use of quasi-experiments. Although the ecological validity and equivalence of experimental stimuli across groups of participants is a general challenge when conducting (quasi-)experimental studies, the problem is more severe when comparing age groups, because the representativeness of experimental stimuli or demands may vary between younger and older individuals. For example, when studying age differences in emotion recognition accuracy, the exclusive use of stimulus material with younger faces could lead to biased age differences as older adults may have less experience with recognizing emotions in younger adults but more experience with recognizing these emotions in older adults. In fact, Fölster et al. (2015) found relative own-age accuracy advantage with respect to recognizing sadness and disgust in the faces of older versus younger persons.

Thus, quasi-experiments must first establish equivalent representativeness of their stimuli for the different assessed age groups (Freund and Isaacowitz 2013). In addition, because chronological age is confounded with cohort differences (such as higher average educational attainment among younger cohorts), researchers must decide whether to make both groups equivalent with regard to these third variables (by oversampling less educated younger adults or highly educated older adults) as this procedure reduces the representativeness of the participants for their age group. Furthermore, many quasi-experiments have compared a younger group of undergraduates who share a similar lifestyle with a more heterogeneous group of older adults. This is likely to lead to an overestimation of group differences due to the restricted variance in the younger group. In contrast, including only healthy participants (as a common way of controlling for confounding effects of health status) is likely to lead to an underestimation of age differences because health problems are more common in older adults. Nonetheless, a well-designed quasi-experiment that eliminates as many threats to internal validity as possible can provide strong circumstantial evidence about causal relationships, although alternative explanations cannot be completely ruled out (Leary 2001).

Examples of Observational and Experimental Studies in Aging Research

Observational studies dominate most fields of aging research because many research questions can be answered with these studies, chronological age cannot be experimentally manipulated (Cavanaugh and Blanchard-Fields 2019), and many social/societal conditions would be difficult to manipulate (Weil 2017). Among the observational studies, large-scale aging surveys are particularly valuable because they collect data from a large, representative sample mostly across several points in time and often adopt an interdisciplinary approach, such as the Health and Retirement Study (HRS; see ▶ "Health and Retirement Study") or the Midlife in the United States (MIDUS) study. For example, since 1992 the HRS has collected data from a nationally representative, longitudinal survey of individuals over age 50 in the USA with 2-year intervals between assessments and about 18–23,000 participants in any given wave. A proxy respondent is sought for respondents who are unwilling or unable to do an interview themselves, which is essential in order to maintain coverage of cognitively impaired individuals. Every 6 years a new 6-year birth cohort is added. The topics addressed include resources for successful aging (e.g., economic, social, physical, psychological), behaviors and choices (e.g., health behaviors, work, residence, transfers, program use), as well as events and transitions (e.g., retirement, widowhood, institutionalization).

Since 2006, data collection has expanded to include biomarkers and the collection of DNA samples. Even experimental modules will be administered to randomly selected subsamples of HRS respondents in the 2020 wave. The goal of the HRS is not only to gain scientific knowledge on aging but also to provide scientific data for studying national-level social and policy changes that may affect individuals. As HRS sister surveys are conducted in a number of western and non-western countries, these studies also provide opportunities for cross-national comparisons. Over 3000 papers that cite HRS data have been published between 1993 and 2017, and the data are available for secondary analyses (Institute for Social Research 2018; Sonnega et al. 2014). As conducting this kind of study requires a lot of financial and personal resources, and not all research questions can be addressed with the HRS and related aging surveys, most researchers have to collect their own data by use of smaller and usually less representative samples.

Experimental studies (RCTs) are commonly used in gerontological and geriatric treatment research for meeting the standards of evidence-based medicine (Oxford Centre for Evidence-Based Medicine 2011) and the requirements of regulatory agencies that control and supervise medications and medical devices (such as the US Food and Drug Administration) (Carpenter 2010). Research on biological aging heavily relies on experiments conducted on animal models because of ethical issues, the long, natural human lifespan, genetic heterogeneity, and other factors that limit the use of human subjects. For example, the Interventions Testing Program was launched in 2000 to systematically study the effects of diets (e.g., caloric restriction), drugs, and other interventions on life span among mice (Mitchell et al. 2015). As many organisms, such as mice, age differently than humans, experimental animal research has to be supplemented by research on humans in order to check whether the results can be generalized. For example, after experiments on caloric restriction among animals showed a life span extension, a natural experiment with humans who were confronted with an unexpected low availability of food (the Biosphere 2 trial) and a randomized, controlled trial on caloric restriction in humans found improvements in some biomarkers of aging and longevity (Redman and Ravussin 2011), although the effect of caloric restriction on the humans' length of life has not yet been tested in these studies. Nonetheless, experiments are still underused in most fields of aging research.

Although age cannot be experimentally manipulated within the individual, some (but not all) psychological mechanisms theorized to cause age-associated change can be experimentally varied (Freund and Isaacowitz 2013). The following examples refer to cognitive and socio-emotional aging.

An experiment by Lindenberger et al. (2001) tested whether decline in sensory acuity that is typically observed in old age would cause a decline in cognitive performance. The authors randomly assigned younger adults to one of three age-simulating conditions, or two control conditions. In the experimental conditions, participants experienced either reduced visual acuity (by partial occlusion filters), reduced auditory acuity (by means of a headphone-shaped noise protector), or both, while no such manipulation happened in the control conditions. Next, the authors tested sensory acuity and cognitive performance by means of visually and acoustically presented cognitive tasks. While visual and auditory tests confirmed the intended reduction in sensory acuity in the experimental conditions (by about 1.5 standard deviation units), cognitive performance was *not* reduced, thus indicating that reduced sensory acuity did not cause a reduced cognitive performance. With regard to the internal validity of the data, the authors had not reduced sensory acuity to an extent that the participants could no longer perceive the cognitive tasks and therefore would fail. Such thresholds could be determined in further experiments. With regard to external validity, it could be questioned whether the time-limited experimental reduction of sensory acuity would have the same effects as a permanent reduction that is commonly associated with old age.

An experimental study by Haslam et al. (2012) manipulated the age-based categorization (older vs. younger) and expectations about general versus domain-specific declines in performance in old age and tested whether these manipulations affected cognitive test scores. In the "Older" condition, the 60- to 70-year-old participants were told that the participant age spectrum ranged from 40 to 70 years and that they were at the older end of this, while those in the "Younger" condition were told the age spectrum ranged from 60 to 90 years and that they were at the younger end of this. In addition, they had to read a paper that either stated that aging negatively affects memory or that aging is associated with general cognitive decline. Following these manipulations, participants were asked to complete a self-categorization manipulation check (to indicate whether they felt younger or older than other study participants) and to perform a memory test, as well as a test of general cognitive performance. As predicted, the categorization as older reduced performance of the participants, and the reduction of performance was either restricted to the memory test or observed in the field of general cognitive ability (depending on the content of the paper which was read). Among the participants in the "Older" condition who were forced to expect general age-associated cognitive decline, the test scores of 70% met the diagnostic criterion for probable dementia, compared to an average of 14% of the participants in other conditions. The experimental study indicates that part of the age-associated decline in cognitive performance could be based on a self-fulfilling prophecy, although there are many empirical data (mostly from non-experimental studies) and good theoretical arguments indicating that there are also other causes of cognitive change in old age (Cavanaugh and Blanchard-Fields 2019).

As a final example, the theory of socio-emotional selectivity posits that age differences in future time perspective (e.g., perceived time left to live) regulate the selection of social partners. In the case of a limited future time, people are expected to search for emotionally close social ties that provide immediate emotional benefits, while in the case of an open future, people would be more interested in contact with new, knowledgeable partners that may be more beneficial in the long run (see ▶ "Socioemotional Selectivity Theory"). Because age and the remaining time one perceives to have left in the future show, on average, a strong negative correlation, the research team of Laura L. Carstensen conducted some experimental studies that manipulated the perception of future time irrespectively of the present age of the participants, such as letting persons imagine that a new medical discovery would extend their life span by at least two decades or that they were soon going to move across the country and leave their family members behind. The limitation of future time perspective simulates old adulthood in younger adults, while the extension of future time perspective simulates young adulthood in older adults (for overview, see Löckenhoff and Carstensen 2004). These experiments lead to a reversal of effects on social selectivity and provided an elegant experimental test of the underlying theory. Although it can be questioned whether the results of the imaginative tasks would generalize to the real world, some quasi-experimental studies with more externally valid conditions (e.g., on people who really plan to leave their country or on patients who have been successfully treated for a life-shortening disease) have confirmed these findings (Löckenhoff and Carstensen 2004; Pinquart and Silbereisen 2006). This example shows that, as different research designs have different benefits and limitations, combining their results can provide more informative and compelling insights.

Future Directions

In general, researchers need to choose the research design that is best suited for answering their research question and take into account methodological as well as ethnical and practical considerations. Even if one study design is, in principle, best suited for answering a particular kind of research question, researchers must implement this design as effectively as possible, for example, by ensuring sufficient statistical power, using representative samples, and making the compared groups in quasi-experiments as similar as possible in regard to potentially confounding variables. Researchers should avoid weak study designs, such as uncontrolled intervention studies, when better designs could be implemented. There is a big discussion about a replication crisis in the field of psychology as larger numbers of results have not been replicated in later studies, in part due to weak study designs such as using small samples with large measurement errors or other uncontrolled variation, as well as selective publication of significant results (Maxwell 2015). Empirical evidence for publication bias and low statistical power have also recently been found in the field of gerontological psychology, and there was mixed evidence for possible questionable research practices (p-hacking – the use of multiple statistical analyses to find statistically significant results; Byrkes and Bielak under review). Increasing the methodological quality of studies can also increase the validity and replicability of the results in the field of gerontology. If results have been found using a weak study design, it should be tested whether they can be replicated with a stronger design. Better training in methods of aging research may help with further increasing the methodological quality of gerontological research.

As experiments allow rigorous testing of the causal hypothesis, we need more experimental studies in the field of aging research, and researchers should be creative when searching for operationalization of relevant variables that can, in principle, be experimentally manipulated. Given the design-specific strengths and weaknesses, more studies and research programs that combine different research designs are also recommended. For example, when combining the longitudinal design and experimental design, researchers can test whether the past development affects behavior during experiments and whether the experimental manipulation has long-term effects on older adults. Similarly, experimental studies should be supplemented with other study designs that test the generalizability of their results under real-life conditions and translate scientific knowledge into programs and policies that promote positive aging.

Summary

This entry describes the main characteristics and different forms of experimental and non-experimental, observational studies and compares their benefits and limitations within the contexts of gerontology and research on population aging. As observational studies dominate many domains of aging research, we discuss reasons for the limited use of experiments. After providing examples of the application of the two study designs in the field of aging research, we provide suggestions for future research practice.

Cross-References

▶ Age-Period-Cohort Models
▶ Big Data
▶ Cross-Sectional and Longitudinal Studies
▶ Dyad/Triad Studies
▶ Mobile Data Collection with Smartphones
▶ Qualitative Research/Quantitative Research
▶ Recruitment and Retention in Aging Research
▶ Repeated Cross-Sectional Design
▶ Selective Bias in Longitudinal Studies
▶ Timeline of Aging Research

References

Atalay K, Barrett GF (2015) The impact of age pension eligibility age on retirement and program dependence: evidence from an Australian experiment. Rev Econ Stat 97:71–87. https://doi.org/10.1162/REST_a_00443

Bergeman L, Boker SM (eds) (2016) Methodological issues in aging research. Psychology Press, Hove

Byrkes CR, Bielak AMA (under review) Evaluation of publication bias and statistical power in gerontological psychology. Manuscript submitted for publication

Campbell DT, Stanley JC (1966) Experimental and quasi-experimental designs for research. Rand-McNally, Chicago

Carpenter D (2010) Reputation and power: organizational image and pharmaceutical regulation at the FDA. Princeton University Press, Princeton

Cavanaugh JC, Blanchard-Fields F (2019) Adult development and aging, 8th edn. Cengage, Boston

Fölster M, Hess U, Hühnel I et al (2015) Age-related response bias in the decoding of sad facial expressions. Behav Sci 5:443–460. https://doi.org/10.3390/bs5040044

Freund AM, Isaacowitz DM (2013) Beyond age comparisons: a plea for the use of a modified Brunswikian approach to experimental designs in the study of adult development and aging. Hum Dev 56:351–371. https://doi.org/10.1159/000357177

Haslam C, Morton TA, Haslam A et al (2012) "When the age is in, the wit is out": age-related self-categorization

and deficit expectations reduce performance on clinical tests used in dementia assessment. Psychol Aging 27:778–784. https://doi.org/10.1037/a0027754
Institute for Social Research (2018) The health and retirement study. Aging in the 21st century: Challenges and opportunities for americans. Survey Research Center, University of Michigan
Jung J (1971) The experimenter's dilemma. Harper & Row, New York
Leary MR (2001) Introduction to behavioral research methods, 3rd edn. Allyn & Bacon, Boston
Lindenberger U, Scherer H, Baltes PB (2001) The strong connection between sensory and cognitive performance in old age: not due to sensory acuity reductions operating during cognitive assessment. Psychol Aging 16:196–205. https://doi.org/10.1037//0882-7974.16.2.196
Löckenhoff CE, Carstensen LL (2004) Socioemotional selectivity theory, aging, and health: the increasingly delicate balance between regulating emotions and making tough choices. J Pers 72:1395–1424. https://doi.org/10.1111/j.1467-6494.2004.00301.x
Maxwell SE (2015) Is psychology suffering from a replication crisis? What does "failure to replicate" really mean? Am Psychol 70:487–498. https://doi.org/10.1037/a0039400
Menard S (2002) Longitudinal research (2nd ed.). Sage, Thousand Oaks, CA
Mitchell SJ, Scheibye-Knudsen M, Longo DL et al (2015) Animal models of aging research: implications for human aging and age-related diseases. Ann Rev Anim Biosci 3:283–303. https://doi.org/10.1146/annurev-animal-022114-110829
Moher D (1998) CONSORT: an evolving tool to help improve the quality of reports of randomized controlled trials. JAMA 279:1489–1491. https://doi.org/10.1001/jama.279.18.1489
Oxford Centre for Evidence-Based Medicine (2011) OCEBM levels of evidence working group. The Oxford Levels of Evidence 2. Available at: https://www.cebm.net/category/ebm-resources/loe/. Retrieved 2018-12-12
Patten ML, Newhart M (2018) Understanding research methods: an overview of the essentials, 10th edn. Routledge, New York
Piccinin AM, Muniz G, Sparks C et al (2011) An evaluation of analytical approaches for understanding change in cognition in the context of aging and health. J Geront 66B(S1):i36–i49. https://doi.org/10.1093/geronb/gbr038
Pinquart M, Silbereisen RK (2006) Socioemotional selectivity in cancer patients. Psychol Aging 21:419–423. https://doi.org/10.1037/0882-7974.21.2.419
Redman LM, Ravussin E (2011) Caloric restriction in humans: impact on physiological, psychological, and behavioral outcomes. Antioxid Redox Signal 14:275–287. https://doi.org/10.1089/ars.2010.3253
Rutter M (2007) Proceeding from observed correlation to causal inference: the use of natural experiments. Perspect Psychol Sci 2:377–395. https://doi.org/10.1111/j.1745-6916.2007.00050.x
Schaie W, Caskle CI (2005) Methodological issues in aging research. In: Teti D (ed) Handbook of research methods in developmental science. Blackwell, Malden, pp 21–39
Shadish WR, Cook TD, Campbell DT (2002) Experimental and quasi-experimental designs for generalized causal inference. Houghton Mifflin, Boston
Sonnega A, Faul JD, Ofstedal MB et al (2014) Cohort profile: the health and retirement study (HRS). Int J Epidemiol 43:576–585. https://doi.org/10.1093/ije/dyu067
Weil J (2017) Research design in aging and social gerontology: quantitative, qualitative, and mixed methods. Routledge, New York

Experimental Studies: Experiments, Randomized Controlled Trials (RCTs)

▶ Experimental Studies and Observational Studies

Expressive Arts

▶ Expressive Arts Therapy

Expressive Arts Therapy

Rainbow Tin Hung Ho[1,2,3] and Adrian H. Y. Wan[1,2]
[1]Department of Social Work and Social Administration, The University of Hong Kong, Hong Kong, China
[2]Centre on Behavioral Health, The University of Hong Kong, Hong Kong, China
[3]Sau Po Centre on Ageing, The University of Hong Kong, Hong Kong, China

Synonyms

Creative arts; Expressive arts; Holistic wellness; Integrative approach to healthcare; Quality of life

Definition

Expressive arts therapy is the practice of multiple creative arts modalities, including visual arts, music, dance movement, drama, poetry, and expressive writing in an integrated way to foster human growth and to facilitate healing (IEATA 2017).

Overview

Expressive arts therapy combines the visual arts, movement, drama, music, writing, and other creative processes to foster deep personal growth (and community development). It is a distinct therapeutic discipline from the realm of expressive psychotherapies, in which the therapist and the client transit freely between different modalities during the therapy sessions. By integrating the arts creation processes and allowing one to flow into another, people gain access to their inner resources for healing, clarity, illumination, and creativity (IEATA 2017). Expressive arts therapists find their grounding not in one art form but in a theoretical orientation rooted in what all the arts have in common. The use of different arts-based modalities thus becomes itself a new modality separate and distinct from the sum of its part. Expressive arts therapy is, therefore, an integrative approach to psychotherapy and counselling as it is characterized of specific features not always present in traditional verbal therapies. Through the process of art-making, expressive arts therapy encourages self-exploration and self-expression, active participation, imagination, as well as mind-body connection (Malchiodi 2003).

There has been a growing number of practitioners in the field of aging showing increased appreciation for the profound impact that creative arts have on elder's quality of life (Patterson 2011). An expanding body of research provides evidence that creative exploration stimulated by the arts is not only good for the psychological well-being but also for general health and cognitive fitness of older adults (Castora-Binkley et al. 2009). Therefore, the application of expressive arts therapy (and its associated creative arts therapy) has been adopted as one of the adjuvant treatment options to enhance quality of life and well-being of older adults over the past decade (Dunphy et al. 2012; Reynolds 2010).

Despite steadfast interest in the use of expressive arts therapies among elderly, research into the benefits of expressive arts therapy is limited. Existing evidence relied upon the studies on modalities that compose of the integrative expressive arts intervention, most notably on the use of visual arts, dance/movement, drama, and music.

Emerging research seems to suggest the use of music, visual arts, drama, and dance/movement therapies is helpful in improving quality of life and interpersonal engagement of people with dementia, despite methodological differences among the studies conducted in the field (Beard 2011). Further, Brauninger (2014) found that the application of dance movement therapy (DMT) improves quality of life, fosters participation, and strengths resilience. Systematic review showed that group-based dance/movement therapy showed potential benefits to older people with psychiatric conditions including depression and dementia (Jiménez et al. 2018). Furthermore, dance/movement therapy was also found effective in improving biopsychosocial well-being of older adults with mild dementia; Ho et al. (2019, in press) found that the DMT group showed significant decreases in depression, loneliness, and negative mood and improved daily functioning, as well as diurnal cortisol slope. Furthermore, the effects on daily functioning and cortisol slope remained at 1-year follow-up, while exercise group of matched intensity showed no significant effects on the outcomes.

Randomized controlled trials assessing the potential benefits of arts among older adults focused primarily on the psychiatric diagnosis of dementia. The use of drama and dance/movement were found effective in improving subjective well-being of older people with dementia. Drama play was beneficial to older adults in expressing ideas and feelings as well as in unveiling of conscious awareness of participants' own well-being and quality of life (Jaaniste et al. 2015). In addition, a cluster, randomized trial on a group-based music intervention showed effectiveness in improving agitation, aberrant motor behavior, and dysphoria in Chinese older adults with dementia (Ho et al. 2018). Music therapy was also found effective in reducing depressive symptoms to some extent (Zhao et al. 2016).

Although a randomized, controlled study showed potential benefits of arts therapy intervention in reducing anxiety and negative emotions and in improving self-esteem among community-dwelling old American-Korean adults (Kim 2013), the benefit of the use of arts in promoting healthy aging is yet to be explored.

One methodological challenge in the research of expressive arts therapy is the limited amount of studies evaluating the effectiveness of expressive arts therapy as an integrative approach; much of the current evidence relies on studies on modalities that are composed of the expressive arts therapy. This indicates an important area for development of expressive arts therapy research. Despite research efforts over the past decades, the underlying mechanism of how expressive arts therapy works is yet to be fully understood by practitioners and researchers; additional research efforts are needed to better understand the process of change for expressive arts therapy.

Summary

Expressive arts therapy is a form of psychotherapy utilizing multiple modalities of creative arts, combining visual arts, movement, drama, music, writing, and other creative processes for personal growth, healing, and community development. Although the use of arts modalities such as music, drama, and dance/movement was found effective in enhancing quality of life and well-being of older adults with dementia and depression, additional research is needed to better inform helping professionals and researchers the benefits of these arts modalities on well-being of older adults who are healthy or who are facing other forms of psychosocial challenges when these arts modalities are delivered as an integrative expressive arts intervention and the underlying mechanism of human thriving.

Cross-References

▶ Affect Regulation
▶ Psychotherapy
▶ Quality of Life

References

Beard RL (2011) Art therapies and dementia care: a systematic review. Dementia 11:633–656

Brauninger I (2014) Dance movement therapy with elderly: an international internet-based survey undertaken with practitioners. Body Mov Dance Psychother 9:138–153

Castora-Binkley M, Noelker L, Prohaska T, Satariano W (2009) Impact of arts participation on health outcomes for older adults. J Aging Humanit Arts 4:352–367

Dunphy K, Mullane S, Jacbosson M (2012) The effectiveness of expressive arts therapies: a review of the literature. Psychother Couns J Aust Second edition, Volume 2, No. 1, July 2014

Ho RTH, Fond TCT, Sing CY, Lee PHT, Leung ABK, Chung KSM, Kwok JKL (2018) Managing behavioral and psychological symptoms in Chinese elderly with dementia via group-based music intervention: a cluster randomized controlled trial. Dementia. https://doi.org/10.1177/1471301218760023

Ho RTH, Fong TCT, Chan WC, Kwan JSK, Chiu PKC, Yau JCY, Lam LCW (2019) Psychophysiological effects of dance movement therapy and physical exercise on older adults with mild dementia: a randomized controlled trial. J Gerontol Psychol Sci (in press). https://doi.org/10.1093/geronb/gby145.

IEATA (2017) What are the expressive arts. Accessed 8 Mar 2017

Jaaniste J, Linnell S, Ollerton RL, Slewa-Younan S (2015) Drama therapy with older people with dementia – does it improve quality of life. Arts Psychother 43:40–48

Jiménez J, Bräuninger I, Meekums B (2018) Dance movement therapy with older people with a psychiatric condition: a systematic review. Arts Psychother. https://doi.org/10.1016/j.aip.2018.11.008.

Kim SH (2013) A randomized, controlled study of the effects of art therapy on older Korean-American's healthy aging. Arts Psychother 40:158

Malchiodi CA (2003) Expressive therapies. Guildford Press, New York

Patterson MCP (2011) Good for the heart, good for the soul: the creative arts and brain health in later life. J Am Soc Aging 35:27–36

Reynolds F (2010) 'Colour and communion': exploring the influences of visual art-making as a leisure activity on older women's subjective well-being. J Aging Stud 24:135

Zhao K, Bai ZH, Bo A, Chi I (2016) A systematic review and meta-analysis of music therapy for the older adults with depression: review of music therapy for elderly depression. Int J Geriatr Psychiatry 31:1188. https://doi.org/10.1002/gps.4494

Extended Care

▶ Skilled Care

Extended Families

▶ Intergenerational Family Structures

Extended Family

▶ Stem Family

Extended Family Structure

▶ Co-residence

Extending Working Lives

▶ Delaying Retirement

Extra-Care, Service Integrated, Sheltered-Care Housing

▶ Supportive Housing

Extreme Heat Event

▶ Heatwaves and Older People

Extreme Longevity

▶ Supercentenarians

Extremely Impoverished Older Adults

▶ "Three-No" and "Five-Guarantee" Older Adults

Extremely Long-Lived Human Beings Database

▶ (World) Supercentenarian Database

F

Fable

▶ Aging in the Short Story

Facilitating Therapeutic R&D

▶ Clinical Translation Acceleration

Faith-Based Organizations

▶ Faith-Based Social Services

Faith-Based Social Services

Beth R. Crisp
School of Health and Social Development, Deakin University, Geelong, VIC, Australia

Synonyms

Faith-based organizations; Religious charities; Religious welfare agencies

Definition

There is no single understanding as to what is meant by the term "faith-based organizations" which range from those which have a very local focus to large international networks with operations in many countries run by hundreds of thousands of staff and volunteers (Davis 2009). Whether religious organizations who receive government funding can be considered faith-based providers of care is questioned by some (Manuel and Glatzer 2019a). In addition to service organizations with professional staff and their own premises, more informal services are often based around places of worship, often without paid staff as an expression of their faith (Jawad 2012). While it is often argued that faith-based social services pertain particularly to religious organizations established to provide professional health and welfare services (Crisp 2014; Leis-Peters 2019), the distinction between formal and informal services is not necessarily straightforward. For example, professional expertise within a religious community may be made available to older adults within the local community, including services not provided elsewhere in the community, such as planning for death (Collicutt 2015). Supports offered by local faith communities range from social events, home visiting, and transport to professional services such as counseling and nursing.

The distinction between faith-based social services and faith-based health organizations is also not necessarily straightforward when it comes to older adults. For instance, in many countries, social services may be responsible for housing, but palliative care or residential nursing homes are usually deemed to be within the health sector. Furthermore, a faith-based organization may provide both social services and health services to older adults (Brault et al. 2018).

Overview

Irrespective of how they are defined, faith-based organizations make an important contribution to the care and support of older adults in many countries, but this contribution has been particularly significant in the areas of palliative care and residential aged care services. However, the contribution of faith-based providers as compared to the state and private providers varies considerably between countries. While many older service users demonstrate a preference for services provided by faith-based providers, some older adults consider faith-based services are unable to meet their needs.

Key Research Findings

Respect and care for older people is a common factor across many religions (Jawad 2012). At the same time, faith is very important to many older people, with its importance highlighted in an analysis of posts to online communities by older adults, with faith being among the seven most frequent subjects of posts (Nimrod 2010). Older people may be attracted to a faith-based service, even one that is provided by a different faith group to their own. They may be drawn to somewhere where there is a general respect for people having a faith or like the ethos of a place which is underpinned by faith values (Doud 2017).

Religious institutions have always cared for the dying, with palliative care services having their origins in faith-based organizations. The principles of hospice care which place strong emphasis on spiritual as well as physical needs have now been adopted much more widely through emerged from the faith-based sector (Pentaris 2019). Despite the emergence of other providers, a reputation for providing high-quality care often results in faith-based providers not having the capacity to accept potential referrals and needing to refer to other providers (Duckett 2017).

Faith communities have also traditionally played a significant role in providing supported residential services, such as retirement villages where people live in their own homes but can access support and services as needed. Historically these aimed to provide support to members, particularly clergy and missionaries, who may not had the financial resources to provide for their accommodation needs as they aged (Doud 2017). Importantly, faith-based services recognize the importance of people's religious and spiritual journey, past and ongoing, and how this contributes to identity (Douek 2015). Hence, access to spiritual support is integral to much faith-based care for older adults. For example, 90% of Catholic care homes in England and Wales have an on-site chapel, which can be accessed by residents and visitors (Ryan et al. 2009), but this would be much less likely in residential care settings in New Zealand, many of which are now operated by the private sector (Perkins 2015).

Older adults from minority religions may believe, often based on available evidence, that generalist services for the aged will not provide them with appropriate care (Ashencaen Crabtree 2017). In contrast, faith-based service providers are more likely to be able to provide services which are congruent to core beliefs and values, such as dietary requirements which have a religious, rather than medical, basis. For older people whose faith places particular obligations on them, for example, Kosher or Halal dietary requirements, having to reside in a facility which cannot meet these needs may be highly inappropriate. At the same time, there may be others from the same faith background for whom such factors would be readily unimportant (Graham et al. 2014).

Older people who believe their faith needs are best met by a faith-based service provider are not the only individuals who might prefer to receive services from faith-based organizations. Older people who are used to being self-reliant, and have never used welfare services, may be more open to using services provided by faith-based organizations than by government providers (Brault et al. 2018). For example, for Spaniards, the trust worthiness of the church is reportedly higher than for government. Moreover, although the church was only perceived positively by 39 percent of the population, 75% had high regard for social services run by the church (Itçaina 2019). Similarly, no matter how Swedes feel about the Church of Sweden in general, they regard it as a suitable provider for services to those who are marginalized, as long as there are no overt displays of religious beliefs (Leis-Peters 2019).

For older adults whose values or lifestyle is in conflict with the values of a faith-based service provider, a faith-based service provider may be problematic, and there are those who would prefer not to receive services from particular faith-based organizations which are underpinned by very different belief systems (Jameson et al. 2012). For example, care providers who are religious more likely to disapprove of same-sex relations, and may deny the presence of LGBTIQ+ residents, or claim to treat them no differently to anyone who is heterosexual and cisgendered, rather than acknowledging specific needs/issues (Simpson et al. 2018).

Some groups of older people, such as those who are LGBTQI+, are much more likely to have abandoned the religious beliefs of their families and report no religion, particularly when they have been subject to religious intolerance. As such, they may have no interest in engaging with faith-based organizations. Nevertheless, there are older LGBTQI+ individuals for whom religion is important and for whom faith-based organizations may be preferred over other service providers (Henrickson 2007).

Communities, in which there are strong faith-based support networks, record lower levels of hospitalization for older adults with complex health needs (Brault et al. 2018). Faith-based services also provide opportunities for older adults to volunteer in their communities which enable them to be connected with others, as well as contributing to their self-worth and sense of belonging (Brault et al. 2018; Douek 2015). An exercise program for frail-aged who were home-bound matched volunteer trainers with elderly residents, through local faith-based organizations. Participants reported high levels of satisfaction with the program, as well as significantly improvements in social interactions (Etkin et al. 2006).

In addition to direct provision of services, faith groups may contribute to services for the elderly by providing funding and resources such as buildings, encouraging members to become volunteers in various other services, and taking an active interest in social policy development concerning services for the elderly (Parsloe 1999). In particular, faith-based services have taken an active interest in legislation and other social policy initiatives around palliative care and assisted dying. While advocating perspectives in accordance with those of the religious groups which own or auspice their activities, the moral viewpoint on which arguments are based will often differ from secular organizations concerned with the same issues (Duckett 2017).

Examples of Application

The role of faith-based providers varies between countries depending on understandings as to what is the role of the welfare state; whether there is an official relationship between a religion and government (e.g., a state church) and national economy; and whether there is a single dominant faith tradition or multiple faith (Manuel and Glatzer 2019b). Differences in policy contexts, including expectations about the role of religious organizations within the civic state, mean that what is possible in one country is not necessarily possible within another (Davis 2009).

Within countries there can also be variations as to the involvement of faith-based organizations, depending on factors such as extent of unmet need from government providers in a region and the capacity of faith-based organizations to provide services within a region (Manuel and Glatzer 2019b). For example, in Italy, the church often provided services for those not considered as making a positive contribution to economic development such as the elderly and sick, whereas the state took responsibility for providing services for those currently or in the future economically active, such as workers and children (Ascoli and Arlotti 2019).

Different groups within a single faith tradition will also have very different orientations as to what the roles of faith-based and government providers of services should be (Itçaina 2019). In faith communities in which there are strong beliefs that offspring should provide any needed care for older adults, the development of faith-based services is less well-developed. Increasing numbers of elderly who have no adult children is providing the impetus for the development of faith-based aged care services (Arjouch 2008).

In many countries, faith-based organizations work in partnership with government authorities, or accept funding from the state to provide services. Whereas organizations with low dependence on government funding are able to dictate their own priorities (Harris et al. 2003), stipulations of state funding may include the requirement that services be made available to anyone who qualifies and that state funding must be used for the provision of care and not for religious purposes (Crisp 2014). However, while politicians may be eager to encourage faith-based organizations, as a cost-cutting measure for governments, religious benefactors may prefer to fund services which meet specific religious needs of their communities rather than organizations which deliver services to all members of a community irrespective of religion. Nevertheless, faith-based organizations and the financial and people resources they can attract, are not necessarily located in areas of greatest need (Harris et al. 2003).

Furthermore, faith-based organizations are not homogenous entities in respect of which values are given precedence. For example, if an older person in care decided they no longer wanted to live and refused food, respect for autonomous choices might guide practitioners working in Protestant or Hindu settings, whereas the principle of sanctity of life would present issues in Catholic, Jewish, and Muslim settings, all of which are religions which might consider such actions to be suicide which is morally reprehensible (Linzer 2006).

Future Directions of Research

Decreasing numbers of older adults who identify with established religious groups and growing numbers who identify with other belief systems or no religion raise questions as to the future role of faith-based service providers in many countries.

Summary

Faith-based providers have been critical in the development of models of social services for older adults in many countries. Although the role of these agencies differs between countries, for many older people, they provide services which are congruent with their beliefs and values, particularly in respect of housing. There are, however, older people for whom receiving services from faith-based providers is problematic. Social service policy which encourages provision of services by both faith-based and other providers is required for meeting the diverse needs of older adults.

Cross-References

▸ Aged Care Homes
▸ LGBT in Old Age
▸ Religion and Spirituality in End-of-Life Care
▸ Retirement Villages

References

Arjouch K (2008) Muslim faith communities: links with the past, bridges to the future. Generations 32 (2):47–50

Ascoli U, Arlotti M (2019) Religiously oriented welfare organizations in Italy before and after the great recession: toward a more relevant role in the provision of social services? In: Manuel PC, Glatzer M (eds) Faith-based organizations and social welfare. Palgrave Macmillan, Cham, pp 47–64

Ashencaen Crabtree S (2017) Social work with Muslim communities: treading a critical path over the crescent moon. In: Crisp BR (ed) The Routledge handbook of religion, spirituality and social work. Routledge, London, pp 118–127

Brault MA, Brewster AL, Bradley EH, Keene D, Tan AX, Curry LA (2018) Links between social environment and health care utilization and costs. J Gerontol Soc Work 61(2):203–220. https://doi.org/10.1080/01634372.2018.1433737

Collicutt J (2015) Living in the end times: a short course addressing end of life issues for older people in an English parish church setting. Working Older People 19(3):140–149. https://doi.org/10.1108/WWOP-11-2014-0034

Crisp BR (2014) Social work and faith-based organizations. Routledge, London

Davis F (2009) Faith of advocacy and the EU anti-poverty process: a case of caritas. Public Money Manag 29(6):379–386. https://doi.org/10.1080/09540960903378241

Doud RE (2017) The way of life in a retirement community. Way 56(2):31–39

Douek S (2015) Faith and spirituality in older people: a Jewish perspective. Working Older People 19(3):114–122. https://doi.org/10.1108/WWOP-03-2015-0005

Duckett S (2017) Arguing in the public square: Christian voices against assisted dying in Victoria. J Academic Study Religion 30(2):165–187. https://doi.org/10.1558/jasr.34354

Etkin CD, Prohaska TR, Harris BA, Latham N, Jette A (2006) Feasibility of implementing the strong for life program in community settings. The Gerontologist 46(2):284–292. https://doi.org/10.1093/geront/46.2.284

Graham D, Staetsky LD, Boyd J (2014) Jews in the United Kingdom in 2013: Preliminary findings from the National Jewish Community Survey. The Institute for Jewish Policy Research, London. https://www.jpr.org.uk/publication?id=3351 Accessed 1 March 2019

Harris M, Halfpenny P, Rochester C (2003) A social policy role for faith-based organisations? Lessons from the UK Jewish voluntary sector. J Soc Policy 32(1):93–112. https://doi.org/10.1017/S0047279402006906

Henrickson M (2007) Lavender faith. J Relig Spiritual Soc Work 26(3):63–80. https://doi.org/10.1300/J377v26n03_04

Itçaina J (2019) Mixed vibrancy and the invisible politics of religion: Catholic third sector, economic crisis, and territorial welfare in Spain. In: Manuel PC, Glatzer M (eds) Faith-based organizations and social welfare. Palgrave Macmillan, Cham, pp 75–102

Jameson JP, Shrestha S, Escamilla M, Clark S, Wilson N, Kunik M, Zeno D, Harris TB, Peters A, Varner IL, Scantlebury C, Scott-Gurnell K, Stanley M (2012) Establishing community partnerships to support late-life anxiety research: lessons learned from the calmer life project. Aging Ment Health 16(7):874–883. https://doi.org/10.1080/13607863.2012.660621

Jawad R (2012) Religion and faith-based welfare: from Well-being to ways of being. Policy Press, Bristol

Leis-Peters A (2019) Combining secular public space and growing diversity? Interactions between religious organizations as welfare providers and the public in Sweden. In: Manuel PC, Glatzer M (eds) Faith-based organizations and social welfare. Palgrave Macmillan, Cham, pp 161–184

Linzer N (2006) Spirituality and ethics in long-term care. J Relig Spiritual Soc Work 25(1):87–106. https://doi.org/10.1300/J377v25n01_06

Manuel PC, Glatzer M (2019a) The state, religious institutions, and welfare delivery: the case of Portugal. In: Manuel PC, Glatzer M (eds) Faith-based organizations and social welfare. Palgrave Macmillan, Cham, pp 103–133

Manuel PC, Glatzer M (2019b) "Only use words if necessary": the strategic silence of organized religion in contemporary Europe. In: Manuel PC, Glatzer M (eds) Faith-based organizations and social welfare. Palgrave Macmillan, Cham, pp 1–18

Nimrod G (2010) The fun culture in seniors' online communities. Gerontologist 51(2):226–237. https://doi.org/10.1093/geront/gnq084

Parsloe P (1999) Some spiritual and ethical issues in community care for frail elderly people: a social work view. In: Jewell A (ed) Spirituality and ageing. Jessica Kingsley Publishers, London, pp 136–145

Pentaris P (2019) Religious literacy in hospice care: challenges and controversies. Routledge, London

Perkins C (2015) Promoting spiritual care for older people in New Zealand: the Selwyn Centre for Ageing and Spirituality. Working Older People 19(3):107–113. https://doi.org/10.1108/WWOP-01-2015-0003

Ryan L, D'Angelo A, Tilki M, Castro-Ayala A (2009) National Mapping of services to older people provided by the Catholic Community. Social Policy Research Centre, Middlesex University, London. http://www.csan.org.uk/wp-content/uploads/2016/10/Mapping-OP-services-summreport.pdf. Accessed 1 Mar 2019

Simpson P, Almack K, Walthery P (2018) "We treat them all the same": the attitudes, knowledge and practices of staff concerning old/er lesbian, gay, bisexual and trans residents in care homes. Ageing Soc 38(5):869–899. https://doi.org/10.1017/S0144686X1600132X

Fallopian Tube Cancer

▶ Tumors: Gynecology

Falls

Edgar Ramos Vieira[1], Rubens A. da Silva[2], Lindy Clemson[3] and Matthew Lee Smith[4,5,6]
[1]Department of Physical Therapy, Florida International University, Miami, FL, USA
[2]Département des Sciences de la Santé, Programme de physiothérapie de l'université McGill offert en extension à l'Université du Québec à Chicoutimi (UQAC), Centre intersectoriel en santé durable, Lab BioNR – UQAC, Saguenay, QC, Canada
[3]Faculty of Health Sciences, The University of Sydney, Sydney, NSW, Australia
[4]Center for Population Health and Aging, Texas A&M University, College Station, TX, USA
[5]Department of Environmental and Occupational Health, School of Public Health, Texas A&M University, College Station, TX, USA
[6]Department of Health Promotion and Public Health, College of Public Health, The University of Georgia, Athens, GA, USA

Synonyms

Accidental falls

Definition

A fall is an event during which a person inadvertently comes to rest on the ground or other lower level (WHO 2007).

Overview

Falls are more common as people age, but the consequences of a fall can dramatically change people's life trajectory, causing disability (See ▶ "Physical Disability"), functional limitation and, for some, early institutionalization. The good news is that evidence suggests many falls can be prevented, which can help older adults maintain quality of life, health, and capacity to participate in home- and community-based activities (See ▶ "Ageing and Dance"). Fall prevention interventions such as exercise programs and home safety checks have been proven to be cost-effective (See ▶ "Aerobic Exercise Training and Healthy Aging"). This entry focuses on falls among community-dwelling older adults ($\geq$65 years); there is a vast literature on falls in other settings (e.g., hospitals, nursing homes, assisted living facilities). Falls are a common cause of disability and functional limitation among older adults (See ▶ "Disability Measurement" and ▶ "Functional Limitation"). According to the World Health Organization (WHO) Global Report on Falls Prevention in Older Age, 28–35% of all older adults fall each year globally with higher rates among older groups (WHO 2007). Approximately 646,000 people die globally due to falls each year; more than 80% of these deaths happen in low- and middle-income countries, and 60% of these happen in the Western Pacific and Southeast Asia (WHO 2018). Even when people do not die, falls can have serious consequences. Injuries occur in 40–60% of the falls, and emergency department or primary care physician visits are required for approximately 25% of all falls (Masud and Morris 2001; Tinetti and Speechley 1989). In a longitudinal study, 68% of fallers had injuries, of which 24% needed health care, and 35% reported functional decline (Stel et al. 2004). Falls are the leading cause of unintentional injury and injury-related disability and deaths among older adults in the USA (29,668 deaths, 61.6/100,000), and the rate of fall-related deaths among older adults increased 31% from 2007 to 2016 (Burns and Kakara 2018). An older adult goes to an emergency room in the USA due to a fall every 11 s (3 million visits per year), and an older adult dies from a fall-related injury every 19 min (National Council on Aging 2016). Falls are responsible for 40% of nursing home admissions (Spoelstra et al. 2012). Close to 95% of hip fractures are falls-related (See ▶ "Hip Fracture"); 95% of the hip fracture patients are discharged to nursing homes, and 20% die within a year (Florida Department of Health 2011; Schnell et al. 2010).

Key Research Findings

Risk Factors

Falls among older adults are multifactorial and include intrinsic and extrinsic risk factors; examples of intrinsic factors, those within the individual include fear of falling, poor balance, gait impairments (See ▶ "Sarcopenia"), lower limb weakness, reduced physical activity, and frailty (Oliveira et al. 2018; Vieira et al. 2016) (See ▶ "Physical Activity, Sedentary Behaviors, and Frailty"). Additional intrinsic risk factors include cognitive, visual, and hearing impairments as well as certain chronic conditions (See ▶ "Chronic Disease Self-Management") (e.g., arthritis, diabetes, stroke, Parkinson's disease, neurological diseases, incontinence) (See ▶ "Parkinson's Disease") and associated medication issues (e.g., polypharmacy, interactions, psychoactives) (Lord et al. 2001). Examples of extrinsic factors include environmental risks that exist within and/or external to the individual's home (Ambrose et al. 2013). Common examples of environmental risk factors within the home include excessive clutter, throw rugs, and exposed cords, lack of handrails or grab bars in the bathroom, and children or pets under foot. Common examples of environmental risk factors external to the home include slippery surfaces due to weather and outdoor conditions (e.g., rain, sleet, snow), uneven walkways and sidewalks, and crowds. Further, risk factors such as dim or insufficient lighting can exist within and external to the home. Table 1 presents common risk factors for falls among older adults (Adapted from Vieira et al. 2016).

In terms of physical functioning, walking slower than 0.8 m/s and being unable to complete at least eight chair rises in 30 s indicate increased risk of falls and frailty (Guccione et al. 2011) (See ▶ "Frailty Screening"). Decreased physical reserve capacity is associated to frailty and increased risk of falling in older people (Fried 2001) (See ▶ "Biology of Frailty"). Therefore, falls are associated with a cycle of aging-related physiological declines, biomechanical/mobility impairments (e.g., gait and balance), falls, fear of

Falls, Table 1 Common risk factors for falls among older adults (Adapted from Vieira et al. 2016)

Risk	Characterization
Previous falls	During the previous 12 months
Fear of falling	Low fall efficacy confidence scale scores
Balance problems	Increased postural center of pressure sway
Gait and mobility problems	Increased variability of step length, shorter single support time during dual task gait, timed up, and go test time >12 s
Pain	Lower limb and foot pain
Drugs	Polypharmacy (≥ 4), psychotropic, antidepressants, benzodiazepine (See ▶ "Benzodiazepines")
Cognitive impairment	Decreased attention, verbal ability, processing speed (executive function), and immediate memory
Urinary incontinence	Rushing to the bathroom at night
Stroke	Decreased paretic limb contribution to standing balance control, increased variability of step length, inability to step with the blocked limb (See ▶ "Stroke")
Diabetes	Peripheral neuropathy, as well as accelerated balance, somatosensory, visual, vestibular, and cognitive function decline

falling, decreased physical activity, functional decline, deconditioning, frailty, social isolation, reduced quality of life, depression, and increased risk of subsequent falls (Vieira et al. 2016; Mikaela et al. 2011) (See ▶ "Mobility and Frailty"). Fallers have 66% chance of suffering a subsequent fall within a year (Nevitt et al. 1989). Figure 1 illustrates the cycle of aging-related changes that increase the risk of falls among older adults.

Gait changes associated with falls include decreased speed, increased variability, decreased step length, increased single (stance) and double support times, and decreased swing time (time with only one foot on the ground) (Toebes et al. 2012; Verghese et al. 2009). Balance impairments associated with falls include increased center of pressure displacement area and velocity during one-leg stance – area under the curve = 0.72,

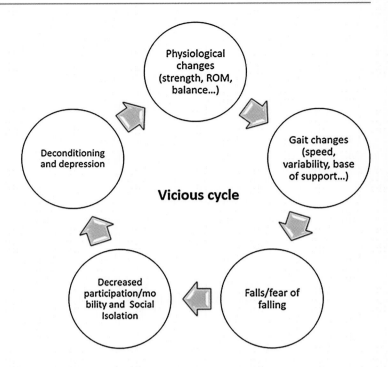

Falls, Fig. 1 Cycle of changes and factors associated with falls among older adults

sensitivity = 78%, and specificity = 68% (Oliveira et al. 2018). Cognitive impairments also affect gait and are associated with falls and therefore should be considered when designing interventions (Zhang et al. 2019) (See ▶ "Prevention of Age-Related Cognitive Impairment, Alzheimer's Disease, and Dementia").

Prevention

Prevention is a term used often to describe efforts to reduce fall occurrences. However, given the multifactorial nature of falls, and the effects of aging-related physiological changes and functional decline, "falls reduction" is more precise because some falls will always happen. With that disclosure, we will discuss some of the evidence-based strategies used to "prevent"/reduce falls among older adults living in the community. Guidelines have been published to guide efforts to reduce falls among older adults (e.g., AGS et al. 2001; Tinetti and Speechley 1989). A vast collection of evidence documents the effectiveness of falls reduction strategies, which can be implemented in a variety of settings (e.g., health care, community, home) and populations (based on risk factors, race/ethnicity, frailty level,

Falls, Table 2 Select online resources for falls reduction

Resource	Website
Centers for Disease Control and Prevention (CDC)	https://www.cdc.gov/homeandrecreationalsafety/falls/community_preventfalls.html
Cochrane library	https://www.cochrane.org
National Council on Aging (NCOA) National Falls Prevention Resource Center	https://www.ncoa.org/center-for-healthy-aging/falls-resource-center
National Institutes on Aging	https://www.nia.nih.gov/health/topics/falls-and-falls-prevention
US Preventive Services Task Force	https://www.uspreventiveservicestaskforce.org/Page/Document/final-evidence-summary24/falls-prevention-in-older-adults-counseling-and-preventive-medication

mobility), are available in the literature. Table 2 contains a select list of online resources that report the effectiveness of interventions based on randomized controlled trials and community-based translations of such interventions.

These resources contain systematic reviews, compendiums, and recommendations for implementing interventions. For example, the CDC website presents "A Guide to Implementing Effective Community-Based Fall Prevention Programs"; the guide lists programs and implementation strategies for community-based organizations. It includes examples and explains the resources need to implement and sustain the programs (CDC 2015). Some of the most commonly used and effective strategies to reduce falls among older adults include exercise, improvement in environmental, and home safety (Vieira et al. 2016).

Exercise (See ▶ "Sport and Healthy Aging")

Exercise is an effective intervention to reduce falls among older adults (Albert and King 2017) (See ▶ "Yoga Practices and Health Among Older Adults"). Gait and balance training and lower limb strengthening exercises and physical therapy reduce falls among community-dwelling older adults from 20% to 30% (Gillespie et al. 2012; Sherrington et al. 2017). Exercise and physical therapy help retrain, recover, and improve balance, strength, and gait, as well as reduce fear of falls and falls in community-dwelling older adults (risk ratio = 0.87, 95%CI = 0.81–0.94) (Michael et al. 2010). To maintain the body in a state of balance, three major postural control strategies are used (i.e., ankle, hip trunk, and step); this control depends on optimum functioning of the neural and locomotor systems (Horak 1987). Therefore, exercises to improve these three strategies are needed to improve balance and prevent falls among older adults with postural control deficits. This fact is further supported by a recently published Cochrane review, which found high-certainty evidence that exercise programs primarily involving balance and functional exercises reduce the number of falls and the number of fallers among community-dwelling older adults (Sherrington et al. 2019). Further, as reported, these programs only present nonserious adverse events, if any. Examples of evidence-based exercise programs to reduce falls are presented next.

The Otago Exercise Program (OEP) was originally developed by the Falls Prevention Research Group at the University of Otago Medical School in New Zealand as an individually tailored exercise program delivered by a nurse or physical therapist within the older adult's home (Campbell et al. 1997). The program was shown to significantly reduce falls by 35% among high-risk older adults (Colligan et al. 2015). Based on its history of success, OEP was adopted by the Centers for Disease Control and Prevention and translated for use in the

USA in 2012 (Shubert et al. 2015). The program includes a series of 17 strength ($n = 5$) and balance ($n = 12$) exercises. The participant receives six face-to-face visits over a 1-year period, four within the first 8 weeks, and then another after 6 and 12 months, respectively. When possible, the program is complemented by telephone sessions, and participants are encouraged to engage in a self-led walking program. Studies have confirmed the effectiveness of the translated version, which is shown to significantly improve participants' actual and perceived functional performance (Shubert et al. 2017a). From that time, to account for the complexities of the health-care system in the USA and increase dissemination, additional group-based and Internet-delivered models have been developed with similar effectiveness (Shubert et al. 2017b, 2019).

The Stepping On fall prevention program is a small group program underpinned by cognitive behavioral theory, adult learning principles, and self-management strategies. This multifaceted program was developed by a group of researchers at the University of Sydney in Australia (Clemson and Swann 2019). It was proven effective in reducing falls by 31% in older people with a history of a fall in the past year or who were concerned about falling. The program runs for seven weekly 2-h sessions with a follow-up phone call or home visit and a 3-month booster session. The program covers a range of issues, including personal fall risk, strength and balance exercises, home hazards, safe footwear, vision and falls, safety in public places, community mobility, coping after a fall, and understanding how to initiate a medication review. It includes specific practical strategies and resources that assist participants to self-regulate changes and taps into social and environmental influences to maintain self-selected actions and lifestyle changes (Clemson and Swann 2019). Training in Stepping On commenced in the USA in 2006 with a series of translational projects which determined key features for delivery (Mahoney et al. 2017) and supported its applicability and effectiveness in practice (Ory et al. 2014; Strommen et al. 2017). The program is supported by the CDC (Stevens and Burns 2015), and ongoing training is offered by the Wisconsin Institute for Healthy Aging enabling widespread distribution (See ▶ "Healthy Aging").

As part of the evidence-based movement in the USA, the Administration for Community Living (ACL) has supported the development of a training and delivery infrastructure to disseminate evidence-based programs for older adults through the aging services network (Boutaugh et al. 2014). Through a series of grants, 37 ACL grantees delivered 1 or more of 8 evidence-based fall prevention programs to 45,812 older adults across 22 states from 2014 to 2017 (Smith et al. 2018). The expanding number of programs supported by ACL and other funding mechanisms enable a diverse set of interventions to be available to older adults with varying levels of fall-related risk and mobility. Because these programs were purposively created for particular populations based on their content, structure, and activities, these interventions can have unique benefits for older adults. Based on the participant's level of fall risk, they may enroll in a particular program, and upon its conclusion, they may enroll in another program that is more challenging or includes higher levels of physical activity (See ▶ "Physical Activities"). Such enrollment can be sequential or concurrent and is intended to sustain or incrementally improve upon the benefits received from the initial program (Lee et al. 2018).

Environmental/Home Safety (See ▶ "Home Modification")

The physical environment at home and in the community may pose risks for falls among older adults. For example, a fall may happen when an older adult gets up from bed at night to go to the bathroom and trips on the shoes or some other piece of closing she/he left on the ground on the way to the bathroom or the person may trip on a cord or hit a poorly placed piece of furniture. We all tend to accumulate things over time; reducing clutter, removing rugs and mats, and properly placing furniture can minimize the risk of slips, trips, and falls. Another common environmental risk factor are stairs, which may have been fine for many years, but start posing a risk as the person gets older and starts to decline physically. Grab bars and rails may also need to be placed strategically around the home and in bathrooms and showers. Other things as simple as the placement of often used pots and pans within reach (not deep under the sink or cabinets) may help reduce falls. Home screens for risk factors (e.g., rugs, clutter, cords) and subsequently making environmental adaptations to reduce hazards as well as behavioral strategies to safely negotiate the environment (such as stairs/steps replacement with ramps) have been found to reduce the risk of falls in older adults (Clemson et al. 2008). This intervention has been shown to be effective when delivered by an occupational therapist. Further, studies have demonstrated that the best investment is when it is provided for older people who are at risk, such as those who had a recent hospitalization and previous fall or have a vision impairment (Pighills et al. 2011).

Future Directions of Research and Practice

Falls reduction relies on the removal, reversal, or compensation for the risk factors. Older adults with a history of falls and/or disabilities should undergo comprehensive physical, functional, and cognitive evaluation and treatment to prevent subsequent falls and related injuries (Colon-Emeric et al. 2013) (See ▶ "Comprehensive Geriatric Assessment"). Targeted interventions including exercise and home safety improvements are effective in reducing the number of falls among older adults by 20–30% (Gillespie et al. 2012). However, in the USA, Medicare does not cover preventative evaluation and treatments, and falls remain a significant problem in both magnitude and costs. The direct medical costs of fall-related injuries in the USA will reach $68 billion by 2020 (CDC 2014); a 20% reduction would represent $13.5 billion in savings.

While much is known about effective interventions and strategies to reduce falls, the persistent high prevalence of falls, related injuries, and death indicates that further advances in research and practice are needed. As with any solution, efforts are needed to ensure what works get into the hands of the people who need it most. As such, additional dissemination and implementation research is needed to better understand the science of diffusing innovation and the factors that foster or impede adoption at the organizational and individual level. Further, assessing organizational capacity to host a single intervention relative to multiple interventions over time could improve our understanding about embedding effective interventions in diverse settings. Better understanding the incremental impact of coordinated sequential evidence-based program enrollment can help define appropriate sequences (order) and cadences (timing) to optimize impact on fall-related risk.

While community-based falls reduction programs are extremely beneficial, and can be offered in a variety of settings (e.g., residential facilities, faith-based organizations, senior centers), they also need to be integrated into health-care systems. Therefore, it is important to develop and integrate fall a prevention framework to identify risks, address the issues that pose immediate risk for falls, refer older adults at risk to ongoing services to address the risks and potential consequences (i.e., falls and related injuries), and follow-up to monitor progress (See ▶ "Geriatric Rehabilitation, Instability, and Falls"). Communication between clinical and community settings is essential when developing prevention frameworks because community programs need to educate clinicians about the value of evidence-based fall prevention programs and who they are appropriate for and inform them when and where they will be occurring in their community (leave behind flyers and other materials, websites, and shared community calendars are often helpful). Clinicians can then recommend and refer older adults to attend an evidence-based program adequate to address their limitations. When older adults arrive at the program, they should share and tell the program deliverer they were referred to attend by a health-care provider. This will help the deliverer coordinate an appropriate channel for feedback. After participating in the program, the participant can return to their health-care provider and inform them about their progress in the program, which should activate assessments, referrals, and plans of care.

Figure 2 illustrates a clinical/community fall prevention collaboration, which links and integrates fall prevention efforts across these sectors. These relationships exist within the context of social determinants of health where not all social or physical environments are similar (nor are those who live within them) and therefore the services offered differ (in availability, access, affordability, and quality). While these social determinants and policies that influence health-care delivery can be protective or harmful, they must be considered in the context of creating and influencing systems change for fall prevention in a given area (Schneider et al. 2015; Smith et al. 2017).

In this example, an older adult goes to their physician who uses the Stopping Elderly Accidents, Deaths, and Injuries (STEADI) toolkit (Stevens and Phelan 2013). This toolkit is based on an algorithm for fall-related decision-making in clinical settings about screening, treatment, referral, and follow-up. The STEADI toolkit and algorithm have been adopted as a basis for implementing fall prevention in Australia (Clemson et al. 2017). Such systems and approaches enable a broader perspective engaging a whole of primary care approach as well as processes to support efficient practices and facilitate referral pathways. As part of patient care, the clinician refers older adults at risk for falls to health-care specialists, like a physical therapist, to improve their lower limb strength and flexibility, for example. After seeing the patient for about 6 weeks, the physical therapist believes she/he is ready to start the Otago Program. After successfully progressing through the exercises over time, the older adult's mobility improves, and the physical therapist recommends that they enroll in

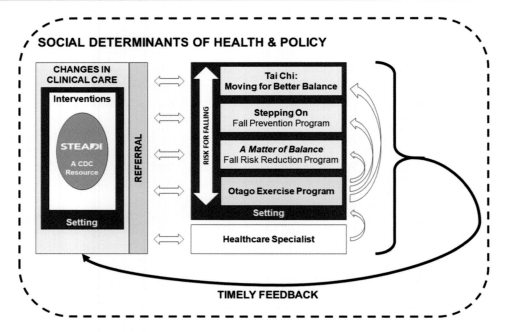

Falls, Fig. 2 Clinical/community fall prevention collaboration example

another evidence-based program such as Stepping On. After completing the 7-week Stepping On workshop, the program deliverer recommends that the client enroll in a longer intervention like Tai Chi: Moving for Better Balance, which meets three times per week for 24+ weeks and can help them maintain their strength, mobility, and functioning. Throughout this older adult's journey, there has been timely feedback (directly or indirectly) between the clinician, the physical therapist, the program deliverer, and the older adult.

In addition to emerging frameworks for falls prevention program delivery and integration into health-care systems, technological advances and translations are becoming readily available. Many Internet-based exercise programs exist to facilitate convenience and overcome obstacles associated with going to an exercise facility. However, these self-driven efforts require older adults to actively participate in the program (See ▶ "Exercise Adherence"); thus, efforts are needed to understand the intrinsic motivators that drive utilization. Similarly, many evidence-based programs have been translated for Internet-based delivery (e.g., OEP). These translated interventions have

great promise to overcome traditional barriers to participation such as time, transportation, and geographical locale (Smith et al. 2018).

Technologies are also emerging that screen for risk, monitor safety, and detect occurrences (See ▶ "Managing Long-Term Conditions: Wearable Sensors and IoT-Based Monitoring Applications"). Many of these advances are in the forms of sensors that can be worn or stationed around the home. Wearables outfitted with accelerometers and gyroscopes can identify falls based on sudden movements and changes in trajectory. Wearables can also be used to detect fall-related risk factors such as hand tremors associated with low blood sugar (diabetes and malnutrition are both risks for falling) and frailty (Vieira et al. 2019) (See ▶ "Nutrition and Aging: Nutrition Balance and Dietary Protein Needs"). Telehealth capabilities can be useful to provide medical-related counseling or assistance for medication adherence, physical activity, diet, or mental health, which can detect and offset fall-related risk. Further, in the instance of a fall, telehealth can be used to assess the situation, and other wearables and sensors can contact authorities, emergency medical personnel, and loved ones as needed.

Summary

Falls are common among community-dwelling older adults. Falls are one of the leading causes of injuries and injury-related deaths among older adults. Falls are not "natural/normal/expected" as people get older; they are the result of aging-related changes and behavioral factors (e.g., low physical activity levels) (See ▶ "Health Literacy and Health Behavior"). Falls are multifactorial; risk factors include physical and cognitive impairments, environmental risks, health conditions, and medications. Many falls can be prevented; effective measures to reduce falls include exercise programs, home safety checks, and medication reviews.

Cross-References

- ▶ Aerobic Exercise Training and Healthy Aging
- ▶ Ageing and Dance
- ▶ Benzodiazepines
- ▶ Biology of Frailty
- ▶ Chronic Disease Self-Management
- ▶ Comprehensive Geriatric Assessment
- ▶ Disability Measurement
- ▶ Exercise Adherence
- ▶ Frailty Screening
- ▶ Functional Limitation
- ▶ Geriatric Rehabilitation, Instability, and Falls
- ▶ Health Literacy and Health Behavior
- ▶ Healthy Aging
- ▶ Hip Fracture
- ▶ Home Modification
- ▶ Managing Long-Term Conditions: Wearable Sensors and IoT-Based Monitoring Applications
- ▶ Mobility and Frailty
- ▶ Nutrition and Aging: Nutrition Balance and Dietary Protein Needs
- ▶ Parkinson's Disease
- ▶ Physical Activities
- ▶ Physical Activity, Sedentary Behaviors, and Frailty
- ▶ Physical Disability
- ▶ Prevention of Age-Related Cognitive Impairment, Alzheimer's Disease, and Dementia
- ▶ Sarcopenia
- ▶ Sport and Healthy Aging
- ▶ Stroke
- ▶ Yoga Practices and Health Among Older Adults

References

Albert SM, King J (2017) Effectiveness of statewide falls prevention efforts with and without group exercise. Prev Med 105:5–9. https://doi.org/10.1016/j.ypmed.2017.08.010

Ambrose AF, Paul G, Hausdorff JM (2013) Risk factors for falls among older adults: a review of the literature. Maturitas 75:51–61. https://doi.org/10.1016/j.maturitas.2013.02.009

American Geriatrics Society (AGS), British Geriatrics Society (BGS) and American Academy of Orthopaedic Surgeons (AAOS) Panel on Falls Prevention (2001) Guideline for the prevention of falls in older persons. J Am Geriatr Soc 49(5):664–672. https://doi.org/10.1046/j.1532-5415.2001.49115.x

Boutaugh ML, Jenkins SM, Kulinski KP, Lorig KL, Ory MG, Smith ML (2014) Closing the disparity: the work of the administration on aging. Generations 38(4):107–118. https://www.asaging.org/blog/closing-disparity-gap-requires-integrated-response-policy-research-and-programs. Accessed 28 June 2019

Burns E, Kakara R (2018) Deaths from falls among persons aged ≥65 years – United States, 2007–2016. MMWR Morb Mortal Wkly Rep 67:509–514. https://doi.org/10.15585/mmwr.mm6718a1

Campbell AJ, Robertson MC, Gardner MM, Norton RN, Tilyard MW, Buchner DM (1997) Randomised controlled trial of a general practice programme of home based exercise to prevent falls in elderly women. BMJ 315(7115):1065–1069. https://doi.org/10.1136/bmj.315.7115.1065

Centers for Disease Control and Prevention (CDC) (2014) Costs of falls among older adults. http://www.cdc.gov/HomeandRecreationalSafety/Falls/fallcost.html. Accessed 24 June 2015

Centers for Disease Control and Prevention (CDC) (2015) Preventing Falls: A Guide to Implementing Effective Community-Based Fall Prevention Programs. Available at: https://www.cdc.gov/homeandrecreationalsafety/falls/community_preventfalls.html. Accessed 25 Aug 2019

Clemson L, Swann M (2019) Stepping on: building confidence and reducing falls. A community based program for older people, 3rd edn. Sydney University Press, Camperdown

Clemson L, Mackenzie L, Ballinger C, Close JCT, Cumming RG (2008) Environmental interventions to prevent falls in community-dwelling older people: a meta-analysis of randomized trials. J Aging Health 20

(8):954–971. https://doi.org/10.1177/0898264308324672

Clemson L, Mackenzie L, Roberts C, Poulos RG, Tan A, Lovarini M, Sherrington C, Simpson JM, Willis K, Lam M et al (2017) Integrated solutions for sustainable fall prevention in primary care, the iSOLVE project: a type 2 hybrid effectiveness-implementation design. Implement Sci 12(1):12. https://doi.org/10.1186/s13012-016-0529-9

Colligan EM, Tomoyasu N, Howell B (2015) Community-based wellness and prevention programs: the role of Medicare. Front Public Health 2:189. https://doi.org/10.3389/fpubh.2014.00189

Colon-Emeric CS, Whitson HE, Pavon J, Hoenig H (2013) Functional decline in older adults. Am Fam Physician 88(6):388–394

Florida Department of Health (FDH) (2011) Florida injury facts: unintentional falls. Injury Prevention Program. www.floridahealth.gov/statistics-and-data/florida-injury-surveillance-system/_documents/data-factsheets/falls-2011.pdf. Accessed 10 Nov 2014

Fried et al (2001) Frailty in older adults: evidence for a phenotype. J Gerontol A Biol Sci Med Sci 56(3):M146–M156. https://doi.org/10.1093/gerona/56.3.M146

Gillespie LD, Robertson MC, Gillespie WJ et al (2012) Interventions for preventing falls in older people living in the community. Cochrane Database Syst Rev (9):CD007146. https://doi.org/10.1002/14651858.CD007146.pub3

Guccione AA, Avers D, Wong R (2011) Geriatric physical therapy. Elsevier Health Sciences, London

Horak FB (1987) Clinical measurement of postural control in adults. Phys Ther 67:1881–1885

Lee S, Smith ML, Towne SD Jr, Ory MG (2018) Effects of sequential participation in evidence-based health and wellness programs among older adults. Innov Aging 2(2):igy016. https://doi.org/10.1093/geroni/igy016

Lord SR, Sherrington C, Menz HB (2001) Falls in older people: risk factors and strategies for prevention. Cambridge University press, New York, p 249

Mahoney J, Clemson L, Schlotthauer A et al (2017) Modified Delphi consensus to suggest key elements of stepping on falls prevention program. Front Public Health 5:21. https://doi.org/10.3389/fpubh.2017.00021

Masud T, Morris RO (2001) Epidemiology of falls. Age Ageing 30(Suppl 4):3–7

Michael YL, Whitlock EP, Lin JS, Fu R, O'Connor EA, Gold R et al (2010) Primary care-relevant interventions to prevent falling in older adults: a systematic evidence review for the U.S. Preventive Services Task Force. Ann Intern Med 153(12):815–825. https://doi.org/10.7326/0003-4819-153-12-201012210-00008

Mikaela B, Bonsdorff V, Rantanen T (2011) Progression of functional limitations in relation to physical activity: a life course approach. Eur Rev Aging Phys Act 8:23–30. https://doi.org/10.1007/s11556-010-0070-9

National Council on Aging (NCA) (2016) Falls prevention facts. https://www.ncoa.org/news/resources-for-reporters/get-the-facts/falls-prevention-facts/. Accessed 01 Sept 2017

Nevitt MC, Cummings SR, Kidd S et al (1989) Risk factors for recurrent nonsyncopal falls: a prospective study. JAMA 261(18):2663–2668. https://doi.org/10.1001/jama.1989.03420180087036

Oliveira MR, Gil AW, Fernandes KBP, Teixeira D, Amorim CF, da Silva RA (2018) One-legged stance sway of older adults with and without falls. PLoS One 13(9):e0203887. https://doi.org/10.1371/journal.pone.0203887. eCollection 2018

Ory MG, Smith ML, Jiang L et al (2014) Fall prevention in community settings: results from implementing stepping on in three states. Front Public Health 2:232. https://doi.org/10.3389/fpubh.2014

Pighills AC, Torgerson DJ, Sheldon TA, Drummond AE, Bland JM (2011) Environmental assessment and modification to prevent falls in older people. J Am Geriatr Soc 59(1):26–33. https://doi.org/10.1111/j.1532-5415.2010.03221.x

Schneider EC, Smith ML, Ory MG, Altpeter M, Beattie BL, Scheirer MA, Shubert TE (2015) State fall prevention coalitions as systems change agents: an emphasis on policy. Health Promot Pract 17(2):244–253. https://doi.org/10.1177/1524839915610317

Schnell S, Friedman SM, Mendelson DA, Bingham KW, Kates SL (2010) The 1-year mortality of patients treated in a hip fracture program for elders. Geriatr Orthop Surg Rehabil 1(1):6–14. https://doi.org/10.1177/2151458510378105

Sherrington C, Michaleff ZA, Fairhall N, Paul SS, Tiedemann A, Whitney J, Cumming RG, Herbert RD, Close JC, Lord SR (2017) Exercise to prevent falls in older adults: an updated systematic review and meta-analysis. Br J Sports Med 51(24):1750–1758. https://doi.org/10.1136/bjsports-2016-096547

Sherrington C, Fairhall NJ, Wallbank GK, Tiedemann A, Michaleff ZA, Howard K, Clemson L, Hopewell S, Lamb SE (2019) Exercise for preventing falls in older people living in the community. Cochrane Database Syst Rev 1:CD012424. https://doi.org/10.1002/14651858.CD012424.pub2

Shubert TE, Smith ML, Ory MG, Clarke C, Bomberger SA, Roberts E, Busby-Whitehead J (2015) Translation of the Otago exercise program for adoption and implementation in the United States. Front Public Health 2:152. https://doi.org/10.3389/fpubh.2014.00152

Shubert TE, Ory MG, Jiang L, Smith ML (2017a) Disseminating the Otago exercise program in the United States: perceived and actual physical performance improvements from participants. J Appl Gerontol 37(1):79–98. https://doi.org/10.1177/0733464816675422

Shubert TE, Ory MG, Jiang L, Smith ML (2017b) The Otago exercise program in the United States: a comparison of two implementation models. Phys Ther 97(2):187–197. https://doi.org/10.2522/ptj.20160236

Shubert TE, Chokshi A, Mendes VM, Grier S, Buchanan H, Basnett J, Smith ML (2019) Stand tall: a virtual translation of the Otago exercise program. J Geriatr Phys Ther. https://doi.org/10.1519/JPT.0000000000000203. (online first)

Smith ML, Schneider EC, Byers IN, Shubert TE, Wilson AD, Towne SD Jr, Ory MG (2017) Reported systems changes and sustainability perceptions of three state departments of health implementing multi-faceted evidence-based fall prevention efforts. Front Public Health 5:120. https://doi.org/10.3389/fpubh.2017.00120

Smith ML, Towne SD Jr, Herrera-Venson A, Cameron K, Horel SA, Ory MG, Gilchrist CL, Schneider EC, DiCocco C, Skowronski S (2018) Delivery of fall prevention interventions for at-risk older adults in rural areas: findings from a national dissemination. Int J Environ Res Public Health 15(12):2798. https://doi.org/10.3390/ijerph15122798

Spoelstra SL, Given B, You M, Given CW (2012) The contribution falls have to increasing risk of nursing home placement in community-dwelling older adults. Clin Nurs Res 21(1):24–42. https://doi.org/10.1177/1054773811431491

Stel VS, Smit JH, Pluijm SMF, Lips P (2004) Consequences of falling in older men and women and risk factors for health service use and functional decline. Age Ageing 33(1):58–65. https://doi.org/10.1093/ageing/afh028

Stevens JA, Burns ER (2015) CDC compendium of effective fall interventions: what works for community-dwelling older adults, 3rd edn. Centres for Disease Control and Prevention, National Center for Injury Prevention and Control, Atlanta

Stevens JA, Phelan EA (2013) Development of STEADI: a fall prevention resource for health care providers. Health Promot Pract 14:706–714. https://doi.org/10.1177/1524839912463576

Strommen J, Brotherson S, Yang Z (2017) Older adult knowledge and behavior change in the stepping on fall prevention program in a community setting. J Hum Sci Ext 5(3):99–121

Tinetti ME, Speechley M (1989) Prevention of falls among the elderly. N Engl J Med 320(16):1055–1059. https://doi.org/10.1056/NEJM198904203201606

Toebes MJ1, Hoozemans MJ, Furrer R, Dekker J, van Dieën JH (2012) Local dynamic stability and variability of gait are associated with fall history in elderly subjects. Gait Posture 36(3):527–531. https://doi.org/10.1016/j.gaitpost.2012.05.016

Verghese J, Holtzer R, Liptom R, Wang C (2009) Quantitative gait markers and incident fall risk in older adults. J Gerontol A 64A(8):896–901. https://doi.org/10.1093/gerona/glp033

Vieira ER, Palmer RC, Chaves PH (2016) Prevention of falls in older people living in the community. BMJ 353:i1419. https://doi.org/10.1136/bmj.i1419

Vieira ER, Menendez N, Amick III BC (2019). The potential to prevent, identify and treat frailty and its consequences in community-dwelling older adults using a digital platform and existing smartphone capabilities. In: 9th International Conference on Frailty & Sarcopenia Research (ICFSR), Miami Beach, FL, USA. Journal of Frailty & Aging, S56

World Health Organization (2007) WHO ageing, & life course unit. WHO global report on falls prevention in older age. World Health Organization. https://www.who.int/ageing/publications/Falls_prevention7March.pdf. Accessed 28 June 2019

World Health Organization (2018) Newsroom facts sheet on falls. https://www.who.int/en/news-room/fact-sheets/detail/falls. Accessed 28 June 2019

Zhang WH, Low LF, Gwynn JD, Clemson L (2019) Interventions to improve gait in older adults with cognitive impairment: a systematic review. J Am Geriatr Soc 67(2):381–391. https://doi.org/10.1111/jgs.15660

False Negative/False Positive

Tirth R. Bhatta
Department of Sociology, University of Nevada, Las Vegas, NV, USA

Synonyms

Type I and Type II errors

Definition

A false positive occurs when a diagnostic instrument erroneously detects a nonexistent disease or condition and presents it as reality (i.e., a positive result). False positives have been encountered in a variety of settings, including medical diagnostics, the criminal justice system, drone warfare, and research studies (Ioannidis 2005; Boyle 2013; Angwin et al. 2016; Morgan 2018). The opposite scenario, labeled as a false negative, occurs when an existing condition is not detected by the diagnostic instrument (i.e., fails to recognize a positive result). Failures to detect diseases, such as lung and breast cancers (Nelson et al. 2016; Davies et al. 2018), due to inadequate or faulty diagnostic tests have been reported in medical settings.

Overview

As probability thresholds used to determine statistical significance, Type I and Type II errors typify the essence of false positives/false negatives in null hypothesis significance testing. Early developments in significance testing led by Francis Ysidro Edgeworth did not establish formal a priori probability to differentiate substantive difference from chance occurrence. The term *statistical significance*, which Edgeworth coined in 1885, was not accompanied by any theory to distinguish between "material" (or substantive) differences and "accidental" occurrences (Perezgonzalez 2015; Ziliak and McCloskey 2008). Although his later writings discouraged the use of a conventional level of significance, Ronald A. Fisher was the first to use a probability threshold known as the level of significance (typically 0.05) as a convention for null hypothesis testing in 1925. Rejection of the null hypothesis required calculation of the *p*-value, which was introduced by Karl Pearson in 1900 (1900). Often erroneously understood as the probability that the null hypothesis is true, the *p*-value refers to the probability of obtaining an effect equal to or greater than the observed effect when the null hypothesis is true (Pollard and Richardson 1987). The probability that a null hypothesis will be rejected substantially increases when the *p*-value is low, which – if lower than the level of significance – offers us evidence against the null hypothesis.

Further formalization of hypothesis testing procedures by Jerzy Neyman and Egon Pearson in 1928 saw the inclusion of alternative hypotheses with an explicit recognition of Type I (i.e., rejecting the null hypothesis when it is true) and Type II (i.e., accepting the null hypothesis when it is false) errors. The inclusion of alternative hypotheses and, hence, incorporation of Type II errors (typically 0.1 or 0.2) and effect size (e.g., expected differences in later life health between Blacks and Whites) set Neyman and Pearson's hypothesis testing framework apart from Fisher's null hypothesis testing. The specification of the two errors allowed them to limit the risk of inferring statistical significance when there is none (i.e., Type I error or α) or not inferring significance when there is one (Type II error or β) while conducting numerous experiments. Under their framework, the null hypothesis is rejected in favor of an alternative hypothesis if the observed effect falls in a critical region delineated by these two errors. Unlike previous approaches, their theory of hypothesis testing requires that the magnitude of Type I and Type II (alternatively, 1-β = power) errors and effect sizes be chosen prior to data collection and factored into sample size calculations (Gigerenzer 2004).

Current research practices tend to deviate from Neyman and Pearson's hypothesis testing framework by exclusively focusing on the Type I error rate to determine statistical significance. Such practices, which stem from a desire to obtain significant findings due to factors such as publication bias and increased pressure to publish, could lead to false positives (Simmons et al. 2011; Ioannidis 2005; Sterne and Smith 2001; Rosenthal 1979). To avoid findings that emanate from what some scholars refer to as "null ritual" (Gigerenzer 2004), proposals such as reducing Type I error range, changing the threshold to $p < 0.005$ (Benjamin et al. 2018), or designing theoretically informed studies (Murayama et al. 2014) have been offered.

Summary

False-positive findings could impede scholarly advancements by initiating debates that lack substantive merit. Further, false positives could lead to inconsistencies in research findings that may run counter to established scientific reality. Ensuing debates that are not theoretically justified run the risk of delegitimizing scientific research, which – in effect – could reduce public trust of research evidence (Simmons et al. 2011). Findings from inherently faulty diagnostic tests or research studies may not only lead to disproportionate and devastating effects of public policies on poor and minority groups but also misguide those policies. Individuals from racial minority groups are more likely to be charged and wrongly convicted based on results from questionable tests (e.g., a $2 roadside drug test). Similarly, courtroom reliance on questionable computer algorithms that are known to (more often than not) falsely label Blacks as future criminals put their lives at higher risk – namely, in the form of restricted participation with regard to employment, housing, and voting rights (Gabrielson and Sanders 2016; Angwin et al. 2016).

False-positive findings could inform deliberations on public policy and are likely to be particularly concerning when the allocation of funding for programs or policies is contingent on their statistically significant effects on individuals (e.g., Investing in Innovation Fund; Haskins and Margolis 2014). The investment of public resources premised on false-positive findings may not exert the intended effects on the issues (e.g., disparities in later life health) being addressed. Due to the failure of hypothesis tests to detect statistically significant results, false-negative findings can obstruct or completely shut down substantively relevant lines of scholarly pursuits and impede policy efforts designed to reduce adverse impacts of structural disadvantages on individuals. For instance, if a research study fails to find significant effects of housing on health outcomes, the policies under consideration to reduce health disparities could exclude housing as an intervention to achieve such a goal.

Cross-References

▶ Cross-Sectional and Longitudinal Studies
▶ Experimental Studies and Observational Studies
▶ Qualitative Research/Quantitative Research
▶ Structural Equation Models
▶ Survival Analysis

References

Angwin J, Larson J, Mattu S et al (2016) Machine bias. https://www.propublica.org/article/machine-bias-risk-assessments-in-criminal-sentencing. Accessed 11 Mar 2019
Benjamin DJ, Berger JO, Johannesson M et al (2018) Redefine statistical significance. Nat Hum Behav 2(1):6
Boyle MJ (2013) The costs and consequences of drone warfare. Int Aff 89(1):1–29
Davies KD, Le AT, Sheren J et al (2018) Comparison of molecular testing modalities for detection of ROS1 rearrangements in a cohort of positive patient samples. J Thorac Oncol 13(10):1474–1482
Gabrielson R, Sanders T (2016, July 7) Busted. https://www.propublica.org/article/common-roadside-drug-test-routinely-produces-false-positives. Accessed 11 Mar 2019
Gigerenzer G (2004) Mindless statistics. J Socio Econ 33(5):587–606
Haskins R, Margolis G (2014) Show me the evidence: Obama's fight for rigor and results in social policy. Brookings Institution Press, Washington, DC
Ioannidis JP (2005) Why most published research findings are false. PLoS Med 2(8):e124
Morgan D (2018) What the tests don't show. https://www.washingtonpost.com/news/posteverything/wp/2018/10/05/feature/doctors-are-surprisingly-bad-at-reading-lab-results-its-putting-us-all-at-risk/?noredirect=on&utm_term=.00f431213979. Accessed 11 Mar 2019
Murayama K, Pekrun R, Fiedler K (2014) Research practices that can prevent an inflation of false-positive rates. Personal Soc Psychol Rev 18(2):107–118
Nelson HD, O'Meara ES, Kerlikowske K et al (2016) Factors associated with rates of false-positive and false-negative results from digital mammography screening: an analysis of registry data. Ann Intern Med 164(4):226–235
Pearson K (1900) X. On the criterion that a given system of deviations from the probable in the case of a correlated system of variables is such that it can be reasonably supposed to have arisen from random sampling. Lond Edinb Dubl Phil Mag 50(302):157–175
Perezgonzalez JD (2015) Fisher, Neyman-Pearson or NHST? A tutorial for teaching data testing. Front Psychol 6:223
Pollard P, Richardson JT (1987) On the probability of making Type I errors. Psychol Bull 102(1):159
Rosenthal R (1979) The file drawer problem and tolerance for null results. Psychol Bull 86(3):638
Simmons JP, Nelson LD, Simonsohn U (2011) False-positive psychology: undisclosed flexibility in data collection and analysis allows presenting anything as significant. Psychol Sci 22(11):1359–1366
Sterne JA, Smith GD (2001) Sifting the evidence – what's wrong with significance tests? Phys Ther 81(8):1464–1469
Ziliak S, McCloskey DN (2008) The cult of statistical significance: how the standard error costs us jobs, justice, and lives. University of Michigan Press, Ann Arbor

Familial Aid

▶ Intergenerational Exchanges

Familial Dyslipidemias

Carlos Alberto Aguilar Salinas
Unidad de Investigación en Enfermedades Metabólicas, Instituto Nacional de Ciencias Médicas y Nutrición Salvador Zubirán, Mexico City, Mexico
Departamento de Endocrinología y Metabolismo, Instituto Nacional de Ciencias Médicas y Nutrición Salvador Zubirán, Mexico City, Mexico
Escuela de Medicina y Ciencias de la Salud, Tecnologico de Monterrey, Monterrey, Nuevo Leon, Mexico

Synonyms

Genetic dyslipidemias; Hypercholesterolemia; Lipid disorders; Primary hyperlipidemias

Definition

Dyslipidemias is a group of asymptomatic disorders identifiable by the presence of abnormal concentrations of cholesterol, triglycerides, and HDL cholesterol caused by an abnormal composition and/or number of plasma lipoproteins. They are a major risk factor for having atherosclerotic cardiovascular disease; its treatment has been highly effective to reduce cardiovascular mortality (Catapano et al. 2016).

Overview

Lipoproteins are complex structures that transport lipids in their interior to make it soluble in the aqueous media of plasma (Durrington 2003). Lipid transport has two destinations. The first route transports lipids from the sites of production or absorption (intestine and liver) to the tissues where they are used or stored (muscle, fat, glands, etc.). In the case of the liver, the lipoprotein synthesis takes 8–12 h and it is highly regulated by insulin and the fatty acid concentration. Apolipoprotein B-100 provides the structure of the particles; also, it contains the region that is recognized by the receptors that mediate its elimination from the plasma (LDL receptors among others) (Morita 2016). The lipoproteins synthesized in the liver are known as very low-density lipoproteins (VLDL), which when metabolized in the plasma undergo lipolysis (due to the action of lipoprotein lipase) and change of their composition becoming the intermediate density lipoproteins (IDL); these particles turn on low-density lipoproteins (LDL) due to the action of hepatic lipase and CETP (Cholesteryl ester transfer protein, an enzyme that transfer cholesterol from other lipoproteins to LDLs). In contrast, lipoprotein synthesis in the intestine is a fast and highly efficient process that allows the absorption of large amounts of fat in few hours. Apolipoprotein B-48 (the edited product of APOB) gives the structure to the particles but lacks the binding site to the LDL receptors. The lipoproteins synthesized in the intestine are known as chylomicrons, which undergo the same changes in the plasma as the VLDL, becoming particles called remnants. Lipolysis makes possible the inclusion of other lipoproteins (i.e., apolipoprotein E) in the surface of the particles, which plays the role of ligands to mediate its elimination from plasma (Marais 2019). The smaller subclasses of VLDL, IDL, and LDL have the capacity to be accumulated in the subendothelial space and therefore contribute to the atherogenic burden. The chylomicrons are large in size, which prevents them from depositing on atheromatous plaques; in contrast, the remnants are atherogenic particles (Tada et al. 2019).

The second route moves lipids from the periphery to the liver to be excreted in the bile. The high density lipoproteins (HDL) are the particles in charge. These particles prevent the ectopic deposit of cholesterol, have anti-inflammatory and antioxidant properties. Therefore, they protect against the development of atherosclerosis (Nicholls and Nelson 2019).

Monogenic or polygenic disorders of lipoprotein metabolism are strong risk factors for cardiovascular morbidity. The best example is familial hypercholesterolemia. It is characterized by extreme elevation of LDL cholesterol (in adults above 190 mg/dl), due to defects in its catabolism explained by abnormalities of LDL receptor function. Its inheritance pattern may be autosomal recessive or dominant. The autosomal dominant variant is caused by mutations in the LDL receptor gene, activating mutations of PCSK9 (protein that participates in the recycling of the LDL receptor) or *APOB* (ligand of the LDL receptor). In the autosomal dominant form, there are two forms of presentation: the heterozygous and the homozygous. The heterozygous form is the most common form; cases usually have tendinous xanthomata, corneal arc, and cholesterol between 300 and 400 mg/dl. The site to look for xanthomas is the Achilles heel. There are several diagnostic criteria, being those proposed by the Dutch Lipid Clinics the most used (Alonso et al. 2018). The prevalence of this condition is from 1: 200 to 1: 500 (although in some populations (i.e., South Africa or Quebec) it is greater due to a founder effect). The recessive variant is related to mutations of the ARH gene. The severity depends on the type of mutation; those that abolish the LDL receptor function have a more serious clinical picture than those that result in dysfunctional forms of the receptor. In cases where *APOB* is mutated, cholesterol values are usually lower than in those caused by LDL receptor defects because a percentage of LDL has another ligand (apolipoprotein E) on its surface to recognize the LDL receptor (Benito-Vicente et al. 2018).

Familial combined hyperlipidemia is the most common of the primary dyslipidemias (Bello-Chavolla et al. 2018). The pattern of inheritance is variable (autosomal dominant in most cases) with variable penetrance. It is expressed by hypercholesterolemia, hypertriglyceridemia, or mixed hyperlipidemia, accompanied by high concentrations of apolipoprotein B (above the 90th percentile of the population, >120 mg/dl in Caucasians) and predominance of small, dense LDL. The severity of dyslipidemia is moderate in most cases; only in coexistence of a secondary hyperlipidemia, cholesterol or triglycerides concentrations are found above 300 mg/dl. To diagnose it, the existence of a relative with hypercholesterolemia, other with hypertriglyceridemia, and another with both abnormalities should be demonstrated. It is associated with the metabolic syndrome and type 2 diabetes. USF1, TCF7L2, and HNF4-alpha are among the genes that are part of its pathophysiology.

Another atherogenic primary dyslipidemia is dysbetalipoproteinemia (Koopal et al. 2017). Its clinical presentation is mixed hyperlipidemia caused by accumulation of remnants and IDLs due to decreased clearance of these and other atherogenic lipoproteins. The E2 isoform of apolipoprotein E is the most common cause; it results in a low affinity for its receptors. To be expressed, it must coexist with a condition that alters the function or quantity of the LDL receptors, since the remnants are also eliminated through this receptor. Some precipitating factors are diabetes, hypothyroidism, obesity, use of beta-blockers, or diuretics. It is suspected if there are simultaneous increases of cholesterol and triglycerides in a range of 300 mg/dl. This diagnosis is confirmed by lipoprotein electrophoresis in which a broad beta pattern (type III) is found, by ultracentrifugation (i.e., the ratio VLDL / triglycerides>0.3) or by ApoE genotyping.

Another common form of primary dyslipidemia is familial hypertriglyceridemia which is characterized by severe hypertriglyceridemia, normal or low LDL cholesterol, and with normal levels of apolipoprotein B. It is a common cause of pancreatitis or eruptive xanthomas. It is diagnosed when the subject and a family member have this lipid pattern. By itself it does not increase atherogenic risk, but it is common among patients with type 2 diabetes. The majority of cases with severe hypertriglyceridemia (>1000 mg/dl) are polygenic in nature; polymorphisms of genes involved in VLDL/chylomicron catabolism (i.e., apolipoprotein AV or lipoprotein lipase) are usually involved. Monogenic forms of severe hypertriglyceridemia (also known as chylomicronemia) result from homozygous or compound heterozygous states of mutations of LPL (lipoprotein lipase), APOC2 (apolipoprotein CII), LMF1 (lipase maturation factor 1) and glycosylphosphatidylinositol-anchored high-density lipoprotein-binding protein 1 (GPIHBP1), apolipoprotein AV, or glycerol-3-phosphate dehydrogenase 1 (G3PDH1). Recurrent pancreatitis is the most common expression of monogenic chylomicronemia (Falko 2018).

Primary hypoalphalipoproteinemia is the most common primary dyslipidemia of the HDL metabolism. It is caused by heterozygous ABC-A1 mutations. HDL cholesterol concentrations are found between 25 and 39 mg/dl; no other lipid abnormalities coexists. It is associated with an increased cardiovascular risk. Extremely low HDL cholesterol concentrations (<20 mg/dl) are caused by mutations in the LCAT (Lecithin cholesterol acyl transferase), Apolipoprotein A1, and ABC-A1 genes. Premature atherosclerosis is found mainly in cases with *APOA1* mutations (Schaefer et al. 2016).

Key Research Findings

Primary dyslipidemias are a strong risk factor for having premature cardiovascular deaths (Sharma and Baliga 2017). As a result, information about these conditions in elders is scant. Expected life expectancy is not reached in 60% of patients with familial hypercholesterolemia. Primary dyslipidemias are associated with increased risk for having stroke, despite, cholesterol concentrations are not independent risk factors for carotid atherosclerosis. On the other hand, there are genetic abnormalities of the lipid metabolism that are associated with longevity. This is the case of apolipoprotein AI Milano, a natural variant identified in cases with low HDL cholesterol levels, caused by a replacement of arginine by cysteine at position 173 (Vallejo-Vaz and Ray 2016). The change of sequence results in the formation of ApoAI homodimers and heterodimers with apolipoprotein AII. These abnormal compounds have an increased affinity to accept cholesterol efflux from cells and improve vascular function. Polymorphisms located in several genes involved in lipoprotein metabolism have been associated with longevity. This is the case of *PON1*, insulin-like growth factor 1 (*IGF1*), and *CETP*. In particular, the substitution of isoleucine by valine in codon 405 (I405V) results in decreased CETP activity, high HDL cholesterol levels, and protection against cognitive decline. However, these findings have not been consistently replicated (Milman and Barzilai 2015).

Prevalence of VV homozygosity is 0.09 in long-lived subjects; the attributable risk fraction of this variant is 18.1. This CETP variant is associated with large LDL and HDL particle sizes. The same lipoprotein pattern is present in a large percentage of individuals older than age 80, even if they are not carriers of the CETP polymorphism. Other CETP polymorphisms (i.e., rs709272) modify HDL cholesterol levels, but the assessment of its impact on atherosclerosis or lifespan has rendered inconsistent results (Milman et al. 2014).

Homozygosity states of the 641C allele of apolipoprotein C-III (*APOC3*) are also associated with longevity; usually, cases have low triglycerides and higher HDL cholesterol concentrations without any other comorbidity. The same trend has been informed for other null mutations of *APOC3* (i.e., R19X). On the other hand, epidemiological studies have linked high HDL cholesterol levels with longevity. As a mean, HDL cholesterol decreases 1% per year during the aging process. Individuals with extended lifespan keep their HDL cholesterol stable, despite the aging process. High HDL cholesterol is associated with lower incidence of cardiovascular disease, hypertension, dementia, and cancer (Walter 2009). The protection could result from the anti-atherogenic, anti-inflammatory actions of some HDL particles. However, not every subject with high HDL cholesterol is protected against aging-related diseases. No laboratory or clinical traits are useful to identify the protective forms of hyperalphalipoproteinemia.

Examples of Application

The information above described has at least two clinical implications. First, treatment of primary dyslipidemias has extended lifespan and reduced cardiovascular morbidity (Orkaby et al. 2017). Consequently, a growing proportion of the affected cases are becoming elders. The best example is the near normal lifespan of FH cases in some European countries. The use of lipid-lowering drugs in older populations has been the matter of review in several recent guidelines. Main position documents recommend applying the same therapeutic approach in older populations than in adult populations (Catapano et al. 2016; Grundy et al. 2018). Exemptions are cases with major functional limitations or comorbidities that severely shorten the life expectancy. Impact of lipid lowering agents on cognitive functions is matter of concern. However, no conclusive evidence has been reached in randomized controlled trials.

Second, new therapeutic approaches have been developed to mimic the natural protection against atherosclerosis that confers defects in the CETP or APOC-III functions. Several CETP inhibitors have reached advanced stages of development (Armitage et al. 2019). This is the case of torcetrapib, evacetrapib, dalcetrapib, and anacetrapib. Torcetrapib studies were stopped due to unanticipated off-target effects that increased risk of death. Dalcetrapib and evacetrapib studies were terminated early for futility. The REVEAL trail (Randomized Evaluation of the Effects of Anacetrapib through Lipid Modification) concluded that anacetrapib reduced major vascular events by 9% over 4 years. The drug accumulates in adipose tissue raising concerns about its long-term safety. As a result, CETP inhibition is not considered as a valid option to reduce cardiovascular events in the near future. On the other hand, several apolipoprotein C-III targeted therapies are being developed. This is the case of Volanesorsen, a second generation antisense complementary to the APOC3 mRNA. This approach blocks apolipoprotein C-III gene expression and reduces triglycerides concentrations by 56–80% (Warden and Duell 2018).

Future Directions of Research

The role of HDL metabolism in healthy aging requires additional studies. Quantitative and qualitative studies are needed to disclose if high HDL cholesterol concentrations are protective against atherosclerosis and/or neurodegenerative conditions. On the other hand, on-going long-term safety studies of available lipid lowering therapies may provide information its effects on several aging-related events (abnormal cognition, dementia, sarcopenia, among others). More studies are needed about the interaction of the factors that modulate plasma lipid levels and healthy aging.

Summary

Genetic factors play major determinants of plasma lipid levels and longevity. The study of monogenic forms of dyslipidemias has made possible the identification of treatment targets (i.e., statins and PCSK9 inhibitors) to prevent cardiovascular events and reduce cardiovascular mortality (Sabatine 2019). The availability of potent lipid lowering agents has expanded the life expectancy of patients with primary dyslipidemias, resulting in new challenges and unsolved questions (i.e., the effect of lipid lowering therapies in healthy aging). New venues for research are under development as natural experiments in which mutations on genes involved in lipoprotein metabolism (i.e., *APOC3*) are associated with longevity.

Cross-References

▶ Healthy Diet
▶ Ischemic Heart Disease
▶ Nutrition Issues in Geriatrics
▶ Vascular Diseases of Ageing

References

Alonso R, Perez de Isla L, Muñiz-Grijalvo O, Diaz-Diaz JL, Mata P (2018) Familial hypercholesterolaemia diagnosis and management. Eur Cardiol 13(1):14–20. https://doi.org/10.15420/ecr.2018:10:2

Armitage J, Holmes MV, Preiss D (2019) Cholesteryl ester transfer protein inhibition for preventing cardiovascular events: JACC review topic of the week. J Am Coll Cardiol 73(4):477–487. https://doi.org/10.1016/j.jacc.2018.10.072

Bello-Chavolla OY, Kuri-García A, Ríos-Ríos M, Vargas-Vázquez A, Cortés-Arroyo JE, Tapia-González G, Cruz-Bautista I, Aguilar-Salinas CA (2018) Familial combined hyperlipidemia: current knowledge, perspectives and controversies. Rev Invest Clin 70(5):224–236. https://doi.org/10.24875/RIC.1800257

Benito-Vicente A, Uribe KB, Jebari S, Galicia-Garcia U, Ostolaza H, Martin C (2018) Familial hypercholesterolemia: the most frequent cholesterol metabolism

disorder caused disease. Int J Mol Sci 19(11):pii: E3426. https://doi.org/10.3390/ijms19113426
Catapano A et al (2016) ESC/EAS guidelines for the management of dyslipidaemias. The task force for the management of dyslipidemias of the European Society of Cardiology (ESC) and the European Atherosclerosis Society (EAS). Eur Heart J 37:2999–3058. https://doi.org/10.1093/eurheartj/ehw272.
Catapano AL, Graham I, De Backer G, Wiklund O, Chapman MJ, Drexel H, Hoes AW, Jennings CS, Landmesser U, Pedersen TR, Reiner Ž, Riccardi G, Taskinen MR, Tokgozoglu L, Monique Verschuren WM, Vlachopoulos C, Wood DA, Luis Zamorano J, Cooney MT (2016) ESC/EAS guidelines for the management of dyslipidaemias. Atherosclerosis 252:207–274
Durrington P (2003) Dyslipidemia. Lancet 362:717–731. https://doi.org/10.1016/S0140-6736(03)14234-1
Falko JM (2018) Familial chylomicronemia syndrome: a clinical guide for endocrinologists. Endocr Pract 24(8):756–763. https://doi.org/10.4158/EP-2018-0157
Grundy SM, Stone NJ, Bailey AL, Beam C, Birtcher KK, Blumenthal RS, Braun LT, de Ferranti S, Faiella-Tommasino J, Forman DE, Goldberg R, Heidenreich PA, Hlatky MA, Jones DW, Lloyd-Jones D, Lopez-Pajares N, Ndumele CE, Orringer CE, Peralta CA, Saseen JJ, Smith SC Jr, Sperling L, Virani SS, Yeboah J (2018) 2018 AHA/ACC/AACVPR/AAPA/ABC/ACPM/ADA/AGS/APhA/ASPC/NLA/PCNA guideline on the management of blood cholesterol. Circulation 10:CIR0000000000000625. https://doi.org/10.1161/CIR.0000000000000625
Koopal C, Marais AD, Visseren FL (2017) Familial dysbetalipoproteinemia: an underdiagnosed lipid disorder. Curr Opin Endocrinol Diabetes Obes 24(2):133–139. https://doi.org/10.1097/MED.0000000000000316
Marais AD (2019) Apolipoprotein E in lipoprotein metabolism, health and cardiovascular disease. Pathology 51(2):165–176. https://doi.org/10.1016/j.pathol.2018.11.002
Milman S, Barzilai N (2015) Dissecting the mechanisms underlying unusually successful human health span and life span. Cold Spring Harb Perspect Med 6(1): a025098. https://doi.org/10.1101/cshperspect.a025098
Milman S, Atzmon G, Crandall J, Barzilai N (2014) Phenotypes and genotypes of high density lipoprotein cholesterol in exceptional longevity. Curr Vasc Pharmacol 12(5):690–697
Morita SY (2016) Metabolism and modification of apolipoprotein B-containing lipoproteins involved in dyslipidemia and atherosclerosis. Biol Pharm Bull 39(1):1–24. https://doi.org/10.1248/bpb.b15-00716
Nicholls SJ, Nelson AJ (2019) HDL and cardiovascular disease. Pathology 51(2):142–147. https://doi.org/10.1016/j.pathol.2018.10.017
Orkaby AR, Gaziano JM, Djousse L, Driver JA (2017) Statins for primary prevention of cardiovascular events and mortality in older men. J Am Geriatr Soc 65(11):2362–2368. https://doi.org/10.1111/jgs.14993
Sabatine MS (2019) PCSK9 inhibitors: clinical evidence and implementation. Nat Rev Cardiol 16(3):155–165. https://doi.org/10.1038/s41569-018-0107-8
Schaefer EJ, Anthanont P, Diffenderfer MR, Polisecki E, Asztalos BF (2016) Diagnosis and treatment of high density lipoprotein deficiency. Prog Cardiovasc Dis 59(2):97–106. https://doi.org/10.1016/j.pcad.2016.08.006
Sharma K, Baliga RR (2017) Genetics of dyslipidemia and ischemic heart disease. Curr Cardiol Rep 19(5):46. https://doi.org/10.1007/s11886-017-0855-9.
Tada H, Nohara A, Inazu A, Mabuchi H, Kawashiri MA (2019) Remnant lipoproteins and atherosclerotic cardiovascular disease. Clin Chim Acta 490:1–5. https://doi.org/10.1016/j.cca.2018.12.014
Vallejo-Vaz AJ, Ray KK (2016) Promoting high-density lipoprotein function via intravenous infusion: the rebirth of apoA-I Milano? Eur Heart J Cardiovasc Pharmacother 2(1):30–31. https://doi.org/10.1093/ehjcvp/pvv042
Walter M (2009) Interrelationships among HDL metabolism, aging, and atherosclerosis. Arterioscler Thromb Vasc Biol 29(9):1244–1250. https://doi.org/10.1161/ATVBAHA.108.181438
Warden BA, Duell PB (2018) Volanesorsen for treatment of patients with familial chylomicronemia syndrome. Drugs Today (Barc) 54(12):721–735. https://doi.org/10.1358/dot.2018.54.12.2899384

Families and Family Life

▶ Intergenerational Solidarity

Familism

Lin Jiang
School of Social Work, University of Texas Rio Grande Valley, Edinburg, TX, USA

Synonyms

Communalism; Filial piety

Definition

The familism, a cultural value, is a "strong identification and attachment of individuals with their families (nuclear and extended) and strong feelings of loyalty, reciprocity and solidarity among members of the same family" (Sabogal et al. 1987, pp. 397–398).

Overview

In most of cases, the familism has been identified to be central to Latino cultural value (Fuller-Iglesias and Antonucci 2016). Nevertheless, it has also been applied to the other population, whose value orientation is collectivism (Ihara, Tompkins and Sonethavilay 2012), such as Asian (filial piety; See Filial Piety and Responsibilities among the Chinese) and African Americans and Caribbean Blacks (communalism) (Campos et al. 2016; Ihara et al. 2012; Schwartz et al. 2010; Wallace and Constantine 2005; Yeung and Fung 2007). According to the familism scale designed by Sabogal et al. (1987), there are three factors: (1) familial obligations, (2) family as a source of support, and (3) family as a referent. The familism plays an essential role in the gerontology field regarding older adults' mental health, service utilization, and caregivers' burden.

Key Research Findings

The familism has been frequently studied related to older adults' caregivers. However, the results are mixed. For example, studies conducted by Shurgot and Knight (2004) and Lai (2007) indicated that caregivers with familism values felt less burden and had lower depressive symptoms (See ▶ Caregiver Stress). Nevertheless, a positive relationship between levels of depression and burden and the familism has been found by Cox and Monk (1993), and Gupta et al. (2009). Additionally, the familism has the protective effects on older adults' mental health (e.g., Fuller-Iglesias and Antonucci 2016) by increasing the family cohesion and social support (Guo et al. 2015). Furthermore, the familism affects care arrangements among Latino and Chinese older adults. For instance, Crist and Speaks (2011) pointed out that the Mexican older adults were comforted to get care from family members instead of outsiders. Song et al. (2017) suggested that Chinese older adults with stronger familism had a higher expectation for home care.

Summary

With the rapidly growing diverse older adult populations, researcher, practitioners, and policymakers should consider the factors, such as the familism, affecting minority older adults' well-being. In the future, the researcher may further explore the following areas, the acculturation and familism level among older adults, the social networks/support (both inside and outside the family) and familism, caregivers' acculturation level, coping skills, familism value, and their stress, etc. The practitioners may consider how to build trust with older adults and make their family members involved during the process of health care decision making. The policymakers may discuss how to support the family caregiving for older adults who are influenced by familism and their caregivers.

Cross-References

▶ Caregiver Stress
▶ Filial Piety and Responsibilities among the Chinese

References

Campos B, Perez OFR, Guardino C (2016) Familism: a cultural value with implications for romantic relationship quality in US Latinos. J Soc Pers Relat 33(1):81–100. https://doi.org/10.1177/0265407514562564
Cox C, Monk A (1993) Hispanic culture and family care of Alzheimer's patients. Health Soc Work 18(2):92–100. https://doi.org/10.1093/hsw/18.2.92
Crist JD, Speaks P (2011) Keeping it in the family: when Mexican American older adults choose not to use home healthcare services. Home Healthc Nurse 29:282–290. https://doi.org/10.1097/NHH.0b013e3182173859
Fuller-Iglesias HR, Antonucci TC (2016) Familism, social network characteristics, and well-being among older adults in Mexico. J Cross Cult Gerontol 31(1):1–17. https://doi.org/10.1007/s10823-015-9278-5

Guo M, Li S, Liu J et al (2015) Family relations, social connections, and mental health among Latino and Asian older adults. Res Aging 37(2):123–147. https://doi.org/10.1177/0164027514523298

Gupta R, Rowe N, Pillai VK (2009) Perceived caregiver burden in India: implications for social services. Affilia 24(1):69–79. https://doi.org/10.1177/0886109908326998

Ihara E S, Tompkins C J, Sonethavilay, H (2012) Culture and familism: An exploratory case study of a grandparent-headed household. J Intergener Relatsh 10(1): 34–47. https://doi.org/10.1080/15350770.2012.645737

Lai DW (2007) Cultural predictors of caregiving burden of Chinese-Canadian family caregivers. Can J Aging 26 (S1):133–147. https://doi.org/10.3138/cja.26.suppl1.133

Sabogal F, Marin G, Otero-Sabogal R et al (1987) Hispanic familism and acculturation: what changes and what doesn't? Hispanic J Behav Sci 9:397–412. https://doi.org/10.1177/07399863870094003

Schwartz SJ, Weisskirch RS, Hurley EA et al (2010) Communalism, familism, and filial piety: are they birds of a collectivist feather? Cultur Divers Ethnic Minor Psychol 16(4):548–560. https://doi.org/10.1037/a0021370

Shurgot GSR, Knight BG (2004) Preliminary study investigating acculturation, cultural values, and psychological distress in Latino caregivers of dementia patients. J Ment Health Aging 10(3):183–194

Song YJ, Yan EC, Sörensen S (2017) The effects of familism on intended care arrangements in the process of preparing for future care among one-child parents in urban China. Ageing Soc 36(7):1416–1434. https://doi.org/10.1017/S0144686X16000349

Wallace BC, Constantine MG (2005) Afrocentric cultural values, psychological help-seeking attitudes, and self-concealment in African American college students. J Black Psychol 31:369–385. https://doi.org/10.1177/0095798405281025

Yeung GT, Fung HH (2007) Social support and life satisfaction among Hong Kong Chinese older adults: family first? Eur J Ageing 4(4):219–227. https://doi.org/10.1007/s10433-007-0065-1

Family and Medical Leave Act

Dorian R. Woods
Nijmegen School of Management, Radboud University, Nijmegen, Gelderland, The Netherlands

Synonyms

US employment protection policy during employees' self-care or family caregiving

Definition

The Family and Medical Leave Act (FMLA) is a US federal law enacted in 1993 that allows a 12-week unpaid employment leave per 12-month period (2014). Employees can take this leave in the case of birth and care of a newborn child, the placement of a child for adoption or foster care, or the care of an immediate family member (spouse, child, or parent) with a serious health condition. The leave also allows employees to take medical leave who are themselves unable to work because of infirmity and long-term illness. Workers are granted a continuation of their health-care insurance while on leave and the same or an equivalent job upon returning to work. The law does not supersede company or state regulations that offer more generous plans (See ▶ "Laboring Work and Healthy Aging"). Employee eligibility is limited by a minimum of 1250 h of work in the last 12 months at the job. The FMLA is confined to nonprofit and private companies that have 50 or more employees within a 75-mile radius, but the law covers all federal, state, and local public agencies, the military, and all public schools, regardless of the 50-employee rule. The US has been one of the last countries to establish employment protection for reasons of illness and care, and it is still one of the last industrial countries in the world that does not have national, comprehensive paid leave. However, this law is unique in its coupling of parental leave with sick leave and with the protection of those who require time off from work while caring for their spouses and older parents with serious health conditions (See ▶ "Long-Term Care"). In international comparison, the law was a forerunner in terms of its employment protection for leaves due to the long-term care of immediate relatives (See ▶ "Laws for Older Adults").

Overview

The Family and Medical Leave Act was a watershed moment in US labor protection law in 1993 because it was the first national policy to provide health insurance and job reinstatement protection for US Americans for employment leave due to illness or familial care (See ▶ "Interventions for Caregivers of Older Adults"). While employment protection laws are widespread among industrial countries, the USA enacted this law relatively late in comparison (See ▶ "Social Security in the United States"). The FMLA, however, is significant, because not only does it cover parents of newborns and small children with maternity and parental leave but it also protects an individual's leave for serious illnesses themselves or for the care of spouses and frail parents. This is consequential for older adults because families tend to be the primary source of care in cases of infirmity, and spouses and adult children tend to do the bulk of this care (FIFARS 2016) (See ▶ "Primary Caregivers"). Also, older adults are projected to be in the workforce for longer periods in the future, so that old age illnesses are likely to be a disruptive factor for individuals and families if health insurance and job projection are not protected (See ▶ "Socioeconomic Status"; ▶ "Quality of Life"; ▶ "Caregivers' Outcomes").

The FMLA covers about 60% of the workforce because of the exemption of employers with less than 50 employees. Other issues that restrict the take-up of FMLA are not widespread knowledge of the law and its lack of compensation. Also, some part-time workers might not be able to accrue the necessary hours that would make them eligible. On the other hand, while most industrial countries have sickness protection leave, few countries have legislation that protects the jobs of individuals who care for their family members. Family members include those in same-sex marriages, and in some cases, other relatives and older adults can be eligible for such care, if they acted in loco parentis at the time the employee turned 18 (See ▶ "Primary Caregivers"). The law defines "serious health conditions" broadly, in which long-term care and regular treatments are necessary: in the case of older adults, this could include Alzheimer's, dementia, a stroke, or other serious conditions impacting aging individuals. Most people take a leave of absence because of severe health conditions of their own.

Historical Background and Legislative Process

The slow process under which the Family and Medical Leave Act became US law stretched over three presidencies and five congressional sessions. Before the law's passage, leave programs had been sporadic and state-administered or company-specific. Advocates for a federal law mobilized after a court ruling in 1984 struck down a state-specific Californian leave law for mothers on the grounds that it was discrimination against men. Soon after, the Women's Legal Defense Fund (now called the National Partnership for Women and Families) drafted the first federal bill with House congressional members. The bill was first drafted to cover maternal and parental leave and guaranteed sick leave for employees. However, in the lobbying process, and in the attempt to garner more support, the bill was expanded to cover leave for workers caring for a seriously ill child, spouse, or parent. The lobby group, AARP (formerly American Association of Retired Persons), was instrumental in supporting this last provision and consolidating legislative support for its passage (See ▶ "AARP"; ▶ "Politics of Aging and Interest Groups"). Proponents of the bill on both sides of the political aisle considered issues of pay too far-reaching and controversial, so ultimately considerations for a paid leave were not included in the bill. Debates in congressional hearings were primarily concerned with the economic drawbacks that an FMLA law might incur, such as undue hardship on small businesses. Therefore, an exemption was adopted for companies with less than 50 employees. After enough votes were mustered for bill passage, President George H.W. Bush vetoed it twice, stating that it would still unduly hurt businesses. In both cases, Congress was not able to mobilize a 2/3 majority to override the vetoes. The new 1992 presidential elections heightened public awareness, and polls at the time showed that the public was overwhelmingly in favor of the bill. After campaigning on this issue, and having assumed the presidency in 1992, Bill Clinton signed the previous existing bill into law (Woods 2012, 2018; Elving 1995) (See ▶ "Older Americans Act").

Changes After 1993 and Take-Up of the Leave

Minimal changes have been made since the law's inception, and eligibility and use have remained relatively stable. An amendment in 2008 authorizes leaves for up to 26 weeks for family members caring for an injured service member (See ▶ "Political Activism"). Military service members can also use 12 weeks to address needs arising from deployment. In 2010, the definition of "son" and "daughter" was clarified: an employee has the right to take leave if he or she assumes a caring role for a child, regardless of a legal or biological relationship. The law applies to all public sector agencies, all public and private elementary and secondary schools, but only to companies and nonprofit organizations with 50 or more employees within a 75-mile radius. Low-wage workers, workers with young children, and working welfare recipients tend to have lower coverage because of their work histories and the employment sectors they work in. Even if older workers might have access to FMLA because of accrued seniority or the types of employment, workers will often opt out if leave is not paid. Studies have shown that unpaid leave has had small effects on leave-taking by new mothers and little effects on leave usage by new fathers (see Kaufman and Gabel 2018). Some low take-up might be accounted for by employees' lack of knowledge about the policy. The availability of paid leave, however, increases the likelihood of take-up from mothers with lower levels of education, unmarried mothers, and Hispanic and African-American mothers (Rossin-Slater et al. 2013). Several studies show that most employers report either positive or no noticeable effect on productivity, turnover, or morale (Klerman et al. 2012; Appelbaum and Milkman 2011).

Extended Unpaid Coverage in Individual US States

While states must comply with the federal FLMA, they are free to set more comprehensive leave policies, and some states have expanded provisions. Longer unpaid leave can be taken in California, Connecticut, DC, Hawaii, Maine, Minnesota, New Jersey, Oregon, Rhode Island, Vermont, Washington, and Wisconsin. Eligibility thresholds have been lowered in seven states (and Washington, DC) to include companies with fewer numbers of employees. Another nine states (and Washington, DC) have expanded the definition of family to include stepparents, grandparents, grandparents-in-law, grandchildren, domestic partners, and domestic partners' children and/or siblings with serious health conditions. Three states have broadened health conditions to include organ/bone marrow transplant donations, and Oregon has broadened care for children to include nonserious illnesses. Nine states (and Washington, DC) have included a limited number of hours in annual leave for parents to attend school-related events and activities for their children or for taking family members to routine medical visits. A few states have explicitly included the effects of domestic violence, stalking, or sexual assault to be included in their definition of a "serious health condition."

Some Compensation in Individual US States

In 2019, four US states have paid family leave (See ▶ "Long-Term Care Financing"). These states are California (in 2004), New Jersey (in 2009), Rhode Island (in 2014), and New York (2018). Washington State and Washington, DC, have passed legislation that will cover workers in 2020, and Massachusetts has passed legislation that is slated to take effect in 2021. More than two dozen other states have introduced legislation related to family or medical leave. Most programs are funded by employee contributions, and private sector workers are covered who pay into state temporary disability funds, with some public sector workers also included (See ▶ "Intellectual Disability (Cognitive Disability)"). In California, the benefits provided are up to 6 weeks within 12 months, and the yearly minimal insurance cost for an employee averages 0.1% of income. The Californian weekly benefit amount was 55%

of earnings up to a maximum of $1,216 a week, and as of 2018, the rate was raised to 60/70% of earnings. Benefit levels are adjusted annually, and the average leave payment was $601 a week in December 2017. New Jersey insurance provides 66% of earnings up to $637 a week for 12 weeks, which can be taken across 24 months. Rhode Island insurance covers 60% of average monthly wages, capped at $637 a week for 4 weeks, the average annual benefit in December 2017 slated at $524. When entirely phased in, New York will provide 12 weeks of paid leave. The benefits start this year with 50% of weekly earnings, capped at $652, but will be extended to 67% of weekly earnings by 2021. Some of these programs are paid in addition to the states' temporary disability insurance (Kaufman and Gabel 2018). An independent financial resource for long-term care in addition to the FMLA is the US Cash and Counseling Program for needy families (See ▶ "Financial Assets"). These measures are usually administered as a Medicaid service, and needy recipients can choose to hire and pay family members as care providers for personal assistance, usually their adult children, but 10 states also allow for spouses to be compensated (See ▶ "Medicare for People with Disabilities").

Future Directions of Research

Work-life balance issues have become pressing for employees because the demographics of the workforce have slowly changed. Women are increasingly in paid employment, and retirement-age workers are staying in the workforce longer (FIFARS 2016, 21) (See ▶ "Gender and Employment in Later Life"). At the same time, the number of older Americans is projected to be about 74 million by 2030, over twice their number in 2000 (FIFARS 2016, 2) (See ▶ "Primary Caregivers"). As the workforce ages and more workers stay in employment for longer periods, families face long-term care responsibilities that formal institutional care facilities cannot offset (See ▶ "Formal and Informal Care"). Also, working adults in their 30s to 60s, in what has been called the sandwich generation (See ▶ "Sandwich Generation"), face pressures where they might be both raising children and caring for frail parents (See ▶ "Intergenerational Relations"). In all very likelihood, these workers' lives become precarious, as they reduce working hours during a time that is usually expected for career development (See ▶ "Employment and Caregiving"). All the while, they expend extra time and accrue extra costs for care in double.

Improvements in the FMLA have been debated on state and federal levels (See ▶ "Caregiver Stress", especially concerning paid leave (See ▶ "Aging Network"). An overwhelming number of people who do not take leave do so because it is not paid, and they cannot afford it (See ▶ "Intergenerational Solidarity"). Indeed, low-income workers are expected to profit the most from a federal paid leave. While only a handful of states have passed legislation for paid leave, the momentum has been gathering for a federal bill for paid leave. One such bill has been pending in Congress since 2013, and it aims to provide participants with up to 12 weeks' paid leave, funded by payroll contributions from employees and employers. Debates on the drawbacks of a paid leave center on economic feasibility and consequences on the economy, although proponents of the bill point to established legislation in other countries. Leaves in other countries for the care of frail adults, however, are generally recent to the policy stage (most beginning after the year 2000), and measures vary widely in countries in terms of duration, payment, levels of care need, circles of claimants, and types of employment protection. For example, Germany provides leave protection for some employees in the form of part-time employment but with a loan top-up in wages that is repaid in working hours at a later date. Other countries, like France and Italy, have different leaves and compensation according to distinct types of employment (Schmidt et al. 2016) (See ▶ "Social Security Around the World"). Generally, the development of employment protection policy based on long-term care needs is new and dynamic for most countries, and regulations around compensation are not standard or across the board (See ▶ "Aging Policy Transfer, Adoption, and Change").

Summary

The FMLA is a significant US labor law for the protection of older adults and their families. While more than half of FMLA leave-takers take time off for the care of their own serious illness (See ▶ "Self, Informal, and Formal Long-Term Care: The Interface"), another 30% take this leave to care for a seriously ill family member, and about 18% take this leave for the care of a new child. Overall, older workers were more likely to take leave (Institute for Women's Policy Research, & IMPAQ International 2017, 3). The FMLA is one of the few and first laws of its kind that protects the employment and health insurance of individuals who take time off to care for seriously ill family members. This is important for older adults because severe illnesses of old age can have devastating effects if health care and employment are not protected (See ▶ "Caregivers' Outcomes"; ▶ "Quality of Life"). Care for serious illness in old age is unlike care for new children in its unpredictability in timing, intensity, and length (See ▶ "Hospice and Caregiving"; ▶ "End-of-Life Care"; ▶ "Post-Acute Care"), and so such protective leaves offer long-term stability for older adults and their families (See ▶ "Aging Policy Analysis and Evaluation").

Cross-References

- ▶ AARP
- ▶ Aging Network
- ▶ Aging Policy Analysis and Evaluation
- ▶ Caregivers' Outcomes
- ▶ Caregiver Stress
- ▶ Employment and Caregiving
- ▶ End-of-Life Care
- ▶ Financial Assets
- ▶ Formal and Informal Care
- ▶ Hospice and Caregiving
- ▶ Intergenerational Relations
- ▶ Intergenerational Solidarity
- ▶ Interventions for Caregivers of Older Adults
- ▶ Laboring Work and Healthy Aging
- ▶ Laws for Older Adults
- ▶ Long-Term Care
- ▶ Long-Term Care Financing
- ▶ Medicare for People with Disabilities
- ▶ Older Americans Act
- ▶ Political Activism
- ▶ Politics of Aging and Interest Groups
- ▶ Post-Acute Care
- ▶ Primary Caregivers
- ▶ Quality of Life
- ▶ Sandwich Generation
- ▶ Self, Informal, and Formal Long-Term Care: The Interface
- ▶ Social Security in the United States
- ▶ Socioeconomic Status
- ▶ Welfare States

References

Appelbaum E, Milkman R (2011) Leaves that pay: employer and worker experiences with paid family leave in California. Center for Economic and Policy Research, Washington, DC

Elving RD (1995) Conflict and compromise: how congress makes the law. Simon & Schuster, New York

Family and Medical Leave Act (2014), Pub.L. 103–3; 29 U.S.C. sec. 2601; 29 CFR 825. http://www.dol.gov/whd/fmla/index.htm. Accessed 01 Nov 2014

FIFARS (Federal Interagency Forum on Aging-Related Statistics) (2016) Older Americans 2016: key indicators of Well-being. U.S. Government Printing Office, Washington, DC. https://agingstats.gov/docs/LatestReport/Older-Americans-2016-Key-Indicators-of-WellBeing.pdf. Accessed 01 Dec 2018

IWPR (Institute for Women's Policy Research, & IMPAQ International) (2017) *Family and Medical Leave-Taking among Older Workers*. https://iwpr.org/publications/family-medical-leave-taking-among-older-workers/. Accessed 01 Jul 2018

Kaufman G, Shirley Gatenio Gabel (2018) United States. In: Blum S, Koslowski A, Macht A, Moss P (eds) 14th international review of leave policies and related research, 2018 pp 443–449. Available via International Network on Leave Policies and Research. http://www.leavenetwork.org/lp_and_r_reports/. Accessed 01 Dec 2018

Klerman J, Daley K, Pozniak A (2012) Family and Medical Leave in 2012. Abt Associates, Cambridge, MA. https://www.dol.gov/asp/evaluation/completed-studies/Family_Medical_Leave_Act_Survey/DATA_DOCUMENTATION_family_medical_leave_act_survey.pdf. Accessed 1 Dec 2018

Rossin-Slater M, Ruhn C, Waldfogel J (2013) The effects of California's paid family leave program on mothers' leave-taking and subsequent labor market outcomes. J Policy Anal Manage 32(2):224–245. https://doi.org/10.1002/pam.21676

Schmidt AE, Fuchs M, Rodrigues R (2016) Juggling family and work – leaves from work to care informally for frail or sick family members – an international perspective. Policy Brief, September 2016. European Centre for Social Welfare Policy and Research, Vienna pp 1–19. http://eurocarers.org/Juggling-family-and-work–Leaves-from-work-to-care-informally-for-frail-or-sick-family-members-an-international-perspective. Accessed 1 Feb 2019

Woods DR (2018) The UK and the US: Liberal models despite family policy expansion? In: Eydal GB, Rostgaard T (eds) Handbook of child and family policy. Edward Elgar, Cheltenham/Northampton, pp 182–194

Woods DR (2012) Family policy in transformation: US and UK policies. Palgrave Macmillan, Basingstoke

Family Care

▶ Formal and Informal Care

Family Care Decisions

▶ Preparation for Future Care: The Role of Family Caregivers

Family Caregiver

▶ Caregivers' Outcomes

Family Caregiver Stress

▶ Caregiver Stress

Family Caregiving

▶ Employment and Caregiving
▶ Hospice and Caregiving

Family Change

▶ Family Formation and Dissolution

Family Conflict

▶ Intergenerational Solidarity

Family Demography

Naomi J. Spence
Department of Sociology, Lehman College, City University of New York, Bronx, NY, USA

Synonyms

Family studies; Household studies

Definition

Family demography is a subfield of demography (the scientific study of populations) focused on family and household formation, composition, change, and function. Families are typically defined as those related through blood, marriage, or adoption. Households may be comprised of family members, unmarried cohabiting adults, and/or unrelated persons. Family has always been central to the field of demography, but the subfield of family demography was not formally recognized until 1961 (Seltzer 2019).

Overview

Demography is the scientific study of human populations that examines how populations grow and change with particular attention to fertility, mortality, migration, and the interrelated topics of marriage, health, and living arrangements. The concerns of demography are foundational to the field of gerontology and population aging. As more people live to the later years of life, gerontology becomes increasingly important. And, while longevity may be an obvious influence on population aging, demography demonstrates that fertility is the strongest contributor to population aging (Agree 2018). When fertility rates are high, the relative age structure of a population will be younger. However, reductions in fertility around the world, coupled with increased longevity, yield populations with significant proportions of older adults. Thus, family formation vis-à-vis fertility underlies the concerns of gerontology and population aging.

The subfield of family demography is well positioned to make ongoing, significant contributions to gerontology. Family demography offers analytic tools commonly used in demography, such as life tables and simulations, and novel ones, such as the family tree (see ▶ "Family Tree"). These tools enable rich description, probability estimates, and forecasting of family changes. Also, family demography is situated at the nexus of theoretically rich social science disciplines from which it draws to generate empirically testable predictions and to explain research findings. With these tools, family demographers are equipped to document characteristics and functions of families and households and to illuminate the implications of these factors for the lives of older adults and for aging populations.

Key Research Findings

In light of the influential role of fertility in shaping population age composition, reproductive aged adults remain a strong focus of family demographers. However, the scope of family change in recent decades yields calls for broader thinking about family relationships and their functions (Seltzer 2019). Improvements in longevity increase the number and importance of multi-generational family connections (Bengtson 2001). Concurrently, more diverse family formation and dissolution patterns create stepfamily and other less institutionalized family relationships. Theoretical developments, data collection efforts, and methodological innovation have advanced the subfield of family demography well-beyond basic concerns with young adult fertility behaviors and the burdensome role of aging parents (Seltzer 2019).

Research documents complexities in the extent to which family and household members receive and/or provide resources and care. For example, in a crowded nest household structure, assumptions may be made about the young adults' needs for continued parental assistance; however, the relationship may be reversed or reciprocal in situations where aging parents rely upon the support of their adult children (see ▶ "Crowded Nest" in this volume). Following a transition to empty nest, the assumption may be made that adult children no longer require the support of their parents. However, they may continue to rely upon their parents and/or provide support for their aging parents despite separate living arrangements (see ▶ "Empty Nest" in this volume). Also, multi-generational families increasingly have an older adult and young adult who depend upon the middle generation for support and care. This so-called sandwich generation may bear the burden of caring for two generations of dependents, but they may also enjoy the support of their adult children and/or their aging parents (see ▶ "Sandwich Generation" in this volume).

As populations age, demographic processes influence family and household formation, composition, and change which in turn influence the allocations of resources to address the needs of older adults in the population. After all, family members are fundamentally connected through linked lives (Elder 1985) and shared obligations to ensuring the well-being of one another. Lower fertility affects the presence and number of adult children in the family of older people, who may provide or demand social support and resources. As health and longevity improve, older adults are no longer simply a generation in need of care and assistance (Margolis and Wright 2017). More families will have actively engaged grandparents who may assist or substitute for parents in the care of grandchildren (Leopold and Skopek 2015). At the same time, increases in childlessness mean that more adults reach the later stages of life without children and grandchildren (Dykstra and Hagestad 2007). Union formation and dissolution trends lead to increases in older adults who are ex-spouses, stepparents, or cohabiting partners (Manning and Brown 2011) (see ▶ "Cohabitation" and

"Family Formation and Dissolution" in this volume). Taken together, family changes yield more diverse relationships, wherein the availability of support and care may be offset by stepfamily, half siblings, and other informal parent-child and sibling bonds (Seltzer 2019; Wachter 1997).

Emerging family forms and living arrangements are the result of social, economic, political, and technological changes (Agree 2018). Social changes, especially women's liberation, influence family formation behavior early in life with lasting implications for older families. Considered within a life course perspective (see ▶ "Life Course Perspective" in this volume), lives are linked through the formation, maintenance, and dissolution of family relationships, which take place in specific historical contexts. High rates of divorce and remarriage among baby boomers make this aging cohort the first to feature some emerging family forms and a generation with relatively a high level of family diversity (see ▶ "Family Diversity" in this volume). Still, the current generation coming of age and making family formation decisions is expected to have even more complex family relationships when they reach older ages (Manning and Brown 2011).

Directions of Future Research

Population aging has myriad implications for the structure of families and households.

In aging populations, a greater share of people will occupy more family roles and experience more changes in family and household composition. Family demography will document these trends and study the consequences of them for the changing needs of older adults. To do so, family demographers must continue to build upon existing methodological tools and data collection to better incorporate broader definitions of families. As norms and laws change around the world, LGBTQ families need to be better understood. Also, future research should delve deeper into understanding family relationships and provisions for support in childless families, stepfamilies, and ex-stepfamily relationships. This research may fruitfully inform programmatic and policy decisions about the allocation of resources to older adults in the population.

The inextricable linkage and need to better understand the complex interconnections of family demography and gerontology are exemplified in one of the stated goals of the US National Institute on Aging (NIA): "Assess the impact of changing family structures on health and caregiving." This national priority is not specific to the USA. The 2018 report on aging to the United Nations General Assembly highlighted the role of unpaid family caregivers and migrants (primarily women) in providing care for older individuals (United Nations 2018). The migration of young, working aged women from developing nations to serve the healthcare and caregiving needs of aging populations in more developed countries is likely to have indirect effects on family formation and composition in those populations who are sending migrants in significant numbers.

Moreover, as family members continue to assume caregiving roles, family demographers must work to document and understand family changes and their consequences. Multigenerational families will continue to increase, leading to more people in the sandwich generation (see ▶ "Sandwich Generation" in this volume) and in bean pole (see ▶ "Beanpole Family Structure") and stem families (see ▶ "Stem Family" in this volume). Family relationships that shape living arrangements warrant further research on the effects of living arrangements on older adults' health, quality of life, and social engagement (see ▶ "Living Arrangements" in this volume). Such research will continue to effectively inform gerontological practice and population-level responses to having substantial proportions of older adults.

Summary

Complexities of family relationships define the scope of the subfield of family demography and create challenges for new approaches to understanding the implications of family for older adults. Research in this burgeoning subfield must continue to expand the traditional focus on reproductive age adults in light of "longer years of shared lives" (Bengtson 2001), linked lives (Elder 1985), and the significance of cohort changes (Agree 2018). Multigenerational families will play an important role in the contexts that shape young adults' fertility decisions, as well as family formation and dissolution at all ages. Nascent research on the range of consequences of greater family diversity and change for older adults provides a strong foundation for necessary advancements. Understanding family change over the past several decades is key to developing approaches to successfully address the needs of older people and of aging populations.

Cross-References

► Beanpole Family Structure
► Crowded Nest
► Empty Nest
► Family Diversity
► Family Formation and Dissolution
► Family Tree
► Intergenerational Family Structures
► Life Course Perspective
► Living Arrangements
► Multigenerational Families
► Primary Caregivers
► Sandwich Generation
► Stem Family

References

Agree EM (2018) Demography of aging and the family. In: Majmundar MK, Hayward MD (eds) Future directions for the demography of aging: proceedings of a workshop. National Academies of Sciences, Engineering, Medicine, & Committee on Population. National Academies Press, Washington, DC, pp 159–186
Bengtson V (2001) Beyond the nuclear family: the increasing importance of multigenerational bonds. J Marriage Fam 63(1):1–16
Dykstra P, Hagestad G (2007) Roads less taken: developing a nuanced view of older adults without children. J Fam Issues 28(10):1275–1310
Elder GH Jr (1985) Perspectives on the life course. In: Elder GH Jr (ed) Life course dynamics: trajectories and transitions, 1968–1980. Cornell University Press, Ithaca, pp 23–49
Leopold T, Skopek J (2015) The demography of grandparenthood: an international profile. Soc Forces 94(2):801–832
Manning WD, Brown SL (2011) The demography of unions among older Americans, 1980–present: a family change approach. In: Handbook of sociology of aging (Handbooks of sociology and social research). Springer, New York, pp 193–210
Margolis R, Wright L (2017) Older adults with three generations of kin: Prevalence, correlates, and transfers. J Gerontol B: Psychol Soc Sci 72(6):1067–1072
Seltzer J (2019) Family change and changing family demography. Demography 56(2):405–426
United Nations (2018) Follow-up to the international year of older persons: second world assembly on ageing. Report of the Secretary General. United Nations General Assembly. [online] pp 1–15. https://undocs.org/A/73/213. Accessed 29 June 2019
Wachter K (1997) Kinship resources for the elderly. Philos Trans R Soc B: Biol Sci 352(1363):1811–1817

Family Diversity

Jaroslava Hasmanová Marhánková
Department of Sociology, University of West Bohemia, Pilsen, Czech Republic
Department of Gender Studies, Charles University in Prague, Prague, Czech Republic
Department of Sociology, Charles University, Prague, Czech Republic

Synonyms

Complex family structures; Complexity of family relationships; Heterogeneity of family life

Definition

The term "family diversity" refers to the plurality of family types and households. It addresses the diversity in family processes (e.g., how people understand what family means and how family relationships are established in daily life). In a broader sense, the term refers to numerous family structures that exist outside the concept of the "traditional family" (a two-parent home with children). This entry considers the impact of current demographical changes on diversity in family formations; it also looks at patterns of intergenerational solidarity and the impact of diverse family structures on intergenerational relationships and the experience of aging.

Population Aging and the Changing Dynamics of Family Life

The process of population aging in western countries, associated with an increase in life expectancy and a decrease in birth rates, significantly changes the structure of multigenerational families and patterns of family reciprocity. Due to the decline in birth rates, the number of bonds within the horizontal structure of the family (between siblings and cousins) is decreasing, and in contrast, vertical bonds (between individual generations) are being strengthened, thanks to the increase in life expectancy. Thus, representatives of different generations spend more time together over the course of their lives (See ▶ "Beanpole family structure"). Harper (2006) notes that current demographic trends in western society will lead to a state in which most people spend at least some of their lives as part of a three- to four-generation family which, at the same time, will comprise fewer members. She argues that the decrease in the number of family members in each generation will strengthen ties across generations. And at the same time, people will spend more time in their roles within the family cycle.

However, this new dynamic in intergenerational bonding patterns gives rise to specific demands on the organization of care within the family, associated, for example, with the phenomenon of the so-called sandwich generation. This term describes the situation for people (mostly women) of a productive age, with not yet fully independent children, who also take on the commitment of caring for their aging relatives. As caregivers, they are exposed to two different types of care demands (See ▶ "Sandwich Generation"). This phenomenon becomes relevant in the context of the previously mentioned demographic changes. The number of children in a family decreases, and they are also born when their parents are at a higher age. Therefore, members of the middle generation more often face the demands of assisting their aging parents while simultaneously providing care for their own children. These people provide two types of care, which are also often combined with employment.

At the same time, an increase in mobility levels and international migration in western and other countries leads to the fragmentation of family bonds, especially between individual generations. In a global context, migration is one of the most significant factors affecting intergenerational relationships and is one of the main barriers to caring for family members (See ▶ "Intergenerational Migration and Relations in the International Context"). The gap between individual generations can occur particularly in cases where the grandchildren live in a different country and grow up in a different cultural and language environment (Chambers 2012).

The context of the mentioned trends also gives rise to discourses on the crisis of the family and the erosion of intergenerational bonds in current developed western societies. Bengtson (2001), however, argues that it has never before been so important for the functioning of a family to have intergenerational bonds. He points out that the current instability of partnerships and the high divorce rate strengthen the importance of intergenerational relationships, to the detriment of the nuclear family. While the bond between partners is characterized by this fragility, parental bonds are stronger. The destabilization of the nuclear family is therefore often accompanied by an increase in the importance of other family bonds which, from the long-term point of view, can be considered more crucial than the structure of the nuclear family itself. Bengtson places special emphasis on the relationship between grandparents and grandchildren, which, according to him, acquires a new intensity and importance. Thanks to the prolonged life expectancy and decrease in the birth rate, grandparents can spend a greater part of their lives with their grandchildren and focus more of their attention on fewer children. In cases where the parents break up, grandparents often substitute in the family functions and roles that have been weakened by the breakup.

Chambers (2012) points out that in the future, among other things, the experience of aging will be characterized by an increase in the diversity of the forms of family bonds. With an increase in the diversity of partner life (e.g., an increase in the number of childless couples and single-person households), different forms of intimate relationships (friendship, neighborly bonds) also come to the fore and could acquire new importance in future aging cohorts. They could even replace some functions that were previously fulfilled within long-lasting marriages and kin ties.

The next section addresses the diversity in family structures and households. It outlines the current trends, focusing in particular on their prevalence and impact on the older cohorts. It specifically concentrates on the impact of family diversity on life in older age and on intergenerational relationships.

Divorce and Stepfamilies

The experience of divorce not only changes the form of partner arrangement, but it also interferes with the family structure and relationships within individual generations. In this regard, the high divorce rate is one of the most significant factors forming current intergenerational bonds and forms of family arrangement. Previous studies have identified the negative impact of divorce on relationships and the amount of contact between fathers and children (Hetherington and Kelly 2003). Parents' divorce, however, also significantly affects relationships between other family members. Great attention has been paid to the impact of parents' divorce on the relationships between grandchildren and their grandparents. In this case, divorce significantly affects the very possibility of fulfilling the grandparenting role and the ways in which grandparenthood is experienced. Parents' divorce has a negative impact on the involvement of paternal grandparents, who also express a subjective feeling of deterioration in their relationships with their grandchildren (Ahrons 2007), while maternal grandparents intensify their contact with their daughters and grandchildren, not to mention the amount of financial support and other resources (Uhlenberg 2005; Cherlin and Fustenberg 1991). The current high divorce rate also contributes to the greater diversity of the grandparental experience. The deterioration of the relationship between parents often also becomes the catalyst for the transition of the relationship between grandchildren and grandparents – whether that occurs in the form of limited contact or in the transformation of the content of the grandparenting role. A study by Nelson (2006) shows that in single-mother families, the roles of parent and grandparent are often redefined. In her research, women stated that grandmothers were more important for their children than biological fathers, whose role within the family was often substituted for by the grandmothers.

The high divorce rate also contributes to the higher diversity of partner trajectories in older age. Over the last two decades, the divorce rate in middle-aged people has doubled, some authors even describing this as the "Silver Divorce Revolution." For example, while only 1 in 10 divorced people was over 50 in the United States in 1990, in 2010, that ratio was 1 in 4. If the divorce rate remains constant over the next 20 years, the number of people over 50 experiencing divorce can be expected to increase by one third (Brown and Lin 2012). In some northern European countries, a new historical trend can be observed where people aged 60–90 have a higher probability of experiencing divorce than widowhood (Bildtgard and Öberg 2017: 32–33).

The increase in the divorce rate in older age is often ascribed to changes in the perception of the institution of marriage and an increased emphasis on individualism, which places new demands on partnerships. A higher life expectancy, a better quality of life, relative economic security, and a reduction in the stigma associated with divorce have all led to a situation in which divorce is increasingly becoming a choice, even for people of older age (See ▶ "Marriage and remarriage"). People also expect more from their later years, not only in terms of the amount of time remaining, but also in terms of the roles of partnership and intimacy in their individual lives. The emphasis on self-fulfillment as a basis for partnerships, together with the weakening of the normative notions of marriage as a lifelong commitment, has led to the reluctance of individuals to maintain empty shell marriages – that is, marriages that are maintained only on the basis of a formal commitment and not a mutual emotional bond (Wu and Schimmele 2007). However, the divorce rate, and not only in older age, obviously reflects patterns in demographic behavior, in the legislative environment, and primarily in social norms, which can differ from country to country. Divorce has been legal for several decades in most countries in the world. There are only a few countries that maintain a low divorce rate. India is one example, where divorce has been legalized, but persisting social stigma keeps the divorce rate at a low level. The number of countries that maintain stable low-divorce societies, however, is decreasing over time (Cherlin 2017).

Stepfamilies in Late Life

A high divorce rate also affects forms of help in old age and is becoming an important catalyst for the increasing diversity of family arrangements, where the foundations of family bonds can seem weaker and the very contours of who is perceived as a family member become blurred. The increase in the number of stepfamilies represents one of the most prominent trends contributing to an increase in the diversity of family arrangements and even to the redefinition of the very idea of family relationships. A high divorce rate and an increase in the number of stepfamilies leads to the fact that within their family circles, people are increasingly connected to more people. These family bonds, however, include a lower level of commitment and are, in fact, unstable. People are therefore forced to actively create a family network from a complex number of bonds on which they must constantly "work" in order for them even to be perceived as bonds.

The presence of children from previous relationships, together with the possible presence of children born into a new partnership, further contributes to the increasing complexity of family bonds, even within stepfamilies. These types of families can be divided into simple stepfamilies, where only one partner has a child from a previous relationship, and complex stepfamilies, where both partners have their own biological children. The term "blended family" then describes a family structure where the new family arrangement includes both children from previous relationships and children conceived in the current relationship. As pointed out by Lin et al. (2018), the more complex the structure of a stepfamily, the less clear the individual roles and commitments are within the family. The increase in the number of stepfamilies also brings with it important questions in relation to the provision of care for older family members. Relationships between stepchildren and stepparents can be characterized by a lower level of commitment and less clearly defined roles. The ambivalent character of these bonds and the lower level of commitment can weaken intergenerational relationships (not only) in older age. For example, relationships between adult non-resident stepchildren and parents over the age of 50 show lower levels of intergenerational solidarity compared to relationships in families with both biological parents (Steinbach and Hank 2016).

Studies note that family relationships and systems of help are significantly affected by the structure of the family arrangement. The increase in the number of stepfamilies also strengthens the diversity of the ways in which individual roles are experienced within the family. A study by Allan et al. (2008) points out the ambivalent character of the relationship between step-grandparents and grandchildren. Although these grandparents often talk about the fact that they actively try to treat all the grandchildren equally, they confess that they perceive the bond with their "own" grandchildren differently.

The Growing Number of Less Institutionalized Forms of Partnership and Family Life

An increase in cohabitation represents another of the most significant changes in family behavior witnessed in the last few decades, contributing to diversity in family arrangements, and not only in older age (See ▶ "Cohabitation"). Although an increase in cohabitation is not an equally significant trend in all countries, in some countries, cohabitation currently represents the most significant alternative to marriage. The study of cohabitation was previously focused primarily on younger age groups where the percentage of cohabitation is the highest. But the cohabitation rate is increasing with time even among older adults. For example, in the United States, older people represent the fastest growing group of cohabitants. Through the first decade after the year 2000, the proportion of cohabitants over 55 increased by a third in the United States (Vespa 2012: 1104). With regard to other demographic trends and changes in partner trajectories, an even higher increase can be expected in the future. The cohabitation rate in older age will probably be fueled by the high divorce rate; the choice of cohabitation is statistically more typical for people who have been through a divorce. Along with the aging of the generation for whom cohabitation is the acceptable and increasingly preferred form of living together, the number of cohabitants among older people will probably increase.

A study by King and Scott (2005) shows that people in different life stages enter into cohabitation for different reasons and that they also experience this type of partnership differently. Older people show greater satisfaction in cohabitation and mention future marriage plans less often. While younger people often perceive cohabitation as an initial step leading to marriage, for older people, it represents, rather, the alternative to marriage. King and Scott's study, among other things, suggests that the conclusions from previous cohabitation studies, which were focused on younger people, may not be applicable to older adults; people at different life stages enter cohabitation with different expectations, and this form of living together plays different roles in their lives.

In fact, marriage still does represent the dominant partnership arrangement in older age (See ▶ "Marital relationships"). However, when looking at new partnerships established in older age, it has been discovered that marriage is certainly not the most preferred form of partnership arrangement. The probability of entering into marriage decreases after 70 years of age, while the probability of choosing a different, "less traditional" arrangement, such as cohabitation or living separately, increases (de Jong Gierveld 2002: 67). Based on data from Sweden, Bildtgard and Öberg (2017) show that while relationships established before 40 years of age most often led to marriage, people who established their relationships between the ages of 40 and 59 more often chose an "alternative" form of partnership arrangement. Relationships established after the age of 60 were dominated by living apart together and by cohabitation, while marriage accounted for the lowest number of such arrangements.

The "postproductive" age and older age as a stage of human biography also opens up a specific space for the establishment and experience of partnerships outside many frameworks defined in previous life stages. This applies to the limiting structure of employment as well as, for example, the presence of small children and the need to care for them, or the (very often unspoken yet still present) assumption that a long-term relationship should lead to its gradual institutionalization and should include reproduction. According to Bildtgard and Öberg (2017), people in older age find themselves at a stage of "postreproductive freedom," which opens up a freer space for choosing partnership arrangements that satisfy individual needs. This also corresponds to the preference for less institutionalized relationships in older age (such as living apart together), which do not place significant demands on new household arrangements or changes in lifestyle.

LGBTQ Families in Later Life

The study of LGBTQ (lesbian, gay, bisexual, transgender, or queer/questioning) families has thus far mostly focused on younger age groups. LGBTQ families in later life are still rather on the periphery of research interest (See ▶ "LGBT in old age"). A study from the USA notes that older LGBTQ people have more than twice as high a probability of living alone and up to four times as low a probability of having children (SAGE 2010). Today's generation of older LGBTQ people have lived most of their lives under conditions in a society in which anything other than a heterosexual orientation has symbolized a fundamental social stigma. Coming out often brought along with it a significant deterioration in relationships with the biological family. Empirical studies show that older LGBTQ people have less "traditional" social networks compared to the majority of the population. They do not rely on help from their families and emphasize the importance of a network of friends instead. A study from the United States points out that older LGBTQ adults mention their friends almost twice as often as their primary sources of help when they are in need. In the case of older LGBTQ people, biological bonds are often replaced by bonds based on a different conceptualization of kinship. The so-called "chosen family" becomes a crucial axis of help. Empirical studies have identified a relatively high incidence of caregiving provided for an adult friend or family among older LGBTQ adults. This study suggests that compared to men overall in the US population, gay and bisexual men may be providing care much more frequently (Metlife 2010).

LGBTQ family and partnership arrangements are significantly affected by legislation in individual countries as well as by societal attitudes toward non-heterosexual forms of expression. The current generation of older LGBTQ people has more often experienced previous heterosexual relationships and marriage. Many of today's adult children of parents who identify themselves as LGBTQ were not raised in same-sex families. Changes in legislation in some countries of the world as well as new developments in reproductive technology (the expanding possibility of in vitro fertilization or surrogate motherhood) have led to an increase in the number of children raised by same-sex couples. This development also brings with it an increased probability of LGBTQ people becoming grandparents. Stelle et al. (2010) estimate that in the United States alone, there are from one to two million grandparents who identify themselves as LGBTQ, while the assumption for the future is that the experience of grandparenthood will increasingly become an expanded part of the lives of older LGBTQ people.

Childlessness in Later Life

Since the second half of the twentieth century, the number of childless people has increased most significantly in developed western countries. The relatively high number of childless people in these societies is often interpreted in the context of a growing emphasis on individualism, a decrease in the amount of pressure from the social norms of parenthood, the prevalence of reliable contraceptive methods, and the associated decrease in birth rates. However, looking at the historical development of the numbers of childless people in developed western countries shows that the relatively high rate of childlessness is not a completely new phenomenon. For example, in some European countries in the nineteenth and the beginning of the twentieth century, more than 20% of women remained childless (Kreyenfeld and Konietzka 2017: 5). In western Europe (mainly in Austria, Switzerland, West Germany, and England), this increase in childlessness was again observed in cohorts born in the 1950s. In southern and eastern Europe, the increase in childless people came later but with comparable intensity. Rates of childlessness are currently high in all European countries. Similarly, the increase in the numbers of childless people can be observed in other countries of the world. For example, in Japan, the number of childless people has gradually increased in recent birth cohorts (Kreyenfeld and Konietzka 2017: 5–6).

Childlessness does not represent a monolithic experience. The paths to childlessness can be many, as can the different levels of importance that people ascribe to parenthood. The cultural norms of parenthood, the potential stigmatization of childlessness, and the associated effects of both can change culturally and historically. The impact of the experience of childlessness can change through different societies and different cohorts over time. Previous studies on childlessness in older age were mainly focused on the negative aspects of the absence of children as an important source of help in older age. Childless people have relatively less help in older age. For example, older people without children enter institutions of formal help earlier than those with their own children (Pezzin et al. 2013).

However, as noted in a study by Albertini and Kohli (2009), at the same time, support networks of childless people tend to be more diverse and more often include strong bonds with other relatives (siblings, nephews, and nieces), friends, and neighbors. Compared to parents, childless older people more often have wider help networks, although these do not necessarily provide the same level of support as children. The study also shows that childless people do not significantly differ from parents in their social support of others, but the core receivers of such support are more often nonrelatives. Some childless people also maintain close bonds with their nephews and nieces, which helps them maintain intergenerational links within the family. Albertini and Kohli (2009: 1272) end their study by emphasizing the need for policymakers and researchers to contemplate not only the specific needs of childless people in older age but also these elders' contributions to those close to them. The study authors point out that childless people can also be perceived as "pioneers of a culture of postfamilial civic engagement." The absence of children does not necessarily lead to the absence of the practice of generativity. However, childless people, when compared to parents, build diverse support networks and invest in volunteer and charity activities more often (See ▶ "Childless older adults").

Summary

The process of population aging in western developed countries raises issues regarding intergenerational relationships and caring responsibilities. We currently experience both benefits and challenges from prolonged life expectancy, which also results in lengthened shared lifetimes of different generations. At the same time, relationships in families are becoming less predictable and more diverse, as documented by the growing diversity of family formations and practices. Over time, researchers have moved from concerns about the erosion of family and intergenerational relationships that may arise from such changes to an emphasis on the continued importance of family bonds and also new ways of organizing family relationships beyond the structure of the "traditional" heterosexual nuclear family and the framework of biological kinship. Our future research on aging families must incorporate the fluidity of the family relationship as well as intimate family-like relationships that are established over the course of people's lives. Policymakers and professionals from a wide range of disciplines must be aware of and take into consideration the wider array of family situations characterizing not only older adulthood. They must do this to ensure the enactment of policies that are sensitive to the growing variability of relationships and living circumstances experienced by older adults as well as being sensitive to new challenges concerning the social ties and support that may arise hand in hand with growing family diversity.

Cross-References

▶ Beanpole Family Structure
▶ Caregiving Among the LGBT Community
▶ Cohabitation
▶ Family Formation and Dissolution
▶ Living Arrangements in Later Life
▶ Marriage and Remarriage

References

Ahrons CR (2007) Family ties after divorce: long-term implications for children. Fam Process 46:53–65
Albertini M, Kohli M (2009) What childless older people give: is the generational link broken? Ageing Soc 29(8):1261–1274
Allan G, Hawker S, Crow G (2008) Kinship in stepfamilies. In: Pryor J (ed) The international handbook of stepfamilies: policy and practice in legal, research and clinical environments. Wiley, New York, pp 323–344
Bengtson VL (2001) Beyond the nuclear family: the increasing importance of multigenerational relationships in American society. J Marriage Fam 63:1–16
Bildtgard T, Öberg P (2017) Intimacy and ageing. New relationships in later life. Policy Press, Bristol
Brown SL, Lin I-F (2012) The gray divorce revolution: rising divorce among middle-aged and older adults,

1990–2010. J Gerontol B Psychol Sci Soc Sci 67 (6):731–741
Chambers D (2012) A sociology of family life. Change and diversity in intimate relations. Polity Press, Cambridge
Cherlin AJ (2017) Introduction to the special collection on separation, divorce, repartnering, and remarriage around the world. Demogr Res 37:1275–1296
Cherlin AJ, Fustenberg F (1991) The new American grandparent: a place in the family, a life part, 2nd edn. Harvard University Press, Cambridge
De Jong Gierveld J (2002) The dilemma of repartnering: consideration of older men and women entering new intimate relationships in later life. Ageing Int 27:61–78
Harper S (2006) The ageing of family life transitions. In: Vincent JA, Phillipson C, Downs M (eds) The future of old age. Sage, London, pp 164–172
Hetherington ME, Kelly J (2003) For better or for worse: divorce reconsidered. W.W. Norton, New York
King V, Scott ME (2005) A comparison of cohabiting relationships among older and younger adults. J Marriage Fam 67(2):271–285
Kreyenfeld M, Konietzka D (2017) Analyzing childlessness. In: Kreyenfeld M, Konietzka D (eds) Childlessness in Europe: contexts, causes, and consequences. Demographic research monographs. Springer, Cham
Lin IF, Brown SL, Cupka CJ, Carr D (2018) A national portrait of stepfamilies in later life. J Gerontol B Psychol Sci Soc Sci 73(6):1043–1054
MetLife (2010) Still out, still aging. The MetLife Study of lesbian, gay, bisexual, and transgender baby boomers. MetLife and The American Society on Aging, New York. https://www.metlife.com/assets/cao/mmi/publications/studies/2010/mmi-still-out-still-aging.pdf. Accessed 10 Nov 2018
Nelson MK (2006) Single mothers 'do' family. J Marriage Fam 68:781–795
Pezzin LE, Pollak RA, Schone BS (2013) Complex families and late-life outcomes among elderly persons: disability, institutionalization, and longevity. J Marriage Fam 75(5):1084–1097
SAGE (2010) Improving the lives of LGBT older adults. Services and Advocacy for Gay, Lesbian, Bisexual and Transgender Elders, New York. http://www.lgbtmap.org/file/improving-the-lives-of-lgbt-older-adults.pdf. Accessed 10 Nov 2018
Steinbach A, Hank K (2016) Intergenerational relations in older stepfamilies: a comparison of France, Germany, and Russia. J Gerontol B Psychol Sci Soc Sci 71(5):880–888
Stelle C, Fruhauf CA, Orel N, Landry-Meyer L (2010) Grandparenting in the 21st century: issues of diversity in grandparent-grandchild relationships. J Gerontol Soc Work 53(8):682–701
Uhlenberg P (2005) Historical forces shaping grandparent-grandchild relationships: demography and beyond. In: Silverstein M (ed) Intergenerational relations across time and place. Springer, New York, pp 77–97
Vespa J (2012) Union formation in later life: economic determinants of cohabitation and remarriage among older adults. Demography 49(3):1103–1125
Wu Z, Schimmele C (2007) Uncoupling in late life. Generations 31(3):41–46

Family Dynamics

▶ Intergenerational Family Dynamics and Relationships

Family Formation and Dissolution

Naomi J. Spence
Department of Sociology, Lehman College, City University of New York, Bronx, NY, USA

Synonyms

Family change; Family life course transitions; Household formation and dissolution; Marriage formation and dissolution

Definition

Family formation and dissolution refer to processes through which people transition between family statuses often with emphasis on marital status. These processes may be characterized by the number, timing, duration, and sequence of transitions, which have all undergone substantial changes in recent decades. A narrow focus on the transition to and consequences of widowhood has been replaced by research on "gray divorce" (i.e., divorce over the age of 50), second (or subsequent) partnerships (i.e., later life marriage, cohabitation, and living apart together), and duration spent in different family statuses. This shift reflects the complexity of family formation and dissolution in aging populations.

Overview

Family changes that have captured the attention of scholars, policy-makers, and the general public in the past 50 years also apply to the experiences of older adults. Remarriage among older adults is not a recent phenomenon (Vinick 1978); however, family formation and dissolution have become more complex than transitions from marriage to widowhood to remarriage. Divorces have reduced the duration of marriages and increased the number of family transitions. Fewer lifelong marriages coupled with greater acceptance of cohabitation before or in lieu of marriage produce a complex and increasingly diverse set of family transitions across the life course.

Diverse transitions are the result of interrelated demographic, cultural, social, and economic factors. Demographically, people are more geographically mobile, having fewer children, and living longer. Lower fertility and increased geographic mobility results in fewer adult children to care for aging parents. This is related to the erosion of filial piety in cultures that have historically expected children to support older parents. Increased longevity yields an extended life course within which to experience multiple family transitions, while cultural shifts in attitudes about partnerships increase the diversity of family formation and dissolution experiences. Socially, a generation of people in countries like the United States, who de-emphasized marriage, are more likely to engage in nonmarital repartnering such as cohabitation or living apart together. The diversity of these transitions affords benefits to and present challenges for older adults especially within social and economic contexts built around the nuclear family ideal with women's dependence on husbands or in societies that value filial piety and maintain multigenerational households.

Key Research Findings

In a series of studies, Brown and colleagues have documented patterns of family formation and dissolution among older adults in the United States. Their work shows a substantial increase in cohabitating older adults and gray divorce, along with a growing number of older adults who were never married (Brown and Lin 2012; Brown and Wright 2017; Lin and Brown 2012). Other research on older Americans shows that widowed women do not commonly remarry although many remain involved in romantic relationships with some reporting a desire to remarry and others being disinterested in involvement with men (Moorman et al. 2006; Talbott 1998).

Outside of the United States, the diversity in family formation and dissolution patterns provides insights into the influences of social and cultural factors. In Singapore, widows and their children tend not to favor remarriage; instead multigenerational households are more common following the death of a spouse (Mehta 2002). As a society transitioning from more traditional ideas about family and romantic relationships, repartnering among older adults in Israel is approached with some reluctance and may be characterized according to physical and/or emotional togetherness (Koren and Lipman-Schiby 2014). Thus remarriage, cohabitation, and living apart together make up the diverse experiences even as couples perceived their repartnering to be socially unacceptable and may keep romances secret (Koren and Eisikovits 2011). Cohabitation may pose a higher risk of dissolution than marriages, as has been demonstrated using recent Canadian data on middle-aged and older adults (Wu and Penning 2018). Although this is generally found among younger adults as well, the factors that predict separation and divorce in middle to later life may be unique from those that influence union dissolution earlier in life (Wu and Penning 2018).

Sociocultural factors feature prominently in research on repartnering in later life, with older Americans endorsing rather favorable attitudes about diverse family formation patterns among older adults (Brown et al. 2016). However, partnerships after the death of a spouse may be culturally repudiated among ethnic groups who do not view death as the end of a marriage or who disregard the sexual desires of older people (Mehta 2002). Perhaps not surprisingly, having more children born within a marriage predicts a lower chance of union dissolution, while separation and divorce are more likely with greater numbers of children born outside of the union (Wu and Penning 2018). Moreover, those who approach midlife with young children are at increased risk of separation or divorce later in life (Wu and Penning 2018). In addition to these social and cultural influences on family formation and dissolution among older adults, economic considerations abound.

Marriage has been a well-established source of economic stability for women. In later life, public pension benefits may protect widows from poverty (McDonald and Robb 2004). Higher entitlement incomes are associated with lower likelihood of marriage for older women in the United States because remarriage can reduce entitlement benefits (Vespa 2013). Relatedly, economic consequences of singlehood among older women may be more severe for separated or divorced women than those who were never married or are widowed (McDonald and Robb 2004). Men may be less likely to reject future marriages following a gray divorce than women (Crowley 2019). Among Chinese widowers (men who lost wives to death), those who are younger or dissatisfied with widowhood are more likely to favor remarriage although attitudes about remarriage may be generally unfavorable (Chiu and Ho 2010). Among older cohabitors in the United States, men are most likely to transition to marriage when they are unhealthy and wealthy, while healthy poor women had a higher likelihood of marriage (Vespa 2013). These patterns parallel those found in research on gender differences in the benefits of marriage showing significant economic benefits for wives and health benefits for husbands (Waite 1995).

In response to increasingly complex family formation and dissolution patterns, researchers have begun to amass evidence of the consequences of these experiences for older adults. For example, marital quality may not be significantly different in remarriages, but gender differences in marital quality persist in higher-order (i.e., second, third, etc.) marriages (Cooney et al. 2016). Moreover, gender differences in the health consequences of marital trajectories show negative health effects of early marriage and divorce for women and more years spent divorced or widowed for men (Dupre and Meadows 2007). The mental health consequences of family dissolution through divorce or widowhood may depend, in part, on the quality of the marriage prior to the terminating event (Ye et al. 2018).

The positive and negative effects of marriage, divorce, and widowhood have been widely researched (Waite 1995). However, much of this work is limited by the availability of survey data for studying trajectories, which less frequently include experiences of cohabitation, living apart together, and other less institutionalized forms of family formation and dissolution. Despite increases in less institutionalized forms of family formation and dissolution around the world, these experiences may remain too uncommon in some societies to be studied with quantitative data and methods. In such cases, qualitative interview data may continue to serve as the best tool for understanding how older adults think about and experience repartnering and relationship dissolution.

Future Directions of Research

Future research should elaborate our understanding of the consequences of complex family formation and dissolution patterns for aging societies to better inform policies and programs to support older populations. The diversity of family life course transitions among older people implicates a need to change the structure of public pension systems that are based on outmoded ideas about family formation and dissolution, as these structures systematically disadvantage members of certain social groups, especially women. Comparative research may illuminate the pitfalls and promises of public and private pensions for ensuring the economic security of older adults. Disentangling financial needs from personal and romantic relationships should enable older adults to enjoy more flexibility in their family formation and dissolution decisions. The resulting consequences are likely to differ by gender and may promote greater equality across gender and sexual preferences.

Summary

The intersection of demographic, cultural, social, and economic changes that have contributed to population aging around the world has also influenced family formation and dissolution. Consequently, people increasingly approach middle and later life with complex and diverse partnership histories. Social science has documented this trend and demonstrated varied consequences of family formation and dissolution later in life. Less institutionalized arrangements such as cohabitation and living apart together shift attention to these important parts of older adults' life course experiences. Research and policies would do well to address family changes and their consequences for older individuals and for aging societies.

Cross-References

▶ Cohabitation
▶ Family Diversity
▶ Widowhood

References

Brown SL, Lin IF (2012) The gray divorce revolution: rising divorce among middle-aged and older adults, 1990–2010. J Gerontol: Ser B 67:731–741
Brown SL, Wright MR (2017) Marriage, cohabitation, and divorce in later life. Innov Aging 1:1–11. https://doi.org/10.1093/geroni/igx015
Brown SL, Lin IF, Hammersmith AM, Wright MR (2016) Later life marital dissolution and repartnership status: a national portrait. J Gerontol: Ser B 73(6):1032–1042
Chiu M, Ho W (2010) Intent to remarry among Chinese elderly widowers: an oasis or an abyss? Am J Mens Health 4(3):258–266
Cooney TM, Proulx CM, Snyder-Rivas LA (2016) A profile of later life marriages: comparisons by gender and marriage order. In: Divorce, separation, and remarriage: the transformation of family. Emerald Group Publishing, Bingley, pp 1–37
Crowley J (2019) Once bitten, twice shy? Gender differences in the remarriage decision after a gray divorce. Sociol Inq 89(1):150–176
Dupre ME, Meadows SO (2007) Disaggregating the effects of marital trajectories on health. J Fam Issues 28(5):623–652
Koren C, Eisikovits Z (2011) Life beyond the planned script: accounts and secrecy of older persons living in second couplehood in old age in a society in transition. J Soc Pers Relat 28(1):44–63
Koren C, Lipman-Schiby S (2014) "Not a replacement": emotional experiences and practical consequences of Israeli second couplehood stepfamilies constructed in old age. J Aging Stud 31:70–82
Lin IF, Brown SL (2012) Unmarried boomers confront old age: a national portrait. Gerontologist 52:153–165
McDonald L, Robb A (2004) The economic legacy of divorce and separation for women in old age. Can J Aging/La Rev Can Du Vieillissement 23(5):S83–S97
Mehta K (2002) Perceptions of remarriage by widowed people in Singapore. Ageing Int 27(4):93–107
Moorman S, Booth A, Fingerman K (2006) Women's romantic relationships after widowhood. J Fam Issues 27(9):1281–1304
Talbott M (1998) Older widows' attitudes towards men and remarriage. J Aging Stud 12(4):429–449
Vespa J (2013) Relationship transitions among older cohabitors: the role of health, wealth, and family ties. J Marriage Fam 75:933–949. https://doi.org/10.1111/jomf.12040
Vinick B (1978) Remarriage in old age. Fam Coord 27(4):359–363
Waite L (1995) Does marriage matter? Demography 32(4):483–507
Wu Z, Penning M (2018) Marital and cohabiting union dissolution in middle and later life. Res Aging 40(4):340–364
Ye M, DeMaris A, Longmore MA (2018) Role of marital quality in explaining depressive symptoms after marital termination among older adults. Marriage Fam Rev 54(1):34–49

Family History

▶ Family Tree

Family Life Course Transitions

▶ Family Formation and Dissolution

Family Medicine

Vivian J. Miller
Department of Social Work, College of Human Services, Bowling Green State University, Bowling Green, OH, USA

Synonyms

Family physician; General practitioner; Primary care; Primary care physician (PCP)

Definition

Family medicine by way of a family physician-patient relationship provides primary care to the whole person taking into account the medical needs of an individual, as well as social determinants and community factors, within the context of their communities (AAFP 2019; Rakel and Rakel 2016). Family medicine is provided to individuals across the lifespan, families, and populations, who are referred to as patients in this setting (Duke Family Medicine and Community Health 2019). Family medicine is provided by physicians, commonly referred to as primary care physicians (PCPs), within clinical and neighborhood community settings. Family medicine is also provided in hospital settings, emergency care settings, and nursing home facilities (University of Minnesota 2019). The training requirements for PCP in family medicine include medical school, medical licensing examinations, and clerkship grants. In US and Canada settings, once conferring the "Doctor of Medicine" (MD) degree, the physician must complete a family medicine residency, which is typically 3 years, at minimum (AAFP 2019). The final certification is by the American Board of Family Medicine (ABFM) that determines where the physician can practice as a licensed expert. In other jurisdictions, including most European countries, a total of 6 years of intensive university courses are taken prior to a residency. The final exam is then delivered following residency, whereby a specialty (e.g., geriatrics, internal medicine, etc.) can be chosen (Martinho 2012). There are additional trainings, fellowships, and certifications for family physicians to advance opportunities in concentrated areas such as preventative medicine, rural medicine, and geriatric medicine.

Overview

Founded as a distinct health science discipline in 1969 (Rakel and Rakel 2016; Tulane University 2019), family medicine encompasses whole healthcare (Rakel and Rakel 2016). This holistic approach includes an initial visit where the PCPs will review medical and family history, develop a list of any existing medical conditions, identify current medical providers and prescription medicines, assess and detect any cognitive limitations or impairment, provide personalized health advice, and, overall, help prevent disease and disabilities based on personal risk and health factors (Fernald et al. 2013). Compared to medical disciplines such as neurology, dermatology, or emergency medicine that have a particularly narrow focus when caring for an individual, family medicine and, more specifically, geriatric-focused family medicine take responsibility for broad healthcare of the patient from first contact to initial assessment and all subsequent ongoing care of chronic health problems and conditions (Rakel and Rakel 2016). It is the role of the geriatric PCP to coordinate, refer, and integrate the patient to any healthcare services and treatments that the individual might need with the utmost continuity in care. Coupled with an assessment of the medical needs of the older adult patient, the family medicine physician takes a comprehensive approach that includes functional, psychological, and social domains, with the goal of maintaining function and attaining the highest quality of life (Rakel and Rakel 2016). Moreover, the family medicine physician addresses social issues that affect the patient (McWinney and Freeman 2009), as understanding the perspective of the person-in-environment is critical as neighborhood, community, and societal characteristics, together with health behaviors, which strongly influence healthy aging (World Health Organization 2018).

Family medicine by way of healthcare delivery through family practitioners and geriatric family physicians is the primary medical professional that controls access to emergency medical services and hospital health services (Fung et al. 2015). Family physicians follow the life cycle of a patient (Cleveland Clinic 2019); they help individuals understand illness, set goals, and navigate complex healthcare systems (Phillips et al. 2014). In family medicine, the physician must have a comprehensive understanding of the healthcare needs of the older adult patients served, in addition to an expertise of specialists and acute care services to respond to the individual's needs and make referrals out, as necessary, keeping the patient at the center of care (Stollenwerk et al. 2019). Examples of subspecialties that the geriatric family physician may make a referral to include surgical management, obstetrician-gynecologists, ophthalmology and optometry, oncologist, and mental health professionals.

Key Research Findings

Broadly, family medicine is the primary medical gatekeeping of all healthcare services such as specialists (e.g., allergists, endocrinologists), acute medical needs (e.g., surgeries), and other healthcare interventions (e.g., mental health). For older adults utilizing family physicians, innovative clinical interventions can be used in routine medical care, which can increase the comprehensiveness of care and "expand health care providers' capacity to facilitate education, counseling, and documentation" (Lum et al. 2017, p. 481). Moreover, the family physician engages the patient in all aspects of care through the life cycle.

The latest family medicine research has found that older adults most often utilize a PCP, compared to younger counterparts. FM utilization rates reflect this. In 2015, older adults age 65 years and above had 658 visits to the family physician per 100 persons, compared to 366 visits for adults aged 45–64 years old, and 204 visits for adults aged 18–44 years old (Centers for Disease Control and Prevention [CDC] 2017b). These visits by Medicare beneficiaries included visits with a primary care physician, a medical specialist, or surgeon (Pham et al. 2007). For older adults with chronic conditions, the number of physician visits was increased, particularly to medical professionals providing specialized care, rather than primary care physicians (Zayas et al. 2016). Although older adults utilize family medicine most often, younger adults more frequently maintain continuity of care with the same practitioner due to a number of reasons (Gerst-Emerson and Jayawardhana 2015; Mattson 2011; Valtorta et al. 2018). Extant literature found that older adult family medicine visits span across seven physicians in four different settings each year (Amjad et al. 2016). These trends in PCP utilization are found to be due in part to geographical access to care in areas with more primary care physicians (Daly et al. 2018), relationship with a practitioner (Daly et al. 2018; Chipidza et al. 2015), and social relationships in the community that may encourage family medicine visits (Bremer et al. 2017). While the PCP aims to assist in continuity of care and successful referrals to services, this is not always achieved and may result in disconnected services. Such disjointed medical visits lead to fragmented and fractured practitioner-patient communication and overall care delivery.

Moreover, such disconnected relationships between the older adult and their healthcare provider within family medicine are extremely costly on the individual and the overall healthcare system. Older adults having a family physician is associated with decreased hospitalizations, better compliance (e.g., medicine, medical advice), lower costs to the individual through the maintenance of chronic health conditions, and fewer emergency department visits (Birch 2010; Phillips et al. 2014), thus lowering the associated costs to the patient and healthcare system. The fees-for-service associated with physician visits under Medicare healthcare coverage are much less than compared to emergency and hospital visits (Shi 2012) and are reported to be the most cost-effective healthcare service (Fung et al. 2015).

Equally so, a strong working patient-provider relationship is necessary for optimal health and wellness of aging adults. This continuity of care (COC) is a defining feature of family medicine, whereby this long-term, team-based partnership that is led by the physician in primary care prevents diseases and improves receipt of preventative care, assists in maintaining health, and lowers mortality rates (Birch 2010; Rakel and Rakel 2016). For the older adult, recognizing the importance of patient-centered care (PCC) is becoming increasingly important in healthcare delivery within primary care. While improving the overall health delivery model through PCC has been at the forefront of policy and care delivery within long-term care and dementia care (Grabowski et al. 2014; Koren 2010), PCC must be especially targeted toward older adults receiving care from their PCP in the community, as they are most likely to have increasingly dynamic and complex healthcare needs. A systematic review of the literature identified that whole-person care, respect and value, dignity, choice, purposeful living, and self-determination were the most prominent principles and values of PCC, published within literature across the past 25 years (Kogan et al. 2016). Next, identifying the barriers to implementing these practices and this care delivery into family medicine and primary healthcare settings, followed by determining a definition that encompasses each feature of this care model, can help support older adults with complex needs (Goodwin 2016).

One PCP appointed task that is of utmost importance for older adults is advanced care planning (ACP). ACP preferences are important to document as adults age. It is the role of the family physician to discuss ACP. ACP includes legal documents and medical orders, which are comprised of living wills, healthcare surrogates, and do-not-resuscitate (DNR) orders. ACP has been associated with well-being outcomes, such as improved quality of life, satisfaction, and receiving medical care in line with the patients' wishes (Carr and Luth 2017; Integrated Healthcare Association 2016; Lum et al. 2017). In one study, an ACP group intervention was conducted within three primary care clinics that were found to be effective in improving ACP documentation among older adults (Lum et al. 2017). Implementing this person-centered intervention within the context of the family physician-patient relationship was associated with high levels of overall patient satisfaction.

Future Directions of Research

As the geriatric population continues to rise, healthcare needs and healthcare utilization will increase (Foley et al. 2017; WHO 2015). Moreover, the population of older adults is becoming increasingly diverse. By the year 2030, ethnic and racial minorities will encompass 28 percent of the total population of persons aged 65 years and older in the USA (Miller and Hamler 2019; Fox-Grage 2016). The subset of minority older adults experiences increased levels of chronic illnesses, disabilities, and preventable diseases at a disproportionately higher rate than nonminority Whites (Administration for Community Living 2017; Centers for Disease Control and Prevention [CDC] 2017a). Thus, family practitioners prepared with a comprehensive understanding of the unique needs of this growing, diverse population and the skillset to treat aging adults will be in the greatest demand. Below are three specific areas for future research targeted at improving patient-provider relationships and older adult health outcomes.

First, the evaluation of efficacy and effectiveness for healthcare assessments informed by ethnicity, race, and culture must be of focus in future research (Miller and Hamler 2019). Existing research states that the current geriatric FM assessment(s) are comprehensive and "is multidimensional, multidisciplinary assessment designed to evaluate an older person's functional ability, physical health, cognition and mental health, and socioenvironmental circumstances" (Elsawy and Higgins 2011, p. 48). While aspects of functional ability, medical condition, and quality of life are all included in the current geriatric assessment, research has found that this patient-provider interaction can be more problem-directed rather than holistic, amidst a busy clinical practice (Elsawy and Higgins 2011). The geriatric assessment must include ethnicity, education, age, and religious and/or spiritual practices and beliefs. Additionally, assessments must take into account diet, tradition, routines, language of origin, medical illness and/or disabilities, sexual orientation, occupation, life experiences, socioeconomic status, functional status, social support systems, and networks. While the family practitioner conducts the medical assessment to screen for common diagnosable disorders (e.g., hearing loss, changes in cognition, impaired mobility), it is important that they do not leave out critical psychological and social assessments, which may be uniquely of interest to older adults of racial and ethnic minority status. Research targeted at critically evaluating geriatric assessments may assist to inform the development of future assessments that focus on the mental, physical, and environmental needs of minority aging adults.

Next, future research ought to include an evaluation of assessments used by family medicine practitioners with a special focus on adverse events (e.g., past trauma, abuse). Adults now aging into older adulthood are at an especially high risk for having experienced adverse events. Physicians must be able to recognize signs of physical, behavioral, and emotional abuse. Through models of care, including trauma-informed care, family practitioners can assess for traumatic events (Rakel and Rakel 2016) and reduce or prevent any re-traumatization (Kusmaul and Anderson 2018) to best understand the unique needs of older adults. Research that is aimed at examining trauma assessments may inform practice within family medicine and improve patient-provider relationships working toward overall positive healthcare outcomes.

Furthermore, exploring "patient voice" as an effort toward quality care in family medicine is an area that warrants continued research, as the population of older adults becomes increasingly diverse and unique. Noted in the seminal work published over 50 years ago by medical leader Donabedian (2005), the patient's voice is imperative to evaluate the effectiveness of the healthcare system, whereby taking into account the voice of a diverse population is necessary (Holt 2018). Evaluations of healthcare delivery and healthcare effectiveness are necessary for patient health outcomes and overall organizational quality. Quality improvement efforts within PCP practices are often relied on to improve the healthcare delivered to older patients; however a recent study (Hsu et al. 2019) reports that these efforts fall short on incorporating the patient's voice. "Patient voice" is an integral feature of patient-centered care (PCC). As such, research must continue to explore best practices to understand the voice of patients. Integrating the patient into family medicine research, as well as asking researchers to engage directly with family medicine physicians and older adult patients, may be a critical next step that takes into account the perspectives of all key stakeholders in the healthcare system.

Lastly, there is a need for expanded research on the role of the family medicine physician in rural communities. In rural areas within the USA, older adults most often rely on visits to their family physician; however healthcare accessibility for this population continues to be a major barrier (AAFP 2014). Rural-based family medicine physicians must take a considered assessment of older adults' access to healthcare that is primarily driven by transportation access and opportunities. Research has noted that transportation is vital to community connectedness and accessing healthcare services (Syed et al. 2014), whereby future efforts ought to more closely explore the first/last mile issues, including access to transportation and travel time, especially in underserved, high-need communities (Miller 2018).

In addition to transportation-focused research efforts for aging adults in rural communities, an exploration of telehealth and telemedicine is warranted. Virtual technologies, such as telehealth and telemedicine, allow older adults to remain at home in the community and have interaction with their family medicine doctor by way of interactive, real-time telecommunications (e.g., two-way video or audio) or asynchronous communications (e.g., transmission of medical information through email, video, or audio). These alternatives are becoming increasingly important as a solution to mitigate issues in accessibility of family medicine as experienced by older adults (HRSA 2015). Continued examinations of the value and quality of these services offered for aging older adults are of importance to reduce the barriers to care.

Summary

The primary feature of family medicine provided to older adults across care settings is comprehensive care. This comprehensive delivery model helps for older adults to attain their highest well-being and is associated with decreased healthcare costs and hospitalization rates for this population (Bazemore et al. 2015). Continued research into retention of FM physicians specialized in geriatrics, culturally sensitive and trauma-informed assessments for this population, and access to healthcare within underserved communities will contribute to an increased health and well-being of FM for aging adults.

Cross-References

► Comprehensive Geriatric Assessment
► Healthcare Utilization
► Preventive Care

References

Administration for Community Living [ACL] (2017) Minority aging. Retrieved from https://www.acl.gov/aging-and-disability-in-america/data-and-research/minority-aging

American Academy of Family Physicians [AAFP] (2014) Rural practice, keeping physicians in (position paper). Retrieved from https://www.aafp.org/about/policies/all/rural-practice-paper.html

American Academy of Family Physicians [AAFP] (2019) Family medicine: comprehensive care for the whole person. Retrieved from https://www.aafp.org/medical-school-residency/choosing-fm/model.html

Amjad H, Carmichael D, Austin AM, Chang C-H, Bynum JPW (2016) Continuity of care and health care utilization in older adults with dementia in fee-for-service Medicare. JAMA Intern Med 176(9):1371–1378

Bazemore A, Petterson S, Peterson LE, Phillips RL (2015) More comprehensive care among family physicians is associated with lower costs and fewer hospitalizations. Ann Fam Med 13(3):206–213

Birch JT (2010) Continuity of care and the geriatric patient. Retrieved from http://classes.kumc.edu/coa/education/AMED900/CONTINUITY%20OF%20CARE.pdf

Bremer D, Inhestern L, von dem Knesebeck O (2017) Social relationships and physician utilization among older adults – a systematic review. PLoS One 12(9): e0185672

Carr D, Luth EA (2017) Advanced care planning: contemporary issues and future directions. Innov Aging 1(1): igx012

Centers for Disease Control and Prevention [CDC] (2017a) Health equity. Retrieved from https://www.cdc.gov/minorityhealth/index.html

Centers for Disease Control and Prevention [CDC] (2017b) Health, United States, 2017 – data finder. Retrieved from https://www.cdc.gov/nchs/hus/contents2017.htm#076

Chipidza FE, Wallwork RS, Stern TA (2015) Impact of the doctorpatient relationship. Prim Care Companion CNS Disord. 17(5):10.4088/PCC.15f01840

Cleveland Clinic (2019) Family medicine. Retrieved from https://my.clevelandclinic.org/departments/medicine/depts/family

Daly MR, Mellor JM, Millones M (2018) Do avoidable hospitalization rates among older adults differ by geographic access to primary care physicians? Health Serv Res 53(4):3245–3264

Donabedian A (2005) Evaluating the quality of medical care; 1966. Milbank Q 83(4):691–729

Duke Family Medicine and Community Health (2019) Family medicine residency. Retrieved from https://fmch.duke.edu/division-family-medicine/family-medicine-residency

Elsawy B, Higgins KE (2011) The geriatric assessment. Am Fam Physician 83(1):48–56

Fernald DH, Tsai AG, Vance B, James KA, Barnard J, Staton EW, Pace WD, West DR (2013) Health assessments in primary care, A how-to guide for clinicians and staff. Retrieved from https://www.ahrq.gov/sites/default/files/publications/files/health-assessments_0.pdf

Foley KT, Luz CC, Hanson KV, Hao Y, Ray EM (2017) A national survey on the effect of the geriatric academic career award in advancing academic geriatric medicine. J Am Geriatr Soc 65(5):896–900

Fox-Grage W (2016) The growing racial and ethnic diversity of older adults. AARP. Retrieved from https://blog.aarp.org/thinking-policy/the-growing-racial-andethnic-diversity-of-older-adults

Fung CSC, Wong CKH, Fong DYT, Lee A, Lam CLK (2015) Having a family doctor was associated with lower utilization of hospital-based health services. BMC Health Serv Res 15(42). https://doi.org/10.1186/s12913-015-0705-7

Gerst-Emerson K, Jayawardhana J (2015) Loneliness as a public health issue: the impact of loneliness on health care utilization among older adults. Am J Public Health 105(5):1013–1019

Goodwin C. from The American Geriatrics Society Expert Panel on Person-Centered Care (2016) Person-centered care: a definition and essential elements. JAGS 64:15–18

Grabowski DC, O'Malley J, Afendulis CC, Caudry DJ, Elliot A, Zimmerman S (2014) Culture change and nursing home quality of care. Gerontologist 54(Suppl 1):S35–S45

Health Resources and Services Administration [HRSA] (2015) Telehealth in rural America. Retrieved from https://www.hrsa.gov/advisorycommittees/rural/publications/telehealthmarch2015.pdf

Holt JM (2018) Patient experiences in primary care: a systematic review of CG-CAHPS surveys. J Patient Exp. https://doi.org/10.1177/2374373518793143

Hsu KY, Contreras VM, Vollrath K, Cuan N, Lin S (2019) Incorporating the patient voice into practice improvement: a role for medical trainees. Fam Med 51(4):348–352

Integrated Healthcare Association (2016) Fact sheet: quality of life conversation on advanced care planning. Retrieved from https://www.iha.org/sites/default/files/resources/fact-sheet-advance-care-planning-2015.pdf

Kogan AC, Wilber K, Mosqueda L (2016) Person-centered care for older adults with chronic conditions and functional impairment: a systematic literature review. J Am Geriatr Soc 64(1):e1–e7

Koren MJ (2010) Person-centered care for nursing home residents: the culture-change movement. Health Aff 29(2):312–317. https://doi.org/10.1377/hlthaff.2009.0966

Kusmaul N, Anderson K (2018) Applying a trauma-informed perspective to loss and change in the lives of older adults. Soc Work Health Care. https://doi.org/10.1080/00981389.2018.1447531

Lum HD, Sudore RL, Matlock DD, Juarez-Colunga E, Jones J, Nowels M, Schwartz RS, Kutner JS, Levy CR (2017) A group visit initiative improves advance care planning documentation among older adults in primary care. J Am Board Fam Med 30(4):480–490

Martinho AM (2012) Becoming a doctor in Europe: objective selection systems. AMA J Ethics 14(12):984–988

Mattson J (2011) Transportation, distance, and health care utilization for older adults in rural and small urban areas. Transp Res Rec 2256(1):192–199

McWinney IR, Freeman T (2009) Textbook of family medicine, 3rd edn. Oxford, New York

Miller VJ (2018) Investigating barrier to family visitation of residents in nursing homes: a systematic review of the literature. J Gerontol Soc Work 62(2):1–18

Miller VJ, Hamler T (2019) A value-critical policy analysis of the nursing home reform act: a focus on care of African American and Latino residents. Soc Work Health Care. https://doi.org/10.1080/00981389.2019.1587660

Pham HH, Schrag D, O'Malley AS, Wu B, Bach PB (2007) Care patterns in Medicare and their implications for pay for performance. N Engl J Med 356(11):1130–9

Phillips RL, Brungardt S, Lesko SE, Kittle N, Marker JE, Tuggy ML, LeFevre ML, Borkan JM, DeGruy FV, Loomis GA, King N (2014) The future role of the family physician in the United States: a rigorous exercise in definition. Ann Fam Med 12(3):250–255. https://doi.org/10.1370/afm.1651

Rakel RE, Rakel DP (2016) Textbook of family medicine, 9th edn. Elsevier, Philadelphia

Shi L (2012) The impact of primary care: a focused review. Scientifica 22. https://doi.org/10.6064/2012/432892

Stollenwerk D, Kennedy LB, Hughes LS, O'Connor M (2019) A systematic approach to understanding and implementing patient-centered care. Fam Med 51(2):173–178. https://doi.org/10.22454/FamMed.2019.320829

Syed ST, Gerber BS, Sharp LK (2014) Traveling towards disease: transportation barriers to health care access. J Community Health 38(5):976–993

Tulane University (2019) Family and community medicine research guide. Retrieved from https://www.libguides.tulane.edu/c.php?g=182531andp=1204239

University of Minnesota (2019) Department of family medicine and community health. Retrieved from https://familymedicine.umn.edu/community-impact

Valtorta NK, Moore DC, Barron L, Stow D, Hanratty B (2018) Older adults' social relationships and health care utilization: a systematic review. Am J Public Health 108(4):e1–e10

World Health Organization (WHO) (2015) World report on ageing and health. WHO Press, Geneva. Retrieved from https://apps.who.int/iris/bitstream/handle/10665/186463/9789240694811_eng.pdf?sequence=1

World Health Organization (WHO) (2018) Ageing and health. Retrieved from https://www.who.int/news-room/fact-sheets/detail/ageing-and-health

Zayas CE, He Z, Yuan J, Maldonado-Molina M, Hogan W, Modave F, Guo Y, Bian J (2016) Examining Healthcare Utilization Patterns of Elderly Middle-Aged Adults in the United States. Proc Int Fla AI Res Soc Conf 361–366. pmid:27430035

Family Network

▶ Kin Availability

Family Networks

▶ Kinship Networks

Family Physician

▶ Family Medicine

Family Poverty Among Older Adults

Duygu Basaran Sahin
Department of Sociology, The Graduate Center, City University of New York, New York, NY, USA

Definition

The concept of poverty "concerns itself with having too few resources or capabilities to participate fully in a society" (Smeeding 2016: 21). Family poverty among the older population refers to poor families whose head of household is 65 years old or older. In the USA, a family is poor when their total household income falls short of the poverty threshold established for the family's size and composition. The official poverty threshold dates back to 1960s when Mollie Orshansky, a staff economist at the Social Security Administration, "multiplied the cost of a minimum food diet by three to account for other expenses" (The U.S. Census Bureau 2017). According to this calculation, households with a householder aged 65 years and older have a lower poverty threshold than other family types as nutritional needs, and thereby costs of living were assumed to decline with age. In 2017, the official poverty threshold for an individual age 65+ living alone was $11,756.

Old age in and of itself is not a cause of poverty. Instead, poverty among older adults is a function of disadvantages that accumulate over the life course (Crystal and Shea 1990).

Overview

Poverty Measures

Conceptual Definitions of Poverty
Two broad measures of poverty are widely used: absolute versus relative poverty. Absolute measures set a monetary value to the "goods deemed to be basics in a given country" (Smeeding 2016: 27) and consider people poor if their income is less than this value. Absolute measures consist of a fixed threshold which does not vary by geography or time except for inflationary adjustments; however, the threshold usually varies by family size and type. The USA is among the high-income countries that uses an absolute measure of poverty. Relative poverty measures refer to one's economic standing relative to some standard, such as other persons in one's nation or community. To calculate this measure, researchers typically take an individual or household's income and compare it to some standard such as the median individual or household income for that country or to a concrete standard of living such as 60% of the nation's median individual or household income. Most scholars in Europe use relative poverty measures because from a theoretical perspective, they argue that measuring poverty through shortfalls of income does not take into account the social norms and behaviors needed to participate fully in a society (Townsend 1979).

In a paper comparing various measures of poverty across different demographics in the USA, Iceland (2005) showed that relative measures yield a higher rate of poverty (13.8%) among the 65+ population compared to the official (absolute) poverty measure (9.3%). He argues that it is because older people often have incomes above the official poverty lines but below the relative ones. Furthermore, others found that (e.g., Bauman 1999) despite having lower incomes, persons age 65+ report fewer material hardships (e.g., lack of food, heat, access to health care). Therefore, low income alone may not necessarily indicate poverty for this age group. Iceland concluded that relative measures may not be useful indicators of poverty because even though older people have incomes below the relative poverty lines, they are more likely to own their homes and may have less variable expenses, which serve as a protection against poverty spells. Instead, he argues for a quasi-relative measure, recommended by the National Academy of Sciences, which motivated the Supplemental Poverty Measure discussed below.

Statistical Indicators of Poverty
The most commonly used poverty measure in the USA is the official poverty measure. As previously mentioned, an individual aged 65 and older is classified as poor if she/he had a yearly income equal to or less than $11,756. Using this measure, a little over 9% of older adults were classified as poor in 2017, a rate considerably below the 17.5% rate documented for youth under ages 18 (Fontenot et al. 2018). This gap has fueled debates about public investments in social problems targeting older adults versus youth (Klein 2013), yet recent evidence suggests that the official poverty rate considerably underestimates late-life poverty and that the recently developed Supplemental Poverty Measure

(SPM) provides a more accurate snapshot. Whereas the official poverty threshold has only used consumer price index (CPI) to adjust for yearly inflation, the SPM introduced flexibility by providing different thresholds based on geography and homeownership status (Short 2013). The SPM also subtracts medical expenses but includes noncash income (e.g., food stamps) in the resource category in contrast to the official poverty measure, which only includes gross before-tax income. These adjustments result in a higher poverty rate for older population due to their high medical needs.

In 2017, the SPM rate for people aged 65 and older was 14.1%, which is 4.9 points higher than the official poverty rate (Fox 2018). This is because the SPM poverty thresholds do not vary by the householder's age, while the official poverty thresholds are set lower for older householders (Mather et al. 2015). Although there is a significant difference between the two rates for the older population, the difference was only 1.6 percentage points for the overall poverty in the USA.

Lastly, the Census Bureau has historically provided detailed analysis of the "near poor" population, which consists of families whose incomes are between 100 and 125% of the official poverty threshold (Hokayem and Heggeness 2014). In 2012, 5.5% of persons age 65+ were living in near poverty, in addition to 9.4% living below the official poverty threshold.

General Trends

According to the official poverty rates, in 2017, persons age 65+ had lower poverty rates (9.2%) than both working age adults ages 18–64 (11.2%) and youth under age 18 (17.5%). Historically, in the USA, older adults experienced a greater decline in their poverty rates than other age groups even though they started with the highest poverty rate (see Fig. 1). More than one-third (35.2%) of persons age 65+ were poor in 1959 compared to 17% of working age adults and 27.3% of children (Iceland 2006: 39). The biggest contributor to this steep decline in late-life poverty was the introduction and expansion of Social Security and other benefit programs targeting the poor.

Poverty rates among older adults vary widely by state, regardless of which measure is used. In 2016, nearly 29% of older adults in the District of Columbia were living at 100% of the poverty line, as defined by the SPM, although rates were just over 5% in Vermont and Iowa. When the official poverty measure was used, the proportion of older adults living at 100% of the poverty line ranged from 5.4% in Minnesota to just over 15% in the District of Columbia. Differences in state-level poverty rates under the official measure versus

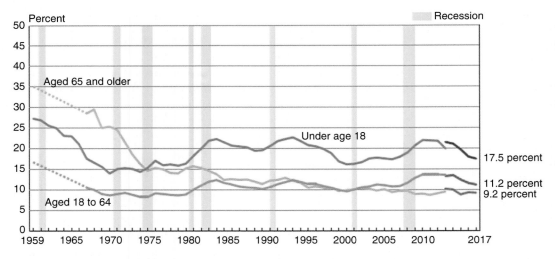

Family Poverty Among Older Adults, Fig. 1 Official poverty rates by age (1959–2017). (Source: Fontenot et al. 2018)

the SPM reflect factors including state income distributions; housing prices, which are calculated into the SPM poverty thresholds; variations in medical use and costs, since such costs are deducted from income under the SPM but not the official measure; and differences in state Medicaid programs expenditures, which affect older adults' out-of-pocket medical expenses (Cubanski et al. 2018).

Cross-national comparisons show that the USA has one of the highest old-age poverty rates among high-income nations. Smeeding (2006) used the data from the Luxembourg Income Study (years 1999–2000) and analyzed the poverty rates of 11 rich countries (Ireland, Italy, the UK, Canada, Germany, Belgium, Austria, Netherlands, Sweden, Finland, and the USA). He considered households poor if they had incomes less than 50% of the median disposable income and found that in households headed by individuals age 65 and older, Ireland had the highest poverty rate (43%) followed by the USA (23%). The lowest rate was in the Netherlands (2%). It is important to note that among these countries, the USA has the smallest share of its resources devoted to antipoverty programs.

Key Research Findings

Race and gender are two of the most powerful correlates of economic well-being at all stages of the life course, and old age is no exception. Employment and family characteristics have been identified as two key mechanisms that account, in part, for these disparities.

Race/Ethnicity

In the USA, the racial/ethnic distribution of poverty among the older population is similar to that of the younger population. Although Social Security has been very effective in reducing the overall number of older adults living in poverty, it is not sufficient to eradicate deep-seated racial gaps. In 2014, only 8% of non-Hispanic Whites lived in poverty compared to 18% for Hispanics and 19% for non-Hispanic Blacks (Mather et al. 2015). As discussed below, racial gaps in late-life poverty are a function of multiple factors dating back to earlier stages of the life course, including more precarious employment, lower rates of marriage, and the implications of work and family arrangements for the Social Security benefits received in later life.

Gender

The interaction of gender, age, and poverty results in having more women living in poverty than men at any age. In 2017, 10.5% of older women lived in poverty compared to 7.5% of older men in the USA. The gender gap of poverty among the old population widens as the age increases. Among the ages 75 and older, women are nearly twice as likely to be poor (15%) compared with men (8%) (Mather et al. 2015).

Consistent with the theme of intersectionality theory that race and gender have multiplicative effects on life chances, recent data powerfully show that women of color have poverty rates that are substantially higher than White men. As Fig. 2 shows, women have higher rates of poverty than men in all racial subgroups, and ethnic minorities have higher rates of poverty than Whites among both men and women in the USA. Yet when intersecting identities are considered, the gaps are most profound. Black and Latino women evidence poverty rates more than three times that of White men (20.5 and 20.9 versus 5.7%, respectively).

Studies focusing on the European Union countries show similar findings. Antczak and Zaidi (2016) used the European Union Survey on Income and Living Conditions data (years 2005–2014) to assess the poverty risk of the older population in 28 E.U. countries and the prevalence of gender gap among the poor. They found that in 2014, 22% of women ages 75+ were at-risk-of-poverty compared to 15% of men in the same age group although the gender gap remained quite stable since the beginning of twenty-first century. The authors concluded that structural assumptions in pension systems, including the assumption that women would leave or take a break from employment after marriage or during childcare period, result in lower pension contributions, which in turn lead to lower income in late life.

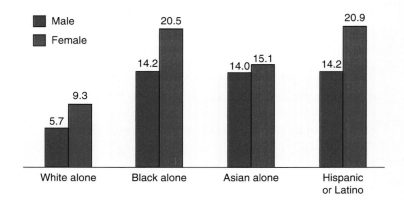

Family Poverty Among Older Adults, Fig. 2 Percent poor among the population 65+ by sex, race, and Hispanic origin (2010). (Source: West et al. 2014)

Pathways to Racial/Ethnic and Gender Differences in Old-Age Poverty

Family structure and employment patterns are two main mechanisms behind race and gender differences in poverty.

Employment Patterns

In the USA, employment history has a direct impact on whether older adults will benefit from Social Security and if so how much they will. The youngest of the current 65+ population entered the labor force in the late 1970s. Even though the Civil Rights movement opened the way for equal employment opportunities for all, Whites were able to move into white-collar occupations much faster than Blacks due to their higher educational attainment (Newman 2003). Whereas Black women closed the income gap by 1980, earning 92% of what their White peers took home, the same was not true for Black men. In 1980, a black male worker's pay was only 63% of his White counterpart. In sum, the gender and racial disparities that we see in today's 65+ population are reflection of their employment histories (Carr 2019).

Changes in employment status, especially for the working-age population, are the most important factor in transitioning in and out of poverty (Duncan et al. 1995). However, it does not have quite the same impact for the older population. On the one hand, it is relatively rare for older people to change jobs to increase their income, due in part to ageism in hiring practices. Therefore, their chance of upward mobility and thereby escaping poverty is more limited compared to the younger population. When possible, older people rely on their assets rather than finding a better paying job when they have unexpected costs to address. On the other hand, despite a recent increase in the labor force participation of older men and women, the majority of older people are retired or about to retire and therefore have at least a steady income, assuming they contributed to Social Security during their employment years. As of June 2018, almost 90% of individuals, age 65 and older, received Social Security benefits (Social Security Administration 2018).

Social Security, private pensions, and earnings are the three main sources of income for older adults in the USA. Social Security has always been an important source of income especially for those in the lowest-income quintile who collect over 80% of their income from Social Security compared to 17% of people in the highest-income quintile (West et al. 2014). Scholars estimate that without Social Security, 40% of older adults rather than 9% would be living in poverty (Romig 2018).

Despite Social Security's effectiveness in reducing late-life poverty, older women, especially women of color, consistently have higher poverty rates than older men for two main reasons: (a) Social Security benefits are based on lifetime earnings, and throughout their working years, women receive lower wages than men; (b) women have more sporadic employment and more part-time work compared to men (Carr 2019). As a result, women have fewer retirement savings, benefit less from private pensions, and acquire "fewer income-producing assets to use in retirement" (Women's Bureau U.S. DOL 2015). A third explanation relates to the structure of Social Security benefits, such that people qualify for benefits as a retired worker or as the spouse, ex-spouse, or widow(er) of a retired worker. This system penalizes women of color in part because in labor force they earned less than White women, resulting in lower Social Security checks, and in part because of their historically lower rates of marriage (people should be married to a qualified worker for at least 10 years in order to receive spousal benefits), higher rates of divorce, and widowhood relative to White women (Carr 2019).

Family Structure
Marital status, including whether one is married, single, or formerly married, is an important risk factor for poverty. This is because married persons are financially more secure than never married, widowed, and divorced persons both because of the earnings (or Social Security benefits) of both spouses receive. Married couples also enjoy economies of scale in sharing costs for rent or mortgage, food, utilities, and the like. At every stage of the life course, Whites are more likely than Blacks to get married and remain married. Similarly, women are much more likely than men to become widowed and are less likely than men to remarry following divorce and widowhood. Consequently, marital status is a key mechanism linking race and gender to poverty status.

In 2014, 53% of older Americans were married and lived with their spouse (Stepler 2016). However, only 58% of women ages 65–74 were married compared to 75% of men of the same age (Carr 2019) as women live longer than men, and they are less likely to remarry (McGarry and Schoeni 2000), resulting in a higher share of women who live alone. Historically, divorce rates have been low in older cohorts; however, it is currently on the rise. For adults aged 65 and older, the divorce rate tripled since 1990, and 6 people out of 1000 married people were divorced in 2015. Unless divorced women remained married longer than 10 years, they will not benefit from Social Security spouse or survivor benefits and must rely on benefits linked to their own earnings, which are significantly lower compared to their male counterparts.

Figure 3 summarizes the impact of living arrangement and marital status on poverty for different racial/ethnic groups. In 2010, regardless of racial/ethnic background, older women living alone had the highest poverty rate, ranging from

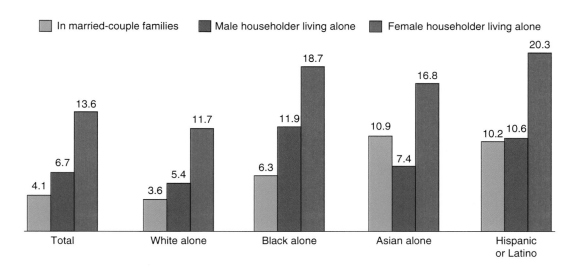

Family Poverty Among Older Adults, Fig. 3 Percent poor among the population 65+ by living arrangement, race, and Hispanic origin (2010). (Source: West et al. 2014)

20% of Hispanic/Latino women living in poverty to 12% of White women. Families made of married couples had the lowest poverty rate for each racial/ethnic group, except for Asians. Asian married couples had higher poverty rates than Asian men living alone.

Future Directions of Research

Racial and ethnic differences are a prominent theme in the late-life poverty literature in the USA Regardless of the type of measure used, whether official poverty threshold or the SPM, older Blacks and Latinos have higher poverty rates than Whites. Demographic projections show that more than half of the 65+ population will be non-Hispanic White by 2060; however, one fifth will be Hispanic/Latino, a little over 10% will be Black alone, and 9% will be Asian (Mather et al. 2015), primarily due to aging of the post-1965 immigrants from Latin American and Asian countries. More importantly, demographers expect the share of foreign-born population among persons age 65+ to double from 13% in 2014 to 26% in 2060 (Colby and Ortman 2015). Given the expected shift in the racial/ethnic composition of the 65+ population, future research can focus on understanding the poverty levels of older immigrants in the USA for three main reasons: (a) immigrants have significantly lower earnings than natives (Sevak and Schmidt 2014), resulting in lower Social Security benefits and putting them at risk of late-life poverty; (b) older immigrants may not qualify for Social Security as Social Security requires 40 quarters or a total of 10 years of earnings (Carr 2019); and (c) those who are eligible may receive lower benefits than natives because they worked for fewer quarters or they have worked "off the books." At international level, by 2061, the share of older persons with a foreign background is projected to be as high as two-thirds of the population in some EU member states such as Luxembourg (Lanzieri 2011). In sum, understanding the sources behind racial, ethnic, and nativity status differentials in poverty is a critical step toward ultimately eradicating them.

Summary

Family poverty among the older population refers to poor families whose head of household is 65 years old or older. The USA uses an absolute measure of poverty and considers people poor based on a calculation introduced in the 1960s, "which multiplied the cost of a minimum food diet by three to account for other expenses" (The Census Bureau 2017). According to the official poverty threshold used for householders aged 65 years and older, approximately 10% of older adults were living in poverty in 2017. However, while aggregate portraits of late-life poverty based on official poverty measures show an optimistic picture, and reveal that older adults are less likely than young persons to live in poverty, these statistics are somewhat misleading. When the SPM is used, older adults are roughly as likely as younger persons to be poor. This entry also showed that poverty in later life does not emerge instantly on one's 65th birthday but is the product of long-standing and cumulative disadvantages linked to race and gender that form earlier in the life course; and when poverty rates are stratified by race and gender, stark disparities emerge (Carr 2019).

Cross-References

▶ Poverty and Gender Issues in Later Life in the UK

References

Antczak R, Zaidi A (2016) Risk of poverty among older people in EU countries. CESifo DICE. https://www.cesifo-group.de/DocDL/dice-report-2016-1-zaidi-antczak-march.pdf. Accessed 18 Dec 2018

Bauman KJ (1999) Extended measures of well-being: meeting basic needs. The U.S. Census Bureau. https://www.census.gov/prod/99pubs/p70-67.pdf. Accessed 28 Mar 2019

Carr D (2019) The golden years: social inequality in late life. Russell Sage, New York

Colby SL, Ortman JM (2015) Projections of the size and the composition of the U.S. population: 2014 to 2060. The U.S. Census Bureau. https://www.census.gov//content/dam/Census/library/publications/2015/demo/p25-1143.pdf. Accessed 27 Jan 2019

Crystal S, Shea DG (1990) Cumulative advantage, cumulative disadvantage, and inequality among elderly people. Gerontologist 30(4):437–443

Cubanski J, Orgera K, Damico A, Neuman T (2018) How many seniors are living in poverty? National and states estimates under the official and supplemental poverty measures in 2016. Kaiser Family Foundation. http://files.kff.org/attachment/Data-Note-How-Many-Seniors-Are-Living-in-Poverty-National-and-State-Estimates-Under-the-Official-and-Supplemental-Poverty-Measures-in-2016. Accessed 19 Jan 2019

Duncan GJ, Gustafsson B, Hauser R, Schmaus G, Jenkins S, Messinger H et al (1995) Poverty and social-assistance dynamics in the United States, Canada, and Europe. In: McFate K, Lawson R, Wilson WJ (eds) Poverty, inequality and the future of social policy: western states and the new world order. Russell Sage, New York

Fontenot K, Semega J, Kollar M (2018) Income and poverty in the United States: 2017. The U.S. Census Bureau. https://www.census.gov/library/publications/2018/demo/p60-263.html. Accessed 21 July 2018

Fox L (2018) The supplemental poverty measure:2017. The U.S. Census Bureau. https://www.census.gov/content/dam/Census/library/publications/2018/demo/p60-265.pdf. Accessed 21 July 2018

Hokayem C, Heggeness ML (2014) Living in near poverty in the United States: 1966–2012. The Census Bureau. https://www.census.gov/prod/2014pubs/p60-248.pdf. Accessed 25 Jan 2019

Iceland J (2005) Measuring poverty: theoretical and empirical considerations. Measurement 3(4):199–235

Iceland J (2006) Poverty in America: a handbook. University of California Press, Berkeley

Klein E (2013) Feds spend $7 on elderly for every $1 on kids. The Washington Post. https://www.washingtonpost.com/news/wonk/wp/2013/02/15/feds-spend-7-on-elderly-for-every-1-on-kids/?noredirect=on&utm_term=.5186f940971a. Accessed 25 Jan 2019

Lanzieri G (2011) Fewer, older and multicultural? Projections of the EU populations by foreign/national background. Eurostat European Commission. https://ec.europa.eu/eurostat/documents/3888793/5850217/KS-RA-11-019-EN.PDF/0345b180-b869-4cb0-907b-d755b699a369. Accessed 27 Mar 2019

Mather M, Jacobsen LA, Pollard KM (2015) Aging in the United States. Population Reference Bureau. https://www.prb.org/wp-content/uploads/2016/01/aging-us-population-bulletin-1.pdf. 21 July 2018

McGarry K, Schoeni R (2000) Social security, economic growth, and the rise in elderly widows' independence in the twentieth century. Demography 37(2):221–236

Newman KS (2003) A different shade of gray midlife and beyond in the inner city. The New Press, New York

Romig K (2018) Social security lifts more Americans above poverty than any other program. Center on Budget and Policy Priorities. https://www.cbpp.org/research/social-security/social-security-lifts-more-americans-above-poverty-than-any-other-program. Accessed 25 Jan 2019

Sevak P, Schmidt L (2014) Perspectives: immigrants and retirement resources. Soc Secur Bull 74(1):27

Short K (2013) The research supplemental poverty measure: 2012. The U.S. Census Bureau. https://www.census.gov/prod/2013pubs/p60-247.pdf. Accessed 25 Jan 2019

Smeeding T (2006) Poor people in rich nations: the United States in comparative perspective. J Econ Perspect 20(1):69–90

Smeeding T (2016) Poverty measurement. In: Brady D, Burton LM (eds) The Oxford handbook of the social science of poverty. Oxford University Press, New York, pp 21–46

Social Security Administration (2018) Fact sheet social security. Social Security Administration. https://www.ssa.gov/news/press/factsheets/basicfact-alt.pdf. Accessed 15 Dec 2018

Stepler R (2016) Smaller share of women ages 65 and older are living alone. Pew Research Center. http://www.pewsocialtrends.org/2016/02/18/smaller-share-of-women-ages-65-and-older-are-living-alone/. Accessed 18 June 2018

The U.S. Census Bureau (2017) The history of the official poverty measure https://www.census.gov/topics/income-poverty/poverty/about/history-of-the-poverty-measure.html. Accessed 3 Dec 2018

Townsend P (1979) Poverty in the United Kingdom. Penguin, Harmondsworth

West LA, Cole S, Goodkind D, He W (2014) 65 + in the United States:2010. The U.S. Census Bureau.https://www.census.gov/content/dam/Census/library/publications/2014/demo/p23-212.pdf. Accessed 3 June 2018

Women's Bureau U.S. DOL (2015) Older women workers and economic security. https://www.dol.gov/wb/resources/older_women_economic_security.pdf. Accessed 11 Dec 2018

Family Relationships

▶ Intergenerational Family Dynamics and Relationships
▶ Kinship Networks

Family Service Credit

▶ Caregiver Credits

Family Service(s)

▶ Social Work

Family Solidarity

▶ Intergenerational Solidarity

Family Studies

▶ Family Demography

Family Tree

Graziella Caselli[1] and Lucia Pozzi[2]
[1]Department of Statistical Sciences, Sapienza University of Rome, Rome, Italy
[2]Department of Economics and Business, University of Sassari, Sassari, Italy

Synonyms

Family history; Genealogies

Definition

A family tree is a diagram used in genealogical studies to represent individuals belonging to a family, showing their ties to one other over many generations (Fig. 1). The individuals are organized into a "tree," choosing one index individual and working backwards in time to lineal ancestors, and/or a lineage working forwards in time from an ancestor to descendants, together with collateral lines (Durie 2017). The family ties between individuals derive from individual-level demographic events, which provide the most relevant information included in the diagram, with the dates and often the place of birth, marriage(s), and death (Smith and Mineau 2003).

Overview

The practice of reconstructing genealogies and family trees goes back to a very ancient past: family trees carved on the walls of ancient Egyptian temples are conserved and genealogies are included in the Bible. The oldest conserved western exemplars are included in the copies of an ancient codex, *Commentary on the Apocalypse*, illustrated by a monk, Beatus of Liébana, in 776. However, only in the first decades of the sixteenth century, with the *Medici family tree*, conserved in Florence, did the trees assume their modern structure, with the progenitor at the top and the descendants branching out vertically down the tree (Klapisch-Zuber 2003).

Over the centuries, the trees have progressed from oral history, to become written texts and then diagrams, the changes in their form corresponding to the cognitive needs of an increasing number of disciplines (Mitchel 2014). Born as purely genealogical tools, nowadays, they address research questions in many fields (anthropology, family history, genetics, demography, etc.) and serve various purposes. One of the main advantages of genealogical data, from a research perspective, derives from the combined information being presented in a format of family relations and descendance, which is extremely useful in the study of the dynamic of population change over various generations (Gellatly 2015; Kaplanis et al. 2018), and it is crucial for studying the family determinants of longevity (Beeton and Pearson 1901; Pearl 1931).

Key Research Findings

In recent decades, there has been a proliferation of studies concerned with the familial components of human longevity. Many of them refer to the historical archives, such as Utah (Kerber et al. 2001), Arles (Robine and Allard 1997), and/or Quebec genealogies (Philippe 1980; Desjardins and Charbonneau 1990). Today, the number of people reaching extreme ages is increasing in developed societies. This fast-growing segment of the population has inspired research into the role played by genetics in longevity. In particular, extreme longevity in a centenarian's family history supports the inheritability of the oldest old longevity (Perls et al. 2000). This would imply increased homogeneity among the oldest persons and could depend on a

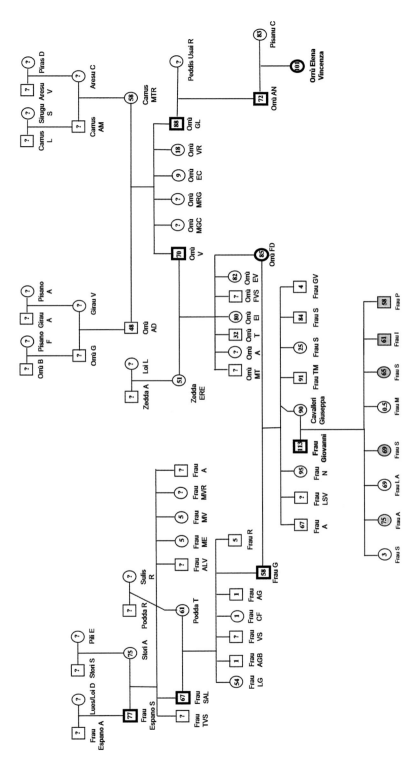

Family Tree, Fig. 1 Family tree of Giovanni Efisio Emanuele Frau (born 19.12.1890–died 19.6.2003). The number inside the symbols (circle for women and square for men) represents the age in years at death (white symbols) and the age reached (gray symbols). Figure published in Caselli et al. 2006. (Note: Reprinted from Experimental Gerontology, vol 41/June 2006, Fig. 2, authors: Graziella Caselli, Lucia Pozzi, James W. Vaupel, Luca Deiana, Gianni Pes, Ciriaco Carru, Claudio Franceschi, Giovannella Baggio: Family clustering in Sardinian Longevity: A genealogical approach, Pages 727–736, Copyright (2006), with permission from Elsevier, Order N° 4542590933171, 5 March 2019, Licensee: University of Sassari)

shared genetic factor or factors (Abbott et al. 1974; Bouquet-Appel and Jakobi 1991). However, the clustering of exceptional survival within families suggests a familial component affecting mortality differences, especially at extreme old ages, but of course, not all familial effects are genetic: the transmission of risky behaviors and/or life style is also important (McGue et al. 1993; Christensen and Herskind 2006).

Demography could help to provide deeper insights into the role inheritability plays in longevity, even in the absence of genetic variables. To this end, validated data are needed to reconstruct the family histories of the older person and, in this instance, of centenarians. With these data, an analysis of genealogies and family life histories – through the family trees – can be performed, using features from individual trajectories to reconstruct the life history of each sibling (Robine and Allard 1997; Cournil and Kirkwood 2001; Motta et al. 2007).

Examples of Application

Combining extensive information contained in the Civil Registers, and in the Parish Registers since 1861, Caselli et al. (2006) reconstructed the family trees of two Sardinian people aged 112 years. As an example, the family tree of Giovanni Efisio Emanuele Frau is presented in Fig. 1 and it shows the age at death of each of the "relatives" for whom validated information is available. An important aspect emerging from the Sardinian Centenarians Study (Poulain et al. 2006) is that longevity was found among the ascendants of a particular branch of the family, and in most cases, along the maternal line, as in other studies (Jalavisto 1951; Abbott et al. 1974; Bouquet-Appel and Jakobi 1991; Caselli et al. 2013).

Future Directions of Research

The analysis of family trees over multiple generations, combined with individual-level demographic databases, could favor a deeper understanding of late-in-life survival, identifying a more adequate definition of longevity and taking into account other essential aspects like birth order, number of children, socioeconomic status and geographical background (van den Berg et al. 2017).

Cross-References

▶ Beanpole Family Structure
▶ Clan Families
▶ Family Formation and Dissolution
▶ Intergenerational Family Structures
▶ Multigenerational Families

References

Abbott MH, Murphy EA, Bolling DR, Abbey H (1974) The familial component in Longevity. A study of offspring of nonagenarians. II. Preliminary analysis of the complete study. Hopkins Med J 134(1):1–16

Beeton M, Pearson K (1901) On the inheritance of the duration of life, and the intensity of natural selection in men. Biometrika 1:50–89

Bouquet-Appel JP, Jakobi L (1991) La transmission familial de la longévité à Arthez d'Asson (1685–1975). Population 46(2):327–347

Caselli G, Pozzi L, Vaupel JW et al (2006) Family clustering in Sardinian longevity: A genealogical approach. Exp Gerontol 41:727–736

Caselli G, Lipsi RM, Lapucci E, Vaupel JW (2013) Exploring Sardinian longevity: women fertility and parental transmission of longevity. In: Luy M, Caswelli G, Butz WP (eds) Determinants of unusual and differential longevity. Vienna Yearbook of Population Research, pp 247–266

Christensen C, Herskind AM (2006) Genetic factors associated with individual life duration: Heritability. In: Robine J-M, Crimmins EM, Horiuchi S, Yi Z (eds) Human longevity. Individual life duration, and the growth of the oldest-old population. Springer, Dordrecht, pp 237–250

Cournil A, Kirkwood TRL (2001) If you would live long, choose your parents well. Trends Genet 17(5): 233–235

Desjardins B, Charbonneau H (1990) L'héritabilité de la longèvité. Population 4–5:755–792

Durie B (2017) What is genealogy? Philosophy, education, motivations and future prospects. Genealogy 1:4. https://doi.org/10.3390/genealogy1010004

Gellatly C (2015) Reconstructing historical populations from genealogical data file. In: Blothooft G, Christen P, Mandemakers K, Schraagen M (eds) Population reconstruction. Springer, pp 111–128

Jalavisto E (1951) Inheritance of longevity according to Finnish and Swedish genealogies. Ann Med Exp Biol Fenn 40:263–274

Kaplanis J, Gordon A et al (2018) Quantitative analysis of population-scale family trees with millions of relatives. Science 360(6385):171–175. https://doi.org/10.1126/science.aam9309

Kerber RA, O'Brien E, Smith KR, Cawthon RM (2001) Familial excess longevity in Utah genealogies. J Gerontol A Biol Sci Med Sci 56(3):B130–B139

Klapisch-Zuber C (2003) L'Arbre des Familles. Editions de La Martinière, Paris

McGue MJ, Vaupel JW, Holm N, Harvald B (1993) Longevity is moderable heritable in a sample of Danish twins born 1870–1880. J Gerontol 48:B237–B244

Mitchel M (2014) Fitting issues: the visual representation of time in family tree diagrams. Sign Systems Studies 42(2–3):241–280. ISSN 1406-4243

Motta M, Malaguarnera M et al (2007) Genealogy of centenarians and their relatives: a study of 12 families. Arch Gerontol Geriatr 45(1):97–102. Online 2006. https://doi.org/10.1016/j.archger.2006.10.004

Pearl R (1931) Studies in human longevity. IV. The inheritance of longevity. Preliminari report. Hum Biol 1468:209–212

Perls T, Shea-Drinkwater M, Bowen-Flynn J et al (2000) Exceptional familial clustering for extreme longevity in humans. J Am Geriatr Soc 48(11):1483–1485

Philippe P (1980) Longevity: some familial correlates. Soc Biol 27(3):211–219

Poulain M, Pes GM et al (2006) The validation of exceptional male longevity in Sardinia. In: Robine J-M, Crimmins EM, Horiuchi S, Yi Z (eds) Human longevity. Individual life duration, and the growth of the oldest-old population. Springer, pp 147–166

Robine J-M, Allard M (1997) Towards a genealogical epidemiology of longevity. In: Robine J-M, Vaupel JW, Jeune B, Allard M (eds) Longevity: to the limits and beyond. Springer, pp 121–129

Smith KS, Mineau GP (2003) Genealogies in demographic research. In: Demeny P, McNicchol G (eds) Encyclopedia of population. Macmillan, New York, pp 448–451

van den Berg N, Beekman M, Smith KR et al (2017) Historical demography and longevity genetics: back to the future. Aging Res Rev. https://doi.org/10.1016/j.arr.2017.06.005

Fasting Mimicking Diet

▶ Slow-Ageing Diets

Fear of Aging

▶ Anxiety About Aging

Fear of Death

▶ Anxiety About Aging
▶ Death Anxiety

Federal Old-Age Security Program

▶ Social Security in the United States

Federal Old-Age, Survivors, and Disability Insurance Program

▶ Social Security in the United States

Fee-for-Service

Yu Guo and Heminxuan Wei
Renmin University of China, Beijing, China

Synonyms

Care costs; Cost-based payment; Health service payments

Definition

Fee-for-service (FFS) is a traditional and widely used payment model for healthcare where services are unbundled and paid for separately. It is a retrospective cost-based payment. The specific method is that the insurance institution reimburses according to the patient's cost of the medical report reported by the institution or doctor. The reimbursement depends on the price and quantity of each service item they provided.

Overview

Regulating FFS payment is more feasible and straightforward to administer than other payment models since it requires less complex mechanisms to ensure appropriate function. So physicians have historically been paid FFS, and it continues to be the dominant method in most countries. In FFS, individual items of care are reimbursed retrospectively, but excessive services and unnecessary or inappropriate care may be encouraged. The drawbacks include up-coding, overtreatment, and excessive readmissions.

Key Research Findings

The early history of the fee-for-service reimbursement system can be traced back to the Title XVIII and Title XIX of the Social Security Act which established the Medicare and Medicaid programs in 1965. This is the foundation of the fee-for-service reimbursement system, where services are paid for separately (Montgomery 2018).

Research has shown that FFS (FFS) has an impact on healthcare behavior and is different from other payment systems; however, evidence of the advantages and disadvantages of FFS is mixed. Joseph et al. (1983) used two sources of data that were available for actual claim costs per month, as well as various client characteristics, to get the conclusion that paying pharmacies on a capitation basis reduced drug prescription costs compared to spending them on an FFS basis. However, it has been hypothesized that capitation payment reduces costs but also lowers quality of care compared with FFS (Gosden et al. 1999). Most scholars believe that there is the incentive to deliver more care to inflate income under FFS payment. This can lead to supplier-induced demand where the patient receives more care than they would have chosen if they had the required knowledge (Gosden et al. 2011).

During the last few years, scholars had reached the conclusion that, rather than to regard FFS payment as an evil practice, it would be more practical and preferable to co-opt physicians into developing a fee schedule that defines and sets the fee of each item. Imposing and monitoring adherence to the regulations can mitigate not only inappropriate volume expansion but also assure basic quality standards be met. It should also facilitate the control of extra billing and balance billing by explicitly setting the terms for billing publicly financed services (Ikegami 2015a). Regulating FFS payment is better than introducing other payment models. For example, electronic medical record would make gaming easier even though it would improve the quality of care and increase efficiency (Ikegami 2015b).

Examples of Application

The FFS payment system is the prevailing method of payment in Chinese public hospitals. The retrospective payment system, which reimburses hospitals based on clinic visits, examinations, and treatment programs, is feasible and simple to administer. However, improper incentives as part of China's dominant FFS payment model are largely responsible for the rising costs of healthcare. Under the Chinese FFS system, the government controls the pricing of medical services, so the prices for advanced care and drugs were set higher than their actual cost, while the rates for primary care were set lower. For hospitals to obtain their funds, physicians were encouraged to prescribe expensive and profitable medications or diagnostic tests that were not always beneficial to patients. Consequently, overtreatment and overprescription caused by the FFS payment system were widespread in China, leading to rising medical costs. In order to contain the continuing growth of health expenditures, the Chinese government was called upon by the World Bank to convert its health system from a purely FFS system to a mixed payment system as early as 1997 (Zhao et al. 2018).

Future Directions for Research

Reimbursement for healthcare has utilized a variety of payment mechanisms with varying degrees of effectiveness. No single method of provider payment is perfect. Each has its own strength and weakness and can induce unintended behavior. Rigorous evidence-based assessment emphasizing the quality of medical care and outcomes and measures to dissociate care providers' profit motives from the incentives of physicians they employ are essential and need to be promptly implemented. Moreover, professional ethics and norms in medicine should be reinstituted to ensure the lasting social benefits of payment reform. Whether these mechanisms are used singly or in combination, it is imperative that the resulting systems remunerate on the basis of the quantity, complexity, and quality of care provided. Expanding the role of the electronic medical record (EMR) to monitor provider practice, patient responsiveness, and functioning of the healthcare organization has the potential to not only enhance the accuracy and efficiency of reimbursement mechanisms but also to improve the quality of medical care (Ginsburg 2012; Ikegami 2015b).

Summary

FFS insurance provides a subscriber with indemnification for expenses incurred in obtaining health services in the open market, and it has been widely used in healthcare systems worldwide. It creates purchasing power without any direct effect on the availability of services; thus, demand may be increased without increasing the supply of services. In FFS, individual items of care are reimbursed retrospectively, but excessive services and unnecessary or inappropriate care may be encouraged. If the government wants to improve the quality of healthcare and the efficiency of the expenditure, it is necessary to regulate the FFS system.

Cross-References

- Adult Day Services
- Adult Protective Services
- Care Coordination
- Care Management
- Faith-Based Social Services
- Formal and Informal Care
- Health Insurance
- Healthcare Utilization
- Home- and Community-Based Services
- Legal Services
- Medical Ethics
- Mental Health Services
- Personal Assistant Services
- Personal Care
- Preventive Care
- Quality of Care
- Respite Care
- Skilled Care
- Social Services Utilization
- Social Work

References

Ginsburg P (2012) Fee-for-service will remain a feature of major payment reforms, requiring more changes in Medicare physician payment. Health Aff 31(9):1977: Payment reform to achieve better heal care. https://doi.org/10.1377/hlthaff.2012.0350

Gosden T, Pedersen L, Torgerson D (1999) How should we pay doctors? A systematic review of salary payments and their effect on doctor behaviour. Q J Med 92(1):47–55. [PUBMED: 10209672]

Gosden T, Forland F, Kristiansen I, Sutton M, Leese B, Giuffrid A, Sergison M, Pedersen L (2011) Capitation, salary, fee-for-service and mixed systems of payment: effects on the behaviour of primary care physicians. Cochrane Database Syst Rev (3). https://doi.org/10.1002/14651858.CD002215

Ikegami N (2015a) Fee-for-service payment–an evil practice that must be stamped out? Int J Health Policy Manag 4(2):57. https://doi.org/10.15171/ijhpm.2015.26

Ikegami N (2015b) In defense of regulated fee-for-service payment: a response to recent commentaries. Int J Health Policy Manag 4(9):635. https://doi.org/10.15171/ijhpm.2015.131

Joseph H, Burmeister L, Fisher W, Lipson DP, Norwood GJ, Standridge CR, Yesalis CE (1983) Pharmacy costs: capitation versus fee-for-service. Q J Bus Econ 22:41–51

Montgomery M (2018) The origin of fee-for-service. American College of Cardiology. Retrieved from https://www.acc.org/membership/sections-and-councils/cardiovascular-management-section/section-updates/2018/07/10/14/42/the-origin-of-fee-for-service

Zhao C, Wang C, Shen C, Wang Q (2018) Diagnosis-related group (DRG)-based case-mix funding system, a promising alternative for fee for service payment in China. Biosci Trends 12(2):109–115. https://doi.org/10.5582/bst.2017.01289

Feeling

▶ Emotion

Felt Age

▶ Subjective Age

Female Stars

▶ Representations of Older Women and White Hegemony

Feminism and Aging in Literature

Roberta Maierhofer
Center for Inter-American Studies, University of Graz, Graz, Austria

Synonyms

Cultural representations; Gender and age/aging; Intersectionality

Definition

Since the 1980s, scholars in the field of humanistic gerontology have turned to cultural manifestations to investigate ideas about the meaning of age and aging and discuss models presented in literature, art, and film. This research is based on an awareness that our understanding of aging has considerable implications for medical ethics and practice, psychotherapy and the education of older people, research on the biology of aging and the prolongation of human life, public policy, and religious and spiritual life. In a 1986 anthology of essays, Thomas R. Cole demanded in his introduction entitled "The Tattered Web of Cultural Meanings" an awareness of the importance of cultural images of aging. Numerous practical decisions in professional ethics and public policy rely on the cultural and theoretical meanings we give to aging (Cole and Gadow 1986, p3f.). Therefore, literary texts and literary studies fulfill an essential role in gerontology, even if literary analysis has been more hesitant to incorporate the aspect of age into its scholarship. However, US-American literary criticism has been an influential force in the humanities and arts through its method of identifying cultural manifestations in the matrix of "race, class, and gender" and questioning the established literary canon of the time (Bercovitch 1994, p1f.). By identifying the traditional canon of literary texts as an expression of power and thus of inclusion and exclusion, texts were reread as providing a new focus for inter- and cross-disciplinary investigation. Gender studies, ethnic studies, and popular culture studies, for example, drew on literary texts to initiate a discussion of social, political, economic, and cultural marginalization and demand recognition of diversity and inclusion of previously marginalized groups. This development was theoretically fueled by postmodernism and poststructuralism offering an understanding of the world rendered through cultural representations and positioning literary texts and literary studies as central in terms of understanding the world as social and historical narratives that are culturally constructed.

While first-wave feminism focused on legal aspects of gender inequality, since the 1960s second-wave feminism expanded the discussion by offering an understanding of detailed mechanisms of discrimination and domination, such as rape and violence, but also institutionalized gender disparities in terms of workplace discrimination and unequal access to education. With the slogan "the personal is political" or "the private is political" (Hanisch 1970), personal experiences were now understood as determined within social, political, and economic contexts. Most importantly, Simone de Beauvoir's pivotal work *The Second Sex* (1949) laid the groundwork for distinguishing between biology (sex) and the cultural associations with that biology (gender) that create a patriarchal system privileging on the basis of sex, i.e., a male-centered ideology as the norm that defines the female as "the other." By defining the feminist cause as a movement and a theoretical approach, it allowed for adaptions including any form of oppression and an understanding of the cultural construction of all identity markers. In 1989, Kimberlé Williams Crenshaw's (1999) term "intersectionality" referred to the interconnectedness of power based on different forms of social categories, thus expanding the previously used slogan of "race, class, and gender." Although introduced during the second wave of feminism, it only later became the term commonly used to express all forms of oppression. In "Becoming the Third Wave" (1992), Rebecca Walker defined being a feminist as integrating an ideology of equality and female empowerment into every fiber of one's life with the demand to turn theoretical outrage into political power and activism by becoming more socially and politically involved. Within the third and fourth waves of feminism, all forms of social role stereotypes were addressed and diversity and individual difference were of central concern. Feminism has had a long tradition of interpreting literary texts as cultural manifestations that both reinforce and document patriarchal society, but also provide a platform for strong but neglected female voices of resistance and subversion. In the feminist tradition of Judith Fetterley's concept of the "resisting reader" and her feminist approach to American fiction (1977), feminist criticism set out to change the world by reassessing the past in order to stimulate a re-vision of fiction. Critical interpretations of the established literary canon as well as the incorporation of previously neglected and forgotten texts into the so-called new canon were one of the important goals of feminist literary scholarship in the 1980s and 1990s. In the tradition of critical theory, literary critics in the field of aging have continued this feminist tradition and have been performing a political act with the aim not simply to interpret the world but to change it by changing the consciousness of those who read and their relationship to what they read. Therefore, literary texts are still at the center of any discussion of the categories of gender as a basis of establishing implicit social, political, and cultural meanings of what age and aging mean.

Overview

The rise of age as a cultural category in the 1990s has very much been indebted to the establishment of gender as a category of analysis in literary and cultural studies in the decade before. Age/aging studies would not have been established as a field without the theoretical and methodological approaches established through feminist theory. Susan Sontag was the first to address the intersection of gender and age, when first in an article in 1972 and then at a conference of the Institute of Gerontology in 1973 she pointed to the "double standard of aging" as applied to men and women and distinguished between old age and growing older. Sontag (1972, p31) defined old age as "a genuine ordeal, one that men and women undergo in a similar way" and growing older as "an ordeal of the imagination – a moral disease, a pathology – intrinsic to which is the fact that it afflicts women much more than men" (Sontag 1972, p31f.). Thus, men reach old age, women grow old, or more precisely they grow "older." In the feminist tradition, Sontag had early on identified aging as a social judgment of women rather than a biological eventuality ordained by the way society limits how women feel free to imagine themselves. It is therefore the narrative (social judgment and imagination) that expresses social boundaries that limit women, not the fact of aging as such. As youth is seen as a metaphor for energy, restless mobility, and the general state of "wanting," all attributes that have traditionally been linked to "masculinity" and age have been associated with incompetence, helplessness, passivity, noncompetitiveness, and being nice, qualities that have stereotypically been defined as "feminine," and agism – the systematic stereotyping of and discrimination against people because they are old – reinscribes women in cultural definitions on the assumption that appearance creates identity. This is part of the ideology which positions women and age as "other" against masculinity and youth, which is regarded by contemporary Western society as the norm of human behavior (Woodward 1991). Second-wave feminists, such as Germaine Greer, have thus demanded an inclusion of the female and old through a recognition of menopause as a feminist rite of passage of empowerment (Greer 1992, p32f). Since the beginning of the 1990s, anocriticism has been the term used to describe the application of Susan Sontag's approach of linking theories of gender and age in search of a specific culture of aging. In the tradition of Elaine Showalter's gynocriticism (1977, 1985) – a study of women writers and of the history, styles, themes, genres, and structures of writing by women – and Germaine Greer's use of the Latin word "anus," old woman, to create the term "anophobia" (1992, p4) to describe the fear of old women, anocriticism defines a method to trace the aspect of aging in cultural representations, the stories we tell ourselves, in order to generate understanding of what it means, in Margaret Morganroth Gullette's (2004) term, to be "aged by culture." As feminist theory distinguishes between sex and gender, a distinction has been made between chronological age and the cultural stereotypes associated with old people, which allows an escape from the confining binary opposition of young and old. Starting with the premise that age – similar to race, class, and gender – does not flow naturally or inevitably from the individual's anatomical body, literary scholarship analyzes the way age identity is constructed in literature and thus in society, for both young and old. By determining in what way "youth" and "age" come to have certain meanings at a particular place and time, and stressing the necessary interrelatedness of these meanings, an understanding can be reached that what is considered typically "young" in a given society depends in part on being different from what is "old" and what is "old" on not being "young." This understanding is based on a feminist approach and leads to the conclusion that what is considered age-neutral, i.e., "universal," is implicitly often male and young and exclusive of the female and old (Maierhofer 1995, 2003).

However, this feminist position concerning age/aging only slowly developed. Although feminist theory became increasingly sensitive to the ways in which gender, race, ethnicity, and class structure affect the lives of individuals, up to the late 1980s, it had virtually ignored age. In the opening lecture of the annual meeting of the National Women's Studies Association in 1985, Barbara Macdonald and Rich (1983) angrily points out that "agism," which she defines as a central feminist issue, has its basis in the patriarchal system. Having reached a certain age, she felt that the message of the women's movement has been that women over 60 were not part of the proclaimed "sisterhood." That the younger women in the movement saw older women the way men do – as "women who used to be women but aren't any more" (Macdonald and Rich 1983, p6), and by ignoring their present lives and not identifying with their issues, they stereotype, exploit, and patronize them. When younger women define their relationship to older women in family relational terms, comparing them to their grandmothers, mothers, or aunts, they are guided by "the man in [their] own heads" (Macdonald and Rich 1983, p11); they ignore their individuality relying on a man's definition of motherhood, which infantilizes younger women and erases older. Already in 1983, Macdonald called for a feminist approach toward agism that goes beyond defining old women as needy, simpleminded, and helpless exploiting them thus for the needs of others. By reducing older women to witnesses of the past with no interest in their present struggles, feminism supports a social system that negates women generally, but is particularly obvious in academia and life history research projects. Macdonald's angry critique is that within the feminist movement, young women do not treat the old equally, "You come to old women who have been serving young women for a lifetime and ask to be served one more time" (Macdonald and Rich 1983, p8). The feminist historian Lois Banner explains the neglect of the subject of women and aging in her book *In Full Flower: Aging Women, Power, and Sexuality; A History* in terms of the scholars in the field as "young women, working out the dynamics of their own lives in their studies". The topic of aging therefore remains obscure, as the older generation is perceived as antifeminist and young women define their struggle in rebellion and their "own youth as the real reality" (Banner 1992, p6).

Key Research Findings

With the graying of feminists in the United States, however, age becomes a concern. An updated version of *Our Bodies, Ourselves* with the title *Ourselves, Growing Older* at the beginning of the 1990s, special issues of such magazines as *Feminist Studies* on women and aging, and books on menopause and old age by prominent second-wave feminists such as *The Change* (1992) by Germaine Greer and Betty Friedan's *The Fountain of Age* (1993) are only a few examples of this newly discovered interest in women and aging at the time. One good example of such changes in attitude was the reception of Simone de Beauvoir's study on age and aging published in 1970 as *La Vieillesse – The Coming of Age* – that moved from a very negative to a very positive evaluation of her work. While humanistic gerontologists recognized early on the importance of Beauvoir's contribution to the discussion of old age and aging, prominent second-wave feminist critics writing in the early 1990s on the topic of women and aging showed very little appreciation for Beauvoir's work and rejected it as too pessimistic, agist, and sexist. Thus, Kathleen Woodward (1999) speaks of the fact that the book, which she refers to as "encyclopedic" and its breadth "stunning," has virtually been ignored by intellectuals and academics in the United States, even those interested in the work of Beauvoir. The first American edition of 1972 went out of print in the early 1980s, but the 1996 Norton paperback edition reflects the newly arisen feminist interest in the subject of aging. The lack of interest in *The Coming of Age* in the 1970s and 1980s reflects the unwillingness of feminists to deal with the topic of age in the women's movement of the time. Whereas Greer (1992) demanded an acceptance of old age as a time of calm, detachment, and tranquility and defined the serene old woman not as a continuation of the submissiveness of the feminine woman, but as a statement of an authentic and autonomous individual, Beauvoir's ferocious demands were criticized. Beauvoir's anger and rage were to Greer evidence that Beauvoir insisted to such a large extent on living through men and never looked to solitude as a means to freedom in old age. But feminism has proven that rage is a powerful force, and speaking out against oppression and breaking the silence is a method of achieving change. When Beauvoir offered a bitter indictment of Western culture for its "criminal" treatment of the elderly, she can be seen as using the method of "ranting" to rail against the atrocities old people endure and to protest against the silencing of all those who are "other" to patriarchy. In the late 1990s, feminist scholars finally integrated Beauvoir's pessimistic stance on women and aging and recognized the importance of her position for age studies in more general terms. Barbara Frey Waxman asserts in the introduction to her book *To Live in the Center of the Moment: Literary Autobiographies of Aging* (1997) that it is no longer important to discuss the negative aspects of Beauvoir's work, but to simply acknowledge her contribution to a heightened understanding of what it means to be woman and old. It is Beauvoir's notion of recognizing the self in the other, however, that informs Waxman's intention to express concerns, values, and feelings that are of importance to all generations reinforcing again Sontag's idea of the imagination and of narrative that was also expressed in the anocritical approach (Maierhofer 2004).

Literary texts exploring the interconnectedness of gender, ethnicity, and race provided earlier examples of narrating different identity categories. In 1982, Adrienne Rich uses the term "split at the root" when referring to the concept of an ethnic author. Looking back on her suppressed Jewish past, Rich recognizes the denial of her own ethnicity and talks about her disappointment in not having learned about "resistance, only about passing" (Rich 1986, p107). Realizing that she can neither be "them" nor "us," that the self and other meet, she feels not a unified whole, but a person "split at the root" (Rich 1986, p100f). Accepting her identity as perpetually in flux, Rich demands resistance and an acceptance of the many facets of one's identity and links this to a "moving into accountability" (Rich 1986, p123). In the course of accounting for her various identities, such as being Jewish and lesbian, politicized by the civil rights movement and the feminist movement of the 1960s, she also refers to the aspect of age: "The woman limping with a cane, the woman who has stopped bleeding are also accountable" (Rich 1986, p123). Maya Angelou's (1994, p170) poem "On Aging" can offer another example of how the feminist movement provided the basis for a recognition of age as an identity marker. Although the speaker has to deal with the limitations of her body due to her age – such as stiff and aching bones, "a little less hair, a little less chin, a lot less lungs and much less wind" – she angrily rejects condescending pity. The speaker finally asks for "understanding if you got it" – referring to an understanding that allows for a recognition of the self in the other, a knowledge that expectations of another person's behavior based on race, class, gender, and age limit and belittle that person: "Don't study and get it wrong" (Angelou 1994 p170). By stating "I am the same person I was back then" (Angelou 1994, p170), she is both asserting self-confidently her own identity as being both self and other, and offering a basis for identification not only of the old person with the young but for the young person with the old, and deconstructing binaries on various levels. Paule Marshall's novel *Praisesong for the Widow* (1983) also positions the female protagonist at the center; a movement from silence to narration, from negating to accepting one's own feelings and needs, and reevaluating a past life lived according to the norm are shown as conscious acts of aging.

In *Images of Women in Fiction: Feminist Perspectives* (1972), one of the first feminist contributions to academic literary scholarship, Susan Koppelman Cornillon writes that "people are beginning to see literature in a new 'perspective,' as the subtitle indicates. Feminist literary scholarship in the late 1970s turned from analyzing female characters in male texts according to male experience and instead chose female experiences and perspectives. In 1978, Nina Baym criticized American literary scholarship for having a bias in favor of things male: whaling ships rather than sewing circles as a symbol of the human community. Baym (1992, p20) speaks of literary critics as "displaying an exquisite compassion for the crises of the adolescent male, but altogether impatient with the parallel crises of the female." Whereas the second wave of feminism in the 1970s and 1980s emphasized issues associated with the earlier years in life, with "the graying of American feminism," it is no longer adolescence but old and middle age that are a concern to feminists, and age has been introduced as a social and cultural marker. Feminist literary criticism reclaimed the female hero of traditional literature and reinterpreted her in the light of feminist analysis. In the 1981 Pearson and Pope study of *The Female Hero in American and British Literature*, the main emphasis of the investigation is on youth, but the text acknowledges the possibility of age as a decisive factor of the heroic: "In many cases, women begin new lives in old age" (Pearson and Pope 1981, p8). The recognition of life as a process led to an acknowledgment of aging as a narrated individual experience. At the end of the 1980s and beginning of the 1990s, early literary studies of age and aging in literature, therefore, emphasized the matrix of time and experience. In her part memoir, autobiography, recollection, and analysis of literary texts, Margaret Morganroth Gullette (1988), for example, uses the term "midlife progress novel," and Barbara Frey Waxman in *A Feminist Study of Aging in Contemporary Literature* (1990) talks of the new fictional genre of the "Reifungsroman or Fiction of Ripening" and links growing old to a journey "from the hearth to the open road." Feminist scholars such as Margaret Cruikshank (2003) have referred to literary texts as informing their scholarship on aging, which offers a way of discussing aging as individual experiences within specific collective social, political, and cultural contexts. Cultural representations, especially fiction and film, challenge the traditional notion of life as a decline narrative, and female characters provide an awareness of the social and cultural dimensions of aging (Maierhofer 1995, 2003; Chivers 2003; Brennan 2004; Worsfold 2005, Marshall 2006, 2009).

Future Directions of Research

Academic traditions demand a constant framing of research in terms of a narrative of progress. Innovation is based on the aspect of the new and overcoming the old. Therefore, literary gerontology is caught up in discovering unchartered territory and claiming unknown academic land for exploration. However, feminist scholarship has shown that it is not the new, but the hidden, the marginalized that contributes to the field. The focus so far has been on scholarship from English language settings. This needs to be expanded in terms of languages and academic cultures. Feminist scholarship aware of the interconnectedness of the private and the political will be reflecting their own embodied aging in terms of the abstract and the concrete, the material and the immaterial, and intergenerational connections of solidarity and support. Cultural representations of material realities will be the basis of such discussions.

Summary

In "The Coming of Age of Literary Gerontology" (1990), Anne M. Wyatt-Brown optimistically declared the possibility of bridging the gap between gerontology and literary criticism, and after the initial phase of literary gerontology in the 1990s, scholarship at the beginning of the twenty-first century provided new points of departure for literary analysis, where aging characters in their own rights were discussed as portrayals of human existence incorporating all stages of life and not merely as a point of reference for the young playing inspirational or redemptive roles by rescuing young protagonists from indecision or despair. Feminist scholarship has provided a recognition of the inherent ambivalence of life and its representations in literary texts to escape the dualistic mode that has traditionally governed our thinking. In feminism, identity has been discussed both as the possibilities and as the restrictions of the individual through social structures. The often quoted "re-vision – the act of looking back, of seeing with fresh eyes, of entering an old text from a new critical direction" (Rich 1972, p18) has led to a new awareness of our own identity and the aging process. Literature has produced a wealth of texts that offer in Fetterly's (1977) resisting reader terms new ways of imagining and narrating the life course. Literary texts exemplify that contrary to popular conceptions of old age as a distinct period in life, old people themselves "narrate" with creativity and imagination their own life course in terms of continuity and change. When Le Guin (1992) speaks of starting to pretend to be an old woman as she ages, she is calling for a recognition of the constructedness of social and cultural categories and more importantly for an understanding of the power of imagination, narration, and creativity and a recognition of literature as a means of resistance and subversion.

In "A Declaration of Grievances" (2017), Margaret Morganroth Gullette demands an end to agism by evoking the US "Declaration of Independence" (1776) by its reference to an earlier version, the "Declaration of Rights and Grievances" (1765), and positions the text in the tradition of the first wave of feminism by employing feminist acts of resistance: consciousness-raising and an imagined and thus narrated road map. Whereas the "Declaration of Sentiments and Resolutions" (1848) demanded the acknowledgment of women as right-bearing individuals, it is now the old whose rights are infringed, where discrimination and marginalization are occurring unnoticed. These declarations take on the role of a radical statement of defining (or declaring) the role of social structures (government) in relation to the rights of the individual. By reversing the hierarchy of society and individual – at least in a declared statement – a cultural narrative is established to measure government not by how well society is regulated, but by how free the individual is from social structures (Urofsky 1994). On the basis of a definition of identity that describes the subject as perpetually in flux, pursuing an illusion of wholeness and selfhood that is ultimately unattainable, the other can be seen as defining the subject because it is the ultimate signifier of everyone the subject is not. Texts in which age is identified as other lead to a repositioning of the self and an understanding of the fluidity of identity, thus representing age as both self and other and emphasizing the aspect of "we are you grown old." In terms of my anocritical approach, knowing one's possibilities as well as one's limitations is a political act of resistance.

Cross-References

▶ Aging Definition
▶ Aging Theories
▶ Critical Gerontology
▶ Feminist Theories and Later Life
▶ Gender Issues in Age Studies
▶ History of Ageism
▶ Poetry and Age
▶ Timeline of Aging Research

References

Angelou M (1994) On aging. In: the complete collected poems of Maya Angelou. Random House, New York, p 170
Banner LW (1992) In full flower: aging, women, power, and sexuality. Knopf, New York
Baym N (1992) Creating a national literature. In: Luedtke LS (ed) Making America: the society and culture of the United States. U of North Carolina P, Chapel Hill, pp 219–235
Bercovitch S (1994–95) The Cambridge history of American literature. Cambridge University Press, Cambridge
Brennan Z (2004) The older woman in recent fiction. McFarland, London
Chivers S (2003) From old woman to older women: contemporary culture and women's narrative. Ohio UP, Columbus
Cole TR, Gadow SA (eds) (1986) What does it mean to grow old? Reflections from the humanities. Duke University Press, Durham
Crenshaw KW (1999) Demarginalizing the intersection of race and sex. A black feminist critique of antidiscrimination doctrine, feminist theory and antiracist politics. In: Phillips A (ed) Feminism and politics. Oxford University Press, Oxford, pp 314–343
Cruikshank M (2003) Learning to be old. Rowman and Littlefield, New York
De Beauvoir S (1949) Le deuxième sexe. [The Second Sex] Gallimard, Paris
Fetterley J (1977) The resisting reader: a feminist approach to American fiction. Indiana University Press, Bloomington
Friedan B (1993) The fountain of age. Simon and Schuster, New York
Greer G (1992) The change: women, aging and the menopause. Knopf, New York
Gullette MM (1988) Safe at last in the middle years: the invention of the midlife progress novel: Saul Bellow, Margaret drabble, Anne Tyler, and John Updike. University of California Press, Berkley
Gullette MM (2004) Aged by culture. University of Chicago, Chicago
Gullette MM (2017) Ending ageism, or how not to shoot old people. (global perspectives on aging). Rutgers University Press, New Jersey
Hanisch C (1970) The personal is political. In: Notes from the second year: women's liberation, major writings of the radical feminists. Radical Feminism, New York. Retrieved from http://www.carolhanisch.org/CHwritings/PIP.html
Le Guin U (1992) Dancing at the edge of the world: thoughts on words, women, places. Paladin, London
Macdonald B, Rich C (1983) Look me in the eye: old women, aging and ageism. Spinster Ink, San Francisco
Maierhofer R (1995) The graying of American feminism. In: Frank T (ed) Values in American society. Eötvös Loránd University Press, Budapest, pp 113–121
Maierhofer R (2003) Salty old women: Frauen, Altern und Identität in der amerikanischen Literatur. Blaue Eule, Essen
Maierhofer R (2004) The old woman as the prototypical American – an anocritical approach to gender, age, and identity. In: Hoelbling W, Rieser K (eds) What is American? New identities in US culture. LIT, Vienna, pp 319–336
Marshall P (1983) Praisesong for the widow. Putnam, New York
Marshall L (2006) Aging: a feminist issue. NWSA J 18(1):vii–xiii
Marshall L (2009) Teaching ripening: including age when teaching the body. Transformation 19(2):55–80
Pearson C, Pope K (1981) The female hero in American and British literature. Bowker, New York
Rich A (1972) When we dead awaken: writing as re-vision. In: College English: Women, Writing and Teaching 34(1):18–30
Rich A (1986) Split at the root: an essay on Jewish identity. In: Blood, Bread, and Poetry. Selected Prose 1979–1985. Norton/New York, pp 100–123
Showalter E (1977) A literature of their own: British women novelists from Bronte to Lessing. Princeton UP, Princeton1985) (ed) New feminist criticism: essays on women, literature, and theory. Pantheon Books, New York
Sontag S (1972) The double standard of aging. Saturday Rev 55:29–38
Urofsky MI (ed) (1994) Basic readings in US democracy. USIA Washington, DC
Walker R (1992) Becoming the third wave. Ms 11(2):39–41
Waxman BF (1990) From the hearth to the open road: a feminist study of aging in contemporary literature. Greenwood, Westport

Waxman BF (1997) To live in the center of the moment: literary autobiographies of aging. University Press of Virginia, Charlottesville

Woodward K (1991) Aging and its discontents: Freud and other fictions. Indiana University Press, Bloomington

Woodward K (ed) (1999) Figuring age: women, bodies, generations. Indiana University Press, Bloomington

Worsfold BJ (ed) (2005) Women ageing through literature and experience. DEDA-LIT, Lleida

Wyatt-Brown AM (1990) The coming of age of literary gerontology. J Aging Stud 4(3):299–315

Feminist Life Course Approach

▶ Feminist Theories and Later Life

Feminist Theories and Later Life

Ann Therese Lotherington
Centre for Women's and Gender Research and Faculty of Humanities, Social Sciences and Education, UiT The Arctic University of Norway, Tromsø, Norway

Synonyms

Critical feminist gerontology; Feminist life course approach; Intersectionality

Definition

Feminist theories are critical, meaning that they explore inequality, oppression, and power dynamics to enable conceptual or social transformation. They recognize gender, race, ethnicity, class, sexual orientation, and other differentiating categories as relational organizing principles in society, and claim that privilege, dis/advantages, and oppression are effects of societal processes, not of individual choice. Age as a differentiating category has gained some momentum over the past two decades but is not commonly found in feminist theories and analysis. However, when feminist theories include age, the outcome is new understandings: Later life is analyzed as dynamically formed over time by intersections of two or more differentiating categories.

Overview

Understanding diversity and heterogeneity among old people, and diversification of the meaning of age along the lines of intersectionality, are important contributions from feminist theory to studies of later life (Hooyman et al. 2002). Intersectionality as a theoretical approach allows for inclusion in the analysis not only gender, but also other differentiating power mechanisms, such as nationality, ethnicity, race, sexuality, ability, class, and age, without privileging one. The categories are not considered isolated from each other but intertwined in different ways according to time and place. The approach aims to demonstrate ways in which differentiating categories, through their interconnectedness, might strengthen oppressive, marginalizing practices, revealing how unequal social regulations form the lives of old women, men, and trans people worldwide.

Feminist intersectional analysis might start with gender but does not a priori privilege gender, or take other structures, positions, or categories for granted (Calasanti 1999, 2004a). The approach is, in line with a feminist life course approach (see "▶ Life Course Perspective") and critical feminist gerontology (see "▶ Critical gerontology"), open to historical, political, and economic contexts, and includes in the analysis long-term effects of intersecting dimensions of power (Utz and Nordmeyer 2007; Freixas et al. 2012). Studies of later life will accordingly not take any understanding of age for granted but inquire into how the meaning of age might shift according to gender, class, race, sexuality, etc., recognizing the effects on old age of earlier dependency and discrimination. Done appropriately intersectional analyses acknowledge differences between women and avoid essentializing and universalizing, providing new, diverse, and complex interpretations of female aging. Critical feminist theories analyze life trajectories in detail, document old women's experiences, and identify potentials for emancipatory social change. The use of such theories sensitizes the researcher and enables understandings of complexity in everyday life, opposing dominating notions of old age as purely a matter of loss, frailty, and decline, but at the same time without completely denying such aspects of later life (Freixas et al. 2012; Latimer 2018). Feminist theories move beyond gendered binaries, such as decline and success, and conceptualize old age differently, for example as "affirmative old age" (Sandberg 2013: 35), accepting old age in all its diversity.

While intersectionality is open for most differentiating social categories, new material feminist theory suggests strengthening the approach with materiality. New material feminisms – in plural, as it is not one coherent, grand theory but a set of theoretical propositions within a shared-onto epistemology – have much to offer studies of later life even though these theories not so much have later life as their interest (Alaimo and Hekman 2008; Barad 2007). New material feminism is a nonrepresentational set of theories, understanding the world as ongoing intra-actions of becoming that cannot be represented, but must be inquired as they unfold. Hence, the world is continuously becoming, making the distinction between ontology and epistemology blurred, if at all possible to draw. A core point of these theoretical propositions is that simplistic binary understandings of gender and other differentiating categories, such as young and old or male and female, are condemned. No category is seen as fixed or clean, but performed or enacted (Mol 2002) differently in different entanglements with human and nonhuman creatures, artefacts, and technologies (Haraway 1996). Being entangled means to lack independent, self-contained existence. Consequently, existence is not a static, individual affair, and individuals are not rational, coherent individualized preexisting agents but emerge through and as part of the entanglements as results of intra-action (Barad 2007). Because "inter-action" refers to actions between separate agents, "intra-action" is used to underscore the intertwined understanding of the individual.

Prospects

With this approach, later life is understood as ongoing intra-actions. What age might be under what circumstances, and how later life might be lived, are not defined a priori or taken for granted. The emphases of the research are rather on how everyday life is recursively and relationally configured, how and when humans and nonhumans take part in the configurations, and what effects the configurations might have for whom. For example, neither age nor the body would be analyzed as properties pertaining to the individual but as effects of the configurations that shape everyday life. Hence, the collective rather than the individual becomes central to the analyses, and the dominating biomedical and essentializing understanding of ageing in gerontology might be avoided (Twigg 2004). The merger of feminist theory and later life approaches into critical feminist gerontology holds major potentials for the development of new understandings of later life, also into deep old age. However, further theorizing of difference and complexity in later life is required (Calasanti 2004b). New understandings of age and embodiment might fortify such theorizing processes (Sandberg 2013).

Cross-References

▶ Critical Gerontology
▶ Life Course Perspective

References

Alaimo S, Hekman S (eds) (2008) Material feminisms. Indiana University Press, Bloomington/Indianapolis
Barad K (2007) Meeting the univers halfway. Quantum physics and the entanglement of matter and meaning. Duke University Press, Durham/London
Calasanti T (1999) Feminism and gerontology: not just for women. HIJA 1(1):44–55
Calasanti T (2004a) Feminist gerontology and old men. J Gerontol B Psychol Sci Soc Sci 59(6):S305–S314. https://doi.org/10.1093/geronb/59.6.S305
Calasanti T (2004b) New directions in feminist gerontology: an introduction. J Aging Stud 18(1):1–8. https://doi.org/10.1016/j.jaging.2003.09.002

Freixas A, Luque B, Reina A (2012) Critical feminist gerontology: in the back room of research. J Women Aging 24(1):44–58. https://doi.org/10.1080/08952841.2012.638891

Haraway D (1996) Modest witness: feminist diffractions in science studies. In: Galison P, Stump DJ (eds) The disunity of science, boundaries, contexts, and power. Stanford University Press, Stanford, pp 428–445

Hooyman N, Browne C, Ray R, Richardson V (2002) Feminist gerontology and the life course. Gerontol Geriatr Educ 22(4):3–26. https://doi.org/10.1300/J021v22n04_02

Latimer J (2018) Repelling neoliberal world-making? How the ageing-dementia relation is reassembling the social. Sociol Rev 66(4):832–856. https://doi.org/10.1177/0038026118777422

Mol A (2002) The body multiple: ontology in medical practice. Duke University Press, Durham/London

Sandberg L (2013) Affirmative old age: the ageing body and feminist theories on difference. IJAL 8(1):11–40. https://doi.org/10.3384/ijal.1652-8670.12197

Twigg J (2004) The body, gender and age: feminist insights in social gerontology. J Aging Stud 18:59–73. https://doi.org/10.1016/j.jaging.2003.09.001

Utz RL, Nordmeyer K (2007) Feminism, aging, and the life course perspective. Gerontologist 47(5):705–716. https://doi.org/10.1093/geront/47.5.705

Feng Shui and Aging in Place

Miyoko Fuse
Portland Community College, Portland, OR, USA

Synonyms

Aging and homes; Aging well at home; Healthy home and aging in place

Definition

In lieu of *Stedman's Medical Dictionary for the Health Professions and Nursing* (2012, p. 625), *feng shui* is "an ancient Chinese belief system used to configure one's living or work environment to promote health, happiness, and prosperity, by enhancement of energy flow (*Chi*); includes use of space, color, and order as factors in satisfaction."

Overview

Aging in Place (see ▶ "Aging in Place")

Until recent decades, *feng shui* has primarily been practiced in Eastern cultures. Now it is gaining widespread practice outside of Eastern countries. *Feng shui* has been popular in the USA since the mid-1970s. Today, *feng shui* can be examined within the Western sciences of quantum physics and environmental psychology in conjunction with practical applications for aging in place.

Aging in place is not limited to living in one's own home but extends to alternative types of homes (World Health Organization 2015) (see ▶ "Aging in the Right Place"). The term "aging in place" is associated with health and independence to maintain older adults' quality of life by adapting the homes to their changing needs (see ▶ "Aging in Place: Maintaining Quality of Life for Older Persons at Home"). Before this happens, whether accessibilities for aging in place have been implemented or not, people personalize their homes. Aging in place should consider homes that are comfortable for the occupants' emotional well-being.

Homes affect the occupants' quality of life in terms of emotions and behaviors. *Feng shui* observes how visible and invisible components in a home environment, such as shapes and sounds, interact with each other, the space, and the people in the space, and makes the desired adjustments (Stehr 2013; Tiller 1997). *Feng shui* is a holistic approach and is a gentle, natural remedy and is affordable for all types of aging in place.

As the aging population increases rapidly (World Health Organization 2015), policy makers, researchers in environmental gerontology, and other industries of interest are probably providing information on aging in place more to professionals than the older people who need the information most. In reality, how much the term aging in place has permeated to older adults who are the center of it is unknown. Wiles et al. (2012) found that most of the older people who participated in their research were not familiar with the term. They pointed out that the word "place" is ambiguous.

The intention of aging in place is retaining independence while experiencing some degrees of difficulty for activities of daily living such as bathing, using the toilet, dressing, and cooking (see ▶ "Independent Aging"). Moreover, many houses are not built for aging in place; door thresholds may be too high, stairs do not have guardrails, hanging windows are too heavy to push up, and doors are too narrow for wheelchairs (see ▶ "Home Modification"). Thus, accessibility, safety, and practicalities are focal points for aging in place (see ▶ "Smart Homes").

Assisted living facilities are increasing (American Senior Communities 2016). These facilities are equipped with the necessary adaptations and have corrected the obstructions found in old houses. Other types of homes such as home sharing (see ▶ "Homeshare") or building an accessory dwelling unit are also alternatives (AARP 2018).

No matter the type of home, occupants make their homes comfortable by personalizing their homes with their personal interests or cultural backgrounds, probably with colors, textures, shapes, or family heirloom furniture. Aging in place is built on the individual's unique, personalized home.

Chi: The Principle of Feng Shui

Feng shui originated in China over 3,000 years ago (Mak and So 2011). *Feng shui* is a holistic approach that observes not only physical settings for accessibility, safety, and usability but also air, smell, sound, colors, and touch. Its goal is to create a harmonious environment to promote physical and mental health (Hale 2002).

The principle of *feng shui* is *Chi* (*qi*), the vibration of subtle energy (Stehr 2013; Tiller 1997). *Chi* is radiated from all matters and things by frequencies of vibration. *Chi* is vibration; therefore, it is not seen by the naked human eye. *Chi* flows everywhere in the universe, our town, home, and into our body and mind (Hale 2002; Loretta 2010).

Quantum physics seems to be in agreement with the notion of *Chi*. Spear (2010, p. 31) stated that "this invisible electromagnetic energy radiates in particular patterns from objects of all shapes and sizes." Hawkins (2012, p. 153) wrote that "everything in the universe constantly gives off an energy pattern of a specific frequency that remains for all time and can be read by those who know how." Loretta (2010, p. 12), referring to the aforementioned statement by Hawkins, describes that "energetic adjustment is the basis for why *feng shui* works and why it can have such an impact as it does on personal energy flows, our thoughts and actions, and hence, the outcomes of both our personal and professional lives."

Chi has two modalities, *yin* and *yang* (Goswami 2008; Loretta 2010). A *yin yang* duality explains a degree of harmony. *Chi* can be either positive or negative. The *yin yang* symbol, also known as the *Tai Chi* (*Tai ji*) symbol, is balanced with equal portions of black and white in the circle. *Yin* and *yang* are opposite yet complementary, such as the qualities of female and male, darkness and brightness, softness and hardness, etc. One does not exist without the opposite one. Darkness needs some light; otherwise darkness cannot be seen. In order to create harmony with *yin* and *yang* in a given situation, the portion of *yin* and *yang* is examined and an appropriate balance of them will be applied. Also, *yin* always contains *yang* and *yang* always contains *yin* within; meaning that one quality has a small portion of the other differing quality.

The five elements – Wood, Fire, Earth, Metal, and Water – are what comprise the *yin yang* duality (Hale 2002; Stehr 2013). Each element has symbolic qualities, shapes, materials, and colors, and interacts constantly, constructively, and destructively to create harmonious environments. The five elements are associated with everything in the universe, in nature, the seasons, in people, and in man-made environments (Hale 2002; Stehr 2013).

In America, except in homes that have had *feng shui* applied, *feng shui* has been seen mainly in alternative medicine clinics, spiritual studios, and hospitality services. Some general hospitals and medical science clinics have also applied *feng shui*. Evans and McCoy (1998) stated that certain types of settings, including hospitals and therapeutic facilities, create a retreat environment that may uplift the spirit and promote healing.

Some *feng shui* examples used in healthcare facilities are hallways with curved patterns, nurse stations with half-moon shaped counters, selections of colors in treatment rooms, or creating a sanctuary in lobbies and gardens. They are purposely designed for specific areas and goals such as calming, healing, happiness, peace, and prosperity.

Key Findings

Environmental psychology is a relatively new field, recognized in the late 1960s (Bonnes and Carrus 2004; Gifford 2007; Bonaiuto et al. 2010). The concept of *feng shui* overlaps with the definition of environmental psychology in architecture. Bonnes and Carrus (2004, p. 801) defined environmental psychology as the study of "the relationship between people and the socio-physical features of the built and natural environment, in order to enhance human well-being and to improve people-environment relations."

Evans and McCoy (1998) discussed that architectural design may affect human health in the form of psychological stress. They defined architectural design dimensions such as restoration, control, stimulation, coherence, and affordance, and their links to stress. Recognizing similarities between *feng shui* and environmental psychology, Bonaiuto et al. (2010) compared both and concluded that there are at least two convergences between them. Those are restoration and control. With respect to healthy emotions and *feng shui*, this entry examines restoration, control, stimulation, coherence, and affordance in terms of aging in place.

Restoration. Steg et al. (2012, p. 58) defined restoration as "an umbrella term that, within environmental psychology, refers to the experience of a psychological and/or physiological recovery process that is triggered by particular environments and environmental configurations, i.e. restorative environments." Evans and McCoy (1998) embraced retreat, fascination, and exposure to nature in the restorative dimension. Examples of these are having a space for resting the mind (Evans and McCoy 1998), seeing a burning fireplace to recuperate (Gregoire 2014), and seeing a view of nature through the window or a picture of nature to heal (Ulrich 2001).

Evans and McCoy (1998, p. 91) stated that intended restorative spaces "may uplift the human spirit and promote healing. Design may offer opportunities to combat stress by providing rest, recovery, or contemplation." For *feng shui*, restoration is one of the key components to make a harmonious environment for balanced emotions and behavior. Individuals, whether consciously or not, create their homes with their personal input. Culturally related environmental components make older adults feel connected and supported, and personal interest components make older adults feel fascinated and involved. *Feng shui* examines *Chi* in the home and adapts for a better flow of *Chi* with colors, shapes, materials, and patterns, as they associate with the five elements of Wood, Fire, Earth, Metal, and Water to create a restorative space.

Garden *feng shui* provides an emotional sanctuary to older adults who are aging in place. Garden *feng shui* uses designs with different shaped leaves, varieties in the height and sizes of bushes, and different colored flowers to harmonize *Chi* by applying the *yin yang* duality and the five elements. In addition, *feng shui* encourages animated *Chi* flow such as by attracting chirping birds or with running water. Hearing birds chirping or running water recuperates the mind. Working in the garden or seeing the garden through the window impacts the emotions of older adults. The effect of a garden on older adults was revealed by Detweiler et al. (2012, p. 101) that horticultural therapy "is a restorative technique to improve memory, attention, sense of responsibility and social interaction with few to no adverse side effects."

For older adults who may not be able to enjoy gardening, *feng shui* brings nature inside the house such as with indoor plants and a tabletop water fountain. Some indoor plants contribute to reducing off gas which is emitted from building materials or adhesives applied to furniture (Wolverton 1997). *Feng shui* for aging in place further analyzes the causes of allergies from poor air quality and works to reduce allergens such as dust mites. Indoor air quality and reducing allergens are critical for people who have asthma. The Asthma and Allergy Foundation of America (2019) has stated that asthma is a commonly found disease in people over age 65.

A tabletop water fountain or wind chime has the effect of creating positive *Chi* flow. The sounds of running water from a tabletop fountain or from a wind chime actively radiate *Chi* around it. The sounds of running water are a retreat for the mind. Sounds from a wind chime create a cheerful yet relaxing atmosphere and at the same time help the mind recuperate.

Control. Control in environmental psychology for architecture means one's control over a space. Examples are control over a door or window to open or close it, or the arrangement, temperature, and lighting of a space (Evans and McCoy 1998; Bonaiuto et al. 2010). Bonaiuto et al. (2010, p. 30) stated that "control is considered one of the most important dimensions of the people-environment relationship and is often associated with a high level of environmental satisfaction." Evans and McCoy (1998, p. 88) stated that it is "mastery or the ability to either alter the physical environment or regulate exposure to one's surroundings. Physical constraints, flexibility, responsiveness, privacy, spatial syntax, defensible space, and certain symbolic elements are key design concepts salient to control."

Probably the most well-known control factor in *feng shui* is the positioning of a bed or a desk. Over thousands of years, predecessors encountered uncomfortable incidents in their living environments and modified them for their safety and security. After countless times of testing and trying, *feng shui* developed its method to control the door, in other words to react before a predator enters, by the position of the furniture. Today, this position is more for safety of the subconscious than for actual predator attacks.

For aging in place, control over the bed position is noteworthy for sleeping well. A bedroom is used for the longest hours. Not only is a bed's position placed subconsciously for control over the door but also for getting into and out of bed without obstruction, which is an important consideration, particularly for going to the bathroom during the night. Further, a bedroom should be quiet, cool, and dark. These are control components, which environmental psychology describes as sound reduction, thermal control, and light control, respectively.

Regarding spatial syntax, flexibility, and responsiveness, *feng shui* arranges space for a harmonious living environment. *Feng shui* evaluates the appropriateness of *Chi* flowing for any given situation and makes adjustments as necessary (Stehr 2013; Tiller 1997). *Feng shui* for aging in place recommends creating space for interpersonal communication and physical touching. Fuse (2016) asserts that no matter the degree of changed needs or socioeconomic class, accessibility for physical touching is most needed for older adults. Touching is healing (Montagu 1971). Knapp and Hall (2002, p. 275) stated that "The use of touch to communicate emotional and relational messages to the elderly may be crucial, particularly as the reliance on verbal/cognitive message wanes." A simple furniture adjustment for allowing touching would contribute to an older adult's emotional responsiveness without any cost.

Stimulation. Stimulation in environmental psychology is described by Gifford (2007, p. 8) as being "relatively simple stimuli such as light, color, sound, noise, heat, and cold, but also more complex stimuli such as buildings, streets, outdoor settings, and other people." Evans and McCoy (1998, p. 85) stated that stimulation is "the amount of information in a setting or object that impinges upon the human user." The interpretation of stimuli differs by personal and cultural background (Gifford et al. 2011; McCunn 2015).

For aging in place, the personal meaning of stimuli varies depending on the older person's physical, mental, and psychological needs, as well as cultural values and personal interests. *Feng shui* suggests applying personal meanings and cultural values that only bring positive emotions to the occupants. For older adults' changes in vision and hearing, *feng shui* adapts colors, shapes, and sounds to stimulate their sensory perceptions while adjusting *Chi* flow harmoniously to promote safety and wellness. *Feng shui* also encourages positive human communication and relationships at home and in the community for the older adults' healthy emotions (see ▶ "Neighborhood Social Environment and Health").

Coherence. Coherence in environmental psychology in a living environment emulates a harmonious *feng shui* environment. *Feng shui* creates coherence with the *yin yang* duality and the five elements. Evans and McCoy (1998, p. 87) stated that "ambiguity, disorganization, and disorientation are major impediments to coherence."

Installing accessible items, modifications, or remodeling for aging in place (see ▶ "Supportive Housing") is new to older adults for their homes. The new devices or technologies should be blended in the home environment rather than predominantly standing out themselves. *Feng shui* observes the material, color, shape, and size of the new items and adapts *Chi* flow by constructing or destructing the five elements, as well as adjusting portions of *yin* and *yang*.

It is necessary to minimize obstacles which may be created by the new accessible devices or ambiguity caused by the newly modified home environment. *Feng shui* asserts coherence by maintaining the occupants' accustomed behaviors or habitual movements as much as possible while keeping learning new body movements or skills to a minimum. *Feng shui* sees keeping the old habits as a way of making an older adult feel respected.

Affordance. Affordance in environmental psychology for architecture is described by Evans and McCoy (1998, p. 87) such that "We utilize interior space according to our understanding or the functions that they provide us. We also rely on information systems to provide feedback about building or equipment performance." A living room with a sunken floor area, for example, without any cue of depth difference could cause someone to keep walking as if it is all on the same level surface (Evans and McCoy 1998).

Feng shui and aging in place also provide alerts for attention to steps. Using different colors on treads and risers, as well as different patterns or materials from adjacent surfaces to signal where levels or surfaces start and end is highly suggested. Further, installing guardrails on both sides of a set of steps in order to not lose balance is necessary. There would be positive *Chi* flowing along stairs that have safety cues for vision and when a person has secure balance by holding the guardrails.

The architectural dimensions of environmental psychology such as restoration, control, stimulation, coherence, and affordance are used to explain the concept of *feng shui* and applications of aging in place. *Feng shui* is a holistic approach, thus each dimension shares and interacts with each other to create a harmonious interrelated home environment.

Prospects

During the past half century, *feng shui* in the USA has been adapted to the trends of the times, such as the New Age movement (Too 1997) followed by influential publications on *feng shui* interior design (Mak and So 2011). In the twenty-first century, with the awareness of environmental sensitivity and energy conservation, as well as the increase of aging in place, *feng shui* could corroborate with these social movements. *Feng shui* will probably provide empirical analyses (Mak and So 2011) to work closely with professionals such as architects, designers, policy makers, and researchers in gerontology. Furthermore, when *feng shui* is looked at with a more scientific approach, it could become a part of the education system in Western cultures for architecture and environmental design (Mak and So 2011).

Examples of challenges in research on *feng shui* and aging in place could be in an assisted living facility where a number of participants are in one place. Finding a facility that has applied *feng shui* design and occupants who would agree to participate in the research could be difficult. Moreover, the research would rely on more subjective responses than objective measurements.

Research on *feng shui* within existing literature may also be limited. Western researchers point out that *feng shui* lacks empirical analyses (Bonaiuto et al. 2010). It probably derives from cultural differences: Western thinking emphasizes clear, linear, analytic, and logical explanations while Eastern thinking is relatively imprecise and is oriented toward spiritual attainment (Kim 1988). In addition, *feng shui* literature is varied from classical to modern and among different *feng shui* schools. Further, the original concept of *feng shui* might have been lost in translation or language barriers may have caused miswording of the Chinese terminology (Mak and So 2011).

In order for *feng shui* to be correctly understood and practiced for aging in place in Western culture, the future direction of research on *feng shui* may accommodate using empirical testing and analyses. Such testing could be an evidence-based design (Brawley 2006) which has been used for hospital environments (Ulrich 2001). The center for health design (2019, para. 3) described evidence-based design as "the process of basing decisions about the built environment on credible research to achieve the best possible outcomes." The population of older adults is increasing and so are their health concerns and needs, hence providing empirical test results on the effects of *feng shui* as a natural remedy for home environment health would benefit aging in place.

Summary

Aging in place builds on the individual's unique, personalized home and is also applied to any alternative of home. Positive *Chi* flowing in the home is essential to healthy emotions and behaviors. The *yin yang* duality and the five elements perform the key roles for creating a harmonious living environment. The concept of *Chi* has been explained by quantum physics that all matters and things in the universe vibrate with a specific frequency. Although the history and terminologies are different between *feng shui* and environmental psychology, they share the purpose of creating a comfortable environment for people. *Feng shui* is a gentle, natural, and affordable remedy for aging in place that can be used in any culture and socio-economic class.

Cross-References

▶ Affordable Housing
▶ Aged Care Homes
▶ Aging Households
▶ Aging in Place
▶ Aging in Place and Quality of Life
▶ Aging in Place: Maintaining Quality of Life for Older Persons at Home
▶ Aging in the Right Place
▶ Aging Well
▶ Congregate Housing
▶ Co-residence
▶ Healthy Aging
▶ Home Health Therapies
▶ Home Modification
▶ Homeshare
▶ Housing
▶ Housing Assistance
▶ Independent Aging
▶ Living Arrangements in Later Life
▶ Longevity Areas and Mass Longevity
▶ Mobile and Manufactured Homes
▶ Neighborhood Social Environment and Health
▶ Resilience: Measures and Models
▶ Smart Homes
▶ Social Housing
▶ Supportive Housing

References

AARP (2018) 2018 Home and community preferences: a national survey of adults age 18-plus. https://www.aarp.org/research/topics/community/info-2018/2018-home-community-preference.html. Accessed 31 Jan 2019

American Senior Communities (2016) Assisted living statistics – a deeper dive into the demographics. https://www.asccare.com/assisted-living-statistics-a-deeper-dive-into-the-demographics. Accessed 31 Jan 2019

Asthma and Allergy Foundation of America (2019) Asthma in older adults. http://asthmaandallergies.org/asthma-allergies/asthma-in-older-adults/. Accessed 31 Jan 2019

Bonaiuto M, Bilotta E, Stolfa A (2010) Feng shui and environmental psychology: a critical comparison. J Archit Plan Res 27(1):23–34. https://www.researchgate.net/publication/289047711_Feng_shui_and_environmental_psychology_A_critical_comparison. Accessed 1 Feb 2019

Bonnes M, Carrus G (2004) Environmental psychology, overview. In: Encyclopedia of applied psychology, pp 801–804. https://doi.org/10.1016/B0-12-657410-3/00252-X

Brawley E (2006) Design innovations for aging and Alzheimer's creating caring environments. Wiley, Hoboken, pp 299–308

Center for Health Design (2019). https://www.healthdesign.org/certification-outreach/edac/about. Accessed 15 Mar 2009

Detweiler MB, Sharma T, Detweiler JG, Murphy PF, Lane S, Carman J, Chudhary AS, Halling MH, Kim KY (2012) What is the evidence to support the use of therapeutic gardens for the elderly? Psychiatry Investig 9(2):100–110. https://doi.org/10.4306/pi.2012.9.2.100

Evans GW, McCoy JM (1998) When buildings don't work: the role of architecture in human health. J Environ Psychol 18(1):85–94. https://doi.org/10.1006/jevp.1998.0089

Feng shui (2012) Stedman's medical dictionary for the health professions and nursing, 7th edn. Wolters Kluwer/Lippincott Williams & Wilkins, Philadelphia

Fuse M (2016) Healthy home for healthy aging with feng shui. CreateSpace, North Charleston, pp 55–78

Gifford R (2007) Environmental psychology: principles and practice, 4th edn. Optimal, Colville, pp 1–21

Gifford R, Steg L, Reser JP (2011) Environmental psychology. https://www.rug.nl/staff/e.m.steg/giffordstegreser2011.pdf. Accessed 31 Jan 2019

Goswami A (2008) Creative evolution: a physicist's resolution between darwinism and intelligent design, pp 221–241. https://ebookcentral.proquest.com. Accessed 13 Mar 2019

Gregoire C (2014) The evolutionary reason why we love sitting by a crackling fire. https://www.huffingtonpost.com/2014/11/18/the-evolutionary-reason-w_n_6171508.html. Accessed 1 Mar 2019

Hale G (2002) The practical encyclopedia of feng shui. Barnes & Noble, New York, pp 6–29

Hawkins D (2012) Power vs. force: the hidden determinants of human behavior (rev. 1st ed). Hay House, Carlsbad, pp 149–153

Kim YY (1988) Intercultural personhood: an integration of Eastern and Western perspectives. In: Samover L, Porter RE (eds) Intercultural communication: a reader, 5th edn. Wadsworth, Belmont, pp 363–373

Knapp ML, Hall JA (2002) Nonverbal communication in human interaction, 5th edn. Wadsworth/Thomson Learning, Belmont, pp 272–304

Loretta J (2010) The hidden energies behind feng shui. Mira Digital, St. Louis, pp 4–24

Mak M, So AT (2011) Scientific feng shui for the built environment: fundamentals and case studies. City University Press of Hong Kong, Kowloon

McCunn L (2015) Environmental stimulation and environmental psychology: returning to the theory of things. https://www.psychologytoday.com/us/blog/ienvironment/201501/environmental-stimulation-and-environmental-psychology. Accessed 1 Feb 2019

Montagu A (1971) Touching: the human significance of the skin. Harper & Row, New York

Spear W (2010) Feng shui made easy. North Atlantic, Berkeley, pp 29–37

Steg L, van den Berg AE, de Groot JIM (eds) (2012) Environmental psychology: an introduction. https://ebookcentral.proquest.com

Stehr G (2013) Feng shui for life: connecting the dots. CreateSpace, North Charleston, pp 15–46, pp 59–86

Tiller W (1997) Science and human transformation: subtle energies, intentionality and consciousness. Pavior, Walnut Creek, pp 41–94

Too L (1997) Lillian Too's basic feng shui: an illustrated reference manual. Oriental, Kuala Lumpur, pp 1–13

Ulrich R (2001) Effects of healthcare environmental design on medical outcomes. http://www.capch.org/wp-content/uploads/2012/10/Roger-Ulrich-WCDH2000.pdf. Accessed 1 Mar 2019

Wiles JL, Leibing A, Guberman N, Reeve J, Allen RES (2012) The meaning of "aging in place" to older people. The Gerontologist 52(3):357–366. https://doi.org/10.1093/geront/gnr098

Wolverton BC (1997) How to grow fresh air. Penguin, New York

World Health Organization (2015) World report on ageing and health. https://www.who.int/ageing/events/world-report-2015-launch/en/. Accessed 15 Feb 2019

Fertility History

▶ Reproduction and Longevity in Humans

Fertility Transition Theories

▶ Demographic Transition Theories

Fictive Kin

Laura M. Funk
Department of Sociology and Criminology,
University of Manitoba, Winnipeg, MB, Canada

Synonyms

Intentional families; Voluntary kin

Definition

The process by which emotionally close and supportive relationships between individuals not related through blood or legal ties are interpreted as being "like family" by those individuals; this can include shifts in identities. In critical gerontology, the broader ideal of fictive kin relationships between older clients and care workers, invoked by organizations to market their services.

Overview

Drawn from family social science scholarship, in gerontology, the fictive kin concept acknowledges the important contributions of nonkin support for older adults without robust, traditionally defined family networks (Jordan-Marsh and Harden 2005; Macrae 1992). More specifically, the term signifies a shift in the meaning of particular relationships by one or both parties. Older adults often use kin terms to reframe supportive relationships involving friends and other nonkin (Barker 2002; Macrae 1992). For instance, in a study of Dutch persons (Voorpostel 2013), older adults – especially those who were widowed, divorced, or never married – were more likely than younger adults to include a nonrelative as part of their defined family. Allen et al. (2011) identified this process of "converting" nonkin to kin as a form of agency that helps older adults adapt to changing networks as well as bolster a sense of mutual support and closeness with others.

Fictive kin relationships may be particularly important for older gay and lesbian persons isolated from traditional family support (Barranti and Cohen 2000; Brotman et al. 2003); however, the role of fictive kin is often unrecognized within health and social service systems. Fictive kin relationships might also develop between step-parents acquired by adult children (e.g., where there are no legal ties); factors precipitating this process (e.g., frequency of interaction) were examined by Ganong et al. (2017). In broader sociological family scholarship, however, fictive kin terminology has been critiqued for its tendency to be applied to the study of ethnic minority populations, perpetuating a largely unfounded assumption that fictive kin processes are more common in these groups (Nelson 2014).

Paid care workers can also become interpreted as fictive kin by older clients. This can facilitate the social integration of workers into clients' private households and be a way for older adults to become more comfortable with formal help (Karner 1998). In one study of Taiwanese and Hong Kong immigrant families in California, Lan (2002) explained how the fictive kin process supports the commodification of care for older adults when adult children become less able to provide family care yet want to maintain the cultural ideal. However, although paid care workers often value family-like relationships with their clients, and such interpretations can promote good care, there is concern that these relationships can inadvertently exploit low wage, often racialized and/or migrant female care workers (Dodson and Zincavage 2007; Karner 1998; Johnson 2015). Workers might feel compelled to go above and beyond to contribute unpaid work, or be less likely to report workplace violence (Johnson 2015; Outcalt 2013). Moreover, many workers wish to maintain boundaries in their relationships with clients (Daly and Armstrong 2016; Piercy 2000).

Future Directions of Research

The fictive kin concept could be explored in intimate partnerships in which older adults "live apart together" (Funk and Kobayashi 2016) as well as among older residents in intentional co-housing communities. The potential for and implications of fictive kin relationships between older adults and different kinds of care workers (e.g., paid companions, volunteers) could be explored, as well as how these ideals are used by organizations to promote emotional labor among their staff.

Summary

The fictive kin concept highlights the role and importance of nontraditional forms of informal support and care for older adults who lack traditionally defined sources of family support. The concept also offers insights into the complex subjective processes involved in "constructing" family as well as potential antecedents and outcomes of these processes. Lastly, critical gerontologists have used the term to expose how close relationships with older clients, alongside organizational imperatives for emotional labor, can disadvantage paid care workers.

Cross-References

▶ Caregiving and Social Support: Similarities and Differences
▶ Familism
▶ Family Diversity

References

Allen KR, Blieszner R, Roberto A (2011) Perspectives on extended family and fictive kin in the later years: strategies and meanings of kin reinterpretation. J Fam Issues 32(9):1156–1177. https://doi.org/10.1177/0192513X11404335

Barker JC (2002) Neighbors, friends, and other nonkin caregivers of community-living dependent elders. J Gerontol Ser B 57(3):S158–S167. https://doi.org/10.1093/geronb/57.3.S158

Barranti C, Cohen H (2000) Lesbian and gay elders: an invisible minority. In: Schneider R, Kropt N, Kisor A (eds) Gerontological social work: knowledge, service settings and special populations, 2nd edn. Wadsworth/Thompson Learning, Belmont, pp 343–367

Brotman S, Ryan B, Cormier R (2003) The health and social service needs of gay and lesbian elders and their families in Canada. Gerontologist 43(2):192–202. https://doi.org/10.1093/geront/43.2.192

Daly T, Armstrong P (2016) Liminal and invisible long-term care labour: precarity in the face of austerity. J Indust Rel 58(4):473–490. https://doi.org/10.1177/0022185616643496

Dodson L, Zincavage RM (2007) 'It's like a family': caring labor, exploitation and race in nursing homes. Gend Soc 21(6):905–928. https://doi.org/10.1177/0891243207309899

Funk LM, Kobayashi K (2016) From motivations to accounts: an interpretive analysis of "living apart together" relationships in mid- to late- life. J Fam Issues 37(8):1101–1122. https://doi.org/10.1177/0192513X14529432

Ganong L, Coleman M, Chapman A et al (2017) Stepchildren claiming stepparents. J Fam Issues 39(6):1712–1736. https://doi.org/10.1177/0192513X17725878

Johnson EK (2015) The business of care: the moral labour of care workers. Sociol Health Ill 37(1):112–126. https://doi.org/10.1111/1467-9566.12184

Jordan-Marsh M, Harden JT (2005) Fictive kin: friends as family supporting older adults as they age. J Gerontol Nurs 31(2):24–31. https://doi.org/10.3928/0098-9134-20050201-07

Karner TX (1998) Professional caring: homecare workers as fictive kin. J Aging Stud 12(1):69–82. https://doi.org/10.1016/S0890-4065(98)90021-4

Lan PC (2002) Subcontracting filial piety: elder care in ethnic Chinese immigrant families in California. J Fam Issues 23(7):812–835. https://doi.org/10.1177/019251302236596

Macrae H (1992) Fictive kin as a component of the social networks of older people. Res Aging 14(2):226–247. https://doi.org/10.1177/0164027592142004

Nelson MK (2014) Whither fictive kin? Or, what's in a name? J Fam Issues 35(2):201–222. https://doi.org/10.1177/0192513X12470621

Outcalt L (2013) Paid companions: a private care option for older adults. Can J Aging 32(1):87–102. https://doi.org/10.1017/S0714980813000093

Piercy K (2000) When it is more than a job: close relationships between home health aides and older clients. J Aging Health 12(3):362–387. https://doi.org/10.1177/089826430001200305

Voorpostel M (2013) Just like family: fictive kin relationships in the Netherlands. J Gerontol Ser B 68(5):816–824. https://doi.org/10.1093/geronb/gbt048

Filial Obligations

▶ Confucian Culture and Filial Piety
▶ Filial Piety and Responsibilities Among the Chinese

Filial Piety

▶ Familism
▶ Filial Piety and Responsibilities Among the Chinese

Filial Piety and Responsibilities Among the Chinese

Daniel W. L. Lai and Gracy B. Y. Fang
Department of Applied Social Sciences, The Hong Kong Polytechnic University, Hong Kong, Hong Kong

Synonyms

Filial obligations; Filial piety; Filial responsibilities

Definitions

Filial piety is a social norm based upon Confucian ideology, referring to the virtue that children should respect and care for their parents due to their parents' love and care for them (Ha et al. 2016). Filial piety is constructed upon the cultural value of familial collectivism, which is defined as the tendency to value the needs of the family and other family members over one's individual needs (Schwartz et al. 2010). When translated into behaviors, this generally means treating parents or older generations of the family well and taking good care of them. This involves providing them not only with daily necessaries, but also care and respect both during the time they are alive, during illness, and after their death (Ho et al. 2015). Obeying their teachings and instructions also constitutes a key component of filial piety. These good deeds should be extended outside the family, by showing love, care, and respect to older people so as to demonstrate the good teachings of one's parents and family (Liu 2013).

Overview

While filial piety has been considered a cultural virtue, it drives motivations and reasons for caring for older people among younger generations and defines roles and responsibilities in Chinese families and communities (Ho 2008). Children are often taught and socialized into filial concepts by parents, grandparents, and clan elders. For centuries, filial piety has played a central role in maintaining family order by promoting responsibility, interdependence, sacrifice, and harmony (Bengtson and Putney 2000).

Key Research Findings

As one of the most influential cultural ethics guiding intergenerational relationships in Chinese culture, the concept of filial piety has received much attention in gerontological research. Due in part to shifts in social norms and ideologies in recent decades, changes have occurred in filial expectations and practices. Contrary to the traditional filial value that prescribes children's unconditional obligation to provide material and nonmaterial support to aging parents (Bell 2010), the realization of filial responsibilities in contemporary social contexts has become increasingly conditional and dependent on adult children's availability and socioeconomic situation (Yeh 2009). For example, paying for the costs of parents' long-term care and related services, rather than performing hands-on caring tasks, has been adopted as one form of realizing filial piety in a contemporary context in which adult children and in-laws are often engaged with employment and caring roles as parents (Yeh 2009; Lum et al. 2016).

Multigenerational co-residence has been an essential element promoting intergenerational support (Laidlaw et al. 2010), but a lack of space and the realities of global employment have contributed to its decline. Higher education attainment, greater desires for individual career pursuits, and decreased familial collectivism are believed to have shaped young people's perspective and manifestation of filial piety (Lum et al. 2016; Wu and Li 2012; Cheung and Kwan 2009).

Gender has also played a role in filial obligation. In traditional Chinese contexts, while the oldest son is expected to carry the responsibilities to shoulder the filial obligations, the daughter-in-law is often the one to provide the actual care (Yi et al. 2016). Gender variations in performing filial responsibilities are reflected in the phenomenon that sons tend to provide financial and emotional support while daughters or daughters-in-law tend to provide instrumental support and hands-on care (Chen and Jordan 2018). However, in most cases, married women are expected to reside with their husbands' families to ease care provision for their parents-in-law (Hu and Scott 2016), which may reduce their physical and emotional contact with their natal families (Zhou et al. 2017).

The increase of women's involvement in the employment markets in the last half century has resulted in women having the multiple roles of earning a family income and providing care to both children and older parents-in-law or parents (Liu et al. 2010). This new emerging situation can be particularly challenging for married daughters who are the only child for their parents in Mainland China (Zhan 2003). As this first cohort of single children reached age 40, both adult sons and daughters are expected to take on filial responsibilities on a more equal basis (Chen and Jordan 2018). For each single-child couple, they are likely to have to take care of four parents and parents-in-law of their own and/or another four grandparents and grandparents-in-law in longer term (Feng et al. 2014). The dual responsibilities of caring for both coresiding parents-in-law and non-coresiding parents means that married women who are only children may be subject to a high level of care demands (Hu 2017).

Although the term "filial piety" is derived from local Chinese culture, the virtue of familial obligation in providing parental or intergenerational care has universal significance for Chinese communities in different parts of the world. Studies indicate that among older Chinese adults in the United Kingdom and Australia, filial piety continues to exert significant influence on relationships with children, reflected in high levels of structural, affectual, and normative solidarity among generations (Laidlaw et al. 2010; Lin et al. 2018). In the United States, older Chinese American adults view filial piety as a key symbol of respect and highly expect filial piety from younger generations (Dong et al. 2014). In Canada, despite its gendered nature (Chappell and Kusch 2007), filial piety serves to reduce caregiving stressors and enhance positive appraisal, thus reducing caregiving burden.

Studies of the health of older adults have described beneficial effects of filial piety. Filial receipt has been linked to improved life satisfaction (Li and Dong 2018) and psychological well-being (Cheng and Chan 2006; Hsu 2017) as well as a decreased risk of loneliness (Dong and Zhang 2016) and suicidal ideation (Simon et al. 2014). Similarly, perceived filial piety has been associated with reduced depression and anxiety (Guo et al. 2017). Children's fulfillment of filial responsibilities has been described as a result of successful parenting and family fortune (Simon et al. 2014), further boosting older adults' self-image and self-worth and thus contributing to positive well-being and psychological outcomes.

Debates on the impact of filial piety on caregiving outcomes have gained much momentum given the growth of the aging population and increasing demands for caregiving (Kim and Kang 2015). Filial piety is internalized in Chinese family caregiving contexts although cross-cultural variations have been reported. For instance, Canadian-Chinese family caregivers with stronger filial beliefs tend to perceive their caregiving experience as positive, beneficial, and less stressful (Lai 2010). However, findings from Mainland China suggest that filial piety is associated with greater psychological stress among family caregivers (Kim and Kang 2015), likely due to the discrepancy between social expectations of caregivers and caregivers' capacity to fulfill increasingly challenging care demands. Thus, while filial values might foster positive appraisal of caregiving experiences, some caregivers might interpret filial piety as a rigorous and rigid moral code predetermining their care obligations and potentially leading to increased stress and psychological burden.

To conclude, industrialization and urbanization processes have resulted in filial piety being realized differently when compared to traditional contexts of absolute parental authority and unconditional submission of children. However, family remains a central and critical unit to social organization in Chinese culture and societies, in which family members share resources to meet physical, psychological, and social needs (Cheng and Chan 2006). The expansion of the current global and transient market economy has further affected the capacity of many individuals to provide adequate and appropriate personal caregiving and support. Without sufficient state interventions and assistance (Tsai et al. 2008), the virtue of filial piety may risk "burnout," particularly when adult children and young relatives are faced with multiple roles and associated demands.

Future Directions of Research

Future research should further examine strategies for maintaining an optimal balance between perceived familial obligations and caregivers' capacity. Additionally, further research on the potentially detrimental impacts of over-reliance on family caregivers for care provision, and the benefits of appropriate supports from governments and service agencies, should be conducted to better meet the emerging needs of older cohorts in contemporary Chinese communities across the world.

References

Bell D (2010) China's new Confucianism: politics and everyday life in a changing society. Princeton University Press, Princeton

Bengtson VL, Putney N (2000) Who will care for tomorrow's elderly? Consequences of population aging east and west. In: Bengtson VL, Kim K-D, Myers GC, Eun K-S (eds) Aging in the east and west: families, states, and the elderly. Springer, New York, pp 263–286

Chappell NL, Kusch K (2007) The gendered nature of filial piety: a study among Chinese Canadians. J Cross Cult Gerontol 22(1):29–45. https://doi.org/10.1007/s10823-006-9011-5

Chen J, Jordan LP (2018) Intergenerational support in one- and multi-child families in China: does child gender still matter? Res Aging 40(2):180–204. https://doi.org/10.1177/0164027517690883

Cheng ST, Chan ACM (2006) Filial piety and psychological well-being in well older Chinese. J Gerontol Ser B Psychol Sci Soc Sci 61(5):262–269. https://doi.org/10.1093/geronb/61.5.P262

Cheung CK, Kwan AYH (2009) The erosion of filial piety by modernisation in Chinese cities. Ageing Soc 29(2):179–198. https://doi.org/10.1017/s0144686x08007836

Dong XQ, Zhang M (2016) The association between filial piety and perceived stress among Chinese older adults in greater Chicago area. J Geriatr Palliat Care 4(1):1–24. https://doi.org/10.13188/2373-1133.1000015

Dong X, Zhang M, Simon MA (2014) The expectation and perceived receipt of filial piety among Chinese older adults in the greater Chicago area. J Aging Health 26(7):1225–1247. https://doi.org/10.1177/0898264314541697

Feng X, Poston DL, Wang X (2014) China's one-child policy and the changing family. J Comp Fam Stud 45(1):17–29

Guo M, Steinberg NS, Dong X et al (2017) A cross-sectional study of coping resources and mental health of Chinese older adults in the United States. Aging Ment Health 2017:1448. https://doi.org/10.1080/13607863.2017.1364345

Ha JH, Yoon H, Lim YO et al (2016) The effect of widowhood on parent-child relationships in Korea: do parents' filial expectations and geographic proximity to children matter? J Cross-Cult Psychol 31(1):73–88. https://doi.org/10.1007/s10823-016-9280-6

Ho DYF (2008) Chinese patterns of socialization: a critical review. In: Bond MH (ed) The psychology of the Chinese people. The Chinese University Press, Hong Kong, pp 1–37

Ho DYF, Xie W, Liang X et al (2015) Filial piety and traditional Chinese values: a study of high and mass cultures. Psy Ch J 1(1):40–55. https://doi.org/10.1002/pchj.6

Hsu H (2017) Parent-child relationship and filial piety affect parental health and Well-being. J Soc Anthropol 5(5):404–411. https://doi.org/10.13189/sa.2017.050504

Hu A (2017) Providing more but receiving less: daughters in intergenerational exchange in mainland China. J Marriage Fam 79:739–757. https://doi.org/10.1111/jomf.12391

Hu Y, Scott J (2016) Family and gender values in China: generational, geographic, and gender differences. J Fam Issues 37(9):1267–1293. https://doi.org/10.1177/0192513X14528710

Kim YJ, Kang HJ (2015) Effect of filial piety and intimacy on caregiving stress among Chinese adult married children living with parents. Indian J Sci Technol 8(S1):434–439. https://doi.org/10.17485/ijst/2015/v8iS1/59361

Lai DWL (2010) Filial piety, caregiving appraisal, and caregiving burden. Res Aging 32(2):200–223. https://doi.org/10.1177/0164027509351475

Laidlaw K, Wang DH, Coelho C et al (2010) Attitudes to ageing and expectations for filial piety across Chinese and British cultures: a pilot exploratory evaluation. Aging Ment Health 14(3):283–292. https://doi.org/10.1080/13607860903483060

Li M, Dong XQ (2018) The association between filial piety and depressive symptoms among U.S. Chinese older adults. J Geriatr Med Gerontol 4:1–7. https://doi.org/10.1177/2333721418778167

Lin X, Dow B, Boldero J et al (2018) Parent–child relationships among older immigrants from mainland China: a descriptive study using the solidarity, conflict, and ambivalence perspectives. J Fam Stud. https://doi.org/10.1080/13229400.2018.1441057

Liu YL (2013) Autonomy, filial piety, and parental authority: a two-year longitudinal investigation. J Genet Psychol 174(5):557–581. https://doi.org/10.1080/00221325.2012.706660

Liu L, Dong XY, Zheng X (2010) Parental care and married Women's labor supply in urban China. Fem Econ 16(3):169–192. https://doi.org/10.1080/13545701.2010.493717

Lum TYS, Yan ECW, Ho AHY et al (2016) Measuring filial piety in the 21st century: development, factor structure, and reliability of the 10-item contemporary filial piety scale. J Appl Gerontol 35(11):1235–1247. https://doi.org/10.1177/0733464815570664

Schwartz SJ, Weisskirch RS, Hurley EA et al (2010) Communalism, familism, and filial piety: are they birds of a collectivist feather? Cult Divers Ethn Min 16(4):548–560. https://doi.org/10.1037/a0021370

Simon MA, Chen R, Chang ES et al (2014) The association between filial piety and suicidal ideation: findings from a community-dwelling Chinese aging population.

J Gerontol A Biol Sci Med Sci 69(S2):S90–S97. https://doi.org/10.1093/gerona/glu142

Tsai HH, Chen MH, Tsai YF (2008) Perceptions of filial piety among Taiwanese university students. J Adv Nurs 63(3):284–290. https://doi.org/10.1111/j.1365-2648.2008.04711.x

Wu F, Li J (2012) Policy approaches to development capacity of family. Popul Rep 4:37–43

Yeh KH (2009) Intergenerational exchange behaviors in Taiwan: the filial piety perspective. Indig Psychol Res Chinese Soc 31:97–141

Yi Z, George L, Sereny M et al (2016) Older parents enjoy better filial piety and care from daughters than sons in China. Am J Med Res 3(1):244–272. https://doi.org/10.22381/AJMR3120169

Zhan HJ (2003) Gender and elder care in China: the influence of filial piety and structural constraints. Gender Soc 17(2):209–229

Zhou T, Onojima M, Kameguchi K et al (2017) Family structures and women's status in rural areas of Xining, China: a family image study in the villages of Qinghai province. Asian J Women Stud 23(1):89–109. https://doi.org/10.1080/12259276.2017.1279887

Filial Responsibilities

▶ Confucian Culture and Filial Piety
▶ Filial Piety and Responsibilities Among the Chinese

Financial Abuse of Older People

▶ Financial Frauds and Scams

Financial Assets

Delali Adjoa Dovie
Department of Sociology, University of Ghana, Accra, Ghana

Synonyms

Available resources; Cash assets; Financial capital; Financial holdings; Financial resources; Monetary assets

Definition

Financial assets refer to income and other assets that individuals own that affect future income and consumption. Vis-à-vis retirement, financial assets enable individuals, particularly retirees, to accumulate funds in aid of retirement income security. It is often considered to be synonymous with financial resources.

Overview

Retirement entails disengagement from active service. It is an inevitable life transition that requires preparation (Dovie 2018a). "In the beginning, there was no retirement. There were no old people. In the Stone Age, everyone was fully employed until age 20, by which time nearly everyone was dead..." (Weisman 1999: 1 cited in Hershenson 2016: 1). Several centuries down the line, people live longer with implication for population aging. From a global perspective, a new historical period emerged in which retirement has raised social awareness of status as well as increased readiness in societal initiatives in a myriad of forms (Ibrahim and Wahat 2015). In contemporary times, retirement represents a key life change. Hence, it requires adequate planning and/or preparation.

Hershenson (2016) conceptualized the phenomenon of retirement to include six distinct statuses, namely, retrenchment, exploration, tryout, involvement, reconsideration, and exiting, which constitute the acronym RETIRE. Retrenchment entails fully or partially cutting down on one's first employment. The exploration status encompasses thinking about and gathering information regarding possible activities and lifestyle to be engaged in during retirement. This status may be affected by energy levels, adventurousness, and anticipations. Tryout status pertains to the selection and trying out of options to retirement activities and lifestyle. Involvement status depicts retirement activities and lifestyle that retirees may expect to participate in, perhaps vetted by tryout. However, the reconsideration status is activated when involvement looses its allure or that new attractive options emerge. The existing status is activated when voluntary or imposed termination of a retirement activities or lifestyle is activated. Essentially, a return to full-time employment entails exiting retirement. By and large, these are induced by social change. Social change has been the factor of transformation in contemporary life worldwide including Africa and Ghana. It is induced by modernization, urbanization, and globalization (Kpessa-Whyte 2018). It is associated with social conditions such as weakened extended family support system (De-Graft Aikins et al. 2016; Dovie 2018a; Van der Geest 2016), inadequate formal support infrastructure (Dovie 2018a), and increasing nucleation of the family.

Retirement preparation is becoming an increasingly enduring and diverse process. Compared to the past, retirement is different today and will be different in the nearest future (Ibrahim and Wahat 2015: 154). This raises awareness of the critical need for retirement planning and assists to outline strategies that encourage early preparation for retirement. In recent times, the decline in the extended family support system, population aging, and increased life expectancy (Dovie 2018a, 2019a) highlights the problem of substantial old-age financial insecurity in post-retirement life.

Planning facilitates in individuals the propensity to hold substantial amounts of wealth and to invest in such wealth holdings (Sulliman 2019) in high-return financial assets, namely, pension contribution, stocks, shares, savings, and bonds, including commodities. Retirement preparation is significant in explaining the savings and investment behaviors of workers and individuals at large. The nonfinancial assets encompass research and development, technologies, patents, and other intellectual properties as well as real assets. Real assets refer to value-generating physical assets that are owned, for example, land, buildings, inventory, precious metals, commodities, real estate, land, and machinery. An income accumulation strategy may require that individuals invest in a mixture of risky assets (mainly equity) and risk-free assets, with the balance of risky and risk-free shifting over time to optimize the likelihood of achieving the investment goal (Dovie 2018a).

Economic security or financial security is the condition of having a stable income or other resources to support a standard of living now and shortly. It ensures the predictability of individuals' cash flow. Snyman et al. (2017) articulated a guideline that suggests that saving 15% income yearly in aid of retirement is enough. However, most people often underestimate what they require for retirement, to be able to save for an independent retirement financially. The importance of starting retirement planning, savings, and investments early in an individual's career has also been underscored by (Dovie 2018a; Snyman et al. 2017) (see ▶ "Financial Gerontology").

Financial Risk-Taking Behavior

Financial risk is classified into four broad categories: market risk, credit risk, liquidity risk, and operational risk. Market risk involves the risk of changing conditions in the specific marketplace in which there is competition for business. Liquidity risk includes asset liquidity and operational funding liquidity risk. Asset liquidity refers to the relative ease with which assets can be converted into cash, should there be a sudden, substantial need for additional cash flow pertains. Operational risks refer to the various risks that can arise from a company's ordinary business activities.

Cash and cash equivalents are cash on hand and in bank accounts and their equivalents in various currencies. Cash equivalents include short-term (up to 3 months) liquid investments into securities, traveler's checks, and other financial assets and/or financial resources (Dovie 2018a) that meet the definition of cash equivalents. Examples of available-for-sale financial assets entail purchased loans or other amounts receivable, other entity's shares, bonds, or other securities purchased by an entity for purposes of trading. Examples of held-to-maturity financial assets include bonds with a fixed interest rate and specified maturity date. Available-for-sale financial assets are classified as held-to-maturity financial assets only when an entity which acquired them has an intention and ability to hold them until maturity date without taking the opportunity to sell.

A pension plan can expect about 40 years of contributions for a full-career worker to finance about 20 years of retirement. The number of years spent in retirement is about half of the number of years spent in the labor force in recent times. Older people more often than not are concentrated in urban cities, where the challenges of aging are compounded by problems in urban environments such as crime, congestion, decaying neighborhoods, and inaccessibility of affordable housing (Dovie et al. 2018). Further, increasing economic hardship and the associated excessive cost of living in Africa including Ghana coupled with the increasing nucleation of the family system as well as the fact that the older persons are no longer employed due to several reasons including retirement from active service impact negatively on living conditions. This suggests that to some extent that there is significant poverty among older adults, for instance, in Ghana. There exist state welfare mechanisms such as the livelihood empowerment against poverty program (LEAP), older people's welfare card, and national health insurance exemptions for older people (Dovie 2019). However, these are insufficient. Hence, the need for workers to harness financial assets for sustenance in old age.

Key Research Findings

Preparing for retirement requires financial education and knowledge, which facilitate the understanding of basic retirement planning concepts, informed decision while prioritizing advancement and achievement of retirement planning goals (Dovie 2018b).

Mobilized financial assets and/or resources can be categorized into savings and investments. The savings category encompasses susu, retirement savings, fixed deposits, and mutual and provident funds. Susu is an informal savings scheme. Midterm savings portfolios such as T-bills and fixed deposits also pertain. The investment component entails financial products such as life insurance, funeral policy, shares, stocks, mfund, and epack.

The other long-term invested financial products comprise farming (i.e., palm plantation), school establishment and management, transportation business, and trading. Others include the acquisition of landed properties, namely, plots of land and houses. These could be rented out and therefore are financial resource generators (Dovie 2017).

Instituting a given array of financial assets is an indication of the diversification of income sources. It brings to the fore the significance of the development of multiple streams of income. Another financial measure is farming, which takes two forms, namely, crop production and livestock farming. From a retirement preparation standpoint, livestock may serve as a "revolving fund." Establishment and management of schools from primary through to junior and senior high school levels are other observations. As in the case of farming, the object of school establishment, ownership, and management constitutes an activity to occupy the planners in postretirement life. This feature plays a significant role in the use of the extended time that can probably be spent in retirement due to increased life expectancy (Dovie 2017). Also, engagement in the transportation business sector provides another angle to the paradigm of financial measures instituted by the workers and its connection with retirement-oriented activities. These are all other ways of creating new income sources as a set of yet another income generator(s) for the individual investor.

The diversification of income and investment sources is imperative. Because, should something happen to the institution the individual is saving or investing with, the only guarantee is up to certain FDIC limits, thus the need to diversify funds across several savings institutions and/or portfolios. The diversification of financial assets encompasses bonds and shares including stocks. Noteworthy is that owning stocks is one way to fight inflation risk. This is consistent with Maverick's (2018) argument that stock market diversification works and is the best way to balance overall portfolio risks and returns. However, while a plethora of assets is amassed in preparation for life in old age, it is prudent to watch out for schemes that sound too good to be true. In other words, it is emphasized that old-age financial asset mobilizers need to avoid get-rich-quick schemes. The essence is due to the upsurge in Ponzi scheme operations, which has been on the increase in recent times, with some people falling victims to such schemes. Should financial asset mobilizers fall prey to such scams, the string of loss can be alleviated with the assistance of plan diversification.

Diversified income has the propensity to reduce vulnerability to income shocks, and this could be a proxy to retiree's ability to respond to economic changes as experienced in old age. Retirement is primarily related to financial preparedness, yet it is also important that the individual's mobilization of financial assets towards retirement should entail health (Dovie, 2018b), social networks, leisure (Dovie 2018a), and the spiritual (Ibrahim and Wahat 2015). Each domain relies on the others, all of which are connected and work together to constitute a whole (Dovie 2018a). Even during retirement, older adults desire to continue working after formal retirement.

In a study conducted in Ghana, Dovie (2017) found that most of the workers (85.5%) were actively planning for retirement. Schofield et al. (2010 cited in Snyman et al. 2017) reported comparable findings among Australian dentists aged 50+, namely, 84% were actively planning for (financial independent) retirement. There might be a concern over those who are not planning for retirement at all albeit not seriously. Dovie (2017, 2018a) agrees that it is imperative to start planning and saving towards retirement as early as practicable, preferably as soon as income is being earned.

Comparable findings show that in Ghana for instance, formal sector workers were found to be well prepared for retirement than their informal sector counterparts (Dovie 2017). Similarly, American dentists acknowledged that they were not prepared for retirement and might have to postpone retirement (Du Molin 2015). Australian dentists were however said to be well prepared for retirement (Schofield et al. 2010 cited in Snyman et al. 2017). South African dentists indicated that they are not confident or well prepared for retirement (Snyman et al. 2017).

The mobilization of financial assets is aimed at a dramatic reduction in old-age poverty. The poverty rate, which refers to income less than 50% of median income for individuals aged 65+, for instance, was close to the rate for the entire population of countries of the Organisation for Economic Co-operation and Development (Anderson 2015). Also, tighter links between employment income and contributions and benefits as well as the more significant role played by private and workplace pensions in complementing reduced public pensions may induce inequality among pensioners. Thus, different financial assets when amassed have the propensity to ensure adequate financial flows in an era of increased life expectancy. Financial assets when mobilized provide retirement income security for retirees (Dovie et al. 2018).

Lower pension incomes imply the necessity to combine pension receipt with paid employment. Most retirees may prefer to work beyond retirement based on the availability of work. However, those constrained by insufficient pension income (Dovie 2018b) may be forced to resort to paid work (Yu and Schömann 2015) in the informal sector economy. As De Wind et al. (2016) note, motivation to work, physical health, and the financial situation were the most relevant aspects regarding working beyond retirement, which supports the idea that the principle of "human agency" of the life course perspective is useful in understanding factors that impact working beyond retirement.

Significantly, how much one requires to be able to retire confidently depends on specific needs: namely, healthcare benefits and expenses; spending habits; how long one would live for; inflation in the future; and returns on income. It is worth noting that the last three variables/factors are unpredictable.

Essentially, the best way to manage money and financial health is to be acutely aware of the relationship between financial risk and monetary return. By managing risks, investors are better equipped with addressing unpredictable scenarios that can wreak havoc with their finances. Such risk needs to be controlled or calculated, keeping policies up to date and enough for their respective circumstances (Maverick 2018).

Examples of Application

Financial asset by application is mobilized, facilitated by financial education and knowledge (Dovie 2017, 2018c; de Scheresberg and Lusardi 2016); this fosters awareness of financial resources to be accumulated. It also boosts the rate of savings and investments. A plethora of financial assets is mobilized, namely, pension contribution, stocks, bonds, treasury bills, shares, savings, and commodities. These financial assets are complemented by non-financial assets such as houses and other properties, which are rent inducing (Dovie et al. 2018). This also includes employers' retirement benefits (Dovie 2017, 2019b). Collectively, these yield consolidated retirement income security outcomes. Paid work beyond pension is however resorted to when necessary.

Future Directions of Research

Globally, the population of older adults is aging exerting pressure on welfare state mechanisms, creating the silver economy in the process. These factors will continually drive financial asset outcomes at individual levels. Future research should, therefore, track strategies adopted to ensure financial asset availability and security. Primarily, the paid work beyond pensions dimension should be explored, especially in the African region. There is articulate the sector of work albeit public/private, type of work, designated work hours including remuneration without sex barriers. Future research should also explore the increase in retirement age and its effects on the economy and the means of adjustment particularly in regions where an upward adjustment in retirement age has not yet being taken into consideration as a retirement policy dimension.

Summary

This chapter emphasizes financial assets' mobilization ascertained via the process of retirement planning at the individual level, as a key phenomenon to obtaining income security at old age. It investigated the financial investments workers mobilized during pre-retirement preparation towards sustainable well-being in post-retirement life. There is evidence that supports workers' strategic investment in formal and informal financial products to amass financial wealth in aid of retirement. This is critical because greater wealth is needed to finance more years of retirement. However, these investment portfolios are more short-term oriented, for instance, savings, investments in the money market, and treasury bills. This presupposes that retirement planners should invest in long-term portfolios such as corporate bonds including venturing into equity-based investments and a host of others, thereby overturning risk aversion into risk-taking exploits. These results were discussed in terms of portfolio diversification, longevity, and pension incomes (in)security. Nevertheless, retirement income need not replace gross pre-retirement earnings. Failing traditional support systems, population aging, and longevity have implications for financial retirement aspirations and preparations, and therefore financial asset mobilization toward retirement adequacy.

Cross-References

▶ Financial Gerontology

References

Anderson KM (2015) Pension reform in Europe: context, drivers, impact. In: Scherger S (ed) Paid work beyond pension age. Palgrave Macmillan, New York, pp 177–197

De Wind A, van der Pas S, Blatter BM, van der Beek AJ (2016) A life course perspective on working beyond retirement-results from a longitudinal study in the Netherlands. BMC Public Health 16:499

De-Graft Aikins A, Kushitor M, Sanuade O, Dakey S, Dovie DA, Kwabena-Adade J (2016) Research on aging in Ghana from the 1950s to 2016: a bibliography and commentary. Ghana Stud J 19:173–189

Dovie DA (2017) Preparations of Ghanaian formal and informal sector workers towards retirement. Unpublished doctoral thesis, University of Ghana, Accra

Dovie DA (2018a) Systematic preparation process and resource mobilisation towards postretirement life in urban Ghana: an exploration. Ghana Soc Sci J 15(1):64–97

Dovie DA (2018b) Leveraging healthcare opportunities for improved access among Ghanaian retirees: the case of active ageing. J Soc Sci 7(92):1–18. https://doi.org/10.3390/socsci7060092

Dovie DA (2018c) Utilization of digital literacy in retirement planning among Ghanaian formal and informal sector workers. Interações: Sociedade E As Novas Modernidades, 34:113-140. https://doi.org/10.31211/interacoes.n34.2018.a6. Special Issue on (In)Equalities and Social (In)Visibilities in the Digital Age.

Dovie DA (2019a) The influence of MIPAA in formal support infrastructure development for the Ghanaian elderly. International Journal of Ageing in Developing Countries, 3: 47–59. Special MIPAA Issue 2: Review and appraisal of Country-level progress under MIPAA

Dovie DA (2019b) Assessment of how house ownership shapes health outcomes in urban Ghana. Societies, 9(43):1–18. https://doi.org/10.3390/soc9020043. Special Issue Families, Work and Well-being

Dovie DA, Ayimey IR, Adodo-Samani P (2018) Pension policy dimension to Ghanaian workers' housing needs provision. Soc Nov Modernidades (35):30–56

Du Molin J [Internet]. Retirement? Dentists may have to keep working. The Wealthy Dentist; [cited 2015 June 29]. Available from: http://thewealthydentist.com/pr/103-dental-retirement.htm

Hershenson DB (2016) Reconceptualising retirement: a status-based approach. J Aging Stud 38:1–5

Ibrahim DKA, Wahat NWA (2015) A new pathway towards retirement preparation: integration of holistic life planning. Eur J Soc Sci Educ Res 5(1):154–160

Kpessa-Whyte M (2018) Aging and demographic transition in Ghana: state of the elderly and emerging issues. Gerontologist 58:403–408

Lusardi A, de Bassa Scheresberg C (2016) Americans' troubling financial capabilities: A profile of pre-retirees. Public policy & Ageing Report, 26(1):23–29

Maverick JB (2018) What are the major categories of financial risk for a company? https://www.investopedia.com/ask/answers/062415/what-are-major-categories-financial-risk-company.asp. Accessed 15 Mar 2019

Snyman L van der Berg-Cloete SE, White JG (2017) Planning for financially independent retirement. SADJ 72(5):204–208.

Sulliman NN (2019) The intertwined relationship between power and patriarchy: examples from resource extractive industries. Societies 9(14):1–11. https://doi.org/10.3390/soc9010014. www.mdpi.com/journal/societies

van der Geest S (2016) Will families in Ghana continue to care for older people? Logic and contradiction in policy. In: Hoffman J, Pype K (eds) Ageing in Sub-Saharan Africa: spaces and practices of care. Policy Press, London, pp 21–42

Yu G, Schömann K (2015) Working pensioners in China: financial necessity or luxury of choice. In: Scherger S (ed) Paid work beyond pension age. Palgrave Macmillan, New York, pp 151–173

Financial Behavior

▶ Financial Socialization

Financial Capital

▶ Financial Assets

Financial Counseling

▶ Financial Therapy

Financial Education

▶ Financial Literacy
▶ Financial Socialization

Financial Exploitation by Family Members

Gordon Alley-Young
Department of Communications and Performing Arts, Kingsborough Community College – City University of New York, Brooklyn, NY, USA

Synonyms

Economic abuse of older adults by family members; Elder family financial exploitation; Elder financial abuse in the family; Elder/dependent financial abuse by family members; Elder/vulnerable financial abuse by family members; Financial exploitation of vulnerable adults by family members; Granny battering

Definitions

Financial exploitation of older adults (FEOA) by family members is the abuse, mistreatment, and/or neglect of older persons by family members who steal or take control of an older person's money and/or assets without permission and/or by way of deception, manipulation, and coercion (See ▶ "Financial Assets"). Gender, race, poverty, health, and relationship status can determine one's susceptibility to exploitation, while perpetrators are propelled by motives ranging from addiction to entitlement. Addressing FEOA may involve maintaining social connectedness, making older persons and their families aware of the phenomenon, educating medical/financial professionals, and improving legislation. Embarrassment, stigma, and fear of losing relationships prevent many victims from reporting perpetrators. Financial exploitation by family members affects people from all levels of society and is expected to become more prevalent as the older adult population is increasing.

Overview, History, and Scope of the Problem

FEOA by family members is part of a larger phenomenon of abuse, mistreatment, and/or neglect of older persons first identified in British journals in 1975 as granny battering (Krug et al. 2002). This broader phenomenon was identified and most studied in the North America and Europe with political action on the issue in the 1980s beginning in Australia, Canada, Hong Kong, Norway, Sweden, and the USA and then in the 1990s in Argentina, Brazil, Chile, India, Israel, Japan, South Africa, and Eastern Europe (Krug et al. 2002). The World Health Organization in 2002 specified that abuse can be a single/repeated act, or be the lack of appropriate action, occurring within any relationship where there is an expectation of trust that causes harm or distress to an older person, taking forms such as physical, psychological/emotional, sexual, financial abuse, and/or intentional/unintentional neglect (Hall 2014) (See ▶ "Older Adults Abuse and Neglect"). This definition specifies financial exploitation as a specific abuse type.

FEOA, also called economic or financial abuse, is defined as theft or fraudulent use of an older person's money/belongings possibly involving undue influence (using one's role/power to exploit another) to gain control over/access to an older person's assets (Gibson and Qualls 2012). A South African workshop in 1992 distinguished mistreatment (e.g., verbal abuse, passive and active neglect, financial exploitation, and overmedication) from abuse (including physical, psychological, and sexual violence and theft) (Krug et al. 2002). FEOA includes both exploitation and theft and thus implicates both mistreatment and abuse. Undue influence is blackmail or threats of violence, suicide, begging, and/or claiming entitlement and/or a need to keep family relationships balanced (Gibson and Qualls 2012). This discussion of FEOA specifies family perpetrators, but research often does not distinguish by the perpetrator.

Six million cases of FEOA occurred in 2012 in the USA, most frequently in California, followed by Florida (Darst 2012). A survey of over 4,000 people found that FEOA had a 1-year prevalence of 2.7% and a lifetime prevalence of 4.7%; however, older persons are reluctant to report exploitation by family (Strouse and Gomulka 2014). Referrals of FEOA cases rose more than 35% from 2010 to 2014 (Idzelis 2017). The victim usually knows the perpetrators (Gleason 2017). FEOA perpetrators are family (68%), friends/neighbors (17%), and healthcare aides (15%) (See ▶ "Abuse and Caregiving").

While an estimated five million older adults experience financial exploitation each year, it is believed that many do not report it for a variety of reasons such as embarrassment or fear of retaliation (Gleason 2017). When they do report it, they might be reluctant to name a loved one as the perpetrator. A MetLife study estimated the annual pecuniary loss from FEOA, by all perpetrators, to be at least $2.9 billion, while the Consumer Financial Protection Bureau (CFPB) suggests the number could be as high as $36.48 billion (Gleason 2017; Idzelis 2017; Office for Older Americans 2016). This problem will grow as the population of older adults increases steadily. The US Census Bureau estimates that, by 2050, Americans over the age of 65 will exceed 20% of the US population (Gleason 2017). Currently, that translates to roughly 10,000 people turning 65 every day over the next decade (Olson 2015).

Identifying Exploitation, Potential Victims, and Perpetrators

FEOA can take many different forms. Older adults might encounter exploitation as stolen property (e.g., stolen prescription drugs, mail, cashing worthless checks for others), identity theft, a family member not contributing or under-contributing toward shared household expenses, not receiving necessary living assistance from family (e.g., thus leaving one destitute or reliant on welfare), and being forced or misled into surrendering rights, property, or signing documents (e.g., changing wills, taking out/over leases/property) (Peterson et al. 2014). Types of financial exploitation include a family member borrowing money and never repaying, taking a portion of an older relative's pension or Social Security check, adding one's name to accounts or opening accounts, taking money or assets directly or as a joint signatory, transferring property titles or selling property/home equity, and/or using another's checks/ATM/debit/credit cards without the older person's permission (BITS). Perpetrators might rationalize taking money as compensation for their caring duties/assistance/companionship.

Often financial professionals recognize FEOA before the older adult or other family members due to unexplained/atypical withdrawals, difficulty contacting the older adult, seeing an older person intimidated/reluctant to speak around specific people, sudden/increased social isolation, unusual checks issued, forged signatures, and misuse of conservatorships, guardianships, or powers of attorney (BITS 2016) (See ▶ "Isolation"). Nonfamily members might become family via marriage scams (online or face-to-face), and once the older adult realizes the fraud, they are too embarrassed or afraid of loneliness to speak out (BITS 2016; Darst 2012) (See ▶ "Financial Frauds and Scams"; See ▶ "Loneliness"). Perpetrators use intimidation, deceit, coercion, emotional manipulation, psychological/physical abuse (isolating the victim from support), and/or false promises (BITS 2016).

Research suggests different potential victim profiles and risk factors for exploitation. One in five Americans 65 or older has been victimized by fraud, and some estimates argue as much as 55% of financial abuse is committed by family, friends, and caregivers (Atkinson 2016). One suggested victim profile identifies white, widowed females 70–89, with cognitive impairment, in isolation, with daily assistance needs as having a higher exploitation risk (Gibson and Qualls 2012) (See ▶ "Widowhood"). Another profile identifies African-Americans, living below the poverty line with multiple non-spousal household members, and disability as being at higher risk (Peterson et al. 2014; Strouse and Gomulka 2014). Other research suggests that older men might be exploited by family members as they have more financial resources than women (Melchiorre et al 2016; Olson 2015). A recent survey found that over a third of men and women suffered financial exploitation (Olson 2015). That said, for every case reported, experts suggest that 10–44 of cases often go unreported with family reporting only 2% and banks/financial institutions most often making referrals (Idzelis 2017). Studies suggest that men were 3% more likely than women to report financial and emotional abuse (Krug et al. 2002). Also, some cultures are based upon providing intergenerational support to struggling family members, so what is considered exploitation in one culture may be an expected familial obligation in another (Olson 2015).

Victims of FEOA by the family can come from all social strata. Actor Mickey Rooney was awarded a 2.8 million judgment against his stepson and his stepson's wife for taking money from Rooney, as Rooney testified before Congress (Khalfani-Cox 2011). Similarly, a New York City jury convicted the son of wealthy philanthropist Brooke Astor of first-degree grand larceny for defrauding his late, older mother, out of millions of dollars while she had Alzheimer's disease (Gleason 2017). Though celebrity cases garner media attention, FEOA by family affects average Americans such as Sina Harris, in a case that encompassed both exploitation and possibly abuse and/or neglect. Harris was left dead and decomposing in a back bedroom of her Louisville, KY, home for 6 months, while her son Julius claimed her social security checks to support his crack addiction. Julius only received a 1-year sentence, a misdemeanor, most of which was served on probation, but a subsequent driving under the influence (DUI) charge actually brought Julius a more substantial sentence than he received for what he did to his mother (Darst 2012). While the law often seeks restitution for victims, some victims (e.g., Rooney) may never be made entirely whole financially.

Older adults fear to report exploitative family, fearing that this could mean them losing their independence/decision-making ability (Idzelis 2017). Older parents may feel emotionally responsible for the economic well-being of adult children (Khalfani-Cox 2011) or feel guilty/responsible for raising a child who would exploit them (Idzelis 2017). Living with a spouse/partner significantly lowered one's relative risk of exploitation (Peterson et al. 2014). Due to the reticence/fear associated with speaking out against exploitative family members, the signs of FEOA could be hard to spot. For instance, an older adult voicing anxiety about affording daily expenses could be an indirect expression of exploitation even though the victim might not even recognize the abuse (Olson 2015).

Perpetrators might feel they are owed money, due to a bad childhood, or for the support they give the older person (Gibson and Qualls 2012). Carers who exploit relatives might be more likely to experience anxiety, depression, overwork, and/or abuse from care recipients; others cite a reaction to being depended upon or feeling entitled to inheritance (Hall 2014; Khalfani-Cox 2011) (See ▶ "Gender and Caring in Later Life"). The economic downturn of 2011–2012 exacerbated FEOA by making more adult children financially dependent on older parents (Khalfani-Cox 2011) (See ▶ "Economics of Aging"). FEOA often ends with the death of the older person and/or the depletion of the account (Gibson and Qualls 2012). Carers/overseers of older relatives must act in the best interest of the individual (e.g., as power of attorney), use the older adult's funds only for caring for, the older person, and never use funds for personal benefit (BITS 2016).

Financial advisors often spot FEOA first, but in an InvestmentNews survey, of the 62% who have seen/suspected FEOA, 56% said they did not report it (e.g., lacking conclusive proof, fearing litigation) (Idzelis 2017). A lack of public awareness about the problem (Idzelis 2017) and the difficulty of distinguishing a legitimate/consensual financial transaction from an exploitive one compound the problem (Olson 2015). InvestmentNews survey found that 16% of financial advisors reported cutting ties with older clients showing mental impairment/cognitive problems (affecting 50% of adults over 80) because of the difficulties in managing family situations, fear of litigation, and client's refusal to seek medical consultation or follow financial advice (Idzelis 2017). In the same InvestmentNews survey, 65% of advisors identified a family member as a suspected perpetrator of FEOA (Idzelis 2017). Bank of America Merrill Lynch surveyed its advisors about perpetrators of FEOA, and 71% cited children of the victim, 32% cited other family members, and only 18% identified anonymous fraudsters (e.g., respondents could identify more than one type of perpetrator) (Idzelis 2017). Financial professionals, as the first to spot FEOA, can be a resource by educating older clients and their families (Darst 2012). Financial advisors can offer advice tailored to older clients, urging them to stay actively connected to others in their community, to regard all uninvited solicitations for money critically, to monitor account statements/credit reports, and to report any unrecognized financial activity (BITS 2016) (See ▶ "Financial Literacy").

General practitioners (GPs) might spot FEOA early when it affects a patient's health (e.g., stolen prescriptions) or when it co-occurs with other abuse (e.g., physical). In an opinion survey of GPs, older carers, independent older adults, and care-receiving older adults, on the seriousness of potential abuse scenarios, findings showed that female GPs and female care-receivers rated financial abuse as being of lower importance than did male caregivers (Olson 2015). Community health resources are also needed as older persons might need to leave homes shared with an exploiter and seek emergency housing such as places such as the Weinberg Center for Elder Abuse Prevention at the Hebrew Home in the Bronx, NY (Olson 2015). Between 2005, when the center opened, and 2015, 14 other such centers opened (Olson 2015).

Addressing Exploitation by Family Members

Financial institutions can flag and report potential abuse, but they fear violating privacy rules as section 502 of the Gramm-Leach-Bliley Act (GLBA) generally prohibits a financial institution from disclosing nonpublic personal information about a consumer to nonaffiliated third parties unless the consumer is notified and can opt out; however, the law exempts actions taken to protect consumers from actual/potential fraud, unauthorized transactions, claims, or other liability (Gleason 2017). Financial planners can have a trusted family member of cognitively impaired clients attend the meeting and receive copies of financial documents (Atkinson 2016). Financial institution Wells Fargo has trained its employees to identify and respond to suspected FEOA (Strouse and Gomulka 2014). Signs of mental changes and/or abuse could include disorientation, confusion with simple concepts, forgetting basic facts or recent conversations or how to do simple tasks, drastic investment changes, unexplained withdraws or wire transfers, erratic behavior/mood swings, memory loss, diminished vision/hearing, inability to make decisions, and/or overreliance on a third party (BITS 2016).

CFPB's recommendations are to coordinate efforts to educate better, use technology to the flag, identify and report suspected abuse, develop a relationship with the state adult protective services agency, and extend the time to report unauthorized transactions when persons are in hospital/adult care (Gleason 2017). Age-friendly banking recommendations from the National Community Reinvestment Coalition (NCRC) include customizing products for older customers and providing them with knowledgeable personnel, offering affordable retirement planning services, ensuring that older adults have access to critical income support programs and electronic benefits, designing online banking training and features to be age-friendly, and creating programs for identifying and reporting suspected elder abuse (Gleason 2017).

Juanita Stone pressed charges against her granddaughter Casey Jarvis who had power of attorney to use her grandmother's money to pay off her nursing home bills but drained her grandmother's finances instead. Stone pressed charges, and Jarvis received 5 years (Darst 2012). The first step is reporting (intervention), but older people hesitate due to self-blame and dependency on the abuser or feelings of sympathy/protection (toward ill/addicted/dependent abusers). The next step is to determine cognition. With cognitively functional victims, they can be given counseling to recognize the dysfunction of the relationship and also to keep themselves safe and to survive (Gibson and Qualls 2012). Crimes can be hard to prove, especially if the victim does not have their mental faculties and when the money is not used for large purchases; restitution is frequently sought as a solution (Darst 2012).

One way to stave off any of these issues is to develop a financial plan as soon as possible that includes making specific plans of action in the cases of increased age and incapacity (Rapacon 2018). Financial experts advise older adults to establish and stick to a budget that allows for all current obligations, to question fees and risks before purchasing/investing in any financial product, to set aside 3 months' worth of emergency funds, and to account for unexpected emergencies when doing estate planning (e.g., creating a durable power of attorney, living will, property will, and healthcare proxy) with a lawyer or through free/low-cost legal services available to older persons (BITS 2016) (See ▶ "Legal Services"). In addition, experts advise storing estate documents in a safe place where only trusted family members could access them (BITS 2016). When choosing a financial power of attorney (FPOA), experts advise picking a trusted individual who understands the responsibilities and limits of the FPOA role and informing family of such financial plans before something happens (BITS 2016). Older adults should also have a team in place that is apprised of and can be trusted to follow the plans and should keep multiple people apprised of one's plans and wishes to thus create a system of checks and balances (Rapacon 2018).

Future Directions of Research

The Elder Justice Act (EJA) was the first comprehensive US legislation to address the abuse of older persons in 2010 but was unfunded until 2015 when it was underfunded (e.g., $4 million vs. reported promises of $1 billion in funding) (Leiber 2018). EJA was passed under the Patient Protection and Affordable Care Act and is said to be the first attempt to address exploitation through US federal law (Finkel 2012). Other such laws include 2018's the National Senior Investor Initiative Act that would require the Securities and Exchange Commission to create an interagency task force on financial exploitation of older Americans (Schoeff 2018) and the Senior $afe Act proposed by Sen. Susan Collins (R-Maine) (Pearson 2018). The S$A had bipartisan support, was modeled on Maine state protections for older citizens, and gave financial institutions immunity to report suspected FEOA to federal/state authorities (Pearson 2018) (See ▶ "Laws for Older Adults"). Internationally, the United Nations (UN) passed the Madrid International Plan of Action on Aging at the Second World Assembly on Aging in 2002, and it was adopted by 159 nations (See ▶ "World Assembly on Ageing"). However, it does not provide specific steps for countries to take in order to outlaw abuse and neglect (Finkel 2012). In December 2010, the UN General Assembly established an open-ended working group on aging (Finkel 2012). The trouble with establishing an international law is that older persons living in nations with so-called fat or robust economies will experience financial exploitation much differently than older persons living in nations in the majority world (i.e., where, as World Bank statistics indicate, 80% of humanity lives on less than US$10 a day) and vice versa (Finkel 2012).

Summary

The American Bar Association states in the first half of this century, people over age 60 globally will increase from 600 million to almost 2 billion (Finkel 2012). There are 45 million older Americans, and 10,000 more will reach 65 every day (Idzelis 2017). As the older adult population grows, so too has greater attention been given to their mistreatment (Krug et al. 2002). Increased attention and passing legislation are just the beginning steps in addressing exploitation. Given the complexity and diversity of the older adults who experience FEOA by family, and the motivating factors for exploitative family members, any solutions must include several stakeholders (e.g., older adults, families of older adults, the community, medical/financial/social services professionals). When FEOA by family members is discovered and addressed and/or remedied, older adults can expect to face a myriad of issues that can be emotional, psychological, financial, and physical in nature.

Cross-References

▶ Abuse and Caregiving
▶ Economics of Aging
▶ Financial Assets
▶ Financial Frauds and Scams
▶ Financial Literacy
▶ Gender and Caring in Later Life
▶ Isolation
▶ Laws for Older Adults
▶ Legal Services
▶ Loneliness
▶ Older Adults Abuse and Neglect
▶ Widowhood
▶ World Assembly on Ageing

References

Atkinson W (2016) Family, friend or foe? Financ Plann 46(2):54–55

BITS (2016) Protecting the elderly from financial fraud and exploitation. In: White paper for the elder justice coordinating council. BITS Financial Services Roundtable. Available via Administration for Community Living. https://acl.gov/sites/default/files/programs/2016-09/Smocer_White_Paper.pdf. Accessed 20 Oct 2018

Darst A (2012) Exploiting the vulnerable. KY Law Enforcement Mag 11(1):48–51

Finkel E (2012) The forgotten demographic: ABA wants an international convention on elder rights. ABA J 98(5):58–59

Gibson S, Qualls SH (2012) A family systems perspective of elder financial abuse. Generations 36(3):26–29

Gleason, L (2017) Combating elder financial abuse. In: Consumer Compliance Outlook. Federal Reserve System. Available via Consumer Compliance Outlook. https://consumercomplianceoutlook.org/2017/first-issue/combating-elder-financial-abuse/. Accessed 20 Oct 2018

Hall M (2014) Elder abuse. In: Davies P, Francis P, Wyatt T (eds) Invisible crimes and social harms: critical criminological perspectives. Palgrave Macmillan, London, pp 102–122

Idzelis C (2017, April 3) Easy prey; Growing numbers of seniors are vulnerable to financial abuse. Investment News. Gale Academic One File, Farmington Hills. https://www.gale.com/c/general-onefile. Accessed 20 Oct 2018

Khalfani-Cox L (2011) Are you a victim of financial abuse? Recognize the signs – and fight back. In: AARP Magazine. American Association of Retired Persons. Available via AARP Online. https://www.aarp.org/money/scams-fraud/info-03-2011/are-you-being-financially-abused-by-a-family-member.html. Accessed 20 Oct 2018

Krug EG, Dahlberg LL, Mercy JA, Zwi AB, Lozano R (eds) (2002) Abuse of the elderly. In: World report on violence and health. World Health Organization. Available via World Health Organization International. https://www.who.int/violence_injury_prevention/violence/global_campaign/en/chap5.pdf. Accessed 20 Oct 2018

Lieber N (2018) How criminals steal $37 billion a year from America's elderly. In: Bloomberg Businessweek. Bloomberg LP. Available via Bloomberg online U.S. Edition. https://www.bloomberg.com/news/features/2018-05-03/america-s-elderly-are-losing-37-billion-a-year-to-fraud. Accessed 20 Oct 2018

Melchiorre MG, Di Rosa M, Lamura G, Torres-Gonzales F, Lindert J, Stankunas M, Ioannidi-Kapolou E, Barros H, Macassa G, Soares JJF (2016) Abuse of older men in seven European countries: a multilevel approach in the framework of an ecological model. PLoS One 11(1). Gale Academic One File, Farmington Hills. https://www.gale.com/c/general-onefile. Accessed 20 Oct 2018

Office for Older Americans (2016) A resource guide for elder financial exploitation prevention and response networks. Consumer Financial Protection Bureau. Available via Consumer Financial Protection Bureau. https://files.consumerfinance.gov/f/documents/082016_cfpb_Network_Resource_Guide.pdf. Accessed 20 Oct 2018

Olson E (2015, November 28) A hidden scourge: preying on the elderly. New York Times. EBSCO host Research Platform, Ipswich. https://www.ebsco.com/products/ebscohost-platform. Accessed 20 Oct 2018

Pearson, KC (2018) The new Senior Safe Act encourages reporting financial abuse. In: Wealth management. Informa. Available via Informa USA. https://www.wealthmanagement.com/high-net-worth/new-senior-safe-act-encourages-reporting-financial-abuse. Accessed 20 Oct 2018

Peterson JC, Burnes DP, Caccamise PL et al (2014) Financial exploitation of older adults: a population-based prevalence study. J Gen Intern Med 29(12):1615–1623. https://doi.org/10.1007/s11606-014-2953-3

Rapacon, S (2018) Beware elder financial abuse in the family: aging boomers are particularly at risk for economic abuse perpetrated by friends, family or neighbors. In: US News and World Report. US News & World Report L.P. Available from Money.USNews.com. https://money.usnews.com/money/personal-finance/family-finance/articles/2018-03-05/beware-elder-financial-abuse-in-the-family. Accessed 20 Oct 2018

Schoeff M (2018, July 18) 3 Bills key to advisers pass house; Measures include SEC easing regulatory burden on small firms. In: Investment News. Crain Communications. Available via General OneFile. http://link.galegroup.com/apps/doc/A547789763/ITOF?u=cuny_kingsboro&sid=ITOF&xid=44ac855c. Accessed 20 Oct 2018

Strouse C, Gomulka J (2014) Capsule commentary on Peterson et al., "Financial exploitation of older adults: a population based prevalence study". J Gen Intern Med 29(12):1688. https://doi.org/10.1007/s11606-014-2953-3

Financial Exploitation of Vulnerable Adults by Family Members

▶ Financial Exploitation by Family Members

Financial Frauds and Scams

Paul Salvin and Nimkit Lepcha
Department of Peace and Conflict Studies and Management, Sikkim University, Gangtok, Sikkim, India

Synonyms

Financial abuse of older people; Frauds and older people; Older financial abuse; Older financial exploitation; Older financial scams; Scams and older people

Definition

The complex and diverse nature of financial abuse of older people makes it difficult to define the financial abuse of older people. However, Wilson and Brown state "Financial abuse can range from failure to access benefits, through inadvertent mismanagement and opportunistic exploitation to deliberate and targeted abuse, often accompanied by threats and intimidation" (Wilson and Brown 2003). Mervyn Eastman (1994) identified financial abuse as situations "when relatives either fraudulent for personal gain or... where a family member actually steals from the purse of the elderly dependent." Action on Elder Abuse (a United Kingdom-based voluntary organization working for protection and prevention of elder abuse) defines financial abuse as stealing from, defrauding someone off, or coercing someone to apart with goods and property. Its definition of elder abuse focuses on any relationship where there is an expectation of trust, excluding abuse perpetrated by strangers (Fitzgerald 2004).

Older Americans Act of 1965 defines older individuals as individuals who are 60 years of age or older and defines old financial frauds as the illegal or improper use of an older person's funds, property, or assets. Some of the examples are cashing an elder adult person's check without authorization or permission; forging an older person's signature; misusing or stealing an older person's money or possessions; coercing or deceiving an older person into signing any documents; and the improper use of conservatorship, guardianship, or power of attorney. Moreover, a more detailed definition refers to "financial exploitation of an older person by another person or entity, that occurs in any setting (e.g., home, community, or facility), either in a relationship where there is an expectation of trust and/or when an older person is targeted based on age or disability" (Connolly et al. 2014). Some definitions distinguish between two types of seniors' financial exploitation: financial abuse, in which a relationship trust has been violated by family members, friends, or others, and old fraud, such as scams perpetrated by strangers (DeLiema and Kendon 2017).

Overview

Financial frauds and scams have become common nomenclature in contemporary society as the scope of fraudulent activities are growing, and to deal with such challenges has become an awful task for the administration. It becomes so appalling when older people are subjected to financial frauds and scams, thinking that they have a significant amount of money in their accounts. Older people, regardless of being wealthy, low income, or middles class, are all becoming the target of fraudulent activities, and the perpetrators are not always the stranger but their own close ones, including family members. As financial exploitation of older people has become an international issue all across the ethnic races, it has gained the attention of financial regulators, firms, and professionals to adopt new measures to overcome and protect especially the vulnerable people such as older people whose health and aged ailments may not support to follow up to get justice. Most of the minor financial scams go unreported, especially if the target is older people; such crimes get the least attention by the judiciary and police as they are considered a low-risk crime. Financial exploitation can be a devastating effect on older people.

Not only their comfortable lifestyle disappear, but they do not have the time or opportunity to recover financially which can be a life-threatening event characterized by fear, lack of trust, and often acute and chronic anxiety (James and Graycar 2000).

Meaning of Fraud

Fraud is a complex and elusive concept ranging from such cursory ones as "the obtaining of goods and or money by deception" (Levi 2009). Fraud is defined as "a human endeavor, involving deception, purposeful intent, the intensity of desire, the risk of apprehension, violation of trust, rationalization, etc." (Ramamoorti and Olsen 2007). In the context of financial market activities such as banking, securities, and insurance, fraud is attributed to a specific meaning and best understood as the unlawful falsification or manipulation of financial information (Fligstein and Roehrkasse 2013).

Some of the examples of fraud include but not limited to corruption include conflicts of interest, bribery, illegal gratuities, and extortion; misappropriation of cash asset including larceny, skimming, check tampering, and fraudulent disbursement and expense reimbursement schemes; fraudulent statement which includes financial reporting, employment credentials, and external reporting; misappropriation of noncash asset which includes larceny, false asset requisitions, destruction, removal or inappropriate use of records and equipment, inappropriate disclosure of confidential information, and document forgery or alteration; and fraudulent practice by customers, vendors, or other parties including bribes or inducements and fraudulent (rather than erroneous) invoices from a supplier or information from customer (Internal Audit 2012). Most of the fraud research has focused on fraud against governments and organizations, which is distinct from fraud against individuals, both in its methods and its players. The lack of a clear definition has allowed only financial fraud to remain relatively overshadowed and overlooked (Beals et al. 2015). The Federal Bureau of Investigation (FBI) in the United States defines financial fraud as "illegal acts characterized by deceit, concealment, violation of trust. They are committed to obtaining money, property, or services" (Deem 2000).

Financial Frauds Related to Older People

Older people become the victims of this through their friends, relatives, and caregivers who financially exploit them taking advantage of their trust to gain control of bank accounts, checkbooks, and payment cards, often under the guise of helping them manage their finances. The abuser may be appointed a power of attorney, a legal guardian, a trustee, or someone else in a fiduciary role or having open access to older people's money through a familial bond. Fraud perpetrators, by contrast, are predatory strangers who earn their target's trust by promising a future benefit or reward in exchange for money or personal information upfront (DeLiema and Deevy 2016).

Older financial frauds include but not limited are taking money or items from an older adult's home or accounts without proper authority or approval; occupying, selling, or transferring property against an older adults wishes or best interests; unauthorized credit or debit card use; opening credit accounts in older adults name using their good credit and older personal information to obtain services such as telephone, cable, and basic utilities, rent lease, or buy properties; use of insurance information to obtain medical services; creating or changing insurance policies to benefit another; changing wills, trusts, or inheritance arrangements for another's benefit without an older adult's knowledge or permission; abusing joint signature authority on a bank account; misappropriation of funds from a pension; getting an older adult to sign a deed, will, contract, or power of attorney through deception; negligently mishandling assets, including misuse by a fiduciary or caregiver; and denying older persons access to their money or preventing them from controlling their assets (Hall et al. 2016). One 2015 report estimated that older Americans lose $36.5 billion each year to financial scams and abuse. Three in ten state security regulators say they have seen an uptick over the past year in cases and complaints involving senior financial fraud and exploitation, according to a new survey from the North American Securities Administrators Association.

The association president, Rothman, said, "it is easier to try to exploit a senior citizen with cognitive or other impairments in financial issues, who are alone than it is to rob a bank, so they are targets" (Grant 2017).

According to the North American Securities Administrators Association (Grant 2017), percentage of cases reported regarding the older financial fraud are such third-party abuse/exploitation 27%; family member, trustee, or power of attorney taking advantage 23%; diminished responsibility 12%; combined diminished capacity and third-party abuse 12%; fraud 6.30%; older exploitation 5.70%; taking advantage from friends, housekeeper, or caretaker <1%; and excessive withdrawals <1%.

According to the United States Senate Special Committee on Aging, seniors' financial abuse was one of the ten consumer scams in 2017 targeting seniors. The committee maintains a Fraud Hotline as a resource for seniors and others affected by scams, and it tracks consumer frauds reported to it. Other top ten scams included the Internal Revenue Service (IRS) in the US impersonation scams (impersonating an IRS agent and falsely accusing seniors of owing back taxes and penalties in order to scam them); sweepstakes scams, including a Jamaican lottery scam; computer scams; romance scams; and grandparent scams (impersonating a grandchild who claims to need money for an emergency). As the list indicates, older people are targeted for a wide variety of scams; a number of scams involve the false promise of goods, services, or financial benefits that do not exist (Santucci 2017).

Types of Financial Frauds

The National Council on Aging (NCAG 2015) has categorized older people financial frauds or scams into ten types.

Healthcare/Medicare/health insurance fraud: in these types of scams, perpetrators may pose as a Medicare representative to get older people to give them their personal information, or they will provide bogus services for older people at makeshift mobile clinics and then use the personal information they provide to bill Medicare and pocket the money.

Counterfeit prescription drugs: most commonly, these types of scams operate on the Internet, where seniors increasingly go to find better prices on specific medications. The danger in that besides paying money for something that will not help a person's medical condition, victims may purchase unsafe substances that can inflict even more harm which can be as hard on the body as it is on the wallet.

Funeral and cemetery scams: the FBI warns two types of funeral and cemetery fraud perpetrated on seniors; in one approach, scammers read obituaries and call or attend the funeral service of a stranger to take advantage of the grieving widow or widower. Claiming the deceased had an outstanding debt with them, scammers will try to extort money from relatives to settle the fake debts. Another tactic is for the disreputable funeral home to capitalize on family member's unfamiliarity with the considerable cost of funeral services to add unnecessary charges to the bill.

Fraudulent antiaging products: society is putting so much emphasis on physical appearance; many individuals feel the need to find treatments or products that claim to help them conceal their age. Scammers advertise antiaging products that are either worthless or harmful. Some products might contain materials that can be harmful yet touted by scammers as being as effective as a brand name product, such as Botox. Scammers might also advertise products as being effective and natural, but in reality, the product has no antiaging effects.

Telemarketing: phone scams are the most common frauds used against older people. Scammers might get seniors to wire or send them money by claiming to be a family member who is in trouble and needs money. They might also solicit money from older people by posing as a fake charity, especially after a natural disaster.

Internet fraud: older people are usually not as savvy with handling emails and surfing the Internet; they are targets for scammers. Victims have been tricked into downloading fake antivirus software that allows scammers access to personal information on their computers. Seniors might also respond to phishing emails sent by scammers asking them to update their bank or credit card information on a phony website.

Investments: many seniors find themselves planning for retirement and managing their savings once they finish working; a number of investment schemes have been targeted at seniors looking to safeguard their cash for their later years. From pyramid schemes such as Bernie Madoff's (which counted a number of senior citizens among its victims) to fables of a Nigerian prince looking for a partner to claim inheritance money to complex financial products that many economists do not even understand, investment schemes have long been a successful way to take advantage of older people.

Homeowner/reverse mortgage scams: older adults who own their homes can be valuable assets to scammers. Property tax scam in San Diego saw fraudster sending personalized letters to different properties apparently on behalf of the country Assessor's Office.

Sweepstakes/lotteries: this scam usually contacts older adults by mail or telephone and inform them that they have won a prize of some sort but must pay a fee to obtain. Scammers send a fake check to the senior to deposit in their bank account.

The grandparent scam: this fraud is exceptionally deceptive because it plays on older peoples' emotions. In this type of scam, the scammer calls an older person and pretends to be their grandchild. They ask them if they know who is calling, and when the grandparent guesses the name of one of their grandchildren, they pretend to be that grandchild. The scammer tells the grandparent that they are in some financial bind and asks if they can send money using Western Union or MoneyGram to help them out. The scammer asks the grandparent not to tell anyone about their situation. Once the scammer receives the money, he continues to contact the grandparent and asks them to send more money.

Causes or Factors Relating to Financial Frauds

The report by David Burnes et al. (2017) estimates that 5.4% of older adults experience some form of fraud or scam each year. The estimate includes only seniors living on their own and excluding those in institutional settings and most who are cognitively impaired. The scammers target an older adult who is lonely and socially isolated and will develop an online relationship and over time will ask them to send money over and over again. There are several factors which are primarily confined to the perception of fraudsters toward older adults. *Firstly*, fraudster thinks that older people are easier to confuse and therefore more likely to fall victims. The scammer believes that the longer he or she can keep an older person talking, the more the victim will become wrapped up into the fraudster's story. *Secondly*, they believe that older adults are more trusting of others and are more likely to make small talk with a stranger, whether in person, over the phone, or by email. *Finally*, they are unlikely to report the crime, some older adults' victims feel embarrassed that they have fallen for the scam, and in some cases, they fear the family members will think they are unfit to manage their own finances.

Stephen Deane (2018) has given three interrelated sets of factors for financial exploitation of older people, which include cognitive impairment, poor physical health, functional impairment, and dependency on others, which are associated with seniors' financial exploitation and other forms of elder abuse. The second set of factors are related to financial and retirement trends. The wealth or assets that older people have accumulated through life can make them a target of financial exploitation. Pension trends increasingly have shifted responsibility to older people themselves to manage their retirement savings and investments. The final growth in the aging population has, to some extent, emerged as the factor for the problem of seniors' financial exploitation (Deane 2018).

Cognition: cognitive decline is a crucial factor that makes older people more susceptible to financial exploitation. In the older age, most of older adults suffer from this cognitive decline which loses their ability to sense that something is just not right or become too trusting and fail to recognize false claims, suspicious intentions, and signs of riskiness (Blazer et al. 2015).

Financial and Retirement Trends: older adults become targets because of the retirement savings and other assets which they have built throughout their life. In a New York study (2014) on older financial exploitation, authors have observed that older adults are more likely to have financial resources than their younger counterparts, and this in combination with the higher prevalence of social isolation, cognitive impairment, and other factors renders them uniquely susceptible to financial exploitation (Deane 2018).

Older people who undergo financial fraud may experience a range of psychological and financial implications as they are potentially on fixed incomes and may not have the resources to recover from an instance of financial fraud. Due to the feelings of embarrassment and shame, older people may not report the incidents, which makes legal intervention difficult if not impossible (Rabiner 2004).

Impact of Fraud upon Older People

According to Deem, the impact that financial fraud may have on older people is profound after being victimized by financial abuse; older adults not only lose their trust in others but also the trust of self and own judgment (Deem 2000). Older people may feel unable to handle their finances any longer following an incident of financial abuse. Deem reported the feelings of the victim who stated, "I felt like I was raped" which shows that it is not just the loss of money but more than that. The older person experiences a chain reaction of symptoms that include depression and feelings of guilt and shame, disbelief, anger, depression, a sense of betrayal, and a loss of trust (Deem 2000).

Legal Remedies

The Securities and Exchange Commission (SEC) and Financial Industry Regulatory Authority (FINRA) actions to protect senior investors: The SEC's efforts to combat older people's financial exploitation mostly fall within the broader goal to protect the senior investor, including those nearing retirement and those who are already retired. The SEC's efforts are based on the three E's: education, exams, and enforcement. Seniors are a key audience for the SEC's education works with the enforcement staff to promote awareness aimed at preventing fraud. It hosted events for World Elder Abuse Awareness Day each year since the inaugural summit in June 2015. Examinations form the second leg of the SEC efforts to protect senior investors. Each year commission staff conducts a National Exam Program of broker-dealers and investment advisers. Enforcement constitutes the third leg of SEC efforts to protect seniors. The SEC's Division of enforcement places a priority on protecting older people in its investigations and actions. FINRA operates a Securities Helpline for seniors, which seniors can call toll-free to receive neutral, knowledgeable assistance on topics such as understanding how to review investment portfolios or account statements, concerns about the handling of a brokerage account, and investor tools and resources from FINRA (Deane 2018).

Future Directions of Research

It is easier to say why older financial exploitation is expected to grow than to determine how big a problem is. The older financial exploitation can be measured through several dimensions, we can compare them to the corresponding rates of other types of elder abuse, and it is also essential to measure the extent that older financial exploitation goes unreported to authorities. It must also be considered the cost of older financial exploitation, both financial and nonfinancial, to the victims and society. As digitalization in the society has become a reality, how older get impacted by such nuances and whether the technology could bring a solution to the financial frauds and scams in such an emerging society need to be delved into.

Summary

Financial abuse of older people has grown rapidly in the present era; along with the older people, their families and all who provide services and financial assistance to them get victimized. Many consider these financial scams as the crime of the twenty-first century. It is assumed that the highest part of the wealth in the country is controlled by older people, but all seniors, regardless of income, are at risk. Abuse may be perpetuated by anyone, a professional con artist, paid caregiver, stranger, casual acquaintance, or even a son, daughter, or another family member. Older people are likely to be the victims of financial scams for various reasons; some of the examples are they are known to possess a large amount of wealth and assumed to be vulnerable and in some cases they are often dependent on others for help and support maybe due to loneliness and desire for companionship. They often do not understand financial matter and being emotional; they tend to trust very easily compared to younger counterparts, which may have trouble spotting fraud. Since older people population is increasing, it is contributing to the source of wealth for financial abusers; in order to eliminate this abuse, the two most effective strategies are education and awareness.

Cross-References

▶ Financial Assets
▶ Financial Exploitation by Family Members
▶ Financial Literacy
▶ Financial Socialization
▶ Older Adults Abuse and Neglect

References

Beals M, DeLiema M, Deevy M (2015) Framework for a taxonomy of fraud. Stanford Longevity Centre, FINRA Financial, Washington, DC. http://longevity3.stanford.edu/wp-content/uploads/2015/11/Full-Taxonomy-report.pdf. Accessed 28 Nov 2018

Blazer DG, Yaffe K, Liverman CT (eds) (2015) Cognitive aging: Progress in understanding and opportunities for action. The. National Academies Press, Washington, DC. https://doi.org/10.17226/21693

Burnes D, Henderson CR, Sheppard C, Zhao R, Pillemer K, Lachs MS (2017) Prevalence of financial fraud and scams among older adults in the United States: a systematic review and meta-analysis. Am J Public Health 107:e13_e21. https://doi.org/10.2105/AJPH.2017.303821

Connolly MT, Brandl B, Breckman R (2014) The elder justice roadmap: a stakeholder initiative to respond to an emerging health, justice, financial and social crisis. US Department of Justice, Washington, DC. https://www.justice.gov/file/852856/download. Accessed 28 Nov 2018

Deane S (2018) U.S. Securities and Exchange Commission, Office of the Investor Advocate, Elder Financial Exploitation. http://www.sec.gov/files/elder-financial-exploitation.pdf

Deem DL (2000) Notes from the field: observations in working with the forgotten victims of personal financial crimes. J Elder Abuse Negl 12(2):33–48

DeLiema M, Deevy M (2016) Aging and financial victimization: how should the financial service industry respond? Pension Research Council, The Wharton School, University of Pennsylvania, Philadelphia. https://pensionresearchcouncil.wharton.upenn.edu/wp-content/uploads/2016/07/WP2016-4-DeLiema-and-Deevy-Text.pdf. Accessed 28 Nov 2018

DeLiema M, Kendon JC (2017) Financial exploitation of older adults. In: Dong XQ (ed) Elder abuse: research, practice and policies. Springer, Heidelberg, pp 141–157

Eastman M (1994) Old age abuse: a new perspective. Springer, New York

Fitzgerald G (2004) Hidden voices: older People's experience of abuse an analysis of calls to the action on elder abuse helpline. Action on Elder Abuse is an Association with Help the Aged, London

Fligstein N, Roehrkasse A (2013) All the incentives were wrong: opportunism and the financial crisis. Conference paper, American sociology annual meeting, New York. https://www.law.berkeley.edu/files/csls/Fligstein_Paper_CSLS_23_Sep13(1).pdf. Accessed 28 Nov 2018

Grant KB (2017) Elder Financial Fraud is $36billion and Growing. CNBC. https://www.cnbc.com/2017/08/25/elder-financial-fraud-is36-billion-andgrowing.html. Accessed 28 Nov 2018

Hall J, Karch DL, Crosby A (2016) Elder abuse surveillance: uniform definitions and recommended core data elements. Center for Disease Control and Prevention and Control Division of Violence Prevention, Atlanta. https://www.cdc.gov/violenceprevention/pdf/EA_Book_Revised_2016.pdf. Accessed 28 Nov 2018

Internal Audit (2012) http://internalaudit.ku.edu/fraud. Accessed on 28 Nov 2018

James M, Graycar A (2000) Preventing Crime Against Older Australasians, Australia Institute of Criminology, Research and Public Policy Series, No. 32, Canberra

Levi M (2009) Financial Crimes. In: Tonry M (ed) Oxford handbook of crime and public policy. Oxford University Press, New York, pp 223–246

NCAG (National Council of Aging) (2015) Top 10 financial scams targeting seniors. https://www.ncoa.org/economic-security/money-management/scams-security/top-scams-targetting-seniors/. Accessed on 30 Nov 2018

Rabiner DJ (2004) Bridging essay to link two financial exploitation. J Elder Abuse Neglect 16(1):51–52

Ramamoorti S, Olsen W (2007) Fraud: the human factor. Financial Exec 23(7):35–53

Santucci L (2017) Can data sharing help financial institutions improve the financial health of older Americans? Payment cards center. Federal Reserve Bank of Philadelphia, Philadelphia. https://www.philadelphiafed.org/-/media/consumer-finance-institute/payment-cards-center/publications/discussion-papers/2017/dp17-01.pdf?la=en. Accessed 28 Nov 2018

Wilson B, Brown H (2003) Introducing special issue on financial abuse, Journal of Adult Protection 5(2):2

Financial Gerontology

Erik Selecky[1] and Andrzej Klimczuk[2]
[1]Center for Continuing Education, Technical University, Zvolen, Slovakia
[2]SGH Warsaw School of Economics, Warsaw, Poland

Definition

Financial gerontology can be defined as investigating relations between finances and aging. Authors such as Neal E. Cutler, Kouhei Komamura, Davis W. Gregg, Shinya Kajitani, Kei Sakata, and Colin McKenzie (Kajitani et al. 2017) affirm that financial literacy is an effect of aging with concern about the issue of finances, as well as stating that it is the effect of longevity and aging on economies or the financial resilience of older people.

Overview

Financial gerontology dates back to the 1980s when Jozef Boettner and Davis Gregg started to deal with the effects of financial education on population aging (Vitt and Siegenthaler 1996; Cutler 2016). They felt the need to educate older generations about finances and financial services in a professional way. In the year 1986, the first research entity focused on the financial gerontology, "Boettner Research Institute" in the United States, was founded and led by Gregg. The first director of the Institute was Neal E. Cutler. The Institute focused not only on professional education but also on research. The American Institute of Financial Gerontology (AIFG) (2019) was later founded and still exists today. Financial gerontology is a scientific discipline that overlaps with other branches of science such as biology, psychology, sociology, economics, and demography. As a scientific discipline, it evolved due to rapidly advancing technology, medicine, and their impacts on the society and economy. Financial gerontology, as a whole, is perceived as a science aimed at supporting a healthy aging, active lifestyle, self-sufficiency in later life, and in general improving the quality of life of older people (Timmermann 2018). Financial gerontology does not solely mean the research of older people but rather focuses on the processes of financial planning for aging and longevity (Komamura 2017).

The Effects of Aging and Old Age on Financial Well-Being

Aging is a natural process involving biological, health, economic, and social changes. It is an irreversible process (effect of genetics, the environment, and lifestyle), connected with psychological age (e.g., individual behavioral changes during the process of aging) and sociological age (e.g., termination of employment and social role transition). The challenge of population aging is discussed in many strategic documents around the world, and it is implemented in many programs and statements. Old age is the last stage of the human life cycle. Being aware of an approaching and inevitable end can result in various difficulties relating to the adaptation and coping with later age and age-related changes as well as demand financial education and socialization and retirement planning (Timmermann 2016a). Coping with aging successfully in terms of financial well-being is an essential condition for achieving satisfaction in later life. Old age is accompanied by physical and mental changes arising due to changes in the external as well as the internal environment. All older people can live with dignity and age gracefully, enjoy life in old age, and prevent medical problems. Many of these factors depend upon their personal choice and the way of living which they prefer. Individual approaches to successful aging affect the quality of life to a substantial extent.

The Effect of the Economy on Retirement

Older citizens can be defined as being citizens who can receive a pension. Terms such as an early pensioner or pensioner with disability are quite common. Pensions received by these people differ from universal basic pensions related to old age. An increase in life expectancy can be seen due to many factors such as the rapid development of technology, electronics, information communication technologies (ICTs), and increasingly effective medicine. The retirement age varies around the world. One of the reasons for this is the various levels of socioeconomic development. For instance, in Germany, the retirement age is 67, whereas in Slovakia, it is 63 and 55 in Ukraine. Concerning the recent world statistics, Japan's retirement age is the highest at 75 (Komamura 2017).

Pension systems in individual countries must reflect a variety of standards of retirement ages. Pension schemes must be flexible because of gradual changes in human life spans. The role of public and private pension systems must be considered, especially concerning how their influence and the rules of investment appraisal. The term "the problem with longevity" can be perceived negatively mainly from an economic point of view (Allenby 2013; Mitra 2017). Everybody is willing to live as long as possible, but, of course, quality of life plays a significant role. Moreover, negative age stereotypes and agism are already also challenging for financial services advisors, firms, and related financial sectors such as banking and insurances (Migliaccio 2019).

The issue of population aging is also affected by migration. The rules in developed countries provide accepting foreign-born people because of social or economic reasons. The economy of a country is also significantly affected by education, and it must be recognized that the percentage of populations with university degrees is getting higher. To a considerable extent, university students continue in their studies and apply for PhD courses. On the one hand, full-time students are a part of the education system. On the other hand, the retirement age has to be increased gradually. It follows that the productive years of people hiring in secondary education job positions and jobs oriented to manual work are extended, and it is beneficial to the state because of the levied taxes.

Financial Issues in the Context of the Silver Economy

The term silver economy is very closely related to financial gerontology. The silver economy is used to describe the consumption, production, and distribution of goods and services for the aging population, especially older people (Klimczuk 2016). As the number of older adults increases, the number of older consumers and voters increases as well. States and their economies will have to react and adapt to such changing situations. The more developed the economy is, the higher the purchasing power of seniors is. In the structure of consumption expenditure, Engel's law (Kindleberger 1990) is valid: the more money someone has, the less money they spend on necessities of life such as food and accommodation, but more money they spend on a healthy lifestyle and higher needs such as healthcare, holidays, culture, and charity. Therefore, the fact that an increase in life expectancy can increase health and social care and related expenses can be expected. The significant role of the economy in individual countries must be flexible enough to look at the different angles of such changes. This is an essential task not only for economists and analysts but also for politicians.

Educational Programs Related to the Financial Gerontology

Systematic education can help enhance theoretical and practical knowledge concerning financial gerontology. Education programs in this area concentrate on people both close to retirement and those already receiving pensions. Other target groups include consultants, lecturers, and teachers. In other words, two groups can be affirmed: those being educated and the educators themselves.

A critical publication in the field entitled *Encyclopedia of Financial Gerontology* was edited by Lois Vitt and Jurg Siegenthaler (1996). The encyclopedia entries were divided into eight main categories: economic and income security; employment, work, and retirement; family and intergenerational issues; financial advice, investments, and consumer services; healthcare and health insurance; housing and housing finance; legal issues and services; and quality of life and well-being. These topics could be included in preparing education courses, seminars, and programs.

The American Institute of Financial Gerontology (AIFG) (2019) is one of the most important institutions that deal with financial gerontology. The AIFG provides education and consulting services for older people and their families. Education is aimed at older generations and understanding their needs. The main focus is on the family and intergenerational aspects, aging, health-related expenses, marketing to people age 50+, and older people's readiness and preparation for later life. Neil E. Cutler, the current president and a principal author in the field of financial gerontology, puts great emphasis on financial planning (Cutler 2008).

In the European environment, there are publications (Gracova et al. 2017; Selecky 2017) that are analyzing seniors' education provided by the Universities of the Third Age. Education concerning finances is oriented toward topics such as how debts arise and how not to become insolvent; financial environment; tips to get finances under control (incomes-expenditures); financial goals; setting money aside; payslips; the tax system; the economy of married couples or partners; personal finances; banking and bank services; how to purchase goods and services effectively; types of loans; savings; investment appraisals; and insurance and insurance products. The term financial literacy is commonly used at the Universities of the Third Age as a synonym to the financial gerontology. However, it cannot be defined precisely when older people are financially literate. It is more about being familiar with finances and financial mechanisms and about the information on how to use them correctly in various life situations.

Future Directions of Research

As has already been mentioned, financial gerontology concerns both people who are going to receive retirement and those who are already retired and are receiving a pension. A third group consists of consultants, lecturers, and teachers – people specialized in education (Migliaccio 2017). Informal education, for example, further education providing training courses and seminars, should be offered to people just before their retirement. The second target group can be educated at the Universities of the Third Age (Selecky 2017; Klimczuk 2013). Education can be provided by universities, nongovernmental organizations, civic associations, community centers, towns, and villages or private companies. The third target group could enhance their knowledge using formal education provided by universities, professional associations, and associations focused on the economy or private companies. A flourishing area of research potential might concern the members of the first target group – people just before retirement who need preventing despair and promoting resilience (Timmermann 2016b). The essential research topics cover questions such as how to think about working life, financial planning for the future, retirement, and how to be prepared for old age.

Summary

Financial gerontology is a young scientific discipline that focuses on the issue of finances, especially the relationships between finances, aging, and older people (Cutler 2008). Financial gerontology covers a range of cutting-edge topics with immense potential not only in the future but also at present. Financial gerontology is necessary to deal with the demographic transition models, life span, and later retirement regarding the economy (Cutler 2016). Financial gerontology is needed to raise awareness of economic and financial issues. Not only developed countries should adapt to changing situations and provide new models for life insurance and pension funds. The transition from the economically active population to the retired population should be supported by various programs, seminars, and courses focused on financial gerontology.

Cross-References

▶ Autonomy and Aging
▶ Financial Assets
▶ Financial Exploitation by Family Members
▶ Financial Frauds and Scams
▶ Financial Literacy
▶ Financial Social Work
▶ Financial Socialization
▶ Financial Therapy
▶ Gray is the New Green: Opportunities of Population Aging
▶ Health-Care Financing
▶ Human Wealth Span
▶ Joint Tenancy and Tenancy in Common
▶ Life Cycle Theories of Savings and Consumption
▶ Long-Term Care Financing
▶ Objectively Measured Financial Resources
▶ Socioeconomic Status
▶ Subjectively Measured Financial Resources

References

Allenby B (2013) Talking 'bout my generation: the real walking dead. Slate. https://slate.com/technology/2013/10/the-big-problem-with-longevity-old-people.html?via=gdpr-consent. Accessed 23 June 2019
Cutler NE (2008) Financial gerontology, family aging, and middle-aged boomers: using the "senior sandwich generation" concept in retirement planning. TIAA Institute, New York, pp 1–14
Cutler NE (2016) Twenty-five for 25: a quarter-century of financial gerontology. J Financ Serv Prof 70:25–33
Gracova D, Spulber D, Selecky E (2017) Universities of the third age in Europe. Carocci, Roma
Kajitani S, Sakata KEI, McKenzie C (2017) Occupation, retirement and cognitive functioning. Ageing Soc 37:1568–1596. https://doi.org/10.1017/S0144686X16000465
Kindleberger CP (1990) Engel's law. In: Kindleberger CP (ed) Economic laws and economic history. Cambridge University Press, Cambridge, UK, pp 3–20. https://doi.org/10.1017/CBO9780511559495.003
Klimczuk A (2013) Universities of the third age in Poland: emerging model for 21st century. J Educ Psychol Soc Sci 1(2):8–14
Klimczuk A (2016) Comparative analysis of national and regional models of the silver economy in the European Union. Int J Ageing Later Life 10(2):31–59. https://doi.org/10.3384/ijal.1652-8670.15286
Komamura K (2017) Longevity across the disciplines: new frontiers in financial gerontology. Keio University, Tokyo. https://www.keio.ac.jp/en/keio-times/features/2017/2/. Accessed 23 June 2019
Migliaccio JN (2017) Planning for the utterly unexpected: advice for the retirement advisor. J Financ Serv Prof 71:32–37
Migliaccio JN (2019) The ageism within and how to counter it. J Financ Serv Prof 73:26–31
Mitra S (2017) Our future – the problem with longevity. HuffPost News. https://www.huffpost.com/entry/our-future-the-problem_b_14003944. Accessed 23 June 2019
Selecky E (2017) Universities of the third age in Europe – best practice. Technical University in Zvolen, Zvolen
Timmermann S (2016a) Life planning and retirement planning: where do they intersect? J Financ Serv Prof 70:33–36
Timmermann S (2016b) Shocks and loss in retirement: preventing despair, promoting resilience. J Financ Serv Prof 70:34–38
Timmermann S (2018) Financial gerontology: what is it? Do we need it? What can we learn? J Financ Serv Prof 72:37–42
Vitt LA, Siegenthaler JK (eds) (1996) Encyclopedia of financial gerontology. Greenwood Press, Westport

Financial Holdings

▶ Financial Assets

Financial Knowledge

▶ Financial Literacy
▶ Financial Socialization

Financial Learning

▶ Financial Literacy

Financial Literacy

Lucie Vidovićová
Research Institute for Labour and Social Affairs, Brno, Czech Republic

Synonyms

Financial education; Financial knowledge; Financial learning; Financial proficiency; Financial skills

Definition

Financial literacy is a group of skills and abilities in the financial area. It includes an understanding of financial products and concepts with the help of information and advice and ability to identify and understand financial risks and make informed decisions (Vidovićová 2013: 191, OECD 2005, 2006, 2016). It can be empirically measured as knowledge (information), behavior (ability), and/or attitudes (values) related to finances. It can affect a broad range of everyday life experiences, such as using the credit cards and managing a monthly family budget, as well as more complex issues such as the stock market orientation or retirement planning, vulnerability to finacial scams and financial abuse.

Financial literacy can include basic numerical skills tested by mathematical tasks (numeracy); values and preferences for a particular investment, savings, and consumer lifestyles; and information and knowledge about the possibilities for ensuring financial stability in old age. Empirical studies may be interested (a) in an ability to calculate and (b) in a developed or advanced (as it is sometimes characterized) ability to navigate in financial markets and to use the information to develop one's own potential, well-being, and (financial) security and hence expect and minimize economic risks that can be related to pre-retirement, postretirement, and old age.

Overview

Financial literacy – that is, the ability to count, to navigate in financial systems, and to react adequately to risks – has three essential characteristics: relativity, variability, and continuity. The term *relativity* here refers to the current state and the expected development of a given society (Rabušicová 2002: 42). More demanding environments will require higher levels of literacy and vice versa, for example, complicated and complex rules for the calculation of pensions will pose higher demands on financial and functional literacy of people in pre-retirement ages to plan their financial future. The historical/generational conditions will exercise its influence on the financial literacy requirements as well. As a result, the financial literacy itself becomes a fluid concept, requiring complex and multidimensional measurements; making empirical comparisons between various groups, such as people of different age groups or people living in different countries; and being subject to different educational, economic, the welfare state, and cultural contexts, rather uneasy.

The concept of *variability* has two components: individual and contextual. Individuals develop and expand various literacy skills through the life course but can also limit or lose these depending on the extent to which they use a given skill. The other component is contextual as demands on the individual change in time and space and depending on the performance of various social roles and their cultural, social, economic, and legislative contexts. It is therefore impossible to characterize an individual in the duality of being financially literate or illiterate, but literacy should be seen as a *continuum* between these two poles, and these positions may differ for the different components of the financial literacy as well. Fourth, a *cumulative* characteristic of financial literacy can be added. The later life attitudes to financial management will be influenced by the layers of earlier experience gained through the lifetime, and it can have both positive and adverse effect on the material situation of the older person.

One of the critical components of financial literacy is quantitative literacy (numeracy)

(Skagerlund et al. 2018; Lusardi and Mitchell 2007a), the skill to count and understand basic as well as more advanced mathematical and logical operations. It can be illustrated by the fact that countries that rank high in Program for International Student Assessment (PISA) mathematical scales test tend to rank also high in financial literacy tests (Brown and Graf 2013). However, as the formal education (measured in years of educational attainment) tends to be generally lower for older respondents, in older adults the lack of formal education may be compensated for by the layers of experience with outcomes of various financial decisions gained over the lifetime. On the other hand, if there was not a reasonable basis for financial literacy, the knowledge deprivation may accumulate financial distress or unseized opportunities over life as well. It seems that later life financial literacy is determined by the early life, including culturally driven systematic differences (Hershey et al. 2007), and early life financial socialization (Brown et al. 2018), (See ▶ "Financial Socialization") fully supporting the idea of cumulative disadvantage (Dannefer 2003). (See ▶ "Theory of Cumulative Disadvantage/Advantage")

In the case of (pre-)retirement planning, there is another crucial, decisive factor related to more general attitudes toward aging and old age. As there are significant levels of ageism motivated by fears of aging, frailty, and dependence (Price et al. 2014), some individuals tend to avoid thinking about future in old age. About 6% in the research by Šlapák et al. (2010) gave as reason for not saving for retirement "I will not live so long to survive into my retirement, saving makes no sense". On the other hand, future time perspective (future orientation) is related to the tendency to plan and save (cf. Hershey et al. (2007) for an overview). Last but not least, the cognitive abilities of the individuals seem to play a significant role in the levels of financial literacy (Skagerlund et al. 2018).

Key Research Findings

Many authors conclude that the level of financial literacy in different countries around the world is low in many individuals (Lusardi and Mitchell 2014). According to Lusardi (2012), the lack of numeracy is not only widespread but is particularly severe among some demographic groups, such as women, older adults, and those with low educational attainment. Hershey et al. (2007) based on their study of retirement planners in the Netherlands and the USA, quite unsurprisingly, add the low-income groups to the list. Brown and Graf (2013) when testing the relation between financial literacy and retirement planning in Switzerland found that female respondents, respondents with low educational attainment, foreign nationals, and respondents with low income and wealth have significantly lower levels of financial literacy. In terms of age, both young and old respondents were less financially literate than middle-aged respondents. As the authors explain: "(t)his result is driven by two countervailing effects: Knowledge about inflation is positively correlated with age, while knowledge about compound interest and risk diversification is negatively correlated with age" (Brown and Graf 2013).

The standard financial literacy measures (Lusardi and Mitchell 2007a) heavily rest on the numeracy and include three aspects: (1) knowledge of interest compounding, (2) knowledge of inflation, and (3) knowledge of risk diversification. These questions involve knowledge of financial concepts, such as inflation, interest rates, and risk diversification as well as a firm grasp of numbers, percentages, and calculation procedures (Skagerlund et al. 2018). It includes the ability to work with the numerical information and to analyze and use it in a proper way. The typical question: *If the chance of getting the disease is 10%, how many people out of 1,000 would be expected to get the disease?* is in some studies also used as a filter and only those who do answer it correctly get to answer also more complex questions such as this one: "*The account earns 10 percent interest per year. How much would you have in the account at the end of the two years?*" For example, in the Czech Republic, the correct answer to the first question (100 persons) gave 87% of respondents 10 years before effective retirement age. For the second question, the correct figure, which would take into account the interest of the first year before adding the next year's 10% interest (e.g., 20,000 deposit would bring 2,000 in the year one, that is 22,000 in total in year 1 of which 10% in year 2 is 2,200, so the final figure is 22,000 + 2,200 = 24,200), gave 25% of the pre-retirees in the Czech Republic (Vidovićová 2013; Šlapák et al. 2010) and 18% in the original study by Lusardi and Mitchell (2007a) in the USA. Additional 37% in the Czech sample and 43% in the US Health and Retirement Survey sample forgot to add the in-between step of adding the interest of the first year before moving to calculate the total result. In these cases, the level of numeracy could be labeled as satisfactory, but the full financial literacy potential is not reached. Individuals with a higher level of financial literacy perform better in retirement planning (Lusardi and Mitchell 2007a; van Rooij et al. 2012) and are less prone to indebtedness (Lusardi and Tufano 2015).

Examples of Application

Latter life financial literacy influences the retirement planning (See ▶ "Pre-retirees' Preparation for Retirement"), savings and assets accumulation, indebtedness, vulnerability to financial scams (See ▶ "Financial Frauds and Scams") (Burnes et al. 2017; Wood and Lichtenberg 2017), and many other areas of everyday experiences of older people. The lack of financial literacy in itself may be both a result and a product of social exclusion. As such, it is unevenly distributed in the societies, affecting some groups of older adults more severally as others. (See ▶ "Financial Gerontology")

While the empirical results show that financial literacy has an extensive and broad social and socioeconomic root, some authors believe that appropriate financial education may increase and/or sustain the levels of financial literacy during the life course and for various ages (Lusardi et al. 2017; Lusardi and Mitchell 2007b). However, in a meta-analysis of 201 studies by Fernandes et al. (2014) in which they analyzed the role of financial literacy and financial education on financial behavior, financial literacy could only explain 0.1% of the variance in financial behavior. Authors suggested that the effect of financial literacy on financial behavior reported in correlation studies may be driven by individual cognitive ability (Skagerlund et al. 2018). Similarly, Wilson et al. (2017) concluded that higher levels of financial as well as health literacy (See ▶ "Health Literacy and Health Behavior") are associated with maintenance of cognitive health in old age (see also Myck et al. 2019).

Future Directions of Research

Overall the amount of research on the financial literacy in older adults is rising, but vast majority of studies concentrate on the students or working-age individuals and/or are related only to retirement planning. The research on retired older adults is almost missing, despite they have either accumulated many assets (See ▶ "Financial Assets") that require cautious management or may be in severe financial situation due to the absence of previous planning, significant economic downturns, individual crises (sickness, disability, family dissolution), low pre-retirement incomes or low replacement rates between pre-retirement and postretirement incomes (See ▶ "Replacement Rate"), and/or because of the increasing expenditures on housing or medical care. So far relatively little is known on extreme cases of financial literacy failures such as in cases of homelessness (See ▶ "Homelessness"), detrains, and financial scam victims (See ▶ "Financial Frauds and Scams").

Future research should also revisit and revise the empirical measures of financial literacy to make them more relevant for later life situations and contexts. Further, as the work with information is an essential part with the financial literacy maintenance in the complex and evolving economic world (Braunstein and Welch 2002), more attention needs to be paid on how the information is presented to and acquired by older adults with a different individual, socioeconomic, and cultural backgrounds and functional capacities.

Summary

It is believed that active and healthy aging involves planning in the form of active interventions in various life areas, and financial security is one of the most crucial factors in determining the quality of the outcomes. The financial literacy as a set of skills and abilities to navigate the world of money is, therefore, a key aspect of ensuring successful aging. The modern states show growing tendencies to shift the responsibilities for financial security in old age to individuals (Macnicol 2015). Under these conditions, the lifelong and repeated (updated) financial education, are becoming a necessity to ensure a reasonable level of orientation in the finances and economic concepts to ensure financial security.

Cross-References

▶ Financial Assets
▶ Financial Frauds and Scams
▶ Financial Gerontology
▶ Financial Socialization
▶ Health and Retirement Study
▶ Health Literacy and Health Behavior
▶ Homelessness
▶ Pre-retirees' Preparation for Retirement
▶ Replacement Rate
▶ Theory of Cumulative Disadvantage/Advantage

Acknowledgment The entry was supported by "Active aging, family, and intergenerational solidarity" grant (MŠMT 2D06004) of the Research Institute for Labour and Social Affairs in the Czech Republic.

References

Braunstein SF, Welch C (2002) Financial literacy: an overview of practice, research, and policy. Fed Reserv Bull 88:445–457

Brown M, Graf R (2013) Financial literacy and retirement planning in Switzerland. Numeracy 6(2):6. https://doi.org/10.5038/1936-4660.6.2.6

Brown M, Henchoz C, Spycher T (2018) Culture and financial literacy: evidence from a within-country language border. J Econ Behav Organ 150:62–85. https://doi.org/10.1016/j.jebo.2018.03.011

Burnes D, Henderson CR, Sheppard C, Zhao R, Pillemer K, Lachs MS (2017) Prevalence of financial fraud and scams among older adults in the United States: a systematic review and meta-analysis. Am J Public Health 107(8):e13–e21. https://doi.org/10.2105/AJPH.2017.303821

Dannefer D (2003) Cumulative advantage/disadvantage and the life course: cross-fertilizing age and social science theory. J Gerontol B Psychol Sci Soc Sci 58:S327–S337. https://doi.org/10.1093/geronb/58.6.S327

Fernandes D, Lynch JG Jr, Netemeyer RG (2014) Financial literacy, financial education, and downstream financial behaviors. Manag Sci 60(8):1861–1883. https://doi.org/10.1287/mnsc.2013.1849

Hershey DA, Henkens K, Van Dalen HP (2007) Mapping the minds of retirement planners: a cross-cultural perspective. J Cross-Cult Psychol 38:361–382. https://doi.org/10.1177/0022022107300280

Lusardi A (2012) Numeracy, financial literacy, and financial decision-making. Numeracy 5(1):2. https://doi.org/10.5038/1936-4660.5.1.2

Lusardi A, Mitchell OS (2007a) Financial literacy and retirement planning: new evidence from the Rand American life panel. SSRN Electron J. https://doi.org/10.2139/ssrn.1095869

Lusardi A, Mitchell OS (2007b) Financial literacy and retirement preparedness: evidence and implications for financial education programs. SSRN Electron J. https://doi.org/10.2139/ssrn.957796

Lusardi A, Mitchell OS (2014) The economic importance of financial literacy: theory and evidence. J Econ Lit 52(1):5–44. https://doi.org/10.1257/jel.52.1.5

Lusardi A, Tufano P (2015) Debt literacy, financial experiences, and overindebtedness. J Pension Econ Financ 14:332–368. https://doi.org/10.1017/S1474747215000232

Lusardi A, Samek A, Kapteyn A et al (2017) Visual tools and narratives: new ways to improve financial literacy. J Pension Econ Financ 16:297–323. https://doi.org/10.1017/S1474747215000323

Macnicol J (2015) Neoliberalising old age. Cambridge University Press, Cambridge

Myck M, Najsztub M, Oczkowska M (2019) Implications of social and material deprivation for changes in health of older people. J Aging Health 1–25. https://doi.org/10.1177/0898264319826417

OECD (Organisation for Economic Co-operation and Development) (2005) Improving financial literacy: analysis of issues and policies. Organisation for Economic Co-operation and Development, Paris. Available via OECD. https://www.oecd.org/daf/fin/financial-education/improvingfinancialliteracyanalysisofissuesandpolicies.htm. Accessed 3 June 2019

OECD (Organisation for Economic Co-operation and Development) (2006) The importance of financial education. The policy brief. Available via OECD. http://www.oecd.org/finance/financial-education/37087833.pdf. Accessed 3 June 2019

OECD (Organisation for Economic Co-operation and Development) (2016) OECD/INFE international survey of adult financial literacy competencies. Available via OECD. http://www.oecd.org/finance/OECD-INFE-International-Survey-of-Adult-Financial-Literacy-Competencies.pdf. Accessed 3 June 2019

Price D, Bisdee D, Daly T, Livsey L, Higgs P (2014) Financial planning for social care in later life: the 'shadow' of fourth age dependency. Ageing Soc 34:388–410. https://doi.org/10.1017/S0144686X1200 1018

Rabušicová M (2002) Gramotnost: staré téma v novém pohledu. (Literacy: an old theme in a new perspective). Masarykova univerzita, Brno

Skagerlund K, Lind T, Strömbäck C, Tinghög G, Västfjäll D (2018) Financial literacy and the role of numeracy–how individuals' attitude and affinity with numbers influence financial literacy. J Behav Exp Econ 74:18–25. https://doi.org/10.1016/j.socec.2018.03.004

Šlapák M, Soukup T, Vidovićová L, Holub M (2010) Finanční příprava na život v důchodu: informovanost, postoje a hodnoty. (Financial preparation for life at retirement: information, attitudes, and values). Research Institute for Labour and Social Affairs, Praha

van Rooij M, Lusardi A, Alessie R (2012) Financial literacy, retirement planning, and household wealth. Econ J 122:449–478. https://doi.org/10.1111/j.1468-0297.2012.02501.x

Vidovićová L (2013) Financial literacy in retirement planning context: the case of Czech older workers. In: Phellas C (ed) Aging in European societies. Healthy aging in Europe. Springer, New York, pp 191–203

Wilson RS, Yu L, James BD, Bennett DA, Boyle PA (2017) Association of financial and health literacy with cognitive health in old age. Aging Neuropsychol Cognit 24:186–197. https://doi.org/10.1080/13825585.2016.1178210

Wood S, Lichtenberg PA (2017) Financial capacity and financial exploitation of older adults: research findings, policy recommendations and clinical implications. Clin Gerontol 40:3–13. https://doi.org/10.1080/07317115.2016.1203382

Financial Planning

▶ Financial Therapy

Financial Proficiency

▶ Financial Literacy

Financial Resources

▶ Financial Assets

Financial Resources over One's Life

▶ Human Wealth Span

Financial Satisfaction

▶ Subjectively Measured Financial Resources

Financial Skills

▶ Financial Literacy

Financial Social Work

Burcu Özdemir Ocaklı
Faculty of Health Sciences, Department of Social Work, Ankara University, Ankara, Turkey

Synonyms

Social work and financial issues of older people

Definitions

Financial social work is defined as "a comprehensive approach to individual and community economic stabilization" by the Financial Social Work Initiative (FSWI) at the University of Maryland School of Social Work (UMBSSW 2018). As one of the most financially vulnerable populations, older adults are one of the main client groups for financial social work. At the micro-level, financial social work employs assessments and interventions that strengthen psychosocial and economic assets to help older adults to build or rebuild their financial security. At the meso-level, financial social work aims at promoting access to affordable credit, appropriate financial products, and available public benefits and economic support for older adults. Moreover, at the macro-level, financial social work pursues policy advocacy for the interests of vulnerable and underserved populations (UMBSSW 2018) such as older citizens. Financial therapy, financial healing, and financial counseling have similar approaches to financial social work since they provide a psychosocial, strength-based behavioral approach that features financial education, motivation, validation, and support (CFSW 2018). However, financial social work is distinct itself due to its multiple operating levels (micro, meso, and macro).

Overview

Financial social work is a relatively new subfield of social work practice which aims at financial empowerment of vulnerable individuals and communities. It first emerged as a transformative learning model that was developed by Reeta Wolfsohn at the Center for Financial Social Work (CFSW) in North Carolina, USA (Wolfsohn 2012). Initially, in 1997, Wolfsohn used the term "femonomics" in order to draw attention to "gender of money" and financial inequities between men and women (CFSW 2018). In 2003, this term has taken a more comprehensive outlook and turned into "financial social work," becoming a significant subfield of social work. This current umbrella term not only addresses the gender inequalities but also age inequalities and financial disadvantage experienced by older adults.

The need for financial social work has always been a topical issue for older people; however, it has started to be emphasized more since the financial system has become more complicated in recent years. Social security, taxing, and billing system along with benefits constitute a very complex set of information, especially for older consumers. Moreover, recent financial crises that many countries face require austerity measures and welfare cuts that end up negatively affecting the pension systems and individual finances. In this respect, senior citizens' ability to manage their own financial resources becomes more crucial at the times of financial instability and precarity. Accompanied by demographic aging, economic instability all around the world is likely to persist in the upcoming years; the need for financial social work will be on the rise (Birkenmaier et al. 2013).

In addition to financial vulnerability due to economic crises, older people are more prone to be a victim of financial abuse/exploitation, which increases the significance of financial social work interventions for older clients. Financial exploitation, which is mainly committed by family members, has negative effects on older people's economic well-being and quality of life (Kemp and Mosqueda 2005; Rabiner et al. 2008). Moreover, increased isolation of older people renders them more prone to fraud scams (Alves and Wilson 2008). In this respect, financial social work interventions work as preventive solutions to alleviate financial abuse situations.

Objectives and Tools

Empowerment is one of the main goals in social work, and it can be described as a process where the clients are expected to take control of their lives and reveal their potential as a human being. The empowerment process in financial social work is tailored according to the needs of older adults where social workers take on various roles as advisors, consultants, counselors, therapists, trainers, educators, advocators, and resource binders. Among all these intervention types, empowerment through training is the most common intervention type for financial social work. The training subjects include a various range of subjects such as mortgage and credit system, taxing and billing system, social security and benefits, employment opportunities, and budgeting unique to the country. Old age jeopardizes social and financial safety nets, which result in an increased need for financial guidance for older people.

In financial social work, one of the ways to empower clients is to increase their financial literacy levels. Even though there is not an agreement on the definition and the scope of financial literacy (Huston 2010), the concept encompasses the knowledge of financial concepts, ability to communicate about financial concepts, aptitude in managing personal finances, skill in making appropriate financial decisions, and confidence in planning effectively for future financial needs (Remund 2010). Research shows that financial literacy is lower among older people (Bennett et al. 2012; Lusardi 2012; MacLeod et al. 2017) and, as individuals get older, their capacity to make financial decisions deteriorates (Finke et al. 2011; Bennett et al. 2012; Gamble et al. 2015). Social, psychological, and physical losses and cognitive declines make it harder for older people to make sound financial decisions. In that respect, the need for financial guidance for older people emerges. Financial social work as a support mechanism aims to equip older people with skills to help them take control of their finances and hence gain control of their lives.

Moreover, financial social work helps to support the welfare state at the macro-level. As conventional defined benefit pension systems are no longer sustainable, emphasis on the accumulation of personal funds increases. The cuts on welfare benefits also call for better financial planning for older people. In that respect, financial literacy provides older clients with the necessary knowledge and lifelong skills to manage their own financial resources instead of temporarily providing them with financial resources or making them dependent on social welfare. Hence, the empowerment of older people through financial social work benefits both older people at the micro-level and welfare state at the macro-level.

Key Research Findings

Current research regarding financial social work takes place in three different areas, including the need, availability, and effectiveness of financial social work programs. As stated above, there is a substantial amount of evidence on the insufficient financial literacy and capability of older people (Bennett et al. 2012; Lusardi 2011, 2012; Lusardi and Mitchell 2011; MacLeod et al. 2017). Moreover, as individuals get older, their ability to make financial decisions declines (Finke et al. 2011; Bennett et al. 2012; Gamble et al. 2015). These low financial literacy levels among older people clearly show the extensive need for financial social work interventions.

In order to practice financial social work, social workers require specific skills which can be acquired through formal and nonformal education. There is also a considerable amount of research that investigates the financial literacy and capability of social workers and advocates the insertion of financial social work courses in the curricula of social work students at universities (Kindle 2009, 2013; Sherraden et al. 2007). Research also suggests that incorporating finance modules in social work curricula is effective in changing attitudes, building confidence in helping clients with basic financial management, increasing knowledge about financial capability, and improving some personal financial behaviors (Sherraden et al. 2017). In order to practice financial social work with older adults, social workers should also be equipped with physical, psychological, and cognitive characteristics of older people.

There is also some research that shows that financial training programs positively influence financial behavior and outcomes (Brown et al. 2016; Lusardi et al. 2017; Totenhagen et al. 2015). A quasi-experimental study conducted by the Center for Financial Social Work also reveals positive outcomes for the trainings provided by the center such as improved understanding and knowledge of financial concepts and terms, improved self-efficacy in performing various financial behaviors, improved understanding of financial matters and ability to apply what is learned, and decreased debt levels (KSRC 2015).

Future Directions of Research

Even though there is a considerable amount of research that demonstrates the need for financial social intervention programs and providers, research focusing on the effectiveness of existing programs for older people is very limited. This limitation occurs due to the lack of existing financial social work programs that target older people. Expansion of financial social work programs, measurement of their effectiveness, and gathering feedback from older clients are the key steps to take in order to develop better tools and training programs. Even though most of the social work schools all around the world do not offer classes or training programs on financial social work, revision of the curricula and incorporation of financial training will be a significant point of discussion in the near future (Gillen and Loeffler 2012).

Summary

Financial social work is a relatively new subfield of social work that aims to empower individuals and communities through economic stabilization. The concept of economic stabilization initially emerged for financial empowerment of women (Wolfsohn 2012). Demographic aging and economic recessions have rendered older people more susceptible to financial instability, and they have become a significant part of the client groups for financial social work. Through mainly education and mentoring, social workers aim to financially empower older clients, and increasing their financial literacy is the main tool that is used toward this end. Even though financial healing and financial therapy have similar objectives (Financial Therapy Association 2018), financial social work stands out as an overarching discipline that operates at many levels (micro, meso, and macro). Though limited, current research provides evidence on the need and effectiveness of financial social work interventions on vulnerable groups such as older clients (FINRA Investor Education Foundation 2009; Lusardi 2011; Lusardi and Mitchell 2011). However, in order to design effective intervention programs for older people, financial social work needs to be incorporated into the social work curricula, and social workers should be equipped with relevant skills.

Cross-References

▶ Financial Literacy
▶ Financial Socialization
▶ Financial Therapy
▶ Geriatric Social Workers
▶ Social Work

References

Alves LM, Wilson SR (2008) The effects of loneliness on telemarketing fraud vulnerability among older adults. J Elder Abuse Negl 20(1):63–85. https://doi.org/10.1300/J084v20n01_04
Bennett JS, Boyle PA, James BD, Bennett DA (2012) Correlates of health and financial literacy in older adults without dementia. BMW Geriatr 12(1):30. https://doi.org/10.1186/1471-2318-12-30
Birkenmaier J et al (2013) The role of social work in financial capability: shaping curricular approaches. In: Birkenmaier J, Sherraden M, Curley J (eds) Financial capability and asset development: research, education, policy and practice. Oxford University Press, New York, pp 278–301
Brown M, Grigsby J, van der Klaauw W, Wen J, Zafar B (2016) Financial education and the debt behavior of the young. Rev Financ Stud 29(9):2490–2522. https://doi.org/10.1093/rfs/hhw006
CFSW (Center for Financial Social Work) (2018) Financial social work: what it is, what it does, why it matters in all economic times. Center for Financial Social Work, Huntersville, North Carolina, US
Financial Therapy Association (2018) FTA information. https://www.financialtherapyassociation.org/. Accessed Dec 2018
Finke MS, Howe JS, Huston, SJ (2011) Old Age and the Decline in Financial Literacy. SSRN Electronic Journal, August. https://dx.doi.org/10.2139/ssrn.1948627
FINRA Investor Education Foundation (2009) Financial capability in the United States national survey: executive summary. FINRA Investor Education Foundation, Washington, DC
Gamble K, Boyle P, Yu L, Bennett D (2015) Aging and financial decision making. Manag Sci 61(11): 2603–2610. https://doi.org/10.1287/mnsc.2014.2010
Gillen M, Loeffler DN (2012) Financial literacy and social work students: knowledge is power. J Financ Ther 3 (2):27–38. https://doi.org/10.4148/jft.v3i2.1692
Huston SJ (2010) Measuring financial literacy. J Consum Aff 44(2):296–316. https://doi.org/10.1111/j.1745-6606.2010.01170.x
Kemp BJ, Mosqueda LA (2005) Elder financial abuse: an evaluation framework and supporting evidence. J Am Geriatr Soc 53(7):1123–1127. https://doi.org/10.1111/j.1532-5415.2005.53353.x

Kindle PA (2009) Financial literacy and social work: questions of competence and relevance. Doctoral dissertation, University of Houston, ProQuest UMI Dissertation Publishing

Kindle PA (2013) The financial literacy of social work students. J Soc Work Educ 49(3):397–407. https://doi.org/10.1080/10437797.2013.796853

KSRC (Keystone Research Corporation) (2015) Financial social work: an evaluation of its effectiveness in changing financial behaviors and improving self-sufficiency, executive summary. Keystone Research Corporation, Erie

Lusardi A (2011) Americans' financial capability. Report prepared for the Financial Crisis Inquiry Commission, and NBER working paper no 17103, National Bureau of Economic Research, Cambridge, Massachusetts, US.

Lusardi A (2012) Financial literacy and financial decision-making in older adults. Generations 2(8):25–32

Lusardi A, Mitchell OS (2011) Financial literacy around the world: an overview. J Pension Econ Finance 10:497–508. https://doi.org/10.1017/S1474747211000448

Lusardi A, Samek A, Kapteyn Glinert LA, Hung A, Heinberg A (2017) Visual tools and narratives: new ways to improve financial literacy. J Pension Econ Finance 16(3):297–323

Macleod S, Musich S, Hawkins K, Armstrong DG (2017) The growing need for resources to help older adults manage their financial and healthcare choices. BMC Geriatr 17(84). https://doi.org/10.1186/s12877-017-0477-5

Rabiner DJ, O'Keeffe J, Brown D (2008) A conceptual framework of financial exploitation of older persons. J Elder Abuse Negl 16(2):53–73. https://doi.org/10.1300/J084v16n02_05

Remund DL (2010) Financial literacy explicated: the case for a clearer definition in an increasingly complex economy. J Consum Aff 44(2):276–2965. https://doi.org/10.1111/j.1745-6606.2010.01169.x

Sherraden MS, Laux S, Kaufman C (2007) Financial education for social workers. J Community Pract 15(3):9–36. https://doi.org/10.1300/J125v15n03_02

Sherraden MS, Birkenmaier J, McClendon GG, Rochelle M (2017) Financial capability and asset building in social work education: is it "the big piece missing?". J Soc Work Educ 53(1):132–148. https://doi.org/10.1080/10437797.2016.1212754

Totenhagen CJ, Casper DM, Faber KM, Bosch LA, Wigg CB, Borden LM (2015) Youth financial literacy: a review of key considerations and promising delivery methods. J Fam Econ Iss 36(2):167–191. https://doi.org/10.1007/s10834-014-9397-0

UMBSSW (University of Maryland School of Social Work) (2018) Financial social work initiative. https://www.ssw.umaryland.edu/fsw/about-fsw/. Accessed Dec 2018

Wolfsohn R (2012) Linking policy and practice. In: Hoffler EF, Clark EJ (eds) Social work matters: power of linking policy and practice. NASW Press, Washington, DC, pp 219–223

Financial Socialization

Md. Mostafijur Rahman
Department of Law, Prime University, Dhaka, Bangladesh

Synonyms

Financial behavior; Financial education; Financial knowledge; Financial literacy

Definition

Financial socialization is a method by which individuals gain and increase financial knowledge, attitudes, and behaviors that develop their financial skills (Fox et al. 2000). Financial socialization is called consumer socialization also, and it is primarily related to the knowledge, skill, and approaches of the children and young consumers who work with finance in the marketplaces (Ward 1974). "Socialization opportunity derives from individual, organizational, or institutional agents with whom children come into contact or maintain a relationship" (Fluellen 2013, p. 11). In a broader sense, financial socialization is the method by which financial knowledge, financial behavior, and financial values, norms, and attitudes are gained and developed by individuals that are necessary for their financial ability and well-being (Senevirathne et al. 2018). According to many, the financial socialization commences in early childhood and continues to the entire time of life of the individual. Thus, adults learned about financial behavior since their childhood by many agents.

Overview

In modern age many individuals are facing complexities in managing their individual finance due to varieties of new and complex financial products on a daily basis which are jeopardized. But nationally or globally, the echelon of financial literacy of youth, children, and even older adults is disappointing. In this milieu, individuals particularly the older adults fall into financial troubles that can exacerbate financial stability hampering economic development of a nation. Considering this issue many literatures have established that financial literacy and financial behavior play a significant role in financial decision-making in aging and older adults. Since financial literacy and financial behavior have vast impacts on financial management, the correlation between these two is upsetting in view of the fact that in many cases moreover, old-aged people are poorly seen equipped in financial decision-making due to their low levels of financial literacy and they are not prepared to keep up with the latest financial background around the world. To measure the significant factors that are influential among the older adults to their financial socialization, a notable example from the American perspective has been examined due to the fact that the American Congress have declared April 2008 as the financial literacy month, realizing financial education as a key module in financial growth of individuals.

Financial Literacy and Financial Knowledge

Generally, literacy is related to a variety of knowledge. A literate person is he who is enriched with knowledge of a specific area. Thus, financial literacy is an individual's capacity to handle, understand, and communicate money affairs (Vitt et al. 2000). Financial literacy means the ability to understand and evaluate the financial problems that arise as a developmental impact on the complexity of global finance (Danes and Haberman 2007). Thus, financial literacy is the ability or knowledge of a person who is indispensable for him to handle financial defies and challenges in daily life. Similarly, financial education edifies persons to be competent to find out financial hardships as well as the ways and means to solve financial challenges, to make right the decision in financial management, to evaluate the ongoing financial situation, and to handle finances in a relevant way to overcome all the hurdles in the path of financial management. A financially educated and civilized individual is capable in money management and is proficient in handling own funds. Financial education empowers people to manage their own finances and gives lifelong financial security for themselves and their families.

Impacts of Financial Socialization Agents on Financial Literacy

Several studies have discussed that financial socialization agents are parents (family), peers, schools, employers, media, and religion. For example, Moschis (1987) examined that parents, schools, media, and peers play an important role in learning children's behavior. Moschis's view is collaborated by Pinto et al. (2005) who mentioned financial habits and values of children are influenced by parents. Knowledge and information on finances given at educational institutions such as schools, colleges, and universities have significant outcomes on the children's financial behavior in terms of their obtaining and determining financial knowledge and skills in marketplaces. This is why; a teacher as parents also plays a role in shaping a child's financial behaviors. According to many, mass media such as radio, television, newspaper, and the Internet can play a vital role in financial socialization of young people. Financial socialization agents have a significant influence on the financial literacy of an individual in financial management in this sense that financial literacy leads the individuals to ethical financial decision-making in their daily life. On the contrary, lack of financial literacy may lead them to serious financial problems. Thus, it will not be exaggerated to say that every individual of a country should be financially literate. Such financial literacy is particularly important to resolve fundamental financial problems rising in modern society.

Financial Literacy and Financial Behavior

Financial literacy and financial behavior are linked to each other. According to Tezel (2015) financial behavior means one's capacity to be aware of impacts, precautions, and opportunities of financial decisions and to be competent to take the right decisions in financial management and budget planning. Besides, in the UK, five behaviors such as keeping track, making ends meet, planning, choosing products, and staying informed (Dolan et al. 2012, pp. 126–142) are regarded as yardsticks for judging individuals' financial capability. Many scholars' studies proved that functional financial literacy controls individual's financial behavior to a constructive approach of financial management, for instance, the payment of a bill on time, having savings and investment, and ability to manage credit cards wisely (Lusardi et al. 2010). Chen and Volpe (1998) examine the linkage between financial literacy and financial decisions of college students. These findings showed that a student who has a low financial literacy tends to the wrong financial choice than students with higher financial literacy. On the other hand, persons with lower financial literacy have financial troubles such as debt and lack of social security. Low literate individuals are more likely to be default and wrong on financial management. For instance, Lusardi and Tufano (2009a, b) found that low financial literate individuals have a propensity to high-cost approaches in managing their finances. The less knowledgeable also reports that their debt loads are excessive or that they are unable to judge their debt position.

Financial Knowledge and Financial Behavior

Since financial behavior means how a financial manager will plan, manage, control finance, and use money at the individual level or in a company, thus financial behavior is connected to financial knowledge. Individual's financial knowledge sways financial behavior. Several studies revealed a positive relationship between financial knowledge and financial behavior. Since financial behavior is positively related to financial literacy, the effects of financial education on financial behavior are also widespread. Comparing the results of *Financial Practices Index* based upon financial behavior in four variables, cash flow management, credit management, savings, and investment practices with scores on the financial literacy quiz, Hilgert et al. (2003) found higher *Financial Practices Index* scores among those who were more financially literate, and this result indicates a relationship between financial knowledge and financial behavior. In retirement seminars, Lusardi and Mitchell (2007) found a positive wealth effect of some retired due to those retired persons who had less wealth and financial education.

The Importance of Financial Literacy and Financial Education in Older Adults

Financial education is more significant due to the rapid changes of contemporary world in finance, growth of the number of financial institutions, incensement of complex financial products, advancement of financial assets, growth economies in GDPs, and several new options and defies in contemporary globe such as annuity, Internet access, cable networks, online business, share market, massive use of debit and credit cards, security exchange, modern banking system, investment, pension fund, share, debenture, securities, the modern global economy, the modern period of science, technology, and so on. Simultaneously, the financial background has turned into trouble owing to complex modern financial markets that offer a variety of new financial services and products that are not easy to understand for the older individual. These accelerations have existed newly in modern finance that make individuals over indebted and unable to make understand in financial decision-making and lead them to financial difficulties. The Minister of Finance of the Russian Federation Siluanov (2013, p. 69) said "A country's economic development largely depends on its financial literacy.... Therefore, the development and introduction of financial literacy strategies is a vital element of our policy. Financial literacy has become an inalienable part of education in the 21st century." Financial education is becoming crucial for a family to settle or balance its budget on buying and funding and to ensure an earning of adults when they retire (OECD 2006). Thus it is important for people of all ages. An individual's daily life is frequently affected in financial decision-making nowadays. Thus, financial literacy is necessary for understanding financial socialization, which makes an individual able to avoid numerous financial problems. Due to the lack of financial literacy, only an individual falls into adverse financial condition. Thus, for better financial management at the individual level, for example, for controlling, managing, planning, and budgeting of finance, financial behavior is almost all the times appreciated. According to Layli (2014), strong financial behavior depends on an individual's approaches how he/she can manage his/her money, loan, and investments also. High quality financial behavior affects everybody's life, especially on the life of older adults. The menace for financial decisions has a major impact on future life of older adults particularly in the case of pension of workers and employers owing to growing life expectancy as they are to enjoy a long episode of retirement (OECD 2006). So, financial education is more likely in all ages to teach how to behave with finance or how to take a right decision toward finance, saving, and spending of money in daily life and other issues relating to finance. There is a need for older people who are not outfitted to keep up with this latest financial background around the world.

Level of Financial Literacy and Financial Education of Older Adults

According to Fabris and Luburić (2016) some recent studies, the level of financial education of older adults is not adequate even in rich countries. In 2005 the Organization for Economic Co-operation and Development (OECD) highlighted that financial illiteracy is common across age groups and geographical areas. OECD (2006) points out that in most of the countries, even so in developed countries, the level of financial literacy is not so high. For example, in Japan 71% of adults have no idea on investment in equities and bonds as OECD revealed (OECD 2006). Several studies have revealed how the level of financial literacy among adults Americans is rather low (Chen and Volpe 1998; Lusardi and Mitchell 2011a, b, c). OECD (2006) mentions that in the year of 2003 in the USA, the rate of bankruptcy for increasing rate of credit card use was almost 1 in 10 and in Austria it was 11%. OECD further revealed that, at the same time, another two countries Korea and Germany have experienced of increasing consumer debts as well as private insolvencies due to more use of credit cards. Thus, financial education is in these days more essential than it was in the past. According to OECD (2006), financial literacy is essential particularly to provide an adequate income in retirement to avoid high levels of debt that may increase personal bankruptcy. This is why it is seen nowadays in the developed countries like the USA and even worldwide that for financial well-being their family agents are more liable (Lusardi 2012). In the 2004 Health and Retirement Study (HRS), Olivia Mitchell and Annamaria Lusardi surveyed respondents who are ages 50 and older for assessing the knowledge of basic financial literacy of the adults. Responses to their three questions (inflation, risk diversification, and the capacity to calculate the interest rates) discovered a very low level of financial literacy among older Americans (Lusardi 2012). They found financial illiteracy of the elderly females was lacking of financial education as they were lean to have skill with credit cards, bank accounts, and mortgage. Examining these three questions, the numerous national surveys in America such as the RAND American Life Panel and the 2009 Financial Capability Study (Lusardi and Mitchell 2011c) also revealed the low rate of financial literacy among older respondents, and it was found to decrease with the age of the adults. Not only that, those questions were surveyed in Germany, the Netherlands, Italy, Sweden, Russia, Japan, and New Zealand (Lusardi and Mitchell 2011b) which show financial literacy is low in those countries as well as the lowest level of financial knowledge of older people. Annamaria Lusardi and Peter Tufano (2009a, b) further examined financial knowledge among elderly respondents, and they found a gap between these two things among the respondents (Lusardi 2012). OECD (2006) and Lusardi and Mitchell (2007) assessed international support on financial literacy, and they found financial illiteracy is common in many developed countries like Australia, Japan, and Korea, as well as developed countries in Europe.

Future Direction of Research

The findings suggest that there is a correlation between financial literacy and financial behavior, and financial literacy is influenced by financial socialization agents such as family, peers, teachers, and mass media. Inadequate financial education of the individuals, in particular, older adults, creates a lack of financial knowledge. But the concern is that an adult, who is not financially literate, will not be competent to choose the right savings or investments for their future. If he/she is financially educated, he/she will be able to save finance and to take any financial challenge. Numerous programs and international institutions are working in many countries to promote and extend financial education such as UNICEF, OECD, the UN Committee on the Rights of the Child, Child and Youth Finance International (CYFI), Program of Social and Financial Education for children, GIZ, USAID, UNCDF, FinMark Trust, the World Bank, etc. These programs and organizations provide and foster financial education to children in the world. Despite these good efforts, the level of financial knowledge of children is not praiseworthy. The findings from numerous studies suggested that students, for example, young people, lack basic financial knowledge which leads them to bad financial decision-making and several financial problems accordingly. Similarly, lower financial educated adults are seen to manage their finance in higher cost and higher fees (Lusardi 2012). Many researchers have discovered that most of the students, the young people, and even the older people are not well-informed of their finance and they do not obtain sufficient financial education, financial education programs need to be strengthened to educate more financial literacy to students, young people, and older people also because financial literacy has a significant effect on students' as well as older people's financial behavior, and thus, financial literacy is an essential constituent in financial behavior for today and the future.

Summary

The above discussion clears that, due to the rampant changing situation of the contemporary globe and at the same time more complexity of vast innovative financial products, individuals are to face some risks in dealing with modern finance owing to their lack of financial literacy or financial education. For example, children or adults who have no sufficient financial knowledge from their financial socialization agents become the victims of financial risks that hamper their personal life and financial stability in the society accordingly. Since financial education and economic development are interrelated, financial education is a prerequisite for smooth economic development of a country. The reason is financial education changes individuals' financial habits and behavior. Moreover, financial management is a particularly important part of financial constancy and growth. Thus, it has a unique value in financial firmness. Internationally, financial management among individuals has turned into a critical issue for financial well-being nowadays. Mainly, this is important among children in educational institutions because the lack of financial knowledge leads the students in the debts. Furthermore, this is important due to the fact that today's children are the representatives of upcoming economic progress. So, financial literacy is necessary to improve their financial behavior.

If the inadequately financially educated adults are increased in the society, then it may create many problems either in individual level or in the society as a whole. For example, in most of the cases, financially educated adults take erroneous financial decision, and thusly they become over indebted either in their family or in the society. Various social problems particularly expenditures from the funds, the economic development of the society, financial stability, and even their living standard are hindered. For this very reason, financial awareness in terms of financial literacy and education has been seen as a basic module of financial well-being. Proper financial education influences proper financial behavior among adults. On the contrary, as a result of improper or inappropriate financial management, adults are affected by unwanted debts. Thus, a comprehensive and well-planned financial plan is required to make sure appropriate financial management among adults. In summary, the older-aged population in the world has grown in the recent years, and it is likely to continue to grow. This entry aims to increase high consciousness about older-aged population and their need of financial knowledge for which some risks they might face.

Cross-References

▶ Developmental Stake Hypothesis
▶ Financial Literacy
▶ Financial Social Work
▶ Financial Therapy
▶ Generativity and Adult Development
▶ Human Wealth Span
▶ Life Course Perspective
▶ Life Cycle Theories of Savings and Consumption
▶ Life-Span Development

References

Chen H, Volpe RP (1998) An analysis of personal financial literacy among college students. Financ Serv Rev 7(2):107–128. http://citeseerx.ist.psu.edu/viewdoc/download?doi=10.1.1.392.4650&rep=rep1&type=pdf

Danes SM, Haberman H (2007) Teen financial knowledge, self-efficacy, and behavior: a gendered view. Journal of Financial Counseling and Planning 18(2). Available at SSRN: https://ssrn.com/abstract=2228406

Dolan P, Elliott A, Metcalfe R, Vlaev I (2012) Influencing financial behavior: from changing minds to changing contexts. J Behav Financ 13:126–142. https://doi.org/10.1080/15427560.2012.680995

Fabris N, Luburić R (2016) Financial education of children and youth. J Cent Bank Theory Pract 5(2):65–79. https://doi.org/10.1515/jcbtp-2016-0011. https://www.researchgate.net/publication/303664105_The_Financial_Education_of_Children_and_Youth. Received 1 Dec 2015; accepted 5 Jan 2016

Fluellen VM (2013) Exploring the relationship between financial behaviors and financial well-being of African American college students at one historically black institution. Graduate theses and dissertations. 12987. https://lib.dr.iastate.edu/etd/12987. Accessed 15 Dec 2018

Fox J, Bartholomae S, Gutter MS (2000) What do we know about socialization? Consum Interes Annu 46:217. https://www.consumerinterests.org/assets/docs/CIA/CIA2000/foxbartholomaegutter.pdf. Accessed 15 Dec 2018

Hilgert MA, Hogarth JM, Beverly SG (2003) Household financial management: the connection between knowledge and behavior. Fed Reserv Bull 89:309–322. Retrieved from https://www.federalreserve.gov/pubs/bulletin/2003/0703lead.pdf. On 02 Aug 2019

Layli N (2014) Pengaruh literasi keuangan terhadap perilaku mahasiswa dalam mengelola keuangan. JPA UM Malang 1(4):277–285

Lusardi A (2012) Financial literacy and financial decision making in older adults. Am Soc Aging Blog, [on line]. https://www.asaging.org/blog/financial-literacy-and-financial-decision-making-older-adults. Accessed 15 Dec 2018

Lusardi A, Mitchell O (2007) Financial literacy and retirement preparedness: evidence and implications for financial education. Bus Econ 42:35–44. Retrieved from https://www.dartmouth.edu/~alusardi/Papers/Financial_Literacy.pdf. On 02 Aug 2019

Lusardi A, Mitchell OS (2011a) Financial literacy and planning: implications for retirement wellbeing. In: Mitchell OS, Lusardi A (eds) Financial literacy: implications for retirement security and the financial marketplace. Oxford University Press, Oxford, UK, pp 17–39

Lusardi A, Mitchell OS (2011b) Financial literacy around the world: an overview. J Pension Econ Finance 10(04):497–508. https://doi.org/10.3386/w17107. Cambridge University Press

Lusardi A, Mitchell OS (2011c) Financial literacy and retirement planning in the United States. J Pension Econ Finance 10(4):509–525. https://doi.org/10.1017/S147474721100045X. Cambridge University Press

Lusardi A, Tufano P (2009a) Debt literacy, financial experiences, and overindebtedness. J Pension Econ Finance 14(04):332–368. https://doi.org/10.3386/w14808. Cambridge University Press

Lusardi A, Tufano P (2009b) Teach workers about the peril of debt. Harv Bus Rev 87(11):22–24

Lusardi A, Mitchell OS, Curto V (2010) Financial literacy among the young. J Consum Aff 44(2):358–380. https://doi.org/10.1111/j.1745-6606.2010.01173.x

Moschis GP (1987) Consumer socialization: a life cycle perspective. Lexington Books, Lexington

OECD (2006) The importance of financial education, the OECD policy briefs. Retrieved from http://www.oecd.org/finance/financial-education/37087833.pdf. On 31 July 2019

Pinto MB, Parente DH, Mansfield PM (2005) Information learned from socialization agents: its relationship to credit card use. Fam Consum Sci Res J 33(4):357–367. https://doi.org/10.1177/1077727X04274113

Senevirathne WAR, Jayendrika WADK, Silva GAJ (2018) Impact of financial socialization agents towards financial literacy among young micro business entrepreneurs in Colombo district. In: Sri-Lanka, Proceedings of the Wayamba University international conference, Sri Lanka, 24–25 August 2018

Sulivanov G (2013) G20 national financial literacy strategies presented. Retrieved from http://en.g20russia.ru/news/20130905/782412777.html on 31.07.2019

Tezel Z (2015) Financial education for children and youth. In: Copur Z (ed) Handbook of research on behavioral finance and investment strategies: decision making in the financial industry. IGI Global, Hershey, pp 69–92. https://doi.org/10.4018/978-1-4666-7484-4.ch005

Vitt LA, Anderson C, Kent J, Lyter DM, Siegenthaler JK, Ward J (2000) Personal finance and the rush to competence: financial literacy education in the U.S. Fannie Mae Foundation, Washington, DC. Retrieved from https://www.isfs.org/documents-pdfs/rep-finliteracy.pdf. On 31 July 2019

Ward S (1974) Consumer socialization. J Consum Res 1(2):1–14. Retrieved from https://takechargetoday.arizona.edu/system/files/ward74jcr.pdf

Financial Therapy

Ella Schwartz
Israel Gerontological Data Center, Paul Baerwald School of Social Work and Social Welfare, The Hebrew University of Jerusalem, Jerusalem, Israel
School of Social Work, Bar-Ilan University, Ramat Gan, Israel

Synonyms

Financial counseling; Financial planning

Definition

Financial therapy is an integration of the financial planning and mental health professions, aimed at helping individuals, couples, and families deal with their financial difficulties through integrative work on the cognitive, emotional, behavioral, relational, and economic issues relating to these problems.

Overview

Financial therapy is an emerging field that encompasses the collaboration and the synthesis among financial planning, financial counseling, marriage and family therapy, and psychology. Formally established in 2008 with the gathering of the "Financial Therapy Forum," the field brings together two distinct disciplines of finances and therapy and focuses on their areas of overlap (Grable et al. 2010). Therapists are experts in helping clients with relationship issues, while financial planners and counselors are experts in helping clients improve their financial literacy and money management (Kim et al. 2011). An integrative approach of these disciplines is needed in relation to late-life issues such as retirement planning, financing healthcare, and coping with limited financial resources postretirement. In such matters, both family therapists and financial planners share challenges when client's relationship and financial needs are intertwined. While many of these practitioners lack cross-professional training, it has become increasingly common that their clients expect them to be helpful in areas that go beyond traditional bounds of training (Sussman and Dubofsky 2009), for example, by encouraging a discussion of financial postretirement investments among older couples (Maton et al. 2010). Moreover, during old age, individuals may face increasing financial hardships that can be a leading cause of stress and familial conflicts (Lincoln 2007; Marshall 2015).

Financial therapy is aimed at helping individuals, couples, and families cope with the stress of financial problems through working on financial as well as therapeutic goals (Smith et al. 2018). Money and the interpersonal and intrapersonal aspects of older adults' lives are considered to be inseparable, such that financial well-being cannot be attained without considering the whole person and their relationships with others around them (Britt et al. 2015). Emotional well-being is also related to financial behaviors; for example, saving toward retirement might be impacted by emotional tendencies such as distrust and anxiety (Hayhoe et al. 2012). It is believed that money affects relationships and overall well-being in old age, while psychological and relational well-being influence one's financial management. In financial therapy, therapeutic interventions are implemented in order to increase rational thinking around money or improve financial behaviors (Goetz and Gale 2014).

Financial Therapy with Older Adults

There is currently little research on financial therapy specifically aimed at older adults. While financial therapy interventions can be applied to this population group, some unique characteristics should be considered when working with older persons, and modifications should be made to their needs. Such a specific focus is warranted since older people are more prone to financial stressors, compared to younger persons, and have little opportunity to regain their money if it is lost (Marshall 2015). They are more likely to develop chronic illnesses, with medical expenses often putting them at risk for debt while also posing new mental health challenges.

Older women are a specifically vulnerable group, as often they have less financial knowledge and resources compared to older men while being more likely to become widowed in old age. Often women have not discussed financial matters with their spouse before his passing and are not prepared financially for the unexpected death of their spouse (Whirl and DeVaney 2006). These women might benefit from help in budgeting and planning their expenses, especially in the face of possible emotional challenges as they become caregivers or lose their spouse (Into 2003).

An additional area in need of attention is planning for retirement, which might seem daunting for some adults. Financial therapists can help their aging clients cope with the mental impediments to planning their future retirement (Kiso and Hershey 2017). Moreover, even the decision to seek the assistance of a financial expert can be affected by emotional factors, as some older adults were found to experience anxiety at the prospect of visiting a financial adviser (Van Dalen et al. 2017). Such adults might benefit from a more therapeutically oriented approach to help ease their anxiety about financial counseling. Future research should develop and assess interventions tailored to older adults' characteristics and needs.

Key Research Findings

Since the formation of the field in 2008, there is increasing academic writing on financial therapy. However, as with any new field, empirical data are still limited (Britt et al. 2015). Several studies have investigated the effectiveness of financial therapy interventions, indicating positive outcomes. One such study assessed 33 participants in an experiential financial therapy program. The treatment focused on resolving unfinished business that might underlie dysfunctional financial behaviors. Following the program, participants showed significant reductions in psychological distress, anxiety, and worries about finance-related issues and improved financial health. These improvements were stable at a 3-month follow-up (Klontz et al. 2008). Another pilot study examined an intervention model utilizing co-therapy teams of marriage and family therapists and financial planners. It employed a five-session treatment approach in working with couples reporting both relationship and financial stress. The participants reported positive relationship changes and improvements in their financial knowledge and behaviors (Kim et al. 2011). Although not specifically targeting the older population, these interventions could be applied to financial matters concerning older adults.

Solution-focused financial therapy was examined in a pilot study of a video-based brief intervention, delivered in conjunction with income tax preparation services. The results showed an increase in both the frequency and amount of self-reported savings at tax time (Palmer et al. 2016). Such an intervention can prove useful in encouraging older adults to save for retirement and save following their retirement.

An intervention that has been explored more often is cognitive behavioral therapy applied to financial disorders. A meta-analysis has indicated that cognitive behavioral therapy can significantly decrease the severity of hoarding disorders (Tolin et al. 2015). The authors suggest that adults of older ages might face a different set of challenges. Older adults with hoarding disorders may be at particular risk for chronic medical illness and executive dysfunction, and they might be more severely psychiatrically ill in general than younger patients (Tolin et al. 2015). Therefore, cognitive behavioral therapy aimed at financial disorders should be adapted for the specific characteristics of the older population.

Examples of Application

Financial therapy is multifaceted, and there is no one right way to deliver interventions, tools, and approaches (Archuleta et al. 2012). Moreover, as this is an emerging field, there are few financial therapy training programs, with some schools offering counseling courses to financial planning students (Britt et al. 2015). Currently, no professional credentials or designations are specific to financial therapy (Goetz and Gale 2014). Moreover, there are no guidelines or training programs explicitly aimed at older adults and their needs. However, there are some interventions and guidelines that can be utilized by practitioners to provide financial therapy for older adults.

The financial therapy literature identifies several strategies of collaborations between finance and mental health professionals. A first primary collaborative strategy involves two professionals in different fields (e.g., a therapist and financial planner or counselor) working with a client simultaneously in the same session (Kahler 2005; Kim et al. 2011; Seay et al. 2015). For example, a woman who has trouble saving for retirement can benefit from a financial planner's help in getting out of debt and begin saving for retirement, while a therapist can help her address the underlying processes behind her financial stress (Jorgensen and Taylor 2014). Another model consists of two professionals in different fields consulting with each other and collaborating on planning the treatment but not meeting with the client at the same time (Falconier and Epstein 2011). Another approach is training financial therapists as a profession, in both mental health and finances (Gudmunson 2011). Financial therapy might also be administered in online settings (Smith et al. 2018). This option can have relevance for frail older adults who might find it difficult to leave their house for financial therapy sessions.

Theoretical perspectives and conceptual frameworks from the mental health fields are beginning to be used to develop interventions to influence financial thinking and behavior. Several models of financial therapy currently exist. These models often incorporate elements from well-established therapy models with elements of financial counseling (Klontz et al. 2015). One model aims to foster communication between older spouses, to help older women prepare financially for the chance of unexpectedly losing their spouse. Whirl and DeVaney (2006) stress the need for older couples to communicate openly about financial matters and especially about the possibility of an unexpected death. Financial advisors can be important in encouraging such communication. They should be aware of the different communication styles of men and women. Several approaches can be used to facilitate spousal discussion and financial preparation for spousal loss, such as the Life Scripts or the Emotionally Focused Approach. The Life Script approach, for example, maintains that a script should be developed before conversations on difficult topics that can facilitate successful communication (Whirl and DeVaney 2006). The financial therapist can understand the couple's goals before the conversation, help them develop plans to accommodate these goals, and maintain communication as they advance toward accomplishing these goals.

Solution-focused financial therapy (SFFT) is a pragmatic approach offering techniques to focus on clients' strengths in order to achieve their desired outcomes in the financial realm. It helps clients focus on future financial goals, identify solutions, and emphasize their strengths. Clients are encouraged to take responsibility for making changes in their lives (Archuleta et al. 2015a, b; Smith et al. 2016). Cognitive behavioral financial therapy uses cognitive behavioral theory and techniques in financial therapy. It utilizes cognitive behavioral therapy concepts, such as automatic thoughts, schemas, underlying beliefs, and behavioral techniques (Mitchell et al. 2006). Financial therapy uses these techniques to identify, challenge, and change problematic money scripts and automatic thoughts (Nabeshima and Klontz 2015). Although cognitive behavioral financial therapy was not examined in an older population, the general effectiveness of cognitive behavioral therapy has been previously shown in older adults, and it was found to be more effective than other interventions in treating late-life depression (Pinquart et al. 2007). Therefore, cognitive behavioral financial therapy might prove to be helpful among the older population as well.

Future Directions of Research

Financial therapy is an emerging field, and there are few clinical or experimental studies into the effectiveness of its interventions and no studies that specifically focus on the older population (Archuleta et al. 2012). As approaches to financial therapy are developed and refined, efforts should be made to measure the impact these approaches have on older clients' financial, emotional, and relational well-being (Britt et al. 2015). They should particularly focus on specific subpopulations of older adults that might be at an increased risk of financial and emotional problems, such as older adults with lower education and lower income (Lincoln 2007).

Models that have been examined empirically should be further examined and adapted to the older population. As mentioned above, this population might have unique needs and can respond differently to interventions designed for the general population. Such studies should be based on large samples and include randomized clinical trials with treatment and control groups (Klontz et al. 2008; Kim et al. 2011; Archuleta et al. 2015a). More research is also needed to validate financial therapy interventions with diverse samples of older people (Archuleta et al. 2015a). For example, the needs of older men and older women might differ and require different interventions (Into 2003).

Approaches to financial therapy vary and include different approaches to collaborations between financial and mental health professionals, as well as different intervention techniques. Future research should compare the various approaches and assess their relative contribution (Archuleta et al. 2012). Guidelines might be formulated to determine older clients' characteristics that might predict better outcomes and help practitioners choose from a variety of theoretical models. Future research should also be conducted to establish the various potential outcomes of financial therapy. For example, possible outcomes may include decreased financial stress, increased financial self-confidence, reduced debt, increased relational stability, and increased overall well-being (Goetz and Gale 2014).

Summary

Financial therapy takes a psychological and relational perspective in understanding and changing financial behavior. It emphasizes the interconnectedness of persons' financial, personal, and relational well-being. Financial therapy has been developed to address this complexity of human behavior and to help individuals, couples, and families make better financial choices and improve their overall well-being. Empirically supported research on approaches and interventions tailored to the needs of older individuals should be further carried out to advance this emerging field and to ensure that older clients receive the best possible services.

Cross-References

▶ Financial Assets
▶ Financial Exploitation by Family Members
▶ Financial Frauds and Scams
▶ Financial Gerontology
▶ Financial Literacy
▶ Financial Social Work
▶ Financial Socialization

References

Archuleta KL, Burr EA, Dale AK et al (2012) What is financial therapy? Discovering mechanisms and aspects of an emerging field. J Financ Ther 3:57–78. https://doi.org/10.4148/jft.v3i2.1807

Archuleta KL, Burr EA, Bell Carlson M et al (2015a) Solution focused financial therapy: a brief report of a pilot study. J Financ Ther 6:1–16. https://doi.org/10.4148/1944-9771.1081

Archuleta KL, Grable JE, Burr E (2015b) Solution-focused financial therapy. In: Klontz BT, Britt SL, Archileta KL (eds) Financial therapy: theory, research, and practice. Springer, New-York, pp 121–141

Britt SL, Klontz BT, Archuleta KL (2015) Financial therapy: establishing an emerging field. In: Klontz BT, Britt SL, Archileta KL (eds) Financial therapy: theory, research, and practice. Springer, New-York, pp 3–13

Falconier MK, Epstein NB (2011) Couples experiencing financial strain: what we know and what we can do. Fam Relat 60:303–317. https://doi.org/10.1111/j.1741-3729.2011.00650.x

Goetz JW, Gale JE (2014) Financial therapy: de-biasing and client behaviors. In: Baker HK, Ricciardi V (eds) Investor behavior: the psychology of financial planning and investing. Wiley, Hoboken, pp 227–244

Grable J, McGill S, Britt S (2010) The financial therapy association: a brief history. J Financ Ther 1:1–6. https://doi.org/10.4148/jft.v1i1.235

Gudmunson CG (2011) Researcher profile: an interview with Clinton G. Gudmunson, Ph.D. J Financ Ther 2:86–91. https://doi.org/10.4148/jft.v2i2.1557

Hayhoe CR, Cho SH, Devaney SA et al (2012) How do distrust and anxiety affect saving behavior? Fam Consum Sci Res J 41:69–85. https://doi.org/10.1111/j.1552-3934.2012.02129.x

Into FH (2003) Older women and financial management: strategies for maintaining independence. Educ Gerontol 29:825–839. https://doi.org/10.1080/716100365

Jorgensen BL, Taylor AC (2014) Understanding financial literacy and competence: considerations for training, collaboration, and referral for MFTs. J Financ Ther 5:1–18. https://doi.org/10.4148/1944-9771.1062

Kahler RS (2005) Financial integration: connecting the client's past, present, and future. J Financ Plan 18:62–71

Kim JH, Gale J, Goetz J, Bermúdez JM (2011) Relational financial therapy: an innovative and collaborative treatment approach. Contemp Fam Ther 33:229–241. https://doi.org/10.1007/s10591-011-9145-7

Kiso H, Hershey DA (2017) Working adults' metacognitions regarding financial planning for retirement. Work Aging Retire 3:77–88. https://doi.org/10.1093/workar/waw021

Klontz BT, Bivens A, Klontz PT et al (2008) The treatment of disordered money behaviors: results of an open clinical trial. Psychol Serv 5:295–308. https://doi.org/10.1037/1541-1559.5.3.295

Klontz BT, Britt SL, Archileta KL (eds) (2015) Financial therapy: theory, research, and practice. Springer, New-York

Lincoln DK (2007) Financial strain, negative interactions, and mastery: pathways to mental health among African Americans. J Black Psychol 33:439–462. https://doi.org/10.1177/0095798407307045

Marshall GL (2015) Financial hardship in later life: social work's challenge or opportunity. Soc Work 60:265–267. https://doi.org/10.1093/sw/swv015

Maton C, Maton M, Martin W (2010) Collaborating with a financial therapist: the why, who, what and how. J Financ Plan 23:62–70

Mitchell JE, Burgard M, Faber R et al (2006) Cognitive behavioral therapy for compulsive buying disorder. Behav Res Ther 44:1859–1865. https://doi.org/10.1016/j.brat.2005.12.009

Nabeshima G, Klontz BT (2015) Cognitive-behavioral financial therapy. In: Klontz BT, Britt SL, Archileta KL (eds) Financial therapy: theory, research, and practice. Springer, New-York, pp 143–160

Palmer L, Pichot T, Kunovskaya I (2016) Promoting savings at tax time through a video-based solution-focused brief coaching intervention. J Financ Ther 7:0–16. https://doi.org/10.4148/1944-9771.1103

Pinquart M, Duberstein PR, Lyness JM (2007) Effects of psychotherapy and other behavioral interventions on clinically depressed older adults: a meta-analysis. Aging Ment Health 11:645–657. https://doi.org/10.1080/13607860701529635

Seay M, Goetz JW, Gale J (2015) Collaborative relational model. In: Klontz BT, Britt SL, Archileta KL (eds) Financial therapy: theory, research, and practice. Springer, New-York, pp 161–172

Smith TE, Shelton VM, Richards KV (2016) Solution-focused financial therapy with couples. J Hum Behav Soc Environ 26:452–460. https://doi.org/10.1080/10911359.2015.1087921

Smith TE, Williams JM, Richards KV, Panisch LS (2018) Online financial therapy. J Fam Psychother 29:106–121. https://doi.org/10.1080/08975353.2017.1368812

Sussman L, Dubofsky D (2009) The changing role of the financial planner part 1: from financial analytics to coaching and life planning. J Financ Plan 22:48–57

Tolin DF, Frost RO, Steketee G, Muroff J (2015) Cognitive behavioral therapy for hoarding disorder: a meta-analysis. Depress Anxiety 32:158–166. https://doi.org/10.1002/da.22327

Van Dalen HP, Henkens K, Hershey DA (2017) Why do older adults avoid seeking financial advice? Adviser anxiety in the Netherlands. Ageing Soc 37:1268–1290. https://doi.org/10.1017/s0144686x16000222

Whirl SP, DeVaney SA (2006) Communication strategies to help prepare for the unexpected loss of a spouse. J Pers Financ 5:42–53

Financial Transfers

▶ Intergenerational Exchanges

Financing Long-Term Care

▶ Long-Term Care Financing

Fin-de-siècle

▶ Aging Research in the Late Nineteenth to Early Twentieth Century

Finite Mixture Models

▶ Latent Class Analysis

First Caregivers

▶ Primary Caregivers

Fiscal Welfare

Ryan D. Edwards
University of California, Berkeley, CA, USA

Synonyms

Public pensions; Senior financial support; Social safety net; Welfare state

Definition

Generally speaking, fiscal welfare is a term that refers to the actions by governments to improve well-being among citizens through transfers of funds and in-kind goods and services financed either by tax revenues, tariffs, or borrowing. Government provision of transfers and services intended to improve the welfare of specific individuals or groups, as opposed to the public at large, is usually referred to as the "welfare state" (Morel et al. 2018a, b).

Fiscal welfare usually refers to government spending and transfers that are aimed at meeting needs, usually but not always those that cannot be met through individuals transacting in private marketplaces. Fiscal welfare is also further understood to mean that portion of the welfare state associated with meeting aging-related needs, which arise due to retirement from working and the loss of earnings, to typical patterns of declining health and increasing disability over the life cycle, and to spousal death in particular and the reduction in human contact and social networks that may arise with advancing mortality rates through age among a birth cohort.

Overview

Governments perform many kinds of functions that enhance welfare and require financing. Protection of legal rights, resolution of criminal and civil complaints, inspection of food and water supplies, regulation of physical borders, police, fire, waste disposal, enumeration, monetary policy, environmental regulation, and national defense are examples of what are typically called "public goods" that benefit all residents and sometimes local nonresidents. Hybrid services include maintenance of roads and other physical infrastructure, which may include public transit systems, and postal systems, where sometimes governments charge user fees usually with substantial subsidies. These myriad functions lie outside the common definition of fiscal welfare (Morel et al. 2018a, b).

In modern market economies, shifts away from traditional family structures, such as geographically dispersed living near new employment centers and specialization among children in tasks unrelated to caregiving, amplify the potential for unmet need related to aging. Although well-functioning markets for annuitizing financial savings and for formal care could also meet these needs, heterogeneity in wealth and access to markets, in idiosyncratic and unpredictable shocks to health, and in market quality often implies that fiscal welfare systems of transfers financed by taxes or borrowing are best equipped to meet these needs.

By far the most common type of program in this category, and arguably the original motivation for establishing systems of fiscal welfare for aging, is pension systems. As discussed elsewhere (see "▶ Pension Systems" in this volume), pensions can take a variety of forms but in their basic incarnation are funds transferred to retired workers with regularity, financed by general tax revenue either from contributions from current workers, from liquid assets, from borrowing, or in some cases from printing money.

The development of modern pension systems is usually traced either to the German system developed in the 1880s and introduced in 1889 (Börsch-Supan and Wilke 2004) or to the US system of Civil War pensions that was developed earlier but significantly expanded around the same time (Costa 1998). Today, public pension systems are almost universal in industrialized countries, and they are present also in many middle-income and developing countries as well.

While fiscal welfare programs began with pensions, direct medical care or insurance is an increasingly large part of support for the elderly in most countries, especially in those where diseases associated with aging rather than external hazards such as infectious disease play a key role in mortality. The epidemiologic revolution that occurred in the earlier part of the twentieth century in advanced countries, owing to advances in medical knowledge, public health initiatives, and improvements in nutrition, set the stage for increasingly greater need by surviving elderly of medical intervention against diseases associated with aging, such as cardiovascular disease, cancer, and dementia (Sinfield 2012; Barrios et al. 2019).

Long-term care (see "▶ Long-Term Care" in this volume), usually consisting of formal direct assistance with the activities of daily living for those with limitations, is often combined with older adults' medical care at least conceptually. Funding and delivery mechanisms are often different. The nature and cost structure of long-term care are much more like a traditional pension than medical care (see "▶ Long-Term Care Financing" in this volume), although components of medical care per se can resemble elements of long-term care. Like pensions, long-term care is much more like the public assumption of roles traditionally performed privately by families. Caring for the sick was also a traditional role, but modern medicine, with microsurgical techniques, the use of other technologies, and pharmaceuticals, has transformed that part of care beyond much recognition.

In terms of total net fiscal transfers, pensions and medical and long-term care are by far the largest components of fiscal welfare for the elderly and are the primary foci of most research and policy interest. Although all citizens typically benefit from public goods like the rule of law and the enforcement of contracts, it could be true that spending on these programs disproportionately benefits the elderly or any other group. Patterns of local public finance alone might produce such patterns, if voters and taxpayers tend to locate in certain municipalities according to their age and then vote and pay taxes differently. Because large countries where this might happen tend also to be countries with strong central or federal government support of older adults, as opposed to greater local government support, this dynamic is not usually researched (Branco and Costa 2019; Morel et al. 2019).

The National Transfer Accounts (Lee and Mason 2011; United Nations 2013; see "▶ National Transfer Accounts Project" in this volume) provides a unified but flexible methodology for categorizing parts of the entire fiscal welfare system, i.e., all taxes, transfers, and other flows between entities, as specific to particular ages. What has emerged from the NTA database and the scientific research it has generated is a more nuanced view that proceeds beyond pension system and old-age medical and long-term support. But pensions and health care remain a central part of the story because they are so large. And a program-specific categorization still remains useful when interpreting other research that approaches issues using that perspective.

Key Research Findings

In most rich countries, fiscal welfare systems are usually relatively large and play an important role in transferring resources to older adults. But there still is heterogeneity among rich countries in the size and significance of fiscal welfare programs, although there are patterns that connect countries with similar histories and geography. As Mason and Lee (2018) show using a sample of 29 countries in the NTA, European countries, except for the UK and possibly Spain, tend to have very large public transfers to older adults relative to other methods of support. Their preferred metric is the share of the life-cycle deficit funded through public transfers, as opposed to private transfers between families or asset-based reallocations, the latter of which represent capital income and the like. The life-cycle deficit is defined as the excess of consumption over labor earnings, which must be funded in one of those ways. Mason and Lee find that for ages 65 and older, public transfers typically fund 86% of the life-cycle deficit for countries like Austria (89%), France (73%), Hungary (99%), and Sweden (108%).

Spain (63%) and especially the UK (46%) look more like the USA (37%) and Australia (47%), with the differences made up using asset-based reallocations. The common history of Australia, the USA, and the UK, with colonial relationships, language, and laws, is a clear connecting pattern, although all of those precede the development of the modern welfare state. Presumably the development of fiscal welfare systems drew in part from demographic necessity, from external shocks like warfare, and from underlying institutions and political preference. In terms of the size of net public transfers, Japan (57%) looks more like the USA than Europe.

Per Mason and Lee (2018), many countries in Central and South America have large public transfer systems similar to those in most of Europe. These include Brazil (108% of that country's life-cycle deficit), Ecuador (114%), and Peru (82%). But other middle-income countries in the Americas like Costa Rica (66%), Mexico (37%), and Uruguay (66%) have public transfer systems that redistribute at rates more like the USA, also relying more on asset-based reallocations. An outlier is El Salvador (18%), where older adults rely even more heavily on their assets.

In parts of Asia and East Asia, public transfers are a very low share of the life-cycle deficit, and the elderly rely almost exclusively on their assets. These include countries like Cambodia (5%), India (3%), Indonesia (2%), the Philippines (-1%), and Thailand (8%). To be sure, this is not to say that the elderly in those countries do not receive pensions or medical care. Rather, the point is about net transfers to those groups or the excess in benefits received over taxes contributed.

This is the essence of the fiscal welfare concept, which concerns how resources can be transferred between rich and poor states, which often but does not always align with life-cycle patterns of wealth and vitality.

Prima facie, fiscal welfare systems ought to be instrumental in reducing elderly poverty rates, and that is ultimately what the literature shows. The subtlety derives from the well-known disincentives to work and save in the presence of distortionary taxation and when progressive social benefits are available. For the USA, Engelhardt and Gruber (2006) provide an overview of how much of the decline in poverty rates among older adults can be attributed to the US Social Security program. They assess the behavioral responses in labor supply using average replacement rates across birth cohorts and the famous Social Security "notch" for those born between 1917 and 1921, which are exogenous to individuals. Their bottom line is that all of the decline in poverty rates among older adults between 1967 and 2000 can be attributed to Social Security.

A subject of considerable debate in the USA is the extent to which gaps in the health insurance system below age 65 may contribute to poverty and destitution via bankruptcy due to medical debt. Although different, this issue speaks to the question of how the fiscal welfare system affects elderly poverty rates. Evidence from recent public insurance expansions suggests they reduce medical debt (Finkelstein et al. 2012), but there is much disagreement over the degree to which medical bills fuel bankruptcies (Dobkin et al. 2018).

Fiscal welfare transfers to older adults improve well-being, but social programs typically include side effects. Gruber and Wise (1999, 2002) were among the first to characterize the problem in a systematic way, drawing from cross-country variation in public pension eligibility and implicit rates of taxation to explain retirement ages. Many public pension programs that supply a "defined benefit" in exchange for a certain amount of working in the system typically levy an implicit tax on continued working. These implicit taxes can be quite high, and Gruber and Wise showed how older workers rationally retire earlier than is probably necessary given their health and disability.

Sometimes high rates of exit from the labor force by older workers may be viewed as beneficial for younger citizens. In this view, the young might otherwise go unemployed unless jobs are vacated for them. Most economists do not view this perspective as persuasive, akin to the "lump of labor" hypothesis sometimes underpinning popular arguments against immigration, but it holds common appeal.

Reduced poverty associated with fiscal welfare transfers is unambiguously good for health. But along the margin, some of the side effects of a fiscal welfare system may be suboptimal. Producing a spiritual successor to the earlier work by Gruber and Wise, Rohwedder and Willis (2010) show that public pension systems that promote earlier retirement may also produce earlier cognitive decline through inactivity.

Still, the overall sense is that fiscal welfare systems are beneficial to health on net. A much more direct effect is the impact of medical insurance on health outcomes. At the aggregate level, life expectancies at older ages have risen robustly across the developed world for most of the past half century, losing only a small amount of their gains recently due to elevated opioid deaths and related issues (Case and Deaton 2017). Card et al. (2009) examine discontinuities around the age of eligibility for the US Medicare system to identify the effect of medical insurance on mortality, which otherwise is difficult to detect in time series. They find a 20% reduction in deaths for the most ill patients, which lasts almost a year.

The benefits of fiscal welfare systems are not costless, and the sustainability of their systems is a key policy issue for most countries. De la Maisonneuve and Martins (2014) discuss how in particular health and long-term care are the key challenges for OECD and middle-income countries alike. Financial markets sometimes punish governments presiding over fiscal policies they perceive as unsustainable. The experience of Canada in the 1990s suggests markets will react to an unsustainable trajectory and also to a correction (Edwards 2013). In that case, pension reform was probably a small to moderate part of the policy change. In the case of the Greek debt crisis of the 2010s, pensions appeared to play a larger role, both literally and figuratively speaking. The long-term fiscal outlook in most countries has not improved since the global financial crisis of 2007–2008. But for the most part, high levels of implicit debt through unfunded pension and medical care obligations seem not to have impeded further borrowing by the US government or other governments of large economies.

Future Directions for Research

How fiscal welfare programs will adapt to population aging around the world, which takes a great variety of forms due to the heterogeneity of demographic transitions, remains a topic of interest for policy and research. The National Transfer Accounts (NTA) project is pushing this agenda forward in a key way by providing the basic measurement and quantification of the large variation in conditions across economies.

Several recent papers illustrate the broad applicability of NTA methods and the richness of data and insights that they produce. Cai et al. (2018) compare the "nascent and fragmented" fiscal welfare system inside China, a rapidly aging country, to select Latin American and OECD countries, and they highlight the challenge for policy that is posed by unsustainable future paths. Gál et al. (2018) examine NTA data on public and private transfers of money and also of time, such as unpaid child care within the household, to compare how working-age populations support the elderly and children across ten European countries. Hammer et al. (2018) model the sustainability of fiscal welfare systems in 24 countries in the European Union by analyzing cohorts, reporting imbalances across all countries that appear to be more associated with the levels of contributions and benefits than with retirement ages and employment rates. An emergent contribution of the NTA program is to assess how fiscal welfare programs have evolved, which requires constructing national transfer accounts across many points in time. As these accounts grow and evolve, more knowledge will emerge that can guide policy responses.

Summary

Fiscal welfare systems that are aimed at meeting the needs of aging vary considerably across geographic regions and levels of development, and historically they vary across time as well. Originally envisioned as income replacement in retirement during the age when infectious diseases and hunger were chief barriers to human well-being, these systems have had to adapt in advanced economies to the massively changed landscape of aging brought about by advances in knowledge and medical technologies. Although they do not yet account for the majority of resources distributed in advanced fiscal welfare systems, the provisions of medical and long-term care to aging populations have grown substantially in size over the past decades and will likely become the dominant components of fiscal welfare systems in countries that have completed their demographic transitions.

Cross-References

▶ Pension Systems
▶ Social Security Around the World
▶ Social Security: History and Operations
▶ Social Security: Long-Term Financing and Reform
▶ Welfare States

References

Barrios S, Coda Moscarola F, Figari F, Gandullia L, Riscado, S. (2019). The fiscal and equity impact of social tax expenditures in the EU. J Eur Soc Policy 0958928719891341. https://doi.org/10.1177/0958928719891341

Börsch-Supan A, Wilke CB (2004) The German public pension system: how it was, how it will be. NBER working paper 10525, May. https://doi.org/10.3386/w10525

Branco R, Costa E (2019) The golden age of tax expenditures: fiscal welfare and inequality in Portugal (1989–2011). New Polit Econ 24(6):780–797. https://doi.org/10.1080/13563467.2018.1526264

Cai Y, Wang F, Shen K (2018) Fiscal implications of population aging and social sector expenditure in China. Popul Dev Rev 44(4):811–831. https://doi.org/10.1111/padr.12206

Card D, Maestas N, Dobkin C (2009) Does Medicare save lives? Q J Econ 124(2):597–636. https://doi.org/10.1162/qjec.2009.124.2.597

Case A, Deaton A (2017) Mortality and morbidity in the 21st century. Brookings Pap Econ Act 2017:397–476

Costa DL (1998) The evolution of retirement: an American economic history, 1880–1990. University of Chicago Press, Chicago

De la Maisonneuve C, Martins JO (2014) The future of health and long-term care spending. OECD J Econ Stud 2014(1):61–96. https://doi.org/10.1787/eco_studies-2014-5jz0v44s66nw

Dobkin C, Finkelstein A, Kluender R, Notowidigdo MJ (2018) Myth and measurement – the case of medical bankruptcies. N Engl J Med 378(12):1076–1078. https://doi.org/10.1056/NEJMp1716604

Edwards C (2013) Canada's fiscal reforms. Cato J 33(2):299–306

Engelhardt GV, Gruber J (2006) Social security and the evolution of elderly poverty, Chapter 6. In: Auerbach AJ, Card D, Quigley JM (eds) Public policy and the income distribution. Russell Sage Foundation, New York, pp 259–287

Finkelstein A, Taubman S, Wright B, Bernstein M, Gruber J, Newhouse JP, Allen H, Baicker K, Oregon Health Study Group (2012) The Oregon health insurance experiment: evidence from the first year. Q J Econ 127(3):1057–1106

Gál RI, Vanhuysse P, Vargha L (2018) Pro-elderly welfare states within child-oriented societies. J Eur Publ Policy 25(6):944–958. https://doi.org/10.1080/13501763.2017.1401112

Gruber J, Wise DA (1999) Social security and retirement around the world. University of Chicago Press, Chicago

Gruber J, Wise DA (2002) An international perspective on policies for an aging society. In: Altman S, Schactman D (eds) Policies for an aging society: confronting the economic and political challenges. Johns Hopkins Press, Baltimore, pp 34–62

Hammer B, Prskawetz A, Gál RI, Vargha L, Istenič T (2018) Human capital investment and the sustainability of public transfer systems across Europe. J Popul Ageing 12:427. https://doi.org/10.1007/s12062-018-9224-8

Lee R, Mason A (2011) Population aging and the generational economy: a global perspective. Edward Elgar Publishing, Northampton

Mason A, Lee R (2018) Intergenerational transfers and the older population, Chapter 7. In: Hayward MD, Majmundar MK (eds) Future directions for the demography of aging. National Academies Press, Washington. https://doi.org/10.17226/25064

Morel N, Touzet C, Zemmour M (2018a) Fiscal welfare in Europe: why should we care and what do we know so far? J Eur Soc Policy 28(5):549–560. https://doi.org/10.1177/0958928718802553

Morel N, Touzet C, Zemmour M (2018b) Conceptualizing fiscal welfare in Europe. Rev Fr Socio-Écon 1:123–141. https://doi.org/10.3917/rfse.020.0123

Morel N, Touzet C, Zemmour M (2019) From the hidden welfare state to the hidden part of welfare state reform: analyzing the uses and effects of fiscal welfare in

France. Soc Policy Adm 53(1):34–48. https://doi.org/10.1111/spol.12416

Rohwedder S, Willis RJ (2010) Mental retirement. J Econ Perspect 24(1):119–138. https://doi.org/10.1257/jep.24.1.119

Sinfield A (2012) Fiscal welfare. In: Bent Greve (ed) The Routledge handbook of the welfare state. Routledge, Abingdon. Accessed 03 Feb 2020, Routledge Handbooks Online. https://doi.org/10.4324/9780203084229.ch3

United Nations Population Division (2013) National transfer accounts manual: measuring and analysing the generational economy. United Nations, New York

Fixed Body Size Limits Lifespan

▶ Cessation of Somatic Growth Aging Theory

Flexible Goal Adjustment

▶ Dual Process Theory of Assimilation and Accommodation

Flooding and Older People

Gary Haq
Stockholm Environment Institute, Department of Environment and Geography, University of York, Heslington, York, UK

Synonyms

Deluge; Downpour; Surge; Torrent; Tsunami

Definition

A flood is the overflowing of the normal confines of a stream or other body of water, or the accumulation of water over areas that are not normally submerged. Floods include river (fluvial) floods, flash floods, urban floods, pluvial floods, sewer floods, coastal floods, and glacial lake outburst floods (IPCC 2012). The rise in the frequency of climate change-related extreme weather events combined with urban expansion in flood prone areas has increased flood risk. Floods and extreme precipitation have a particular detrimental impact on the health and well-being of vulnerable older people.

Overview

A rise in the global mean surface temperature due to the release of carbon dioxide and other greenhouse gases is changing precipitation patterns, melting snow and ice and affecting the quantity and quality of water resources. According to the Intergovernmental Panel on Climate Change (IPPC), global heating has resulted in an increase in the frequency, intensity and amount of heavy precipitation events worldwide (IPCC 2018). A hotter climate would further increase the risk of floods (IPCC 2012). This is occurring at time when the world's population is aging. By 2050, the number of older people aged over 60 is predicted to reach 2.1 billion people with the older population of developing regions growing faster than in developed regions (DESA 2017). By 2050, 79% of the world's older population will live in developing countries who may not be equipped to deal with the impacts of a changing climate such as flooding (See ▶ "Climate Change, Vulnerability, and Older People").

Floods have several potential negative effects on the health of older people. As well as immediate injury and death from floodwater, longer-term impacts on health include the spread of infectious disease and mental illness, both are exacerbated by the destruction of infrastructure, homes, and livelihoods (Watts et al. 2018).

Key Research Findings

In 2017, there was 126 reported flood events worldwide affecting 55 million people and resulting in 3,331 deaths and economic losses of USD 20 billion (CRED 2018). Although the number of flood events were lower compared to the annual average of the previous decade 2007–2016, the 2017 flood events had the highest economic cost. Flooding in Asian countries such as Bangladesh, India, and Nepal affected approximately 27 million people, with 450 million people living in flood risk areas (CRED 2018).

Higher levels of global heating mean that the number of people exposed to river floods and the effects of sea level rise is expected to increase (IPCC 2014). In particular, populations in low-income countries who inhabit high-risk flood plains and coastal zones are more vulnerable because of poor public health infrastructure, poverty, and lack of insurance coverage. There is increasing evidence to suggest that older people in these areas have more difficulty than other groups when leaving their place of residence in an emergency (Oven et al. 2012).

Older people are vulnerable to the effects of flooding because of preexisting health conditions such as disabilities, poor health, dementia, and other mental health problems. Mobility restrictions means they are less likely to respond to flood warnings (Fielding et al. 2007). This demographic group therefore has high mortality rates due to immediate injury and death from floodwater via drowning.

Flooding can have adverse psychological impacts on older people with some experiencing symptoms of posttraumatic stress disorder (PTSD) such as post-flood anxiety and mood changes (Bei et al. 2013). This, together with preexisting health conditions, can have a significant impact on the health and well-being of an older flood victim.

Secondary health impacts from flooding include hypothermia, heart problems fecal-oral disease, vector-borne disease, rodent-borne disease, and mental disorders (Ahern et al. 2005). Flood-related water contamination is more likely to have a greater impact on older people due to their weakened immune systems (Vardoulakis and Heaviside 2012). Flooding may also restrict older people's access to vital emergency medical services and routine nonurgent appointments. If an older person lives alone, they be less prepared to protect themselves or able to recover after a flood (Fernandez et al. 2002).

National level adaption planning needs to include disaster risk reduction and flood event preparedness. There should also be strategies to engage older people in river basin management and urban planning to better manage and protect against flood risk (HelpAge International 2015).

Future Research Challenges

In order to meet Sustainable Development Goal (SDG13) on climate action, further research is needed on strengthening the resilience and adaptive capacity of older people to flood events. While research has been devoted to characterizing and quantifying the health risks associated with natural disasters such as flooding, more can be done to understand the impact on older people. Older people may find it hard to recover from the effects of flooding, especially those living with mental disorders such as dementia. A greater understanding is required of how older people can be better prepared for, cope with, and recover from flood events. In particular, determining the effectiveness of intervention strategies and how they can address the needs of an aging population (Bukvic et al. 2018). Research should facilitate the better design and delivery of support services such as primary and secondary mental health care by providing an understanding of the short-, medium- and long-term impact of flooding on the psychosocial experience of older people (Stanke et al. 2012).

Summary

Flooding has direct and indirect detrimental effects on the health and well-being of older people. With a changing climate, the frequency and intensity of floods pose a threat to older people living in flood risk areas. Greater planning to manage and reduce flood risk is necessary that meets the needs and vulnerabilities of older people. This requires a better of understanding of the impact of flood events on older people.

Cross-References

▶ Neighborhood Social Environment and Health

References

Ahern M, Kovats RS, Wilkinson P, Few R, Matthies F (2005) Global health impacts of floods: epidemiologic evidence. Epidemiol Rev 27(1):36–46. https://doi.org/10.1093/epirev/mxi004

Bei B, Bryand C, Gilson KM, Koh J, Gibson P, Komiti A, Jackson H, Judd F (2013) A prospective study of the impact of floods on the mental and physical health of older adults. Aging Ment Health 17(8):992–1002. https://doi.org/10.1080/13607863.2013.799119

Bukvic A, Gohlke J, Borate A, Suggs J (2018) Aging in flood-prone coastal areas: discerning the health and Well-being risk for older residents. Int J Environ Res Public Health 15:2900. https://doi.org/10.3390/ijerph15122900

CRED (2018) Natural disasters 2017. Centre for Research on Epidemiology of Disasters, Brussels

DESA (2017) World population ageing – highlights. United Nations, New York

Fernandez LS, Byard D, Lin CC, Benson S, Barbera JA (2002) Frail Elderly as Disaster Victims: emergency Management Strategies. Prehosp Disaster Med J 17 (2):67–74. https://doi.org/10.1017/S1049023X00000200

Fielding J, Burningham K, Thrush D, Catt R (2007) Public response to flood warning. R&D Technical Report SC020116. Environment Agency, Bristol

HelpAge International (2015) Climate change in an ageing world. HelpAge International, London

IPCC (2012) Managing the risks of extreme events and disasters to advance climate change adaptation. A special report of working Groups I and II. of the intergovernmental panel on climate change. Cambridge University Press, Cambridge

IPCC (2014) Climate Change 2014: synthesis report. Contribution of working groups I, II and III to the fifth assessment report. Intergovernmental Panel on Climate Change, Geneva, Switzerland

IPCC (2018) Summary for Policymakers. In: Global warming of 1.5 °C. An IPCC Special Report on the impacts of global warming of 1.5 °C above pre-industrial levels and related global greenhouse gas emission pathways, in the context of strengthening the global response to the threat of climate change, sustainable development, and efforts to eradicate poverty. World Meteorological Organization, Geneva

Oven KJ, Curtis SE, Reaney S, Riva M, Stewart MG, Ohlemüller R, Dunn CE, Nowell S, Dominelli L, Holden R (2012) Climate change and health and social care: defining future hazard, vulnerability and risk for infrastructure systems supporting older people's health care in England. Appl Geogr 33:16–24. https://doi.org/10.1016/j.apgeog.2011.05.012

Stanke C, Murray V, Amlôt R, Nurse J, Willams R (2012) The effects of flooding on mental health outcomes and recommendations from a review of the literature. Version 1. PLoS Curr 4. https://doi.org/10.1371/4f9f1fa9c3cae

Vardoulakis S, Heaviside C (2012) Health effects of climate change in the UK. Health Protection Agency, Didcot

Watts N, Amann M, Aveb-Karlsson S, Belesova K, Bouley T, Bovkoff M et al (2018) The *Lancet* countdown on health and climate change: from 25 years of inaction to a global transformation for public health. Lancet 391 (10120):581–630. https://doi.org/10.1016/S0140-6736(17)32464-9

Fluid Deficit

▶ Dehydration in Older Adults

Fluid Imbalance

▶ Dehydration in Older Adults

Fluid Reasoning

▶ Intelligence (Crystallized/Fluid)

Fluid/Crystallized Knowledge

▶ Intelligence (Crystallized/Fluid)

Follicular Dendritic Cells

Péter Balogh
Department of Immunology and Biotechnology, University of Pécs Clinical Center, Pécs, Hungary

Synonyms

Antigen-retaining reticulum

Definition

Follicular dendritic cells (FDCs) are non-hematopoietic cells of the follicles (B-cell rich zones) of the secondary lymphoid tissues in mammalians (including humans) that form the scaffolding of follicles and promote the follicular colonization, expansion, and survival of B cells during immune responses associated with antibody production upon antigenic stimulation.

Overview

Identification, Functions, and Cellular Characteristics of FDCs

The ability of the mammalian immune system to mount effective humoral immune responses with high-affinity antibody production coupled with the establishment of immunological memory (the capacity to elicit prompt reaction upon secondary antigenic encounter) critically depends on the organized microstructure of peripheral lymphoid tissues. These include the spleen, a unique singular lymphoid organ to combat blood-borne pathogens; lymph nodes, forming an extensive network of encapsulated organs arranged in a chain-like fashion filtering interstitial fluid throughout the body; and various forms of lymphoid aggregates in the gastrointestinal and genitourinary mucosa and in the airways.

After their production in the bone marrow, B cells expressing cell surface immunoglobulin recirculate between various peripheral lymphoid tissues where they accumulate within the follicles, separately from the bulk of T cells. Although they constitute the vast majority of cells dwelling in these regions, other hematopoietic cells, including some T cells, macrophages, and dendritic cells that can take up or recognize antigens, can also be found here. Pivotal studies in the 1960s using radioisotope-labeled antigens performed on rodents demonstrated that, following their entry, antigens accumulate within the follicles (Miller and Nossal 1964) in a clumped arrangement (Nossal et al. 1968; Szakal and Hanna 1968). According to ultrastructural studies, this deposition of antigens represents their retention mediated by non-phagocytic (i.e., macrophage) cells. These cells possess extensive filiform membrane extensions (henceforth named as follicular dendritic cells; Chen et al. 1978) or dendrites with bead-like microsphere structures (Schnizlein et al. 1985; Szakal et al. 1985), which can be transferred to B cells as immune complex containing the antigens together with antibody and complement fragments (Szakal et al. 1988). The general structure of lymph node follicles is illustrated schematically in Fig. 1a.

The FDCs are rather uniformly distributed within the primary follicles, whereas following antigenic stimulation they are typically confined to the light zone of secondary follicles containing germinal centers (Szakal et al. 1989; MacLennan 1994).

Given the difficulties of selectively locating FDCs within the follicles overwhelmed by B lymphocytes using traditional staining, the identification of FDCs by immunohistological labeling represented an important advance for their subsequent analyses. In mice, they were found to express FcγRIII/FcγRIIB (CD16/32; Schnizlein et al. 1985; Kosco et al. 1986), followed by the demonstration of complement receptor 1/2 (CD21/35) display (Yoshida et al. 1993), similarly to humans (Reynes et al. 1985; Johnson et al. 1986). Although these markers are not specific for FDCs, their increased production compared to that of B cells has further strengthened the hypothesis of their involvement in converting and presenting immune complexes into a highly immunogenic form by FDCs as a critical element for supporting recall immune responses (Qin et al. 1998; Qin et al. 2000). Figure 1b shows a typical FDC cluster in a primary follicle using multicolor immunofluorescence with specific antibodies against T cells, B cells, and FDCs.

To strengthen antigen-specific interactions with antigen-stimulated B cells during immune responses, FDCs also demonstrate enhanced adhesion with B cells by increased expression of several adhesion molecules (including VCAM-1, ICAM-1 and MAdCAM-1), which further increase during immune responses, demonstrating the inducible capacities of FDCs (Koopman et al. 1991; Kosco et al. 1992; Szabo et al. 1997; Balogh et al. 2002).

Development of FDCs and their Pre-immune Role in the Formation of Follicles

The origin of FDCs remained a contested issue for decades. Their restricted presence in the follicles of secondary lymphoid tissues suggested

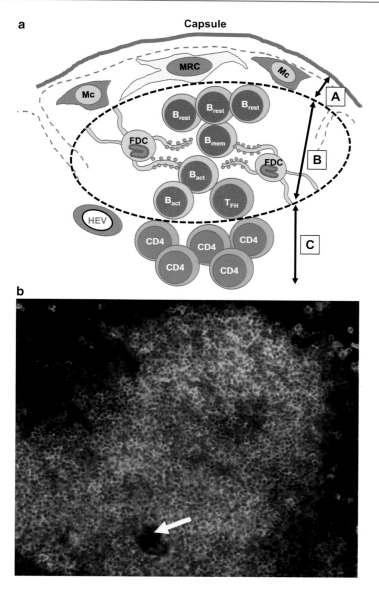

Follicular Dendritic Cells, Fig. 1 (a) Schematic diagram of a segment of a lymph node. **A: Subcapsular area**, containing macrophages (Mc) and marginal reticular cells (MRC; putative FDC precursors), which area serves as antigen entry route via the afferent lymphatics (indicated as brown dashed line). **B: Follicle** (black encircled), arranged around FDC meshwork retaining the antigen in immune complex form (depicted as small grey dots), surrounded mostly by B cells (resting/rest or activated/act or memory/mem) as well as follicular T helper cells (Tfh). **C: Extrafollicular (paracortical) region** comprising mainly of T cells, majority belonging to the CD4 subsets. Their stromal scaffolding as well as possible other stromal elements at the T/B boundary are not depicted. The blood-borne lymphocytes enter through the high endothelial venules (HEV). (**b**) Splenic follicular dendritic cells (green fluorescence staining of CR 1/2 antigen) with arborized extensions are located in B-cell rich regions (red fluorescence staining of B220 antigen) adjacent to T-cell zone (blue fluorescence staining of Thy-1.2) forming the periarteriolar lymphoid sheath (PALS) around the central arteriole (arrow). 40 × magnification hematopoietic origin, in a close dependence with the presence of B cells, as the elimination of B cells resulted in the loss of FDCs (Cerny et al. 1988), although unlike the hematopoietic cells, FDCs are radioresistant (Humphrey et al. 1984). Subsequently, congenitally severe immunodeficient (SCID) mice lacking B-and-T-cells were found to lack FDCs, but capable of generating FDCs upon transfer of mature B cells with increased expression of FcγR and complement receptors involved in long-term immune complex retention (Kapasi et al. 1993; Yoshida et al. 1994). Thus, the absence of FDCs in SCID mice offered a feasible in vivo experimental system to study the emergence of FDCs. In this approach, the origin of FDCs was defined either as local host-derived (Yoshida et al. 1994) or could also at least partly be supplied by the bone marrow (Kapasi et al. 1998). Using several advanced cell-lineage tracing in vivo experimental systems, the tissue origin of FDCs is now generally accepted as local mesenchymal derivatives from undifferentiated perivascular precursors (Krautler et al. 2012) also present at the periphery of lymph nodes as marginal reticular cells (Jarjour et al. 2014), also serving as expansion pools for FDCs in immune responses. However, possible peripheral lymphoid tissue-specific features distinguishing between the spleen and different types of peripheral lymph nodes may exist (Wang et al. 2011; Castagnaro et al. 2013).

The discovery of the role of lymphotoxin α (LTα) represented a seminal finding, opening the road to define critical molecular elements of lymphoid organogenesis generally, and the development of FDCs particularly (De Togni et al. 1994). LTα-deficient mice lack lymph nodes and Peyer's patches, and their spleen demonstrated impaired T/B distribution, coupled with the lack of FDCs, also resulting in defective germinal center formation upon immunization (Matsumoto et al. 1996). Analyzing similar roles for other members of the expanding tumor necrosis-lymphotoxin (TNF-LT), it turned out that, in addition to LTα that can be generated as a soluble LTα3homotrimer, the appearance of FDCs also requires the related member LTβ, complexed in an LTαβ2 heterotrimeric form that binds to a separate receptor LTβR (Androlewicz et al. 1992; Crowe et al. 1994). Several subsequent studies have established that splenic FDC development requires LTβR engagement via its heterotrimeric LTαβ2 ligand (Koni et al. 1997; Fütterer et al. 1998), whereas their follicular maturation is dependent on TNF, both supplied by B cells (Pasparakis et al. 1996; Endres et al. 1999; Pasparakis et al. 2000). Importantly, even fully developed FDC network can be dissolved by disrupting LTβR-mediated signaling using a soluble decoy receptor analogue (Mackay and Browning 1998), indicating the continued need for LTαβ2 ligands also for the maintenance of FDCs.

The general feature of lymphoid cell segregation within the peripheral tissues into T- and B-cell zones raised the issue of the mechanism of this separation. This segregation is sensitive to pertussis toxin blocking G-protein-coupled receptor-mediated (GPCR) signalization (Lyons and Parish 1995; Cyster and Goodnow 1995). Subsequent discovery of the GPCR member chemokine receptor CXCR5 (originally denoted as BLR1) on B cells as a key molecule for their follicular recruitment represented a major progress in defining the mechanism of follicular build-up (Förster et al. 1996). As its ligand, CXCL13 (originally named B-lymphocyte chemoattractant/BLC) produced by FDCs, was identified (Gunn et al. 1998). After its discovery, CXCL13 was demonstrated to be able to induce the upregulation of LTαβ2 heterotrimer on B cells, thus suggesting the existence of a positive feedback relationship between the FDCs (or their precursors) expressing LTβR and producing CXCL13, and the B cells generating the pair of complementary ligand and receptor LTαβ2 and CXCR5, respectively (Ansel et al. 2000). The process of CXCL13-driven follicular movement of B cells leading to the follicular segregation is linked to the follicular conduit as a physical platform, a nonvascular drainage system, formed by follicular stromal cells, possibly FDCs (Nolte et al. 2003).

An important aspect for follicular organization and B-cell responsiveness is the B-cell survival within follicles, promoted by another member of TNF family BAFF (B-cell activating factor; Schneider et al. 1999) and its analogues TACI and BCMA (Gross et al. 2000). This function is probably performed by a subset of non-hematopoietic cells (possibly related to T-zone fibroblastic cells/FRCs) producing BAFF within the follicles distinct from the FDCs, and these two cell types jointly control follicular B-cell recruitment (via FDC-derived CXCL13) and survival (via follicular FRC-derived BAFF; Cremasco et al. 2014).

Age-Associated Deviations of FDC Functions in Supporting Humoral Immune Reactions and Memory Responses

Aging has been characterized by declining immune responsiveness, and enhanced frequency of autoimmunity and malignancies ▶ "Human Immune System in Aging". As long-term preservation of FDC-associated antigen is necessary for the maintenance of memory, the aging-related decline of FDCs to sustain germinal center reactions has been known to impair humoral immune responsiveness, including reduced preservation of immunological memory and the capacity to mount recall responses (Szakal et al. 2002; McElhaney and Effros 2009; Ciabattini et al. 2018) ▶ "Cytomegalovirus and Human Immune System Aging" and ▶ "Influenza Vaccination in Older Adults". Unlike the studies performed on isolated aged T and B cell subsets, however, similar analyses addressing the impact of aging on purified FDCs are notoriously difficult to perform in mice, owing to the technical difficulties in their isolation. Moreover, such cells are beyond availability in humans unless some invasive procedure (in most cases lymph node biopsy or tonsillectomy) is employed. Nevertheless, histological analyses and in vivo cell murine experiments via transfer of young lymphocytes into aged recipients have been informative in revealing significant functional impairment of FDCs in old animals.

Initial observations indicated substantially reduced germinal center formation in aged mice (Hanna et al. 1967; Kosco et al. 1989). Subsequent immunohistochemical analyses demonstrated that, although the size of FDC reticulum was not significantly different between aged and young mice, the time-course for upregulating FcγR (inhibitory type FcγRIIB) by FDCs was substantially reduced, associated with the FDCs' reduced capacity to retain antigen. Using in vitro experimental approaches to combine young or old B cells and young and old FDCs, a significant reduction in the co-stimulatory capacity of old FDCs was noted, and explored in details (reviewed by Aydar et al. 2004). According to the hypothesis based on these findings, FDCs in young mice efficiently upregulate and dominantly grab immune complexes via *inhibitory-type FcγRIIB receptors*, precluding their suppressive effect prevailing on antigen-stimulated B cells which, in turn, will preferentially utilize *complement receptors (CR1/2)*, thus delivering potent co-stimulatory signals. In contrast, *in aged FDCs such upregulation of FcγRIIB is defective*, thereby antigen-stimulated B cells will not be rescued from having their FcγRIIB engaged by immune complexes (Szakal et al. 2002; Aydar et al. 2002; Aydar et al. 2003). Using immune complexes for immunization resulted in an enhanced in vivo responsiveness, increased antibody formation coupled with long-lived plasma cell differentiation, and accumulation in the bone marrow (Zheng et al. 2007).

In the spleen, the acquisition of immune complexes requires the shuttling of marginal zone (MZ) B cells into the follicles, delivering the antigen in immune complex form on the surface of FDCs (Cinamon et al. 2008). It was found that this shuttling of MZ B cells is impaired, probably resulting in a reduced amount of available immune complex to be deposited onto FDCs (Turner and Mabbott 2017). In aged mice, the splenic distribution of CXCL13 chemokine production was also notably different from that of the young mice, which may be implicated in the reduced mobility of MZ B cells transporting antigens into the follicles (Wols et al. 2010).

Of the key morphogenic members of the TNF/LT family for FDC differentiation, up to date no significant alteration associated with aging has been reported either in mice or humans. On the other hand, various autoimmune diseases affecting synovium, kidney, and other target organs have been described to be associated with ectopic (tertiary) lymphoid neogenesis, including the local appearance of FDCs within the affected tissues (Aloisi and Pujol-Borrell 2006; Bombardieri et al. 2017). Thus while the fitness of differentiated FDCs within secondary lymphoid tissues may deteriorate locally, the organism's general capacity to produce such cells is retained. This uncertainty (possible reduction in secondary lymphoid tissues, and increased appearance ectopically) may also hinder their laboratory analysis in patients through measuring soluble mediators (like BAFF or CXCL13) due to their undeterminable origin. Using recent advances in lineage-related cell tracing combined with multiparameter analysis of purified stromal cells, including FDCs (Rodda et al. 2018), the eventual changes of mRNA profile of purified FDCs from old mice can shed light on vital alterations associated with aging. Figure 2 illustrates the possible elements contributing to the decline of FDC functions in aging.

While the overwhelming majority of findings point to a negative impact of aging on FDC functions involved in humoral immune responsiveness, the decline of FDCs appear to reduce the transmission and neurodegeneration induced by scrapie agents (Brown et al. 2009). FDCs are necessary for initial prion replication (Brown et al. 1999), and their inactivation through the blockade of LTβR by soluble decoy receptor treatment resulted in blocked prion propagation (Mabbott et al. 2000; Montrasio et al. 2000), before the translocation of prion agent(s) to the central nervous system via the vegetative innervation of peripheral lymphoid tissues (Glatzel et al. 2001). Although this blunted spreading may be perceived as a positive consequence of FDC impairment associated with aging, it also warrants caution for the possibility of more frequent occurrence of subclinical transmissible spongioform encephalopathy amongst aged individuals.

Key Research Findings

FDCs develop from local non-hematopoietic cells of the peripheral lymphoid tissues that create suitable microenvironment for B cells and promote their follicular clustering by chemotactic stimuli. To perform their functions throughout life, the

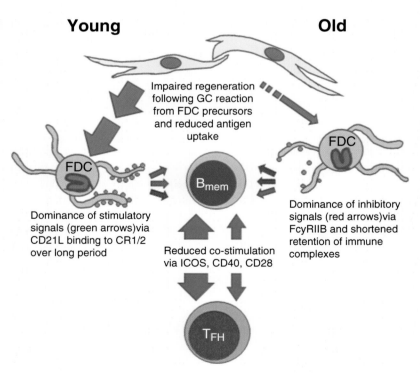

Follicular Dendritic Cells, Fig. 2 Possibilities contributing to impaired humoral immune responses involving FDCs in old individuals (right) compared to young (left). FDC-associated immune complexes are depicted as grey dots. FDC precursors likely originate from local tissue precursors

FDCs' persistence requires several members of TNF/LT family. In their interactions with B cells, FDCs utilize different types of receptors for antigen retention, which also determines the resulting signal preference (activating or inhibitory) of the partnering B cells, which alters during aging.

Future Directions of Research

As FDCs may derive from different precursors in various peripheral lymphoid tissues, the identification of tissue-specific factors affecting the formation of FDCs in different peripheral lymphoid tissues and at ectopic location may reveal possible means for their manipulation in a tissue-specific manner. Further studies should reveal whether aging-related impairment of FDCs can be reversed, by enhancing their encounter with B cells to promote the stimulatory effects or by overcoming inhibitory signals. As FDCs are usually inaccessible for in vitro analyses, monitoring their functionalities necessitates the development of novel laboratory diagnostic approaches.

Summary

FDCs are non-hematopoietic cells with significant role for, and dependence upon, B cells, in promoting high-affinity antibody responses and establishing a highly ordered lymphoid tissue architecture. Their development requires recognition of several members of TNF/LT family produced by B cells or other lymphoid cells. The FDC:B cell-clustering B cells involves the production of CXCL13 chemokine by FDCs. Upon antigenic encounter FDCs retain antigen for activated B cells as immune complexes in a favorable proportion of receptor binding to circumvent B-cell suppression. In contrast, in aged FDCs this preference is altered, so aged FDCs can no longer shield B cells from receiving inhibitory signals, thus leading to premature termination of B-cell expansion and antibody production. While aging causes a progressive loss of FDC support capacity, autoimmune disease often manifest in the ectopic appearance of FDCs, which may potentially perpetuate tissue damage of the affected organ.

Cross-References

▶ Cytomegalovirus and Human Immune System Aging
▶ Human Immune System in Aging
▶ Influenza Vaccination in Older Adults

References

Aloisi F, Pujol-Borrell R (2006) Lymphoid neogenesis in chronic inflammatory diseases. Nat Rev Immunol 6(3):205–217

Androlewicz MJ, Browning JL, Ware CF (1992) Lymphotoxin is expressed as a heteromeric complex with a distinct 33-kDa glycoprotein on the surface of an activated human T cell hybridoma. J Biol Chem 267(4):2542–2547

Ansel KM, Ngo VN, Hyman PL, Luther SA, Förster R, Sedgwick JD, Browning JL, Lipp M, Cyster JG (2000) A chemokine-driven positive feedback loop organizes lymphoid follicles. Nature 406(6793):309–314

Aydar Y, Balogh P, Tew JG, Szakal AK (2002) Age-related depression of FDC accessory functions and CD21 ligand-mediated repair of co-stimulation. Eur J Immunol 32(10):2817–2826

Aydar Y, Balogh P, Tew JG, Szakal AK (2003) Altered regulation of Fc gamma RII on aged follicular dendritic cells correlates with immunoreceptor tyrosine-based inhibition motif signaling in B cells and reduced germinal center formation. J Immunol 171(11):5975–5987

Aydar Y, Balogh P, Tew JG, Szakal AK (2004) Follicular dendritic cells in aging, a "bottle-neck" in the humoral immune response. Ageing Res Rev 3(1):15–29

Balogh P, Aydar Y, Tew JG, Szakal AK (2002) Appearance and phenotype of murine follicular dendritic cells expressing VCAM-1. Anat Rec 268(2):160–168

Bombardieri M, Lewis M, Pitzalis C (2017) Ectopic lymphoid neogenesis in rheumatic autoimmune diseases. Nat Rev Rheumatol 13(3):141–154

Brown KL, Stewart K, Ritchie DL, Mabbott NA, Williams A, Fraser H, Morrison WI, Bruce ME (1999) Scrapie replication in lymphoid tissues depends on prion protein-expressing follicular dendritic cells. Nat Med 5(11):1308–1312

Brown KL, Wathne GJ, Sales J, Bruce ME, Mabbott NA (2009) The effects of host age on follicular dendritic cell status dramatically impair scrapie agent neuroinvasion in aged mice. J Immunol 183(8):5199–5207

Castagnaro L, Lenti E, Maruzzelli S, Spinardi L, Migliori E, Farinello D, Sitia G, Harrelson Z, Evans SM, Guidotti LG, Harvey RP, Brendolan A (2013) Nkx2-5(+)islet1(+) mesenchymal precursors generate distinct spleen stromal cell subsets and participate in restoring stromal network integrity. Immunity 38(4):782–791

Cerny A, Zinkernagel RM, Groscurth P (1988) Development of follicular dendritic cells in lymph nodes of B-cell-depleted mice. Cell Tissue Res 254(2):449–454

Chen LL, Adams JC, Steinman RM (1978) Anatomy of germinal centers in mouse spleen, with special reference to "follicular dendritic cells". J Cell Biol 77(1):148–164

Ciabattini A, Nardini C, Santoro F, Garagnani P, Franceschi C, Medaglini D (2018) Vaccination in the elderly: The challenge of immune changes with aging. Semin Immunol 40:83–94

Cinamon G, Zachariah MA, Lam OM, Foss FW Jr, Cyster JG (2008) Follicular shuttling of marginal zone B cells facilitates antigen transport. Nat Immunol 9(1):54–62

Cremasco V, Woodruff MC, Onder L, Cupovic J, Nieves-Bonilla JM, Schildberg FA, Chang J, Cremasco F, Harvey CJ, Wucherpfennig K, Ludewig B, Carroll MC, Turley SJ (2014) B cell homeostasis and follicle confines are governed by fibroblastic reticular cells. Nat Immunol 15(10):973–981

Crowe PD, VanArsdale TL, Walter BN, Ware CF, Hession C, Ehrenfels B, Browning JL, Din WS, Goodwin RG, Smith CA (1994) A lymphotoxin-beta-specific receptor. Science 264(5159):707–710

Cyster JG, Goodnow CC (1995) Pertussis toxin inhibits migration of B and T lymphocytes into splenic white pulp cords. J Exp Med 182(2):581–586

De Togni P, Goellner J, Ruddle NH, Streeter PR, Fick A, Mariathasan S, Smith SC, Carlson R, Shornick LP, Strauss-Schoenberger J et al (1994) Abnormal development of peripheral lymphoid organs in mice deficient in lymphotoxin. Science 264(5159):703–707

Endres R, Alimzhanov MB, Plitz T, Fütterer A, Kosco-Vilbois MH, Nedospasov SA, Rajewsky K, Pfeffer K (1999) Mature follicular dendritic cell networks depend on expression of lymphotoxin beta receptor by radioresistant stromal cells and of lymphotoxin beta and tumor necrosis factor by B cells. J Exp Med 189(1):159–168

Förster R, Mattis AE, Kremmer E, Wolf E, Brem G, Lipp M (1996) A putative chemokine receptor, BLR1, directs B cell migration to defined lymphoid organs and specific anatomic compartments of the spleen. Cell 87(6):1037–1047

Fütterer A, Mink K, Luz A, Kosco-Vilbois MH, Pfeffer K (1998) The lymphotoxin beta receptor controls organogenesis and affinity maturation in peripheral lymphoid tissues. Immunity 9(1):59–70

Glatzel M, Heppner FL, Albers KM, Aguzzi A (2001) Sympathetic innervation of lymphoreticular organs is rate limiting for prion neuroinvasion. Neuron 31(1):25–34

Gross JA, Johnston J, Mudri S, Enselman R, Dillon SR, Madden K, Xu W, Parrish-Novak J, Foster D, Lofton-Day C, Moore M, Littau A, Grossman A, Haugen H, Foley K, Blumberg H, Harrison K, Kindsvogel W, Clegg CH (2000) TACI and BCMA are receptors for a TNF homologue implicated in B-cell autoimmune disease. Nature 404(6781):995–999

Gunn MD, Ngo VN, Ansel KM, Ekland EH, Cyster JG, Williams LT (1998) A B-cell-homing chemokine made in lymphoid follicles activates Burkitt's lymphoma receptor-1. Nature 391(6669):799–803

Hanna MG Jr, Nettesheim P, Ogden L, Makinodan T (1967) Reduced immune potential of aged mice: significance of morphologic changes in lymphatic tissue. Proc Soc Exp Biol Med 125(3): 882–886

Humphrey JH, Grennan D, Sundaram V (1984) The origin of follicular dendritic cells in the mouse and the mechanism of trapping of immune complexes on them. Eur J Immunol 14(9):859–864

Jarjour M, Jorquera A, Mondor I, Wienert S, Narang P, Coles MC, Klauschen F, Bajénoff M (2014) Fate mapping reveals origin and dynamics of lymph node follicular dendritic cells. J Exp Med 211(6): 1109–1122

Johnson GD, Hardie DL, Ling NR, Maclennan IC (1986) Human follicular dendritic cells (FDC): a study with monoclonal antibodies (MoAb). Clin Exp Immunol 64(1):205–213

Kapasi ZF, Burton GF, Shultz LD, Tew JG, Szakal AK (1993) Induction of functional follicular dendritic cell development in severe combined immunodeficiency mice. Influence of B and T cells. J Immunol 150(7):2648–2658

Kapasi ZF, Qin D, Kerr WG, Kosco-Vilbois MH, Shultz LD, Tew JG, Szakal AK (1998) Follicular dendritic cell (FDC) precursors in primary lymphoid tissues. J Immunol 160(3):1078–1084

Koni PA, Sacca R, Lawton P, Browning JL, Ruddle NH, Flavell RA (1997) Distinct roles in lymphoid organogenesis for lymphotoxins alpha and beta revealed in lymphotoxin beta-deficient mice. Immunity 6(4): 491–500

Koopman G, Parmentier HK, Schuurman HJ, Newman W, Meijer CJ, Pals ST (1991) Adhesion of human B cells to follicular dendritic cells involves both the lymphocyte function-associated antigen 1/intercellular adhesion molecule 1 and very late antigen 4/vascular cell adhesion molecule 1 pathways. J Exp Med 173(6):1297–1304

Kosco MH, Tew JG, Szakal AK (1986) Antigenic phenotyping of isolated and in situ rodent follicular dendritic cells (FDC) with emphasis on the ultrastructural demonstration of Ia antigens. Anat Rec 215(3):201–213, 219–225

Kosco MH, Burton GF, Kapasi ZF, Szakal AK, Tew JG (1989) Antibody-forming cell induction during an early phase of germinal centre development and its delay with ageing. Immunology 68(3):312–318

Kosco MH, Pflugfelder E, Gray D (1992) Follicular dendritic cell-dependent adhesion and proliferation of B cells in vitro. J Immunol 148(8):2331–2339

Krautler NJ, Kana V, Kranich J, Tian Y, Perera D, Lemm D, Schwarz P, Armulik A, Browning JL, Tallquist M, Buch T, Oliveira-Martins JB, Zhu C, Hermann M, Wagner U, Brink R, Heikenwalder M, Aguzzi A (2012) Follicular dendritic cells emerge from ubiquitous perivascular precursors. Cell 150(1): 194–206

Lyons AB, Parish CR (1995) Are murine marginal-zone macrophages the splenic white pulp analog of high endothelial venules? Eur J Immunol 25(11):3165–3172

Mabbott NA, Mackay F, Minns F, Bruce ME (2000) Temporary inactivation of follicular dendritic cells delays neuroinvasion of scrapie. Nat Med 6(7):719–20

Mackay F, Browning JL (1998) Turning off follicular dendritic cells. Nature 395(6697):26–27

MacLennan IC (1994) Germinal centers. Annu Rev Immunol 12:117–139

Matsumoto M, Mariathasan S, Nahm MH, Baranyay F, Peschon JJ, Chaplin DD (1996) Role of lymphotoxin and the type I TNF receptor in the formation of germinal centers. Science 271(5253):1289–1291

McElhaney JE, Effros RB (2009) Immunosenescence: what does it mean to health outcomes in older adults? Curr Opin Immunol 21(4):418–424

Miller JJ, Nossal GJ (1964) Antigens in immunity. VI. The phagocytic reticulum of lymph node follicles. J Exp Med 120:1075–1086

Montrasio F, Frigg R, Glatzel M, Klein MA, Mackay F, Aguzzi A, Weissmann C (2000) Impaired prion replication in spleens of mice lacking functional follicular dendritic cells. Science 288(5469):1257–1259

Nolte MA, Beliën JA, Schadee-Eestermans I, Jansen W, Unger WW, van Rooijen N, Kraal G, Mebius RE (2003) A conduit system distributes chemokines and small blood-borne molecules through the splenic white pulp. J Exp Med 198(3):505–512

Nossal GJ, Abbot A, Mitchell J, Lummus Z (1968) Antigens in immunity. XV. Ultrastructural features of antigen capture in primary and secondary lymphoid follicles. J Exp Med 127(2):277–290

Pasparakis M, Alexopoulou L, Episkopou V, Kollias G (1996) Immune and inflammatory responses in TNF alpha-deficient mice: a critical requirement for TNF alpha in the formation of primary B cell follicles, follicular dendritic cell networks and germinal centers, and in the maturation of the humoral immune response. J Exp Med 184(4):1397–1411

Pasparakis M, Kousteni S, Peschon J, Kollias G (2000) Tumor necrosis factor and the p55TNF receptor are required for optimal development of the marginal sinus and for migration of follicular dendritic cell precursors into splenic follicles. Cell Immunol 201(1):33–41

Qin D, Wu J, Carroll MC, Burton GF, Szakal AK, Tew JG (1998) Evidence for an important interaction between a complement-derived CD21 ligand on follicular dendritic cells and CD21 on B cells in the initiation of IgG responses. J Immunol 161(9):4549–4554

Qin D, Wu J, Vora KA, Ravetch JV, Szakal AK, Manser T, Tew JG (2000) Fc gamma receptor IIB on follicular dendritic cells regulates the B cell recall response. J Immunol 164(12):6268–6275

Reynes M, Aubert JP, Cohen JH, Audouin J, Tricottet V, Diebold J, Kazatchkine MD (1985) Human follicular dendritic cells express CR1, CR2, and CR3 complement receptor antigens. J Immunol 135(4):2687–2694

Rodda LB, Lu E, Bennett ML, Sokol CL, Wang X, Luther SA, Barres BA, Luster AD, Ye CJ, Cyster JG (2018) Single-Cell RNA Sequencing of Lymph Node Stromal Cells Reveals Niche-Associated Heterogeneity. Immunity 48(5):1014–1028

Schneider P, MacKay F, Steiner V, Hofmann K, Bodmer JL, Holler N, Ambrose C, Lawton P, Bixler S, Acha-Orbea H, Valmori D, Romero P, Werner-Favre C, Zubler RH, Browning JL, Tschopp J (1999) BAFF, a novel ligand of the tumor necrosis factor family, stimulates B cell growth. J Exp Med 189(11):1747–1756

Schnizlein CT, Kosco MH, Szakal AK, Tew JG (1985) Follicular dendritic cells in suspension: identification, enrichment, and initial characterization indicating immune complex trapping and lack of adherence and phagocytic activity. J Immunol 134(3):1360–1368

Szabo MC, Butcher EC, McEvoy LM (1997) Specialization of mucosal follicular dendritic cells revealed by mucosal addressin-cell adhesion molecule-1 display. J Immunol 158(12):5584–5588

Szakal AK, Hanna MG Jr (1968) The ultrastructure of antigen localization and viruslike particles in mouse spleen germinal centers. Exp Mol Pathol 8(1):75–89

Szakal AK, Gieringer RL, Kosco MH, Tew JG (1985) Isolated follicular dendritic cells: cytochemical antigen localization, Nomarski, SEM, and TEM morphology. J Immunol 134(3):1349–59

Szakal AK, Kosco MH, Tew JG (1988) A novel in vivo follicular dendritic cell-dependent iccosome-mediated mechanism for delivery of antigen to antigen-processing cells. J Immunol 140(2):341–353

Szakal AK, Kosco MH, Tew JG (1989) Microanatomy of lymphoid tissue during humoral immune responses: structure function relationships. Annu Rev Immunol 7:91–109

Szakal AK, Aydar Y, Balogh P, Tew JG (2002) Molecular interactions of FDCs with B cells in aging. Semin Immunol 14(4):267–274

Turner VM, Mabbott NA (2017) Ageing adversely affects the migration and function of marginal zone B cells. Immunology 151(3):349–362

Wang X, Cho B, Suzuki K, Xu Y, Green JA, An J, Cyster JG (2011) Follicular dendritic cells help establish follicle identity and promote B cell retention in germinal centers. J Exp Med 208(12):2497–2510

Wols HA, Johnson KM, Ippolito JA, Birjandi SZ, Su Y, Le PT, Witte PL (2010) Migration of immature and mature B cells in the aged microenvironment. Immunology 129(2):278–290

Yoshida K, van den Berg TK, Dijkstra CD (1993) Two functionally different follicular dendritic cells in secondary lymphoid follicles of mouse spleen, as revealed by CR1/2 and FcR gamma II-mediated immune-complex trapping. Immunology 80(1):34–39

Yoshida K, van den Berg TK, Dijkstra CD (1994) The functional state of follicular dendritic cells in severe combined immunodeficient (SCID) mice: role of the lymphocytes. Eur J Immunol 24(2):464–468

Zheng B, Switzer K, Marinova E, Wansley D, Han S (2007) Correction of age-associated deficiency in germinal center response by immunization with immune complexes. Clin Immunol 124(2):131–137

Foot Health

▶ Podiatric Medicine

Force of Mortality

Marius D. Pascariu[1,2] and Catalina Torres[3]
[1]Biometric Risk Modelling Chapter, SCOR Global Life SE, Paris, France
[2]University of Southern Denmark, Odense, Denmark
[3]Interdisciplinary Centre on Population Dynamics (CPOP), University of Southern Denmark, Odense, Denmark

Synonyms

Hazard rate mortality intensity; Instantaneous death rate; Instantaneous risk of mortality

Definitions

The force of mortality is a continuous function of age and can be defined as the instantaneous effect of mortality at a certain age. If we denote by $\ell(x)$ the number of survivors at age x in a population, the force of mortality can be specifically described as the ratio of the rate of change of $\ell(x)$ to the value of $\ell(x)$. Mathematically, then,

$$\mu(x) = -\frac{1}{\ell(x)} \frac{d\ell(x)}{dx}, \quad (1)$$

where $d\ell(x)/dx$ represents differentiation with respect to age x and indicates the rate of decrease of the survivorship over an infinitesimally small increment of age. In formal demography and actuarial science, the force of mortality is denoted by the Greek letter μ and is a positive measure over the entire age range. Alternatively, the force of mortality can be regarded as the behavior of a death rate $_nm_x$, over a small duration n converging to zero,

$$\mu(x) = \lim_{n \to 0} {_nm_x}. \quad (2)$$

It is natural to think about the force of mortality in terms of the expected number of deaths in a population (Preston et al. 2001). The difference, within a tiny time frame of length n, in the number of survivors in a cohort aged exactly x years old, is given by the expected number of deaths, which can be written algebraically as follows:

$$\ell(x) - \ell(x+n) = n \cdot \mu(x) \cdot \ell(x), \quad n \to 0. \quad (3)$$

Following eq. (1), it can be shown that the ratio of survivors at any age $x + n$ to the survivors at age x is determined solely by the aggregate mortality between those two ages: a relationship of great importance in the theory of stable populations and the life table framework. This is done by writing the survival probability at age x over the period n, $_np_x$, as a function of μ:

$$_np_x = \exp\left(-\int_0^n \mu(x+t)dt\right). \quad (4)$$

Age-specific mortality data are usually available in discrete intervals (e.g. for 1, 5 or 10-year age classes). As a consequence, in practice, the force of mortality for noninteger ages can be derived from age-specific death rates by assuming either (1) a constant force of mortality within the age interval, or (2) a uniform distribution of deaths (i.e. increasing hazard rate), or (3) a decreasing failure rate also known as the Balducci assumption (Pitacco et al. 2009).

Overview

The age-pattern of human mortality has a characteristic shape, as the force of mortality decreases quickly after birth, reaches its lowest point during childhood around ages 8–14, and then increases with age, with a *hump* around young adult ages marking accidental and maternal mortality (see examples in Fig. 1). At the oldest ages, for

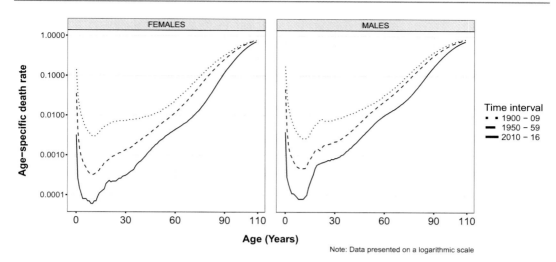

Force of Mortality, Fig. 1 Mortality decline and postponement over age in France, based on age-specific death rates for selected historical periods, by sex. Source: Life tables from the Human Mortality Database (2019)

example, after age 100, the force of mortality starts to level off as the risk of dying remains approximately constant, forming the so-called *mortality plateau*. Nevertheless, such behavior of the mortality curve at advanced ages is based on scarce data, as the number of people reaching very old ages was relatively small until recently. Population aging is providing new opportunities for the improvement of knowledge about the force of human mortality beyond age 100.

Given the variation of mortality by age, one of the main roles of the force of mortality is to provide a tool for a fundamental statement of assumptions about the behavior of individual mortality as a function of the attained age. This was of such importance in the past, that researchers and actuaries tried to represent mortality by means of mathematical functions (Smith, 1948) in their efforts to discover the universal law governing human mortality (analogous to, for instance, the law of gravitation in physics). For this reason, it is common to find the term *mortality laws* in the scientific literature, which refers to parametric models of mortality. Although a perfect fit to the entire mortality experience over age has never been achieved, some of these attempts are of particular interest.

The Gompertz model provides an excellent example. In the early nineteenth century, Gompertz (1825) made the first noteworthy proposal to describe mortality, by speculating that "it is possible that death may be the consequence of two generally co-existing causes; the one, chance, without previous disposition to death or deterioration; the other, a deterioration, or an increased inability to withstand destruction." He showed that, for various human populations, arithmetic increases in age were consistently accompanied by geometric increases in mortality between ages 20 and 60 (Olshansky and Carnes, 1997). He concluded that as the ageing process takes place, the inability to slow down physiological deterioration increases nearly exponentially.

An important modification of the model proposed by Gompertz was made by Makeham (1860). He assumed that in addition to the part of the human age-range where the force of mortality increases geometrically as the individuals grow old, a baseline mortality level operating with equal intensity at all ages can be identified. Both Benjamin Gompertz and Matthew Makeham were British actuaries trying to provide solutions to the practical work involving mortality statistics, such as smoothing data, eliminating or reducing errors, aiding inferences from incomplete data, facilitating comparisons of mortality between various groups and populations, forecasting, and life annuity pricing.

In addition to the two models mentioned above, numerous other mortality laws have been proposed during the past two centuries, attempting to cover either the entire experience from infancy to old age or distinct parts of it, such as life spent in retirement. Tabeau (2001) provides an overview of the most notable contributions.

The concept of the force of mortality plays an important role in the theory of life contingencies and actuarial science. For instance, information about changes in mortality intensities is essential for determining feasible ages of retirement and public policies needed to maintain a viable social security system in the context of transition to an ageing society. Insurance companies and private pension funds rely on mortality data for accurately determining the present value of their future liabilities and the adequate level of capital required to ensure their solvency under stressed market conditions, characterized among other factors by significant short and long-term demographic changes.

Moreover, information on mortality levels by age is fundamental for the identification of needs in intervention, for instance, in the public health domain (see "Mortality leveling"). Since the start of the *statistical revolution* in the eighteenth and the nineteenth centuries – when increased efforts began to be made for improving the collection and quality of population data and statistics – the availability of accurate estimates of the force of mortality in a given population is essential for the implementation of efficient health measures and policies. For example, given the consideration of ageing as something problematic for both individuals and societies (Victor 2004), an understanding of the dynamics in the force of mortality at the oldest ages is essential for informed social policies.

Key Research Findings

Most populations around the world have experienced impressive reductions in death rates at most ages during the last centuries and decades (see examples in Fig. 1). Those reductions have triggered unprecedented improvements in human survival: the rise in life expectancy, referred to as the "crowning achievement of the modern era" by Riley (2001), is a major demographic phenomenon with deep individual and social implications.

In the developed countries, where the demographic transition started earlier and proceeded slowly, various historical phases of change in the force of mortality can usually be distinguished, based on the ages and causes of death contributing the most to reductions in overall mortality (i.e. to gains in life expectancy). During the early stages in the 19th and early twentieth century, important reductions in infectious mortality at the youngest ages (0 to 14) were achieved, generating rapid gains in life expectancy. This decline was accompanied by reductions in young adult mortality in the first half of the twentieth century, especially from tuberculosis. From about the middle of the twentieth century onwards, important reductions in old age mortality (65+) have contributed the most to further (albeit slower) gains in life expectancy. The *cardiovascular revolution* has been key in the later development (Vallin and Meslé 2001). The Epidemiologic Transition (Omran 1971) and the Health Transition (e.g., Vallin and Meslé 2009) formalize and illustrate such trajectories, although the former theory does not include the current phase of improvements in old age mortality. In the developing countries, where the demographic transition started later and has proceeded at a fast pace, very rapid gains in life expectancy have been achieved, as large reductions in mortality at young and old ages are taking place simultaneously.

The sustained reductions in mortality referred to above, especially the ongoing improvements at old ages, have sparked academic discussion about the *biological limits to human lifespan (see "maximum lifespan")*. For instance, various studies that make hypotheses about future mortality improvements conclude that such a limit exists, and it will soon be reached (e.g., Olshansky et al. 1990). On the other hand, Oeppen and Vaupel (2002) showed that the idea of being close to such a ceiling is not supported by empirical evidence, as asserted limits to life expectancy have repeatedly been broken since the 1920s and maximum recorded life expectancy levels have continued to increase at a constant pace of "2.5 years per decade for a century and a half."

In addition to variation by age and over time, research has shown that levels of death rates may differ considerably between population subgroups according to certain attributes such as sex (women experience lower mortality and live longer than men, e.g., Zarulli et al. 2018), marital status (married men tend to have lower mortality than their nonmarried counterparts, e.g., Rendall et al. 2011), education and socio-economic status (mortality is generally lower among highly educated and wealthy individuals, e.g., Lleras-Muney 2005; Valkonen 2006), urban/rural place of residence (e.g., higher urban than rural mortality in the past vs. lower urban mortality in the present, e.g., Woods 2003; Van De Poel et al. 2009; Vierboom and Preston 2020), lifestyle (e.g., impact of smoking and obesity on health and survival, e.g., Mehta and Chang 2009; Lindahl-Jacobsen et al. 2016), and migration status (e.g., the *healthy migrant* effect and the *salmon bias* hypothesis, see Abraido-Lanza et al. 1999) among other characteristics.

Future Directions of Research

Many aspects of health and human mortality at advanced ages remain unclear due to the limitations imposed by the data available (e.g., Grundy 1997). However, with the rapid increase in the share of the population reaching older and older ages, new opportunities are emerging for improving understanding in those areas. For example, new models to fit, explain, and forecast the force of mortality after ages 90 or 100 are of great interest, motivated by the desire to accurately reflect past experience while also providing plausible future mortality trajectories (Heuveline and Clark, 2011). Additionally, thanks to advances in computational capacity for the storage, linkage, and analysis of large databases, future research will shed light on the characteristics (genetic and environmental) associated with longer lifespans.

Furthermore, age-specific mortality improvements are not constant over time, as they depend on a multitude of factors (see Cutler et al. 2006). Considering that current gains in life expectancy are increasingly driven by reductions in old-age mortality, postponement of senescence will have to take place in order for life expectancy to continue rising at the same rate (Vaupel, 2010).

Finally, developments in age-specific mortality, especially at old ages, and their impact on life expectancy will most likely continue, fueling the debate on the biological limits to human lifespan.

Summary

The force of mortality is a continuous function representing the instantaneous death rate. Given the characteristic age-pattern of human mortality, numerous mathematical models have been proposed for the estimation of mortality levels as a function of attained age. Impressive reductions in the force of mortality at most ages have been achieved during the last centuries and decades. As more and more people reach advanced ages, current gains in life expectancy are increasingly driven by reductions in old-age mortality. Understanding past and present trends in the force of mortality is essential for taking informed decisions on social and health policies. Sustained reductions in old-age mortality and population ageing are bringing new opportunities for improving understanding on health and mortality at advanced ages.

Cross-References

► Demographic Transition Theories
► Maximum Lifespan
► Mortality Leveling
► Mortality Modeling
► Probability of Dying

References

Abraido-Lanza AF, Dohrenwend BP, Ng-Mak DS, Turner JB (1999) The Latino mortality paradox: a test of the "salmon bias" and healthy migrant hypotheses. Am J Public Health 89(10):1543–1548. https://doi.org/10.2105/AJPH.89.10.1543

Cutler D, Deaton A, Lleras-Muney A (2006) The determinants of mortality. J Econ Perspect 20(3):97–120. https://doi.org/10.1257/jep.20.3.97

Gompertz B (1825) On the nature of the function expressive of the law of human mortality, and on a new mode of determining the value of life contingencies. Philos Trans R Soc Lond 115:513–583. https://doi.org/10.1098/rstl.1825.0026

GRUNDY E (1997) Demography and gerontology: mortality trends among the oldest old. Ageing Soc 17(6):713–725. https://doi.org/10.1017/S0144686X97006715

Heuveline P, Clark SJ (2011) Model schedules of mortality. In: International handbook of adult mortality. Springer, pp 511–532. https://doi.org/10.1007/978-90-481-9996-9_24

Human Mortality Database (2019). University of California, Berkeley (USA), and Max Planck Institute for Demographic Research (Germany). Data downloaded on 13/03/2019

Lindahl-Jacobsen R, Rau R, Jeune B, Canudas-Romo V, Lenart A, Christensen K, Vaupel JW (2016) Rise, stagnation, and rise of Danish women's life expectancy. Proc Natl Acad Sci 113(15):4015–4020. https://doi.org/10.1073/pnas.1602783113

Lleras-Muney A (2005) The relationship between education and adult mortality in the United States. Rev Econ Stud 72(1):189–221. https://doi.org/10.1111/0034-6527.00329

Makeham WM (1860) On the law of mortality and the construction of annuity tables. J Inst Actuaries 8(6):301–310. https://doi.org/10.1017/S2046165800000126X

Mehta NK, Chang VW (2009) Mortality attributable to obesity among middle-aged adults in the United States. Demography 46(4):851–872. https://doi.org/10.1353/dem.0.0077

Oeppen J, Vaupel JW (2002) Broken limits to life expectancy. Science 296(5570):1029–1031. https://doi.org/10.1126/science.1069675

Olshansky SJ, Carnes BA (1997) Ever since Gompertz. Demography 34(1):1–15. https://doi.org/10.2307/2061656

Olshansky SJ, Carnes BA, Cassel C (1990) In search of methuselah: estimating the upper limits to human longevity. Science 250(4981):634–640. https://doi.org/10.1126/science.2237414

Omran AR (1971) The epidemiologic transition: a theory of the epidemiology of population change. Milbank Mem Fund Q 49(4):509–538

Pitacco E, Denuit M, Haberman S, Olivieri A (2009) Modelling longevity dynamics for pensions and annuity business. Oxford University Press

Preston S, Heuveline P, Guillot M (2001) Demography: measuring and modeling population processes. Wiley-Blackwell

Rendall MS, Weden MM, Favreault MM, Waldron H (2011) The protective effect of marriage for survival: a review and update. Demography 48(2):481–506. https://doi.org/10.1007/s13524-011-0032-5

Riley JC (2001) Rising life expectancy: a global history. Cambridge University Press

Smith FC (1948) The force of mortality function. Am Math Mon 55(5):277–284. https://doi.org/10.1080/00029890.1948.11999234

Tabeau E (2001) A review of demographic forecasting models for mortality. Springer, pp 5–11. https://doi.org/10.1007/0-306-47562-6_1

Valkonen T (2006) Social inequalities in mortality. In: Caselli G, Vallin J, Wunsch G (eds) Demography: analysis and synthesis, vol 2. Elsevier, pp 195–206

Vallin J, Meslé F (2001) Trends in mortality in Europe since 1950: age-, sex-and cause-specific mortality. In Vallin J, Meslé F, Valkonen T (eds) Trends in mortality and differential mortality, number 36. Council of Europe Publishing, Population Studies, pp 33–188

Vallin J, Meslé F (2009) The segmented trend line of highest life expectancies. Popul Dev Rev 35(1):159–187. https://doi.org/10.1111/j.1728-4457.2009.00264.x

Van De Poel E, O'donnell O, Van Doorslaer E (2009) What explains the rural-urban gap in infant mortality: household or community characteristics? Demography 46(4):827–850. https://doi.org/10.1353/dem.0.0074

Vaupel JW (2010) Biodemography of human ageing. Nature 464(7288):536–542. https://doi.org/10.1038/nature08984

Victor C (2004) The social context of ageing: a textbook of gerontology. Routledge

Vierboom YC, Preston SH (2020) Life beyond 65: changing spatial patterns of survival at older ages in the United States, 2000–2016. J Gerontol Ser B 75(5):1093–1103. https://doi.org/10.1093/geronb/gbz160

Woods R (2003) Urban-rural mortality differentials: an unresolved debate. Popul Dev Rev 29(1):29–46. https://doi.org/10.1111/j.1728-4457.2003.00029.x

Zarulli V, Jones JAB, Oksuzyan A, Lindahl-Jacobsen R, Christensen K, Vaupel JW (2018) Women live longer than men even during severe famines and epidemics. Proc Natl Acad Sci 115(4):E832–E840. https://doi.org/10.1073/pnas.1701535115

Forced Migrants

▶ Aging Refugees

Forcibly Displaced People

▶ Aging Refugees

Forest Fires

▶ Wildfires and Older People

Formal and Informal Care

Jia Li[1] and Yajun Song[2]
[1]Department of Applied Social Sciences, The Hong Kong Polytechnic University, Hong Kong, China
[2]Department of Social Work, East China University of Science and Technology, Shanghai, China

Synonyms

Care of older adults; Family care

Definition

Formal care for older people usually refers to paid care services provided by a healthcare institution or individual for a person in need. Informal care refers to unpaid care provided by family, close relatives, friends, and neighbors. Both forms of caregiving involve a spectrum of tasks, but informal caregivers seldom receive enough training for these tasks. Formal caregivers are trained in the field, but the depth of their training varies.

Overview

Formal Care

Formal care can be organized into three different categories: (1) home-based care; (2) community-based care (such as daycare centers with trained staff); and (3) residential care in the form of nursing homes. Residential care is the most traditional and predominant type of formal care. While informal home-based caregivers still take the predominant role, some older adults also receive formal home care, which includes assistance in personal care (such as dressing and bathing), homemaking (e.g., laundry and cleaning), and clinical care (e.g., wound care) (Lee et al. 2018).

Nowadays, formal care is usually provided by the government and private enterprises. The government used to be the sole provider of formal care. Over the recent decades, due to a neoliberal shift in social care provision, private sector organizations are increasingly involved in providing formal care. The role of the government in formal care for older adults has been shifting from that of "supplier and provider" to that of "purchaser and regulator." For instance, in Germany, formal care providers are predominantly private so that the quality can be mostly assured due to the competition, but the prices of services are determined by the government (Campbell et al. 2010).

Deinstitutionalization of formal care for older people is also underway. A global trend (evident in nations including Germany, Japan, and the USA) is that the government reallocates more resources to home-based and community-based care from residential care due to the benefits of aging in place and, more importantly, the low cost-efficiency of traditional modes of residential care (see ▶ "Aging in place"). Moreover, increased support for home-based formal care can effectively reduce the burden of informal caregivers (Murphy and Turner 2017). Many older adults prefer to receive care at home, which can help them age in place (Campbell et al. 2010). This is even true in Asian countries, where home-based care aligns with traditional values suggesting that it is unfilial to send older adults to residential care (Huang et al. 2012).

However, the privatization, marketization, and deinstitutionalization of formal care face a variety of challenges. For instance, in China, there are inconsistent policies and relatively weak regulatory frameworks for the accreditation and oversight of private-sector care. Currently, Chinese initiatives promoting home-based and community-based services remain largely confined to megacities such as Shanghai. Privatization encounters the greatest difficulties in rural areas due to barriers in the physical environment; in these areas, formal care remains highly centralized and operationalized by the local government (Feng et al. 2012).

Formal care is not universally provided across different countries, and there are differences between regional and municipal services, as well as differences in the quantity and quality of formal care provided. Older people from different countries do not have equal access to formal care. Only 5.6% of the global population of older adults (including residents of Germany, Japan, and Scandinavian countries) have universal rights to long-term formal care, according to the statistics provided by the World Bank in 2015. In these countries, eligibility is determined via an objective need assessment process which is extended equally to everyone. Globally, 48% of older people have no right to any long-term care rights (e.g., residents of Brazil and India), and 46.3% have to be poor to be eligible for long-term care based on the mean-tested approach used in countries such as China, Russia, and the USA (Scheil-Adlung 2015) (see ▶ "Long-term care").

Formal caregivers tend to have more professional experience providing care for older adults. These caregivers include licensed professionals, such as social workers, registered nurses, medical doctors, occupational therapists, physiotherapists, and so forth. In addition, formal care is also provided by unlicensed direct caregivers who received short-term training; these caregivers provide services to older people in institutions such as nursing homes, assisted living facilities, community-based facilities, and private residences (Stone and Harahan 2010). However, the standards for the formal caregivers, especially those frontline workers, vary across countries. For instance, within Europe, Scandinavian countries such as Sweden have the highest professional standards and requirements for formal caregivers. In these countries, the wages and job satisfaction of formal caregivers are higher than those in some other European countries such as the UK, where no professional training is required to become a formal caregiver. In the middle, between these extremes, lie countries such as Germany, where formal caregivers are required to have attained at least secondary school graduation (Murphy and Turner 2017).

Informal Care

Informal care comprises assistance in four main areas: (1) routine activities of daily living (e.g., bathing, toileting, and eating); (2) instrumental activities of daily living (e.g., housework, transportation, and managing finances); (3) companionship and emotional support; and (4) medical and nursing tasks, such as injections and colostomy care (Reinhad et al. 2012).

Although recent years have witnessed an increase in male caregivers, the majority of caregivers are still female. While women tend to take care of emotional needs and daily living activities, men are more likely to provide financial and legal support (Barbabella et al. 2018). According to the hierarchical compensatory model, married older adults turn first to their spouses for primary care, then to their children, relatives, friends, and neighbors (Cantor 1979). In fact, informal caregivers are mainly children and children-in-law. For western families with multiple children, mothers' preferred caregivers tend to be daughters who share their values and live in close proximity (Pillemer and Suitor 2014). For eastern families with Confucian heritage, sons are still preferred as the primary caregivers of their aging parents (Cong and Silverstein 2012).

Informal caregivers play a vital but often invisible role in welfare systems (Barbabella et al. 2018). In the USA, informal caregivers created an estimated economic value of about $350 billion in 2006 (Gibson and Houser 2007), and approximately 34.2 million adults provided informal care to an adult aged 50 and over in 2014 (National Alliance for Caregiving and AARP 2015). In Europe, informal caregivers greatly outnumbered formal caregivers and created an indirect financial contribution ranging from 40% to 90% of the overall costs of long-term care (Triantafillou et al. 2010). In 2015, Australian informal caregivers were estimated to create a replacement value equivalent to 3.8% of the gross domestic product and 60% of the health and social work industry (Deloitte Access Economics 2015).

This significant economic value indicates the high intensity and long duration of informal caregiving. On average, a US spousal caregiver spends 44.6 h a week providing care (National Alliance for Caregiving and AARP 2015). In Europe, caregivers spend a mean of 46 h a week providing care, and the mean length of caregiving is 60 months (Triantafillou et al.

2010). In the long run, informal caregivers may experience burdens and need various forms of support as they become deprived of time and energy.

Key Research Findings

Existing studies of formal and informal care mostly cover the following areas: (1) factors associated with the usage of formal care; (2) working conditions, job satisfaction, and related psychosocial outcomes of formal caregivers; (3) care-related challenges including health outcomes and work conflicts for informal caregivers; (4) health outcomes and experiences of older adults living in formal care settings; (5) support and services for informal caregivers; and (6) the relationship between formal and informal care and related governmental policies.

Factors Influencing the Usage of Formal Care

Three main factors can influence people's choice to utilize formal care. Previous studies have examined the different characteristics of older adults receiving formal and informal care. The first of these factors is sociodemographic characteristics: formal care recipients tend to be older and have higher socioeconomic status compared to older adults receiving informal care (Coley et al. 2015). There is an inconsistency regarding whether men or women are more likely to receive formal care (Kuzuya et al. 2010). The second factor is health status and functional limitation. Those who have limitations in daily functional activities and cognitive functioning (Coley et al. 2015) are more likely to depend upon formal care. The third factor is the availability and affordability of formal and informal care. Those who do not have informal care (for instance, "empty nesters" living alone and widowed older adults) are more likely to use formal care (Li et al. 2017). These three categories of factors are in accordance with Andersen's behavioral model of health services use, which suggests the role of predisposing factors (e.g., sociodemographics), enabling factors (e.g., resources), and need (e.g., illness and mobility level) in influencing people's choices of health services (Zhu 2015).

Challenges in Formal Care Provision

The industry of formal care for older adults faces various challenges, such as workforce shortage and deficiency in governmental funding (Scheil-Adlung 2015). One of the most frequently discussed challenges is the unsatisfying working conditions of formal caregivers, including high turnover due to low wages, low job satisfaction, and substantial physical and emotional demand (Gao et al. 2017). Care is usually perceived as "feminized, dirty and migrant work" (Huang et al. 2012, p. 198). Women, immigrants, and individuals with low education are more likely to work the care industry, and these characteristics contribute to their low wages despite the high job demand. This is true in western countries, such as the USA and Canada, as well as some high-income nations in Asia such as Singapore, Taiwan, and Hong Kong, where domestic workers play an important role in both informal and formal care. Even caregivers who are professional healthcare workers, such as registered nurses, still experience devaluation by their surrounding society (Huang et al. 2012).

Due to agism in society, senior care settings are not typically regarded as a critical part of the healthcare system. The frontline nurses and direct care workers are neither respected by the public nor by managerial teams, who rarely consider frontline workers' opinions during decision-making processes. The marketization of formal care also contributes to the low wages of formal caregivers due to competition. The working conditions in private formal care settings tend to be even worse than in public institutions (Murphy and Turner 2017). Another challenge faced by formal care sectors is related to the negative image and reputation of formal care usually portrayed by the media (e.g., abuse of older people) and the misunderstanding that those who seek formal services are lazy or selfish, as care for older adults is viewed more as a family obligation. This stigmatization may exacerbate the unfavorable working conditions of formal caregivers.

Challenges for Informal Caregivers

It is widely acknowledged that caregiving can have negative influence on caregivers' physical and psychological health. Previous studies indicate that providing informal care may lead to the following health consequences for caregivers: (1) negative changes in lifestyle behaviors, such as sleep problems (Happe 2002), nonachievement of heart-healthy diet, and insufficient physical activity (Mochari-Greenberger and Mosca 2012); (2) physical strain and musculoskeletal discomfort (Darragh et al. 2015); and (3) psychological stress such as depression (Covinsky et al. 2003) (see ▶ "Caregiver Stress").

Caregivers of higher age, lower socioeconomic status, and less informal support are more likely to have poor health (Pinquart and Sörensen 2007). Compared with non-caregivers, caregivers were found to have worse outcomes with respect to stress, depression, subjective well-being, self-efficacy, and physical health. Though statistically significant, most differences were small to medium. However, large differences were found between non-caregivers and dementia caregivers, the latter of whom had higher stress levels and higher risks for poor health (Pinquart and Sörensen 2003). Compared with children and children-in-law, spouse caregivers also reported higher levels of depression and caregiving burden and worse subjective well-being (Pinquart and Sörensen 2011).

Informal caregivers sometimes feel that they are under house arrest, especially when they live with and take care of a dependent older person. Hence, getting out to work gives caregivers a break from care responsibilities and a chance to rejuvenate. This may explain why, compared to non-caregivers, women older than 50 years were more likely to remain employed when they started providing low-intensity care (King and Pickard 2013). However, both time and energy are finite resources. Each additional effort in caregiving comes at the potential cost of paid work (i.e., employment) and leisure activities (i.e., social life). In the USA, 52.4% of employed caregivers found it difficult to reconcile work and caregiving roles, and 39.8% of non-working caregivers reported that they had left work because of caregiving responsibilities (Longacre et al. 2017). In the long run, women taking care of older adults may have early withdrawal from employment and higher risks of living in poverty (Wakabayashi and Donato 2006) (see ▶ "Employment and caregiving").

Besides work-care conflicts, informal caregivers may also experience care-family conflicts. An adult daughter caregiver may also be a wife and a mother at the same time. These individuals are likely to divide their efforts between taking care of aging parents and providing support to their own children. Families with adolescent children, especially those with fewer socioeconomic resources, may experience heightened strains which could intensify conflicts between the roles of caregiver and mother (Stephens et al. 2001).

Experiences and Health Outcomes of Older People Using Formal Care

Older adults report both negative and positive feelings and health outcomes related to residential care use. Being a resident in nursing homes is usually associated with loss of identity and independency, less time seeing children and friends, and fewer opportunities for social interaction (Riedl et al. 2013). However, positive outcomes of using formal care have also been reported. For instance, Lee et al. (2018) found that due to more opportunities for social interaction within formal care settings, Canadian older adults receiving formal care report a higher level of life satisfaction and a lower level of loneliness compared to their peers receiving solely informal home care or a blend of the two. Morris et al. (2019) reported the effectiveness of a formal home care program in promoting older adults' functional recovery. Whether the experience of using formal care is positive or negative may be influenced by a number of factors, such as older adults' preference to receive home-based formal care instead of institutionalization and the quality and professionalism of care providers.

Support and Services for Informal Caregivers

There are four main kinds of support and services for informal caregivers: (1) formal support, such as respite and paid care leaves; (2) intervention at the individual level, such as psychoeducational programs; (3) group support; and (4) integrated services. On average, support and services had significant but small effects on caregivers' well-being (Pinquart and Sörensen 2006). Respite services help reduce depression, caregiving burden, and anger (Lopez-Hartmann et al. 2012) (see ▶ "Effectiveness of Respite Care for Caregivers of Older Adults"). Australian caregivers may receive a Carer Allowance if they give extra care to frail older people and meet eligibility requirements. In China, adult children in certain provinces (e.g., Henan) get 20 days' paid leave to care for older adults. Interventions at the individual level aim at improving caregivers' coping ability by teaching strategies to manage care and reduce stress. The last decades have witnessed the benefits of information and communication technology (ICT) for caregivers. ICT provides not only online support for caregivers but also remote telecare for care recipients at home (Topo 2009). Group support increases social support enabling caregivers to share both positive and negative experiences. Support groups for dementia caregivers had higher effects when they were designed with theoretical models, were composed of 6–10 members, lasted longer with follow-ups, and had interdisciplinary leaders (Chien et al. 2011). Integrated services, or multicomponent interventions, refer to the combination of all the above support and services. Integrated services are the only effective intervention for delaying dementia patients' institutionalization (Pinquart and Sörensen 2006).

Integrating Formal and Informal Care

There are two models used to describe the relationship between formal and informal care: (1) a complementary task-specific model, referring to formal care providing services beyond informal caregiver's expertise and capability, and (2) a supplementary or substitutional model, where formal and informal care provide similar services and thus are replaceable (Rogero-García and Rosenberg 2011). The aging population worldwide has been rising, and the family structure has been evolving at the same time. Informal care may not be able to meet the growing care needs for older adults, and formal care will be in greater demand (Li et al. 2017) (see ▶ "Self, Informal, and Formal Long-Term Care: The Interface").

Both substitutive and complementary models of the relationship between formal and informal care have received popular support, but the substitutive model has relatively less evidence to support its effectiveness (Garcia et al. 2008). For the complementary model, previous studies in countries such as the USA, Germany, and the Netherlands show that older adults prefer informal caregivers to take care of short-term needs, household tasks, and emotional support, but they prefer formal care for their long-term needs. In line with the substitutive model, Garcia et al. (2008) found that informal care can substitute for some functions of health professionals; for example, after controlling for health status and other sociodemographics, those with informal care have a significantly lower frequency of doctor visits.

Government policies influence the model of care provision. Following the trend of neoliberalism, governments have recently begun to encourage individuals and families to seek informal care for older adults using their own networks, thereby departing from the traditional universal welfare state model of care provision. While formal care has become increasingly marketized, some governments still provide subsidies to caregivers using a strict mean test approach (Murphy and Turner 2017). There are two ways to provide support for informal caregivers: monetary allowances and direct provision of services such as respite care, home modification, and visiting nurses. In Germany, most informal caregivers use a cash allowance, but in Japan, most prefer direct provision of services (Campbell et al. 2010).

Future Directions of Research

Future research should further examine the factors at different levels influencing the provision and intersection of both informal and formal care. The factors influencing the job exhaustion and satisfaction of formal caregivers need to be further investigated at intrapersonal, interpersonal, and organizational levels. These factors should also be examined in relation to social policies, such as employment policies advocating for more female labor participation, delayed retirement, and benefits for informal caregivers (Murphy and Turner 2017). Solutions to tackle the barriers in collaboration between informal and formal care for older adults need to be addressed in future studies. For example, informal caregivers should be delivered more knowledge about formal care, including residential care (which is the most well-known type) as well as other types of home-based and community-based formal care. Future studies should also examine the effectiveness of and provide evidence for more comprehensive and efficient care systems in comparison to the traditional segmented, inefficient, and costly approach (Campbell et al. 2010).

Summary

Care for older adults has gone through a process of marketization, privatization, and deinstitutionalization. Utilization of formal care is related to factors at multiple levels and can both complement and substitute for informal care. However, no matter how challenging care can be, informal caregivers are unlikely to give up their care responsibilities even when their care recipients have moved to institutions. On one hand, more professional training should be offered to formal caregivers. On the other hand, informal caregivers could benefit from more government support. A comprehensive model of integrating informal and formal care should be explored and implemented to better address the emerging caregiving burden.

Cross-References

► Aging in Place
► Caregiver Stress
► Effectiveness of Respite Care for Caregivers of Older Adults
► Employment and Caregiving
► Long-Term Care
► Self, Informal, and Formal Long-Term Care: The Interface

References

Barbabella F, Poli A, Santini S et al (2018) The role of informal caregivers in long-term care for older people. In: Boll T, Ferring D, Valsiner J (eds) Cultures of care in aging. Information Age Publishing, Charlotte, pp 193–212
Campbell JC, Ikegami N, Gibson MJ (2010) Lessons from public long-term care insurance in Germany and Japan. Health Aff 29(1):87–95
Cantor MH (1979) Neighbors and friends: an overlooked resource in the informal support system. Res Aging 1(4):434–463
Chien LY, Chu H, Guo JL et al (2011) Caregiver support groups in patients with dementia: a meta-analysis. Int J Ger Psy 26(10):1089–1098
Coley N, Gallini A, Gares V et al (2015) A longitudinal study of transitions between informal and formal care in Alzheimer disease using multistate models in the European ICTUS cohort. J Am Med Dir Assoc 16(12):1104–11e1
Cong Z, Silverstein M (2012) Parents' preferred care-givers in rural China: gender, migration and intergenerational exchanges. Ageing Soc 34(05):727–752
Covinsky KE, Newcomer R, Fox P et al (2003) Patient and caregiver characteristics associated with depression in caregivers of patients with dementia. J Gen Int Med 18(12):1006–1014
Darragh AR, Sommerich CM, Lavender SA et al (2015) Musculoskeletal discomfort, physical demand, and caregiving activities in informal caregivers. J Appl Gerontol 34(6):734–760
Deloitte Access Economics (2015) The economic value of informal care in Australia in 2015. Carers Australia, Canberra
Feng Z, Liu C, Guan X et al (2012) China's rapidly aging population creates policy challenges in shaping a viable long-term care system. Health Aff 31(12):2764–2773
Gao F, Newcombe P, Tilse C et al (2017) Challenge-related stress and felt challenge: predictors of turnover and psychological health in aged care nurses. Collegian 24(4):361–369
Garcia JR, Prieto-Flores ME, Rosenberg MW (2008) Health services use by older people with disabilities in Spain: do formal and informal care matter? Ageing Soc 28(7):959–978
Gibson M, Houser A (2007) Valuing the invaluable: a new look at the economic value of family caregiving. AARP Public Policy Institute, Washington, DC
Happe S (2002) The association between caregiver burden and sleep disturbances in partners of patients with Parkinson's disease. Age Ageing 31(5):349–354

Huang S, Yeoh BS, Toyota M (2012) Caring for the elderly: the embodied labour of migrant care workers in Singapore. Global Netw 12(2):195–215

King D, Pickard L (2013) When is a carer's employment at risk? Longitudinal analysis of unpaid care and employment in midlife in England. Health Soc Care Community 21(3):303–314

Kuzuya M, Hasegawa J, Enoki H et al (2010) Gender difference characteristics in the sociodemographic background of care recipients. Nihon Ronen Igakkai zasshi/Jpn J Geriatr 47(5):461–467

Lee Y, Barken R, Gonzales E (2018) Utilization of formal and informal home care: how do older Canadians' experiences vary by care arrangements?. J Appl Gerontol 0733464817750274

Li F, Fang X, Gao J et al (2017) Determinants of formal care use and expenses among in-home elderly in Jing'an district, Shanghai, China. PLoS One 12(4): e0176548

Longacre ML, Valdmanis VG, Handorf EA et al (2017) Work impact and emotional stress among informal caregivers for older adults. J Gerontol Ser B Psychol Sci Soc Sci 72(3):522–531

Lopez-Hartmann M, Wens J, Verhoeven V et al (2012) The effect of caregiver support interventions for informal caregivers of community-dwelling frail elderly: a systematic review. Int J Integr Care 12(10):e133

Mochari-Greenberger H, Mosca L (2012) Caregiver burden and nonachievement of healthy lifestyle behaviors among family caregivers of cardiovascular disease patients. Am J Health Promot 27(2):84–89

Morris JN, Berg K, Howard EP et al (2019) Functional recovery within a formal home care program. J Am Med Dir Assoc. (in press) 30:1–6

Murphy C, Turner T (2017) Formal and informal long term care work: policy conflict in a liberal welfare state. Int J Sociol Soc Policy 37(3/4):134–147

National Alliance for Caregiving, AARP (2015) Caregiving in the U.S. AARP, Washington, DC

Pillemer K, Suitor JJ (2014) Who provides care? A prospective study of caregiving among adult siblings. The Gerontologist 54(4):589–598

Pinquart M, Sörensen S (2003) Differences between caregivers and noncaregivers in psychological health and physical health: a meta-analysis. Psychol Aging 18(2):250–267

Pinquart M, Sörensen S (2006) Helping caregivers of persons with dementia: which interventions work and how large are their effects? Int Psychogeriatr 18(4):577–595

Pinquart M, Sörensen S (2007) Correlates of physical health of informal caregivers: a meta-analysis. J Gerontol Ser B Psychol Sci Soc Sci 62(2):126–137

Pinquart M, Sörensen S (2011) Spouses, adult children, and children-in-law as caregivers of older adults: a meta-analytic comparison. Psychol Aging 26(1):1–14

Reinhad S, Levine C, Samis S (2012) Home along: family caregivers providing complex chronic care. AARP Public Policy Institute, Washington, D.C

Riedl M, Mantovan F, Them C (2013) Being a nursing home resident: a challenge to one's identity. Nursing Res Pract 2013:1–9

Rogero-García J, Rosenberg MW (2011) Paid and unpaid support received by co-resident informal caregivers attending to community-dwelling older adults in Spain. Eur JAgeing 8(2):95–107

Scheil-Adlung X (2015) Long-term care protection for older persons: a review of coverage deficits in 46 - countries. ILO, Geneva

Stephens MAP, Townsend AL, Martire LM et al (2001) Balancing parent care with other roles: Interrole conflict of adult daughter caregivers. J Gerontol Ser B Psychol Sci Soc Sci 56(1):24–34

Stone R, Harahan MF (2010) Improving the long-term care workforce serving older adults. Health Aff 29(1):109–115

Topo P (2009) Technology studies to meet the needs of people with dementia and their caregivers: a literature review. J Appl Gerontol 28(1):5–37

Triantafillou J, Naiditch M, Repkova K et al (2010) Informal care in the long-term care system. Interlinks, Athens/Vienna

Wakabayashi C, Donato KM (2006) Does caregiving increase poverty among women in later life? Evidence from the health and retirement survey. J Health Soc Beh 47(3):258–274

Zhu H (2015) Unmet needs in long-term care and their associated factors among the oldest old in China. BMC Geriatr 15(1):46–56

Formal Care

▶ Self, Informal, and Formal Long-Term Care: The Interface
▶ Social Services Utilization

Formal Service Use

▶ Social Services Utilization

Formal Volunteering

▶ Volunteering and Health Outcomes Among Older Adults

Formal Worship

▶ Religiosity

Former Caregivers

▶ Former Carers

Former Carers

Mary Larkin
Faculty of Wellbeing, Education and Language Studies, The Open University, Milton Keynes, UK

Synonyms

Ex-caregivers; Ex-carers; Former caregivers; Past caregivers

Definition

Former carers are variously defined as such when the person they were caring for dies, is admitted to a long-term care setting, is admitted to a hospital or hospice, goes into remission (e.g., for cancer patients), or recovers from their health problem (e.g., as with substance misusers or those who have undergone major surgery) (Cavaye and Watts 2016; Larkin and Milne 2017). The term former carer is used throughout; "caring" and "caregiving" are used interchangeably as are "carer" and "caregiver" (the terms "caregiving" and "caregiver" tend to be favored more in US-based research).

Overview

That population aging has led to the global increase in the number of family carers (caregivers) is now well-recognized (Humphries et al. 2016; International Labour Organization 2018; OECD 2011). Much less recognized is the fact that those factors leading to the growth in the number of family carers also lead to a growth in the number of those who "were" carers – namely former carers (Carers UK 2014a; Larkin and Milne 2017). It is estimated that each year, around a third of carers cease "active" caring (Carers UK 2014a) for a variety of reasons.

Some of the routes into former caring in the definition above are less clear than others. For instance, research shows that caregiving merely takes a different form as opposed to ending when the care recipient enters long-term care (Davies and Nolan 2004, 2006; Moore and Dow 2015; Roland and Chappell 2015). Furthermore, being a former carer may not be a static state. Carers may transition in and out of caring roles across their adult life course (Department of Health 2014; Larkin 2009), being a carer and a former carer more than once and simultaneously. For example, a midlife woman may care for her parent(s), upon whose death she is a former carer. She may then care for her husband and become a former carer once again when he dies.

Although former carers are marginal to carer-related research, discourse, and services (Larkin 2009; Orzeck and Silverman 2008; Scrutton and Creighton 2015), there are signs that they are beginning to have a more visible public and policy profile. In addition, a growing international body of research is developing (Hirst 2014; Larkin and Milne 2017, 2018). While it is very small in comparison to the many thousands of papers about carers, it has been the subject of a number of reviews (Cavaye and Watts 2018; Cronin et al. 2015; Henwood Larkin and Milne 2017; Larkin 2009; Larkin and Milne 2017). These reviews show that research consistently evidences the disadvantages from which former carers suffer as a consequence of their caring experiences. Larkin and Milne (2017) use the term "Legacies of caring" when referring to these disadvantages.

One of the most significant of these legacies is the impact on former carers' health and wellbeing. The well-documented physical and psychological health problems linked with long-term (defined as more than 6 months) and intensive caring (defined as caring for at least 20 h per week) often persist into, and may worsen, post-caring. These include back problems, e.g., caused by regular lifting, exhaustion, skin disorders and infections (linked to suppressed immunity), arthritis, sciatica, high blood pressure, and heart problems. In some cases, new health problems start after caring ends – these relate particularly to sleep, exercise, eating, and alcohol consumption. Many former carers also experience significant distress post-caring, especially in relation to depression (Cavaye and Watts 2016, 2018; Larkin et al. 2017; McLaughlin and Ritchie 1994).

Another legacy is social isolation; long-term and intensive caring also has a well-established link with increased risk of loneliness and social isolation. This is amplified post-caring, particularly when contact with the agencies involved with the (ex) cared for person ceases, e.g., home care services, health services (Larkin 2009; Larkin and Milne 2017).

Many former carers also experience financial hardship because care-related expenses (e.g., additional laundry linked to incontinence, extra heating) have depleted their savings and increased their levels of debt. This financial hardship is deepened by the withdrawal of any carer-related benefits once active caring ends. Income reduction can also result from an inability to undertake paid work; long-term carers are at particular risk (Black et al. 2010; Carers UK 2014a, b). As pension contributions are reduced because of time out of the workforce, post-retirement income is also adversely affected. Furthermore, skills lost during caring can damage prospects of returning to pre-caring jobs or embarking on a new career post-caring. Not only does this mean that opportunities to address financial losses incurred during caring are jeopardized, but nonfinancial benefits linked to work (such as status and identity, and membership in social networks) are also lost (Carmichael and Ercolani 2016; Cronin et al. 2015).

Whatever legacies former carers face, few receive support from services; once they are no longer actively providing care, they are not regarded as carers by agencies with a statutory duty to assess the needs of, or provide support to, carers. Where carers are not assessed, they lose the support which had been provided to the care recipient and indirectly to them (Larkin 2007, 2009; Larkin and Milne 2017).

There is also evidence that the degree to which these legacies are experienced depends on a number of factors. For instance, problematic caring experiences (almost inevitably) lead to lower levels of wellbeing among former carers. Specifically, intensive caring, lower levels of social activity, smaller social networks, dissatisfaction with support received, and a strained relationship with the cared-for person are all correlated with higher levels of post-caring depression. In addition, while there are feelings of relief, bereavement can be more challenging for carers than noncarers and have longer term emotional consequences (Aneshensel et al. 2004; Mullan 1992). One explanation is that carers – unlike those whose relatives die without having required any family care – often experience many years of "anticipatory loss" (e.g., loss of freedom, hopes for the future) linked to a deterioration in the care recipient's health. This process can give rise to a multitude of complex emotions, both during caring and after death (Orzeck and Silverman 2008; Wuest 2000). A number of potential influences on bereaved older carers' emotional wellbeing have been identified. A close relationship with the care recipient, higher levels of self-esteem, socio-emotional support, and higher levels of education and income are all protective of the bereaved older carers' emotional wellbeing (Aneshensel et al. 2004; Burton et al. 2008; Guerra et al. 2016). In contrast, higher levels of distress among bereaved carers appear to be linked to specific negative care-related experiences such as dissatisfaction with caregiving and emotional strain, role overload, and lack of support during caring (Boerner et al. 2004; Pruchno et al. 2009).

Often overlooked is the fact that former carers may move into another caring role. Former carers can feel they have "little control over their resumption of the role of carer; somebody who was closely related to them had needed care" (Larkin 2009, p. 1039) with "habit" and familial obligations playing an important role as drivers.

Concepts such as "vocation carer" (Lewis and Meredith 1988) and "serial carer" (Larkin 2009) have been used to describe this group reflecting the sequential nature of caregiving roles across the life course. If a former carer takes on another caring role soon after ending one caring role, there is no time to recover from the legacies of caring and some legacies, such as those associated with their health, may be compounded by further caring.

Future Directions of Research

While existing research provides invaluable insights into former carers in terms of showing that they can have significant levels of need, identifying some of the factors that can shape the extent of their needs, and that they have limited access to support, it suffers from a number of weaknesses. Many of these are methodology related. Most studies are small scale, short term, and focus on a single group of former carers, e.g., bereaved former carers. The evidence base is also fragmented and atomized. More fundamentally, there is an almost total absence of any conceptual or theoretical analysis of former carers as a group, the process of *becoming* a former carer or about former carers lives and experiences (Cavaye and Watts 2016; Kelleher and McGrath 2016; Larkin and Milne 2017).

These sorts of flaws constrain opportunities to generate further knowledge about former carers and develop ways of supporting them. However, there will be a growing number of former carers who, because of their caring experiences, will suffer a range of complex financial, social, and emotional and health needs. Importantly too, it is likely the legacies of caring will worsen; the move to self-management of long-term conditions and health and social care funding shortfalls mean that more demands are being placed on former carers.

Lessening the "legacies of caring" for this internationally growing group is a matter of social justice. There is also a social and economic rationale for doing so. The extent to which former carers' needs impact their lives can be costly, both in terms of individual life experiences and ultimately public spending because of services they require. Moreover, in relation to changing demographics, lessening the legacies of caring can help address the ever-increasing demand for care by increasing the likelihood that former carers can re-enter the pool of carers.

Further knowledge generation about former carers is therefore not simply a question of more research; the weaknesses of existing research also need to be addressed. For example, there needs to be a more systematic and coordinated approach to research into former carers that explores both the negative and positive aspects of life post-caring, more longitudinal studies, and exploration of discipline-based constructs and theories to deepen understanding. Progressing such research is timely and can provide the future direction for more effective policy and practice around the growing numbers of former carers – supporting them in ways during caring that will help them post-caring and supporting them post-caring. In so doing, it can help address some of the many challenges of population aging.

Summary

While the status of the growing population of former carers is ambiguous and research about them remains limited, the evidence that they experience a range of disadvantages because of their caring experiences is unambiguous. In order to provide a robust evidence base to underpin policy and practice that improves outcomes for former carers, research activity needs to be both increased and strengthened by rectifying the flaws in existing knowledge generation.

Cross-References

▶ Formal and Informal Care
▶ Respite Care

References

Aneshensel CS, Botticello AL, Yamamoto-Mitani N (2004) When caregiving ends: the course of depressive symptoms after bereavement. J Health Soc Behav

45:422–440. https://doi.org/10.1177/002214650404500405

Black E, Gauthier S, Dalziel W et al (2010) Canadian Alzheimer's disease caregiver survey: baby-boomer caregivers and burden of care. Psychol Aging 25:807–813. https://doi.org/10.1002/gps.2421

Boerner K, Schulz R, Horowitz A (2004) Positive aspects of caregiving and adaptation to bereavement. Psychol Aging 19(4):668–675. https://doi.org/10.1037/0882-7974.19.4.668

Burton AM, Haley WE, Small B et al (2008) Predictors of well-being in bereaved former hospice caregivers: the role of caregiving stressors, appraisals, and social resources. Pall Support Care 6(2):149158. https://doi.org/10.1017/S1478951508000230

Carers UK (2014a) Need to know: transitions in and out of caring: the information challenge. Carers UK, London

Carers UK (2014b) Caring and family finances inquiry, UK report. Carers UK, London

Carmichael F, Ercolani MG (2016) Unpaid caregiving and paid work over life-courses: different pathways, diverging outcomes. Soc Sci Med 156:1–11. https://doi.org/10.1016/j.socscimed.2016.03.020

Cavaye J, Watts JH (2016) Experiences of bereaved carers: insights from the literature. Europ J Pall Care 23(4):200–203

Cavaye J, Watts JH (2018) Former carers: issues from the literature. Fam, Rel Soc 7(1):141–157. https://doi.org/10.1332/204674316X14676464160831

Cronin P, Hynes G, Breen M et al (2015) 'Between worlds': the experiences and needs of former family carers. Health Soc Care Comm 23(1):88–96. https://doi.org/10.1111/hsc.12149

Davies S, Nolan M (2004) 'Making the move': relatives' experiences of the transition to a care home. Health Soc Care Comm 12:517–526. https://doi.org/10.1111/j.1365-2524.2004.00535.x

Davies S, Nolan M (2006) 'Making it better': self-perceived roles of family caregivers of older people living in care homes: a qualitative study. Int J Nurs Stud 43:281–291. https://doi.org/10.1016/j.ijnurstu.2005.04.009.b

Department of Health (2014) Carers strategy: second national action plan 2014–2016. Department of Health, London

Guerra S, Figueiredo D, Patrão M et al (2016) Family integrity among older caregivers of relatives with dementia. Paideia 26(63):15–23. https://doi.org/10.1590/1982-43272663201603

Henwood, M, Larkin, M, Milne A (2017) Seeing the wood for the trees. Carer related research and knowledge: a scoping review. Available at https://www.scie-socialcareonline.org.uk/seeing-the-wood-for-the-trees-carer-related-research-and-knowledge-a-scoping-review/r/a110f00000RCtCnAAL

Hirst M (2014) Transitions into and out of unpaid care. Working Paper No. CUK 2644, York

Humphries R, Thorlby R, Holder H (2016) Social care for older people: home truths. The King's Fund, London

International Labour Organisation (2018) Care work and care jobs for the future of decent work. International Labour Organisation, Geneva

Kelleher C, McGrath H (2016) What comes next? Family carers' experiences of role and identity transition on cessation of the caring role. Adv Consum Res 44:508–509

Larkin M (2007) Group support during caring and post-caring – the role of carers groups. Group 17(2):28–51

Larkin M (2009) Life after caring: the post-caring experiences of former carers. Brit J Soc Work 3(6):1026–1042. https://doi.org/10.1093/bjsw/bcn030

Larkin M, Milne A (2017) What do we know about older former carers? Key issues and themes. Health Soc Care Comm 25(4):1396–1403. https://doi.org/10.1111/hsc.12437

Larkin M, Milne A (2018) The case for a carer-related knowledge exchange network: enhancing the relationship between research and evidence and policy and practice. Evidence and Policy. https://doi.org/10.1332/174426418X15299594689303

Lewis J, Meredith B (1988) Daughters who care. Routledge, London

McLaughlin E, Ritchie J (1994) Legacies of caring: the experiences and circumstances of ex-carers. Health Soc Care Comm 2(4):241–253

Moore KJ, Dow B (2015) Carers continuing to care after residential care placement. Internat Psychogeri 27(6):877–880. https://doi.org/10.1017/S1041610214002774

Mullan JT (1992) The bereaved caregiver: a prospective study of changes in wellbeing. The Gerontologist 32:673–683

OECD (2011) Informal carers. In Health at a Glance 2011: OECD Indicators. OECD Publishing. https://doi.org/10.1787/health_glance-2011-en

Orzeck P, Silverman M (2008) Recognizing post-caregiving as part of the caregiving career: implications for practice. J Soc Work Pract 22(2):211–220. https://doi.org/10.1080/02650530802099866

Pruchno R, Cartwright P, Wilson-Genderson M (2009) Effects of marital closeness on the transition from caregiving to widowhood. Ag Ment Health 13(6):808–817. https://doi.org/10.1080/13607860903046503

Roland KP, Chappell NL (2015) A typology of care-giving across neurodegenerative diseases presenting with dementia. Ag Soc 35:1905–1927. https://doi.org/10.1017/S0144686X1400066X

Scrutton J, Creighton H (2015) The emotional wellbeing of older carers. International Longevity Centre, London

Wuest J (2000) Negotiating with helping systems: an example of grounded theory evolving through emergent fit. Qual Health Res 10(1):51–70. https://doi.org/10.1177/104973200129118246

Fourth Age Imaginary

▶ Representations of Older Women and White Hegemony

Fragile X Syndrome and Premutation Aging Disorders

Maria Jimena Salcedo-Arellano, Hazel Maridith Barlahan Biag, Sumra Afzal and Randi J. Hagerman
Medical Investigation of Neurodevelopmental Disorders (MIND) Institute, University of California Davis, Sacramento, CA, USA
Department of Pediatrics, University of California Davis School of Medicine, Sacramento, CA, USA

Definition

Fragile X-associated disorders (FXD) are a family of genetic conditions that are all caused by changes in the *FMR1* gene. Fragile X syndrome (FXS) is among this family and is the most common inherited cause of intellectual disability (ID) or autism spectrum disorder (ASD). FXS is caused by a CGG expansion of greater than 200 repeats in the 5′ end of the promoter region of the fragile X mental retardation 1 (*FMR1*) gene. The premutation is the carrier state, and it is caused by 55 to 200 repeats in the *FMR1* gene, whereas the normal range of repeats is from 5 to 45.

Overview

FXS is predominantly caused by the absence of *FMR1* gene activity; the full mutation is usually methylated so the *FMR1* protein product FMRP is absent or diminished (Hagerman et al. 2017). The mental impairment related to FXS can range from mild to severe and is typically diagnosed between 2 and 3 years of age. Males are more frequently affected than females and generally with greater severity. The fragile X-associated tremor ataxia syndrome (FXTAS) is caused by the premutation, and it is a late adult-onset neurological disorder that affects older adult carriers – usually occurring after age 50. This condition affects more males then females with symptoms worsening with age. The characteristics of FXTAS are intention tremor, which typically develops first, followed later with problems in balance, also called ataxia (Hagerman and Hagerman 2016). The fragile X-associated primary ovarian insufficiency (FXPOI) is also caused by the premutation and refers to a spectrum of impaired ovarian function that includes infertility and early menopause in women prior to the age of 40 years. FXPOI occurs in about 20% of adult female *FMR1* premutation carriers; however it has less commonly been reported in teenagers who are carriers (Sullivan et al. 2005). The fragile X-associated neurodevelopmental disorders (FXAND) is an umbrella term that covers several neuropsychiatric disorders that are more common in carriers of the premutation compared to the general population. FXAND includes depression, anxiety, obsessive compulsive behavior, chronic fatigue, chronic pain syndrome, restless legs syndrome, and migraine headaches, and one or more of these symptoms can occur in up to 50% of carriers of the premutation (Hagerman et al. 2018).

Although no effective cure exists for these disorders, there are treatments available to help improve the quality of life of affected individuals and their families. This entry reviews the research that has been conducted relevant to explaining the differences between these disorders as issues regarding aging.

Key Research Findings

Though many studies involving individuals with FXS have been done to date, a majority of them focus on the cognitive and behavioral issues in children. The only study known to focus on the problems involved in those aging with FXS highlights several medical problems in those that are 40 years and older (Utari et al. 2010). The retrospective study collected data from past clinical and research appointments and found that the five most common medical problems were neurological problems, gastrointestinal problems, obesity, hypertension, and heart problems (Utari et al. 2010). Among the 62 individuals involved in the study, 44 were male and 18 were female. Between the two groups, one medical issue that significantly differed was movement disorders (Utari et al. 2010) which included tremor, Parkinson's disease (PD), bradykinesia without PD, tardive dyskinesia, and tics. Since mood instability and aggression are present in approximately 75% of those with FXS (Hessl et al. 2008), treatment with atypical antipsychotic medications is common (Hagerman et al. 2009). Prolonged use of this drug class puts these individuals at risk for antipsychotic-induced movement disorder which can include tardive dyskinesia and parkinsonism (Caligiuri et al. 2009; Miller et al. 2005). The Utari et al. study found that 17% of aging people with FXS had PD or Parkinsonism. While there had been no previous report of those with FXS having PD or loss of dopamine in the substantia nigra, there is evidence of dysregulation of dopamine in the hippocampal region (Wang et al. 2008) which may put these individuals at risk for PD. Due to the documented negative side effects of atypical medications, it is best to avoid these medications in older individuals with FXS. However, if treatment is necessary, the safest atypical in regard to parkinsonian symptoms is quetiapine.

Older adults with FXS may present with signs of dementia: irritability, mood disorders, sleep disturbances, and binge-eating related to anxiety and OCD leading to obesity in about 30% of the patients. However, FXS seems to have lower risk for developing dementia when compared to Down and Williams syndromes (Sauna-Aho et al. 2018).

In addition to there being a statistically significant difference between movement disorders in males and females such was the case when it came to obesity (Utari et al. 2010). For those with FXS, this may be exacerbated by reclusive behavior or autism like symptoms increasing with age (Hatton et al. 2006). In addition, the use of atypical antipsychotics increases appetite and the presence of the Prader-Willi phenotype (PWP), a subtype of FXS that is associated with hyperphagia, severe obesity, and lack of satiation after a meal. The PWP is seen in less than 10% of patients with FXS (Nowicki et al. 2007; McLennan et al. 2011). In the Utari study (Utari et al. 2010), the mean BMI in females was significantly higher than that of males ($p < 0.001$), but overall 82.6% of the patients were either obese or overweight.

Notably, hypertension in men with FXS is common but may be under-recognized due to anxiety and tactile defensiveness in the clinical setting which can lead to a transient increase of blood pressure. In the Utari study (Utari et al. 2010), 24% were diagnosed with hypertension, which is roughly the same as the general population (Cutler et al. 2008) but possibly an underestimate. Those with FXS may be at increased risk for hypertension because of documented sympathetic hyperarousal (Miller et al. 1999) and possibly because of elastic abnormalities in the vessel walls (Waldstein and Hagerman 1988).

Premutation carriers have a different clinical pathway. Throughout their life, carriers mainly experience mood disorders, particularly anxiety and depression; however, specific phobias, panic disorder, and social phobias are also found to be more prevalent in carriers when compared to the general population (Bourgeois et al. 2011). In most cases carriers have a normal IQ; nevertheless, some may encounter learning challenges. Additionally, they often experience chronic pain, fibromyalgia, OCD, ADHD, sleep disturbances, and autoimmune disorders. These disorders have been recently grouped to be recognized as FXAND (Hagerman et al. 2018). Female carriers may experience early signs of ovarian insufficiency (Cronister et al. 1991), commonly identified as fertility problems, and approximately 20% will present with menopause before age 40. This is recognized as FXPOI (Sherman 2000).

FXTAS is a neurodegenerative disease associated with the neurotoxic effect of a two- to eight-fold increase of *FMR1* mRNA leading to sequestration of proteins that are important to neuronal function. In addition, there is formation of inclusions in neurons and astrocytes throughout the central nervous system (CNS) and in the peripheral nervous system and in some organs (Hunsaker et al. 2011; Hagerman and Hagerman 2016). This disease develops in ~40% of male and 16% of female carriers as they age. Symptoms start at a mean age range between 55 and 65 years old; patients usually present with an onset of mild intention tremor and cognitive decline in memory and executive functioning abilities. The presentation is variable and can include executive dysfunction, parkinsonism, cerebellar ataxia, peripheral neuropathy, and lower-limb proximal muscle weakness (Jacquemont et al. 2004). Both tremor and ataxia are considered the major clinical criteria for the diagnosis of FXTAS (Tassone and Hall 2016). Progression to severe manifestations of the syndrome may take between 5 and 25 years from the time of diagnosis. A slower rate of progression and milder symptomatology is experienced by female carriers and those with a smaller number of CGG repeats (Leehey et al. 2008). An earlier onset and exacerbation of the symptoms and a faster rate of progression of FXTAS have been reported with the chronic use of alcohol or addictive substances (Muzar et al. 2014; Martinez-Cerdeno et al. 2015). The use of volatile anesthetics during general anesthesia has also been reported to increase the severity of FXTAS as well as accelerate disease in the CNS including the decline in cognition (Ligsay et al. 2018). The findings of FXTAS on MRI include global brain atrophy, dilated ventricles and white matter disease in the middle cerebral peduncles, periventricular areas, the insula, and the splenium of the corpus callosum (Hagerman and Hagerman 2016). During the final stages of FXTAS, 40% of the affected male carriers experience dementia (Seritan et al. 2008) and severe motor impairment.

The prevalence of FXTAS increases with age such that FXTAS occurs in 75% of male carriers in their 70s. However, for carriers who do not have FXTAS, due to their neuronal vulnerability especially to environmental toxins (Saldarriaga et al. 2019), they have a greater risk for experiencing working memory deficits, major depression disorder, increased rate of seizures, hypertension, and FXAND symptoms.

Ricaurte, Colombia, a rural town, is home of an aging cluster of FXS and premutation carriers (Saldarriaga et al. 2018); special social and economic circumstances have led to the exposure of the population to additional environmental factors. In the case of FXS, frequent alcohol consumption is found to be a triggering factor for increased impulsivity and aggressive behavior, as well as increased frequency in seizure episodes (Salcedo-Arellano et al. 2016). Premutation carriers, on the other hand, have experienced an increased frequency and severity of FXD including FXTAS, FXPOI, and FXAND. This increase is hypothesized to be associated with a chronic neurotoxic exposure, such as exposure to pesticides (Saldarriaga et al. 2019).

Future Directions of Research

Many of the aging problems in premutation carriers can begin in childhood, particularly in boys, and the psychiatric symptoms often include anxiety and ADHD (Farzin et al. 2006). The premutation problems associated with FXAND are considered developmental, although the symptoms can worsen with aging such as the anxiety and attention problems. However, the symptoms of FXTAS are definitely neurodegenerative, and once the symptoms of tremor, ataxia, and CNS changes of atrophy and white matter disease begin, the progression of symptoms accelerate. The RNA toxicity secondary to the elevated mRNA in carriers first leads to calcium dysregulation with elevated calcium in the cytoplasm of the neuron, then subsequently mitochondrial dysfunction, and eventually neuronal cell death (Napoli et al. 2016). We recommend treatment with exercise and antioxidants which can help to maintain or improve mitochondrial function in the CNS and perhaps elsewhere in the body (Polussa et al. 2014). In the near future, agents that are super antioxidants which also improve mitochondrial function will help with the accelerated aging process of FXTAS. An example of one of these agents is Anavex 2–73 which is a sigma-1 agonist that works in the area between the endoplasmic reticulum and the mitochondria. Preliminary studies suggest benefits in the cognitive decline in Alzheimer's disease and also improvements in the tremor of PD, which may suggest likelihood of beneficial effects in FXTAS.

FXPOI represents premature aging of the ovaries in female carriers, and while a super antioxidant may help with this problem, currently only hormonal replacement has been used for treatment of this condition (Sullivan et al. 2016). The presence of an abnormal protein called FMRpolyG which occurs because of repeat-associated non-AUG (RAN) translation of the mRNA of the premutation. The use of this non-AUG start site leads to a poly-Glycine track in the FMRP that has been named FMRpolyG. This protein is toxic to the cell, and it has been found in the ovaries of individuals with FXPOI and in the CNS of those with FXTAS (Sellier et al. 2017). Although it may not occur in all individuals with FXTAS or FXPOI, research is also taking place to block the formation of FMRpolyG. The main aim of current research is to elucidate a more comprehensive understanding of the underlying pathophysiology of FXTAS with the objective of implementing targeted treatments that can help with prevention and delay progression of the syndrome to final stages.

In FXS a few new targeted treatments are also currently being studied. Metformin, a common medication used as first-line treatment for type 2 diabetes mellitus and obesity, has demonstrated benefit for the management of weight loss in overweight and obese patients with FXS (McLennan et al. 2011; Dy et al. 2018). In FXS, there is upregulation of many pathways in the absence of FMRP that usually inhibits the translation of many proteins (Hagerman et al. 2017). The mTOR pathway, MEK-ERK pathways, and the insulin receptor are upregulated in FXS, and metformin can downregulate these pathways. Metformin has also rescued FXS symptoms in the animal models of FXS (Gantois et al. 2018). Therefore, metformin appears to be a targeted treatment in FXS since clinical benefit has been shown in behavior and the structure of verbal communication as reported by caregivers of those treated clinically (Dy et al. 2018). Therefore, we have initiated a double-blind, randomized, controlled trial of metformin in those 6–25 with FXS, in the USA and in Canada (NCT03479476). There are several additional new targeted treatment trials in FXS including cannabidiol (CBD) and a GABA agonist, Gaboxidol (Berry-Kravis et al. 2018; Erickson et al. 2017). The future looks bright for new medications for both FXS and FXTAS that may improve development and aging in these disorders.

Summary

Those with the premutation and the full mutation of the *FMR1* gene are at risk for aging disorders. The premutation causes neurotoxicity because of the elevated levels of *FMR1* mRNA that sequesters proteins needed for neuronal functioning. This can lead to neurodevelopmental problems such as anxiety and ADHD in addition to aging problems including premature aging of the ovary causing FXPOI and premature aging in the CNS associated with FXTAS. Those with FXS caused by the full mutation are also at risk for early aging because FMRP is important for neurogenesis throughout life, and without FMRP there is a higher risk for PD and in some cases cognitive decline. The use of new targeted treatments such as metformin may improve symptoms in FXS, and research is also focused on finding targeted treatments of premutation disorders.

Cross-References

► Aging Pathology
► Alzheimer's Disease
► Brain Atrophy
► Cognitive Disorders in Older Patients with Cancer
► Comprehensive Geriatric Assessment
► Dementia
► Genetics: Gene Expression
► Geriatric Rehabilitation, Instability, and Falls
► Human Aging, Mitochondrial and Metabolic Defects (The Novel Protective Role of Glutathione)
► Intellectual Disability (Cognitive Disability)
► Mutation
► Neuroscience of Aging
► Parkinson's Disease

- Prevention of Age-Related Cognitive Impairment, Alzheimer's Disease, and Dementia
- Rheumatic Diseases Among Older Adults
- Vascular Diseases of Ageing
- White Matter Hyper-Intensities

References

Berry-Kravis EM, Lindemann L, Jonch AE et al (2018) Drug development for neurodevelopmental disorders: lessons learned from fragile X syndrome. Nat Rev Drug Discov 17(4):280–299. https://doi.org/10.1038/nrd.2017.221

Bourgeois JA, Seritan AL, Casillas EM et al (2011) Lifetime prevalence of mood and anxiety disorders in fragile X premutation carriers. J Clin Psychiatry 72 (2):175–182. https://doi.org/10.4088/JCP.09m05407blu

Caligiuri MP, Teulings HL, Dean CE et al (2009) Handwriting movement analyses for monitoring drug-induced motor side effects in schizophrenia patients treated with risperidone. Hum Mov Sci 28(5):633–642. https://doi.org/10.1016/j.humov.2009.07.007

Cronister A, Schreiner R, Wittenberger M et al (1991) Heterozygous fragile X female: historical, physical, cognitive, and cytogenetic features. Am J Med Genet 38(2–3):269–274

Cutler JA, Sorlie PD, Wolz M et al (2008) Trends in hypertension prevalence, awareness, treatment, and control rates in United States adults between 1988–1994 and 1999–2004. Hypertension 52(5):818–827. https://doi.org/10.1161/HYPERTENSIONAHA.108.113357

Dy ABC, Tassone F, Eldeeb M et al (2018) Metformin as targeted treatment in fragile X syndrome. Clin Genet 93 (2):216–222. https://doi.org/10.1111/cge.13039

Erickson CA, Davenport MH, Schaefer TL et al (2017) Fragile X targeted pharmacotherapy: lessons learned and future directions. J Neurodev Disord 9:7. https://doi.org/10.1186/s11689-017-9186-9

Farzin F, Perry H, Hessl D et al (2006) Autism spectrum disorders and attention-deficit/hyperactivity disorder in boys with the fragile X premutation. J Dev Behav Pediatr 27(2 Suppl):S137–S144

Gantois I, Popic J, Khoutorsky A et al (2018) Metformin for treatment of fragile X syndrome and other neurological disorders. Annu Rev Med. https://doi.org/10.1146/annurev-med-081117-041238

Hagerman RJ, Hagerman P (2016) Fragile X-associated tremor/ataxia syndrome – features, mechanisms and management. Nat Rev Neurol 12(7):403–412. https://doi.org/10.1038/nrneurol.2016.82

Hagerman RJ, Berry-Kravis E, Kaufmann WE et al (2009) Advances in the treatment of fragile X syndrome. Pediatrics 123(1):378–390. https://doi.org/10.1542/peds.2008-0317

Hagerman RJ, Berry-Kravis E, Hazlett HC et al (2017) Fragile X syndrome. Nat Rev Dis Primers 3:17065. https://doi.org/10.1038/nrdp.2017.65

Hagerman RJ, Protic D, Rajaratnam A et al (2018) Fragile X-associated neuropsychiatric disorders (FXAND). Front Psych 9. https://doi.org/10.3389/fpsyt.2018.00564

Hatton DD, Sideris J, Skinner M et al (2006) Autistic behavior in children with fragile X syndrome: prevalence, stability, and the impact of FMRP. Am J Med Genet A 140A(17):1804–1813. https://doi.org/10.1002/ajmg.a.31286

Hessl D, Tassone F, Cordeiro L et al (2008) Brief report: aggression and stereotypic behavior in males with fragile X syndrome – moderating secondary genes in a "single gene" disorder. J Autism Dev Disord 38(1):184–189. https://doi.org/10.1007/s10803-007-0365-5

Hunsaker MR, Greco CM, Tassone F et al (2011) Rare intranuclear inclusions in the brains of 3 older adult males with fragile X syndrome: implications for the spectrum of fragile X-associated disorders. J Neuropathol Exp Neurol 70(6):462–469. https://doi.org/10.1097/NEN.0b013e31821d3194

Jacquemont S, Hagerman RJ, Leehey MA et al (2004) Penetrance of the fragile X-associated tremor/ataxia-syndrome in a premutation carrier population. JAMA 291(4):460–469. https://doi.org/10.1001/jama.291.4.460

Leehey MA, Berry-Kravis E, Goetz CG et al (2008) FMR1 CGG repeat length predicts motor dysfunction in premutation carriers. Neurology 70(16 Pt 2):1397–1402. https://doi.org/10.1212/01.wnl.0000281692.98200.f5.

Ligsay A, El-Deeb M, Salcedo-Arellano MJ et al (2018) General anesthetic use in fragile X spectrum disorders. J Neurosurg Anesthesiol. https://doi.org/10.1097/ANA.0000000000000508

Martinez-Cerdeno V, Lechpammer M, Lott A et al (2015) Fragile X-associated tremor/ataxia syndrome in a man in his 30s. JAMA Neurol 72(9):1070–1073. https://doi.org/10.1001/jamaneurol.2015.1138

McLennan Y, Polussa J, Tassone F et al (2011) Fragile X syndrome. Curr Genomics 12(3):216–224. https://doi.org/10.2174/138920211795677886

Miller LJ, McIntosh DN, McGrath J et al (1999) Electrodermal responses to sensory stimuli in individuals with fragile X syndrome: a preliminary report. Am J Med Genet 83(4):268–279

Miller DD, McEvoy JP, Davis SM et al (2005) Clinical correlates of tardive dyskinesia in schizophrenia: baseline data from the CATIE schizophrenia trial. Schizophr Res 80(1):33–43. https://doi.org/10.1016/j.schres.2005.07.034

Muzar Z, Adams PE, Schneider A et al (2014) Addictive substances may induce a rapid neurological deterioration in fragile X-associated tremor ataxia syndrome: a report of two cases. Intractable Rare Dis Res 3(4):162–165. https://doi.org/10.5582/irdr.2014.01023.

Napoli E, Song G, Schneider A et al (2016) Warburg effect linked to cognitive-executive deficits in FMR1 premutation. FASEB J 30(10):3334–3351. https://doi.org/10.1096/fj.201600315R

Nowicki ST, Tassone F, Ono MY et al (2007) The Prader–Willi phenotype of fragile X syndrome. J Dev Behav Pediatr 28(2):133–138. https://doi.org/10.1097/01.DBP.0000267563.18952.c9

Polussa J, Schneider A, Hagerman R (2014) Molecular advances leading to treatment implications for fragile X premutation carriers. Brain Disord Ther 3. https://doi.org/10.4172/2168-975X.1000119.

Salcedo-Arellano MJ, Lozano R, Tassone F et al (2016) Alcohol use dependence in fragile X syndrome. Intractable Rare Dis Res 5(3):207–213. https://doi.org/10.5582/irdr.2016.01046

Saldarriaga W, Forero-Forero JV, Gonzalez-Teshima LY et al (2018) Genetic cluster of fragile X syndrome in a Colombian district. J Hum Genet 63(4):509–516. https://doi.org/10.1038/s10038-017-0407-6

Saldarriaga W, Salcedo-Arellano MJ, Rodriguez-Guerrero T et al (2019) Increased severity of fragile X spectrum disorders in the agricultural community of Ricaurte, Colombia. Int J Dev Neurosci 72:1–5. https://doi.org/10.1016/j.ijdevneu.2018.10.002

Sauna-Aho O, Bjelogrlic-Laakso N, Siren A et al (2018) Signs indicating dementia in Down, Williams and Fragile X syndromes. Mol Genet Genomic Med 6(5):855–860. https://doi.org/10.1002/mgg3.430

Sellier C, Buijsen RAM, He F et al (2017) Translation of expanded CGG repeats into FMRpolyG is pathogenic and may contribute to fragile X tremor ataxia syndrome. Neuron 93(2):331–347. https://doi.org/10.1016/j.neuron.2016.12.016.

Seritan AL, Nguyen DV, Farias ST et al (2008) Dementia in fragile X-associated tremor/ataxia syndrome (FXTAS): comparison with Alzheimer's disease. Am J Med Genet B Neuropsychiatr Genet 147B(7):1138–1144. https://doi.org/10.1002/ajmg.b.30732

Sherman SL (2000) Premature ovarian failure in the fragile X syndrome. Am J Med Genet 97(3):189–194. https://doi.org/10.1002/1096-8628(200023)97:3<189::AID-AJMG1036>3.0.CO;2-J

Sullivan AK, Marcus M, Epstein MP et al (2005) Association of FMR1 repeat size with ovarian dysfunction. Hum Reprod 20(2):402–412. https://doi.org/10.1093/humrep/deh635

Sullivan SD, Sarrel PM, Nelson LM (2016) Hormone replacement therapy in young women with primary ovarian insufficiency and early menopause. Fertil Steril 106(7):1588–1599. https://doi.org/10.1016/j.fertnstert.2016.09.046

Tassone F, Hall D (2016) FXTAS, FXPOI, and other premutation disorders, 2nd edn. Springer, New York

Utari A, Adams E, Berry-Kravis E et al (2010) Aging in fragile X syndrome. J Neurodev Disord 2(2):70–76. https://doi.org/10.1007/s11689-010-9047-2

Waldstein G, Hagerman R (1988) Aortic hypoplasia and cardiac valvular abnormalities in a boy with fragile X syndrome. Am J Med Genet 30(1–2):83–98

Wang H, Wu LJ, Kim SS et al (2008) FMRP acts as a key messenger for dopamine modulation in the forebrain. Neuron 59(4):634–647. https://doi.org/10.1016/j.neuron.2008.06.027

Fragility

▶ Frailty in Clinical Care

Frailty

▶ Psychological Fatigue

Frailty and Social Vulnerability

Judith Godin and Melissa K. Andrew
Division of Geriatric Medicine, Dalhousie University and Nova Scotia Health Authority, Halifax, NS, Canada

Definition

Researchers from multiple disciplines have demonstrated connections between a variety of social factors and health and well-being (Andrew 2015). Traditionally, only a single social factor or a small group of related social factors were examined in a single study. Social vulnerability, on the other hand, simultaneously encompasses a broad and comprehensive range of social factor stemming from different levels of influence (e.g., individual, family, peer group, neighborhood, society) (Andrew and Keefe 2014). Social vulnerability is the extent to which a person's overall social circumstances leave them susceptible to the development of health issues, such as frailty. The concept of social vulnerability is closely aligned with the concept of social frailty, which also considers a broad range of social circumstances and provides a holistic understanding of how social factors may be influencing health (Bunt et al. 2017).

Overview

Similar to social vulnerability, frailty can be considered a state of increased susceptibility to further health insults (Searle et al. 2008). Both social vulnerability and frailty can be measured using a deficit accumulation approach: a count of the number of deficits a person has divided by the number of deficits considered. A Frailty Index (FI) needs to include a number of deficits covering a range of bodily functions (Searle et al. 2008), whereas a Social Vulnerability Index (SVI) needs to include a variety of social deficits such as socioeconomic status (SES), living situation, social supports and engagement, and neighborhood context, relevant functional abilities (e.g., using the telephone), and psychological aspects (e.g., sense of mastery, self-efficacy, and self-esteem) (Andrew et al. 2008). Despite the similar approaches, the FI and SVI are only moderately correlated and represent related but distinct concepts (Andrew et al. 2008).

Operationalizing these holistic constructs using a deficit accumulation approach allows for the simultaneous consideration of a large number of variables. Although examining the association between individual social factors and health and well-being independently provides useful information, it precludes the possibility that a single factor may not be associated with health and well-being in the absence of other social deficits. For instance, living alone may not exert any influence on health for an individual with a strong social network and social support outside the home; however, living alone in the context of a limited social network and social support could be detrimental. Examining social vulnerability in a single index provides a holistic approach to understand how social factors and circumstances are associated with health and well-being.

Key Research Findings

Women are more socially vulnerable than men and social vulnerability is more strongly correlated with frailty in women (Andrew et al. 2008). Social vulnerability tends to increase with age, making it an important construct to consider when investigating determinants of health of older adults, as even in the fittest older adults, social vulnerability is associated with mortality (Andrew et al. 2012).

Most of the research examining frailty and social vulnerability has posited that social vulnerability predicts frailty. In non-frail adults aged 65 years and older, baseline social frailty was associated with physical frailty 4 years later (Makizako et al. 2018). Cognition and frailty are reciprocally related (Godin et al. 2017) and social vulnerability has also been found to be predictive of cognitive decline even after controlling for frailty and demographic variables (Andrew and Rockwood 2010; Armstrong et al. 2015b). On the other hand, it is also plausible that frailty can lead to increased social vulnerability. For instance, in adults aged 65 years and older, frailty was associated with increased loneliness over 3 years after adjusting for age, sex, partner status, education, and chronic disease (Hoogendijk et al. 2016).

Social vulnerability has been found to interact with frailty and other health and well-being outcomes. Frailty was negatively associated with life satisfaction more strongly in young-old compared to old-old in participants with moderate (middle tertile) or high (third tertile) levels of social vulnerability but not in those with low levels of social vulnerability (Yang et al. 2016). This suggests that although the young-old see a greater impact of frailty on life satisfaction compared to the old-old, low levels of social vulnerability reduce this effect. Social vulnerability was associated with an increased risk of mortality in older men who had low levels of frailty but not in those with higher levels of frailty (Armstrong et al. 2015a). On the other hand, social vulnerability was a consistent predictor of quality of life in adults aged 50 years and older across all levels of frailty (Godin et al. 2019).

Evidence exists that the inclusion of social factors from multiple levels of influence is important to understand the association between social vulnerability and health. Deprivation at both the neighborhood and individual level are independently associated with frailty (Lang et al. 2009). In an analysis that examined the association between social vulnerability and mortality and disability after controlling for frailty and demographic variables, social vulnerability was associated with mortality in Mediterranean and Continental countries but not in Nordic countries (Wallace et al. 2015). As Nordic countries tend to have stronger institutional supports available through the welfare state (versus emphasizing family responsibility in Mediterranean cultures) (Pichler and Wallace 2007), this finding highlights the importance of higher levels of influence, such as government and policy. These findings taken together demonstrate the importance of considering multiple levels of influence.

When a range of social factors are combined into a single index, it is relevant to wonder and test whether a single social factor could be the driving force in any significant associations. Two techniques have been used to test whether this may be the case: jackknifing and bootstrapping. With these techniques, multiple indices are built but an individual deficit (jackknifing) or a percentage of deficits (e.g., 20%; bootstrapping) are randomly left out of each SVI. Analyses are run using each SVI to investigate whether results change. Research involving these techniques has demonstrated that findings are generally not dependent on any single item or group of items, with the exception of women who are the least socially vulnerable in whom the single factors of widowhood and living alone may be importantly driving social vulnerability (Andrew et al. 2008).

Future Directions of Research

A recent systematic review reported on 15 studies examining the connection between the social environment and frailty; however, the vast majority of these studies examined single domains of the social environment such as social support or neighborhood characteristics (Duppen et al. 2017). In order to capture the complex way these individual factors interact, more research is need using holistic measures, such as the SVI.

The majority of studies examining social vulnerability have examined it at a single point in time; however, it is likely that an individual's level of social vulnerability varies somewhat across the life course. Examining how changes in social vulnerability are associated with frailty and other health outcomes could shed light on whether the effects of social vulnerability can easily be minimized with an improvement in social circumstances or if the effects are more long-lasting.

Cross-References

▶ Epidemiology of Frailty
▶ Measurement of Frailty
▶ Neighborhood Coherence
▶ Neighborhood Social Environment and Health
▶ Social Support
▶ Socioeconomic Status

References

Andrew MK (2015) Frailty and social vulnerability. In: Theou O, Rockwood K (eds) Frailty in aging: biological, clinical, and social Implications. Karger, Basel, pp 188–195

Andrew MK, Keefe JM (2014) Social vulnerability from a social ecology perspective: a cohort study of older adults from the National Population Health Survey of Canada. BMC Geriatr 14:90

Andrew MK, Rockwood K (2010) Social vulnerability predicts cognitive decline in a prospective cohort of older Canadians. Alzheimers Dement 6:319–325. https://doi.org/10.1016/j.jalz.2009.11.001

Andrew MK, Mitnitski AB, Rockwood K (2008) Social vulnerability, frailty and mortality in elderly people. PLoS One 3:e2232. https://doi.org/10.1371/journal.pone.0002232

Andrew MK, Mitnitski AB, Kirkland SA, Rockwood K (2012) The impact of social vulnerability on the survival of the fittest older adults. Age Ageing 41:161–165. https://doi.org/10.1093/ageing/afr176

Armstrong JJ, Andrew MK, Mitnitski A et al (2015a) Social vulnerability and survival across levels of frailty in the Honolulu-Asia aging study. Age Ageing 44:709–712. https://doi.org/10.1093/ageing/afv016

Armstrong JJ, Mitnitski AB, Andrew MK et al (2015b) Cumulative impact of health deficits, social vulnerabilities, and protective factors on cognitive dynamics in late life: a multistate modeling approach. Alzheimers Res Ther 7:1–9. https://doi.org/10.1186/s13195-015-0120-7

Bunt S, Steverink N, Olthof J et al (2017) Social frailty in older adults: a scoping review. Eur J Ageing 14:323–334. https://doi.org/10.1007/s10433-017-0414-7

Duppen D, Van der Elst MCJ, Dury S et al (2017) The social environment's relationship with frailty. J Appl Gerontol:073346481668831. https://doi.org/10.1177/0733464816688310

Godin J, Armstrong JJ, Rockwood K, Andrew MK (2017) Dynamics of frailty and cognition after age 50: why it matters that cognitive decline is mostly seen in old age. J Alzheimers Dis 58:231–242

Godin J, Armstrong JJ, Wallace L, Rockwood K, Andrew M (2019) The impact of frailty and cognitive impairment on quality of life: employment and social context matter. Int Psychogeriatr 31(6):789–797. https://doi.org/10.1017/S1041610218001710

Hoogendijk EO, Suanet B, Dent E et al (2016) Adverse effects of frailty on social functioning in older adults: results from the longitudinal aging study Amsterdam. Maturitas 83:45–50. https://doi.org/10.1016/j.maturitas.2015.09.002

Lang IA, Hubbard RE et al (2009) Neighborhood deprivation, individual socioeconomic status, and frailty in older adults. J Am Geriatr Soc 57:1776–1780. https://doi.org/10.1111/j.1532-5415.2009.02480.x

Makizako H, Shimada H, Doi T et al (2018) Social frailty leads to the development of physical frailty among physically non-frail adults: a four-year follow-up longitudinal cohort study. Int J Environ Res Public Health 15. https://doi.org/10.3390/ijerph15030490

Pichler F, Wallace C (2007) Patterns of formal and informal social Capital in Europe. Eur Sociol Rev 23:423–435

Searle SD, Mitnitski AB, Gahbauer EA et al (2008) A standard procedure for creating a frailty index. BMC Geriatr 8:24. https://doi.org/10.1186/1471-2318-8-24

Wallace LMK, Theou O, Pena F et al (2015) Social vulnerability as a predictor of mortality and disability: cross-country differences in the survey of health, aging, and retirement in Europe (SHARE). Aging Clin Exp Res 27:365–372. https://doi.org/10.1007/s40520-014-0271-6

Yang F, Gu D, Mitnitski A (2016) Frailty and life satisfaction in Shanghai older adults: the roles of age and social vulnerability. Arch Gerontol Geriatr 67:68–73. https://doi.org/10.1016/j.archger.2016.07.001

Frailty Assessment

▶ Frailty in Clinical Care
▶ Measurement of Frailty

Frailty Assessment Instruments

▶ Frailty Screening

Frailty Assessment Tools

▶ Frailty Screening

Frailty in Clinical Care

Darryl B. Rolfson
Geriatric Medicine, University of Alberta, Edmonton, AB, Canada

Synonyms

Fragility; Frailty assessment; Frailty management; Frailty measurement; Geriatric medicine; Older adults; Overview

The health care challenges of an aging population cannot be adequately addressed by health care systems and providers without fully integrating the paradigm of frailty. The construct of frailty is both robust and complex, perfectly suited to evolving societal needs in clinical care. A family of frailty measures are available for both case-finding and assessment. Judgment-based measures, physical performance measures, self-report, and the electronic frailty index are well-suited for case-finding. Multidimensional measures are ideal for the further assessment of frailty. The assessing teams should recommend physical exercise, ideally group-based and multimodal. Individuals who live with frailty should be guided with respectful language to develop an integrated plan of care and to consider therapies that are most appropriate to their circumstances and wishes. A multidimensional assessment of frailty should identify the specific components that require attention in areas such as cognition, mood, balance and mobility, nutrition, medications, continence, functional independence, and social support.

With an unprecedented rise in life expectancy over the past century, health care systems worldwide are struggling to realign clinical care to meet the needs of an aging population (Cesari et al. 2016). Clinical care a century ago targeted a younger population comprised of individuals with single diseases and classic presentations of illness. In the twenty-first century, clinical care strategies in both ambulatory and acute care settings must account for the needs of an older population who live with multiple chronic illnesses (multi-morbidity), take multiple medications (polypharmacy), and whose presentations of acute illness is sometimes atypical (Jarrett et al. 1995; Yarnall et al. 2017). Systems of care that focus primarily on discrete diseases in otherwise healthy individuals and ignore presentations of illness in frailty will miss the mark in the twenty-first century (Muscedere et al. 2016).

Perhaps no single paradigm better captures the essence of population aging than frailty. Frailty is a state of exaggerated vulnerability, manifest as a multidimensional syndrome, involving a dynamic interaction between intrinsic capacity, external resources, and stress (Clegg et al. 2013; Abbasi et al. 2018). This definition captures the essential characteristics of frailty (the state of vulnerability, the multidimensional syndrome, and the dynamic nature), and it resonates with the daily realities of clinical care.

The construct of frailty is a useful alternative to the ageism that creeps so easily into care practices and protocols. Past practices that have used age criteria to deny eligibility for surgeries, interventions, and chemotherapy have been rightfully challenged and abandoned in favor of risk indices and decision aids that account for the continuum from fitness to frailty. For many frontline clinicians, the term frailty has become synonymous with risk stratification. For others, frailty is framed as an "illness script," a presentation of illness to be recognized by clinicians, representing underlying pathophysiology, and prompting investigations and treatment. Still others, see frailty as a clinically recognizable syndrome that can inform clinical decisions with multimodal strategies.

Over the past two decades, implementation of frailty measures into health care systems has been stalled by a perceived lack of consensus on its definition and the absence of a universally recognized criterion standard. Still, there is a large body of evidence demonstrating relationships between frailty, however defined and operationalized, and outcomes in a wide variety of clinical settings. By comparison, evidence for the benefits of therapies and interventions in frailty is only now emerging. As consensus grows on the definition of frailty, and the recognition of the unique contributions of different frailty measures, working sometimes together as a suite of measures, we have an opportunity to close the evidence gap for treatment and care strategies (Rodríguez-Mañas 2013).

With the impact of population aging at the doorstep, the urgent need to continue the work of bridging gaps in the definition, operationalization, and management of frailty has never been greater. Understanding frailty in real-life clinical contexts can link up an ontological divide between the conceptualization of frailty as a discretely defined clinical syndrome and frailty as a stochastic and multidimensional state of risk that is highly interactive with external context and acute stress (Kuchel 2018). The world of clinical care is naturally placed between these two shorelines. Research which understands frailty as narrowly defined phenotype will seek to discover physiologic, molecular, and genetic alterations as a basis for diagnostic and therapeutic approaches to clinically recognizable syndromes (Vina et al. 2016). Research that defines frailty as a multidimensional state leads naturally to illness scripts that can be recognized by clinicians as a multidimensional frailty measure (Rodríguez-Mañas et al. 2013). The particular dimensions or illness scripts that comprise frailty can then be lined up with already established evidence-based practice for geriatric syndromes. Care pathways will already exist in many care environments for dementia, delirium, multi-morbidity, polypharmacy, falls and immobility, incontinence syndromes, malnutrition, and social isolation.

Clinical care is defined broadly as the full range of settings where medical assessment and treatment is directly provided to real patients to address the acute and chronic aspects of illness and injury. This includes, but is not limited to, hospitals, including emergency care and preadmission clinics; intermediate care settings such as rehabilitation and subacute care; ambulatory settings such as the care provided in clinics and in homes; and continuing care such as home care, supportive living, and long-term care; and transition services that bridge the movement between these settings. Multidisciplinary and interdisciplinary teams, offering a coordinated range of services, are increasingly found in all of these settings, and ideally these teams are joined up with other community, voluntary, and family partners. The way that these health and social care services are organized, coordinated, and person-centered can make all the difference for individuals living with frailty.

What Frailty Means in Clinical Care

Clinicians must first understand that frailty is a state of exaggerated vulnerability. In a robust health state, individuals who encounter stress have a repertoire of responses, allowing them to resolve the threat, retain independence, and thrive. When intrinsic or extrinsic stress is sufficiently noxious to overwhelm these assets, even the most robust individuals may need to depend on others to cope or recover. The particular predisposition to lose functional independence in the context of stress falls on a continuum between fitness and frailty. This is a powerful way to understand acute illness presentations of individuals living with frailty (Campbell and Buchner 1997; Rockwood et al. 2004, Romero-Ortuno and O'Shea 2013). This is especially relevant in clinical contexts, where stress is manifest as acute health concerns, and illness presentation in older adults living with frailty is best understood in a dynamic model where a minor stressor "tips the balance" with profound functional consequences (Rockwood et al. 1994).

This concept of vulnerability underpins all geriatric syndromes (Inouye et al. 2007) and is well illustrated with the multicomponent model of delirium (Inouye and Charpentier 1996). It helps to understand what it takes to cause delirium by acknowledging the reciprocal relationship between the predisposing and precipitating variables. When the vulnerability (number or impact of predisposing variables) is low, as in the robust state, a very noxious insult (precipitant) is required to cause delirium. However, with increasingly higher levels of vulnerability, less noxious insults will be sufficient to cause and sustain delirium. A similar application of predisposing and precipitating variables can be used to understand other acute geriatric syndromes such as failure to thrive, falls, urinary incontinence, and nutritional crises. Preoperative frailty has been independently associated with postoperative delirium, independent of other known risk factors (Leung et al. 2011). A multidimensional assessment might be better suited to define the underlying pattern of frailty, with its unique components of vulnerability, then focus on prevention using a multicomponent approach.

The tension between frailty as a state of accumulated deficits and frailty as a narrowly defined phenotype (Fried et al. 2001) has been at an impasse for almost two decades (Sternberg et al. 2011). However, in 2013, important consensus was achieved on the notion that frailty is a multidimensional syndrome with decreased reserve and diminished resistance to stressors, and it is expressed in functional terms (Rodríguez-Mañas et al. 2013). If indeed the syndrome of frailty is multidimensional, then its expression must make sense to clinicians. Acute and chronic geriatric syndromes are the language of frailty and can be anticipated by its early recognition.

The way forward for clinicians is to understand that frailty is by nature complex but by no means outside their grasp to recognize and address. A useful analogy compares frailty to the Golden Gate Bridge (Kuchel 2018). Here, it becomes apparent how a number of frailty models help to explain an integrated whole. The phenotype model of frailty is understood by the structural integrity of a tower in the bridge. Other towers should represent the multidimensionality of the syndrome. Stochastic aspects, as represented by the deficits in the many vertical cables, capture the state of vulnerability. External forces such as wind, water, and traffic bring to mind the dynamic aspects of frailty, and important reminder that frailty may be silent and unknown until placed under conditions of stress.

Prevalence and Outcomes of Frailty in Clinical Care

Frailty identification in clinical care may be used for risk assessment, to inform care planning, and possibly to track the impact of an intervention. For risk modelling, frailty certainly is more meaningful than chronological age, and rivals or exceeds other risk indices when applied to at-risk clinical populations.

Though clearly common in late life, the prevalence of frailty in the community will vary by the community population with weighted prevalence estimated to be 10–24% depending on the measure used and the particular population of interest (Collard et al. 2012; Shamliyan et al. 2013). Frailty is strongly associated with poor survival in older community-dwelling populations with a dose-response reduction per increasing number of frailty criteria (Shamliyan et al. 2013), and it is inversely associated with quality of life (Kojima et al. 2016).

The impact of frailty on risk prediction has been examined in the context of selected chronic diseases and is influencing revisions in disease-based risk modelling. In ischemic heart disease, there is an association between frailty and adverse outcomes in the short- (Kang et al. 2015) and medium-term (Blanco et al. 2017). In advanced heart failure, frailty is associated with higher rates of comorbidity, hospitalization, mortality, and other adverse outcomes (Jha et al. 2016). Frailty confers a higher risk of mortality in chronic kidney disease (Walker et al. 2013) and hemodialysis (McAdams-DeMarco et al. 2013). Frailty and sarcopenia appear to be important in predicting waitlist mortality in liver transplant candidates (Kahn et al. 2018). There is an important reciprocal interaction between cognitive impairment and frailty (Robertson et al. 2013), resulting in accelerated functional decline, and the effect might be further amplified in contexts of social vulnerability (de Jesus et al. 2018). The prevalence of frailty in cancer patients is high and frailty raises the risk of chemotherapy intolerance, postoperative complications, and mortality (Handforth et al. 2015).

For a comprehensive review of frailty in the acute care setting, readers are referred to two excellent and recent scoping reviews (Hogan et al. 2017; Theou et al. 2018). There is a vast array of articles, with a wide variety of operational definitions used for frailty, and many of these are not well-established frailty instruments. However, as a general rule, frailty appears to be a good predictor of adverse outcomes in acute settings, and there is still a need to better define how these measures can be used to inform the care plan within defined settings (Fig. 1).

The predictive validity of frailty for outcomes after acute illness has shown mixed results (Wou et al. 2013; Basic and Shanley 2015). Frailty raises the risk of postoperative complications and institutionalization (Dasgupta et al. 2009). There is an increased risk of postsurgical discharge institutionalization (Robinson et al. 2011) and a higher chance of readmission within 30 days of discharge from hospital (Kahlon et al. 2015). It does appear that preoperative frailty assessment predicts distal outcomes in cardiac surgery, including mortality, morbidity, functional decline, and major adverse cardiac and cerebrovascular events (Sepehri et al. 2014; Kim et al. 2016) In critical care settings, frailty is common and is clearly associated with worse outcomes (Muscedere et al. 2017; Koizia et al. 2019).

Frailty measures, regardless of the measure used, appears to predict mortality and hospitalization in assisted living environment (Hogan et al. 2012). Frailty is a major trajectory in the last year of life, more common than any of advanced dementia, organ failure, cancer, and sudden death (Gill et al. 2010). Despite this, it is rarely included to inform goals of care discussions, as is common for the other patterns.

The clinical approach to frailty can be clustered into three steps: case-finding, assessment, and management as shown in Fig. 2. This approach is most relevant when applied by frontline clinical teams and emphasizes the approach to individual patients over more healthy populations.

Case Finding of Frailty

Once clinicians are persuaded that frailty is a meaningful and valid way to anticipate outcomes, they will seek practical ways to identify and act on it. Even armed with valid, reliable, and acceptable measures, *screening* of entire populations is unlikely, but *case-finding* in populations at risk is more realistic. Although chronological age would seem to be a logical way to define this population at risk, no particular age cut point has been accepted. The diagnostic accuracy and predictive validity of frailty measures in adults aged 60 and older from a broad range of clinical settings was summarized in a recent scoping review (Apostolo et al. 2017), and only a few were deemed to be adequately valid, reliable, diagnostically accurate, and had good predictive ability. Although gaps are still present in this literature, the frailty index and gait speed were highlighted for routine care in community settings.

To be adopted, case finding in clinical settings must be both valid and acceptable to clinicians. First, these measures must demonstrate good diagnostic test accuracy with high sensitivity and specificity at a defined cut point. This is challenging because no single frailty measure, including the two that are best known (Cardiovascular Health Study Physical Phenotype (Fried et al. 2001) and Canadian Study on Health and Aging Frailty Index (Mitnitski et al. 2001)), has emerged as the undisputed criterion standard. Second, an ideal candidate measure should have good predictive validity for the intended purpose, and this has been established, as already described, primarily on the basis of cross-sectional studies. Measures acceptable to clinicians should offer rapid administration and interpretation, and the ability to function independently as a stand-alone tool.

Frailty in Clinical Care, Fig. 1 Integration of frailty models using the Golden Gate Bridge analogy. (Used with permission from the Journal of the American Geriatric Society)

Frailty in Clinical Care, Fig. 2 Clinical approach to frailty

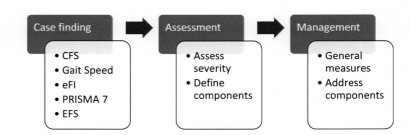

Regardless of setting, clinicians tend not to use measures that require additional steps such as the prior population of a special database, special equipment, or special skills (see Table 1). Compromises in these ideals can be justified depending on setting and purpose (Rolfson et al. 2018).

In community and primary care settings, gait speed meets many of these criteria and is becoming a popular starting point for opportunistic case finding (Pamoukdjian et al. 2015). Likewise, an electronic version of the frailty index (Clegg et al. 2016) was developed, validated, and implemented broadly in primary care throughout the United Kingdom, and demonstrates how frailty can be detected systematically in the right health system. Self-report measures such as PRISMA 7 (Raiche et al. 2008) and multidimensional measures such as the Edmonton Frail Scale (Rolfson et al. 2006) can be completed in under 5 minutes and require no prior assessment. Multidimensional measures provide additional information as will be described below.

When an individual is acutely unwell, it is challenging to apply and interpret measures that assume baseline cognitive or physical performance. Thus, in acute settings where the operator assumes the ability to make inferences based on previously known clinical data, judgment-based measures, such as the Clinical Frailty Scale, are often employed (Rockwood et al. 2005). A number of candidate biological markers have been tested for frailty, but none have been proven to be ready for use in clinical settings (Vina et al. 2016).

A classification of frailty measures, as shown in Table 2, with single examples of each illustrates the broad range frailty measures that can be employed, both alone and in combination with other measures.

Frailty in Clinical Care, Table 1 Characteristics of frailty measures needed to be acceptable for case finding

Good diagnostic test accuracy
Good predictive validity
No prior assessment or pre-populated database
Does not require an expert operator
Does not require special equipment
Rapid administration and interpretation
Functions as a stand-alone bedside tool

Assessment of Frailty

Frailty, once recognized, becomes part of a dialogue conducted by clinicians both with their patients and with their teams. That dialogue must be evidence-informed and respectful, as there is a relationship between the use of the word frailty and an older adults own perception

Frailty in Clinical Care, Table 2 A description of different categories of frailty measures with examples

Category of measure	Description	Example
Physical frailty	Classification based on observable physical traits	Frailty phenotype (Fried et al. 2001)
Biological markers	Blood tests and imaging	Blood tests, imaging
Cumulative deficit	Quotient of deficits present from a predetermined list of at least 30 available in the health record.	Frailty index (Mitnitski et al. 2001)
Judgment-based	A judgment-based scale that supports clinicians to assign a frailty status.	Clinical Frailty Scale (Rockwood et al. 2005)
Self-report	A set of discrete questions related to frailty that can be answered by a patient or family member	PRISMA 7 (Raiche et al. 2008)
Performance-based	Measured physical performance	Gait speed (Pamoukdjian et al. 2015)
Multidimensional	A set of items that comprise multiple frailty domains, presented in clinical terms. May be a combination of self-report, actual performance, and judgment.	Edmonton Frail Scale (Rolfson et al. 2006)

of aging (Warmoth et al. 2018). It is more accurate and acceptable to refer to frailty as a condition that a person lives with, or to start with colloquial language that is culturally acceptable.

Moving beyond general risk prediction, the individual characteristics of frailty need to be understood, thus opening new doors of clinical inquiry and care. A referral to another specialist or team for comprehensive geriatric assessment is one way to make this happen. However, depending on available resources, interdisciplinary teams might prefer to first draw from their own resources. This is where a multidimensional frailty measure, such as the Edmonton Frail Scale (Rolfson et al. 2006), can guide a team to address what is most needed (Perna et al. 2017; Rolfson et al. 2018).

The first question regards the assessment of frailty severity, as this will inform goals and strategies of care. When less severe, or with fewer or less burdensome concurrent illnesses, the approach emphasizes those strategies that encourage healthy and independent community-based living. Prevention strategies are more population based, including an emphasis on activity, nutrition, social support, and harm reduction such as medication reviews. With greater frailty, the approach to prevent further worsening becomes more individualized and clinically oriented. Comprehensive geriatric assessment (CGA) (Parker et al. 2018) can be applied by identifying and addressing the particular components of frailty (i.e., geriatric syndromes). Frailty *case finding* has less relevance in populations with very high prevalence of frailty, such as institutional long-term care. However, the *multidimensional assessment* of frailty remains highly relevant when it is severe in an individual or highly prevalent in a population (Cesari et al. 2018).

A closer assessment of frailty reveals that its expression is heterogeneous and clinically meaningful. In some individuals with frailty, the downstream consequences of a single illness, such as dementia, may have a multidimensional impact, leading to problems with nutrition, functional independence, deconditioning, and social isolation. In other individuals, there may be a large burden of other illnesses resulting in a very different constellation of frailty components such as polypharmacy, poor overall health status, and low mood. When these components of frailty are discovered through CGA, primary care and interdisciplinary teams are empowered to conduct further assessments and make recommendations to improve the overall status of their patients. This is how multidimensional frailty assessment empowers frontline care teams to assess and act upon obvious issues without delay, reserving the more complex assessment needs to those with greater expertise.

Management of Frailty

There is clearly a need for high quality research to better understand how to best manage frailty in the community setting (Karunananthan and Bergman 2018), but several important strategies have already been established. A recent scoping review searching for interventions to prevent or reduce the level of frailty in community dwelling adults found a significant impact of physical activity and pre-habilitation (Puts et al. 2017). Physical exercise programs are effective in reducing or postponing frailty (Silva et al. 2017; Apostolo et al. 2018). Favorable effects on frailty have also been shown with different combinations of exercise, nutritional supplementations, and cognitive training (Apostolo et al. 2018), and multi-domain compared to mono-domain interventions appear to be superior in terms of frailty status, muscle mass, and physical functioning (Dedeyne et al. 2017). Therefore, for older adults living with frailty in community settings, clinicians should recommend group-based exercise programs, ideally with a multi-domain focus that includes an emphasis on nutrition and cognition. If frailty severity or complexity warrants it, CGA should also be performed as it has been shown to increase the likelihood of an older adult living with frailty being alive and in their home up to 12 months (Ellis et al. 2011).

Because frailty anticipates functional decline, and the accompanying need for changes in care models or transitions in care settings, all older adults living with frailty should be assisted to obtain and develop an integrated, portable plan of care. This, combined with clinical judgment, will help determine how best to apply disease-based guidelines and how to proceed in times of acute illness and care transitions. Likewise, the recognition of frailty can be used in anticipation of upcoming interventions, such as surgery or therapies known to have high toxicity. In these circumstances, the recognition of frailty motivates optimization prior to the procedure, modified therapeutic approaches, greater specialized attention before and after procedures, or avoidance of the proposed stressor altogether. Indeed, hospital-wide interventions will prevent adverse outcomes in older adults with frailty (Bakker et al. 2011).

Whether in the community or hospital settings, busy clinicians need compelling reasons to assess and act on frailty. If available, a simple referral to a specialist in geriatric medicine has been the traditional approach. However, in many places and for the foreseeable future, the need for geriatricians outstrips supply. Clinical teams in primary care, community care, assisted living, institutional care, and hospitals are being empowered to assess and manage frailty. The particular components of frailty discovered in those being assessed warrants involvement of different members of these interdisciplinary teams. As one example, the British Geriatric Society offers a CGA Toolkit in 2019 for primary care practitioners that spells out general and specific approaches that encompass the needs of individuals living with frailty.

Future Directions

Whether frailty measurement is opportunistic or systematic, greater consensus on definition and operational characteristics of instruments within different categories will be helpful in research, quality improvement, and clinical care. In the status quo, frailty measures are being used opportunistically, especially by physicians in primary and acute care. Future frailty research should help inform clinicians and patients regarding both prognosis and individualized care planning.

If a more coordinated system-wide approach needs to be implemented, then a number of conditions must be met. These include standardizing instrumentation, specifying data standards and coding rules, providing training, establishing reporting standards, negotiating cross-sector consistency, and other considerations (Rolfson et al. 2018). In a system-wide approach, access to suite of measures that are valid, reliable, and acceptable across sectors would enhance efficiency for frontline assessors, reduce assessment fatigue of the individuals who are in constant transition, and enable quality improvement and research through data sharing and comparability.

Summary

Frailty is a highly relevant construct in clinical care, especially in anticipation of a massive increase in the number of older adults over the next generation. Frailty is both a state of risk and a multidimensional clinical syndrome. It is a strong predictor of mortality, institutionalization, and medical complications and as such is highly relevant in both community and acute care settings. There are a number of categories of case-finding options that can be employed, followed by respectful and patient-centered dialogue regarding further assessment and care planning. Case-finding measures can be used with multidimensional measures to better understand the characteristics that comprise frailty, thus motivating an interdisciplinary approach to address these components. All individuals living with frailty should be encouraged to be physically active, ideally through multimodal interventions that include attention to nutrition, cognitive stimulation, socialization, and where needed, comprehensive geriatric assessment.

Cross-References

▶ Delirium
▶ Dementia
▶ Falls
▶ Formal and Informal Care
▶ Malnutrition
▶ Multidimensional Views of Aging and Old Age
▶ Phenotype

▶ Physical Activity, Sedentary Behaviors, and Frailty
▶ Polypharmacy and Frailty
▶ Sarcopenia
▶ Self-Perceptions of Aging
▶ Social Isolation

References

Abbasi M, Rolfson D, Khera AS, Dabravolskaj J, Dent E, Xia L (2018) Identification and management of frailty in the primary care setting. CMAJ 190(38):E1134–e1140

Apostolo J, Cooke R, Bobrowicz-Campos E, Santana S, Marcucci M, Cano A, Vollenbroek-Hutten M, Germini F, Holland C (2017) Predicting risk and outcomes for frail older adults: an umbrella review of frailty screening tools. JBI Database System Rev Implement Rep 15(4):1154–1208

Apostolo J, Cooke R, Bobrowicz-Campos E, Santana S, Marcucci M, Cano A, Vollenbroek-Hutten M, Germini F, D'Avanzo B, Gwyther H, Holland C (2018) Effectiveness of interventions to prevent pre-frailty and frailty progression in older adults: a systematic review. JBI Database System Rev Implement Rep 16(1):140–232

Bakker FC, Robben SH, Olde Rikkert MG (2011) Effects of hospital-wide interventions to improve care for frail older inpatients: a systematic review. BMJ Qual Saf 20(8):680–691

Basic D, Shanley C (2015) Frailty in an older inpatient population: using the clinical frailty scale to predict patient outcomes. J Aging Health 27(4):670–685

Blanco S, Ferrieres J, Bongard V, Toulza O, Sebai F, Billet S, Biendel C, Lairez O, Lhermusier T, Boudou N, Campelo-Parada F, Roncalli J, Galinier M, Carrie D, Elbaz M, Bouisset F (2017) Prognosis impact of frailty assessed by the Edmonton frail scale in the setting of acute coronary syndrome in the elderly. Can J Cardiol 33(7):933–939

Campbell AJ, Buchner DM (1997) Unstable disability and the fluctuations of frailty. Age Ageing 26(4):315–318

Cesari M, Prince M, Thiyagarajan JA, De Carvalho IA, Bernabei R, Chan P, Gutierrez-Robledo LM, Michel JP, Morley JE, Ong P, Rodriguez Manas L, Sinclair A, Won CW, Beard J, Vellas B (2016) Frailty: an emerging public health priority. J Am Med Dir Assoc 17(3):188–192

Cesari M, Araujo de Carvalho I, Amuthavalli Thiyagarajan J, Cooper C, Martin FC, Reginster JY, Vellas B, Beard JR (2018) Evidence for the domains supporting the construct of intrinsic capacity. J Gerontol A Biol Sci Med Sci 73(12):1653–1660

Clegg A, Young J, Iliffe S, Rikkert MO, Rockwood K (2013) Frailty in elderly people. Lancet 381(9868):752–762

Clegg A, Bates C, Young J, Ryan R, Nichols L, Ann Teale E, Mohammed MA, Parry J, Marshall T (2016) Development and validation of an electronic frailty index using routine primary care electronic health record data. Age Ageing 45(3):353–360

Collard RM, Boter H, Schoevers RA, Oude Voshaar RC (2012) Prevalence of frailty in community-dwelling older persons: a systematic review. J Am Geriatr Soc 60(8):1487–1492

Dasgupta M, Rolfson DB, Stolee P, Borrie MJ, Speechley M (2009) Frailty is associated with postoperative complications in older adults with medical problems. Arch Gerontol Geriatr 48(1):78–83

de Jesus ITM, Orlando FS, Zazzetta MS (2018) Frailty and cognitive performance of elderly in the context of social vulnerability. Dement Neuropsychol 12(2):173–180

Dedeyne L, Deschodt M, Verschueren S, Tournoy J, Gielen E (2017) Effects of multi-domain interventions in (pre) frail elderly on frailty, functional, and cognitive status: a systematic review. Clin Interv Aging 12:873–896

Ellis G, Whitehead MA, Robinson D, O'Neill D, Langhorne P (2011) Comprehensive geriatric assessment for older adults admitted to hospital: meta-analysis of randomized controlled trials. BMJ 343:d6553

Fried LP, Tangen CM, Walston J, Newman AB, Hirsch C, Gottdiener J, Seeman T, Tracy R, Kop WJ, Burke G, McBurnie MA (2001) Frailty in older adults: evidence for a phenotype. J Gerontol A Biol Sci Med Sci 56(3):M146–M156

Gill TM, Gahbauer EA, Han L, Allore HG (2010) Trajectories of disability in the last year of life. N Engl J Med 362(13):1173–1180

Handforth C, Clegg A, Young C, Simpkins S, Seymour MT, Selby PJ, Young J (2015) The prevalence and outcomes of frailty in older cancer patients: a systematic review. Ann Oncol 26(6):1091–1101

Hogan DB, Freiheit EA, Strain LA, Patten SB, Schmaltz HN, Rolfson D, Maxwell CJ (2012) Comparing frailty measures in their ability to predict adverse outcome among older residents of assisted living. BMC Geriatr 12(1):56

Hogan DB, Maxwell CJ, Afilalo J, Arora RC, Bagshaw SM, Basran J, Bergman H, Bronskill SE, Carter CA, Dixon E, Hemmelgarn B, Madden K, Mitnitski A, Rolfson D, Stelfox HT, Tam-Tham H, Wunsch H (2017) A scoping review of frailty and acute Care in Middle-Aged and Older Individuals with recommendations for future research. Can Geriatr J 20(1):22–37

Inouye SK, Charpentier PA (1996) Precipitating factors for delirium in hospitalized elderly persons. Predictive model and interrelationship with baseline vulnerability. JAMA 275(11):852–857

Inouye SK, Studenski S, Tinetti ME, Kuchel GA (2007) Geriatric syndromes: clinical, research, and policy implications of a core geriatric concept. J Am Geriatr Soc 55(5):780–791

Jarrett PG, Rockwood K, Carver D, Stolee P, Cosway S (1995) Illness presentation in elderly patients. Arch Intern Med 155(10):1060–1064

Jha SR, Hannu MK, Chang S, Montgomery E, Harkess M, Wilhelm K, Hayward CS, Jabbour A, Spratt PM, Newton P, Davidson PM, Macdonald PS (2016) The prevalence and prognostic significance of frailty in patients with advanced heart failure referred for heart transplantation. Transplantation 100(2):429–436

Kahlon S, Pederson J, Majumdar SR, Belga S, Lau D, Fradette M, Boyko D, Bakal JA, Johnston C, Padwal RS, McAlister FA (2015) Association between frailty and 30-day outcomes after discharge from hospital. CMAJ 187(11):799–804

Kahn J, Wagner D, Homfeld N, Muller H, Kniepeiss D, Schemmer P (2018) Both sarcopenia and frailty determine suitability of patients for liver transplantation-a systematic review and meta-analysis of the literature. Clin Transpl 32(4):e13226

Kang L, Zhang SY, Zhu WL, Pang HY, Zhang L, Zhu ML, Liu XH, Liu YT (2015) Is frailty associated with short-term outcomes for elderly patients with acute coronary syndrome? J Geriatr Cardiol 12(6):662–667

Karunananthan S, Bergman H (2018) Managing frailty in primary care: evidence gaps cannot be ignored. CMAJ 190(38):E1122–e1123

Kim DH, Kim CA, Placide S, Lipsitz LA, Marcantonio ER (2016) Preoperative frailty assessment and outcomes at 6 months or later in older adults undergoing cardiac surgical procedures: a systematic review. Ann Intern Med 165(9):650–660

Koizia L, Kings R, Koizia A, Peck G, Wilson M, Hettiaratchy S, Fertleman MB (2019) Major trauma in the elderly: frailty decline and patient experience after injury. Trauma 21(1):21–26

Kojima G, Iliffe S, Jivraj S, Walters K (2016) Association between frailty and quality of life among community-dwelling older people: a systematic review and meta-analysis. J Epidemiol Community Health 70(7):716–721

Kuchel GA (2018) Frailty and resilience as outcome measures in clinical trials and geriatric care: are we getting any closer? J Am Geriatr Soc 66(8):1451–1454

Leung JM, Tsai TL, Sands LP (2011) Preoperative frailty in older surgical patients is associated with early postoperative delirium. Anesth Analg 112(5):1199–1201

McAdams-DeMarco MA, Law A, Salter ML, Boyarsky B, Gimenez L, Jaar BG, Walston JD, Segev DL (2013) Frailty as a novel predictor of mortality and hospitalization in individuals of all ages undergoing hemodialysis. J Am Geriatr Soc 61(6):896–901

Mitnitski AB, Mogilner AJ, Rockwood K (2001) Accumulation of deficits as a proxy measure of aging. ScientificWorldJournal 1:323–336

Muscedere J, Andrew MK, Bagshaw SM, Estabrooks C, Hogan D, Holroyd-Leduc J, Howlett S, Lahey W, Maxwell C, McNally M, Moorhouse P, Rockwood K, Rolfson D, Sinha S, Tholl B (2016) Screening for frailty in Canada's health care system: a time for action. Can J Aging 35(3):281–297

Muscedere J, Waters B, Varambally A, Bagshaw SM, Boyd JG, Maslove D, Sibley S, Rockwood K (2017) The impact of frailty on intensive care unit outcomes: a systematic review and meta-analysis. Intensive Care Med 43(8):1105–1122

Pamoukdjian F, Paillaud E, Zelek L, Laurent M, Levy V, Landre T, Sebbane G (2015) Measurement of gait speed in older adults to identify complications associated with frailty: a systematic review. J Geriatr Oncol 6(6):484–496

Parker SG, McCue P, Phelps K, McCleod A, Arora S, Nockels K, Kennedy S, Roberts H, Conroy S (2018) What is comprehensive geriatric assessment (CGA)? An umbrella review. Age Ageing 47(1):149–155

Perna S, Francis MD, Bologna C, Moncaglieri F, Riva A, Morazzoni P, Allegrini P, Isu A, Vigo B, Guerriero F, Rondanelli M (2017) Performance of Edmonton frail scale on frailty assessment: its association with multi-dimensional geriatric conditions assessed with specific screening tools. BMC Geriatr 17(1):2

Puts MTE, Toubasi S, Andrew MK, Ashe MC, Ploeg J, Atkinson E, Ayala AP, Roy A, Rodriguez Monforte M, Bergman H, McGilton K (2017) Interventions to prevent or reduce the level of frailty in community-dwelling older adults: a scoping review of the literature and international policies. Age Ageing 46(3):383–392

Raiche M, Hebert R, Dubois MF (2008) PRISMA-7: a case-finding tool to identify older adults with moderate to severe disabilities. Arch Gerontol Geriatr 47(1):9–18

Robertson DA, Savva GM, Kenny RA (2013) Frailty and cognitive impairment–a review of the evidence and causal mechanisms. Ageing Res Rev 12(4):840–851

Robinson TN, Wallace JI, Wu DS, Wiktor A, Pointer LF, Pfister SM, Sharp TJ, Buckley MJ, Moss M (2011) Accumulated frailty characteristics predict postoperative discharge institutionalization in the geriatric patient. J Am Coll Surg 213(1):37–42; discussion 42–34

Rockwood K, Fox RA, Stolee P, Robertson D, Beattie BL (1994) Frailty in elderly people: an evolving concept. CMAJ 150(4):489–495

Rockwood K, Howlett SE, MacKnight C, Beattie BL, Bergman H, Hebert R, Hogan DB, Wolfson C, McDowell I (2004) Prevalence, attributes, and outcomes of fitness and frailty in community-dwelling older adults: report from the Canadian study of health and aging. J Gerontol A Biol Sci Med Sci 59(12):1310–1317

Rockwood K, Song X, MacKnight C, Bergman H, Hogan DB, McDowell I, Mitnitski A (2005) A global clinical measure of fitness and frailty in elderly people. CMAJ 173(5):489–495

Rodríguez-Mañas L, Féart C, Mann G, Viña J, Chatterji S, Chodzko-Zajko W, Gonzalez-Colaço Harmand M,

Bergman H, Carcaillon L, Nicholson C, Scuteri A, Sinclair A, Pelaez M, Van der Cammen T, Beland F, Bickenbach J, Delamarche P, Ferrucci L, Fried LP, Gutiérrez-Robledo LM, Rockwood K, Rodríguez Artalejo F, Serviddio G, Vega E (2013) Searching for an operational definition of frailty: a Delphi method based consensus statement. The frailty operative definition-consensus conference project. J Gerontol A Biol Sci Med Sci 68(1):62–67

Rolfson DB, Majumdar SR, Tsuyuki RT, Tahir A, Rockwood K (2006) Validity and reliability of the Edmonton frail scale. Age Ageing 35(5):526–529

Rolfson DB, Heckman GA, Bagshaw SM, Robertson D, Hirdes JP (2018) Implementing frailty measures in the Canadian healthcare system. J Frailty Aging 7(4):208–216

Romero-Ortuno R, O'Shea D (2013) Fitness and frailty: opposite ends of a challenging continuum! Will the end of age discrimination make frailty assessments an imperative? Age Ageing 42(3):279–280

Sepehri A, Beggs T, Hassan A, Rigatto C, Shaw-Daigle C, Tangri N, Arora RC (2014) The impact of frailty on outcomes after cardiac surgery: a systematic review. J Thorac Cardiovasc Surg 148(6):3110–3117

Shamliyan T, Talley KM, Ramakrishnan R, Kane RL (2013) Association of frailty with survival: a systematic literature review. Ageing Res Rev 12(2):719–736

Silva RB, Aldoradin-Cabeza H, Eslick GD, Phu S, Duque G (2017) The effect of physical exercise on frail older persons: a systematic review. J Frailty Aging 6(2):91–96

Sternberg SA, Wershof Schwartz A, Karunananthan S, Bergman H, Mark Clarfield A (2011) The identification of frailty: a systematic literature review. J Am Geriatr Soc 59(11):2129–2138

Theou O, Squires E, Mallery K, Lee JS, Fay S, Goldstein J, Armstrong JJ, Rockwood K (2018) What do we know about frailty in the acute care setting? A scoping review. BMC Geriatr 18(1):139

Vina J, Tarazona-Santabalbina FJ, Perez-Ros P, Martinez-Arnau FM, Borras C, Olaso-Gonzalez G, Salvador-Pascual A, Gomez-Cabrera MC (2016) Biology of frailty: modulation of ageing genes and its importance to prevent age-associated loss of function. Mol Asp Med 50:88–108

Walker SR, Gill K, Macdonald K, Komenda P, Rigatto C, Sood MM, Bohm CJ, Storsley LJ, Tangri N (2013) Association of frailty and physical function in patients with non-dialysis CKD: a systematic review. BMC Nephrol 14:228

Warmoth K, Tarrant M, Abraham C, Lang IA (2018) Relationship between perceptions of ageing and frailty in English older adults. Psychol Health Med 23(4):465–474

Wou F, Gladman JR, Bradshaw L, Franklin M, Edmans J, Conroy SP (2013) The predictive properties of frailty-rating scales in the acute medical unit. Age Ageing 42(6):776–781

Yarnall AJ, Sayer AA, Clegg A, Rockwood K, Parker S, Hindle JV (2017) New horizons in multimorbidity in older adults. Age Ageing 46(6):882–888

Frailty Index

▶ Measurement of Frailty

Frailty Indicator(s)

▶ Measurement of Frailty

Frailty Instruments

▶ Measurement of Frailty

Frailty Management

▶ Frailty in Clinical Care

Frailty Measurement

▶ Frailty in Clinical Care
▶ Measurement of Frailty

Frailty Measures

▶ Measurement of Frailty

Frailty Phenotype

▶ Measurement of Frailty

Frailty Scale(s)

▶ Measurement of Frailty

Frailty Screening

M. E. Hamaker
Department of Geriatric Medicine,
Diakonessenhuis Utrecht, Utrecht/Zeist/Doorn,
The Netherlands

Synonyms

Frailty assessment instruments; Frailty assessment tools; Frailty screening instruments; Frailty screening tools

Definitions

| Frailty screening | The process in which a short screening instrument (a frailty screening tool) is used to identify patients who may potentially be vulnerable to adverse health outcomes and who may benefit from a more elaborate assessment (Decoster et al. 2015). |

Overview

The number of older cancer patients will increase substantially in the coming decades as a result of increasing life expectancy and aging of the population (www.cbs.nl). Cancer specialists are faced with the challenge of determining the optimal treatment for this patient population, with its wide individual variety in comorbidity, physiological reserve, disability, and geriatric conditions (Caillet et al. 2014; Jolly et al. 2015).

To capture this heterogeneity, the concept of frailty has been adopted by geriatric oncology. Frailty is considered a state of decreased physiological reserves, arising from cumulative decline across multiple physiological systems, resulting in a diminished resistance to stressors (Ferrucci et al. 2014). As cancer and its treatment can both be significant stressors, requiring patients to encroach on their reserves, the concept of frailty appears particularly relevant for older cancer patients. The original definition of frailty as formulated by Fried et al. focuses primarily on physical weakness and wasting (Fried et al. 2001), but many other definitions and criteria have been postulated, incorporating different aspects of ageing that contribute to diminishing reserves (Decoster et al. 2015).

Geriatric assessment (GA) is considered the gold standard for detecting frailty. This is a systematic procedure used to objectively appraise the health status of older people, focusing on somatic, functional, and psychosocial domains (Extermann et al. 2015), ideally followed by multidisciplinary interventions aimed at optimizing comorbidity and impairments. Its value in geriatric medicine has been proven extensively (Ellis et al. 2017). In geriatric oncology, evidence regarding the value of GA in prognostication, predicting feasibility of treatment and the likelihood of treatment-related complications is rapidly accumulating (Hamaker et al. 2017). Therefore, results of GA are increasingly used in clinical practice to support treatment decisions.

However, GA is resource-consuming and does not appear necessary for all older cancer patients. Therefore, a two-stepped approach is often implemented, in which a frailty screening tool is used to separate fit older cancer patients – who do not need a more elaborate assessment and are probably able to receive standard cancer treatment – from vulnerable patients who should subsequently undergo GA to guide tailoring of their treatment regimen (Decoster et al. 2015). In addition, some of these frailty screening tools themselves have shown association with prognosis and course of treatment and thus can help to inform patients and cancer specialists in the decision-making process.

Many operationalizations of frailty exist (Collard and Oude Voshaar 2012), as well as a great variety of frailty screening tools. For this entry, only brief screening instruments (15 items or less) were included, provided they encompassed more than one geriatric domain, and at least two studies were performed within a geriatric oncology setting that assessed their association with either geriatric assessment or with the outcomes of cancer treatment. Ultimately, seven frailty screening tools are included: Geriatric8 or G8 (Bellera et al. 2012), Vulnerable Elders Survey-13 or VES-13 (Saliba et al. 2000), Groningen Frailty Indicator or GFI (Steverink et al. 2001), Triage Risk Screening Tool or TRST (Meldon et al. 2003), Fried frailty criteria (Fried et al. 2001), and Barber (Molina-Garrido and Guillen-Ponce 2011).

Table 1 provides an overview of the frailty screening tools most commonly studied in geriatric oncology. Of these seven, only the G8 was designed specifically for cancer patients. Other instruments focused on patients in primary care or the community (VES-13, Barber), hospitalized or emergency room patients (TRST, ISAR-HP), or did not have one specific setting (GFI, Fried frailty criteria). Three can be self-administered (VES-13, GFI, ISAR-HP) while the other four are administered by a health-care professional (G8, Fried, Barber, TRST). All except the Fried frailty criteria can be completed in just a few minutes and do not require any additional equipment.

The screening tools vary significantly in the geriatric domains they address and the weight these domains are given within the screening (Table 2). All seven include functional impairments such as (instrumental) activities of daily living, mobility, or falls, and most include self-reported health; for other geriatric impairments, there is much more variation.

Key Research Findings

Frailty Screening Tools in Relation to Geriatric Assessment

When using a frailty screening tool to select vulnerable patients that should subsequently receive a GA to guide tailoring of their treatment regimen, the sensitivity of a frailty screening tool is of primary importance. This allows the treating physician to trust that frail patients will correctly be identified by the screening tool. However, to optimize the resource-saving potential of a two-stepped approach, a good specificity is also required to ensure that the number of fit patients incorrectly identified as frail on the screening tool and who will thus unnecessarily receive a GA will be limited.

None of the currently available screening instruments have a perfect sensitivity. In a

Frailty Screening, Table 1 Overview of commonly used frailty screening tools

	Patient population for which tool was developed	Number of items	Range of scores	Cut-off for frailty	Time needed to complete	Administered by
G8	Cancer patients	8	0–17	≤ 14	5 min	Health-care professional
VES-13	Community dwelling older people	13	0–10	≥ 3	5 min	Self-administered
GFI	Validated in various patient populations	15	0–5	≥ 4	5 min	Self-administered
Fried	No specific population	5	0–5	≥ 3	8 min	Health-care professional
ISAR-HP	Hospitalized patients	5	0–5	≥ 2	2 min	Self-administered
TRST	Patients in the emergency room	5	0–5	≥ 2	2 min	Health-care professional
Barber	Patients in primary care	9	0–9	≥ 1	5 min	Health-care professional

G8 geriatric 8, *VES* Vulnerable Elders Survey, *GFI* Groningen Frailty Indicator, *ISAR-HP* Identification of seniors at risk-hospitalized patients, *TRST* Triage Risk Screening Tool

Frailty Screening, Table 2 Domains addressed per frailty screening tool. Proportions represent the percentage of items in each screening tool addressing each domain

	Functional status	Psychosocial domain	Neuro-sensory deficits	Nutritional status	Polypharmacy	Recent hospitalization	Geriatric syndromes	Self-reported health	Education	Age
G8	11%	11%		46%	6%			11%		11%
VES-13	60%							10%		30%
GFI	27%	40%	13%	7%	7%			7%		
Fried	60%			20%				20%		
ISAR-HP	75%								25%	
TRST	20%	20%			20%	20%	20%			
Barber	44%	11%	22%			11%		11%		

G8 geriatric 8, *VES* Vulnerable Elders Survey, *GFI* Groningen Frailty Indicator, *ISAR-HP* identification of seniors at risk-hospitalized patients, *TRST* Triage Risk Screening Tool

systematic review of the association between frailty screening tools and GA, G8 showed the most consistently high sensitivity for impairment (Decoster et al. 2015; Hamaker et al. 2012), ranging from 65 to 92%. This means that 8 to 35% of potentially frail patients are still incorrectly identified as being fit. Specificity of G8 is around 60% meaning that two-fifth of fit patients are unnecessarily referred for geriatric assessment.

VES-13 has also been studied elaborately in relation to GA. Given its strong focus on functioning (Table 2), it is not surprising that the sensitivity for a GA, which encompasses many other domains in addition to functioning, is poorer, with a recent systematic review reporting a sensitivity over 80% in only two out of 11 studies (Decoster et al. 2015). Specificity was better, ranging from 62 to 100%.

In three studies using the GFI, sensitivity was 39–66% but specificity was high (86–87%, Decoster et al. 2015). The discriminative power of the other frailty screening instruments has only been assessed in two studies or less and generally showed little consistency in results (Decoster et al. 2015).

Frailty Screening Tools in Relation to Course of Treatment and Prognosis

For G8, VES-13, and GFI, one or more studies have demonstrated that being screened as potentially frail with the screening tool was associated with a higher risk of toxicity of chemotherapy (Decoster et al. 2015) and G8 was also associated with a higher risk of side effects from endocrine treatment (Dottorini et al. 2019). Most screening instruments appear to have little association with surgery-related complications. A systematic review reported negative results for VES-13, GFI, and the Fried criteria (Decoster et al. 2015), while two recent studies reported G8 was also not predictive (Fagard et al. 2017; Souwer et al. 2018). However, in a small cohort of colorectal cancer patients, the Identification of Seniors At Risk (ISAR) tool showed a significant association with adverse outcomes (Souwer et al. 2018) while another study addressing emergency abdominal surgery found VES-13 to have a high sensitivity and negative predictive value for postoperative complications and mortality (Kenig et al. 2014).

With regard to prognosis, G8 and TRST have both been found to have an association with functional decline after cancer treatment (Decoster et al. 2015; Martinez-Tapia et al. 2017; Schulkes et al. 2017; Takahashi et al. 2017) For mortality, G8 has been demonstrated to predictive in multiple studies (Decoster et al. 2015). Two studies found an association for ISAR in patients undergoing chemotherapy (Schulkes et al. 2017) and surgery (Souwer et al. 2018); one study each found an association with overall survival for TRST and Fried criteria (Decoster et al. 2015). For VES-13 and GFI, an association with mortality was seen in patients receiving chemotherapy but not those undergoing oncologic surgery (Decoster et al. 2015).

In summary, G8 has been studied most often and is associated with chemotherapy toxicity, side effects of endocrine treatment, functional decline, and mortality, but not surgery-related complications. For patients receiving chemotherapy, VES-13 and GFI are associated with toxicity and mortality. In surgical patients, the limited available evidence appears to favor ISAR and VES-13.

Application of Tools in Clinical Practice

The 2014 International Society of Geriatric Oncology recommendation on screening tools for multidimensional health problems warranting geriatric assessment in older cancer patients (Decoster et al. 2015) concluded that data for G8 were the most robust, because this tool has been extensively studied, showed a high sensitivity for geriatric impairment with acceptable specificity and also appeared to have prognostic value in itself. However, the performance of different screening tools may depend on the setting and goal for which they are used. For instance, geriatric impairments which are most relevant to decision-making regarding chemotherapy may not be equally relevant when considering surgery. Similarly, there is a significant variation between countries and between centers within countries in the availability of geriatric resources. If access is limited, a high specificity may be felt to be more relevant, as this will ensure that the patients receiving geriatric care are those that will almost certainly benefit.

Thus, it is not possible to select one tool that will have an optimal performance for all treatment decisions and irrespective of the local situation. However, some recommendations can be made. Over the past few years a large range of new frailty screening tools or predictive models have been developed for older cancer patients, many of which are never adequately validated. Ultimately, this proliferation of little used and non-validated screening tools is unlikely to further the collective knowledge of how these tools can be optimally used in the decision-making process. It is therefore highly recommended to choose one of the existing frailty screening tools – based on the setting you are most interested in and the domains most relevant to you – rather than trying to develop a new instrument in hopes of finding one that will be ideal. Gaining experience with a specific tool will lead the multidisciplinary team to a better understanding of its strengths and weaknesses, and it is this knowledge that will ultimately lead to optimizing its implementation and usefulness within a specific treatment setting.

Future Directions of Research

One of the biggest challenges in geriatric oncology is translating the outcome of a frailty assessment (with a screening tool or GA) to a treatment decision. Multiple studies have assessed the impact of frailty assessment on multidisciplinary decision-making, for example, by assessing whether the presence of frailty resulted in a change in treatment plan (Hamaker et al. 2018). However, few studies have provided sufficient follow-up to determine whether these changes have in fact benefited the patient, for example, because they experience less toxicity or better functional trajectories. This is an important topic for future research.

The value and performance of frailty screening tools for patients undergoing oncologic surgery appears to be different from those receiving chemotherapy, and has thus far been assessed much less extensively. This is another relevant direction for future studies.

Summary

Frailty screening tool can be used to separate fit older cancer patients – who do not need a more elaborate assessment and are probably able to receive standard cancer treatment – from vulnerable patients who should subsequently undergo geriatric assessment to guide tailoring of their treatment regimen. Data for the G8 screening tool appear the most robust, as it has been extensively studied, shows a high sensitivity for geriatric impairment with acceptable specificity, and also has prognostic value in itself. However, the performance of different screening tools may depend on the setting and goal for which they are used.

Cross-References

▶ Comprehensive Geriatric Assessment
▶ Epidemiology of Frailty
▶ Geriatric Assessment for Older Adults with Cancer
▶ Measurement of Frailty

References

Bellera CA, Rainfray M, Mathoulin-Pelissier S et al (2012) Screening older cancer patients: first evaluation of the G-8 geriatric screening tool. Ann Oncol 23:2166–2172

Caillet P, Laurent M, Bastuji-Garin S, Liuu E, Culine S, Lagrange JL, Canoui-Poitrine F, Paillaud E (2014) Optimal management of elderly cancer patients: usefulness of the comprehensive geriatric assessment. Clin Interv Aging 9:1645–1660

Collard RM, Oude Voshaar RC (2012) Frailty; a fragile concept. Tijdschr Psychiatr 54:59–69

Decoster L, Van Puyvelde PK, Mohile S et al (2015) Screening tools for multidimensional health problems warranting a geriatric assessment in older cancer patients: an update on SIOG recommendations dagger. Ann Oncol 26:288–300

Dottorini L, Catena L, Sarno I et al (2019) The role of geriatric screening tool (G8) in predicting side effect in older patients during therapy with aromatase inhibitor. J Geriatr Oncol 10(2):356–358

Ellis G, Gardner M, Tsiachristas A et al (2017) Comprehensive geriatric assessment for older adults admitted to hospital. Cochrane Database Syst Rev (9): CD006211

Extermann M, Aapro M, Bernabei R et al (2015) Use of comprehensive geriatric assessment in older cancer patients: recommendations from the task force on CGA of the International Society of Geriatric Oncology (SIOG). Crit Rev Oncol Hematol 55:241–252

Fagard K, Casaer J, Wolthuis A et al (2017) Value of geriatric screening and assessment in predicting postoperative complications in patients older than 70 years undergoing surgery for colorectal cancer. J Geriatr Oncol 8:320–327

Ferrucci L, Guralnik JM, Studenski S et al (2014) Designing randomized, controlled trials aimed at preventing or delaying functional decline and disability in frail, older persons: a consensus report. J Am Geriatr Soc 52:625–634

Fried LP, Tangen CM, Walston J et al (2001) Frailty in older adults: evidence for a phenotype. J Gerontol A Biol Sci Med Sci 56:M146–M156

Hamaker ME, Jonker JM, Rooij d et al (2012) Frailty screening methods for predicting outcome of a comprehensive geriatric assessment in elderly patients with cancer: a systematic review. Lancet Oncol 13:e437–e444

Hamaker ME, Wildes TM, Rostoft S (2017) Time to stop saying geriatric assessment is too time consuming. J Clin Oncol 35:2871–2874

Hamaker ME, Te Molder MM, Thielen N et al (2018) The effect of a geriatric evaluation on treatment decisions and outcome for older cancer patients – a systematic review. J Geriatr Oncol 9:430–440

Jolly TA, Deal AM, Nyrop KA et al (2015) Geriatric assessment-identified deficits in older cancer patients with normal performance status. Oncologist 20:379–385

Kenig J, Zychiewicz B, Olszewska U et al (2014) Six screening instruments for frailty in older patients qualified for emergency abdominal surgery. Arch Gerontol Geriatr 61:437–442

Martinez-Tapia C, Paillaud E, Liuu E et al (2017) Prognostic value of the G8 and modified-G8 screening tools for multidimensional health problems in older patients with cancer. Eur J Cancer 83:211–219

Meldon SW, Mion LC, Palmer RM et al (2003) A brief risk-stratification tool to predict repeat emergency department visits and hospitalizations in older patients discharged from the emergency department. Acad Emerg Med 10:224–232

Molina-Garrido MJ, Guillen-Ponce C (2011) Comparison of two frailty screening tools in older women with early breast cancer. Crit Rev Oncol Hematol 79:51–64

Saliba D, Orlando M, Wenger NS et al (2000) Identifying a short functional disability screen for older persons. J Gerontol A Biol Sci Med Sci 55:M750–M756

Schulkes KJG, Souwer ETD, van Elden LJR et al (2017) Prognostic value of geriatric 8 and identification of seniors at risk for hospitalized patients screening tools for patients with lung cancer. Clin Lung Cancer 18:660–666

Souwer ETD, Verweij NM, van den Bos F et al (2018) Risk stratification for surgical outcomes in older colorectal cancer patients using ISAR-HP and G8 screening tools. J Geriatr Oncol 9:110–114

Steverink N, Slaets JP, Schuurmans H et al (2001) Measuring frailty: developing and testing the GFI (Groningen Frailty Indicator). Gerontologist 41:236

Takahashi M, Takahashi M, Komine K et al (2017) The G8 screening tool enhances prognostic value to ECOG performance status in elderly cancer patients. PLoS One 12:e0179694

Frailty Screening Instruments

▶ Frailty Screening

Frailty Screening Tools

▶ Frailty Screening

Frailty Tools

▶ Measurement of Frailty

Frauds and Older People

▶ Financial Frauds and Scams

Free Radical Theory of Aging

▶ Mitochondrial Reactive Oxygen Species Aging Theory
▶ Oxidation Damage Accumulation Aging Theory (The Novel Role of Glutathione)

Freehold Property

▶ Home Ownership

Freelance Worker

▶ Self-Employment Among Older Adults

FRTA Theory

▶ Antiaging Strategies

Fuller Working Lives

▶ Delaying Retirement

Functional Impairment

▶ Functional Limitation

Functional Limitation

Ke Shen
School of Social Development and Public Policy, Fudan University, Shanghai, China

Synonyms

Disability in instrumental activities of daily living; Functional impairment; Physical inactivity

Definition

Functional limitation is defined as restrictions in performing vital situation-free physical and mental actions needed in everyday life (Verbrugge and Jette 1994; Nagi 1976). Nagi (1976) identified three dimensions of limitations in the functioning that are conceptually and analytically separable: physical, emotional, and mental. Physical functioning refers to sensory-motor functioning of the organism as indicated by limitations in such activities as walking, climbing, bending, reaching, and hearing. Emotional functioning refers to a person's effectiveness in psychological coping with life stress and can be manifested through levels of anxiety, restlessness, and a variety of psychophysiological symptoms. Mental functioning denotes the intellectual and reasoning capabilities of individuals which have been most commonly measured through problem-solving tests such as the IQ. An important point to be made in connection with these three dimensions of functioning is that their indicators can be found in the characteristics of the human organism itself.

Overview

Nagi (1976) conceptualize the pathway to disability that begins with pathology and impairments, and then functional limitations are the intermediate steps leading to disability (see also Disability Measurement in this volume). It is thus important to distinguish functional limitation and disability in activities of daily living. Functional limitation means restrictions in basic physical and mental actions, and its indicators can be found in the characteristics of the human organism itself (Verbrugge and Jette 1994). Disability, on the other hand, means inability or limitations in performing social roles and activities such as in relation to work, to family, or to independent living. In contrast to functional limitation, indicators of disability can be found in both the characteristics of individuals and in the requirements of the social roles in question. In this sense, the same types and degrees of functional limitations of the organism can lead to varying dimensions and degrees of disability. Accumulating research has adopted this pathway as a working framework to elucidate the dynamics of disability onset and progression. For example, in a study of American older adults, it was demonstrated that impairments had a curvilinear relationship with functional limitations, which were, in turn, related to an index of disability in daily life tasks (Jette et al. 1998).

Functional limitations may be assessed either through self-report or proxy report or through physical performance tests that target this domain. As for the first type of assessment, the original developmental work of Nagi (1976) included 15 functional limitation questions for which answers constituted a 4-point scale. The first seven items, including "difficulty in standing for long periods," "difficulty in lifting or carrying weights of about ten pounds," "difficulty in going up and down stairs," "difficulty in walking," "difficulty in stooping, bending or kneeling," "difficulty in using hands and fingers," and "difficulty in reaching with either or both arms," addressed physical functioning. The following three items, including "nervousness, tension, anxiety and depression," "trouble getting to sleep and staying asleep," and "troubled with hands sweating and feeling damp and clammy," addressed psychophysiological reactions as indicators of emotional functioning. The last four items, including "heart beating hard even when not exercising or working hard," "pains, aches, or swelling in parts of the body," "weakness, tiring easily, no energy," "fainting spells, dizziness, sick feelings," and "shortness of breath, trouble breathing even when not exercising or working hard," related to generalized symptoms which were manifestations of both physical and emotional limitations.

More recently, a much more comprehensive scale of self-report functional limitations has been developed, which includes three dimensions, basic lower extremity function, advanced lower extremity function, and upper extremity function (Haley et al. 2002). The individual items of each dimension are depicted in Table 1. The scales developed by Nagi (1976) and Haley et al. (2002) show that without resorting to physical performance tests, it is possible to assess the functional limitation domain very comprehensively.

Meanwhile, a large number of physical performance tests using objective and standardized procedures have been developed, and many of them assess some aspects of functional limitations. The commonly used performance tests include pegboard test, opening and closing fasteners and latches, picking up object, lifting 10 pounds, gait speed, rise from chair, walk and return to chair, and stair climb. Many performance tests are timed, and people who are unable to accomplish can be assigned the worst time on the test observed in those who completed the test (Guralnik and Ferrucci 2003).

Given the wide choice of measures of functional limitation, it is a challenge to determine the best single measure or battery to use in a particular research study. In general, the measures are quite

Functional Limitation, Table 1 Scale of functional limitation developed by Haley et al. (2002)

Basic lower extremity function	Advanced lower extremity function	Upper extremity function
Walk around one floor of home	Hike a few miles including hills	Remove wrapping with hands only
Pick up a kitchen chair	Carry while climbing stairs	Unscrew lid without assistive device
Get into and out of car	Walk a brisk mile	Pour from a large pitcher
Reach overhead while standing	Go up and down one flight, no rails	Hold full glass of water in one hand
Wash dishes while standing	Walk 1 mile with rests	Put on and take off pants
Up and down from a curb	Walk on slippery surface	Reach behind back
Put on and take off coat	Go up and down three flights indoors	Use common utensils
Open heavy outside door	Walk several blocks	
On and off bus	Run one half mile	
Make bed	Get up from floor	
Bend over from standing position		
Go up and down a flight of stairs		
On and off a step stool		
Stand up from a low, soft couch		

Source: Haley et al. (2002)

robust in capturing the functional limitations of older people, and many different measures can serve the purpose of documenting the level of function (Guralnik and Ferrucci 2003). In a study comparing over 100 alternative scales of functional limitations, Long and Pavalko (2004) used the Bayesian information criterion in regression models of disability to compare the performance of measures. They revealed that including more activities in a scale is preferable to fewer and combing activities into a single scale is more effective than including each activity as a separate variable.

Key Research Findings

With population aging, studies over trends of functional limitation among older adults make up a major line of research. The findings are mixed so far, varying by country and time period. For instance, Freedman and Martin (2000) found that, even as the prevalence of many chronic diseases was increasing, the prevalence of functional limitations was declining among older Americans in the 1980s and 1990s. Possible explanations include earlier diagnosis and better disease management. By contrast, Grundy et al. (1999) found an increasing prevalence of mild disability during the late 1980s and late 1990s in Great Britain. Zimmer et al. (2002) also showed that the proportions of Taiwanese aged 65 and over reporting functional limitations significantly increased for both men and women during 1993 and 1999. Another study in five Asian settings, Indonesia, the Philippines, Singapore, Taiwan, and the Beijing Municipality, confirmed that every setting except for Singapore experienced significant increases in functional limitations among older adults in the 1990s (Ofstedal et al. 2007). The recent improvements in survival at older ages that these five settings have experienced may have disproportionately benefited those in poorer health and with functional limitations, leading to an overall increase in functional limitations. Improvement in functional limitations was also documented in the city of Shanghai of China in the 2000s and was more prominent than the improvement in ADL disability (Feng et al. 2013). Gu et al. (2015) reviewed substantial literature and found mixed patterns of changes in functional limitation across countries. For instance, functional limitation declined in Belgium, Germany, Sweden, China, and Thailand, while increased in Italy, Norway, and Brazil between 2005 and 2009 (see also Disability Trend in this volume).

Numerous studies have confirmed that functional limitations are strong predictors of disability among African-American population, Hispanic population, and Japanese and Dutch populations (Guralnik et al. 2000; Ostir et al. 1998; Furuna et al. 1999; Hoeymans et al. 1996; de Rekeneire and Volpato 2015). These studies have revealed a gradient of risk, with progressively lower rates of disability incidence associated with increasingly better performance scores. These findings highlight the value of clinical trials that focus on prevention or treatment of functional limitations as a means of preventing disability. Moreover, functional limitations are predictive of nursing home admission (Severson et al. 1994), use of hospital services (Mor et al. 1994), primary and preventive care utilization (Beatty and Dhont 2001; Chan et al. 1999), and mortality (Reuben et al. 1992; Keller and Potter 1994; Zimmer et al. 2012; Feng et al. 2010; Jassal et al. 2016).

In addition to research that use functional limitations as predictors, extensive studies also examine risk factors of functional limitations, ranging from sociodemographic characteristics to health factors. It is well established that the incidence of functional limitation increases with age (Crimmins et al. 1994; Guralnik and Simonsick 1993). Many studies show a higher prevalence of functional limitations for women than men, while others reach the opposite conclusion (Boult et al. 1994; Zimmer et al. 2014). Socioeconomic status is demonstrated to have a robust association with the onset and progression of functional limitations. For instance, in a study of American adults, Zimmer and House (2003) showed that those with higher income and education were less likely to experience an onset of functional limitation and those with the highest income were most likely to improve and least likely to get worse in comparison to those with the lowest income. Chronic conditions, such as arthritis, CVD, diabetes, and vision impairment, are also prominent risk factors for future functional limitation among elderly people (Dunlop et al. 2002; Arvidson et al. 2002). In a study of the impact of depression, it is shown that the greater the number of depressive symptoms, the larger the increase in functional limitations over a 4-year follow-up period (Penninx et al. 1998). Behavioral factors also matter. A recent study shows that driving cessation leads to significantly higher incidence of functional limitation for older Japanese adults over a 24-month period (Shimada et al. 2016). Verbrugge (2016) also summarized various kinds of buffers for functional limitations, ranging from medical care and rehabilitation, lifestyle changes, activity modifications, and environmental improvement to equipment and personal assistance.

Future Directions of Research

Since the 1990s, although there has been substantial progress in the assessment of function in older populations due to both the clarification of the concept of functional limitations and the development and use of specific measures of functional limitation in a variety of research settings, there are some key areas that require more attention (Guralnik and Ferrucci 2003). First, further work is required to improve standardization of measurement of functional limitations. Second, it is essential to develop criteria for clinically meaningful change in levels of function over time, particularly for self-report or proxy report measures. For example, it is valuable to know whether a change from a little difficulty to some difficulty is meaningful. Combination of self-report and performance-based measures thus would be essential to better identify the clinical needs (Feng et al. 2010; also see Disability Types in this volume). Third, the challenge in the future is to decide which measures are most relevant for public health applications and whether performance measures can be practically incorporated into public health surveys.

Summary

In sum, functional limitation, as a precursor to ADL disability, reflects restrictions in performing situation-free basic physical and mental actions. With the rising levels of population aging, research on functional limitation has been gaining momentum, ranging from the risk factors of functional limitation to its consequences. Future clinical studies that could better monitor the progression of functional limitation and identify intervening measures are especially warranted.

Cross-References

▶ Biosocial Model of Disability
▶ Disability Measurement
▶ Disability Types
▶ Handicap
▶ Impairment
▶ Physical Disability
▶ Social Disability

References

Arvidson NG, Larsson A, Larsen A (2002) Simple function tests, but not the modified HAQ, correlate with radiological joint damage in rheumatoid arthritis. Scand J Rheumatol 31(3):146–150. https://doi.org/10.1080/rhe.31.3.146.150

Beatty PW, Dhont KR (2001) Medicare health maintenance organizations and traditional coverage: perceptions of health care among beneficiaries with disabilities. Arch Phys Med Rehabil 82(8): 1009–1017. https://doi.org/10.1053/apmr.2001.25135

Boult C, Kane RL, Louis TA et al (1994) Chronic conditions that lead to functional limitation in the elderly. J Gerontol 49(1):M28–M36. https://doi.org/10.1093/geronj/49.1.M28

Chan L, Doctor JN, MacLehose RF et al (1999) Do Medicare patients with disabilities receive preventive services? A population based study. Arch Phys Med Rehabil 80(6):642–646. https://doi.org/10.1016/S0003-9993(99)90166-1

Crimmins EM, Hayward MD, Saito Y (1994) Changing mortality and morbidity rates and the health status and life expectancy of the older population. Demography 31(1):159–175. https://doi.org/10.2307/2061913

De Rekeneire N, Volpato S (2015) Physical function and disability in older adults with diabetes. Clin

Geriatr Med 31(1):51–65. https://doi.org/10.1016/j.cger.2014.08.018

Dunlop DD, Manheim LM, Sohn M et al (2002) Incidence of functional limitation in older adults: the impact of gender, race, and chronic conditions. Arch Phys Med Rehabil 83(7):964–971. https://doi.org/10.1053/apmr.2002.32817

Feng Q, Hoenig HM, Gu D et al (2010) Effect of new disability subtype on 3-year mortality in Chinese older adults. J Am Geriatr Soc 58(10):1952–1958. https://doi.org/10.1111/j.1532-5415.2010.03013.x

Feng Q, Zhen Z, Gu D et al (2013) Trends in ADL and IADL disability in community-dwelling older adults in Shanghai, China, 1998-2008. J Gerontol Ser B Psychol Sci Soc Sci 68(3):476–485. https://doi.org/10.1093/geronb/gbt012

Freedman VA, Martin LG (2000) Contribution of chronic conditions to aggregate changes in old–age functioning. Am J Public Health 90(11):1755–1760. https://doi.org/10.2105/AJPH.90.11.1755

Furuna T, Nagasaki H, Nishizawa S et al (1999) Longitudinal change in the physical performance of older adults in the community. J Jpn Phys Ther Assoc 1(1):1–5. https://doi.org/10.1298/jjpta.1.1

Grundy E, Ahlburg D, Ali M et al (1999) Disability in Great Britain: results from the 1996/97 disability survey. Department of Social Security Research Report. The Stationery Office, London

Gu D, Gomez-Redondo R, Dupre ME (2015) Studying disability trends in aging populations. J Cross Cult Gerontol 30(1):21–49. https://doi.org/10.1007/s10823-014-9245-6

Guralnik JM, Simonsick EM (1993) Physical disability in older Americans. J Gerontol S48:3–10. https://doi.org/10.1093/geronj/48.Special_Issue.3

Guralnik JM, Ferrucci L (2003) Assessing the building blocks of function: Utilizing measures of functional limitation, Am. J. Prev. Med. 25(3): 112–121. https://doi.org/10.1016/S0749-3797(03)00174-0

Guralnik JM, Ferrucci F, Pieper CF, et al (2000) Lower extremity function and subsequent disability: consistency across studies, predictive models, and value of gait speed alone compared to the short physical performance battery. J Gerontol Med Sci 55(4):221–231. https://doi.org/10.1093/gerona/55.4.M221

Haley SM, Jette AM, Coster WJ et al (2002) Late life function and disability instrument: II. Development and evaluation of the function component. J Gerontol Ser A 57(4):M217–M222. https://doi.org/10.1093/gerona/57.4.M217

Hoeymans N, Feskens EJM, van den Bos GAM et al (1996) Measuring functional status: cross sectional and longitudinal associations between performance and selfreport (Zutphen Elderly Study 1990–1993). J Clin Epidemiol 49(10):1103–1110. https://doi.org/10.1016/0895-4356(96)00210-7

Jassal SV, Karaboyas A, Comment LA et al (2016) Functional dependence and mortality in the international dialysis outcomes and practice patterns study (DOPPS). Am J Kidney Dis 67(2):283–292. https://doi.org/10.1053/j.ajkd.2015.09.024

Jette AM, Assmann SF, Rooks D et al (1998) Interrelationships among disablement concepts. J Gerontol Ser A 53(5):M395–M404. https://doi.org/10.1093/gerona/53A.5.M395

Keller BK, Potter JF (1994) Predictors of mortality in outpatient geriatric evaluation and management clinic patients. J Gerontol 49(6):M246–M251. https://doi.org/10.1093/geronj/49.6.M246

Long JS, Pavalko E (2004) Comparing alternative measures of functional limitation. Med Care 42(1):19–27. https://doi.org/10.1097/01.mlr.0000102293.37107.c5

Mor VM, Wilcox V, Rakowski W et al (1994) Functional transitions among the elderly: patterns, predictors, and related hospital use. Am J Public Health 84(8):1274–1280. https://doi.org/10.2105/AJPH.84.8.1274

Nagi SZ (1976) An epidemiology of disability among adults in the United States. Milbank Mem Fund Q Health Soc 54(4):439–467. https://doi.org/10.2307/3349677

Ofstedal MB, Zimmer Z, Hermalin AI et al (2007) Short-term trends in functional limitation and disability among older Asians: a comparison of five Asian settings. J Cross Cult Gerontol 22(3):243–261. https://doi.org/10.1007/s10823-006-9025-z

Ostir GV, Markides KS, Black SA et al (1998) Lower body functioning as a predictor of subsequent disability among older Mexican Americans. The Journals of Gerontology: Series A 53A(6):M491–M495. https://doi.org/10.1093/gerona/53A.6.M491

Penninx BW, Guralnik JM, Ferrucci L et al (1998) Depressive symptoms and physical decline in community- dwelling older persons. JAMA, J Am Med Assoc 279(21):1720–1726. https://doi.org/10.1001/jama.279.21.1720

Reuben DB, Rubenstein LV, Hirsch SH et al (1992) Value of functional status as a predictor of mortality: results of a prospective study. Am J Med 93(6):663–669. https://doi.org/10.1016/0002-9343(92)90200-U

Severson MA, Smith GE, Tangalos EG et al (1994) Patterns and predictors of institutionalization in community-based dementia patients. J Am Geriatr Soc 42(2):181–185. https://doi.org/10.1111/j.1532-5415.1994.tb04949.x

Shimada H, Makizako H, Tsutsumimoto K et al (2016) Driving and incidence of functional limitation in older people: a prospective population-based study. Gerontology 62(6):636–643. https://doi.org/10.1159/000448036

Verbrugge LM (2016) Disability experience and measurement. J Aging Health 28(7):1124–1158. https://doi.org/10.1177/0898264316656519

Verbrugge LM, Jette AM (1994) The disablement process. Soc Sci Med 38(1):1–14. https://doi.org/10.1016/0277-9536(94)90294-1

Zimmer Z, House JS (2003) Education, income, and functional limitation transitions among American adults:

contrasting onset and progression. Int J Epidemiol 32(6):1089–1097. https://doi.org/10.1093/ije/dyg254

Zimmer Z, Martin LG, Chang MC (2002) Changes in functional limitation and survival among older Taiwanese, 1993, 1996, and 1999. Population Studies 56(3):265–276. https://doi.org/10.1080/00324720215931

Zimmer Z, Martin LG, Nagin DS et al (2012) Modeling disability trajectories and mortality of the oldest-old in China. Demography 49(1):291–314. https://doi.org/10.2307/41408229

Zimmer Z, Martin LG, Jones BL et al (2014) Examining late-life functional limitation trajectories and their associations with underlying onset, recovery, and mortality. J Gerontol B Psychol Sci Soc Sci 69(2):275–286. https://doi.org/10.1093/geronb/gbt099

Functionality

▶ Independent Aging

Funeral Ceremonies, Funeral Memorialization

▶ Funerals and Memorial Practices

Funerals and Memorial Practices

William G. Hoy
Medical Humanities Program, Baylor University, Waco, TX, USA

Synonyms

Funeral ceremonies, Funeral memorialization; Memorial ceremonies

Definition

Funerals and memorial practices refer to the various religious, cultural, and kinship rituals that mark the death of an individual and provide social support to the bereaved individuals.

Overview

From time immemorial, humans have gathered in the face of death to ritualize the passage from life to death. First credited with coining the term, "rites of passage," in the early twentieth century, van Gennep (1960) studied tribal ceremonies to better understand how people groups utilize rituals to express emotion following a community or family member's death. Anecdotally, traditional funeral practices are more widely embraced by older members of a society while younger members of a group seem open to experimentation and new styles of memorialization.

Funerals and memorial practices exist for two reasons and individuals and groups may hold one or some combination of these beliefs. Among many people groups, funerals are thought to hold significance primarily for appeasement of the dead or are a necessary component to assure the deceased person's "safe passage" to the next plane of existence. Among others, funerals are seen as commemorative ceremonies primarily designed for the comfort of bereaved family and/or community, including perhaps, the expiation of guilt over words and actions between bereaved survivors and deceased (Pine 1972) (See ▶ "Social support in bereavement"). Nearly all communities studied have one or more ritual "experts" who take their cues from the primary interests of the community according to one or some combination of these two broad patterns.

Key Research Findings

Funerals, as part of a wider interest in ceremony and ritual, have been widely studied in academic and popular literature, especially since the beginning of the twentieth century but organized methods for death rituals are found at least dating to the Cerny (4600–4300 BCE) who buried their dead with grave goods such as arrowheads, pottery, and shell jewelry (Thomas et al. 2011). The utilization of "gifts" to accompany the deceased is seen throughout history including trinkets placed in the coffin or cremation container of a contemporary deceased person or the libations poured into a grave before burial of a fallen military comrade.

Van Gennep's (1960) early work created a three-part structure for understanding funeral rituals: rituals of separation (preliminal), transition (liminal), and incorporation (postliminal). Rites of *separation* provide mourners with an opportunity to say goodbye to the relationship, as previously known, while the rites of *transition* reference the "between and betwixt," liminal time. Finally, van Gennep suggested, ritual carries the participant into a period of *incorporation* where the new roles and responsibilities are taken on, with rituals demarcating the new life that begins for mourners.

Malinowski (1948) declared that death creates a paradox in which humans are both drawn to the dead in an attempt to stay connected and repulsed by the transformation that death brings to the body. Memorial practices, Malinowski believed, attempt to reconcile these two contradictory purposes of keeping the relationship alive while severing the physical bond. Geertz (1973) reflected on Malinowski's premise by suggesting that "Mortuary rituals maintain the continuity of human life by preventing the survivors from yielding either to the impulse to flee panic-stricken from the scene or to the contrary impulse to follow the deceased into the grave" (p. 163).

Communities employ various systems to care for their dead. In many communities, deceased individuals are cared for by the family and community members as an act of filial devotion. In some societies, this familial-based care eventually gives way to a system of death care managed by variously trained-and-certified professionals such as funeral directors, the "ceremonial tender" of the funeral (Pine 1972). In the United States, for example, this second era began in earnest with the popularization of arterial embalming in the late nineteenth century making possible the return of some military dead to their hometowns during the American Civil War as well as the public funeral rituals for President Abraham Lincoln. Many common citizens embraced the notion of temporary preservation of their dead when they saw the option employed in the war dead's return and the president's funeral (Laderman 2003).

In a few western societies, a minority of individuals have embraced the idea of returning funeral rituals to the hands of community and away from professionals; this can be seen in the North American movement of providing care for one's own dead (Slocum and Carlson 2011).

Fulton (1994) posited that the funeral is more important than simply a means to dispose of the corpse; its purpose extends to gathering the community, providing a "socializing experience for the participants, particularly the young ... [and] serves as an important vehicle of cultural transmission," (p. 309).

Memorial practices seem to serve at least four purposes. Funerals make real the fact of the death. Funeral rituals also provide stability in the chaos of early loss. Community funeral rituals remind mourners that in a real sense, the community has been here before and knows the way through. Whether it is the highly prescribed funeral ritual of Roman Catholicism or the beating of a tribal drum on the African continent to notify the entire village of a death, ritual gives order to the chaos. Pine (1972) wrote that funerals are more than religious ceremonies and that their purposes extend to helping community members cope with their loss while dealing with the reality of their own future deaths.

Third, funeral rituals help consolidate the legacy of the dead. In the face of a loved one's death, bereaved individuals tend to find in the deceased person's life one or more character qualities worth imitating, and those qualities become the values talked about in the funeral eulogy, sermon, or remembrance speech. Tribute-speakers often use stories and anecdotes to articulate the positive values evidenced in the deceased's life: compassion, courage, respect, generosity, enthusiasm, humor, positivity, warmth, politeness, peacefulness, and heroism. Funeral rituals provide socially sanctioned places for mourners to honor the legacy of the deceased person.

Fourth, funeral rituals remind mourners of social continuation. The re-incorporative task of rites of passage (van Gennep 1960/1908) remind mourners that while death changes the landscape of relationships, the death – even of a community leader – does not end the social order. State funerals are thought to help provide psychological reassurance to citizens that their government is safe and will be peacefully transferred to the next administration. Funerals after the assassinations of world leaders such as the US President John F. Kennedy in 1963, the Israeli Prime Minister Yitzhak Rabin in 1995, and the former Pakistani Prime Minister Benazir Bhutto in 2007 are examples of how the order of funeral rituals provides psychological reassurance to citizens.

Much recent scholarship has pointed to the efficacy of memorial practices for the healthy integration of the loss experience and funerals are viewed as positive in contemporary theories and models of the bereavement process (Hoy 2013, 2016; Worden 2018). Romanoff and Terenzino (1998), for example, noted, "Rituals can serve moderating, mediating, and connecting functions within the bereavement process, and thus facilitate intrapsychic transformation, the psychosocial status transition, and the continuation of communal and symbolic connections" (p. 697).

Hoy (2013) articulated five attributes in funeral rituals, indicating that in grounded theory research with more than 135 people groups across the historical periods and contemporary cultures, these "anchors" or ritual attributes were found to be present in the death rituals of all groups. *Significant symbols*, the first of these anchors reminds observers that while words are important, images and material objects are paramount. Ritual symbols such as the earth, water, fire, and air accompany the well-known symbols of death such as caskets, flowers, and hearse. In contemporary memorial gatherings, these traditional symbols are mixed – and even sometimes supplanted by – "life symbols" such as the deceased's golf clubs, doll collection, and toolbox.

Gathered community create the second anchor of funeral rituals, characterized by people from varied communities important to the deceased and the bereaved. In contemporary western communities, it is not uncommon for individuals to attend the memorial gatherings for the parents of colleagues as a show of support even though the deceased was not known to the attendee. Also present in communities around the world and throughout history, Hoy identified the anchor of *ritual action*, demonstrating that grief includes behaviors as well as feelings. Rather than standing by as casual observers, many groups of mourners become actively involved in carrying out the duties of mourning and memorialization.

A fourth anchor is that of *cultural heritage*, reminding observers that funerals and memorial gatherings are created within a cultural framework, often with ideas, rituals, and concepts handed down over many generations. The incensing of the coffined body at a Roman Catholic funeral mass belongs to a traditional ritual hundreds of years old and that carries theological and emotional meaning for mourners. *Presence of the dead*, the fifth anchor identified by Hoy seems has lost some prominence in recent North American funeral rituals (Long 2009) even though this seems to be a largely western, Caucasian phenomenon.

A mid-twentieth century movement by some to remove the corpse from the funeral and replace that increasingly elaborate ritual with a simple cremation and memorial service seemed at least temporarily quelled when, as historian Gary Laderman (2003) wrote the assassination of John F. Kennedy brought the body back to the center of the funeral: "(Kennedy's funeral) did momentarily reveal the complex set of relations that exist between the living and the dead. In this case, as with (President Abraham) Lincoln roughly 100 years before, private and public attachments to the dead body required its presence and accessibility to mourners participating in the final ceremonies..." (p. xli).

Though not necessarily central to the body's presence at funeral rites, the technical process of embalming often accompanies funerals with the deceased present. Mayer (1990) summarized *embalming* procedures as chemical treatment of the dead human body to reduce microorganism growth, slow decomposition, and to restore "an acceptable physical appearance" (p. 14). Mayer continued, "The goal of restoration of the dead human body is not so much to make the deceased look lifelike, but rather to try and remove from the body the devastation caused by many long-term diseases and illnesses" (p. 17). Whether employing embalming, other preservation techniques, or no artificial preservation strategy at all, it appears that the majority of the contemporary societies employ the dead body's presence at the funeral rites.

Future Research Directions

Starting in the first decades of the twentieth century, calls for reform and alleviation of elaborate funeral practices became louder in the years following World War II. These criticisms questioned the social and emotional value of funeral rituals as well as the economic cost born by the bereaved families.

Global development workers have long expressed concern over the large parts of family finances dedicated to funeral costs in much of the developing world, frequently levying strong criticism over such practices. In a work in Uganda, however, Jones (2009) found that funeral rituals without bodies were seen by the Katine people as proof the west had lost its way in caring for its elderly. Sometimes even in a quest to protect public health, European and American aid workers have resorted to mass burials during public health crises like the 2010 Haitian earthquake and the west African Ebola crisis, leading to significant distress on the part of grieving family members and communities (Delva 2010; Maxmen 2015). Clearly, more research is needed to discover the perceived value versus economic cost of various kinds of funeral options chosen.

Funeral rituals are also often the impetus to ignite passion for sweeping social change. In addition to the public funeral for Rev. Martin Luther King, Jr., in 1968, the funeral rituals for Medgar Evers in 1963 and Jimmie Lee Jackson in 1965 provided impetus to the growing Civil Rights movement in the United States as have 21st century funerals that utilize eulogy to call for social justice. Sociological researchers will be interested to discover how rituals can inspire and encourage social change.

Summary

That funeral customs will continue to change seems obvious, but if past trends predict future activity, these changes will come slowly. *Cremation*, and other alternative disposal methods take the place of burial but do not impede the use of funeral rituals. The physical act of corpse-disposal in no way erases the tendency of humans to engage in diverse memorial practices, even when the body is not available to be included in the rites. Older populations may or may not readily embrace these changes.

Cross-References

▶ Professional Grief and Burnout
▶ Rumination in Bereavement
▶ Social Support in Bereavement

References

Delva JG (2010) Haiti's voodoo priests object to mass burials. Reuters World News. Retrieved from http://www.reuters.com

Fulton R (1994) The funeral in contemporary society. In: Fulton R, Bendicksen R (eds) Death and identity, 3rd edn. Charles Press, Philadelphia, pp 288–312

Geertz (1973) The interpretation of culture: selected essays. Basic Books, New York

Hoy WG (2013) Do funerals matter? The purposes and practices of death rituals in global perspective. Routledge, New York

Hoy WG (2016) Bereavement groups and the role of social support: integrating theory, research, and practice. Routledge, New York

Jones B (2009) Beyond the state in rural Uganda. University of Edinburgh Press, Edinburgh

Laderman G (2003) Rest in peace: a cultural history of death and the funeral home in twentieth-century America. Oxford University Press, New York

Long TG (2009) Accompany them with singing: the Christian funeral. Westminster John Knox Press, Louisville

Malinowski B (1948) Magic, science and religion and other essays. Anchor Doubleday, New York

Maxmen A (2015) How the fight against Ebola tested a culture's traditions. Natl Geogr. Retrieved from http://www.nationalgeographic.com

Mayer RG (1990) Embalming: history, theory and practice. Appleton and Lange, New York

Pine VR (1972) Social meanings of the funeral. In: Pine VR et al (eds) Acute grief and the funeral. Charles Press, Philadelphia, pp 115–125

Romanoff BD, Terenzino M (1998) Rituals and the grieving process. Death Stud 22:697–711

Slocum J, Carlson L (2011) Final rights: reclaiming the American way of death. Upper Access, Hinesburg

Thomas A, Chambon P, Murail P (2011) Unpacking burial and rank: the role of children in the first monumental cemeteries of Western Europe (4600–4300 BC). Antiquity 85:772–786 van Gennep A (1960) The rites of passage. Routledge and Kegan Paul, London. Originally published as *Rites de passage,* 1908

Worden JW (2018) Grief counseling and grief therapy: a handbook for the mental health practitioner, 5th edn. Springer, New York

Future Care Needs Planning

▶ Preparation for Future Care: The Role of Family Caregivers

Future Care Planning

▶ Preparation for Future Care: The Role of Family Caregivers

Future Selves

▶ Possible Selves Theory

G

Gaining Strength

▶ Women Empowerment

Gaps in Internet Access

▶ Digital Divide and Robotics Divide

Gastrointestinal Malignant Neoplasms

▶ Tumors: Gastrointestinal Cancers

Gateway to Global Aging Data

Jinkook Lee[1,2,3], Drystan Phillips[1] and Jenny Wilkens[1]
[1]Center for Economic and Social Research, University of Southern California, Los Angeles, CA, USA
[2]Department of Economics, University of Southern California, Los Angeles, CA, USA
[3]RAND Corporation, Santa Monica, CA, USA

Overview

The Gateway to Global Aging Data (https://g2aging.org/) is a data and information platform developed to facilitate longitudinal and cross-country analyses on aging, especially those using the family of Health and Retirement Studies (HRS) around the world. The Gateway indexes and extensively documents survey questionnaires, including the order of modules and questions, also referred to as survey metadata. The Gateway has indexed metadata from 17 surveys in 45 countries from 1992 to 2017. Additionally, for many of these surveys, the Gateway also creates and releases files that use the survey's individual-level data, or microdata, to produce harmonized datasets, which are comprised of variables representing a subset of the original data and that have been defined to be as comparable as possible across different surveys and over time. Currently, more than 25,518 harmonized variables have been created in 14 datasets, representing 29 countries, 261,545 individuals, and 832,895 observations. The Gateway is funded by the National Institutes on Aging (R01AG030153), and all data and documentation are available to researchers without cost.

Development and Scope of Gateway to Global Aging Data Collections

The Health and Retirement Study (HRS) began in 1992 as a nationally representative, longitudinal, multidisciplinary survey of individuals over age 50 in the United States (See ▶ "Health and Retirement Study"). Since then, it has been fielded every

© Springer Nature Switzerland AG 2021
D. Gu, M. E. Dupre (eds.), *Encyclopedia of Gerontology and Population Aging*,
https://doi.org/10.1007/978-3-030-22009-9

2 years with periodic refresher samples to maintain a nationally representative sample over time (Sonnega et al. 2014). As population aging has progressed in every region of the world, the success of the HRS generated substantial interest in collecting similar data in a similar manner, leading to the development of a number of surveys designed to be comparable with the HRS.

Like the HRS, the following surveys are nationally representative, longitudinal, multidisciplinary surveys of older individuals. The Mexican Health and Aging Study (MHAS) began in 2001 with respondents aged 50 and older and has had additional waves in 2003 and triennial interviews beginning in 2012 (Wong et al. 2017) (See ▶ "Mexican Health and Aging Study"). The English Longitudinal Study of Ageing (ELSA) has conducted biennial interviews since 2002 with respondents 50 and older (Steptoe et al. 2013) (See ▶ "English Longitudinal Study of Ageing"). The Survey of Health, Ageing and Retirement in Europe (SHARE) has also conducted biennial interviews on respondents 50 and older since 2004 in a growing list of over 20 countries and Israel (Börsch-Supan et al. 2013) (See ▶ "The Survey of Health, Ageing and Retirement in Europe"). The Costa Rican Longevity and Healthy Aging Study (CRELES) conducted interviews with two distinct cohorts, the first including respondents aged 60 and older in 2004, 2006, and 2008 and the second including respondents aged 55 to 65 in 2010 and 2012 (Rosero-Bixby et al. 2013) (See ▶ "Costa Rican Longevity and Healthy Aging Study"). The Korean Longitudinal Study of Aging (KLoSA) has interviewed South Korean respondents aged 45 and older biennially since 2006 (KLoSA team 2018) (See ▶ "Korean Longitudinal Study of Ageing"). The Japanese Study of Aging and Retirement (JSTAR) has conducted biennial interviews of respondents aged 50 to 75 who are municipality-based, rather than nationally representative, since 2009 (Ichimura et al. 2009). The Irish Longitudinal Study on Ageing (TILDA) has conducted biennial interviews since 2010 for respondents aged 50 and older (Kearney et al. 2011) (See ▶ "The Irish Longitudinal Study on Ageing"). The China Health and Retirement Longitudinal Study (CHARLS) began conducting biennial interviews on respondents aged 45 and older in 2011 (Zhao et al. 2014) (See ▶ "China Health and Retirement Longitudinal Study (CHARLS)"). More recent additions to the HRS family of surveys include Health and Aging in Africa: A Longitudinal Study of an INDEPTH Community in South Africa (HAALSI) (See ▶ "Health and Aging in Africa: A Longitudinal Study of an INDEPTH Community in South Africa"), the Brazilian Longitudinal Study of Aging (ELSI-Brazil) (See ▶ "Brazilian Longitudinal Study of Aging (ELSI-Brazil)"), the Healthy Ageing in Scotland (HAGIS) (See ▶ "Healthy Ageing in Scotland (HAGIS)"), the Northern Ireland Cohort for the Longitudinal Study of Ageing (NICOLA) (See ▶ "The Northern Ireland Cohort for the Longitudinal Study of Ageing (NICOLA)"), and the Longitudinal Aging Study in India (LASI) (See ▶ "Longitudinal Aging Study in India"). See Table 1 for more detailed information.

The Gateway to Global Aging Data originally began as a search portal for the HRS family of surveys in 2007, with its website launch in 2011 under the RAND Meta Data Repository. In 2013, the team moved to the University of Southern California and re-branded the site as the Gateway to Global Aging Data. Prior to the Gateway, several barriers limited the use of HRS family surveys for cross-wave and cross-country research, including difficulty identifying concordance information, the need to merge multiple data files, dispersed documentation, and a lack of knowledge about what is available. The upfront costs of identifying and understanding complex, longitudinal datasets have overwhelmed even experienced researchers. With the Gateway's extensive documentation and harmonized data files, all found within a single website, the analysis of HRS family surveys across time and countries became significantly easier, lowering the costs of entry for new researchers and saving time and effort for more experienced researchers.

Documentation

The extensive documentation available on the Gateway to Global Aging website, https://g2aging.org/, begins with the inclusion of all

Gateway to Global Aging Data, Table 1 Number of HRS family interviews by year

Survey	Age eligibility	Country	2000–2001	2002–2003	2004–2005	2006–2007	2008–2009	2010–2011	2012–2013	2014–2015	2016–2017
HRS[a]	51+	USA	19,579	18,165	20,129	18,469	17,217	22,034	20,554	18,747	20,912
MHAS	50+	Mexico	15,186	13,704					15,723	14,779	
ELSA	50+	England		12,099	9,432	9,771	11,050	10,274	10,601	9,666	8,445
SHARE[b]	50+	Austria			1,563	1,197		5,247	4,378	3,397	3,206
	50+	Belgium			3,810	3,227		5,322	5,637	5,815	4,902
	50+	Denmark			1,706	2,630		2,287	4,146	3,733	3,246
	50+	France			3,122	2,990		5,851	4,506	3,947	3,331
	50+	Germany			2,995	2,628		1,619	5,751	4,412	3,821
	50+	Greece			2,897	3,412				4,928	3,072
	50+	Israel			2,449	2,447			2,599	2,035	2,132
	50+	Italy			2,552	2,984		3,570	4,745	5,311	4,571
	50+	Netherlands			2,968	2,683		2,789	4,168		
	50+	Spain			2,316	2,423		3,727	6,693	5,623	4,711
	50+	Sweden			3,049	2,796		1,969	4,556	3,906	3,197
	50+	Switzerland			997	1,498		3,786	3,049	2,803	2,402
	50+	Czech Republic				2,736		5,526	5,640	4,856	4,219
	50+	Poland				2,466		1,733		1,826	4,704
	50+	Ireland				1,035					
	50+	Estonia						6,863	5,752	5,638	5,117
	50+	Hungary						3,072			1,538
	50+	Portugal						2,020		1,675	508
	50+	Slovenia						2,748	2,958	4,224	3,692
	50+	Luxembourg							1,610	1,564	1,254
	50+	Croatia								2,495	2,408
	50+	Bulgaria									2,006
	50+	Cyprus									1,233
	50+	Finland									2,007
	50+	Latvia									1,756
	50+	Lithuania									2,035
	50+	Malta									1,261

(continued)

Gateway to Global Aging Data, Table 1 (continued)

Survey	Age eligibility	Country	2000–2001	2002–2003	2004–2005	2006–2007	2008–2009	2010–2011	2012–2013	2014–2015	2016–2017
	50+	Romania									2,114
	50+	Slovakia									2,077
CRELES[c]	60+/55–65	Costa Rica			2,827	2,364	1,855	2,798	2,430		
KLoSA	45+	South Korea				10,254	8,688	7,920	7,486	7,949	7,490
JSTAR	50–75	Japan				3,742	4,122	5,138	4,937		
TILDA	50+	Ireland						8,504	7,207	6,400	
CHARLS	45+	China						17,708	18,612	21,097	
LASI	45+	India									71,071

Sources: RAND HRS 1992–2016 ver1, H_MHAS verB, H_ELSA verF, H_SHARE verD5, H_CRELES verB, H_KLoSA verC, H_JSTAR verC, H_TILDA verB, H_CHARLS verC, H_LASI verA

[a]HRS number of interviews: 12,652 in 1992; 8,222 in 1993; 11,527 in 1994; 7,027 in 1995; 10,964 in 1996; 21,384 in 1998

[b]SHARE 2008–2009 is their life history questionnaire and is not considered to be a longitudinal wave

[c]CRELES survey contains two separate cohorts: pre-1945 cohort (waves 1–3) and 1945–1955 retirement cohort (waves 4–5)

core and supplemental questionnaires for the HRS family of surveys. The core interview is the main interview either given in person or over the phone by trained interviewers and often using computer-assisted personal interviewing (CAPI) technology. There are additional self-completion questionnaires given to the respondent, often used for more sensitive topics, along with health assessments, life history interviews, and end-of-life interviews given to a proxy once the respondent has passed away. Each of these interviews for each HRS family study is available on the Gateway in list, codebook, or flowchart formats for the researcher's examination. Additionally, the search-enabled website allows researchers to easily and quickly find and identify survey questions and harmonized variables of interest by survey and year using keywords or subtopics.

Concordance tables for a multitude of subjects have been created to allow the identification of changes to survey questions over time and of differences between studies. To further promote cross-country analyses, user guides, written by researchers in the given field, detailing study-specific differences with suggestions for appropriate methods of analysis are available for health and cognitive functioning (Hu and Lee 2011, Shih et al. 2011), health behaviors (Wang et al. 2014), physical and anthropometric measurements (Kwon and Hu 2018), healthcare utilization and expenditure (Angrisani et al. 2017), economic status (Angrisani and Lee 2011a, Angrisani and Lee 2011b, Sachedeva et al. 2016), family transfers (Jain et al. 2016, Zissimopoulos et al. 2011), stress (Gruenewald et al. 2019), work and retirement (Zamarro and Lee 2011), and expectations about uncertain future events (Delavande et al. 2011).

Using the data provided in the harmonized data files, the Gateway website also includes tools to visualize the data and to find publications based on the various datasets. The interactive graph and table tool allows researchers to select from over 100 topics to produce population-weighted estimates for countries and years of interest and to filter these results by individual-level or household-level characteristics, which can be viewed as a graph and table or in a global map. Additionally, these survey variables can be viewed next to macro-level variables on the general economy, workforce, public pension, health, and long-term care to provide context. Lastly, the Gateway provides a catalog of publications based on the HRS family of surveys or the harmonized datasets, enabling researchers to easily find articles from their study or topic of interest.

Core Harmonized Datasets

Because the HRS contains a vast amount of data per wave, the need for a user-friendly version of the HRS data has risen. This need was initially met by the RAND Corporation and their release of the RAND HRS, a user-friendly version of a subset of the longitudinal HRS data. In order to facilitate cross-country analyses on aging, the Gateway to Global Aging Data began to create harmonized datasets for the other HRS family studies using the RAND HRS as a model and subsequently expanding on those offerings. The Gateway has currently released harmonized datasets comparable to the RAND HRS for MHAS, ELSA, SHARE, CRELES, KLoSA, JSTAR, TILDA, CHARLS, and LASI and a dataset for the HRS containing additional variables which supplement the RAND HRS.

When first created, the harmonized datasets released variables covering demographics, health, healthcare utilization and insurance, cognition, financial and housing wealth, income and consumption, family structure, employment history, and retirement. Overtime, additional variables have been or planned to be added in new areas, including pension, physical measures and biomarkers, caregiving, stress, and end-of-life planning, and existing areas have been expanded with new variables detailing medication use, retirement expectations, job characteristics and job search, incentives to retire, pension wealth, social connectedness and isolation, and psychosocial variables. See Table 2 for more information for the variables currently available.

Similar to the RAND HRS, the Gateway's harmonized data files are created as a single user-friendly, respondent-level file per survey and include all longitudinal waves. The data are stored so that a single observation represents one

Gateway to Global Aging Data, Table 2 Number of harmonized variables released by dataset

Module	HRS[a] w1-12	MHAS w1-3	ELSA w1-8	SHARE w1,2,4-6	CRELES w1-5	KLoSA w1-6	JSTAR w1-3	TILDA w1-2	CHARLS w1,2,4	LASI Pilot
Demographics	384	149	660	331	126	267	83	76	237	55
Health	1,920	638	1,829	906	268	1,502	345	316	586	80
Healthcare utilization and insurance	0	90	48	175	82	268	33	50	100	0
Cognition	0	72	252	132	53	168	30	50	66	6
Financial and housing wealth	0	85	225	258	130	446	72	50	82	0
Income and consumption	0	132	363	262	110	242	36	50	105	24
Family structure	65	127	436	335	47	331	15	95	201	3
Employment history	0	54	352	168	60	444	76	36	42	2
Retirement and expectations	0	8	180	82	8	144	12	26	36	0
Pension[b]	0	34	118	30	12	72	0	18	38	0
Stress[b]	1,374	72	1,362	216	11	114	86	154	0	0
Physical measures[b]	582	0	0	0	0	132	0	0	0	0
Assistance and caregiving[b]	678	0	0	0	0	350	0	0	0	0
End-of-life planning[b]	568	0	0	0	0	54	0	0	0	0
Total number variables	5,571	1,461	5,825	2,895	907	4,534	788	921	1,493	170
Total number observations	37,495	21,369	18,529	119,988	5,625	11,174	7,116	8,504	25,504	1,683

Sources: H_HRS verB, H_MHAS verB, H_ELSA verF, H_SHARE verD5, H_CRELES verB, H_KLoSA verC, H_JSTAR verC, H_TILDA verB, H_CHARLS verC, H_LASI verA

[a] Harmonized HRS contains limited modules because it serves as a supplement to the existing RAND HRS longitudinal file
[b] Variable expansion has not yet been incorporated into all harmonized datasets

respondent and matched spouse-level variables are created, along with corresponding household-level variables, such as housing, financial wealth, income, and consumption variables. Harmonized variables, essentially standardized variables, are constructed to be as identically defined across studies as possible, a process which entails thoughtful consideration of question wording and survey skip patterns to ensure that identical concepts are measured and require careful effort to implement. Further, the cause of missing values is identified to the extent possible when values cannot be safely assigned. A codebook detailing the variable creation and the code used to create it are provided with each harmonized data file to ensure total transparency. The creation code and codebooks for every harmonized dataset are available on the Gateway's website, and most harmonized datasets can be downloaded from the original study's website in Stata, SAS, or SPSS formats. The harmonized datasets for SHARE and KLoSA are required to be created by the user using the creation code from the Gateway and the study's original data files.

Additional Datasets

In 2019, the Gateway released two new sets of data products describing the life history and end-of-life for survey respondents in multiple studies. Life history questionnaires are in-depth interviews collecting retrospective data on multiple domains and have been conducted by ELSA, SHARE, CHARLS, and the HRS so far. These interviews are increasingly popular for allowing researchers to take a life-course approach, bridging multiple disciplines and connecting events during earlier life stages with outcomes in later life. Harmonized life history datasets are structured in a user-friendly sequence data format with annual information for each year of age from age 15 to 80 for five key domains: children, partnership, housing, work, and health. The Gateway has released harmonized life history data for ELSA and SHARE, with plans to create these datasets for CHARLS and the HRS and to add additional domains in the near future.

Because these studies are conducted longitudinally, an end-of-life interview is attempted with a knowledgeable proxy once the respondent has passed away. These interviews are conducted in order to obtain information on key core content areas for the period between the last core interview and the respondent's death, as well as to obtain information on the respondent's death. To better facilitate the study of the end-of-life and its relation to the lifespan, the Gateway has produced harmonized end-of-life datasets covering demographics, health, caregiving, cognition, healthcare utilization and insurance, family, employment, and end-of-life planning for the HRS, SHARE, and ELSA, with plans to create similar datasets for MHAS, KLoSA, CHARLS, and JSTAR as well.

Applications of Gateway to Global Aging Data for Research on Aging

While researchers could go to individual study websites to search questionnaires and download data in its original format, there is no other place that researchers can go to access such a wealth of comparative information or user-friendly and comparable datasets for the HRS family of surveys. Further, the Gateway is continually expanding its offerings by adding new survey questionnaires, improving search methods, providing additional concordance tables and user guides, including more individual-level and macro-level variables, enlarging current harmonized datasets, and creating brand new harmonized datasets. The Gateway's dedication to growth and to helping researchers is evident with more than 4,800 registered users. Indeed, as the Gateway is a public resource, even more people use the site than are registered users. In the past 12 months, the Gateway has had 25,000 unique visitors on the site, including over 46,000 sessions with more than 220,000 page views from 146 countries. Since October 2015, users have downloaded over 7,000 harmonized data creation codes, codebooks, or data files for use in published articles, as well as unpublished presentations and class reports. As of summer 2019, there have been more than 7,656 journal publications written about one or more of the HRS family of surveys, all of which can be found on the

Gateway's website. Additionally, the Gateway offers training to users through in-person user workshops and online webinars, which are available for review on the website, training 725 researchers in 2018 alone.

Future Directions of Research

Due to the increasing global burden of Alzheimer's disease and other dementias, there has been increased interest in research into the risk factors of dementias around the world. In answer to these interests, the HRS developed the Harmonized Cognitive Assessment Protocol (HCAP) to obtain high-quality data on late-life cognition and dementia, and other HRS family studies have adapted the HCAP for their own study populations (MHAS, ELSA, SHARE, LASI, CHARLS, KLoSA, HAALSI). Upon release of this newly collected data, the Gateway plans to create HCAP datasets, enabling in-depth cross-country analysis of late-life cognition and dementia.

Further, the Gateway to Global Aging Data plans to incorporate individual-level estimates of exposure to pollution, specifically $PM_{2.5}$, using a spatiotemporal model for the samples in the United States (HRS), England (ELSA), and India (LASI). These pollution estimates will be de-identified, linked to the survey data, and released to the research community through the Gateway, allowing for the epidemiologic analysis of $PM_{2.5}$ exposures, cognition, and dementia in each of these cohorts.

Summary

The ever-expanding core datasets and numerous tools for studying cross-country comparability, together with continual innovations designed to increase data usability and enable sophisticated analyses, make the Gateway to Global Aging Data a tremendous resource for the study of aging populations across the world.

Cross-References

▶ Brazilian Longitudinal Study of Aging (ELSI-Brazil)
▶ China Health and Retirement Longitudinal Study (CHARLS)
▶ Costa Rican Longevity and Healthy Aging Study
▶ English Longitudinal Study of Ageing
▶ Health and Aging in Africa: a Longitudinal Study of an INDEPTH community in South Africa
▶ Health and Retirement Study
▶ Healthy Ageing in Scotland (HAGIS)
▶ Indonesia Family Life Survey
▶ Korean Longitudinal Study of Ageing
▶ Longitudinal Aging Study in India
▶ Malaysia Ageing and Retirement Survey
▶ Mexican Health and Aging Study
▶ Panel Survey and Study on Health, Aging, and Retirement in Thailand
▶ The Irish Longitudinal Study on Ageing
▶ The Northern Ireland Cohort for the Longitudinal Study of Ageing (NICOLA)
▶ The Survey of Health, Ageing and Retirement in Europe
▶ WHO's Study on Global AGEing and Adult Health (SAGE)

References

Angrisani M, Lee J (2011a) Harmonization of cross-national studies of aging to the Health and Retirement Study: income measures. RAND Corporation working paper WR-861/5

Angrisani M, Lee J (2011b) Harmonization of cross-national studies of aging to the Health and Retirement Study: wealth measures. RAND Corporation working paper WR 861/6

Angrisani M, Park S, Hu P, Lee J (2017) Harmonization of cross-national studies of aging to the Health and Retirement Study user guide: health care utilization and expenditure. University of Southern California, CESR report no. 2017-001

Börsch-Supan A, Brandt M, Hunkler C et al (2013) Data resource profile: the survey of health, ageing and retirement in Europe (SHARE). Int J Epidemiol 42(4):992–1001. https://doi.org/10.1093/ije/dyt088

Delavande A, Yoong J, Lee J (2011) Harmonization of cross-national studies of aging to the Health and

Retirement Study: expectations. Rand Corporation, working paper WR-861/3
Gruenewald TL, Crosswell AD, Mayer S, Lee J (2019) Measures of stress in the Health and Retirement Study (HRS) and the HRS family of studies: user guide. University of California, San Francisco, user guide.
Hu P, Lee J (2011) Harmonization of cross-national studies of aging to the Health and Retirement Study: chronic medical conditions. Rand Corporation working paper WR-861/1
Ichimura H, Shimizutani S, Hashimoto H (2009) JSTAR first results 2009 report. The Research Institute of Economy, Trade and Industry, discussion paper series 09-E-047
Jain U, Min J, Lee J (2016) Harmonization of cross-national studies of aging to the Health and Retirement Study – user guide: family transfer – informal care. University of Southern California, CESR-Schaeffer working paper series no. 2016-008
Kearney PM, Cronin H, O'Regan C et al (2011) Cohort profile: the Irish longitudinal study on ageing. Int J Epidemiol 40(4):877–884. https://doi.org/10.1093/ije/dyr116
KLoSA Team (2018) User's guide for the 2018 Korean Longitudinal Study of Ageing (KLoSA). Available via KEIS. https://survey.keis.or.kr/. Accessed 24 July 2019
Kwon E, Hu P (2018) Harmonization of cross-national studies of aging to the Health and Retirement Study user guide: physical and anthropometric measurement. University of Southern California, CESR report no. 2018-001
Rosero-Bixby L, Fernández X, Dow WH (2013) CRELES: Costa Rican longevity and healthy aging study, 2005, Costa Rica Estudio de Longevidad y Envejecimiento Saludable. Inter-university Consortium for Political and Social Research, Ann Arbor. https://doi.org/10.3886/ICPSR26681.v2
Sachedeva A, Jung D, Angrisani M, Lee J (2016) Harmonization of cross-national studies of aging to the Health and Retirement Study: household expenditure. Center for Economic and Social Research, University of Southern California, CESR report no. 2016-002
Shih R, Lee J, Das L (2011) Harmonization of cross-national studies of aging to the Health and Retirement Study: cognition. RAND Corporation working paper WR 861/7
Sonnega A, Faul JD, Ofstedal MB et al (2014) Cohort profile: the health and retirement study (HRS). Int J Epidemiol 43(2):576–585. https://doi.org/10.1093/ije/dyu067
Steptoe A, Breeze E, Banks J, Nazroo J (2013) Cohort profile: the English longitudinal study of ageing. Int J Epidemiol 42(6):1640–1648. https://doi.org/10.1093/ije/dys168
Wang S, Min J, Lee J (2014) Harmonization of cross-national studies of aging to the Health and Retirement Study: health behavior. RAND Corporation working paper WR 861/8
Wong R, Michaels-Obregon A, Palloni A (2017) Cohort profile: the Mexican health and aging study (MHAS). Int J Epidemiol 46(2):e2(1-10). https://doi.org/10.1093/ije/dyu/263
Zamarro G, Lee J (2011) Harmonization of cross-national studies of aging to the Health and Retirement Study: employment and retirement measures. RAND Corporation, working paper WR-861/4
Zhao Y, Hu Y, Smith JP et al (2014) Cohort profile: the China health and retirement longitudinal study (CHARLS). Int J Epidemiol 43(1):61–68. https://doi.org/10.1093/ije/dys203
Zissimopoulos J, Lee J, Carroll J (2011) Harmonization of cross-national studies of aging to the Health and Retirement Study: financial transfer. RAND Corporation, working paper WR-861/3

Gc

▶ Intelligence (Crystallized/Fluid)

GCS

▶ The Georgia Centenarian Study

Gender

▶ Representations of Older Women and White Hegemony

Gender and Age in Film

▶ Cougars and Silver Foxes in Film and TV

Gender and Age/Aging

▶ Feminism and Aging in Literature

Gender and Caregiving

▶ Gender and Caring in Later Life

Gender and Caring in Later Life

Athina Vlachantoni and Maja Fuglsang Palmer
Centre for Research on Ageing and ESRC Centre for Population Change, Faculty of Social Sciences, University of Southampton, Southampton, UK

Synonyms

Gender and caregiving; Gender role in caregiving; Women and informal care provision

Overview

The provision of informal care is an increasingly common experience for individuals globally, as a result of demographic trends resulting in population ageing and an increase in the demand for care; and socioeconomic trends, such as greater economic migration, which can affect the supply of care to individuals within and outside the household (Janus and Doty 2018; Pickard 2015; Zhou and Walker 2015) (See ▶ "Demographic transition theories"). Although both men and women can find themselves providing informal care at different stages of the life course, historically research into informal care provision has predominantly focused on the experiences of women providing informal care (Barnett 2013; Leinonen 2011; Williams et al. 2017). Nevertheless, men's improving life expectancy alongside women's, combined with the increasing entry of women into the formal labor market has had an impact on patterns of informal care provision in different country contexts and placed increasing emphasis on men's caring roles (Greenwood and Smith 2015; Milligan and Morbey 2016) (see ▶ "Gender issues in age studies"). Indeed, the renegotiation of traditional gender roles, combined with demographic trends, suggests a changing balance between men's and women's contribution to informal care provision over time, resulting in a narrowing gender gap in this respect (Grigoryeva 2017) (see ▶ "Feminist Theories and Later Life").

Key Research Findings

(a) Cross-sectional evidence: gender differences in informal care provision

Cross-sectional analysis has consistently found women to provide a disproportionate share of unpaid care, in fact women contribute to 71% of the global hours of informal care, with the highest proportion in low-income countries (Wimo et al. 2018). The 2011 United Kingdom (UK) Census data found that 58% of carers were women; however, this gender division was reverse among carers aged over 85, with 12% of men and 5% of women in this age group providing mostly spousal care (ONS 2013). Women are more likely to care for parents/parents-in-law, other kin and nonkin, while men mainly provide care for their spouse or partner (Glauber 2016; Vlachantoni 2010). Women also tend to provide more demanding and intensive forms of daily care, such as personal care (e.g., help with bathing), whereas men are more likely to provide support of an instrumental nature (e.g., help with shopping) (Eurocarers 2017). However, gender differences become more nuanced later in the life course; while women overall provide more hours of care, most studies found that men committed more time to caregiving after the age of 70 (Arber and Ginn 1995; ONS 2013; Vlachantoni 2010). Part of the reason for this pattern relates to men's greater likelihood to provide spousal care and their greater likelihood to be married in later life, thereby "catching-up" with women in care provision in later age (Glauber 2016).

Gender differences are more pronounced in care provided outside the household.

Using the American Health and Retirement study, Grigoryeva (2017) found that daughters on average provided 13.6 h of care per month to a parent, compared to sons' 5.9 h. The author also noted that brothers often pass on caregiving responsibilities to their sisters, whereas sisters tend to share caregiving among themselves. Gender norms are also evident in respect to the gender of the parents, and daughters provide relatively more care to mothers and sons provide relatively more care to fathers (Arber and Ginn 1995; Grigoryeva 2017; Pillemer and Suitor 2014). However, such gender differences are less noticeable among spousal and partner caregivers (Pinquart and Sörensen 2006; Silverstein and Giarrusso 2010). Del Bono et al. (2009) used individual records from the 2001 UK Census to show that after adjusting for marital status and household type, gender differences in care provision disappear (Del Bono et al. 2009). Some research has found that husbands are less likely to provide care for their wives, should their difficulties with Instrumental Activities of Daily Living (IADL) increase (Feld et al. 2010). However, this has been disputed by Langner and Furstenberg (2018) who noted that men increased their care provision equal to women in time of need, resulting in men and women providing similar care hours.

(b) Longitudinal evidence: the gendered dynamics and impact of care provision

Using longitudinal analysis has allowed researchers to examine the dynamics of informal care provision over time, as well as the impact of care provision on the carer's circumstances, for example, their health or economic activity status (Brown et al. 2003; Fredman et al. 2010; Jenkins et al. 2009; Vlachantoni et al. 2013). This body of work has often been permeated by gender differences. For example, focusing on the impact of care provision on the carer's health, O'Reilly et al. (2008) analyzed registration data and data from the 2001 Northern Ireland Census on informal care provision alongside registration data on individuals' mortality risk 4 years later. The study showed broadly that, although male and female caregivers had lower mortality than non-caregivers, nevertheless the mortality risk increased among caregivers as the number of hours of care provision per week increased. However, there were key gender differences in this respect, even after controlling for age, general health status, and other characteristics. For example, men providing 50 h or more of care per week were 25% less likely than male non-caregivers, and women providing the same number of hours of care were 14% less likely than female non-caregivers, to die in the four subsequent years (Ibid) (see ▶ "Benefits of caregiving"). The gender differences in the health impact of moving in and out of the caring role has been examined to a lesser extent (Ross et al. 2008). For example, Vlachantoni et al. (2016) analyzed data from the UK's Office for National Statistics Longitudinal Study linking 2001 and 2011, and showed that individuals who provided more than 20 h of care per week in 2001 but were not caring in 2011 were more likely to report poor health than non-carers, while those who provided more than 20 h of care per week in both 2001 and 2011 were one-third less likely to report poor health in 2011 compared to non-carers at both time points (see ▶ "Caregivers' Outcomes").

The evidence on the impact of care provision on individuals' economic activity status presents a more complex and gendered picture, with female carers typically faring worse in terms of their socioeconomic status than male carers (Carmichael and Charles 2003; Proulx and Le Bourdais 2014). For example, using US data, Van Houtven et al. (2013) found that women care providers who continued working decreased their employment by 3–10 h per week (Van Houtven et al. 2013). Similar results have been found in the British context. For example, Gomez-Leon et al. (2017) analyzed data from the National Child Development Study (1958 birth cohort) to examine among carers (a) the likelihood of exiting the labor force versus continuing work and (b) among those continuing in work, the likelihood of reducing their hours of employment. This research found that providing care for more personal tasks, and for a higher number of hours, are associated with exiting employment for both men and women carers; however, the negative impact of more intense care-giving on reducing one's working hours was significant only for men, which the authors suggested that it meant that women may juggle intensive care commitments alongside their work or leave work altogether. (see ▶ "Employment and caregiving").

Future Directions of Research

Existing research into gender differences in patterns and dynamics of care provision has gone a long way towards improving our understanding of an activity which is becoming increasingly important from a social policy perspective in the context of population ageing. From a gender perspective, there are at least three areas which could guide future research endeavors. Firstly, the vast majority of surveys incorporating questions on informal care provision use the indicator of hours of care provided per week in order to measure the intensity of care provision (Carmichael and Ercolani 2016; Ramsay et al. 2013). However, a more widespread use of time use indicators on the part of informal carers, which can spread over a number of days within each week, would help provide a more detailed picture of male and female carers' time commitment, as well as their perceptions of activities which can or cannot be defined as informal care provision. A second area of future research could focus on the under-researched gender differences within the so-called "care dyads" that is the examination of gender differences among both carers and the care recipients (Rutherford and Bu 2017) (see ▶ "dyad/triad studies"). Such research could contribute to a more nuanced understanding of cultural differences and perceptions of filial obligation in terms of informal care provision, as well as the preferences of care recipients. A final research direction could explore conceptualizations of male carers globally, aiming to unravel instances of "hidden carers," particularly in country contexts where informal care provision has been synonymous with women carers (Hughes et al. 2017; Knowles et al. 2015) (see ▶ "Caregiving and ethnicity" and ▶ "Rural-Urban Comparisons in Caregiving for Older Adults").

Summary

Although informal care provision is an activity which is becoming increasingly common for men and women alike, nevertheless the vast majority of literature in this area has emphasized women's caring roles across the life course. However, men are more likely than women to provide informal care in older ages, and also to provide more intense care, largely to their greater likelihood of being spousal carers. The cross-sectional literature highlights important gender differentials in the caring activity, with women being more likely to provide care across most of the life course (except in older ages), and being more likely to provide personal care and to more than one care recipient. An emerging strand of cross-sectional research relates to the gender of both the care provider and the care recipient, and the extent to which sons/daughters are expected to provide care to their fathers/mothers respectively in particular country contexts. The longitudinal evidence on the other hand has highlighted the importance of examining the characteristics of care provision (e.g., relationship to care recipient, intensity of care provided) when exploring care provision over time (see ▶ "Cross-Sectional and Longitudinal Studies"). Notwithstanding such distinctions, this body of work shows that the provision of informal care does not necessarily result in poor health on the part of the carer; nevertheless, it does appear to result in more adverse effects for women's economic activity and socioeconomic status compared to men's. Future research examining the provision of informal care in greater detail, by both men and women, can contribute to our better understanding of how the needs of care recipients can be met in the future, and what the role for formal systems of social care is.

Cross-References

- ▶ Benefits of Caregiving
- ▶ Caregivers' Outcomes
- ▶ Caregiving and Ethnicity
- ▶ Cross-Sectional and Longitudinal Studies
- ▶ Demographic Transition Theories
- ▶ Dyad/Triad Studies
- ▶ Employment and Caregiving
- ▶ Feminist Theories and Later Life
- ▶ Gender Issues in Age Studies
- ▶ Rural-Urban Comparisons in Caregiving for Older Adults

References

Arber S, Ginn J (1995) Gender differences in informal caring. Health Soc Care Community 3:19–31. https://doi.org/10.1111/j.1365-2524.1995.tb00003.x

Barnett AE (2013) Pathways of adult children providing care to older parents. J Marriage Fam 75:178–190. https://doi.org/10.1111/j.1741-3737.2012.01022.x

Brown SL, Nesse RM, Vinokur AD, Smith DM (2003) Providing social support may be more beneficial than receiving it: results from a prospective study of mortality. Psychol Sci 14:320–327. https://doi.org/10.1111/1467-9280.14461

Carmichael F, Charles S (2003) The opportunity costs of informal care: does gender matter? J Health Econ 22:781–803. https://doi.org/10.1016/s0167-6296(03)00044-4

Carmichael F, Ercolani MG (2016) Unpaid caregiving and paid work over life-courses: different pathways, diverging outcomes. Soc Sci Med 156:1–11. https://doi.org/10.1016/j.socscimed.2016.03.020

Del Bono E, Sala E, Hancock R (2009) Older carers in the UK: are there really gender differences? New analysis of the individual sample of anonymised records from the 2001 UK census. Health Soc Care Community 17:267–273. https://doi.org/10.1111/j.1365-2524.2008.00826.x

Eurocarers (2017) The gender dimension of informal care – 2017. Eurocarers. https://eurocarers.org/userfiles/files/The%20gender%20dimension%20of%20informal%20care.pdf. Accessed on 26 Sept 2018

Feld S, Dunkle RE, Schroepfer T, Shen H-W (2010) Does gender moderate factors associated with whether spouses are the sole providers of IADL care to their partners? Res Aging 32:499–526. https://doi.org/10.1177/0164027510361461

Fredman L, Cauley JA, Hochberg M, Ensrud KE, Doros G (2010) Mortality associated with caregiving, general stress, and caregiving-related stress in elderly women: results of caregiver-study of osteoporotic fractures. J Am Geriatr Soc 58:937–943. https://doi.org/10.1111/j.1532-5415.2010.02808.x

Glauber R (2016) Gender differences in spousal care across the later life course. Res Aging:1–26. https://doi.org/10.1177/0164027516644503

Gomez-Leon M, Evandrou M, Falkingham J, Vlachantoni A (2017) The dynamics of social care and employment in mid-life. Ageing Soc:1–28. https://doi.org/10.1017/S0144686X17000964

Greenwood N, Smith R (2015) Barriers and facilitators for male carers in accessing formal and informal support: a systematic review. Maturitas 82:162–169. https://doi.org/10.1016/j.maturitas.2015.07.013

Grigoryeva A (2017) Own gender, sibling's gender, parent's gender: the division of elderly parent care among adult children. Am Sociological Rev 82:116–146. https://doi.org/10.1177/0003122416686521

Hughes M, McKay J, Atkins P, Warren A, Ryden J (2017) Chief cook and bottle washer: life as an older male carer. J Community Nursing 31:63–66

Janus AL, Doty P (2018) Trends in informal care for disabled older Americans, 1982–2012. Gerontologist 58:863–871. https://doi.org/10.1093/geront/gnx076

Jenkins KR, Kabeto MU, Langa KM (2009) Does caring for your spouse harm one's health? Evidence from a United States nationally-representative sample of older adults. Ageing Soc 29:277–293. https://doi.org/10.1017/s0144686x08007824

Knowles S, Combs R, Kirk S, Griffiths M, Patel N, Sanders C (2015) Hidden caring, hidden carers? Exploring the experience of carers for people with long-term conditions. Health Soc Care Community 24:203–213. https://doi.org/10.1111/hsc.12207

Langner LA, Furstenberg FF (2018). Gender differences in spousal caregivers' care and housework: fact or fiction? J Gerontol: Ser B:gby087-gby087. https://doi.org/10.1093/geronb/gby087

Leinonen AM (2011) Adult children and parental caregiving: making sense of participation patterns among siblings. Ageing Soc 31:308–327. https://doi.org/10.1017/s0144686x10001042

Milligan C, Morbey H (2016) Care, coping and identity: older men's experiences of spousal care-giving. J Aging Stud 38:105–114. https://doi.org/10.1016/j.jaging.2016.05.002

ONS (2013) 2011 census analysis: unpaid care in England and Wales, 2011 and comparison with 2001 vol 2013. Office for National Statistics, London

O'Reilly D, Connolly S, Rosato M, Patterson C (2008) Is caring associated with an increased risk of mortality? A longitudinal study. Soc Sci Med 1982(67):1282–1290. https://doi.org/10.1016/j.socscimed.2008.06.025

Pickard L (2015) A growing care gap? The supply of unpaid care for older people by their adult children in England to 2032. Ageing Soc 35:96–123. https://doi.org/10.1017/S0144686X13000512

Pillemer K, Suitor JJ (2014) Who provides care? A prospective study of caregiving among adult siblings. Gerontologist 54:589–598. https://doi.org/10.1093/geront/gnt066

Pinquart M, Sörensen S (2006) Gender differences in caregiver stressors, social resources, and health: an updated meta-analysis. J Gerontol Seri B: Psy Sci Soc Sci 61: P33–P45. https://doi.org/10.1093/geronb/61.1.P33

Proulx C, Le Bourdais C (2014) Impact of providing care on the risk of leaving employment in Canada. Can J Aging 33:488–503. https://doi.org/10.1017/s0714980814000452

Ramsay S, Grundy E, O'Reilly D (2013) The relationship between informal caregiving and mortality: an analysis using the ONS Longitudinal Study of England and Wales. J Epidemiol Community Health 67:655–660. https://doi.org/10.1136/jech-2012-202237

Ross A, Lloyd J, Weinhardt M, Cheshire H (2008) Living and caring? An investigation of the experiences of older carers. International Longevity Centre, London

Rutherford AC, Bu F (2017) Issues with the measurement of informal care in social surveys: evidence from the English Longitudinal Study of Ageing. Ageing Soc:1–19. https://doi.org/10.1017/S0144686X17000757

Silverstein M, Giarrusso R (2010) Aging and family life: a decade review. J Marriage Fam 72:1039–1058. https://doi.org/10.1111/j.1741-3737.2010.00749.x

Van Houtven CH, Coe NB, Skira MM (2013) The effect of informal care on work and wages. J Health Econ 32:240–252. https://doi.org/10.1016/j.jhealeco.2012.10.006

Vlachantoni A (2010) The demographic characteristics and economic activity patterns of carers over 50: evidence from the English Longitudinal Study of Ageing. Popul Trends 141:54. https://doi.org/10.1057/pt.2010.21

Vlachantoni A, Evandrou M, Falkingham J, Robards J (2013) Informal care, health and mortality. Maturitas 74:114–118. https://doi.org/10.1016/j.maturitas.2012.10.013

Vlachantoni A, Robards J, Falkingham J, Evandrou M (2016) Trajectories of informal care and health. SSM – Popul Health 2:495–501. https://doi.org/10.1016/j.ssmph.2016.05.009

Williams LA, Giddings LS, Bellamy G, Gott M (2017) 'Because it's the wife who has to look after the man': a descriptive qualitative study of older women and the intersection of gender and the provision of family caregiving at the end of life. J Palliat Med 31:223–230. https://doi.org/10.1177/0269216316653275

Wimo A, Gauthier S, Prince M (2018) Global estimates of informal care. Alzheimer's Disease International (ADI), London

Zhou J, Walker A (2015) The need for community care among older people in China. Ageing Soc 36:1312–1332. https://doi.org/10.1017/S0144686X15000343

Gender and Employment in Later Life

Holly Birkett, Fiona Carmichael and Joanne Duberley
Birmingham Business School, University of Birmingham, Birmingham, UK

Synonyms

Gender and empowerment; Gender and reemployment

Overview

With current retirement literature increasingly focusing upon the diversity of experiences of later-life working, there is a suggestion that a fundamental change has taken place in the meaning of retirement for workers, organizations, and society. This variety of experiences, faced by workers, is closely connected to their career pathways throughout their working lives and to the complex network of resources they have had access to, which they take with them into retirement (Birkett et al. 2017). These are often highly gendered. Implicit within this argument is the fact that these networks of resources are diversifying and becoming more complex, as career patterns, as well as social norms around work and families, have changed.

As career patterns have been changing significantly in the past 20 years, so too has research on careers and retirement. In the 1990s, research remained predominantly focused on traditional, male-oriented, career pathways managed by the organization. A clear critique developed at the time which began to emphasize the role of agency in the development of career trajectories. These critiques led to new concepts, such as the protean career – a career managed by the individual – and the boundaryless career – careers existing beyond the boundaries of one organization. More recently, there has also been recognition of increased fragmentation and precariousness of careers, for both men and women, and careers research is moving to mirror the current state of careers. Consequently, research explores the roles and impacts of: the agents undertaking the careers, employing organizations, and the broader structural and contextual environment the agent inhabits throughout her entire career.

Research Findings

Factors Impacting Employment in Later Life

Although employment rates for people in their 50s and 60s have been rising (particularly among older women in the 60–64 age group, who have been most affected by rising statutory pension ages), they remain lower in comparison with younger adults. In 2018, the average employment rate of people in OECD countries aged 55–64 was 61.27%, compared with 78.4% for people aged 25–54. In 2005, the comparable figures were 51.65% and 75.86% (OECD 2018a). Although female employment rates are lower than those of men (at every age), age-employment gaps by gender are similar. In OECD countries in 2017 (OECD.Stat 2019), employment rates for women and men aged 55–64 were 52.2% and 69.1%, respectively. This compares with figures of 68.9% and 86.8% for women and men aged 25–54. As such, there are female and male age-employment gaps of 16.7 pp and 17.7 pp (compared with 23.1 pp and 24.6 pp in 2005).

Age-employment gaps vary considerably by country. In the European Union, the gap is larger than for the OECD as a whole, as the employment rate of older people is lower (58.41%) and that of younger people higher (80.43%). The gap is narrowest in New Zealand (6.36 pp) and widest in Luxembourg (44.31 pp). Part of the explanation for these gaps is that older workers exit into retirement and this can happen early, i.e., prior to statutory pension ages (Schmitt and Starke 2015). Differences in these gaps between countries reflect, among other factors, different structural and contextual environments including different statutory pension ages and differences in other institutional arrangements relating to: pension and welfare eligibility, pension replacement rates, mandatory retirement ages, and access to flexible working arrangements (Blossfeld et al. 2011). Notably, New Zealand abolished the compulsory age of retirement in 1999 and flexible employment has been supported for all workers since 2015. It is important to recognize that much discussion of gender and employment in later life focuses on the wealthier countries of the Global North. The relationship between gender, ageing, health, and work are less well understood in the Global South. Labor force participation among the over 65s is on average much higher in low-income countries, particularly where there is no public income support. In many low-income countries, older adults cannot afford retirement and therefore must continue to work. Often this work may be in the informal economy – for example, this accounts for over 70% of nonagricultural employment in Sub-Saharan Africa (Jütting and de Laiglesia 2009). According to Staudinger et al. (2016), this means that the poorest face high economic risks as they age due to an accumulation of adverse life experiences. This is particularly the case for women who have lower labor participation rates, lower incomes, and are more likely to live to an older age.

In addition to macro institutional factors that may encourage or facilitate (early) exit into retirement, the lower employment rates of older people reflect a wide range of barriers and challenges that make it difficult for some older workers to access, or maintain, secure, full-time employment, particularly in the event of displacement (Harris et al. 2018). Some of these barriers are gendered, others less so. In the extensive body of research investigating these issues, the perspectives of both employers (Irving et al. 2005; Loretto and White 2006b; McNair 2006; Rego et al. 2017) and older workers (Loretto and White 2006a; Walker et al. 2007; Porcellato et al. 2010) have been considered. Following Neumark et al. (2017), it is convenient to subdivide influences into demand-side effects, which reduce employers' demand for older workers and restrict opportunities for employment and reemployment, and supply-side factors, which reduce incentives for older workers to remain in employment. However, the categorization of these influences as either supply-side or demand-side effects is not always clear cut (Porcellato et al. 2010).

On the demand side, organizational policies and practices, such as mandatory retirement (where legal), seniority wage systems, and the take-up by older workers (often women) of particular types of employment practices, such as flexible working (where available), can make older workers costly, or more difficult to employ. Furthermore, if employers expect younger workers to stay with them for longer, human capital theory (Becker 1964) predicts that there will be less incentive to employ or train older workers, as employers will expect the net returns from hiring or training costs to be lower for older workers, given they are recouped over a shorter period.

Lower levels of educational attainment and lack of, or inappropriate, skills further constrain demand for older workers. In all countries, employment rates are lower for the less-educated and lower-skilled, and, for all age groups, employment rates vary considerably by educational level (Sonnet et al. 2014). Furthermore, in most countries, the proportion of older people who have completed tertiary levels of education is substantially lower than for the younger population, while the proportion of older people who have studied to only secondary levels or less is higher (OECD 2018b). Extant research demonstrates that lower levels of educational attainment have a negative impact on the employability of older people (van Zon et al. 2017; Rutledge et al. 2017). Deindustrialization and technological change have also reduced demand for traditional skills, leading to skill gaps and redundancies among the older population. Furthermore, the intangible intellectual and career capital (Tempest and Coupland 2017) that older people bring to the workplace through experience and seniority (e.g., their knowledge and expertise of context management) is not always highly valued by employers and can also deteriorate over time.

Ageist perceptions further constrain demand for older workers, particularly when linked to discriminatory workplace practices (House of Commons 2018), and there is evidence that older women experience worse age discrimination than men (Neumark et al. 2017). Negative perceptions and stereotypes that characterize traditional concepts of old-age underpin negative societal attitudes towards older workers and assumptions about their abilities (McGregor and Gray 2002; Harris et al. 2018). These are manifested in a preference for young workers (Loretto and White 2006a, b; Neumark et al. 2017), who are expected to fit in better and be around for longer. Aesthetics in labor hiring, discrimination against workers on the "basis of needing to look good or sound right" (Warhurst and Nickson 2007) may also be problematic for older workers (and particularly older women), as it potentially excludes them from the service economy, one of the biggest growth areas in most economies.

In addition to institutional rules that reduce incentives for older workers to continue working beyond statutory pension ages (e.g., policies that raise state retirement benefits or lower ages for eligibility), supply-side factors include declining health and negative perceptions of self in the workplace. While good health is an enabler for longer working lives, poor health is an important reason for labor market exit and early retirement among older men and women (Carmichael et al. 2013; Reeuwijk et al. 2017). Older people, particularly working age women, are also more likely to be caring for a relative or friend who is in need of support (Carmichael and Ercolani 2016). Health issues and caring responsibilities can lead to unemployment, or early retirement, by limiting the type of work an older worker can do, or how well they can do it and by influencing an employer's attitude to their employability. In this respect, physically demanding jobs, more likely to be undertaken by men, are likely to present particular problems for older workers (McLaughlin and Neumark 2017). However, chronological age is not a reliable marker for reduction in physical and mental health (Waddell and Burton 2006, p. 15), while health conditions and disability may also be caused by previous employment, for example, through accident, disease, or mental health issues, including stress (Carmichael et al. 2013). Informal care responsibilities and a greater emphasis on work/life balance (Blau and Shvydko 2007) may also explain why older workers may be more willing to extend their working lives when flexible working arrangements are available (OECD 2004).

While age discrimination has a demand-side effect on the employment of older workers, negative stereotypes of older people, negative attitudes towards them, or prejudice against them may be internalized (Loretto and White 2006b; Yang 2012). This internalization can foster negative perceptions of oneself as a worker, or negative attitudes to work, thus imposing supply-side dispositional or psychological barriers to employment. For example, older people may lack confidence in the workplace because they worry about their ability to cope with the demands of the job in the context of change. The perception that older people are resistant to change can additionally lead to age discrimination by employers (Loretto and White 2006b). Older workers can also worry that, by continuing in employment, they are displacing younger workers, so should therefore consider stepping aside (Porcellato et al. 2010).

The challenges faced by older workers in their later-life careers may lead some to exit early into retirement and others to choose not to extend their careers. However, people move into retirement for a range of reasons, both positive and negative and their experiences of retirement vary considerably.

What This Means for Retirement

It has been argued that our understanding of retirement is underpinned by an implicit, stereotypical, masculine model of career, which means that retirement is conceptualized as a point of transition between a state of continuous employment and a time of leisure. Authors have challenged the extent to which this takes account of women's experiences in the labor market (see, for example, Duberley et al. 2014). Research has also suggested that women have more diverse experiences of retirement than men, with older women being more likely to engage in paid work (see, for example, Kinsella and Velkoff 2001), while continuing to balance this with a variety of nonpaid activities, including voluntary, domestic, and caring work. Recent research has suggested a variety of factors that may be responsible for this higher level of diversity and complexity in older women's work/nonwork activities, including: the impact of divorce and family circumstances (Miller 2013); social class and pension differences (Hank and Korbmacher 2013; Radl 2013); caring responsibilities (Jacobs et al. 2014); and ethnicity (Butrica and Smith 2012). Clearly, these factors should not be viewed in isolation. Studies which give a greater understanding of the individual context of career, and its embeddedness in family and community life, are needed.

Future Directions of Research

Life expectancy is increasing and, with the removal of a mandatory retirement-age in many countries, people are working in paid employment for longer. Increasingly, people are retiring and reentering the workforce or reducing working hours in preparation for retirement, sometimes even taking on bridging roles. Work is becoming more precarious, more women are entering and returning to the workforce following childbirth or adoption, and careers are becoming more fragmented. Simultaneously, workplace pensions continue to become less generous. These structural changes mean a career, even for white, middle-class men, is no longer a linear progression of roles in one or two companies, culminating in a worker's most senior and well-remunerated position shortly before retirement in one's 60s. Consequently, the meaning of retirement is changing. Retirement can no longer be seen as a time in later life when a person leaves paid employment to pursue other interests and hobbies. This raises the question of the extent to which retirement itself is still a useful concept at all, or whether it should it be replaced by alternative concepts more in keeping with contemporary work and careers. This is not to say that traditional linear careers no longer exist, rather they are less common, and even employees who do move up the career ladder, undertaking a progression of roles and achieving their final, and most lucrative, role in their 60s, have often had numerous jobs across multiple organizations along the way. Such broad structural changes around work and retirement mean that contemporary and future retirees have access to a significantly different set of resources than their predecessors, access which is significantly gendered. As such, future research on gender and work in later life could usefully focus on a number of areas. In terms of future research on employment in later life, it is clear that it is less relevant to conceptualize retirement as an age-related life stage at the end to paid employment. Instead, future research on later-life working needs to focus on how people combine and navigate their paid employment, unpaid work, and leisure activities. Existing research, and our arguments here, suggests that, in order effectively to support those wanting to exit paid employment in later life, organization and government policies need to be flexible enough to provide differential support for people with different career histories and different resource portfolios. Some of those exiting the workforce, particularly those with discontinuous work histories and small pension pots, will need support to manage their finances, or help with retraining, while those with more traditional linear carer pathways are likely to need more support around developing nonwork related social resources to support their retirement.

Summary

Employment rates in later life are rising but remain stubbornly low in comparison with younger adults. This age employment gap varies significantly by country, related to the spread of demand-side effects, which reduce employers' demand for older workers and restrict opportunities for employment and reemployment, and supply-side factors, which reduce incentives for older workers to remain in employment. This has implications for how older people (and women in particularly) experience employment in later life.

Historically, men and women globally have followed very different career paths, due to the differential impact of family and domestic responsibilities and the disparity in access to significant sectors of the labor market. Thus, women's careers rarely followed the conventional male career norms, often implicit in the literature, where retirement is seen as marking a neat ending to continuous employment. These traditional linear careers, particularly in the Global North, historically resulted in access to significant economic resources throughout life and in retirement. However, as these career pathways become less common, even white, middle-class males are increasingly having to rely on other resources in retirement, including broad-ranging social networks and opportunities to retrain so they can reenter the workforce in roles which better fit their ambitions and opportunities in later life.

Cross-References

▶ Age Discrimination in the Workplace
▶ Delaying Retirement
▶ Gender Disparities in Health in Later Life
▶ Gender Issues in Age Studies
▶ Health and Retirement Study
▶ Retirement Patterns
▶ Retirement Transition
▶ Self-Employment Among Older Adults

References

Becker G (1964) Human capital: a theoretical and empirical analysis, with special reference to education. Columbia University Press, New York
Birkett H, Carmichael F, Duberley J (2017) Activity in the third age: examining the relationship between careers and retirement experiences. J Vocat Behav 103:52–65
Blau D, Shvydko T (2007) Labor market rigidities and the employment behaviour of older workers. IZA discussion paper 2996. Institute for the Study of Labour, Bonn
Blossfeld H-P, Buchholz S, Kurz K (2011) Aging populations, globalization and the labor market. Comparing late working life and retirement in modern societies. Edward Elgar, Cheltenham
Butrica BA, Smith KE (2012) The retirement prospects of divorced women. Soc Secur Bull 72:11–22
Carmichael F, Ercolani M (2016) Unpaid caring and paid employment: different life-histories and divergent outcomes. Soc Sci Med 156:1–11
Carmichael F, Hulme C, Porcellato L (2013) Older age and ill-health: links to work and worklessness. Int J Workplace Health Manag 6(1):54–65
Duberley J, Carmichael F, Szmigin I (2014) Exploring women's retirement: continuity, context and career transition. Gend Work Organ 21(1):71–90
Hank K, Korbmacher J (2013) Parenthood and retirement. Eur Soc 15(3):446–461

Harris K, Krygsman S, Waschenko J, Rudman DL (2018) Ageism and the older worker: a scoping review. The Gerontologist 58(2):e1–e14. https://doi.org/10.1093/geront/gnw194

House of Common (2018) Older people and employment. Fourth report of session 2017–19 Women and Equalities Committee HC359. www.parliament.uk/womenandequalities. Accessed 27 Dec 2018

Irving P, Steels J, Hall N (2005) Factors affecting the labour market participation of older workers: qualitative research. Research report no 281. Department of Work and Pensions, London

Jacobs JC, Laporte A, Van Houtven CH, Coyte PC (2014) Caregiving intensity and retirement status in Canada. Soc Sci Med 102:74–82

Jütting J, de Laiglesia JR (2009) Is informal normal?: towards more and better jobs in developing countries. In: Jütting J, De Laiglesia (Eds) Development Centre of the Organisation for Economic Co-operation and Development, Paris

Kinsella K, Velkoff VA (2001) An aging world. US Government Printing Office, Washington

Loretto W, White P (2006a) Work, more work and retirement: older workers' perspectives. Soc Policy Soc 5:495–506

Loretto W, White P (2006b) Employers' attitudes, practices and policies towards older workers. Hum Resour Manag J 16(3):313–330

McGregor J, Gray L (2002) Stereotypes and older workers: the New Zealand experience. Soc Policy J N Z 18:163–177

McLaughlin JS, Neumark D (2017) Barriers to later retirement for men: physical challenges of work and increases in the full retirement age. Res Aging 40(3):232–256

McNair S (2006) How different is the older labour market? Attitudes to work and retirement among older people in Britain. Soc Policy Soc 5(4):485–494

Miller AR (2013) Marriage timing, motherhood timing and women's wellbeing in retirement in lifecycle events and their consequences: job loss, family change, and declines in health. In: Couch KA, Daly MC, Zissimopoulos J (eds) Stanford University Press, Stanford, pp 109–132

Neumark D, Burn I, Button P (2017) Age discrimination and hiring of older workers. Federal Reserve Bank of San Francisco Economic Letter February. http://www.frbsf.org/economic-research/publications/economic-letter/. Accessed 24 Dec 18

OECD (2004) Ageing and employment policies United Kingdom. Organisation for Economic Co-operation and Development, Paris

OECD (2018a) Employment rate by age group *(indicator)*. https://doi.org/10.1787/084f32c7-en. https://data.oecd.org/emp/employment-rate-by-age-group.htm. Accessed 24 Dec 2018

OECD (2018b) Population with tertiary education *(indicator)*. https://doi.org/10.1787/0b8f90e9-en. Accessed 27 Dec 2018

OECD.stat (2019) LFS by sex and age – indicators, OECD. https://stats.oecd.org/Index.aspx?DataSetCode=LFS_SEXAGE_I_R. Accessed 7 Jan 2019

Porcellato L, Carmichael F, Hulme C, Ingham B, Prashar A (2010) Giving older workers a voice; constraints on employment. Work Employ Soc 24(1):85–103

Radl J (2013) Labour market exit and social stratification in Western Europe: the effects of social class and gender on the timing of retirement. Eur Sociol Rev 29(3):654–668

Reeuwijk KG, van Klaveren D, van Rijn RM, Burdorf A, Robroek SJW (2017) The influence of poor health on competing exit routes from paid employment among older workers in 11 European countries. Scand J Work Environ Health 43(1):24–33

Rego A, Vitória A, Pina e Cunha M, Tupinambá A, Leal S (2017) Developing and validating an instrument for measuring managers' attitudes toward older workers. Int J Hum Resour Manag 28(13):1866–1899

Rutledge MS, Sass SA, Ramos-Mercado JD (2017) How does occupational access for older workers differ by education? J Lab Res 38(3):283–305. https://doi.org/10.1007/s12122-017-9250-y

Schmitt C, Starke P (2015) The political economy of early exit: the politics of cost-shifting. Eur J Ind Relat 22(4):391–407

Sonnet A, Olsen H, Manfredi T (2014) Towards more inclusive ageing and employment policies: the lessons from France, The Netherlands, Norway and Switzerland. De Economist 162:315–339

Staudinger UM, Finkelstein R, Calvo E, Sivaramakrishnan K (2016) A global view on the effects of work on health in later life. The Gerontologist 56(Suppl_2):S281–S292

Tempest S, Coupland C (2017) Lost in time and space: temporal and spatial challenges facing older workers in a global economy from a career capital perspective. Int J Hum Resour Manag 28(15):2159–2183

van Zon S, Reijneveld SA, Mendes de Leon CF, Bültmann U (2017) The impact of low education and poor health on unemployment varies by work life stage. Int J Public Health 62(9):997–1006

Waddell G, Burton AK (2006) Is working good for your health and wellbeing? The Stationery Office, London. https://www.gov.uk/government/publications/is-work-good-for-your-health-and-well-being. Accessed 27 Dec 2018

Walker H, Grant D, Meadows M, Cook I (2007) Women's experiences and perceptions of age discrimination in employment: implications for research. Soc Policy Soc 6:37–48

Warhurst C, Nickson D (2007) Employee experience of aesthetic labour in retail and hospitality. Work Employ Soc 21:103–120

Yang Y (2012) Is adjustment to retirement an individual responsibility? Socio-contextual conditions and options available to retired persons: the Korean perspective. Ageing Soc 32:177–195

Gender and Empowerment

▶ Gender and Employment in Later Life

Gender and Reemployment

▶ Gender and Employment in Later Life

Gender Differences

▶ Gender Inequity

Gender Disparities in Health in Later Life

Stefanie Doebler
Department of Sociology, Social Policy and Criminology, University of Liverpool, Liverpool, UK

Synonyms

Aging and gender; Gender paradox in health; Gendered health disparities; Sex differentials in health

Definition

Gender disparities in health in later life capture health inequalities during the life-course period from age 65 onward – the retirement age in most Western countries. Although there is no fixed consensus about what age cutoff defines later life, the retirement age is the most commonly used operationalization in the empirical literature. Research on gender disparities in later life acknowledges that health outcomes are often gendered and at the same time age dependent.

Research in this field views gender and age as socially constructed categories that intersect and interact to produce health inequalities. Gender is seen as the ascription of female and male social identities to individuals based on socioculturally constructed role expectations. While earlier second wave feminist scholarship theorized "gender" as socially and culturally constructed and distinguished it from (biological) "sex" (Pickard 2014, pp. 1281–1282; Griffin 2017), postmodern feminist theory (Pickard 2016, pp. 48–55; Butler 2006) qualified this distinction further by pointing out that "sex" itself is influenced by culture and constructed through gender norms. Recent feminist accounts also critically reflect on categorizations such as "later life," "old age," "oldest old," and "deep old" (Pickard 2014).

Overview

Gender disparities in health in later life are subject to a growing body of literature, particularly in the fields of Epidemiology, Gerontology, and Social Policy, while there are fewer contributions in Sociology. A simple literature search on academic search engines for entries containing a combination of the terms "health disparity," "gender," and "age" in publications between 1996 and 2018 yielded 3,646 hits on the ISI Web of Science, 2,460 hits on Scopus, 775,039 on Medline, and 34,700 hits on Google Scholar on the 22nd March 2019. The majority of contributions so far have made use of large-scale population data to examine predictors of disparities in the physical and mental health of men and women at older ages. There is also a growing number of qualitative studies that use interviews and focus groups to uncover the health and social care needs of older men and women. Critical sociological theory approaches in a feminist and gender studies tradition (e.g., Backes et al. 2006; Pickard 2018) have queried the socially constructed nature of gendered concepts of age and aging and have focused on the role of gendered discourses and gender-age expectations in shaping health inequalities at older ages.

Key Theory Approaches

Epidemiological and Medical Accounts

The majority of epidemiological contributions on gendered health disparities is rooted in a positivist background and has mainly been concerned with the physical and mental health of men and women (Arber and Ginn 1993; Pickard 2014). Medical accounts on sex health differentials of the nineteenth and early twentieth centuries ignored gender inequality as an important explanatory variable of sex health differentials; instead, they often uncritically used gender stereotypes to sexualize disease and illness – the Victorian narrative of "hysteria" as a disease of a "weaker," "fair" female sex is a famous example of such stereotyping in the gender studies literature (Butler 2006; Bradby 2008; Pickard 2018).

Epidemiological studies on gender and health prior to the 1980s have largely focused on sex differentials in health, which have often been viewed as being conditional on biological differences between the sexes. Conditions such as congestive heart failure and stroke were found in many studies to be more prevalent among men (Schocken et al. 1992; Ho et al. 1993; Gustafsson et al. 2004), while other illnesses, particularly mental health conditions such as anxiety and depression, were found to be more prevalent among women, albeit findings are not always consistent and vary by the national context (Gu et al. 2009).

Empirical findings of sex differentials of various health conditions highlight the importance of gender as an influential, albeit historically neglected variable. Researchers (Gove 1984; Arber and Ginn 1993; Raleigh and Kiri 1997; Williams and Umberson 2004) found shifts in the sex distribution of physical and mental health patterns in Western countries coinciding with female emancipation and a shift in gender roles lifting sexist social taboos on male displays of emotion.

The Cultural Turn and the Influence of Gender Studies

The cultural turn in the social sciences since the 1970s has brought a new critical awareness of gender as a cultural category (Jacobs and Spillman 2005; Chen et al. 2006, pp. 268–269). Feminist accounts since the 1980s, inspired especially by the works of Foucault (Foucault 1998, 2001) and Judith Butler (Butler 2006, 2011), gained momentum in the study of gender and health and criticized earlier medical accounts for conflating gender and sex, for an often uncritical use of gender stereotypes (Courtenay 2000; Schofield 2002, p. 29; Connell 2012; Pickard 2014), and for largely omitting (gender) discrimination as an important predictor of poor health (Arber and Ginn 1991, 1993; Bradby 2008). Gender discourse, discursive power, and its manifestation in sociocultural perceptions of the (female and male) body are analytical categories that were coined by post-structuralist and deconstructivist approaches. Such approaches increasingly focus on intersections between gender, sex, and age. The physical (bodily) process of aging is judged differently for women than men in Western mainstream societies (Backes et al. 2006; Reimann and Backes 2006, pp. 59–66; Pickard 2018, pp. 16–20). The male gaze at the aging female body, social taboos, and censorship of the (gendered) aging body in mainstream mediatized discourses are emerging fields of study (Pickard 2016, 2018). This is increasingly relevant to individuals at older ages.

Other social scientists in a critical realist tradition have focused on empirical analyses of gender health differentials based on surveys and population health data. Studies in this tradition found that gender role expectations and discrimination are indeed very influential and lead to gendered health effects at all ages (Verbrugge 1985; Bird and Rieker 1999; McDonough and Walters 2001; Idler 2003). Contemporary gerontological accounts accept that it is inappropriate to explain sex differences in health merely by referencing biologistic determinism (Bird and Rieker 1999). A recent literature also focuses on gender differences in mental health at older ages (Moriarty et al. 2015; Doebler et al. 2017). Studies in this area increasingly take a life-course approach looking at how exposure to adverse conditions earlier in life affects mental well-being, mental health, and mortality later life.

The Life-Course Approach

Accounts on health and aging increasingly take a life-course approach (Kim and Moen 2002; Dewilde 2003; McMunn et al. 2009; Pickard 2018), studying individuals not only by taking static snapshots at different ages, but aiming to capture their wider life-span from childhood into adulthood and later life (Arber et al. 2003a). Events that affect later-life health and well-being often occur at earlier stages of the life-course; later-life well-being is influenced by processes that span over longer periods of time as well as single shock events.

Since the 1980s, the cumulative disadvantage approach (Allison et al. 1982; Dannefer 1987, 2003; DiPrete and Eirich 2006; Willson et al. 2007) has inspired the analysis of life-course inequalities.

New longitudinal data sources on aging and computational developments facilitating longitudinal analysis methods have enabled researchers to follow individuals over the course of their lives. This approach is particularly well suited for empirical studies of gendered health disparities in later life.

Measurement

The question of measurement is crucial when analyzing health disparities – decisions over what is considered "old age," what distinguishes young bodies from older bodies, and how health and disease are distinguished and operationalized are crucial for the design of empirical research on health disparities (Grundy and Holt 2001). The problem of measurement effects – that decisions over the measurement of social phenomena influence the results of a given study and can thus lead to bias (Braveman 2006; Billiet and Matsuo 2012; Lynn et al. 2012; Walsh et al. 2012) – is a well-known and important one to consider when designing any research on aging and health.

Measurement approaches to aging and health vary by epistemological stance and academic discipline. Many survey-based studies have used self-reported, subjective ratings of health using Likert scale-type survey instruments, since measures of "objective" and physical health are not always available (Streiner et al. 2015, pp. 100–122). Subjective reported health is a very different measure than objective measures, such as blood pressure, or biomarkers. Different measures operationalize different dimensions of health and well-being. Interpretations of findings therefore need to carefully consider the implications and limitations of the measures used and should cautiously consider what conclusions a study can legitimately draw based on the method it used (Carr-Hill and Chalmers-Dixon 2005).

Key Empirical Findings

Discrimination and Sexism

Gender discrimination and sexism are important factors contributing to gender disparities in health. Research found experiences of sexism to be associated with a higher likelihood of victim blaming and a lower acceptance of gender wage equality in women (Connor and Fiske 2018). Similar patterns of internalizing discrimination were also found with regard to ageism (Chrisler et al. 2016). Pickard (2014) suggests using a deconstructivist analytical framework to examine both sexism and ageism – viewing both gender and age as socially constructed categories.

Sexism and discrimination are confirmed causes for inequalities such as the gender pay gap, unequal access to education, healthcare (Travis et al. 2012), and human rights (Bourassa et al. 2004) in many countries. Several studies found that sexism and gender discrimination cause reduced well-being, poorer health, and mental health (Klonoff et al. 2000) over the life-course and thus a lower life expectancy (Idler 2003; Bourassa et al. 2004).

The Gender Pay Gap and Later-Life Inequalities

An established tradition of research on socioeconomic gender disparities over the life-course has taken a political economy perspective (Price and Ginn 2003; Vlachantoni 2012; Blau 2012). A sizeable literature exists on the gender pay gap across the world. Findings in this literature indicate that in most countries, there still exists a sizeable gender pay gap (Newell and Reilly 2001; Blau and Kahn 2003, 2007; Vlachantoni

2012; Minkus and Busch-Heizmann 2018) affecting women's well-being and later-life health. Research also found that over the life-course, women have been disadvantaged when it comes to accumulating pensions (Price and Ginn 2003; Frericks et al. 2009, Pickard 2018). Such life-course disadvantages accumulate at older ages.

Gender, Unpaid Work, and Informal Caregiving

Unpaid house- and care work is known to be gendered – women still shoulder the main burden. A large literature on the health effects of providing informal care found gender disparities both in the amount of care delivered and in the effect of caregiver burden on health (Olsen and Mehta 2006; Burholt and Dobbs 2010; Hwang et al. 2010; Vlachantoni 2012; Vlachantoni et al. 2013; Lee and Tang 2013; Robards et al. 2015; Doebler et al. 2017; Falkingham et al. 2016). Informal caregiving was found to be associated with experiences of caregiver strain and burden (Iecovich 2008; Etters et al. 2008; Chang et al. 2010; Shahly et al. 2013) and a higher likelihood of mental health conditions, such as depression (Livingston et al. 1996; Moriarty et al. 2015; Doebler et al. 2017). Unpaid housework and caregiving occur at defined stages of the life-course – spells of stay-at-home parenting in young to mid-adulthood are often followed by caregiving for older relatives in mid-life. This has implications for the (predominantly female) caregivers' chances of pursuing a lucrative career and of accumulating pension earnings (Backes et al. 2006, pp. 30–32, Pickard 2018). Such financial losses and lack of professional self-fulfillment due to informal caregiving work have direct consequences for later-life quality of life and well-being. Due to their longer life expectancy, older women are often caregivers of their partners at late and latest stages of the life-course, thus further affecting their late-life well-being (Backes et al. 2006; Falkingham et al. 2016).

Since informal caregiving is highly gendered, women being the main caregivers, caregiver strain and its adverse effect on health and mental health were found to affect mainly women; however, some recent studies emphasize that male caregivers are often neglected both in research and in the provision of support by policy-makers. Furthermore, male caregivers may face different support needs than female caregivers (Ryan et al. 2014; Doebler et al. 2017). The importance of considering male experiences and masculinities more generally in research on aging has been emphasized by Arber et al. (2003b, pp. 4–5) and Calasanti (2003).

National and Regional Contexts Matter

Country and area levels of poverty and wealth, income inequality, access to resources such as clean water and food, but also cultural factors, such as the extent to which women's rights and gender equality are implemented in a country's legal system, all matter greatly for population health. Access to resources and basic human rights are often gendered (Richie and Kanuha 1993; Bourassa et al. 2004; Kabeer and Mahmud 2004; Neyhouser et al. 2018) and so is labor force participation (Olsen and Mehta 2006) with women often being disadvantaged. On the macro-level, structural conditions, i.e., policies, are important in that they can both cause and ameliorate gender disparities in later-life health (Palència et al. 2014).

Studies found that in countries with high levels of poverty, women are often disproportionally affected (Kabeer and Mahmud 2004; Goldberg 2009; Lastrapes and Rajaram 2016; Jacobson 2018) and as a consequence suffer adverse effects on their health. Gu et al. (2009) found that modernization in China has improved population health for both sexes, but women benefited more than men.

Adverse contextual conditions, such as poverty, affect later-life well-being and life expectancy. Gendered health disparities can be prevented by implementing policies that ensure gender equality and distributional fairness across all social strata.

Prospects

The empirical literature on gender inequalities over the life-course is still young and growing.

Health inequalities by sex have been found to be related with gender roles and gender inequalities, both were found to vary by country and geographical area, and they are influenced by the policy context. The differential impact of the national and local context is subject to an emerging literature, but more research is needed to disentangle these area-level effects on gendered health trajectories over the life-course. One shortcoming of the literature so far is the scarcity of accounts on the impact of gender discrimination on male health and well-being. Several studies found that the health and social care needs of older men are under-researched, as the majority of the literature on gender inequalities focuses on women. Future studies on social policy effects on masculinities and the later-life health of men would greatly contribute to the field.

Cross-References

▶ Age Stereotypes
▶ Ageism Around the World
▶ Aging and Health Disparities
▶ Feminist Theories and Later Life
▶ Gender Equality in Later Life
▶ Gender Issues in Age Studies
▶ Heterosexism and Ageism
▶ Life Course Perspective
▶ Sexism and Ageism

References

Allison PD, Long JS, Krauze TK (1982) Cumulative advantage and inequality in science. Am Sociol Rev 47:615–625. https://doi.org/10.2307/2095162

Arber S, Ginn J (1991) The invisibility of age: gender and class in later life. Sociol Rev 39:260–291. https://doi.org/10.1111/j.1467-954X.1991.tb02981.x

Arber S, Ginn J (1993) Gender and inequalities in health in later life. Soc Sci Med 36:33–46. https://doi.org/10.1016/0277-9536(93)90303-L

Arber S, Davidson K, Ginn J (2003a) Gender and aging: changing roles and relationships. Open University Press, Maidenhead, p 2003

Arber S, Davidson K, Ginn J (2003b) Changing approaches to gender and later life. In: Gender and aging: changing roles and relationships. Open University Press, Maidenhead

Backes GM, Amrhein L, Lasch V, Reimann K (2006) Gendered life course and ageing – implications on "Lebenslagen" of ageing women and men. In: Backes GM, Lasch V, Reimann K (eds) Gender, health and ageing. European perspectives on life course, health issues and social challenges. VS Verlag fuer Sozialwissenschaften, Wiesbaden, pp 29–56

Billiet J, Matsuo H (2012) Chapter 10: non-response and measurement error. In: Lior G (ed) The handbook of survey methodology in social sciences. Springer, New York, pp 149–178

Bird CE, Rieker PP (1999) Gender matters: an integrated model for understanding men's and women's health. Soc Sci Med 48:745–755. https://doi.org/10.1016/S0277-9536(98)00402-X

Blau FD (2012) Gender, inequality, and wages. Oxford University Press, Oxford

Blau FD, Kahn LM (2003) Understanding international differences in the gender pay gap. J Labor Econ 21:106–144. https://doi.org/10.1086/344125

Blau FD, Kahn LM (2007) The gender pay gap. Acad Manag Perspect 21:7–23. https://doi.org/10.5465/amp.2007.24286161

Bourassa C, McKay-McNabb K, Hampton M (2004) Racism, sexism and colonialism: the impact on the health of Aboriginal women in Canada. Can Woman Stud 24:23

Bradby H (2008) Virtual Special Issue on feminism and the sociology of gender, health and illness. Sociol Health Illn. https://onlinelibrary.wiley.com/page/journal/14679566/homepage/feminism_and_the_sociology_of_gender__health_and_illness.htm

Braveman P (2006) Health disparities and health equity: concepts and measurement. Annu Rev Public Health 27:167–194. https://doi.org/10.1146/annurev.publhealth.27.021405.102103

Burholt V, Dobbs C (2010) Caregiving and carereceiving relationships of older South Asians: functional exchange and emotional closeness. GeroPsych J Gerontopsychol Geriatr Psychiatry 23:215–225. https://doi.org/10.1024/1662-9647/a000023

Butler J (2006) Gender trouble, 1st edn. Routledge, New York

Butler J (2011) Bodies that matter: on the discursive limits of sex. Routledge, London

Calasanti T (2003) Masculinities and care work in old age. In: Arber S, Davidson K, Ginn J (eds) Gender and aging: changing roles and relationships. Open University Press, Buckingham, pp 15–31

Carr-Hill R, Chalmers-Dixon P (2005) In: Lin J (ed) The public health observatory handbook of health inequalities measurement. University of York and South East Public Health Observatory, York/Oxford

Chang H-Y, Chiou C-J, Chen N-S (2010) Impact of mental health and caregiver burden on family caregivers' physical health. Arch Gerontol Geriatr 50:267–271. https://doi.org/10.1016/j.archger.2009.04.006

Chen K-H, Morley D, Morley FP of TCHD (2006) Stuart hall: critical dialogues in cultural studies. Routledge, London

Chrisler JC, Barney A, Palatino B (2016) Ageism can be hazardous to women's health: ageism, sexism, and stereotypes of older women in the healthcare system. J Soc Issues 72:86–104. https://doi.org/10.1111/josi.12157

Connell R (2012) Gender, health and theory: conceptualizing the issue, in local and world perspective. Soc Sci Med 74:1675–1683. https://doi.org/10.1016/j.socscimed.2011.06.006

Connor RA, Fiske ST (2018) Not minding the gap: how hostile sexism encourages choice explanations for the gender income gap. Psychol Women Q. 0361684318815468. https://doi.org/10.1177/0361684318815468

Courtenay WH (2000) Constructions of masculinity and their influence on men's well-being: a theory of gender and health. Soc Sci Med 50:1385–1401. https://doi.org/10.1016/S0277-9536(99)00390-1

Dannefer D (1987) Aging as intracohort differentiation: accentuation, the Matthew effect, and the life course. Sociol Forum 2:211–236. https://doi.org/10.1007/BF01124164

Dannefer D (2003) Cumulative advantage/disadvantage and the life course: cross-fertilizing age and social science theory. J Gerontol B Psychol Sci Soc Sci 58:S327–S337. https://doi.org/10.1093/geronb/58.6.S327

Dewilde C (2003) A life-course perspective on social exclusion and poverty. Br J Sociol 54:109–128. https://doi.org/10.1080/0007131032000045923

DiPrete TA, Eirich GM (2006) Cumulative advantage as a mechanism for inequality: a review of theoretical and empirical developments. Annu Rev Sociol 32:271–297

Doebler S, Ryan A, Shortall S, Maguire A (2017) Informal care-giving and mental ill-health – differential relationships by workload, gender, age and area-remoteness in a UK region. Health Soc Care Community: 987–999. https://doi.org/10.1111/hsc.12395

Etters L, Goodall D, Harrison BE (2008) Caregiver burden among dementia patient caregivers: a review of the literature. J Am Acad Nurse Pract 20:423–428. https://doi.org/10.1111/j.1745-7599.2008.00342.x

Falkingham J, Sage J, Stone J, Vlachantoni A (2016) Residential mobility across the life course: continuity and change across three cohorts in Britain. Adv Life Course Res. https://doi.org/10.1016/j.alcr.2016.06.001

Foucault M (1998) The history of sexuality: the will to knowledge: the will to knowledge, vol 1, New edn. Penguin, London

Foucault M (2001) Madness and civilization, 2nd edn. Routledge, London

Fredricks P, Knijn T, Maier R (2009) Pension reforms, working patterns and gender pension gaps in Europe. Gend Work Organ 16:710–730. https://doi.org/10.1111/j.1468-0432.2009.00457.x

Goldberg GS (2009) Poor women in rich countries: the feminization of poverty over the life course. Oxford University Press, Oxford

Gove W (1984) Gender differences in mental and physical illness: the effects of fixed roles and nurturant roles. Soc Sci Med 19:77–84. https://doi.org/10.1016/0277-9536(84)90273-9

Griffin G (2017) Gender. In: A dictionary of gender studies. Oxford University Press, Oxford

Grundy E, Holt G (2001) The socioeconomic status of older adults: how should we measure it in studies of health inequalities? J Epidemiol Community Health 55:895–904. https://doi.org/10.1136/jech.55.12.895

Gu D, Dupre ME, Warner DF, Zeng Y (2009) Changing health status and health expectancies among older adults in China: gender differences from 1992 to 2002. Soc Sci Med 68:2170–2179. https://doi.org/10.1016/j.socscimed.2009.03.031

Gustafsson F, Torp-Pedersen C, Burchardt H et al (2004) Female sex is associated with a better long-term survival in patients hospitalized with congestive heart failure. Eur Heart J 25:129–135. https://doi.org/10.1016/j.ehj.2003.10.003

Ho KK, Pinsky JL, Kannel WB, Levy D (1993) The epidemiology of heart failure: the Framingham Study. J Am Coll Cardiol 22:A6–A13

Hwang B, Luttik ML, Dracup K, Jaarsma T (2010) Family caregiving for patients with heart failure: types of care provided and gender differences. J Card Fail 16:398–403. https://doi.org/10.1016/j.cardfail.2009.12.019

Idler EL (2003) Discussion: gender differences in self-rated health, in mortality, and in the relationship between the two. The Gerontologist 43:372–375. https://doi.org/10.1093/geront/43.3.372

Iecovich E (2008) Caregiving burden, community services, and quality of life of primary caregivers of frail elderly persons. J Appl Gerontol 27:309–330. https://doi.org/10.1177/0733464808315289

Jacobs MD, Spillman L (2005) Cultural sociology at the crossroads of the discipline. Poetics 33:1–14. https://doi.org/10.1016/j.poetic.2005.01.001

Jacobson JL (2018) Women's health: the price of poverty. In: Gay, J (Ed) The health of women. Routledge, London, pp 3–32

Kabeer N, Mahmud S (2004) Globalization, gender and poverty: Bangladeshi women workers in export and local markets. J Int Dev 16:93–109. https://doi.org/10.1002/jid.1065

Kim JE, Moen P (2002) Retirement transitions, gender, and psychological well-being a life-course, ecological model. J Gerontol B Psychol Sci Soc Sci 57:P212–P222. https://doi.org/10.1093/geronb/57.3.P212

Klonoff EA, Landrine H, Campbell R (2000) Sexist discrimination may account for well-known gender differences in psychiatric symptoms. Psychol Women Q 24:93–99. https://doi.org/10.1111/j.1471-6402.2000.tb01025.x

Lastrapes WD, Rajaram R (2016) Gender, caste and poverty in India: evidence from the National Family Health Survey. Eur Econ Rev 6:153–171

Lee Y, Tang F (2013) More caregiving, less working caregiving roles and gender difference. J Appl Gerontol:0733464813508649. https://doi.org/10.1177/0733464813508649

Livingston G, Manela M, Katona C (1996) Depression and other psychiatric morbidity in carers of elderly people living at home. BMJ 312:153–156. https://doi.org/10.1136/bmj.312.7024.153

Lynn P, Jaeckle A, Jenkins SP, Sala E (2012) The impact of questioning method on measurement error in panel survey measures of benefit receipt: evidence from a validation study. J R Stat Soc 175:289–308

McDonough P, Walters V (2001) Gender and health: reassessing patterns and explanations. Soc Sci Med 52:547–559. https://doi.org/10.1016/S0277-9536(00)00159-3

McMunn A, Nazroo J, Breeze E (2009) Inequalities in health at older ages: a longitudinal investigation of the onset of illness and survival effects in England. Age Aging 38:181–187. https://doi.org/10.1093/aging/afn236

Minkus L, Busch-Heizmann A (2018) Gender wage inequalities between historical heritage and structural adjustments: a German–German comparison over time. Soc Polit Int Stud Gend State Soc, jxy032, https://doi.org/10.1093/sp/jxy032

Moriarty J, Maguire A, O'Reilly D, McCann M (2015) Bereavement after informal caregiving: assessing mental health burden using linked population data. Am J Public Health 105:1630–1637. https://doi.org/10.2105/AJPH.2015.302597

Newell A, Reilly B (2001) The gender pay gap in the transition from communism: some empirical evidence. Econ Syst 25:287–304. https://doi.org/10.1016/S0939-3625(01)00028-0

Neyhouser C, Quinn I, Hillgrove T et al (2018) A qualitative study on gender barriers to eye care access in Cambodia. BMC Ophthalmol 18:217. https://doi.org/10.1186/s12886-018-0890-3

Olsen W, Mehta S (2006) Female labour participation in rural and Urban India: does housewives' work count? Radical Stat 93:57

Palència L, Malmusi D, De Moortel D et al (2014) The influence of gender equality policies on gender inequalities in health in Europe. Soc Sci Med 117:25–33. https://doi.org/10.1016/j.socscimed.2014.07.018

Pickard S (2014) Biology as destiny? Rethinking embodiment in 'deep' old age. Aging Soc 34:1279–1291. https://doi.org/10.1017/S0144686X13000196

Pickard S (2016) Age studies: a sociological examination of how we age and are aged through the life course. Sage, London

Pickard S (2018) Age, gender and sexuality through the life course: the girl in time. Routledge, Abingdon

Price D, Ginn J (2003) Sharing the crust? Gender, partnership status and inequalities in pension accumulation. In: Arber S, Davidson K, Ginn J (eds) Gender and aging: changing roles and relationships. Open University Press, Maidenhead, pp 127–147

Raleigh VS, Kiri VA (1997) Life expectancy in England: variations and trends by gender, health authority, and level of deprivation. J Epidemiol Community Health 51:649–658. https://doi.org/10.1136/jech.51.6.649

Reimann K, Backes GM (2006) Men in later life: perspectives on gender, health, and embodiment. In: Backes GM, Lasch V, Reimann K (eds) Gender, health and ageing. European perspectives on life course, health issues and social challenges. VS Verlag fuer Sozialwissenschaften, Wiesbaden, pp 57–70

Richie BE, Kanuha V (1993) Battered women of color in public health care systems: racism, sexism, and violence. In: Wings of gauze: women of color and the experience of health and illness. Wayne State University Press, Detroit, pp 288–299

Robards J, Vlachantoni A, Evandrou M, Falkingham J (2015) Informal caring in England and Wales – Stability and transition between 2001 and 2011. Adv Life Course Res 24:21–33. https://doi.org/10.1016/j.alcr.2015.04.003

Ryan A, Taggart L, Truesdale-Kennedy M, Slevin E (2014) Issues in caregiving for older people with intellectual disabilities and their aging family carers: a review and commentary. Int J Older People Nursing 9:217–226. https://doi.org/10.1111/opn.12021

Schocken DD, Arrieta MI, Leaverton PE, Ross EA (1992) Prevalence and mortality rate of congestive heart failure in the United States. J Am Coll Cardiol 20:301–306. https://doi.org/10.1016/0735-1097(92)90094-4

Schofield T (2002) What does 'gender and health' mean? Health Sociol Rev 11:29–38. https://doi.org/10.5172/hesr.2002.11.1-2.29

Shahly V, Chatterji S, Gruber MJ et al (2013) Cross-national differences in the prevalence and correlates of burden among older family caregivers in the World Health Organization World Mental Health (WMH) Surveys. Psychol Med 43:865–879. https://doi.org/10.1017/S0033291712001468

Streiner DL, Norman GR, Cairney J (2015) Health measurement scales: a practical guide to their development and use. Oxford University Press, Oxford

Travis CB, Howerton DM, Szymanski DM (2012) Risk, uncertainty, and gender stereotypes in healthcare decisions. Women Ther 35:207–220. https://doi.org/10.1080/02703149.2012.684589

Verbrugge LM (1985) Gender and health: an update on hypotheses and evidence. J Health Soc Behav 26:156–182. https://doi.org/10.2307/2136750

Vlachantoni A (2012) Financial inequality and gender in older people. Maturitas 72:104–107. https://doi.org/10.1016/j.maturitas.2012.02.015

Vlachantoni A, Evandrou M, Falkingham J, Robards J (2013) Informal care, health and mortality. Maturitas 74:114–118. https://doi.org/10.1016/j.maturitas.2012.10.013

Walsh K, Scharf T, Cullinan J, Finn C (2012) Deprivation and its measurement in later life: findings from a mixed-methods study in Ireland. ICSG, Galway

Williams K, Umberson D (2004) Marital status, marital transitions, and health: a gendered life course perspective. J Health Soc Behav 45:81–98. https://doi.org/10.1177/002214650404500106

Willson AE, Shuey KM, Elder J, Glen H (2007) Cumulative advantage processes as mechanisms of inequality in life course health. Am J Sociol 112:1886–1924. https://doi.org/10.1086/512712

Gender Equality

▶ Gender Inequity

Gender Equality in Later Life

Siobhan Austen
School of Economics, Finance and Property, Curtin Business School, Perth, WA, Australia

Synonyms

Gender equity; Gender inequality; Gender inequity; Gender issues; Gender paradox; Genderism

Definition

Gender equality refers to the equal treatment of individuals of different genders. In the context of older communities, the concept refers to an allocation of economic and other resources that does not discriminate between individuals of different genders on the basis of their different forms of contribution across the life course and which is sensitive to the different needs and circumstances of older men and women.

Overview

In most contexts imbalanced sex ratios are a source of policy and research focus. However, the severely imbalanced sex ratio in many older communities, where older women outnumber older men by almost 2 to 1, attracts surprisingly little attention. In an effort to redress this oversight, this entry calls attention to how the issues of an aging population are in many ways the issues of older women and how, as a result, there is an important need for policies, technologies, and services to be designed and implemented in ways that respond to the particular economic, physical, and social circumstances of older women.

The entry also calls attention to gender inequality in older communities. The different economic and social positions of men and women pre-retirement have ongoing and somewhat cumulative effects on the gender distribution of financial resources in old age. The typical gendered pattern of care within working-age households, where men are heavily involved in paid work while many women contribute more unpaid work, produces a substantial gender gap, favoring men, in the distribution of financial resources at retirement.

This gender-based inequality is particularly consequential because women outlive men in most communities and often live the latter part of their life without a spouse. Thus, women's need for economic resources in retirement is relatively large. A focus on gender equity in policy making can help to ensure that policies on retirement income, aged care, and housing, as well as policies that target earlier life stages, both minimize gender gaps in resources in old age and ensure that women are not penalized in later life for the care roles they have performed across the life course.

While most of the specific data points in this entry are based on the Australian experience, Australia is a midsized western industrialized nation with an aging population, and many if not most of the challenges of gender inequality in this community are similar to those currently being encountered in countries around the globe (see Stark 2005 for an earlier account based on the European experience).

Key Research Findings

Gender Gaps in Life Expectancy, Widowhood, Health and Aged Care Service Utilization, and Unpaid Care Roles

The rapid growth in the older population throughout the world has featured particularly strong growth in the number of older women in our communities. In Australia, there are currently *1.7 times* as many women over the age of 85 as there are men, and across the total population aged over 50 years, there are 9.6% more women than men (Australian Bureau of Statistics (ABS) 2018a). This pattern reflects the different average longevities of men and women: at age 65, the (remaining) life expectancy is, on average, 19.6 years for Australian men and 22.3 years for women, a 13.8% gap (ABS 2016a). The gender gap in the older population is likely to narrow in coming decades as the life expectancy of men and women merges and as the influence of events such as major wars on the population profile abates. However, for the foreseeable future, women are likely to continue to comprise the majority of the older population in most communities.

A related demographic feature of many older communities is the high proportion of widows (see ▶ "Widowhood"): In Australia, in the 85+ age group, there are 3.8 times as many widows as there are widowers (ABS 2018a). This partly results from women's great average longevity but is also the product of an age gap between most marriage partners (see ▶ "Marriage and Remarriage"). The median age at marriage for Australia men who married for the first time in 1984 (and, thus, would be close to 60 years of age now) was 25.1, while for women this age was 22.9 (a 2.2-year age gap) (ABS 1994). Among divorcees and widowers who remarry, the age gap at marriage is even larger (at 3.8 years and 8.5 years, respectively), reflecting the tendency of men to remarry to younger women (ABS 2016b).

These gender gaps in longevity and widowhood have a number of important consequences. On the one hand, the gender gap in longevity points to a significant health advantage that favors women. However, the age gap in marriage, combined with women's greater longevity, contributes to many more men than women having a surviving spouse at the end of their life. Men are more likely than women to benefit from the care and support of a partner in later life, while women, more than men, are likely to spend at least part of their later years caring for a frail partner. Importantly, women, more than men, are also likely to be dependent on the provision resources from *outside* their home for their own care in later life, and this is a significant vulnerability.

Data from the most recent Australian census, conducted in 2016, cast light on these differences in the provision and receipt of care in later life (see ▶ "Gender and Caring in Later Life"). They show that in the group of married individuals over the age of 60, 21.2% of women and 14.8% of men are involved in the provision of unpaid assistance to a person with a disability, including to those with a long-term health condition or problems related to old age. Speaking to the greater reliance of women on care provided outside the home, the census data also show how women in each age group are relatively more likely to live in a residential aged care facility (or nursing home) than men. Across the population aged 80+, 35.8% of women, as compared to 28.4% of men, live in residential aged care facilities. Together with the higher representation of women in the older community, their greater need for out-of-home care leads to severely imbalanced sex ratios in most aged care facilities. In Australia there are currently 26 women for each 10 men over the age of 85 living in either a residential aged care facility or in hospital.

Gender Gaps in Retirement Income, Savings, and Wealth

Contrasting the above patterns of aged care needs, the gendered distribution of financial resources is one that strongly favors men (see ▶ "Poverty and Gender Issues in Later Life in the UK"):

> ... the people who receive the most unpaid care from their spouses are men, who as a group also have higher retirement incomes. The people who receive much less unpaid care from spouses are women, who as a group have lower retirement incomes. (Stark 2005: 19)

This is certainly true in Australia where the gender gap in superannuation wealth is large. The median superannuation balance of Australian women who are retired and have at least some superannuation is currently only about 1/3 that of male retirees ($110,952 to $325,000), and a substantially higher proportion of women (34.6%) than men (26.1%) have no superannuation at all (Wilkins 2017: 73).

These gender gaps in superannuation wealth have arisen because women's average wage rate and paid work hours are lower and their exposure to career interruptions is higher than men's. In line with other earnings-based retirement income systems, the Australian *superannuation guarantee charge* (SGC) scheme is structured in such a way that retirement income is proportional to the individual's labor market earnings. As such, it produces larger retirement incomes for men than women, who are, on average, disadvantaged by lower wage rates, lower paid work hours, and career interruptions (often brought about by unpaid caring roles) (see ▶ "Pension Systems").

In Australia, the gender pay gap favoring men (between male and female weekly full-time earnings) currently stands at 15%, the workforce participation gap favoring men is about 10 percentage points, and the rate of part-time employment among men is only about 1/3 the rate for women (ABS 2018b). Together these aspects combine to produce a very substantial gap between the labor market earnings of women and men over the life course. Data from the Housing, Income and Labour Dynamics in Australia (HILDA) survey show, for example, that at mean values, men's earnings over the 15 years from 2001 to 2015 exceed those of women by 74.1% (Austen and Mavisakalyan 2018).

The strong link between the gender gap in labor market earnings over the life course and the gendered division of the paid and unpaid caring roles is revealed in additional analysis of the HILDA data by Austen and Mavisakalyan (2018). Applying regression methods to 2001–2015 HILDA data, they found that, at mean values, women who had a child under the age of 2 in 2001 (as compared to no children) recorded 77.5% less earnings over the subsequent 15 years. For men, in contrast, becoming a parent was not a statistically significant source of variation in their long-term earnings. This study also identified strong links between men and women's long-term earnings and economic resources at retirement, adding to evidence from a raft of previous studies (see, e.g., Preston and Austen 2001; Jefferson and Preston 2005; Rake 2000) and reinforcing the point that earnings-based retirement income systems both penalize the performance of unpaid work and contribute to greater gender inequality in later life.

The Design of Government Policies and Retirement Incomes

In Australia we have witnessed a shift toward an earnings-based system – the SGC – and away from a public age pension, which is increasingly being assigned a residual, safety net role. The pension, which is supported through general tax revenue and expenditure and allocated on the basis of citizenship and assessed means, has been an important mechanism for equalizing men's and women's incomes in old age. In contrast, the growth in superannuation (supported by generous tax concessions) has redirected economic resources toward men because the value of the tax concessions is proportional to an individual's marginal tax rate and men are much more highly represented in the highest tax brackets than women (Australian Tax Office 2013) (see ▶ "Tax Policies and Older Adults").

A number of other policy settings, including those relating to the taxes on earned income and child care, have also been consequential for gender inequality in later life by influencing the incentive and capacity for individuals to participate in paid work. In Australia, a form of income support known as family tax benefits is available to households with young children. Similarly, financial subsidies for the cost of child care are made available by the Australian government to relevant households. These benefits are valuable and well-motivated; however, because individuals' entitlement to them is governed by a means test conducted at the *household* level, second earners in many households (typically women) confront extraordinarily large marginal effective tax rates if they return to work or increase their work hours (Austen, Sharp and Hodgson 2015:

777; Productivity Commission 2014: 892–893). By discouraging workforce participation and full-time work among women, these policy settings contribute negatively to gender inequality in superannuation wealth and, thus, in retirement income in later life.

The results of recent economic modelling of the effect on old-age poverty rates of the shift to an earning-based retirement income system also reveal gender impacts. On a positive note, the modelling predicts an overall fall in incidence of old-age poverty as the Australian SGC scheme matures (Gallagher 2016). The results for a baseline scenario (where the proportion of each worker's wage that is contributed to superannuation reaches 12%; retirement occurs at 65 years; superannuation funds generate conservative rates of return; and superannuants fully draw down their accounts by life expectancy) indicate retirement incomes in excess of OECD poverty line levels for men and women from most earnings groups. However, in scenarios where retirement occurs before age 65 years, which is the norm for Australian women (ABS 2017), the retirement income of women on low wages is predicted to remain below the OECD poverty line (Gallagher 2016: 12, 16). Thus, while the growth in superannuation assets appears likely to reduce the risk of poverty for many older Australians, the benefits for women will be less than those for men, and women will remain more vulnerable to economic hardship in old age than men, even under relatively optimistic policy and economic scenarios.

Intra-Household Issues
The importance of the gender gap in retirement assets and income is downplayed by some on the grounds that women who live in a couple household can benefit from their partner's economic resources. These commentators, however, overlook the issues affecting the substantial number of older single women (at age 65 close to one third of Australian women are not married, and by age 75 half of all women are single (Gallagher 2016: chart 4)). They also ignore the impacts that an unequal distribution of economic resources *within households* is likely to have on the well-being and other outcomes of individual household members. An extensive empirical literature on working-age households shows how women's financial independence is enhanced and the proportion of household resources spent on children is improved when women have ownership and control of at least some of their households' economic resources (see Browning et al. 2011, Chapter 5.5 for a comprehensive overview).

In older couple households, there are also good reasons to expect that the ownership of financial resources within the household will be consequential for the distribution of outcomes. For one, the fact that women typically outlive their partners implies that they have a particular interest in ensuring that sufficient economic resources are set aside to cater for the health and aged care needs of later life. While the international evidence base on these issues is less comprehensive, a number of studies of older couple households find that the allocation of economic assets in older couple contexts is influenced by the distribution of bargaining power (see Browning 1995; Lundberg and Ward-Batts 2000; Gibson et al. 2006; Keese 2011).

In the light of this evidence, it is of concern that, in Australia at least, with the shift toward an earnings-based retirement income system, the outcomes of many older women have become increasingly vulnerable to decisions made by their partners about the use of superannuation wealth. These include decisions about levels of life and unemployment insurance, the drawdown of account balances, lump-sum withdrawals, and the choice of annuities or pension streams – and whether these reflect the life expectancy of both the account holder and his/her spouse.

In Australia, spouses currently have no legal right to information about the details of their partners' superannuation wealth and no legal authority to participate in decisions made about these funds. There is no regulatory requirement in the Australian system for survivor benefits (a form of life insurance for spouses) to be incorporated into individuals' superannuation accounts. This appears to be an important regulatory oversight as there are international precedents for this type of requirement. The US 1984 Retirement Equity Act (REA) altered the regulated procedures for changing the survivor benefits, requiring the written permission of the account holder's spouse (usually a woman) before the survivor benefits could be waived. Perhaps not surprisingly, following the REA, when women were provided with a voice into decisions over retirement annuities, household life insurance holdings and income security for widows increased. Most retirement income systems around the world incorporate spouse benefits, reflecting both an understanding that many women are economically dependent on their partner's superannuation in later life and that unregulated decision-making places these interests at risk.

Of course, a well-resourced age pension in combination with publicly funded health and aged care is further a critical element of the policy mix required for gender equity in later life, both within and across households. In intra-household contexts, the age pension acts to equalize the distribution of economic resources. Thus, it both promotes women's financial independence in retirement and enhances their ability to influence decisions within their households.

An age pension that has *universal* elements is especially important in ensuring women can achieve financial independence in retirement and in protecting them from the effects of imbalance and poor decision-making within their households. However, the Australian age pension is currently means-tested on household, rather than individual income and wealth. Thus, as men's superannuation wealth grows, an increasing number of older women will become ineligible for the age pension and find themselves in a situation of financial dependence on their spouses. This will be negative for gender equality in later life.

Future Directions for Policy and Research

In the future, if the weight of numbers of older women translates into political pressure for action on gender inequality in later life, a number of pathways for policy development work should be followed. Some of these pathways focus squarely on the situation of older people, while others address earlier life course stages that are relevant to outcomes in later life. Each of the pathways requires a stronger evidence base than currently exists – and this is where the opportunities and challenges for researchers lie.

The pathways for improving gender equity within the current group of older Australians include bolstering the age pension and reducing tax expenditures on superannuation. Recent moves to reduce the tax concessions available on very high superannuation account balances are a positive step. However, this needs to be matched by an increased commitment to the age pension to reflect its key role in equalizing the retirement incomes of men and women and in minimizing the effect of prior care roles on individuals' economic outcomes in later life. The "citizenship dividend" aspects of the age pension need to be reasserted, and the pension needs to be set at a level that will ensure that all Australians are protected from the risks of poverty in old age. Combining a basic universal pension entitlement with taxation of retirement income and additional means-tested pension payments appears to offer promise as a revenue-neutral option for reforming the Australian retirement income system to increase gender equality.

There is significant scope for economic modelling of alternative retirement income systems and their effects on gender and other inequalities in later life. This type of research can build on cross-country comparisons of retirement income systems and also utilize new micro-simulation methods and data to generate important, policy-relevant results.

Other policy options for directly improving gender equality in later life include improving the quality and affordability of aged care, health care, and housing. Such measures are required to counterbalance the maldistribution of private economic resources in later life. Within the superannuation system, in couple household contexts, mechanisms that improve women's access to information on – and control over – decisions on superannuation wealth would help to better ensure that women who survive their spouse have access to financial resources at the end of their life. These measures involve some intrusion of the state into household decisions that have previously been seen as private matters and beyond the legitimate concerns of government. However, the intervention can be justified both by the significant potential effects on women's well-being in later life of their current lack of information and lack of voice on the use of superannuation wealth and by the large tax concessions provided by the state to superannuation (partly on the grounds that superannuation assets will be used to support the needs of both partners in later life).

The design of such policies will also benefit from a strong research base. In Australia a project is currently underway on how the ownership, control, and management of resources is distributed and negotiated in older couple households and how this affects the well-being of older Australians (Austen et al. 2015). Additional international studies are warranted in this research space, given the likely importance of institutional variables, including those related to gendered social norms, retirement income systems, and legal frameworks on outcomes.

Divorce laws are an important backstop protection for many women's well-being in later life. These laws need to ensure that individuals are able to make a claim on their partners' current superannuation wealth. They also need to ensure that the division of assets on divorce takes account of how the intra-household distribution of paid and unpaid work might have impacted the *future* superannuation chances of each partner (Dewar et al. 1999). Longitudinal and/or cross-jurisdictional studies that can track the effect of changing/alternative divorce laws on gender equity within and outside of marriage would make an important contribution to research and policy development in this field.

If, as seems likely, earnings-based retirement incomes system will persist into the future, policy measures that target earlier life stages will be important for gender equity in later life. The negative impacts of care and other domestic roles on long-term earnings (and thus superannuation wealth) can be reduced by policy settings that minimize the financial penalties faced by parents when they return to work or increase their work hours. Additional measures are required to address age and sex discrimination in the workplace and perhaps especially in hiring practices, to ensure that women and men of all ages have an equal ability to participate in paid work if they wish to do so. Addressing the gender pay gap – and its impact on the distribution of retirement income – will require other policy interventions, including measures that address the undervaluing of women's work. This is likely to be supported by the large and ongoing effort to research the sources and solutions to gender pay and participation gaps.

Summary

This entry has highlighted how the issues of an aging population are in many ways the issues of older women. It has identified the large imbalance in the sex ratio in the older population, the prevalence of widows over widowers, and the predominance of women in residential aged care facilities. The entry has contrasted the overrepresentation of women in the older population with their relatively small share of the economic resources needed to support well-being in retirement. It has linked this to the typical gendered pattern of care within working-age households by showing how, in an earnings-based retirement income system, the distribution of financial resources at retirement strongly favors men. The extract argues that new policies on retirement income, aged care, and housing, as well as policies that target earlier life stages, are needed to achieve gender equality in later life and to ensure vital care roles across the life course are protected and supported.

Cross-References

► Gender Equity
► Gender Issues in Age Studies

References

Austen S, Mavisakalyan A (2018) Gender gaps in long-term earnings and retirement wealth: the effects of education and parenthood. J Ind Relat 60(4):492–516

Austen S, Sharp R, Hodgson H (2015) Gender impact analysis and the taxation of retirement savings in Australia. Aust Tax Forum 30:763–781

Australian Bureau of Statistics (2016a) Life tables states territories and Australia 2014–2016. ABS, Canberra

Australian Bureau of Statistics (2016b) Marriages and divorces Australia 2016. Cat. no. 3310.0. ABS, Canberra

Australian Bureau of Statistics (2017) Retirement and retirement intentions Australia July 2016 to June 2017. Cat. no. 6238.0. ABS, Canberra

Australian Bureau of Statistics (2018a) Census of population and housing 2016 TableBuilder. ABS, Canberra

Australian Bureau of Statistics (2018b) Gender indicators Australia Feb 2016. Cat. no. 4125.0. ABS, Canberra

Australian Bureau of Statistics (ABS) (1994) Marriages and divorces Australia 1994. Cat. no. 3310.0. ABS, Canberra

Australian Taxation Office (2013) Taxation statistics 2011–12. ATO, Canberra

Browning M (1995) Saving and the intra-household allocation of income. Ric Econ 48(3):277–292

Browning M, Chiappori P, Wise Y (2011) Family economics. https://www.tau.ac.il/~weiss/fam_econ/BCW_Book_index_07_09_2011_MB.pdf. Accessed 8 Oct 2018

Dewar J, Sheehan G, Hughes J (1999) Superannuation and divorce in Australia working paper 18. Australian Institute of Family Studies, Melbourne, pp 1–4

Gallagher P (2016) Modelling adequacy for the broad retiree population paper presented to Committee for Sustainable Retirement Incomes Canberra, 6 April

Gibson J, Trinh L, Scobie G (2006) Household bargaining over wealth and the adequacy of womens retirement incomes in New Zealand. Fem Econ 12(1–2):226–246

Jefferson T, Preston A (2005) Australia's "other" gender wage gap: baby boomers and compulsory superannuation accounts. Fem Econ 11(2):79–101

Keese M (2011) Thrifty wives and lavish husbands? Ruhr Econ Pap 258

Lundberg S, Ward-Batts J (2000) Saving for retirement: household bargaining and household net worth. Working paper WP2000-004, University of Michigan Retirement Research Centre

Preston A, Austen S (2001) Women superannuation and the SGC. Aust Bull Lab 27(4):272–295

Productivity Commission (2014) Childcare and early childhood learning. Inquiry Report 73. Productivity Commission, Canberra

Rake K (2000) Women's incomes over the lifetime. The Stationery Office, London

Stark A (2005) Warm hands in cold age – on the need of a new world order of care. Fem Econ 11(2):7–36

Wilkins R (2017) The household income and labour dynamics in Australia survey: selected findings from waves 1 to 15. Melbourne Institute for Applied Economic and Social Research, Melbourne

Gender Equity

▶ Gender Equality in Later Life
▶ Gender Inequity

Gender Fairness

▶ Gender Inequity

Gender Inequality

▶ Gender Equality in Later Life

Gender Inequity

Ka Ki Chan[1] and Francisca Yuen-ki Lai[2]
[1]Department of Social Work, Hong Kong Baptist University, Hong Kong, China
[2]Center for General Education, National Tsing Hua University, Hsinchu, Taiwan

Synonyms

Gender differences; Gender equality; Gender equity; Gender fairness

Definition

In general, the word *equity* is commonly used to refer to inequalities on different bases (United Nations 2014). The term *gender inequity* specifically refers to disparities and inequalities between treatment of men and women (Mencarini 2014).

This concept is a subtle term because it may not refer to the equality of outcomes for both men and women but refer to fair outcomes between men and women. Therefore, the matter is not only whether treatment of each gender is equal or not but also which gender is rendered unable to access to access the full range of opportunities, rights, or benefits. Therefore, the pursuit of gender equity emphasizes the fairness of the process of resource allocation and decision-making to both men and women, allowing both a full range of opportunities to achieve their social, psychological, and physical needs (United Nations 2010). In this entry, *gender* not only includes people who are *cisgender* (whose gender attributes and expressions match their sex at birth) but also people of different gender identifications and gender expressions that do not align with heteronormative expectations. Since the late 1990s, studies of gender inequity are no more limited to the two sexes, men and women, but include the experiences of lesbian, gay, bisexual, and transgender (LGBT) people (Richardson 1998).

Overall: Aging and Gender

Aging is never a homogenizing experience, as it is deeply affected by the social locations occupied by individuals. In a society where the social locations of men and women are different, the aging experiences of men and women are also very different. Gender and age are not merely discrete demographic variables; rather, they are in an interdependent relationship. Gender shapes one's aging experience – that is, gender shapes how men, women, and non-gender-conforming individuals behave differently when they are aging – while aging shapes gendered bodies, that is, what different meanings of aging are imposed on gendered bodies.

There are three primary approaches to gender in gerontology: (i) the "adding women" approach that simply adds women into existing research models, which use men as the main reference group; (ii) taking "gender as a variable" that treats gender as a fixed demographic variable without examining the experiences of men and women or the complicated impact of gendered social structures; and (iii) taking gender as being relevant only for women, as if only women experience gender relations, not men (Calasanti and Slevin 2001). To remedy the problems of the previous approaches, scholars turn to adopt a feminist perspective in gerontology emphasizing gender as an organizing principle that shapes individual interactions, social structures, and institutional processes. Such a feminist exploration leads researchers to examine both oppressions and privileges experienced by different social groups (not only of gender but also of race, class, sexual orientation, migration, and disability) by gaining insights into the power relationships among groups at many different levels (Westwood 2019a). Connecting feminism and gerontology, this approach lays important foundations for recognizing that the aging experience is never a homogenizing experience but is diversely experienced among genders and groups.

Active Aging

With the increase of longevity in European countries, *active aging* has become a leading policy that has responded to such challenges. In 2002, the World Health Organization (WHO 2002) provided a definition in a policy frame for active aging. The guiding principles of active aging provided by the World Health Organization further stated that gender perspective in active aging should also be addressed, particularly regarding the differences in gendered roles and experiences between two sexes in old age (WHO 2012). Tracing the development of active aging reveals that more comprehensive views on the concept have emerged since the 1990s. Active aging began to develop and emphasize the link between activity and health and the importance of healthy aging. The word *active* is also assumed developed a new meaning as "continuing participation in social, economic, cultural, spiritual, and civic affairs, no longer put the focus on the ability to be physically active or participation in the labor force" (WHO

2002:12). Therefore, "active" is about the optimization of activities involving employment, politics, education, the arts, and religion and the contribution to society regardless of paid or unpaid status. Moreover, the WHO (2002) also provided a more holistic approach to the examination of well-being that links to policy terms such as *quality of life*, mental and physical well-being, and participation. Gender is utilized in examining the effectiveness and the impact of various policies on the differences in well-being of men and of women (WHO 2002). The European Union designated 2012 as the European Year for Active Ageing and Solidarity between Generations. This action took a further step toward changing culturally ingrained attitudes about older people and strived to provide better opportunities for them to fully actively participate in societies. The ultimate vision of this policy aims to facilitate rights of older people who will be able to remain healthy, stay in employment longer, and have fully participation in community life (Walker and Zaidi 2016).

Several scholars stated that one of the aspects of active aging that has long been neglected is the impact of gender on aging experiences, particularly how the gender differences and experiences in aging are further connected with the problems of social exclusion and poverty. They noted that EU member states overlooked a coherent and integrated approach to and understanding of the implication of gender on active aging strategies (Walker 2009). Indeed, some changes did occur when the EU designated 2012 as the European Year for Active Ageing and Solidarity. According to Chang et al. (2014), gender has become one of the key clusters in active aging research. The first change is the launch of the Active Ageing Index (AAI), a project that accesses and monitors developing active and healthy older population across European countries. One of the measurements includes a breakdown by gender of inequalities in experiences of active and healthy aging. Bennett and Zaidi (2016) also stated the potentials of taking a gendered approach to aging and development. They pointed out that the new post-2015 Sustainable Development Goal (SDG) specifically mentioned older people and aging as constituting the cornerstone of sustainable development. The work of the Global AgeWatch Index provided substantial evidence that cumulative gendered disadvantages have occurred in developing countries and the Active Ageing Index placed more emphasis on gender inequality affecting older women in low income within EU countries (Bennett and Zaidi 2016; Zaidi and Stanton 2015). They found that the low level of completed secondary education, healthy life, and diminished life expectancy are explicitly shown when comparing with those of female counterparts in high-income countries (HelpAge International 2015). Based on the above trend on the global level, they recommended that the emerging discourse on aging and development should include a gender dimension of active aging on the national development agenda in order to achieve a global commitment to tackling gender inequality everywhere.

The first recommendation is to provide a critical view on the productivist approaches to the notion of active aging (Foster and Walker 2013). For example, the existing measurement on this area is predominantly based on male trajectories of work and retirement but largely ignore the related needs of women and gender-based differences. At least several key features of women's employment such as career breaks, part-time work, flexible employment, and low-paid status are not sufficiently addressed (Ginn 2003; Foster 2011).

The second recommendation is to identify and review any consequences of gender inequality that are raised to allow for the reconciliation of work and care responsibilities through family-friendly strategies (Foster and Walker 2013). The third recommendation is to interpret *activity* that can include all meaningful pursuits that contribute to individual well-being. Instead of paid work or production, volunteer activities should also be valued. This focus is also important for older women who are taking up unpaid roles in society including taking a role of carer, particularly providing caring for their grandchildren (Foster and Walker 2015) (See ▶ "Gender Disparities in Health in Later Life").

Further Directions of Research

Economic Resources

According to Foster and Walker (2013), there are several phenomena that have emerged in different aging experiences of men and women in European countries. First, poverty rates for women aged over 65 are twice as high as those of men. Second, high poverty rates among women aged over 65 lead to other issues such as deterioration in living conditions, material deprivation, unequal provision of care, and inadequate pensions.

Apart from the disadvantaged economic position of older women, there is research showing that older women have enjoyed advantages, for instance, better relationships with kin and friends than their male counterpart (Arber et al. 2003). For older men, despite their better economic situation than their female counterpart, the conventional discourse of masculinity (i.e., being the breadwinner) has subjected them to a peripheral position when they retire from a paid and professional position (Arber et al. 2003).

The gendered institutions of marriage and family shape the later life of LGBT people. A recent study in the UK shows that heterosexual people over 55 are less likely to be in paid employment than LGBT people (Westwood 2019a). It supports the view that older LGBT people have a more need to get a paid job to secure their financial situation when their heterosexual counterpart has enjoyed spousal benefits of the insurance system, tax exemptions, and rights of inheritance (Calasanti and Slevin 2001). Some older LGBT people have entered a same-sex marriage and become eligible to obtain spousal benefits. There are 26 countries, mostly in Europe and the Americas, allowing same-sex couples to legally marry (Pew Research Center 2015). However, it is worth noting that same-sex marriage does not appeal to all LGBT older people. A survey in the USA shows that only a half of older people in same-sex relationships opted to legally marry, with an average age of 61.82, which is slightly younger than those remained single whose average age is 63.98 (Goldsen et al. 2017). This research also found that legally married gay men and lesbians have enjoyed a higher socioeconomic status, including education, income, ownership of assets and property, etc. compared to their (i) unmarried partnered and (ii) single counterparts in USA. Another study on older transgender people has addressed how prejudice and discrimination in both school and workplace have generated negative impact on their employment, including the limited kind of work they choose to work and the reduced opportunity of job promotion (Bishop and Westwood 2019). This economic disadvantage continues in their older age (See ▶ "Gender and Employment in Later Life").

Bodies

One of the most important topics under the theme of gender inequality concerns aging bodies. Since the dominant discourses and images of aging are in line with the biomedical discourses that emphasize medical accounts of old age, such as growing old, decline, onset of ill-health, and loss of function (Tulle 2015), another line of research has shifted to emphasize the significance of the cultural and social construction of aging body and has challenged the above medical discourse of body by developing an analytic engagement with the corporeality and materiality of aging bodies (Calasanti and Slevin 2001; Gilleard and Higgs 2018; Katz 2010; Martin and Twigg 2018). A cultural and social analysis can also provide more reflexive accounts on identities and lived experiences of old age.

An early study conducted by Sinnott (1984) raised concern about the construction of aging bodies. She argued that the polarities of masculinity and femininity became blurred with age over the life course of individuals, and the bodies of older men and older women were more likely to experience no differences in the sex-role attributes. However, some research demonstrated that there are significant movements toward inequity in old age, even in the face of bodily decline. For example, Davidson et al. (2000) found that older men who received care would find ways to exert power over their wives because older men lacked their former physical abilities. Such older men attempted to seek ways to maintain their dominant positions in their marriage, for example, controlling finances or becoming overly demanding of their wives. Later, Silver (2003) further argued that this phenomenon may have the effect of "de-gendering" later life toward greater equality. Being "old" is signified as the loss of power and authority to maintain control over one's body and then of experienced economic marginalization, social stigmatization, and social exclusion from full adulthood and citizenship (Boyle 2017; Sandberg 2018). In the works of Calasanti and King (2018) as well as Katz and Marshall (2004), both investigated the ways that people do gender from the influence of aging bodies and examined to what extent masculinity and femininity become less relevant with age and whether people feel themselves to be less gendered in later life. Even though older people worried less about being "sexy" but put more focus on maintaining self-sufficiency in old age, the strategies that older people took were gendered. Men continued to be hardworking and maintain the gendered role of breadwinner against aging, while women preferred to engage in antiaging industry such as beauty against aging. The above phenomenon indicated by Calasanti and King reflected that people continue to strive to achieve masculinity and femininity in their later life. Many scholars utilized different themes in discussing the intersection between aging bodies and gender, including the role of the body in feminist insights and subjective experiences of aging and receiving personal care (Twigg 2004), the design of residential care (Buse et al. 2016), fashion and dress (Twigg 2018), and embodied masculinities and cultural discourse of aging (Lodge and Umberson 2013; Clarke and Lefkowich 2018) (See ▶ "Gender and Caring in Later Life").

Sexuality

Countering the stereotype of an asexual old age, research findings suggest that older people do not stop desiring sex or engaging in sexual behavior. Since the 1990s, a trend of research has concerned the potential benefits of sex for older people – e.g., prolonged life expectancy (Hinchliff and Merryn 2016). While the number of research studies about older people's sex life has been increasing, there has also emerged a feminist critique of the "normal sex" model, which defines sex as male-dominated, heterosexual, and defined by penetration and orgasms. This model seriously overlooks female notions of sexuality as well as the diverse forms of sexual behaviors practiced by older people (Calasanti and Slevin 2001; Westwood 2019b). Recent studies have begun to examine the relationship between aging, gender, and sexuality. Older people of different genders have different expectations and experiences of sex. Older men typically equated sex with sexual intercourse and place primacy upon their erectile ability; older women, on the contrary, articulated a sense of liberation through a sexuality that is no longer defined by the duties of being a wife or of reproduction (Gott and Hinchliff 2003).

Social Well-Being

The social well-being of older men and women is affected by their living arrangements and social networks. In de Jong Gierveld's (2003) study of older men and women living alone, findings show that men have smaller networks and less weekly contact with kin and non-kin. Older men living alone are more likely to be lonely than older women living alone. The findings also suggest that lone women have more social resources and frequent contacts with network members, which help them alleviate loneliness.

Older transgender people are more likely to suffer from a range of mental health issues due to a lifetime of discrimination with low levels of social support from family, friends, colleagues, and religious organizations (Bishop and Westwood 2019). Studies highlight the diverse experiences among older trans people. For example, Bailey et al. (2019) indicate that trans women suffer poorer mental health than trans men or other non-binary identified individuals. An alternative angle is provided by the Trans Mental Health Study in the UK, which is the first comprehensive study about trans people's mental health. This study argues that trans people's mental health improves following transition (Bailey et al. 2019). Another research project in Australia provides evidence to suggest that older trans people have better mental health than younger trans people (Bailey et al. 2019).

Cross-References

▶ Caregiving among the LGBT Community
▶ Gender and Caring in Later Life
▶ Gender and Employment in Later Life
▶ Gender Disparities in Health in Later Life
▶ Gender Equality in Later Life
▶ LGBT in Old Age

References

Arber S, Davidson K, Ginn J (2003) Changing approaches to gender and later life. In: Arber S, Davidson K, Ginn J (eds) Gender and ageing: changing roles and relationships, 1st edn. Open University Press, Maidenhead, pp 1–14

Bailey L, McNeil J, Ellis S (2019) Mental health and Wellbeing among older trans people. In: King A, Almack K, Suen Y, Westwood S (eds) Older lesbian, gay, bisexual and trans people minding the knowledge gaps, 1st edn. Routledge, New York, pp 44–60

Bennett R, Zaidi A (2016) Ageing and development: putting gender back on the agenda. Int J Ageing Dev Countries 1(1):5–19

Bishop J, Westwood S (2019) Trans(gender)/gender-diverse ageing. In: Westwood S (ed) Ageing, diversity and equality: socials justice perspectives, 1st edn. Routledge, London, pp 82–97

Boyle G (2017) Revealing gendered identity and agency in dementia. Health Soc Care Community 25(6):1787–1793. https://doi.org/10.1111/hsc.12452

Buse C, Nettleton S, Martin D, Twigg J (2016) Imagined bodies: architects and their constructions of later life. Ageing Soc 37(07):1435–1457. https://doi.org/10.1017/s0144686x16000362

Calasanti T, King N (2018) The dynamic nature of gender and aging bodies. J Aging Stud 45:11–17. https://doi.org/10.1016/j.jaging.2018.01.002

Calasanti T, Slevin K (2001) Gender, social inequalities, and aging. AltaMira Press, Oxford

Clarke HL, Lefkowich M (2018) 'I don't really have any issue with masculinity': older Canadian men's perceptions and experiences of embodied masculinity. J Aging Stud 45:18–24. https://doi.org/10.1016/j.jaging.2018.01.003

Chang C, Lu H, Luor T, Yang P (2014) Active aging: a systematic literature review of 2000–2014 SSCI Journal Articles. Retrieved from https://doi.org/10.6171%2fntuswr2015.32.05

Davidson K, Arber S, Ginn J (2000) Gendered meanings of care work within late life marital relationships. Can J Aging 19(04):536–553. https://doi.org/10.1017/s0714980800012502

de Jong Gierveld J (2003) Social networks and social well-being of older men and women living alone. In: Arber S, Davidson K, Ginn J (eds) Gender and ageing: changing roles and relationships, 1st edn. Open University Press, Maidenhead, pp 95–110

Foster L (2011) Privatization and pensions: what does this mean for women? J Poverty Soc Justice 19(2):103–115. https://doi.org/10.1332/175982711x573978

Foster L, Walker A (2013) Gender and active ageing in Europe. Eur J Ageing 10(1):3–10. https://doi.org/10.1007/s10433-013-0261-0

Foster L, Walker A (2015) Active and successful aging: a European policy perspective. The Gerontologist 55(1):83–90. https://doi.org/10.1093/geront/gnu028

Gilleard C, Higgs P (2018) Unacknowledged distinctions: corporeality versus embodiment in later life. J Aging Stud 45:5–10. https://doi.org/10.1016/j.jaging.2018.01.001

Ginn J (2003) Gender, pensions and the life course, 1st edn. Policy Press, Bristol

Goldsen J, Bryan AEB, Kim H-J, Muraco A, Jen S, Fredriksen-Goldsen KI (2017) Who says I do: the changing context of marriage and health and quality of life for LGBT older adults. Gerontologist 57(S1): S50–S62. https://doi.org/10.1093/geront/gnw174

Gott M, Hinchliff S (2003) Sex and ageing: a gendered issue. In: Arber S, Davidson K, Ginn J (eds) Gender and ageing: changing roles and relationships, 1st edn. Open University Press, Maidenhead, pp 63–78

HelpAge International (2015) Global AgeWatch index 2015: insight report. HelpAge International, London

Hinchliff S, Merryn G (2016) Ageing and sexuality in Western societies: changing perspectives on sexual activity, sexual expression and the "Sexy" older body. In: Peel E, Harding R (eds) Ageing and sexualities: interdisciplinary perspectives, 1st edn. Ashgate, Surrey

Katz S (2010) Sociocultural perspectives on ageing bodies. In: The SAGE handbook of social gerontology. Sage, Los Angeles, pp 357–366. https://doi.org/10.4135/9781446200933.n27

Katz S, Marshall B (2004) Is the functional 'Normal'? Aging, sexuality and the bio-marking of successful living. Hist Hum Sci 17(1):53–75. https://doi.org/10.1177/0952695104043584

Lodge A, Umberson D (2013) Age and embodied masculinities: midlife gay and heterosexual men talk about their bodies. J Aging Stud 27(3):225–232. https://doi.org/10.1016/j.jaging.2013.03.004

Martin W, Twigg J (2018) Editorial for special issue ageing, body and society: key themes, critical perspectives. J Aging Stud 45:1–4. https://doi.org/10.1016/j.jaging.2018.01.011

Mencarini L (2014) Gender equity. In: Encyclopedia of quality of life and well-being research. Springer Netherlands, Dordrecht, pp 2437–2438. https://doi.org/10.1007/978-94-007-0753-5_1131

Pew Research Center (2015) Gay marriage around the world. Retrieved from https://www.pewforum.org/2017/08/08/gay-marriage-around-the-world-2013/

Richardson D (1998) Sexuality and citizenship. Sociology 32(1):83–100

Sandberg L (2018) Dementia and the gender trouble?: Theorising dementia, gendered subjectivity and embodiment. J Aging Stud 45:25–31. https://doi.org/10.1016/j.jaging.2018.01.004

Silver C (2003) Gendered identities in old age: toward (de)gendering? J Aging Stud 17(4):379–397. https://doi.org/10.1016/s0890-4065(03)00059-8

Sinnott J (1984) Older men, older women: are their perceived sex roles similar? Sex Roles 10(11–12):847–856. https://doi.org/10.1007/bf00288508

Tulle E (2015) Theorizing embodiment and ageing. In: Twigg J, Martin W (eds) The Routledge handbook of cultural gerontology. Routledge, London, pp 125–132

Twigg J (2004) The body, gender, and age: feminist insights in social gerontology. J Aging Stud 18(1):59–73. https://doi.org/10.1016/j.jaging.2003.09.001

Twigg J (2018) Dress, gender and the embodiment of age: men and masculinities. Ageing Soc 1–21. https://doi.org/10.1017/s0144686x18000892

UN Committee on the Elimination of Discrimination against Women (CEDAW) (2010) General recommendation no. 28 on the core obligations of states parties under Article 2 of the convention on the elimination of all forms of discrimination against women, 16 December 2010, CEDAW/C/GC/28

United Nations (2014) Women's rights are human rights. New York and Geneva: United Nations

Walker A (2009) The emergence and application of active aging in Europe. J Aging Soc Policy 21(1):75–93. https://doi.org/10.1080/08959420802529986

Walker A, Zaidi A (2016) New evidence on active ageing in Europe. Intereconomics 51(3):139–144. https://doi.org/10.1007/s10272-016-0592-0

Westwood S (2019a) Ageing, diversity and equality: social justice perspectives, 1st edn. Routledge, London

Westwood S (2019b) Heterosexual ageing: interrogating the taken-for-granted norm. In: Westwood S (ed) Ageing, diversity and equality: social justice perspectives, 1st edn. Routledge, London, pp 147–164

World Health Organization (WHO) (2002) Active ageing: a policy framework. World Health Organization, Geneva

World Health Organization (WHO) (2012) Strategy and action plan for healthy ageing in Europe, 2012–2020, 2012–2020. WHO Regional Office for Europe, World Health Organization, Copenhagen

Zaidi A, Stanton D (2015) Active ageing index 2014: analytical report. European Commission, Directorate General for Employment, Social Affairs and Inclusion, Geneva

Gender Issues

▶ Gender Equality in Later Life

Gender Issues in Age Studies

Susan Pickard
Department of Sociology, Social Policy and Criminology, School of Law and Social Justice, University of Liverpool, Liverpool, UK

Synonyms

Age system; Gender regime; Health inequalities; Intersectionality; Life course studies

Overview

Turning a conceptual and practical gaze on gender issues in later life is important for increasing our understanding of how both age and gender operate as hierarchical systems that furthermore work together to naturalize and obscure inequality. This requires understanding of the age system, including the life course framework, age ideology, and the age relations that comprise it. Women are disadvantaged in, and through, both the age and gender systems, and the effects of such deep-rooted and interlocking disadvantage comprise material and symbolic dimensions, including consequences for health and well-being through the life course.

What is Age Studies?

Age studies is the name given to the study of age as a system, which, like gender, is hierarchical and associated with inequality. As a system, it recognizes the importance of the part each stage plays in relation to the whole and the fact that, while specific inequalities are associated with each life stage, this cannot be disconnected from the ways in which each stage works together to shape inequality both synchronically and longitudinally, with material and ideological dimensions (Calasanti and Slevin 2006; Pickard 2016). Pioneers in this field include Margaret Morganroth Gullette (1997, 2004, 2011, 2018) and Kathleen Woodward (1999). Gullette points out, "age studies handles *age* in the same self-reflexive ways that have slowed down and enriched the analysis of cultural critics who use concepts of *gender* and *race*" (Gullette 2018: 255; original emphasis) recognizing its multi-stranded dimension. However, when compared to gender and critical race studies, age studies is a fledgling discipline, and more work is needed to theorize its relationship to "body-mind and narrative" as well as to neoliberal structures and ideology (Gullette 2004).

Key Research Findings

The Age System

Throughout western modernity, the age system has both regulated populations in line with the needs of capitalism, disciplining the workforce and latterly shaping consumer identities, and has served to inject meaning into an increasingly disenchanted existence (Cole 1992). The age system refers to this structure, which consists of a hierarchically constituted mode of governance, operating through the framework of the life course, in which the role of particular ages and their relationship to each other underpins and legitimizes an assortment of material and other inequalities (Calasanti and Slevin 2006; Gullette 1997; Pickard 2016). Age relations privilege prime of life adults, and the perspective is that of a jointly youthful and male gaze. Certain ideologies are associated with age relations including norms and practices. For example, both the younger and older populations have served as a reserve army of labor; unwillingness to invest in an older labor force either in terms of training or by providing the flexibility that would support (particularly female) workers' caring obligations in turn justifies poor conditions and underpins claims regarding their "natural" obsolescence.

The Life Course Framework

The central vehicle through which the age system works is that of the life course which plays a key role in organizing individuals and groups into seemingly natural age-based categories. A historical gaze reveals the degree to which the life course is not "natural" but rather a social construction that, since modernity, has shaped and regulated populations in particular ways. For example, the "invention" of youth as a stage of life defined in relation to education, symbolically constituted in terms of the "future of the nation," only began to take the shape it possesses today in the course of the late nineteenth and early twentieth centuries (Cunningham 2006). Similarly, within the same period, old age as a category was given meaning and substance in relation to the institution of retirement as well as the discourses of geriatric medicine that established a different ontology for older bodies (Katz 1996). More recently the latter part of the life course has been reconstituted and old age given an entirely different meaning (see section "Late Modern Life Course and Later Life: The Third and Fourth Ages").

In late modernity, the institutional life course has been replaced by a more fluid concept, and education and retirement as the "boxes" containing youth and old age, respectively, no longer apply with such determinacy. There are still norms operating which signal the "appropriate" timing for transitions and stages, including retirement (Kohli 2007). Furthermore, the life course is inherently gendered, with men's life courses comprising the "standard" (e.g., in employment terms) from which women are seen to "deviate." In addition, beneath the apparent fluidity lies a new structure, namely, that of neoliberalism, with an emphasis on individual productivity, success, and entrepreneurial qualities expected at all ages and stages and in all contexts, including unemployment, ill health, and retirement.

Late Modern Life Course and Later Life: The Third and Fourth Ages

The resultant "neoliberalization" of old age (Macnicol 2015) has been accompanied in later life by the construction of the "third" and "fourth" age categories, defined according to a distinction between biological (amenable to lifestyle and self-care practices) and chronological dimensions, with successful aging displaying biological and other characteristics of youthfulness (Rowe and Kahn 1987). Although in contemporary times these distinct categories have been shaped by medical discourses, they can be traced back as far as the nineteenth century where a split between a "good" and "bad" old age associates "virtue" with aging well (Cole 1992). In contemporary times such moral underpinnings remain alongside medical approaches, and the third age is characterized by levels of autonomy, youthfulness, consumerism/productivity, and health. By comparison the fourth age is defined as a loss of all these attributes. In a discourse shaped by age ideology, "choice" and agency determine third age or fourth age classification in which the nature and degree to which one ages become a matter of individual responsibility (Kirkwood 2008). In turn this justifies the end of mandatory retirement and the associated raising of the pension age together with individual responsibility for savings and pensions. This ideological distinction glosses over the fact that men and women in later life are positioned in very different ways (Lain et al. 2018). Indeed, where older people are increasingly taking low-level, low-paid jobs out of necessity to supplement their inadequate pensions, this situation of "precarity" particularly applies to divorced and single women (Lain et al. 2018).

Age Ideology

Age ideology, as an intrinsic element of the age system developed in association with the individualizing and civilizing processes of modernity and the Enlightenment, which emphasize autonomy, separation, and privatization and conversely express horror and disgust at their opposite qualities. It is based on a temporal narrative of growth, stability/stasis, and decline. This has both an intellectual/philosophical dimension in the Cartesian association of knowledge as "objective" and "transcendent" and a psychological dimension. The former embeds the perspective of normative gender (male) and age (youthful prime of life) in the definition of knowledge; the latter associates personhood with clear boundaries, autonomy, and transcendence of time, with aging threatening the loss of these attributes.

Moreover, age ideology is fundamentally gendered. The ideology works at all societal levels: so the "master narrative of decline" (Gullette 1997) which defines the process of moving through the life course, seeing aging as loss and deficit, is also reflected in the bildungsroman, the main form taken by the novel which itself sprung up with modernity. Expressing the symbolic dimension of the capitalist revolution, the *Bildungsroman* rejects the old and the past, focusing on the future and featuring a young (male) hero's social and developmental journey to (self) knowledge, stopping a point just before maturity. Youth becomes a synonym for progress, productivity, and value. Conversely, it is the gendered image of the reviled Hag that distils the negative essence of the age system, representing all that is feared and dreaded (Gullette 2017; Pickard 2016). This image crystallizes the double standard of aging, wherein women are considered to age faster than men and with loss of their sexual value and femininity. Indeed, it can be argued that all of western modernity's conceptual systems, from psychology and medicine to aesthetics, and literature, embed and naturalize a problematization not just of old age but of old women in particular. In its symbolic form of the hag, an image with a long genealogy in literature, art, and religious iconography (Creed 1993), aging femininity is depicted with an essentially dual nature: good or evil, pitiful or terrifying, as well as woman-as-monster and woman-as-victim (Creed 1993). This figure also appears as the Phallic Mother in Freudian thought, sometimes represented with both a penis and a vagina, the omnipotent mother who is the "'whole' in relation to which man is lacking" (Gallop 1982: 22). In Jungian philosophy of the archetypes, drawn from examples of ancient myth, the Great Mother has two aspects: the good nurturing mother and the terrible mother who "draws the life of the individual back into herself" (Neumann 1963: 71), the opposite of the regenerative principle, leading to sterility and death. This representation underpins "age war" discourses in which, in a wide array of media, older generations (and older women in particular) are represented as obstructing the progress and possibility of younger generations through their politics, their ideas, and their greed (Pickard 2018). However, the degree to which gender discrimination at this latter stage is fed by such images and tropes in the representational regime is as yet not fully recognized.

Gender Inequalities, Age, and Stage

Despite multiple social changes giving women political and legal equality, sociocultural change has been uneven. In global terms the fact that modernization does not necessarily lead to progress in gender equality is clear in the recent history of China, South Korea, and India (Gosh 2009; Campbell 2013). For example, in China the gender pay gap has widened significantly in the past decade, and the double standard of aging is used in a particularly violent way to underscore inequality, where urban professional women who are unmarried at the age of 27 and above are labelled "leftover women" – a term carrying considerable social stigma (Fincher 2016). In broader global terms, certain aspects of gender equality are more evident than others. For example, educational attainment is relatively advanced: on average 65% of girls and 66% of boys are in secondary education with 39% and 34%, respectively, in tertiary education. By contrast, political empowerment and economic participation and opportunity have the highest gender gaps, respectively; 58% of the economic participation and opportunity gap has been closed globally, while the figure for educational attainment is 95% (World Economic Forum (WEF) 2018).

Turning to the global north, gains made in recent times coexist with both old and new inequalities in a confusing mix. For example, girls' educational advantage does not continue on into the workplace. While the "mommy pay gap" is one of the most significant factors underpinning this (Gardiner 2017), there are broader cultural norms at work, and these are evident from the start. For example, high-achieving female graduates are at a direct disadvantage when it comes to hiring policies, and Quadlin found that "employers value competence and commitment among men applicants, but instead privilege women applicants who are perceived as likeable" – a fact which helps "moderate" female achievers over and above their high-earning female peers (Quadlin 2018). Looking to the very top of the gendered hierarchy, only 1 in 20 of the highest earners, or 1% households, are underpinned by women's earnings, and as the authors of this research state, "marrying a man with good income prospects is a woman's main route to the one per cent" (Yavorksky et al. 2019). Meanwhile, further down in the socioeconomic hierarchy, women remain hugely overrepresented in lower-paid jobs, including part-time work, with consequences in terms of poverty throughout the life course, and women are still predominantly responsible for both childcare and other adult care, as well as domestic work, including in more progressive regimes (Aboim 2010; Pfau-Effinger 1998). For women across the socioeconomic hierarchy, the gender pay gap widens with each child, indicating how particular practices and norms are entrenched in workplace and policy practices, as well as gendered norms and expectations between couples (Aboim 2010).

Age itself brings a number of specific inequalities to women. Women are working in greater numbers than ever before and are also now working later in life than in previous decades: for example, involvement in the labor market by UK women aged 50 and over increased by 3.2 percentage points between 1994 and 2014 (ONS 2015) accounting for 72% of women's employment growth during that time, while the biggest increase in employment rates over the past three decades has been for women aged 60–64 and 55–59 (ONS 2015; Brewis et al. 2017). This picture applies more generally both to Europe and Australia (Brewis et al. 2017). However, women over 50 are also most disadvantaged in terms of finding, retaining, and progressing in work as compared to both men their own age and younger women. Using UK evidence, the gender pay gap is twice as large for women in their 50s as it is for women overall (TUC 2013). This age group also has the highest proportion of carers with almost 1 in 4 women in this age group caring (Department for Work and Pensions 2015), and a relatively high proportion of workers in Sweden, Ireland, Germany, and the UK, among others, work part-time for this reason (Ní Léime and

Loretto 2017). Gendered norms similarly contribute to this. For example, looking at universities, far more men than women have chairs by the age of 50, and relatedly women are more likely to gain chairs over the age of 50 than below this age (HESA 2015). As well as achieving career success later than men, when they do finally achieve this, they are likely to find themselves limited by sexist and ageist attitudes: for example, interviews found a general managerial emphasis for women not on developing their own careers but on being more collegiate and taking on more academic housework, including mentoring more junior female colleagues and generally "passing the baton on" or making things "easier for future generations of women," looking behind themselves, as it were, rather than to the future (MacFarlane and Burg 2018).

Women are encouraged to blame "menopause" for many of their difficulties at mid-life, including in employment (Gullette 1997). However, there is evidence that older female workers are often not offered the training and support they need to negotiate technical change (Commission on Older Women 2013), while even in public sector organizations more likely to offer training, women are often not in jobs that qualify for such training. Additionally, older women managers are more likely to give older men training opportunities (Lossbroek and Radl 2018).

However, the "life course" as a framework for analyzing gendered experience partly obscures the overarching structures and everyday practices that disadvantage women; I discuss this next.

The Life Course Framework and Gender

Women bear most of the risk in terms of the late modern "flexibility" of the life course. Widmer and Ritschard (2009) found that it is women who bear the main burden of the destandardization of the life course; in terms of occupational trajectories, this relates to a fluctuating back and forth between part-time work and family care, the impact of which is greater later on in the life course. While for men it largely concerns the transition from education to paid work and is complete by around the age of 30, for women it increases after that point: "In other words," they note, "uncertainty has become a permanent state in women's occupational trajectories, while it is only transitional in men's occupational trajectories" (2009: 23). Four points can be made in relation to the gendered nature of the contemporary life course. Firstly, change has been mostly in one direction. For women, the changes have brought them closer in line with a "male" pattern, in terms of the centrality of employment, where previously they had defined themselves with regard to the home and caregiving roles and any work was seen as "supplementing" the breadwinner's main income. Men have not taken on caregiving or domestic roles to any significant extent, and while expanding their role in the labor economy, women have retained responsibility for caregiving. Indeed, the male breadwinner model remains embedded in workplace practices, policy norms, and cultural expectations around caring (Aboim 2010; Krüger and Levy 2001; Moen and Spencer 2006; Ní Léime and Loretto 2017). Moreover, both the values enshrined in the third age and the means to achieve it are more associated with male experience (Katz and Calasanti 2015). As a result of all these factors, the vast majority of part-time workers are female, and they are also more likely to leave the work force early, including in countries with gender-progressive norms and values (Aboim 2010). Secondly, and resultantly, women's structural disadvantages persist and indeed grow over the life course, regardless of their involvement in the labor economy, culminating in a significant pension gap across Europe and other developed states (Tinios et al. 2015). Thirdly, such gender differences continue to be present in the third age and indeed are increased at this point, where women (and men) moving into the third age "bring with them existing disparities and outmoded scripts about age and gender, producing asymmetries in power, resources, needs and preferences" (Moen and Spencer 2006: 140). This results in a greater likelihood of women experiencing both structural and ontological precarity in later life (Lain et al. 2018). Moreover, in that a greater proportion of sufferers of both Alzheimer's disease (AD) and frailty are women, the fourth age is predominantly a feminine life stage. Yet the "life course" as it is used in both policy and academic discourse is still the "male" life course, from which women are seen to deviate, a perspective that entirely overlooks the fact that the life course as a mechanism embeds these gender-based differences in its very fabric. For example, in employment terms, the narrative of linear progress to a peak followed by a clear-cut retirement process both obscures the temporal quality of the gender disadvantage and encourages further discrimination (e.g., underpinning the fact that women are less likely to be offered ongoing training and promotion than men). The assumptions contained in the framework also impact on the way data itself is captured: for example, the World Economic Forum's Global Gender Gap Report (2018), a compendious 355-page report, nowhere presents statistics that trace equality with age and stage but are rather "static" in their approach (an a-temporal bias that is embedded in social science disciplines more generally: see Pickard (2016)).

As noted, the nature of gender inequality is complex and, for multiple reasons, is hard to capture. However, one way through which the complexity of this inequality can be captured is through the lens of health, which offers a unique insight into the consequences and expressions of the gender regime (Annandale 2009; Pickard 2019b). At the same time, research into gendered health inequality often does not take into account temporality, age, and stage. In the next section, I trace some of the effects on health and well-being of gendered inequality, as it unfolds over the life course.

Gender and Health Inequalities Through the Life Course

Recent reports in the UK of a significant increase in anxiety and depression among teenage girls (Nuffield Foundation 2012) found that rates have doubled in the past 30 years, and more than a third of 14-year-old girls are reporting symptoms of distress (Lessof et al. 2016). This "slow-growing epidemic" (Kennedy 2016) has been linked to the high expectations of girls' educational success combined with lower instrumental coping skills (West and Sweeting 2003), pressures in family and personal life, the desire to "look good," peer pressures, sexual harassment at school, and the effect of social media, which continue into university (Darlington et al. 2011). Even as many girls excel at examinations, entering higher education at higher rates and performing better than men, overall there is ethnographic evidence that girls manage their school performance so that their ambition to achieve is tempered by their performance of femininity, such as not raising their hand in class and "looking" feminine (Renold and Allan 2006). Eating disorders and self-harm among young women are also on the rise, sometimes requiring hospitalization (Campbell 2016). Girls experience mental health problems at double the rate of teenage boys; white girls or mixed-race backgrounds are particularly likely to suffer, and there is evidence that middle-class girls and working-class girls suffer equal, if contrasting, symptoms (Patalay and Fitzsimmons, 2017). Although this is maintained throughout the life course, the gender gap in depression and anxiety is at its greatest during the reproductive years (Bird and Rieker 2008).

Health problems also cluster around mid-life. At this life stage, a wide variety of symptoms are attributed to menopause, from vasomotor symptoms, mood changes and fatigue, insomnia, anxiety, depression, and memory problems (Cheung et al. 2004). There are large individual differences in the nature, extent, and duration of such symptoms; many of these differences relate to prior experience of ill health in addition to socioeconomic and other disadvantages, with those suffering worst and longest having a string of other disadvantages, including low levels of education, previous anxiety and stress, poverty, and nonwhite ethnicity (Avis et al. 2015). Negative effects on mid-life women's quality of working life and performance at work include reduced engagement with work, reduced job satisfaction, reduced commitment to the organization, higher sickness absence, and an increased desire to leave work altogether (Brewis et al. 2017).

Gullette (1997) sees menopause as a "cultural consolidation," namely, an ideological framework within which women's bodies and lives are understood and as a result of which all kinds of (unrelated) symptoms and experiences are explained. These include, she notes, life events that are the effect of various discriminatory attitudes and which, culminating in ill health or unemployment at mid-life, are explained away by the label of "menopause." Here, not only workplace discrimination but the effect of the unequal division of labor at home will play its part too. For example, studies have found that women experiencing an unequal division of labor in terms of housework had significantly lower well-being and physical/psychosomatic symptoms than women with more equal relationships (Eek and Axmon 2015). This is the case even for Sweden (positioned at no. 1 in the Global Gender Gap Index, see World Economic Forum 2018) and indeed has been noted as exacerbated in the latter country where the divergence between official rhetoric and reality leads to frustration (Strandh and Nordenmark 2006).

In the latter part of the life course, the long-term results of inequality result in a significant gendered health gap. Indeed, health and poverty are directly linked in this stage with lower income associated with disability and lower self-assessed health (Arber and Cooper 1999). Older women have a higher number of limiting and non-limiting long-term conditions than men, spend a greater proportion of their lives in disability as compared to men, and are more likely to experience both physical ill health and poor mental health and cognitive decline (Allen and Sesti 2018). They are further disadvantaged by being more likely to live alone and thus not have a spousal carer on hand to help (Arber and Cooper 1999). Life expectancy for women is not increasing in the UK, and the trend is worse for women than for men. Not only are men closing the gap on women's life expectancy, they are adding healthy life years, thereby increasing the health gap. Women represent two-thirds of all persons with dementia (PWD) (Allen and Sesti 2018), and significantly more women than men are frail in all age groups over 65 (Age UK 2018). Women who have other disadvantages are more likely to suffer worse health, including frailty; they also fluctuate more between varied states of frailty than men (Kojima et al. 2019). Yet despite conditions such as frailty appearing unique to the "fourth age," they can be said to have their origin in structures, practices, and dispositions contained in gendered inequality throughout the life course (see Pickard 2019b, for a further discussion of the fundamental linking of frailty and femininity).

Future Directions of Research

The above discussion suggests that there are several lacunae in the field regarding gender issues in age studies which will need further exploration. More research globally is required on the uneven nature of progress as women gain equality in some areas but not others, together with the consequences of this in multiple dimensions of life. This includes the way inequality accumulates over time and with specific origins in discrete ages and stages, especially later life, together with the interplay between both. Research on alternative ways of conceptualizing the life course will also be very important in capturing inequality, particularly the gendered differences that are currently structured by the life course itself as it has evolved in late modernity. This research will also form the basis of further enquiry as to how gender and age work with other ascribed characteristics, such as race, disability, sexuality, and so on, both longitudinally and at specific points in the life course.

Summary

Foregrounding gender in age studies is important in order to demonstrate how age and gender work together to naturalize and obscure a range of inequalities. Furthermore, a temporal dimension is crucial to understanding gender as a structure, just as a gendered dimension is crucial to understanding the age system.

Cross-References

▶ Decline and Progress Narrative
▶ Feminist Theories and Later Life
▶ Gender Disparities in Health in Later Life
▶ Gender Equality in Later Life
▶ Gendered Aging and Sexuality in Audiovisual Culture

References

Aboim S (2010) Gender cultures and the division of labour in contemporary Europe: a cross-national perspective. Sociol Rev 58(2):171–196. https://doi.org/10.1111/j.1467-954X.2010.01899.x

Age UK (2018) Later life in the United Kingdom. https://www.ageuk.org.uk/globalassets/age-uk/documents/reports-and-publications/later_life_uk_factsheet.pdf

Allen J, Sesti F (2018) Health inequalities and women – addressing unmet needs. BMA, London

Annandale E (2009) Women's health and social change. Routledge, London

Arber S, Cooper H (1999) Gender differences in health in later life: the new paradox? Soc Sci Med 48:61–76. https://doi.org/10.1016/S0277-9536(98)00289-5

Avis NE et al (2015) Duration of menopausal vasomotor symptoms over the menopause transition. JAMA Intern Med 175(4):531–539. https://doi.org/10.1001/jamainternmed.2014.8063

Bird CE, Rieker PP (2008) Gender and health: the effects of constrained choices and social policies. Cambridge University Press, Cambridge

Brewis J, Beck V, Davies A, Mattheson J (2017) The effects of menopause transition on women's economic participation in the UK. Research Report. Department for Education. https://www.gov.uk/government/publications/menopause-transition-effects-on-womens-economic-participation

Calasanti TM, Slevin KF (eds) (2006) Age matters: realigning feminist thinking. Routledge, London

Campbell B (2013) End of equality: the only way is women's liberation. Seagull, London

Campbell D (2016) NHS figures show 'shocking' rise in self-harm among young. The Guardian, 23 October

Cheung AM, Chaudhry R, Kapral M, Jackevicius C, Robinson G (2004) Perimenopausal and postmenopausal health. BMC Womens Health 4(1):S23. https://bmcwomenshealth.biomedcentral.com/articles/10.1186/1472-6874-4-S1-S23

Cole TR (1992) The journey of life. Cambridge University Press, Cambridge

Commission on Older Women (2013) Interim report. London, House of Commons

Creed B (1993) The monstrous-feminine: film, feminism, psychoanalysis. Routledge, London

Cunningham H (2006) The invention of childhood. BBC, London

Darlington R, Margo J, Sternberg S with Burks BK (2011) Through the looking glass: teenage girls' self-esteem is more than skin-deep. Demos, London

Department of Work and Pensions and Baroness Altmann, CBE (2015) Older women see a dramatic rise in rate of employment over the past 30 years. Press Release. Available at https://www.gov.uk/government/news/older-women-see-a-dramatic-rise-in-employment-rate-over-past-30-years. Accessed 5 Jan 2016

Eek F, Axmon A (2015) Gender inequality at home is associated with poorer health for women. Scand J Public Health 43:176–182. https://doi.org/10.1177/1403494814562598

Fincher LH (2016) Leftover women: the resurgence of gender inequality in China. Zed Books, London

Gallop J (1982) Feminism and psychoanalysis: the daughter's seduction. Macmillan, London

Gardiner L (2017) Is the gender pay gap on the brink of closure for young women today? Available at http://www.resolutionfoundation.org/media/blog/the-genderpay-gap-has-almost-closed-for-millennial-women-but-it-comes-shooting-back-when-they-turn-30/. Accessed 5 Jan 2017

Gosh J (2009) Never done and poorly paid: women's work in globalising India. Women Unlimited, New Delhi

Gullette MM (1997) Declining to decline. University of Virginia Press, Charlottesville

Gullette MM (2004) Aged by culture. University of Chicago Press, Chicago

Gullette MM (2011) Agewise. University of Chicago Press, Chicago

Gullette MM (2017) Ending Ageism, or how not to shoot old people, Rutgers University Press, New Brunswick

Gullette MM (2018) Against 'aging' – how to talk about growing older. Theory Cult Soc 35(7–8):251–270

HESA (2015) Age and gender statistics for HE staff, February 26. Available at https://www.hesa.ac.uk/news/26-02-2015/age-and-gender-of-staff

Katz S (1996) Disciplining old age. University of Virginia Press, Charlottesville

Katz S, Calasanti T (2015) Critical perspectives on successful ageing: does it 'appeal' more than it 'illuminates. The Gerontologist 55(1):26–33

Kennedy M (2016) More than a third of teenage girls in England suffer depression and anxiety. The Guardian, 22 August

Kirkwood T (2008) Understanding ageing from an evolutionary perspective. J Intern Med 263:117–127. https://doi.org/10.1111/j.1365-2796.2007.01901.x

Kohli M (2007) The institutionalization of the life course: looking back to look ahead. Res Hum Dev 4(3–4):253–271. https://doi.org/10.1080/15427600701663122

Kojima G, Taniguchi Y, Illife S et al (2019) Transitions between frailty states among community-dwelling older people: a systematic review and meta-analysis. Ageing Res Rev 50:81–88. https://doi.org/10.1016/j.arr.2019.01.010

Krüger H, Levy R (2001) Linking life courses, work, and the family: theorizing a not so visible nexus between women and men. Can J Sociol 26(2):145–166. https://www.jstor.org/stable/3341676

Lain D, Airey L, Loretto W, Vickerstaff S (2018) Understanding older worker precarity: the intersecting domains of jobs, households and the welfare state. Ageing Soc:1–23. https://doi.org/10.1017/S0144686X18001253

Lessof C, Ross A, Brind R, Bell C, Newton S - TNS BRMB (2016) Longitudinal Study of Young People in England cohort 2: health and wellbeing at wave 2, Department for Education, UK. Available at https://assets.publishing.service.gov.uk/government/uploads/system/uploads/attachment_data/file/599871/LSYPE2_w2-research_report.pdf; Accessed 2 Jan 2019

Lossbroek J, Radl J (2018) Teaching older workers new tricks: workplace practices for gender differences in 9 European countries. Ageing Soc. https://doi.org/10.1017/S0144686X1800079X

MacFarlane B, Burg D (2018) Women professors as intellectual leaders. Leadership Foundation for Higher Education, London

Macnicol J (2015) Neoliberalising old age. Cambridge University Press, Cambridge

Moen P, Spencer D (2006) The gendered life course. In: Binstock R, George L, Cutler S, Hendricks J, Schulz J (eds) Handbook of ageing and the social sciences. Academic, San Diego, pp 127–144

Neumann E (1963) The Great Mother: an analysis of the archetype. Princeton University Press, Princeton

Ní Léime A, Loretto W (2017) Gender perspectives on extended working policies. In: Ní Léime A, Street D, Vickerstaff S, Krekula C, Loretto W (eds) Gender and extended working life: cross-national perspectives. Policy Press, Bristol, pp 53–75

Nuffield Foundation (2012) Changing adolescence briefing paper social trends and mental health: introducing the main findings. Available at www.nuffieldfoundation.org

Office for National Statistics (2015) Participation rates in the UK labour market: 2014, 19 March

Patalay P, Fitzsimons E (2017) Mental ill-health among children of the new century: trends across childhood with a focus on age 14. Centre for Longitudinal Studies, London

Pfau-Effinger (1998) Gender cultures and the gender arrangement – a theoretical framework for cross-national gender research. Innov Eur J Soc Sci Res 11(2):147–166

Pickard S (2016) Age studies: a sociological examination of how we age through the life course. Sage, London

Pickard S (2018) Age war as the new class war? Contemporary representations of intergenerational inequity. J Soc Policy. Available at https://doi.org/10.1017/S0047279418000521

Pickard S (2019a) Introduction. Embodying the gender regime: health, illness and disease across the life course. In: Pickard S Robinson J (eds) Ageing, the body and the gender regime: health, illness and disease across the life course. Routledge, London, to be published July 2019. (in press)

Pickard S (2019b) Femininity and frailty through the life course. In: Pickard S, Robinson J (eds) Ageing, the body and the gender regime: health, illness and disease across the life course. Routledge, London, to be published July 2019. (in press)

Quadlin N (2018) The mark of a woman's record: gender and academic performance in hiring. Am Sociol Rev 83(2):331–360. https://doi.org/10.1177/0003122418762291

Renold E, Allan A (2006) Bright and Beautiful: high-achieving girls, ambivalent femininities and the feminisation of success. Discourse 27(4):547–573. https://doi.org/10.1080/01596300600988606

Rowe JW, Kahn RL (1987) Human ageing: usual and successful. Science 237:143–149. https://doi.org/10.1126/science.3299702

Strandh M, Nordenmark M (2006) The interference of paid work with household demands in different social policy contexts: perceived work–household conflict in Sweden, the UK, the Netherlands, Hungary, and the Czech Republic. Br J Sociol 57(4):597–617. https://doi.org/10.1111/j.1468-4446.2006.00127.x

Tinios P, Bettio F, Betti G (2015) Men, women and pensions. Available online at http://ec.europa.eu/justice/gender-equality/files/documents/vision_report_en.pdf. Accessed 16 Jun 2018

TUC Economic and Social Affairs Department/Equality and Employment Rights Department (EERD) (2013) Older women and the labour market. TUC, London

West P, Sweeting H (2003) Fifteen, female and stressed: changing patterns of psychological distress over time. J Child Psychol Psychiatry 44(3):399–411. https://doi.org/10.1111/1469-7610.00130

Widmer E, Ritschard G (2009) The de-standardization of the life course: are men and women equal? Adv Life Course Res 14(1/2):28–39. https://doi.org/10.1016/j.alcr.2009.04.001

Woodward K (ed) (1999) Figuring age: women, bodies, generations. Indiana University Press, Bloomington

World Economic Forum (WEF) (2018) The global gender gap report 2018. WEF, Geneva. Available at https://www.weforum.org/reports/the-global-gender-gap-report-2018. Accessed 1 Feb 2019

Yavorksky JE, Keister LA, Qian Y, Nau M (2019) Women in the one per cent: gender dynamics in top income positions. Am Sociol Rev 84(1):54–81. https://doi.org/10.1177/0003122418820702

Gender Paradox

▶ Gender Equality in Later Life

Gender Paradox in Health

▶ Gender Disparities in Health in Later Life

Gender Pay Gap

▶ Wage Gap

Gender Regime

▶ Gender Issues in Age Studies

Gender Role in Caregiving

▶ Gender and Caring in Later Life

Gendered Aging and Sexuality in Audiovisual Culture

Barbara Zecchi
University of Massachusetts Amherst,
Amherst, MA, USA

Definition

Western audiovisual culture is male-dominated and youth-centric. In the context of the obsession with physical perfection and eternal youth of film and television productions, little or no visibility is given to the sex life of older characters, in particular of women, LGBTIQ subjects, racial minorities, and people with disabilities. The representation of sexuality in the years of midlife and beyond is rare and poses a thematic and aesthetic challenge.

Gender is one of the key factors in audiovisual culture's age-based discrimination. The entertainment industry marks asymmetrically the way male actors and female actors age. Male celebrities remain in the spotlight longer than women: "Men are usually credited with a longer plateau at their prime, whereas women climb the slope of social desirability more swiftly and are more rapidly thrown from its peak" (Gardiner 2001, p. 98). While older male actors continue to find parts that depict them as sexually active and appealing – and their much younger female counterparts are essential instruments to substantiate their desirability – older women are less visible and erotically uninteresting. If their sexuality is represented, it becomes the focus of criticism. However, exceptions to this norm abound in alternative film productions.

Overview

Audiovisual culture capitalizes on the representation of young white straight men who are sexually active. Older men's sexuality is also frequently depicted, but, in general, their age is made invisible and irrelevant by the plot. In Hollywood, silver foxes (interpreted by de-aged men such as George Clooney, Harrison Ford, Warren Beatty, Liam Neeson, or Tom Cruise, just to mention a few) are often paired with female partners who are half their age or even younger. Such an age discrepancy is not addressed, let alone problematized by the story. When the reverse occurs and older women romance younger men, the age gap plays a central role in the narrative and is often magnified and treated as pathology or immorality, such as, for example, in *The Mother* (2004).

Women of "a certain age" are generally represented as secondary figures "in relation to" the main characters – they are shadows without an identity, let alone a sexuality. In the rare instances in which they are instrumental to the plot, their sexuality becomes the object of compassion, derision, or disdain. Aging women's sexuality is either linked to their past or memories and exists only in their flashbacks (see, for instance, *Titanic* 1997); presented as wrong, out of place, and punishable (such as in *The Graduate* (1967), in *The Mother* (2004), or in *Ladies in Lavender* (2004), among others); or embodies the abject in films about Alzheimer's disease or other forms of dementia (as illustrated by Medina 2018), that is, films that profit from the spectacularization and morbidity of old-age related suffering (for instance, in British-American film *Iris* (2001)). In all these cases, women's sexuality is "aged by culture" and represented according to the "dominant narrative of decline" (see Gullette 2004). Older female characters are denied the pleasure of their bodies and are reduced to sexless human beings (who live in their past), nymphomaniacs (who live off excesses in the present), or quasi-corpses (who do not live at all).

Less frequent and under-examined initiatives in audiovisual practice disrupt these canonical modes of representing aging characters and pose alternatives to the hegemonic narratives that capitalize on the invisibility or stigmatization of later life sexuality. These alternative productions give agency to mature female characters and visibility to sexuality in later life. By departing from the abovementioned sexless, sex-addicted, or abject paradigms, they create powerful female figures who, despite old age and memory loss – and in many cases precisely because of them – grant depth and meaning to the aging female body. These narratives reclaim what Michelle Fine (1988, p. 54) has called "the missing discourse of desire" of the hegemonic audiovisual discourse.

There are at least four different paradigms of representation at work in these productions. A first group of independent audiovisual products attempt to eroticize and "spectacularize" the uncommon view of the sexually active mature woman. Despite age, the older female actor's body is represented as the glamorous and erotically charged object of heterosexual desire, usually the desire of a younger man. These depictions eroticize the mature woman by intentionally adopting the very same strategies that commercial cinema implements to represent and fetishize the younger female's body. The age issue is ignored by the plot. By making use of mainstream discourses, these films suggest that nothing changes in a woman's sex life after she has turned 50, 60, or even older. Conversely, these works aim to underscore that cultural norms on aging obscure a biological reality that predominantly affects men's sexual performance. A second group of films represents the mature woman's body as unglamorous as a means of denouncing the very stereotypes that have condemned it to scorn. In these films, the woman is not the object of a traditional male protagonist's scopophilia. The gaze of the camera neither embellishes nor conceals the older body; rather, it reveals and exhibits it as it is, often using a style of representation that is graphically and intentionally over-realistic. In spite of our antiaging culture, these texts argue that there is nothing wrong with becoming older. In a third group of films, the act of forgetting plays an important role in the recovery and rediscovery of sexual pleasure. Only when women can forget and overcome their past (personal memories of unhappy marriages, sexual violence, compulsory heterosexuality, that is, patriarchal norms that suppressed their sexual freedom) can they learn to live their sexuality and enjoy erotic pleasure. Either through a deliberate attempt to forget, or through a process of memory loss due to a disease that erases their past for them, women enter a new reality of disregarding conventions and ignoring restraint. Finally, in a fourth group of films, older women's sexuality is displaced and empathically recovered – and experienced – by younger women's imaginaries. These movies are not the product of what Marianne Hirsch has called a process of "post-memory" (i.e., "the relationship that the 'generation after' bears to the personal, collective, and cultural trauma of those who came before, to experiences they 'remember' only by means of the stories, images, and behaviors among which they grew up" Hirsch 2012, p. 5) but rather the result of the empathic projection of the bodily experiences of a younger woman onto an older one, without oversimplifying the differences between youth and old age, but rather dipping into the question of their interaction: towards a discourse on what Sandberg (2013) has defined as "affirmative aging" (the older body is different, but not in a way that depicts it as degraded by decline or loss).

Key Research Findings

A 2015 study conducted by *Time* magazine on the top 5,000 grossing movies since the beginning of Hollywood cinema reveals that male actors attain their career peak at the age of 46, while women reach their professional pinnacles at age 30, and that the age gap between the sexes is becoming wider every year.

The Agent 007 film series is a good example of how deeply ingrained and normalized this trend is and not only in Hollywood. Since his first appearance in *Dr. No* (1962), the British spy James Bond has aged and de-aged according to the different actors that embodied him: he was 58 when Roger Moore interpreted him in *A view to a Kill* (1985), 49 with Pierce Brosnan in *Die Another Day* (2002), and he will be 52 with Daniel Craig in *Bond 25* (2020). Throughout his 25 films, Agent 007 has interacted romantically with numerous women, often two or sometimes even three decades younger than him, namely, the "Bond girls." The character interpreted by the 50-year-old Italian star Monica Bellucci in *Spectre* (2015) became his oldest love interest. The fact that 007 could be attracted to a woman of a "certain age" was broadly reviewed by critics as a positive reversal of the sexist age gap and applauded as evidence that mature women could finally play strong and sexy roles in European film production. Bellucci called herself a "Bond woman," rather than a "Bond girl." However, Bellucci's very limited screen presence and her rather degrading role in *Spectre* proved quite the opposite. Her last appearance in the film in particular is humiliating and stereotypical. After a night of sex, crawling still half naked on the bed, she pleads with James Bond not to abandon her, but the womanizer leaves her behind and moves on to his next adventure with a woman 20 years younger. Despite all the publicity that proclaimed the opposite, *Spectre* maintains the sexist norm. The fate of the Italian beauty – a disposable object after a one-night stand with Bond – is similar to that of many other characters in hegemonic audiovisual productions who dare to think romantically about younger men. Even if men aren't much younger, the productions suggest that they are, and women's desire is condemned and/or pathologized. If Bellucci's character is depicted as a dependent and needy woman, in The *Graduate* (1967), the sexually unsatisfied Mrs. Robinson, interpreted by 36-year-old Anne Bancroft, is presented as an immoral and dissolute seductress who takes advantage of the innocence and inexperience of a young man, who could be her son. Like Bellucci, who is only 3 years older than Craig, Anne Bancroft is only 6 years older than Dustin Hoffman, the graduate. Almost 50 years separate *Spectre* from *The Graduate*. Clearly, not much has changed.

A recent trend of comedies with casts of all-star veterans constitutes only an apparent alternative to this status quo of gender/age inequality and an obsession with youth: TV series such as *Grace and Frankie* (2015) (featuring the for-ever-young Jane Fonda and Lily Tomlin); action films such as *Just Getting Started* (2017) (with Morgan Freeman, Tommy Lee Jones, and Rene Russo); *The Bucket List* (2007) (with Jack Nicholson and Morgan Freeman); and *Red* (2010) and its sequel *Red 2* (2013) (both starring Bruce Willis, John Malkovich, Helen Mirren, and, once again, Morgan Freeman) represent aging retirees who are still very active. Age is a central topic of these productions, and age-related goofiness often becomes the origin of the comedy puns. However, the exceptionality of the characters' lifestyles and of their actions does not challenge what Margaret Gullette has called the dominant narrative of decline but rather celebrates the ability of the older characters to maintain a youthful body and independent existence. These films follow the successful/active aging paradigm, in opposition to Sandberg's affirmative aging, mentioned earlier: old age is still inscribed as an undesirable stage of human life that the protagonists manage to avert and drive off.

The double standard of age/gender inequity is reversed in several different ways by numerous independent and international productions. For instance, some films feature older women having sex with younger men without presenting the reversal of traditional age difference as a central element to the plot. These films correspond to the first paradigm indicated above. One example is the Spanish road movie ¡Vámonos, Bárbara! (1978), a groundbreaking and pioneering work directed by Cecilia Bartolomé that focuses on a middle-aged woman's experience after her matrimonial separation, 3 years before the introduction of the divorce law in Spain. As in all road movies, the protagonist embarks on a journey that will have a transformative effect. But in this case, the driver is an older woman. Throughout her trip, Ana encounters several men – oftentimes younger than her – accidental partners whose presence has no other relevance than that of satisfying her sexual pleasure. In the sex scenes, by eroticizing the older female body and presenting it as "spectacle," Cecilia Bartolomé subverts traditional modes of representation that conventionally keep the mature woman out of sight. Such an exhibition has at least two functions: it exposes the body of a mature woman as an object of voyeurism, and, on the other hand, from the protagonist's subjective perspective, it offers a comment on the pleasure of such a display as the site of narcissism. This spectacle corresponds only apparently to the traditional male gaze's object denounced by Laura Mulvey (1975), since its objective is precisely that of giving visibility to the traditionally invisible woman. Active scopophilia – male voyeurism – is displaced into passive scopophilia, feminine narcissism, a pleasure of being-looked-at that could empower women's spectatorial gaze. Bartolomé reverses the traditional parameters of narration in commercial cinema (i.e., the male character leads the story, while the woman's presence halts the story) by using Ana's lovers as a pause in the action. These men fail as trip companions, let alone as life partners. Eventually Ana decides that she does not want to "return to be dependent on men" and that she "prefers to be alone." Conversely, Spanish films Nosotras (2000) and Flores de otro mundo (1999), among other examples of this first approach, ridicule older men wooing much younger women. In Nosotras, for instance, a middle-aged male character is depicted through non-flattering, nearly nude shots during a sex scene in which he believes he is pleasing a very young prostitute. A close-up of her facial expression, however, reveals that the woman is bored by his poor sexual performance. In Flores de otro mundo, on the other hand, the older male body remains covered. Carmelo brings his attractive Cuban fiancée, Milady, to the aging community of his village. When they first appear on the screen together, the medium shot places Carmelo's short and stout body in comparison with Milady's tall and slender figure. Her exuberant youthfulness differs dramatically from the vintage local atmosphere, emphasized by Carmelo's gray hair and the wrinkled and toothless faces of three old men who gaze in awe at the newcomer. When Carmelo takes her to his bedroom for the first time, it becomes apparent that Milady does not want to have sex with him, nor does she want him to be naked. Thus, she manages to make him reach a premature orgasm with the simple stimulation of his genitals through his pants, which is, ultimately, a statement about Carmelo's lack of ejaculatory control. These films can be seen as part of the trend that Molly Haskell has called "The revenge of the older woman" (2002) – that is films used to highlight or critique the fact that male and female aging are culturally marked in highly asymmetrical ways.

A completely different approach to the representation of sexuality in midlife and beyond is found in another group of films that depict the older body graphically, without concealing its age, with all its "defects." This group corresponds to the second paradigm highlighted above. In diverse national productions, such as La cama (Argentina, 2019), La vida empieza hoy (Spain, 2010), Wolke 9 (Germany, 2009), Mejor que nunca (Spain, 2009), and Magnolia (Colombia, 2012), contrary to its treatment in commercial cinema, such an imperfect body is not the object of scorn but rather the site of a new pleasure. If middle age means, according to patriarchal discourse, the loss of sex appeal and if menopause marks the "beginning of the end," in these audiovisual products, they represent, as Anna Freixas (et al.) puts it, the beginning of a "more to come tomorrow" (2012, p. 119). These films challenge the status quo in forms that are still strictly anchored to a heteronormative imaginary; nevertheless, they envision a female eroticism that often dispenses with men and the erect phallus. The protagonists are seen to enjoy sex after menopause through intercourse but also through masturbation, the most obvious manifestation of the separation of sex from procreation. For example, Juanita's gigantic strawberry-shaped dildo in *La vida empieza hoy* is an irreverent substitute for a penis. These films also attempt to counterbalance the inequity of cultural norms by stressing that biology has in fact favored women, who do not need an erection for intercourse and whose age is never an impediment to reaching an orgasm, with or without the need for a penis.

Memory is seen as an impediment to the complete sexual satisfaction of these female characters. If in mainstream audiovisual productions, as Medina (2018) has argued, people with Alzheimer's disease and other forms of dementia are represented as sexless zombies, in another group of films that illustrate the third paradigm indicated above, only when women can free themselves from their personal memories of gender-based violence that suppressed their sexual freedom can they learn how to live their sexuality and enjoy erotic pleasure. In *La vida empieza hoy*, the new widow Juanita has never experienced an orgasm; she does not even know *what* her clitoris is. After understanding that she has been sexually repressed during her entire life by her late husband, she requests a sort of "divorce postmortem": she informs the funeral home of her firm determination not to be buried near him in their family grave. Only then will she finally feel free from her past and be ready to learn how to enjoy her own newly discovered sexuality. In a similar fashion, in *Away from Her* (Canada, 2006), a woman with Alzheimer's disease disregards conventions and starts an adulterous relationship with a patient in the same nursing home, thus leaving her unfaithful husband wondering whether she is taking revenge against him by faking – or exaggerating – the symptoms of her disease.

Finally, the fourth paradigm is represented by Mexican films *No quiero dormir sola* (2012) and *Las buenas hierbas* (2010), among other works, that is, films that engage in a strategy of evoking a tactile eroticism that centers on the mature female body as an escape from the prominently heterosexual visual economy of commercial audiovisual productions. Even though they cannot be considered as lesbian films, these works dispense with the phallocentric and heteronormative paradigm of pleasure and challenge the myth that identifies female sexuality with genitality and vaginal penetration. Through the empathic projection of the bodily experiences of younger women onto older ones, without oversimplifying the differences between youth and old age, but rather dipping into the question of their interaction, these movies represent "affirmative aging bodies," that is, they "aim to acknowledge the material specificities of the ageing body [...] in terms of difference, but without understanding it as a body marked by decline, lack or negation" (Sandberg 2013, p. 12).

Future Directions of Research

This entry highlights the fact that unlike mainstream cinema, where glamorous girls are in the spotlight and aging female actors are virtually invisible, a considerable number of independent films put powerful (and often erotic) mature women at the forefront. However, lack of research in this area on films beyond Hollywood and British cinema is noticeable.

Summary

Unlike men, female film stars find it increasingly more difficult to be offered leading roles as they get older. The hegemonic audiovisual discourse is youth-centric and male-dominated, and aging women's sexuality is rarely addressed. When it is, it belongs to the successful aging regime, the neoliberal imperative to maintain a youthful body and remain sexually active as long as possible.

By resisting the hegemonic narratives of aging, despite their different approaches, many independent or international productions coincide in representing the older female body as the site/sight of sexual pleasure: these works reveal that, as Margaret Gullette (2004) has indicated, we are aged by culture. In essence, alternative filmmakers give voice to the unspoken and visibility to the invisible, thus contributing to the fight to free the female aging body from mainstream cultural impositions.

Cross-References

- Aging and The Road Movie
- Aging, Stardom, and "The Economy of Celebrity"
- Cougars and Silver Foxes in Film and TV
- Decline and Progress Narrative
- Dementia Narratives
- Silvering Screen
- The Youthful Structure of the Look

Acknowledgments In these pages, I elaborate and summarize some ideas that I have previously published in Fouz-Hernández, S., Spanish Erotic Cinema (Edinburgh University Press, Edinburgh 2017) and in Studies in Hispanic Cinema (2007).

References

Fine M (1988) Sexuality, schooling, and adolescent females: the missing discourse of desire. Harv Educ Rev 58(1):54–63
Freixas A, Luque B, Reina A (2012) Secretos y silencios en torno a la sexualidad de las mujeres mayores. In: Salinas A (ed) Estudios Etarios y relaciones intergeneracionales. MIC Género, Mexico, pp 117–127
Gardiner JK (2001) Masculinity studies and feminist theory. University of Columbia University, New York
Gullette MM (2004) Aged by culture. University of Chicago Press, Chicago
Haskell M (2002) Revenge of the Older Woman. The Guardian. https://www.theguardian.com/film/2002/jun/28/mollyhaskell.culture
Hirsch M (2012) The generation of postmemory: writing and visual culture after the holocaust. University of Columbia, New York
Medina R (2018) Cinematic representations of Alzheimer's disease. Palgrave Macmillan, London
Mulvey L (2009 [1975]) Visual pleasure and narrative cinema. In: Mulvey L (ed) Visual and other pleasures. Palgrave Macmillan, London, pp 14–27
Sandberg L (2013) Affirmative old age: the ageing body and feminist theories on difference. Int J Ageing Later Life 8(1):11–40

Filmography

A view to a kill (1985) Directed by John Glen, written by Richard Maibaum and Michael G. Wilson. UK
Away from her (2007) Directed and written by Sarah Polley. Canada
Bond 25 (2020) Directed and written by Cari Joji Fukunaga. USA
Die another day (2002) Directed by Lee Tamahori, written by Neal Purvis and Robert Wade. UK
Flores de otro mundo (1999) Directed by Iciar Bollaín, written by Iciar Bollaín and Julio Llamazares. Spain
Grace and Frankie (2015–) Written by: Marta Kauffman, David Budin, Brendan McCarthy. USA
Iris (2001) Directed by Richard Eyre, written by Richard Eyre and Charles Wood. UK
Just getting started (2017) Directed and written by Ron Shelton. USA
La cama (2018) Directed and written by Mónica Lairana. Argentina
La vida empieza hoy (2010) Directed by Laura Maña, written by Alicia Luna and Laura Mañá. Spain
Ladies in lavender (2004) Directed and written by Charles Dance. UK
Las buenas hierbas (2010) Directed and written by María Novaro. Mexico
Magnolia (2012) Directed and written by Diana Montenegro. Colombia
Mejor que nunca (2008) Directed and written by Dolores Payás. Spain
No quiero dormir sola (2012) Directed and written by Natalia Berinstáin. Mexico
Nosotras (2000) Directed by Judith Colell, written by Jordi Cadenas and Isabel-Clara Simó. Spain
The bucket list (2008) Directed by Rob Reiner, written by Justin Zackham. USA
The graduate (1967) Directed by Mike Nichols, Calder Willingham and Buck Henry. USA
The mother (2004) Directed by Roger Mitchell, written by Hanif Kureishi. UK
¡Vámonos, Bárbara! (1978) Directed by Cecilia Bartolomé, written by Cecilia Bartolomé, Sara Azcárate, and Concha Romero. Spain
Wolke 9 (2009) Directed by Andreas Dresen, written by Andreas Dresen, Jörg Hauschild, Laila Stieler and Cooky Ziesche. Germany
Zecchi B (2007) Women filming the male body: Subversions, inversions and identifications. Studies in Hispanic Cinema 3(3):187–204
Zecchi B (2017) Sex after fifty: The 'Invisible' female ageing body in spanish women-authored cinema, Spanish Erotic Cinema, Santiago Fouz-Hernandez, ed.,

Gendered Health Disparities

▶ Gender Disparities in Health in Later Life

Genderism

▶ Gender Equality in Later Life

Gene

▶ Alleles
▶ Chromosome

Gene Chip

▶ DNA Chip

Genealogies

▶ Family Tree

General Fluid/Crystallized Ability

▶ Intelligence (Crystallized/Fluid)

General Practitioner

▶ Family Medicine

Generalized Linear Mixed Effects Model

▶ Hierarchical Models

Generational Differences

▶ Intergenerational Solidarity

Generational Family

▶ Stem Family

Generational Migration

▶ Aging Refugees

Generational Resources

▶ Intergenerational Resource Tensions

Generational Stake

▶ Intergenerational Stake

Generativity and Adult Development

Holger Busch and Jan Hofer
Department of Developmental Psychology, Trier University, Trier, Germany

Definition

Generativity is defined as "the concern in establishing and guiding the next generation" (Erikson 1963, p. 240). It thus subsumes all goals and actions that intend to have a positive impact on future generations (McAdams and de

St. Aubin 1992), such as parenting, teaching, and mentoring. As this impact is ego-transcending in nature, Kotre (1996, p. 10) defines generativity as "a desire to invest one's substance in forms of life and work that will outlive the self."

Overview

The concept of generativity originates from Erik Erikson's (1963) theory of psychosocial development. This theory postulates eight developmental tasks, each of which characterizes a specific period in the life span. After having run through four childhood stages, in adolescence, people figure out who they are and want to be (i.e., develop an identity), then form lasting interpersonal commitments to a romantic partner (i.e., develop intimacy) in young adulthood, subsequently widen their radius of care to younger generations and the world at large (i.e., generativity), and finally review their lives in old age to come to a positive acceptance of the life they have lived (i.e., ego-integrity). Thus, according to Erikson, generativity is the defining developmental task of middle adulthood, which is often defined as beginning at 35 and ending at 65 years of age (e.g., Vaughan and Rodriguez 2013). However, the view of generativity as limited to this – albeit long – period has received a lot of criticism. For example, Vaillant and Milofsky (1980) have argued that due to the wide variety of potential generative behaviors, some forms of generativity precede others and can be found before middle adulthood while others become relevant in later periods in life and thus are shown in old age. Currently, generativity is rather seen as increasing in importance in midlife but no decline in importance is assumed in old age (Villar 2012).

The quality that develops with generativity is the care for the coming generations and for the legacy one leaves behind; in Erikson's words, generativity thus entails "'to care to do something', 'to care for' somebody or something, 'to take care of' that what needs protection and attention, and 'to take care not to' do something destructive" (Evans 1967, p. 53). Those who fail to develop such care remain in a state of stagnation or self-absorption which means they care only for themselves. Thus, they are at risk of not aging successfully and productively in that they do not engage in social and societal commitments (Erikson 1963).

In its beginning, research has employed various empirical approaches to studying generativity (such as interviews, single-case studies), with each approach highlighting various aspects of generativity. To systemize generativity research, McAdams and de St. Aubin (1992) introduced a model which distinguishes seven facets of generativity: *Cultural demand* and *inner desire* represent motivational sources of generativity. Whereas cultural demand refers to societal expectations concerning appropriate timing and outlets of generativity, inner desire is often seen in terms of implicit, that is unconscious, motives. Consisting of a need to be needed by others and a desire for symbolic immortality, the inner desire combines the motivational trends of communion and agency. These motivational sources feed a *generative concern*, which is the extent to which an individual is willing to care for future generations. Provided the individual holds an optimistic view of humanity's potential for positive development which Erikson called *belief in the species*, this investment is channeled into specific generative goals (*commitment*). In turn, generative goals energize *generative action*. Finally, an individual's life *narration* serves to provide generative efforts with a sense of meaning and embed generativity into a coherent narrative identity.

The model has sparked a lot of studies which have generally confirmed the predictions it makes (for an overview, see McAdams 2013). Moreover, the Loyola Generativity Scale (e.g., I try to pass along the knowledge I have gained through my experiences) and the Generative Behavior Checklist (e.g., Over the past 2 months, how often have you taught somebody about right and wrong, good and bad?), which McAdams and de St. Aubin (1992) have introduced along with their model, have become the gold standard self-report questionnaires for assessing generative concern and action, respectively.

Key Research Findings

The key research findings presented in the following will focus on the consequences of generativity. More specifically, they concern the behaviors that generativity can be expressed through. Here, parenting and grandparenting will be highlighted along with political and environmentalist activities. Furthermore, research will be reviewed that examines how generativity relates to well-being and the way individuals think about their past. Finally, some examples of how generativity might be applied in various contexts will be discussed.

Behavioral Expressions of Generativity

As has been stated previously, generativity can be expressed in a host of generative actions. In fact, various authors have suggested taxonomies of generative behaviors. Schoklitsch and Baumann (2011) distinguish six classes of generative actions: biological (i.e., the interest in passing on one's genes), parental (i.e., the interest in raising and educating one's children or grandchildren), technical (i.e., the interest in passing on specific skills), social (i.e., the interest in guiding younger people by passing on knowledge and values), cultural (e.g., the interest in making political contributions), and ecological (i.e. the interest in preserving the environment for future generations) generativity. The authors also present questionnaires to capture these generative aspects, which they validated specifically for older adults.

Parenting and Grandparenting

For Erikson, parenting is the prototypical generative behavior. It is, however, not the only behavioral expression of generativity so that also people whose desire for parenthood is thwarted can achieve generativity as Snarey et al. (1987) demonstrated in a sample of involuntarily childless men. In line with this finding, generativity is associated with well-being in parents as well as childless adults (Rothrauff and Cooney 2008). That is, parenting is an important but certainly not the only way to achieving generativity and can be compensated by other forms of generative behavior if need be.

Among parents, generativity relates to the way they behave towards their children; with more generative concern, parents tend toward a more authoritative parenting style (Peterson et al. 1997). That is, they blend demandingness, which is a combination of maturity demands and monitoring of children's behavior, with responsiveness, which is a combination of emotional warmth and autonomy support. The findings further suggest that authoritative parenting is the mechanism that connects parents' generative concern and their success in transmitting their attitudes to their children. Indeed, parents' generative concern predicts their children's positive affect, political interest, and attitude similarity with their parents (Peterson 2006; Peterson et al. 1997). Moreover, with more generative concern, parents also encourage a stronger involvement of grandparents in the socialization of their children, possibly because they view grandparents as important agents in the transmission of family traditions and lore (Pratt et al. 2008).

This is in line with Erikson's later writings (Erikson et al. 1986) in which he acknowledges the importance of generativity for old age, stating that grandparenthood provides a new opportunity for generativity. Correspondingly, older adults state that as grandparents they can pass on traditions, experiences, and knowledge to their grandchildren in a more conflict-free relationship than with their children (Warburton et al. 2006; cf. Erikson et al. 1986) and that they feel this exchange is beneficial for their grandchildren as well as themselves (Hebblethwaite and Norris 2011). Indeed, in a sample of grandmothers, Moore and Rosenthal (2014) found grandparental engagement to contribute to a sense of generative achievement. In turn, sense of generative achievement predicts women's successful aging composite score of life satisfaction, societal involvement, and general health over a 10-year interval (Versey et al. 2013). Similarly, generative concern predicts older people's satisfaction with their role as grandparent (Thiele and Whelan 2008). It thus seems that grandparenting is a suitable area for older people to express their generativity and experience it as rewarding.

Political and Environmentalist Activities

Political and environmentalist activities represent promising behavioral outlets for generativity because they allow individuals to leave future generations a world worth living in. It is thus not surprising that generative concern is positively associated with political interest and political contributions (Peterson 2006; Peterson and Duncan 1999). It is important to note, however, that the specific political causes that persons commit to are dependent on their general political orientation. More recently, it has been argued that preserving natural resources serves the well-being of future generations and hence benefits from people's motivation for generativity. Consistent with this argument, generative concern predicts the extent to which people save water and energy both at home and at their workplace (Wells et al. 2016). Research findings are thus in line with Schoklitsch and Baumann's (2011) notion of cultural and ecological generativity categories.

Well-Being

Generally, scholars agree that generativity should contribute to well-being; indeed, many psychological but also laypersons' (Fisher 1995) theories cite generativity as an integral part of successful aging. However, when specific aspects of the McAdams and de St. Aubin (1992) generativity model are investigated, findings consistently show that generative concern relates positively to well-being whereas generative behavior does not necessarily do so (e.g., Grossbaum and Bates 2002). An explanation may be found in the fact that generative actions require resources (Hofer et al. 2016): You have to invest time, energy, and sometimes also material resources. As McAdams (2013, p. 200) puts it: "Generativity is tough work." Similarly, examining generativity in the context of need satisfaction, Hofer et al. (2016) demonstrated that some needs might be more readily satisfied by generativity than others: Unlike the needs for relatedness and competence which benefit from generative action in that contributing to the thriving of future generations serves to connect generative individuals with others and provide them with a sense of achievement, the need for autonomy relates negatively with generative action. Oftentimes, generative action might be enacted in direct response to some request by juniors, so that generative individuals might feel they are compelled to engage in a form of generativity that is not of their own choice. Findings such as the one by Hofer et al. (2016) suggest that generativity does not come without costs.

Moreover, generative individuals cannot be sure if their generative efforts are appreciated by their intended recipients. The more people view their generativity as not valued, the less does their well-being benefit from generativity and the less they are inclined towards generativity in the future (Cheng 2009). That is, recipients' reactions to generative endeavors play a decisive role in whether generativity contributes to well-being. Similarly, the lack of what Erikson termed a belief in the species decreases the well-being people can derive from their generativity: In individuals with Machiavellian attitudes, i.e., the conviction that others cannot be trusted because they would readily exploit one's efforts, generativity is unrelated to purpose in life, whereas there is a positive relation in those low in Machiavellianism (Busch and Hofer 2012). Taken together, research thus suggests that although generativity is generally associated with well-being, there are internal and external obstacles that can lessen the well-being benefits that generative individuals reap.

Despite such restrictions, research converges towards the conclusion that, generally speaking, generativity does indeed contribute to well-being in older adults. For instance, generative behavior is linked to meaning in life in older adults from Cameroon, the Czech Republic, Germany, and Hong Kong (Hofer et al. 2018). Moreover, as has been stated before, Versey et al. (2013) found a sense of generative achievement to be positively associated with successful aging. Similarly, Gruenewald et al. (2012) showed that more self-perceived generativity is associated with a smaller increase in activities of daily living disabilities over a 10-year period. Also, older individuals' failure to meet their self-set generativity standards across this 10-year period is predictive of lower life satisfaction, with decreased positive affect, self-worth, and social connectedness mediating the relationship (Grossman and Gruenewald 2018).

Finally, an indicator of well-being that is argued to be specific to old age, however, is (a relative absence of) the fear of death. From the perspective of terror management theory, it has been argued that generativity is motivated by the attempt at warding off the fear of death. An empirical test of this hypothesis had indeed found that generative concern was increased in older adults when they had been reminded of their mortality compared to older adult who had not as well as to younger adults in both experimental conditions (Maxfield et al. 2014). This finding is compatible with the inner desire for symbolic immortality which is a motivational source in McAdams and de St. Aubin's (1992) generativity model. More recently, however, Busch et al. (2018) demonstrated that generativity does not have a direct effect on the fear of death. Generativity is related to a decreased fear of death in that it is associated with ego-integrity which in turn predicts a reduced fear of death. Hence, it is ego-integrity which is defined as the acceptance of one's one and only life-course that is directly linked to the absence of fear of death, and generativity serves to develop ego-integrity.

How and Why Generative People Think About Their Lives

In their generativity model, McAdams and de St. Aubin (1992) postulated that generativity would eventually be incorporated into individuals' narrative identity. Such narrative identities are the corpus of stories that people tell about themselves and that serve to integrate memories and life-episodes into a meaningful whole. McAdams (e.g., 2013) upholds that, specifically, compared to less generative adults, highly generative adults remember more (a) having been privileged during childhood, (b) having been sensitive to the suffering of other people, (c) having established firm moral values by adolescence which have guided their lives from then on, and (d) seeing how even negative life-events can eventually have positive outcomes. Moreover, these stories are characterized by (e) a balance between agency and communion and (f) prosocial goals that the person wants to act upon in the future. In sum, generative adults remember having been advantaged in comparison to others and want to give something back to society. Their moral values and optimism that the generative effort will come to a good end eventually support their generative endeavors.

Generativity is, however, related not only to the structure of autobiographical memories as illustrated by the commitment stories delineated above, it reflects also in the reasons why people think back to their past. Reasons for remembering the past, called reminiscence functions in the corresponding research literature, can be manifold. Two of these reminiscence functions, namely thinking back to one's past to, first, instruct others and, second, prepare for death, motivate generative behavior in older adults (Hofer et al. 2018). That is, older individuals might use the reminiscence function teach/inform to respond to generative demands such as requests for specific advice. The reminiscence function of death preparation might be an attempt at creating a self-transcending legacy by sharing the memories of one's life with others. Thus, generative individuals seem to remember their lives in a specific way and also use these memories to initiate generative behavior.

Applied Generativity Research

Research on applications of generativity is a relatively recent phenomenon. Nevertheless, generativity has been found to be relevant in various contexts. As can be seen, the majority of these proposed applications ties in with the major research findings delineated above.

Considering clinical settings, generativity is an integral aspect of dignity therapy (Chochinov et al. 2005) which was designed to alleviate psychological distress in terminally ill patients: Based on patients' autobiographical narrations, a generativity document is written which patients can hand on as a self-transcending legacy. Moreover, to foster older adults' well-being, programs have been developed that seek to provide older people with opportunities to engage in generative intergenerational interactions. Research shows that such programs do indeed increase generative concern (e.g., Ehlman et al. 2014) or a sense of generative achievement in their participants (e.g., Gruenewald et al. 2016).

Apart from such clinical interventions, generativity can also be applied in an organizational context. For example, Zacher et al. (2011) showed that team members judged older leaders as less effective leaders when their leader generativity was low, with leader generativity defined as the willingness to renounce one's own career benefits to support members of the younger generation at work in establishing their own careers. That is, to help older superiors to maintain leader effectiveness, organizations might benefit from fostering leader generativity in older team leaders.

Future Directions of Research

As outlined above, for a long time, generativity has not been examined systematically and has received much less attention in psychological research than other developmental stages in Erikson's theory. It is, thus, not surprising that there are some open questions that have not yet been addressed. Among these are questions that concern further applications of generativity, if and how generativity can be experimentally induced, and how generativity relates to individual differences in the perception of time, such as, e.g., time orientation (i.e., how much one thinks of the past, present, or future). In the following, two specific future direction of generativity research will be elaborated upon: first, generativity and culture because cross-cultural insight into generativity not only broadens the generalizability of scientific statements about generativity but might also be useful in fostering intercultural intergenerational contact in a globalized world, and second, the role of the recipient in the generative process because the recipient's response is crucial in determining if generative endeavors are successful or not, thus determining the consequences for the generative individual.

Generativity and Culture

All cultures rely on individuals to pass on the spirit of their culture: their values, convictions, codes, and stories. Generativity can be assumed to play a pivotal role in motivating the corresponding behavior. Consequently, one would expect generativity to be cherished in all cultures (McAdams 2013). Accordingly, the McAdams and de St. Aubin (1992) generativity model assumes cultural demand as a motivational source of generativity. That is, individuals are expected to behave generatively in culturally adequate way at some culturally appropriate point in their lives.

Seeing generativity as a desirable developmental outcome across cultures implies that generativity also has comparable consequences across cultures. In fact, some of the studies reviewed above investigating the relation between generative concern and well-being have done so cross-culturally and found comparable effects (e.g., Busch et al. 2018; Hofer et al. 2018); as of now, effects of the generative recipient's reactions have exclusively been studied in Asian contexts (e.g., Cheng 2009). In a rare attempt at examining this specific aspect of the generativity model, cultural demand indeed positively affected generative concern in older adults from Cameroon, China (Hong Kong), the Czech Republic, and Germany (Hofer et al. 2016). In this study, cultural demand was assessed in the form of self-transcendence values, which subsume values concerning the welfare of close interaction partners, people in general, and nature.

However, the mere fact that all cultures require generativity and that cross-cultural equivalence in the correlates of generativity has been established for some variables do not preclude cross-cultural differences concerning generativity. For example, generative concern seems to be higher in more collectivist compared to individualistic cultures (Hofer et al. 2008). Moreover, which specific acts are considered appropriate in a given culture might well differ; in individualist cultures, creating a legacy to achieve symbolic immortality might be viewed more favorably than in collectivist cultures. Cross-cultural generativity research is thus needed to generate more knowledge on which behaviors are seen generative, who is allowed to be generative in specific ways, and

The Role of the Recipient

So far, generativity research has been a rather one-sided affair: Although the beneficial effect of generativity on well-being should be seen for both the recipient of the generative effort and the generative individual (cf. Hebblethwaite and Norris 2011), to date, only little is known about generativity's effect on recipients; the extant research, however, indicates that adult children experience more positive affect the more generative concern their parents express (Peterson 2006) and that via authoritative parenting, more generative concern in parents is associated with a stronger perception in adult children that they have similar attitudes as their parents (Peterson et al. 1997).

Still, the generative individual has been at the center of research, whereas the recipients of generative efforts have rarely been considered (e.g., Peterson 2006; Peterson et al. 1997) despite clear evidence that their reaction is crucial: As stated above, more perceived respect for generative action facilitates generative concern over a 1-year interval (Cheng 2009). On the other hand, more perceived rejection of generative efforts seems to hamper generative concern 12 months later (Tabuchi et al. 2015).

However, generativity research to date has been mute on what characteristics such reactions depend on; certainly, sometimes what well-meaning seniors intend to be generative is perceived as irrelevant or outdated by their juniors, which bears the potential for intergenerational conflicts. How can such conflicts be resolved without damaging adults' willingness to act generatively or prevented in the first place? Furthermore, are there some characteristics of recipients that make them more likely to appreciate the generative efforts of others? For example, adolescents who are currently on the search for their identity and have not made firm identity commitments yet might be particularly grateful for others' generativity.

Additionally, it is not always clear who the intended recipient of generative efforts is. Biological and parental generativity clearly are directed towards the generative individual's child but who is chosen as pupil or mentee? As Kai Erikson (2004) points out, generative efforts aimed at benefitting one group of recipients might have detrimental effects on others. Erikson (1982) described the exclusion of others from one's care as rejectivity and went on to discuss the danger that lies in rejecting others as unworthy of one's generativity due to prejudice. Thus, in the choice of the recipients of generativity, in- vs. out-group effects might come to bear. Future generativity research might want to empirically address this issue.

Summary

Taken together, generativity describes the desire to pass something on to future generations, creating a self-transcending legacy from which future generations might benefit. It is a developmental task that is relevant across the entire adulthood: not, as originally assumed, only in midlife but in old age, too. It can find behavioral expression in a wide array of behaviors. Research has shown that – with certain restrictions – generativity pays off for generative individuals in that it is associated with well-being. An important limitation is the generative recipient's reaction; the benefit for the generative individual seems to increase with the recipient's appreciation of the generative effort. A major challenge for future generativity research will be to learn more about how culture shapes generativity and the effects of generativity on the recipient. More knowledge in this area might make practical applications of generativity more fruitful.

Cross-References

- ▶ Aging Well
- ▶ Death Anxiety
- ▶ Indigenous Cultural Generativity: Teaching Future Generations to Improve Our Quality of Life
- ▶ Intergenerational Programs
- ▶ Productive Aging
- ▶ Terror Management Theory and Its Implications for Older Adults

References

Busch H, Hofer J (2012) Self-regulation and milestones of adult development: intimacy and generativity. Dev Psychol 48:282–293. https://doi.org/10.1037/a0025521

Busch H, Hofer J, Poláčková Šolcová I et al (2018) Generativity affects fear of death through ego integrity in German, Czech, and Cameroonian older adults. Arch Gerontol Geriatr 77:89–95. https://doi.org/10.1016/j.archger.2018.04.001

Cheng T-S (2009) Generativity in later life: perceived respect from younger generations as a determinant of goal disengagement and psychological well-being. J Gerontol Psychol Sci 64:45–54. https://doi.org/10.1093/geronb/gbn027

Chochinov HM, Hack T, Hassard T et al (2005) Dignity therapy: a novel psychotherapeutic interventions for patients near the end of life. J Clin Oncol 23:5520–5525. https://doi.org/10.1200/JCO.2005.08.391

Ehlman K, Ligon M, Moriello G (2014) The impact of intergenerational oral history on perceived generativity in older adults. J Intergener Relatsh 12:40–53. https://doi.org/10.1080/15350770.2014.870865

Erikson EH (1963) Childhood and society, 2nd edn. Norton, New York

Erikson EH (1982) The life cycle completed. Norton, New York

Erikson K (2004) Reflections on generativity and society: a sociologist's perspective. In: de St. Aubin E, DP MA, Kim T-C (eds) The generative society: caring for future generations. American Psychological Association, Washington, DC, pp 51–61

Erikson EH, Erikson JM, Kivnick HQ (1986) Vital involvement in old age. Norton, New York

Evans RI (1967) Dialogue with Erik Erikson. Harper & Row, New York

Fisher BJ (1995) Successful aging, life satisfaction, and generativity in later life. Int J Aging Hum Dev 41:239–250. https://doi.org/10.2190/HA9X-H48D-9GYB-85XW

Grossbaum MF, Bates GW (2002) Correlates of psychological well-being at midlife: the role of generativity, agency and communion, and narrative themes. Int J Behav Dev 26:120–127. https://doi.org/10.1080/01650250042000654

Grossman MR, Gruenewald TL (2018) Failure to meet generative self-expectations is linked to poorer cognitive-affective well-being. J Gerontol: Psychol Sci. https://doi.org/10.1093/geronb/gby069

Gruenewald TL, Liao DH, Seeman TE (2012) Contributing to others, contributing to oneself: perceptions of generativity and health in later life. J Gerontol: B 67:660–665. https://doi.org/10.1093/geronb/gbs034

Gruenewald TL, Tanner EK, Fried LP et al (2016) The Baltimore experience corps trial: enhancing generativity via intergenerational activity engagement in later life. J Gerontol: B 71:661–670. https://doi.org/10.1093/geronb/gbv005

Hebblethwaite S, Norris J (2011) Expressions of generativity through family leisure: experiences of grandparents and adult grandchildren. Fam Relat 60:121–133. https://doi.org/10.1111/j.1741-3729.2010.00637.x

Hofer J, Busch H, Chasiotis A et al (2008) Concern for generativity and its relation to implicit pro-social power motivation, generative goals, and satisfaction with life: a cross-cultural investigation. J Pers 76:1–30. https://doi.org/10.1111/j.1467-6494.2007.00478.x

Hofer J, Busch H, Au A et al (2016) Generativity does not necessarily satisfy all your needs: associations among cultural demand for generativity, generative concern, generative action, and need satisfaction in the elderly in four cultures. Dev Psychol 52:509–519. https://doi.org/10.1037/dev0000078

Hofer J, Busch H, Au A et al (2018) Reminiscing to teach others and to prepare for death is associated with meaning in life through generative behavior in elderlies from four cultures. Aging Ment Health. https://doi.org/10.1080/13607863.2018.1548568

Kotre J (1996) Outliving the self: how we live on in future generations, 2nd edn. Norton, New York

Maxfield M, Greenberg J, Pyszczynski T et al (2014) Increases in generative concern among older adults following reminders of mortality. Int J Aging Hum Dev 79:1–21. https://doi.org/10.2190/AG.79.1.a

McAdams DP (2013) The positive psychology of adult generativity: caring for the next generation and constructing a redemptive life. In: Sinnott JD (ed) Positive psychology: advances in understanding adult motivation. Springer, New York, pp 191–205

McAdams DP, de St. Aubin E (1992) A theory of generativity and its assessment through self-report, behavioral acts, and narrative themes in autobiography. J Pers Soc Psychol 62:1003–1015. https://doi.org/10.1037/0022-3514.62.6.1003

Moore SM, Rosenthal DA (2014) Personal growth, grandmother engagement and satisfaction among non-custodial grandmothers. Aging Ment Health 19:136–143. https://doi.org/10.1080/13607863.2014.920302

Peterson BE (2006) Generativity and successful parenting: an analysis of young adult outcomes. J Pers 74:847–869. https://doi.org/10.1111/j.1467-6494.2006.00394.x

Peterson BE, Duncan LE (1999) Generative concern, political commitment, and charitable actions. J Adult Dev 6:105–118. https://doi.org/10.1023/A:1021620824878

Peterson BE, Smirles KA, Wentworth PA (1997) Generativity and authoritarianism: implications for personality, political involvement, and parenting. J Pers Soc Psychol 72:1202–1216. https://doi.org/10.1037/0022-3514.72.5.1202

Pratt MW, Norris JE, Hebblewaith S et al (2008) Intergenerational transmission of values: family generativity and adolescents' narratives of parent and grandparent value teaching. J Pers 76:171–198. https://doi.org/10.1111/j.1467-6494.2007.00483.x

Rothrauff T, Cooney TM (2008) The role of generativity in psychological well-being: does it differ for childless adults and parents? J Adult Dev 15:148–159. https://doi.org/10.1007/s10804-008-9046-7

Schoklitsch A, Baumann U (2011) Measuring generativity in older adults: the development of new scales. GeroPsych 24:31–43. https://doi.org/10.1024/1662-9647/a000030

Snarey J, Son L, Kuehne VS et al (1987) The role of parenting in men's psychosocial development: a longitudinal study of early adulthood infertility and midlife generativity. Dev Psychol 23:593–603. https://doi.org/10.1037/0012-1649.23.4.593

Tabuchi M, Nakagawa T, Miura A et al (2015) Generativity and interaction between the old and young: the role of perceived respect and perceived rejection. The Gerontologist 55:537–547. https://doi.org/10.1093/geront/gnt135

Thiele DM, Whelan TA (2008) The relationship between grandparent satisfaction, meaning, and generativity. Int J Aging Hum Dev 66:21–48. https://doi.org/10.2190/AG.66.1.b

Vaillant GE, Milofsky E (1980) Natural history of male psychological health: IX. Empirical evidence for Erikson's model of the life cycle. Am J Psychiatr 137:1348–1359. https://doi.org/10.1176/ajp.137.11.1348

Vaughan MD, Rodriguez EM (2013) The influence of Erik Erikson on positive psychology theory and research. In: Sinnott JD (ed) Positive psychology: advances in understanding adult motivation. Springer, New York, pp 231–245

Versey HS, Stewart AJ, Duncan LE (2013) Successful aging in late midlife: the role of personality among college-educated women. J Adult Dev 75:63–75. https://doi.org/10.1007/s10804-013-9157-7

Villar F (2012) Successful ageing and development: the contribution of generativity in older age. Ageing Soc 32:1087–1105. https://doi.org/10.1017/S0144686X11000973

Warburton J, McLaughlin D, Pinsker D (2006) Generative acts: family and community involvement in older Australians. Int J Aging Hum Dev 63:115–137. https://doi.org/10.2190/9TE3-T1G1-333V-3DT8

Wells VK, Taheri B, Gregory-Smith D et al (2016) The role of generativity and attitudes on employees home and workplace water and energy saving behaviours. Tour Manag 56:63–74. https://doi.org/10.1016/j.tourman.2016.03.027

Zacher H, Rosing K, Henning T et al (2011) Establishing the next generation at work: leader generativity as a moderator of the relationship between leader age, leader-member exchange, and leadership success. Psychol Aging 26:241–252. https://doi.org/10.1037/a0021429

Generosity

▶ Altruism

Genes that Delay Aging

Helena M. Hinterding and Joris Deelen
Max Planck Institute for Biology of Ageing, Cologne, Germany

Synonyms

Age-related genes; Longevity genes; Longevity-related genes

Definition

Genes that delay aging are genes that have shown to be associated with lifespan and/or longevity in animal models and/or humans when altered (e.g., overexpressed or knocked down).

Overview

Aging is to a minor, yet significant, extent genetically determined. Moreover, the genetic component of aging appears, at least partially, to be evolutionary conserved across species. Hence, to unravel this genetic component, research in model organisms, i.e., yeast (*Saccharomyces cerevisiae*), worms (*Caenorhabditis elegans*), fruit flies (*Drosophila melanogaster*), and mice (*Mus musculus*), has proven to be highly valuable due to the relatively easy genetic manipulation of these organisms in comparison to humans. The overexpression or deletion of genes (knockout) in these model organisms has resulted in the identification of several signaling pathways involved in nutrient sensing that seem to play a central role in aging. However, the complexity of aging is reflected by the fact that hundreds of genes seem to influence lifespan in yeast, worms, and flies, but most of them do not show any effect in mice, let alone humans. In this entry, we will provide an overview of the most important genes and pathways that have been shown to influence aging across species.

Key Research Findings

Evolutionary Conserved Aging-Associated Pathways

Over the past three decades, research in model organisms has highlighted a role for evolutionary conserved nutrient-sensing pathways in aging (Fontana et al. 2010). The most important of which are (1) the insulin/insulin-like growth factor 1 (IGF-1) signaling (IIS) pathway, (2) the mammalian target of rapamycin (mTOR) signaling pathway, and (3) the mitogen-activated protein kinase (MAPK)/extracellular signal-regulated kinase (ERK) signaling pathway.

Insulin/Insulin-Like Growth Factor 1 Signaling Pathway
Undoubtedly, the IIS pathway constitutes a crucial piece in the yet to be fully comprehended puzzle of the genetics of aging. In fact, most of the thus far identified genes involved in aging are part of this complex pathway that consists of several sub-branches (see below). This hormonally regulated pathway responds to insulin and insulin-like hormones that bind to their corresponding receptors and consists of various downstream targets and effectors, mainly involved in cell growth and metabolism. Reducing the physiological and anabolic functions of the IIS pathway in cell growth and nutrient regulation through gene mutations or dietary restriction has been shown to consistently extend lifespan in different model organisms.

The first ever gene to be discovered in relation to increased lifespan in any species, *age-1* (Klass et al. 1983), is one of the main effectors in this pathway and encodes the mammalian orthologue of phosphatidylinositol-3-OH kinase (PI(3) K) catalytic subunits (Morris et al. 1996). The role of IIS in the lifespan regulation of worms was further supported by the discovery that *daf-2* mutant worms lived twice as long as their wildtype counterparts (Kenyon et al. 1993). *Daf-2* encodes the insulin-like growth factor 1 receptor (IGF1R) that is located upstream in the IIS pathway and is well conserved throughout different species. The observed lifespan extension of *daf-2* mutant worms was, in turn, shown to be dependent on the activity of *daf-16*, the orthologue of the mammalian forkhead box protein O (FOXO) transcription factors. Both *daf-2* and *daf-16* are also known for their protective role against oxidative and thermal stress, which might be at least partially underlying this longevity phenotype. After its initial discovery in the worm, it turned out that genetic manipulation of the IIS pathway was also able to extend lifespan in flies (by knockout of IIS-related genes, such as *chico* (encoding the insulin receptor substrate), *Inr* (encoding the insulin-like receptor), *Ilp1-7* (encoding insulin-like ligands), and *foxo* (Partridge et al. 2011)) and even in mice (i.e., by means of knockout of *Irs-1* and *Igf1r* (Holzenberger et al. 2003; Selman et al. 2008) (Table 1). Genetic studies in humans have shown that genetic variation in *FOXO3*, one of the downstream targets within the IIS pathway, is one of the few genes associated with human aging (Flachsbart et al. 2009; Willcox et al. 2008). Moreover, the combined genetic variation in the IIS pathway has also been associated with human aging (Deelen et al. 2013).

Mammalian Target of Rapamycin Signaling Pathway
The mTOR signaling pathway occupies an indispensable role in today's biomedical research and is widely studied, spanning, among others, the fields of cancer, immunological, neurological, and aging research. mTOR belongs to the PI3K-related kinase family and comes in the form of two protein complexes, mTORC1 and mTORC2, that regulate cell growth and are thereby crucial for overall physiology in many organisms: nowadays we know that mTORC1 mediates mostly anabolic processes such as protein, lipid, and nucleotide synthesis by phosphorylating RPS6K1 and other downstream effectors while at the same time blocking catabolic processes such as autophagy.

Orthologues of mTORC-related genes and RPS6K1 were shown to extend lifespan across species (Johnson et al. 2013) (Table 1). Moreover, rapamycin, which was named after the island Rapa Nui on which it was discovered in 1964 (Vézina et al. 1975) and is a natural inhibitor of mTORC1, is one of the few compounds that is known to extend lifespan across species (Fontana et al. 2010). Interestingly, the combined genetic

Genes that Delay Aging, Table 1 Lifespan-extending genes involved in the IIS, mTOR, and MAPK/ERK signaling pathways, grouped per species

	Yeast	Worm	Fly	Mouse	Human
IIS	GPR1	age-1	Ilp2,3,5	Irs1	FOXO3
		daf-2	Inr	Igf1r	
		akt-1	Lnk		
		daf-16	chico		
			14-3-3		
			foxo		
mTOR	SCH9	let-363	Tsc1	mTOR	
	TOR1	daf-15	gig	Rictor	
		rict-1	Tor	Rps6k1	
		sinh-1	S6k		
		rsks-1	Thor		
		rps-6			
MAPK/ERK	RAS1	let-60	ras	Rasgrf1	
	RAS2	ets-4	ras85D		
			rl		
			pnt		
			aop		

variation in the mTOR signaling pathway has also been linked to human aging, although the specific genes responsible for this association have not yet been identified (Passtoors et al. 2013).

Mitogen-Activated Protein Kinase/Extracellular Signal-Regulated Kinase Signaling Pathway
The MAPK/ERK signaling pathway should not be neglected in this context, not only because of its well-established role in cell growth and cancer but also because of its lesser, yet emerging role in metabolism and aging. This pathway transmits extracellular cues, mostly in the form of growth factors, to the cytoplasm and nucleus through a three-tier cascade of MAPKs. RAS proteins are located upstream in this cascade and cycle between inactive, GDP-bound, and active, GTP-bound, states, which in turn leads to phosphorylation of their downstream targets RAF, MEK, and ERK.

Research coming from yeast, flies, and mice provides evidence for a role of the MAPK/ERK signaling pathway in the aging process of these species (Slack 2017) (Table 1). Moreover, *HRAS1* was shown to interact with the strongest human longevity-associated gene, *APOE*, indicating that the MAPK/ERK signaling pathway may also be involved in human aging (Jazwinski et al. 2010).

Genes Influencing Aging in Model Organisms and Humans

In addition to the pathways described above, we would like to highlight some of the genes that have shown major effects on lifespan in the most studied model organisms (*Saccharomyces cerevisiae*, *Caenorhabditis elegans*, *Drosophila melanogaster*, and *Mus musculus*) and humans. An overview of all aging-associated genes in model organisms is provided in the GenAge database (http://genomics.senescence.info/genes/) (Tacutu et al. 2018).

Saccharomyces cerevisiae
Among the first discoveries that lead to an increase in replicative lifespan in yeast are sirtuins, NAD+-dependent deacylases, named after the initial member silent mating-type information regulation 2 (*SIR2*). The deletion of this gene was found to decrease mean lifespan by as much as 50%, while overexpression extends lifespan by 30% through transcriptional silencing (Kaeberlein et al. 1999). However, the translation

of these findings to worms and flies quickly became the platform of controversy and was subsequently repelled as the effects attributed to sirtuins were abolished after controlling for the appropriate genetic background (Burnett et al. 2011). With more recent technical advances, large-scale analyses have found a vast number of genes to be involved in the aging process in yeast. The genes that show the most prominent effects on lifespan are involved in the signaling pathways outlined above and additionally include (ribosomal) genes involved in mitochondrial translation and the TCA cycle (Janssens and Veenhoff 2016; McCormick et al. 2015; Powers et al. 2006; Smith et al. 2007).

Caenorhabditis elegans
Besides the prominent role of the IIS pathway in the aging process of worms (see Table 1), RNA interference longevity screens additionally identified several other genes with substantial effects on lifespan, such as the mitochondrial gene *lrs-2* (Lee et al. 2003) and components of the electron transport chain and TCA cycle, like the cytochrome C oxidase (*cco-1*), NADH ubiquinone oxidoreductases (*nuo-2*, *nuo-3*, *nuo-4*) (Hansen et al. 2005), and iron sulfur-protein of complex 3 (*isp-1*) (Feng et al. 2001).

Drosophila melanogaster
One of the genes whose functional compromise has been linked to a vast increase in lifespan in flies is Methuselah (*Mth*), a G protein-coupled transmembrane protein, which has been shown to function in an mTOR-dependent manner (Lin et al. 1998). Another gene worth mentioning is integrin-linked kinase (*Ilk*), whose (slightly) reduced expression was shown to increase lifespan in both males and females and specifically prevent cardiac aging phenotypes such as arrhythmias. Interestingly, the complete knockdown of this gene resulted in the loss of cardiac integrity, which implies a closely regulated balance of this kinase in heart physiology (Nishimura et al. 2014). Other interesting findings came from studies of the evolutionary conserved link between the olfactory system and appetite/food intake. These studies showed that the positive effects of dietary restriction on lifespan are reversible by exposing flies to nutrient odors. However, lifespan extension can also be achieved through mutagenesis of the odorant receptor *Orco* that additionally leads to improved stress resistance (Libert et al. 2007). The concept of dietary restriction and lifespan extension also seems to involve serotonin signaling, as the disruptions of serotonin signaling-related genes, such as tryptophan hydroxylase (*Trh*), serotonin receptor 2a (*5HT2A*), and amino acid transporter *JhI-21*, made flies more attracted to protein-rich diets, resulting in an increased lifespan of these flies (Ro et al. 2016).

Mus musculus
In mice, one of the first antiaging genes to be discovered was *Klotho* (Kuro-o et al. 1997). Defects in this gene result in an overall aging-like phenotype that includes a reduction in lifespan and fertility as well as the occurrence of arteriosclerosis, osteoporosis, and emphysema. Subsequent studies of *Klotho* showed that the function of this gene can be traced back to the inhibition of IIS and, moreover, that the gene plays a role in kidney and vascular disease in humans. Generally speaking, the inhibition or knockout of growth factors dominates the aging field in the mouse: besides insulin-like growth factors, knockout of the growth hormone-releasing hormone and its receptor, *Ghrh* and *Ghrhr* (Flurkey et al. 2001; Sun et al. 2013), the epidermal growth factor receptor pathway substrate 8 (*Eps8*) (Tocchetti et al. 2010), and growth differentiation factor 15 (*Gdf15*) (Wang et al. 2014) as well as overexpression of fibroblast growth factor 21 (*Fgf21*) (Zhang et al. 2012) were found to increase the average lifespan of mice by 14–40%. The downregulation of somatic growth and the subsequent prevention of age-related diseases and increase in lifespan appears to be a highly conserved feature that may also exist in humans (Bartke and Quainoo 2018).

Homo sapiens
In humans, the identification of genetic variants that associate with longevity has proven to be challenging. Taking into account the strongly

limiting factor of replicability, the only consistent gene to show up in genetic studies across the globe has been apolipoprotein E (*APOE*) (Partridge et al. 2018). The protein encoded by this gene is a major player in cholesterol/lipoprotein homeostasis in various tissues, including the brain. The most common isoforms of APOE are ε2, ε3, and ε4, which are defined by the two common genetic variants rs429358 and rs7412. While the ε2 isoform seem to exert an overall protective function, ε4 has been associated with many different diseases, including cognitive decline and Alzheimer's disease and coronary artery disease (Mahley and Rall 2000).

Future Directions of Research

Correlations between reduced IIS/mTOR signaling and lifespan that have been observed in animal models may also be present in humans. Nonetheless, it remains largely unknown how certain individuals seem to escape age-related diseases and become exceptionally long-lived. Future research in validated as well as emerging model organisms, like the exceptionally long-lived naked-mole rat (lifespan >30 years as compared to <10 years for its genetic relatives) and short-lived African turquoise killifish (lifespan of 4–9 months), will hopefully continue to reveal insights into conserved aging mechanisms and consolidate existing knowledge (Kim et al. 2016; Lewis et al. 2018). Moreover, the continuation of research in model organisms will likely result in improved cross-species comparisons, which may ultimately facilitate the translation of findings to humans, with the overall goal of ameliorating the burden of age-related diseases in the future.

Summary

Research in model organisms has proven to be very helpful in unraveling the genetic component of aging, by revealing evolutionary conserved signaling pathways that appear to underlie the aging process. Likewise, these studies have shown that most aging-associated genes act in a pleiotropic manner and lead to an overall improved stress resistance on a cellular level, which ultimately results in protection against age-related traits and diseases. Future studies using animal models, including newly emerging ones, will hopefully continue to provide insights that could be used to tackle the increasing burden of age-related diseases.

Cross-References

▶ Animal Models of Aging
▶ Genetic Control of Aging
▶ Genetic Theories of Aging
▶ Genomics of Aging and Longevity
▶ Progeria: Model Organisms

References

Bartke A, Quainoo N (2018) Impact of growth hormone-related mutations on mammalian aging. Front Genet 9(November):1–11. https://doi.org/10.3389/fgene.2018.00586

Burnett C, Valentini S, Cabreiro F, Goss M, Somogyvari M, Piper MD, ... Gems D (2011) Absence of effects of Sir2 overexpression on lifespan in *C. elegans* and Drosophila. Nature 477(7365):482–485. https://doi.org/10.1038/nature10296

Deelen J, Uh H, Monajemi R, van Heemst D, Thijssen P, Boehringer S, ... Beekman M (2013) Gene set analysis of GWAS data for human longevity highlights the relevance of the insulin/IGF-1 signaling and telomere maintenance pathways. Age (Dordr) 35(1):235–249

Feng J, Bussière F, Hekimi S (2001) Mitochondrial electron transport is a key determinant of life span in *Caenorhabditis elegans*. Dev Cell 1(5):633–644. https://doi.org/10.1016/S1534-5807(01)00071-5

Flachsbart F, Caliebe A, Kleindorp R, Blanche H, von Eller-Eberstein H, Nikolaus S, ... Nebel A (2009) Association of FOXO3A variation with human longevity confirmed in German centenarians. Proc Natl Acad Sci 106(8):2700–2705

Flurkey K, Papaconstantinou J, Miller RA, Harrison DE (2001) Lifespan extension and delayed immune and collagen aging in mutant mice with defects in growth hormone production. PNAS 98(12):6736–6741. https://doi.org/10.1073/pnas.111158898

Fontana L, Partridge L, Longo V (2010) Extending healthy life span – from yeast to humans. Science 328(5976):321–326

Hansen M, Hsu AL, Dillin A, Kenyon C (2005) New genes tied to endocrine, metabolic, and dietary regulation of lifespan from a *Caenorhabditis elegans* genomic RNAi screen. PLoS Genet 1(1):0119–0128. https://doi.org/10.1371/journal.pgen.0010017

Holzenberger M, Dupont J, Ducos B, Leneuve P, Geloen A, Even PC, ... Le Bouc Y (2003) IGF-1 receptor regulates lifespan and resistance to oxidative stress in mice. Nature 421(6919):182–187

Janssens G, Veenhoff L (2016) Evidence for the hallmarks of human aging in replicatively aging yeast. Microbial Cell 3(7):263–274. https://doi.org/10.15698/mic2016.07.510

Jazwinski SM, Kim S, Dai J, Li L, Bi X, Jiang JC et al (2010) HRAS1 and LASS1 with APOE are associated with human longevity and healthy aging. Aging Cell 9(5):698–708. https://doi.org/10.1161/CIRCULATIONAHA.111.040337.Wnt

Johnson SC, Rabinovitch PS, Kaeberlein M (2013) mTOR is a key modulator of ageing and age-related disease. Nature 493(7432):338–345

Kaeberlein M, Mcvey M, Guarente L (1999) The SIR2/3/4 complex and SIR2 alone promote longevity in *Saccharomyces cerevisiae* by two different mechanisms. Genes Dev 13(19):2570–2580. Retrieved from http://www.ncbi.nlm.nih.gov/entrez/query.fcgi?cmd=Retrieve&db=PubMed&dopt=Citation&list_uids=10521401

Kenyon C, Chang J, Gensch E (1993) A *C. elegans* mutant that lives twice as long as wild type. Nature 366(December):461–464. https://doi.org/10.1038/366461a0

Kim Y, Nam HG, Valenzano DR (2016) The short-lived African turquoise killifish: an emerging experimental model for ageing. Dis Model Mech 9(2):115–129. https://doi.org/10.1242/dmm.023226

Klass M, Nguyen PN, Dechavigny A (1983) Age-correlated changes in the DNA template in the nematode. Mech Ageing Dev 22(3–4):253–263

Kuro-o M, Matsumura Y, Aizawa H, Kawaguchi H, Suga T, Utsugi T, ... Nabeshima Y (1997) Mutation of the mouse klotho gene leads to a syndrome resembling ageing. Nature 390:45–51. https://doi.org/10.1002/3527607307.ch14

Lee SS, Lee RYN, Fraser AG, Kamath RS, Ahringer J, Ruvkun G (2003) A systematic RNAi screen identifies a critical role for mitochondria in *C. elegans* longevity. Nat Genet 33(1):40–48. https://doi.org/10.1038/ng1056

Lewis KN, Rubinstein ND, Buffenstein R (2018) A window into extreme longevity; the circulating metabolomic signature of the naked mole-rat, a mammal that shows negligible senescence. GeroScience 40(2):105–121. https://doi.org/10.1007/s11357-018-0014-2

Libert S, Zwiener J, Chu X, VanVoorhies W, Roman G, Pletcher SD (2007) Regulation of Drosophila life span by olfaction and food-derived odors. Science 315(5815):1133–1137. https://doi.org/10.1126/science.1136610

Lin Y, Seroude L, Benzer S (1998) Extended life-span and stress resistance in the Drosophila mutant methuselah. Science 282(5390):943–946

Mahley R, Rall SJ (2000) Apolipoprotein E: far more than a lipid transport protein. Annu Rev Genomics Hum Genet 1:507–537

McCormick MA, Delaney JR, Tsuchiya M, Tsuchiyama S, Shemorry A, Sim S, ... Kennedy BK (2015) A comprehensive analysis of replicative lifespan in 4,698 single-gene deletion strains uncovers conserved mechanisms of aging. Cell Metab 22(5):895–906. https://doi.org/10.1007/s11065-015-9294-9.Functional

Morris JZ, Tissenbaum HA, Ruvkun G (1996) A phosphatidylinositol-3-OH kinase family member regulating longevity and diapause in Caenorhabditis elegans. Nature 382(August):536–539. Retrieved from http://www.wormbase.org/db/misc/paper?name=WBPaper00002514

Nishimura M, Kumsta C, Kaushik G, Diop SB, Ding Y, Bisharat-Kernizan J, ... Ocorr K (2014) A dual role for integrin-linked kinase and β1-integrin in modulating cardiac aging. Aging Cell 13(3):431–440. https://doi.org/10.1111/acel.12193

Partridge L, Alic N, Bjedov I, Piper MD (2011) Ageing in Drosophila: the role of the insulin/Igf and TOR signalling network. Exp Gerontol 46(5):376–381

Partridge L, Deelen J, Slagboom P (2018) Facing up to the global challenges of ageing. Nature 561(7721):45–56

Passtoors W, Beekman M, Deelen J, van der Breggen R, Maier A, Guigas B, ... Slagboom P (2013) Gene expression analysis of mTOR pathway: association with human longevity. Aging Cell 12(1):24–31

Powers RW, Kaeberlein M, Caldwell SD, Kennedy BK, Fields S (2006) Extension of chronological life span in yeast by decreased TOR pathway signaling. Genes Dev 20(2):174–184. https://doi.org/10.1101/gad.1381406

Ro J, Pak G, Malec PA, Lyu Y, Allison DB, Kennedy RT, Pletcher SD (2016) Serotonin signaling mediates protein valuation and aging. eLife 5:1–21. https://doi.org/10.7554/elife.16843

Selman C, Lingard S, Choudhury A, Batterham R, Claret M, Clements M, ... Withers DJ (2008) Evidence for lifespan extension and delayed age-related biomarkers in insulin receptor substrate 1 null mice. FASEB J 22(3):807–813

Slack C (2017) Ras signaling in aging and metabolic regulation. Nutr Healthy Aging 4(3):195–205

Smith ED, Kennedy BK, Kaeberlein M (2007) Genome-wide identification of conserved longevity genes in yeast and worms. Mech Ageing Dev 128(1):106–111. https://doi.org/10.1016/j.mad.2006.11.017

Sun LY, Spong A, Swindell WR, Fang Y, Hill C, Huber JA, ... Bartke A (2013) Growth hormone-releasing hormone disruption extends lifespan and regulates response to caloric restriction in mice. eLife 2:1–24. https://doi.org/10.7554/elife.01098

Tacutu R, Thornton D, Johnson E, Budovsky A, Barardo D, Craig T, ... de Magalhaes J (2018) Human Ageing

Genomic Resources: new and updated databases. Nucleic Acids Res 46:D1083–D1090

Tocchetti A, Ekalle Soppo CB, Zani F, Bianchi F, Gagliani MC, Pozzi B, ... Offenhäuser N (2010) Loss of the Actin remodeler Eps8 causes intestinal defects and improved metabolic status in mice. PLoS ONE 5(3). https://doi.org/10.1371/journal.pone.0009468

Vézina C, Kudelski A, Sehgal SN (1975) Rapamycin (AY-22, 989) a new antifungal antibiotic. J Antibiot 28:721–726. https://doi.org/10.7164/antibiotics.28.727

Wang X, Chrysovergis K, Kosak J, Kissling G, Streicker M, Moser G, ... Eling TE (2014) hNAG – 1 increases lifespan by regulating energy metabolism and insulin/IGF-1/mTOR signaling. Aging 6(8):690–704. https://doi.org/10.18632/aging.100687

Willcox B, Donlon T, He Q, Chen R, Grove J, Yano K, ... Curb J (2008) FOXO3A genotype is strongly associated with human longevity. Proc Natl Acad Sci 105 (37):13987–13992

Zhang Y, Xie Y, Berglund ED, Coate KC, He TT, Katafuchi T, ... Mangelsdorf DJ (2012) The starvation hormone, fibroblast growth factor-21, extends lifespan in mice. eLife 1:1–14. https://doi.org/10.7554/elife.00065

Genetic Alternation

▶ Mutation

Genetic Ancestry

▶ Genetics: Ethnicity

Genetic Code

▶ Alleles
▶ Chromosome

Genetic Constitution

▶ Genotype

Genetic Control of Aging

Maarouf Baghdadi[1], David Karasik[2,3] and Joris Deelen[1]

[1]Max Planck Institute for Biology of Ageing, Cologne, Germany
[2]Azrieli Faculty of Medicine, Bar Ilan University, Safed, Israel
[3]Marcus Institute for Aging Research, Hebrew SeniorLife, Boston, MA, USA

Synonyms

Candidate-gene studies; Endophenotypes; Genome-wide association studies; Heritability; Lifespan; Linkage studies; Longevity; Whole-genome/exome sequencing

Definition

The genetic control of aging is the contribution of genetic variation to lifespan and longevity.

Overview

The present day lifespan of individuals is higher than it has ever been before. Concomitant with the increase in global lifespan is the increase in the burden of late-life diseases, since advanced age is the greatest common risk factor for most chronic debilitating diseases.

Heritability studies have convincingly demonstrated that aging is subject to genetic control, with additive effects comprising a small fraction of total lifespan heritability. This opens the door for genetic studies of aging, which aim to understand the pathways underlying this process and, using that knowledge, to develop strategies for prevention, intervention, and therapy. Thus far, these studies have shown an inverse effect of multiple age-related disease-associated variants on lifespan regulation, while the number genes containing well-replicated longevity-conferring variants is limited to *APOE* (ApoE ε2) and *FOXO3*.

Key Research Findings

Use of Different Definitions for Genetic Studies of Aging

Aging is a highly complex and heterogeneous process. Biologists usually define aging in terms of the gradual and progressive decline in structure and function of multiple organs and systems, which culminate in death of the individual. Although recent developments in advanced molecular genetic techniques and a considerable reduction in costs have dramatically improved our understanding of the genetic underpinnings of human aging, genetic studies of aging have been hindered by an inconsistent use of definitions. The two main ways of conducting research on the genetics of aging are by studying (1) the continuous trait lifespan and (2) the dichotomous trait longevity (i.e., belonging to the longest-lived individuals within a specific population). In addition to the two main definitions used to study the genetics of aging, there are other, less well-established, definitions that are used in the field, such as frailty, health span, and successful aging, which are not reviewed here. Despite awareness in the field about the problem with the inconsistent use of definitions (Sebastiani et al. 2016; Partridge et al. 2018), the adherence to these two main definitions has not been widely adopted. Throughout this chapter, we will therefore discriminate between studies that used one of the two main definitions, to provide a comprehensive picture of the current state of the field.

Heritability of Aging

Many efforts have been undertaken to estimate the heritability of aging, but this has proven to be difficult. In human studies, unlike model organism studies in laboratory settings, the effect of environmental variance cannot be controlled. Given the complex polygenic nature of aging, the fine delineation between different sources of variance is essential. The two main study designs that have been used to investigate the heritability of aging are twin and genealogical (e.g., pedigree) studies. The twin study design has been widely used to measure heritability in a variety of phenotypic traits. The main advantage of this design is the ability to discriminate between the effect of genetic, shared, and non-shared early-life environmental influences on a trait (Veale and Falconer 1960). However, as aging has a delayed onset, the effect of adult environment can have a significant effect on the phenotype as well (McGue et al. 2014) and this is not captured by this design. On the other hand, the pedigree/genealogical study design has the advantage of having access to a much larger sample size, especially for the older members of a population. When pedigrees get large and multigenerational, one can investigate a traits' genetic inheritance pattern and delineate between additive and nonadditive components of heritability.

Heritability of Lifespan

Altogether, twin studies have shown that the heritability of lifespan ranges between 0.01 and 0.27 in various European populations (van den Berg et al. 2017). Although most studies estimated the heritability based on the differences between mono- and dizygotic twins, one study was able to consider early environmental effects by comparing twins reared apart and twins reared together, showing a heritability of 0.01 for males and 0.15 for females (Ljungquist et al. 1998). Unfortunately, the low sample size for twins reared apart (<100 pairs) prohibited making any confident conclusions. The notoriously high variation in estimated heritability between studies may result from the different designs used across studies, such as variable minimum age thresholds for inclusion. These thresholds are typically used to remove the influence of early life mortality. Moreover, many twin studies suffer from small sample sizes. Finally, it is important to note an inherent limitation of twin studies, namely, geographical confinement of participants, which leads to population-specific estimates. Large genealogical studies have enough power to address unanswered questions in twin studies, such as to what extent nonadditive genetic variance contributes to the heritability of lifespan. Kaplanis and colleagues used more than three million pairs of relatives and found that the additive component of heritability of lifespan was 0.16 (comparable to twin studies), while there was only a mild effect of the nonadditive component of heritability (~0.04) (Kaplanis et al. 2018). Ruby and colleagues used an even more impressive dataset, consisting of hundreds of millions of historical individuals, and showed a similar heritability of lifespan. However, after correcting for the effects of assortative mating, the remaining heritability of lifespan was below 7%, indicating that other heritability studies may have overestimated the true heritability of lifespan (Ruby et al. 2018).

Altogether, the twin and genealogical studies have shown that lifespan is heritable, but only to a small extent, which may explain why genetic studies of lifespan have proven to be challenging.

Heritability of Longevity

The only study on the heritability of longevity that has thus far been performed was a twin study by Ljungquist and colleagues (1998), in which they found that the heritability of longevity increases with advancing age. However, the age thresholds used to define longevity were relatively low (i.e., reaching an age above 80 or 85 years) and the sample size was limited, which prevents making confident conclusions about the heritability of longevity. However, several genealogical studies of long-lived families have consistently found that siblings of long-lived individuals have a high probability to also become long-lived. Hence, parental longevity could be considered a valid proxy for life expectance. Moreover, long-lived parents have a high probability to bear long-lived offspring, which gives an indication that longevity is indeed heritable (van den Berg et al. 2017). Interestingly, members of long-lived families have an even more important phenotype than extended lifespan; they seem to be escaping or delaying age-related disease and show a compression of late life morbidity. The genetic makeup of these individuals can thus hold key insights into the mechanisms or pathways that result in delay of or even escape from age-associated diseases (Andersen et al. 2012). An important drawback of longevity research is the variability in the age threshold that is used to signify an extreme age. Recently, van den Berg and colleagues used two independent multigenerational genealogical datasets to determine the most optimal definition of longevity (van den Berg et al. 2019). They found that the strongest heritable component of longevity is present in individuals belonging to the top 10% survivors of their birth cohort with equally long-lived family members.

In conclusion, future genetic studies of longevity may thus benefit from only using the 10% longest lived individuals within a population, although the sample size of such studies will likely be small due to limited availability of genetic data on such individuals worldwide.

Genetic Studies of Aging

People have tried to study the genetics of aging using multiple different approaches, including candidate gene, linkage, and genome-wide association studies (GWAS), the results of which are summarized below.

Candidate Gene Studies

Before the rise of gene array-based technologies, the genetics of aging was studied using the candidate gene approach, in which single or groups of genes that were implicated by studies on animal models and age-related diseases in humans were studied. These candidate gene studies, an overview of which can be found in the LongevityMap database (http://genomics.senescence.info/longevity/) (Budovsky et al. 2013), revealed only two loci that have withstood replication and validation. The first locus is *APOE*, which has been identified already more than two decades ago (Schächter et al. 1994) and has two aging-associated alleles, ApoE ε2, which has shown to be protective, and ApoE ε4, which has shown to be deleterious. The second locus is *FOXO3*, which has consistently been associated with a protective effect on lifespan/longevity across a variety of populations (Willcox et al. 2008; Flachsbart et al. 2009) and has recently been functionally validated (Flachsbart et al. 2017; Grossi et al. 2018). The candidate gene studies, that studied the combined effect of genetic variants in groups of genes, have highlighted a role for genetic variation in insulin/insulin-like growth factor 1 signaling, mammalian target of rapamycin signaling, telomere maintenance, and DNA damage/repair in aging (Passtoors et al. 2012; Deelen et al. 2013; Debrabant et al. 2014).

Linkage Studies

The first hypothesis-free genome-wide studies used a linkage approach to investigate the genetics of aging. These studies implemented a family-based design to identify regions in the genome associated with lifespan/longevity and were thus prone to identification of population-specific, rather than generalizable, loci. This likely explains the inconsistency between the results of these studies in relatively small populations. The largest and most comprehensive linkage study on aging to date is the GEHA study which looked at long-lived sibling pairs from 11 European countries (Beekman et al. 2013). They found four chromosomal regions that exhibit linkage with longevity, of which one could be explained by the presence of the ApoE ε2 and ApoE ε4 alleles. The genetic variants responsible for the linkage at the remaining three regions have thus far not been identified by fine-mapping or sequencing of these regions.

Genome-Wide Association Studies

The GWAS approach is suitable to detect the effect of common genetic variation (i.e., minor allele frequency >1%) on a phenotypic trait. To date, this approach has resulted in the identification of tens of thousands of genomic loci underlying individual human traits (Buniello et al. 2019). Interestingly, some biological candidate genes (sometimes described as the "usual suspects") proposed at earlier stages of genetic exploration have generally not been confirmed by subsequent large GWAS of the same phenotypes.

GWAS of Lifespan The first lifespan-related GWAS, based on age at death (Walter et al. 2011), did not identify any genome-wide significant loci, which was likely due to the relatively low sample size. However, the recent emergence of the UK Biobank has significantly enhanced research on the genetics of lifespan. The most recent effort using parental lifespan data from this study, as well as several additional studies that took a part in the LifeGen initiative, has resulted in the identification of 12 loci that passed genome-wide significance. These loci are located in or near *MAGI3*, *KCNK3*, *HTT*, *HLA-DQA1*, *LPA*, *CDKN2B-AS1*, *ATXN2/BRAP*, *CHRNA3/5*, *FURIN/FES*, *HP*, *LDLR*, and *APOE*. Many of the loci have previously been associated with age-related diseases, including cardiometabolic, auto-immune, and neuropsychiatric diseases, which likely explains their association with lifespan (Timmers et al. 2019).

GWAS of Longevity GWAS of longevity to date have provided only a handful of longevity-associated loci. The most consistent evidence has been obtained for variants in the *APOE* gene (Partridge et al. 2018), which was already reported to be associated with longevity in the candidate gene studies. Generally, GWAS have been less successful for diseases in which phenotypes have been more difficult to quantify and to standardize (homogenize) among collaborating cohorts, such as cognitive and behavioral traits and mental-health-related diseases. Not surprisingly, in the aging field, the progress is slowed down by a vague definition of the phenotype (Sebastiani et al. 2016; Partridge et al. 2018). Hence, more objective definitions, such as restricting oneself to the 10% longest lived individuals within a population (van den Berg et al. 2019), should be both promising for discovery and bear potential to be replicable.

GWAS of Aging Using Alternative Phenotypes In order to identify and quantify the genetic contributions to a complex disease or trait, such as osteoporosis, osteoarthritis, or aging, there is a need to identify and validate an "endophenotype" (Karasik and Kiel 2015). Endophenotypes are defined as intermediate components of a phenotype of interest proximal to a biological process (Schulze and McMahon 2004; Pan et al. 2006), as opposing to a final ("visible") exophenotype, such as fracture, spondylosis, or death (Karasik and Kiel 2015). There is currently no consensus on the ultimate set of (multi-)biomarkers that should be used as endophenotypes in genetic studies of aging. Several studies have used composite scores to classify individuals as healthy agers. However, the subsequent use of such score in GWAS has thus far not resulted in identification of novel replicated genome-wide significant loci (Singh et al. 2017). The recent development of indices of biological aging may help to overcome this problem. For the purposes of genetic exploration, it is important that these endophenotypes are heritable. Moreover, it would be valuable if the endophenotypes are translatable to animal models so they can be used to perform more in-depth functional studies.

In addition to the search for endophenotypes of aging, there is an ongoing effort to systematically dissect the genetics of physical function phenotypes that are present among older adults, including hand grip strength (Willems et al. 2017; Tikkanen et al. 2018) and gait speed (Ben-Avraham et al. 2017), since the loci emerging from these GWAS could give insight into healthy aging. Moreover, genetic research into aging may benefit from the ongoing efforts to study the genetics of health span and age-related disease-informed GWAS (McDaid et al. 2017; Zenin et al. 2019), which could lead to identification of loci that are not picked up in regular GWAS of lifespan and longevity.

Whole-Genome/Exome Sequencing of Long-Lived Individuals

One disadvantage of the GWAS approach is that it is only capable of detecting the effect of common genetic variants on aging. However, most of the regions identified through linkage cannot be explained by common genetic variants (Beekman et al. 2013), indicating a potential involvement of rare variants in aging. Rare variants can be detected through whole-genome or exome sequencing and several studies have used this approach to investigate the genome of long-lived individuals. However, all of the sequencing studies of long-lived individuals that have been performed so far were based on a very small number of samples (the largest study contains data on 17 supercentenarians (Gierman et al. 2014)), which makes it impossible to draw any solid conclusions about the involvement of rare variants in aging.

Other Types of Genome-Wide Studies

In addition to the study of single nucleotide polymorphisms, one could also investigate the effect of structural variation, such as copy number variations (CNVs). CNVs can shed light on missing heritability as they contribute to human genetic variation to a larger extent than SNPs (Manolio et al. 2009) and have been associated with several aging-related traits (Owen et al. 2018). Two studies of CNVs in relation to lifespan showed that carriers of a higher burden of deletions had an increased mortality risk compared to noncarriers (Kuningas et al. 2011; Nygaard et al. 2016), supporting a role for genome instability in aging. However, these and other studies were unable to find specific CNVs associated with lifespan or longevity at a genome-wide significant level, although a duplication on 7p11.2 was found to be associated with longevity in both European and Asian populations (Zhao et al. 2018).

Hence, these results indicate that CNVs may contribute to the genetic component of aging, but investigation in larger studies is required to confirm these findings.

Future Directions of Research

Genetic studies on aging have proven to be challenging, which is likely due to the limited heritability of this process, especially when looking at the trait lifespan. Given that longevity seems to be more heritable than lifespan, it will be interesting to look at the heritability of this trait in more detail. This can be done by performing twin and genealogical studies using a well-defined threshold of longevity, such as the previously proposed population-specific survival percentile (i.e., 10%). The recent availability of large genealogical datasets, such as the ones used by Kaplanis and Ruby, should make these kinds of studies possible.

The newly proposed longevity phenotype can also be used for genetic studies on aging and a worldwide GWAS using this phenotype is already ongoing. However, given the limited number of identified aging-associated genetic variants through GWAS and the availability of affordable exome and whole-genome sequencing, future genetic studies of aging may also start focusing on rare genetic variants with more potent effects on aging that are not detectable through GWAS. The application of this approach, at a candidate-gene level, has already resulted in the identification of functional genetic variants in *IGF1R* and *FOXO3* (Tazearslan et al. 2011; Flachsbart et al. 2017). The most interesting individuals to sequence would originate from multigenerational long-lived families, since such families are not only long-lived but also seem to evade morbidity. Unraveling the genetics of these individuals will potentially help identify novel mechanisms involved in healthy aging that can subsequently guide therapeutic interventions.

One major drawback of the genetic studies on aging that have thus far been performed is that they have mostly focused on individuals of European ancestry, which constitute only a minor fraction of the global population. Given that the majority of genetic variants in the genome are not shared or show differences in frequency between diverse populations (Gibbs et al. 2015), most of the findings cannot be directly translated between populations. Hence, future genetic studies of aging should try to include individuals from diverse lineages, to benefit the increasingly globalized world. Moreover, expanding studies to include populations of additional ancestries may lead to identification of novel genetic variants associated with aging that shed light on the molecular mechanisms underlying this process.

Large-scale GWAS have identified numerous loci associated with age-related traits (Buniello et al. 2019). Given that aging is a complex process, it may benefit from exploration of pleiotropy across these age-related traits to (1) refine the phenotypic definition of aging and its proxies (endophenotypes), (2) identify novel (pleiotropic) genetic variants involved in aging, and (3) increase the power of genetic studies on aging. Moreover, gene-by-environment interactions should be investigated.

Summary

Aging is a complex process, which has been explored using two distinct definitions, lifespan and longevity. Heritability studies using these definitions have shown that genetics likely explains only a minor fraction of aging. Genetic studies of lifespan and longevity have thus far only provided a limited number of loci involved in aging, with *APOE* and *FOXO3* being the most consistently replicated. This is likely due to the inconsistent use of phenotypic definitions of longevity and the relatively small sample size of these studies. Genetic studies of aging may thus benefit from (1) a better definition of the longevity phenotype and (2) the use of endophenotypes of aging that can be analyzed in much larger sample sizes than currently available for lifespan or longevity. Moreover, genetic studies of aging may have to shift their focus to rare protective genetic variants that are present in multigenerational long-lived families, since such families seem to hold the key to healthy aging.

Cross-References

- Centenarians
- Genes that Delay Aging
- Genome-Wide Association Study
- Genomics of Aging and Longevity
- Maximum Lifespan
- Mortality at Older Ages and Mean Age at Death
- Somatic Mutations and Genome
- Supercentenarians
- Whole Genome Sequencing

References

Andersen SL, Sebastiani P, Dworkis DA et al (2012) Health span approximates life span among many supercentenarians: compression of morbidity at the approximate limit of life span. J Gerontol A Biol Sci Med Sci 67A:395–405. https://doi.org/10.1093/gerona/glr223

Beekman M, Blanché H, Perola M et al (2013) Genome-wide linkage analysis for human longevity: genetics of healthy aging study. Aging Cell 12:184–193. https://doi.org/10.1111/acel.12039

Ben-Avraham D, Karasik D, Verghese J et al (2017) The complex genetics of gait speed: genome-wide meta-analysis approach. Aging (Albany NY) 9:209–246. https://doi.org/10.18632/aging.101151

Budovsky A, Craig T, Wang J et al (2013) LongevityMap: a database of human genetic variants associated with longevity. Trends Genet 29:559–560. https://doi.org/10.1016/j.tig.2013.08.003

Buniello A, MacArthur JAL, Cerezo M et al (2019) The NHGRI-EBI GWAS Catalog of published genome-wide association studies, targeted arrays and summary statistics 2019. Nucleic Acids Res 47:D1005–D1012. https://doi.org/10.1093/nar/gky1120

Debrabant B, Soerensen M, Flachsbart F et al (2014) Human longevity and variation in DNA damage response and repair: study of the contribution of sub-processes using competitive gene-set analysis. Eur J Hum Genet 22:1131–1136. https://doi.org/10.1038/ejhg.2013.299

Deelen J, Uh H-W, Monajemi R et al (2013) Gene set analysis of GWAS data for human longevity highlights the relevance of the insulin/IGF-1 signaling and telomere maintenance pathways. Age (Omaha) 35:235–249. https://doi.org/10.1007/s11357-011-9340-3

Flachsbart F, Caliebe A, Kleindorp R et al (2009) Association of FOXO3A variation with human longevity confirmed in German centenarians. Proc Natl Acad Sci U S A 106:2700–2705. https://doi.org/10.1073/pnas.0809594106

Flachsbart F, Dose J, Gentschew L et al (2017) Identification and characterization of two functional variants in the human longevity gene FOXO3. Nat Commun 8:2063. https://doi.org/10.1038/s41467-017-02183-y

Gibbs RA, Boerwinkle E, Doddapaneni H et al (2015) A global reference for human genetic variation. Nature 526:68–74. https://doi.org/10.1038/nature15393

Gierman HJ, Fortney K, Roach JC et al (2014) Whole-genome sequencing of the world's oldest people. PLoS One 9:e112430. https://doi.org/10.1371/journal.pone.0112430

Grossi V, Forte G, Sanese P et al (2018) The longevity SNP rs2802292 uncovered: HSF1 activates stress-dependent expression of FOXO3 through an intronic enhancer. Nucleic Acids Res 46:5587–5600. https://doi.org/10.1093/nar/gky331

Kaplanis J, Gordon A, Shor T et al (2018) Quantitative analysis of population-scale family trees with millions of relatives. Science 360:171–175. https://doi.org/10.1126/science.aam9309

Karasik D, Kiel DP (2015) Genetics of osteoporosis in older age, 2 edn. 141–155. Springer International Publishing Switzerland https://doi.org/10.1007/978-3-319-25976-5

Kuningas M, Estrada K, Hsu YH et al (2011) Large common deletions associate with mortality at old age. Hum Mol Genet 20:4290–4296. https://doi.org/10.1093/hmg/ddr340

Ljungquist B, Berg S, Lanke J et al (1998) The effect of genetic factors for longevity: a comparison of identical and fraternal twins in the Swedish Twin Registry. J Gerontol A Biol Sci Med Sci 53:441–446. https://doi.org/10.1093/gerona/53A.6.M441

Manolio TA, Collins FS, Cox NJ et al (2009) Finding the missing heritability of complex diseases. Nature 461:747–753. https://doi.org/10.1038/nature08494

McDaid AF, Joshi PK, Porcu E et al (2017) Bayesian association scan reveals loci associated with human lifespan and linked biomarkers. Nat Commun 8. https://doi.org/10.1038/ncomms15842

McGue M, Skytthe A, Christensen K (2014) The nature of behavioural correlates of healthy ageing: a twin study of lifestyle in mid to late life. Int J Epidemiol 43:775–782. https://doi.org/10.1093/ije/dyt210

Nygaard M, Debrabant B, Tan Q et al (2016) Copy number variation associates with mortality in long-lived individuals: a genome-wide assessment. Aging Cell 15:49–55. https://doi.org/10.1111/acel.12407

Owen D, Bracher-Smith M, Kendall KM et al (2018) Effects of pathogenic CNVs on physical traits in participants of the UK Biobank. BMC Genomics 19:867. https://doi.org/10.1186/s12864-018-5292-7

Pan W-H, Lynn K-S, Chen C-H et al (2006) Using endophenotypes for pathway clusters to map complex disease genes. Genet Epidemiol 30:143–154. https://doi.org/10.1002/gepi.20136

Partridge L, Deelen J, Slagboom PE (2018) Facing up to the global challenges of ageing. Nature 561:45–56. https://doi.org/10.1038/s41586-018-0457-8

Passtoors WM, Boer JM, Goeman JJ et al (2012) Transcriptional profiling of human familial longevity indicates a role for ASF1A and IL7R. PLoS One 7:e27759. https://doi.org/10.1371/journal.pone.0027759

Ruby JG, Wright KM, Rand KA et al (2018) Estimates of the heritability of human longevity are substantially inflated due to assortative mating. Genetics 210:1109–1124. https://doi.org/10.1534/genetics.118.301613

Schächter F, Faure-Delanef L, Guénot F et al (1994) Genetic associations with human longevity at the APOE and ACE loci. Nat Genet 6:29–32. https://doi.org/10.1038/ng0194-29

Schulze TG, McMahon FJ (2004) Defining the phenotype in human genetic studies: forward genetics and reverse phenotyping. Hum Hered 58:131–138. https://doi.org/10.1159/000083539

Sebastiani P, Nussbaum L, Andersen SL et al (2016) Increasing sibling relative risk of survival to older and older ages and the importance of precise definitions of "Aging," "Life Span," and "Longevity". J Gerontol A Biol Sci Med Sci 71:340–346. https://doi.org/10.1093/gerona/glv020

Singh J, Minster RL, Schupf N et al (2017) Genomewide association scan of a mortality associated endophenotype for a long and healthy life in the long life family study. J Gerontol A Biol Sci Med Sci 72:1411–1416. https://doi.org/10.1093/gerona/glx011

Tazearslan C, Huang J, Barzilai N, Suh Y (2011) Impaired IGF1R signaling in cells expressing longevity-associated human IGF1R alleles. Aging Cell 10:551–554. https://doi.org/10.1111/j.1474-9726.2011.00697.x

Tikkanen E, Gustafsson S, Amar D et al (2018) Biological insights into muscular strength: genetic findings in the UK Biobank. Sci Rep 8:6451. https://doi.org/10.1038/s41598-018-24735-y

Timmers PRHJ, Mounier N, Lall K et al (2019) Genomics of 1 million parent lifespans implicates novel pathways and common diseases and distinguishes survival chances. 1–40. https://doi.org/10.7554/eLife.39856.001

van den Berg N, Beekman M, Smith KR et al (2017) Historical demography and longevity genetics: back to the future. Ageing Res Rev 38:28–39. https://doi.org/10.1016/j.arr.2017.06.005

van den Berg N, Rodríguez-Girondo M, van Dijk IK et al (2019) Longevity defined as top 10% survivors and beyond is transmitted as a quantitative genetic trait. Nat Commun 10:35. https://doi.org/10.1038/s41467-018-07925-0

Veale AMO, Falconer DS (1960) Introduction to quantitative genetics. Appl Stat 9:202. https://doi.org/10.2307/2985722

Walter S, Atzmon G, Demerath EW et al (2011) A genome-wide association study of aging. Neurobiol Aging 32. https://doi.org/10.1016/j.neurobiolaging.2011.05.026

Willcox BJ, Donlon TA, He Q et al (2008) FOXO3A genotype is strongly associated with human longevity. Proc Natl Acad Sci 105:13987–13992. https://doi.org/10.1073/pnas.0801030105

Willems SM, Wright DJ, Day FR et al (2017) Large-scale GWAS identifies multiple loci for hand grip strength providing biological insights into muscular fitness. Nat Commun 8:16015. https://doi.org/10.1038/ncomms16015

Zenin A, Tsepilov Y, Sharapov S et al (2019) Identification of 12 genetic loci associated with human healthspan. Commun Biol 2:41. https://doi.org/10.1038/s42003-019-0290-0

Zhao X, Liu X, Zhang A et al (2018) The correlation of copy number variations with longevity in a genome-wide association study of Han Chinese. Aging (Albany NY) 10:1206–1222. https://doi.org/10.18632/aging.101461

Genetic Dyslipidemias

▶ Familial Dyslipidemias

Genetic Factor

▶ Alleles

Genetic Instability

▶ Genome Instability

Genetic Makeup

▶ Genotype

Genetic Sequence

▶ Nucleotides

Genetic Testing for Cancer

▶ Cancer Screening

Genetic Theories of Aging

Cristina Giuliani[1,2], Paolo Garagnani[3] and Claudio Franceschi[4]
[1]Laboratory of Molecular Anthropology and Centre for Genome Biology, Department of Biological, Geological and Environmental Sciences (BiGeA), University of Bologna, Bologna, Italy
[2]School of Anthropology and Museum Ethnography, University of Oxford, Oxford, UK
[3]Department of Experimental, Diagnostic and Specialty Medicine (DIMES), University of Bologna, Bologna, Italy
[4]IRCCS, Institute of Neurological Sciences of Bologna, Bologna, Italy

Definition

The aging theories are a very high number, and also many scientists proposed different ways of dividing and categorizing them. As a major subset of these theories, the genetic theories of aging include three main concepts: (1) the genetics of aging can be interpreted in the light of the evolutionary theories; (2) the genetics of aging and longevity can be informative if ecological and anthropological views are considered; (3) the genetics components underpin all the theories of aging even if not specifically stated.

Overview

Aging and longevity (as well as each phenotype and complex trait) are the results of complex gene-environment interactions (GxE), and in modern humans, the dimension of the environment is very heterogeneous as it includes the cultural dimension and the ecological one (Giuliani et al. 2017). Moreover if we consider the human body as the reference point, the term environment refers to the past and present environment and includes two dimensions: (1) the external environment (climate, toxicants, water, soil, pollutants, diet, technology, socioeconomic background, etc.) and (2) the internal environment (the internal conditions in which each cell of the body is located termed also the milieu) (Giuliani et al. 2018). Aging itself is characterized by remodeling of the body and of the internal environment, and this process includes changes in metabolisms, immune functions, sensorial perception, body composition, and hormonal profile among others. Changes in both internal and external environment shape the milieu in which each gene is located with the probability of impacting the final phenotype (Ukraintseva et al. 2016).

Moreover, it is to note that the genetic background observed in a given time is the result of past evolutionary dynamics such as migration and local selective pressures indicating that also the environment experienced by past population plays a role in determining the genetic variability observed in today's populations. Thus, the genetic theories of aging are elucidated in the light of evolutionary biology and human ecology (Fig. 1).

Key Research Findings

Genetic Theories of Aging: Evolutionary Considerations

Evolutionary considerations can be divided in two main groups; the first opened by Alfred Russel Wallace supports that aging is a phenomenon fundamental for the benefit of a population as it makes possible the availability of resources for the new generation. Thus, according to this thesis, the good of the species/population is favored if compared with the good of individuals. Also Josh Mitteldorf proposed that aging has evolved in order to stabilize population dynamics (Mitteldorf 2012; Goldsmith 2017).

The second view is supported by three evolutionary biologists (J.B.S. Haldane, Peter B. Medawar, and George C. Williams) that argue that aging is not a mechanism fixed for the "good of the species" as previously suggested, but aging evolves because the strength of natural selection becomes inefficient after the reproductive period and at old age. These theories are born from the observation of an age-related pathology, i.e., Huntington disease, a genetic disease whose outcome (mental illness) occurs after the reproductive ages, and thus the genes that cause this pathology have already been transmitted to the offspring. For a lot of time, the vast majority of consideration focused on this second view (that in our opinion explains why genes involved in age-related pathology are still present in the general populations but do not explain mechanisms of physiological aging or in other words why people die of old age (Rando 2006)).

After these first considerations influenced by the Darwinian thinking, in the 1980s two new theories (supported also by experimental data mainly in animal models) spread in the scientific communities (Partridge and Gems 2002). These are (1) mutation accumulation and (2) trade-off mechanisms of aging (that is closely link to antagonistic pleiotropy and disposable soma).

Mutation Accumulation

Mutation accumulation theory suggests that a genetic mutation can reach a high frequency in

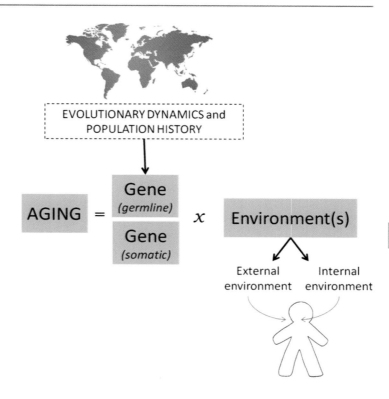

Genetic Theories of Aging, Fig. 1 Aging is the result of gene-environment interactions and thus is highly population-specific

the population because its effects occur late in life, after reproduction. The mutation accumulation (MA) hypothesis proposed by Medawar suggests that aging is the consequence of the declining force of natural selection with age: slightly deleterious germline mutations that are functional in old age are not effectively eliminated by selection and therefore lead to aging-related phenotypes. A data that supports these theories come from a study of >2,500 human genetic variants associated with 120 genetic diseases that demonstrated that variants associated with late-onset disease segregate at higher frequencies than those associated with early-onset disease (Rodríguez et al. 2017).

Trade-Off Theory and Antagonistic Pleiotropy

The concept of trade-off is based on the "resource allocation model" (Lemaitre et al. 2015; Beirne et al. 2015) in which each organism needs to allocate a finite amount of energy across different tissues including growth, reproduction, and "somatic maintenance." The concept of trade-off in economics is very clear: money spent for something is not available for buying something else, and it follows the Y model in which for a given amount of resource (e.g., energy, space, time), it is not possible to increase two traits at once (Garland 2014).

Given a fixed energy "budget," differences in allocation among individuals may result in a negative correlation or trade-offs, among traits at the individual level. From here it is easier to understand the concept of antagonistic pleiotropy. According to the antagonistic pleiotropy, the same allele can act on a trait beneficial for survival/reproductive success in the first stage of life, and the same allele may be associated with a different trait deleterious for survival late in life (Gavrilov and Gavrilova 2002) (e.g., one allele may be associated with reproductive success during pathogen infections, and the same allele may be associated with neurodegeneration late in life as suggested for APOE (van Exel et al. 2017; Abondio et al. 2019)). The term pleiotropy in genetics usually indicates those genes whose encoding protein has multiple qualitatively distinct functions (target). The human genome has around 23,000 genes and encodes a phenotype with more than 23,000 aspects, and this demonstrates that many genes must have many functions. Thus, given the pleiotropic nature of genes, the term antagonistic refers to the fact that the same gene may have different effects (beneficial or detrimental).

It is to note that a number of alleles associated with a range of other diseases may have possessed similar antagonistically pleiotropic effects and led to specific aspects of our senescence. One example is the carcinoma risk (Crespi 2011), and many other examples of antagonistic pleiotropy are reported by Carter and Nguyen (2011).

Following this line of reasoning, T. Kirkwood proposed the disposable soma theory (Kirkwood 1977; Kirkwood and Holliday 1979). The theories are based on the observation that strong evolutionary pressure is necessary to maintain the integrity of germ cells, but very weak pressure is essential for maintaining somatic tissues. As a result, somatic tissues are not maintained well enough, which results in time-dependent deterioration, i.e., aging. This theory can be considered the most inclusive one as it gives reason of many features of aging (including why mutations accumulate during aging). He argued of the hypothetical existence of a class of genetic mutations that save energy for reproduction by reducing molecular proofreading in somatic cells (some examples are reported in animal models but it lacks example in humans because of the difficulties to associate genetics to fertility) (Gavrilov and Gavrilova 2002).

Genetic Theories of Aging: Considerations from Anthropology and Human Ecology

After many studies on the role of genetics in longevity and the discovery of many genes involved in the process (described in other sections), recently some theories on human ecology have been included in the genetic theories of aging and in particular the concept of niche construction.

An example regards the theories of Govindaraju and colleagues in 2015 (Govindaraju et al. 2015) that considered longevity as a complex trait where individuals, families, and populations play a major role. In their paper longevity is proposed as the result of the interactions between genotype-epigenetic and phenotype (called GEP). They suggested that the analysis of the genetics of centenarians and the approach of population ecology and evolutionary biology may provide new insight for extending health span. The authors proposed a "genotype space," a "epigenetic space," and a "phenotype space." The first is influenced by nuclear and mitochondrial genomes and the interactions among them. This genotype space may include also the genomes of microbiota (Garagnani et al. 2014). The second includes all the mechanisms between genotype and the phenotype, and the third includes all the nongenetic factors.

A second example reported the study of genetics of aging and longevity in relation to the **ecological space** (Giuliani et al. 2018) a term that represents and grasps (in a concentric way) the different timescales of interactions between genes, culture, and ecology. In particular genetics should be interpreted not only at individual level but considering family (GxE1), community (GxE1xE2), population (GxE1xE2xE3), and cultural, social, and economic factors that are horizontal across populations (GxE1xE2xE3xE4). For extensive review and examples, see Alvergne et al. (2016) and Giuliani et al. (2017, 2018).

Genetic Variability in the Seven Pillars of Aging

The genetics of aging and longevity is a very complex field impossible to summarize in few pages, but many reviews have been published (Capri et al. 2013; Garagnani et al. 2014; Giuliani et al. 2018).

There is not one unifying theory for the genetics of aging, but the genetics is involved in all the main theories of aging. For this chapter we will consider the seven pillars (i.e., metabolism, macromolecular damage, epigenetics, inflammation, adaptation to stress, proteostasis, stem cells and regeneration) of the recent theories of aging identified by Kennedy and colleagues (2014), and we will describe one example of how genes/network of genes are involved in each theory.

Metabolism

Aging is characterized by metabolic changes, including modulation of mitochondrial function, a decline in insulin sensitivity, and alterations in substrate utilization (Riera and Dillin 2015), and many recent data on animal models showed that reduction of food intake correlates with an extension in lifespan (Finkel 2015). Many molecular mechanisms are involved in this process, and in particular the major ones are (1) nutrient sensing (through mTOR, sirtuins and AMPK are molecular mechanisms that link nutrients and lifespan) and (2) IGF and insulin signaling. Genetics have been studies for all the molecular mechanisms sited above; in particular in centenarians, some common genetic variants located in IGF-1R have been associated with low levels of plasma levels of IGF-1 (Bonafè et al. 2003), and in 2008 Suh and colleagues demonstrated that Ashkenazi Jewish centenarians carry genetic variants in IGF-1R that reduce the efficiency of the receptor generating a sort of IGF-1 resistance (Suh et al. 2008).

Macromolecular Damage

The decline in DNA repair efficiency, the accumulation of unrepaired DNA damages, and replicative senescence are processes that occur during aging (Gorbunova et al. 2007; Vaidya et al. 2014). The role of the genes involved in DNA repair is crucial for the aging process and demonstrated by progeroid syndromes that are characterized by an impairment of the mechanisms involved in DNA repair (Martin and Oshima 2000). Impairment of these genes leads to accelerated aging as a consequence of DNA damage accumulation.

Moreover epistatic interactions between SNPs located in different genes (TP53-DNA repair pathway/TXNRD1-pro-antioxidant pathway and TP53-DNA repair pathway/ERCC2-DNA repair pathway) have been demonstrated to contribute to human longevity (Dato et al. 2018).

Epigenetics

During aging, the epigenetic information changes sporadically in response to exogenous and endogenous factors (for an extensive review, see Pal and Tyler (2016)). Old cells in different tissues are characterized by an abnormal chromatin state (such as altered DNA methylation profiles, epimutations). Moreover, in aging the abnormal chromatin state is associated with genomic instability and to the activation of transposable elements (epigenetic mechanisms and somatic mutations are strictly interconnected). To this purpose recent data showed an accumulation of somatic mutation in hematopoietic cells of people with cardiovascular diseases (and thus mortality) and an increase of these mutations with aging. The genes involved in this process are DNMT3A, TET2, and ASXL1 (Jaiswal et al. 2014).

Inflammation

The theory of inflammaging was proposed by Franceschi and colleagues in 2000 and highlights the role of inflammation during aging and in particular on the chronic low-grade inflammation (Franceschi et al. 2000a). Even in the inflammatory process, genetics play a role. Interleukin 6 (IL6) is a pro-inflammatory cytokine that has been associated in conditions of chronic inflammation as well as in advanced age. Genetic variants located in IL-6 gene have been found negative associated with longevity in the Han Chinese population (rs2069837) (Zeng et al. 2016), and a different genetic variant (rs1800795) located in the same gene has been found associated with longevity in Italian centenarian males (Bonafè et al. 2001). This is an example of genetic factor that may increase susceptibility to high level of inflammation that reduces longevity.

Adaptation to Stress

Some theories of aging are based on stress response (Franceschi et al. 2000b), and in particular aging therefore can be considered a process of progressively falling homeostasis in response to the stresses encountered by the organisms itself. In this view longevity is directly linked with the capability to cope with a variety of stressors. The remodeling theory of aging (proposed in 1995 after experiment on the immune system of centenarians) (Franceschi and Cossarizza 1995; Franceschi et al. 1995) is based on the concept of adaptation in terms of the continuous reshaping of the body to environmental (external and internal) stimuli. Healthy centenarians are "those who have the better capacity to adapt to damaging agents and in particular to immunological stressors." It is thus clear that stress is context-dependent both for animal models and even more for humans (Parsons 1995; Zeng et al. 2016; Giuliani et al. 2017, 2018). This has got many implications also for genetics as the same alleles can be associated with different phenotypes in different human populations and in different environments. Evolutionary medicine approaches and population genetics studies showed that frequency of allele involved in diseases may reach high frequency in certain populations because of different local selective pressures and adaptation to past environment (Quintana-Murci 2016). The above-mentioned evolutionary dynamics and others (such as demographic history, migration, and drift) have a main impact on aging and longevity since there is a strong link between age-related diseases and longevity (Fortney et al. 2015; Giuliani et al. 2017). It is interesting to mention the example of TCF7L2 gene (rs7903146-TT) a SNP associated with type 2 diabetes and to complications of diabetes that are at low frequency in centenarians (Garagnani et al. 2013); interestingly a study of Corella and colleagues showed that the association of this variant to cardiovascular event change according to nutrition and in particular the Mediterranean diet seems to counterbalance the effect of the risk genotype (Corella et al. 2013; Corella and Ordovás 2014).

A second example of how genetics may be involved in this theory is the response to genotoxic stress. Recent study showed that SIRT6 (sirtuin 6) a gene involved in stress responses during the course of aging is depleted from L1 loci (long interspersed element 1), resulting in the upregulation of L1 activity indicating that while mobilization of chromatin-modifying elements in response to stress-stimuli is adaptive in young organisms, constitutive redistribution of these elements may drive age-related phenotypes (Van Meter et al. 2014).

Proteostasis

Protein homeostasis, or proteostasis, is assured through the coordinated action of intricate cellular systems in order to maintain cellular proteins in a state that allows optimum biological activity (Powers and Balch 2013). Proteostasis mechanisms lead to primary, secondary, ternary, and quaternary structural states of a protein, thus conferring more functions to a single protein. Moreover the term proteostasis includes processes that stabilize folded proteins (heat-shock family of proteins) as well as processes for the degradation of proteins (proteasome or the lysosome) (Hartl et al. 2011; López-Otín et al. 2013). Loss of proteostasis is also linked to aging, and it has been associated with many diseases (and age-related disease) such as Parkinson's disease and Alzheimer's disease (Koga et al. 2011). The endoplasmic reticulum (ER) is the cellular compartment crucial for protein folding and trafficking, and the impairment of the ER's adaptive capacity results in activation of the unfolded protein response (UPR) which intersect with many inflammatory signals (for an extensive review, see Hotamisligil (2010)), and it is thus crucial for aging process. Many genetic variants located in genes that encode for ER have been associated with aging and longevity indicating that genetics play a role even in the theories based on proteostasis. Even if the study regarding proteostasis in aging is obviously focused on proteins, genetics may play also in this case a role. In yeast the deletion of several UPR target genes (IRE1 is one of them) significantly increases replicative lifespan, and in mice the genetic ablation of Ire1α in intestinal epithelial cells generated short-lived mice with impaired intestinal epithelial barrier function (Zhang et al. 2015; Luis et al. 2016). In humans a study on Italian centenarians showed that genetic variants located in NACAD gene (a gene involved in the inappropriate targeting of nonsecretory polypeptides to the ER) were among the most significant signals of association with longevity.

Stem Cells and Regeneration

A theory that provides a more comprehensive framework for considering the influences of genetics on the aging process is the "disposable soma theory" (Kirkwood 1977; Kirkwood and Holliday 1979). As introduced before, the basic premise is that species have evolved with genetic programs that optimize the utilization of resources for survival and reproduction. There is overwhelming evidence that there are strong genetic influences on the rate of aging. The disposable soma theory also provides an interesting framework for considering adult stem cells in relation to the disposable soma in which they reside. Adult stem cells are located in many tissues in many organisms, but an overall decline in tissue regenerative potential with age has been observed (Rando 2006) probably because of intrinsic events (such as DNA damage) as well as of extrinsic causes (such as changes in their supporting niches). The decreased regenerative capacity appears to contribute to some aspects of mammalian aging opening to theories such as "stem cell hypothesis" at the basis of age-related diseases. Genetics is highly implicated in this theory as unrepaired genetic lesions in stem cells are passed on to their self-renewing daughters and accumulate during aging and because a stem cell that experiences a DNA damage may be at high risk of malignancy. Mitochondrial DNA mutations play also a role in stem cell aging, and mtDNA acquired mutations expand clonally with tissue-specific patterns during aging (in the colon, small intestine, and stomach among others), thus affecting stem cell homeostasis (for reviews, see Baines et al. (2014) and Su et al. (2018)).

Prospects and Summary

Given the constant reduction of the costs of genetic analyses and the technological possibility of producing more and more genome-wide data on the single individual, the future challenge will be to find models able to integrate these data in the modern theories of aging. In this regard the term phenomics has been introduced to indicate the detailed characterization of all the characteristics of the individual that are urgent to include in genetic studies (Houle et al. 2010). Furthermore, the ability to collect more and more data that characterize the ecological niche of each individual and to build mathematical models that include interdisciplinary data on historical, socioeconomic, and cultural aspects will be fundamental to grasp the role of the genetic variability in human aging.

Aging and longevity (as well as each phenotype and complex trait) are the results of complex gene-environment interactions (GxE), and in modern humans, the dimension of the environment is very heterogeneous as it includes the cultural dimension and the ecological one (Giuliani et al. 2017). We need to start from the genetic history of populations (migrations, selection events, isolated, inbreeding, etc.) and from the study of ecological niche (termed also ecological space in Giuliani et al. (2018)) of each individual to grasp the real weight of the genetic component in longevity. Moreover, genetic changes observed in aging may be understood in the light of the theories of aging such as mutation accumulations, trade-offs, and antagonistic pleiotropy and the disposable soma. The genetic background is a "constant layer" that, even if not explicitly stated, plays a role in all the theories of aging and in particular plays a role also in the seven pillars of aging (i.e., metabolism, macromolecular damage, epigenetics, inflammation, adaptation to stress, proteostasis, stem cells and regeneration) (Kennedy et al. 2014).

Cross-References

▶ Aging Mechanisms
▶ Aging Theories
▶ Biogerontology
▶ Centenarians
▶ DNA Damage Theory
▶ Genes That Delay Aging
▶ Genetics: Gene Expression
▶ Genomics of Aging and Longevity
▶ Immunological Theory of Aging
▶ Supercentenarians
▶ Whole Genome Sequencing

References

Abondio P, Sazzini M, Garagnani P et al (2019) The genetic variability of APOE in different human populations and its implications for longevity. Genes 10:222. https://doi.org/10.3390/genes10030222

Alvergne A, Jenkinson C, Faurie C (eds) (2016) Evolutionary thinking in medicine. Springer International Publishing, Cham

Baines HL, Turnbull DM, Greaves LC (2014) Human stem cell aging: do mitochondrial DNA mutations have a causal role? Aging Cell 13:201–205. https://doi.org/10.1111/acel.12199

Beirne C, Delahay R, Young A (2015) Sex differences in senescence: the role of intra-sexual competition in early adulthood. Proc R Soc B Biol Sci 282:20151086. https://doi.org/10.1098/rspb.2015.1086

Bonafè M, Olivieri F, Cavallone L et al (2001) A gender – dependent genetic predisposition to produce high levels of IL-6 is detrimental for longevity. Eur J Immunol 31:2357–2361. https://doi.org/10.1002/1521-4141(20 0108)31:8<2357::AID-IMMU2357>3.0.CO ;2-X

Bonafè M, Barbieri M, Marchegiani F et al (2003) Polymorphic variants of insulin-like growth factor I (IGF-I) receptor and phosphoinositide 3-kinase genes affect IGF-I plasma levels and human longevity: cues for an evolutionarily conserved mechanism of life span control. J Clin Endocrinol Metab 88:3299–3304. https://doi.org/10.1210/jc.2002-021810

Capri M, Santoro A, Garagnani P et al (2013) Genes of human longevity: an endless quest? Curr Vasc Pharmacol 12:707

Carter AJ, Nguyen AQ (2011) Antagonistic pleiotropy as a widespread mechanism for the maintenance of polymorphic disease alleles. BMC Med Genet 12:160. https://doi.org/10.1186/1471-2350-12-160

Corella D, Ordovás JM (2014) Aging and cardiovascular diseases: the role of gene-diet interactions. Ageing Res Rev 18:53–73. https://doi.org/10.1016/j.arr.2014.08.002

Corella D, Carrasco P, Sorlí JV et al (2013) Mediterranean diet reduces the adverse effect of the TCF7L2-rs7903146 polymorphism on cardiovascular risk factors and stroke incidence: a randomized controlled trial in a high-cardiovascular-risk population. Diabetes Care 36:3803–3811. https://doi.org/10.2337/dc13-0955

Crespi BJ (2011) The emergence of human-evolutionary medical genomics. Evol Appl 4:292–314. https://doi.org/10.1111/j.1752-4571.2010.00156.x

Dato S, Soerensen M, De Rango F et al (2018) The genetic component of human longevity: new insights from the analysis of pathway-based SNP-SNP interactions. Aging Cell 17:e12755. https://doi.org/10.1111/acel.12755

Finkel T (2015) The metabolic regulation of aging. Nat Med 21:1416–1423. https://doi.org/10.1038/nm.3998

Fortney K, Dobriban E, Garagnani P et al (2015) Genome-wide scan informed by age-related disease identifies loci for exceptional human longevity. PLoS Genet 11:e1005728. https://doi.org/10.1371/journal.pgen.1005728

Franceschi C, Cossarizza A (1995) Introduction: the reshaping of the immune system with age. Int Rev Immunol 12:1–4. https://doi.org/10.3109/08830189509056697

Franceschi C, Monti D, Sansoni P, Cossarizza A (1995) The immunology of exceptional individuals: the lesson of centenarians. Immunol Today 16:12–16

Franceschi C, Bonafè M, Valensin S et al (2000a) Inflamm-aging. An evolutionary perspective on immunosenescence. Ann N Y Acad Sci 908:244–254

Franceschi C, Valensin S, Bonafè M et al (2000b) The network and the remodeling theories of aging: historical background and new perspectives. Exp Gerontol 35:879–896

Garagnani P, Giuliani C, Pirazzini C et al (2013) Centenarians as super-controls to assess the biological relevance of genetic risk factors for common age-related diseases: a proof of principle on type 2 diabetes. Aging 5:373–385

Garagnani P, Pirazzini C, Giuliani C et al (2014) The three genetics (nuclear DNA, mitochondrial DNA, and gut microbiome) of longevity in humans considered as metaorganisms. Biomed Res Int 2014:e560340. https://doi.org/10.1155/2014/560340

Garland T (2014) Trade-offs. Curr Biol 24:R60–R61. https://doi.org/10.1016/j.cub.2013.11.036

Gavrilov LA, Gavrilova NS (2002) Evolutionary theories of aging and longevity. Sci World J 2:339–356. https://doi.org/10.1100/tsw.2002.96

Giuliani C, Pirazzini C, Delledonne M et al (2017) Centenarians as extreme phenotypes: an ecological perspective to get insight into the relationship between the genetics of longevity and age-associated diseases. Mech Ageing Dev. https://doi.org/10.1016/j.mad.2017.02.007

Giuliani C, Garagnani P, Franceschi C (2018) Genetics of human longevity within an eco-evolutionary nature-nurture framework. Circ Res 123:745–772. https://doi.org/10.1161/CIRCRESAHA.118.312562

Goldsmith TC (2017) Evolvability, population benefit, and the evolution of programmed aging in mammals. Biochem Mosc 82:1423–1429. https://doi.org/10.1134/S0006297917120021

Gorbunova V, Seluanov A, Mao Z, Hine C (2007) Changes in DNA repair during aging. Nucleic Acids Res 35:7466–7474. https://doi.org/10.1093/nar/gkm756

Govindaraju D, Atzmon G, Barzilai N (2015) Genetics, lifestyle and longevity: lessons from centenarians. Appl Transl Genom 4:23–32. https://doi.org/10.1016/j.atg.2015.01.001

Hartl FU, Bracher A, Hayer-Hartl M (2011) Molecular chaperones in protein folding and proteostasis. Nature 475:324–332. https://doi.org/10.1038/nature10317

Hotamisligil GS (2010) Endoplasmic reticulum stress and the inflammatory basis of metabolic disease. Cell 140:900–917. https://doi.org/10.1016/j.cell.2010.02.034

Houle D, Govindaraju DR, Omholt S (2010) Phenomics: the next challenge. Nat Rev Genet 11:855–866. https://doi.org/10.1038/nrg2897

Jaiswal S, Fontanillas P, Flannick J et al (2014) Age-related clonal hematopoiesis associated with adverse outcomes. N Engl J Med 371:2488–2498. https://doi.org/10.1056/NEJMoa1408617

Kennedy BK, Berger SL, Brunet A et al (2014) Geroscience: linking aging to chronic disease. Cell 159:709–713. https://doi.org/10.1016/j.cell.2014.10.039

Kirkwood TBL (1977) Evolution of ageing. Nature 270:301–304. https://doi.org/10.1038/270301a0

Kirkwood TBL, Holliday R (1979) The evolution of ageing and longevity. Proc R Soc Lond B Biol Sci 205:531–546. https://doi.org/10.1098/rspb.1979.0083

Koga H, Kaushik S, Cuervo AM (2011) Protein homeostasis and aging: the importance of exquisite quality control. Ageing Res Rev 10:205–215. https://doi.org/10.1016/j.arr.2010.02.001

Lemaitre J-F, Berger V, Bonenfant C et al (2015) Early-late life trade-offs and the evolution of ageing in the wild. Proc R Soc B Biol Sci 282:20150209. https://doi.org/10.1098/rspb.2015.0209

López-Otín C, Blasco MA, Partridge L et al (2013) The hallmarks of aging. Cell 153:1194–1217. https://doi.org/10.1016/j.cell.2013.05.039

Luis NM, Wang L, Ortega M et al (2016) Intestinal IRE1 is required for increased triglyceride metabolism and longer lifespan under dietary restriction. Cell Rep 17:1207–1216. https://doi.org/10.1016/j.celrep.2016.10.003

Martin GM, Oshima J (2000) Lessons from human progeroid syndromes. Nature 408:263–266. https://doi.org/10.1038/35041705

Mitteldorf JJ (2012) Adaptive aging in the context of evolutionary theory. Biochem Mosc 77:716–725. https://doi.org/10.1134/S0006297912070036

Pal S, Tyler JK (2016) Epigenetics and aging. Sci Adv 2:e1600584. https://doi.org/10.1126/sciadv.1600584

Parsons PA (1995) Inherited stress resistance and longevity: a stress theory of ageing. Heredity 75:216–221. https://doi.org/10.1038/hdy.1995.126

Partridge L, Gems D (2002) Mechanisms of ageing: public or private? Nat Rev Genet 3:165–175. https://doi.org/10.1038/nrg753

Powers ET, Balch WE (2013) Diversity in the origins of proteostasis networks – a driver for protein function in evolution. Nat Rev Mol Cell Biol 14:237–248. https://doi.org/10.1038/nrm3542

Quintana-Murci L (2016) Understanding rare and common diseases in the context of human evolution. Genome Biol 17:225. https://doi.org/10.1186/s13059-016-1093-y

Rando TA (2006) Stem cells, ageing and the quest for immortality. Nature 441:1080–1086. https://doi.org/10.1038/nature04958

Riera CE, Dillin A (2015) Tipping the metabolic scales towards increased longevity in mammals. Nat Cell Biol 17:196–203. https://doi.org/10.1038/ncb3107

Rodríguez JA, Marigorta UM, Hughes DA et al (2017) Antagonistic pleiotropy and mutation accumulation influence human senescence and disease. Nat Ecol Evol 1:0055. https://doi.org/10.1038/s41559-016-0055

Su T, Turnbull D, Greaves L (2018) Roles of mitochondrial DNA mutations in stem cell ageing. Genes 9:182. https://doi.org/10.3390/genes9040182

Suh Y, Atzmon G, Cho M-O et al (2008) Functionally significant insulin-like growth factor I receptor mutations in centenarians. Proc Natl Acad Sci 105:3438–3442. https://doi.org/10.1073/pnas.0705467105

Ukraintseva S, Yashin A, Arbeev K et al (2016) Puzzling role of genetic risk factors in human longevity: "risk alleles" as pro-longevity variants. Biogerontology 17:109–127. https://doi.org/10.1007/s10522-015-9600-1

Vaidya A, Mao Z, Tian X et al (2014) Knock-in reporter mice demonstrate that DNA repair by non-homologous end joining declines with age. PLoS Genet 10:e1004511. https://doi.org/10.1371/journal.pgen.1004511

van Exel E, Koopman JJE, Bodegom DV et al (2017) Effect of APOE ε4 allele on survival and fertility in an adverse environment. PLoS One e0179497:12. https://doi.org/10.1371/journal.pone.0179497

Van Meter M, Kashyap M, Rezazadeh S et al (2014) SIRT6 represses LINE1 retrotransposons by ribosylating KAP1 but this repression fails with stress and age.

Nat Commun 5:5011. https://doi.org/10.1038/ncomms6011

Zeng Y, Nie C, Min J et al (2016) Novel loci and pathways significantly associated with longevity. Sci Rep 6. https://doi.org/10.1038/srep21243

Zhang H-S, Chen Y, Fan L et al (2015) The endoplasmic reticulum stress sensor IRE1α in intestinal epithelial cells is essential for protecting against colitis. J Biol Chem 290:15327–15336. https://doi.org/10.1074/jbc.M114.633560

Genetic Transformation

▶ Mutation

Genetics and Epigenetics of Aging and Longevity

▶ Genomics of Aging and Longevity

Genetics: Ethnicity

Hoi Shan Kwan
School of Life Sciences, The Chinese University of Hong Kong, Shatin, NT, Hong Kong

Synonyms

Genetic ancestry; Heredity by ethnicity

Definition

Ethnicity does not have a clear definition. It is usually referred to race or ancestry in population genetics. Race and ancestry are more related to genetics of the population. Ethnicity can be more related to cultural background. However, people of the same ethnicity can have no sharing in cultural background. Ethnicity is now a controversial term to use in genetics (Ossorio and Duster 2005).

Overview

Ethnicity is a complicated term without a clear definition. It is commonly used interchangeably with race (Ossorio and Duster 2005). People assign themselves and others to specific ethnic groups by phenotypic traits rather than the genotypes. The phenotypic traits are difficult to define and representing only a small fraction of the genome, determined by only a few genes. Different ethnic groups have different genomic variants which can be more precisely assessed with the advancement of next-generation sequencing (NGS). Genetic variations among different populations can be clearly shown. Population groups could be clustered according to their genetic distances. However, the population groups are usually not corresponding to the generally assigned ethnic groups. Individuals within a population have higher genetic variations than those between populations in an ethnic group. Simply assigning ethnicity of an individual would not reveal the genetic variants of the individual. Compounding to the problem is ethnicity, assigned or self-identified, is complicated by social factors. The non-biological nature is considered to be the result of disruption by demographic changes in recent human evolution events.

Race and ethnicity are controversial and misunderstood terms within the scientific community. When these terms are used as parameters for health care and treatment of diseases and aging-related health issues (Ferraro et al. 2017), they are limited in reflecting the genetic characteristics of the individual. Ethnic ancestry has been used in place of ethnicity for more precise representations but is also limited in tracing the ancestry.

When ethnicity and ancestry can be more precisely determined, genetic disorders can be shown to have different frequencies in different populations. With caution, in specified uses, knowledge obtained on genomic variants in each population would assist the understanding of the role of genomics in health, aging, and diseases. The current approach is to increase the understanding of the genetic determinants related to aging and diseases with large-scale multi-ethnic biobanks and population health resources on a global scale. Genome-wide association studies (GWAS) within small ethnic population would be an important approach to relate genomic loci to key human aging traits (Melzer et al. 2020).

Key Research Findings

Genetic Variation Between Ethnic Groups

Genetic variation between ethnic groups causes phenotypic trait differences. Different ethnic groups are commonly distinguished by their phenotypic traits including more superficial ones such as color of skin, eye, and hair. Only a few genes determine these traits. For example, ~11 genes affect skin color (Sturm and Duffy 2012), and ~ 16 genes affect eye color with 2 genes having the most impact (White and Rabago-Smith 2011). Most people would take these traits as the basis for determining ethnicity. These traits represent only a very small fraction of the genetic variation among ethnic groups.

Now ethnicity is more precisely assessed with genetic testing that analyzes hundreds to thousands of genetic markers or more recently the entire genome. The advancement of DNA sequencing technology lowers the time and cost to the extent that genome sequencing is quite common in developed countries (Turbitt et al. 2019).

Analyzing an extensive number of genes of many individuals reveals the genetic variations among different human populations. Clusters of populations could be identified by cluster analysis or principal component analysis (Li et al. 2008). The greater the geographical distance between two populations, the more genetic variations in general. But that is not absolute. Geologically isolated populations would have high genetic variation from populations outside (Lorusso and Bacchini 2015). Clusters may or may not coincide with ethnicities.

The clusters can be regarded as subgroups of the human population and correspond roughly to ethnicities. Genetic distance between the subgroups is calculated to reveal the genetic relatedness of the subgroups. Fixation index (F_{st}) is used to measure genetic distance between two subgroups. It is interesting to note that genetic distances between populations in Africa are greater than between populations outside Africa. Measurement of European populations returns a low Fst.

So it would be difficult to define ethnicity as a subgroup of the human populations. Genetic distances would be quite variable. In other words, human population subgroups may not have unique genetic variations that are not shared by another subgroup. Ethnic groups are even more ill-defined as population subgroups.

Lewontin (1972) used 17 markers in a Fst analysis of human populations and found that the genetic variations between individuals within a population are 85.4%, between populations in an ethnic group 8.3%, and 6.3% between two ethnic groups. He studied Caucasian, African, Mongoloid, South Asian Aborigines, Amerinds, Oceanians, and Australian Aborigines populations. Later analyses using many more markers generally supported Lewontin's analyses (Risch et al. 2002; Ossorio and Duster 2005; Long and Kittles 2009). However, it has been controversial about this observation of non-biological nature of ethnic groups.

Van Arsdale (2019) reviewed the biological concept of race and concluded that race as a biological concept has been disrupted by demographic changes in recent events in human evolution. Genetic variations do not correspond to our usual ethnicity classification.

"Race" and "Ethnicity" Are Controversial and Misunderstood Terms Within the Scientific Community

Ethnicity is a complex term that reflects biological characteristics such as skin color and cultural characteristics such as religion, history, ancestral geographical origins, socioeconomic status, education, and access to health care. It has been used as a parameter for health care and treatment of diseases. The assumption is that ethnicity would be a surrogate for multiple genetic and environmental factors in disease and health care. This assumption has been controversial (Stryjecki et al. 2018) and is sometimes considered to be flawed (Collins 2004). Indeed, ethnicity per se has been difficult to use to reflect genetic and environmental characteristics of individuals. Human population has a complex migration, colonization, and cultural history that makes self-reported ethnicity unreliable to use in diseases and health care.

Genetics Professionals' Views on Ethnicity and Genetics

Nonetheless, ethnicity, when used to refer to race or ancestry, has been used in genetic studies on human health, aging, and disease. It was considered to have a role in research and clinical care of ethnic groups. There are many studies relating to aging, health, and ethnic groups (Ferraro et al. 2017; Ganapathy et al. 2019). However, in a developed community such as the USA or European Union, ethnicity or race is difficult to define, and its association with genetics is marginal at best. Many studies showed that association between ethnicity and genetics is smaller than between population and genetics which is in turn smaller than that between individuals. Genetic professionals' views on ethnicity and genetics are divided. When more and more genomes of humans are sequenced, it is evident that ethnicity is not a precise parameter to use in research and clinical care (Nelson et al. 2018).

Ethnic Ancestry

Owing to the difficulties in using ethnicity to represent the genetic characteristics of an individual, it is proposed that the ethnic ancestry of the individual should be identified rather than a simple assignment of ethnicity of the individual. In this way, the genetic characteristics of the individual can be more accurately predicted. For health, aging, and diseases, the genetic factor of ethnicity would be the most important factor. Unfortunately, ethnic ancestry may be self-identified by the individual and thus may not be precise and may be influenced by social factors (de Andrade Ramos et al. 2016; Liebler et al. 2017). Ethnic ancestry of an individual can be very complicated for communities with high degree of genetic admixture.

Genetic markers can be used to more precisely assign ethnic ancestry. Ancestry informative markers (AIMs) are single-nucleotide polymorphism markers developed to estimate the ancestry profile from different ethnicities of an individual. AIMs are those with different frequencies among distinct populations. They are discovered and developed with the advancement of next-generation sequencing (NGS). Panels of AIMs have been developed for efficient ethnic ancestry determination (Bauchet et al. 2007; Santos et al. 2010). AIMs are useful but are limited to trace along only one maternal and one paternal line (Duster 2015). The uses of AIMs in aging studies are rare.

Examples of Application

Genetic Disorders with Different Frequency in Different Populations Related to Ethnicity and Ancestry

Ethnicity may be assumed to associate with the genetic disorders. However, since ethnicity is very complex and is only weakly related to the genome of individuals, it is not a useful marker for the diseases. Ethnic ancestry, especially confined to small and coherent populations, may be more precise as marker for associating with generic disorders.

Single gene disorders can have better associations with the population. Some examples of this type are the following:

1. Cystic fibrosis: a common autosomal recessive disease in Northern European group (O'Sullivan and Freedman 2009)
2. Sickle cell anemia: most prevalent in populations with sub-Saharan African ancestry but also common among Latin American, Middle Eastern, and South European populations (Rees et al. 2010.
3. Tay–Sachs disease: an autosomal recessive disorder more frequent among Ashkenazi Jews, French Canadians of Southeastern Quebec, and Cajuns of southern Louisiana (Myrianthopoulos and Aronson 1966; NORD 2017)

For multifactorial polygenic diseases, including age-related diseases, individuals from different ethnic ancestry may be considered high-risk or low-risk based on the estimated probability of the ethnic ancestry population known to expose to genetic factors. However, risk factors other than genetic factors, including environmental, social, and cultural factors, would contribute to the overall risk and not easy to delimit.

One example of multifactorial polygenic gerontological studies related to ethnicity was reported by Zeng et al. (2016). That study reported a GWAS of Han Chinese that identified 11 genetic loci associated with longevity. Eight of them occurred in Han Chinese, European, and US populations.

How Ethnicity and Ancestry Data Are Used in Biomedical Research (Bonham et al. 2018)

Ethnicity and ancestry data are commonly used as population descriptors and ancestral background in human genomic studies. Aiming to be more precise, ancestral informative markers (AIMs) are used to infer disease risk. AIMs have been developed to estimate admixture and genetic ancestral proportions. They are limited in value when used to infer ethnicity.

Future Directions of Research

Genome-wide association studies (GWAS) (Dehghan 2018) investigate association among common genomic variants and complex diseases. Population ancestry data including ethnicity are used in many GWAS. However, most GWAS used only European ancestral populations and lacked ethnicity diversity (Popejoy and Fullerton 2016). The GWAS results are biased and limited in clinical applications to the general public. US National Institutes of Health implemented measures to enhance GWAS to include diverse populations and admixed populations (Hindorff et al. 2018). Large-scale multi-ethnic biobanks and population health resources are built for the studies of genetic determinants associating with diseases on a global scale (Gurdasani et al. 2019). Use of GWAS in identifying genetic loci related to human aging traits has been reviewed by Melzer et al. (2020).

Knowledge obtained on genomic variants in each population would help us to understand the role of genomic research in fundamental and clinical uses in health, aging, and diseases. Studies or clinical applications would need to go beyond ethnicity and ethnic ancestry and target at understanding the genomic variants in each population. Population and ethnicity are highly associated with genetic ancestry at the population level but not at the individual level.

We need to re-visit what Kaplan and Bennett (2003) proposed as guidelines for race and ethnicity which are used in genetics research:

1. "When race/ethnicity is used as a study variable, the reason for its use should be specified."
2. "In the interpretation of racial and ethnic differences, all conceptually relevant factors should be considered."

The advancement of genomic technology allows detailed analysis of individual genomes and population genomic diversity. The use of race or ethnicity in health and aging studies should be useful if the genomic details are considered without regarding as discrete population groups and distinguished clearly from the social components of ethnicity.

Summary

Race and ethnicity are controversial and misunderstood terms without clear definition. It is common that individuals identify their ethnicity by themselves or by others based on few phenotypic traits such as skin, eye, and hair colors. The ethnicity reported as such is often considered to reflect genetic variants for the individuals. The result may be used as references for health or disease decisions. Recent advancement of next-generation sequencing technology and population genetics allows better understanding of the genetic variations in different populations. The populations, however, do not necessarily correspond to estimated ethnicity. Ethnic ancestry can be identified now to more precisely relate individuals with their ancestral population but still has only limited value to trace back.

When ethnicity and ancestry can be more precisely determined, genetic variants can be shown to have different frequency in different populations. We can use this knowledge to assist the understanding of the role of genomics in health, aging, and diseases. The scientific community is developing large-scale multi-ethnic biobanks and population health resources on a global scale to understand the genetic determinants related to diseases in different populations as small coherent ethnic groups.

Cross-References

- ▶ Genetic Theories of Aging
- ▶ Genetics: Gene Expression
- ▶ Genome-Wide Association Study
- ▶ Genomics of Aging and Longevity
- ▶ Genotype
- ▶ Human Genome Project
- ▶ Phenotype
- ▶ Race Crossover in Longevity
- ▶ Racial and Ethnic Disparities in Health
- ▶ Whole Genome Sequencing

References

Bauchet M, McEvoy B, Pearson LN et al (2007) Measuring European population stratification with microarray genotype data. Am J Hum Genet 80:948–956. https://doi.org/10.1086/513477

Bonham VL, Green ED, Pérez-Stable EJ (2018) Examining how race, ethnicity, and ancestry data are used in biomedical research. JAMA 320:1533–1534. https://doi.org/10.1001/jama.2018.13609

Collins FS (2004) What we do and don't know about "race", "ethnicity", genetics and health at the dawn of the genome era. Nat Genet 36:S13–S15. https://doi.org/10.1038/ng1436

de Andrade Ramos BR, D'Elia MPB, Amador MAT et al (2016) Neither self-reported ethnicity nor declared family origin are reliable indicators of genomic ancestry. Genetica 144:259–265. https://doi.org/10.1007/s10709-016-9894-1

Dehghan A (2018) Genome-wide association studies. Humana Press, New York, pp 37–49

Duster T (2015) A post-genomic surprise. The molecular reinscription of race in science, law and medicine. Br J Sociol 66:1–27. https://doi.org/10.1111/1468-4446.12118

Ferraro KF, Kemp BR, Williams MM (2017) Diverse aging and health inequality by race and ethnicity. Innov Aging. 1:igx002. https://doi.org/10.1093/geroni/igx002

Ganapathy V, Jacob ME, Short MI et al (2019) The healthy aging index and its association with mortality in older Mexican and European Americans. Innov Aging 3(Supplement_1):S960. https://doi.org/10.1093/geroni/igz038.3481

Gurdasani D, Barroso I, Zeggini E, Sandhu MS (2019) Genomics of disease risk in globally diverse populations. Nat Rev Genet:1–16. https://doi.org/10.1038/s41576-019-0144-0

Hindorff LA, Bonham VL, Brody LC et al (2018) Prioritizing diversity in human genomics research. Nat Rev Genet 19:175–185. https://doi.org/10.1038/nrg.2017.89

Kaplan JB, Bennett T (2003) Use of race and ethnicity in biomedical publication. JAMA 289:2709. https://doi.org/10.1001/jama.289.20.2709

Lewontin RC (1972) The apportionment of human diversity. In: Evolutionary biology. Springer US, New York, pp 381–398

Li JZ, Absher DM, Tang H et al (2008) Worldwide human relationships inferred from genome-wide patterns of variation. Science 319:1100–1104. https://doi.org/10.1126/science.1153717

Liebler CA, Porter SR, Fernandez LE et al (2017) America's churning races: race and ethnicity response changes between census 2000 and the 2010 census. Demography 54:259–284. https://doi.org/10.1007/s13524-016-0544-0

Long JC, Kittles RA (2009) Human genetic diversity and the nonexistence of biological races. Hum Biol 81:777–798. https://doi.org/10.3378/027.081.0621

Lorusso L, Bacchini F (2015) A reconsideration of the role of self-identified races in epidemiology and biomedical research. Stud Hist Phil Biol Biomed Sci 52:56–64. https://doi.org/10.1016/J.SHPSC.2015.02.004

Melzer D, Pilling LC, Ferrucci L (2020) The genetics of human ageing. Nat Rev Genet 21:88–101. https://doi.org/10.1038/s41576-019-0183-6

Myrianthopoulos NC, Aronson SM (1966) Population dynamics of Tay-Sachs disease. I. Reproductive fitness and selection. Am J Hum Genet 18:313–327

Nelson SC, Yu J-H, Wagner JK et al (2018) A content analysis of the views of genetics professionals on race, ancestry, and genetics. AJOB Empir Bioeth 9:222–234. https://doi.org/10.1080/23294515.2018.1544177

NORD (National Organization for Rare Disorders) (2017) Tay Sachs Disease. https://rarediseases.org/rare-diseases/tay-sachs-disease. Accessed 12 Jul 2019

O'Sullivan BP, Freedman SD (2009) Cystic fibrosis. Lancet 373:1891–1904. https://doi.org/10.1016/S0140-6736(09)60327-5

Ossorio P, Duster T (2005) Race and genetics: controversies in biomedical, behavioral, and forensic sciences. Am Psychol 60:115–128. https://doi.org/10.1037/0003-066X.60.1.115

Popejoy AB, Fullerton SM (2016) Genomics is failing on diversity. Nature 538:161–164. https://doi.org/10.1038/538161a

Rees DC, Williams TN, Gladwin MT (2010) Sickle-cell disease. Lancet 376:2018–2031. https://doi.org/10.1016/S0140-6736(10)61029-X

Risch N, Burchard E, Ziv E, Tang H (2002) Categorization of humans in biomedical research: genes, race and disease. Genome Biol 3:comment2007.1. https://doi.org/10.1186/gb-2002-3-7-comment2007

Santos NPC, Ribeiro-Rodrigues EM, Ribeiro-dos-Santos ÂKC et al (2010) Assessing individual interethnic admixture and population substructure using a 48-insertion-deletion (INSEL) ancestry-informative marker (AIM) panel. Hum Mutat 31:184–190. https://doi.org/10.1002/humu.21159

Stryjecki C, Alyass A, Meyre D (2018) Ethnic and population differences in the genetic predisposition to human obesity. Obes Rev 19:62–80. https://doi.org/10.1111/obr.12604

Sturm RA, Duffy DL (2012) Human pigmentation genes under environmental selection. Genome Biol 13:248. https://doi.org/10.1186/gb-2012-13-9-248

Turbitt E, Roberts MC, Hollister BM et al (2019) Ethnic identity and engagement with genome sequencing research. Genet Med 21:1735–1743. https://doi.org/10.1038/s41436-018-0410-0

Van Arsdale AP (2019) Population demography, ancestry, and the biological concept of race. Annu Rev Anthropol 48:1. https://doi.org/10.1146/annurev-anthro-102218-011154

White D, Rabago-Smith M (2011) Genotype–phenotype associations and human eye color. J Hum Genet 56:5–7. https://doi.org/10.1038/jhg.2010.126

Zeng Y, Nie C, Min J, Liu X, Li M, Chen H, Xu H, Wang M, Ni T, Li Y, Yan H (2016) Novel loci and pathways significantly associated with longevity. Sci Rep 6:21243. https://doi.org/10.1038/srep21243

Genetics: Gender

Terence Kin Wah Lee
Department of Applied Biology and Chemical Technology, The Hong Kong Polytechnic University, Kowloon, Hong Kong

Synonyms

Sex hereditary

Definition

Gender genetics refer to the hereditary factors that distinguish organisms on the basis of their reproductive roles.

Overview

Gender is genetically controlled, which determines the health and fate of an organism. Gender determination results in the evolution of individuals with particular characteristics that allow them to be identified as males, females, or hermaphrodites. However, in certain species such as solid nematode *C. elegans*, the difference in sex characteristics can be very little. The only difference is the presence of a testis versus an ovotestis. The first major breakthrough in understanding gender determination was the discovery of sex chromosomes in the early 1900s. From scrupulous analyses of male and female insect chromosomes, scientists uncovered that there were one or two additional chromosomes that were unevenly represented in the two genders although same numbers of chromosomes are found in both males and females. Analyses of other species through the years have disclosed that chromosomal differences are major determinants for genders in most animals. Among many different species on earth, insects are the most diverse class of creature. Because of this reason, it is expected that they exhibit substantial diversity in their gender determination mechanism (Saccone et al. 2002). Having said that, most insects have dimorphic sex chromosomes that can be recognized cytologically. The sex chromosomes of the fruit fly *Drosophila melanogaster* provided a mechanistic insight in understanding of gender heredity. The gender of *Drosophila* is principally controlled by the X:A ratio or the ratio of the number of X chromosomes to the number of sets of autosomes (Cline and Meyer 1996).

The equilibrium between female-determining factors encoded on the X chromosome and male-determining factors encoded on the autosomes determines which sex-specific mode of transcription will be started off. Therefore, XO and XY flies are males while XX, XXY, and XXYY flies are females. Fruit flies are impotent to live with more than two copies of X chromosome due to the effect of dosage compensation. *Sex determination system of Drosophila* also varies from mammalian one in several ways. First, sex determination begins immediately during the process of fertilization. Second, hormones are not involved for sex-specific characteristics. Third, each cell in the embryo senses the X:A ratio, triggering either the female- or male-specific pattern of transcription instead. Microarray analyses revealed that the sex-specific differences in gene expression are quite substantial. Approximately 30% of genes in *Drosophila* were found to exhibit sex-specific favoritism in expression (Parisi et al. 2004). In birds, some insects, and many reptiles in which heterogametic sex is female, ZW system was identified which is in contrast with the XY system for sex determination. It had been a belief that ZW system was also applied to snakes, but there had been unpredicted effects in the genetics of species in the families Boidae and Pythonidae which drew attention. For instance, parthenogenic reproduction produced only females but not males, and this observation is in contrast with the expected outcome in the ZW system. In mammals, Y chromosome regulates sex. Generally, cells from males contain an X and a Y chromosome, while females contain two X chromosomes. Although the role of the Y chromosome in sex determination in mammals has been uncovered in the early twentieth century, scientists were able to identify the particular region in Y chromosome that regulated this crucial process in 1959 (McLaren 1991). Using DNA hybridization technique with probes specific to different regions of the Y chromosome, scientist discovered that sex-reversed males carried genes from a 140-kilobase (kb) region on the short arm of the Y chromosome (Page et al. 1985). Subsequent studies have restricted and refined this region and identified the sex-determining region of the Y (SRY), to be the master regulator of sex determination (McLaren 1991). The presence of this region in the Y chromosome is adequate to develop to a male individual (Koopman et al. 1991). At week 7 of embryonic development, SRY gene encodes a crucial transcription factor that activates DMRT1, a critical gene responsible for formation of testis. Prior to this period, the embryonic gonad can be developed into either a testis or an ovary. After the action of SRY gene to develop into a testis, two male sex hormones including testosterone and anti-Müllerian hormone are subsequently produced, which further stimulates the formation of other organs in male reproductive system. Since SRY protein is absent in females, pathways related to ovary formation is activated by distinct sets of female sex-related proteins. The completely developed ovary subsequently produces estrogen hormone, which triggers development of other female reproductive organs including the cervix, oviducts, and uterus from the Müllerian duct. However, there are some exceptional cases in which the testes can develop without the presence of an SRY gene. For these cases, SOX9, another gene critically involved in development of the testis, is capable in inducing testis development without the assistance of SRY. Without the presence of both SRY and SOX9, the testes will not be developed. A recent report revealed that a pro-female gene FOXL2 is crucial for ovary development and perpetuation (Pannetier et al. 2016).

Key Research Findings

Gender Determination by Genetics in Human Beings

Human beings contain 46 chromosomes in 23 pairs in which the X and Y chromosomes regulate the gender of a person. The majority of female are 46XX, while 46XY is mostly found in men. In some rare cases, some individuals carry a single sex chromosome (45X or 45Y) and some with three or more sex chromosomes (47XXX, 47XYY, 47XXY, etc.). Furthermore, some males carry 46XX because of the translocation of a small region of the sex-determining region (SDR) of the Y chromosome. Likewise, some females carry 46XY because of the presence of mutations in Y chromosome. In addition to the involvement of XX and XY in sex determination, there are other factors determining the gender of individuals including hormone imbalances, variations in chromosome complements, and phenotypic differences for sex determination.

The biological distinction between men and women is associated with two processes, namely, sex determination and differentiation (Goodfellow and Lovell-Badge 1993). The biological process of sex determination regulates whether sexual differentiation pathway of the male or female will be followed. The procedures of biological sex differentiation are tightly regulated by genetic factors followed by multistep hierarchical development. In general, more than 95% of the Y chromosome is specific to male

(Willard 2003), and a copy of specific gene segment in Y chromosome is capable to induce formation of the testis upon differentiation of the gonad. Male phenotype is mainly determined by Y chromosome. Even individuals with four X chromosomes are phenotypically male if they carry one Y chromosome (49XXXXY) (Arnold et al. 2016). Without the presence of Y chromosome, the ovaries will be developed in the absence of a testis-determining factor (TDF) (Pannetier et al. 2016).

Gender is ordinarily depicted in terms of femininity and masculinity. However, its definition varies across different cultures and may be changed over time (Wood 1997). Some countries may be tidily divided along paired lines like male and female of homosexual and heterosexual. In certain countries, however, a greater gender diversity exists at a particular time. For instance, the berdache in North America, the fa'afafine in the Pacific, and the kathoey in Thailand are all cases of different gender categorization, which is different from the traditional thoughts of Western people who only possess the concept of males and females. In addition, gender is regarded to be a continuum than distinct groupings among certain native communities in North America. Currently, it becomes obvious that different cultures have adopted different points of view to create gender discrimination.

Sex Chromosome Abnormalities

Chromosomes are structures that carry genetic materials passing hereditary information from parents to their offspring. Humans carry 23 pairs of chromosome in which one half of each pair is inherited from each parent. Compared to other chromosomes, Y chromosome is very small which carries limited amount of genes with quite a number of repetitive sequences, while the X chromosome is more autosome-like in structure and content (Lahn et al. 2001). Aneuploidy refers to the condition of having less than (monosomy) or more than (polysomy) the usual diploid number of chromosomes. Aneuploidy was found in around 5% of all pregnancies and is regarded as the most common chromosomal abnormality in human beings (Hassold and Hunt 2001). Deviation from the normal number of X and Y chromosomes is considered sex chromosome aneuploidy (SCA), which occurs at the frequency of 1:400 in the general human populations (Pannetier et al. 2016). Determination for sex chromosome abnormalities has usually been achieved during pregnancy by traditional amniocentesis and chorionic villi sampling (CVS). With the recent advances in medical technology, the incidence of SCA diagnosis is in the rise. The high frequency of SCA cases is due to the fact that their outcomes are not as critical as autosomal abnormalities and are rarely deadly. The life expectancy of individual with SCA is similar to that of normal healthy people. Because of this reason, the abortion rate for SCA has constantly dropped from 100% in the 1970s to 69% in the 1980s. In the 1990s, the number of SCA cases dropped further to 49% (Christianson 2000). Such drastic decrease in number may be attributed to the improved knowledge for SCA disease after genetic counseling. The diseases related to SCA include Turner syndrome, XXX females, Klinefelter syndrome, XYY males, hermaphroditism, congenital adrenal hyperplasia, androgen insensitivity syndrome, and others. For example, women with three X chromosomes (47XXX) show normal development of sexual characteristics and are fertile. However, affected individuals are often taller than average and have slim builds. The frequency of women carrying an extra X chromosome is approximately about 0.1% of female population (Lauritsen 1982). Likewise, men who carry an additional Y chromosome (XYY males) are generally taller than average and with acne-prone skin. The frequency of this disease is similar to women with an extra X chromosome.

Implications and Prospects

The sex heredity has important implications in aging research. For instance, there is difference between the sexes in the actual rates, trends, and particular types of diseases in human beings. Social and health factors such as poor nutrition and poor education are considered to be the general disadvantages of women during their lifetime (see ▶ "Gender Disparities in Health in Later Life", in this volume). Cardiovascular diseases are the leading cause of modality of both sexes in the world. It is generally considered to be more prevalent in male than female. It is because men are less likely than women to seek medical help, at least until a disease has advanced (see ▶ "Aging and Health Disparities", in this volume). As a whole, men's life expectancy is shorter than woman (see ▶ "The Male-Female Health-Mortality Paradox", in this volume). Due to the lack of social skills and familial ties, men were found be more isolated than women at their old ages, resulting in lower quality toward the later part of their life (Steptoe et al. 2013). Apart from cardiovascular diseases, cancer is one of the major causes for death of both sexes. Men are more prone to develop liver cancer, while cancers of colon and breast are more common in female (Bray et al. 2018). In addition, the mortality of various cancers is associated with sex-specific disparities. In the future, researchers are investigating the underlying mechanism to understand the gender disparity in cancer risk. Understanding the genetics of sex difference leads to a new concept of evidence-based sex and gender medicine, which includes the fundamental differences of biology and behavior between women and men.

Summary

Hereditary factors distinguish organisms on the basis of their reproductive roles, which is defined as gender. In many species, sex determination is genetically controlled. Males and females have different alleles or different genes that define their sexual morphology. In general, different animals have their own gender-determining system, which is accompanied by chromosomal distinctions, which are generally determined through matchings of XY, ZW, XO, and ZO chromosomes or haplodiploidy. In mammals, Y chromosome determines sex. Human beings contain 46 chromosomes in 23 pairs in which X and Y chromosomes determine the gender of a person. Male cells carry an X and a Y chromosome, while females contain two X chromosomes. In Y chromosome, SRY was found to be the master regulator of sex determination by inducing the development of testis via inducing DMRT1 expression. However, there are some cases that the testis can still be developed in an individual without the presence of SRY. This is possible due to the presence of SOX9. For ovary development, FOXL2 was found to play crucial role. Gender diversity varies in different countries and differs over time. Defects in sex chromosomes will lead to development of many diseases. Gender differences are critical determinant of health and illness.

Cross-References

▶ Biogerontology
▶ Gender Disparities in Health in Later Life
▶ Gender Issues in Age Studies
▶ Gendered Aging and Sexuality in Audiovisual Culture
▶ Genetics: Gene Expression
▶ Genotype
▶ Heterosexism and Ageism

References

Arnold AP, Reue K, Eghbali M et al (2016) The importance of having two X chromosomes. Philos Trans R Soc Lond B Bio Sci 371:20150113
Bray F, Ferlay J, Soeriomataram I et al (2018) Global cancer statistics 2018: GLOBOCAN estimates of incidence and mortality worldwide for 36 cancers in 185 countries. CA Cancer J Clin 68:394–424. https://doi.org/10.3322/caac.21492
Christianson SM (2000) Parental decisions following prenatal diagnosis of sex chromosome aneuploidy: a trend over time. Prenat Diagn 20:37–40
Cline TW, Meyer BJ (1996) Vive la difference: males vs. females in flies vs. worms. Annu Rev Genet 30:637–702
Goodfellow PN, Lovell-Badge R (1993) SRY and sex determination in mammals. Annu Rev Genet 27:71–92
Hassold T, Hunt P (2001) To err (meiotically) is human: the genesis of human aneuploidy. Nat Rev Genet 2:280–291
Koopman P, Gubbay J, Vivian N et al (1991) Male development of chromosomally female mice transgenic for Sry. Nature 351:117–121
Lahn BT, Person NM, Jegalian K (2001) The human Y chromosome, in the light of evolution. Nat Rev Genet 2:207–216
Lauritsen JG (1982) The cytogenetics of spontaneous abortion. Res Reprod 14:3–4
McLaren A (1991) The making of male mice. Nature 351:96. https://doi.org/10.1038/351096a0

Page DC, De la Chapelle A, Weissenbach J (1985) Chromosome Y-specific DNA in related human XX males. Nature 315:224–226. https://doi.org/10.1038/315224a0

Pannetier M, Chassot AA, Chaboissier MC et al (2016) Involvement of FOXL2 and RSPO1 in ovarian determination, development, and maintenance in mammals. Sex Dev 10:167–184

Parisi M, Nuttall R, Edwards et al (2004) A survey of ovary-, testis-, and some-biased gene expression in *Drosophila melanogaster* adults. Genome Biol 5: R40. https://doi.org/10.1038/6769

Saccone G, Pane A, Polito LC (2002) Sex determination in flies, fruit flies and butterflies. Genetica 116:15–23

Steptoe A, Shankar A, Demakakos P, Wardle J (2013) Social isolation, loneliness, and all-cause mortality in older men and women. Proc Natl Acad Sci 110:5797–5801. https://doi.org/10.1073/pnas.1219686110

Willard HF (2003) Tales of the Y chromosome. Nature 423:810–813

Wood JT (1997) Gendered lives: communication, gender, and culture, 2nd edn. Wadsworth Publishing Company, Belmont

Genetics: Gene Expression

Martina M. L. LEI[1] and Terence Kin Wah Lee[2]
[1]The Hong Kong Polytechnic University, Hong Kong, China
[2]Department of Applied Biology and Chemical Technology, The Hong Kong Polytechnic University, Kowloon, Hong Kong

Synonyms

Age-associated gene expression; Aging gene expression

Definition

Gene expression refers to the process by which the genetic code in DNA is translated into functional products that dictate cell function. The nucleotide sequence in DNA is first transcribed into mRNA (transcription) and then translated into protein or RNA (translation) which gives rise to the phenotype.

Overview

Gene expression is generally defined as the process of protein synthesis from genetic information, producing functional signals in order to control the cell functions and give rise to phenotype. It involves two stages: transcription and translation. During transcription, the DNA strands unwind, and RNA polymerase binds to the promoter region of the template and moves along the strand to synthesize an mRNA with the recruitment of RNA nucleotides. Transcription terminates in a Rho-dependent or Rho-independent manner, followed by posttranscriptional modification, including addition of $5'$ cap, $3'$ polyadenylated tail, and splicing. The processed mature RNA then binds to ribosome and initiates translation. tRNAs with complementary codon bind to mRNA with peptidyl transferase, linking the peptide bonds and extending the peptide strand. Translation stops when the ribosomal complex meets one or more stop codons, and the released peptide strand undergoes posttranslation modification to form a protein.

Aging is a progressive deterioration of physiological functions, as well as reduction in resistance to stress and damage, resulting in increased risks of morbidity and mortality (Richardson and Schadt 2014). Aging-related phenotypic traits such as declines in immune system, cognitive, and organ functions increase aged individual's vulnerability to certain age-related diseases including cardiovascular diseases, degenerative diseases, Parkinson's disease, Alzheimer's disease, arthritis, and cancers (Rodríguez-Rodero et al. 2011). Interestingly, some people tend to have longer lifespans and are less vulnerable from diseases brought along with aging. Researches indicate that phenotypic traits and gene expressions become heterogeneous with age among somatic cells and individuals (Somel et al. 2006). The appearance of this phenotypic variation is hypothesized to be a connection between aging and morbidity due to a loss in regulatory capacities in aging organisms (Viñuela et al. 2018). To study the reasons behind age-dependent phenotypes, aging research has been focusing on finding corresponding gene expressions.

At cellular level, aging is a multifactorial process of which accumulative damages impair the defense and repair mechanisms (Kirkwood 2005). Since defense and repair systems involve various enzymatic mechanisms, any aberrant gene expression essential for enzyme production and activity can bring about cell damage (Rodríguez-Rodero et al. 2011). Malfunctioned repair system can cause mutations in genomic and mitochondrial DNA and thus distorts the somatic stem cell function. Accumulative damages can be attained through genetic and epigenetic modifications influencing the age-associated alterations, with multiple factors in genetic, environmental, and stochastic aspects involved (Martino et al. 2013).

Key Research Findings

Aging was traditionally perceived as a stochastic course of damage accumulation; aging can also be induced by genetics and environment (Raamsdonk 2018). Factors influencing aging are categorized into damage-related theory and programmed theory.

The free radical theory of aging (Harman 1956) first attributed aging as a consequence of accumulated oxidative stress, generated by reactive oxygen species (ROS) in normal cellular mechanisms. Nevertheless, the aging process is later characterized with nine metabolic hallmarks including genomic instability, telomere shortening, protein aggregation, aberrant nutrient sensing, mitochondrial dysregulation, cellular senescence, stem cell exhaustion, transformed intercellular communication, and epigenetic alterations (López-Otín et al. 2013). These aging hallmarks are interactive and have reciprocal effects on each other.

There are three major metabolic cascades influencing the lifespan: growth hormone (GH)/insulin-like growth factor 1 (IGF-1) signalling pathway (Junnila et al. 2013), FOXO3/sirtuin pathway (Mouchiroud et al. 2014), and mitochondrial electron transport chain (Chistiakov et al. 2014). These pathways are likely to influence aging in separate manners, as simultaneously targeting of these pathways results in additive prolongation of lifespan.

With the development of genome- and epigenome-wide association studies, researches have made use of the human gene expression profile to investigate the underlying mechanisms of aging, filling the limited relevance in studies using model organisms, e.g., yeast, *Drosophila*, and *C. elegans* (Piper and Partridge 2018; Olsen et al. 2006; Heiderstadt and Kennett 2011). Human studies identified thousands of age-related transcripts in diverse tissues, yet those age effects are minimal compared to research in model organism. Transcriptome-wide analyses of human genome compare the transcript alteration at different chronological ages. 1573 of 20,408 protein-coding genes from Genotype-Tissue Expression (GTEx) project (Jia et al. 2018) are identified to have age-dependent expressions. CDC42, a gene with strong correlation of expression level with age, is nominated to be a predictor of longevity (Kerber et al. 2009). It plays a crucial role in premature aging phenotypes including senescence-associated inflammation, osteoporosis, muscle deterioration, and defective wound healing.

A genome-wide RNA-seq study shows that skin possesses the highest association of gene expression change with age (1672 genes), compared to adipose tissue, blood, and the brain (Glass et al. 2013). Those identified genes mostly participate in fatty acid metabolism, mitochondrial function, splicing, and cancer-associated pathways such as p53, Wnt, or Notch pathways. However, only one age-linked gene is found common across four tissues. Another twin cohort study shows similar result, while only 5 out of 5631 age-related genes can be found common in fat, skin, and blood (Viñuela et al. 2018). The tissue-specific gene expressions indicate that aging is incurred by differential expression patterns and signalling in tissues. It is not controlled by a specific gene, neither a universal gene expression program.

A set of intrinsic alterations in gene expression modulating the cellular aging process was proposed to recapitulate the common signature changes, which includes downregulation of genes encoding proteins in mitochondria; downregulation of protein synthesis machinery; dysregulation of immune system genes; reduction in growth factor signalling; constitutive responses to stress and DNA damage; and dysregulation of gene expression and mRNA processing (Frenk and Houseley 2018).

Change in mean expression level is not the mere outcome of gene expression modification under aging influence; most of the aging effects can also be reproducibly found in heterogeneity of gene expression, RNA splicing, and epigenetic regulation. Correlation analysis using the database of Genotype-Tissue Expression (GTEx) (Jia et al. 2018) identified 1573 genes with differential expressions in accordance with chronological age. Age-related gene set is divided into two clusters on the basis of upregulated or downregulated mean expression level under age effect. The downregulated gene cluster ($n = 863$) is more evolutionarily conserved and has higher expression level and lower tissue specificity, implying their importance in global essential functions in fundamental metabolic or catabolic mechanisms for human survival. The enrichment of mitochondrial and innate immunity among those downregulated genes is in parallel with the aging hallmarks mitochondrial dysfunction and immunity decline, while the upregulated age-associated genes are more tissue-specific and consist of more disease genes and single-nucleotide polymorphic genes. Meta-analysis using eight microarray results of human and rat confirms that age-correlated heterogeneity of expression (ACHE) (Somel et al. 2006) has weak impact on single genes but generally acts throughout the transcriptome. Compared to age effect, heritability has a greater power on the proportion of genetic variation on expression.

Environment can influence gene expression via epigenetic DNA methylation (Ciccarone et al. 2018). DNA methylation changes in promoter regions started early in infants during the first year of life, with accumulative changes with age, creating diverging patterns even between monozygous twins as a result of "epigenetic drift" (Teschendorff et al. 2013). On the other side, DNA methylation changes at distinct loci are repeatedly found across the elderly, regardless of gender and tissue specificity. This locus-specific methylation change is proposed to be monitored by unified "epigenetic clock" program (Horvath 2013).

Gene expression modifications and regulation of genes related to age-associated diseases are identified. APOE (apolipoprotein E) is a group of proteins involved in lipid mechanism. Genetic polymorphisms in APOE gene have dissociable effects on age-related physiological declines such as dementia and Alzheimer's disease, inflammation, and atherosclerosis (Fortney et al. 2015). Transcript variants at TOMM40/APOE/APOC1 locus near APOE gene are reported to have correlation with longevity (Beekman et al. 2013). Differential splicing is observed in APOE gene, where age-dependent upregulation of reads connecting exons 2 and 4 is observed. Isoform with APOE exon 3 omitted tends to be more abundant in older individuals (Viñuela et al. 2018).

Example of Application

Hutchinson-Gilford Progeria Syndrome (HGPS)

HGPS is a disease of which premature aging phenotypic traits appear at very early age of life. Early symptoms of progeria include scleroderma, limited growth, and hair loss. Degenerative symptoms aggravate as the child ages, causing wrinkles, atherosclerosis, renal failure, and blindness. It is a genetic disorder caused by a subset of rarely inherited mutated variants. A mutation at G608G in LMNA gene impedes the posttranslational processing of lamin A protein which is crucial for maintaining intact shape of nucleus and nuclear stability (Piekarowicz et al. 2019). Expression change in LMNA exons and links between them are noticed with age association, suggesting the altered production of LMNA isoforms during aging process. Mutation in RNA splicing results in a defective transcript with truncated exon 11 and production of progerin which is a mutated lamin A with 50 amino acids lacked (Ragnauth et al. 2010). In order to interrogate how these gene expression changes lead to the extreme aging disease, gene expression profiles of three progeria fibroblast cell strains with heterozygous LMNA mutations are compared with normal aged cell strain. Significant gene expression changes are observed in 361 genes; most of them encode transcription factors and extracellular matrix proteins which play important roles in the severely impaired tissues in HGPS (Csoka et al. 2004). HGPS is characterized with an aberrant development affect mesodermal and mesenchymal dystrophies, highlighted with the abnormal expression of MEOX2/GAX transcription factor.

Future Direction of Application

Development of New Treatment

No adequate treatment is now available to cure HPGS; most of the treatments focus on relieving the symptoms. With the discoveries from gene expression analysis, such as switchable progerin expression, gene therapy is now under development for HGPS treatment (Piekarowicz et al. 2019). Strategies targeting gene correction, splicing dysregulation, progerin expression knockdown, and selective lamin A knockdown are suggested for gene therapy. Silencing of transcript variant 7 encoding progerin has been found successful in HGPS patients' fibroblasts.

Another strategy targets at blocking the enigmatic splice site at pre-mRNA of LMNA gene using a morpholino oligonucleotide (Ex11) (Scaffidi and Misteli 2005). In this approach, progerin protein levels are largely reduced with the morphology of HGPS cell rescued. Moreover, the expression level of affected genes such as matrix metalloproteinases 3 and 14 and chemokines is restored to normal level. The hope of curing HPGS will lie on the development of gene therapy.

With the development of high-throughput transcriptome analysis, the molecular mechanisms and pathologies underlying aging and age-associated disease are now getting clearer. The future direction of aging research will aim at identifying biomarkers with significant gene expression alteration or gene regulation under aging process. Yet, some challenges still remained to be solved. As gene expression being regulated with age may have a feedback effect on aging process, making the causality relationship between gene expression and aging complex to be interrogated. More sophisticated experimental design and an increase in statistical power, e.g., increase in sample number and the diversity of sample, will help to avoid the insignificant result due to biological or experimental noises.

Future studies can focus on identifying age-related pathways as a single common variant is difficult to be identified under aging effect. Moreover, interpretation of gene expression is not enough to draw the whole picture of aging as it just represents the regulation of age on transcription level. Measurement of protein levels of target gene should also be taken into account as synthesis and degradation rates of proteins are also crucial in studying aging and its related diseases.

Summary

With the development of high-throughput screening, genome and transcriptome analyses provide better understandings of aging effect on transcriptional level and age-associated pathologies. Genetic studies in human help to resolve the limited relevance of animal models in studying aging. Identifying genes with significant expression changes during aging process can help to understand the molecular mechanisms engaged in aging and possibly to find the biomarkers to measure aging rate and predict longevity. Age-related genes such as CDC24, APOE, and LMNA are highlighted with their gene expression change under age effect. To conclude, age-dependent gene expressions are tissue-specific with an increase in variance with age. The high specificity of age-related gene expression profiles implies that aging is controlled by more than one general program. The impact of aging is less powerful than heritability on the genetic variation. Direction of aging modification can be categorized into six features: downregulation of mitochondrial proteins; protein synthesis dysregulation; reduced growth factor signalling; abnormal immune system response; accumulative stress and DNA damage; and abnormal gene expression and mRNA processing. Modification on gene expression can be observed from changes in mean gene expression level, variance in expression, RNA splicing, and epigenetic regulation.

Cross-References

▶ DNA Damage Theory

References

Beekman M, Blanché H, Perola M et al (2013) Genome-wide linkage analysis for human longevity: genetics of healthy aging study. Aging Cell 12:184–193. https://doi.org/10.1111/acel.12039

Chistiakov DA, Sobenin IA, Revin VV et al (2014) Mitochondrial aging and age-related dysfunction of mitochondria. Biomed Res Int 2014:238463. https://doi.org/10.1155/2014/238463

Ciccarone F, Tagliatesta S, Caiafa P et al (2018) DNA methylation dynamics in aging: how far are we from understanding the mechanisms ? Mech Ageing Dev 174:3–17. https://doi.org/10.1016/j.mad.2017.12.002

Csoka AB, English SB, Simkevich CP et al (2004) Genome-scale expression profiling of Hutchinson – Gilford progeria syndrome reveals widespread transcriptional misregulation leading to mesodermal/mesenchymal defects and accelerated atherosclerosis. Aging Cell 3:235–243. https://doi.org/10.1111/j.1474-9728.2004.00105.x

Fortney K, Dobriban E, Garagnani P et al (2015) Genome-wide scan informed by age-related disease identifies loci for exceptional human longevity. PLoS Genet 11: e1005728. https://doi.org/10.1371/journal.pgen.1005728

Frenk S, Houseley J (2018) Gene expression hallmarks of cellular ageing. Biogerontology 19:547–566. https://doi.org/10.1007/s10522-018-9750-z

Glass D, Viñuela A, Davies MN et al (2013) Gene expression changes with age in skin, adipose tissue, blood and brain. Genome Biol 14:R75. https://doi.org/10.1186/gb-2013-14-7-r75

Harman D (1956) Aging – a theory based on free radical and radiation chemistry. J Gerontol 11:298–300

Heiderstadt KM, Kennett MJ (2011) IACUC issues related to animal models of aging. ILAR J 52:106–109. https://doi.org/10.1093/ilar.52.1.106

Horvath S (2013) DNA methylation age of human tissues and cell types. Genome Biol 14:R115. https://doi.org/10.1186/gb-2013-14-10-r115

Jia K, Cui C, Gao Y et al (2018) An analysis of aging-related genes derived from the Genotype-Tissue Expression project (GTEx). Cell Death Discov 4:26. https://doi.org/10.1038/s41420-018-0093-y

Junnila RK, List EO, Berryman DE et al (2013) The GH/IGF-1 axis in ageing and longevity. Nat Rev Endocrinol 9:366–376. https://doi.org/10.1038/nrendo.2013.67

Kerber RA, Brien EO, Cawthon RM (2009) Gene expression profiles associated with aging and mortality in humans. Aging Cell 8:239–250. https://doi.org/10.1111/j.1474-9726.2009.00467.x

Kirkwood TB (2005) Time of our lives. What controls the length of life? EMBO Rep 6 Spec No:S4–S8

López-Otín C, Blasco MA, Partridge L et al (2013) The hallmarks of aging. Cell 153:1194–1217. https://doi.org/10.1016/j.cell.2013.05.039

Martino D, Loke YJ, Ollikainen M et al (2013) Longitudinal, genome-scale analysis of DNA methylation in twins from birth to 18 months of age reveals rapid epigenetic change in early life and pair-specific effects of discordance. Genome Biol 14:R42. https://doi.org/10.1186/gb-2013-14-5-r42

Mouchiroud L, Houtkooper RH, Moullan N et al (2014) The NAD$^+$/sirtuin pathway modulates longevity through activation of mitochondrial UPR and FOXO signaling. Cell 154:430–441. https://doi.org/10.1016/j.cell.2013.06.016

Olsen A, Vantipalli MC, Lithgow GJ (2006) Using *Caenorhabditis elegans* as a model for aging and age-related diseases. Ann N Y Acad Sci 1067:120–128. https://doi.org/10.1196/annals.1354.015

Piekarowicz K, Machowska M, Dzianisava V (2019) Hutchinson-Gilford progeria syndrome – current status and prospects for gene therapy treatment. Cells 8(2): E88. https://doi.org/10.3390/cells8020088

Piper MDW, Partridge L (2018) Drosophila as a model for ageing. Biochim Biophys Acta Mol Basis Dis 1864:2707–2717. https://doi.org/10.1016/j.bbadis.2017.09.016

Ragnauth CD, Warren DT, Liu Y et al (2010) Prelamin A acts to accelerate smooth muscle cell senescence and is a novel biomarker of human vascular aging. Circulation 121:2200–2210. https://doi.org/10.1161/CIRCULATIONAHA.109.902056

Richardson AG, Schadt EE (2014) The role of macromolecular damage in aging and age-related disease. J Gerontol A Biol Sci Med Sci 69(Suppl 1):S28–S32. https://doi.org/10.1093/gerona/glu056

Rodríguez-Rodero S, Fernández-Morera JL, Menéndez-Torre E et al (2011) Aging genetics and aging. Aging Dis 2:186–195

Scaffidi P, Misteli T (2005) Reversal of the cellular phenotype in the premature aging disease Hutchinson-Gilford progeria syndrome. Nat Med 11:440–445. https://doi.org/10.1038/nm1204

Somel M, Bahn S, Pääbo S (2006) Gene expression becomes heterogeneous with age. Curr Biol 16:R359–R360. https://doi.org/10.1016/j.cub.2006.04.024

Teschendorff AE, West J, Beck S (2013) Age-associated epigenetic drift: implications, and a case of epigenetic thrift ? Hum Mol Genet 22:7–15. https://doi.org/10.1093/hmg/ddt375

Van Raamsdonk JM (2018) Mechanisms underlying longevity: a genetic switch model of aging. Exp Gerontol 107:136–139. https://doi.org/10.1016/j.exger.2017.08.005

Viñuela A, Brown AA, Buil A et al (2018) Age-dependent changes in mean and variance of gene expression across tissues in a twin cohort. Hum Mol Genet 27:732–741. https://doi.org/10.1101/063883

Genetics: Parental Influence

Teresa Chung[1] and Lok Ting Lau[1,2]
[1]Department of Applied Biology and Chemical Technology, The Hong Kong Polytechnic University, Hung Hom, Kowloon, Hong Kong
[2]Innovation and Technology Development Office, The Hong Kong Polytechnic University, Hung Hom, Kowloon, Hong Kong

Synonyms

Biological heredity; Parental epigenetics; Parental genetics; Parental mitochondrial genome influences

Definition

The genetic influences of parents onto their offspring can be divided into genetic and epigenetic and mitochondrial levels. Genetic factors include the inheritance of nuclear genomes from the parents and the consequences of any chromosomal aberrations. Parental epigenetics, sometimes with transgenerational effects, involve non-mutational changes in chromatin structure and can be influences by maternal and paternal periconceptional development conditioning, nutritional status, obesity, and genomic imprinting.

Overview

Parents influence child development in infinite number of ways, from genetic factors to a wide range of environmental factors such as nutrition, parenting, experiences, friends, family, education, etc., with certain factors playing a more important role in certain aspects of child development.

Parental Genetic Influences

From the earliest moment of life, a child inherits from parents two complete sets of genetic instructions per nucleus. The sperm and the ova each contain only 23 chromosomes, and when these 2 cells fuse during fertilization, the resulting zygote has 46 chromosomes, which is the same number of chromosomes as in all somatic cells. Within these chromosomes are genes that are made up of DNA (deoxyribonucleic acid). Genes contain the genetic code that acts as a blueprint for human life. Collectively, these genes constitute the genotype of the individual, and the actual expression of these genes, or phenotype, can result in physical and nonphysical traits.

The expression of phenotypes can be influenced by many factors. Gene, or more precisely allele, can interact with each other and follow a dominant-recessive, codominant, or incomplete dominant patterns depending on the ultimate phenotype of the heterozygote. If the phenotype of the heterozygote is identical to the phenotype of homozygous of one of the alleles, that allele whose phenotype is evident in a heterozygote is said to be dominant. The alternative allele, whose phenotype is not evident and is masked in the heterozygote, is the recessive allele. Only if both alleles are recessive does the phenotype of the recessive allele becomes evident. For example, the gene for brown eyes is dominant, and the gene for blue eyes is recessive. If a child inherits a dominant brown eye allele from one parent and a recessive blue eye allele from the other parent, the child will still have brown eyes.

In fact, there are exceptions to these rules, and many traits defy this simple recessive and dominant classification and are modeled by incomplete dominance and codominance instead. Incomplete dominance is when both alleles are neither dominant nor recessive but are instead blended together, expressed, and produce an intermediate phenotypic trait. Codominance is similar to incomplete dominance where neither dominant nor recessive alleles are expressed alone, but instead of a blend of the two alleles, both alleles get expressed and are mixed in the phenotypic trait, and neither is dominant.

Chromosomal Aberrations

Sometimes when a sperm or an ovum is formed, the chromosomes may divide unevenly, which results in the production of gametes with abnormal numbers of chromosomes or aneuploidy. Fertilization with a normal gamete will result in the formation of zygote with aberrant number of chromosomes. In humans, most aneuploidies are lethal because of the ensuing imbalance in gene expression, spontaneously aborted, and are never developed into a full-term baby. In some cases, babies are born with an abnormal number of chromosomes, resulting in various syndromes with distinguish characteristics.

Boys have one X chromosome and one Y chromosome each, and girls have two X chromosomes. Aneuploidies on sex chromosome are mostly tolerated because X inactivation maintains normal gene dosage. Common aneuploidies on sex chromosomes include the Klinefelter syndrome (extra X chromosome(s) in males), Turner syndrome (the presence of only one X chromosome in females), triple X syndrome (trisomy X in females), and XYY syndrome (an extra Y chromosome in males). Klinefelter syndrome is characterized by a lack of development of the secondary sex characteristics and learning disabilities. Turner syndrome is characterized by a short stature, webbed neck, a lack of development of secondary sex characteristics, and psychological impairments.

Aneuploidies on some non-sex small chromosomes can sometimes be tolerated, but they usually result in severe developmental disorders. Common aneuploidies include Down syndrome (trisomy 21), Patau syndrome (trisomy 13), and Edwards syndrome (trisomy 18). Down syndrome is a common genetic disorder and is characterized by facial characteristics such as round face, slanted eyes, and intellectual impairment.

Although it is no doubt that genetics play an important part in child development, nevertheless, genetics is only one piece of the intricate puzzle that can affect the outcome. Other than the inheritance of nuclear genome from the parents and the interaction in between genes, phenotypes/genetic traits can also be largely influenced by environmental factors. The environment a child is exposed to periconceptionally, pre-, and postnatally can impact whether to express or repress these heredity instructions, as well as the level of expression, which eventually help shaping the outcome. For example, the onset of puberty is largely dependent on heredity, but the availability of nutrition can also have a huge impact. In addition, although height is largely determined by heredity, it can be suppressed if the child has experienced poor nutrition or chronic illness. The effects of environmental factors on phenotypes, often mediated by epigenetic mechanisms, involve non-mutational changes in chromatin structure that lead to the expression of various phenotypes from a single genotype. In certain cases, parental epigenetics acquired in one generation can also be transmitted to the next generation, affecting transgenerational development.

Parental Epigenetic Influences: Periconceptional Development Conditioning

During the periconceptional period, the embryonic cell is sensitive to a wide range of parental environmental cues such as maternal lifestyle factors (e.g., diet, alcohol consumption), maternal body composition and physiological status (obesity, BMI, and fatty acid concentration), advanced maternal age, maternal stress, maternal systemic inflammation status as well as paternal lifestyle. The above parental environmental cues alter the embryonic epigenetic, gene expression profiles, and mitochondrial activity which ultimately drive an altered growth trajectory on offspring's long-term health. This periconceptional developmental plasticity provides a key opportunistic window for the embryo to biologically embed early life exposures, to optimize subsequent development to best suit survival and fitness when the offspring is exposed to a similar environment postnatally (Fleming et al. 2018).

Maternal Overnutrition and Obesity

Maternal overnutrition at conception alters the embryonic epigenome during early stage of embryonic development and affects offspring lifelong immune function and obesity status. For example, in maternal obese mice with hyperlipidemia, the combination of metabolic, mitochondrial, epigenome, and chromosomal alterations in oocytes and embryos reduces the developmental competence of offspring with smaller fetuses and pups but with an overgrowth, development of adiposity, and glucose intolerance after birth (Sunde et al. 2016; Luzzo et al. 2012; Yang et al. 2012; Igosheva et al. 2010). Transfer of mouse blastocysts from obese mothers to normal recipients produces similar growth-restricted fetuses (Luzzo et al. 2012). In sheep, similar experiments produce offspring with increased adiposity and dysregulated insulin signaling due to an upregulation of hepatic microRNAs (Nicholas et al. 2013).

Maternal Undernutrition

Similar to maternal overnutrition, the effects of periconceptional maternal undernutrition can also persist into adulthood of the offspring. For example, offspring of Dutch famine have a differential DNA methylation pattern of the imprinted growth- and metabolism-regulating genes such as IGF2 (Tobi et al. 2014). In response to maternal low-protein dietary restriction during embryonic period, the nutrient-sensing ribosome factor, Rrn3, modulates DNA methylation at the rDNA promoter, suppresses ribosome biogenesis, and affects the growth of different somatic organs (e.g., liver and kidney). Interestingly, rRNA expression is reversed once the dietary challenge is removed (Denisenko et al. 2016). Periconceptional maternal diet deficient in one-carbon (1-C) metabolites (including substrates and cofactors vitamin B12, folate, methionine) in sheep leads to offspring with adverse cardiometabolic and immune dysfunctions (Sinclair et al. 2007). Similarly, mouse offspring of paternal low folate diet are born with craniofacial and musculoskeletal malformations (Lambrot et al. 2013). In contrast, folate addition to rodent maternal LPD rescues DNA and histone methylation in gametes and embryos and restores normal expression of metabolic regulators in offspring with prevailing cardiovascular dysfunction (Lillycrop et al. 2005).

Paternal Periconceptional Programming

In addition to maternal factors, paternal physiological and lifestyle factors, such as obesity, BMI, and diet high in sugar and fat, can also affect offspring health and development through epigenetic dysfunction. Direct mechanisms affect sperm quality such as reduced sperm motility, increased sperm abnormality, reduced sperm DNA integrity, and increased sperm reactive oxygen species levels. Indirect mechanisms alter seminal fluid composition such as the granulocyte-macrophage colony-stimulating factors (Chocron et al. 2019). For example, diet of paternal rats high in fat have offspring with increased body weight, glucose intolerance, and fasting hyperglycemia (Ng et al. 2010). Low-protein diet in paternal mice triggered a dysregulated hepatic DNA methylation pattern of lipid regulator gene PPARα, which lead to enhanced hepatic lipid and cholesterol biosynthesis, and offspring with higher birth weight, increased adult adiposity, glucose intolerance, and elevated serum TNF-α levels (Watkins and Sinclair 2014; Carone et al. 2010). Nevertheless, sperm-mediated effects can sometimes be transient and even reversible through paternal dietary and exercise interventions in mice (Palmer et al. 2012).

Transgenerational Epigenetic Inheritance of Longevity

Other than parental periconceptional programming, deficiencies in the H3K4me3 (histone H3 lysine 4 trimethylation) regulatory complex ASH-2, WDR-5, or SET-2 in the parental generation were shown to induce an epigenetic memory of longevity and extend the life span of descendants up until the third generation. This transgenerational inheritance of longevity was associated with chromatin changes at the conserved H3K4me3 chromatin modifier loci in parents, which might be incompletely reprogrammed during the generation of gametes and subsequently lead to changes in the gene expression (Greer et al. 2011).

Other Genetic Components of Exceptional Longevity

With the strong emphasis on healthy aging nowadays, the exceptional longevity of centenarians is being studied. Evidences showed an increasing heritability component at greater ages. For example, centenarians tended to cluster in families (Perls et al. 2002). The heritability of surviving to at least 100 was estimated at 0.33 for women and 0.48 for men (Sebastiani and Perls 2012). In a study of 20,000 Scandinavian twins, it was demonstrated that the heritability of longevity was negligible below the age of 60 but increased thereafter (Hjelmborg et al. 2006). Two more studies demonstrated that longevity was heritable beyond 70 and 55 years in an Icelander study and an Okinawan centenarian sibling studies, respectively (Gudmundsson et al. 2000; Willcox et al. 2006). Centenarians also have significantly different genetic profiles as demonstrated in several meta-analyses and genome-wide association studies (GWAS). For example, a genetic variant rs2070325 of the bactericidal/permeability-increasing fold-containing family B member 4 (BPIFB4) gene is associated with exceptional longevity under a recessive genetic model by reducing the blood pressure level (Villa et al. 2015). A rare mutation of ELOVL6 is associated with longevity as this accumulated protective palmitoleic acid (C16:1) (Sebastiani et al. 2017). Several genetic variants of IGF-1 receptor and FOXO3A are associated with longevity (Di Bona et al. 2014). Apart from the inheritance of the right combination of genetic variants from the parents, longevity may also be subjected to the acquisition of epigenetic modifications induced by the environment and lifestyle, which may be the cause of increased survival rate in industrialized countries. In the future, there is a hope of identifying particular DNA chemical signatures as prognostic markers for developing pathologies and healthy aging (Puca et al. 2018).

Parental Mitochondrial Genome Influences

In addition to nuclear genome, the inheritance of maternal mitochondrial genome (mtDNA) can also affect offspring development. mtDNA, comprised of 16.6 kilobase DNA which encodes 37 genes involved in oxidative phosphorylation and mitochondrial translation machinery, is inherited maternally in mammals. Mitochondrial dysfunction is a hallmark of aging and deteriorates with age in different tissues. The declines in mtDNA copy number, epigenetic modifications, and mtDNA mutations with age are all linked with age-associated functional decline of mitochondria and increased susceptibility to a range of age-associated diseases including Alzheimer's disease, Parkinson's disease, sarcopenia, heart failure, and cancer. Fertility declines with age, and it is now demonstrated that oocytes in aged mothers contain a higher level of mtDNA mutations than younger mothers. For example, Barritt, Cohen, and Brenner found that oocytes in older mother aged 36–42 contained more mtDNA T414G transversion point mutations in D-loop than younger mothers aged 26 to 35 and that oocyte quality was negatively influenced by BMI and smoking (Chocron et al. 2019; Barritt et al. 2000).

Genomic Imprinting

Another important factor that parental influences have on fetal growth and child development is genomic imprinting. Endogenous genomic imprinting determines which of the parental allele to express, where mono-allelic expression of imprinted genes in embryos is retained as memory of parent-of-origin specific inheritance of epigenetic marks in the germline and leads to differential gene expression. Heritable epigenetic marks are usually brought about by DNA methylation, where addition of a methyl (CH_3) group to cytosines on regulatory sequences such as promoters is associated with transcriptional repression (Tucci et al. 2019; Reik and Walter 2001).

During the life cycle of the imprints, imprints are generally erased epigenetically in between generations during germ cell development into sperm or egg, followed by establishment and reprogramming at a later stage of germ cell development. Imprinted genes are usually clustered with other imprinted genes at imprinting control regions (ICRs) with a coordinated regulation of multiple imprinted genes in a chromosomal domain. A well-known example is the X-inactivation center which controls the inactivation of the entire X chromosome. Other imprinted genes are differentially methylated: some are maintained in all developmental stages and tissues; some acquire tissue-specific methylation patterns; in somatic cells, imprints are maintained and are changed in methylation during development; some are methylated in the inactive or active gene copy; different epigenetic modifications can also be displayed, such as histone acetylation. The brain appears to be an interesting organ where parental regulation of gene dosage (differential imprinting) across functionally distinct brain regions may be required for normal brain development and function. For example, deletion of the paternal allele of the paternally biased gene Bcl2l results in the loss of specific neuron types (Tucci et al. 2019; Perez et al. 2016; Reik and Walter 2001).

Erroneous genomic imprinting, or epimutation in imprinted genes, alters the DNA methylation or chromatin patterns and can sometimes lead to biallelic expression or silencing (Reik and Walter 2001) and imprinting diseases. Neurodevelopmental disorders such as AS and PWS are due to reciprocal deletion of 15q11-q13. AS is due to the lack of maternal UBE3A gene expression, point mutations in UBE3A, or paternal uniparental disomy (pUPD15). PWS can be caused by the deletion of the paternal 15q11-q13, chromosomal maternal uniparental disomy (mUPD15), or ICR deletion. In PWS, all deletions lead to the loss of paternal gene expression, but mUPD15 and ICR deletion are more often associated with psychotic illness than single paternal 15q11-q13 deletion as the prior doubles the expression of maternal gene UBE3A. In contrast, duplication of the paternal expressed gene does not increase the risk of psychotic illness. Some other imprinting diseases include autism (maternal duplications of PWACR in 15q11-q13), bipolar affective disorder, epilepsy, schizophrenia, Tourette syndrome, Turner syndrome, Wilms tumor (H19 methylation), and BWS (H19 methylation, KvDMR1 demethylation), whose disease susceptibility and severity largely depend on which of parent's allele is expressed (Tucci et al. 2019; Reik and Walter 2001).

Future Directions of Research

Scientists have made a significant step toward our understanding of inherited genetic effects on offspring health and development due to recent advances in technologies such as next-generation sequencing and bioinformatic analysis. However, identification of susceptible genes is only the first step; the next crucial stage is to investigate how each of these genes and their associated mutations affect disease pathogenesis.

The study of how epigenetically and genetically mutated mitochondria influence offspring development is currently hindered by technological obstacles such as the identification, isolation, and characterization of enough mutated mitochondria for further analysis. In addition, further insights on how different epigenetic inheritance such as periconceptional developmental conditioning, transgenerational epigenetic inheritance, and genomic imprinting affecting child development are needed, as well as a deepened understanding on the underlying mechanisms that drive them. Challenges remain to enhance technological advances to study different epigenetic modifications such as methylation, histone modification, microRNA, and other small RNAs on child development.

Summary

Child development is a highly heterogenous and dynamic phenotype which involves a complicated interaction between a wide range of genetic, epigenetic, and environmental factors. Overall, understanding how different genetic and epigenetic regulatory processes, such as genomic imprinting and periconceptional development conditioning, is critical in understanding how our health is programmed during early pregnancy. This enabled us to design interventions during early development to combat different lifelong diseases.

Cross-References

- ▶ Alleles
- ▶ Chromosome
- ▶ DNA Damage Theory
- ▶ Genetics: Ethnicity
- ▶ Genetics: Gender
- ▶ Genetics: Gene Expression
- ▶ Genotype
- ▶ Mitochondrial DNA Mutations
- ▶ Nucleotides

References

Barritt JA, Cohen J, Brenner CA (2000) Mitochondrial DNA point mutation in human oocytes is associated with maternal age. Reprod Biomed Online 1(2000):96–100

Carone BR, Fauquier L, Habib N, Shea JM, Hart CE, Li R, Bock C, Li C, Gu H, Zamore PD, Meissner A, Weng Z, Hofmann HA, Friedman N, Rando OJ (2010) Paternally induced transgenerational environmental reprogramming of metabolic gene expression in mammals. Cell 143(7):1084–1096

Chocron ES, Munkácsy E, Pickering AM (2019) Cause or casualty: the role of mitochondrial DNA in aging and age-associated disease. Biochim Biophys Acta Mol basis Dis 1865(2):285–297

Denisenko O, Lucas ES, Sun C, Watkins AJ, Mar D, Bomsztyk K, Fleming TP (2016) Regulation of ribosomal RNA expression across the lifespan is fine-tuned by maternal diet before implantation. Biochim Biophys Acta 1859(7):906–913

Di Bona D, Accardi G, Virruso C, Candore G, Caruso C (2014) Association between genetic variations in the insulin/insulin-like growth factor (Igf-1) signaling pathway and longevity: a systematic review and meta-analysis. Curr Vasc Pharmacol 12:674–681

Fleming TP, Watkins AJ, Velazquez MA, Mathers JC, Prentice AM, Stephenson J, Barker M, Saffery R, Yajnik CS, Eckert JJ, Hanson MA, Forrester T, Gluckman PD, Godfrey KM (2018) Origins of lifetime health around the time of conception: causes and consequences. Lancet 391(10132):1842–1852

Greer EL, Maures TJ, Ucar D, Hauswirth AG, Mancini E, Lim JP, Benayoun BA, Shi Y, Brunet A (2011) Transgenerational epigenetic inheritance of longevity in Caenorhabditis elegans. Nature 479(7373):365–371

Gudmundsson H, Gudbjartsson DF, Frigge M, Gulcher JR, Stefansson K (2000) Inheritance of human longevity in Iceland. Eur J Human Genet 8:743–749

Hjelmborg JV, Iachine I, Skytthe A, Vaupel JW, McGue M, Koskenvuo M, Kaprio J, Pedersen NL, Christensen K (2006) Genetic influence on human lifespan and longevity. Hum Genet 119:312–321

Igosheva N, Abramov AY, Poston L, Eckert JJ, Fleming TP, Duchen MR, McConnell J (2010) Maternal diet-induced obesity alters mitochondrial activity and redox status in mouse oocytes and zygotes. PLoS One 5(4): e10074

Lambrot R, Xu C, Saint-Phar S, Chountalos G, Cohen T, Paquet M, Suderman M, Hallett M, Kimminsa S (2013) Low paternal dietary folate alters the mouse sperm epigenome and is associated with negative pregnancy outcomes. Nat Commun 4:2889

Lillycrop KA, Phillips ES, Jackson AA, Hanson MA, Burdge GC (2005) Dietary protein restriction of pregnant rats induces and folic acid supplementation prevents epigenetic modification of hepatic gene expression in the offspring. J Nutr 135(6):1382–1386

Luzzo KM, Wang Q, Purcell SH, Chi M, Jimenez PT, Grindler N, Schedl T, Moley KH (2012) High fat diet induced developmental defects in the mouse: oocyte meiotic aneuploidy and fetal growth retardation/brain defects. PLoS One 7(11):e49217

Ng SF, Lin RC, Laybutt DR, Barres R, Owens JA, Morris MJ (2010) Chronic high-fat diet in fathers programs beta-cell dysfunction in female rat offspring. Nature 467(7318):963–966

Nicholas LM, Rattanatray L, MacLaughlin SM, Ozanne SE, Kleemann DO, Walker SK, Morrison JL, Zhang S, Muhlhäusler BS, Martin-Gronert MS, McMillen IC (2013) Differential effects of maternal obesity and weight loss in the periconceptional period on the epigenetic regulation of hepatic insulin-signaling pathways in the offspring. FASEB J 27(9):3786–3796

Palmer NO, Bakos HW, Owens JA, Setchell BP, Lane M (2012) Diet and exercise in an obese mouse fed a high-fat diet improve metabolic health and reverse perturbed sperm function. Am J Physiol Endocrinol Metab 302(7):E768–E780

Perez JD, Rubinstein ND, Dulac C (2016) New perspectives on genomic imprinting, an essential and multifaceted mode of epigenetic control in the developing and adult brain. Annu Rev Neurosci 39:347–384

Perls TT, Wilmoth J, Levenson R, Drinkwater M, Cohen M, Bogan H, Joyce E, Brewster S, Kunkel L, Puca A (2002) Life-long sustained mortality advantage of siblings of centenarians. Proc Natl Acad Sci USA 99:8442–8447

Puca AA, Spinelli C, Accardi G, Villa F, Caruso C (2018) Centenarians as a model to discover genetic and epigenetic signatures of healthy ageing. Mech Ageing Dev 174:95–102

Reik W, Walter J (2001) Genomic imprinting: parental influence on the genome. Nat Rev Genet 2(1):21–32

Sebastiani P, Perls TT (2012) The genetics of extreme longevity: lessons from the New England centenarian study. Front Genet 3:277

Sebastiani P, Gurinovich A, Bae H, Andersen S, Malovini A, Atzmon G, Villa F, Kraja AT, Ben-Avraham D, Barzilai N, Puca A, Perls TT (2017) Four genome-Wide association studies identify new extreme longevity variants. J Gerontol Sers A Biol Sci Med Sci. https://doi.org/10.1093/gerona/glx027. (Mar 15. [Epub ahead of print])

Sinclair KD, Allegrucci C, Singh R, Gardner DS, Sebastian S, Bispham J, Thurston A, Huntley JF, Rees WD, Maloney CA, Lea RG, Craigon J, McEvoy TG, Young LE (2007) DNA methylation, insulin resistance, and blood pressure in offspring determined by maternal periconceptional B vitamin and methionine status. Proc Natl Acad Sci U S A 104(49):19351–19356

Sunde A, Brison D, Dumoulin J, Harper J, Lundin K, Magli MC, Van den Abbeel E, Veiga A (2016) Time to take human embryo culture seriously. Hum Reprod 31(10):2174–2182

Tobi EW, Goeman JJ, Monajemi R, Gu H, Putter H, Zhang Y, Slieker RC, Stok AP, Thijssen PE, Müller F, van Zwet EW, Bock C, Meissner A, Lumey LH, Eline Slagboom P, Heijmans BT (2014) DNA methylation signatures link prenatal famine exposure to growth and metabolism. Nat Commun 5:5592

Tucci V, Isles AR, Kelsey G, Ferguson-Smith AC, Erice Imprinting Group (2019) Genomic imprinting and physiological processes in mammals. Cell 176(5):952–965

Villa F, Carrizzo A, Spinelli CC, Ferrario A, Malovini A, Maciąg A, Damato A, Auricchio A, Spinetti G, Sangalli E, Dang Z, Madonna M, Ambrosio M, Sitia L, Bigini P, Calì G, Schreiber S, Perls T, Fucile S, Mulas F, Nebel A, Bellazzi R, Madeddu P, Vecchione C, Puca AA (2015) Genetic analysis reveals a longevity-associated protein modulating endothelial function and angiogenesis. Circ Res 117:333–345

Watkins AJ, Sinclair KD (2014) Paternal low protein diet affects adult offspring cardiovascular and metabolic function in mice. Am J Physiol Heart Circ Physiol 306(10):H1444–H1452

Willcox BJ, Willcox DC, He Q, Curb JD, Suzuki M (2006) Siblings of Okinawan centenarians share lifelong mortality advantages. J Gerontol Ser A Biol Sci Med Sci 61:345–354

Yang X, Wu LL, Chura LR, Liang X, Lane M, Norman RJ, Robker RL (2012) Exposure to lipid-rich follicular fluid is associated with endoplasmic reticulum stress and impaired oocyte maturation in cumulus-oocyte complexes. Fertil Steril 97(6):1438–1443

Genome Instability

Hoi Shan Kwan
School of Life Sciences, The Chinese University of Hong Kong, Shatin, NT, Hong Kong

Synonyms

DNA damage; Genetic instability; Genomic instability

Definition

Genome instability refers to occurrence of mutations within the genome. These mutations include base changes, chromosomal rearrangements, or aneuploidy.

Overview

Genome instability refers to the increased chance to have mutations caused by deficiencies in DNA repair and replication or chromosome segregation. DNA can be damaged by reactive oxidative species (ROS), UV irradiation, or environmental mutagens (Cui et al. 2012).

It is estimated that each cell in our body can have over 10,000 damages per day. Fortunately, our cells have genome maintenance systems to repair the damages and restore the original base sequence. Sometimes, repair systems make mistakes, or the DNA fails to replicate, resulting in damages that are not repaired thus becoming irreversible as DNA mutations.

DNA mutations can occur in the germline or in somatic cells. Germline mutations are easily recognized, and their frequencies can be calculated. Mutations in the germline are passed to a child who carries the mutation in every cell. Somatic cell mutations, on the other hand, are difficult to detect because they are carried by individual cells. Recent advances of sequencing of single cells allowed the detection of genome mutations (Sanders et al. 2016).

Genome instability is considered to cause aging. It is clear that genome instability giving rise to somatic cell mutations can accumulate and cause changes in the fitness of the cells. Questions are how frequent the mutations occur and whether the mutations would cause decline in cellular function, which in turn cause recognized health issues accompanying aging. Indeed, somatic mutations accumulating during a person's life span are associated with cancer which has higher frequencies in older people (Roos et al. 2016). Genomic instability caused by DNA damage are linked to aging. The various aspects of this linkage are described in this entry.

Key Research Findings

Genome instability refers to the increased chance to have mutations caused by deficiencies in DNA repair and replication or chromosome segregation. DNA can be damaged by ROS, UV irradiation, or environmental mutagens. It is estimated that each cell in our body can have over 10,000 damages per day. DNA damages trigger DNA damage response (DDR) (Jackson and Bartek 2009). Fortunately, in DDR, our cells have genome maintenance systems to repair the damages and restore the original base sequence. Sometimes, repair systems make mistakes, or the DNA fails to replicate, resulting in damages that are not repaired and thus becoming irreversible as DNA mutations. Occasionally, the DNA replication system makes mistakes and caused changes in the DNA which are also mutations.

DNA mutations can occur at single bases causing point mutations, deletions, and insertions. They can also involve several to many bases, causing larger-scale insertions and deletions, transposition, and inversions. Larger pieces of DNA can also be mutated, even chromosomal aberrations. DNA damages and mutations are now clearly shown as causes of cancer (Hanahan and Weinberg 2011; Roos et al. 2016). The mutations may occur in the germline as heritable mutations that predispose the person to high risk of cancer. Later life occurrence of other mutations in the somatic cells would cause cancer.

There are numerous reports showing that mutations accumulate in human during the life span, associated to aging (López-Otín et al. 2013). The question is whether mutations are the cause or the consequence of aging. Evidences are accumulating that mutations are the cause. Many mutations have been identified that can cause progeroid syndromes, that is, accelerated aging in humans. Some of them are systemic, for example, WRN (RecQ DNA helicase gene) mutation. WRN mutation causes growth retardation, premature graying of hair, lipodystrophy, and multiple age-related diseases including arteriosclerosis, type 2 diabetes, and cancer. Moreover, most genome instability disorders are characterized by accelerated aging of multiple organ systems with or without cancer predisposition.

Another hint suggesting that DNA damage causes aging is from cancer studies. Cancer treatments with genotoxic agents are not specific for the cancer cells and damage DNA of the normal cells. The outcomes are apoptosis, senescence, and mutagenesis, associated with acceleration of aging. One example is that survivors of childhood cancer when get older would have a 2.5-fold higher chance in suffering severe age-related diseases than their siblings (Armstrong et al. 2014). These hints show that DNA damages accelerated aging.

In the normal aging process, several lines of observations provide evidences that DNA damage increases with age. First, many studies showed that oxidative damage to all cellular macromolecules increases with age. The main oxidative damaging agent is ROS. Indeed, overexpression of enzymes that reduce ROS, such as copper and zinc superoxide dismutase (SOD), manganese SOD, and catalase, can extend the life of the fruit fly *Drosophila melanogaster*. Many studies showed that ROS and oxidative damage increase during aging (Cui et al. 2012).

Second, DNA damage levels increase with age. For example, oxidative DNA damages as measured with the levels of 8-oxo-7,8-dihydro-2′-deoxyguanosine (8-oxodG) have been shown to increase with age in the heart, liver, kidney, and lung in mice (Wong et al. 2009). Jacob et al. (2013) showed that more 8-oxodG occurs in older people than younger people. Older women were shown to have high level of urinary 8-oxodG (Hou et al. 2016). A meta-analysis of 36 studies showed a significant correlation between age and DNA damage occurrence (Soares et al. 2014).

Third, DNA repair capacity seems to decrease with age. Vaidya et al. (2014) used a transgenic reporter in mice to show that nonhomologous end joining repair decreased from mid-age to old age. However, there have been contradictory reports on this point. The in vivo tools to test DNA report are not satisfactory in providing conclusive evidence for its decline with age.

While DNA damages are usually repaired, DNA mutations are not reversible. Mutations occur in somatic cells and accumulate over time, as physical indicator of genome instability. Somatic mutations occur in individual cells and are very difficult to identify because the cells are just one of many in the tissue or organ. Now we can use high-resolution methods to identify somatic mutations. We can identify chromosomal alterations with fluorescence in situ hybridization (FISH). FISH analysis showed that more lymphocytes have chromosomal aberrations in human and mice when they get older (Ramsey et al. 1995). However, FISH cannot detect small genome structural variations (SV) or point mutations which are the major types of spontaneous mutations constituting genome instability in somatic cells when they age.

Clonal lineages of some cell types provide the materials to show genome instability. In human lymphocytes, clonally expanded copy number variants accumulation is related to aging (Jacobs et al. 2012; Laurie et al. 2012). Recent advances of sequencing of single cells allowed the detection of genome SVs (Falconer et al. 2012; Sanders et al. 2016), but they are not used to show the accumulation of mutations with aging.

Cells keep genome instability in check by the DNA repairing systems. Therefore, we can expect that long-lived individuals (centenarians) would have better DNA repair systems than their normal life span peers. Centenarians have similar proportions of spontaneous DNA strand breaks as youngsters, implying better DNA repairs (Chevanne et al. 2003; Franzke et al. 2015). Other studies showed that cells from centenarians seem to tolerate damaging oxidants better (Franzke et al. 2015). The relationship between genome maintenance and aging is indirectly supported by genome-wide association studies (GWAS) in which many genome maintenance genes are found to associate with longevity (Cho and Suh 2014). A study on the age of natural menopause identified 44 genetic variants related to early or late menopause, a biomarker of healthy aging (Laven et al. 2016). Most of these variants are in genes related to genome maintenance.

There are at least four possible ways that genome instability may impact aging (Terence: are the following paragraphs describing these 4 ways? If so, they need to be revised into a clearer manner. If not, they may need to be linked up somehow):

1. DNA damage can make cells to be senescent by causing mitochondrial dysfunction and telomere dysfunction. Higher levels of senescence markers can be detected in tissues of older mice and human. Senescent cells have senescence-associated secretory phenotype (SASP) with induction of autocrine and paracrine, including interleukin-6 (IL-6), transforming growth factor-β (TGF-β), IL-1α, tumor necrosis factor-α, matrix metalloproteinases, insulin-like growth factor-1 binding proteins (IGFBPs), plasminogen activator inhibitor-1 (PAI-1), and monocyte chemoattractant protein-1 (MCP-1) (Acosta et al. 2013). SASP is a response to persistent DNA damage (Rodier et al. 2009).
2. Cellular senescence can be activated and maintained by many signaling pathways. Some of these pathways are linked to DNA damage. These include DNA damage response (DDR), insulin/IGF-1 signaling (Ock et al. 2016), NF-KB (Tilstra et al. 2012), and sirtuins (Langley et al. 2002).
3. DNA damage can promote aging by affecting the number or function of stem cells such as hematopoietic stem cells (Rossi et al. 2007; Ermolaeva et al. 2018).
4. Mutations in nuclear and mitochondrial genes may cause mitochondrial autophagy

(Scheibye-Knudsen et al. 2012). DNA damage response (DDR) activates autophagy (Eliopoulos et al. 2016), which is required for cell senescence, such as that reported in cardiovascular aging (Abdellatif et al. 2018).

Examples of Application

Genome instability is mainly DNA damages caused by genotoxic events. It is now clear that DNA damages are abundant, happen all the time, and cause aging. Our current understanding of genome instability is focused on DNA damage and mutations impacting aging, and our knowledge have been applied to achieve the following:

Detection includes knowing the occurrence of DNA damage in somatic cells linked to aging; appropriate high-throughput and high-resolution methods can be used to detect the damaged DNA. The recently developed single-cell DNA sequencing technology serves the purpose. Identifying the genes related to aging provides us the target genes to detect DNA damages and mutations.

Assessment for genome instability and the agents responsible includes the following:

(a) Genotoxins types – The type of genotoxins that can cause DNA damages is now quite well-established with ROS as an example. Identifying their occurrence would help to prevent people from the harmful effects they bring about.
(b) Genotoxin exposure – For individuals or populations, the level of exposure can be assessed, and precautions can be taken to lower the risk of genome instability. The types of tissues and cells would need to be taken into consideration.
(c) Genetic predisposition – Other than somatic cell DNA damages and mutations, germline can be affected too. Germline genome instability is mainly mutations and can be detected more easily by DNA sequencing because they occur in almost all the cells of an individual. Knowing the kind of mutation an individual has would provide information on the aging trend of the individual (Turkez et al. 2017).

Genes analyses are as follows:

(a) GWAS – GWAS can be carried out to validate the implicated aging-related genes and deepen our knowledge about them. The current knowledge provides the basis to assess GWAS results.
(b) Functional roles – Knowing the genes which would cause aging when damaged, we can try to deduce their functional roles and in turn better understand the aging process.
(c) Antiaging medicine aims at genome stability maintenance to know more about the relations between genome instability and aging, it is possible to develop antiaging medicine by aiming at maintaining genome stability. The antiaging drug metformin is a promising one (Valencia et al. 2017).

Future Directions of Research

Detection – current detection method of single-cell genome sequencing relies mainly on next-generation sequencing (NGS). NGS is very useful but still limited by high setup capital and running cost and also yielding short-read sequences. More recently developed single-molecule DNA sequencing methods are useful technologies in detecting and quantifying DNA damages in cells (van Dijk et al. 2018).

DNA damages in signaling pathways have been shown to link to aging of cells, but the detail mechanisms still need to be elucidated. The following are some of the issues:

Assessment aims at genotoxins types and exposure – Based on our better understanding of genotoxins, we need to explore and identify potential genotoxins in our environment (Brinke and Buchinger 2016). It would also be useful to avoid generation of genotoxins in our technology development in many areas (Turkez et al. 2017). A toxicogenomics database would help to associate chemicals with aging (Davis et al. 2019).
Genes – GWAS needs to be expanded to include more people from different ethnicities and regions in order for us to better understand

the variations in genome instabilities in different individuals.

Antiaging measures aiming at genome stability maintenance – Antiaging by antigenotoxic substances can be explored as a good strategy to slow down aging. Natural products such as fruits, vegetables, natural resins, and polysaccharides have been shown to have antigenotoxic effects (Izquierdo-Vega et al. 2017; López-Romero et al. 2018). Their antigenotoxic mechanisms need to be explored. Drugs that can improve genome stability need to be developed (Vaiserman et al. 2017). Gene therapy to repair gene mutations related to aging would help to tackle some problems caused by genome instability. Recently, gut microbiota has been shown to relate to numerous health issues; among them is antiaging (Vaiserman et al. 2017). Gut microbiota dysfunction may be genotoxic (Grasso and Frisan 2015; Yu and Schwabe 2017), so maintenance for a health gut microbiota is necessary in preventing genotoxic impacts.

Summary

Genome instability refers to DNA damages which escape repairing or are repaired with mistakes. Examples of genotoxins include ROS, UV irradiation, and environmental mutagens. DNA mutations can occur in both somatic cells and germline cells. Accumulated evidences showed that DNA damages and mutations are abundant, are happening all the time, and cause aging. Recent advances in single-cell DNA sequencing technology allowed the detection of somatic mutations. Some mutated genes related to aging have been identified and become the target in characterizing the aging process. Many agents responsible for genome stability have also been identified and assessed. Knowing the roles of genome instability would allow us to design targeted antiaging measures.

Cross-References

▶ Aging Definition
▶ Aging Mechanisms
▶ Aging Theories
▶ Disposable Soma Aging Theory
▶ DNA Damage Theory
▶ Genes that Delay Aging
▶ Genetic Theories of Aging
▶ Genetics: Gene Expression
▶ Genome-Wide Association Study
▶ Genomics of Aging and Longevity
▶ Human Genome Project
▶ Mitochondrial DNA Mutations
▶ Mitochondrial Reactive Oxygen Species Aging Theory
▶ Molecular and Epigenetic Clocks of Aging
▶ Mutation
▶ Mutation Accumulation Aging Theory
▶ Mutation Load and Aging
▶ Oxidation Damage Accumulation Aging Theory (The Novel Role of Glutathione)
▶ Programmed Cell Death
▶ Whole Genome Sequencing

References

Abdellatif M, Sedej S, Carmona-Gutierrez D et al (2018) Autophagy in cardiovascular aging. Circ Res 123:803–824. https://doi.org/10.1161/CIRCRESAHA.118.312208

Acosta JC, Banito A, Wuestefeld T et al (2013) A complex secretory program orchestrated by the inflammasome controls paracrine senescence. Nat Cell Biol 15:978–990. https://doi.org/10.1038/ncb2784

Armstrong GT, Kawashima T, Leisenring W et al (2014) Aging and risk of severe, disabling, life-threatening, and fatal events in the childhood cancer survivor study. J Clin Oncol 32:1218–1227. https://doi.org/10.1200/JCO.2013.51.1055

Brinke A, Buchinger S (2016) Toxicogenomics in environmental science. In: Reifferscheid G, Buchinger S (eds) In vitro environmental toxicology – concepts, application and assessment. Advances in biochemical engineering/biotechnology. Springer, Cham, pp 159–186

Chevanne M, Caldini R, Tombaccini D et al (2003) Comparative levels of DNA breaks and sensitivity to oxidative stress in aged and senescent human fibroblasts: a distinctive pattern for centenarians. Biogerontology 4:97–104. https://doi.org/10.1023/A:1023399820770

Cho M, Suh Y (2014) Genome maintenance and human longevity. Curr Opin Genet Dev 26:105–115. https://doi.org/10.1016/J.GDE.2014.07.002

Cui H, Kong Y, Zhang H (2012) Oxidative stress, mitochondrial dysfunction, and aging. J Signal

Transduction 2012:646354. https://doi.org/10.1155/2012/646354

Davis AP, Grondin CJ, Johnson RJ et al (2019) The comparative toxicogenomics database: update 2019. Nucleic Acids Res 47:D948–D954. https://doi.org/10.1093/nar/gky868

Eliopoulos AG, Havaki S, Gorgoulis VG (2016) DNA damage response and autophagy: a meaningful partnership. Front Genet 7:204. https://doi.org/10.3389/fgene.2016.00204

Ermolaeva M, Neri F, Ori A, Rudolph KL (2018) Cellular and epigenetic drivers of stem cell ageing. Nat Rev Mol Cell Biol 19:594–610. https://doi.org/10.1038/s41580-018-0020-3

Falconer E, Hills M, Naumann U et al (2012) DNA template strand sequencing of single-cells maps genomic rearrangements at high resolution. Nat Methods 9:1107–1112. https://doi.org/10.1038/nmeth.2206

Franzke B, Neubauer O, Wagner KH (2015) Super DNAging – new insights into DNA integrity, genome stability and telomeres in the oldest old. Mutat Res 766:48–57. https://doi.org/10.1016/J.MRREV.2015.08.001

Grasso F, Frisan T (2015) Bacterial genotoxins: merging the DNA damage response into infection biology. Biomol Ther 5:1762–1782. https://doi.org/10.3390/biom5031762

Hanahan D, Weinberg RA (2011) Hallmarks of cancer: the next generation. Cell 144:646–674. https://doi.org/10.1016/J.CELL.2011.02.013

Hou J, Yang Y, Huang X, Song Y, Sun H, Wang J, Hou F, Liu C, Chen W, Yuan J (2016) Aging with higher fractional exhaled nitric oxide levels are associated with increased urinary 8-oxo-7, 8-dihydro-2′-deoxyguanosine concentrations in elder females. Environ Sci Pollut Res 23(23):23815–23824. https://doi.org/10.1007/s11356-016-7491-6

Izquierdo-Vega J, Morales-González J, SánchezGutiérrez M et al (2017) Evidence of some natural products with antigenotoxic effects. Part 1: fruits and polysaccharides. Nutrients 9:102. https://doi.org/10.3390/nu9020102

Jackson SP, Bartek J (2009) The DNA-damage response in human biology and disease. Nature 461:1071–1078. https://doi.org/10.1038/nature08467

Jacob KD, Noren Hooten N, Trzeciak AR, Evans MK (2013) Markers of oxidant stress that are clinically relevant in aging and age-related disease. Mech Ageing Dev 134:139–157. https://doi.org/10.1016/J.MAD.2013.02.008

Jacobs KB, Yeager M, Zhou W et al (2012) Detectable clonal mosaicism and its relationship to aging and cancer. Nat Genet 44:651–658. https://doi.org/10.1038/ng.2270

Langley E, Pearson M, Faretta M et al (2002) Human SIR2 deacetylates p53 and antagonizes PML/p53-induced cellular senescence. EMBO J 21:2383–2396. https://doi.org/10.1093/emboj/21.10.2383

Laurie CC, Laurie CA, Rice K et al (2012) Detectable clonal mosaicism from birth to old age and its relationship to cancer. Nat Genet 44:642–650. https://doi.org/10.1038/ng.2271

Laven JSE, Visser JA, Uiterlinden AG et al (2016) Menopause: genome stability as new paradigm. Maturitas 92:15–23. https://doi.org/10.1016/J.MATURITAS.2016.07.006

López-Otín C, Blasco MA, Partridge L et al (2013) The hallmarks of aging. Cell 153:1194–1217. https://doi.org/10.1016/J.CELL.2013.05.039

López-Romero D, Izquierdo-Vega J, Morales-González J et al (2018) Evidence of some natural products with antigenotoxic effects. Part 2: plants, vegetables, and natural resin. Nutrients 10:1954. https://doi.org/10.3390/nu10121954

Ock S, Lee WS, Ahn J et al (2016) Deletion of IGF-1 receptors in cardiomyocytes attenuates cardiac aging in male mice. Endocrinology 157:336–345. https://doi.org/10.1210/en.2015-1709

Ramsey MJ, Moore DH, Briner JF et al (1995) The effects of age and lifestyle factors on the accumulation of cytogenetic damage as measured by chromosome painting. Mutat Res 338:95–106. https://doi.org/10.1016/0921-8734(95)00015-X

Rodier F, Coppé JP, Patil CK et al (2009) Persistent DNA damage signalling triggers senescence-associated inflammatory cytokine secretion. Nat Cell Biol 11:973–979. https://doi.org/10.1038/ncb1909

Roos WP, Thomas AD, Kaina B (2016) DNA damage and the balance between survival and death in cancer biology. Nat Rev Cancer 16:20. https://doi.org/10.1038/nrc.2015.2

Rossi DJ, Bryder D, Seita J et al (2007) Deficiencies in DNA damage repair limit the function of haematopoietic stem cells with age. Nature 447:725–729. https://doi.org/10.1038/nature05862

Sanders AD, Hills M, Porubský D et al (2016) Characterizing polymorphic inversions in human genomes by single-cell sequencing. Genome Res 26:1575–1587. https://doi.org/10.1101/gr.201160.115

Scheibye-Knudsen M, Ramamoorthy M, Sykora P et al (2012) Cockayne syndrome group B protein prevents the accumulation of damaged mitochondria by promoting mitochondrial autophagy. J Exp Med 209:855–869. https://doi.org/10.1084/jem.20111721

Soares JP, Cortinhas A, Bento T et al (2014) Aging and DNA damage in humans: a meta-analysis study. Aging 6:432–439. https://doi.org/10.18632/aging.100667

Tilstra JS, Robinson AR, Wang J et al (2012) NF-κB inhibition delays DNA damage-induced senescence and aging in mice. J Clin Invest 122:2601–2612. https://doi.org/10.1172/JCI45785

Turkez H, Arslan ME, Ozdemir O (2017) Genotoxicity testing: progress and prospects for the next decade. Expert Opin Drug Metab Toxicol 13:1089–1098. https://doi.org/10.1080/17425255.2017.1375097

Vaidya A, Mao Z, Tian X et al (2014) Knock-in reporter mice demonstrate that DNA repair by non-homologous end joining declines with age. PLoS Genet 10:e1004511. https://doi.org/10.1371/journal.pgen.1004511

Vaiserman AM, Koliada AK, Marotta F (2017) Gut microbiota: a player in aging and a target for anti-aging intervention. Ageing Res Rev 35:36–45. https://doi.org/10.1016/J.ARR.2017.01.001

Valencia WM, Palacio A, Tamariz L, Florez H (2017) Metformin and ageing: improving ageing outcomes beyond glycaemic control. Diabetologia 60:1630–1638. https://doi.org/10.1007/s00125-017-4349-5

van Dijk EL, Jaszczyszyn Y, Naquin D, Thermes C (2018) The third revolution in sequencing technology. Trends Genet 34:666–681. https://doi.org/10.1016/J.TIG.2018.05.008

Wong YT, Gruber J, Jenner AM et al (2009) Elevation of oxidative-damage biomarkers during aging in F2 hybrid mice: protection by chronic oral intake of resveratrol. Free Radic Biol Med 46:799–809. https://doi.org/10.1016/J.FREERADBIOMED.2008.12.016

Yu LX, Schwabe RF (2017) The gut microbiome and liver cancer: mechanisms and clinical translation. Nat Rev Gastroenterol Hepatol 14:527–539. https://doi.org/10.1038/nrgastro.2017.72

Genome-Wide Association Studies

▶ Genetic Control of Aging

Genome-Wide Association Study

Teresa Chung and Lok Ting Lau
Department of Applied Biology and Chemical Technology, The Hong Kong Polytechnic University, Hung Hom, Kowloon, Hong Kong

Synonyms

GWA; GWAS; WGA study; Whole genome association study

Definition

In genetic epidemiology, a genome-wide association study (GWAS) is a type of biological study that is commonly used to identify genetic variants associated with phenotypic traits, usually in the form of single nucleotide polymorphisms (SNPs). GWAS enhances our understanding on the etiology and pathophysiology of the underlying trait being investigated.

Overview

No two human genomes are the same. Human genome is about three billion base pair long and contains around 30,000 genes. An allele is one of two or more different forms on the same genetic site or a locus (loci for plural). Allele has DNA sequence(s) in the form of nucleotides that carry information about observable physical characteristics or traits, and different alleles contribute to genetic variations that result in different phenotypic traits. For a diploid organism such as human, each of the two alleles is inherited by the offspring from each parent. Genotype is an organism's combination of alleles at genetic loci. Therefore, at a particular locus, there are usually two reads in GWAS for a binary trait, one inherited from the mother and another inherited from the father.

Ever since the publication of the first GWAS in 2000s, the number of GWAS, the respective sample size, as well as the number of statistically significant SNPs detected have increased exponentially. Up until April 2019, more than 3923 GWAS have been published on GWAS Catalog records identifying genetic variants across more than 3500 unique traits. With subjects from the USA, the UK, and Iceland representing 72% of all GWAS discoveries (Mills and Rahal 2019), GWAS still shows a predominant European ancestry, and there is an imbalanced representation from ethics minority.

Design and Analysis

Before GWAS becoming popular, pedigree linkage analysis and candidate gene studies were mainly used. Linkage analysis evaluates the inheritance patterns of widely spaced genetic variants along family members, initially used to find genetic variants of single gene disorders with large effect sizes and high allele frequencies.

When it is applied on the common complex diseases such as diabetes or coronary heart disease, which rely on the complex interactions between multiple genetic variants and environmental factors, results are often non-replicative. In addition, linkage analysis has a low power to detect such genetic differences due to the limited generations of family members. An alternative method, candidate gene study can overcome this problem by using a gene-specific approach that focuses on a small number of pre-specified genetic regions and tests the association of these variants with the trait of interest. However, because candidate gene study ignores the rest of the genome, it is very likely to miss true causal genetic variants.

In contrast, GWAS adopts an unbiased, non-hypothesis-driven approach (Jeck et al. 2012) to study genetic variants without the need to pre-specify a particular gene region. Genetic markers on the array panel are scattered throughout the entire genome, which allow the comparison of genetic variants between the affected and unaffected individuals to test for associations with the trait of interest and to determine the allele frequencies. Recent advances in array-based genotyping of SNPs have allowed researchers to assay millions of SNPs simultaneously, which give GWAS a higher coverage, a higher throughput, and a greater power to detect genetic variants. The steadily decrease in array price has also enabled the widespread use of GWAS (Witte 2010).

In order to minimize false-positive results, GWAS is usually organized in multistage study design. A proportion of the samples are first genotyped in the discovery cohort using genome-wide SNP array, then the most significant SNPs are then genotyped in an additional replication/follow-up cohort(s) with the remaining study samples. The exact split in sample size between discovery and replication cohorts depends on the tolerance of false-negative SNPs that would be missed if the sample size of the discovery cohort is too small. Hence, the usual practice would be to ensure as many SNPs are carried over from one stage to another, splitting around one-third to one-half of the study sample for the discovery cohort (Witte 2010). Follow-up data are sometimes treated as separate replication cohort with the goal of confirming the findings from the discovery cohort. But depending on the situations, e.g., the need to obtain sufficient power due to SNPs with modest effects and to avoid multiple comparison issues, it is sometimes common to combine follow-up data as joint analysis. Afterwards, it is common to combine multiple GWAS dataset in the form of meta-analysis to further validate the results. Hence, careful multistage design is vital to the success of GWAS (Witte 2010).

Most GWAS compare case subjects bearing a particular trait with unaffected controls or can study a continuous trait among subjects. Subjects selected should be representative of the population. To increase the power of the study, subjects with extreme phenotype should be selected. Power calculation is required to determine a minimum sample size in order to have sufficient power to detect statistically significant SNPs, and usually at least thousands cases and controls are needed. Unrelated cases and controls should be well matched to each other in terms of ethnicity, age, and sex. Family-based design directly controls for confounding by ethnicity, geographical origin, and environmental exposures as increasing amount of genetic information are shared between family members. To reduce the cost of recruitment and genotyping, it is also common to use readily available control data that were recruited to previous studies to increase the sample size. In this case, it is important to control for confounding due to the different geographically origins and ethnicity between case and control groups (Witte 2010), and population stratification should be examined using principle component analysis (PCA). Gender is another confounding factor to take into account as genetic variations differ enormously between genders.

During quality control, the proportion of successfully genotyped SNPs are being tested for Hardy-Weinberg equilibrium, where actual and expected gene frequencies are compared and tested for significance using Pearson's Chi-square and genes whose frequencies are significantly deviated from such equilibrium are removed for further analysis. To overcome the limitation of SNP restriction on array chip, fine mapping imputation is also routinely done to increase the number of SNPs and the power of the study as well as facilitate the meta-analysis of GWAS with comparable SNPs. Imputation is based on linkage disequilibrium (LD), merely the nonrandom association of neighboring alleles at different loci in a given population. With genotyped SNPs and LD haplotype reference from HapMap, GWAS data can be imputed with neighboring, untyped common genetic variants (Witte 2010). Fine mapping is also commonly employed to genotype on a denser coverage all variants within the associated locus to identify the causal variants.

The minor allele of each SNP is then tested for association with the trait of interest, taking into account different genetic models such as allelic/additive, recessive, and dominant (Witte 2010). Once the genetic variants are identified, the effect sizes or risks of these disease-predisposing variants are determined. The most common method is to calculate the odds ratio (OR), that is, the ratio of the odds of the observed risk allele frequency between cases and controls. The higher than 1 the odd ratio is, the stronger the association the SNP is with the trait of interest. Effect sizes of SNPs of most common diseases are small with a median odd ratio of only 1.28 (Witte 2010). There is also a risk of overestimating the true genetic effect size as a result of the winner's curse issue (Lohmueller et al. 2003). If the power calculations for subsequent cohorts are based on this overestimated OR, then all subsequent replication studies are likely to be underpowered.

To determine the overall statistical significance of GWAS results, p-value of the odd ratio is calculated using chi-squared test. In most GWAS, up to one million independent statistical tests are being performed simultaneously, and the issue of multiple comparisons arises. To overcome this problem, correction to Bonferroni is used, in which the p-value 0.05 is divided by the number of statistical tests being performed simultaneously, i.e., $0.05/1,000,000 = 5 \times 10^{-8}$. It is important to note that such cutoff point is arbitrary and does not reflect actual biological association (Witte 2010). To visualize the GWAS results, Manhattan plots are often used to graphically represent the results. Y-axis is represented by the negative logarithm of the P-value/association level as a function of genomic location on the x-axis. With each SNP being represented by a dot, significantly associated SNPs often stand out as stacks of points and can easily be recognized on the plot.

Key Research Findings

Longevity, a type of healthy ageing, is a polygenic trait that is influenced by a complex interaction of multiple gene variants and the environment (Shadyab and LaCroix 2015). Clustering of longevity in families suggests that genetics accounts for 20–35% of longevity (Shadyab and LaCroix 2015) with greater importance at extreme ages (Beekman et al. 2013). Although numerous GWAS have been performed, only genetic variants of the apolipoprotein E (APOE), FOXO3A, and the AKT1 loci have been robustly replicated in GWAS (Shadyab and LaCroix 2015), and many more variants are still to be characterized. For example, in a GWAS meta-analysis with replication and subset analysis that comprised more than 26,000 cases ($\geq$85 years) and 92,000 controls (<65 years) of European ancestry, Deelen et al. identified that rs4420638 (TOMM40/APOE/APOC1 locus) (OR 0.72, $p = 3.4 \times 10-36$) and a novel locus rs2149954 (T) (OR 1.10, $p = 1.74 \times 10-8$) were significantly associated with longevity (Deelen et al. 2014). Reasons for not identifying additional longevity loci may be due to the need to harmonize the phenotypic criteria and insufficient sample size of subjects with extreme phenotypes.

One of the ageing-related late-onset diseases is the Alzheimer's disease (AD), which is further classified into late-onset AD (or sporadic AD) and early-onset AD (EOAD) (or familial AD) based on the age of onset and inheritance of genetic factors. LOAD is by far the most common form of AD with an average age of onset of around 65 years of age and is characterized by the pathological presence of (1) neurofibrillary tangles which are composed of phosphorylated tau proteins and (2) senile plaques which are composed of amyloid β (Aβ)-protein. Although the different isoforms of the polymorphic apolipoprotein (ApoE) genes are by far the only well-established risk factor for LOAD, several meta-analyses on GWAS have identified additional genetic risk variants on LOAD which are suggested to act via the following physiological pathways:

1. Immune response and inflammation: complement receptor 1 gene (CR1), membrane-spanning 4-domains, subfamily A (MS4A) gene cluster, ephrin type-A receptor 1 gene (EPHA1), CD33, triggering receptor expressed on myeloid cells gene 2 (TREM2)
2. Lipid (cholesterol) metabolism: clusterin gene (CLU) and ABCA7
3. Endocytosis and synaptic function: phosphatidylinositol binding clathrin assembly protein gene (PICALM), bridging integrator 1 gene (BIN1), CD2AP, and EPHA1.

With the above and the still-to-be-discovered genes to be identified, the exact pathogenesis of LOAD remains unclear, and the links of these genetic variants to both the amyloidogenic and the tau hyperphosphorylation pathways remain to be elucidated (Misra et al. 2018).

Prospects

Although several robustly associated SNPs have been identified using GWAS, most of them are novel, with unknown functions, are located outside gene regions (around 30%), and require further fine mapping and functional studies to demonstrate causality (Witte 2010). Common diseases can be caused by common genetic variants each with a small effect size. But despite the clustering in families, genetic variants that are identified from GWAS still only account for little heritability. An alternate hypothesis is that common diseases may be due to rare alleles each with a large effect (Atzmon 2015). With the tagged variants on the genotyping array, GWAS are only designed to measure common variants at a minor allele frequency >5%. Therefore, the heritability issue may be due to the fact that GWAS is unable to determine measure such rare variants. Alternative methods are needed to be sought. Fine mapping such as deep sequencing may be required to unearth such rare variants. Other than SNPs, genetic variants in the form of copy number variants (CNVs) may also be another reason for the small heritability accounted (Witte 2010). With the continued scientific and technological advances, our ability to study complex phenotypes that are caused by the less common, different genetic variations with smaller effect sizes will become easier (Witte 2010). One major implication of GWAS is the availability of genetic testing based on the SNPs identified from GWAS, and the area under the receiver operating characteristic curve (AUC) is used to measure the discriminatory ability of such test.

Summary

GWAS is a very useful technology to identify risk-SNP biomarkers to help elucidate the pathophysiology, improve the accuracy of prognosis and diagnosis, assess drug responses, and design new therapeutic drug targets based on the patient's genotype. However, successful application of GWAS requires stringent quality control as there are multiple sources of errors, confounding factors and bias, compounded by the modest effect sizes for most common traits of interest. Therefore, careful study design, stringent threshold, result analysis, and interpretation are required to ensure robust GWAS results.

Cross-References

- ► Alleles
- ► DNA Chip
- ► Genetics: Gene Expression
- ► Genotype
- ► Nucleotides
- ► Whole Genome Sequencing

References

Gil Atzmon (ed) (2015) Longevity genes: a blueprint for aging. Advances in experimental medicine and biology. Springer, New York

Beekman M, Blanché H, Perola M et al (2013) Genome-wide linkage analysis for human longevity: Genetics of Healthy Aging Study. Aging Cell 12(2):184–193. https://doi.org/10.1111/acel.12039

Deelen J, Beekman M, Uh HW et al (2014) Genome-wide association meta-analysis of human longevity identifies a novel locus conferring survival beyond 90 years of age. Hum Mol Genet 23(16):4420–4432. https://doi.org/10.1093/hmg/ddu139

Jeck WR, Siebold AP, Sharpless NE (2012) Review: a meta-analysis of GWAS and age-associated diseases. Aging Cell 11(5):727–731. https://doi.org/10.1111/j.1474-9726.2012.00871.x

Lohmueller KE, Pearce CL, Pike M et al (2003) Meta-analysis of genetic association studies supports a contribution of common variants to susceptibility to common disease. Nat Genet 33(2):177–182. https://doi.org/10.1038/ng1071

Mills MC, Rahal C (2019) A scientometric review of genome-wide association studies. Commun Biol 2:9. Advances in experimental medicine and biology. https://doi.org/10.1038/s42003-018-0261-x

Misra A, Chakrabarti SS, Gambhir IS (2018) New genetic players in late-onset Alzheimer's disease: findings of genome-wide association studies. Indian J Med Res 148(2):135–144. https://doi.org/10.4103/ijmr.IJMR_473_17

Shadyab AH, LaCroix AZ (2015) Genetic factors associated with longevity: a review of recent findings. Ageing Res Rev 19:1–7. https://doi.org/10.1016/j.arr.2014.10.005

Witte JS (2010) Genome-wide association studies and beyond. Annu Rev Public Health 31:9–20, 4 p following 20. https://doi.org/10.1146/annurev.publhealth.012809.103723

Genomic Alterations

▶ Mutation Load and Aging

Genomic Instability

▶ Genome Instability

Genomics of Aging and Longevity

Ghadeer Falah, Danielle Gutman and Gil Atzmon
Faculty of Natural Science, University of Haifa, Haifa, Israel

Synonyms

Genetics and epigenetics of aging and longevity; Genomics of centenarians; Genomics of healthy aging; Genomics of healthy lifespan

Definition

Genomics of aging and longevity encompasses the fields of genomics and epigenomics implemented towards the understanding of a healthy aging phenotype. Exceptionally long-lived individuals (centenarians) are used as a model for longevity alongside animal and cell models that are used for mechanism and pathway illustration and elucidation.

Overview

Aging can be defined as a progressive decline in organ function and is considered the main risk factor for chronic disease, weakened health, and increased risk of morality (Macedo et al. 2017). Aging research has progressed in recent years, especially since it has been found to be mediated, at least to some extent, by genetic pathways and biochemical processes (López-Otín et al. 2013). Further, longevity is one of the most complex phenotypes (Brooks-Wilson 2013) with studies centering on long-lived individuals (LLI) who are considered a model for healthy aging (Murabito et al. 2012). According to López-Otín, there are nine tentative hallmarks which drive aging and aging characteristics of organs especially in mammalian aging. These hallmarks are genomic instability, telomere attrition, epigenetic alterations, loss of proteostasis, deregulated nutrient sensing, mitochondrial dysfunction, cellular senescence, stem cell exhaustion, and altered intercellular communication (López-Otín et al. 2013).

Many research efforts are invested in attempts to find the connection between these hallmarks and their contribution to aging, aiming to improve human health during aging and thereby increase healthy lifespan. Each hallmark is affected by many factors, such as DNA methylation, telomere attrition, and chromatin remodeling, and any disruption in them may lead to cellular senescence. Cellular senescence (CS) is involved in many physiologic and pathological processes, such as cancer protection, aging, and disease (Sturmlechner et al. 2017), and is also a causal factor in aging playing an important role in age-related diseases, stem cell exhaustion, and tissue damage.

Key Research Findings

Genomics of Aging and Longevity

Genomic instability and mitochondrial dysfunction are considered to be among the primary drivers to cellular aging in mammalian cells (Vijg et al. 2017). Cells that have defects in one or more of DNA repair pathways are likely to accumulate continuous DNA damage that will result with deleterious genetic mutations and will promote aging. As a result, many mutated alleles in DNA repair genes are the genetic basis of several heritable human disorders that characterized by accelerated aging (Lombard et al. 2005).

The knowledge about the genes and pathways that are impaired in the process of aging and in age-related diseases has increased dramatically in recent years. According to many studies, there is a list of genes that were identified as associated with aging and multi morbidity, placing aging as a malleable process that can be strongly influenced by genetics (Wegman et al. 2014; Cardoso et al. 2018). Only two decades ago, the tedious work to explore the genetic predisposed mechanisms associated with aging was built on the candidate gene approach, a procedure that is much less common today, since the completion of the human genome project in 2003 brought forth vast development and more efficient strategies to detect these genetic elements. Genomic technologies and availability of human genomic data provide a powerful tool for high-throughput, automatic cheap screening. Genomic-wide association studies (GWAS) identify genes associated with biological performance, and in particular, genetic contributors associated with aging-related phenotypes have flourished (Kronenberg 2008).

In parallel, Human Aging Genomic Resources (HAGR), a collection of many resources for studying the biology of human aging, was established. This database includes (a) GenAge, a database of genes associated with human aging, and (b) AnAge, composed of longevity and aging genes found in animal species (Warner 2005) and provided the aging research community a resourceful component for the aging studies.

Another support for measuring the genetic component of healthy human aging came from the project of GEnetics of Healthy Aging (GEHA) which began on May 2004 that identifies genes involved in longevity and aging (Heshmati 2018). After sample collection from 2,800 pairs of European long-lived siblings (90+ years old (YO)) and controls, the final results were published by Beekman et al. in 2013 and included 4 large chromosomal regions of association with longevity, 1 SNP within the TOMM40/APOE/APOC1 locus including the APOEe4 and APOEe2 alleles, also associated with longevity. This study reconfirmed the previously established association of the APOE gene with the longevity phenotype. Additionally, three chromosomal regions were found to be associated with longevity in a gender-specific manner (Beekman et al. 2013).

In 2014, a meta-analysis of genome-wide association studies was conducted on 6,036 long-lived individuals (age >90) and 3,757 participants who died between ages 55 and 80 used as a control group, in an attempt to identify SNPs associated with longevity. Broer et al. revealed significant evidence for the involvement of SNPs near the genes CADM2 and GRIK2, both involved in neuronal pathways, particularly in neuron cell-cell adhesion and regulation of glutamatergic synaptic transmission. A significant association with longevity was found at rs1416280 located 369 kb upstream the GRIK2 gene. APOE and FOXO3 were also validated as genes association of with longevity (Broer et al. 2015).

These efforts resulted in a handful of candidates of aging associated genes. Yet, APOE shines in almost every report (Zhao et al. 2018; Theendakara et al. 2018). APOE is a well-studied gene with multiple effects on aging and longevity. APOE have three alleles (ε2, ε3, ε4), and the prevalence of its genotypes ε2/ε3, ε3/ε3, and ε3/ε4 is commonly associated with Alzheimer's disease (AD). For example, Ferri et al. reported differential frequency of APOE genotypes in centenarians versus AD patients (Ferri et al. 2019). Another study found APOE ε2 to be associated with healthy aging and longevity (Gurinovich et al. 2018).

The immense use of genomic chips and GWAS in the last decade rapidly declined due to the dramatic reduction in sequence operations' cost. Whole-exome and whole-genome sequencing have become exponentially abundant in general, and in longevity and aging studies in particular. For example, in China a study in 502 women confirmed candidate genes that were previously reported as responsible for signs of skin aging. Significant association between the SNPs rs2066853 in exon 10 of the aryl hydrocarbon receptor gene AHR and rs10733310 in intron 5 of BNC2 and pigment spots on the arms was observed. Also significant association between the SNP rs11979919, 3kb downstream of COL1A2, and laxity of eyelids was reported (Gao et al. 2017). Another study examined whether SNPs within the REST gene are linked with cognitive aging; they analyzed 634 Taiwanese aged over 60 and found that the REST rs1277306 SNP was significantly associated with cognitive aging. Moreover, this association remained significant after removal of individuals with APOE ε4 homozygote allele, but not for individuals with at least one APOE ε4 allele. In addition, interactions between the rs1713985 and rs1277306 (variants of the REST gene), SNPs' effect on cognitive aging was observed (Lin et al. 2017).

In the same line, premortem data and postmortem tissue samples were collected from 155 PLWH patients (persons living with HIV) displaying evidence of "accelerated aging." According to the result, IL-6 and IL-10 genes were found significantly associated with greater accelerated aging, an association that was depleted when taking into account the TNF genotypes (Sundermann et al. 2018). Further, a study conducted by Sahba et al. revealed that SPARCL1 accelerates AD pathogenesis and links neuroinflammation with common changes in brain structure and function during aging (Seddighi et al. 2018).

Studies of long-lived individuals have revealed few genetic mechanisms for protection against age-associated disease. Erikson et al. pursued a whole-genome study of a healthy aging phenotype in order to find the genetic factors that protect from age-associated diseases; the study was performed on individuals ranging from 80 to 105 years old with no chronic diseases and who are not taking chronic medications. Their efforts did not prevail, and they were unable to highlight association to the healthy aging phenotype; however they were able to find a variant that was associated with preserved cognition (Erikson et al. 2016). The idea of "protective" variants gained further support from a whole-exome sequencing screen of 44 centenarians that identified 130 rare variants that have been previously reported as cause of numerous inherited diseases. Surprisingly, one of the centenarians who was not cognitively impaired at all was found to be homozygous for the known risk variant of APOE ε4 (Freudenberg-Hua et al. 2014). These studies support an earlier hypothesis suggested by (Bergman et al. 2007) stating that healthy aging and longevity could be mediated through protective variants or possibly a different mechanism serving as a "buffer" and enabling a healthy phenotype (Bergman et al. 2007). The effect of the "buffering" and protective variants remains elusive as are the longevity and healthy aging phenotypes. None of the genetic mechanisms can represent the phenotype, leading the research community towards consideration of the epigenetic interaction of the genome with the environment.

Aging Epigenomics

Epigenetic Alterations and Aging

The field of epigenetics in aging- and age-related diseases (such as Alzheimer's disease (AD), rheumatoid arthritis (RA), type 2 diabetes (TD2), cardiovascular, and cancer) attracted the attention of researcher in molecular physiology and medicine (Lipman and Tiedje 2006). The term epigenetics first arose to describe heritable changes in gene expression that did not involve changes to the coding sequence of DNA. Epigenetic mechanisms are the major contributors to genome structure changes and the aging process. According to Susana Gonzalo (Gonzalo 2010), epigenetic regulation includes DNA methylation (Schübeler 2015; Byun et al. 2009), histone modifications (Chen and Zhao 2015), and noncoding (nc) RNA species (Holoch and Moazed 2015); changes in these factors affect nuclear processes, DNA replication and repair, cell cycle progression, and telomere and centromere structure and function (Gonzalo 2010).

DNA Methylation and the Epigenetic Clock

DNA methylation is an important epigenetic mechanism that is involved in development, differentiation, and cell fate diversity (Vijg et al. 2018). In addition, it is one of the most intensely studied epigenetic modifications in mammals (Marta Kulis 2010). DNA methylation occurs by the formation of a covalent bond between a methyl group and the cytosine residues in the context of cytosine-guanine dinucleotides (so-called CpG sites) by a family of enzymes called DNA methyltransferases (DNMT) (Aquino et al. 2018). Methylation in mammals is achieved through the operation of three DNA methyltransferases: DNMT1, which has a maintenance role (Jin and Li 2011), and DNMT3a and 3b, that are de novo methylases (Okano et al. 1999). The marks of methylation are established during embryological development by de novo methyltransferases DNMT3a and DNMT3b that play an important role in normal development and disease in addition to gene expression in adult somatic cells (Okano et al. 1999); another DNMT is DNMT2 that methylates cytosine-38 in the anticodon loop of the tRNA for aspartic acid, but does not methylate DNA (Goll et al. 2006).

The general decrease in DNA methylation together with the increase in variability of methylation patterns with age is termed "epigenetic drift" (Pérez et al. 2018). Methylation of the DNA was used as a biomarker of healthy versus unhealthy aging and disease risk (Field et al. 2018), as well as a predictor of age in "aging clock" (Aquino et al. 2018). Two clocks are vastly used, the Hannum et al. (2013) and Horvath (2013) models, independently developed using different methods for calculating the epigenetic age of the individual. The Hannum model was developed on the basis of DNA methylation measurements using Illumina 450 K array data from blood samples of more than 650 people from the general population. The model identified 71 sites that predicted the chronological age (Hannum et al. 2013). In comparison, Horvath's model involved over 7,800 samples from 51 different cell and tissue types. This model identified 353 CpG sites with strong correlations with chronological age (Horvath 2013). A comparison between the two clocks demonstrated that both underestimate DNAm age with respect to chronological age but are opposed in estimating the rate of age acceleration (Armstrong et al. 2017). These accumulated reports suggest a shift in the aging field towards focus on the association between DNA methylation and age. Primary human fibroblasts grown in culture (Wilson and Jones 1983), human lymphocytes (Drinkwater et al. 1989), and peripheral blood cells (Bjornsson et al. 2008; Fuke et al. 2004) also support the observation that during aging, there is a decrease in DNAm.

A study by Manel Esteller's group on monozygotic twins (MZ) shows that although the twins share nearly identical DNA sequences and the same methylation profile at birth, when they mature, they exhibit differences in methylation patterns in addition to significant epigenetic differences (Lipman and Tiedje 2006); thus twin studies can serve as a model for studying the effect of the epigenome on phenotype and gene expression. Another thorough whole-genome sequencing study performed by Ye et al. (2013) on two monozygotic twin pairs (aged 100 and 40 years) demonstrated a low number of somatic mutations accumulated in a century of life, thus weakening the notion that aging is a result of accumulation of damaging mutations over time (Ye et al. 2013).

Hypomethylation of repeating sequence-like elements (SINEs) and LTR retrotransposons gradually declines with age and is linked to organ function suppression (Jintaridth and Mutirangura 2010). Methylation drift (differential site-specific hypo- or hypermethylation) occurs in all individuals past a certain age (Mcclay et al. 2014; Schübeler 2015), but the rates of methylation drift differ depending on local transcriptional activity (Takeshima et al. 2009), the methylation state (Sharma et al. 2011), and histone tail modifications (Ooi et al. 2007).

The epigenetic field went through the same developmental process of identifying genes associated with selected traits (e.g., candidate gene approach to unbiased genome approach using DNA chip and whole-genome sequencing), starting with tracing epiloci to epigenome-wide association studies (EWAS). EWAS replaced the candidate epiloci approach and was used to reveal differential epigenomic changes, mostly differentially methylated regions (DMRs), which occur in the genome (Birney et al. 2016). The decrease in costs of whole-genome sequencing and whole-genome bisulfite sequencing (WGBS) has become more prominent and now serves as the gold standard method for mapping DNA methylation status (Pidsley et al. 2016).

A WGBS study performed by Heyn H. on newborn and centenarian genomes found that the centenarian DNA had low levels of DNA methylation and more hypomethylated CpG islands compared with the more homogeneously methylated newborn DNA (Heyn et al. 2012). In conclusion, over the last years, there is a growing body of research that led to the progressive and sharp description of the impact of methylation on aging.

Histone Modifications

Histones are a protein family that share a typical 3D structure, are the basic components of chromatin, and are wrapped by DNA forming a nucleosome. A nucleosome is an octamer usually formed by the histones H2A, H2B, H3, and H4. Binding of the octamer with the linker histone H1 promotes higher-order chromatin organization. The histones affect the dynamic structure of the chromatin via posttranslational histone modifications (PTMs) (Biterge and Schneider 2014). Those modifications most commonly occur on the N-terminal tails of histones including methylation, acetylation, phosphorylation, and more (Strahl and Allis 2000). In addition, histone PTMs are important for DNA replication, cell differentiation, and tissue specification during development of numerous diseases (Cao and Dang 2018). Likewise, there are three classes of proteins that interact with histones either increasing or repressing the transcription of target genes: writers, enzymes that add PTMs to histones; erasers, enzymes that remove specific PTMs from histone substrates; and readers, enzymes that recognize specific posttranslational marks on histones and direct a particular transcriptional outcome (Gillette and Hill 2015).

According to a study performed by Wang, an association between aging and histone lose was detected; in this study they elaborate on the connection between aging and histone modifications (Wang et al. 2018). A study performed by Shumaker analyzed the expression of histones and histone chaperones in HGPS (Hutchinson-Gilford progeria syndrome) patients. They found a loss of heterochromatin along with a loss of histone-3 lysine-27 tri-methylation (H3K27me3) and H3K9me3 during replicative aging (Shumaker et al. 2006). The model of "loss of heterochromatin" has been proposed and confirmed by studies linking changes in the heterochromatin to aging phenotypes in humans (Shumaker et al. 2006; Zhang et al. 2015). Studies in premature aging-related diseases have also shown that there is indeed a connection between changes in heterochromatin and aging (Zhang et al. 2015; Shumaker et al. 2006). Loss of H3K9me3, HP1, and the H3K9 methyltransferase, SUV39H1, has been observed in both Werner syndrome model cells and cells derived from old individuals, suggesting a relevant mechanism of physiological aging (Zhang et al. 2015).

H3K27me3 seems to suppress gene groups during development and differentiation (Bracken and Helin 2009), while the H3K4me3 (trimethylation of position 4 lysine on histone 3) histone mark activates transcription since these histone modifications block methylation of DNA (Hödl and Basler 2012). The simultaneous appearance of these two markers on the promoters of numerous developmental genes in multipotent adult stem cells is defined as bivalent domains which presumably act as a dynamic "switch" towards activation or repression of the downstream gene, as a function of the cell fate (Bernstein et al. 2006). Another study in humans shows that a germline mutation in SIRT6 causes fetal demise, thus defining SIRT6 as a key factor in human development (Mcclay et al. 2014). SIRT6 is a member of the sirtuin family of NAD+-dependent enzymes that has a role in chromatin signaling and genome maintenance. Through these functions, SIRT6 protects against aging-associated pathologies (Tasselli et al. 2017).

In conclusion, histone posttranslational modifications play a crucial role in epigenetic alteration during aging, among which histone methylation is emerging as a prominent modification method.

Noncoding RNAs

As histone modifications and DNA methylation have been established as the majority of epigenetic modifiers, recently, noncoding RNAs (ncRNAs) were demonstrated as players in chromatin remodeling and in transcriptional or posttranscriptional gene silencing. Three types of ncRNAs were found in mammalian cells regulating numerous molecular pathways (Mello and Conte 2004). The best characterized ncRNAs are miRNAs (microRNAs); these ncRNAs consist of approximately 22 nucleotides (Unda and Villegas 2017), responsible for multiple functions such as cell cycle regulation, differentiation, apoptosis, tumor suppression, and mechanisms of aging. In addition, miRNAs degrade the mRNA target and inhibit their translation (Grillari and Grillari-Voglauer 2010). A study that surveys mRNA, miRNA, and protein expression changes in the prefrontal cortex of humans supports the idea of miRNA modulation of the aging process and many expression changes that had been previously associated with aging, such as downregulation of genes involved in neural functions (Somel et al. 2010). Another study published by Huan et al. identified 127 age-differentially expressed miRNAs. Furthermore, they established a model predicting miRNA age which was suggested by them as a biomarker for accelerated aging by using the difference between chronological age and miRNA age, which was found associated with all-cause mortality, as an indicator (Huan et al. 2018). Analysis of miRNA expression in human serum from young and old individuals found that the expression of miR-151a-5p, miR-181a-5p, and miR-1248 is significantly decreased in older individuals (Noren Hooten et al. 2013). In conclusion, miRNA expression may change the prognosis of patients and may affect their quality of life; however, further studies are needed to validate their association with target genes (Huang et al. 2017).

Age-Associated Chromatin Remodeling

The structure of the chromatin affects the accessibility of DNA to transcription, repair, and replication. As mentioned above many of the epigenetic alterations and chromatin changes are hallmarks of aging. Studies that examine cells from patients with the premature aging disorder HGPS reveal loss of heterochromatin, loss of some of the key architectural chromatin proteins such as HP1, and changes in the levels of heterochromatin-associated histone PTMs including H3K9me3 and H3K27me3 (Shumaker et al. 2006; Goldman et al. 2004). Observations from Pegoraro G show that suppression of the activity of chromatin modifiers increased levels of endogenous DNA damage, one of the key players in human aging (Pegoraro et al. 2009). Cellular response to DNA damage leads to chromatin defects resulting in gene misregulation (Oberdoerffer et al. 2008).

The development of chromatin formation capture technologies, such as 3C, 4C, and Hi-C, revealed the regulatory elements that may serve as modulators of the aging phenotype in vitro (Fraser et al. 2015; Jin et al. 2013). For example, CTCF (a DNA-binding protein that is necessary for regulation of chromatin architecture by mediating both short- and long-range chromosomal contacts (Ghirlando 2016)) was significantly associated with frailty (Sathyan et al. 2018). This study adds support to the crucial role of these binding sites in the process of aging. Another recent study performed on three types of primary human cells found that upon senescence, there are nuclear reorganizations and factors such as high-mobility group B1 and B2 which are removed from the nucleus prior to the appearance of senescence markers. This change leads to the marking of certain topologically associating domains via clustering of binding sites and thus chromatin reorganization (Zirkel et al. 2018).

In conclusion, according to studies in humans (Ishimi et al. 1987), the decreased levels of histone proteins and changes in histone modifications lead to alteration of the chromatin during aging; these alterations are likely to result in a more open chromatin structure during aging, presumably allowing inappropriate transcription.

Mitochondrial Dysfunction

The mitochondria is central to all basic and advanced cellular and organismal functions (see mitochondria in this volume). It produces most of the cellular ATP through OXPHOS and the ETC and supplies cells with several factors such as NAD^+, NADH, and other intermediate metabolites that are essential for cellular function. Over time, mitochondrial damage accumulates, leading to reduced quality and activity of mitochondria in each cell. This is correlated with the fact that some age-related diseases are characterized by a decline in energy homeostasis and mitochondrial function (Fang et al. 2016); thus in mammalian cells mitochondrial dysfunction is thought to be among the primary drivers of cellular aging (Huang et al. 2017).

The health of the mitochondria is maintained by fission and fusion of individual mitochondrion and autophagic removal of damaged mitochondria by mitophagy (Pickles et al. 2018). Mitochondrial fission proteins are regulated by a range of protein modifications, including phosphorylation, ubiquitination, sumoylation, and nitrosylation, and the fusion promotes maintenance of the mitochondrial membrane potential (MMP), increases ATP production, decreases mitophagy, and prevents cell death (Westermann 2012; Mariño et al. 2014).

In response to DDR (DNA damage response), there is a failure to reproduce NAD, and this affects the cellular redox balance which worsens oxidative and metabolic stress (Massudi et al. 2012; Blacker and Duchen 2016).

In cells that sustain high levels of DNA damage, continuous activation of DDR creates a high-energy demand, triggering a cascade of events that leads to metabolic stress and mitochondrial dysfunction (Wang et al. 2017). The mtDNA (mitochondrial DNA) encodes for some critical proteins that are essential for electron transport (Kauppila et al. 2017), increasing in mutated mtDNA the drive to mammalian aging (Singh 2004).

In general, relative to young organisms, mitochondria from aged people show decreased ATP production, increased free radical production, and depolarization of the mitochondrial (Green et al. 2011).

Example of Application/Future Directions

The genetic contribution to the aging phenotype is low based on recent estimation (Ruby et al. 2018; Axelrad and Atzmon 2013), suggesting a much larger environmental input. However, people age differently, pointing to the fact that genetic-environmental interactions are more flexible in this trait than what was traditionally believed. The future of this field should focus on tailoring the best environment for a given genome in order to fulfill its own life expectancy potential.

Hallmark studies such as the methylation age clocks presented by Horvath or Hannum set the stage to finding the DMRs that are associated with the longevity phenotype. One recent study reports differential methylation in the PCDHGA3 gene promotor with age, placing it as a key mediator of healthy longevity (Kim et al. 2018). Another study conducted by Slieker et al. shows the complexity and dynamics of methylation within and between tissues, highlighting the importance of finding the aging methylation biomarkers of each tissue (Slieker et al. 2018). Understanding the nature of those changes in methylation over lifespan will enable the tailoring of a controlled environment that will maximize lifespan potential.

Summary

Recent advances in aging research have continued to raise our understanding and knowledge about the process of aging and longevity. We illustrated here the progression from the specific and limited candidate gene approach for the study of aging, to the unbiased, large-scale screening in genomic and epigenomics studies in the field. Though this progress has been substantial and impressive, it raised many more questions and only begun the process of elucidating the enigmatic aging phenotype. As we mentioned above, many biological pathways are associated and affect aging and longevity. Despite of all the information and the tools that we have, it is still unclear which biological pathway causes aging because every pathway is influenced by environmental stimuli, nutrient signaling, and metabolic state.

Epigenetics adds a great amount of complexity, as it translates the interaction between the environment and the DNA. Flexible DNA (in terms of epiloci opportunities) can aid with personal adaptation to multiple environments and achieving longevity. Thus, the environment is better for those who can adapt (e.g., flexible genome) and less beneficial to those who do not.

Cross-References

▶ Aging as Phenoptotic Phenomenon
▶ Aging Definition
▶ Aging Mechanisms
▶ Aging Pathology
▶ Aging Theories
▶ Animal Models of Aging
▶ Cell Senescence
▶ Molecular and Epigenetic Clocks of Aging
▶ Timeline of Aging Research

References

Aquino EM, Benton MC, Haupt LM, Sutherland HG, Griffiths LR, Lea RA (2018) Current understanding of DNA methylation and age-related disease. OBM Genet 2:1–1

Armstrong NJ, Mather KA, Thalamuthu A et al (2017) Aging, exceptional longevity and comparisons of the Hannum and Horvath epigenetic clocks. Epigenomics. https://doi.org/10.2217/epi-2016-0179

Axelrad MA, Atzmon G (2013) Epigenomic of aging. Genetics 2(1):e106. https://doi.org/10.4172/2161-1041.1000e106

Beekman M, Blanché H, Perola M et al (2013) Genome-wide linkage analysis for human longevity: genetics of healthy aging study. Aging Cell. https://doi.org/10.1111/acel.12039

Bergman A, Atzmon G, Ye K et al (2007) Buffering mechanisms in aging: a systems approach toward uncovering the genetic component of aging. PLoS Comput Biol 3(8):e170. https://doi.org/10.1371/journal.pcbi.0030170

Bernstein BE, Mikkelsen TS, Xie X et al (2006) A bivalent chromatin structure marks key developmental genes in embryonic stem cells. Cell. https://doi.org/10.1016/j.cell.2006.02.041

Birney E, Smith GD, Greally JM (2016) Epigenome-wide association studies and the interpretation of disease - omics. PLoS Genet 12:e1006105

Biterge B, Schneider R (2014) Histone variants: key players of chromatin. Cell Tissue Res 356:457

Bjornsson HT, Sigurdsson MI, Fallin MD et al (2008) Intra-individual change over time in DNA methylation with familial clustering. JAMA. https://doi.org/10.1001/jama.299.24.2877

Blacker TS, Duchen MR (2016) Investigating mitochondrial redox state using NADH and NADPH autofluorescence. Free Radic Biol Med 100:53

Bracken AP, Helin K (2009) Polycomb group proteins: navigators of lineage pathways led astray in cancer. Nat Rev Cancer 9:773

Broer L, Buchman AS, Deelen J et al (2015) GWAS of longevity in CHARGE consortium confirms APOE and FOXO3 candidacy. J Gerontol A Biol Sci Med Sci. https://doi.org/10.1093/gerona/glu166

Brooks-Wilson AR (2013) Genetics of healthy aging and longevity. Hum Genet 132:1323

Byun HM, Siegmund KD, Pan F et al (2009) Epigenetic profiling of somatic tissues from human autopsy specimens identifies tissue- and individual-specific DNA methylation patterns. Hum Mol Genet. https://doi.org/10.1093/hmg/ddp445

Cao X, Dang W (2018) Histone modification changes during aging: cause or consequence? – what we have learned about epigenetic regulation of aging from model organisms. In: Epigenetics of aging and longevity. Academic, London

Cardoso AL, Fernandes A, Aguilar-Pimentel JA et al (2018) Towards frailty biomarkers: candidates from genes and pathways regulated in aging and age-related diseases. Ageing Res Rev 47:214

Chen HP, Zhao YTZT (2015) Histone deacetylases and mechanisms of regulation of gene expression. Crit Rev Oncog 20:35

Drinkwater RD, Blake TJ, Morley AA, Turner DR (1989) Human lymphocytes aged in vivo have reduced levels of methylation in transcriptionally active and inactive DNA. Mutat Res DNAging. https://doi.org/10.1016/0921-8734(89)90038-6

Erikson GA, Bodian DL, Rueda M et al (2016) Whole-Genome sequencing of a healthy aging cohort. Cell. https://doi.org/10.1016/j.cell.2016.03.022

Fang EF, Scheibye-Knudsen M, Chua KF et al (2016) Nuclear DNA damage signalling to mitochondria in ageing. Nat Rev Mol Cell Biol 17:308

Ferri E, Gussago C, Casati M et al (2019) Apolipoprotein E gene in physiological and pathological aging. Mech Ageing Dev. https://doi.org/10.1016/j.mad.2019.01.005

Field AE, Robertson NA, Wang T et al (2018) DNA methylation clocks in aging: categories, causes, and consequences. Mol Cell 71:882

Fraser J, Williamson I, Bickmore WA, Dostie J (2015) An overview of genome organization and how we got there: from FISH to Hi-C. Microbiol Mol Biol Rev. https://doi.org/10.1128/mmbr.00006-15

Freudenberg-Hua Y, Freudenberg J, Vacic V et al (2014) Disease variants in genomes of 44 centenarians. Mol Genet Genomic Med. https://doi.org/10.1002/mgg3.86

Fuke C, Shimabukuro M, Petronis A et al (2004) Age related changes in 5-methylcytosine content in human peripheral leukocytes and placentas: an HPLC-based study. Ann Hum Genet. https://doi.org/10.1046/j.1529-8817.2004.00081.x

Gao W, Tan J, Hüls A et al (2017) Genetic variants associated with skin aging in the Chinese Han population. J Dermatol Sci. https://doi.org/10.1016/j.jdermsci.2016.12.017

Ghirlando R, Felsenfeld G (2016) CTCF: making the right connections. Genes Dev 30:881

Gillette TG, Hill JA (2015) Readers, writers, and erasers: Chromatin as the whiteboard of heart disease. Circ Res 116:1245

Goldman RD, Shumaker DK, Erdos MR et al (2004) Accumulation of mutant lamin A causes progressive changes in nuclear architecture in Hutchinson–Gilford progeria syndrome. Proc Natl Acad Sci. https://doi.org/10.1073/pnas.0402943101

Goll MG, Kirpekar F, Maggert KA, et al (2006) Methylation of tRNAAsp by the DNA methyltransferase homolog Dnmt2. Science (80-). https://doi.org/10.1126/science.1120976

Gonzalo S (2010) Epigenetic alterations in aging. J Appl Physiol. https://doi.org/10.1152/japplphysiol.00238.2010

Green DR, Galluzzi L, Kroemer G (2011) Mitochondria and the autophagy-inflammation-cell death axis in organismal aging. Science (80-) 333:1109

Grillari J, Grillari-Voglauer R (2010) Novel modulators of senescence, aging, and longevity: small non-coding RNAs enter the stage. Exp Gerontol 45:302

Gurinovich A, Bae H, Andersen S et al (2018) Ethnic-specific effect of Apoe alleles on extreme longevity. Innov Aging. https://doi.org/10.1093/geroni/igy023.373

Hannum G, Guinney J, Zhao L et al (2013) Genome-wide methylation profiles reveal quantitative views of human aging rates. Mol Cell. https://doi.org/10.1016/j.molcel.2012.10.016

Heshmati A (2018) Healthy aging as a solution to the 'ticking time bomb': dealing with aging population in urban china. Sociol Int J. https://doi.org/10.15406/sij.2018.02.00038

Heyn H, Li N, Ferreira HJ et al (2012) Distinct DNA methylomes of newborns and centenarians. Proc Natl Acad Sci. https://doi.org/10.1073/pnas.1120658109

Hödl M, Basler K (2012) Transcription in the absence of histone H3.2 and H3K4 methylation. Curr Biol. https://doi.org/10.1016/j.cub.2012.10.008

Holoch D, Moazed D (2015) RNA-mediated epigenetic regulation of gene expression. Nat Rev Genet 16:71

Horvath S (2013) DNA methylation age of human tissues and cell types. Genome Biol. https://doi.org/10.1186/gb-2013-14-10-r115

Huan T, Chen G, Liu C et al (2018) Age-associated microRNA expression in human peripheral blood is associated with all-cause mortality and age-related traits. Aging Cell. https://doi.org/10.1111/acel.12687

Huang F, Yi J, Zhou T et al (2017) Toward understanding non-coding RNA roles in intracranial aneurysms and subarachnoid hemorrhage. Transl Neurosci. https://doi.org/10.1515/tnsci-2017-0010

Ishimi Y, Masatoyo Kojima, Fujio Takeuchi, Terumasa Miyamoto, Masa-Atsu Yamada, Fumio Hanaoka (1987) Changes in chromatin structure during aging of human skin fibroblasts. Experimental Cell Research 169(2):458–467

Jin B, Li Y, Robertson KD (2011) DNA methylation: superior or subordinate in the epigenetic hierarchy? Genes Cancer 2:607

Jin F, Li Y, Dixon JR et al (2013) A high-resolution map of the three-dimensional chromatin interactome in human cells. Nature. https://doi.org/10.1038/nature12644

Jintaridth P, Mutirangura A (2010) Distinctive patterns of age-dependent hypomethylation in interspersed repetitive sequences. Physiol Genomics. https://doi.org/10.1152/physiolgenomics.00146.2009

Kauppila TES, Kauppila JHK, Larsson NG (2017) Mammalian mitochondria and aging: an update. Cell Metab 25:57

Kim S, Wyckoff J, Morris AT et al (2018) DNA methylation associated with healthy aging of elderly twins. GeroScience. https://doi.org/10.1007/s11357-018-0040-0

Kronenberg F (2008) Genome-wide association studies in aging-related processes such as diabetes mellitus, atherosclerosis and cancer. Exp Gerontol 43:39

Lin E, Tsai SJ, Kuo PH et al (2017) The rs1277306 variant of the REST gene confers susceptibility to cognitive aging in an elderly Taiwanese population. Dement Geriatr Cogn Disord. https://doi.org/10.1159/000455833

Lipman T, Tiedje LB (2006) Epigenetic differences arise during the lifetime of monozygotic twins. MCN Am J Matern Child Nurs. https://doi.org/10.1097/00005721-200605000-00016

Lombard DB, Chua KF, Mostoslavsky R et al (2005) DNA repair, genome stability, and aging. Cell 120:497

López-Otín C, Blasco MA, Partridge L et al (2013) The hallmarks of aging. Cell. https://doi.org/10.1016/j.cell.2013.05.039

Macedo JC, Vaz S, Logarinho E (2017) Mitotic dysfunction associated with aging hallmarks. Adv Exp Med Biol 1002:153

Mariño G, Niso-Santano M, Baehrecke EH, Kroemer G (2014) Self-consumption: the interplay of autophagy and apoptosis. Nat Rev Mol Cell Biol 15:81

Marta Kulis ME (2010) 2 – DNA methylation and cancer. ScienceDirect 70:27–56

Massudi H, Grant R, Guillemin GJ, Braidy N (2012) NAD+ metabolism and oxidative stress: the golden nucleotide on a crown of thorns. Redox Rep. https://doi.org/10.1179/1351000212y.0000000001

Mcclay JL, Aberg KA, Clark SL et al (2014) A methylome-wide study of aging using massively parallel sequencing of the methyl-CpG-enriched genomic fraction from blood in over 700 subjects. Hum Mol Genet. https://doi.org/10.1093/hmg/ddt511

Mello CC, Conte D (2004) Revealing the world of RNA interference. Nature 431:338

Murabito JM, Yuan R, Lunetta KL (2012) The search for longevity and healthy aging genes: Insights from epidemiological studies and samples of long-lived individuals. J Gerontol Ser A Biol Sci Med Sci. https://doi.org/10.1093/gerona/gls089

Noren Hooten N, Fitzpatrick M, Wood WH et al (2013) Age-related changes in microRNA levels in serum. Aging (Albany NY) 5:725

Oberdoerffer P, Michan S, McVay M, Mostoslavsky R, Vann J, Park SK, Hartlerode A, Stegmuller J, Hafner A, Loerch P, Wright SM, Mills KD, Bonni A, Yankner BA, Scully R, Prolla TA, Alt FW, Sinclair DA (2008) SIRT1 redistribution on chromatin promotes genomic stability but alters gene expression during aging. Cell 135:907

Okano M, Bell DW, Haber DA, Li E (1999) DNA methyltransferases Dnmt3a and Dnmt3b are essential for de novo methylation and mammalian development. Cell. https://doi.org/10.1016/S0092-8674(00)81656-6

Ooi SKT, Qiu C, Bernstein E et al (2007) DNMT3L connects unmethylated lysine 4 of histone H3 to de novo methylation of DNA. Nature. https://doi.org/10.1038/nature05987

Pegoraro G, Kubben N, Wickert U, Gohler H, Hoffmann K, Misteli T (2009) Ageing-related chromatin defects through loss of the NURD complex. Nat Cell Biol:1261–1267

Pérez RF, Tejedor JR, Bayón GF et al (2018) Distinct chromatin signatures of DNA hypomethylation in aging and cancer. Aging Cell. https://doi.org/10.1111/acel.12744

Pickles S, Vigié P, Youle RJ (2018) Mitophagy and quality control mechanisms in mitochondrial maintenance. Curr Biol 28:R170

Pidsley R, Zotenko E, Peters TJ et al (2016) Critical evaluation of the Illumina MethylationEPIC BeadChip microarray for whole-genome DNA methylation profiling. Genome Biol. https://doi.org/10.1186/s13059-016-1066-1

Ruby JG, Wright KM, Rand KA et al (2018) Estimates of the heritability of human longevity are substantially inflated due to assortative mating. Genetics. https://doi.org/10.1534/genetics.118.301613

Sathyan S, Barzilai N, Atzmon G et al (2018) Genetic insights into frailty: association of 9p21-23 locus with frailty. Front Med. https://doi.org/10.3389/fmed.2018.00105

Schübeler D (2015) Function and information content of DNA methylation. Nature

Seddighi S, Varma VR, An Y et al (2018) SPARCL1 accelerates symptom onset in Alzheimer's disease and influences brain structure and function during aging. J Alzheimers Dis. https://doi.org/10.3233/JAD-170557

Sharma S, De Carvalho DD, Jeong S, Jones PA, Liang G (2011) Nucleosomes containing methylated DNA stabilize DNA methyltransferases 3A/3B and ensure faithful epigenetic inheritance. PLoS Genet 7:e1001286

Shumaker DK, Dechat T, Kohlmaier A et al (2006) Mutant nuclear lamin A leads to progressive alterations of epigenetic control in premature aging. Proc Natl Acad Sci. https://doi.org/10.1073/pnas.0602569103

Singh KK (2004) Mitochondrial dysfunction is a common phenotype in aging and cancer. Ann N Y Acad Sci 1019:260

Slieker RC, Relton CL, Gaunt TR et al (2018) Age-related DNA methylation changes are tissue-specific with ELOVL2 promoter methylation as exception. Epigenetics Chromatin. https://doi.org/10.1186/s13072-018-0191-3

Somel M, Guo S, Fu N et al (2010) MicroRNA, mRNA, and protein expression link development and aging in human and macaque brain. Genome Res. https://doi.org/10.1101/gr.106849.110

Strahl BD, Allis CD (2000) The language of covalent histone modifications. Nature 403:41

Sturmlechner I, Durik M, Sieben CJ et al (2017) Cellular senescence in renal ageing and disease. Nat Rev Nephrol. https://doi.org/10.1038/nrneph.2016.183

Sundermann E, Levine A, Horvath S, Moore D (2018) Inflammation-related genes are associated with accelerated aging in HIV. Am J Geriatr Psychiatry 26:S118

Takeshima H, Yamashita S, Shimazu T et al (2009) The presence of RNA polymerase II, active or stalled, predicts epigenetic fate of promoter CpG islands. Genome Res. https://doi.org/10.1101/gr.093310.109

Tasselli L, Zheng W, Chua KF (2017) SIRT6: novel mechanisms and links to aging and disease. Trends Endocrinol Metab 28:168

Theendakara V, Peters-Libeu CA, Bredesen DE, Rao RV (2018) Transcriptional effects of ApoE4: relevance to Alzheimer's disease. Mol Neurobiol 55:5243

Unda SR, Villegas EA (2017) MicroRNA: a major key in pain neurobiology. Int J Cell Sci Mol Biol 3(5):555621. https://doi.org/10.19080/ijcsmb.2017.03.555621

Vijg J, Dong X, Milholland B, Zhang L (2017) Genome instability: a conserved mechanism of ageing? Essays Biochem:305–315

Vijg J, Gravina S, Dong X (2018) Chapter 9 – Intratissue DNA methylation heterogeneity in aging. ScienceDirect 4:201–209

Wang T, Zhang M, Jiang ZSE (2017) Mitochondrial dysfunction and ovarian aging. Am J Reprod Immunol 77:e12651

Wang Y, Yuan Q, Xie L (2018) Histone modifications in aging: the underlying mechanisms and implications. Curr Stem Cell Res Ther. https://doi.org/10.2174/1574888x12666170817141921

Warner HR (2005) Longevity genes: from primitive organisms to humans. Mech Ageing Dev 126(2):235

Wegman MP, Guo MH, Bennion DM et al (2014) Practicality of intermittent fasting in humans and its effect on oxidative stress and genes related to aging and metabolism. Rejuvenation Res. https://doi.org/10.1089/rej.2014.1624

Westermann B (2012) Bioenergetic role of mitochondrial fusion and fission. Biochim Biophys Acta Bioenerg 1817:1833

Wilson VL, Jones PA (1983) DNA methylation decreases in aging but not in immortal cells. Science (80-). https://doi.org/10.1126/science.6844925

Ye K et al (2013) Aging as accelerated accumulation of somatic variants: whole-genome sequencing of centenarian and middle-aged monozygotic twin pairs. Twin Res Hum Genet 16:1026–1032

Zhang W, Li J, Suzuki K, et al (2015) A Werner syndrome stem cell model unveils heterochromatin alterations as a driver of human aging. Science (80-). https://doi.org/10.1126/science.aaa1356

Zhao N, Liu CC, Qiao W, Bu G (2018) Apolipoprotein E, receptors, and modulation of Alzheimer's disease. Biol Psychiatry 83:347

Zirkel A, Nikolic M, Sofiadis K et al (2018) HMGB2 loss upon senescence entry disrupts genomic organization and induces CTCF clustering across cell types. Mol Cell. https://doi.org/10.1016/j.molcel.2018.03.030

Genomics of Centenarians

▶ Genomics of Aging and Longevity

Genomics of Healthy Aging

▶ Genomics of Aging and Longevity

Genomics of Healthy Lifespan

▶ Genomics of Aging and Longevity

Genotype

Nancy B. Y. Tsui[1] and Johnson Y. N. Lau[1,2]
[1]Avalon Genomics (Hong Kong) Limited, Shatin, Hong Kong
[2]Department of Applied Biology and Chemical Technology, The Hong Kong Polytechnic University, Hung Hom, Hong Kong

Synonyms

Allelic combination; Genetic constitution; Genetic makeup

Definition

Genotype refers to the genetic constitution at each locus, i.e., a particular position within the genome, of an individual. In human, a genotype comprises two alleles at each genetic locus, with one allele inherited from each parent. Genotype is one of the factors that determine phenotype, which is the outward appearance of an individual.

Overview

Human are diploid organisms. Each human cell contains 22 pairs of autosomes and 1 pair of sex chromosomes. For all autosomal pairs and X chromosome pair of female individuals, each genetic locus comprises a pair of alleles that are inherited from each parent. Alleles are variant forms of the same gene occupying the same locus on homologous chromosomes. They produce variants of the same gene product and give rise to human variation. The composition of the two alleles in each locus is described as genotype. Typically, one refers to an individual's genotype with regard to a particular gene of interest. A locus typically has three possible combinations of alleles, generically denoted as AA, Aa, and aa, whereas "A" is the reference allele that is the allele appears in the human reference genome (International Human Genome Sequencing Consortium 2004) and "a" is the variant allele that is the allele different from that in the reference genome. "A" and "a" may also refer to wild-type and mutant alleles, respectively, in a mutation. More than one variant allele can be present in some loci. A genotype is described as homozygous if it contains two identical alleles, i.e., AA or aa, and as heterozygous if the two alleles are different, i.e., Aa. Homozygote and heterozygote are individuals who carry homozygous and heterozygous genotypes, respectively.

There are scenarios in which a genotype is not made up of exactly two copies of alleles. Human male contains one X chromosome and one Y chromosome in each cell. Hence, there is only one allele at a given X or Y genetic locus, giving rise to the genotype of A or a. This kind of genotype is also called hemizygous. Besides X and Y chromosomal loci, the loss of one allele in autosomal loci can occur through a mutation, e.g., deletion mutation, resulting in hemizygous genotype. Other mutations and chromosomal aneuploidies can also change the number of allele at a locus. For example, in a trisomy-21 (Down syndrome) individual, a locus on chromosome 21 would have the genotypes of AAA, AAa, Aaa, or aaa.

The genotype frequencies of a population can be estimated from allele frequencies by the Hardy-Weinberg principle, with the assumptions that the allele frequencies will not change from generation to generation, and any disturbing factor is absent (Mayo 2008). When the frequency of allele A is p and the frequency of allele B is q, the expected frequency of genotype AA is p^2, the frequency of genotype Aa is $2pq$, and the frequency of genotype aa is q^2. The Hardy-Weinberg principle is fundamental to the study of evolution biology. Recently, as genome-wide association study (GWAS) has become a popular research tool, the Hardy-Weinberg test has been utilized for quality control of genotyping result (Ryckman and Williams 2008).

Genotyping is the experimental process to determine a genotype. DNA sequencing by Sanger's method is widely accepted as the gold standard of genotyping, as it determines the exact nucleotide sequence of a certain length of DNA spanning the locus of interest (Shendure et al. 2017). In the past decades, a wide variety of genotyping techniques have been developed for different applications. For examples, polymerase chain reaction (PCR), mass spectrometry, and automatic DNA chip methods have been applied for low-to-medium throughput genotyping applications (Ragoussis 2006). On the other hand, microarrays (Marzancola et al. 2016) and next-generation sequencing (Goodwin et al. 2016) allow high-throughput genotyping in a genome-wide scale.

The difference of genotypes between two individuals can lead to difference in phenotypes such as aging and, more importantly, disease susceptibility. The genotyping of mutations, single nucleotide polymorphisms (SNPs), and other genetic variations is a common tool for genetic research and genetic test development. While both mutations and SNPs are allelic variations in particular loci, they are defined as the alternative alleles being present in <1% or >1% of the population, respectively (Karki et al. 2015). By comparing the genotype differences between patients and healthy controls, disease-associated alleles and genotypes can be identified. The understanding of genotype-phenotype relationships would be valuable in predicting, diagnosing, and risk assessing of gerontological diseases.

Key Research Findings

Alzheimer's disease (AD) is the most common cause of dementia. Genetic research on AD has revealed that disease-causing mutations of AD are rare and constitute less than 1% of all AD cases. These cases are mostly familial early-onset AD. The pathogenic mutations causing early-onset AD have been found on the genes encoding *presenilin-1 (PSEN1)*, *presenilin-2 (PSEN2)*, and *amyloid protein precursor (APP)* (Lanoiselee et al. 2017). Recently, the pathogenic mutational genotypes of these genes have been comprehensively studied in large patient cohorts (Lanoiselee et al. 2017). The information is valuable for identifying individuals at risk for early-onset AD (Tsao 2013). Since not all families with early-onset AD have identifiable mutations in *PSEN*, *PSEN2*, or *APP*, additional genes that influence the manifestation of early-onset AD have been suggested (Raux et al. 2005).

SNP is the dominant source of sequence variation in the human genome. Most SNPs do not directly cause AD. Some of them, however, may increase the likelihood of developing the disease. The genetic variations on *apolipoprotein E* (*APOE*) gene are the most well-studied risk factor for late-onset AD. The *APOE* gene comprises three alleles that are defined by two SNPs, i.e., rs429358 with C and T alleles and rs7412 with C and T alleles. The three *APOE* alleles are named ε2 (the combination of rs429358[T] and rs7412[T]), ε3 (rs429358[T] and rs7412[C]), and ε4 (rs429358[C] and rs7412[C]). The ε4 allele has been consistently demonstrated to be associated with an increased risk of AD among research studies focused on different regions and populations (Ward et al. 2012). An individual homozygous for ε4, i.e., with genotype ε4/ε4, has an odds ratio of 11.8 in developing AD, while an individual heterozygous for ε4 has an odds ratio of 2.8 (Bertram et al. 2007). Approximately 8% and 42% of AD patients were homozygote and heterozygote for ε4, respectively (Ward et al. 2012).

Despite the well-characterization of *APOE* ε4, around 50% of AD patients did not carry the risk-associated ε4 allele (Ward et al. 2012). Hence, additional genetic factors have to be identified in order to better understand the cause of the disease. Over the past decades, researchers have used different methods, such as GWAS, to search for genetic variations that contribute to the risk of AD. GWAS detects the genotypes of all or a representative set of SNPs within the whole genome. It allows the identification of disease-associated variants without a prior knowledge of potential candidates. By comparing the genotype and/or allele frequencies of the analyzed SNPs between patients and healthy controls, the disease-associated SNPs and their risk-contributing genotypes/alleles can be identified. Thus far, a considerable number of GWAS reports on AD have been published. According to a meta-analysis of 875 GWAS reports on AD, 20 SNPs in 13 genes that were significantly associated with the risk of AD were identified (Bertram et al. 2007). While each of these SNPs typically conferred a small effect size with odds ratios around 1.2, they may cumulatively explain a substantial proportion of genetic risk of the disease.

Recently, next-generation sequencing has substantially put forward the throughput and quality of DNA analysis. It facilitates the genotyping of rare variants in a genome-wide scale and single-base resolution. For example, using this technology, a variant in *triggering receptor expressed on myeloid cells 2* gene, which has an allele frequency of 0.64%, has been found to confer a significant risk of AD with odds ratio of 2.9 (Jonsson et al. 2013). Along with the exponential growing of the volume of genotype data, more discoveries are expected in the future.

Examples of Application

Following the rapid discovery of relationships between genotypes and diseases, researchers have begun to study their value in healthy aging (Manolio 2013). One actively investigating application is disease prediction and prevention. In general, the genetically contributed risk of a disease of an individual can be assessed based on the presence or absence of the risk association genotypes. By genotyping the disease-associated mutations or SNPs, individuals at a high genetic risk of a disease could be identify early before symptoms onset. Lifestyle improvement or medical intervention may then be considered in order to reduce the risk.

Genotype testing for AD represents a clinical advance in being able to examine one's genetic risk of the disease. For early-onset AD, genotype testing is now clinically available to screen for rare causative mutations in *PSEN1*, *PSEN2*, and *APP* genes. The test can be used for assisting diagnosis in symptomatic individual. It has also been used for predictive or presymptomatic test for high-risk individuals who have a strong family history of early-onset AD. For late-onset AD, disease susceptibility genetic test for *APOE* ε4 allele is also available. However, it should be noted that *APOE* ε4 allele represents a risk factor for AD, rather than a disease-causing variant. Thus far, *APOE* ε4 genotype test has not yet been adopted in the clinical setting. However, the test has already gained attention in general public due to the availability of direct-to-customer genetic testing products.

With the aging of global population, the number of patients with AD and other gerontological diseases is projected to be increased significantly. Since many gerontological diseases are not curable, the society should evaluate the disease burdens of its population and prepare for diseases with increased burden for the future. Population genomics is a potentially useful tool for assessing the genetic risk burden of diseases predisposed in a society. The population risk profiles of diseases can be estimated by recruiting a representative proportion of population from the community and determining the frequencies of risk-associated genotypes carried by the population. In a pilot study focused on Chinese population in Macau community, the top three diseases with the highest percentages of population having elevated genetic risks were rheumatoid arthritis (22%), age-related macular degeneration (17%), and AD (17%) (Tsui et al. 2018). Notably, age-related macular degeneration and AD are diseases of the older adults. Such the information may facilitate healthcare workers and policy-makers to identify gerontological health burdens of the community, so that public health policies and practice can be planned accordingly.

Future Directions of Research

Late-onset AD and many other age-related diseases are complex diseases. Genetics is estimated to contribute 60–80% of the risk of AD. However, genetic loci that contribute to the disease risk are still largely unknown. With advanced DNA analyzing technologies, genotyping can be readily performed among large cohorts of patients and in a whole-genome scale. The continual accumulation of the vast amount of genotype data (Bertram et al. 2007) would facilitate the identification of genetic causes of the disease in the future. In addition to individual genotypes, genotype sets that are closely linked together, namely, haplotypes (International HapMap Consortium 2005), may provide another level of genetic analysis of the disease. All these analyses may also pave the way for therapeutic target development.

Further effort is also required to catalyze the clinical implementation of genotype testing for screening high-risk individuals. For examples, guidelines should be established for both clinicians and general public in order to allow them to understand the benefit and impact of the genetic test (Goldman et al. 2011). As genotype frequencies of mutations and SNPs are often different among ethnical groups (Ward et al. 2012), in-depth studies of genotype-disease relationship in specific populations would be necessary in order to validate the applicability of the genotype testing in a particular population (Tsui et al. 2018).

Summary

Genotype describes the genetic makeup of an individual. It denotes which version of a gene that an individual has been inherited from the parents. The determination of one's genotype is a useful tool for predicting his/her phenotype, such as disease susceptibility. The rapid development of DNA analyzing technologies has allowed genotype detection to be performed at high accuracy and reduced cost. It greatly advances genetic research discoveries and revolutionizes medical practice toward personalized medicine (Katsanis and Katsanis 2013).

Cross-References

▶ Alleles
▶ DNA Chip
▶ Genome-Wide Association Study
▶ Mutation
▶ Phenotype

References

Bertram L, McQueen MB, Mullin K, Blacker D, Tanzi RE (2007) Systematic meta-analyses of Alzheimer disease genetic association studies: the AlzGene database. Nat Genet 39:17–23. https://doi.org/10.1038/ng1934

Goldman JS et al (2011) Genetic counseling and testing for Alzheimer disease: joint practice guidelines of the American College of Medical Genetics and the National Society of Genetic Counselors. Genet Med 13:597–605. https://doi.org/10.1097/GIM.0b013e31821d69b8

Goodwin S, McPherson JD, McCombie WR (2016) Coming of age: ten years of next-generation sequencing technologies. Nat Rev Genet 17:333–351. https://doi.org/10.1038/nrg.2016.49

International HapMap Consortium (2005) A haplotype map of the human genome. Nature 437:1299–1320. https://doi.org/10.1038/nature04226

International Human Genome Sequencing Consortium (2004) Finishing the euchromatic sequence of the human genome. Nature 431:931–945. https://doi.org/10.1038/nature03001

Jonsson T et al (2013) Variant of TREM2 associated with the risk of Alzheimer's disease. N Engl J Med 368:107–116. https://doi.org/10.1056/NEJMoa1211103

Karki R, Pandya D, Elston RC, Ferlini C (2015) Defining "mutation" and "polymorphism" in the era of personal genomics. BMC Med Genet 8:37. https://doi.org/10.1186/s12920-015-0115-z

Katsanis SH, Katsanis N (2013) Molecular genetic testing and the future of clinical genomics. Nat Rev Genet 14:415–426. https://doi.org/10.1038/nrg3493

Lanoiselee HM et al (2017) APP, PSEN1, and PSEN2 mutations in early-onset Alzheimer disease: a genetic screening study of familial and sporadic cases. PLoS Med 14:e1002270. https://doi.org/10.1371/journal.pmed.1002270

Manolio TA (2013) Bringing genome-wide association findings into clinical use. Nat Rev Genet 14:549–558. https://doi.org/10.1038/nrg3523

Marzancola MG, Sedighi A, Li PC (2016) DNA microarray-based diagnostics. Methods Mol Biol 1368:161–178. https://doi.org/10.1007/978-1-4939-3136-1_12

Mayo O (2008) A century of Hardy-Weinberg equilibrium. Twin Res Hum Genet 11:249–256. https://doi.org/10.1375/twin.11.3.249

Ragoussis J (2006) Genotyping technologies for all. Drug Discov Today Technol 3:115–122. https://doi.org/10.1016/j.ddtec.2006.06.013

Raux G et al (2005) Molecular diagnosis of autosomal dominant early onset Alzheimer's disease: an update. J Med Genet 42:793–795. https://doi.org/10.1136/jmg.2005.033456

Ryckman K, Williams SM (2008) Calculation and use of the Hardy-Weinberg model in association studies. Curr Protoc Hum Genet 57(1):1.18.1. https://doi.org/10.1002/0471142905.hg0118s57. Chapter 1: Unit

Shendure J, Balasubramanian S, Church GM, Gilbert W, Rogers J, Schloss JA, Waterston RH (2017) DNA sequencing at 40: past, present and future. Nature 550:345–353. https://doi.org/10.1038/nature24286

Tsao JW (2013) Genetic testing for early-onset Alzheimer disease. Continuum (Minneap Minn) 19:475–479. https://doi.org/10.1212/01.CON.0000429170.50686.4e

Tsui NBY et al (2018) Population-wide genetic risk prediction of complex diseases: a pilot feasibility study in Macau population for precision public healthcare planning. Sci Rep 8:1853. https://doi.org/10.1038/s41598-017-19017-y

Ward A, Crean S, Mercaldi CJ, Collins JM, Boyd D, Cook MN, Arrighi HM (2012) Prevalence of apolipoprotein E4 genotype and homozygotes (APOE e4/4) among patients diagnosed with Alzheimer's disease: a systematic review and meta-analysis. Neuroepidemiology 38:1–17. https://doi.org/10.1159/000334607

Geographical Gerontology

Zhixin Feng[1] and David R. Phillips[2]
[1]Primary Care and Population Sciences, Faculty of Medicine, University of Southampton, Southampton, Hampshire, UK
[2]Department of Sociology and Social Policy, Lingnan University, Hong Kong, China

Synonyms

Geographies of ageing; Spatial analysis of ageing

Definition

Geographical gerontology (GG) refers to a burgeoning multidisciplinary subject that encompasses the application of geographical perspectives, concepts, and approaches to the study of ageing, old age, and older populations (Skinner et al. 2018). Human geography and social gerontology have together influenced the development of the field of geographical gerontology, and the thematic scopes of geographical gerontology are presented in Fig. 1. This body of work includes gerontological work by geographers which involves examination and explanation of how geographical approaches can be used to research and understand gerontologically related issues such as demography and population geography (issues such as the spatial patterning of demographic ageing, patterns of migration, and movements in and of ageing populations), health geography (i.e., health-care services and infrastructure), and social geography (such as living arrangements and environments of older people and their families), and geographically orientated work by gerontologists from different disciplines including social work, social policy, public health, nursing, planning, and a full range of social and health sciences (Andrews et al. 2007, 2009; Skinner et al. 2015, 2018).

Overview

Origins and History of Geographical Gerontology

Many Western countries in Europe entered the phase of being "ageing societies" in the 1800s and have been demographically ageing for over two centuries. It is sometimes held to be unavoidable that older people will have to face increasing chronic conditions (noncommunicable diseases and mental health problems) and financial and social issues related to their longer lives and particular in the "oldest-old" years. Against

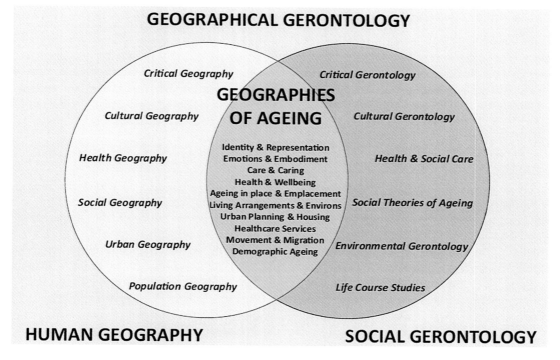

Geographical Gerontology, Fig. 1 A thematic diagram of geographic gerontology. (Source: Adapted from Skinner et al. (2018))

widespread backgrounds of problems among population, resources, and environmental sustainability, an ageing population can add pressures to health and social care systems, welfare systems, and governments (Andrews et al. 2009). With ever limited resources, this provides challenges for governments and older people and their families. The ageing of societies is however rather controversial. The ageing population is often positioned as a potential catastrophe by "apocalyptic demography" that treats people living longer as a burden to societies with negative long-term effects (Byteway and Johnson 2010; Gee 2002; Longino 2005). This can be referred to as the "moral panic" perspective, in which concern is aroused over a social issue (Scott 2014). On the other hand, a much more positive view of ageing can be found embedded in the policy objectives such as "active ageing" from WHO and the European Commission, WHO's (2015) new definition of "healthy ageing," a replacement of "active ageing," or "successful ageing" from academics (i.e., Rowe and Kahn 1987; Bulow and Soderqvist 2014; WHO 2015; European Commission 2016). In addition, the world's largest ageing population country, China, has an idiom: "an old person is like a treasure to the family" to indicate that many people (used to) show respect to their seniors. Ageing is a complex, diverse, and social and geographically influenced concept.

The early concerns of gerontology often focused on physiological and psychological features of ageing with gerontology as an initial development as a health science that closely aligned to medical and clinical geriatrics (Kontos 2005). Over time, gerontology has developed simultaneously as a social science; even the tradition of gerontology has remained strong. "Social gerontology," as a widely known subfield of gerontology, has perhaps become the larger focus involving older people in their social and family contexts, using sociological, psychosocial, economic, anthropological, geographical, political, social theory, and other perspectives to explore a broad range of issues in older peoples' health, health care, welfare, and social and cultural life (Enright 1994; Andrews et al. 2009). The perspectives have helped shape social gerontology in dynamic ways that make gerontology today a multidisciplinary subject.

How do geography and gerontology engage to form geographical gerontology? Many quasi-spatial/geographical perspectives have been employed in different disciplines including but not confined to epidemiology, social medicine, and public health (Diez-Roux 2002). The two have long individual histories, but GG has an established and growing body of research on geographical aspects of ageing by academics from the social sciences (Skinner et al. 2015, 2018). Geography as a dedicated academic subject has many fields, and, within human geography, Harper and Laws (1995) provided a landmark review to highlight how advances within human geography could create new possibilities for geographers involved in ageing and older people. The place-embedded implications and uneven spatial distribution of population ageing among different geography scales (household, group, community, country, global levels) and settings (rural-urban or metropolitan) have become increasingly evident to policy makers and academics as Skinner et al. (2015) discuss in their review two decades later. The core of geographical gerontology is the relationships between older people and the spaces and places in which and through which age and ageing occur (Cutchin 2009). These very perspectives have also made geographical gerontology a multidisciplinary subject.

The development of geographical gerontology has close relations and parallels with the development of human geography as a discipline. Many features of geographical gerontology could be dated to the 1960s, alongside the quantitative revolution and the shift from more descriptive geography to an empirical scientific geography (Burton 1963; Warnes 1981; Rowles 1986). Until the 1990s, geographical gerontology tended to focus on the spatial perspective of population ageing, the location and movement of older people, retirement migration, and the associations with their environment and services at different scales using descriptive analysis (Warnes 1981; Rowles 1986). It was a first phase of empirical accumulation of geographical gerontology as other emerging field of research (Rowles 1986; Andrews and Phillips 2005). During this period, the objectives of geographical gerontology were debated by scholars. In Rowles' (1986) review, a focused was detected on the relationship between older people and their living environments at different scales. He called for research on the meaning of place for older people, the spatial distribution of older populations, and the perspectives that could reflect those in social gerontology regrading to appreciate different ages and cultures of older persons (Rowles 1986; Andrews and Phillips 2005). There had also been a growing body of work on provision for older persons, for example, the development of private residential care, their emergence as businesses, locational concentrations and their associations with changes in government policies, and the risks associated with some trends, especially in the UK (see, e.g., Phillips et al. 1987; Phillips and Vincent 1988) and elsewhere (Rosenberg and Everitt 2001). As Andrews et al. (2018) note, a common hope or aim of much of this research was that policy makers and service planners might note the trends analyzed and better tune and target future policies and programs. Warnes also pointed out that the global evolution in population ageing (and its implications), locational dimensions in the circumstances of older people's lives, and temporal change in the interaction between older people and the environment had not been given enough attention by geographers; and he suggested that geographers needed to change their priorities and objectives from the theoretical needs of human geography to the needs of older people (Warnes 1990; Andrews et al. 2007). In later progress, Harper and Laws (1995) recognized the growing contributions of geographical gerontology and the advances within human geography that could create new possibilities for geographical research on ageing and older people. They argued that geographical gerontology required to learned theoretical and methodological lessons from the cultural turn in human geography and great adoption of social theory and the possibilities within the field for closer engagement with postmodern perspectives (such as feminism, postmodernism, and political economy) to underpin geographical gerontology research should be considered.

By the mid-1990s, the analysis of spatial patterns in population ageing and the movement of older populations, thematic concentrations in the areas of health, health care and caregiving, and the different settings and environments of ageing among geographical gerontology were well-established (Skinner et al. 2015). Some 10 years after Harper and Laws' (1995) review, Andrews and his colleagues (2007) reviewed the process of geographical gerontology between 1995 and 2006; they found that geographical gerontology was constituted of multiple fields of empirical interest studies working with multiple academic disciplines during this period. Indeed, geographical gerontology's focus was on dynamics, distributions, and movements in older populations' health, and postmodern perspectives and qualitative approaches were developed to explore the complex relationships between older people and the varied places within which they live and are cared for (Andrews et al. 2007, 2009). A more formal recognition of geographical gerontology has broadened and deepened the scope of geographical interests in ageing, particularly in social and cultural geography (Del Casino 2009; Skinner et al. 2015). It has enriched theoretical and methodological pluralism, particularly in the discipline within feminism, postmodernism, and post-structuralism (Andrews et al. 2007, 2009).

Subsequently, in the twenty-first century, the microscales of the subjective experiences of older persons in a wide variety of health and care settings have been further explored by human geographers, for example, finer microscale to reflect human experience in places, which transcend space, place, and scale to obtain a picture of older people's life courses and different concepts of cross- or multiscale issues that geographers could pursue in particular locations (Skinner et al. 2015). In addition, the contemporary theoretical orientations, like relational geographies of ageing and nonrepresentational geographies of ageing, are the growing interest of geographers (Skinner et al. 2015). Interests in therapeutic landscapes as relating to older persons are also growing (Winterton 2018).

Significance of Geographical Gerontology

The twenty-first century is the era of both stabilizing and booming population ageing, depending on the different areas of the globe and sometimes among different social groups. It is estimated that there will be 3.14 billion people aged 60 and above in 2100, and the number of the oldest-old, defined as 80+, will increase to 909 million (United Nations DESA Population Division (UNDESA) 2017). From a geographical gerontology perspective, place, space, scale, landscape, territory, and other factors of geographical constructs can shape the experiences of older people. Patterns of ageing are varied at different spatial/geographical levels across countries around the world. Populations have been ageing in more developed countries for over a century, and the ageing process started recently in less developed and developing countries, especially LAMICs (Kinsella and Phillips 2005). Economic, social, cultural, and political from the geography perspective could be reflected by varied ageing process. Why and how place and space matter are key questions for many scholars (particular geographers) in the field of geographical gerontology.

While geographical gerontology may be broad, the data from UNDESA (2017) can be used to illustrate how geographical gerontology can be applied in real-world settings. From a geographical gerontology perspective, it is clear that population ageing is varied around the world (returned to in the Policy section below). The old-age dependency ratio, measured by the ratio of population aged 65+ per 100 population aged 15–64, can be an indicator, if imperfect, of likely pressure on "productive" population (the "productive" ages 15–64). A lower ratio could reflect better pensions and better health care for residents, and a higher ratio could indicate more financial stress between working people and dependents. Table 1 shows the old-age dependency ratio by region and subregion between 1950 and 2100. In general, the old-age dependency ratio in the world increased from 8.37 to 12.64 between 1950 and 2015, and it is predicted that old-age dependency ratio will reach 39.75 in 2100. This indicates that there will be 2.5 adults aged 15–64 taking financial responsibility for one older people in 2100, comparing to 10 adults aged 15–64 taking financial responsibility for one older people in 2015. That will be a huge potential financial stress in the

Geographical Gerontology, Table 1 Old-age dependency ratios by region and subregion 1950–2100

Region and subregion	1950	1970	1990	2010	2015	2030	2050	2070	2090	2100
World	8.37	9.29	10.10	11.65	12.64	18.03	26.17	33.04	37.85	39.76
More developed regions	11.87	15.46	18.74	23.74	26.68	37.43	42.97	41.80	45.10	47.24
Less developed regions	6.54	6.57	7.43	8.95	9.75	14.80	23.53	31.60	36.60	38.45
Least developed countries	5.91	5.46	5.97	6.23	6.29	7.20	13.19	25.02	31.68	33.52
Less developed regions, excluding least developed countries	6.62	6.70	7.61	9.34	10.28	16.34	25.90	33.18	37.79	39.63
Less developed regions, excluding China	6.23	6.50	6.87	8.00	8.47	11.95	19.86	29.42	34.75	36.72
High-income countries	12.29	15.66	18.30	22.81	25.69	36.34	43.37	43.58	46.38	48.86
Middle-income countries	6.92	7.25	8.12	9.64	10.49	16.17	25.27	32.58	37.29	39.06
Upper-middle-income countries	7.24	7.68	9.01	11.41	12.89	22.25	34.79	39.52	43.84	45.23
Lower-middle-income countries	6.50	6.72	7.06	7.83	8.18	11.37	18.66	27.89	32.92	34.92
Low-income countries	5.65	5.38	5.95	6.23	6.22	6.41	11.41	23.31	30.64	32.50

Sources: United Nations DESA/Population Division (2017). (https://population.un.org/wpp/Download/Standard/Population/)

future. In Table 1, geographical gerontologists may focus various features and raise many research questions. These could be on space patterns of old-age dependency ratios across different regions; do the geographical patterns of different regions increase in the same direction? Why do more developed regions have higher old-age dependency ratios than the less or least developed regions? What factors are associated with increasing old-age dependency ratios in different regions; and are these factors the same or different in different regions (a geographic perspective); from the gerontological perspective, are more developed regions age-friendly regions? Will health-care services be sufficient for the coming ageing of populations in different regions? Will older people age in place as different regions grow older? What policies could be implied for high old-age dependency ratios? In order to answer these questions, theoretical developments in geographical gerontology could be applied. Readers could develop their own interesting questions from Table 1.

This example has illustrated what geographical gerontologists consider in a real world, and it also leads to the important questions of geographical gerontology: space and place. Why and how space and place matter? People's physical and mental capacities may vary across his/her life course and into older age, also strongly determined by the environments in which people live; environments also determine how well people adjust to loss of function and other forms of stress that people may experience at different stages of life and in particular in his/her later life (World Health Organization (WHO) 2018a). This has great implications for the technical definitions such as old-age dependency ratios and shows the limitations of using age cutoffs to define such ratios. In many cases, persons aged 65+ will still be active and will hardly be dependent. In many countries, especially the low- and middle-income group, a lower age dependency cutoff may be used, and in many countries in sub-Saharan Africa, for example, ages between 50 and 55 are used to represent older age, to reflect local epidemiological and demographic circumstances.

How space and place matter to older people and reflect on older people can also be influenced by length of residence. Many continue to value and enjoy their homes; they value the natural environment and would like to participate in outdoor leisure activities. Many value the place of which they have many memories and where they may have lived in for many years; they may have established close relationships in a retirement community, but how far are their requirements for physical and emotional supports met in residential care settings? (Andrews et al. 2007, 2009, 2018). In comparison to younger adults, older people are sometimes (erroneously) felt to be less able to adapt and more reliant on resources available in and around their living residence and dependent on the support of others locally (Muramatsu 2003; Robert and Li 2001; Feng et al. 2012; Phillips and Yeh 1999). A place with accessible wide range of health services and social support is very important to one's sense of security and belonging, particularly for people in later life (Hanlon 2018; Menec et al. 2011). The quality of space and place that older people live could contribute to differential health, well-being, and welfare of older people, which becomes a major concern for geographical gerontologists.

Key Findings

In geographical gerontology both historic and contemporary, three main focuses have been the spatial concentration and distribution of ageing population, spatial patterns of health outcomes among older adults, and health service availability, access, and utilization.

The Spatial Concentration and Distribution of Older Populations

The spatial concentration and distribution of older population have long been interested by geographers, demographers, and social gerontologist. As previous statement, the patterns of ageing vary at different scales including local, national, and global levels. The global pattern of ageing has been estimated and described in UNDESA, and many countries have their own census datasets to illustrate the ageing pattern at national levels (e.g., the United States Census Bureau in the USA, Office for National Statistics in the UK, National Bureau of Statistics of China in China) (see, e.g., Kinsella and Phillips 2005). In addition, several countries operate longitudinal studies of ageing which allow academics to access individual dataset and conduct ageing studies at different levels (individual, household, or community levels).

Illustrating global patterns of ageing, Figs. 2 and 3 show the proportions (usually in percentages) of persons aged 60+ in 2015 and 2050 (projected), respectively, from the UNDESA using the GIS mapping. It is apparent that only Japan currently has more than 30% of its population aged 60+ years, and almost all African countries, most countries in Central America, South, and Southeast Asia, have fewer than 10% of population in this age group (Fig. 2); looking forward, however, in 2050, Japan will still have more than 30% of its population aged 60+ years, joined by many countries in East and Southeast Asia, Europe, and some in Latin America (Chile) (Fig. 3). There are to be huge geographical changes in distribution of the 60+ population globally in 2015 and 2050, especially impacting the current low- and middle-income countries with concomitant challenges of health and social care and support.

The country borders do not reflect the endorsement or the view of the publisher, the editor, or the author.

At a smaller geographical scale (within a nation), spatial patterns of ageing population are also varied. Taking the UK as an example, life expectancy at birth has nearly doubled over the last 100 years in the UK (Office for National Statistics 2015). Figure 4 shows the number of people aged 100 years and over (centenarians) per 100,000 population for England, Scotland, Wales, and Northern Ireland between 1987 and 2017. In 1987, the proportions of centenarians in England, Wales, and Northern Ireland were 7 per 100,000, and the proportion of centenarians in Northern Ireland was 5 per 100,000. In 2017, Wales has the highest proportion at 26 per 100,000, followed by England at 22, and Scotland at 17, and Northern Ireland has the lowest at 15 (Office for National Statistics 2018).

Global ageing is the result of the continued decline in fertility rates and increased life expectancy (Kinsella and Phillips 2005; Phillips and

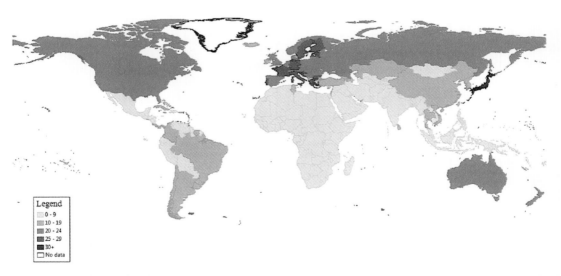

Geographical Gerontology, Fig. 2 Percentage of population aged 60+ in 2015. Source: authors' calculations, Probabilistic Population Projections based on data in UNDESA (2017) World Population Prospects: The 2017 Revision World. Population Prospects: The 2017 Revision (https://esa.un.org/unpd/wpp/Download/Probabilistic/Population/). The country borders do not reflect the endorsement or the view of the publisher, the editor, or the author

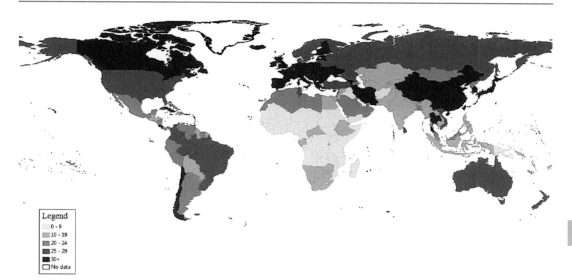

Geographical Gerontology, Fig. 3 Percentage of population aged 60+ in 2050. Source: authors' calculations, Probabilistic Population Projections based on data in UNDESA (2017) World Population Prospects: The 2017 Revision World. Population Prospects: The 2017 Revision (https://esa.un.org/unpd/wpp/Download/Probabilistic/Population/)

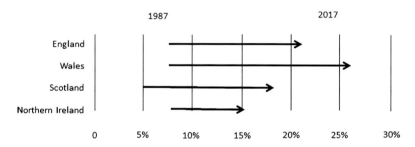

Geographical Gerontology, Fig. 4 Number of people aged 100 years and over per 100,000 population, for England, Scotland, Wales, and Northern Ireland, 1987 and 2017 in the UK. Sources: adapted from Office for National Statistics, National Records of Scotland, Northern Ireland Statistics and Research Agency https://www.ons.gov.uk/peoplepopulationandcommunity/birthsdeathsandmarriages/ageing/bulletins/estimatesoftheveryoldincludingcentenarians/2002to2017

Feng 2018). Improved medical technology and healthier behaviors have also influenced the life course which could result in longer life expectancy, although current evidence on continuing increases in life expectancy is mixed. The varied patterns of ageing in the world and the proportion of centenarians in the UK will reflect different epidemiological factors, medical services and technology, and especially socioeconomic and environmental conditions, plus diets and health behaviors, excluded and included populations, the very crucial "social determinants of health" (Marmot 2015). All of these are squarely within the research interest of health geographers and especially the socio-spatial variations.

The Spatial Patterns of Health Outcomes Among Older Adults

Although global ageing is generally increasing (life expectancy may be static or even decreasing in some countries (i.e., Russia)), it is generally of more interest to know whether older people's later life is in good or less good health (Gatrell and Elliott 2015). It is recognized that older people are at higher risk of having chronic and disabling conditions than younger adults, and older people's health varies in different population groups as well as at different geographical scales (Wiles 2018). In addition, health outcomes are not only affected by individual characteristics but also by the surrounding environment where individuals live and work (Jones et al. 2000). The "area effect" which refer to either there is an independent effect for a place-based variable (or so-called geographical effect) such as per capita GDP or a difference between places that is not reducible to individual characteristics (or so-called geographical differentials) has been well studied (Jones et al. 2000). They also reflect the social determinants of health as mentioned above.

Chinese studies using multilevel models provide some examples. Zeng et al. (2010) analyzed the 2002 and 2005 waves of the Chinese Longitudinal Health Longevity Study (CLHLS) to examine the impacts of environmental factors (at the community level) on older people's health outcomes (activities of daily living (ADLs), cognitive impairment, and mortality). The community level information included labor force participation rate, per capita GDP, illiteracy rate, average temperature in January and July, yearly rainfall, hills or mountains covering, and air pollution. Generally speaking, their findings found that communities' GDP per capita, adult labor force participation rate, and illiteracy rate were associated with physical, mental, and overall health and mortality among older persons in China. More specifically, they found that higher per capita GDP and lower community rates of illiteracy decreased the odds of cognitive impairment; higher per capita GDP increased the rate of ADL disability. Higher labor force participation among persons aged 15–64 years in the community reduced older persons' risk of ADL disabilities, cognitive impairment, health deficits, and death over a 3-year follow-up. Air pollution increased the odds of disability in ADLs, cognitive impairment, and health deficits. More rainfall reducing the odds of ADL disability and cognitive impairment and low seasonal temperatures increased the odds of ADL disability and mortality; high seasonal temperatures increased the odds of cognitive impairment and deficits. All these clearly indicate the importance of spatial differences in socio-environmental factors for older persons health locally.

Feng et al. (2012) used the same CLHLS dataset (CLHLS 2008) and found that higher income inequality at province level was associated with poorer self-rated health among older people. Feng et al. (2015) also found that higher provincial levels of economic development have a negative influence on the survival of the rural older people (CLHLS 2002–2008). Feng et al.'s studies also explored the geographical differentials of health in China. They found that province with the best health was Zhejiang whose residents had the lowest risk of reporting poor health, while Hainan had the highest risk of reporting poor health nationally (Feng et al. 2012, 2013) (see Fig. 5). However, there are no geographical differentials in survival status among older people in China (Feng et al. 2015). Evandrou et al.'s 2014 study used the Chinese Health and Retirement Longitudinal Study (CHARLS 2011) to study the individual and province inequalities in health in China. Their results indicated that persons who lived in economically developed provinces, albeit with lower health expenditures and less developed health-care institutions, were less likely to report difficulties with ADLs, and older people living in a province with a higher proportion of old people are more likely to report difficulty with ADLs. In terms of geographical differentials, Zhejiang province again has the lowest risk of reporting poor health in their study even with the different datasets (Evandrou et al. 2014). These geographical findings have gerontological policy implications as to how to improve older people's living environments (such as narrowing income inequalities, reducing air pollution, developing social security) to improve their health outcomes in later life.

Health Services, Access, and Utilization

The interconnections between older people and the spaces and places are one of the central tenets of geographical gerontology (Hanlon 2018), and a

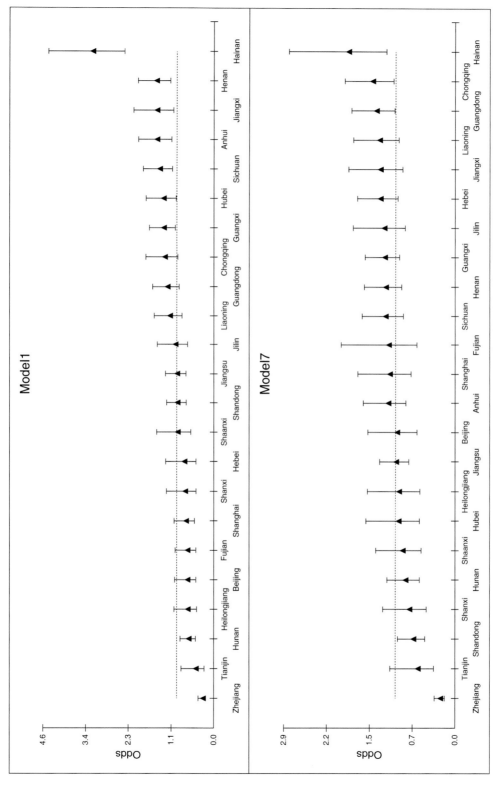

Geographical Gerontology, Fig. 5 Differential relative odds of poor self-reported health for provinces derived from models 1 and 7 compared to the national average set at 1 (Source: adapted from Feng et al. (2012))

place with accessible wide range of health services and social support is very important to one's sense of security and belonging, particularly for people at their later life (Hanlon 2018; Menec et al. 2011). Older persons, in general, have higher demands for health care than younger age groups, and they also face various distinct disadvantages in accessing affordable, appropriate, and quality care (United Nations 2018). The WHO (2015) in the first *World Report on Ageing and Health* emphasizes the need for health systems to be structured, so they are much better aligned to the health and social care needs of older persons. Currently, this is the exception rather than the rule, especially for the growing older populations in low- and middle-income countries.

Older people who have good physical functioning are generally able to retain better control over their lives and are more likely to remain living in their own homes than in residential or institutional care. This refers to "ageing in place" ("the ability to live in one's own home and community safely, independently, and comfortably, regardless of age, income, or ability level" (Centers for Disease Control and Prevention 2009). "Ageing in place" requires older people's homes or communities to have the health and social supports and services older people need to live safely and independently as long as they wish and are able. "Ageing in place" has somewhat different meanings in different parts of the world. In Eastern Asian countries, for example, it is part of cultural beliefs for older people to age in place. Children take the responsibility to support and look after of their parents as they age, and children will move in with their parents when their assistance is needed (i.e., filial piety, the "code" of intergenerational duty in China and similar societies (Phillips and Feng 2015; WHO 2015: Box 1.4); in Western countries, there has been greater reliance on governments, charities, or other organizations to provide the support and the services to help people to remain in their own homes. Increasingly, the global view is that policies should support "ageing in the *right* place" (Golant 2015), in other words, "the ability to live in the place with the closest fit with the person's needs and preferences" (WHO 2015, p. 225). This very important extension of the concept recognizes how being in the correct environment can strongly affect how comfortably and successfully ageing may progress.

From a GG perspective, health services, access, and utilization have been traditionally been concerned with regard to distance to doctors and hospitals (Skinner et al. 2015; Joseph and Phillips 1984). In addition, the neighborhoods in which older people live also determine whether their needs for basic services may be met. Based on the latest Health Survey for England data, Savage (2017) reported that older people in the most deprived areas are twice as likely to lack the basic help they need compared with those in the richest neighborhoods, which reflect inequalities in access to social care in England, sometimes called a "post-code lottery." Such socio-spatial variations in accessibility to services are replicated in many communities worldwide.

The importance of local climate and environmental conditions for older person's health was mentioned previously. The findings indicate that local planning and design standards as well as social conditions can be very influential in older persons' health. Indeed, differences in temperatures, local pollution, and many other aspects potentially related to climate change have been associated with differences on older persons' health and well-being (Phillips and Feng 2018). This may relate to severe weather events seen in almost all areas, flooding, earthquakes, resultant tsunamis, and longer-term risks through changed conditions favoring diseases such as the spread of malarial risk zones (McCracken and Phillips 2016; WHO 2018b). A well-known example of the effects of extreme temperatures was the estimated 70,000 excess deaths among older persons in the very hot summer in Europe in August 2003 (WHO 2018b; Robine et al. 2003). Immobile older persons with pre-existing health problems, persons aged 75+, and especially those living in poorly insulated accommodation were at greatest risk of dying in these extreme temperatures (Vandentorren et al. 2006; Poumadère et al. 2005). All these risks point to the importance of intersectoral planning in accommodation design, health services accessibility, and responses for the needs of older persons.

GG can be a central platform in helping these policy areas. Moreover, many aspects of geographical gerontology are seen in the activities of NGOs such as Global AgeWatch Insights. Their *Global AgeWatch Index* is a good example of analysis of 96 countries, compared and ranked in terms of income security, health status, capability, enabling environment, and also given an overall ranking. A related example is *Pension Watch*, a knowledge hub focusing especially on tax-financed pensions. It is hosted by Global AgeWatch Insights and provides information via a global pension database, showing comparative data for many countries based on UN and World Bank data. This is a very good resource for comparing nations policies and practices in terms of public social pensions which can give a good perspective on the resources available to older persons internationally. Both these indexes and databases provide very useful resources for comparative geographical gerontology studies.

A further policy-related area involving GG is comparative epidemiological transition (McCracken and Phillips 2017). Epidemiological and demographic changes can be studied at regional, national, and local scales. Indeed, at the national/regional scale, GG has been showing how the traditional ageing areas such as Europe and Japan are reaching more steady states in demographic terms and even face population declines. The complex interactions with falling birth rates and apparent levelling off increases in life expectancy provide rather different ageing profiles for countries in this group. Many countries and territories in the Asia-Pacific region are also following suit, including China, Singapore, Taiwan, Hong Kong, and Thailand. By contrast, the low- and middle-income countries are sometimes quite rapidly showing themselves as the locales of current and future ageing in this century, including many large countries such as India and several in sub-Saharan Africa. Their changing demographic and epidemiological/health profiles bring new and often unanswered challenges in terms of resources and provision of care for older persons. Research on changing patterns of demographic ageing and short-, medium-, and long-term changes provide essential information for policy makers in many areas, including health, welfare, housing, transport, and a range of public and private services. Information on changing needs, attitudes, and demands of older consumers is also increasingly being valued in terms of retail provision and planning for older consumers as population profiles change.

GG has a role in many other areas especially in comparative study of older people's well-being and use of services. This can be local or national level studies and can involve matters such as the role of various forms of social support, social exclusion, loss of support, social protection, and the like (Scharf and Keating 2012; Feng et al. 2018, 2019; Gyasi et al. 2018a, b; Gyasi and Phillips 2019). Factors such as variations in health literacy can be important and how these impinge on the success or may limit global policies advocated by WHO such as universal health coverage (Amoah and Phillips 2018). Other policy-related areas to which GG has contributed include the realities of ageing in place, identity, and place attachment in later years, ageing landscapes, and therapeutic landscapes (see Skinner et al. 2018).

Prospects

Looking to the future, it is clear that geographical gerontology is an increasingly multidisciplinary subject that relates to a majority of disciplines to research on ageing and older people. In order to achieve an even more fully transdisciplinary contribution of this field, new frameworks, theories, models, or applications could be developed by researchers, to integrate and transcend disciplinarily as a goal (Cooke and Hilton 2015; Cutchin et al. 2018). Geographical gerontology requires and is achieving inputs from an increasing range of contributors from within and outside geography. As its contributions grow, shared language and concepts could unite scholars in visions, collaboration, and the production of valuable insights about ageing (Cutchin et al. 2018). In particular, the ability to harness increasingly sophisticated forms of spatial analysis and mapping using complex datasets at different scales is likely to be an ever more important facet of this subdiscipline and one which will also contribute to policy and social planning. Geographical gerontology has long contributed to planning both physical and social policy. Findings such as those from the above China studies clearly indicate the value of identifying targetable socio-spatial variables that appear to influence older persons' health. Analysis of spatial differences in access to and use of health services is also a traditionally strong area in this respect (see, e.g., Andrews and Phillips 2005; Hanlon 2018), and this will continue as an important future focus. Other key areas will include quality of life, well-being, social care, housing, and an increasing concern with socio-spatial differences in resource allocation. In all these areas, there is a growing trend for interdisciplinary perspectives and multidisciplinary team research. Increasing interest is being paid to matters such as environment, climate, and climate change as they impinge on older populations and their families (see Haq in this encyclopedia).

Cross-References

▶ Climate Resilience and Older People
▶ Flooding and Older People
▶ Heatwaves and Older People

References

Amoah PA, Phillips DR (2018) Health literacy and health: rethinking the strategies for universal health coverage in Ghana. Public Health 159:40–49. https://doi.org/10.1016/j.puhe.2018.03.002
Andrews GJ, Phillips DR (2005) Geographical studies in ageing: progress and connections to social gerontology. In: Andrews GJ, Phillips DR (eds) Ageing and place: perspectives, policy practice. Routledge, London, pp 7–12
Andrews GJ, Cutchin M, McCracken K, Phillips DR, Wiles J (2007) Geographical gerontology: the constitution of a discipline. Soc Sci Med 65(1):151–168. https://doi.org/10.1016/j.socscimed.2007.02.047
Andrews GJ, Milligan C, Phillips DR, Skinner MW (2009) Geographical gerontology: mapping a disciplinary intersection. Geogr Compass 3(5):1641–1659
Andrews GJ, Cutchin MP, Skinner MW (2018) Space and place in geographical gerontology. In: Skinner MW, Andrews GJ, Cutchin MP (eds) Geographical gerontology: perspectives, concepts, approaches. Routledge, Oxford, pp 11–28
Bulow MH, Soderqvist T (2014) Successful ageing: a historical overview and critical analysis of a successful concept. J Ageing Stud 31:139–149. https://doi.org/10.1016/j.jaging.2014.08.009
Burton I (1963) The quantitative revolution and theoretical geography. Can Geogr (4):151–162
Bytheway B, Johnson J (2010) An ageing population and apocalyptic demography. Radical Statistics 100: 4–10
Centre for the Study of Agingand Human Develeopment. Chinese Longitudinal Health Longevity Study (CLHLS) (2008). https://sites.duke.edu/centerforaging/programs/chinese-longitudinal-healthy-longevity-surveyclhls/project-goals/
Centers for Disease Control and Prevention (2009) Healthy places terminology. Retrieved 28 Jan 2018, from https://www.cdc.gov/healthyplaces/terminology.htm
CHARLS homepage. Chinese Health and Retirement Longitudinal Study (CHARLS) (2011). http://charls.pku.edu.cn/en
Cooke NJ, Hilton ML (eds) (2015) Enhancing the effectiveness of team science. Committee on the Science of Team Science, Behavioral, Cognitive, and Sensory Sciences, Division of Behavioral and Social Sciences and Education, National Research Council. The National Academies Press, Washington, DC
Cutchin MP (2009) The geography of ageing: preparing communities for the surge in seniors. Gerontologist 49 (3):440–444
Cutchin MP, Skinner MW, Andrews GJ (2018) Geographical gerontology-process and possibilities. In: Skinner MW, Andrews GJ, Cutchin MP (eds) Geographical gerontology: perspectives, concepts, approaches. Routledge, Oxford, pp 11–28
Del Casino JV (2009) Ageing and the 'new' social geographies of older people. Wiley-Blackwell, Chichester
Diez-Roux AV (2002) Invited commentary: places, people, and health. Am J Epidemiol 155(6):516–519. https://doi.org/10.1093/aje/155.6.516
Enright RB Jr (1994) Perspectives in social gerontology. Allyn and Bacon, Boston
European Commission (2016) Employment, social affairs & inclusion. https://translate.google.com/#view=home&op=translate&sl=en&tl=zh-CN&text=Employment%2C%20Social%20Affairs%20%26%20Inclusion.
Evandrou M, Fakingham J, Feng Z, Vlachantoni A (2014) Individual and province inequalities in health among older people in China: evidence and policy implications. Health Place 30:134–144. https://doi.org/10.1016/j.healthplace.2014.08.009
Feng Z, Wang W, Jones K, Li Y (2012) An exploratory multilevel analysis of income, income inequality and self-rated health of the elderly in China. Soc Sci Med 75(12):2481–2492. https://doi.org/10.1016/j.socscimed.2012.09.028
Feng Z, Wang W, Jones K (2013) A multilevel analysis of the role of the family and the state in self-rated health of

elderly Chinese. Health Place 23:148–156. https://doi.org/10.1016/j.healthplace.2013.07.001

Feng Z, Jones K, Wang W (2015) An exploratory discrete-time multilevel analysis of the effect of social support on the survival of elderly people in China. Soc Sci Med 130:181–189. https://doi.org/10.1016/j.socscimed.2015.02.020

Feng Z, Phillips DR, Jones K (2018) A geographical multivariable multilevel analysis of social exclusion among older people in China: evidence from the CLASS ageing study. Geogr J 184(4):413–428. https://doi.org/10.1111/geoj.12274

Feng Z, Jones K, Phillips DR (2019) Social exclusion, self-rated health and depression among older people in China: evidence from a national survey of older persons'. Arch Gerontol Geriatr 82:238–244

Gatrell A, Elliott S (2015) Ageing and place. In: Gatrell A, Elliott S (eds) Geographies of health: an introduction, 3rd edn. Wiley, Chichester, pp 244–272

Gee EM (2002) Misconceptions and misapprehensions about population ageing. Int J Epidemiol 31:750–753

Golant SM (2015) Ageing in the right place. Health Professions Press, Baltimore

Gyasi RM, Phillips DR (2019) Risk of psychological distress among community-dwelling older adults experiencing spousal loss in Ghana. Gerontologist (in press). https://academic.oup.com/gerontologist/advance-article-abstract/doi/10.1093/geront/gnz052/5490141?redirectedFrom=fulltext

Gyasi RM, Phillips DR, Amoah PA (2018a) Multi-dimensional social support and health services utilization among non-institutionalized older persons in Ghana. J Ageing Health. https://doi.org/10.1177/0898264318816217. Published online Dec 3 2018

Gyasi RM, Phillips DR, Buor D (2018b) The role of a health protection scheme in healthcare utilisation among community-dwelling older persons in Ghana. J Gerontol B Psychol Sci Soc Sci. https://doi.org/10.1093/geronb/gby082

Hanlon N (2018) Oder persons, place and health care accessibility. In: Skinner M, Andrews G, Cutchin M (eds) Geographical gerontology: perspectives, concepts, approaches. Routledge, Oxford, pp 229–240

Harper S, Laws G (1995) Rethinking the geography of ageing. Prog Hum Geogr 19(2):199–221. https://doi.org/10.1177/030913259501900203

Jones K, Gould MI, Duncan C (2000) Death and deprivation: an exploratory analysis of deaths in the health and lifestyle survey. Soc Sci Med 50(7-8):1059–1079

Joseph AE, Phillips DR (1984) Accessibility and utilization: geographical perspectives on health care delivery. Harper and Row, London/New York

Kinsella K, Phillips DR (2005) Global ageing: the challenge of success. Population Reference Bureau, Washington, DC. www.prb.org

Kontos P (2005) Multi-disciplinary configurations in gerontology. In: Andrews GJ, Phillips DR (eds) Ageing and place: perspectives, policy, practice. Routledge, London, pp 24–36

Longino CF (2005) The future of ageism: baby boomers at the doorstep. Gener-J Am Soc Ageing 29(3):79–83

Marmot M (2015) The health gap: the challenge of an unequal world. Bloomsbury, London

McCracken K, Phillips DR (2016) Climate change and the health of older people in South-East Asia'. In: Akhtar R (ed) Climate change and human health scenario in South and Southeast Asia. Springer, New York, pp 29–52

McCracken K, Phillips DR (2017) Demographic and epidemiological transition. In: Richardson D (Editor-in-chief) The international encyclopedia of geography. Wiley-Blackwell, New York, pp 1–8. https://doi.org/10.1002/9781118786352.wbieg0063

Menec VH, Means R, Keating N, Parkhurst G, Eales J (2011) Conceptualizing age-friendly communities. Can J Ageing-Revue Canadienne Du Vieillissement 30(3):479–493. https://doi.org/10.1017/S0714980811000237

Muramatsu N (2003) County-level income inequality and depression among older Americans. Health Serv Res 38(6):1863–1883

Office for National Statistics (2015) How has life expectancy changed over time?. https://www.ons.gov.uk/peoplepopulationandcommunity/birthsdeathsandmarriages/lifeexpectancies/articles/howhaslifeexpectancychangedovertime/2015-09-09. Accessed 15 Feb 2019

Office for National Statistics (2018) Estimates of the very old, including centenarians, UK: 2002 to 2017–Annual mid-year population estimates for people aged 90 years and over by sex and single year of age (90 to 104) and 105 years and over, and comparisons between UK countries. https://www.ons.gov.uk/peoplepopulationandcommunity/birthsdeathsandmarriages/ageing/bulletins/estimatesoftheveryoldincludingcentenarians/2002to2017. Accessed 15 Feb 2019

Phillips DR, Feng Z (2015) Challenges for the ageing family in the People's Republic of China. Can J Ageing-Revue Canadienne Du Vieillissement 34(3):290–304. https://doi.org/10.1017/S0714980815000203

Phillips DR, Feng Z (2018) Global ageing. In: Skinner MW, Andrews GJ, Cutchin MP (eds) Geographical gerontology: perspectives, concepts and approaches. Routledge, London/New York, pp 93–109

Phillips DR, Vincent J (1988) Privatising residential care for the elderly: the geography of developments in Devon. Soc Sci Med 26(1):37–47

Phillips DR, Yeh G (1999) Environment and ageing: environmental policy, planning and design for elderly people in Hong Kong. Centre of Urban Planning and Environmental Management, University of Hong Kong, Hong Kong

Phillips DR, Vincent J, Blacksell S (1987) Spatial concentration of residential homes for the elderly: planning responses and dilemmas. Trans Inst Brit Geogr NS 12:73–83

Poumadère M, Mays C, Sophie Le Mer S, Blong R (2005) The 2003 heatwave in France: dangerous climate change here and now. Risk Anal 25(6):1483–1494

Robert SA, Li LW (2001) Age variation in the relationship between community socioeconomic status and adult health. Res Ageing 23(2):233–258

Robine JM, Cheung SLK, Le Roy S, Van Oyen H, Griffiths C, Michel J-P, Herrmann FR (2003) Death toll exceeded 70,000 in Europe during the summer of 2003. Les Comptes Rendus/Série Biologies 331 (2):171–178

Rosenberg MW, Everitt J (2001) Planning for ageing populations: inside or outside the walls. Prog Plan 56 (3):119–168. https://www.infona.pl/resource/bwmeta1.element.elsevier-c5b5a812-fdfe-399f-a0f4-5dc60e8630b9

Rowe JW, Kahn RL (1987) Human ageing – usual and successful. Science 237(4811):143–149. https://doi.org/10.1126/science.3299702

Rowles G (1986) The geography of ageing and aged: towards an integrated perspective. Prog Hum Geogr 10(4):511–539

Savage M (2017) Social care postcode gap widens for older people, The Guardian, pp. https://www.theguardian.com/society/2017/dec/16/social-care-for-elderly-postcode-gap-grows. Accessed 15 Feb 2019

Scharf T, Keating NC (eds) (2012) From exclusion to inclusion in old age: a global challenge. Policy Press, Bristol

Scott J (ed) (2014) Moral panic. In: A dictionary of sociology, 4th edn. Oxford University Press, Oxford/New York

Skinner MW, Cloutier D, Andrews GJ (2015) Geographies of ageing: progress and possibilities after two decades of change. Prog Hum Geogr 39(6):776–799. https://doi.org/10.1177/0309132514558444

Skinner MW, Andrews GJ, Cutchin M (2018) Introducing geographical gerontology. In: Skinner MW, Andrews GJ, Cutchin M (eds) Geographical gerontology: perspectives, concepts, approaches. Routledge, Oxford, pp 3–10

United Nations (2018) Health inequalities in old age. In: the United Nations (ed) Department of Economic and Social Affairs, New York

United Nations DESA Population Division (UNDESA) (2017) World Population Prospects 2017. United Nations, New York. Available from: https://esa.un.org/unpd/wpp/Download/Probabilistic/Population/. Accessed 15 Feb 2019

Vandentorren S, Bretin P, Zeghnoun A, Mandereau-Bruno L, Crosier A, Cochet C, Ribéron J, Siberan I, Declercq B, Ledrans M (2006) August 2003 heat wave in France: risk factors for death of elderly people living at home. Eur J Public Health 16(6):583–591. https://doi.org/10.1093/eurpub/ck1063

Warnes AM (1981) Towards a geographical contribution to gerontology. Prog Hum Geogr 5(2):317–341

Warnes AM (1990) Geographical questions in gerontology – needed directions for research. Prog Hum Geogr 14(1):24–56. https://doi.org/10.1177/030913259001400103

Wiles J (2018) Health geographies of ageing. In: Skinner M, Andrews G, Cutchin M (eds) Geographical gerontology: perspectives, concepts, approaches. Routledge, Oxford, pp 31–42

Winterton R (2018) Therapeutic landscapes of ageing. In: Skinner MW, Andrews GJ, Cutchin MP (eds) Geographical gerontology: perspectives, concepts, approaches. Routledge, Oxford, pp 292–303

World Health Organization (WHO) (2015) World report on ageing and health. WHO, Geneva. https://www.who.int/ageing/events/world-report-2015-launch/en/. Accessed 15 Feb 2019

World Health Organization (WHO) (2018a) Age-friendly environments ageing and life-course. https://www.who.int/ageing/age-friendly-environments/en/. Accessed 15 Feb 2019

World Health Organization (WHO) (2018b) Climate change and health. Fact sheet. WHO, Geneva. https://www.who.int/en/news-room/fact-sheets/detail/climate-change-and-health. Accessed 15 Feb 2019

Zeng Y, Gu DN, Purser J, Hoenig H, Christakis N (2010) Associations of environmental factors with elderly health and mortality in China. Am J Public Health 100(2):298–305. https://doi.org/10.2105/Ajph.2008.154971

Geographies of Ageing

▶ Geographical Gerontology

Geriatric Anxiety Inventory

Nancy A. Pachana
School of Psychology, The University of Queensland, Brisbane, QLD, Australia

Definition

The Geriatric Anxiety Inventory (GAI) is a 20-item self-report inventory designed for and validated on older adults to measure anxiety symptoms in later life. There is a five-item short version of the scale. The GAI has been translated into over 20 languages and is in wide use globally (Molde et al. in press).

Overview

Anxiety is prevalent globally in those over age 60 (e.g., Byers et al. 2010). Anxiety in later life is highly comorbid with both depressive disorders and chronic health conditions (Tampi and Tampi 2014). A challenging diagnostic issue, late-life anxiety has been widely discussed, particularly in light of changes in Diagnostic and Statistical Manual of Mental Disorders, fifth edition (DSM5) (Bryant et al. 2013). Reviews of anxiety treatment in older adults suggest both pharmacological and psychotherapy treatments are appropriate (Gonçalves and Byrne 2012). Nevertheless, anxiety is often underdiagnosed and undertreated in older populations (Bower et al. 2015). This is despite the fact that anxiety symptoms and diagnoses are associated with increased morbidity, loss of function, and poorer quality of life (Miloyan et al. 2014). A reliable, valid instrument, simple to administer and score, is essential to accurately detect anxiety in older populations. The GAI fills this role.

Key Research Findings

The Geriatric Anxiety Inventory (GAI; Pachana et al. 2007) was developed and validated on an older cohort, designed to be sensitive to issues relevant to older adults, including the brevity of its items and an agree-disagree response format for ease of use in the context of poor education or mild cognitive impairment. It is suitable for administration by a range of health care professionals across numerous health care settings (e.g., inpatient, outpatient, community settings; Pachana et al. 2007). Numerous clinical trials have used the GAI (e.g., Ball et al. 2015).

The original 20-item GAI scale had a Cronbach's alpha of 0.91 in normal older adults and 0.93 in a psychogeriatric sample, with robust inter-rater and test-retest reliability; a score of 9 or above correctly classified 78% of patients (ROC analysis) with a sensitivity of 73% and a specificity of 80% (Pachana et al. 2007). A short five-item version of the scale (GAI-SF; Byrne and Pachana 2011) had similar psychometric properties, with a Cronbach's alpha of 0.91 in normal older adults and 0.93 in a psychogeriatric sample, with robust inter-rater and test-retest reliability; a score of 9 or above correctly classified 78% of patients (ROC analysis) with a sensitivity of 73% and a specificity of 80%. A bifactor analysis by Molde et al. (2017) found a single general factor for both the GAI and GAI-SF.

A range of studies globally have offered cross-cultural validation of the GAI, which has been translated into over two dozen languages (Pachana and Byrne 2012). For example, Ribeiro et al. (2011) found good concurrent validity and reliability in a Portuguese sample; similar support has been found for Spanish (Márquez-González et al. 2012), Brazilian (Massena et al. 2015), and French Canadian (Champagne et al. 2018) versions of the GAI. Support for the use of the GAI with specific populations has been forthcoming, such as nursing home residents (Gerolimatos et al. 2013) and patients with Parkinson's disease (Matheson et al. 2012).

Most recently, a large international study evaluated the "fitness" of individual items of the GAI based on the common variance for each item across pooled data from 10 countries (Australia, Brazil, Canada, The Netherlands, Norway, Portugal, Spain, Singapore, Thailand, and the USA) was undertaken (Molde et al. in press). The data support a strong general (G) factor and good comparability of GAI scores across countries. The GAI also compares favorably with the Geriatric Anxiety Scale (Gould et al. 2014).

Summary

The GAI has sound psychometric properties, is easy to administer and score, and has a growing number of translation and validation studies emerging. Increased awareness and assessment of anxiety symptoms in older adults by health professionals leads to increased chances of treatment.

Cross-References

▶ Hamilton Rating Scale for Depression

References

Ball SG, Lipsius S, Escobar R (2015) Validation of the geriatric anxiety inventory in a duloxetine clinical trial for elderly adults with generalized anxiety disorder. Int Psychogeriatr 27:1533–1539. https://doi.org/10.1017/S1041610215000381

Bower ES, Wetherell JL, Mon T et al (2015) Treating anxiety disorders in older adults. Harv Rev Psychiatry 23:329–342. https://doi.org/10.1097/HRP.0000000000000064

Bryant C, Mohlman J, Gum A et al (2013) Anxiety disorders in older adults: looking to DSM5 and beyond.... Am J Geriatr Psychiatry 21:872–876. https://doi.org/10.1016/j.jagp.2013.01.011

Byers AL, Yaffe K, Covinsky KE et al (2010) High occurrence of mood and anxiety disorders among older adults: the national comorbidity survey replication. Arch Gen Psychiatry 67:489–496. https://doi.org/10.1001/archgenpsychiatry.2010.35

Byrne GJA, Pachana NA (2011) Development and validation of a short form of the Geriatric Anxiety Inventory – the GAI-SF. Int Psychogeriatr 23:125–131. https://doi.org/10.1017/S1041610210001237

Champagne A, Landreville P, Gosselin P et al (2018)- Psychometric properties of the French Canadian version of the Geriatric Anxiety Inventory. Aging Ment Health 22:40–45. https://doi.org/10.1080/13607863.2016.1226767

Gerolimatos LA, Gregg JJ, Edelstein BA (2013) Assessment of anxiety in long-term care: examination of the geriatric anxiety inventory (GAI) and its short form. Int Psychogeriatr 25:1533–1542. https://doi.org/10.1017/S1041610213000847

Gonçalves DC, Byrne GJ (2012) Interventions for generalized anxiety disorder in older adults: systematic review and meta-analysis. J Anxiety Disord 26:1–11. https://doi.org/10.1016/j.janxdis.2011.08.010

Gould CE, Segal DL, Yochim BP et al (2014) Measuring anxiety in late life: a psychometric examination of the geriatric anxiety inventory and geriatric anxiety scale. J Anxiety Disord 28:804–811. https://doi.org/10.1016/j.janxdis.2014.08.001

Márquez-González M, Losada A, Fernández-Fernández V et al (2012) Psychometric properties of the Spanish version of the Geriatric Anxiety Inventory. Int Psychogeriatr 24:137–144. https://doi.org/10.1017/S1041610211001505

Massena PN, de Araújo NB, Pachana N et al (2015) Validation of the Brazilian Portuguese version of Geriatric Anxiety Inventory–GAI-BR. Int Psychogeriatr 27:1113–1119. https://doi.org/10.1017/S1041610214001021

Matheson SF, Byrne GJ, Dissanayaka NN et al (2012) Validity and reliability of the Geriatric Anxiety Inventory in Parkinson's disease. Australas J Ageing 31:13–16. https://doi.org/10.1111/j.1741-6612.2010.00487.x

Miloyan B, Byrne GJ, Pachana NA (2014) Late-life anxiety. In: Pachana NA Laidlaw K (ed) Oxford handbook of clinical geropsychology: international perspectives. Oxford University Press, New York, pp 470–489. https://doi.org/10.1093/oxfordhb/9780199663170.013.049

Molde H, Hynninen KM, Torsheim T et al (2017) A bifactor and item response analysis of the geriatric anxiety inventory. Int Psychogeriatr 29:1647–1656. https://doi.org/10.1017/S1041610217001004

Molde H, Nordhus IH, Torsheim T et al (in press) A cross-national analysis of the psychometric properties of the Geriatric Anxiety Inventory (GAI). J Gerontol B Psychol Sci Soc Sci. https://doi.org/10.1093/geronb/gbz002

Pachana NA, Byrne GJ (2012) The Geriatric Anxiety Inventory: international use and future directions. Aust Psychol 47:33–38. https://doi.org/10.1111/j.1742-9544.2011.00052.x

Pachana NA, Byrne GJ, Siddle H et al (2007) Development and validation of the Geriatric Anxiety Inventory. Int Psychogeriatr 19:103–114. https://doi.org/10.1017/S1041610206003504

Ribeiro O, Paúl C, Simões MR et al (2011) Portuguese version of the Geriatric Anxiety Inventory: transcultural adaptation and psychometric validation. Aging Ment Health 15:742–748. https://doi.org/10.1080/13607863.2011.562177

Tampi RR, Tampi DJ (2014) Anxiety disorders in late life: a comprehensive review. Healthy Aging Res 3:1–9. https://doi.org/10.12715/har.2014.3.14

Geriatric Anxiety Scale

Daniel L. Segal, Katie L. Granier, Marissa A. Pifer and Lisa E. Stone
Department of Psychology,
University of Colorado at Colorado Springs,
Colorado Springs, CO, USA

Definition

The Geriatric Anxiety Scale (GAS; Segal et al. 2010) is a 30-item self-report measure designed to assess, screen, and quantify severity of anxiety symptoms among older adults. The GAS is an increasingly popular and widely used screening measure for anxiety which has a wealth of psychometric data supporting its use in diverse community, medical, and clinical samples.

Overview

Development
Development of the GAS was spurred by the urgent need for brief, self-report screening measures of anxiety designed and validated specifically for use with older adult populations. A specific aspiration was to develop a measure that covers three common domains of anxiety symptoms among older adults (i.e., somatic, cognitive, and affective symptoms). Indeed, these three domains are hallmarks of anxiety that are commonly assessed during a thorough clinical evaluation of anxiety (Segal et al. 2018). Initially, items were derived from the full range of anxiety disorder symptoms in the *DSM-IV-TR*, most of which remained unchanged in *DSM-5*. Then, preliminary analyses identified items that were most heavily endorsed by older adults, items that were endorsed more by older adults than younger adults, and items that improved internal consistency of the measure. The final 30-item measure includes 25 scorable items and 5 content items (described below).

Scoring
During administration, respondents are asked to rate symptoms of anxiety by indicating how often they have experienced each symptom during the last week, answering on a 4-point Likert-type scale ranging from 0 (*not at all*) to 3 (*all of the time*). Scoring provides a total score and three subscale scores (somatic symptoms, cognitive symptoms, and affective symptoms) for which normative and interpretive guidelines are available. The GAS total score is based on the first 25 items. Possible total scores range from 0 to 75, with higher scores indicating the presence of more severe anxiety. The somatic subscale contains nine items, the cognitive subscale contains eight items, and the affective subscale contains eight items. Each scorable item loads on only one subscale. The additional 5 items on the GAS (items 26–30) are not scorable but rather assess areas of anxiety commonly reported to be of concern for older adults: concern about finances, concern about one's health, concern about children, fear of dying, and fear of becoming a burden to others. These content items are used clinically and thus do not load on any score.

Key Research Findings

Internal Consistency
As depicted in Table 1, the GAS demonstrates excellent internal consistency of scale scores across six samples of community-dwelling older adults ($\alpha = 0.90$–0.95). The GAS also demonstrated excellent internal consistency among older adults with one or more chronic medical condition ($\alpha = 0.94$), and older adults meeting criteria for one or more anxiety disorder or seeking outpatient mental health treatment ($\alpha = 0.91$–0.93), making it a well-rounded measure for use in different settings. Across the aforementioned samples, the total GAS averaged an excellent Cronbach's α of 0.92. The cognitive (mean $\alpha = 0.85$), somatic (mean $\alpha = 0.77$), and affective (mean $\alpha = 0.83$) subscales demonstrated good or acceptable average internal consistency across the samples. As depicted in Table 2, the cognitive (mean $r = 0.92$), somatic (mean $r = 0.89$), and affective (mean $r = 0.92$) subscales were also strongly correlated with the total GAS in two community and two clinical samples. Item response theory analyses performed by Mueller et al. (2015) on the entire GAS showed that each item reliably contributes to the assessment of anxiety, further supporting these findings.

Construct Validity
Older Adults. The GAS has demonstrated robust convergent validity through its strong correlations with other well-validated measures of anxiety and worry, including the State-Trait Anxiety Inventory, Beck Anxiety Inventory (BAI), Adult Manifest Anxiety Scale-Geriatric Version, Geriatric Anxiety Inventory (GAI), Hamilton Anxiety Scale, Anxiety Sensitivity Index-3, and Penn State Worry Questionnaire with medium to large effect sizes (see Table 3). Yochim et al. (2011) found that the GAS was most highly correlated with the Geriatric Depression Scale (GDS) as opposed to the BAI or GAI. However, the GAS also correlated more strongly with both the BAI

Geriatric Anxiety Scale, Table 1 Summary of internal consistency of the GAS in older adults across multiple studies

Study	Total GAS	Cognitive	Somatic	Affective
Segal et al. (2010)	0.93	0.90	0.80	0.82
Segal et al. (2010)[a]	0.93	0.85	0.80	0.82
Yochim et al. (2011)	0.90	–	–	–
Yochim et al. (2013)	0.91	0.74	0.74	0.83
Gould et al. (2014)	0.90	0.84	0.68	0.80
Mahoney et al. (2015)	0.95	0.90	0.87	0.89
Gould et al. (2018)	0.90	0.84	0.68	0.80
Gould et al. (2019)[b]	0.91	–	–	–
Segal and Mueller (2019)[c]	0.94	0.89	0.79	0.87
Mean alphas	0.92	0.85	0.77	0.83

Note: Dashes indicate unreported data. All data is reported as Cronbach's α. Gould et al. (2019) did not report individual alphas for each subscale. However, the subscales were reported to have acceptable to good reliability (0.75–0.88)
[a]Clinical sample of outpatients seeking psychological treatment
[b]Community sample who met criteria for one or more anxiety disorder
[c]Clinical sample of older adults with at least one chronic health condition

Geriatric Anxiety Scale, Table 2 Subscale-total correlations in older adult samples across three studies

Scale	Yochim et al. (2013)	Segal and Mueller (2019)	Segal et al. (2010)	Segal et al. (2010)	Mean r
Cognitive	0.90*	*0.94**	0.91*	*0.91**	0.92
Somatic	0.91*	*0.87**	0.86*	*0.91**	0.89
Affective	0.92*	*0.93**	0.92*	*0.91**	0.92

Note: *Italic* text indicates data from a clinical sample
*$p < 0.01$

and GAI than the two measures correlated with each other, indicating that the GAS may more comprehensively assess the symptoms of anxiety covered by both of these measures.

The GAS was closely related to measures of depression including the Beck Depression Inventory-II, GDS, and Patient Health Questionnaire-9 as well as a measure of distress management (Acceptance and Action Questionnaire-II), with large effect sizes. The lack of discriminant validity with measures of depression has been critiqued as a weakness of the GAS. However, due to depression and anxiety being highly related constructs, especially in later life, as well as the high rates of comorbidity and use of the *DSM* criteria within the GAS, this relationship is to be expected. The GAS subscales also differ in the strength of their relations to measures of depression, with the somatic subscale demonstrating lower correlation coefficients than the cognitive and affective subscales (see Table 3), indicating that specific domains may have greater discriminant value and can be used to better assess differential diagnoses.

The GAS also demonstrated good convergent validity with measures of health and medical burden including the 36-item Short Form Survey, Comorbidity Index, and general health ratings as expected with medium to large effect sizes. The somatic subscale of the GAS had higher correlation coefficients with these measures compared to the cognitive and affective scales, indicating good construct validity for the domains they were specifically designed to assess. The GAS also demonstrated strong convergent validity with measures of PTSD (i.e., Posttraumatic Stress Disorder Checklist-Civilian version), mindfulness (i.e., Mindful Attention Awareness Scale,

Geriatric Anxiety Scale, Table 3 Summary of convergent and discriminant validity across seven studies in older adults

Measure	Total GAS	Cognitive	Somatic	Affective
STAI-trait total	0.79**	0.81**	0.57**	0.75**
STAI-state total	0.74**	0.78**	0.50**	0.71**
STAI-Y1	0.77**	–	–	–
BAI	**0.60**–0.82****	0.53**–0.79**	0.61**–0.70**	0.45**–0.76**
AMAS	0.77**	0.74**	0.65**	0.69**
Worry	0.76**	0.75**	0.62**	0.67**
Physiological	0.65**	0.65**	0.54**	0.55**
Fear of aging	0.46**	0.40**	0.44**	0.39**
GAI	**0.60**–0.82****	0.74**–0.85**	0.43**–0.61**	0.65**–0.80**
HAM-A	**0.60****	–	–	–
ASI-3	0.37**	–	–	–
PSWQ	**0.57****	–	–	–
BDI-II	**0.59**–0.73****	**0.81****	**0.65****	**0.65****
GDS	*0.73**–0.78***	*0.67**–0.82***	*0.53**–0.68***	*0.72**–0.76***
PHQ-9	*0.84***	*0.80***	*0.67***	*0.83***
SF-36	*−0.68***	*0.60***	*−0.66***	*−0.62***
CMI	0.34**	0.28**	0.38**	0.22*
GAF	*−0.39***	*−0.38***	*−0.34***	*−0.37***
Health rating	**0.29****	–	–	–
PCL-C	**0.60****	–	–	–
WTAR	−0.36**	–	–	–
AAQ-II	0.70**	–	–	–
MAAS	−0.69**	–	–	–
KIMS	−0.54**	–	–	–
Education	*−0.01–0.20**	*−0.10*	*0.01*	*0.06*
Sex	0.06–0.15	–	–	–
Age	*−0.08–0.18*	–	–	–

Bolded text indicates use of Spearman correlations to account for positive skewness
Italic text indicates data from a clinical sample
STAI, State-Trait Anxiety Inventory; *STAI-Y1*, State-Trait Anxiety Inventory Form Y1; *BAI*, Beck Anxiety Inventory; *AMAS*, Adult Manifest Anxiety Scale-Geriatric Version; *GAI*, Geriatric Anxiety Inventory; *HAM-A*, Hamilton Anxiety Scale; *ASI-3*, Anxiety Sensitivity Index-3; *PSWQ*, Penn State Worry Questionnaire; *BDI-II*, Beck Depression Inventory-II; *GDS*, Geriatric Depression Scale; *PHQ-9*, Patient Health Questionnaire-9; *SF-36*, 36-item Short Form Survey; *CMI*, Comorbidity Index; *GAF*, Global Assessment of Functioning; *PCL-C*, Posttraumatic Stress Disorder Checklist-Civilian version; *WTAR*, Wechsler Test of Adult Reading. *AAQ-II*, Acceptance and Action Questionnaire-II; *MAAS*, Mindful Attention Awareness Scale; *KIMS*, Kentucky Inventory of Mindfulness Skills
*p < 0.05; **p < 0.01

Kentucky Inventory of Mindfulness Skills), and overall functioning (i.e., Global Assessment of Functioning). The GAS showed good discriminant validity with demographic variables (i.e., education, sex, age). The GAS and its subscales also had small to no association with cognitive measures such as the California Verbal Learning Test-II, trail making 20 Question Initial Abstraction, letter fluency, category fluency, Wechsler Adult Intelligence Scale-III Digit-Symbol Coding, Rey Auditory Verbal Learning Test, and visual reproduction, thus demonstrating good discriminant validity (Gould et al. 2014; Yochim et al. 2011, 2013).

Impact of Memory. Gould et al. (2014) observed minor shifts in the psychometric properties of the GAS in samples of different cognitive functioning. There was reduced but still good internal consistency on the GAS scale scores for average compared to superior performers on verbal and visual memory tasks. The convergent validity of the GAS was stable when assessing the impact of varying memory abilities on the measure. Some discrepancies in discriminant validity were found between average and superior memory groups. These findings suggest that slight changes in the GAS' reliability and validity may be observed as a result of respondents' cognitive abilities.

Other Properties

Mueller et al. (2015) found that the GAS is most reliable in distinguishing individuals at the higher end of the anxiety spectrum, as opposed to the very low end. However, these researchers emphasize that this is not problematic because clinicians are typically interested in elevated levels of anxiety. Additionally, their results suggested that each item on the GAS is a reliable indicator of anxiety. Somatic items provided the least amount of information compared to cognitive and affective, likely because of the physical ailments commonly endorsed by older adults. However, they were still informative, and their inclusion distinguishes the GAS from other measures of anxiety. Indeed, evidence suggests that the subscales have clinical and practical utility; however, they should be used with appropriate caution.

Mueller et al. (2015) also used item response theory to explore potential age (under 80 vs. over 80 years) and sex (male vs. female) bias on the GAS. They found that two items were biased, one for age and one for sex. However, they argue that the amount of bias was extremely minimal, and they recommend retaining the items on the measure. The minimal age and sex bias on the GAS indicate that it is useful among all older adults rather than just the young-old or women. The GAS is best used to distinguish individuals at the high end of the anxiety spectrum, which is logical given its development as a clinical instrument.

Based on an efficiency of 89%, Gould et al. (2014) identified a cutoff score of >16 with a sensitivity of 0.40 and specificity of 0.94 as optimal at the $p < 0.01$ level, indicating that the >16 cutoff score correctly classified participants 89% of the time in this study. However, a less stringent cutoff score of >9 was also identified with an efficiency of 73% and more balanced sensitivity (0.60) and specificity (0.75). These findings suggest that a narrower cutoff score of >16 is recommended to attain the highest percentage of correct classification. However, a broader cutoff score of >9 is recommended for a more balanced chance of correctly identifying both those with and without clinical levels of anxiety.

Examples of Application

Alternate Forms

There are two alternative versions available for use by researchers and clinicians when the full GAS may not be appropriate. These include the Geriatric Anxiety Scale-10 (GAS-10), a 10-item short form, and the Geriatric Anxiety Scale-Long Term Care (GAS-LTC), a 10-item version created for use within long-term care settings. The GAS-10 was created from the full GAS, as short forms are often preferred in busy clinical and research settings where time is limited. The GAS-10 was created using item response theory, which identified items on the GAS that provided the most information about a person's anxiety and that were best able to discriminate endorsement of the item based on actual anxiety, rather than a certain personality or character trait (Mueller et al. 2015). Items from the full GAS with the highest discrimination parameters and information curve peaks were selected from each subscale in order to ensure that the GAS-10, although shorter, still captured cognitive, affective, and somatic components of anxiety. Three items were selected from the somatic subscale, three were selected from the affective subscale, and four were selected from the cognitive subscale. Mueller et al. (2015) suggested that the GAS-10 best assesses anxiety for people with average anxiety, up to 2.5 standard deviations above the mean and that the reduced items of GAS-10 did not significantly reduce precision of anxiety detection. The GAS-10 had excellent internal consistency ($\alpha = 0.89$) and was significantly, positively correlated with the full GAS ($r = 0.96$, $p < 0.001$). Results of this study provide strong support for the use of the GAS-10, especially when time is limited or when individuals being assessed may fatigue easily and not be able to tolerate longer measures. Despite having fewer items, the GAS-10 still provides good amounts of symptom information and is still precise in detecting anxiety for older adults.

The GAS-LTC was created to address the gap in assessment measures available and validated for use in long-term care settings. The GAS-LTC was created by modifying the GAS-10 with simpler language and replacing the Likert-type response format with a more simple yes-no response format. Preliminary psychometric validation for the GAS-LTC by Pifer et al. (2019) was conducted with 66 older adult long-term care residents. Results revealed good internal consistency for the GAS-LTC ($\alpha = 0.80$). Item-total correlations ranged from moderate to strong positive correlations (range r: 0.35–0.65) between each item and the total GAS-LTC score, indicating that all items on the GAS-10 meaningfully contribute to the total score. Tests of convergent validity for the GAS-LTC revealed a strong positive correlation between the GAS-LTC and the GAI ($r = 0.70, p < 0.01$), indicating that both are measuring similar constructs but are not identical (Pifer et al. 2019). The GAS-LTC was also strongly positively correlated with a measure of depression, the Geriatric Depression Scale-15 (GDS-15; $r = 0.67, p < 0.01$). As noted above, depression has regularly been shown to be related to anxiety, with both disorders having significant symptom overlap and strong correlations between depression and anxiety measures (Cairney et al. 2008). Therefore, it is expected that the GAS-LTC would relate strongly to a measure of depression. Results from this study showed strong support for the use of the GAS-LTC in long-term care settings. However, this was the first study to date to use the GAS-LTC, and further studies examining the psychometric properties of the GAS-LTC are needed, with larger and more diverse samples.

Translated and International Versions

At present, the GAS has been formally translated into six languages: Persian (Bolghan-Abadi et al. 2013), German (Gottschling et al. 2016), Chinese (Xiao-Ling et al. 2017), Arabic (Hallit et al. 2017), Turkish (Karahan et al. 2018), and Italian (Gatti et al. 2018). Each of these versions underwent a rigorous translation process to ensure that the translated measure was culturally and semantically appropriate. Then, the psychometric properties of each translated measure were carefully evaluated, with these measures generally demonstrating strong evidence of reliability and validity, typically comparable to the original English version. Other informal translations of the GAS are also available in several additional languages (e.g., Spanish, Korean, Vietnamese, Japanese, and Croatian), but these measures have not yet been formally evaluated.

Research Applications

The GAS has been used as a measure of anxiety in numerous empirical studies with older adults. For example, Gould et al. (2019) used the GAS as a treatment outcome measure to assess effectiveness of a video-delivered relaxation intervention among older adults. The intervention was not designed to target one specific anxiety diagnosis, so the GAS was selected as the outcome measure because of its ability to assess a diverse array of anxiety symptoms. Results demonstrated that the GAS was sensitive to clinical change as an outcome measure. Specifically, the GAS was able to detect treatment changes in individuals with generalized anxiety disorder, social anxiety disorder, panic disorder, agoraphobia, and anxiety disorder unspecified. The GAS has also been utilized to examine relationships between late-life anxiety and other domains, including cognitive impairment (Yochim et al. 2013), sleeping difficulties (Gould et al. 2018), and loneliness (Khademi et al. 2015). Mahoney et al. (2015) examined age differences in experiential avoidance, anxiety sensitivity, and mindfulness, using the GAS as a measure of state anxiety symptoms. This study illustrates that the GAS can be used in cross-sectional age studies in order to adequately capture state anxiety among both younger and older samples. Overall, these studies suggest that the GAS appears to be an excellent measure of late-life anxiety and should be considered for use in research studies aiming to examine a wide variety of anxiety symptoms among older adults.

Future Directions of Research

New avenues for research on the GAS may examine alternative methods of administration for those unable to complete the self-report measure independently, for example, through computer-assisted administration, oral administration, or by gathering data from collateral reports. In addition, further studies should explore the psychometric properties of the GAS among the oldest-old populations (those 85 years old and older), since this is the fastest growing segment of older people. Finally, research is needed to explore the psychometric properties and clinical utility of the GAS in culturally and ethnically diverse older adult populations in the USA and internationally, with extensive opportunities for the further development of culturally and linguistically appropriate translated versions of the measure.

Summary

The GAS is an increasingly popular self-report measure of anxiety that has been successfully used in diverse community, psychiatric, and medical samples of older adults. The available psychometric data strongly supports its use in diverse clinical and research endeavors.

Cross-References

▶ Anxiety About Aging
▶ Geriatric Anxiety Inventory
▶ Geriatric Mental Health
▶ Psychopathology

References

Bolghan-Abadi M, Segal DL, Coolidge FL et al (2013) Persian version of the Geriatric Anxiety Scale: translation and preliminary psychometric properties among Iranian older adults. Aging Mental Health 17:896–900. https://doi.org/10.1080/13607863.2013.788999

Cairney J, Corna LM, Veldhuizen S et al (2008) Comorbid depression and anxiety in later life: patterns of association, subjective well-being, and impairment. Am J Geriatr Psychiatry 16:201–208. https://doi.org/10.1097/JGP.0b013e3181602a4a

Gatti A, Gottschling J, Brugnera A et al (2018) An investigation of the psychometric properties of the Geriatric Anxiety Scale (GAS) in an Italian sample of community-dwelling older adults. Aging Mental Health 22:1170–1178. https://doi.org/10.1080/13607863.2017.1347141

Gottschling J, Segal DL, Häusele C et al (2016) Assessment of anxiety in older adults: translation and psychometric evaluation of the German version of the Geriatric Anxiety Scale (GAS). J Psychopathol Behav Assess 38:136–148. https://doi.org/10.1007/s10862-015-9504-z

Gould C, Segal DL, Yochim BP et al (2014) Measuring anxiety in late life: a psychometric examination of the Geriatric Anxiety Inventory and Geriatric Anxiety Scale. J Anxiety Disord 28:804–811. https://doi.org/10.1016/j.janxdis.2014.08.001

Gould CE, Spira AP, Liou-Johnson V et al (2018) Associations of anxiety symptom clusters with sleep quality and daytime sleepiness. J Gerontol B Psychol Sci Soc Sci 73:413–420. https://doi.org/10.1093/geronb/gbx020

Gould CE, Kok BC, Ma VK et al (2019) Video-delivered relaxation intervention reduces late-life anxiety: a pilot randomized controlled trial. Am J Geriatr Psychiatry 27:514–525. https://doi.org/10.1016/j.jagp.2018.12.018

Hallit S, Hallit R, Hachem D et al (2017) Validation of the Arabic version of the Geriatric Anxiety Scale among Lebanese population of older adults. J Psychopathol 23:26–34

Karahan FS, Hamarta E, Karahan AY (2018) The Turkish adaptation and psychometric properties of the Geriatric Anxiety Scale. Ment Illn 10:1–5. https://doi.org/10.4081/mi.2018.7580

Khademi MJ, Rashedi V, Sajadi S et al (2015) Anxiety and loneliness in the Iranian older adults. Int J Psychol Behav Sci 5:49–52

Mahoney CT, Segal DL, Coolidge FL (2015) Anxiety sensitivity, experiential avoidance, and mindfulness among younger and older adults: age differences in risk factors for anxiety symptoms. Int J Aging Hum Dev 81:217–240. https://doi.org/10.1177/0091415015621309

Mueller AE, Segal DL, Gavett B et al (2015) Geriatric anxiety scale: item response theory analysis, differential item functioning, and creation of a ten-item short form (GAS-10). Int Psychogeriatr 27:1099–1111. https://doi.org/10.1017/S1041610214000210

Pifer MA, Segal DL, Noel OR et al (2019) Development and validation of the geriatric anxiety scale-long term care version. Poster presented at the 127th American Psychological Association Annual Convention, Chicago

Segal DL, Mueller AE (2019) Evidence of validity of the geriatric anxiety scale for use among medically ill older adults. J Depress Anxiety Forecast 2:1010

Segal DL, June A, Payne M et al (2010) Development and initial validation of a self-report assessment tool for anxiety among older adults: the geriatric anxiety scale. J Anxiety Disord 24:709–714. https://doi.org/10.1016/j.janxdis.2010.05.002

Segal DL, Qualls SH, Smyer MA (2018) Aging and mental health, 3rd edn. Wiley/Blackwell, Hoboken

Xiao-Ling L, Lu D, Gottschling J et al (2017) Validation of a Chinese version of the geriatric anxiety scale among community-dwelling older adults in mainland China. J Cross Cult Gerontol 32:57–70. https://doi.org/10.1007/s10823-016-9302-4

Yochim BP, Mueller AE, June A et al (2011) Psychometric properties of the geriatric anxiety scale: comparison to the Beck anxiety inventory and geriatric anxiety inventory. Clin Gerontol 34:21–33. https://doi.org/10.1080/07317115.2011.524600

Yochim BP, Mueller A, Segal DL (2013) Late life anxiety is associated with decreased memory and executive functioning in community dwelling older adults. J Anxiety Disord 27:567–575. https://doi.org/10.101 6/j.janxdis.2012.10.010

Geriatric Assessment for Older Adults with Cancer

Nikesha Gilmore[1], Sindhuja Kadambi[2], Allison Magnuson[2] and Supriya G. Mohile[2]
[1]Department of Surgery, Division of Cancer Control, University of Rochester Medical Center, Rochester, NY, USA
[2]Wilmot Cancer Institute, University of Rochester Medical Center, Rochester, NY, USA

Synonyms

Geriatric Assessment in Oncology

Definition

The comprehensive geriatric assessment has been used in geriatric medicine since the 1980s and is defined as "a multi-dimensional, interdisciplinary, diagnostic process to identify care needs, plan care, and improve outcomes of frail older people" (Rubenstein 2004; Solomon 1988; Puts and Alibhai 2018). In older adults with cancer, the comprehensive geriatric assessment is guided by a set of tools that are referred to as the geriatric assessment (GA). In oncology settings the GA is used to determine the general health status of older adults with cancer by assessing impairments in a variety of domains that are associated with negative outcomes, such as morbidity and mortality (Mohile et al. 2018; Loh et al. 2018).

Overview

Older adults, aged 65+, account for approximately 70% of patients with cancer (Smith et al. 2009; Schulkes et al. 2017). With advances in health care and the progressive increase in life expectancy, the number of older adults with cancer is expected to rapidly escalate in the next two decades (Smith et al. 2009; Hamaker et al. 2018a). Older adults will therefore comprise an increasing proportion of patients seen by oncologists. Thus, some have advocated that all oncologists receive formal training in geriatrics. Even in the absence of such training, evidence-based GA measures specific to the care of older adults with cancer are accessible and feasible for all oncology teams. The GA can aid in treatment decision-making and facilitate the development of geriatric-specific supportive care interventions for older patients with cancer.

The aging process is heterogeneous, and older adults of the same chronologic age may have very different physiologic ages. As individuals age, they may develop impairments, such as functional or cognitive declines, which ultimately manifest as frailty. However, these impairments are not always apparent during routine physical examinations. Performance status measures traditionally used in oncology (e.g., the Eastern Cooperative Oncology Group (ECOG), Performance Status (PS), and Karnofsky Performance Status (KPS)) fail to accurately predict which older adults with cancer are at greatest risk for treatment-related adverse events (Mohile et al. 2011).

Mounting evidence indicates that the GA can guide oncology teams through their shared decisions for the treatment of vulnerable older adults with cancer. The GA predicts adverse events related to cancer treatment and early mortality and detects age-related problems not commonly found during routine medical examinations (Mohile et al. 2015, 2018). Further, it guides interventions known to improve outcomes important for older patients such as emotional health, quality of life, pain control, and functional status (Mohile et al. 2015; Puts et al. 2014b). Unfortunately, the GA is not routinely used in oncology practice due to the perception of the GA being time- and resource-intensive. Furthermore, older adults are frequently underrepresented in oncology clinical trials, making it challenging for oncologists to apply evidence-based medicine while making treatment decision for older adults with cancer (Ludmir et al. 2019a, b). Recently, based on increasing and robust evidence demonstrating the value of GA, an American Society of Clinical Oncology (ASCO) guideline recommended that older adults be given the opportunity to participate in oncology clinical trials and that all adults with cancer aged 65+ should undergo a GA prior to initiating chemotherapy (Mohile et al. 2018). Modifications to the GA in oncology settings have also been suggested (Hamaker et al. 2017).

Key Research Findings

Domains and Measurement of Geriatric Assessments

The GA, in oncology practice, is largely self-administered and has been shown to be feasible in the clinical care of older adults with cancer as well as in oncology clinical trials. In an initial study of 45 older patients with cancer recruited from an academic center, Hurria et al. demonstrated that a majority of older patients with cancer were able to complete a primarily self-administered cancer-specific GA without assistance (Hurria et al. 2005). Feasibility was further evaluated in a larger population of older patients with cancer at an academic tertiary care center and a community-based satellite clinic (Hurria et al. 2007). In a cohort of 250 patients (mean age 76, range 65–95), most patients were able to complete the cancer-specific GA questionnaires without assistance (78%), in a mean time of 15 min (range 20 to 60 min), and most patients reported satisfaction with the length of the questionnaire (91%). Similarly, Williams et al. demonstrated the feasibility of the GA in community oncology clinics by administering the cancer-specific GA to 1088 patients including 339 from 7 community clinics (Williams et al. 2014). The median time to complete the GA was 23 min in the academic centers and 30 min in the community. Although most patients did not require assistance (76%), more patients in the community required assistance than patients in academic centers (24% vs 14%). Additional studies have also demonstrated the feasibility of electronic capture of the GA and found that it is preferred over paper and pencil (Hurria et al. 2016a; McCleary et al. 2013). The Cancer and Leukemia Group (CALGB) trial 360401 demonstrated the feasibility of the GA in older patients with cancer enrolled in cooperative group clinical trials (Hurria et al. 2011a).

The seven domains routinely captured by the GA are physical function including mobility, cognition, comorbidity, polypharmacy, psychological status, nutrition, and social support. *Functional Status* is a measure of a patient's ability to perform routine activities of daily living (ADLs) in order to maintain their health and well-being. Impaired functional status is common in older adults with cancer and is associated with adverse outcomes including decreased survival and increased risk of toxicity due to chemotherapy (Hurria et al. 2011b; Wildes et al. 2013). ADLs and Instrumental Activities of Daily Living (IADL) are patient-reported measures that give a comprehensive overview of a patient's ability to maintain skills necessary for basic living (Table 1). Physical function including mobility can also be objectively measured using the Timed Up and Go (TUG), Short Physical Performance Battery (SPPB; which includes the gait speed, chair stand, and balance tests), and fall history to assess patients' mobility and risk of falls (Table 1) (Hernandez Torres and Hsu 2017; Loh et al. 2018; Pamoukdjian et al. 2020; Soubeyran et al. 2012).

Aging is also associated with changes in *cognition*; 15–48% of older adults with cancer screen positive for cognitive impairments, depending on

Geriatric Assessment for Older Adults with Cancer, Table 1 Review of geriatric assessment domains and tools to assess impairments

Domain	Tools	Description	Management recommendations
Function	IADL (Lawton scale)[a]	Self-reported 8-item measure to assess the degree to which activities can be performed independently	Modification of cancer treatment regimen Improving function prior to treatment initiation Referral to physical therapy to prescribe strength and balance training and home exercise programs Referral to occupational therapy for assistive device and home safety evaluations Fall prevention discussion
	Fall history[a]	Self-reported measure of history of falls within the last 6 months	
	ADL (Katz index)	Self-reported measure of patients' ability to maintain skill necessary for basic living independently. Patients rank their ability to perform 6 functions bathing, dressing, toileting, transferring, continence, and feeding	
	TUG	An objective measure of performance and function that measures how long it takes to stand up for a chair and walk 3 meters, turn, walk back to chair, and sit back down	
	SPPB	An objective measure of physical performance that evaluates lower extremity function by combining the results from the gait speed, chair stand, and balance tests	
Cognition	Mini-Cog[a]	A test to screen for cognitive problems that includes a word recall and clock drawing test	Assess decision-making capacity and ability to consent for treatment Identify health-care proxy and involve in decision-making Delirium risk counseling for patient and caregivers Medication review to minimize medications with higher risk of delirium Referral to geriatrician or cognitive specialist
	BOMC	A 6-item measure that evaluates orientation, attention, and memory	
	MMSE	An objective 11-item measure that assesses mental status by testing 5 areas of cognitive function: orientation, registration, attention and calculation, recall, and language	
	MOCA	An objective measure to rapidly screen for mild cognitive dysfunction by assessing the following domains: attention and concentration, executive functions, memory, language, visuo-constructional skills, conceptual thinking, calculations, and orientation	
Comorbidities	Chart review[a]	Robust review of chronic medical conditions through routine history	Assess for medication adherence Involve primary care physician and/or geriatrician in decision-making for treatment and management of comorbidities Consider referral to geriatrician
	Physical Health (subscale of OARS)	Self-reported 14-item measure of pre-existing medical conditions rated on the degree to which the comorbidities impair daily activities	
	Charlson Comorbidity Index	A self-reported measure weighted on severity that categorizes patient's comorbidities into 19 different categories	
	CIRS-G	A self-reported 14-item measure that quantifies disease burden in older adults. Higher scores indicating higher severity of disease burden	

(continued)

Geriatric Assessment for Older Adults with Cancer, Table 1 (continued)

Domain	Tools	Description	Management recommendations
Polypharmacy	Beers Criteria	A list that identifies potentially inappropriate medications that should be avoided in older adults	Minimize medications as much as possible; consider involving a pharmacist
	STOPP Criteria	Identifies prescription medications that are potentially inappropriate for older adults	
	START Criteria	Treatments for older adults that should be considered in the absence of contraindications	
Psychological status	GDS[a]	Self-reported 15-item screening tool for depression in older adults	Consider referral to psychotherapy/psychiatry
	GAD-7	Self-reported 15-item measure used to screen for and determine the severity of generalized anxiety disorder	Consider cognitive-behavioral therapy
	HADS	A self-reported 14-item measure to assess symptoms associated with anxiety (7 items) and depression (7 items)	Referral to social work
	Distress Thermometer	A self-reported measure to screen for psychological distress in patients with cancer	Consider pharmacologic therapy
Nutrition	Unintentional weight loss[a]	Measures patients weight loss over the last 6 months that was not done intentionally	Referral to nutritionist/dietician for nutrition counseling
	Mini Nutritional Assessment	A 6-item screening measure to determine nutritional impairments and patients at risk of malnutrition	Evaluate for drug tolerance. Assess for support with meal preparation and recommend support interventions (caregiver, meals-on-wheels)
	Body mass index	A measure of body fat based on height and weight	
Social support	Medical Outcomes Survey Social Support Survey	A self-reported measure of 19-items that assesses 4 social domains: emotional support, tangible support, affectionate support, and medical outcomes	Consider modifying treatment. Referral to social work and home health. Assistance with transportation. Recommend caregiver support services. Home safety evaluation
	Medical Social Support Section (subscale of OARS)	Self-reported measure of the number of support individuals involved in the patient's medical care and the degree of involvement of the support individuals	

ADL Katz Index of Independence in Activities of Daily Living, *IADL* Instrumental Activities of Daily Living, *TUG* Timed Up and Go, *SPPB* Short Physical Performance Battery, *BOMC* Blessed Orientation Memory Concentration Test, *MMSE* Mini-Mental State Examination, *MOCA* Montreal Cognitive Assessment, *OARS* Older Americans Resources and Services, *CIRS-G* Cumulative Index Rating Scale for Geriatrics, *STOPP* Screening Tool of Older Persons' Prescriptions, *START* Screening Tool to Alert Doctors to Right Treatment, *GDS* Geriatric Depression Scale, *GAD-7* Generalized Anxiety Disorder 7, *HADS* Hospital Anxiety and Depression Scale, *OARS* Older Americans Resources and Services
[a]ASCO guideline recommendations of parts of the GA that should be performed on all oncology patients

the type of cancer and cognitive measure used (Corre et al. 2016; Hshieh et al. 2018; Klepin et al. 2011; Loh et al. 2017). Cognitive impairment in older adults with cancer is associated with decreased survival and may be a risk factor for increased chemotherapy toxicity (Magnuson et al. 2016b; Robb et al. 2010). Several screening tests, using an objective approach, can identify cognitive impairment in older adults with cancer and can be easily performed during routine clinical encounters. These include the Mini-Cog, Blessed Orientation Memory Concentration (BOMC) test, Mini-Mental State Examination (MMSE), and the Montreal Cognitive Assessment (MOCA) (Table 1) (Hernandez Torres and Hsu 2017; Loh et al. 2018).

Comorbidities: the presence of at least one significant medical problem in addition to cancer can complicate management of cancer treatment regimens. More than 50% of older adults with cancer have at least one other chronic health condition. The presence of comorbidities and their severity are associated with decreased survival, increased treatment-related toxicities, and increased hospitalizations (Williams et al. 2016; Ferrat et al. 2016). The ASCO guidelines for geriatric oncology recommend robust chart reviews for the presence of chronic medical conditions (Mohile et al. 2011). Other measures included in the GA for screening for comorbidities include the Physical Health subscale of the Older Americans Resources and Services (OARS) measure, Charlson Comorbidity Index and its updated version, and the Cumulative Index Rating Scale for Geriatrics (CIRS-G). These measures allow patients to report the presence of comorbidities and to rate their severity (Table 1) (Hernandez Torres and Hsu 2017; Loh et al. 2018).

Polypharmacy is defined as the use of multiple drugs or more medications than are medically necessary. Many studies define polypharmacy as a medication count of five or more medications. Polypharmacy is widespread among older adults. More than 15% of older adults report taking five or more prescription medications per week (Kantor et al. 2015). Polypharmacy decreases patients' compliance with medications, increases the risk of falls, and increases the risk of drug interactions with other prescriptions, herbal supplements, and chemotherapy agents (Dhalwani et al. 2017; Maggiore et al. 2010; Popa et al. 2014). Patient medications should be evaluated using Beers Criteria to identify potentially inappropriate medications. Screening tool of older people's prescriptions (STOPP) and screening tool to alert to right treatment (START) criteria are other tools that can be used to manage appropriate prescription of medications for older adults with cancer (Table 1) (Hernandez Torres and Hsu 2017; Loh et al. 2018).

Psychological Status, a measure of depression and anxiety, is common in older adults with cancer and associated with emotional distress. Studies have shown that older adults with cancer who screen positive for depression and anxiety have increased problems with making treatment decisions, worse treatment outcomes, and longer hospital stays (Weiss Wiesel et al. 2015). The Geriatric Depression Scale (GDS), Generalized Anxiety Disorder-7 (GAD-7), Hospital Anxiety and Depression Scale (HADS), and the Distress Thermometer (Table 1) are all tools that can be used to assess psychological status (Table 1) (Hernandez Torres and Hsu 2017; Loh et al. 2018).

Nutrition management is critical for the optimal care of older adults with cancer undergoing treatment. Poor nutritional status is highly prevalent in older adults with cancer (Ahmed and Haboubi 2010; Dewys et al. 1980; Argiles 2005). Malnutrition is associated with longer hospital stays, and unintentional weight loss is associated with mortality in older adults with cancer (Lis et al. 2012; Sorensen et al. 2008; Boulahssass et al. 2018; Caillet et al. 2017). Nutritional deficiencies can be assessed by identifying unintentional weight loss, calculating body mass index (BMI), and using the Mini Nutritional Assessment (MNA) (Table 1) (Hernandez Torres and Hsu 2017; Loh et al. 2018).

Social Support is an important factor for older adults with cancer and is essential for maintaining quality of life. Lack of social support predicts mortality and is associated with morbidity (Reblin and Uchino 2008; White et al. 2009). Screening for adequate social support can be accomplished using the Older Americans Resources and Services (OARS) Medical Social Support subscale or the Medical Outcomes Survey Social Support Survey to identify patients with limited or diminished social support systems (Table 1) (Hernandez Torres and Hsu 2017).

Predictive Power of Geriatric Assessments on Health Outcomes

Tools that incorporate elements of the GA have been shown to be better predictors of mortality risk, treatment toxicity, and functional decline than common oncology assessment tools such as the KPS and the ECOG PS. In a systematic review by Puts et al., impairments in the following GA variables – older age, inadequate finances, mental health, comorbidity, high medication use, low MNA score, and impairments in IADLs – were associated with increased mortality across multiple studies involving patients with various cancer types and stages (Puts et al. 2014a). Studies have shown that the GA can be used to stratify patients by frailty and that it predicts mortality (Palumbo et al. 2015; Koroukian et al. 2010; Clough-Gorr et al. 2012; Ferrat et al. 2017). For example, Palombo et al., using a scoring system based on age, comorbidity, and functional status, were able to characterize older adults with multiple myeloma as fit, vulnerable, or frail and demonstrate that vulnerable and frail patients had worse overall 3-year survival.

The GA has been shown to identify older adults with cancer at increased risk for experiencing chemotherapy toxicity and functional decline. Impairments in ADLs, IADLs, mental health, social support, cognitive function, and comorbidity have been associated with treatment-related adverse events (Puts et al. 2014a). These include development of chemotherapy toxicity, postoperative complications, hospitalizations, and early discontinuation of cancer treatment. Functional status and geriatric syndromes, such as hearing impairment, were the most significant predictors of toxicity risk. Gait speed, SPPB scores, and grip strength have also been shown to predict functional decline (Hoppe et al. 2013; Owusu et al. 2017).

The Cancer and Aging Research Group (CARG) toxicity tool and the Chemotherapy Risk Assessment Scale for High-Age Patients (CRASH) tool both incorporate GA variables to identify older adults with increased risk of chemotherapy toxicity. The CARG toxicity tool developed by Hurria et al. consists of 11 items including 5 GA variables: history of falls, assistance with daily medications, hearing impairment, limitations in walking 1 block, and limitations in social activities (Hurria et al. 2016b). The model was shown to discriminate chemotherapy toxicity in older adults with solid tumors better than the KPS. Similarly, the CRASH tool also incorporates GA variables including function, nutrition, and cognition to predict the risk of hematologic and non-hematologic toxicity in patients aged ≥ 70 years (Extermann et al. 2012).

Geriatric Assessments to Guide Decision-Making

Despite a high prevalence of cancer in older adults, this population is frequently underrepresented in oncology clinical trials (Mohile et al. 2018; Hamaker et al. 2018b; Ludmir et al. 2019a; Ludmir et al. 2019b). This underrepresentation creates a knowledge gap about the ideal treatment approach for older adults with cancer. Improved understanding of the potential risks and benefits of cancer treatment in older adults will enhance the decision-making process.

Cancer treatment decisions based on chronological age alone frequently result in overtreatment of vulnerable and frail patients (with treatments that have low likelihood of benefit or high likelihood of complications) and undertreatment of fit/robust patients (by withholding evidence-based treatment). The GA assists oncologists in tailoring decision-making by distinguishing vulnerable and frail patients from fit patients and by identifying non-oncologic vulnerabilities amenable to intervention prior to treatment. Multiple factors should be considered when making treatment decisions for older adults with cancer. These factors include (1) the safety and tolerability of treatments for patients with comorbidities, physical performance impairments, poor nutritional status, and limited social support, (2) the risk of falling for patients with functional impairments, and (3) the complexity of treatment for patients with cognitive impairments and extent of the caregiver's involvement (Mohile et al. 2018).

In a systematic review by Hamaker et al., results of the GA on patient outcomes from 35 studies involving older adults with various cancer types were examined. Of these studies, 11 evaluated oncologic treatment choice before and after the GA (Hamaker et al. 2018b). After the GA, the oncologic treatment plan was altered in 8–54% (median 28%) of patients, with the majority choosing less intensive treatment options: for example, different treatment modality or regimen, dose reduction, or best supportive care (no oncologic treatment). Two studies have demonstrated that the GA was able to identify older adults with cancer who were eligible for multi-regimen chemotherapy (Chaibi et al. 2011; Corre et al. 2016). In another study, although approximately half of the patients had an ECOG PS of 0 or 1, 78% of patients overall were found to have geriatric impairments, with 43% having impairments in three or more (out of eight) geriatric domains. In addition, the GA helped clarify patient preferences and expectations and assisted physicians with initiating advanced care planning in 69% of patients (Schulkes et al. 2017).

Similarly, the ELCAPA (ELderly CAncer PAtient) cohort, a prospective study of 375 older adults with various cancers, found that the GA resulted in adjustment of initial treatment plans in 20% of patients, primarily resulting in decreased treatment intensity (Caillet et al. 2011). Treatment changes were associated with an ECOG PS ≥ 2, dependency on one or more ADLs, malnutrition, cognitive impairment, depression, and greater number of comorbidities. Lower ADL scores and malnutrition were both independently associated with treatment changes.

The GA has also been shown to be helpful in surgical decision-making. It can be used preoperatively to determine fitness and identify risk factors for postsurgical complications and extended hospital stay (Audisio et al. 2008). The PACE (Preoperative Assessment in Elderly Cancer Patients) study demonstrated that the GA could be used to assess frailty in patients preoperatively and accurately predict postoperative mortality and morbidity.

Geriatric Assessments to Guide Management
"GA with management" involves tailoring medical decision-making as well as implementing interventions in response to vulnerabilities identified by the GA. The goal is to improve treatment tolerability and safety, allow vulnerable patients to receive standard of care treatments, optimize functional status, and improve survival and quality of life during treatment and in survivorship.

In the ELCAPA study, the GA led to changes in prescribed medications (31%), social support assistance (46%), physiotherapy (42%), nutritional care (70%), psychological care (36%), and memory evaluation (21%) (Corre et al. 2016). Similarly, a systematic review by Hamaker et al., evaluating the effect of the GA on treatment decision-making in older adults with cancer, found that in all but one study, interventions were recommended to >70% of patients, with social interventions and changes in medication the most frequent recommendations (Hamaker et al. 2014).

Although GA-guided management has been shown to decrease risk of death and nursing home placement and improve physical and mental function in older adults without cancer, there are limited studies demonstrating that GA with management and follow-up improves outcomes of older adults with cancer (Magnuson et al. 2016a). Kalsi et al. demonstrated that high-risk older adults with cancer (with one or more comorbidities and/or significant functional impairment and quality of life difficulties) who received geriatrician-led GA with management (average of 6.2 ± 2.6 interventions) were more likely to complete cancer treatment and required fewer treatment modifications than the observational cohort who received standard oncology care (Kalsi et al. 2015). McCorkle et al. demonstrated that implementing a 4-week specialized program of home visits and telephone calls delivered by advanced practice nurses to postsurgical older adults with cancer resulted in a survival benefit, particularly in those with advanced disease (McCorkle et al. 2000). Goodwin et al. demonstrated that older adults with breast cancer who received nurse case management interventions based on the GA were significantly more likely to return to normal functioning, particularly women with poor social support (Goodwin et al.

2003). Other studies have also demonstrated that GA with management improves nutritional status, pain management, and quality of life (Bourdel-Marchasson et al. 2014; Rao et al. 2005; Nipp et al. 2012).

Based on clinical experience and results of consensus studies, the Delphi consensus panels of experts have established an algorithm to implement GA-guided interventions. These recommendations include the following: (1) modifying cancer treatment regimen based on detected impairments in physical and cognitive function and comorbidities, (2) improving function prior to initiating treatment and considering physical therapy and occupational therapy referral for impaired physical function and history of falls, (3) engaging caregivers and minimizing medications with increased delirium risk for impaired cognition, (4) co-managing with the patient's primary care physician of comorbidities and social work and home health referrals for poor social support, (5) involving social work and recommending counseling for depression and/or anxiety, and (6) recommending a nutrition consult and oral care for poor nutrition (Mohile et al. 2018). This algorithm was evaluated in a large cluster randomized study in the United States which demonstrated that providing oncologists with a summary of GA and management recommendations increased conversations about aging and recommendations for management; the GA intervention also improved the satisfaction with communication and care for older patients with advanced cancer and their caregivers (Mohile et al. 2019).

Future Directions of Research

In order to continue providing the best care possible for older adults with cancer, more research is necessary. Given the fact that the majority of patients with cancer are aged 65+ resulting in the fact that this age group represents the majority of the recipients of cancer treatments, it is imperative that oncology treatment trials stop excluding participants based on age alone. Additional research examining the effect of GA-driven interventions on clinical outcomes, such as mortality and toxicities, in older adults with cancer are also needed. Furthermore, given the increasing evidence of the value of the GA for older adults with cancer, more dissemination and implementation studies should be conducted to determine the factors that influence the successful implementation of the GA into community and academic oncology clinics.

Summary

In summary, utilizing the GA in a multidisciplinary team approach can lead to improved quality of care of older adults with cancer. Up-front treatment selection is based not only on age and disease characteristics, but it is a shared decision based on aging-related conditions and patient preference. The GA allows for early intervention, leading to improved treatment tolerability and completion, improved physical and emotional health, and better quality of life. For these reasons, the GA is recommended for all patients with cancer aged ≥65 prior to initiating treatment.

Cross-References

▶ Aging and Cancer
▶ Aging and Cancer: Concepts and Prospects
▶ Benefits of Caregiving
▶ Cancer Diagnosis
▶ Cancer Screening
▶ Caregiving and Social Support: Similarities and Differences
▶ Cognition and Frailty
▶ Comprehensive Geriatric Assessment
▶ Falls
▶ Frailty and Social Vulnerability
▶ Frailty in Clinical Care
▶ Measurement of Frailty
▶ Mini-Mental State Examination (MMSE)
▶ Mobility and Frailty
▶ Polypharmacy and Frailty
▶ Social Support

References

Ahmed T, Haboubi N (2010) Assessment and management of nutrition in older people and its importance to health. Clin Interv Aging 5:207–216

Argiles JM (2005) Cancer-associated malnutrition. Eur J Oncol Nurs 9(Suppl 2):S39–S50. https://doi.org/10.1016/j.ejon.2005.09.006

Audisio RA, Pope D, Ramesh HS, Gennari R, van Leeuwen BL, West C, Corsini G, Maffezzini M, Hoekstra HJ, Mobarak D, Bozzetti F, Colledan M, Wildiers H, Stotter A, Capewell A, Marshall E (2008) Shall we operate? Preoperative assessment in elderly cancer patients (PACE) can help. A SIOG surgical task force prospective study. Crit Rev Oncol Hematol 65(2):156–163. https://doi.org/10.1016/j.critrevonc.2007.11.001

Boulahssass R, Gonfrier S, Ferrero JM, Sanchez M, Mari V, Moranne O, Rambaud C, Auben F, Levi JMH, Bereder JM, Bereder I, Baque P, Turpin JM, Frin AC, Ouvrier D, Borchiellini D, Largillier R, Sacco G, Delotte J, Arlaud C, Benchimol D, Durand M, Evesque L, Mahamat A, Poissonnet G, Mouroux J, Barriere J, Benizri E, Piche T, Guigay J, Francois E, Guerin O (2018) Predicting early death in older adults with cancer. Eur J Cancer 100:65–74. https://doi.org/10.1016/j.ejca.2018.04.013

Bourdel-Marchasson I, Blanc-Bisson C, Doussau A, Germain C, Blanc JF, Dauba J, Lahmar C, Terrebonne E, Lecaille C, Ceccaldi J, Cany L, Lavau-Denes S, Houede N, Chomy F, Durrieu J, Soubeyran P, Senesse P, Chene G, Fonck M (2014) Nutritional advice in older patients at risk of malnutrition during treatment for chemotherapy: a two-year randomized controlled trial. PLoS One 9(9):e108687. https://doi.org/10.1371/journal.pone.0108687

Caillet P, Canoui-Poitrine F, Vouriot J, Berle M, Reinald N, Krypciak S, Bastuji-Garin S, Culine S, Paillaud E (2011) Comprehensive geriatric assessment in the decision-making process in elderly patients with cancer: ELCAPA study. J Clin Oncol 29(27):3636–3642. https://doi.org/10.1200/jco.2010.31.0664

Caillet P, Liuu E, Simon AR, Bonnefoy M, Guerin O, Berrut G, Lesourd B, Jeandel C, Ferry M, Rolland Y, Paillaud E (2017) Association between cachexia, chemotherapy and outcomes in older cancer patients: a systematic review. Clin Nutr 36(6):1473–1482. https://doi.org/10.1016/j.clnu.2016.12.003

Chaibi P, Magne N, Breton S, Chebib A, Watson S, Duron JJ, Hannoun L, Lefranc JP, Piette F, Menegaux F, Spano JP (2011) Influence of geriatric consultation with comprehensive geriatric assessment on final therapeutic decision in elderly cancer patients. Crit Rev Oncol Hematol 79(3):302–307. https://doi.org/10.1016/j.critrevonc.2010.08.004

Clough-Gorr KM, Thwin SS, Stuck AE, Silliman RA (2012) Examining five- and ten-year survival in older women with breast cancer using cancer-specific geriatric assessment. Eur J Cancer 48(6):805–812. https://doi.org/10.1016/j.ejca.2011.06.016

Corre R, Greillier L, Le Caer H, Audigier-Valette C, Baize N, Berard H, Falchero L, Monnet I, Dansin E, Vergnenegre A, Marcq M, Decroisette C, Auliac JB, Bota S, Lamy R, Massuti B, Dujon C, Perol M, Daures JP, Descourt R, Lena H, Plassot C, Chouaid C (2016) Use of a comprehensive geriatric assessment for the management of elderly patients with advanced non-small-cell lung cancer: the phase III randomized ESOGIA-GFPC-GECP 08-02 study. J Clin Oncol 34(13):1476–1483. https://doi.org/10.1200/JCO.2015.63.5839

Dewys WD, Begg C, Lavin PT, Band PR, Bennett JM, Bertino JR, Cohen MH, Douglass HO Jr, Engstrom PF, Ezdinli EZ, Horton J, Johnson GJ, Moertel CG, Oken MM, Perlia C, Rosenbaum C, Silverstein MN, Skeel RT, Sponzo RW, Tormey DC (1980) Prognostic effect of weight loss prior to chemotherapy in cancer patients. Eastern Cooperative Oncology Group. Am J Med 69(4):491–497

Dhalwani NN, Fahami R, Sathanapally H, Seidu S, Davies MJ, Khunti K (2017) Association between polypharmacy and falls in older adults: a longitudinal study from England. BMJ Open 7(10): e016358. https://doi.org/10.1136/bmjopen-2017-016358

Extermann M, Boler I, Reich RR, Lyman GH, Brown RH, DeFelice J, Levine RM, Lubiner ET, Reyes P, Schreiber FJ 3rd, Balducci L (2012) Predicting the risk of chemotherapy toxicity in older patients: the Chemotherapy Risk Assessment Scale for High-Age Patients (CRASH) score. Cancer 118(13):3377–3386. https://doi.org/10.1002/cncr.26646

Ferrat E, Audureau E, Paillaud E, Liuu E, Tournigand C, Lagrange JL, Canoui-Poitrine F, Caillet P, Bastuji-Garin S, Group ES (2016) Four distinct health profiles in older patients with cancer: latent class analysis of the prospective ELCAPA cohort. J Gerontol A Biol Sci Med Sci 71(12):1653–1660. https://doi.org/10.1093/gerona/glw052

Ferrat E, Paillaud E, Caillet P, Laurent M, Tournigand C, Lagrange JL, Droz JP, Balducci L, Audureau E, Canoui-Poitrine F, Bastuji-Garin S (2017) Performance of four frailty classifications in older patients with cancer: prospective elderly cancer patients cohort study. J Clin Oncol 35(7):766. https://doi.org/10.1200/Jco.2016.69.3143

Goodwin JS, Satish S, Anderson ET, Nattinger AB, Freeman JL (2003) Effect of nurse case management on the treatment of older women with breast cancer. J Am Geriatr Soc 51(9):1252–1259

Hamaker ME, Schiphorst AH, ten Bokkel HD, Schaar C, van Munster BC (2014) The effect of a geriatric evaluation on treatment decisions for older cancer patients – a systematic review. Acta Oncol 53(3):289–296. https://doi.org/10.3109/0284186X.2013.840741

Hamaker ME, Wildes TM, Rostoft S (2017) Time to stop saying geriatric assessment is too time consuming.

J Clin Oncol 35(25):2871. https://doi.org/10.1200/Jco.2017.72.8170

Hamaker ME, Prins M, van Huis LH (2018a) Update in geriatrics: what geriatric oncology can learn from general geriatric research. J Geriatr Oncol 9(4):393–397. https://doi.org/10.1016/j.jgo.2018.01.005

Hamaker ME, Te Molder M, Thielen N, van Munster BC, Schiphorst AH, van Huis LH (2018b) The effect of a geriatric evaluation on treatment decisions and outcome for older cancer patients – a systematic review. J Geriatr Oncol 9(5):430–440. https://doi.org/10.1016/j.jgo.2018.03.014

Hernandez Torres C, Hsu T (2017) Comprehensive geriatric assessment in the older adult with cancer: a review. Eur Urol Focus 3(4–5):330–339. https://doi.org/10.1016/j.euf.2017.10.010

Hoppe S, Rainfray M, Fonck M, Hoppenreys L, Blanc JF, Ceccaldi J, Mertens C, Blanc-Bisson C, Imbert Y, Cany L, Vogt L, Dauba J, Houede N, Bellera CA, Floquet A, Fabry MN, Ravaud A, Chakiba C, Mathoulin-Pelissier S, Soubeyran P (2013) Functional decline in older patients with cancer receiving first-line chemotherapy. J Clin Oncol 31(31):3877–3882. https://doi.org/10.1200/JCO.2012.47.7430

Hshieh TT, Jung WF, Grande LJ, Chen J, Stone RM, Soiffer RJ, Driver JA, Abel GA (2018) Prevalence of cognitive impairment and association with survival among older patients with hematologic cancers. JAMA Oncol 4(5):686–693. https://doi.org/10.1001/jamaoncol.2017.5674

Hurria A, Gupta S, Zauderer M, Zuckerman EL, Cohen HJ, Muss H, Rodin M, Panageas KS, Holland JC, Saltz L, Kris MG, Noy A, Gomez J, Jakubowski A, Hudis C, Kornblith AB (2005) Developing a cancer-specific geriatric assessment: a feasibility study. Cancer 104(9):1998–2005. https://doi.org/10.1002/cncr.21422

Hurria A, Lichtman SM, Gardes J, Li D, Limaye S, Patil S, Zuckerman E, Tew W, Hamlin P, Abou-Alfa GK, Lachs M, Kelly E (2007) Identifying vulnerable older adults with cancer: integrating geriatric assessment into oncology practice. J Am Geriatr Soc 55(10):1604–1608. https://doi.org/10.1111/j.1532-5415.2007.01367.x

Hurria A, Cirrincione CT, Muss HB, Kornblith AB, Barry W, Artz AS, Schmieder L, Ansari R, Tew WP, Weckstein D, Kirshner J, Togawa K, Hansen K, Katheria V, Stone R, Galinsky I, Postiglione J, Cohen HJ (2011a) Implementing a geriatric assessment in cooperative group clinical cancer trials: CALGB 360401. J Clin Oncol 29(10):1290–1296. https://doi.org/10.1200/JCO.2010.30.6985

Hurria A, Togawa K, Mohile SG, Owusu C, Klepin HD, Gross CP, Lichtman SM, Gajra A, Bhatia S, Katheria V, Klapper S, Hansen K, Ramani R, Lachs M, Wong FL, Tew WP (2011b) Predicting chemotherapy toxicity in older adults with cancer: a prospective multicenter study. J Clin Oncol 29(25):3457–3465. https://doi.org/10.1200/JCO.2011.34.7625

Hurria A, Akiba C, Kim J, Mitani D, Loscalzo M, Katheria V, Koczywas M, Pal S, Chung V, Forman S, Nathwani N, Fakih M, Karanes C, Lim D, Popplewell L, Cohen H, Canin B, Cella D, Ferrell B, Goldstein L (2016a) Reliability, validity, and feasibility of a computer-based geriatric assessment for older adults with cancer. J Oncol Pract 12(12):e1025–e1034. https://doi.org/10.1200/JOP.2016.013136

Hurria A, Mohile S, Gajra A, Klepin H, Muss H, Chapman A, Feng T, Smith D, Sun CL, De Glas N, Cohen HJ, Katheria V, Doan C, Zavala L, Levi A, Akiba C, Tew WP (2016b) Validation of a prediction tool for chemotherapy toxicity in older adults with cancer. J Clin Oncol 34(20):2366–2371. https://doi.org/10.1200/JCO.2015.65.4327

Kalsi T, Babic-Illman G, Ross PJ, Maisey NR, Hughes S, Fields P, Martin FC, Wang Y, Harari D (2015) The impact of comprehensive geriatric assessment interventions on tolerance to chemotherapy in older people. Br J Cancer 112(9):1435–1444. https://doi.org/10.1038/bjc.2015.120

Kantor ED, Rehm CD, Haas JS, Chan AT, Giovannucci EL (2015) Trends in prescription drug use among adults in the United States from 1999–2012. JAMA 314(17):1818–1831. https://doi.org/10.1001/jama.2015.13766

Klepin HD, Geiger AM, Tooze JA, Kritchevsky SB, Williamson JD, Ellis LR, Levitan D, Pardee TS, Isom S, Powell BL (2011) The feasibility of inpatient geriatric assessment for older adults receiving induction chemotherapy for acute myelogenous leukemia. J Am Geriatr Soc 59(10):1837–1846. https://doi.org/10.1111/j.1532-5415.2011.03614.x

Koroukian SM, Xu F, Bakaki PM, Diaz-Insua M, Towe TP, Owusu C (2010) Comorbidities, functional limitations, and geriatric syndromes in relation to treatment and survival patterns among elders with colorectal cancer. J Gerontol A Biol Sci Med Sci 65(3):322–329. https://doi.org/10.1093/gerona/glp180

Lis CG, Gupta D, Lammersfeld CA, Markman M, Vashi PG (2012) Role of nutritional status in predicting quality of life outcomes in cancer – a systematic review of the epidemiological literature. Nutr J 11:27. https://doi.org/10.1186/1475-2891-11-27

Loh KP, Pandya C, Zittel J, Kadambi S, Flannery M, Reizine N, Magnuson A, Braganza G, Mustian K, Dale W, Duberstein P, Mohile SG (2017) Associations of sleep disturbance with physical function and cognition in older adults with cancer. Support Care Cancer 25(10):3161–3169. https://doi.org/10.1007/s00520-017-3724-6

Loh KP, Soto-Perez-de-Celis E, Hsu T, de Glas NA, Battisti NML, Baldini C, Rodrigues M, Lichtman SM, Wildiers H (2018) What every oncologist should know about geriatric assessment for older patients with cancer: young international society of geriatric oncology position paper. J Oncol Pract 14(2):85–94. https://doi.org/10.1200/JOP.2017.026435

Ludmir EB, Mainwaring W, Lin TA, Miller AB, Jethanandani A, Espinoza AF, Mandel JJ, Lin SH, Smith BD, Smith GL, VanderWalde NA, Minsky BD, Koong AC, Stinchcombe TE, Jagsi R, Gomez DR, Thomas CR Jr, Fuller CD (2019a) Factors associated with age disparities among cancer clinical trial participants. JAMA Oncol. https://doi.org/10.1001/jamaoncol.2019.2055

Ludmir EB, Subbiah IM, Mainwaring W, Miller AB, Lin TA, Jethanandani A, Espinoza AF, Mandel JJ, Fang P, Smith BD, Smith GL, Pinnix CC, Sedrak MS, Kimmick GG, Stinchcombe TE, Jagsi R, Thomas CR Jr, Fuller CD, VanderWalde NA (2019b) Decreasing incidence of upper age restriction enrollment criteria among cancer clinical trials. J Geriatr Oncol. https://doi.org/10.1016/j.jgo.2019.11.001

Maggiore RJ, Gross CP, Hurria A (2010) Polypharmacy in older adults with cancer. Oncologist 15(5):507–522. https://doi.org/10.1634/theoncologist.2009-0290

Magnuson A, Allore H, Cohen HJ, Mohile SG, Williams GR, Chapman A, Extermann M, Olin RL, Targia V, Mackenzie A, Holmes HM, Hurria A (2016a) Geriatric assessment with management in cancer care: current evidence and potential mechanisms for future research. J Geriatr Oncol 7(4):242–248. https://doi.org/10.1016/j.jgo.2016.02.007

Magnuson A, Mohile S, Janelsins M (2016b) Cognition and cognitive impairment in older adults with cancer. Curr Geriatr Rep 5(3):213–219. https://doi.org/10.1007/s13670-016-0182-9

McCleary NJ, Wigler D, Berry D, Sato K, Abrams T, Chan J, Enzinger P, Ng K, Wolpin B, Schrag D, Fuchs CS, Hurria A, Meyerhardt JA (2013) Feasibility of computer-based self-administered cancer-specific geriatric assessment in older patients with gastrointestinal malignancy. Oncologist 18(1):64–72. https://doi.org/10.1634/theoncologist.2012-0241

McCorkle R, Strumpf NE, Nuamah IF, Adler DC, Cooley ME, Jepson C, Lusk EJ, Torosian M (2000) A specialized home care intervention improves survival among older post-surgical cancer patients. J Am Geriatr Soc 48(12):1707–1713

Mohile SG, Fan L, Reeve E, Jean-Pierre P, Mustian K, Peppone L, Janelsins M, Morrow G, Hall W, Dale W (2011) Association of cancer with geriatric syndromes in older Medicare beneficiaries. J Clin Oncol 29(11):1458–1464. https://doi.org/10.1200/JCO.2010.31.6695

Mohile SG, Velarde C, Hurria A, Magnuson A, Lowenstein L, Pandya C, O'Donovan A, Gorawara-Bhat R, Dale W (2015) Geriatric assessment-guided care processes for older adults: a delphi consensus of geriatric oncology experts. J Natl Compr Cancer Netw 13(9):1120–1130

Mohile SG, Dale W, Somerfield MR, Schonberg MA, Boyd CM, Burhenn PS, Canin B, Cohen HJ, Holmes HM, Hopkins JO, Janelsins MC, Khorana AA, Klepin HD, Lichtman SM, Mustian KM, Tew WP, Hurria A (2018) Practical assessment and management of vulnerabilities in older patients receiving chemotherapy: ASCO guideline for geriatric oncology. J Clin Oncol 36(22):2326–2347. https://doi.org/10.1200/JCO.2018.78.8687

Mohile SG, Epstein RM, Hurria A, Heckler CE, Canin B, Culakova E, Duberstein P, Gilmore N, Xu H, Plumb S, Wells M, Lowenstein LM, Flannery MA, Janelsins M, Magnuson A, Loh KP, Kleckner AS, Mustian KM, Hopkins JO, Liu JJ, Geer J, Gorawara-Bhat R, Morrow GR, Dale W (2019) Communication with older patients with cancer using geriatric assessment: a cluster-randomized clinical trial from the National Cancer Institute Community Oncology Research Program. JAMA Oncol:1–9. https://doi.org/10.1001/jamaoncol.2019.4728

Nipp R, Sloane R, Rao AV, Schmader KE, Cohen HJ (2012) Role of pain medications, consultants, and other services in improved pain control of elderly adults with cancer in geriatric evaluation and management units. J Am Geriatr Soc 60(10):1912–1917. https://doi.org/10.1111/j.1532-5415.2012.04143.x

Owusu C, Margevicius S, Schluchter M, Koroukian SM, Berger NA (2017) Short physical performance battery, usual gait speed, grip strength and vulnerable elders survey each predict functional decline among older women with breast cancer. J Geriatr Oncol 8(5):356–362. https://doi.org/10.1016/j.jgo.2017.07.004

Palumbo A, Bringhen S, Mateos MV, Larocca A, Facon T, Kumar SK, Offidani M, McCarthy P, Evangelista A, Lonial S, Zweegman S, Musto P, Terpos E, Belch A, Hajek R, Ludwig H, Stewart AK, Moreau P, Anderson K, Einsele H, Durie BG, Dimopoulos MA, Landgren O, San Miguel JF, Richardson P, Sonneveld P, Rajkumar SV (2015) Geriatric assessment predicts survival and toxicities in elderly myeloma patients: an international myeloma working group report. Blood 125(13):2068–2074. https://doi.org/10.1182/blood-2014-12-615187

Pamoukdjian F, Aparicio T, Zebachi S, Zelek L, Paillaud E, Canoui-Poitrine F (2020) Comparison of mobility indices for predicting early death in older patients with cancer: the physical frailty in elderly cancer cohort study. J Gerontol A Biol Sci Med Sci 75(1):189–196. https://doi.org/10.1093/gerona/glz024

Popa MA, Wallace KJ, Brunello A, Extermann M, Balducci L (2014) Potential drug interactions and chemotoxicity in older patients with cancer receiving chemotherapy. J Geriatr Oncol 5(3):307–314. https://doi.org/10.1016/j.jgo.2014.04.002

Puts MTE, Alibhai SMH (2018) Fighting back against the dilution of the comprehensive geriatric assessment. J Geriatr Oncol 9(1):3–5. https://doi.org/10.1016/j.jgo.2017.08.009

Puts MT, Santos B, Hardt J, Monette J, Girre V, Atenafu EG, Springall E, Alibhai SM (2014a) An update on a systematic review of the use of geriatric assessment for older adults in oncology. Ann Oncol 25(2):307–315. https://doi.org/10.1093/annonc/mdt386

Puts MTE, Tu HA, Tourangeau A, Howell D, Fitch M, Springall E, Alibhai SMH (2014b) Factors influencing adherence to cancer treatment in older adults with cancer: a systematic review. Ann Oncol 25(3):564–577. https://doi.org/10.1093/annonc/mdt433

Rao AV, Hsieh F, Feussner JR, Cohen HJ (2005) Geriatric evaluation and management units in the care of the frail elderly cancer patient. J Gerontol A Biol Sci Med Sci 60(6):798–803

Reblin M, Uchino BN (2008) Social and emotional support and its implication for health. Curr Opin Psychiatry 21(2):201–205. https://doi.org/10.1097/YCO.0b013e3282f3ad89

Robb C, Boulware D, Overcash J, Extermann M (2010) Patterns of care and survival in cancer patients with cognitive impairment. Crit Rev Oncol Hematol 74(3):218–224. https://doi.org/10.1016/j.critrevonc.2009.07.002

Rubenstein LZ (2004) Joseph T. Freeman award lecture – comprehensive geriatric assessment: from miracle to reality. J Gerontol a-Biol 59(5):473–477

Schulkes KJG, Souwer ETD, Hamaker ME, Codrington H, van der Sar-van der Brugge S, Lammers J-WJ, Portielje JEA, van Elden LJR, van den Bos F (2017) The effect of a geriatric assessment on treatment decisions for patients with lung cancer. Lung 195(2):225–231. https://doi.org/10.1007/s00408-017-9983-7

Smith BD, Smith GL, Hurria A, Hortobagyi GN, Buchholz TA (2009) Future of cancer incidence in the United States: burdens upon an aging, changing nation. J Clin Oncol 27(17):2758–2765. https://doi.org/10.1200/JCO.2008.20.8983

Solomon DH (1988) Geriatric assessment – methods for clinical decision-making. Jama-J Am Med Assoc 259(16):2450–2452. https://doi.org/10.1001/jama.259.16.2450

Sorensen J, Kondrup J, Prokopowicz J, Schiesser M, Krahenbuhl L, Meier R, Liberda M, Osg E (2008) EuroOOPS: an international, multicentre study to implement nutritional risk screening and evaluate clinical outcome. Clin Nutr 27(3):340–349. https://doi.org/10.1016/j.clnu.2008.03.012

Soubeyran P, Fonck M, Blanc-Bisson C, Blanc JF, Ceccaldi J, Mertens C, Imbert Y, Cany L, Vogt L, Dauba J, Andriamampionona F, Houede N, Floquet A, Chomy F, Brouste V, Ravaud A, Bellera C, Rainfray M (2012) Predictors of early death risk in older patients treated with first-line chemotherapy for cancer. J Clin Oncol 30(15):1829–1834. https://doi.org/10.1200/JCO.2011.35.7442

Weiss Wiesel TR, Nelson CJ, Tew WP, Hardt M, Mohile SG, Owusu C, Klepin HD, Gross CP, Gajra A, Lichtman SM, Ramani R, Katheria V, Zavala L, Hurria A, Cancer Aging Research G (2015) The relationship between age, anxiety, and depression in older adults with cancer. Psychooncology 24(6):712–717. https://doi.org/10.1002/pon.3638

White AM, Philogene GS, Fine L, Sinha S (2009) Social support and self-reported health status of older adults in the United States. Am J Public Health 99(10):1872–1878. https://doi.org/10.2105/AJPH.2008.146894

Wildes TM, Ruwe AP, Fournier C, Gao F, Carson KR, Piccirillo JF, Tan B, Colditz GA (2013) Geriatric assessment is associated with completion of chemotherapy, toxicity, and survival in older adults with cancer. J Geriatr Oncol 4(3):227–234. https://doi.org/10.1016/j.jgo.2013.02.002

Williams GR, Deal AM, Jolly TA, Alston SM, Gordon BB, Dixon SA, Olajide OA, Chris Taylor W, Messino MJ, Muss HB (2014) Feasibility of geriatric assessment in community oncology clinics. J Geriatr Oncol 5(3):245–251. https://doi.org/10.1016/j.jgo.2014.03.001

Williams GR, Mackenzie A, Magnuson A, Olin R, Chapman A, Mohile S, Allore H, Somerfield MR, Targia V, Extermann M, Cohen HJ, Hurria A, Holmes H (2016) Comorbidity in older adults with cancer. J Geriatr Oncol 7(4):249–257. https://doi.org/10.1016/j.jgo.2015.12.002

Geriatric Assessment in Oncology

▶ Geriatric Assessment for Older Adults with Cancer

Geriatric Care Coordination

▶ Care Coordination

Geriatric Care Manager/Management

▶ Care Management

Geriatric Depression Scale

Lisa E. Stone, Katie L. Granier and
Daniel L. Segal
Department of Psychology, University of Colorado at Colorado Springs, Colorado Springs, CO, USA

Definition

The Geriatric Depression Scale (GDS; Yesavage et al. 1983) is a 30-item self-report measure designed to assess and screen depressive symptoms among older adults. It is the most popular and widely used screening measure for depression, and it has a wealth of psychometric data supporting its use.

Overview

Development

The GDS was developed in response to concerns of poor applicability of pre-existing depression screeners in assessing older adults (Yesavage et al. 1983). A common deficit of measures not designed specifically for use with older adults is a lack of item relevance to older adults, which skews scores and leads to potential misdiagnosis. As an example, many depression measures include somatic symptoms, which are often over-reported by older adults due to outside factors (e.g., medical conditions), resulting in inaccurate results and high rates of false-positive diagnoses. To address these problems, the GDS includes relatively few somatic items and contains items specifically tailored to the older adult population.

In addition to increased relevance of item content, the GDS is formatted to be straightforward and brief to avoid respondent fatigue and confusion, as well as to decrease administrator burden (Yesavage et al. 1983). Subsequently, respondents typically require less than 10 min to complete the measure. Item development of the GDS began with 100 preliminary items selected by clinicians and researchers with geriatric experience. The original 100 items were administered to samples of older adults with and without diagnoses of depression. The 30 items included in the final version of the GDS were determined by selecting the items with the highest correlation with the total score.

Scoring

The GDS includes a simple dichotomous, forced-choice format prompting respondents to select either *yes* or *no* for each item. Some respondents have difficulty with the yes/no options, instead opting to write the word "sometimes," in circling the space between the two choices, or failing to select a response. None of these are scored responses, and improvement can sometimes be seen with explicit instructions to respond with the answer that *best* reflects their response. Each item on the GDS is scored with one point for a depressive response. Items are summed to determine the total score with a maximum of 30 points. General interpretive guidelines are as follows: scores ranging from 0 to 9 indicate normal mood, scores of 10 to 19 indicate mild depressive symptoms, and scores of 20 to 30 indicate severe depressive symptoms. The GDS is free to use and available in the public domain. It can be accessed at www.stanford.edu/~yesavage/GDS.html.

Key Research Findings

Psychometric Properties

Reliability. Research has indicated that the GDS has excellent reliability of scale scores, with the original study reporting split-half and alpha coefficients of 0.94 (Yesavage et al. 1983). Combining results from 338 published studies on the GDS, a reliability generalization study found an α level of 0.85, indicating that the GDS is a highly reliable measurement across diverse populations (Kieffer and Reese 2002). Some studies have analyzed the GDS over time, finding test-retest correlations between 0.68 and 0.94 (Parmelee et al. 1989; Pedraza et al. 2009).

Validity. Criterion validity is the extent to which a measure relates to a known outcome. The GDS's sensitivity is its ability to correctly identify depressed individuals, whereas specificity refers to the measure's ability to correctly identify those who are not depressed. Among community-dwelling older adults, Yesavage et al. (1983) found an 84% sensitivity rate and a 95% specificity rate. In a review utilizing data from 42 published papers, Wancata et al. (2006) identified a sensitivity rate of 75% and a specificity rate of 77%, indicating that the criterion validity of the GDS is good across various samples.

Convergent validity of the GDS is the degree to which it correlates with other assessments of depression. In the original study (Yesavage et al. 1983), the GDS was validated against the Hamilton Rating Scale for Depression (HRS-D) and the Zung Self-Rating Depression Scale (SDS). High correlations were found among the measures, as expected, suggesting that the GDS adequately overlaps with previous measures of the construct. Researchers have also found significant correlations between the GDS and other well-validated depression questionnaires, such as the Center for Epidemiological Studies Depression Scale (CES-D; Brink and Niemeyer 1992) and the Beck Depression Inventory (BDI; Norris et al. 1987). Overlap between depression measures and the GDS are high, suggesting they are measuring similar constructs, but not perfectly correlated, indicating that the GDS assesses unique aspects of depression in late life. Additionally, the GDS fails to correlate with different constructs such as cognition (Feher et al. 1992) and pain (Parmelee et al. 1991), providing evidence of discriminant validity.

Alternate Forms

Brief Forms. Given cognitive and physical problems experienced by some older adults, the full 30-item version of the GDS has been shortened in a number of ways. Sheikh and Yesavage (1986) developed the GDS Short Form (GDS-SF) by taking the 15 items that had the highest correlations with depressive symptoms in previous studies. Evidence indicated that the two versions are highly correlated with one another ($r = 0.84$) and have acceptable reliability and validity. Alden et al. (1989) recommended the following cutoff points for the GDS-SF: 0–4 (*normal*), 5–9 (*mild*), and 10–15 (*moderate to severe*). Research has indicated that the GDS-SF is also valid for use among young and middle-aged adults, making it a useful research tool (Guerin et al. 2018).

Evidence for validity for the GDS-15 is mixed. Some studies indicate it is not an acceptable substitute for the full GDS in community-dwelling older adults (Alden et al. 1989; Pedraza et al. 2009) and cognitively impaired individuals (Burke et al. 1991). However, it may be as useful as the full GDS among cognitively intact and medical populations (Burke et al. 1991), older adults with affective disorders (Herrmann et al. 1996), and those in primary care settings (Bijl et al. 2005). Overall, Chiang et al. (2009) suggest that the GDS-15 is perhaps not a useful tool for screening depression but may be effective in detecting changes in moderate levels of depression.

Increasingly shorter versions of the GDS have been developed to create efficient and valid methods of screening depression in older adults. D'Ath et al. (1994) created three versions of the GDS, containing 10 items, 4 items, and 1 item, based on the GDS-15. They found that the 10-item version (GDS-10) was acceptable, but the 4- and 1-item versions had low criterion validity. Hoyl et al. (1999) created the GDS-5 by taking the 5 items on the GDS-15 that were most correlated with clinical depression diagnoses. They found the resulting measure to have a sensitivity of 97%, specificity of 95%, and a Cronbach's α of 0.80. Other researchers have found it to be a valid measure among diverse populations (Marquez et al. 2006).

However, Izal et al. (2010) argue that the GDS-10 and GDS-5 may not be valid because they were derived from the GDS-15, not the full GDS. They contend that research is too mixed to assume the GDS-15 contains the most important items from the full GDS and therefore should not be used as the starting point. Subsequently, they started with the GDS and identified 10 items as the most optimal solution to shorten the measure. They created the GDS-R, a 10-item assessment using a separate set of items that comprise the GDS-10, and evidence indicates that it has comparable diagnostic accuracy to the full GDS while increasing sensitivity and specificity relative to other shortened versions.

Nursing Home Forms. Sutcliffe et al. (2000) developed a 12-item version of the GDS for use in residential care facilities (GDS-12R), with presumably frail respondents. Residents of various long-term care facilities were administered the GDS-15, and researchers determined that three items lacked reliability among the residents because the items were irrelevant or ambiguous (e.g., "Do you prefer staying in, rather than going and doing new things?"). Removal of these three items improved internal reliability scores (Cronbach's α increased from 0.76 to 0.81).

Sutcliffe et al. (2000) recommend a cutoff value of 4/5 for the GDS-12R for research purposes, which maximized both sensitivity (79%) and specificity (67%), and a cutoff of 3/4 for clinical purposes, which maximized sensitivity (83%) over specificity (52%).

Jongenelis et al. (2007) developed another version of the GDS that removed additional items deemed inappropriate for frail nursing home populations. Three health professionals rated the appropriateness of each item on the GDS-15, and agreement was independently reached on seven items deemed to be inappropriate. Subsequently, these items were removed from the measure, creating the GDS-8. Using a cutoff point of 2/3, the GDS-8 had an equal sensitivity rate to the GDS-15 (96%), an improved specificity (72% for the GDS-8 vs. 63% for the GDS-15), and a high internal consistency (Cronbach's α of 0.80). Jongenelis et al. (2007) argue that the GDS-8 is more valid for frail and dependent older adults than either the GDS or GDS-15. Allgaier et al. (2011) further validated the GDS-8 as the most effective and least time-consuming alternative for nursing home residents.

Collateral Forms. Nitcher et al. (1993) created the Collateral Source Geriatric Depression Scale (CS-GDS), a 30-item version of the GDS that was designed to be completed by informants, by simply changing the pronoun "you" to "them." In the study, older adults took the GDS, informants took the CS-GDS, and mental healthcare professionals made clinical depression diagnoses. Nitcher et al. (1993) found that on 28 of 30 items, informants reported the presence of symptoms significantly more than the participants, but both versions were congruent with clinical diagnoses. Nitcher et al. (1993) recommend a cutoff point of 21 for the CS-GDS, higher than for the GDS, which showed sensitivity and specificity rates of 68%. Furthermore, the CS-GDS retains its reliability and validity when administered over the telephone, increasing its utility (Burke et al. 1997). Brown and Schinka (2005) created a 15-item informant version of the GDS (GDSI-15), which was as effective and valid as the full CS-GDS. Similarly, Chang et al. (2011) compared the validity of the 30-item, 15-item, and 5-item informant versions of the GDS and found that the shorter versions also have acceptable psychometric properties but that higher cutoff points should be used than for their self-report counterparts.

International Versions. The GDS-30 and its shortened versions have been translated and validated into 14 languages: Arabic, Chinese, Dutch, French, German, Hebrew, Iranian, Japanese, Korean, Romanian, Taiwanese, and Turkish. Two Spanish versions have been developed, with one designed for a Spanish population and another based on a Mexican-American population. Additionally, two Portuguese versions have been developed: one for a Brazilian population and another for a Portuguese population.

Examples of Application

Long-Term Care

In addition to being a reliable and valid tool for screening of depression in the overall older adult population, the GDS has also been validated for use in long-term care settings (Soon and Levine 2002). The GDS has a six-factor structure among long-term care residents: life dissatisfaction, dysphoria, hopelessness/decreased self-attitude, rumination/anxiety, social withdrawal and decreased motivation, and impaired cognition (Abraham et al. 1994). However, accuracy of the GDS and the CS-GDS in nursing homes depends on cognitive functioning, with higher functioning associated with higher congruence between GDS and CS-GDS scores and clinician interviews (Li et al. 2015).

Compared to other measures used in nursing home settings, the GDS has been shown to be a solid screening measure for depression. A modified form of the GDS provided more accurate predictions of psychologist-rated depression compared to a visual analogue scale, which required participants to mark their perceived level of depression on a line, especially when used with participants with higher cognitive functioning (Snowdon and

Lane 1999). Similarly, the GDS-15 was compared to the Minimum Data Set and was found to be a stronger detector of depression in nursing home residents, with increased sensitivity and specificity (when compared to a semistructured interview) and showing a higher percentage of positive screens for depression (Heiser 2004). Researchers also evaluated and compared the effectiveness of the GDS and the Montgomery Asberg Depression Rating Scale (MADRS) among nursing home residents (Smalbrugge et al. 2008). Using multiple forms of the GDS (30-, 15-, and 8-items), results indicated that the GDS' simpler administration was preferable for use with nursing home residents. However, the MADRS, due to higher specificity obtained with a larger item base, showed better detection of depression than each of the three GDS versions.

Jongenelis et al. (2005) also assessed the diagnostic accuracy of multiple version of the GDS with various item counts (30-, 15-, 12-, 10-, 5-, 4-, and 1-item) to assess the ideal length for accurate assessment in an effort to counter concerns regarding the ability of nursing home residents to properly complete self-report measures due to testing fatigue. Findings indicated that as the number of items increased, sensitivity and ability of the measure to detect varying levels of depression also increased. More specifically, it was suggested that scales with four or more items were able to detect major depression, but at least 10 items were necessary for the reliable detection of milder forms of depression.

Cognitive Impairment

Data from several studies have indicated that as cognitive functioning decreases, accuracy of the GDS decreases (Burke et al. 1991; Snowdon and Lane 1999; Børner et al. 2006). Such findings are consistent with literature on other self-report measures, indicating that the format of the self-report items may be responsible for these results as opposed to the nature of the GDS (Miller et al. 2013). Data examining the functioning of the GDS among older adults categorized by level of cognitive impairment (i.e., normal, mild, and impaired) showed minimal differential item functioning between groups, implying the GDS is a valid screener and suggesting that modified interpretation of total scores based on cognitive functioning is unnecessary (Chiesi et al. 2018a).

Though the majority of researchers (with the exception of Burke et al. 1989) suggest that use of the GDS with older adults with mild-to-moderate cognitive impairment can be effective, alternative options may be preferable. For example, clinicians may use the GDS as a supplementary measure with other screening strategies or use collateral sources (e.g., Cornell Scale for Depression in Dementia or CS-GDS) to shift responsibility of reporting to competent informants. In addition, some authors suggest adjusting the cutoff score when administering the GDS to individuals with cognitive impairments (Børner et al. 2006).

Primary Care

A common concern regarding the assessment of mental health in older adults is that many older adults are first seen in primary care settings where psychological screening and assessment is often minimal. As such, it is important to evaluate the utility of measures, like the GDS, in medical settings. Psychometric properties of the GDS have been assessed among older adult primary care patients, showing adequate reliability and validity (Friedman et al. 2005). Systematic review of nine studies evaluating use of the GDS in primary care settings indicated high sensitivity and adequate specificity (Watson and Pignone 2003). The cutoff for GDS scores was also evaluated in primary care settings, with a cutoff of five showing adequate sensitivity and specificity (Bijl et al. 2005). These results imply that use of the GDS in primary care settings can increase detection of depressive symptoms early on in the treatment process (Peach et al. 2001).

Oldest-Old

Among older adults, concern exists regarding the comparability of various stages of later life. A common distinction between age groups in late life is the division of older adults into three subgroups, with the "oldest-old" subgroup defined as adults aged 85 years old and older. Research has shown minimal (one item) differential item functioning between young-old, old-old, and oldest-old groups of older adults, implying minimal differences in the GDS' assessment across age groups, independent of additional factors (e.g., cognitive functioning; Chiesi et al. 2018b). Sensitivity and specificity of the GDS-15 in a sample of oldest-old adults were evaluated, indicating that a low cutoff point (2/3) had high sensitivity and adequate specificity (De Craen et al. 2003).

Compared to the General Health Questionnaire-12 (GHQ-12), the GDS was shown to have higher reliability and sensitivity when used to assess oldest-old Chinese adults (Boey and Chiu 1998). Another study examining the utility of the GDS among the oldest-old found that accuracy varied based on cognitive functioning, as measured by the Mini-Mental State Examination (MMSE). The GDS showed solid detection in oldest-old individuals with higher functioning but lacked detection among those with lower functioning (<10 on the MMSE; Conradsson et al. 2013). Another study with oldest-old participants showed the ability of the GDS to detect changes in depressive symptoms following negative life events, suggesting utility in evaluating longitudinal changes in depression in oldest-old adults (Vinkers et al. 2004).

Medical Populations

Research has also evaluated the utility of the GDS in specific medical populations. For example, a study assessing patients with end-stage renal disease showed potential for the GDS to be used as a predictor of mortality in terminally ill patients (Balogun et al. 2011). Another study has also shown potential for the GDS to be used as a supplementary tool in distinguishing between dementia with Lewy bodies (DLB) and Alzheimer's disease (AD). Results suggested that the GDS is more useful in determining distinctions between DLB and AD than clinician observations of depressive symptoms. These findings were attributed to the detection of specific symptom profiles commonly found in patients with DLB that present independently of other symptoms of the disorder (Yamane et al. 2011). The use of the GDS in assessing specific patient populations is still a relatively new area of research but has shown potential to improve outcome predictions and diagnostic specificity in addition to detection of depression.

Future Directions of Research

Future research on the GDS may examine alternative methods of administration for those unable to complete the self-report measure independently. In addition, the growing use of the GDS in medical contexts has the potential to be expanded upon with further research, introducing new clinical applications in the assessment of specific patient populations. Further research on the factor structure of the GDS may also be beneficial to future understanding and use of the measure, especially in culturally and ethnically diverse older adult populations.

Summary

Overall, the GDS has a tremendous amount of evidence in support of its reliability and validity as an assessment measure of depressive symptoms among diverse groups of older adults. Its translation into many languages and the development of population-specific variations have vastly increased its utility. Test performance depends on respondents' individual characteristics (e.g., cognitive functioning), which is common for many self-report measures used with older adults. Brief versions of the measure increase its potential use in medical and long-term care settings, which may increase the detection of depression in older adults. It is expected that the GDS and its variants will continue to be highly popular and effective tools for the screening and assessment of depressive symptoms in older adults, both for clinical and research purposes.

Cross-References

- ▶ Cognitive Behavioral Therapy (I)
- ▶ Depression and Antidepressants
- ▶ The Center for Epidemiologic Studies Depression (CES-D) Scale

References

Abraham IL, Wofford AB, Lichtenberg PA et al (1994) Factor structure of the Geriatric Depression Scale in a cohort of depressed nursing home residents. Int J Geriatr Psychiatry 9:611–617. https://doi.org/10.1002/gps.930090804

Alden D, Austin CN, Sturgeon R (1989) A correlation between the Geriatric Depression Scale Long and Short Forms. J Gerontol 44:124–125. https://doi.org/10.1093/geronj/44.4.P124

Allgaier AK, Kramer D, Mergl R et al (2011) Validity of the Geriatric Depression Scale in nursing home residents: comparison of the GDS-15, GDS-8, and GDS-4. Psychiatr Prax 38:280–286. https://doi.org/10.1055/s-0030-1266105

Balogun RA, Balogun SA, Kepple AL et al (2011) The 15-item Geriatric Depression Scale as a predictor of mortality in older adults undergoing hemodialysis. J Am Geriatr Soc 59:1563–1565. https://doi.org/10.1111/j.1532-5415.2011.03533.x

Bijl D, van Marwijk HWJ, Adèr HJ et al (2005) Test-characteristics of the GDS-15 in screening for major depression in elderly patients in general practice. Clin Gerontol 29:1–9. https://doi.org/10.1300/J018v29n01_01

Boey KW, Chiu HFK (1998) Assessing psychological well-being of the old-old: a comparative study of GDS-15 and GHQ-12. Clin Gerontol 19:65–75. https://doi.org/10.1300/J018v19n01_06

Brink TL, Niemeyer L (1992) Assessment of depression in college students: Geriatric Depression Scale versus Center for Epidemiological Studies Depression Scale. Psychol Rep 71:163–166. https://doi.org/10.2466/PR0.71.5.163-166

Brown LM, Schinka JA (2005) Development and initial validation of a 15-item informant version of the Geriatric Depression Scale. Int J Geriatr Psychiatry 20:911–918. https://doi.org/10.1002/gps.1375

Burke WJ, Houston MJ, Boust SJ et al (1989) Use of the Geriatric Depression Scale in dementia of the Alzheimer type. J Am Geriatr Soc 37:856–860. https://doi.org/10.1111/j.1532-5415.1989.tb02266.x

Burke WJ, Roccaforte WH, Wengel SP (1991) The short form of the Geriatric Depression Scale: a comparison with the 30-item form. J Geriatr Psychiatry Neurol 4:173–178. https://doi.org/10.1177/089198879100400310

Burke WJ, Rangwani S, Roccaforte WH et al (1997) The reliability and validity of the collateral source version of the Geriatric Depression Rating Scale administered by telephone. Int J Geriatr Psychiatry 12:288–294. https://doi.org/10.1037/t00930-000

Chang YP, Edwards DF, Lach HW (2011) The Collateral Source version of the Geriatric Depression Scale: evaluation of psychometric properties and discrepancy between collateral sources and patients with dementia in reporting depression. Int Psychogeriatr 23:961–968. https://doi.org/10.1017/S1041610211000147

Chiang KS, Green KE, Cox EO (2009) Rasch analysis of the Geriatric Depression Scale-Short Form. The Gerontologist 49:262–275. https://doi.org/10.1093/geront/gnp018

Chiesi F, Primi C, Pigliautile M et al (2018a) Does the 15-item Geriatric Depression Scale function differently in old people with different levels of cognitive functioning? J Affect Disord 227:471–476. https://doi.org/10.1016/j.jad.2017.11.045

Chiesi F, Primi C, Pigliautile M et al (2018b) Is the 15-item Geriatric Depression Scale a fair screening tool? A differential item functioning analysis across gender and age. Psychol Rep 121:1167–1182. https://doi.org/10.1177/0033294117745561

Conradsson M, Rosendahl E, Littbrand H et al (2013) Usefulness of the Geriatric Depression Scale 15-item version among very old people with and without cognitive impairment. Aging Ment Health 17:638–645. https://doi.org/10.1080/13607863.2012.758231

D'Ath P, Katona P, Mullan E et al (1994) Screening, detection, and management of depression in elderly primary care attenders: the acceptability and performance of the 15 item Geriatric Depression Scale (GDS-15) and the development of short versions. Fam Pract 11:260–266. https://doi.org/10.1093/fampra/11.3.260

de Craen AJM, Heeren TJ, Gussekloo J (2003) Accuracy of the 15-item Geriatric Depression Scale (GDS-15) in a community sample of the oldest old. Int J Geriatr Psychiatry 18:63–66. https://doi.org/10.1002/gps.773

Feher EP, Larrabee GJ, Crook TH (1992) Factors attenuating the validity of the Geriatric Depression Scale in a dementia population. J Am Geriatr Soc 40:906–909. https://doi.org/10.1111/j.1532-5415.1992.tb01988.x

Friedman B, Heisel MJ, Delavan RL (2005) Psychometric properties of the 15-item Geriatric Depression Scale in functionally impaired, cognitively intact, community-dwelling elderly primary care patients. J Am Geriatr Soc 53:1570–1576. https://doi.org/10.1111/j.1532-5415.2005.53461.x

Guerin JM, Copersino ML, Schretlen DJ (2018) Clinical utility of the 15-item geriatric depression scale (GDS-15) for use with young and middle-aged adults. J Affect Disord 241:59–62. https://doi.org/10.1016/j.jad.2018.07.038

Heiser D (2004) Depression identification in the long-term care setting: the GDS vs the MDS. Clin Gerontol 27:3–18. https://doi.org/10.1300/J018v27n04_02

Herrmann N, Mittmann N, Silver SL et al (1996) A validation study of the Geriatric Depression Scale short form. Int J Geriatr Psychiatry 11:457–460. https://doi.org/10.1002/(SICI)1099-1166(199605)11:5<457::AID-GPS325>3.0.CO;2-2

Hoyl M, Alessi CA, Harker JO et al (1999) Development and testing of a five-item version of the Geriatric Depression Scale. J Am Geriatr Soci 47:873–878. https://doi.org/10.1111/j.1532-5415.1999.tb03848.x

Izal M, Montorio I, Nuevo R et al (2010) Optimizing the diagnostic performance of the Geriatric Depression

Scale. Psychiatry Res 178:142–146. https://doi.org/10.1016/j.psychres.2009.02.018

Jongenelis K, Pot AM, Eisses AMH et al (2005) Diagnostic accuracy of the original 30-item and shortened versions of the Geriatric Depression Scale in nursing home patients. Int J Geriatr Psychiatry 20:1067–1074. https://doi.org/10.1002/gps.1398

Jongenelis K, Gerritsen DL, Pot AM et al (2007) Construction and validation of a patient- and user-friendly nursing home version of the Geriatric Depression Scale. Int J Geriatr Psychiatry 22:837–842. https://doi.org/10.1002/gps.1748

Kieffer KM, Reese RJ (2002) A reliability generalization of the Geriatric Depression Scale. Educ Psychol Meas 62:969–994. https://doi.org/10.1177/0013164402238085

Kørner A, Lauritzen L, Abelskov K et al (2006) The Geriatric Depression Scale and the Cornell Scale for Depression in Dementia A validity study. Nord J Psychiatry 60:360–364. https://doi.org/10.1080/08039480600937066

Li Z, Jeon YH, Low LF et al (2015) Validity of the Geriatric Depression Scale and the collateral source version of the Geriatric Depression Scale in nursing homes. Int Psychogeriatr 27:1495–1504. https://doi.org/10.1017/S1041610215000721

Marquez DX, McAuley E, Motl RW et al (2006) Validation of Geriatric Depression Scale-5 scores among sedentary older adults. Educ Psychol Meas 66:667–675. https://doi.org/10.1177/0013164405282464

Miller LS, Brown CL, Mitchell MB et al (2013) Activities of daily living are associated with older adult cognitive status: caregiver versus self-reports. J Appl Gerontol 32:3–30. https://doi.org/10.1177/0733464811405495

Nitcher RL, Burke WJ, Roccaforte WH et al (1993) A collateral source version of the Geriatric Depression Rating Scale. Am J Geriatr Psychiatr 1:143–152. https://doi.org/10.1037/t00930-000

Norris JT, Gallagher DE, Wilson A et al (1987) Assessment of depression in geriatric medical outpatients: the validity of two screening measures. J Am Geriatr Soc 35:989–995. https://doi.org/10.1111/j.1532-5415.1987.tb04001.x

Parmelee PA, Lawton MP, Katz IR (1989) Psychometric properties of the Geriatric Depression Scale among the institutionalized aged. Psychol Assess 1:331–338. https://doi.org/10.1037/1040-3590.1.4.331

Parmelee PA, Katz IR, Lawton MP (1991) The relation of pain to depression among institutionalized age. J Gerontol 46:15–21. https://doi.org/10.1093/geronj/46.1.P15

Peach JU, Koob JJ, Kraus MJ (2001) Psychometric evaluation of the Geriatric Depression Scale (GDS): supporting its use in health care settings. Clin Gerontol 23:57–68. https://doi.org/10.1300/J018v23n03_06

Pedraza O, Dotson VM, Willis FB et al (2009) Internal consistency and test-retest stability of the Geriatric Depression Scale-Short Form in African American older adults. J Psychopathol Behav Assess 31:412–416. https://doi.org/10.1007/s10862-008-9123-z

Sheikh JI, Yesavage JA (1986) Geriatric Depression Scale: recent evidence and development of a shorter version. Clin Gerontol 5:165–173. https://doi.org/10.1300/J018v05n01_09

Smalbrugge M, Jongenelis L, Pot AM et al (2008) Screening for depression and assessing change in severity of depression. Is the Geriatric Depression Scale (30-, 15-, and 8-item versions) useful for both purposes in nursing home patients? Aging Ment Health 12:244–248. https://doi.org/10.1080/13607860801987238

Snowdon J, Lane F (1999) Use of the Geriatric Depression Scale by nurses. Aging Ment Health 3:227–233. https://doi.org/10.1080/13607869956181

Soon JA, Levine M (2002) Screening for depression in patients in long-term care facilities: a randomized controlled trial of physician response. J Am Geriatr Soc 50:1092–1099. https://doi.org/10.1046/j.1532-5415.2002.50266.x

Sutcliffe C, Cordingley L, Burns A et al (2000) A new version of the Geriatric Depression Scale for nursing and home populations: the Geriatric Depression Scale (residential) (GDS-12R). Int Psychogeriatr 12:173–181. https://doi.org/10.1037/t00930-000

Vinkers DJ, Gussekloo J, Stek ML et al (2004) The 15-item Geriatric Depression Scale (GDS-15) detects changes in depressive symptoms after a major negative life event. The Leiden 85-plus Study. Int J Geriatr Psychiatry 19:80–84. https://doi.org/10.1002/gps.1043

Wancata J, Alexandrowic R, Marquart B et al (2006) The criterion validity of the Geriatric Depression Scale: a systematic review. Acta Psychiatr Scand 114:398–410. https://doi.org/10.1111/j.1600-0447.2006.00888.x

Watson LC, Pignone MP (2003) Screening accuracy for late-life depression in primary care: a systematic review. J Fam Pract 52:956–964

Yamane Y, Sakai K, Maeda K (2011) Dementia with Lewy bodies is associated with higher scores on the Geriatric Depression Scale than is Alzheimer's disease. Psychogeriatrics 11:157–165. https://doi.org/10.1111/j.1479-8301.2011.00368.x

Yesavage JA, Brink TL, Rose TL et al (1983) Development and validation of a Geriatric Depression Screening Scale: a preliminary report. J Psychiatr Res 17:37–49. https://doi.org/10.1016/0022-3956(82)90033-4

Geriatric Diseases

▶ Aging Pathology

Geriatric Emergency Medicine

▶ Emergency Care

Geriatric Interventions

Oncology and Treatment

Kenis Cindy[1] and Puts Martine[2]
[1]Department of General Medical Oncology and Geriatric Medicine, University Hospitals Leuven, Leuven, Belgium
[2]Canada Research Chair in the Care for Frail Older persons, Lawrence S. Bloomberg Faculty of Nursing, University of Toronto, Toronto, Canada

Synonyms

Comprehensive geriatrics assessment; Geriatric management; Geriatric screening

Definition

Using a screening tool to screen for geriatric issues such as issues with activities of daily living or falls. Geriatric interventions could also be intervention to address any issues identified in the geriatric assessment.

Overview

The world population is aging. Cancer is a disease that predominantly affects older persons, and due to the aging of the world population, there will be a significant increase in the older persons diagnosed with cancer. Older persons experience age-related decline in many body functions, such as kidney function, lung function, cardiac function, etc. In addition, many older persons have additional chronic conditions such as diabetes, cardiovascular diseases, etc. that impact the remaining life expectancy. With older age, the risk of treatment complications may increase and the treatment efficacy may change.

Unfortunately, the older population has been dramatically underrepresented in clinical trials resulting in a lack of evidence on how to best treat them and clinicians have to make treatment decisions based on evidence obtained in younger and more fit older persons (Al-Refaie and Vickers 2012; Al-Refaie et al. 2011; 2013; Hori et al. 2007; Scher and Hurria 2012; Townsley et al. 2006). Research has previously shown that cancer specialists can find it challenging to select the best cancer treatment when older persons have other diseases that also affect life expectancy, quality of life (QoL), and treatment tolerability (Puts et al. 2010; Wan-Chow-Wah et al. 2011). There is ample evidence showing over- and undertreatment of older persons, which can have detrimental effects on the health of an older person as well as leading to suboptimal use of scarce health care resources (Bouchardy et al. 2007; Daskivich et al. 2011; Enewold et al. 2015; Meresse et al. 2017; Meyers and Samson 2014; Monroe et al. 2013; Ng et al. 2005; Noon et al. 2013; Shumway and Hamstra 2015; Van Leeuwen et al. 2011; Weiss et al. 2013).

In order to enhance the quality of the cancer treatment decision-making process, the American Society of Clinical Oncology (ASCO) (Mohile et al. 2018a) and the International Society of Geriatric Oncology (SIOG) (Wildiers et al. 2014) have both recommended that clinicians conduct a comprehensive geriatric assessment (CGA) to help select appropriate treatment(s) and identify health and functional status issues that may affect cancer treatment.

What Is a Comprehensive Geriatric Assessment (CGA)?

CGA has been defined by Rubenstein in 2004 as: *"A multidimensional, interdisciplinary, diagnostic process to identify care needs, plan care, and improve outcomes of frail older people"* (Rubenstein and Joseph 2004). In 1996, Reuben et al. described six components of the CGA process: Step 1: Data gathering (i.e., conducting the CGA); Step 2: Discussion among team members of the multidisciplinary team; Step 3: Development of a treatment plan to address the issues identified in the CGA; Step 4: Implementation of the treatment plan by the team members; Step 5: Monitoring response to the treatment plan; and Step 6: Revising the treatment plan as necessary (Reuben et al. 1996). The ultimate goal of the CGA process is the development of a coordinated and integrated plan for treatment and long-term follow-up.

CGA is the cornerstone of geriatric medicine and the standard of care for older persons, and evidence shows that its implementation leads to improved outcomes such as functional status, improved overall survival, and fewer nursing home admissions ((Deschodt et al. 2011; Ellis et al. 2011) see "Comprehensive Geriatric Assessment"). A CGA has four core domains: physical health (this includes other health conditions and a medication review), functional status (including the basic activities of daily living (ADLs)(Katz et al. 1963) and the instrumental activities of daily living (IADLs)(Lawton and Brody 1969), psychosocial health (including mood (depression/anxiety and cognitive function), and the socioenvironmental domain (including the social support network of older persons) (Rubenstein and Joseph 2004; Reuben et al. 1996; National Institutes of Health 1988). Using validated geriatric scales, this assessment identifies older persons with cancer who are at risk.

In the oncology setting, however, due to the greater volume of older persons seen in daily oncology practice compared to traditional geriatric medicine settings, an additional step of geriatric screening has been implemented (see Table 1). In this step, first a geriatric screening tool is utilized to identify patients who can benefit from a comprehensive geriatric assessment (Decoster et al. 2015).

Geriatric Interventions, Table 1 Components of the comprehensive geriatric assessment process in oncology

1.	Data gathering: Performance of a geriatric screening to identify who can have benefit from the performance of a CGA
2.	Data gathering: Conducting the comprehensive geriatric assessment
3.	Discussion among team members of the multidisciplinary team
4.	Development of a treatment plan to address the issues identified in the CGA
5.	Implementation of the treatment plan by the team members
6.	Monitoring response to the treatment plan
7.	Revising the treatment plan as necessary

Abbreviations: *CGA* comprehensive geriatric assessment
^aAdapted from Reuben et al. (1994)

What Do We Know About the Benefits of CGA in the Care for Older Persons with Cancer?

Multiple reasons are identified to state why the performance of a CGA is recommended. The first reason is the identification of previously unknown geriatric problems that may be missed in the treatment decision-making process in daily oncology practice. Second, to identify the optimal cancer treatment, treating physicians face the difficulty of the heterogeneity of this specific patient population. Furthermore, as other comorbid conditions may limit life expectancy and the ability to tolerate treatment, treatment decisions in older patients with cancer can be challenging. To assist the shared-decision-making process, the predictive capacity of CGA for overall survival and postoperative complications/treatment toxicity can be helpful. More specific, the CGA results can be used by the oncology team to predict who is at high risk of treatment toxicity, enabling the multidisciplinary team to modify treatment and/or to provide additional supportive care (Hurria et al. 2011; 2016). Previous systematic reviews of CGA have demonstrated that the conduct of a CGA can change treatment decisions in 30–40% of patients and thus it helps to reduce over- and undertreatment of older persons with cancer. Therefore, the implementation of a CGA allows for personalization of cancer treatment and provision of the right care to the right patient (Hamaker et al. 2018; Kenis et al. 2013; Puts et al. 2014; Schulkes et al. 2016). Third, CGA is important for individualized care planning by identifying areas to set up targeted geriatric interventions. These interventions can include referrals to other health care facilities (e.g., geriatric day clinic, fall clinic, memory clinic) or the involvement of additional health care professionals (e.g., social worker, occupational therapist, physical therapist, dietician, and psychologist). These interventions can help patients to tolerate and complete the cancer treatment and to maintain functional status, quality of life, and overall survival. In addition, a recent study by Mohile et al. (2018) shows that the implementation of CGA improves communication between treating physicians and their older patients with cancer within the shared decision-making process (Mohile et al. 2018b). An improvement of patients' satisfaction and an increase of the number and quality of discussions about age-related concerns was identified. These reasons show that the implementation of CGA delivers meaningful information for treating physicians and other health care professionals in daily oncology practice. Instead of time-consuming, the integration of the CGA process in the care for older patients with cancer is time well spent.

Key Research Findings

There are many studies in geriatric oncology that have studied the implementation of geriatric screening and CGA in the care for older patients with cancer. One of the largest studies to date was carried out in Belgium (period 2009–2011; n = 1967) and described the relevance of the implementation of geriatric screening and CGA in the care for older patients with cancer focusing on the detection of previously unknown geriatric problems, the impact on geriatric interventions, and the influence on cancer treatment decisions (Kenis et al. 2013). Based on the result of the geriatric screening, 71% of older patients with cancer were in need of the performance of a CGA. The CGA was able to reveal unknown geriatric problems in more than half of the patients and led to subsequent geriatric interventions in a quarter of patients when left at the discretion of the treating physician. The detected problems and interventions occurred in all previously mention core domains confirming their overall importance. In this study, treating physicians stated that geriatric screening and CGA influenced treatment decisions also in a quarter of their patients with the remark that the information of the geriatric screening and CGA didn't always reach the treating physician in time to integrate it in the treatment decision-making process.

In order to optimize the CGA process in the care for older patients with cancer, more specific the comprehensive character, it should be acknowledged that the main benefit relies on geriatric recommendations and interventions based on the CGA results. Discussion within the multidisciplinary team is the main practice to establish these recommendations. Following on the previous observational study, a second part was performed and focused on geriatric recommendations based on the CGA-results and on the interventions that were carried out (Baitar et al. 2015). This Belgian study (period 2011–2012; n = 1550) showed that a median of two geriatric recommendations (range: 1–6) was given, that one-third (35.3%) of all the geriatric recommendations were applied and that geriatric recommendations most frequently consisted of referrals to the dietician (60.4%), social worker (40.3%), and psychologist (28.9%). The results of this study already gave a first insight into geriatric recommendations and interventions which is crucial to optimize the effectiveness of implementing the CGA process in older population with cancer.

In North America, two smaller pilot randomized controlled studies of geriatric assessment and management were completed. In the study by Puts et al. (2018), 58 older persons aged 70 years and over with advanced breast, genitourinary, or gastrointestinal cancer starting chemotherapy for adjuvant or palliative intent were included in the study (Puts et al. 2018) The intervention comprised of a CGA by a nurse and geriatrician at baseline and then clinical follow-up at 6 weeks and 3 months after the CGA. All patients in the intervention arm received an intervention, most commonly provided interventions included referral to dietician, patient education, and changes in the medication. The second randomized study by Magnusson et al. (2018) included 71 older persons with patients with stage III/IV solid tumor malignancies (Magnuson et al. 2018). In this study, a trained coordinator conducted and scored a baseline CGA with predetermined cutoffs for impairment. Patients that were randomized to the intervention arm received interventions based on an algorithm, and these recommendations were provided to the treating physician (in this case the primary oncologist) who was responsible for implementing the recommendations. Both studies that were not designed to be adequately powered showed positive impact of the intervention on quality of life and functional status (Puts et al. 2018; Magnuson et al. 2018).

Another recent observational study (Kenis et al., 2018) investigated the adherence to geriatric recommendations based on the CGA results and subsequent actions undertaken in older patients with cancer (Kenis et al. 2018). In this large study, patients were screened with a geriatric screening tool and a CGA was performed when necessary. At least one geriatric recommendation was made in approximately 80% of patients and at least one recommendation was adhered to in 70%. When taking the total number of geriatric recommendations into account, 46% of them were adhered to. The highest level of adherence was noticed for referral to the dietician, followed by the geriatrician, the social worker, the occupational therapist, the psychologist, and the physiotherapist. The most frequent actions undertaken were nutritional support and supplements, extended home care, and psychological support. Adherence was lowest for the memory clinic, the geriatric day clinic, and the fall clinic. This was the first study to examine in detail geriatric interventions and subsequent actions undertaken for problems detected by CGA in older patients with cancer. As such, the study gave increased insight into the adherence of geriatric recommendations and identified which health care workers and facilities are essential in the optimal management of older patients with cancer.

Next to these studies, international guidelines were developed to support the research field in geriatric oncology related to geriatric interventions. Both the ASCO guideline for geriatric oncology as well as the consensus guideline for Geriatric Assessment-Guided Care Processes for Older persons: A Delphi Consensus of Geriatric Oncology Experts describe interventions that can be considered for each CGA domain by health care professionals (Mohile et al. 2015, 2018a;). However, older persons with cancer may not be interested in following up on all geriatric recommendations provided by the multidisciplinary team to address the issues identified in the CGA. Patient preferences need to be taken into account.

Examples of Application

The recent Delphi consensus framework for interventions guided by the geriatric assessment has been published (43) and the ASCO geriatric oncology guideline (Mohile et al. 2018a) both outline interventions that can be used and that nurses and other members of the multidisciplinary team can take to support older persons with cancer before, during and after treatment based on the geriatric assessment findings. It is important that there is one team member responsible for the monitoring of the treatment plan, e.g., are the tests/referrals implemented and what is the outcome of the interventions, does the plan needs adjustment? In addition, new health status and functional status issues may arise during the treatment as well and ongoing monitoring would allow for identification of these new functional status issues to be addressed by the multidisciplinary team (see Table 2). Nurses are well positioned to be the team member responsible for follow-up and care coordination of the treatment plan with the older patient after the CGA and to communicate with the broader multidisciplinary team.

Depending on the organization of geriatric and oncology care and the community care system, either the multidisciplinary team can implement the interventions or involve the patient's family physician to implement the interventions. It is important to involve the older person and their caregiver in the development of the treatment plan to improve adherence to the treatment plan. For example, if the older person has difficulty coming to the hospital and they are dependent on others to bring them to the hospital, in case they have been experiencing falls instead of coming to the exercise program in the cancer center for multiple extra visits per week, the multidisciplinary team can work with the patient to identify local exercise programs in the community close to where the older person is living which will increase the likelihood of the older person attending the exercise classes or their many be an opportunity to get home visits from the community physical therapist. Another example

Geriatric Interventions, Table 2 Issues identified during the CGA and possible interventions to consider

Domain assessed	Possible interventions to be considered by the multidisciplinary team
- Falls/mobility limitations	- Consider referral to the occupation therapist for home safety assessment and use of assistive devices such as gait aids - To assess fall history, circumstances, symptoms, fear of falling and gait disturbances. - Consider a comprehensive fall assessment if there have been multiple falls and dependent on the etiology and risk factors for future falls implementation of additional interventions. - Consider referral to physical therapy for strength and balance training - Consider vitamin D and calcium supplementation Physician to review medications to identify if there are any medications that increase the fall risk
Weight loss/poor nutritional status	- Consider a referral to the dietician - Nurses should review with the older person the intake, and the weight loss history - Assess the patient for any symptoms that can interfere with adequate food intake (pain, inability to swallow, nausea, vomiting, constipation and address these symptoms) - Patient education on oral care and adequate intake and staying hydrated - Consider referral to Meals on Wheels program
Cognitive impairment	- Referral to the geriatric (day) clinic for further diagnostic workup if the older person is interested - Consider a referral to the social worker for supports at home for ongoing monitoring and for the caregiver - Consider a referral to the psychiatrist for in depth assessment - Physician to review medications to minimize/simplify the medication regime to promote adherence - Delirium prevention - Assess and optimize hearing and vision
Mood impairment	- Referral to psychosocial support services for support and counseling - Physician to review if pharmacological treatment is needed - Assess the social support network to mobilize support - Referral to community support programs - Nurses to do patient education on the importance of sleep hygiene and physical activity
Medication suboptimal adherence	- Assess and identify reasons for suboptimal adherence (side effects of medication, forgetting the medication or not believing that the medication is working) and address/intervene on the identified reasons - Consider dosette for reasons of forgetfulness - Consider referral to the occupational therapist to help with memory aids for medication adherence - Consider strategies such as smart phone alarms etc. for older persons who forget taking the medication - Physicians to review and simplify the medication regime to reduce the chance of forgetting medication
Comorbidities	- To review the symptoms and self-management strategies that the older person is using to manage their comorbidities. Depending on the issues identified, implement interventions to address the issues and collaboration with the patient's family physician. - Monitor symptoms and self-management over time

(continued)

Geriatric Interventions, Table 2 (continued)

Domain assessed	Possible interventions to be considered by the multidisciplinary team
Disability in functional status (Instrumental and Basic Activities of Daily Living)	- Review all supports available to the older person to help with instrumental and basic activities of daily living - Referral to community nursing for support with basic activities of daily living - Referral to social work to identify community supports available - Depending on the needs identified, consider a referral to occupational therapy for assistive devices, home adjustments, physical therapy to increase ability to walk stairs etc.
Lack of social support/isolation	- Referral to social work to identify community supports available - Assess need to support with transportation from and too the cancer center for appointments - Referral to psychosocial oncology for support as needed - Consider referral to community nursing/home care to provide support and ongoing monitoring - Referral to peer support program/ cancer patient support services available in the hospital

would be if the team during the CGA identifies inappropriate medication or no longer beneficial, but the older person is reluctant to stop these as there is not yet an trusting therapeutic relationship between the team and the older person, a team member can connect with the older person's family physician to update the family physician on the CGA findings and explain the rationale for changing the medication regime and that way the family physician can address it again with the older person which may be a more successful approach as many older persons have their family physician for a lengthy period before they enter in the cancer system.

Future Directions of Research

In geriatric medicine, randomized clinical trials in have already demonstrated the benefits of a CGA-based approach. Previous research demonstrated in a non-cancer population that geriatric intervention in a specialized geriatric unit significantly improved survival and return to own home (Ellis et al. 2011; Rubenstein et al. 1991). Recently randomized controlled trials have been initiated in the field of geriatric oncology in order to explore this research area and to demonstrate the clear benefit of CGA in older patients with cancer (see Table 3).

The ongoing randomized clinical trials have included a variety of geriatric intervention delivery models ranging from an nurse practitioner and the treating physician to a geriatrician with nurse follow-up which in the future will provide more evidence whether the intervention is beneficial on cancer treatment outcomes such as chemotherapy-related toxicity as well as important patient reported outcomes such as functional status, quality of life, and patient satisfaction. In addition, the ongoing randomized trials will provide evidence whether it makes a difference how this intervention is delivered to older persons with cancer which is important for the generalizability of the findings and for sustainability of the intervention in the various health care settings.

One could argue that studies could be focused on specific subgroups of patients (for example, specific tumor types or specific treatment groups such as chemotherapy or immunotherapy) in order to decrease heterogeneity of the older population as much as possible. On the other hand, it is virtually impossible to rule out heterogeneity with regards to characteristics like functional status, social status, comorbidity, and polypharmacy. And this heterogeneity is a main character for the older population. Therefore, the performance of a CGA can be similar to each other for all the patients that would have benefit from CGA and thus independent of the cancer diagnosis or the

Geriatric Interventions, Table 3 Examples of ongoing randomized clinical studies in geriatric oncology

Author	Design	Population	Intervention	Outcomes
Mohile et al. ClinicalTrials.gov identifier: NCT02107443	Cluster randomization	70 years and older, advanced solid malignancies	CGA results and geriatric recommendations given to the oncology team	Chemotherapy-related toxicity, overall survival, functional status, communication, satisfaction, quality of life, health care utilization
Puts et al. ClinicalTrials.gov identifier: NCT03154671	Patient randomization	70 years and older, solid malignancies starting first- and second-line chemotherapy	CGA by geriatrician and nurse with nurse follow-up	Quality of life, cost-effectiveness, chemotoxicity, satisfaction, overall survival, cancer treatment plan changes
Soubeyran et al. ClinicalTrials.gov identifier: NCT02704832	Patient randomization	70 years and older, solid tumor malignancies candidate for first- and second-line treatment	Geriatrician with nurse follow-up	Co-primary endpoint: Overall survival and quality of life response; other endpoints: Progression free survival, chemotherapy-related toxicity, health care utilization
Hurria et al. ClinicalTrials.gov identifier: NCT02054741	Patient randomization	65 years and older, solid tumor malignancies start new chemo	Study nurse practitioner in collaboration with treating physician and clinic nurse to follow-up	Primary endpoints: Chemotherapy-related toxicity, rate of hospitalization, change in functional status, change in psychosocial status
Lund et al. ClinicalTrials.gov identifier: NCT03719573	Patient randomization	70 years and older, frail patients who have undergone surgery for stage III /high-risk stage II colorectal cancer or patients with unresectable or metastatic disease who will receive adjuvant chemotherapy or first-line chemotherapy	Geriatrician will perform CGA and conduct the clinical follow-up	Number of patients completing planned treatment without dose reductions, occurrence of dose reductions, delay of treatment, treatment toxicity, time to recurrence, overall survival, cancer-specific mortality
Wildiers et al. ClinicalTrials.gov identifier: NCT04069962	Patient randomization	70 years and older, starting first- or second-line systemic therapy	All patients will undergo CGA and are then randomized to geriatric interventions coordinated by their treating physician or interventions coordinated by a geriatric team including coaching of the patient.	Primary endpoint: Quality of life at 6 months; secondary endpoints: Quality of life at 3 and 12 months, evolution in quality of life over 1 year follow-up, patient satisfaction, evolution of functional status, falls, systemic therapy-related adverse events, premature interruption of treatment

cancer treatment. Specific measuring instruments can be added when necessary and in case of specific points of attention related to the cancer diagnosis or the cancer treatment.

Next to this, studies should focus on longer follow-up in order to determine the long-term effects of treatment on functional status and quality of life. Resilience of patients is an important factor to study and may influence our treatment decisions in the future.

Summary

In this entry, possible geriatric interventions that could be implemented after the conduct of the CGA were highlighted as well as the latest research evidence in CGA studies. It is important to keep in mind that based on the CGA a treatment plan needs to be developed with the older patient to address the issues identified and should be based on the priorities and preferences of the older person. One team member should be responsible for monitoring the implementation of the treatment plan to evaluate the effectiveness and to address any newly identified issues that may arise during the cancer treatment. While currently there have been few completed randomized controlled trials demonstrating the benefit of implementing CGA on important patient centered outcomes such as functional status and quality of life, several studies are ongoing and will provide much needed evidence in the next few years. Until then, there are several consensus guidelines to help with the selection of geriatric interventions based on the best available evidence in traditional geriatric medicine settings.

Cross-References

▶ Care Coordination
▶ Chronic Disease Self-Management.
▶ Comprehensive Geriatric Assessment
▶ Frailty in Clinical Care
▶ Frailty Screening
▶ Geriatric Assessment for Older Adults with Cancer
▶ Gerontological Nursing

References

Al-Refaie WB, Vickers SM (2012) Are cancer trials valid and useful for the general surgeon and surgical oncologist? Adv Surg 46:269–281. https://doi.org/10.1016/j.yasu.2012.05.001

Al-Refaie WB, Vickers SM, Zhong W, Parsons H, Rothenberger D, Habermann EB (2011) Cancer trials versus the real world in the United States. Ann Surg 254(3):438–442. https://doi.org/10.1097/SLA.0b013e31822a7047

Al-Refaie WB, Weinberg A, Nelson H (2013) Are older adults adequately represented in surgical oncology trials? Bull Am Coll Surg 98(5):52–53

Baitar A, Kenis C, Moor R, Decoster L, Luce S, Bron D et al (2015) Implementation of geriatric assessment-based recommendations in older patients with cancer: a multicentre prospective study. J Geriatric Oncol 6(5): 401–410. https://doi.org/10.1016/j.jgo.2015.07.005

Bouchardy C, Rapiti E, Blagojevic S, Vlastos AT, Vlastos G (2007) Older female cancer patients: importance, causes, and consequences of undertreatment. J Clin Oncol 25(14):1858–1869. https://doi.org/10.1200/JCO.2006.10.4208

Daskivich TJ, Chamie K, Kwan L, Labo J, Palvolgyi R, Dash A et al (2011) Overtreatment of men with low-risk prostate cancer and significant comorbidity. Cancer 117 (10):2058–2066. https://doi.org/10.1002/cncr.25751

Decoster L, Van Puyvelde K, Mohile S, Wedding U, Basso U, Colloca G et al (2015) Screening tools for multidimensional health problems warranting a geriatric assessment in older cancer patients: an update on SIOG recommendations. Ann Oncol Off J Eur Soc Med Oncol 26(2):288–300. https://doi.org/10.1093/annonc/mdu210

Deschodt M, Braes T, Broos P, Sermon A, Boonen S, Flamaing J et al (2011) Effect of an inpatient geriatric consultation team on functional outcome, mortality, institutionalization, and readmission rate in older adults with hip fracture: a controlled trial. J Am Geriatr Soc 59(7):1299–1308. https://doi.org/10.1111/j.1532-5415.2011.03488.x

Ellis G, Whitehead MA, Robinson D, O'Neill D, Langhorne P (2011) Comprehensive geriatric assessment for older adults admitted to hospital: meta-analysis of randomised controlled trials. BMJ (Clinical research ed) 343:d6553. https://doi.org/10.1136/bmj.d6553

Enewold L, Harlan LC, Tucker T, McKenzie S (2015) Pancreatic cancer in the USA: persistence of undertreatment and poor outcome. J Gastrointest Cancer 46(1):9–20. https://doi.org/10.1007/s12029-014-9668-x

Hamaker ME, Te Molder M, Thielen N, van Munster BC, Schiphorst AH, van Huis LH (2018) The effect of a geriatric evaluation on treatment decisions and outcome for older cancer patients - a systematic review. J Geriatric Oncol 9(5):430–440. https://doi.org/10.1016/j.jgo.2018.03.014

Hori A, Shibata T, Kami M, Kusumi E, Narimatsu H, Kishi Y et al (2007) Age disparity between a cancer population and participants in clinical trials submitted as a new drug application of anticancer drugs in Japan. Cancer 109(12):2541–2546. https://doi.org/10.1002/cncr.22721

Hurria A, Cirrincione CT, Muss HB, Kornblith AB, Barry W, Artz AS et al (2011) Implementing a geriatric assessment in cooperative group clinical cancer trials: CALGB 360401. J Clin Oncol 29(10):1290–1296. https://doi.org/10.1200/JCO.2010.30.6985

Hurria A, Mohile S, Gajra A, Klepin H, Muss H, Chapman A et al (2016) Validation of a prediction tool for chemotherapy toxicity in older adults with Cancer. J Clin Oncol 34(20):2366–2371. https://doi.org/10.1200/JCO.2015.65.4327

Katz S, Ford AB, Moskowitz RW, Jackson BA, Jaffe MW (1963) Studies of illness in the aged - the index of Adl - a standardized measure of biological and psychosocial function. Jama-J Am Med Assoc 185(12):914–919. https://doi.org/10.1001/jama.1963.03060120024016

Kenis C, Bron D, Libert Y, Decoster L, Van Puyvelde K, Scalliet P et al (2013) Relevance of a systematic geriatric screening and assessment in older patients with cancer: results of a prospective multicentric study. Ann Oncol Off J Eur Soc Med Oncol 24(5):1306–1312. https://doi.org/10.1093/annonc/mds619

Kenis C, Decoster L, Flamaing J, Debruyne PR, De Groof I, Focan C et al (2018) Adherence to geriatric assessment-based recommendations in older patients with cancer: a multicenter prospective cohort study in Belgium. Ann Oncol Off J Eur Soc Med Oncol. https://doi.org/10.1093/annonc/mdy210

Lawton MP, Brody EM (1969) Assessment of older people - self-maintaining and instrumental activities of daily living. Gerontologist 9(3P1):179

Magnuson A, Lemelman T, Pandya C, Goodman M, Noel M, Tejani M et al (2018) Geriatric assessment with management intervention in older adults with cancer: a randomized pilot study. Support Care Cancer Off J Multinat Assoc Suppor Care Cancer 26(2):605–613. https://doi.org/10.1007/s00520-017-3874-6

Meresse M, Bouhnik AD, Bendiane MK, Retornaz F, Rousseau F, Rey D et al (2017) Chemotherapy in old women with breast Cancer: is age still a predictor for under treatment? Breast J 23(3):256–266. https://doi.org/10.1111/tbj.12726

Meyers BF, Samson PP (2014) EGJ and esophageal cancers: choosing induction therapy so as to err on the side of overtreatment rather than undertreatment when staging is imperfect. Oncology (Williston Park) 28(6):512, 9-20

Mohile SG, Velarde C, Hurria A, Magnuson A, Lowenstein L, Pandya C et al (2015) Geriatric assessment-guided care processes for older adults: a Delphi consensus of geriatric oncology experts. J Nat Comprehensive Cancer Network JNCCN 13(9):1120–1130. https://doi.org/10.6004/jnccn.2015.0137

Mohile SG, Dale W, Somerfield MR, Schonberg MA, Boyd CM, Burhenn PS et al (2018a) Practical assessment and Management of Vulnerabilities in older patients receiving chemotherapy: ASCO guideline for geriatric oncology. J Clin Oncol 36(22):2326–2347. https://doi.org/10.1200/JCO.2018.78.8687

Mohile SG, Epstein RM, Hurria A, Heckler CE, Duberstein P, Canin BE et al (2018b) Improving communication with older patients with cancer using geriatric assessment (GA): a University of Rochester NCI Community Oncology Research Program (NCORP) cluster randomized controlled trial (CRCT). J Clin Oncol 36(18_suppl):LBA10003-LBA. https://doi.org/10.1001/jamaoncol.2019.4728

Monroe MM, Myers JN, Kupferman ME (2013) Undertreatment of thick head and neck melanomas: an age-based analysis. Ann Surg Oncol 20(13):4362–4369. https://doi.org/10.1245/s10434-013-3160-x

National Institutes of Health (1988) Consensus development conference statement: geriatric assessment methods for clinical decision-making. J Am Geriatr Soc 36(4):342–347. https://doi.org/10.1111/j.1532-5415.1988.tb02362.x

Ng R, de Boer R, Green MD (2005) Undertreatment of elderly patients with non-small-cell lung cancer. Clin Lung Cancer 7(3):168–174. https://doi.org/10.3816/CLC.2005.n.031

Noon AP, Albertsen PC, Thomas F, Rosario DJ, Catto JW (2013) Competing mortality in patients diagnosed with bladder cancer: evidence of undertreatment in the elderly and female patients. Br J Cancer 108(7):1534–1540. https://doi.org/10.1038/bjc.2013.106

Puts MT, Girre V, Monette J, Wolfson C, Monette M, Batist G et al (2010) Clinical experience of cancer specialists and geriatricians involved in cancer care of older patients: a qualitative study. Crit Rev Oncol Hematol 74(2):87–96. https://doi.org/10.1016/j.critrevonc.2009.04.005

Puts MT, Santos B, Hardt J, Monette J, Girre V, Atenafu EG et al (2014) An update on a systematic review of the use of geriatric assessment for older adults in oncology. Ann Oncol Off J Eur Soc Med Oncol 25(2):307–315. https://doi.org/10.1093/annonc/mdt386

Puts MTE, Sattar S, Kulik M, MacDonald ME, McWatters K, Lee K et al (2018) A randomized phase II trial of geriatric assessment and management for older cancer patients. Support Care Cancer Off J Multinat Assoc Suppor Care Cancer 26(1):109–117. https://doi.org/10.1007/s00520-017-3820-7

Reuben DB, Fishman LK, McNabney M, Wolde-Tsadik G (1996) Looking inside the black box of comprehensive geriatric assessment: a classification system for problems, recommendations, and implementation strategies. J Am Geriatr Soc 44(7):835–838. https://doi.org/10.1111/j.1532-5415.1996.tb03744.x

Rubenstein LZ, Joseph T (2004) Freeman award lecture: comprehensive geriatric assessment: from miracle to reality. J Gerontol A Biol Sci Med Sci 59(5):473–477. https://doi.org/10.1093/gerona/59.5.m473

Rubenstein LZ, Stuck AE, Siu AL, Wieland D (1991) Impacts of geriatric evaluation and management programs on defined outcomes: overview of the evidence. J Am Geriatr Soc 39(9 Pt 2):8S–16S; discussion 7S-8S. https://doi.org/10.1111/j.1532-5415.1991.tb05927.x

Scher KS, Hurria A (2012) Under-representation of older adults in cancer registration trials: known problem, little progress. J Clin Oncol 30(17):2036–2038. https://doi.org/10.1002/cncr.22721

Schulkes KJ, Hamaker ME, van den Bos F, van Elden LJ (2016) Relevance of a geriatric assessment for elderly patients with lung Cancer-a systematic review. Clin Lung Cancer 17(5):341–9.e3. https://doi.org/10.1016/j.cllc.2016.05.007

Shumway DA, Hamstra DA (2015) Ageism in the undertreatment of high-risk prostate cancer: how long will clinical practice patterns resist the weight of evidence? J Clin Oncol 33(7):676–678. https://doi.org/10.1200/JCO.2014.59.4093

Townsley CA, Chan KK, Pond GR, Marquez C, Siu LL, Straus SE (2006) Understanding the attitudes of the elderly towards enrolment into cancer clinical trials. BMC Cancer 6:34. https://doi.org/10.1186/1471-2407-6-34

Van Leeuwen BL, Rosenkranz KM, Feng LL, Bedrosian I, Hartmann K, Hunt KK et al (2011) The effect of undertreatment of breast cancer in women 80 years of age and older. Crit Rev Oncol Hematol 79(3):315–320. https://doi.org/10.1016/j.critrevonc.2010.05.010

Wan-Chow-Wah D, Monette J, Monette M, Sourial N, Retornaz F, Batist G et al (2011) Difficulties in decision making regarding chemotherapy for older cancer patients: a census of cancer physicians. Crit Rev Oncol Hematol 78(1):45–58. https://doi.org/10.1016/j.critrevonc.2010.02.010

Weiss A, Noorbakhsh A, Tokin C, Chang D, Blair SL (2013) Hormone receptor-negative breast cancer: undertreatment of patients over 80. Ann Surg Oncol 20(10):3274–3278

Wildiers H, Heeren P, Puts M, Topinkova E, Janssen-Heijnen ML, Extermann M et al (2014) International Society of Geriatric Oncology consensus on geriatric assessment in older patients with cancer. J Clin Oncol 32(24):2595–2603. https://doi.org/10.1200/JCO.2013.54.8347

Geriatric Management

▶ Geriatric Interventions

Geriatric Management in Persian Medicine

Arman Zargaran[1] and Mohammad M. Zarshenas[2,3]
[1]Department of Traditional Pharmacy, School of Persian Medicine, Tehran University of Medical Sciences, Tehran, Iran
[2]Department of Phytopharmaceuticals (Traditional Pharmacy), School of Pharmacy, Shiraz University of Medical Sciences, Shiraz, Iran
[3]Research Office for the History of Persian Medicine, Shiraz University of Medical Sciences, Shiraz, Iran

Synonyms

Iranian traditional medicine; Persian medicine; Traditional Iranian medicine; Traditional Persian medicine

Definition

Persian medicine is one of the oldest traditional systems of medicine dating back to at least 7000 years ago. This traditional system of medicine has a holistic approach to human and is based on humoral theory. Persian medicine was the main paradigm of medical sciences at least until the seventeenth century in Western Asia, Northern Africa, and Europe and some other traditional systems of medicine like Unani medicine derived from it. Geriatric management in Persian medicine covers all aspects of geriatric management to keep health of older people via lifestyle modification and the special recommendations and treatment approach embedded in Persian medicine for diseases and disorders.

Overview

Aging and geriatric management is a global health challenge (Onder et al. 2013), and finding any strategies to reduce the risk factors of disorders and to improve the quality of life for them is the key for healthy aging (Salas-Crisostomo et al. 2018). One of the sources to find such new strategies and recommendations based on historical perspective and generation by generation experience is traditional systems of medicine (Emami et al. 2013). These traditional systems of medicine like Persian medicine, Chinese medicine, Ayurveda, homeopathy, etc. have a root in history with their own philosophy and also centuries of experiences of physicians during history (Zarshenas et al. 2017).

Persian medicine is one of the traditional systems of medicine that dates back to at least 7000 years ago (Zargaran 2014). It was the main paradigm of medicine in the main parts of Western Asia, Northern Africa, and Europe until the seventeenth century. There were great Persian physicians like Akhawayni, Rhazes, Avicenna, Jorjani, etc. who have flourished this paradigm of medical sciences, in particular during the Islamic Golden Age (ninth to thirteenth century CE) (Ghaffari et al. 2017). One of the most important parts of Persian medicine is its contribution to health care, more than treatment (Nikaein et al. 2012) and in particular to geriatrics (Emami et al. 2013), for example, for their dietary regimen (Emami et al. 2014), and to reduce cardiovascular disease risks (Zarshenas et al. 2016).

Persian Medicine

This traditional system of medicine follows holistic paradigm and is based on four elements: earth, fire, air, and water with dry and cold, hot and wet, hot and dry, and cold and wet qualities, respectively (Zargaran et al. 2013). It is believed to special *mizaj* (temperament) for any person and any health care or treatment is personalized. It means that the physician had to treat patients, based on his/her own condition. It seems that they used phenotypes of the people to categorize them in different temperaments that currently we know that it is rooted in their genotypes (Moeini et al. 2017). Furthermore, current investigations show logical relations between these claims with evidences based on proteomics (Rezadoost et al. 2016). Also, it is believed that there is an age temperament that affects the temperament of each person (Table 1).

However, the wetness of older persons is abnormal and is not good. Therefore, this temperament can affect the temperament of the person and health-care services are based on the specific condition and temperament of older adults (Naseri 2018).

General Considerations for Older Adults in Persian Medicine

The term of geriatric management is under the title of *Tadbeer-e-mashayekh* in Persian medicine. *Tadbeer* means strategy and *mashayekh* means

Geriatric Management in Persian Medicine, Table 1 Age temperament in Persian medicine

Age	Childhood	Youth	Middle age	Old age
Temperament	Hot and wet	Hot and dry	Cold and dry	Cold and more dryness (sometimes abnormal wetness)

Geriatric Management in Persian Medicine, Table 2 Health-care parameters for geriatrics

General considerations	Some useful foods	Some harmful foods
Foods should be in small amount, light, easily digestible at frequent intervals in geriatric regimen	Laxative foods, fruits (like fig, grapes, citruses, and plum), vegetables (like carrot and cabbage), chicken broth, honey, boiled milk with honey, or rock candy	Heavy foods and additives such as eggplant (*Solanum melongena* L.), lentil (*Lens culinaris* Medik.) and watermelon (*Citrullus lanatus* Thunb.), beef, and vinegar

Geriatric Management in Persian Medicine, Table 3 Medicinal herbs mentioned in Persian medicine for geriatric management

Persian name	Scientific name	Part use	Mentioned effects in Persian medicine	Effects based on current evidences
Ain-ol-deek	*Abrus precatorius* L.	Seed	Antiaging	Immunopotentiating activity (Ramnath et al. 2002: 910–913)
Vaj	*Acorus calamus* L.	Root	Health improver	Antioxidant (Manikandan et al. 2005: 2327–2330)
Soom	*Allium sativum* L.	Root	Antiaging	Antioxidant, antiaging effect (Svendsen et al. 1994: 125–133; Moriguchi et al. 1997: 235–242)
Azaaddrakht	*Azadirachta indica* A. Juss	Flower	Health improver	Immunomodulatory effect (Baral and Chattopadhyay 2004: 355–366)
Foshagh	*Bryonia dioica* Jacq.	Fruit, leaf	Antiaging	Antioxidant (Morales et al. 2012: 851–863)
Kommoon	*Bunium persicum* (Boiss.) B. Fedtsch.	Seed	Health improver	Antioxidant (Shahsavari et al. 2008: 183–188)
Ghortom	*Carthamus tinctorius* L.	Seed	Health improver	Antioxidant, neuroprotective (Hiramatsu et al. 2009: 795–805)
Hemmas	*Cicer arietinum* L.	Seed	Appetizer, general tonic	Growth enhancer (Nestares et al. 1996: 2760–2765)
Narjeel	*Cocos nucifera* L.	Fruit	Antiaging, general tonic	Antioxidant (Mantena et al. 2003: 126–131)
Aftimoon	*Cuscuta epithymum* L.	Aerial part	Health improver	–
Zarringiah	*Dracocephalum kotschyi* Boiss	Root	General tonic	Nutritional value (Goli et al. 2013: 188–193)
Lesan-al-sour	*Echium amoenum* Fisch. & Mey.	Flower	General tonic	Antioxidant (Ranjbar et al. 2006: 469–473)
Jowz	*Juglans regia* L.	Fruit	Memory enhancer	Neuroprotective (Orhan et al. 2011: 781–786)
Termes	*Lupinus termis* L.	Seed	Laxative	Having dietary fiber (Písaříková and Zralý, Písaříková and Zralý 2010: 211–216)
Anbaj	*Mangifera indica* L.	Fruit	General tonic	Antioxidant, Immunomodulator (Sánchez et al. 2000: 565–573; García et al. 2003: 1182–1187)
Baboonaj	*Matricaria chamomilla* L.	Aerial part	Sleep improver	Sleep enhancer (Shinomiya et al. 2005: 808–810)
Badrajbooye	*Melissa officinalis* L.	Leaf	Mood enhancer	Mood modulatory effect (Kennedy et al. 2002: 953–964)
Azanolfar	*Myosotis scorpioides* L.	Aerial part	Health improver	–
Reyhan	*Ocimum basilicum* L.	Leaf	Mood enhancer	Serotoninergic antidepressant effect (Abdoly et al. 2012: 211–215)
Aghhovan	*Tanacetum parthenium* (L.) Sch. Bip.	Flower	Sleep improver	–
Ahlilaj	*Terminalia chebula* Retz.	Fruit	Memory enhancer	Neuroprotective, cholinesterase inhibitor (Chang and Lin 2012: 1; Nag and De 2011: 121–124)

older adults. This part covers all aspects of geriatric management including Persian medicine fundamentals, nutrition, health-care strategies, and treatment for older adults (Rezaeizadeh et al. 2009). Fat older adults have cold and wet and thin older adults have cold and dry temperaments. Therefore, health-care services are based on their temperaments.

Health-Care Services for Geriatrics

Geriatrics need special managements including lifestyle modification like light exercise; body massage at morning with olive, almond, lily, and sesame oils; and adequate sleep during the day and night and good nutrition (Avicenna 1988; Jorjani 2006; Rhazes 2005) (Table 2). In general, foods with hot and wet temperament and hot and dry temperament are beneficial for thin and fat older adults, respectively (Emami et al. 2013).

The aim of these recommendations is avoiding some common geriatric complications like vertigo, constipation, and insomnia (Hussain et al. 2002). Furthermore, Persian scholars recommended using some medicinal herbs in geriatric regimen to improve their health and decrease their aging disorders conditions (Avicenna 1988; Jorjani 2006; Rhazes 2005). Some of these herbs and their claimed effects in Persian medicine in comparison with current evidences are summarized in Table 3 (Emami et al. 2013).

Summary and Future Directions of Research

Persian medicine is one of the oldest and most comprehensive traditional systems of medicine with a long historical background. Persian scholars paid attention to the geriatric problems and tried to manage their conditions with lifestyle modification as well as nutritional regimens and medicinal herbs. It seems that they focused on the common problems of older adults and prescribed medicinal plants as antiaging, health improver, appetizer, general tonic, memory enhancer, laxative, and sleep improver. Also, current investigations support their use and ancient claims about many of them. Therefore, with integrative approach to use traditional systems of medicine, Persian medicine can be considered to find new approaches for geriatric management in current medicine.

Cross-References

▶ Ayurveda, Longevity, and Aging
▶ Gerocomia in the History of Aging Care
▶ Life-Extensionism in a Historical Perspective
▶ Siddha Practice and Management of Geriatrics
▶ Unani Medicine and Healthy Living

References

Abdoly M, Farnam A, Fathiazad F, Khaki A, Khaki AA, Ibrahimi A et al (2012) Antidepressant-like activities of *Ocimum basilicum* (sweet Basil) in the forced swimming test of rats exposed to electromagnetic field (EMF). Afr J Pharm Pharmacol 6:211–215

Avicenna (1988) Canon of medicine (trans: Sharafkandi A). Soroosh Press, Tehran, 1025

Baral R, Chattopadhyay U (2004) Neem (*Azadirachta indica*) leaf mediated immune activation causes prophylactic growth inhibition of murine Ehrlich carcinoma and B16 melanoma. Int Immunopharmacol 4:355–366

Chang CL, Lin CS (2012) Phytochemical composition, antioxidant activity, and neuroprotective effect of *Terminalia chebula* Retzius extracts. Evid Based Complement Alternat Med 2012. https://doi.org/10.1155/2012/125247

Emami M, Sadeghpour O, Zarshenas MM (2013) Geriatric management in medieval Persian medicine. J Midlife Health 4(4):210–215

Emami M, Nazarinia MA, Rezaeizadeh H, Zarshenas MM (2014) Standpoints of traditional Persian physicians on geriatric nutrition. J Evid Based Complement Altern Med 19(4):287–291

García D, Leiro J, Delgado R, Sanmartín ML, Ubeira FM (2003) *Mangifera indica* L. extract (Vimang) and mangiferin modulate mouse humoral immune responses. Phytother Res 17:1182–1187

Ghaffari F, Naseri M, Jafari Hajati R, Zargaran A (2017) Rhazes, a pioneer in contribution to trials in medical practice. Acta Med Hist Adriat 15(2):261–270

Goli SAH, Sahafi SM, Rashidi B, Rahimmalek M (2013) Novel oilseed of Dracocephalum kotschyi with high n-3 to n-6 polyunsaturated fatty acid ratio. Ind Crop Prod 43:188–193

Hiramatsu M, Takahashi T, Komatsu M, Kido T, Kasahara Y (2009) Antioxidant and neuroprotective activities of Mogami-benibana (Safflower, *Carthamus tinctorius* Linne). Neurochem Res 34:795–805

Hussain SA, Khan AB, Siddiqui MY, Latafat T, Kidwai T (2002) Geriatrics and unani medicine – a critical review. Anc Sci Life 22:13–16

Jorjani S (2006) Al- Aghraz al- Tebbieh va al- Mabahes al-Alayieh. Tehran University Press, Tehran, 11th century

Kennedy DO, Scholey AB, Tildesley NT, Perry EK, Wesnes KA (2002) Modulation of mood and cognitive performance following acute administration of *Melissa officinalis* (lemon balm). Pharmacol Biochem Behav 72:953–964

Manikandan S, Srikumar R, Jeya Parthasarathy N, Devi RS (2005) Protective effect of *Acorus calamus* L. on free radical scavengers and lipid peroxidation in discrete regions of brain against noise stress exposed rat. Biol Pharm Bull 28:2327–2330

Mantena SK, Jagadish, Badduri SR, Siripurapu KB, Unnikrishnan MK (2003) In vitro evaluation of antioxidant properties of *Cocos nucifera* Linn. Water. Nahrung 47:126–131

Moeini R, Memariani Z, Pasalar P, Gorji N (2017) Historical root of precision medicine: an ancient concept concordant with the modern pharmacotherapy. Daru 25(1):7

Morales P, Carvalho A, Sánchez-Mata M, Cámara M, Molina M, Ferreira I (2012) Tocopherol composition and antioxidant activity of Spanish wild vegetables. Genet Resour Crop Evol 59:851–863

Moriguchi T, Saito H, Nishiyama N (1997) Anti-ageing effect of aged garlic extract in the inbred brain atrophy mouse model. Clin Exp Pharmacol Physiol 24:235–242

Nag G, De B (2011) Acetylcholinesterase inhibitory activity of *Terminalia chebula*, *Terminalia bellerica* and *Emblica officinalis* and some phenolic compounds. Int J Pharm Pharm Sci 13:121–124

Naseri M (2018) Review on traditional Persian medicine. Traditional Persian Medicine Press, Tehran

Nestares T, López-Frías M, Barrionuevo M, Urbano G (1996) Nutritional assessment of raw and processed chickpea (*Cicer arietinum* L.) protein in growing rats. J Agric Food Chem 44:2760–2765

Nikaein F, Zargaran A, Mehdizadeh A (2012) Rhazes' concepts and manuscripts on nutrition in treatment and health care. Anc Sci Life 31(4):160–163

Onder G, van der Cammen TJ, Petrovic M, Somers A, Rajkumar C (2013) Strategies to reduce the risk of iatrogenic illness in complex older adults. Age Ageing 42(3):284–291

Orhan IE, Suntar IP, Akkol EK (2011) In vitro neuroprotective effects of the leaf and fruit extracts of *Juglans regia* L. (walnut) through enzymes linked to Alzheimer's disease and antioxidant activity. Int J Food Sci Nutr 62:781–786

Písaříková B, Zralý Z (2010) Dietary fibre content in lupine (*Lupinus albus* L.) and soya (*Glycine max* L.) seeds. Acta Vet Brno 79:211–216

Ramnath V, Kuttan G, Kuttan R (2002) Immunopotentiating activity of abrin, a lectin from *Abrus precatorius* Linn. Indian J Exp Biol 40:910–913

Ranjbar A, Khorami S, Safarabadi M, Shahmoradi A, Malekirad AA, Vakilian K (2006) Antioxidant activity of Iranian Echium amoenum Fisch & C.A. Mey flower decoction in humans: a cross-sectional before/after clinical trial. Evid Based Complement Alternat Med 3:469–473

Rezadoost H, Karimi M, Jafari M (2016) Proteomics of hot-wet and cold-dry temperaments proposed in Iranian traditional medicine: a network-based study. Sci Rep 6:30133

Rezaeizadeh H, Alizadeh M, Naseri M, Ardakani MS (2009) The traditional Iranian medicine point of view on health and disease. Iran J Public Health 38: 169–172

Rhazes (2005) Al Havi (Liber Continent). Academy of Medical Sciences, Tehran, 10th century

Salas-Crisostomo M, Torterolo P, Veras AB, Rocha NB, Machado S, Murillo-Rodríguez E (2018) Therapeutic approaches for the management of sleep disorders in geriatric population. Curr Med Chem. https://doi.org/10.2174/0929867325666180904113115. [Epub ahead of print]

Sánchez GM, Re L, Giuliani A, Núñez-Sellés AJ, Davison GP, León-Fernández OS (2000) Protective effects of *Mangifera indica* L. extract, mangiferin and selected antioxidants against TPA-induced biomolecules oxidation and peritoneal macrophage activation in mice. Pharmacol Res 42:565–573

Shahsavari N, Barzegar M, Sahari M, Naghdibadi H (2008) Antioxidant activity and chemical characterization of essential oil of *Bunium persicum*. Plant Foods Hum Nutr 63:183–188

Shinomiya K, Inoue T, Utsu Y, Tokunaga S, Masuoka T, Ohmori A (2005) Hypnotic activities of chamomile and passiflora extracts in sleep-disturbed rats. Biol Pharm Bull 28:808–810

Svendsen L, Rattan SIS, Clark BFC (1994) Testing garlic for possible anti-ageing effects on long-term growth characteristics, morphology and macromolecular synthesis of human fibroblasts in culture. J Ethnopharmacol 43:125–133

Zargaran A (2014) Ancient Persian medical views on the heart and blood in the Sassanid era (224–637 AD). Int J Cardiol 172(2):307–312

Zargaran A, Zarshenas MM, Karimi A, Yarmohammadi H, Borhani-Haghighi A (2013) Management of stroke as described by Ibn Sina (Avicenna) in the Canon of Medicine. Int J Cardiol 169(4):233–237

Zarshenas MM, Jamshidi S, Zargaran A (2016) Cardiovascular aspects of geriatric medicines in traditional Persian medicine; a review of phytochemistry and pharmacology. Phytomedicine 23(11):1182–1189

Zarshenas MM, Zargaran A, Blaschke M (2017) Convenient, traditional and alternative therapies for cardiovascular disorders. Curr Pharm Des 23(7): 1112–1118

Geriatric Medicine

- Exercise Adherence
- Frailty in Clinical Care
- Gerocomia in the History of Aging Care

Geriatric Medicines Strategy

- Regulation of Geroprotective Medications

Geriatric Mental Disorder

- Geriatric Mental Health

Geriatric Mental Health

Jessica Khan[1], Richard Holbert[1], Robert Averbuch[1] and Uma Suryadevara[1,2]
[1]Department of Psychiatry, University of Florida, Gainesville, FL, USA
[2]Malcom Randall VA Medical Center, Gainesville, FL, USA

Synonyms

Geriatric mental disorder; Neurocognitive disorder

Definition

Geriatric mental health is the field of medicine that focuses on the comprehensive evaluation and treatment of mental disorders in older adults.

Overview

With the world's population aging rapidly, there is a greater focus on emotional wellbeing in the later stages of life. Older adults face unique challenges with regard to mental health. This age range experiences significant transitions in levels of independence and life roles, which can take a toll on emotional and mental health.

With the advent of industrialization, along with urbanization and profound advances in medical technologies, the world has seen a significant increase in the geriatric population. During these later years, people experience significant changes, as well as challenges, in their life roles. Many older adults spend their later years productively: working jobs, raising children, engaging in hobbies, and contributing to the community at large. But others may face the challenge of not being able to do these things to the extent they could when they were younger. According to Erik Erickson's stages of psychosocial development, people of this age work through the stage "integrity versus despair," marked by reflection over life's accomplishments or regrets.

Key Research Findings

Psychosocial Factors Affecting Mental Health

Generally speaking, healthy aging is subject to the norms by which the individual societies make sense of old age. The perception of how one fits in with these norms can have a profound effect on the mental health of that individual. In addition, psychological factors such as loss of purpose in life, the passing of loved ones/depleted relationships, drop in socioeconomic status with retirement, and loneliness have a huge impact on geriatric mental health (Fukutomi et al. 2013). Along with functional decline in multiple organs, systemic disorders, and sensory impairments, the geriatric population is also more susceptible to emotional stress. Additional factors include a decline in functional ability, chronic pain, and loss of capacities. Physical illness can lead to dependence on others and financial strain, exposing emotional and physical vulnerabilities and increasing the potential for abuse. According to the World Health Organization, about one in six older adults experience elder abuse, which can take the form of physical, verbal, psychological, and financial exploitation or sexual abuse. Abandonment or neglect, in which the designated caregiver fails to meet the needs of the older person, is considered another form of abuse. The negative sequelae of traumatic injury suffered through abuse are obvious, but also worth consideration are the psychological effects, including increased rates of depression and anxiety (Lachs and Pillemer 2004).

Chronic pain in older adults is one of the largest health-related burdens on our society. Both diagnosis and management are complicated by the need to carefully balance a number of factors. Among these are preexisting conditions that can complicate treatment (neurocognitive disorders, cardiovascular disease, renal or hepatic failure) and age-related physiological changes (gastrointestinal drug absorption, distribution, liver metabolism, and renal clearance). An overly cautious approach can lead to untreated or undertreated pain which in turn negatively impacts mental health. The results can include depression, anxiety, social isolation, and sleep disturbances.

Declines in functional ability, decreased mobility, chronic pain, frailty, or other health problems in older adults sometimes warrant a move into a new environment such as long-term care facilities. This transition can take its toll on the individual and families alike. The relative loss of independence can be a particularly difficult adjustment, along with the perceived change in social status and contacts, loss of a "place to call home," and change in habitual activities. Literature shows that there is an increase in rates of depression, anxiety, and other conditions at this stage of life.

Another major stressor in this age group is the sequential loss of family and friends around them. The morbidity and mortality that accompany bereavement is significant. Consequences may include depression, worsening cognition, anxiety, and even suicide, especially when they have been in a long relationship or a caring role for several years. Re-establishing social support can be a particular challenge, further impacting overall mental health.

Neurocognitive Disorders

Among the most studied and commonly occurring mental illnesses in older adults are a group of degenerative neurocognitive disorders, often referred to as dementia. The underlying disease process destroys neurons, which results in a decline in memory, language, speech, reasoning, judgment, problem-solving, and other cognitive abilities. With progression, activities of daily living are affected, and assistance is needed to do things like pay bills, cook, bathe, eat, and even toilet. At advanced stages, the degenerative process impacts even the most basic functions including walking and swallowing.

According to the World Health Organization (2017), neurocognitive disorders affect about 5% of adults over the age of 60 worldwide. There are several different types of neurocognitive disorders that present with unique symptoms and affect different parts of the brain (WHO 2017 and Alzheimer's Association 2016). The most common type is Alzheimer's disease, accounting for 60–80% of all cases of neurocognitive disorders (Alzheimer's Association 2016). The hallmark of Alzheimer's disease is difficulty with recall of new information, due to destruction of the pathways involved in forming new memories. The disease progresses to include symptoms of confusion, difficulty planning or solving problems, poor judgment, reduced social activity, and increased apathy or depression. In severe cases, people lose their ability to walk, swallow, or care for themselves altogether and require full-time care. They commonly forget where they are and who their caregivers are, even losing ability to recognize loved ones.

Alzheimer's disease is ultimately fatal, with those affected dying from the complications of physical deficits, such as aspiration pneumonia from loss of swallowing ability. The exact causes and mechanisms of the disease are not fully understood, but research has shown that two processes, namely, the accumulation of beta-amyloid between neurons and the accumulation of tau proteins within neurons, lead to breakdown and ultimate destruction of neurons in Alzheimer's disease (Alzheimer's Association 2016).

The other common types of neurocognitive disorders include Parkinson's disease, frontotemporal variant (Pick's disease), Lewy body type, and vascular type. Similar symptoms of memory disturbance, personality changes, and decline in executive function exist to varying degrees in all these types of neurocognitive disorders, though the underlying pathophysiology is different.

Depression

Depression affects about 7% of the worldwide geriatric population (WHO 2017). It is far more common among nursing home residents, with one study showing 54% of residents diagnosed with depression (Hoover et al. 2010). Estimates show that about one-fourth of all deaths committed by suicide are among this age group (WHO 2017). Though it is a very common illness in older adults, it often goes undiagnosed and untreated or undertreated (Lebowitz et al. 1997). Because symptoms are often underreported and/or present atypically in this population, caregivers would be wise to have a higher index of suspicion for depression. Geriatric depression is usually characterized by feelings of sadness, loss of interest or pleasure in activities, fatigue, changes in sleeping and eating patterns, and sometimes thoughts of death or suicide. In contrast to younger patients, older adults have more comorbid medical conditions and cognitive disorders and may present additionally with multiple unexplained medical symptoms, social withdrawal, memory complaints, and problems with self-care (Gallo et al. 1997). Comprehensive geriatric evaluations help to inform good coordinated and integrated treatment plans. Research shows medications and therapy can be effective treatments in this age group, though they are more prone to side effects (Taylor 2014). Clinicians should also be alert to issues with compliance due to impaired cognitive status, medical comorbidities, and often complex treatment plans (Zivin et al. 2008). Family members or other home services help facilitate adherence to treatments and improve outcomes.

Treatments for geriatric depression may vary from exercise, cognitive behavioral or other kinds of psychotherapy, and antidepressants to electroconvulsive therapy or other neuromodulation treatments (Kok and Reynolds 2017). Exercise can be difficult for some older patients due to physical limitations. Psychotherapy is effective, but at times limited by lack of health insurance coverage, patient's cognitive status, and availability of psychotherapists. Pharmacotherapy is the mainstay of treatment for geriatric depression and includes mostly antidepressants. Antipsychotics and other medication classes can be used as adjunct treatment options. Electroconvulsive therapy is a safe and effective treatment for geriatric patients with severe depression (Kok et al. 2017).

Anxiety Disorders

Anxiety disorders affect 3.8% of the older population according to the WHO (2017). Of the anxiety disorders, generalized anxiety disorder is the most common among older adults, with prevalence rates of 2.4%–6.3% (Bower et al. 2015). It is often associated with increased disability, poor quality of life, increased use of health services, and cognitive impairment. Generalized anxiety is characterized by uncontrollable and persistent worry accompanied by symptoms such as irritability, feeling on edge, fatigue, muscle tension, restlessness, and insomnia (Bower et al. 2015). Other psychiatric disorders like depressive disorders and cognitive disorders are the frequently encountered comorbidities. Cardiovascular disease, hyperthyroidism, diabetes, chronic pain, lung disease, and gastrointestinal disorders are medical comorbidities most often associated with anxiety disorders. Older patients tend to have a specific phobia of falling, and when excessive, this leads to loss of independence, functional limitations, and compromised quality of life. In addition to completing a comprehensive assessment, clinicians should obtain collateral information about severity, prior treatments, treatment response, and cognition. Understanding comorbid medical conditions and medications helps to individualize treatment for anxiety disorders. Common treatments approved for use in anxiety disorders in older adults include selective serotonin reuptake inhibitors, serotonin-norepinephrine reuptake inhibitors, relaxation training, and cognitive behavioral therapy (Lenze et al. 2005).

Posttraumatic Stress Disorder (PTSD)

People can be exposed to traumatic events at any time across their life span. Most return to their state prior to the trauma within 3 months. But a subset of those exposed continue to experience symptoms like hypervigilance, nightmares, avoidance, flashbacks, distressing memories, and negative alterations in cognitions even after 3 months. Older adults may report somatic complaints more than the classic psychological symptoms (Kaiser et al. 2016). Depressive disorder, generalized anxiety disorder, and substance use disorders are the most prevalent psychiatric comorbidities in older patients with PTSD (citation). Comorbid neurocognitive disorders complicate the assessment and treatment of PTSD and are associated with higher rates of healthcare utilization (Kang et al. 2018). Symptoms of PTSD can have a long-lasting impact in older adults and may be a risk factor for developing neurocognitive disorders (Jakel 2018). Treatment modalities vary from cognitive behavioral therapies to cognitive processing therapy and eye movement desensitization reprocessing techniques (Forbes et al. 2013). Selective serotonin reuptake inhibitors and serotonin-norepinephrine reuptake inhibitors are considered first-line options, and prazosin has shown some promise, especially for nightmares (add a citation).

Sleep Disorders

As people age, the prevalence of sleep disorders increases as do a number of predictable changes in sleep architecture. The most common sleep disorder is insomnia (citation). Other sleep disorders include sleep apnea, periodic limb movement disorder, restless leg syndrome, and circadian rhythm disorders. Multiple factors contribute to insomnia, ranging from psychiatric or neurological disorders, medical conditions, medication side effects, and substance use to environmental changes (home, hospital, nursing home), unrealistic expectations of sleep, psychosocial stressors, and others (citation). Ideally, treatment should focus on finding and treating the underlying cause of the sleep disturbance. Careful and thorough assessment of medical conditions and medications is warranted as it may help individualize the treatment plan. The sleep log is a helpful tool to collect additional data and to educate those affected about good sleep hygiene (citation) if it is not (Schroeck et al. 2016). Cognitive behavioral therapy for insomnia is a beneficial first-line option followed by or in addition to medications (Schroeck et al. 2016). Biological therapies should ideally be limited to short-term use and may include melatonin-based drugs, non-benzodiazepine receptor agonists, and some antidepressants. Whenever possible, benzodiazepines and antihistamines should be avoided in older adults (Roszkowska et al. 2010).

Substance Use Disorders

Although the rates of substance misuse are lower in older adults, they are more susceptible to harmful effects due to changes in pharmacokinetics such as slower metabolism and excretion as well as pharmacodynamic drug interactions. To complicate diagnosis, older adults are not subject to the classic markers of impaired function because they are often retired, may not drive, and can have multiple medical conditions that may explain symptoms. They may tend to isolate more, avoid family activities or have increased family discord, have cognitive changes, may present with unexplained injuries, falls, or malnutrition. They may also present with psychiatric symptoms like sleep problems, mood swings, irritability, anxiety, or depression (Wu et al. 2013). The use of prescription drug monitoring programs may identify older adults who try to fill prescriptions sooner or go to multiple doctors seeking medications. Substance use disorder treatments should include care for people with coexisting medical and psychiatric conditions, best achieved by coordination of services in integrated healthcare settings (Kuerbis et al. 2014).

Future Directions

The increasing aging population, healthcare costs, and the new healthcare policies being implemented to meet these demands have been the driving forces for geriatric mental health trends recently. To enhance integration of the aging and comprehensive mental health services at a community level, a coordinated and seamless set of resources that include in-home psychiatric assessment, medication management, follow-up, individual or group counseling, transportation, and family caregiver support is being developed. Telepsychiatry is another model of care that is being further developed to help increase access to geriatric mental healthcare.

Summary

Older adults face multiple, complex challenges with regard to mental health. This age range is prone to experience significant transitions in levels of independence and life roles, which can impact mental health. The most common geriatric mental disorders include depressive disorders, anxiety disorders, posttraumatic stress disorders, substance use disorders, and neurocognitive disorders. Older patients do not present with the same symptoms as their younger counterparts due to various reasons, one of which is stigma and the other being the medical and cognitive comorbidities. Hence, a comprehensive assessment with the help of a skilled multidisciplinary team along with obtaining collateral information helps the physician develop a solid individualized treatment plan. Medication management of the geriatric mental disorders could be complicated by cognitive deficits, medical comorbidities, and physiological changes seen with aging making them more prone to side effects.

Cross-References

▶ Alzheimer's Disease
▶ Anxiety About Aging
▶ Autonomy and Aging
▶ Death Anxiety
▶ Domestic Violence
▶ Insomnia, Sleep Disorders, and Healthy Aging
▶ Length of Stay in Long-Term Care Settings
▶ Mental Disorders
▶ Mental Health Services
▶ Neurotrophic Factors Link to Alzheimer's Disease
▶ Pain in Older Persons
▶ Parkinson's Disease

References

Alzheimer's Association (2016) 2016 Alzheimer's disease facts and figures. Alzheimers Dement 12(4):459–509

Bower ES, Wetherell JL, Mon T, Lenze EJ (2015) Treating anxiety disorders in older adults: current treatments and future directions. Harv Rev Psychiatry 23(5):329–342. https://doi.org/10.1097/HRP.0000000000000064

Forbes D, Creamer M, Bisson JI, Cohen JA, Crow BE, Foa EB, . . ., Ursano RJ (2013) A guide to guidelines for the treatment of PTSD and related conditions. Focus 11(3):414–427. https://doi.org/10.1080/20008198.2017.1403257

Fukutomi E, Kimura Y, Wada T, Okumiya K, Matsubayashi K (2013) Long-term care prevention project in Japan. Lancet 381(9861):116. https://doi.org/10.1016/S0140-6736(13)60049-5

Gallo JJ, Rabins PV, Lyketsos CG, Tien AY, Anthony JC (1997) Depression without sadness: functional outcomes of nondysphoric depression in later life. J Am Geriatr Soc 45(5):570–578. https://doi.org/10.1111/j.1532-5415.1997.tb03089.x

Hoover DR, Siegel M, Lucas J, Kalay E, Gaboda D, Devanand DP, Crystal S (2010) Depression in the first year of stay for elderly long-term nursing home residents in the U.S.A. Int Psychogeriatr 22(7):1161–1171. https://doi.org/10.1017/S1041610210000578

Jakel RJ (2018) Posttraumatic stress disorder in older adults. Psychiatr Clin 41(1):165–175. https://doi.org/10.1002/gps.5055

Kang B, Xu H, McConnell ES (2018) Neurocognitive and psychiatric comorbidities of posttraumatic stress disorder among older veterans: a systematic review. Int J Geriatr Psychiatry. https://doi.org/10.1002/gps.5055

Kok RM, Reynolds CF (2017) Management of depression in older adults: a review. JAMA 317(20):2114–2122. https://doi.org/10.1001/jama.2017.5706

Kuerbis A, Sacco P, Blazer DG, Moore AA (2014) Substance abuse among older adults. Clin Geriatr Med 30(3):629–654. https://doi.org/10.1016/j.cger.2014.04.008

Lachs MS, Pillemer K (2004) Elder abuse. The Lancet 364 (9441):1263–1272.

Lebowitz BD, Pearson JL, Schneider LS et al (1997) Diagnosis and treatment of depression in late life, consensus statement update. JAMA 278(14):1186–1190. https://doi.org/10.1001/jama.1997.03550140078045

Lenze EJ, Mulsant BH, Mohlman J et al (2005) Generalized anxiety disorder in late life: lifetime course and comorbidity with major depressive disorder. Am J Geriatr Psychiatry 13:77–80. https://doi.org/10.1176/appi.ajgp.13.1.77

Pless Kaiser A, Seligowski A, Spiro A 3rd, Chopra M (2016) Health status and treatment-seeking stigma in older adults with trauma and posttraumatic stress disorder. J Rehabil Res Dev 53:391–402. https://doi.org/10.1682/JRRD.2015.03.0039

Roszkowska J, Geraci SA (2010) Management of insomnia in the geriatric patient. Am J Med 123(12):1087–1090. https://doi.org/10.1016/j.amjmed.2010.04.006

Schroeck JL, Ford J, Conway EL, Kurtzhalts KE, Gee ME, Vollmer KA, Mergenhagen KA (2016) Review of safety and efficacy of sleep medicines in older adults. Clin Ther 38(11):2340–2372. https://doi.org/10.1016/j.clinthera.2016.09.010

Taylor WD (2014) Depression in the elderly. N Engl J Med 371(13):1228–1236. https://doi.org/10.1056/NEJMcp1402180

World Health Organization (2017) Mental health of older adults. Retrieved from https://www.who.int/news-room/fact-sheets/detail/mental-health-of-older-adults. Accessed on July 2019

Wu LT, Blazer DG (2013) Substance use disorders and psychiatric comorbidity in mid and later life: a review. Int J Epidemiol 43(2):304–317. https://doi.org/10.1093/ije/dyt173

Zivin K, Kales HC (2008) Adherence to depression treatment in older adults: a narrative review. Drugs Aging 25(7):559–571. https://doi.org/10.2165/00002512-200825070-00003

Geriatric Nurse Care Manager/Management

▶ Care Management

Geriatric Nursing

▶ Gerontological Nursing

Geriatric Psychology

▶ Psychogeriatrics

Geriatric Rehabilitation, Instability, and Falls

James Fleet[1], Helen Wear[2] and Finbarr C. Martin[3]
[1]Department of Ageing and Health, Guy's and St Thomas' NHS Foundation Trust, London, UK
[2]London Deanery Postgraduate Training Programme, London, UK
[3]Population Health Sciences, King's College London, London, UK

Definitions

Geriatric rehabilitation is the process of enabling well-being in older age through maintenance or restoration of functional ability. Functional ability is "the health-related attributes that enable people to be and to do what they have reason to value" (WHO 2015, p. 227). A *fall* is defined as an event which results in a person coming to rest inadvertently on the ground or floor or other lower level (WHO 2018; see Falls in this volume). Instability is the inability to maintain safe posture or mobility and therefore avoid falls or injury while performing daily activities.

Overview

Rehabilitation is about promoting the restoration of an individual's ability to function as fully as possible in society. This general notion of rehabilitation covers a wide scope including those withdrawn from society through imprisonment or war, etc. In the context of healthcare, it is about promoting healthy aging – in the words of the World Health Organization (WHO 2015, p. 230), rehabilitation consists of "a set of measures aimed at individuals who have experienced or are likely to experience disability to assist them in achieving and maintaining optimal functioning when interacting with their environments." In that sense, rehabilitation is a component of most clinical care of older patients. The goals and techniques differ in different settings, but the building blocks and principles remain the same, including (i) multidimensional assessment, usually by a multidisciplinary team (MDT); (ii) individualized goal setting based on the principles of consent and patient-centered conversations; (iii) design of treatment plans, including self-management where feasible; and (iv) scheduled follow-up, which should include evaluation of the acceptability and adherence to the treatment plan and monitoring of intermediate indicators of progress and of goals.

The Scope of Geriatric Rehabilitation

There are many ways of describing those aspects of health which impact functional ability. The International Classification of Functioning, Disability and Health (ICF) (WHO 2001) provides a conceptual framework to describe dimensions of health and a shared language which can facilitate communication and understanding across disciplines with a focus on health and functioning rather than disease and disability. Its key concepts are used within geriatric rehabilitation.

The Report on Ageing and Health (WHO 2015) proposed that functional ability is a composite of an individual's intrinsic capacity (IC), "the composite of all the physical and mental capacities that an individual can draw on" (p. 229), the environment in which they live, and the interactions between them. Both physical and sociocultural factors contribute to this environment. The focus of rehabilitation, therefore, may include intrinsic capacities such as balance, external factors such as domestic or environmental structures, and social factors such as practical help or facilitating social activity. Individuals' resilience requires them to draw on resources in any of these dimensions to maintain functional ability in the face of adversity; it is influenced by internal traits such as self-efficacy, individuals' belief in their innate ability to achieve their goals, as well as external factors.

The ICF framework and these new healthy aging definitions enable a move forward from the unhelpful dichotomy of the biomedical versus the social model of care. A holistic approach to rehabilitation embraces a multiplicity of factors without a hierarchy. Promoting change in any of the dimensions may help, but a comprehensive assessment helps to target interventions where they are most likely to be effective in terms of promoting and then maintaining functional abilities. For older people, comprehensive geriatric assessment (CGA) is an evidence-based approach to this process: It consists of a multidimensional assessment to identify the medical conditions, physical and mental capacities, functional abilities, and personal and social resources available to the individual, followed by the development of a "problem list," person-centered goals, and a plan of action including treatments and follow-up.

Who Needs Rehabilitation?

Rehabilitation aims to prevent a trajectory of decline and to restore healthy aging. With aging comes a marked and increasing variability of intrinsic capacity within populations, reflecting both personal (e.g., sex, ethnicity, and socioeconomic status) and health-related factors (e.g., health behaviors, multimorbidity, and geriatric syndromes). This variability makes predicting who will benefit from rehabilitation, and when, more difficult. Some older adults will suffer a gradual decline in their intrinsic capacity to which they will adapt and adjust their level of functioning; this may represent a missed opportunity for intervention. For others, an acute event (e.g., illness or hospital admission) will cause a sudden, unexpected decline in capacity with the potential for either recovery to their original trajectory or the possibility of further deterioration (see Fig. 1).

General Considerations on the Organization and Practice of Geriatric Rehabilitation

The overall aim of rehabilitation in practice can be divided into the restoration of function, adaptation to functional losses, and prevention of secondary complications.

Goal Setting in Rehabilitation

Appropriate goal setting requires inputs from individuals about their priorities, beliefs, and attitudes to potentially burdensome interventions and from

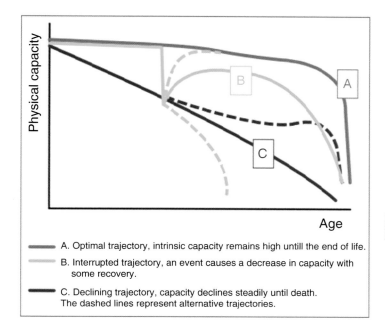

Geriatric Rehabilitation, Instability, and Falls, Fig. 1 Trajectories of physical capacity, impacted by aging, acute events, and variable recovery

professionals in the MDT about identified IC declines, preserved abilities, potentially remediable factors, and likely trajectories with and without interventions. Goals may be general and aspirational, but accompanied by objectives which should be SMART, i.e., specific, measurable, achievable and mutually agreed, relevant, and time defined. These objectives may vary according to the phase of care and stretch over time. In an intensive or prolonged rehabilitation program, it is usual to identify measurable interim objectives, which serve not only to motivate but also to prompt reassessment of the goals or the techniques of intervention.

The Multidisciplinary Team

Rehabilitation may be aimed at any of the following: understanding and modifying disease processes, increasing IC, changing individuals' technique, enhancing confidence, altering the environment, or providing human, mechanical, or computer-based assistance. It is clear therefore that a multiskilled team is needed. Core members in clinical settings are nurses, physical and occupational therapists, and doctors plus when necessary, additional contributions from speech and language therapists, psychologists, social workers, nutritionists, podiatrists, etc. The autonomy of each member varies according to his/her professional standing, availability, the local reimbursement arrangements, and healthcare cultures in different countries. Effective team-working may be helped or hindered by these factors. Sharing information, maintaining consistency of goals along with harmonious divisions of labor, and communicating effectively within the team are critical success factors. This is rarely achieved without team meetings and explicit professional standards to clarify mutual expectations.

Measuring Outcomes from Rehabilitation

Accurate assessment and outcome measurement in rehabilitation allows clear monitoring of progress, which can motivate the patient and also enhance communication among professionals. Standardized measures are imperative for this, and they need to be clinically feasible as well as having good validity, reliability, and responsiveness; those chosen need to be suitable for the individual and the domain that is being assessed. Multiple outcome measures are available within the varying physical domains, including gait assessment (e.g., 4-min or 6-min walk test), balance (e.g., Berg Balance Scale, Performance

Oriented Mobility Assessment), and mobility (e.g., Short Physical Performance Battery, Elderly Mobility Scale). While these measures are objective and reliant on an external observer, another such as the "Falls Efficacy Scale" is a self-reported measure of fear of falling.

Cognition (e.g., Mini-Mental State Examination, Montreal Cognitive Assessment) and depression (e.g., Geriatric Depression Scale, General Health Questionnaire-28) have their own frequently used measures. On a functional level, common measures of activities of daily living are the Functional Independence Measure or Barthel Activities of Daily Living Index. For a more personalized approach, the Goal Attainment Scale allows individualized goal selection and scaling.

Rehabilitation Interventions

The approach, intensity, and duration of a rehabilitation intervention will depend upon the place of the patient in the trajectories illustrated above. For relatively stable individuals with adequate self-efficacy, the emphasis may be on self-management such as an activity program built into daily ongoing life routines. The activity can incorporate physical, cognitive, and social elements, depending upon the results of the CGA. In other situations, the emphasis may need to be professionally delivered and/or supervised treatment, such as specific functional tasks or a progressive strength and balance exercise program. The choices will depend upon the baseline health and expected trajectory of the individual.

Rehabilitation Interventions to Address the Trajectory of Gradual Decline

Losses of intrinsic capacity can be mitigated by healthy behaviors such as exercise and good nutrition and hydration. Higher levels of routine physical activity are associated with a wide range of physical, mental, and social benefits. In all four countries of the United Kingdom and elsewhere, there are national programs aimed at increasing levels of physical activity among the general population and specifically among those aged 65 and older. Within community settings, functional ability should be maximized by reducing environmental barriers in order to promote independent activity.

Assuming that individuals with gradual declines of IC can be identified, then rehabilitation aimed at slowing, stopping, or reversing these declines generally requires generic interventions, not disease-specific approaches. Overall, there is limited evidence about how to design and deliver effective community-based programs targeting populations identified by screening rather than clinical populations. Most involve exercise interventions with encouragement to increase routine physical and social activity levels and optimization of nutrition, including protein intake, which is often suboptimal in frail older people. There is however good evidence for interventions to prevent falls (see below).

Rehabilitation After Acute Illness of Injury

To address acute decompensation in hospitals, rehabilitation objectives include prevention of complications of immobility such as joint contractures and pressure damage to the skin, maintenance of muscle mass and function, and reduction of falls risk. Modification of disease is likely to be helpful, but usually insufficient. In the postacute phase, the interventions are more likely generic, addressing the additional losses of IC incurred through the processes of reduced protein and calorie intake, catabolism, and reduced activity. These acute insults will add to the preexisting declines associated with age-related changes at cellular, organ, and physiological systems level. This broad approach can be achieved through CGA by a multidisciplinary team.

From individuals' perspective, the primary goal in the early stages may be a return to their own environment; therefore, rehabilitation may focus on task-specific activities such as mobilizing to the toilet or modification of the task by assistive devices or environmental adaptions. Motivation, physical environment, and social support networks will impact on their ability to achieve these early goals, and this interaction is often particularly complex in older adults. A focus on self-efficacy and assistive devices can maximize functional recovery within their environment.

During postacute rehabilitation and subsequent community-based programs, the emphasis is to restore function and to rebuild IC reserves lost through deconditioning. The opportunity to focus on optimizing IC after achieving manageable functional recovery is not always available in resource-limited services, and this probably renders the individual more vulnerable to future stressors. There is also inadequate research to clarify the most effective interventions and approaches.

The new respiratory syndrome coronavirus 2 (Sars-CoV-2 / COVID-19) has caused considerable mortality and morbidity throughout the world and will probably continue to do so, with new variants and inadequate vaccination programs continuing to frustrate efforts to contain the pandemic. Older people, those with disabilities and with poorer access to economic or healthcare resources have poorer outcomes and slower recovery. Completeness of recovery is variable. There is ongoing clinical and pathophysiological investigation into the disabling multisystem effects giving rise to a new syndrome termed "Long COVID," or "Post-COVID-19 syndrome," defined as signs and symptoms that develop during or after an infection consistent with COVID-19, continue for more than 12 weeks, and are not explained by an alternative diagnosis (National Institute of Health and Care Excellence 2020).

Features may include those previously associated with critical illness such as polyneuropathy (CIP), critical illness myopathy, and post-intensive care syndrome (PICS) as well as severe physiological deconditioning, cognitive, emotional, and psychosocial problems, which may result in prolonged functional decline and reduced quality of life. Therefore, many people recovering from the acute phase of COVID will need medium or longer-term rehabilitation. Some features are reminiscent of chronic fatigue syndrome. The optimal rehabilitation interventions and models of care are the subject of research programs internationally. Information on progress is available from the US National Library of Medicine (Australian and New Zealand Intensive Care Research Centre 2020).

The Challenge of Postural Stability

Balance is a complex, biological system that maintains bipedal ambulation. To maintain postural control, an individual must coordinate sensory input and task-specific motor output to stabilize the body's center of mass within the base of support (Horak 2006). Falls (like immobility) can be conceptualized as a failure of this balance system. In common with other "geriatric syndromes," impairment of balance often results from the cumulative effects of many small insults. The loss of compensating redundancy of intrinsic balance control factors results in progressive inability to maintain stability faced with balance-stressing activities, resulting in a fall. The contextual environment will impact the degree to which the activity being performed is balance-stressing.

Safe mobility depends on complex sets of integrated functions. Low physiological reserve in one or more systems renders the individual vulnerable to reduced mobility in the face of stressors. Age-related and disease-related declines of IC combine to reduce homeostatic balance reserve, so that relatively minor additional losses result in even low-balance tasks overcoming the reserves. Thus, most falls affecting older people occur while performing apparently routine simple activities. Figure 2 illustrates these points. This vulnerability varies between individuals of the same age, and *frailty* is the term used for this state of vulnerability more generally. Mobility impairment resulting in associated difficulties with everyday activities is a frequent feature of acute illness or hospitalization of frail older people and is thus a major focus of rehabilitation.

What Changes Are Important for Balance and Falls?

Several sensory and orientation modalities are important. Proprioception may be relatively more important than vestibular or visual information for dynamic balance (Peterka 2002). Deficits in one may be compensated by others with no apparent reduction in balance, particularly if the deficits develop gradually. For example, proprioceptive deficit may increase reliance on visual

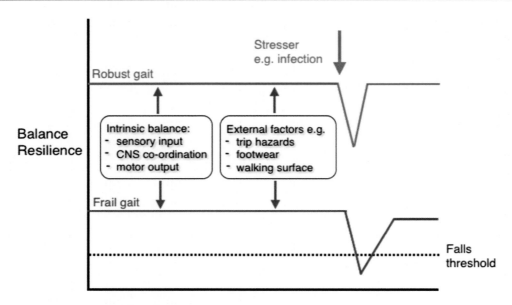

Geriatric Rehabilitation, Instability, and Falls, Fig. 2 Schematic diagram illustrating balance homeostatic reserve and falls (*CNS* central nervous system)

cues so that balance may only be impacted in the dark.

Motor output to maintain balance is mediated by nerves, muscles, tendons, bones, and joints, particularly of the lower limb. Sarcopenia, or age-related loss of muscle function, is a key risk factor for falls. Fracture-related deformity and joint degeneration from arthritis affect proprioception and weight distribution and, in the longer run, reduce muscle strength. Structural abnormalities of the ankle and load-bearing foot are also implicated (Menz et al. 2005).

Falls may also result from gait disorders associated with neurological impairments. Deficits in the peripheral nerves may result in "lower-order" gait disturbance. "Middle-level" gait disorders result from poor coordination due to damage in pyramidal pathways, basal ganglia disorders, and cerebellar syndromes. "Higher-level" disorders, associated with cerebral hemisphere deficits, are often accompanied by cognitive impairment (Snijders et al. 2007): They manifest as impairments of gait initiation and modulation; short, shuffling steps; and ataxia. Even mild cognitive impairment can result in deterioration of gait and falls risk when the individual is challenged to combine several activities simultaneously, e.g., walking while talking (Li et al. 2018). Higher-level gait disorders are a common cause of falls in older people (Thompson 2007). A faster rate of decline in gait speed with aging is associated with earlier mortality (White et al. 2013).

The Impact of Acute Illness on Stability and Falls Risk

Almost any acute illness, including infection and metabolic disturbance (e.g., hyponatremia), can act as a balance-stressing event. Particularly relevant are reductions in levels of consciousness, disruption of postural tone, and/or orthostatic hypotension. In delirium, gait dyspraxia, impaired judgment, and agitation may combine with the deconditioning from acute illness to result in falls during the postacute phase of recovery. Dementia and delirium not only increase the risk of falls severalfold but are also a challenge to successful rehabilitation.

The association of polypharmacy with falls operates particularly when the drug kinetics or dynamics are altered acutely and physiological reserves are reduced. Medication modification with temporary or permanent de-prescribing is an essential component of falls reduction.

Rehabilitation to Improve Stability and Reduce Falls

Falls and fear of falling are relevant in several aspects of geriatric rehabilitation. Firstly, after a fall, rehabilitation is necessary to avoid maladaptive behavioral modification or limitation which the older person may perceive as protective but is in the longer term counterproductive because it leads to deconditioning and consequently an increased risk of falling. Secondly, for individuals with declining strength, balance, sight, or cognition, then rehabilitation can reduce the likelihood of future falls. Finally, after any acute illness or hospitalization, the frequent loss of mobility that results can be reversed or at least minimized by rehabilitation, but during this process, there is a balance to be struck between encouraging increased physical activity for recovery with the temporary increase in falls risk and deconditioning. More specific information on interventions to reduce falls is discussed below.

Key Research Findings

Rehabilitation After Acute Illness or Injury

CGA applied has been shown to deliver better short- and longer-term functional outcomes for older, hospitalized patients than routine medical care (Ellis et al. 2017). Rehabilitation with a focus on exercise training can be successful during an acute hospital episode (Kosse et al. 2013). There is convincing evidence from randomized clinical trials for the effectiveness of rehabilitation after joint replacement or fracture (Prestmo et al. 2015; Perracini et al. 2018) and stroke (Stroke Unit Trialists' Collaboration 2013). Rehabilitation has additional benefits at the level of health systems. It has been associated with an increased likelihood of independent living and reduced admission to 24-h care facilities (Bachmann et al. 2010; Baztán et al. 2009; Ellis et al. 2017) with further positive outcomes including a reduction in hospital admissions and decreased healthcare costs (Pitkala et al. 2009; Taylor et al. 2014).

Rehabilitation as Part of Long-Term Condition Management

It is increasingly recognized that functional ability and quality of life for people living with long-term conditions (LTC) are impacted not only by the organ or system impairments directly consequent on the LTC severity, such as breathlessness, but also by the secondary effects of LTCs such as reduced physical activity, deconditioning, depression, and over- or undernutrition. Therefore, optimizing functional ability requires both disease-specific and generic approaches. Most of the effective interventions include elements specific to the target organ or system plus generic aspects addressing cardiorespiratory fitness and endurance or social participation such as group activities. Space precludes detailed description of the specific features. Exercise and task-specific functional training are common features of both the specific and generic interventions. Good evidence exists for the benefits for patients (improved symptoms and quality of life) and for healthcare systems (reduced frequency of hospital admissions) for older people with heart failure (Taylor et al. 2014) and chronic pulmonary disease (Goldstein et al. 2012; McCarthy et al. 2015).

Community-Based Rehabilitation to Enhance Stability and Reduce Falls

Exercise therapy represents an intervention at the final common pathway of multiple processes resulting in loss of mobility and balance. Exercise therapy incorporating graduated strength and balance training for a total of at least 50 h can reduce falls of older people, including those living with frailty (Sherrington et al. 2017). Treatment programs must be graduated, challenge both strength and balance, and be of sufficient intensity (at least 3 h per week) and duration (6 months); people near the falls threshold of dynamic balance reserve may derive significant functional and fall rate reduction at smaller increments in strength and balance. Long-term behavior changes should be promoted, with effects of exercise regressing after treatment cessation (Sherrington et al. 2017). Supported exercise may also overcome the maladaptive activity limitation response taken by older people who fall (Kumar et al. 2016).

Individuals with only low or moderate impairment of balance may benefit from Tai-Chi (Huang et al. 2017).

The ProAct65+ trial (Iliffe et al. 2015) showed that strength and balance training can result in increased routine physical activity for up to 1 year as well as reducing falls, but uptake to the program was limited. Those who adhere most closely to exercise prescriptions are likely to gain the most benefit, and therefore promoting adherence is essential with individual and program factors being important considerations. Framing exercise positively (promoting function as opposed to fall prevention) may be of benefit (Robinson et al. 2014). Developing social support structures around exercise has been advocated. However, interventions using telecommunication and the integration of exercise into activities of daily living appear most promising when delivering exercise at home (Hughes et al. 2018).

Home-based environmental hazard reduction interventions after hospitalization or falls have been shown to reduce rates of falls, particularly in moderate-risk groups (Lord 2006; Hopewell et al. 2018). Given the complex, multimodal etiology of falls described above, it is not surprising that CGA, with multifaceted interventions to identified risk factors, has been a common approach employed in national and international guidelines (WHO, AGS/BGS 2015; NICE 2013). While efficacious in reducing rates of falls, multifaceted interventions have not been shown to be more effective than exercise alone (Campbell and Robertson 2007; Hopewell et al. 2018) and are also more resource-intensive and, therefore, less cost-effective (Tiedemann et al. 2013). The reasons for this counterintuitive finding are unclear; some have postulated interacting interventions affecting individual efficacy, e.g., reduced focus on the more complex but effective interventions; conflicting advice, e.g., home exercise versus home hazard intervention; the presence of a single dominant risk factor; or the final common pathway of strength and balance deficits leading to ceiling effects.

Patients with cognitive impairment present a particular challenge with most guidelines not providing specific recommendations in this group (Fernando et al. 2017). However, preliminary evidence both inside and outside care homes has been encouraging and is the subject of ongoing trials (Burton et al. 2015). Psychological factors are important. Falls are a traumatic event, with withdrawal from activity an important maladaptive, psychological, compensation mechanism that in itself increases isolation and dependency, which begets further falls (Friedman et al. 2002).

In most population health systems, there is no ongoing systematic monitoring of IC or functional ability, so the prompt for rehabilitation is either an acute destabilizing event or when a more gradual decline reaches a point where functional ability transitions into a degree of dependency. Better prediction of where someone lies on a trajectory of decline will hopefully lead to intervention before a clinical threshold or crisis is reached. Some policy leaders advocate for routine screening, or at least opportunistic case finding, to anticipate such decline and instigate rehabilitative measures (WHO 2017; BGS 2015).

Prevention of Falls During Rehabilitation During and After Acute Illness

NICE guidance (NICE 2013) recommends identification and amelioration of falls risk factors to be initiated soon after a hospital admission with acute illness. But for hospital inpatients and for care home residents, evidence for benefit from an approach based on targeted multifactorial interventions including exercise training and medication modification is poor (Cameron et al. 2018). Anecdotal and some research evidence suggests that awareness of the relevant risk factors, anticipation of risk behaviors, delirium prevention, careful surveillance, and prompt remedial actions may be more effective (Healey et al. 2004; NICE 2013).

Future Directions of Research

There is a substantial research effort currently to address the uncertainties of how best to identify those who may benefit most from general and tailored interventions to modify the trajectory of loss of intrinsic capacity and subsequently functional ability. Screening at population level does not have evidence of cost-effectiveness. The same is true for falls prevention specifically.

Those who adhere most closely to multifactorial interventions or exercise prescriptions are likely to gain the most benefit, and therefore promoting adherence is essential with individual and program factors being important considerations. Research continues to develop the best approaches to maximize uptake and adherence to the interventions and how then to sustain the gains made. A limitation of many of the commonly used measures of functional ability is that they are only sensitive to change once significant losses have already occurred. Research suggests biomarkers in domains including cognition, mood, sensation, and locomotion may be used to identify subclinical physiological changes (Stuck et al. 1999). The implication of these developments is the recognition of preclinical decline and the potential to implement measures (e.g., exercise) to slow or prevent the onset of functional limitation.

Optimum nutrition for health aging remains a subject of increasing research with positive evidence emerging for the Mediterranean diet. In view of the difficulties in adherence and the variable responses to exercise, pharmacological and nutritional treatments for sarcopenia, a central feature of physical frailty, are also needed. People with cognitive impairment and care home residents have particularly high falls rates and so far have proved less able to benefit from current interventions (Taylor et al. 2017). Novel approaches are needed.

Summary

Geriatric rehabilitation aims to enable well-being of older people through maintenance or restoration of functional ability. The clinical context will vary and determine the most appropriate approaches to assessment and the interventions applied, but the broad principles remain the same – comprehensive assessment leading to individualized goal setting based on the principles of consent and patient-centered conversations, followed by agreed treatment plans and follow-up arrangements. The focus of the interventions may include one of more of the following dimensions: the intrinsic capacities of the individual such as mobility or cognition, functional tasks, environmental adaptations, or social supports. The mode of interventions may be advice, facilitated self-management, professionally supervised or delivered therapies, and practical help. Where relevant, enabling the contribution of informal carers can be encouraged. The outcomes of rehabilitation can be measured across several domains and at various levels such as individual functioning, dependency, social participation, and healthcare use, with gains at both individual and societal levels being the most desirable. Enabling safe mobility is a common feature of rehabilitation after acute illness or hospital-induced deconditioning and for preventative community-based approaches. Reducing the risk of falls is intrinsic to this task while avoiding maladaptive limitation of independent activities.

Cross-References

▶ Aerobic Exercise Training and Healthy Aging
▶ Comprehensive Geriatric Assessment
▶ Disability Measurement
▶ Exercise Adherence
▶ Falls
▶ Frailty in Clinical Care
▶ Functional Limitation
▶ Geriatric Depression Scale
▶ Goal Setting
▶ Hip Fracture
▶ Mobility and Frailty
▶ Montreal Cognitive Assessment
▶ Multimorbidity in Aging
▶ Post-acute Care
▶ Sarcopenia

References

Australian and New Zealand Intensive Care Research Centre (2020) Recovery of patients from COVID-19 after critical illness (COVID-Recovery). https://clinicaltrials.gov/ct2/show/NCT04401254. Accessed 22 March 2021

Bachmann S, Finger C, Huss A, Egger M, Stuck AE, Clough-Gorr AM (2010) Inpatient rehabilitation specifically designed for geriatric patients: systematic review and meta-analysis of randomised controlled trials. BMJ 340:c1718. https://doi.org/10.1136/bmj.c1718

Baztán JJ, Suárez-García FM, López-Arrieta J, Rodriguez-Manas L, Rodriguez-Artalejo F (2009) Effectiveness of acute geriatric units on functional decline, living at home, and case fatality among older patients admitted to hospital for acute medical disorders: meta-analysis. BMJ 338:b50. https://doi.org/10.1136/bmj.b50

British Geriatrics Society (2015) Fit for Frailty – developing, commissioning and managing services for people living with frailty in community settings – a report from the British Geriatrics Society and the Royal College of General Practitioners (2015). Available via https://www.bgs.org.uk/sites/default/files/content/resources/files/2018-05-23/fff_full.pdf. Accessed 3 Mar 2019

Burton E, Cavalheri V, Adams R, Browne CO, Bovery-Spencer P, Fenton AM, Campbell BW, Hill KD (2015) Effectiveness of exercise programs to reduce falls in older people with dementia living in the community: a systematic review and meta-analysis. Clin Interv Aging 10:421–434. https://doi.org/10.2147/CIA.S71691

Cameron ID, Dyer SM, Panagoda CE, Murray GR, Hill KD, Cumming RG, Kerse N (2018) Interventions for preventing falls in older people in care facilities and hospitals. Cochrane Database Syst Rev (9):CD005465. https://doi.org/10.1002/14651858.CD005465.pub4

Campbell AJ, Robertson MC (2007) Rethinking individual and community fall prevention strategies: a meta-regression comparing single and multifactorial interventions. Age Ageing 36(6):656–662

Ellis G, Gardner M, Tsiachristas A, Langhorne P, Burke O, Harwood RH, Conroy SP, Kircher T, Somme D, Saltvedt I, Wald H, O'Neill D, Robinson D, Shepperd S (2017) Comprehensive geriatric assessment for older adults admitted to hospital. Cochrane Database Syst Rev (9):CD006211. https://doi.org/10.1002/14651858.CD006211.pub3

Fernando E, Fraser M, Hendriksen J, Kim CH, Muir-Hunter SW (2017) Risk factors associated with falls in older adults with dementia: a systematic review. Physiother Can 69(2):161–170

Friedman SM, Munoz B, West SK et al (2002) Falls and fear of falling: which comes first? A longitudinal prediction model suggests strategies for primary and secondary prevention. J Am Geriatr Soc 50:1329–1335

Goldstein RS, Hill K, Brooks D, Dolmage TE (2012) Pulmonary rehabilitation: a review of the recent literature. Chest 142(3):738–749. https://doi.org/10.1378/chest.12-0188

Healey F, Monro A, Cockram A, Adams V, Heseltine D (2004) Using targeted risk factor reduction to prevent falls in older in-patients: a randomised controlled trial. Age Ageing 33(4):390–395

Hopewell S, Adedire O, Copsey BJ, Boniface GJ, Sherrington C, Clemson L, Close JCT, Lamb SE (2018) Multifactorial and multiple component interventions for preventing falls in older people living in the community. Cochrane Database Syst Rev (7): CD012221. https://doi.org/10.1002/14651858.CD012221.pub2

Horak FB (2006) Postural orientation and equilibrium: what do we need to know about neural control of balance to prevent falls? Age Ageing 35(suppl_2):ii7–ii11

Huang ZG, Feng YH, Li YH, Lv CS (2017) Systematic review and meta-analysis: tai chi for preventing falls in older adults. BMJ Open 7(2):e013661

Hughes KJ, Salmon N, Galvin R, Casey B, Clifford AM (2018) Interventions to improve adherence to exercise therapy for falls prevention in community-dwelling older adults: systematic review and meta-analysis. Age Ageing. https://doi.org/10.1093/ageing/afy164

Iliffe S, Kendrick D, Morris R, Griffin M, Haworth D, Carpenter H, Masud T, Skelton DA, Dinan-Young S, Bowling A, Gage H (2015) Promoting physical activity in older people in general practice: ProAct65+ cluster randomised controlled trial. Br J Gen Pract 65(640): e731–e738

Kosse NM, Dutmer AL, Dasenbrock L, Bauer JM, Lamoth CJC (2013) Effectiveness and feasibility of early physical rehabilitation programs for geriatric hospitalized patients: a systematic review. BMC Geriatr 13:107. http://www.biomedcentral.com/1471-2318/13/107

Kumar A, Delbaere K, Zijlstra GAR, Carpenter H, Iliffe S, Masud T, Skelton D, Morris R, Kendrick D (2016) Exercise for reducing fear of falling in older people living in the community: Cochrane systematic review and meta-analysis. Age Ageing 45(3): 345–352

Li KZH, Bherer L, Mirelman A, Maidan I, Hausdorff JM (2018) Cognitive involvement in balance, gait and dual-tasking in aging: a focused review from a neuroscience of aging perspective. Front Neurol 9:913. Published online 2018 Oct 29. https://doi.org/10.3389/fneur.2018.00913

Lord SR (2006) Visual risk factors for falls in older people. Age Ageing 35(suppl_2):ii42–ii45

McCarthy B, Casey D, Devane D, Murphy K, Murphy E, Lacasse Y (2015) Pulmonary rehabilitation for chronic obstructive pulmonary disease. Cochrane Database Syst Rev (2):CD003793. https://doi.org/10.1002/14651858.CD003793.pub3

Menz HB, Morris ME, Lord SR (2005) Foot and ankle characteristics associated with impaired balance and functional ability in older people. J Gerontol Ser A Biol Sci Med Sci 60A:1546–1552

National Institute of Health and Care Excellence (2013) NICE clinical guideline 161, falls: assessment and prevention of falls in older people. NICE, London. Available via www.nice.org.uk/guidance/CG161. Accessed 20 Feb 2019

National Institute of Health and Care Excellence (2020) COVID-19 rapid guideline: managing the long-term effects of COVID-19. NICE guideline [NG188] December 2020. Available from https://www.nice.org.uk/guidance/NG188. Accessed 22 March 2021

Perracini MR, Kristensenb MT, Cunningham C, Sherrington C (2018) Physiotherapy following fragility fractures. Injury. Int J Care Injured 49:1413–1417

Peterka RJ (2002) Sensorimotor integration in human postural control. J Neurophysiol 88:1097–1118

Pitkala KH, Routasalo P, Kautiainen H, Tilvis RS (2009) Effects of psychosocial group rehabilitation on health, use of health care services, and mortality of older persons suffering from loneliness: a randomized, controlled trial. J Gerontol A Biol Sci Med Sci 64(7): 792–800. https://doi.org/10.1093/gerona/glp011

Prestmo A, Hagen G, Sletvold O, Helbostad JL, Thingstad P, Taraldsen K et al (2015) Comprehensive geriatric care for patients with hip fractures: a prospective, randomised, controlled trial. Lancet 385(9978): 1623–1633. https://doi.org/10.1016/S0140-6736(14) 62409-0

Robinson L, Newton JL, Jones D, Dawson P (2014) Self-management and adherence with exercise-based falls prevention programmes: a qualitative study to explore the views and experiences of older people and physiotherapists. Disabil Rehabil 36(5):379–386. https://doi.org/10.3109/09638288.2013.797507

Sherrington C, Michaleff ZA, Fairhall N, Paul SS, Tiedemann A, Whitney J, Cumming RG, Herbert RD, Close JC, Lord SR (2017) Exercise to prevent falls in older adults: an updated systematic review and meta-analysis. Br J Sports Med 51(24):1750–1758

Snijders AH, van de Warrenburg BP, Giladi N, Bloem BR (2007) Neurological gait disorders in elderly people: clinical approach and classification. Lancet Neurol 6(1):63–74

Stroke Unit Trialists' Collaboration (2013) Organised inpatient (stroke unit) care for stroke. Cochrane Database Syst Rev 11(9):CD000197. https://doi.org/10.1002/14651858.CD000197.pub3

Stuck AE, Walthert JM, Nikolaus T, Bula CJ, Hohmann C, Beck JC (1999) Risk factors for functional status decline in community-living elderly people: a systematic literature review. Soc Sci Med 48(4):445–469

Taylor RS, Sagar VA, Davies EJ, Briscoe S, Coats AJ, Dalal H et al (2014) Exercise-based rehabilitation for heart failure. Cochrane Database Syst Rev 4: CD003331. https://doi.org/10.1002/14651858.CD003331.pub4

Taylor ME, Lord SR, Brodaty H, Kurrle SE, Hamilton S, Ramsay E, Webster L, Payne NL, Close JC (2017) A home-based, carer-enhanced exercise program improves balance and falls efficacy in community-dwelling older people with dementia. Int Psychogeriatr 29(1):81–91

Thompson PD (2007) Higher level gait disorders. Curr Neurol Neurosci Rep 7(4):290–294

Tiedemann A, Sherrington C, Lord SR (2013) The role of exercise for fall prevention in older age. Motriz: Revista de Educação Física 19(3):541–547

White DK, Neogi T, Nevitt MC, Peloquin CE, Zhu Y, Boudreau RM et al (2013) Trajectories of gait speed predict mortality in well-functioning older adults: the health, aging and body composition study. J Gerontol Ser A Biol Sci Med Sci 68(4):456–464

World Health Organization (2001) International classification of functioning, Disability and Health (ICF) World Health Organization, Geneva. Available via http://www.who.int/classifications/icf/en/. Accessed 23 Feb 2019

World Health Organization (2015) Report on ageing and health. Available via https://www.who.int/ageing/events/world-report-2015-launch/en/. Accessed 23 Feb 2019

World Health Organization (2017) Integrated care for older people: guidelines on community-level interventions to manage declines in intrinsic capacity. World Health Organization, Geneva. Available via https://apps.who.int/iris/handle/10665/258981. Accessed 08 Mar 2019

World Health Organization (2018) Falls-key facts, Geneva. Available via https://www.who.int/ageing/projects/falls_prevention_older_age/en/. Accessed 01 Feb 2020

Geriatric Screening

▶ Geriatric Interventions

Geriatric Social Workers

Daniel B. Kaplan
School of Social Work, Adelphi University, Garden City, NY, USA

Synonyms

Gerontological social workers; Social services providers

Definition

Social workers, often with relevant education and/or certification in geriatric or gerontological practice, serving older adults and their families in adjusting to biopsychosocial and environmental challenges associated with aging into the latter decades of the life course, ongoing challenges which persist from earlier stages of life, or their intersections.

Overview

Today's older adults have lived through a transformational era where population aging radically altered the landscape of families, communities, politics, economics, and health-care systems. In the year 1900, older adults accounted for only 4% of the US population. This proportion grew to 8% by 1946 – the year when the baby boomer generation began – and then continued to increase to 13% by 2010. The US Census Bureau projects that older adults will account for over 20% of the population by 2050 (Vincent and Velkoff 2010). Longer lifespans are the welcomed result of significant achievements in public health, medicine, and community and organizational safety. Yet, aging into very advanced years of life also increases the likelihood of confronting a number of challenges which threaten quality of life and emotional well-being for individuals and families (Hooyman and Kiyak 2018) (See ▶ "Multidimensional Views of Aging and Old Age").

Social workers and other helping professionals have both the opportunity and responsibility to support and assist older adults in facing their unique challenges. Social workers are trained in distinct areas of competence to utilize assessment and intervention skills and relevant knowledge of human behavior and systems change to empower clients to achieve personal goals that will optimize quality of life (See ▶ "Social Work"). This requires simultaneous work to overcome individual and familial challenges, community and service system barriers, and problematic policies which govern service delivery, health care and insurance, and income and housing.

Key Roles and Settings

Geriatric social workers respond to the biopsychosocial, environmental, economic, and spiritual impacts of aging for older adults, their families, and society (Kaplan and Berkman 2015). Their clients include healthy older adults who lead fulfilling lives, as well as older adults who are challenged by chronic physical and mental health conditions, intellectual and developmental disabilities, neurological or substance-use disorders, incarceration, isolation, poverty and homelessness, and mistreatment and neglect, as well as their family and other caregivers. They practice in community-based health and social service settings, long-term care facilities, palliative and end-of-life care agencies, educational programs, justice systems, individual and family counseling centers, and specialized health settings such as neurology clinics. They develop policies and programs affecting retirement, employment and voluntarism, housing, transportation, health, mental health, and caregiving for older adults.

As such, geriatric social workers provide comprehensive biopsychosocial assessments, cognitive-behavioral interventions, crisis mitigation, family counseling, supportive psycho-education, program planning and evaluation, community organizing and advocacy work, and case management and care coordination services. They are driven by unique values (National Association of Social Workers 2017) which guide strengths-based and solution-focused practice with older adults and the use of approaches to empower individuals and communities (See ▶ "Person-Centered Care for Older Adults"). They attend to significant historical and contemporary issues facing older adults with particular attention to the intersectionality of aging with gender, race, ethnicity, nationality, class, sexual orientation, religion, and physical and mental disability. Their commitments to social justice demand continual work to celebrate human diversity while fighting against ageism, building knowledge about the harmful effects of the cumulative disadvantages which result from oppression and discrimination, and creating innovative models of inclusive and equitable care and community (See ▶ "Theory of Cumulative Disadvantage/Advantage").

Example: Geriatric Social Workers and Dementia Care

Neurocognitive disorder, or dementia, is a syndrome creating devastation in nearly every domain of existence, with serious negative impacts which reverberate through families, communities, and society. Dementia is the 6th leading cause of death in the USA but the only 1 of 10 leading causes of death which cannot be prevented or cured (Alzheimer's Association 2017). It is a syndrome caused by nearly 60 distinct diseases and conditions (Qui et al. 2007). Dementia is present in 1/10th of older adults, and any setting serving older adults and their families routinely interacts with people who are struggling with dementia.

In order to understand and to respond effectively to the needs of their clients with dementia, geriatric social workers study the complex interactions of disease pathology, individual strengths, environmental conditions, informal supports, formal resources, and societal influences such as health-care policies. They are able to identify evidence-informed interventions for clients throughout the unpredictable fluctuations and evolving needs across stages of dementia. Importantly, geriatric social workers also work to explain the multidimensional needs of families confronting dementia in order to educate interdisciplinary care teams about the importance of addressing psychosocial factors.

While there are considerable needs to increase the overall quality and quantity of providers who can meet the mental health needs of older adults (Shah and Kaplan 2015) and a significant need for enhanced supervision in gerontological care (Kaplan et al. 2018), geriatric social workers are already situated throughout the continuum of care and are appropriately trained to tackle the multidimensional impacts of the dementia syndrome. They are skillful in delivering clinical interventions and coordinating complementary supportive services with individuals with dementia and their families. Social workers are driven by values which inform the ethical application of person-centered care to promote human and community well-being, which is essential in the context of dementia care (Kaplan and Andersen 2013).

Summary

Geriatric social workers serve older adults in any number of health, mental health, social service, or community-based care settings. They strive to create equitable access to effective programs that will respond to the complex needs of diverse older adults and their families and communities. The interventions of geriatric social workers promote personhood and dignity, self-determination, safety, emotional well-being and fulfillment, optimal functioning, and a high quality of life.

Comprehensive reference materials and educational guidance on geriatric social work can be found in the following sources: (1) *Oxford Handbook of Social Work and Aging* (Kaplan and Berkman 2015) and (2) *Oxford Bibliographies* (if accessible through university or public library subscriptions), offering annotated lists of readings on social work and aging, geriatric health and mental health, gerontological assessments and practice interventions, family caregiving, public policies, and health insurance programs. In addition, the National Association of Social Workers offers geriatric social work credentials, a specialty practice section on aging, and published standards and guidelines on social work practice with family caregivers of older adults, palliative and end-of-life care, and services in long-term care facilities.

Cross-References

▶ Adult Protective Services
▶ Aging and Health Disparities
▶ Alzheimer's Disease
▶ Care Needs

- ▶ Caregiving and Ethnicity
- ▶ Case Management
- ▶ Dementia
- ▶ End-of-Life Care
- ▶ Gender Inequity
- ▶ Geriatric Mental Health
- ▶ Healthy Aging
- ▶ Home- and Community-Based Services (HCBS)
- ▶ LGBT in Old Age
- ▶ Mental Health Services
- ▶ Multidimensional Views of Aging and Old Age
- ▶ Person-Centered Care for Older Adults
- ▶ Psychological Theories of Health and Aging
- ▶ Religion and Spirituality in End-of-Life Care
- ▶ Sexism and Ageism
- ▶ Social Services Utilization
- ▶ Social Support
- ▶ Social Work
- ▶ Socioeconomic Status
- ▶ Theory of Cumulative Disadvantage/Advantage

References

Alzheimer's Association (2017) Alzheimer's disease facts and figures. Alzheimer's & Dementia 13(4):325–373

Hooyman NR, Kiyak HA (2018) Social gerontology: a multidisciplinary perspective, 10th edn. Allyn & Bacon, Boston

Kaplan DB, Andersen TA (2013) The transformative potential of social work's evolving practice in dementia care. J Gerontol Soc Work 56(2):164–176. https://doi.org/10.1080/01634372.2012.753652

Kaplan DB, Berkman B (eds) (2015) The Oxford handbook of social work in health and aging, 2nd edn. Oxford University Press, New York

Kaplan DB, Silverstone B, Zlotnik JL et al (2018) NASW's Supervisory Leaders in Aging: an effective and acceptable model for training and supporting social work supervisors. Clin Soc Work J, Special Issue: Clinical Supervision and Field Education in Social Work 46(4):321–330. https://doi.org/10.1007/s10615-018-0673-6

National Association of Social Workers (2017) Code of ethics of the National Association of Social Workers. Retrieved 14 Feb 2019, from https://www.socialworkers.org/About/Ethics/Code-of-Ethics/Code-of-Ethics-English

Qui C, de Ronchi D, Fratiglioni L (2007) The epidemiology of the dementias: an update. Curr Opin Psychiatry 20:380–385. https://doi.org/10.1097/YCO.0b013e32816ebc7b

Shah A, Kaplan DB (2015) Hard truth, with dire consequences: America lacks trained mental health workers to care for an aging population. Aging Today 36(2):8–14

Vincent GK, Velkoff VA (2010) The next four decades: the older population in the United States: 2010 to 2050. Current population reports number P25-1138. U.S. Census Bureau, Washington, DC

Geriatrics

- ▶ Biogerontology

Geriatrics Curriculum

- ▶ Teaching Aging Medicine

German Ageing Survey (DEAS)

Claudia Vogel, Daniela Klaus, Markus Wettstein, Julia Simonson and Clemens Tesch-Römer
German Centre of Gerontology (DZA), Berlin, Germany

Overview

Aging research needs longitudinal data to adequately describe and study age-related trajectories. Consequently, many longitudinal aging studies have been conducted worldwide and are still ongoing. In addition, aging research needs to monitor social change and societal trends in order to keep track of the changing living situations of older people. Sequential cross-sectional designs are suitable for this task, and aging research also abounds with these types of studies. However, neither longitudinal nor sequential cross-sectional designs alone allow for the disentanglement of age and cohort effects, as well as of secular trends in individual age-related changes. Cohort-sequential designs are necessary to reach this goal. This type of design combines sequential cross-sectional samples with longitudinal data collection. Because of their expansive design, cohort-sequential studies are costly and, hence, rare in aging research.

The German Ageing Survey (Deutscher Alterssurvey – DEAS) is an example of a cohort-sequential study that has been ongoing for nearly a quarter century (Kohli 2000). The DEAS combines large cross-sectional samples with longitudinal data collection (Klaus et al. 2017). Starting in 1996, representative samples of the German population in the age range between 40 and 85 years have been drawn every 6 years. There are currently four independent samples (1996, 2002, 2008, and 2014) and a fifth cross-sectional sample has been drawn for ongoing data collection in 2020. Each of these samples is followed over time. Longitudinal data have been collected in 2002, 2008, 2011, 2014, and 2017. To date, nearly 40,000 interviews have been conducted with more than 20,000 participants. Observations cover a time span of up to 24 years, from 1996 to 2020. Since its start in 1996, the DEAS has been funded by the German Federal Ministry for Family Affairs, Senior Citizens, Women, and Youth. The design of the DEAS allows researchers to investigate: (a) intra-individual change by comparing the same individuals at different points in time, (b) social change by comparing different birth cohorts at the same ages, and (c) social change of aging processes by comparing intra-individual change between different birth cohorts.

Study Design

Study Population

The major aim of the DEAS is the monitoring of the living conditions of middle-aged and older individuals. Hence, each DEAS cross-sectional sample is nationwide representative for community-dwelling adults aged 40–85 years. The target population in 1996 and in 2002 was defined as community-dwelling German citizens aged 40–85 years. Since 2008, the definition was widened to include individuals without German citizenship. The population share of individuals without German citizenship is increasing – according to the Federal Statistical Office – from about 9% of the total population living in Germany in the year 2008 to about 12% in the year 2019 (Statistisches Bundesamt 2019: 19).

Baseline Samples

Independent cross-sectional samples (baseline samples) of individuals aged 40 to 85 years have been drawn since 1996 every 6 years (in 1996, 2002, 2008, 2014, and – currently running – in 2020). The cross-sectional samples are drawn in a two-step sampling methodology:

- In the first step, in 1996, a random sample of 290 municipalities was drawn from the total of around 15,000 municipalities that existed in Germany at that time, providing the basis for the baseline samples in 1996, 2002, 2008, and 2014. For the wave 2020, a new random sample of 200 municipalities was drawn from the total of about 11,000 municipalities that exist in Germany nowadays. The local population registries of these municipalities provide information for the sampling procedure (such as name, sex, year of birth, and address).
- In the second step, within the selected municipalities, samples are drawn from the population of individuals living in private households, aged between 40 and 85 years. These samples have been disproportionally stratified into three age groups (40–54, 55–69, and 70–85 years), gender (men and women), and region (East and West Germany). Individuals in the oldest age group, along with men and individuals living in East Germany, were oversampled to ensure that there would be a sufficient number of participants in subgroups of interest for differentiated analyses, such as descriptions of the living situation of older men living in East Germany. In addition, this oversampling increases the number of participants taking part at follow-up measurement occasions so that all relevant population subgroups provide sufficient observations for longitudinal analyses.

Up to 2017, a total of 20,129 individuals aged between 40 and 85 years at the first interview have participated at least once in the DEAS. By the end of 2020, it will be more than 26,000 individuals. The sample sizes range from about 3,000 to about 6,000: $n_{1996} = 4,838$; $n_{2002} = 3,084$; $n_{2008} = 6,205$; $n_{2014} = 6,002$; $n_{2020} = 6,000$ (expected). DEAS participants were born between 1911 and 1980. Case numbers are highest for the birth cohorts 1930 to 1969.

Longitudinal Samples

Each baseline sample is re-assessed every 3 years (before 2008 every 6 years); thus, longitudinal data have been collected in the years 2002, 2008, 2011, 2014, 2017, and (currently running) 2020. The next panel data collection is planned to take place in 2023. The long-time span starting from 1996 allows researchers to analyze individual trajectories starting from age 40 years to old and very old ages. So far, the oldest participants are in their 90s (theoretically there could be participants aged 100 years and above, but none have participated so far) (Klaus et al. 2019: 21–22). Due to the cohort-sequential design, with a recruitment of a new baseline sample every 6 years, samples sizes of participants providing repeated observations have been largely increasing over time: $n_{2002} = 1,524$; $n_{2008} = 1,991$; $n_{2011} = 4,854$; $n_{2014} = 4,322$; $n_{2017} = 6,626$; $n_{2020} = 5,750$ (expected). Until 2017, 310 individuals have participated in all six waves (Klaus et al. 2019: 21). As life expectancy is increasing, the DEAS also allows continuously more panel interviews with respondents aged 90 years and above, providing a representative and growing data base of old and very old individuals living in private households.

The assessment instrument consists of two components. Face-to-face interviews (CAPI) are conducted in each wave with an average duration of about 90 min. Additionally, respondents are asked to fill in a self-administered questionnaire (paper-pencil version or alternatively (since 2014) as online version (CAWI)). About 85% of the interviewees do so. Since 2017, interviews with proxy persons are additionally conducted in case that panel participants may not be re-interviewed due to health issues, to reduce selectivity. The documentation of the survey instruments in German and English is available here: https://www.dza.de/en/fdz.html.

Data Weights

Weights are provided to permit an unbiased generalization of descriptive findings calculated from the stratified sample for the population living in Germany. Weights are computed for each wave. They adjust for the disproportional stratification in the sample against the population distribution, which is obtained from the national micro census (provided by the Federal Statistical Office – Statistisches Bundesamt). And they model attrition processes to adjust for differential nonresponse in the successive panel waves. These combined weights allow joint cross-sectional analyses of baseline and panel data up to the age of 85 years, and from 2014 onwards even up to the age of 90 years.

Data Quality

The response rates for the baseline samples are defined as the proportion of respondents with valid interviews against the gross sample of eligible individuals. The response rates were 50.3% in 1996, 37.8% in 2002, 35.7% in 2008, and 27.1% in 2014 (2020 is currently running, 25% are expected). This trend of decreasing response rates is in line with surveys all over the Western world. Response rates in the DEAS are low as compared with other European surveys on aging (See ▶ "The Survey of Health, Ageing and Retirement in Europe"), but are similar to other surveys conducted in Germany (e.g., Bergmann et al. 2017; Brüderl et al. 2015).

As in most surveys, sample selectivity can be observed in the DEAS (Klaus and Engstler 2017): Response rates tend to be lower in large cities, among women and in the age groups 40–54 years as well as 70–85 years. Such selectivity effects are, however, small, and the distribution of sociodemographic characteristics such as marital status, household size, and employment status in the weighted baseline samples is very close to the distribution within the population. However, the share of individuals without German citizenship is somewhat lower, which might be a result of conducting the interviews in German language only.

Panel attrition is high in the first re-interview but attenuates in subsequent follow-ups, a phenomenon that is common in panel studies. For example, from the baseline 2008, 46% participated in the next wave 2011, 41% in 2014, and still 34% in 2017 (retention rate defined as valid re-interviews as a proportion of the number of valid interviews in the baseline). The retention rate has increased, however, over the waves, after the interval between panel waves has been reduced from 6 to 3 years from 2008 onwards. For example, the retention rate based on the 2008 baseline was considerably higher in 2014 than the retention rate based on the 1996 baseline in 2002 (41–32%; Klaus et al. 2019: 21). Retention rates are associated with a variety of demographic and socioeconomic characteristics. In the DEAS, panel participants tend to be somewhat younger, healthier, and better educated, and they tend to have larger networks than respondents who drop out of the study (Klaus et al. 2019: 26).

Data Availability

Data of the German Ageing Survey (DEAS) are available free of charge to the scientific community for nonprofit purposes. The research data center of the German Centre of Gerontology provides access and support to scholars interested in using DEAS data for their research. Information on how to access the data is available here: www.fdz-dza.de.

Topics

The overarching aim of the survey is social monitoring of the living conditions of middle-aged and older individuals. Hence, a broad range of topics concerning age and aging is covered in the DEAS: beyond sociodemographic characteristics, data are collected on various aspects of quality of life such as household composition, provisions for old age and housing, employment, work and volunteering, family relationships and social support, as well as health and subjective well-being. Additionally, a pulmonary function test measuring the lung capacity and a digit number test measuring the cognitive capacity of the participants are conducted.

Major Findings

Provision for Old Age and Housing

How pronounced is social inequality in terms of the income and housing situation among older people in Germany, and how has social inequality changed over the last decades? Findings of the DEAS show that income poverty levels are on the rise, but not for everybody (Lejeune et al. 2017). While the poverty rates for women aged 40–85 years are larger than those of men, they have increased only slightly (from 12% to 14%) from 1996 to 2014 (See ▶ "Poverty and Gender Issues in Later Life in the UK"). In contrast, the increase in poverty rates for men was larger, from 8% to 11%, but the rates for men are still lower than those for women.

Regarding educational differences, poverty rates among individuals aged 40–85 years with high education have remained about the same (less than 5%), whereas poverty rates among those with low education have increased strongly from 22% to 35%. While the majority of older people is doing well in financial terms, and many parents support their adult children financially (Künemund et al. 2005), the situation of older individuals with low education has become more precarious in recent years. Furthermore, such disadvantages might accumulate during the second half of life and jeopardize the quality of life, when costs, e.g., for health care or housing, additionally increase, while economic resources decrease (Wetzel et al. 2019).

Indeed, tenants aged 40–85 years spend today on average 35% of their income on housing (in 1996 it was only 28%; Nowossadeck and Engstler 2017). Especially those tenants aged 65 and above who belong to the quintile of the population with the lowest incomes have experienced a strong increase in housing cost burden to 41% in 2016, which highly limits the financial scope in old age (Romeu Gordo et al. 2019). In addition, entering income poverty increases the risk for social isolation (Hajek and König 2020).

Employment, Work, and Volunteering

Active aging is linked to rising participation levels in both paid and unpaid work such as volunteering. The DEAS shows similar trends in paid and unpaid work: individuals in their 50s, 60s, and 70s in Germany nowadays are more active and tend to participate longer both in paid and in unpaid positions than the birth cohorts before them.

Although the rate of employment of older workers has increased since the mid-1990s, a majority continues to retire before the standard retirement age. Individual plans and expectations in relation to retirement tend to lag behind changes in the regulations set out for retirement. A considerable minority of older employees in Germany still plan to retire before the age of 65 years or even before the age of 60 years (Engstler and Romeu Gordo 2017). At the same time, remaining in employment after completing the transition into retirement has become increasingly common over the past few years (Franke and Wetzel 2017). Using data from the DEAS and the English Longitudinal Study of Ageing (ELSA), it was shown that fewer people pursue post-retirement employment in Germany than in England, mainly for institutional and/or structural reasons (Scherger et al. 2012; Hokema and Scherger 2016).

In line with findings on volunteering in other European countries (Erlinghagen and Hank 2006), the DEAS shows that volunteering is more widespread among the highly educated than among individuals with low-educational status (Wetzel and Simonson 2017). In addition, there are pronounced regional differences in rates of volunteering. In economically stronger districts, the rate of volunteering is higher than in economically weaker ones, even when controlling for social inequality at an individual level (Simonson et al. 2013). Between 1996 and 2014, not only did the rates of volunteering in organizations increase in general, but also the rates for people volunteering in organizations with a particular focus on older people (Wetzel and Simonson 2017).

While the likelihood to volunteer reaches a maximum around 65 years of age, it slightly decreases afterwards with increasing age. Also, individuals born in the 1950s are not only much more likely to volunteer during the second half of life than individuals born in the 1930s. The time point when volunteering is given up seems to be delayed (Vogel and Romeu Gordo 2019). Findings from the DEAS suggest that volunteering affects subjective well-being differentially in the second half of life. Whereas volunteering affects well-being directly for people aged 45–84 years, it is only in the age groups around retirement (55–74 years) that volunteering turns out to be favorable for subjective well-being not just by its direct effects, but also indirectly via its beneficial effects on self-efficacy (Müller et al. 2014).

Family Relationships and Social Support

Whether and to what extent older adults live in close companionships with others and can rely on family members and non-kin for comfort and help in case of need is a classical question in research on aging (See ▶ "Intergenerational Exchanges"). The DEAS provides a great variety of information on the structure and quality of the social network in general and of family relationships in particular.

With respect to the relationship between older adults and their grown-up children, a remarkable stability in the frequency of contact and emotional closeness is observed between 1996 and 2014 despite a growing geographical distance (Steinbach et al. 2019). The pattern of intergenerational support is changing over time but seems to adapt to social and economic opportunities, constraints, and needs the familial generations face (Steinbach et al. 2019; Klaus and Mahne 2017). Also, well-known gender differences show, for instance, women are found to be more involved in informal support and caregiving to those who are frail and limited in their health than men (Klaus and Vogel 2019).

Changing family structures are a major challenge for old age. For instance, grandparenthood is both being delayed and becoming less likely as a consequence of declining fertility as well as the postponement of family formation (Leopold and Skopek 2015). Thus, later-born cohorts look after their grandchildren later in life than earlier-born cohorts (Klaus and Vogel 2019). Grandchild care has no impact on grandparents' health (Ates 2017), but a high quality of grandparents' relationship to their adolescent and adult grandchildren promotes their life satisfaction and allows a potential for support provided when in need of such support (Mahne and Huxhold 2014; Wetzel and Hank 2020). For the majority of adults in later life the grandparent role is still highly important (Mahne and Motel-Klingebiel 2012).

Growing childlessness not only reduces the chance for grandparenthood but also changes the conditions for receiving help and care in old age. The majority of older adults without children is, however, not isolated or suffers severe lack in support. Instead, most of them are successful in compensating missing (grand-)children by friends, collateral kin and efficient support ties (Klaus and Schnettler 2016). As a general trend, non-family ties gain in importance. For instance, later-born cohorts have a larger network of friends and spend more time in activities with them than earlier-born cohorts (Huxhold 2019). Also, there is no empirical support for the widespread assumption of a "loneliness epidemic" in old age (Huxhold and Engstler 2019) (See ▶ "Loneliness").

Health and Subjective Well-Being

How do health and well-being (See ▶ "Subjective Well-being") change over the life course into old age? Is there a secular trend in favor of later-born cohorts, indicating that more individuals reach old and very old age in good health nowadays than the generations before them? And is there evidence in favor of "compression of morbidity" (Fries 2005), with time spent in poor health being delayed and "compressed" to the very last years of life? The DEAS provides the opportunity to investigate a number of indicators of health and well-being in a cross-sectional as well as longitudinal perspective.

Generally, there is evidence for declines in functional health with advancing age (Spuling et al. 2019) and particularly with increasing proximity to death (Wettstein et al. 2019). Recent findings suggest that later-born cohorts reach old and very old age with better functional health (i.e., less restrictions in everyday activities such as walking or lifting groceries) than earlier-born cohorts, though their self-reported health is not different from earlier-born cohorts at any age within the second half of life (Spuling et al. 2019).

With regard to secular trends in age-related changes of mental health outcomes, findings from the DEAS suggest that individuals from more recent birth cohorts will reach old age with a higher life satisfaction than the birth cohorts before them, whereas age-related changes in depressive symptoms are not different for subsequent birth cohorts (Wettstein and Spuling 2019). Mental, functional, and subjective health are interrelated, that is, individuals with better mental health also report better functional and subjective health, and this association seems to get closer when individuals approach the end of life (Wettstein et al. 2019) as well as across subsequent birth cohorts (Spuling et al. 2015).

Various resources and protective factors are associated with functional health and subjective well-being. For instance, higher cognitive abilities are associated with better functional health, particularly among individuals who are older or closer to death and who are affected by pain (Wettstein et al. 2019; Wettstein et al. 2020). In addition, individuals with more favorable self-perceptions of aging report a better physical and mental health (Wurm and Benyamini 2014).

Future Research Directions

The main strengths of the DEAS are: (a) its combined cohort-sequential and longitudinal design, now providing an observational period of up to 24 years, which enables researchers to analyze both social change on the macro-level as well as intraindividual trajectories on the micro-level (embedded in social change) and to disentangle age effects from cohort effects respective period effects; (b) its broad variety of measures, which cover major aspects of life conditions and quality of life of individuals in the second half of life; and

(c) the very large, age-heterogeneous and representative samples of the community-dwelling population in Germany.

The DEAS has been and will be continuously optimized as a data base for longitudinal social and behavioral aging research as well as for social reporting and monitoring social and demographic change. New issues arising in societies characterized by demographic change and increasing longevity, such as caring for an increasing number of individuals with dementia or for frail individuals for more years, provide new challenges for both, the families in aging societies and for aging research. New indicators for attitudes towards individuals with dementia, as well as for digitization and technological change will enlarge the scope of the German Ageing Survey for future research.

Summary

The German Ageing Survey (Deutscher Alterssurvey – DEAS) is a cohort-sequential study which has been ongoing for nearly a quarter century. The DEAS combines large cross-sectional baseline samples with longitudinal data collection. Since 1996, representative samples of the population in the age range between 40 and 85 years have been drawn every 6 years. There are currently four independent samples (1996, 2002, 2008, and 2014) and a fifth cross-sectional sample has been drawn for ongoing data collection in 2020. Each of these samples is followed over time: Longitudinal data have been collected in the years 2002, 2008, 2011, 2014, 2017, and 2020 is currently running. To date, nearly 40,000 interviews have been conducted with more than 20,000 participants. The cohort-sequential design enables researchers to analyze social change as well as intraindividual age-related trajectories (embedded in social change) and to disentangle age, cohort, and period effects. The DEAS covers a broad range of topics concerning age and aging in the second half of life, such as provision for old age and housing, employment, work and volunteering, family relationships, and social support, as well as health and subjective well-being.

Cross-References

▶ Intergenerational Exchanges
▶ Loneliness
▶ Pain in Older Persons
▶ Subjective Well-being
▶ The Survey of Health, Ageing and Retirement in Europe

References

Ates M (2017) Does grandchild care influence grandparents' self-rated health? Evidence from a fixed effects approach. Soc Sci Med 190:67–74. https://doi.org/10.1016/j.socscimed.2017.08.021

Bergmann M, Kneip T, De Luca G, Scherpenzeel A (2017) Survey participation in the Survey of Health, Ageing and Retirement in Europe (SHARE), Wave 1–6. SHARE working paper series 31

Brüderl J, Schmiedeberg C, Castiglioni L, Arránz Becker O, Buhr P, Fuß D, Ludwig V, Schröder J, Schumann N (2015) The German family panel: study design and cumulated field report (waves 1 to 6). Pairfam, s.l

Engstler H, Romeu Gordo L (2017) Der Übergang in den Ruhestand: Alter, Pfade und Ausstiegspläne [Retirement age, pathways into retirement and exit plans of older workers]. In: Mahne K, Wolff JK, Simonson J, Tesch-Römer C (eds) Altern im Wandel. Zwei Jahrzehnte Deutscher Alterssurvey [Ageing in change: two decades German Ageing Survey]. Springer VS, Wiesbaden, pp 65–80. https://doi.org/10.1007/978-3-658-12502-8_4

Erlinghagen M, Hank K (2006) The participation of older europeans in volunteer work. Ageing Soc 26(4):567–584. https://doi.org/10.1017/S0144686X06004818

Franke J, Wetzel M (2017) Länger zufrieden arbeiten? Qualität und Ausgestaltung der Erwerbstätigkeit in der zweiten Lebenshälfte [Working longer satisfied? Quality and formation of employment in the second half of life]. In: Mahne K, Wolff JK, Simonson J, Tesch-Römer C (eds) Altern im Wandel. Zwei Jahrzehnte Deutscher Alterssurvey [Ageing in change: two decades German Ageing Survey]. Springer VS, Wiesbaden, pp 47–63. https://doi.org/10.1007/978-3-658-12502-8_3

Fries JF (2005) The compression of morbidity. Milbank Q 83(4):801–823. https://doi.org/10.1111/j.1468-0009.2005.00401.x

Hajek A, König H-H (2020) Does the beginning and the end of income poverty affect psychosocial factors among middle-aged and older adults? Findings based on nationally representative longitudinal data. Aging Ment Health 24:1–7. https://doi.org/10.1080/13607863.2020.1725740

Hokema A, Scherger S (2016) Working pensioners in Germany and the UK: quantitative and qualitative evidence on gender, marital status, and the reasons for working. J Popul Ageing 9(1-2):91–111. https://doi.org/10.1007/s12062-015-9131-1

Huxhold O (2019) Gauging effects of historical differences on aging trajectories: the increasing importance of friendships. Psychol Aging 34(8):1170–1184. https://doi.org/10.1037/pag0000390

Huxhold O, Engstler H (2019) Soziale Isolation und Einsamkeit bei Frauen und Männern im Verlauf der zweiten Lebenshälfte [Social isolation and loneliness among women and men during the second half of life]. In: Vogel C, Wettstein M, Tesch-Römer C (eds) Frauen und Männer in der zweiten Lebenshälfte [Women and men in the second half of life: social change of ageing]. Springer VS, Wiesbaden, pp 71–89. https://doi.org/10.1007/978-3-658-25079-9_5

Klaus D, Engstler H (2017) Daten und Methoden des Deutschen Alterssurveys [Data and methods of the German Ageing Survey]. In: Mahne K, Wolff JK, Simonson J, Tesch-Römer C (eds) Altern im Wandel: Zwei Jahrzehnte Deutscher Alterssurvey [Ageing in change: two decades German Ageing Survey]. Springer VS, Wiesbaden, pp 29–45. https://doi.org/10.1007/978-3-658-12502-8_2

Klaus D, Mahne K (2017) Zeit gegen Geld? Der Austausch von Unterstützung zwischen den Generationen [Time for money? The exchange of support between the generations]. In: Mahne K, Wolff JK, Simonson J, Tesch-Römer C (eds) Altern im Wandel. Zwei Jahrzehnte Deutscher Alterssurvey [Ageing in change: two decades German Ageing Survey]. Springer VS, Wiesbaden, pp 247–256. https://doi.org/10.1007/978-3-658-12502-8_16

Klaus D, Schnettler S (2016) Social networks and support for parents and childless adults in the second half of life: convergence, divergence, or stability? Adv Life Course Res 29:95–105. https://doi.org/10.1016/j.alcr.2015.12.004

Klaus D, Vogel C (2019) Unbezahlte Sorgetätigkeiten von Frauen und Männern im Verlauf der zweiten Lebenshälfte [Care of women and men during the second half of life]. In: Vogel C, Wettstein M, Tesch-Römer C (eds) Frauen und Männer in der zweiten Lebenshälfte: Älterwerden im sozialen Wandel [Women and men in the second half of life: social change of ageing]. Springer VS, Wiesbaden, pp 91–112. https://doi.org/10.1007/978-3-658-25079-9_6

Klaus D, Engstler H, Mahne K, Wolff JK, Simonson J, Wurm S, Tesch-Römer C (2017) Cohort profile: the German Ageing Survey (DEAS). Int J Epidemiol 46(4):1105–1105g. https://doi.org/10.1093/ije/dyw326

Klaus D, Engstler H, Vogel C (2019) Längsschnittliches Design, Inhalte und Methodik des Deutschen Alterssurveys (DEAS) [Longitudinal design, topics and methods of the German Ageing Survey (DEAS)]. In: Vogel C, Wettstein M, Tesch-Römer C (eds) Frauen und Männer in der zweiten Lebenshälfte: Älterwerden im sozialen Wandel [Women and men in the second half of life: social change of ageing]. Springer VS, Wiesbaden, pp 17–34. https://doi.org/10.1007/978-3-658-25079-9_2

Kohli M (2000) Der Alters-Survey als Instrument wissenschaftlicher Beobachtung [Ageing-Survey as instrument of scientific monitoring]. In: Kohli M, Künemund H (eds) Die zweite Lebenshälfte. Gesellschaftliche Lage und Partizipation im Spiegel des Alters-Survey [The second half of life: state of society and participation on the basis of the ageing-survey]. Leske + Budrich, Opladen, pp 10–32

Künemund H, Motel-Klingebiel A, Kohli M (2005) Do intergenerational transfers from elderly parents increase social inequality among their middle-aged children? Evidence from the German aging survey. J Gerontol Soc Sci 60(1):S30–S36. https://doi.org/10.1093/geronb/60.1.S30

Lejeune C, Romeu Gordo L, Simonson J (2017) Einkommen und Armut in Deutschland. Objektive Einkommenssituation und deren subjektive Bewertung [Income and poverty in Germany: objective income situation and subjective assessment]. In: Mahne K, Wolff JK, Simonson J, Tesch-Römer C (eds) Altern im Wandel. Zwei Jahrzehnte Deutscher Alterssurvey [Ageing in change: two decades German Ageing Survey]. Springer VS, Wiesbaden, pp 97–110. https://doi.org/10.1007/978-3-658-12502-8_6

Leopold T, Skopek J (2015) The delay of grandparenthood: a cohort comparison in east and west Germany. J Marriage Fam 77(2):441–460. https://doi.org/10.1111/jomf.12169

Mahne K, Huxhold O (2014) Grandparenthood and subjective Well-being: moderating effects of educational level. J Gerontol Soc Sci 70(5):782–792. https://doi.org/10.1093/geronb/gbu147

Mahne K, Motel-Klingebiel A (2012) The importance of the grandparent role – a class specific phenomenon? Evidence from Germany. Adv Life Course Res 17(3):145–155. https://doi.org/10.1016/j.alcr.2012.06.001

Müller D, Ziegelmann JP, Simonson J, Tesch-Römer C, Huxhold O (2014) Volunteering and subjective well-being in later adulthood: is self-efficacy the key? Int J Dev Sci 8(3–4):125–135. https://doi.org/10.3233/DEV-14140

Nowossadeck S, Engstler H (2017) Wohnung und Wohnkosten im Alter [Housing and housing costs in old age]. In: Mahne K, Wolff JK, Simonson J, Tesch-Römer C (eds) Altern im Wandel. Zwei Jahrzehnte Deutscher Alterssurvey [Ageing in change: two decades German Ageing Survey]. Springer VS, Wiesbaden, pp 287–300. https://doi.org/10.1007/978-3-658-12502-8_19

Romeu Gordo L, Grabka MM, Lozano Alcántara A, Engstler H, Vogel C (2019) Immer mehr ältere Haushalte sind von steigenden Wohnkosten schwer belastet [More and more older households are burdened with rising housing costs]. DIW-Wochenbericht 86(27):468–476. https://doi.org/10.18723/diw_wb:2019-27-1

Scherger S, Lux T, Hagemann S, Hokema A (2012) Between privilege and burden. Work past retirement age in Germany and the UK. Zentrum für Sozialpolitik, Bremen

Simonson J, Hagen C, Vogel C, Motel-Klingebiel A (2013) Ungleichheit sozialer Teilhabe im Alter [Unequal social participation in old age]. Z Gerontol Geriatr 46(5):410–416. https://doi.org/10.1007/s00391-013-0498-4

Spuling SM, Wurm S, Tesch-Römer C, Huxhold O (2015) Changing predictors of self-rated health: disentangling age and cohort effects. Psychol Aging 30(2):462–474. https://doi.org/10.1037/a0039111

Spuling SM, Cengia A, Wettstein M (2019) Funktionale und subjektive Gesundheit bei Frauen und Männern im Verlauf der zweiten Lebenshälfte [Functional and subjective health of women and men during the second half of life]. In: Vogel C, Wettstein M, Tesch-Römer C (eds) Frauen und Männer in der zweiten Lebenshälfte [Women and men in the second half of life: social change of ageing]. Springer VS, Wiesbaden, pp 35–52. https://doi.org/10.1007/978-3-658-25079-9_3

Statistisches Bundesamt (2019) Bevölkerung und Erwerbstätigkeit. Ausländische Bevölkerung. [Population and employment. Population without German citizenship]. Statistisches Bundesamt, Wiesbaden

Steinbach A, Mahne K, Klaus D, Hank K (2019) Stability and change in intergenerational family relations across two decades: findings from the German Ageing Survey, 1996–2014. J Gerontol Soc Sci 75(4):899–906. https://doi.org/10.1093/geronb/gbz027

Vogel C, Romeu Gordo L (2019) Ehrenamtliches Engagement von Frauen und Männern im Verlauf der zweiten Lebenshälfte [Volunteering of women and men during the second half of life]. In: Vogel C, Wettstein M, Tesch-Römer C (eds) Frauen und Männer in der zweiten Lebenshälfte [Women and men in the second half of life: social change of ageing]. Springer VS, Wiesbaden, pp 113–132. https://doi.org/10.1007/978-3-658-25079-9_7

Wettstein M, Spuling SM (2019) Lebenszufriedenheit und depressive Symptome bei Frauen und Männern im Verlauf der Zweiten Lebenshälfte [Satisfaction and depression of women and men during the second half of life]. In: Vogel C, Wettstein M, Tesch-Römer C (eds) Frauen und Männer in der zweiten Lebenshälfte [Women and men in the second half of life: social change of ageing]. Springer VS, Wiesbaden, pp 53–70. https://doi.org/10.1007/978-3-658-25079-9_4

Wettstein M, Spuling SM, Cengia A (2019) Trajectories of functional health and its associations with information processing speed and subjective well-being: the role of age versus time to death. Psychol Aging 35:190. https://doi.org/10.1037/pag0000418

Wettstein M, Spuling SM, Cengia A, Nowossadeck S, Tesarz J (2020) Associations of age and pain with 9-year functional health trajectories. J Gerontopsychol Geriatr Psychiatry. https://doi.org/10.1024/1662-9647/a000221

Wetzel M, Hank K (2020) Grandparents' relationship to grandchildren in the transition to adulthood. J Fam Issues. https://doi.org/10.1177/0192513X19894355

Wetzel M, Simonson J (2017) Engagiert bis ins hohe Alter? [Volunteering until old age?]. In: Mahne K, Wolff JK, Simonson J, Tesch-Römer C (eds) Altern im Wandel. Zwei Jahrzehnte Deutscher Alterssurvey [Ageing in change: two decades German Ageing Survey]. Springer VS, Wiesbaden, pp 81–95. https://doi.org/10.1007/978-3-658-12502-8_5

Wetzel M, Bowen CE, Huxhold O (2019) Level and change in economic, social, and personal resources for people retiring from paid work and other labour market statuses. Eur J Ageing 16(4):439–453. https://doi.org/10.1007/s10433-019-00516-y

Wurm S, Benyamini Y (2014) Optimism buffers the detrimental effect of negative self-perceptions of ageing on physical and mental health. Psychol Health 29(7):832–848. https://doi.org/10.1080/08870446.2014.891737

Gerocardiology

▶ Vascular Diseases of Ageing

Gerocomia

▶ Life-Extensionism in a Historical Perspective

Gerocomia in the History of Aging Care

Ilia Stambler
Science, Technology and Society, Bar Ilan University, Ramat Gan, Israel

Synonyms

Care for the aged; Geriatric medicine; Gerocomica; Gerontocomia; Longevity hygiene; Medicine for the aged

Definition

The term "gerocomia" derives from the Greek phrase "care for the aged." Gerocomia, as an early form of healthcare for old age, was practiced since Greco-Roman antiquity through the middle ages until the early modern period. The concepts and practices of gerocomia influenced the emergence of early modern longevity hygiene and modern geriatric medicine.

Overview

In the historical pursuit of healthy longevity and amelioration of aging-related ill health, many fields have been explored with diverse titles and emphases (Gruman 1966). One of the most ancient and influential relevant fields in the medical care of aging was mainly concerned with lifestyle improvements to achieve healthy longevity – the so-called gerocomia (Freeman 1938; Stambler 2014).

The term "gerocomia" ("gerocomica" or "gerontocomia" derived from the Greek phrase "care for the aged") was used at least since the time of the Greco-Roman physician Galen (Aelius/Claudius Galenus, c. 129–217 CE). Galen used this term in his *De tuenda Sanitate. Gerontocomia* (5th book *On the Preservation of Health. Gerontocomia*). Galen's authority was still recognized as late as the eighteenth century by another prominent proponent of gerocomia, Sir John Floyer (1649–1734), in *Medicina gerocomica or The Galenic Art of Preserving Old Men's Health* (1725) (Galen 1725; Floyer 1725; Johnston 2018). Indeed, the term "gerocomia" was commonly used from antiquity through the eighteenth century. Even without an explicit association with the field of "gerocomia," principles of the "care of the aged" were expounded by ancient Greek physicians and scholars (Grmek 1958). The "father of medicine" – Hippocrates (c. 460–c. 370 BCE) – gave prescriptions for healthy longevity, such as "exertion, food, drink, sleep, sexual activity, in moderation" (Smith 1994). Aristotle (384–322 BCE) advised on moderation for the preservation of health in old age, for example, in his treatises *On Length and Shortness of Life* and *On Youth, Old Age, Life and Death, and Respiration* (Barnes 1984). The Romans carried this tradition. Thus, *On Old Age* (*De Senectute*) by Cicero (106–43 BCE) called to "fight" the infirmities and feebleness produced by aging. "We must stand up against old age and make up for its drawbacks by taking pains," Cicero wrote. "We must fight it as we should an illness. We must look after our health, use moderate exercise, take just enough food and drink to recruit, but not to overload, our strength. Nor is it the body alone that must be supported, but the intellect and soul much more" (1900). This statement may provide a short summary of the advice given by most works on the "care of old age" in Greco-Roman antiquity.

The tradition continued through the middle ages to the early modern period. One of the more influential works on gerocomia was written by the Italian professor Gabriele Zerbi (1445–1505) entitled *Gerontocomia, scilicet de senium cura atque victu* (1489, "Gerontocomia, or, care and nutrition for old age," written in Rome upon the request of Pope Innocent VIII, 1432–1492). Further prominent proponents of this art were the Italian long-lived writer Luigi Cornaro (1467–1566), the author of *Discorso sulla vita sobria* (*Discourse on the Sober Life*, 1566), the Flemish Jesuit priest Leonardus Lessius (1554–1623, the author of *A Treatise of Health and Long Life – Hygiasticon*, 1613) (Lessius and Cornaro 1743), and the German scholar Johann Heinrich Cohausen (1665–1750) who wrote *Tentaminum physico-medicorum curiosa decas de vita humana theoretice et practice per pharmaciam prolonganda* (Curious physico-medical theoretical and practical attempts to prolong the decades of human life by pharmacological means, 1699) and *Hermippus Redivivus or the Sage's Triumph over Old Age and the Grave* (1742, including the advice about chaste proximity to young maidens) (Cohausen 1748). Indeed, the main defining recommendation for the attainment of healthy longevity, throughout the entire gerocomia tradition, was moderation, especially moderation in food and sexual moderation (Shapin and Martyn 2000). This tradition has parallels in oriental traditions. In the words of Lao-Tse, the great teacher of Taoism (c. sixth century BCE), "For regulating the human in our constitution and rendering the proper service to the heavenly, there is nothing like moderation" (Lao-Tse 1891).

These concepts of gerocomia formed the basis for the ideology and practice of early modern hygienists who strove to extend longevity and preserve health in old age. Perhaps the most notable in this hygienic tradition was Christoph Wilhelm Hufeland (1762–1836), the renowned German hygienist, physician to the King of Prussia, Friedrich Wilhelm III, and to Goethe and Schiller. In his book *Macrobiotics or the Art of Prolonging Human Life* (1796), Hufeland coined a particular term for the pursuit of healthy longevity – "macrobiotics" which has survived to the present (Wilson 1867). Hufeland specifically distinguished the art of longevity from the general medical art that commonly aims to treat individual diseases and symptoms and mainly considers short term effects. As Hufeland wrote: "The object of medical art is health; that of the macrobiotic, long life. The means employed in the medical art are regulated according to the present state of the body and its variations; those of the macrobiotic, by general principles." Some of the general principles that determine human longevity, according to Hufeland, include "the innate quantity of vital power," "firmness of organization of the vital organs," and the rates of "consumption" versus "renovation" ("restoration" or "regeneration") of the vital force and of the organs. Moderation, in Hufeland, is an absolutely vital means for the conservation of vital power: "the more intensively a being lives, the more will its life lose its extension"; "strengthening, carried too far, may tend to accelerate life, and consequently, to shorten its duration." Similar principles were professed by several European hygienists of the eighteenth century, such as the German-Latvian proponent of healthy longevity, Johann Bernhard Fischer (1685–1772), who served as "Archiatrus" (head of the ministry of medicine) of the Russian Empire and published the book *On Old Age, its Degrees and Diseases* (*De Senior Eiusque Gradibus et Morbis*, 1754, republished in 1760).

The traditions of longevity hygiene of the eighteenth century laid the foundations for the emergence of "medicine for the aged" (*médecine de vieillards*) in France in the nineteenth century and "Geriatrics" in the early twentieth century in the United States.

Summary and Future Directions of Research

In our time when new health regimens and health correlations are being proposed continuously, it may be helpful to recall the health regimens suggested in the past in the gerocomia tradition and discover that not much has changed in the basic recommendations of moderation, exercise, and mental stimulation. At the same time, many of the old issues of gerocomia and allied hygienic schools of thought still persist today, such as the generality of recommendations, the difficulty of quantitative prediction and individualization, as well as the temporal limitations of the health benefits. It may be hoped and expected that the ancient tradition of gerocomia will be better studied and appreciated by modern scientists who will also be able to enhance the old recommendations with more quantitative precision and achieve health benefits more substantial than those that were achieved in the past.

Cross-References

▶ Aging Research in the Late Nineteenth to Early Twentieth Century
▶ Homeostasis in the History of Aging Research
▶ Life-Extensionism in a Historical Perspective
▶ Medical Alchemy in History

References

Barnes J (ed) (1984) The complete works of Aristotle. The revised Oxford translation. Princeton University Press, Princeton

Cicero MT (1900) Two essays on old age and friendship. Translated from the Latin of Cicero by E. S. Shuckburgh. McMillan and Co., New York

Cohausen JH (1748) Hermippus Redivivus or the sage's triumph over old age and the grave, wherein a method is laid down for prolonging the life and vigour of man,

including a commentary upon an ancient inscription, in which this great secret is revealed, supported by numerous authorities. The whole is interspersed with a great variety of remarkable and well attested relations. J. Nourse, London (first published in Latin in 1742)
Floyer J (1725) Medicina gerocomica, or, the Galenic art of preserving old men's healths. J. Isted, London
Freeman JT (1938) The history of geriatrics. Ann Med Hist 10:324–335
Galen (1725) De tuenda Sanitate Gerontocomia. Quoted in: Floyer J. Medicina gerocomica, or, the Galenic art of preserving old men's healths. J. Isted, London, p 107
Grmek MD (1958) On ageing and old age. Basic problems and historic aspects of gerontology and geriatrics. Monogr Biol Den Haag 5(2):1–106
Gruman GJ (1966) A history of ideas about the prolongation of life. The evolution of prolongevity hypotheses to 1800. Trans Am Philos Soc 56:1–102
Johnston I (2018) Galen hygiene, vols I & II. Loeb classical library. Harvard University Press, Cambridge, MA
Lao-Tse (1891) The Tao Teh King. The Tao and its characteristics, (trans: Legge J). In: Müller FM (ed) Sacred books of the East. Clarendon Press, Oxford
Lessius L, Cornaro L (1743) A treatise of health and long life, with the sure means of attaining it. In: 2 books. The first by Leonard Lessius. The second by Lewis Cornaro. Translated into English by Timothy Smith. C. Hitch, London
Shapin S, Martyn C (2000) How to live forever: lessons of history. Br Med J 321:1580–1582
Smith WD (ed) (1994) Hippocrates, vol 7, Epidemics. Loeb classical library. Harvard University Press, Cambridge, MA
Stambler I (2014) A history of life-extensionism in the twentieth century. Longevity History, Rishon Lezion. http://www.longevityhistory.com/. Accessed 15 Feb 2019
Wilson E (ed) (1867) Hufeland's art of prolonging life. Lindsay & Blakiston, Philadelphia (originally C.W. Hufeland, Makrobiotik; oder, Die Kunst das menschliche Leben zu verlängern, Jena, 1796)

Gerocomica

▶ Gerocomia in the History of Aging Care

Gerogogy

▶ Senior Learning

Gerontechnology

Angela Y. M. Leung and Xin Yi Xu
Centre for Gerontological Nursing, School of Nursing, The Hong Kong Polytechnic University, Hong Kong, SAR, China

Synonyms

Technogeriatrics; Technogerontology

Definition

The word "gerontechnology" is derived from words "gerontology" and "technology" (Jansson and Kupiainen 2017). Gerontechnology refers to an interdisciplinary field of science that combines gerontology and technology, designing and using technology to improve physical, mental, social, economic, and cultural environment for aged and aging individuals (Kwon 2016; Millán-Calenti and Maseda 2014). Another definition of gerontechnology shows the connection between technology and environment for independent living and social participation, and gerontechnology is developed for the importance of good health, comfort, and safety in old age (The International Society for Gerontechnology 2018).

Overview

The goal of gerontechnology mainly focuses on three areas: (1) delaying the age-associated changes in health and daily functioning; (2) preventing common functional and perceptual declines of aging; and (3) compensating physical disabilities for aging population (Graafmans 2016; Millán-Calenti and Maseda 2014).

Before 1990, many scientific disciplines have already paid attention to the aged and aging society. Because of the relative increase in aged population in the society and the advancement in technology, gerontechnology gradually became popular. The term "gerontechnology" was invented in 1988 by a small team of researchers, who integrated engineering science with aging research, at the Eindhoven University of Technology in Eindhoven, Netherlands. They targeted to use technology to achieve high standard of physical, mental, and social health for aged population. Gerontechnology can be applied in five human-related domains, (a) health and self-esteem, (b) housing and everyday functioning, (c) communication, (d) transportation and mobility, and (e) work and leisure, and this aimed to improve well-being (Bouma et al. 2009; Graafmans 2016). The textbook *Gerontechnology: Why and How?* described the way how we could pave toward a healthy and independent life in old age with the support by technology (Harrington and Harrington 2000).

With the development of technology, robotics is being developed quickly nowadays, and they have the potentials to play important roles in nursing care (Pollack 2005). There are three main types of robots: robots for assistance, robots as companion, and robots for health and behavior monitoring. Firstly, the robots are used as assistive devices, and, in some occasions, they are called "service robots." These robots usually support independent living by assisting older adults to perform basic activities including eating, bathing, toileting and getting dressed, maintaining safety, providing household maintenance, and supporting mobility. "Nursebot" Pearl is a good example: it provides functional assistance to older adults and assists them to navigate in the nursing facility and provide cognitive training (Pollack et al. 2002). Secondly, some robots are developed to provide companionship. The main function of this kind of robots is to improve psychological health of older adults by its presence, serving as the companions for older adults. For example, Aibo, an entertainment robot dog developed by Sony, is programmed to play and interact with humans (Fujita 2001). Since its development, Aibo has been used extensively in various studies with older adults, and it has effects on quality of life and stress reduction (Broekens et al. 2009). Another systematic review and meta-analysis investigated the effects of social robots on agitation and anxiety, but the existing evidence showed that these effects were not significant (Pu et al. 2018). Nonetheless, in a recent narrative review, social robots were found to reduce loneliness and improve interaction and engagement among older adults (Pu et al. 2018). Thirdly, some robots are developed for monitoring health and behavior for older adults for keeping them healthy and safe (Hosseini and Goher 2017). For example, robots will send alert signals to doctors or other family members when some unusual behaviors, like spending much time in the bathroom, leaving bed in unexpected period for long period of time, medication non-compliance, are captured (Hosseini and Goher 2017).

Mobile devices (i.e., smartphones, tablets) have been steadily accepted by the general public as the development of technology becomes mature and user-friendly (Bulloch et al. 2014). Mobile devices are portable and convenient for users and thus make them popular in elderly healthcare research (BinDhim et al. 2014). Indeed, older adults are more likely to use mobile devices instead of computers, and the application of smartphones grows quickly among this population (Fig. 1) (Riley et al. 2011).

There are billions of mobile app downloads each year with the increase in mobile device usage (Stoyanov et al. 2015). Health apps have been developed for many health purposes, including carrying out simple health interventions (e.g., diet, exercise), conducting quick health screes, and providing health information. Because of the availability and accessibility of mobile apps, they gradually become a promising approach of health promotion for older adults. mHealth becomes a specific term to capture the use of mobile app in relation to health.

Key Research Findings

The use of technology to promote health among older adults has been rapidly developed as "digital health interventions," and these interventions support patients with different kinds of illness, such as cardiovascular disease (Jiang et al. 2019), pre-diabetes (Leung et al. 2018), and stress urinary

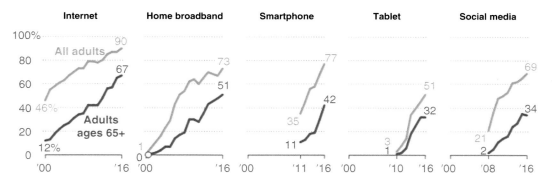

Gerontechnology, Fig. 1 The use of smartphone among American older adults. (Pew Research Center 2017)

incontinence (Sjöström et al. 2017). A variety of strategies have been adopted in digital health interventions, and they are short message service, telephone support, mobile app, video conferencing system, telemonitoring (i.e., physiological data is transmitted), and wearable medical device (Jiang et al. 2019; Leung et al. 2018). Among these, mobile app was the most common strategy that its use has been growing tremendously in health arena. For example, a mobile app was tailor-made to provide instructions for pelvic floor muscle training as the first-line treatment to women who suffered from stress urinary incontinence, and these strategies were found to be both clinical effective and cost-effective (Sjöström et al. 2017). Another example is the Diabetes Risk Score (DRS) app, which provides a useful platform for Chinese people to self-assess their risk of diabetes in a non-traumatizing manner (with no blood glucose testing) (Leung et al. 2018). This DRS app has been shown to be a feasible and reliable tool to identify people with undiagnosed diabetes and prediabetes, and it has also encouraged the at-risk people to modify their diet habits and increase physical activity level (Leung et al. 2018). Other than clinical implications of digital health interventions, attention has been given to cost-effectiveness of these technology-related interventions. A recent systematic review of 14 primary studies using digital health interventions to manage cardiovascular diseases showed that these interventions were cost-effective, gaining quality of life years (QALYs) with cost saving (Jiang et al. 2019). Similar finding in cost-effectiveness was noted in the use of mobile app for stress urinary incontinence (Sjöström et al. 2017).

Even though there are thousand types of technologies for older adults, questions still raised readily among community-dwelling older adults to accept and use these technologies (Peek et al. 2014). A recent systematic review identified the factors influencing the acceptance of electronic technologies that support aging (Peek et al. 2014). The influencing factors of acceptance can be divided into six themes, including concerns about technology such as financial and privacy, expected benefits of technology (e.g., increased safety and perceived usefulness), need for technology (e.g., perceived need and subjective health status), alternatives to technology (e.g., help by family or spouse), social influence (e.g., influence of family, friends, and professional caregivers), and characteristics of older adults (e.g., desire to age at their own homes) (Peek et al. 2014). With the understanding of these technology and communicative concerns, product developers, policymakers, professional caregivers, and family members can work out strategies to support older adults to make good use of gerontechnology to better their living.

Examples of Application

Gerontechnology has been applied to compensate age-related changes, such as declines in sensation and perception. Technical devices were developed to overcome the functional limitation of older adults, such as the anatomical changes in visual and hearing system. In order to overcome the functional limitation related to the visual impairment of aging, a wide variety of devices was used. For example, closed-circuit television magnifiers or computer magnification or screen reading software assist elders to overcome the functional limitations of visual impairment (Corn and Erin 2010). In addition, severe complications (such as diabetic retinopathy) may be resulted from poor blood glucose control. Therefore, it is critical for people with diabetes to monitor and control blood glucose by using blood glucose meters (Uslan et al. 2008), administering accurate amount of insulin by using insulin pens (Williams 2012) or insulin pumps (Marom 2010). Older adults may suffer from hearing loss, which can be compensated by hearing assistive technology (Jenstad and Moon 2011).

An attention has been paid to improve social communication and participation in old age through technology. Older Adults Technology Services (OATS), a nonprofit organization based in New York City, has provided many training programs for older adults. This organization manages 24 technology labs in "senior centers," operates a technology-themed community center for older adults, and produces "seniorplanet.org" website for older adults. OATS provided many courses on computer basics and iPad basics for older adults, and the participants in these courses could learn how to use emails, navigate websites to plan trips, and evaluate websites providing health information online. After attending the courses, 93% of the participants still use computers on daily basis, 94% of them increase their confidence in using computers, and 64% reported to spend substantial time to communicate with friends and families through technology (Gardner et al. 2012). To support older adults to get health information from the Internet, health web-navigating workshops have been developed and implemented in the community (Leung et al. 2007). Evidence has showed that older adults' confidence in seeking health information via the Internet was significantly associated with their satisfaction with the workshops, but not their age, educational level, or experience of using computers before the workshop (Leung et al. 2007). The participants of the workshops continued to use Internet to search for health information after the workshop, and this becomes an alternative source for seeking health information (Leung et al. 2007).

Future Directions of Research

Future research in gerontechnology could be focused on several perspectives: (1) effects on social health, (2) cost implications to healthcare systems, and (3) inclusion of older adults in the design and adaptation of gerontechnology.

By now, a large amount of research studies has been carried out to investigate the effects of gerontechnology on physical health. Limited studies have been focused on its effect on social and psychological health, and there is a recent trend paying attention to this arena (Medtech Boston 2018). A systematic review showed that the positive effect of the information and communication technology (ICT) on social connectedness and social support only had short-term effect (the effect did not last for more than 6 months) and the effects on self-esteem and control were nonsignificant (Chen and Schulz 2016). This implies that more research is needed to further investigate whether technology can be used to improve social health in some specific groups of older adults (older adults are heterogenous group, and they are different in cultures and social background) and which types of ICT (e.g., text-based messages versus audio-based messages) can bring concrete benefits to older adults, as we originally aim to develop different kinds of social robots and senior-oriented websites and devices.

Some attention has been given to cost-effectiveness evaluation of gerontechnology, but limited studies were carried out in the last decade. Further investigation is warranted to assess the direct effect of the use of technology in elderly care on the costs of health services used, for example, how many hospitalization days are reduced with the use of technology in monitoring medication compliance and how many emergency room services are reduced with the fall prevention and monitoring system.

Another key area for future research is the inclusion of older adults in the design and adaptation of gerontechnology. As the users of gerontechnology, older adults should be invited to be involved in all the process when we develop the technology. Individuals are different from others in terms of their demographic characteristics, personality traits, and functional limitations, and therefore their adaptivity to technology could be different (Kamin et al. 2017). This technology adaptivity is subjective, and it is a significant predictor of technology use in old age (Kamin et al. 2017). Thus, further investigation should be carried out to identify factors that could enhance technology adaptivity or identify facilitators that could enhance the use of technology for those who have low subjective technology adaptivity.

Summary

Gerontechnology is a new area of interest in elderly care, and it should be considered as investment rather than expenses because it is expected to reduce government's expenditure in healthcare system, improving individuals' functional capacity and delaying institutionalization (Wong et al. 2017). Gerontechnology should also be used to facilitate caregiving or increasing caregiving efficiency or reducing/alleviating caregivers' burden and workload (Wong et al. 2017). Another significant contribution of gerontechnology goes to health monitoring and support to improve health, leading to a healthy and independent life in old age. Gerontechnology has high potential to be further developed in the two decades, and product providers, policymakers, and health professionals are trying to use technologies to support our senior citizens to lead to a better life in old age.

Cross-References

▶ Comprehensive Digital Self-Care Support System (CDSSS)
▶ Digital Divide and Robotics Divide
▶ Digital Health
▶ Digital Participation
▶ Home Health Technology
▶ mHealth

References

BinDhim NF, Freeman B, Trevena L (2014) Pro-smoking apps for smartphones: the latest vehicle for the tobacco industry? Tob Control 23(1):e4

Bouma H, Fozard JL, Van Bronswijk J (2009) Gerontechnology as a field of Endeavour. Geron 8 (2):68–75

Broekens J, Heerink M, Rosendal H (2009) Assistive social robots in elderly care: a review. Geron 8(2):94–103

Bulloch LR, Andrews C, Dennison T, Elder J, Mitchell A, Rivenbank MT, . . . Gallicchio VS (2014) The impact of mobile devices on healthcare in rehabilitation medicine. Int J Rehabil Res 3:144–148

Chen YRR, Schulz PJ (2016) The effect of information communication technology interventions on reducing social isolation in the elderly: a systematic review. J Med Int Res 18(1):e18. https://doi.org/10.2196/jmir.4596

Corn AL, Erin JN (2010) Optics and low vision devices. In: Zimmerman GJ (ed) Foundations of low vision: clinical and functional perspectives. American Foundation for the Blind, New York

Fujita M (2001) AIBO: toward the era of digital creatures. Int J Robot Res 20(10):781–794

Gardner PJ, Kamber T, Netherland J (2012) "Getting turned on": using ICT training to promote active ageing in New York City. J Commun Inf 8:1

Graafmans JAM (2016) The history and incubation of gerontechnology. In: Kwon S (ed) Gerontechnology: research, practice, and principles in the field of technology and aging. Springer Publishing Company, New York

Harrington TL, Harrington MK (2000) Gerontechnology: why and how? International society for gerontechnology

Hosseini SH, Goher KM (2017) Personal care robots for older adults: An overview. ASS 13(1):11

International Society for Gerontechnology (2018) About us. http://www.gerontechnology.org/ Accessed on 23 Jun 2019

Jansson T, Kupiainen T (2017) Aged people's experiences of gerontechnology used at home: a narrative literature review. https://www.theseus.fi/bitstream/handle/10024/129279/Jansson_Kupiainen_ONT_21.4.17.pdf?sequence=1. Accessed on 25 Jun 2019

Jenstad L, Moon J (2011) Systematic review of barriers and facilitators to hearing aid uptake in older adults. Audiol Res 1(1):e25

Jiang X, Ming WK, You JH (2019) The cost-effectiveness of digital health interventions on the management of cardiovascular diseases: systematic review. J Med Internet Res 21(6):e13166. https://doi.org/10.2196/13166

Kamin ST, Lang FR, Beyer A (2017) Subjective technology adaptivity predicts technology use in old age. Gerontology 63(4):385–392. https://doi.org/10.1159/000471802

Kwon S (2016) Gerontechnology: research, practice, and principles in the field of technology and aging. Springer Publishing Company, New York

Leung A, Ko P, Chan KS, Chi I, Chow N (2007) Searching health information via web: Hong Kong Chinese older adults' experience. Public Health Nurs 24(2):169–175

Leung AY, Xu XY, Chau PH, Yu YTE, Cheung MK, Wong CK, ... Lam CL (2018) A mobile app for identifying individuals with undiagnosed diabetes and prediabetes and for promoting behavior change: 2-year prospective study. JMIR Mhealth Uhealth 6(5):e10662

Marom L (2010) Insulin pump access issues for visually impaired people with type 1 diabetes. Diabetes Res Clin Pract 89(1):e13–e15

Medtech Boston (2018 May) Top 5 gerontechnology products (to increase social engagement for seniors). https://medtechboston.medstro.com/blog/2018/05/22/top-5-gerontechnology-products-to-increase-social-engagement-for-seniors/. Accessed 24 Jun 2019

Millán-Calenti JC, Maseda A (2014) Telegerontology: A new technological resource for elderly support in Handbook of research on personal autonomy technologies and disability informatics. USA: Medical Information Science Reference-IGI Global 705–719

Peek ST, Wouters EJ, van Hoof J, Luijkx KG, Boeije HR, Vrijhoef HJ (2014) Factors influencing acceptance of technology for aging in place: a systematic review. Int J Med Inform 83(4):235–248

Pollack ME, Brown L, Colbry D, Orosz C, Peintner B, Ramakrishnan S, McCarthy CE (2002) Pearl: A mobile robotic assistant for the elderly. Paper presented at the AAAI workshop on automation as eldercare

Pollack ME (2005) Intelligent technology for an aging population: the use of AI to assist elders with cognitive impairment. AI Mag 26(2):9–9

Pu L, Moyle W, Jones C, Todorovic M (2018) The effectiveness of social robots for older adults: a systematic review and meta-analysis of randomized controlled studies. The Gerontologist 59(1):e37–e51

Riley WT, Rivera DE, Atienza AA, Nilsen W, Allison SM, Mermelstein R (2011) Health behavior models in the age of mobile interventions: are our theories up to the task? Trans Behav Med 1(1):53–71

Sjöström M, Lindholm L, Samuelsson E (2017) Mobile app for treatment of stress urinary incontinence: a cost-effectiveness analysis. J Med Internet Res 19(5):e154. https://doi.org/10.2196/jmir.7383

Smartphone adoption among seniors has nearly quadrupled in the last five years. Pew Research Center, Washington, D.C. (May 15, 2017) https://www.pewinternet.org/2017/05/17/tech-adoption-climbs-among-older-adults/pi_2017-05-17_older-americans-tech_0-01/

Stoyanov SR, Hides L, Kavanagh DJ, Zelenko O, Tjondronegoro D, Mani M (2015) Mobile app rating scale: a new tool for assessing the quality of health mobile apps. JMIR Mhealth Uhealth 3(1):e27

Uslan MM, Burton DM, Clements CW (2008) Blood glucose meters that are accessible to blind and visually impaired persons. J Diabetes Sci Technol 2(2):284–287

Williams A (2012) Creating low vision and nonvisual instructions for diabetes technology: an empirically validated process. J Diabetes Sci Technol 6(2):252–259

Wong SYS, Chu KCW, Tsang NSY, Cheng MMC, Wang SWY (2017) Our Hong Kong Foundation – Gerontechnology landscape report. Our Hong Kong Foundation Limited, Hong Kong

Gerontocomia

▶ Gerocomia in the History of Aging Care

Gerontocracy

Raul Magni-Berton[1] and Sophie Panel[2]
[1]Sciences Po Grenoble, Université Grenoble-Alpes, PACTE, Grenoble, France
[2]Sciences Po Bordeaux/Centre Emile Durkheim, PACTE, Bordeaux, France

Definition

A gerontocracy is a political system in which older people have a disproportionate influence on the political process or have higher chances of attaining leadership positions. The term "gerontocracy" (literally, "older people power") was coined by a nineteenth-century French pamphleteer in reaction to the aging of legislators (Eisele 1979).

Gerontocracy can be understood in a narrow or broad sense. In the narrow sense, the concept refers to direct rule by older people. In the broad sense, gerontocracy applies to situations in which older citizens exert more influence on public policies than other age groups. Interestingly, while the former form of gerontocracy is declining in industrialized democracies (see ▶ "Modernization theory"), the latter may be increasing.

This contribution mainly focuses on gerontocracy in the narrow sense. The last section, however, briefly reviews the debates surrounding the rising influence of older people in contemporary democracies.

Overview

Gerontocracy is very widespread. Its existence has been documented in very different types of systems, such as Melanesian societies (Allen 1984), Renaissance Venice (Finlay 1978), the Benin kingdom at the village level (Bradbury 1969), and among indigenous Amazonian peoples (Werner 1981). Gerontocracy is a prevalent feature of age-set societies (e.g., the Samburu of Kenya) in which elderhood entails specific social privileges and older people typically have control over issues such as marriage, the distribution of land and cattle, warfare, and judicial and religious matters (Spencer 1976; Brantley 1978). According to Spencer (2004 [1965]), gerontocracy is a corollary of polygamous societies, because polygyny is unsustainable unless young males are prevented from marrying in order to generate a surplus of brides. Sheehan (1976) puts forward that gerontocracy is typical of sedentary societies, in contrast to nomadic tribes that put a premium on youth. Finally, Cowgill and Holmes (1972) argue that gerontocracy is most prevalent in preliterate societies in which emphasis is placed on the preservation of cultural heritage and mastery of religious rites: in the absence of writings, people's memory becomes the only repository for knowledge, but the diffusion of literacy makes this knowledge less relevant and thus threatens the status of older people.

Gerontocracy can be formal or informal. In the former case, leadership positions are explicitly reserved for the oldest individuals: a well-known historical example is the Spartan *Gerousia*, the powerful council composed of men over 60. Nowadays, some formal rules still survive in Western democracies, such as "seniority" systems that give the oldest MPs committee appointments and chairs (in US senate), or rules according to which, in case of an equal number of votes, the oldest is elected (in many European countries). Some age of candidacy requirements can be considered mild forms of formal gerontocracy: the minimum age to be elected as president is 45 in Singapore and Pakistan, and 50 in Italy. However, formal gerontocratic rules are progressively disappearing, while, on the contrary, age limits are becoming increasingly widespread. For example, mayors in the German region of Baden-Württemberg cannot be older than 68 at the day of their election; and during the 2000s, the Chinese communist party introduced the same mandatory retirement age for members of the Politburo. Yet, these corrective measures are still fairly rare (contrast, for example, measures designed to address gender imbalance among elected officials).

In the case of informal gerontocracy, there is no formal institution reserving leadership positions to elderlies, but the political system is still de facto ruled by old men. This may happen due to some leaders' exceptionally lengthy tenure: for example, Robert Mugabe was at age 93 years when he left office after ruling Zimbabwe for almost 30 years; and Cameroon's president Paul Biya (in office since 1982) recently got reelected at age 86. Yet, some systems are structurally gerontocratic: leaders tend to enter office at an advanced age. The Vatican is a typical example: the ten last popes (since 1903) were elected at age 68 on average, which is about 15 years older than the mean age at which political leaders take office (around 53; see Baturo 2016). Many dictatorships – for example, single-party regimes such as China or the USSR after Stalin – also tend to produce older leaders (Tullock 1987).

Key-Research Findings

The factors behind the emergence or decline of gerontocratic rules are still debated. There are three potential explanations, labeled the wisdom effect, the longevity effect, and the incumbency-limiting effect.

Wisdom effect. The first explanation lies in the experience and knowledge accumulated by older people. They have experienced a large variety of situations and are able to provide solutions for new issues. Older people' influence is at its height when they constitute a rare resource for society – that is, when life expectancy is low (see ▶ "Active life expectancy") – and in periods of slow technological change, which explains why the status accorded to older people has declined with industrialization. Rapid technological changes decrease the social utility of experience and makes elderlies' accumulated knowledge obsolete; meanwhile, the increase in life expectancy reduces the rarity of experienced individuals (Cowgill and Holmes 1972).

The main weakness of this explanation is that it assumes that experience helps leaders take wiser decisions, or, at least, that people believe it (see ▶ "Leadership, politics, and ageism"). Empirical research, however, reports a negative relationship between leaders' age and political performance in a broad range of domains. Older leaders produce less economic growth in democracies (Atella and Carbonari 2017) and in autocracies (Jong-A-Pin and Mierau 2011); with regard to international politics, they are more likely to be involved in an international conflict, either as initiators or targets (Horowitz et al. 2005; Bak and Palmer 2010). A study on 25 European monarchs over the period from 1072 to 1780 finds that their age is negatively correlated with several indicators of military and diplomatic success, with a performance peak occurring around age 40 (Simonton 1984). The results of these macro-level studies echo the findings of the literature analyzing the relationship between age and task performance at the firm or organization level (Walter and Scheibe 2013). Moreover, laboratory experiments indicate that voters' taste for older leaders is not systematic (Spisak 2012; Spisak et al. 2014). Observational studies also strongly suggest that old (or old-looking) candidates do not enjoy any particular electoral advantage (Banducci et al. 2008; Poutvaara et al. 2009; Baltrunaite et al. 2015).

There are two common explanations for the negative correlation between leaders' age and policy performance. The first one lies in leaders' time horizons: as they approach retirement age, they start discounting the future more heavily and prioritizing short-term payoffs over long-term goals. This leads them to invest less time and energy into their political career or (alternatively) to engage in reckless behavior to attain their policy objectives before leaving office (Horowitz et al. 2005). Shorter time horizons may also cause leaders to underinvest in growth-enhancing sectors (Atella and Carbonari 2017) or to prey on the economy in order to maximize their own wealth (Jong-A-Pin and Mierau 2011).

The second possible explanation lies in biological factors, as energy and cognitive abilities may decline among older leaders (see ▶ "Prevention of Age-Related Cognitive Impairment, Alzheimer's Disease, and Dementia"). Schubert (1988) finds that US mayors in their sixties and seventies devote less energy to their job, measured as the frequency and length of oral interventions during municipal council meetings. In a study on the Vermont state legislature, Fengler (1980) finds that legislators aged 65 or older have higher absenteeism rates and initiate fewer bills. However, this study also finds that bills they propose are more likely to become law – an effect that is independent from length of service or committee chairmanship. It points to yet another behavioral difference between younger and older leaders, namely risk aversion: older legislators display a greater propensity to self-select into legislative initiatives that have more chances of success.

Longevity effect. The second explanation for the prevalence of gerontocracy is based on the resources and social ties (friends, professional networks, and offspring) acquired during life (Werner 1981). It is a mechanical effect: since the process of becoming a political leader often requires a long-term investment, older individuals are automatically advantaged. Moreover, older people' effectiveness in the productive sector declines so they hold a comparative advantage in political activity (Mulligan and Sala-i-Martin 1999).

This explanation is consistent with the theory that gerontocracy spreads with the complexification of society (Sheehan 1976) and constitutes another explanation for why polygyny and gerontocracy often go together: in small communities, the age premium increases with the number of potential offspring (Werner 1981). To some extent, it is also compatible with recent data on the career paths of contemporary leaders. Since World War II, the average political leader took office at 53 years after spending about 15.5 years in politics (Baturo 2016): the age of the first job experience and the time invested to obtain the position are probably higher than for any other professional activity. Incumbency advantage in elections also suggests that voters tend to reward experience. However, while the longevity effect explains the rarity of leaders under 35 years of age, it does not account for the prevalence of leaders who are very close to life expectancy. This explanation is thus probably true to some extent, but incomplete.

Incumbency-limiting effect. The third explanation for the prevalence of gerontocracy is that it is often the sole guarantee of regular leadership turnover. In the absence of information on future leaders' performance and of constitutional procedures to remove incompetent ones, selecting the oldest candidates is the safest way to limit their expected tenure length. This strategy is also a way to prevent violent conflict, since it gives potential challengers incentives to wait for the leader's death instead of taking up arms. Conceiving gerontocracy as a strategic choice of people in charge with selecting leaders helps explain why gerontocracy tends to decline in contemporary societies: democracies have introduced fixed terms in office, which solves the alternation problem without having to resort to gerontocracy.

From an historical perspective, the fact that popes are elected for life may explain the Vatican's tendency to produce aging leaders. Between 1400 and 1600, Venetian doges were on average 72 at election (Finlay 1978). These two political figures have in common the fact that they were elected (contrary to kings in the same period, which were much younger) but not dismissible (contrary to elected officials in contemporary democracies). Finlay's (1978) account of the Venetian system suggests that selecting old doges served several purposes, beyond preventing bad leaders from ruling for decades: since the position was given as a reward for long state service, it was a way to select leaders with extensive political experience while keeping younger contenders patient and incite them to invest in the regime.

Thus, the propagation of gerontocracy in predemocratic societies may be explained in two ways: on the one hand, people in charge with selecting leaders have an incentive to choose the oldest one when there is no constitutional way of dismissing governments. On the other hand, to the extent that gerontocracy makes nondemocratic regimes more stable, systems based on such a rule are more likely to survive and develop. Recent empirical data on leaders in the postwar era support both explanations. First, leaders who are elected or appointed constitutionally but cannot be dismissed enter office at a significantly more advanced age than either democratic leaders (who can be removed) or autocratic leaders who self-select into office by illegal means such as coups or revolutions (Magni-Berton and Panel 2017). Second, nondemocratic regimes experience significantly fewer violent rebellions when they are ruled by aging dictators, while there is no relationship between leaders' age and rebellions when leaders can be removed by nonviolent means – that is, in democracies (Magni-Berton and Panel 2018).

Prospects

While gerontocracy in a strict sense (i.e., government by old men) is on the decline in contemporary democracies, it might have been replaced by gerontocracy in a broader sense (see ▶ "Political gerontology"). In the past decades, life expectancy in wealthy societies has risen considerably while the birth rate has fallen, leading to concerns over a growing political influence of elderlies at the expense of other age groups (see ▶ "Intergenerational solidarity"). This may happen through several channels (Mulligan and Sala-i-Martin 1999). First, consistently with the longevity effect, elderlies have accumulated more economic resources and political networks over the life course. Second, they tend to vote more than other age groups. Third, retirees benefit from more leisure time, which they can invest in political activity. Finally, elderlies make for a particularly effective interest group for two reasons: retirees have more homogenous interests than the active population; and young people are likely to become old while the reverse is not true (see ▶ "Politics of aging and interest groups").

Are these concerns borne out? First, survey data on industrial democracies tend to demonstrate that older people prefer significantly less spending in education and more in healthcare and pensions, but the difference is modest (Goerres and Tepe 2010; Sørensen 2013). In addition, there is no clear relationship between aging and pensions or healthcare spending (see, for example, Tepe and Vanhuysse 2009). There are two possible explanations. First, when fertility rates decrease, the young expect to receive less from the succeeding generation than they currently pay: therefore, they oppose spending in pensions and healthcare more than elderlies oppose spending in education (Tepe and Vanhuysse 2009). The second reason is that, while older people' sedentariness allows for higher turnout rates, the young's mobility allows them to "vote with their feet": this leads to a kind of fiscal competition that prevents exploitation of the young by the old (Montén and Thum 2010).

On the other hand, there is some evidence suggesting that public policies in industrialized democracies are becoming more congruent with the preferences of aging people. Elderlies, being less concerned by future investments, should favor short-term policies – such as reducing inflation or increasing public spending. Recent studies find evidence for inflation, but not for public spending (Afflatet 2018; Vlandas 2018). However, more research needs to be done to precisely assess the impact of ageing on public policies.

Cross-References

▶ Active Life Expectancy
▶ Intergenerational Solidarity
▶ Leadership, Politics, and Ageism
▶ Modernization Theory
▶ Political Gerontology
▶ Politics of Aging and Interest Groups
▶ Prevention of Age-Related Cognitive Impairment, Alzheimer's Disease, and Dementia

References

Afflatet N (2018) The impact of population ageing on public debt: a panel data analysis for 18 European countries. Int J Econ Financ Issues 8(4):68

Allen M (1984) Elders, chiefs, and Big Men: authority legitimation and political evolution in Melanesia. Am Ethnol 11:20–41. https://doi.org/10.1525/ae.1984.11.1.02a00020

Atella V, Carbonari L (2017) Is gerontocracy harmful for growth? A comparative study of seven European countries. J Appl Econ 20(1):141–168. https://doi.org/10.1016/S1514-0326(17)30007-7

Bak D, Palmer G (2010) Testing the Biden hypothesis: leader tenure, age, and international conflict. Foreign Policy Anal 6(3):257–273. https://doi.org/10.1111/j.1743-8594.2010.00111.x

Baltrunaite A, Casarico A, Profeta P (2015) Affirmative action and the power of older people. CESifo Econ Stud 61(1):148–164. https://doi.org/10.1093/cesifo/ifu032

Banducci SA, Karp JA, Thrasher M, Rallings C (2008) Ballot Photographs as Cues in Low–Information Elections. *Political Psychology* 29(6):903–17. https://doi.org/10.1111/j.1467-9221.2008.00672.x

Baturo A (2016) Cursus honorum: personal background, careers and experience of political leaders in democracy and dictatorship – new data and analyses. Polit Gov 4(2):138–157. https://doi.org/10.17645/pag.v4i2.602

Bradbury RE (2001 [1969]) Patrimonialism and gerontocracy in Benin political culture. In: Douglas M, Kaberry PM (eds) Man in Africa. Routledge: London, p 17

Brantley C (1978) Gerontocratic government: age-sets in pre-colonial Giriama. Africa 48(3):248–264. https://doi.org/10.2307/1158467

Cowgill DO, Holmes LD (1972) Aging and modernization. Appleton-Century-Crofts and Fleschner Publishing Company, East Norwalk

Eisele F (1979) Origins of "gerontocracy". The Gerontologist 19(4):403. https://doi.org/10.1093/geront/19.4.403

Fengler AP (1980) Legislative productivity of elderly legislators. Polity 13(2):327–333. https://doi.org/10.2307/3234587

Finlay R (1978) The Venetian Republic as a gerontocracy: age and politics in the Renaissance. J Mediev Renaiss Stud 8(1):157–178

Goerres A, Tepe M (2010) Age-based self-interest, intergenerational solidarity and the welfare state: A comparative analysis of older people's attitudes towards public childcare in 12 OECD countries. Eur J Polit Res 49(6):818–851. https://doi.org/10.1111/j.1475-6765.2010.01920.x

Horowitz M, McDermott R, Stam A (2005) Leader age, regime type and violent international relations. J Confl Resolut 49(5):661–685. https://doi.org/10.1177/0022002705279469

Jong-A-Pin R, Mierau J (2011) No Country for old men: aging dictators and economic growth. KOF working papers: KOF Swiss Economic Institute, ETH Zurich No. 289. https://doi.org/10.2139/ssrn.1944146

Magni-Berton R, Panel S (2017) Strategic gerontocracy: why nondemocratic systems produce older leaders. Public Choice 171(3/4):409–427. https://doi.org/10.1007/s11127-017-0449-5

Magni-Berton R, Panel S (2018) Alternation through death: is gerontocracy an equilibrium? Polit Res Q 71(4):975–988. https://doi.org/10.1177/1065912918775251

Montén A, Thum M (2010) Ageing municipalities, gerontocracy and fiscal competition. Eur J Polit Econ 26(2):235–247. https://doi.org/10.1016/j.ejpoleco.2009.11.004

Mulligan C, Sala-i-Martin X (1999) Gerontocracy, retirement, and social security. NWP No. 7117. https://doi.org/10.3386/w7117

Poutvaara P, Jordahl H, Berggren N (2009) Faces of politicians: babyfacedness predicts inferred competence but not electoral success. J Exp Soc Psychol 45(5):1132–1135. https://doi.org/10.1016/j.jesp.2009.06.007

Schubert J (1988) Age and active-passive leadership style. Am Polit Sci Rev 82(3):763–772. https://doi.org/10.2307/1962489

Sheehan T (1976) Senior esteem as a factor of socioeconomic complexity. The Gerontologist 16(5):433–440. https://doi.org/10.1093/geront/16.5.433

Simonton DK (1984) Leader age and national condition: a longitudinal analysis of 25 European monarchs. Soc Behav Pers 12(2):111–114. https://doi.org/10.2224/sbp.1984.12.2.111

Sørensen RJ (2013) Does aging affect preferences for welfare spending? A study of peoples' spending preferences in 22 countries, 1985–2006. Eur J Polit Econ 29:259–271. https://doi.org/10.1016/j.ejpoleco.2012.09.004

Spencer P (1976) Opposing Streams and the Gerontocratic Ladder: Two Models of Age Organisation in East Africa. Man 11(2):153–75. https://doi.org/10.2307/2800202

Spencer P (2004 [1965]) The Samburu: a study of gerontocracy. Routledge, London/New York

Spisak B (2012) The general age of leadership: older-looking presidential candidates win elections during war. PLoS One 7(5):e36945. https://doi.org/10.1371/journal.pone.0036945

Spisak B, Grabo A, Arvey R et al (2014) The age of exploration and exploitation: younger-looking leaders endorsed for change and older-looking leaders endorsed for stability. Leadersh Q 25(5):805–816. https://doi.org/10.1016/j.leaqua.2014.06.001

Tepe M, Vanhuysse P (2009) Are aging OECD welfare states on the path to gerontocracy? Evidence from 18 democracies, 1980–2002. J Publ Policy 29(1):1–28. https://doi.org/10.1017/S0143814X0900097X

Tullock G (1987) *Autocracy*. Boston: Kluwer Academic Publishers

Vlandas T (2018) Grey power and the economy: aging and inflation across advanced economies. Comp Pol Stud 51(4):514–552. https://doi.org/10.1177/0010414017710261

Walter F, Scheibe S (2013) A literature review and emotion-based model of age and leadership: new directions for the trait approach. Leadersh Q 24(6):882–901. https://doi.org/10.1016/j.leaqua.2013.10.003

Werner D (1981) Gerontocracy among the Mekranoti of Central Brazil. Anthropol Q 54(1):15. https://doi.org/10.2307/3317482

Gerontological Nursing

Mary Jo Vetter
NYU Rory Meyers College of Nursing,
New York, NY, USA

Synonyms

Geriatric nursing

Definition

Gerontological nursing is an evidence-based nursing specialty that addresses the unique physiological, social, psychological, developmental, economic, cultural, spiritual, and advocacy needs of older adults. The nursing practice focuses on the process of aging and the protection, promotion, restoration, and optimization of health and generalized functions; prevention of illness and injury; facilitation of healing; alleviation of suffering through the diagnosis and treatment of human responses; and advocacy in the care of older adults, caregivers, families, groups, communities, and populations (Bickford 2018). Gerontologic nursing care is delivered by nurses at all levels of academic preparation in a variety of practice settings that include hospitals, institutional skilled nursing facilities, ambulatory care, the home, and community.

Overview

In response to a growing older adult population, a gerontology specialty focus in nursing began to gain momentum in the early 1950s with the publication of the first geriatric nursing textbook (Newton Shafer 1950). The professionalization movement continued when the American Nurses Association formed a geriatric nursing group in 1962 followed by the establishment of specific standards of practice in 1970 and specialty certification by examination in 1974. In recognition of the holistic focus of the specialty, the term geriatric nursing was replaced by gerontological nursing in 1975 to reflect a health promotion emphasis. Academic programs responded to the need for education of nurses as the registered nurse as well as specialist advanced practice nurses by the late 1970s. Funding support from foundations such as Robert Wood Johnson, Kellogg, and John A. Hartford helped stimulate nursing education curriculum reform, the development of academic centers of excellence, and the promotion of practice and research specialization throughout the 1980s. Early in the evolution of gerontological nursing, the educational emphasis in the 1990s was to prepare advanced practice nurse specialists to work with interprofessional teams to influence positive health outcomes. Despite the availability of federal and private funds and the efforts of academic and professional nursing organizations, the proportion of gerontological nursing specialists remained low, mirroring the workforce of other professions such as medicine and social work. In 2008, the Institute of Medicine (IOM) charged an ad hoc committee on the Future of Health Care Workforce for Older Americans to determine the healthcare needs of individuals over the age of 65. The resulting report concluded that efforts were needed to explore ways to broaden the duties and responsibilities of workers at various levels of training, better prepare informal caregivers to tend to the needs of aging family members, and develop new models of health care delivery and payment that are effective and efficient (IOM 2008). Nursing education responded by shifting the focus of education from specialization to integrating basic gerontological competencies in the generalist nursing role. These efforts were intended to exert maximal impact in meeting the health care needs of older adults who consume the majority of health care services in ambulatory, acute care, home care, and nursing home settings by ensuring that older adults experience appropriate care along the continuum of prevention and health promotion, managing chronic illness, dealing with mental and physical frailty, and facilitating a peaceful death.

As a supplement to the Essentials of Baccalaureate Education for Professional Nursing Practice (AACN 2008), the American Academy of Colleges of Nursing (AACN) published Recommended Competencies and Curricular Guidelines for Nursing Care of Older Adults (AACN 2010). Competency in gerontological nursing encompasses the ability to recognize how one's own attitude and values regarding aging impact care of older adults and their families. The nurse is expected to appreciate the influence of culture, race, roles, language, religion, gender, lifestyle, and other cross-cultural influences in the context of differing international models of geriatric care. The competent gerontological nurse communicates effectively, respectfully, and compassionately during all care encounters with a unique perspective on the impact of transitions between settings on care. Nurses are required to have the ability to conduct an age-specific functional and cognitive assessment using evidence-based tools and manage the full scope of clinical phenomena such as depression, delirium, and dementia, prevention and management of pressure ulcers, maintaining proper nutrition and hydration, managing pain, urinary incontinence, falls, polypharmacy, and physical and chemical restraints. Nursing practice in the specialty involves having expertise in applying ethical and legal principles to complex issues of older adults, understanding principles of caregiving, intergenerational relationships, and the indicators of mistreatment or abuse and associated implications for autonomous decision-making. Gerontological nursing care entails comprehending the role of family, friends, and paid caregivers and the best way to educate and support dealing with the functional, physical, psychological, and social changes that commonly occur with aging. Community resources are essential to maintain independence and social connectivity in the least restrictive, safe environment while empowering endurance capacities of the older adult. Gerontological nurses are called upon to coordinate care and understand health insurance implications for access to and affordability of both preventive and treatment-based care in light of chronic and comorbid diseases. Competency entails working in an interprofessional manner to manage symptoms and ensure individual participation in care as well as promote a desirable, high quality approach at the end of life.

The first edition of *The Scope and Standards of Gerontological Nursing* was published by the American Nurses Association in 1995, followed by the second edition in 2001 and the third edition in 2010 (ANA 1995, 2001, 2010). This document has evolved over time to reflect the evolution of the practice of gerontological nursing. The document provides information about the current context of the specialty as well as guidance for educational preparation, practice setting opportunities, recommendations for professional development, and issues affecting the older population at the time of publication. Gerontological practice standards and professional performance expectations are updated and clearly articulated for the nursing profession to meet the continually changing needs of the older adult population. The next edition, expected to be published in late 2018 after a period of public comment, will link the 2015 Code of Ethics with Interpretive Statements to the newest version of the Scope and Standards to provide examples of how all gerontological nurses integrate ethical behaviors in practice settings (Bickford 2018).

Geriatric Nursing Workforce

Despite the increased demand for nursing roles to care for older adults in all practice settings, it remains difficult to recruit and retain a gerontological workforce (Houde and Deverauz-Millilo 2009). By 2050, the number of older people age 65–85 is expected to quadruple placing a strain on the healthcare system to meet the demands of this population. Nurses represent the largest segment of the workforce providing care to older adults, but only a small percentage is formally prepared with specialized gerontological nursing knowledge. Efforts have been underway for many years to address this gap to ensure competent, high quality care (Harden and Watman 2015).

Globally, nursing education has worked to integrate gerontological content in curricula over the last few decades, yet many practicing nurses have not received formal training to care for this population resulting in a gap between the expected and actual competencies of working nurses (Deschodt et al. 2010). Nursing students may be exposed to clinical environments where negative beliefs and practices toward older adults are prevalent making them less interested in working with this population (Baumbusch et al. 2012). The role of academic faculty as subject matter experts and influencers of attitudes about caring for older adults has been cited as a contributing factor in nursing practice setting preferences (Purfzad et al. 2019). Other contributing factors that negatively influence the choice of gerontological nursing as a practice specialty include challenging physical demands and environments, poor staffing, lack of equipment, and insufficient continuing education about the care of aging adults (Chai et al. 2018). To overcome undesirable perceptions, students need to be exposed to positive nursing role models with a specialized knowledge of gerontology and an empathetic manner (Garbrah et al. 2017). A recent strategy to address workforce issues in the work environment is the Age Friendly Health Systems approach supported by the Institute for Healthcare Improvement (IHI) in partnership with the John A. Hartford Foundation and the American Hospital Association (IHI 2019). The initiative espouses a philosophy that every older adult gets the best care possible, experiences no health care related harms, and is satisfied with the care received. Systems that adopt this model of care optimize value for patients, families, caregivers, healthcare providers, and the system itself. The four essential elements of care comprise the "4Ms" framework of an Age Friendly Health System: What Matters; Medication; Mentation; and Mobility. The system works in an organized manner with all staff to know and align care with each older adult's specific health outcome goals and care preferences, and use medications that do not interfere with what matters to the older adults and their mentation and mobility. System defined strategies seek to prevent, identify, and treat and manage dementia, depression, and delirium across settings and ensure that the older adults move safely every day to maintain function and do what matters to them. The goal is to implement this framework in 20% of US hospitals and medical practices by 2020.

Advanced Practice Gerontological Nursing

Gerontological advanced practice nurses (APRNs) have been prepared in gerontology to function as nurse practitioners and clinical nurse specialists since the 1970s. Like the registered nurse workforce, predictions of increased demand for geriatric population focused care did not result in an increased graduation rate of gerontological specialists including both gerontological clinical nurse specialists (GCNSs) and gerontological nurse practitioners (GNPs). Initial response was to integrate gerontological content into graduate-level curriculum for students in nongerontological programs by educating more competent providers to care for the predicted increasing numbers of older adults (Pencak Murphy et al. 2013). To address this issue, the Consensus Model for APRN education of 2008 (NCSBN 2008) set forth requirements for future licensure, certification, and accreditation of APRN education programs that combined adult and gerontology as one population focus. This mandate was intended to address the problem of declining numbers of graduates applying for certification as gerontological CNSs or NPs intended to increase the number of APRNs with basic competencies in the care of older adults by educating APRN students to provide care across the continuum of care from young to older adults. An unintended consequence was a decrease in APRNs taking the specialty certification in gerontological advanced practice nursing which resulted in the retirement of the Gerontological Nurse Practitioner and Gerontological Clinical Nurse Specialist certification exams in December 2013.

Gerontological nursing leaders recognize that the generalist preparation of adult gerontology APRNs is insufficient to care for the growing number of the oldest-old population (GAPNA 2015). The Gerontological Advanced Practice Nurses Association (GAPNA) engaged in a series of surveys of experienced specialists to capture the unique body of knowledge and skills required to care for this population. A consensus was achieved regarding the role which encompasses multiple proficiencies including the ability to conduct a comprehensive physical, social, cognitive, and functional assessment that considers normal and abnormal changes associated with aging; apply current evidence to inform decision-making for appropriate screening, diagnostic testing, treatment, and planning of care; manage plans of care consistent with regulatory guidelines; engage in prescribing practices that recognize risks and benefits of complex pharmacotherapy regimens; utilize a system-based approach to assess, design, implement, and evaluate effective educational strategies; provide gender inclusive care; coordinate timely palliative and end-of-life care congruent with individualized goals and values; anticipate and manage transitions of care between sites and providers; apply a systems-based approach in deploying resources to optimize health-related outcomes; utilize quality improvement and/or research activities to enhance healthcare; and use data to inform practice and policy development. Beginning June 1, 2018, to affirm specialist level practice proficiencies,

GAPNA began to offer a new APRN Gerontological Specialist Certification Exam.

Gerontological Nursing Research and Policy

With a rich history of commitment to providing interprofessional, team-based health care in a wide variety of settings, gerontological nurses and nursing scientists have been leaders in defining quality care for the older adult. Nursing researchers have contributed to a significant, diverse body of knowledge to improve patient care outcomes and influence policy decisions that impact the structure and process of care delivery (Perez et al. 2018). Studies have been designed utilizing an array of conceptual theories and frameworks producing knowledge that has progressed in sophistication and breadth over time, evolving in response to population needs and the changing profile of the older adult. Nurse scientists conduct research on a wide assortment of topics that inform evidence-based practice across the continuum of health and illness. Areas of focus include but are not limited to geriatric conditions and syndromes, health promotion and prevention, quality of life and well-being, psychosocial issues, genetics, caregiving phenomena, and technology use for population engagement. Research settings encompass any location where older adults receive health services including older adults' home, local community organizations, clinics, hospitals, nursing home, and end of life care settings. Nursing research methods commonly used include qualitative, mixed methods, and large data sets for secondary data analysis (Kovach 2018).

As a profession, nursing has led or collaborated with numerous disciplines in advancing the care of older adults with dementia, incontinence, pain, sleep disorders, and functional mobility issues to impact strategies related to fall and pressure injury prevention, medication and symptom management, care of the frail older adults, and individualized end of life care. Gerontological nursing researchers also lead research on improving long-term care policy and service delivery for older adults with disability (Kovach 2017).

A major source of federal funding for nurse-driven gerontological research is the National Institute of Nursing Research. The ability to conduct research and disseminate results has also been supported by the John A. Hartford Foundation since the establishment of the Building Academic Geriatric Nursing Capacity (BAGNC) in 2000 with four major overlapping goals to increase the number of geriatric nursing faculty through a predoctoral scholarship and postdoctoral fellowship award program; develop the next generation of geriatric nurse leaders; facilitate collaboration among nine Centers of Geriatric Nursing Excellence (CGNE) located at research-intensive universities; and disseminate outcomes that inform the field, policy, and ultimately improve care (Franklin et al. 2011). The program was further developed under the direction of the Gerontological Society of America in 2012 to establish strategic partnerships across all disciplines with the purpose of delivering quality care to older adults. The momentum continues as the National Hartford Center of Gerontological Nursing Excellence supports faculty and leadership development and seeks to impact public policy with interprofessional collaboration. This specialty organization is dedicated to optimal health and quality of life for older adults and works with national and international nursing schools and institutions that demonstrate a high level of commitment to the field of gerontological nursing and are poised for success by accessing best practices in faculty and curriculum development (Franklin et al. 2011).

Future Directions of Nursing Research

The future of gerontological nursing science lies in generating evidence to support the redesign of existing care models and the creation of innovative solutions to improve healthy living among diverse populations while simultaneously attending to decreasing costs (Nannini 2011). Nurse scientists and interprofessional partners have been encouraged by the National Institutes of Health (NIH) to use common data elements across studies to facilitate comparison and synthesis. Additional areas of emphasis include the interplay of multiple factors between older adults and family members or caregivers as an influence on health outcomes, self-management of multiple chronic medical and behavioral health problems with complex healthcare needs, and ensuring health equity among racially/ethnically diverse older adults, palliative and end of life care Wilson and Low 2016). Affiliates across the globe with shared interests in advancing gerontological nursing research must work together to ensure an expert, high quality workforce to provide accessible care to an increasing number of older adults. Professional research initiatives are encouraged among traditional and nontraditional partners that include but are not limited to: Gerontological Society of America (https://www.geron.org/); AARP (https://www.aarp.org/research/), Robert Wood Johnson Foundation (https://www.rwjf.org/), World Health Organization (https://www.who.int/), American Geriatrics Society (https://www.americangeriatrics.org/), American Society on Aging (https://www.asaging.org/), International Association of Gerontology and Geriatrics (https://www.ifa-fiv.org/), and the Diverse Elders Coalition (https://www.diverseelders.org/).

Summary

Gerontological nursing has made significant progress in ensuring the capacity to care for an increasing older adult population with complex health and wellness needs. Organized and coordinated efforts across academia, research, clinical practice, and care delivery systems predict continued emphasis on interprofessional collaboration to achieve concomitant goals of quality improvement, cost effectiveness, and accessibility to health care for all older adult populations. Continued emphasis on multipronged strategies to educate, recruit, and retain an expert gerontological workforce is indicated as the variety of healthcare settings providing services to the older adult and their support network grows more diverse.

Cross-References

▶ Critical Care Nursing
▶ Electronic Nursing Documentation
▶ Nursing Home Policy and Regulations in the United States
▶ Person-Centered Care for Older Adults
▶ Telenursing

References

American Association of Colleges of Nursing (2008) The essentials of baccalaureate education for professional nursing practice. http://www.aacnnursing.org/portals/42/publications/baccessentials08.pdf. Accessed 19 Apr 2019

American Association of Colleges of Nursing (2010) Recommended baccalaureate competencies and curricular guidelines for geriatric nursing care. https://www.aacnnursing.org/Portals/42/AcademicNursing/CurriculumGuidelines/AACN-Gero-Competencies-2010.pdf. Accessed 19 Apr 2019

American Nurses Association (ANA) (1995) Scope and standards of gerontological nursing. ANA, Silver Spring

American Nurses Association (ANA) (2001) Scope and standards of gerontological nursing. ANA, Silver Spring

American Nurses Association (ANA) (2010) Scope and standards of gerontological nursing. ANA, Silver Spring

Baumbusch J, Dahlki S, Phinney A (2012) Nursing students' knowledge and beliefs about care of older adults in a shifting context of nursing education. J Adv Nurs 68(11):2550–2558. https://doi.org/10.1111/j.1365-2648.2012.05958.x

Bickford C (2018) A contemporary look at gerontological nursing. Am Nurse Today 13(6):48

Chai X, Cheng C, Mei J et al (2018) Student nurses' career motivation toward gerontological nursing: longitudinal study. Nurse Educ Today 76:165–171. https://doi.org/10.1016/j.nedt.2019.01.028

Deschodt M, de Casterle B, Milisen K (2010) Gerontological care in nursing programmes. J Adv Nurs 66:139–148. https://doi.org/10.1111/j.1365-2648.2009.05160.x

Franklin P, Archbold P, Fagin C et al (2011) Building academic geriatric capacity: results after the first 10 years and implications for the future. Nurs Outlook 59:198–206. https://doi.org/10.1016/j.outlook.2011.05.01

Garbrah W, Valimaki T, Palovaara M et al (2017) Nursing curriculums may hinder a career in gerontological nursing: an integrative review. Int J Older People Nursing. https://doi.org/10.1111/opn.12152

Gerontological Advanced Practice Nursing Association (GAPNA) (2015) Consensus statement on proficiencies for the APRN gerontological specialist. https://www.gapna.org/sites/default/files/documents/GAPNA_Consensus_Statement_on_Proficiencies_for_the_APRN_Gerontological_Specialist.pdf. Accessed 19 Apr 2019

Harden J, Watman R (2015) The National Hartford Center of gerontological nursing excellence: an evolution of a nursing initiative to improve care of older adults. Gerontologist 55(SI):S1–S12. https://doi.org/10.1093/geront/gnv056

Houde S, Deverauz-Millilo K (2009) Caring for an aging population: review of policy initiatives. J Gerontol Nurs 15(12):8–13. https://doi.org/10.3928/00989134-20091103-04

Institute for Healthcare Improvement (IHI) (2019) The business case for becoming an age-friendly health system. http://www.ihi.org/Engage/Initiatives/Age-Friendly-Health-Systems/Documents/IHI_Business_Case_for_Becoming_Age_Friendly_Health_System.pdf. Accessed 19 Apr 2019

Institute of Medicine (2008) Retooling for an aging America: building the healthcare workforce. National Academies Press, Washington, DC

Kovach C (2017) 10-Year milestone for research in gerontological nursing: trends affecting scientific publishing. Res Gerontol Nurs 10(1):3–4. https://doi.org/10.3928/19404921-20161209-06

Kovach C (2018) Research in gerontological nursing: how are we doing. Res Gerontol Nurs 11(5):227–229. https://doi.org/10.3928/19404921-20180810-03

Nannini A (2011) The future of gerontological nursing. J Gerontol Nurs 37(9):11–15. https://doi.org/10.3928/00989134-20110802-01

National Council of State Boards of Nursing (NCSBN) APRN Consensus Workgroup (2008) Consensus model for APRN regulation: licensure, accreditation, certification & education. https://www.ncsbn.org/Consensus_Model_for_APRN_Regulation_July_2008.pdf. Accessed 19 Apr 2019

Newton Shafer K (1950) Geriatric nursing. Mosby, St. Louis

Pencak Murphy M, Miller J, Siomos M et al (2013) Integrating gerontological content across advanced practice registered nurse programs. J Am Assoc Nurse Pract 26:77–84. https://doi.org/10.1002/2327-6924.12074

Perez G, Mason D, Harden J et al (2018) The growth and development of gerontological nurse leaders in policy. Nurs Outlook 66:168–179. https://doi.org/10.1016/j.outlook.2017.10.005

Purfazad Z, Bahrami M, Keshvari M et al (2019) Effective factors for development of gerontological nursing competence: a qualitative study. J Contin Educ Nurs 50(3):127–133. https://doi.org/10.3928/00220124-20190218-08

Wilson D, Low G (2016) Lifelong health and health services use: a new focus for gerontological nursing research and practice. J Gerontol Nurs 43(2):28–32. https://doi.org/10.3928/00989134-20160727-01

Gerontological Social Workers

▶ Geriatric Social Workers

Gerontology and Political Science

▶ Political Gerontology

Geroprotective Medication

▶ Targeting Aging with Metformin (TAME)

Geropsychology

Hans-Werner Wahl[1,2] and Eva-Luisa Schnabel[2]
[1]Institute of Psychology, Heidelberg University, Heidelberg, Germany
[2]Network Aging Research, Heidelberg University, Heidelberg, Germany

Synonyms

Behavioral gerontology; Psychogerontology; Psychological gerontology; Psychology of aging

Definition

Geropsychology is the area within psychology devoted to the study of aging (APA 2019). Although the focus of geropsychology is on later life, a life-span perspective is fundamental, hence old age is seen as inextricably linked with lifespan development as a whole (Schaie 2016).

Overview

Geropsychology addresses a broad range of domains such as cognitive abilities, personality, social-emotional functioning, as well as the mental health of older adults. Seen against other key disciplines, geropsychology concentrates more on individual behavior than on societal and biological processes as compared to the sociology of aging and biogerontology. Nevertheless, an increasing orientation toward combining biomarkers with traditional measures of geropsychology such as questionnaires has emerged with the goal to better understand the connection between biological and behavioral variables as people age (Kotter-Grühn et al. 2016).

Key Research Findings

Cognitive aging has likely received the most research attention so far in geropsychology. Robust findings mostly based on longitudinal studies underscore that cognitive abilities such as information processing speed, inductive reasoning, working memory, and verbal skills remain on average rather stable until the age of 65–70 years and then decline (Schaie 2013). However, large interindividual differences in cognitive aging exist depending on socio-structural, health, and lifestyle factors.

Further, trait personality development as reflected in the Big Five (neuroticism, extraversion, openness, conscientiousness, agreeableness), which has long been assumed as stable across the adult lifespan after the age of 30 years (McCrae and Costa 1994), should be reinterpreted as rather plastic according to more recent longitudinal research. Indeed, rank-order stability of the Big Five personality traits is decreasing as people age, particularly beyond the age of 80 years (Specht et al. 2011).

The so-called "well-being paradox" of aging describes the phenomenon that positive and negative affect, as well as satisfaction with life, remain on average rather stable until advanced old age despite substantial age-related losses. It has been shown that older adults have a broad range of developmental regulatory strategies at their disposal that allow the maintenance of well-being. For example, older adults adjust their life goals by disengaging from goals that are no longer attainable (Heckhausen et al. 2010).

Driven by socio-emotional selectivity theory, geropsychological research has also found that older adults are active and successful shapers of their social relations. The assumption that a shortened future time perspective is associated with an increased investment into less but more intimate social relations has found much empirical support (English and Carstensen 2017).

Finally, the subjective aging experience has seen intensive research in geropsychology. For example, Westerhof et al.'s (2014) meta-analysis has demonstrated that more positive attitudes toward one's own aging are longitudinally linked with better health outcomes and lowered mortality rates.

Future Directions for Research

Longitudinal research as well as experimental study designs can be seen as the gold standard for geropsychological research. However, ecological momentary assessments (EMA), hence the intensive measurement of behavior across shorter time periods such as a week and frequently with several measurement points per day, have seen a tremendous increase during recent decades (Diehl et al. 2015). Thus, *combining* EMA with long-term study designs running across years and lab-based experimental studies that allow isolating certain mechanisms of interest (e.g., the time flow of emotional down-regulation after a stress occurrence) may become the new gold standard of geropsychology's future.

Another emerging trend in geropsychology is to include a contextual perspective (Wahl and

Gerstorf 2018). The basic idea is that analyzing the behavior of individuals without the empirical consideration of the respective context is oversimplifying and of low ecological validity (Diehl et al. 2017).

Cross-References

▶ Life-Span Development
▶ Personality in Later Life
▶ Plasticity of Aging
▶ Psychogeriatrics
▶ Self-Perceptions of Aging

References

American Psychological Association (2019) Psychology and aging: addressing mental health needs of older adults. https://www.apa.org/pi/aging/resources/guides/aging.pdf. Accessed 18 Feb 2019
Diehl M, Hooker K, Sliwinski MJ (2015) Handbook of intraindividual variability across the life span. Routledge, New York
Diehl M, Wahl H-W, Freund A (2017) Ecological validity as a key feature of external validity in research on human development. Res Hum Dev 14:177–181. https://doi.org/10.1080/15427609.2017.1340053
English T, Carstensen LL (2017) Socioemotional selectivity theory. In: Pachana NA (ed) Encyclopedia of geropsychology. Springer, Singapore, pp 2222–2227. https://doi.org/10.1007/978-981-287-082-7_110
Heckhausen J, Wrosch C, Schulz R (2010) A motivational theory of life-span development. Psychol Rev 117:32–60. https://doi.org/10.1037/a0017668
Kotter-Grühn D, Kornadt AE, Stephan Y (2016) Looking beyond chronological age: current knowledge and future directions in the study of subjective age. Gerontology 62:86–93. https://doi.org/10.1159/000438671
McCrae RR, Costa PT (1994) The stability of personality: observations and evaluations. Curr Dir Psychol Sci 3:173–175. https://doi.org/10.1111/1467-8721.ep10770693
Schaie KW (2013) Developmental influences on adult intelligence: the Seattle Longitudinal Study, 2nd edn. Oxford University Press, New York
Schaie KW (2016) Theoretical perspectives for the psychology of aging in a lifespan context. In: Schaie KW, Willis SL (eds) Handbook of the psychology of aging, 8th edn. Academic, San Diego, pp 3–13. https://doi.org/10.1016/B978-0-12-411469-2.00001-7
Specht J, Egloff B, Schmukle SC (2011) Stability and change of personality across the life course: the impact of age and major life events on mean-level and rank-order stability of the Big Five. J Pers Soc Psychol 101:862–882. https://doi.org/10.1037/a0024950
Wahl H-W, Gerstorf D (2018) A conceptual framework for studying context dynamics in aging (CODA). Dev Rev 50:155–176. https://doi.org/10.1016/j.dr.2018.09.003
Westerhof GJ, Miche M, Brothers AF, Barrett AE, Diehl M, Montepare JM, Wahl H-W, Wurm S (2014) The influence of subjective aging on health and longevity: a meta-analysis of longitudinal data. Psychol Aging 29:793–802. https://doi.org/10.1037/a0038016

Geroscience

Brian K. Kennedy[1,2,3], Jorming Goh[1,2] and Esther Wong[1,2]
[1]NUSMed Healthy Longevity Translational Research Programme, Yong Loo Lin School of Medicine, National University of Singapore, Singapore, Singapore
[2]Centre for Healthy Longevity, National University Health System, Singapore, Singapore
[3]Singapore Institute of Clinical Sciences, A*STAR, Singapore, Singapore

Synonyms

Aging studies; Hallmarks of aging; Pillars of aging; Science of aging; Science of longevity interventions; Science of senescence

Definition

Geroscience is a field of biomedical research which seeks to understand how aging processes drive chronic diseases, in order to develop interventions to extend human healthspan and prevent disease onset with age (Kennedy et al. 2014). Nine hallmarks and seven pillars of aging, which are overlapping and interconnected, have been identified and shown to be modifiable to slow aging and for treating multiple chronic diseases simultaneously (Lopez-Otin et al. 2013; Kennedy et al. 2014).

Overview

Decades of research on the biology of aging have dramatically improved our understanding of the molecular events that drive the process, and the underlying evolutionary explanation for the existence of the phenomenon. From the molecular perspective, one of the main advances has been to escape the mindset that one process, for instance, oxidative stress (Harman 1956), causes aging and embraces the concept of many pathways acting in concert to modulate systemic aging. The initial idea by de Grey et al. (2002) broadly segments the aging process into metabolism, damage, and pathology – three broad stages. de Grey et al. propose interventions that remove the seven major age-related damages, namely, cell loss-atrophy, division-obsessed cells, death-resistant cells, mitochondrial dysfunction/mutations, intracellular junk, extracellular junk and extracellular matrix stiffening, all of which would have the potential to realize "engineered negligible senescence" (de Grey et al. 2002). Subsequently, two studies, both published in *Cell* several years ago, further elaborate a set of pathways that modulate aging and make the important point that aging pathways are highly interconnected, implying that modifying one pathway will likely influence the others. In 2013, Lopez-Otin et al. published *"The Hallmarks of Aging"* (Lopez-Otin et al. 2013), which was modeled after *"The Hallmarks of Cancer,"* published in 2000 by Hanahan and Weinberg (2000). Nine hallmarks of aging were described that met several criteria, including (1) manifestation during aging, (2) experimentation aggravation accelerates aging, (3) experimental amelioration should retard normal aging and extend healthspan. Shortly thereafter, as a result of a National Institute of Aging (NIA)-sponsored meeting, "Geroscience: Linking Aging to Chronic Disease" was published (Kennedy et al. 2014). This review, which involved a number of investigators who presented findings, served to define the aging pillars. Seven pillars were identified and discussed extensively, leading to a consensus in the review.

When comparing pillars and hallmarks, it is clear that there is a strong overlap (Fig. 1). While it is certainly possible that a certain hallmark/pillar may be over-represented or that one or two are yet to be identified, the emerging consensus in the aging field is a mark of (1) the successful model organisms ranging from yeast to mammals and comparative biology studies and (2) the maturation of the field. Most scientists in the field embrace the notion that multiple pathways coordinate aging, although they may weigh pathways differently, and that aging is now modifiable through one or many of the diverse intervention strategies. One note of caution is that studies of human aging still need to be expanded and, while there is a strong likelihood (and significant evidence) that aging pathways will be conserved, surprises are likely yet to be uncovered in the human realm.

Both reviews agreed on at least one other major concept, that aging pathways do not work in isolation. Instead, they are connected in a network that has extensive crosstalk and inter-regulation. Lopez-Otin et al. state that "systems biology approaches will be required to understand the mechanistic links among the processes that accompany, and lead to aging" (Lopez-Otin et al. 2013). Kennedy et al. made a more explicit statement: While "specific recommendations for each pillar (in the meeting) emerged, their connectedness was striking. The themes were not seven independent factors driving aging; rather they were highly intertwined processes, and understanding the interplay between these seven pillars is critical" (Kennedy et al. 2014).

There are a number of interventions described that can delay aging. It is nearly impossible to connect any intervention specifically to one pillar of aging. For examples, the mTOR pathway can be regulated by the highly specific inhibitor, rapamycin (Kennedy and Lamming 2016), and Sirtuins are activated by small molecules that increase NAD(+) levels (Imai and Guarente 2016). Literature searches uncover dozens of papers linking either the mTOR pathway or one or another Sirtuin to each aging pillar (Bonkowski and Sinclair 2016; Kennedy and Lamming 2016). Moreover, rapamycin-mediated lifespan extension is linked to all seven pillars. It is reported to (1) enhance proteostasis, (2) improve adult stem

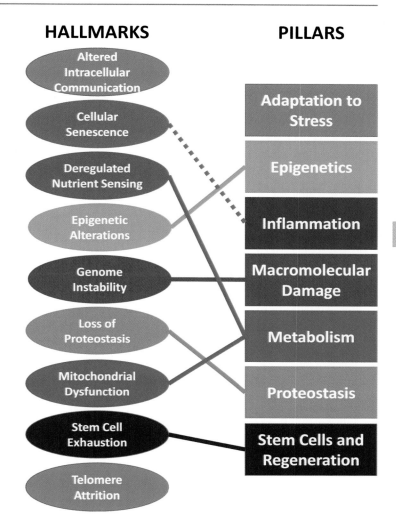

Geroscience, Fig. 1 Relationships between hallmarks and pillars of aging. Solid lines connecting hallmarks and pillars reflect either direct correspondences or cases where they are highly related. The dotted line between cellular senescence and inflammation indicates a partial relationship since senescent cells secrete a set of inflammatory cytokines that contribute to chronic inflammation (Lasry and Ben-Neriah 2015)

cell function, (3) enhance stress responses, (4) reduce chronic inflammation, (5) delay age-associated epigenetic changes, (6) improve the response to macromolecular damage, and (7) favorably alter metabolism (Kennedy and Lamming 2016; Saxton and Sabatini 2017). This is just one example. Similar arguments could be made for dietary restriction and exercise, although these are more complex, whole-organism interventions. These findings suggest that healthy lifespan extension is conferred not by affecting one pathway but by targeting nodes or connectivity points, fostering a homeostatic network during aging that maintains function and prevents disease. The outcomes of these interventions can be read in most or all pillars.

These interventions, if successful, may take on even more relevance, given that aging is the biggest risk factor for mortality due to Covid-19, as well as influenza and many other infectious diseases (Barzilai et al. 2020; see ▶ "COVID-19 Pandemic and Healthy Aging"). If interventions that extend healthspan can be employed successfully prior to the next pandemic, accompanying morbidity and mortality may be dramatically mitigated. Potential aging interventions, which are now being rapidly identified (Partridge et al. 2020; Moskalev et al. 2015), are largely predicated on knowledge of the biological pathways that modulate aging.

Still, one wonders if a hierarchy between hallmarks or pillars can be established, reflecting either a chronological ordering by which aging disrupts homeostasis or a definition of which pathways become dysregulated first. Lopez-Otin et al. propose such a hierarchy (Lopez-Otin et al. 2013), whereby four hallmarks reflect causes of damage, another three respond to damage and a final two that have a more integrative role. This is an interesting analysis, but other interpretations are possible. For instance, deregulated nutrient signaling is considered a response to damage, but in sedentary individuals with chronic overnutrition, it is likely that persistent stimulation of nutrient signaling pathways leads to their dysregulation in a manner that may not require damage. The point is not to disagree with the proposed hierarchy, which is a starting point for discussion, but rather to raise the possibility that dysregulation of aging pathways may be triggered in a personalized manner, depending on the genetics and lifestyle of the individual.

Key Research Findings

With the pathways of aging defined, at least to a reasonable approximation, two major areas of advancement have been achieved in the last decade. First, a large number of interventions to slow aging have been discovered, including lifestyle modifications such as dietary modification ("Diet and Calorie Restriction," "Healthy Diet," "Slow Aging Diet") and exercise (see ▶ "Aerobic Exercise Training and Healthy Aging" and ▶ "Exercise and Healthy Cardiovascular Aging"); dietary supplements including NAD+ precursors, alpha keto-glutarate, and spermidine, and others (Asadi Shahmirzadi et al. 2020; Chin et al. 2014; Madeo et al. 2018; Zhang et al. 2016); and drugs including metformin and rapamycin (and derivatives termed rapalogs). In addition, stem cell- and gene therapy-based interventions are being developed (Beyret et al. 2019; Neves et al. 2017); see ▶ "Stem Cells Aging" and ▶ "Whole Genome Sequencing"). The fact that so many possible interventions exist increases the likelihood that some will translate to humans. Detailed coverage of these interventions is beyond the scope of this manuscript, and readers are referred to recent reviews (Moskalev et al. 2015; Partridge et al. 2020).

A second major step has been the advent of aging biomarkers that may accurately reflect an individual's biologic age (Kudryashova et al. 2020; Zhavoronkov and Mamoshina 2019; Bell et al. 2019). These biomarkers, which appear to be responsive to interventions (Fahy et al. 2019; Gensous et al. 2019, 2020; Pavanello et al. 2019; Sae-Lee et al. 2018; Sandoval-Sierra et al. 2020; Wang et al. 2017) and can be assessed with minimally invasive techniques, have been made possible largely by the generation of omics-based datasets and the use of artificial intelligence-enabled analysis (Chen et al. 2015; Hannum et al. 2013; Horvath 2013; Putin et al. 2016; Pyrkov et al. 2018; Solovev et al. 2020; Tarkhov et al. 2019). Several such biologic age markers now exist, including those based on epigenetic clocks (D'Aquila et al. 2019; Field et al. 2018). More recently, improved biologic age assessment has been proposed through the combination of epigenetic age and a set of clinical markers (Bürkle et al. 2015; Levine et al. 2018; Lu et al. 2019; Benayoun et al. 2019). If biologic age can be accurately measured, then biologic age can be used to stratify individuals at high risk of disease prior to the onset of symptoms and these markers can also be used to assess the efficacy of aging interventions. Detailed reviews of aging biomarkers were cited above.

Examples of Application

If interventions that slow human aging can be validated, then a major new area of medicine will become possible, whereby healthspan can be extended and possibly even compressed morbidity can be achieved. The estimated healthcare cost benefits are massive and, more importantly, the quality of human life can be dramatically enhanced. The challenge is how can these interventions be validated?

There are three standard approaches to validating longevity interventions. The first, and most pursued, involves something of a sleight of hand. Since aging is not recognized as a disease, making FDA approval and reimbursement seemingly impossible, biotechnology and pharmaceutical companies have taken interventions that target aging pathways and tried to treat age-associated diseases, or their complications. These have largely been unsuccessful. The most recent failure involved a study by Unity Biotechnology attempting to reduce pain in osteoarthritis patients using an identified senolytic drug (UNITY Biotechnology 2020). Senolytic drugs target the selective removal of senescent cells that accumulate with age (Robbins et al. 2020). This approach is understandable, particularly with a new chemical entity, but that does not mean it will be successful and continued failure with clinical studies may have an adverse impact on the aging research field. Seemingly, another approach is needed.

A second approach is the most direct: if the interventions were developed to slow aging, we should try to treat aging. At least from the standpoint of validating the potential of geroprotectors, a variety of drugs and supplements reported to sustain health and slow ageing, this seems like an essential step for the field. Two approaches have been proposed to achieve this, both of which involve disease prevention. This is critical, since most geroprotectors have been identified based on their ability to slow aging in experimental animals when initiated at early-to-middle age. The first approach is embodied by the TAME trial (Justice et al. 2018), which will provide metformin to nondiabetic individuals in their 60s for a period of 3 years, with the goal of preventing multiple chronic diseases simultaneously. This approach is most likely to provide data that will extend healthspan in a format acceptable to the FDA. The disadvantage is that it is a long-term trial with thousands of individuals, costing tens of millions of dollars. It remains unclear which interventions will work the best in humans, making comparison of multiple interventions ideal. Therefore, a costly and time-consuming process like the TAME trial has limitations.

A third approach involves combining the two major advances in the aging field – testing whether interventions affect aging biomarkers. These studies may be particularly amenable to lifestyle interventions and natural products, since they are under less regulation. However, they should be tested with drugs as well, at least for drugs that are clinically approved for other indications. These studies are still in early days but already there are suggestions that they might be successful and that public interest in longevity interventions will increase dramatically if they are (Quach et al. 2017; Fahy et al. 2019). It is unclear how long interventions will have to be implemented and whether biologic age can be reversed, or just slowed. Nevertheless, these studies have great promise and should be accelerated as much as possible.

Future Directions of Research

People have dreamed of mitigating aging for millennia, and this may be the generation that witnesses this seismic shift. Much is known about aging; a wide range of possible geroprotectors have been developed, and it is becoming possible to measure the actual rate that a person ages. But the field is far from the finish line. Many of the key questions have been unanswered, including (1) is there a hierarchical process that links aging hallmarks and pillars? (2) to what extent is aging personalized? (3) are there aging pathways in humans that have eluded discovery in animal models? (4) which longevity interventions will translate best to humans? (5) how do we validate interventions most effectively in human clinical studies? and (6) how do these interventions enter widespread use?

While government-based funding for research on the biological basis of aging (the biggest risk factor for the majority of chronic conditions that are driving worldwide disability and mortality) remains paltry, the good news is that the private sector has entered the fray. Venture capitalists are investing in longevity companies and a wide array of new startups are developing interventions and diagnostics. The last decade has seen major advances in geroscience and the next decade looks to be at least as promising.

Summary

Geroscience research has identified a homeostatic network comprising of interconnected mechanisms that preside over healthy aging. When this network is disrupted, chronic diseases occur. Repurposing drugs and natural products to preserve or augment the functionalilty of this homeostatic network may define a new "health" care-based strategy to enhance healthspan and longevity.

Cross-References

- Aerobic Exercise Training and Healthy Aging
- Aging
- Biogerontology
- Diet and Calorie Restriction
- Life-Extensionism in a Historical Perspective
- Nicotinamide Adenine Dinucleotide (NAD+) in Aging
- Senolytic Drugs
- Targeting Aging with Metformin (TAME)

References

Asadi Shahmirzadi A, Edgar D, Liao CY et al (2020) Alpha-ketoglutarate, an endogenous metabolite, extends lifespan and compresses morbidity in aging mice. Cell Metab 32:447–456.e6. https://doi.org/10.1016/j.cmet.2020.08.004

Barzilai N, Appleby JC, Austad SN et al (2020) Geroscience in the age of COVID-19. Aging Dis 11 (4):725–729. https://doi.org/10.14336/AD.2020.0629

Bell CG, Lowe R, Adams PD et al (2019) DNA methylation aging clocks: challenges and recommendations. Genome Biol 20:249. https://doi.org/10.1186/s13059-019-1824-y

Benayoun BA, Pollina EA, Singh PP et al (2019) Remodeling of epigenome and transcriptome landscapes with aging in mice reveals widespread induction of inflammatory responses. Genome Res 29:697–709. https://doi.org/10.1101/gr.240093.118

Beyret E, Liao HK, Yamaoto M et al (2019) Single-dose CRISPR-Cas9 therapy extends lifespan of mice with Hutchinson-Gilford progeria syndrome. Nat Med 25:419–422. https://doi.org/10.1038/s41591-019-0343-4

Bonkowski MS, Sinclair DA (2016) Slowing ageing by design: the rise of NAD(+) and sirtuin-activating compounds. Nat Rev Mol Cell Biol 17:679–690. https://doi.org/10.1038/nrm.2016.93

Bürkle A, Moreno-Villanueva M, Bernhard J et al (2015) MARK-AGE biomarkers of ageing. Mech Ageing Dev 151:2–12. https://doi.org/10.1038/s41591-019-0343-4

Chen W, Qian W, Wu G et al (2015) Three-dimensional human facial morphologies as robust aging markers. Cell Res 25:574–587. https://doi.org/10.1038/cr.2015.36

Chin RM, Fu X, Pai MY et al (2014) The metabolite α-ketoglutarate extends lifespan by inhibiting ATP synthase and TOR. Nature 510:397–401. https://doi.org/10.1038/nature13264

D'Aquila P, Montesanto A, De Rango F et al (2019) Epigenetic signature: implications for mitochondrial quality control in human aging. Aging (Albany NY) 11:1240–1251. https://doi.org/10.18632/aging.101832

De Grey AD, Ames BN, Andersen JK et al (2002) Time to talk SENS: critiquing the immutability of human aging. Ann N Y Acad Sci 959:452–462; discussion 463–5. https://doi.org/10.1111/j.1749-6632.2002.tb02115.x

Fahy GM, Brooke RT, Watson JP et al (2019) Reversal of epigenetic aging and immunosenescent trends in humans. Aging Cell 18:e13028. https://doi.org/10.1111/acel.13028

Field AE, Robertson NA, Wang T et al (2018) DNA methylation clocks in aging: categories, causes, and consequences. Mol Cell 71:882–895. https://doi.org/10.1016/j.molcel.2018.08.008

Gensous N, Franceschi C, Santoro A et al (2019) The impact of caloric restriction on the epigenetic signatures of aging. Int J Mol Sci 20. https://doi.org/10.3390/ijms20082022

Gensous N, Garagnani P, Santoro A et al (2020) One-year Mediterranean diet promotes epigenetic rejuvenation with country- and sex-specific effects: a pilot study from the NU-AGE project. Geroscience 42:687–701. https://doi.org/10.1007/s11357-019-00149-0

Hanahan D, Weinberg RA (2000) The hallmarks of cancer. Cell 100:57–70. https://doi.org/10.1016/S0092-8674(00)81683-9

Hannum G, Guinney J, Zhao L et al (2013) Genome-wide methylation profiles reveal quantitative views of human aging rates. Mol Cell 49:359–367. https://doi.org/10.1016/j.molcel.2012.10.016

Harman D (1956) Aging: a theory based on free radical and radiation chemistry. J Gerontol 11:298–300. https://doi.org/10.1093/geronj/11.3.298

Horvath S (2013) DNA methylation age of human tissues and cell types. Genome Biol 14:R115

Imai SI, Guarente L (2016) It takes two to tango: NAD(+) and sirtuins in aging/longevity control. NPJ Aging Mech Dis 2:16017. https://doi.org/10.1038/npjamd.2016.17

Justice JN, Niedernhofer L, Robbins PD et al (2018) Development of clinical trials to extend healthy lifespan. Cardiovasc Endocrinol Metab 7:80–83. https://doi.org/10.1097/XCE.0000000000000159

Kennedy BK, Lamming DW (2016) The mechanistic target of rapamycin: the grand conducTOR of metabolism and aging. Cell Metab 23:990–1003. https://doi.org/10.1016/j.cmet.2016.05.009

Kennedy BK, Berger SL, Brunet A et al (2014) Geroscience: linking aging to chronic disease. Cell 159:709–713. https://doi.org/10.1016/j.cell.2014.10.039

Kudryashova KS, Burka K, Kulaga AY et al (2020) Aging biomarkers: from functional tests to multi-Omics approaches. Proteomics 20:e1900408. https://doi.org/10.1002/pmic.201900408

Lasry A, Ben-Neriah Y (2015) Senescence-associated inflammatory responses: aging and cancer perspectives. Trends Immunol 36:217–228. https://doi.org/10.1016/j.it.2015.02.009

Levine ME, Lu AT, Quach A et al (2018) An epigenetic biomarker of aging for lifespan and healthspan. Aging (Albany NY) 10:573–591. https://doi.org/10.18632/aging.101414

Lopez-Otin C, Blasco MA, Partridge L et al (2013) The hallmarks of aging. Cell 153:1194–1217. https://doi.org/10.1016/j.cell.2013.05.039

Lu AT, Quach A, Wilson JG et al (2019) DNA methylation GrimAge strongly predicts lifespan and healthspan. Aging (Albany NY) 11:303–327. https://doi.org/10.18632/aging.101684

Madeo F, Eisenberg T, Pietrocola F et al (2018) Spermidine in health and disease. Science 359. https://doi.org/10.1126/science.aan2788

Moskalev A, Chernyagina E, De Magalhaes JP (2015) Geroprotectors.org: a new, structured and curated database of current therapeutic interventions in aging and age-related disease. Aging (Albany NY) 7:616–628. https://doi.org/10.18632/aging.100799

Neves J, Sousa-Victor P, Jasper H (2017) Rejuvenating strategies for stem cell-based therapies in aging. Cell Stem Cell 20:161–175. https://doi.org/10.1016/j.stem.2017.01.008

Partridge L, Fuentealba M, Kennedy BK (2020) The quest to slow ageing through drug discovery. Nat Rev Drug Discov 19:513–532. https://doi.org/10.1038/s41573-020-0067-7

Pavanello S, Campisi M, Tona F et al (2019) Exploring epigenetic age in response to intensive relaxing training: a pilot study to slow down biological age. Int J Environ Res Public Health 16. https://doi.org/10.3390/ijerph16173074

Putin E, Mamoshina P, Aliper A et al (2016) Deep biomarkers of human aging: application of deep neural networks to biomarker development. Aging (Albany NY) 8:1021–1033. https://doi.org/10.18632/aging.100968

Pyrkov TV, Getmantsev E, Zhurov B et al (2018) Quantitative characterization of biological age and frailty based on locomotor activity records. Aging (Albany NY) 10:2973–2990. https://doi.org/10.18632/aging.101603

Quach A, Levine ME, Tanaka T et al (2017) Epigenetic clock analysis of diet, exercise, education, and lifestyle factors. Aging (Albany NY) 9:419–446. https://doi.org/10.18632/aging.101168

Robbins PD, Jurk D, Khosla S et al (2020) Senolytic drugs: reducing senescent cell viability to extend health span. Annu Rev Pharmacol Toxicol. https://doi.org/10.1146/annurev-pharmtox-050120-105018

Sae-Lee C, Corsi S, Barrow TM et al (2018) Dietary intervention modifies DNA methylation age assessed by the epigenetic clock. Mol Nutr Food Res 62: e1800092. https://doi.org/10.1002/mnfr.201800092

Sandoval-Sierra JV, Helbing AHB, Williams EG et al (2020) Body weight and high-fat diet are associated with epigenetic aging in female members of the BXD murine family. Aging Cell 19:e13207. https://doi.org/10.1111/acel.13207

Saxton RA, Sabatini DM (2017) mTOR signaling in growth, metabolism, and disease. Cell 168:960–976. https://doi.org/10.1016/j.cell.2017.02.004

Solovev I, Shaposhnikov M, Moskalev A (2020) Multi-omics approaches to human biological age estimation. Mech Ageing Dev 185:111192. https://doi.org/10.1016/j.mad.2019.111192

Tarkhov AE, Alla R, Ayyadevara S et al (2019) A universal transcriptomic signature of age reveals the temporal scaling of Caenorhabditis elegans aging trajectories. Sci Rep 9:7368. https://doi.org/10.1038/s41598-019-43075-z

UNITY Biotechnology (2020) UNITY Biotechnology announces 12-week data from UBX0101 Phase 2 Clinical Study in Patients with Painful Osteoarthritis of the Knee. https://ir.unitybiotechnology.com/news-releases/news-release-details/unity-biotechnology-announces-12-week-data-ubx0101-phase-2. Accessed 30 Aug 2020

Wang T, Tsui B, Kreisberg JF et al (2017) Epigenetic aging signatures in mice livers are slowed by dwarfism, calorie restriction and rapamycin treatment. Genome Biol 18:57. https://doi.org/10.1186/s13059-017-1186-2

Zhang H, Ryu D, Wu Y et al (2016) NAD^+ repletion improves mitochondrial and stem cell function and enhances life span in mice. Science 352:1436–1443. https://doi.org/10.1126/science.aaf2693

Zhavoronkov A, Mamoshina P (2019) Deep aging clocks: the emergence of AI-based biomarkers of aging and longevity. Trends Pharmacol Sci 40:546–549. https://doi.org/10.1016/j.tips.2019.05.004

Geroscience Funding Frameworks

▶ Crowdsourcing and Crowdfunding in Aging Research

Gero-transcendence

▶ Aging: An Islamic Perspective

Gf

▶ Intelligence (Crystallized/Fluid)

GLBT

▶ LGBT in Old Age

Glioma

▶ Tumors: Brain

Global AgeWatch Index and Insights

Gibran Cruz-Martinez[1] and Gokce Cerev[2]
[1]Institute of Public Goods and Policies, Consejo Superior de Investigaciones Científicas (CSIC), Madrid, Spain
[2]Department of Labor Economics and Industrial Relations, Fırat University, Elazığ, Turkey

Synonyms

Global AgeWatch Insights; Global Aging Data; Global Aging Index; International aging indicators

Definition

Global AgeWatch Index and Insights by HelpAge International aim to contribute to the achievement of long-term transformative change in respect to aging and the lives of older people by advocating for better production of timely and good quality data to inform policy and program response.

The Global AgeWatch Index is a composite index that measures quality of life of older people and ranks countries based on four domains – income security, health status, enabling environment, and capability. The index was developed in partnership with Professor Asghar Zaidi. The index was published during 2013–2015.

The Global AgeWatch Insights is a research-based advocacy tool that examines the situation of older people in low- and middle-income countries, assesses availability of relevant data and evidence, and identifies policy actions. The Insights are produced in partnership with AARP. The reports that were launched in 2018 are planned to be released every 3 years with a different thematic focus. The first report focused on the inequities of the health systems in 12 low- and middle-income countries.

Overview

Life-cycle risks are inherent to every human being. As humans, we desire to go through every stage of life, from childhood to older age. Every stage has social risks associated with it, but older age could be considered as the most sensitive period of human life (Caspari and Lee 2004).

The global population is aging at unprecedented rates. The number of older-age adults in the world has almost quintupled in the last 65 years (see Fig. 1). This exponential growth is especially evident, and above the global average, in the geographical regions of Latin America and the Caribbean, Africa, and Asia (UNPD 2017). The 906 million older adults above 60 years of age represent 12.3% of the world population. See Table 1 for data disaggregated by region.

The main determinants for this exponential aging are the increase in the quality of human life, the decrease of mortality rates, the decrease of birth rates, and the increase in the life expectancy (Stallard 2006). Living longer and healthier lives is a success story. However, policymakers and governments need to consider this when

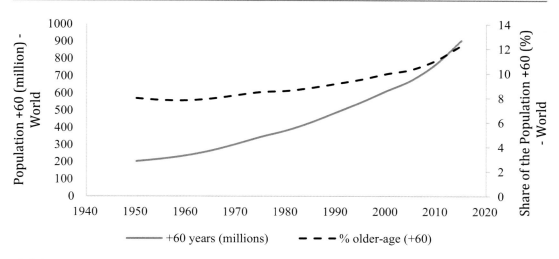

Global AgeWatch Index and Insights, Fig. 1 World population aging from 1950 to 2015. (Source: Calculated by the authors with data from UNPD (2017))

Global AgeWatch Index and Insights, Table 1 Older-age population by geographical region (2015)

Location	Millions of older-age adults	The share of older-age adults (%)
World	906.00	12.27
Africa	64.34	5.39
Asia	513.84	11.63
Europe	176.78	23.86
Latin America and the Caribbean	70.68	11.18
Northern America	73.84	20.74
Oceania	6.52	16.49

Source: Calculated by the authors with data from UNPD (2017)

planning and implementing public policy. Systems, practices, and norms need to adapt to aging. Government policymaking, planning, and implementation must recognize population aging and needs of older people, address barriers to the realization of rights of older people, and ensure that quality evidence and data on older people are collected to inform policy.

The Global AgeWatch Index was developed to: (i) measure the quality of life of older people, (ii) spotlight successes and shortcomings of country responses to population aging, and (iii) stimulate demand for and supply of age-disaggregated data (Mihnovits and Zaidi 2015).

The conceptual framework of the Index draws on the literature review of well-being of older people and the capabilities approach. It further takes into account priority areas identified in the 2012 United Nations Population Fund (UNFPA) and HelpAge report, *Ageing in the Twenty-First Century: A Celebration and a Challenge*, and the Madrid International Plan of Action on Ageing.

Global AgeWatch Index is multidimensional as it considers four areas required for the economic and social well-being of the older-age population: income security, health status, enabling environment, and capability. HelpAge International argues that these four dimensions represent four key domains for older-age people, "covering the most crucial aspects of their wellbeing, experience and opportunities" (HelpAge International 2015, p. 7). The latest available data is from 2015, and it covers 96 countries, which includes 60% of the world's population at age 60 and over.

Table 2 shows domains and indicators of the Global AgeWatch Index. All indicators are outcome indicators, and most of them are absolute-level indicators (Mihnovits and Zaidi 2015).

The income security dimension highlights the importance of adequate income for sustaining

Global AgeWatch Index and Insights, Table 2 Global AgeWatch Index domains, indicators, and indicator weights

Global AgeWatch Index			
Income security	Health status	Capability	Enabling environment
Pension income coverage (1)	Life expectancy at 60 (1)	Employment of older people (1)	Social connections (1)
The poverty rate in old age (0.5)	Healthy life expectancy at 60 (1)	Educational status of older people (1)	Physical safety (1)
The relative welfare of older people (0.5)	Psychological well-being (0.5)		Civic freedom (1)
Gross national income (GNI) per capita (0.5)			Access to public transport (1)

Note: The weight given to each indicator is in parenthesis. It refers to the unit weight used to construct the dimension index. In percentage terms, the HelpAge team assigned the following weights for the income security dimension: 40% for pension income coverage, 20% for poverty rate, 20% for the relative welfare, and 20% for GNI per capita. These are the weights of the three indicators making the health status dimension: 40% for life expectancy, 40% for healthy life expectancy, and 20% for psychological well-being. Each of the two indicators of the capability dimension contributes 50% to the dimension index. Finally, each of the four indicators of the enabling environment dimension contributes 25% to the dimension index
Source: HelpAge International (2015)

well-being in later life. According to the International Labour Organization, over 70% of the world's population is not adequately covered by social protection. In other words, "only 27% of the global population enjoys access to comprehensive social security" (ILO 2014b). Forty-eight percent of older adults over pensionable age do not receive a pension, and for many of the pensioners, the benefit levels are inadequate (ILO 2014a). Low participation rates in the formal economy and under-institutionalized regressive taxation systems are used to justify the implementation of targeted social pensions throughout the globe (Bastagli 2013; Cruz-Martínez 2019). However, recent research demonstrates that universal social policy is a politically and economically viable policy output because there are multiple options to increase fiscal space and implement a basic universal cash transfer in the Global South (Cruz-Martinez 2018).

The health status domain measures the physical and psychological conditions of older people and highlights the importance of accessing quality health and care services. The capability domain measures the employment levels and education status of older adults. HelpAge International considers these two indicators are good proxies for engagement in the labor market and in the society, as well as for human capital in older people.

Finally, the enabling environment domain includes four indicators chosen by older adults themselves. This domain captures some of the social and physical factors of the environment enabling older adults to be and do what they desire and value.

The Index draws on a mix of data sources, most of which are openly available (i.e., the World Bank, World Health Organization, International Labour Organization, Eurostat, Organisation for Economic Co-operation and Development), and one, Gallup WorldView, is proprietary. Data covers latest available observations for population aged 50 or 60 and over.

The methodology used to construct the Global AgeWatch Index is based on the Human Development Index and is inspired by the Active Ageing Index. First, all 13 indicators are expressed as positive values (e.g., a higher value represents a better outcome). Second, all indicators are normalized (e.g., maximum equals 100 and minimum equals 0). Third, the four individual domains' indices are constructed using the geometric mean of the individual indicators. Indicators' weights (see Table 2) are assigned based on the judgment of researchers. Fourth, the domains are aggregated into the composite index using the geometric mean. All domains have equal weight (25%) in the composite index.

Results and Impact: The Global AgeWatch Index

The Index was published annually from 2013 to 2015. During this time the number of countries increased from 91 to 96. In the last edition of the Index, Switzerland ranked first while Afghanistan placed last. Mauritius (42) leads the African region, Japan (8) the Asia-Pacific region, Panama (20) the Latin America and the Caribbean region, and the Czech Republic (22) the Eastern Europe region. The top 7 countries and 18 of the top 20 are located in Western Europe, North America, and Australasia. In contrast, Africa is home to 10 of the bottom 20 countries with the Global AgeWatch Index scores. The ranking is generally in line with the income level of countries.

Global AgeWatch index became a useful advocacy tool as interest among international newspapers as well as national and local outlets helped to amplify key messages, engage new audiences, and create opportunities for more in-depth conversations on aging and development (Gladstone 2015; Anderson 2015). A number of researchers have used the Global AgeWatch Index in their investigations: from determinants of healthy aging (Sadana et al. 2016) to exploring the Index results within the national context (Vidyasova and Grigoryeva 2016), to proposals to strengthen the composite index with population dynamics (da Silva Francisco 2017), and as a complementary measure of well-being and quality of life to existing indices (e.g., Active Ageing Index) (Zaidi 2015). Additionally, the process of developing the Index contributed to building greater understanding about the extent of data gaps which in turn helped to strengthen the call for the establishment of the Titchfield City Group on Ageing and Age-disaggregated data.

Reflecting on lessons from publishing the Index, HelpAge acknowledged that general awareness of data gaps on aging and older people has increased and there is a need for a more in-depth national and regional analysis – more nuanced understanding of diversity and inequality in aging experiences, assessment of what data is available and what is missing at the country level, and more specific policy recommendations on aging and data (Conboy 2017).

Global AgeWatch Insights 2018

In 2017 HelpAge International and AARP partnered to redesign the index to become the Global AgeWatch Insights. The first edition of Global AgeWatch Insights was launched in December 2018. Rather than presenting an updated and strengthened version of the composite Global AgeWatch Index, the Global AgeWatch Insights 2018 focuses on health, examining the extent to which older people's right to health is realized, and the barriers that limit individuals' access to health systems and services. The Insights consist of a global report profiling 12 low- and middle-income countries (LMICs) – Argentina, Colombia, El Salvador, Kenya, Lebanon, Myanmar, Pakistan, Moldova, Serbia, Tanzania, Vietnam, and Zimbabwe – and 2 in-depth country studies (Tanzania and Vietnam).

The report evidences the inequities of the health systems worldwide and the lack of quality data about aging and health in LMICs. Among the barriers to older-age population's right to health are age discrimination, monetary poverty, costs of health services, low health literacy, and lack of institutional outreach to the older-age population in isolated communities (HelpAge International 2018).

The report also considers that health systems around the globe have not been able to keep pace with, what the HelpAge International considers, two major and interconnected global transitions: an epidemiological and a demographic transition. The epidemiological transition refers to the worldwide pattern of disease that has been shifting from communicable toward noncommunicable diseases (NCDs), and the older-age population is disproportionally more affected than younger cohorts. Three-quarter of deaths among people aged 60 and over in LMICs were from NCDs.

According to the World Health Organization, noncommunicable diseases are more commonly known as chronic diseases. Meanwhile, communicable diseases are spread through physical contact, bites from insects, or through the air.

The demographic transition refers to rapid global aging, which is, in part, the result of a decline in the mortality and fertility rates. HelpAge International argues that global health systems need to adapt to these two transitions and develop a more integrated healthcare system with coordinated services responding holistically instead of targeting specific diseases. Moreover, they argue that the Sustainable Development Goals and global commitment to achieve universal health coverage provide an opportunity to advance the realization of human rights for people of all ages, and to provide a pathway for addressing the demographic and epidemiological transitions.

The Global AgeWatch Insights highlights the imperative need for good quality, timely, and individual data that is disaggregated by gender, age, disability, and other socioeconomic characters. Without it governments, civil society, and researchers are not able to accurately measure, understand, and respond to the diverse and particular needs of older people and to monitor realization of the right to health for all. The authors of the report cite a number of challenges when they scoped for data on older people across the 12 profile countries. For example, (a) data is difficult to access, not available or inexistent, (b) metadata is missing for several datasets, (c) the wording of the survey questions varied considerably affecting the comparability of data, (d) age caps in international surveys, and (e) paucity of surveys on aging.

The report identifies a number of actions that should be taken by governments and national statistical agencies to make data, and health and care systems inclusive of aging, and to realize older people's right to health.

In May 2019, two companion country studies were released: *Vietnam and Tanzania Insights: The Right to Health* and *Access to Universal Health Coverage for Older People* (HelpAge International 2019a, b). Despite differences in how health services and financial protection are provided in each country, both countries face similar issues: (i) many older people remain excluded from accessing health services and financial protection; (ii) older men and women struggle to access affordable, appropriate, and good quality care due to low level of awareness of their right to health and lack of recognition of those rights by health and social protection systems; and (iii) data systems designed to inform the planning and delivery of health services exclude older people (HelpAge International 2019c).

Future Directions of Research

Future editions of the Global AgeWatch Insights are expected to examine aging and situations of older men and women in LMICs from global and national perspectives. Unfortunately, the lack of gender-disaggregated data prevented HelpAge International to break down the three Global AgeWatch Index editions (e.g., 2013, 2014, and 2015) nor the latest Global AgeWatch Insights by gender. According to Albone et al. (2014, p. 1), gender disaggregated data for older adults is often not collected, and when it does exist, it is not fully utilized or analyzed. The four domains could be strengthened by including additional indicators *operationalizing* additional facets of older adults' well-being. The unavailability of comparable cross-country disaggregated data is still a significant limitation.

The Global AgeWatch Index and the Global AgeWatch Insights are useful tools to advance awareness of gaps in the post-2015 development agenda regarding aging and older people and to highlight the need for more comprehensive data disaggregated by geographical region, age, and gender (Beales et al. 2014).

Cross-References

▶ Active Aging and Active Aging Index
▶ John A. Hartford Index of Societal Aging
▶ Sustainable Development Goals and Population Aging

Acknowledgments Aleksandr Mihnovits from HelpAge International provided comments for this article.

References

Albone R, Beales S, Mihnovits A (2014) Older people count: making data fit for purpose. In: Global AgeWatch Policy Brief 4. HelpAge International, London, pp 1–6

Anderson M (2015) Poor monitoring renders millions of elderly people worldwide "invisible." The Guardian, September 9. Retrieved from: https://www.theguardian.com/global-development/datablog/2015/sep/09/global-age-watch-index-2015-elderly-people-invisible-helpage-international

Bastagli F (2013) Feasibility of social protection schemes in developing countries. European Union, Brussels

Beales S, Russell-Moyle L, Mikkonen-Jeanneret E (2014) A post-2015 framework for all ages: transforming the future for youth and older people. In: Global AgeWatch Policy Brief 5. HelpAge International, London

Caspari R, Lee SH (2004) Older age becomes common late in human evolution. Proc Natl Acad Sci U S A 101 (30):10895–10900. https://doi.org/10.1073/pnas.0402857101

Conboy P (2017) From Global AgeWatch Index to Global AgeWatch Insights. HelpAge, August 15, 2017. Retrieved from: http://www.helpage.org/global-agewatch/blogs/patricia-conboy-29352/from-global-agewatch-index-to-global-agewatch-insights-1047/

Cruz-Martínez G (2018) Revenue-generating potential of taxation for older-age social pensions. Ageing Int 43 (4):415–437. https://doi.org/10.1007/s12126-017-9298-2

Cruz-Martínez G (2019) Older-age social pensions and poverty: revisiting assumptions on targeting and universalism. Poverty Public Policy 11(1–2):31–56. https://doi.org/10.1002/pop4.243

da Silva Francisco AA (2017) "Gerontogrowth" and population ageing in Africa and the Global AgeWatch Index. J Econ Ageing 9:78–89. https://doi.org/10.1016/j.jeoa.2016.08.003

Gladstone R (2015) Older people are invisible in key data, study warns. The New York Times, September 9, 2015. Retrieved from: https://www.nytimes.com/2015/09/09/world/older-people-are-invisible-in-key-data-study-warns.html

HelpAge International (2015) Global AgeWatch Index 2015: insight report. HelpAge International, London

HelpAge International (2018) Global AgeWatch Insights: the right to health for older people, the right to be counted. HelpAge International & AARP, London. Retrieved from: http://www.globalagewatch.org/

HelpAge International (2019a) Vietnam Insights: the right to health and access to universal health coverage for older people. HelpAge International & AARP, London. Retrieved from: http://www.globalagewatch.org/

HelpAge International (2019b) Tanzania Inisghts: the right to health and access to universal health coverage for older people. HelpAge International & AARP, London. Retrieved from: http://www.globalagewatch.org/

HelpAge International (2019c) Universal health coverage is fundamental for addressing health inequalities in Tanzania and Vietnam, new Global AgeWatch reports reveal. Retrieved from: https://www.helpage.org

ILO (International Labour Organization) (2014a) Social protection for older persons: key policy trends and statistics. International Labour Office, Geneva

ILO (International Labour Organization) (2014b) World Social Protection Report 2014–15: building economic recovery, inclusive development and social justice. International Labour Office, Geneva

Mihnovits A, Zaidi A (2015) Global AgeWatch Index 2015: methodology update. HelpAge International, London

Sadana R, Blas E, Budhwani S, Koller T, Paraje G (2016) Healthy Ageing: Raising Awareness of Inequalities, Determinants, and What Could Be Done to Improve Health Equity. The Gerontologist 56(Suppl 2):S178–S193

Stallard E (2006) Demographic issues in longevity risk analysis. J Risk Insur 73(4):575–609

UNPD (United Nations Development Programme) (2017) World population prospects: the 2017 revision. Percentage of total population by broad age groups. https://esa.un.org/unpd/wpp/Download/Standard/Population/

Vidyasova L, Grigoryeva I (2016) Russia in the international quality of ageing indices. J Sociol Soc Anthropol XIX(1):181–193

Zaidi A (2015) Creating and using the evidence base: the case of the Active Ageing Index. Contemporary Social Science 10(2):148–159

Global AgeWatch Insights

▶ Global AgeWatch Index and Insights

Global Aging Data

▶ Global AgeWatch Index and Insights

Global Aging Index

▶ Global AgeWatch Index and Insights

Global Coalition on Aging

Melissa Mitchell, Susan Wile Schwarz and Charles Pennell
Global Coalition on Aging, New York, NY, USA

Mission

The Global Coalition on Aging (GCOA) (www.globalcoalitiononaging.com) is one of the world's leading business voices on aging-related policy and strategy. GCOA uniquely brings together global corporations across industry sectors with common strategic interests in aging populations, a comprehensive and systemic understanding of aging, and an optimistic view of its impact. Through research, public policy analysis, advocacy, and strategic communications, GCOA is advancing innovative solutions and working to ensure global aging is a path to health, productivity, and economic growth.

History, Goals, and Membership

History and Goals
GCOA was founded in 2011 with the aim to reshape how global leaders approach and prepare for the twenty-first century's profound shift in population aging. At the time of its founding, GCOA's goal was to create an organization that would respond to several needs in the global aging conversation:

- **Recognize the opportunities of aging**. Aging and longevity open a number of unprecedented opportunities. However, too often, discussions are focused narrowly on the challenges, risks, and negative impacts of aging. GCOA was founded to reframe aging as an opportunity to improve the health, productivity, and social engagement of the aging population around the world and thereby enhance individuals' path to healthy aging and health systems' and economies' fiscal sustainability.
- **Catalyze private-sector leadership**. The private sector can and must play a leadership role in how our world understands, discusses, and responds to population aging and longevity. GCOA was founded to address this need – helping businesses to engage in and contribute to the global dialogues on aging, as well as evolve workforce and market strategies for a world defined by the demographic transformation.
- **Collaborate across boundaries**. Effective aging solutions require engagement across sectors, industries, disciplines, and national boundaries. GCOA was founded to facilitate this collaboration, providing platforms for dialogue between and among businesses, governments, global institutions, academia, research groups, NGOs, and advocacy groups. The result is a richer conversation that leads to nuanced perspectives, unique partnerships, and powerful responses.

Membership
GCOA's membership includes global business leaders across all industries and geographies. From health and pharmaceutical companies to technology and financial services and from consumer products to new industries such as home care, all GCOA members share one common vision: they are inspired by the opportunities longevity presents and are committed to leading the global conversation about aging. A full roster of our members can be found at https://globalcoalitiononaging.com/about/#brands.

Approach and Activities

GCOA advances policy and market solutions that ensure aging can be a path for economic growth, winning business strategies, and social well-being. To achieve this goal, GCOA leads the global conversation on aging through:

- **Leadership and insights**: GCOA informs and drives change among policymakers, thought leaders, and the general public. Through its white papers, roundtables, webinars, presentations to third parties, and other communications and engagements, GCOA is leading the global aging dialogue and providing public education designed to enable healthier and more active aging.
- **Partnerships**: GCOA partners with top organizations from around the world and across every sector to connect innovators, researchers, and policymakers to develop lasting solutions for business, government, and society as the global population ages. This diverse network enables GCOA to continuously identify and reach new stakeholders that can play a pivotal role in initiatives driving changes related to aging and longevity.
- **Business strategies**: GCOA works with some of the world's leading companies to align business strategies and workforce policies with aging market and talent opportunities. GCOA combines a unique market understanding with tailored engagement strategies that drive corporate leadership and lead to innovative solutions.
- **Powerful voice**: GCOA creates platforms to demonstrate thought leadership on a global scale. Through its powerful network of influencers, which spans the business and policy elite across all sectors, GCOA works to advance a positive and action-oriented longevity dialogue.

Focus Areas and Key Initiatives

GCOA's initiatives focus on three core areas: the silver economy, healthy and active aging, and elder care.

Silver Economy

Aging populations are too often assumed to be drains on society, when in fact older citizens can be drivers of productivity and wealth creation by remaining active, engaged, and working (Norman 2013; Hodin 2019). GCOA believes that capturing this workforce engine and spending power will necessitate workplace adaptations, updated definitions of retirement and savings, new market strategies, and investments in lifelong training and education (Deloitte Insights 2018).

Healthy and Active Aging

The gift of longevity must be accompanied by policies and private-market solutions that promote a life course of healthy aging. GCOA works with its members and partners to strategize about new products, services, medicines, and technologies – combined with instilling a new mindset about the positive opportunities that come with longer lives – to keep all members of society physically active, socially engaged, and mentally fit as they age.

Elder Care

Rapid aging creates the need for new care models to address the rising demand for quality caregiving as populations age. Declining birthrates, changing family structures, and growing strains on healthcare systems bring about the need for new caregiving solutions that keep older people as independent and engaged as possible for as long as possible, combining new technologies and with a focus on human relationships (Global Coalition on Aging 2018a).

Initiatives

GCOA's key initiatives include:

- **Silver Economy Forum**: The inaugural *High-Level Forum on the Silver Economy* was held in 2019 in Helsinki in partnership with the Finnish Presidency of the European Commission (The Global Coalition on Aging 2019). It convened hundreds of global, national, and local business and government leaders to drive action toward co-created innovation in products, services, and public policies that help enable and meet the needs of a global aging society. The Forum focuses on how worldwide aging and longevity are

transforming the landscape for business and government alike and brings forth new approaches to health, caregiving, financial planning, work, retirement, and technology in light of these shifts.
- **Guiding Principles for the Multi-Generational Workplace**: GCOA's *Guiding Principles for the Multi-Generational Workplace* (Global Coalition on Aging 2018b) help companies create workplaces that embrace and support workers of all ages – key to success in a rapidly aging world. By embracing these principles, companies can better position themselves to realize tangible gains in productivity, competitiveness, and worker satisfaction while also helping to combat ageism in the workplace and across society.
- **Global Institution and National Government Engagement**: GCOA advocates for key priorities related to aging, including combatting ageism, person-centered, integrated care, a focus on the conditions that increase with aging, use of remote care technology and artificial intelligence as twenty-first-century standards of care, and changes in the workforce and workplace. GCOA leads discussions with global and national leaders at forums informing global actions on aging, including the WHO's Decade of Healthy Ageing, the G20 Summit, the Asia-Pacific Economic Cooperation (APEC) High-Level Meeting on Health and the Economy, and the OECD Forum.
- **Healthier Aging Through Adult Vaccines**: GCOA brings together cross-sector experts to prioritize adult vaccines on the global health agenda, engaging with high-level forums like the G20 Finance and Health Ministers' Meetings. As the foundation for this work, GCOA partnered with the Tokyo-based Health and Global Policy Institute to publish a report, *Measures to Ensure Healthy Ageing* (Global Coalition on Aging and the Health and Global Policy Institute 2019), which connects economic growth and fiscal sustainability to healthy aging and preventive healthcare measures – calling for greater focus on adult vaccine policy and uptake in the use of vaccines among older adults.
- **Global Alliance on Heart Failure and Healthy Aging**: Heart failure dramatically increases in prevalence with age, and too often the symptoms of heart failure (e.g., tiredness, weakness, shortness of breath) are conflated with normal aging or other comorbidities associated with aging. As a result, heart failure often goes undiagnosed and under-addressed, with negative impacts on patients' quality of life and health system costs. In 2018, GCOA held a series of roundtables on heart failure and aging that culminated in the creation of the Global Alliance on Heart Failure and Healthy Aging, a platform for cross-sector and cross-disciplinary action to reframe heart failure through the aging lens. The Alliance promotes better monitoring, early detection and diagnosis, and treatment optimization and integrated care for more urgent and effective responses to the global challenge of heart failure (Global Coalition on Aging 2018c).
- **Life Course of Healthy Vision**: GCOA calls for efforts to address widespread vision loss and decline and barriers to eye health in the aging population, enabling older adults to remain productive, active, and healthy. GCOA published a research report in 2019, *A Life Course of Healthy Vision: A Critical Priority for the 21st Century* (Global Coalition on Aging 2019), as a foundation and roadmap to build awareness and galvanize action toward this goal.
- **Remote Care and AI Initiative**: Launched in 2017, GCOA's Remote Care and AI Initiative aims to leverage and mobilize diverse expertise to advance remote care and artificial intelligence as twenty-first-century standards of care. This cross-sector initiative calls for wider adoption of remote care monitoring and AI to improve detection, diagnosis, and monitoring of diseases and health conditions as we age, thereby improving individual health outcomes while reducing costs across healthcare systems.
- **Financial Wellness**: In a 2018 report titled *Financial Wellness for Longer Lives: New Approaches to Working and Saving* (Global Coalition on Aging, the New York Academy

of Medicine, AARP New York 2018), GCOA, the New York Academy of Medicine, and AARP New York defined the dynamic between demographic and financial trends associated with a growing older population, examined strategies for age-friendly business practices to better promote retirement savings and financial management, and explored the importance of financial institutions in helping to identify and manage the risks associated with cognitive decline. GCOA believes that physical wellness and financial wellness must be examined and pursued as interconnected goals on the healthy aging agenda.

- **Dementia Innovation Readiness Index**: GCOA promotes awareness and action to bring innovation to dementia treatment, prevention, and care around the globe through its groundbreaking *Dementia Innovation Readiness Index*. The first Index was co-published with Alzheimer's Disease International (ADI) in 2017, analyzing the readiness of G7 countries to integrate innovative dementia solutions into their healthcare systems and policy frameworks (Global Coalition on Aging, Alzheimer's Disease International 2017). In 2018, GCOA and ADI expanded the Index to cover additional G20 countries (Global Coalition on Aging, Alzheimer's Disease International 2018). In 2020, GCOA plans to publish an Index that will analyze the dementia readiness of 30 cities around the globe.
- **Fragility Fractures**: Urgent action is needed to address the increasing risk, burden, and costs of fragility fractures. Action on fragility fractures and osteoporosis can ensure a more comprehensive and effective approach to frailty and falls. Through greater awareness, new monitoring and care delivery models, and public policy changes, health systems can better improve outcomes and costs for older adults, their caregivers, and their own budgets.
- **Conditions of Aging**: GCOA leads a number of activities and convenes global dialogues to increase awareness of the conditions that increase in prevalence with age and the associated impact on quality of life, social engagement, workplace productivity, and health system costs. From vision and hearing loss to declining bone and muscle mass and from overactive bladder to malnutrition and undernutrition, these conditions do increase in prevalence with age, but are not inevitable results of the aging process. With awareness and continued innovation, policy and individual actions can ensure that longer health spans accompany twenty-first-century longer lifespans.

Please visit www.globalcoalitiononaging.com for more information and details about GCOA's ongoing initiatives.

Cross-References

▶ AARP
▶ Abuse and Caregiving
▶ Age Discrimination in the Workplace
▶ Ageism in Healthcare
▶ Aging Analytics Agency
▶ Aging in Place
▶ Aging Policy Ideas
▶ Alliance for Aging Research
▶ American Federation for Aging Research
▶ Benefits of Caregiving
▶ Cardiovascular System
▶ Corporate Social Responsibility and Creating Shared Value
▶ Delaying Retirement
▶ Dementia
▶ Economics of Aging
▶ Employment and Caregiving
▶ Financial Literacy
▶ Formal and Informal Care
▶ HelpAge International
▶ HelpAge USA
▶ Improving Nutrition in Older Adults
▶ International Federation on Ageing
▶ Malnutrition
▶ Multigenerational Workforce
▶ Telehealth
▶ Telemedicine
▶ The International Association of Gerontology and Geriatrics

References

Deloitte Insights (2018) The rise of the social enterprise. 2018 Deloitte Global Human Capital Trends. https://globalcoalitiononaging.com/wp-content/uploads/2018/06/2018-HCtrends_Rise-of-the-social-enterprise-2.pdf. Accessed 19 Dec 2019

Global Coalition on Aging (2018a) Relationship-based home care: a sustainable solution for Europe's elder care crisis. https://globalcoalitiononaging.com/wp-content/uploads/2018/06/RHBC_Report_DIGITAL.pdf. Accessed 19 Dec 2019

Global Coalition on Aging (2018b) Guiding principles for the multi-generational workplace: harnessing the power of five generations. https://globalcoalitiononaging.com/wp-content/uploads/2019/07/HLG_GCOA_GP_MultigenerationalWorkspace_FINAL.pdf. Accessed 19 Dec 2019

Global Coalition on Aging (2018c) Reimagining and reframing heart failure as we age. https://globalcoalitiononaging.com/wp-content/uploads/2019/08/Reimagining-and-Reframing-Heart-Failure-2018-GCOA-Think-Tank-Series-Report.pdf. Accessed 19 Dec 2019

Global Coalition on Aging (2019) A life course of healthy vision: a critical priority for the 21st century. https://globalcoalitiononaging.com/wp-content/uploads/2019/05/GCOA_Healthy-Vision_FINAL.pdf. Accessed 19 Dec 2019

Global Coalition on Aging, Alzheimer's Disease International (2017) Dementia innovation readiness index. https://globalcoalitiononaging.com/initiatives/page/3/#initiative-152. Accessed 19 Dec 2019

Global Coalition on Aging, Alzheimer's Disease International (2018) Dementia innovation readiness index. https://globalcoalitiononaging.com/initiatives/page/2/#initiative-560. Accessed 19 Dec 2019

Global Coalition on Aging and the Health and Global Policy Institute (2019) Measures to ensure healthy ageing: recommendations for the G20 leaders and health agenda. https://globalcoalitiononaging.com/wp-content/uploads/2019/08/HGPI-GCOA-Roundtable-on-Vaccine-Policy_ENG_20190729.pdf. Accessed 19 Dec 2019

Global Coalition on Aging, the New York Academy of Medicine, AARP New York (2018) Financial wellness for longer lives: new approaches to working and saving. https://globalcoalitiononaging.com/wp-content/uploads/2018/06/financial_report_19.pdf. Accessed 19 Dec 2019

Hodin M (2019) Working longer and differently: the 21st century innovation model. OECD Forum. https://www.oecd-forum.org/users/257659-michael-hodin/posts/48655-working-longer-and-differently-the-21st-century-innovation-model. Accessed 9 Dec 2019

Norman J (2013) The aging population: a crisis in plain sight. CQ Magazine. https://www.commonwealthfund.org/publications/newsletter-article/aging-population-crisis-plain-sight. Accessed 27 Sept 2019

The Global Coalition on Aging (2019) World leaders gather in Helsinki to lead the Silver Economy transformation. https://globalcoalitiononaging.com/2019/07/16/world-leaders-gather-in-helsinki-to-lead-the-silver-economy-transformation/. Accessed 19 Dec 2019

Global Deterioration Scale

Cheshire Hardcastle[1,2], Brad Taylor[1,2] and Catherine Price[3]
[1]Department of Clinical and Health Psychology, University of Florida, Gainesville, FL, USA
[2]Department of Public Health and Health Professions, University of Florida, Gainesville, FL, USA
[3]Department of Clinical and Health Psychology, College of Public Health and Health Professions, University of Florida, Gainesville, FL, USA

Synonyms

Reisberg scale

Definition

The Global Deterioration Scale (GDS) is a clinical rating instrument created to assess stages of primary degenerative dementia, namely, Alzheimer's disease (AD). Medical and scientific colleagues can use the GDS interchangeably for collaboration and clinical research purposes.

Overview

Developed in 1982 by Dr. Barry Reisberg and colleagues, the GDS was among the first of its kind to stage the course of dementia (See ▶ "Alzheimer's Disease" and ▶ "Dementia"). The authors built off of a previously identified model of three progressive disease stages (forgetfulness, confusional, and late dementia). Symptoms of decline are within areas of activities of daily living, instrumental activities of daily living (e.g., managing money, meal preparation), cognitive loss, and psychiatric morbidity (See ▶ "Behavioral and Psychological Symptoms of Dementia"). Psychometric assessments used to define the stages were the Guild Memory subtests from the Wechsler Adult Intelligence Scale (Wechsler 1939), the Guild Memory Test (Crook et al. 1980), as well as the Mental Status Questionnaire (Kahn et al. 1960). This scale can be helpful to appreciate the progression of a disease and rule out dementia; i.e., an individual could meet stage 2, not progress to more advanced stages.

The GDS classified seven stages:

Stage 1 (No Cognitive Decline) – No subjective or objective memory deficits.
Stage 2 (Very Mild Cognitive Decline) – Subjective complaints of memory deficit, but no objective measurements of memory deficit.
Stage 3 (Mild Cognitive Decline) – The individual now meets criteria for mild cognitive impairment (See ▶ "Mild Cognitive Impairment").
Stage 4 (Moderate Cognitive Decline) – The individual is now classified as being mildly demented. This could manifest as a clear deficit on concentration, handling finances, orientation, and recognition of time and place. Symptoms such as flattening of affect and anxiety start to occur.
Stage 5 (Moderately Severe Cognitive Decline) – The individual now meets criteria for moderate dementia and can no longer function without some assistance but can toilet and eat on their own.
Stage 6 (Severe Cognitive Decline) – The individual meets criteria for moderately severe dementia. The individual is entirely dependent on someone else for survival (See ▶ "Caregiver Stress") and are generally unaware of their surroundings, year, season, etc. Personality and emotional changes occur.
Stage 7 (Very Severe Cognitive Decline) – The individual is now severely demented. The individual has lost all verbal abilities and is incontinent, as well as basic psychomotor skills.

The test has been validated with cognitive profiles and radiological approaches. Initial validation was published in 1982 via a retrospective analysis of GDS scores and psychometric assessments of individuals with very mild to moderately severe dementia (Reisberg et al. 1982). GDS scores significantly correlated with 13 of the 19 cognitive items in the Inventory of Psychic and Somatic Complaints in the Elderly and with 25 of 26 other psychometric measures. In 1980, the GDS was validated against computerized tomography (CT) brain metrics (ventricular dilation, sulci enlargement) (de Leon et al. 1980) and positron emission tomography (PET) via the caudate nucleus, thalamus, and temporal regions of basal ganglia, as measured by an ^{18}F-2-deoxy-2-fluoro-D-glucose tracer (Ferris et al. 1980). The GDS has since been validated in test-retest reliability in many other studies (e.g., Gottlieb et al. 1988; Foster et al. 1988).

Key Research Findings

The GDS stages may not be sequential. Eisdorfer et al. (1992) present and demonstrate that individuals with AD can show psychiatric symptoms between stages 3 and 4 instead of 5 and 6. Additionally, the authors found functional impairment across all stages with an increase between stages 3 and 4. While these findings do support a progression in severity of AD, as identified by the GDS, it is possible that deficits of AD can be seen as early as stage 2 and that the progression of stages is homogeneous.

The GDS can be used in related dementias, not just AD. For example, it has been shown that in vascular dementia (VaD) (See ▶ "Vascular Dementia"), GDS stages are associated with cognitive function but also activities of daily living (ADL; e.g., dressing, eating, bathing) and instrumental activities of daily living (IADL; e.g., handling finances, medication management; Paul et al. 2002). In Parkinson's disease (PD), researchers have used the GDS to demonstrate that individuals with subjective cognitive impairment have decreased perfusion in the left angular gyrus compared to individuals with PD without subjective cognitive impairment (Joeng et al. 2016). Another study describing their PD-MCI with GDS scale range between 2 and 3 showed significantly less gray matter volume in the right temporal pole, left precuneus, medial frontal, and posterior cingulate gyrus relative to cognitively well PD (Noh et al. 2014). Also, the GDS has been applied within studies comparing severity of dementia between frontotemporal dementia (FTD), AD, and Diffuse Lewy body disease (Aries et al. 2010).

Examples of Application

The GDS has been used to study brain markers of cognitive abilities as well. In 1994, Prichep and colleagues aimed to determine electroencephalogram (EEG) differences in the spectrum of cognitive impairment, as determined by the GDS. The authors found that even at GDS score of 2, there are significant widespread changes in theta, and these changes increase as severity of scale increases. Only until later stages of the GDS (4–6) are changes in delta observed. These findings indicate that staging of the GDS can differentiate brain changes, suggesting biological brain differences between GDS stages.

While Prichep and colleagues (1994) sought to determine brain differences in the early stages of AD, clear criteria for a prodromal stage of dementia had not yet been determined. Petersen et al. (1999) published a seminal paper identifying cognitive boundaries of mild cognitive impairment (MCI). MCI is currently understood as a stage of cognitive decline that is more severe than cognitive decline seen with typical aging and prodromal to dementia. However, we know that the prevalence of MCI individuals that transition to dementia can range from 3 to 13%, depending on the severity of deficits (Farias et al. 2009). Petersen and colleagues (1999) tracked cognitive functioning (See ▶ "Mini-Mental State Examination (MMSE)" and ▶ "Montreal Cognitive Assessment") at baseline and 12–18 months later in 76 individuals with MCI, compared to to 234 healthy participants and 106 individuals with mild AD. Among the measures used, they compared groups on the GDS. They found that AD and MCI individuals who have the same Clinical Dementia Rating score are distinguished using the GDS because of functional impairment seen in AD. Additionally, MCI individuals changed more rapidly on this more global assessment than healthy participants, but not as rapidly as AD individuals. This study demonstrated that the GDS has value in distinguishing individuals with MCI from normal controls and AD.

Future Directions of Research

Researchers are now using the GDS to explore progression from subjective cognitive complaints or impairment (SCI) to MCI and dementia. Prichep and colleagues (2006) followed community-dwelling older adults who met criteria for GDS stage 2 (SCI) from 1980 to 1997 to assess the timeline of MCI conversion from SCI. The authors were able to show that after 8.9 years, 59.33% of participants had significant cognitive decline. GDS stage 2 appears to be an important indicator of cognitive vulnerability. In a more recent longitudinal study, Reisberg et al. (2019) reported individuals who meet GDS stage 2 significantly declined over a period of 1 to 3 years on measures of remote memory. Future researchers now need to assess how GDS stage 2 may differ by race. It is also unclear if dementia prevalence differences are due to educational differences and geographical region (Sharp and Gatz 2011).

Summary

Since its creation in 1986, the Global Deterioration Scale has been widely used and can now be found on many organizational websites for caregivers to track progression of dementia. The GDS continues to be used in the scientific community to stage dementia progression; however, it should be noted that the GDS is not a diagnostic scale and was developed purely for the use of qualitative severity rating (Hartmaier et al. 1994; Brooke and Bullock 1999; Petersen et al. 1999).

Cross-References

▶ Alzheimer's Disease
▶ Behavioral and Psychological Symptoms of Dementia
▶ Caregiver Stress
▶ Dementia
▶ Mild Cognitive Impairment
▶ Mini-Mental State Examination (MMSE)
▶ Montreal Cognitive Assessment
▶ Vascular Dementia

References

Aries MJ, Le Bastard N, Debruyne H, Van Buggenhout M, Nagels G, De Deyn PP, Engelborghs S (2010) Relation between frontal lobe symptoms and dementia severity within and across diagnostic dementia categories. Int J Geriatr Psychiatr 25(11):1186–1195

Brooke P, Bullock R (1999) Validation of a 6 item cognitive impairment test with a view to primary care usage. Int J Geriatr Psychiatr 14(11):936–940

Crook T, Gilbert JG, Ferris S (1980) Operationalizing memory impairment for elderly persons: the guild memory test. Psychol Rep 47(3_suppl):1315–1318

de Leon MJ, Ferris SH, George AE, Reisberg B, Kricheff II, Gershon S (1980) Computed tomography evaluations of brain-behavior relationships in senile d dementia of the Alzheimer's type. Neurobiol Aging 1:69–79

Eisdorfer C, Cohen D, Paveza GJ, Ashford JW, Luchins DJ, Gorelick PB, . . ., Shaw HA (1992) An empirical evaluation of the global deterioration scale for staging Alzheimer's disease. Am J Psychiatr 149:190–194

Farias ST, Mungas D, Reed BR, Harvey D, DeCarli C (2009) Progression of mild cognitive impairment to dementia in clinic- vs community-based cohorts. Arch Neurol 66:1151–1157

Ferris SH, de Leon MJ, Wolf AP, Farkas T, Christman DR, Reisberg B, . . ., Rampal S (1980) Positron emission tomography in the study of aging and senile dementia. Neurobiol Aging 1:127–131

Foster JR, Sclan S, Welkowitz J, Boksay I, Seeland I (1988) Psychiatric assessment in medical long-term care facilities: reliability of commonly used rating scales. Int J Geriatr Psychiatr 3:229–233

Gottlieb GL, Gur RE, Gur RC (1988) Reliability of psychiatric scales in individuals with dementia of the Alzheimer type. Am J Psychiatr 145:857

Hartmaier SL, Sloane PD, Guess HA, Koch GG (1994) The MDS cognition scale: a valid instrument for identifying and staging nursing home residents with dementia using the minimum data set. J Am Geriatr Soc 42(11):1173–1179

Joeng HS, Oh E, Park J-S, Chung Y-A, Park S, Yang Y, Song I-U (2016) Longitudinal cerebral perfusion changes in Parkinson's disease with subjective cognitive impairment. Dementia Neurocogn Disord 15(4):147–152

Kahn RL, Goldfarb AI, Pollack M, Peck A (1960) Brief objective measures for the determination of mental status in the aged. Am J Psychiatr 117(4):326–328

Noh W, Han H, Mun C, Kim HJ, Seo H, Kim J (2014) Analysis of cognitive profiles and gray matter volume in newly diagnosed Parkinson's disease with mild cognitive impairment. J Neurol Sci 347:210–213

Paul RH, Cohen RA, Moser DJ, Zawacki T, Ott BR, Gordon N, Stone W (2002) The global deterioration scale: relationships to neuropsychological performance and activities of daily living in individuals with vascular dementia. J Geriatr Psychiatr Neurol 15:50–54

Petersen RC, Smith GE, Waring SC, Ivnik RJ, Tangalos EG, Kokmen E (1999) Mild cognitive impairment: clinical characterization and outcome. Arch Neurol 56:303–308

Prichep LS, John ER, Ferris SH, Reisberg B, Almas M, Alper K, Cancro R (1994) Quantitative EEG correlates of cognitive deterioration in the elderly. Neurobiol Aging 15(1):85–90

Prichep LS, John ER, Ferris SH, Rausch L, Fang Z, Cancro R et al (2006) Prediction of longitudinal cognitive decline in normal elderly with subjective complaints using electrophysiological imaging. Neurobiol Aging 27(3):471–481

Reisberg B, Ferris SH, de Leon MJ, Crook T (1982) The global deterioration scale for assessment of primary degenerative dementia. Am J Psychiatr 139:1136–1139

Reisberg B, Torossian C, Shulman MB, Monteiro I, Boksay I, Golomb J, . . ., Rao JA (2019) Two year outcomes, cognitive and behavioral markers of decline in healthy, cognitively normal older persons with global deterioration scale stage 2 (subjective cognitive decline with impairment). J Alzheimers Dis 67(2):685–705

Sharp ES, Gatz M (2011) Relationship between education and dementia: an updated systematic review. Alzheimer Dis Assoc Disord 25:289–304

Wechsler D (1939) The measurement of adult intelligence. Williams & Wilkins, Baltimore

Global End Diastolic Volume Index

▶ End-Diastolic Volume Index

Global Goals and Population Aging

▶ Sustainable Development Goals and Population Aging

Global Public Policy on Aging

▶ Madrid International Plan of Action on Ageing

Global Supercentenarian Database

▶ (World) Supercentenarian Database

Global Warming

▶ Climate Change, Vulnerability, and Older People

Globalized Production

▶ Religion and Spirituality: Older Inmates

Goal Adjustment

▶ Disengagement Theory
▶ Dual Process Theory of Assimilation and Accommodation

Goal Pursuit

▶ Dual Process Theory of Assimilation and Accommodation

Goal Setting

Marie Hennecke
Department of Psychology, University of Siegen, Siegen, Germany

Definition

Goal setting can be understood as the process by which individuals commit to pursuing personal goals. Goals can be defined as personally desirable states to be attained or undesirable states to be avoided through action (Emmons 1996). Such goals are cognitively represented in people's minds and they impact evaluations, emotions, and behaviors (Fishbach and Ferguson 2007). By comparing their current state with regard to the goal state, people are able to gauge their progress and, in turn, self-regulate behavioral adaptations, e.g., by selecting strategies and strategically allocating resources (Carver and Scheier 1982).

Overview

Human behavior is to a large degree goal-driven. Across the human lifespan, people set multiple goals in various domains and on various levels of abstraction (Carver and Scheier 1982; Emmons 1992). For example, "passing an exam" or "picking up one's grandchild from kindergarten" are relatively narrow and specific goals. In contrast, "establishing a meaningful career" or "being a good grandparent" are relatively broad and abstract goals. Whereas abstract goals provide people's lives with purpose and meaning, specific goals provide clear guidelines for action (Emmons 1992; Little 1989). Moreover, whereas abstract goals usually require sustained perseverance over longer periods of time, specific goals may be attained with fewer action episodes.

From a developmental perspective, goals play a double role: On the one hand, they *drive* development through guiding where people invest their time and effort (e.g., Locke and Latham 2006).

Indeed, their consequences are well-documented in various domains, including health (e.g., Mann et al. 2013), work (e.g., Locke and Latham 2006), and relationships (e.g., Impett et al. 2010).

On the other hand, goals are also *subject to development*. As people age, they experience changing social and societal expectations, norms, and opportunities (Freund, 2003). On a more personal level, factors like people's own age stereotypes and ► Self-Perceptions of Aging (e.g., Westerhof et al. 2014; Wurm et al. 2017), changing subjective control beliefs (e.g., Brandtstädter and Rothermund 1994; Lachman 2006, see also ► Locus of Control), or age-related changes in the availability of cognitive, physical, and other resources (e.g., Goodpaster et al. 2006; Levy 1994) have developmental impact. Through the selection and adaptation of personal goals, people are able to respond to such factors and in turn, support their own successful development (e.g., Baltes and Baltes 1990; Freund 2008). In turn, not only the content but also the motivational orientation of personal goals change across adulthood.

Key Research Findings

Goal Content Changes Across Adulthood

Whereas younger and middle-aged adults' goals are typically concerned with their education and career, older adults' goals often refer to health maintenance, their children's lives, as well as politics and global issues (Nurmi et al. 1992; Saajanaho et al. 2016). In addition, healthier older adults are more likely to report goals related to leisure, social, and physical activities, but older adults with only poor social resources are at higher risk for not reporting any personal goals in their lives at all (Nurmi et al. 1992; Saajanaho et al. 2016). The goals of younger adults moreover often refer to extrinsic values like money, attractiveness, and popularity, whereas the goals of older adults more often refer to intrinsic values like self-acceptance, emotional intimacy, and community contribution, reflecting increased maturity and personality integration in old age (Sheldon and Kasser 2001).

A Limited Future Time Perspective Leads to More Present-Focused Goals

Research in the realm of socio-emotional selectivity theory has considered the role of adults' future ► Time Perspective on the content of their goals (e.g., Carstensen et al. 2003). The theory predicts that younger adults, with their extended future time perspective, tend to pursue so-called expansive goals, such as acquiring knowledge and making new social contacts. Older adults, in contrast, with their more restricted future time perspective, should tend to pursue goals with the more immediate payoff of regulating their feelings (see also Penningroth and Scott 2012). As the two types of goals can be attained through different social behavior, younger adults prefer meeting new people, whereas older adults prefer being surrounded by close others like family members and friends (Carstensen et al. 2003; Carstensen et al. 1999).

Goal Selection Is Associated with Successful Aging

Another, much broader, theory of developmental regulation, the model of selection, optimization and compensation (see ► The Model of Selection, Optimization, and Compensation), refers to how people deal with gains and losses across the lifespan (see also ► Life-Span Theory of Control and ► Dual Process Theory of Assimilation and Accommodation). In this model, processes of goal setting are subsumed under the term selection, which is further differentiated into elective and loss-based selection. The former entails the specification of and commitment to selected goals under the ubiquitous condition that resources are always limited (► Selective Engagement of Resources), whereas the latter includes focusing on and restriction to the most important goals, reconstruction of one's goal hierarchy, adaptation of standards, and search for new goals in the face of resource losses (Freund and Baltes 2002). Old-old adults use elective and loss-based selection less than young-old adults. The use of both strategies correlates with satisfaction with one's own aging, the use of elective selection furthermore with the absence of loneliness, and the use of loss-based selection with the experience of positive emotions (Freund and Baltes 1998). Focusing on central and similar goals predicts goal involvement and may help older adults to maintain their goal engagement even in the face of resource limitations (Riediger and Freund 2006). Indeed, older adults report more strongly than younger adults that their goals mutually facilitate each other (Riediger et al. 2005).

Older Adults Are More Prevention-Focused

In addition, older and younger adults' goals also differ with regard to their orientation. Whereas younger adults tend to set goals directed at attaining and improving gains in various domains, older adults tend to set goals directed at preventing gains and maintaining their level of functioning in old age (Ebner et al. 2006; Penningroth and Scott 2012). Interestingly, an orientation toward maintenance is positively related to well-being in older but not younger adults (Ebner et al. 2006). With becoming more prevention/maintenance-oriented, older adults probably also respond to increasing resource restrictions. Indeed, when growth goals are described as requiring the investment of more energy and resources, both younger and older adults become less growth-oriented (Ebner et al. 2006).

Older Adults Are More Process-Focused

Lastly, older adults also become more process-focused. Whereas for younger adults the outcomes of goal pursuit are in the foreground, older adults focus equally on the outcomes and on the processes of goal pursuit. This may also help them to deal effectively with resource losses through better means selection and to more strongly value the here and now rather than potentially distal outcomes (Freund et al. 2010).

Goal Setting Interventions

Goal setting interventions are common and tend to be effective in motivating people to engage in desirable behaviors (e.g., in the health domain, see for example, Evans and Hardy 2002; MacLeod et al. 2008; Pearson 2012) but only little research has applied them for older adults. A few interventions, mostly targeting physical activity, focused on considering older adults' goals and/or explicitly setting goals for and with them, have been shown to be effective. One intervention has combined goal setting, behavioral monitoring, comparative feedback, and action planning to reduce older adults' sedentary time. While it is not clear which of these aspects of the intervention drove the effect, their combination reduced sedentary time effectively from before to after the intervention (Gardiner et al. 2011). Similarly, another study found that, relative to a waiting-list control group, older adults with disabilities who participated in an intervention that included individually tailored goal setting were successful in improving their physical ability (mainly with regard to an objective measures of muscle strength and self-reports of basic functional activities like walking, climbing scares, house cleaning, Jette et al. 1999). In addition, a study with older adults on referral for homecare showed that a group for which individualized goal setting was implemented in the rehabilitation process attained greater increases in their physical ability than a group of older adults who received only standard needs assessment (Parsons et al. 2012).

Another example of application of goal setting, albeit from the lab, is a study showing the effect of goal setting on memory performance, motivation, and memory-related self-efficacy. This study showed improvement in these outcomes in both younger and older adults when they were provided with task goals such as "I will work to remember 7 out of every 10 items." (West et al. 2001).

Given the relative ease by which goal setting can be implemented, it seems like more intervention-focused research should test its applicability and effectiveness in additional domains (e.g., medication adherence or social participation).

Future Directions of Research

There is still a lot to discover about how older adults can adapt the contents and motivational orientations of their personal goals to stay engaged and effective in the face of age-related transitions (e.g., into retirement), new developmental tasks (e.g., generativity), and resource restrictions (e.g., in physical and cognitive capacities). It is possible that different factors affect processes of goal setting in old age than in younger age and it should be tested if and how interventions can be tailored to help older adults set feasible and desirable goals.

One potential approach would be to test the effectiveness of interventions that capitalize on mental contrasting (e.g., Fritzsche et al. 2016). In mental contrasting, people first positively daydream about the future and then contrast it with the here and now, imagining critical obstacles that may keep them from fulfilling their wish. If combined with specific "if-then" action plans (e.g., "if I eat out at night, then I order a salad"; Gollwitzer 1999), mental contrasting may be a useful tool to increase the commitment to and implementation of realistic and desirable goals. Whereas its effectiveness has not been tested specifically in samples of older adults, the approach is nevertheless promising, given that (1) older adults' resource restrictions may increase the necessity of considering obstacles that may stand in the way of goal attainment and that (2) planning interventions have been shown to be effective for older adults (Ziegelmann et al. 2006).

Lastly, given that many studies in fact show that processes of goal setting and pursuit still show gains in older adulthood (see Hennecke and Freund 2010; Riediger et al. 2005), it seems fruitful to identify the processes and strategies that enable older adults to maintain such a high level of functioning despite age-related resource losses.

Summary

Goals are a result of but also *guide* developmental processes across adulthood. As a consequence of age-related changes, such as resource restrictions (e.g., with regard to physical and cognitive resources, but also a restriction in future time perspective), older adults set different goals than their younger counterparts. Goals of older adults differ from goals of younger adults with regard to their goal content (e.g., they become more health-oriented and focused on generativity, intrinsic values, and the present), goal orientation (they become more maintenance/prevention-oriented), and goal focus (they become less outcome-focused), and their relative positions within goal systems (with more mutual facilitation). Future research could focus more strongly on goal setting interventions in various important domains in old age (e.g., maintaining physical and mental ability through regular training, medication adherence, social participation) and should consider how older adults maintain their level of functioning with regard to processes of goal setting and goal pursuit, despite looming resource restrictions in old age.

Cross-References

▶ Dual Process Theory of Assimilation and Accommodation
▶ Life-Span Theory of Control
▶ Locus of Control
▶ Selective Engagement of Resources
▶ Self-Perceptions of Aging
▶ The Model of Selection, Optimization, and Compensation
▶ Time Perspective Across Adulthood

References

Baltes PB, Baltes MM (1990) Psychological perspectives on successful aging: the model of selective optimization with compensation. In: Baltes PB, Baltes MM (eds) Successful aging: perspectives from the behavioral sciences. Cambridge University Press, New York

Brandtstädter J, Rothermund K (1994) Self-percepts of control in middle and later adulthood: buffering losses by rescaling goals. Psychol Aging 9(2):265–273. https://doi.org/10.1037/0882-7974.9.2.265

Carstensen LL, Isaacowitz DM, Charles ST (1999) Taking time seriously: a theory of socioemotional selectivity.

Am Psychol 54(3):165–181. https://doi.org/10.1037/0003-066X.54.3.165

Carstensen LL, Fung HH, Charles ST (2003) Socioemotional selectivity theory and the regulation of emotion in the second half of life. Motiv Emot 27(2):103–123

Carver CS, Scheier MF (1982) Control theory: a useful conceptual framework for personality–social, clinical, and health psychology. Psychol Bull 92(1):111–135. https://doi.org/10.1037//0033-2909.92.1.111

Ebner NC, Freund AM, Baltes PB (2006) Developmental changes in personal goal orientation from young to late adulthood: from striving for gains to maintenance and prevention of losses. Psychol Aging 21(4):664–678. https://doi.org/10.1037/0882-7974.21.4.664

Emmons RA (1992) Abstract versus concrete goals: personal striving level, physical illness, and psychological well-being. J Pers Soc Psychol 62(2):292–300. https://doi.org/10.1037//0022-3514.62.2.292

Emmons RA (1996) Striving and feeling. Personal goals and subjective well-being. In: Gollwitzer PM, Bargh JA (eds) The psychology of action: linking cognition and motivation to behavior. Guilford Press, New York, pp 292–300

Evans L, Hardy L (2002) Injury rehabilitation: a goal-setting intervention study. Res Q Exerc Sport 73(3):310–319. https://doi.org/10.1080/02701367.2002.10609025

Fishbach A, Ferguson MJ (2007) The goal construct in social psychology. In: Kruglanski AW, Higgins ET (eds) Social psychology: handbook of basic principles. Guilford Press, New York, pp 490–515

Freund AM (2003) Die Rolle von Zielen für die Entwicklung. [The role of goals for development]. Psychol Rundsch 54(4):233–242. https://doi.org/10.1026//0033-3042.54.4.233

Freund AM (2008) Successful aging as management of resources: the role of selection, optimization, and compensation. Res Hum Dev 5(2):94–106. https://doi.org/10.1080/15427600802034827

Freund AM, Baltes PB (1998) Selection, optimization, and compensation as strategies of life management: correlations with subjective indicators of successful aging. Psychol Aging 13(4):531–543. https://doi.org/10.1037/0882-7974.13.4.531

Freund AM, Baltes PB (2002) Life-management strategies of selection, optimization and compensation: measurement by self-report and construct validity. J Pers Soc Psychol 82(4):642–662. https://doi.org/10.1037//0022-3514.82.4.642

Freund AM, Hennecke M, Riediger M (2010) Age-related differences in outcome and process goal focus. Eur J Dev Psychol 7(2):198–222. https://doi.org/10.1080/17405620801969585

Fritzsche A, Schlier B, Oettingen G, Lincoln TM (2016) Mental contrasting with implementation intentions increases goal-attainment in individuals with mild to moderate depression. Cogn Ther Res 40(4):557–564. https://doi.org/10.1007/s10608-015-9749-6

Gardiner PA, Eakin EG, Healy GN, Owen N (2011) Feasibility of reducing older adults' sedentary time. Am J Prev Med 41(2):174–177. https://doi.org/10.1016/j.amepre.2011.03.020

Gollwitzer PM (1999) Implementation intentions: strong effects of simple plans. Am Psychol 54(7):493–503. https://doi.org/10.1037//0003-066x.54.7.493

Goodpaster BH, Park SW, Harris TB, Kritchevsky SB, Nevitt M, Schwartz AV, ... Newman AB (2006) The loss of skeletal muscle strength, mass, and quality in older adults: the health, aging and body composition study. J Gerontol A Biol Sci Med Sci 61(10):1059–1064. https://doi.org/10.1093/gerona/61.10.1059

Hennecke M, Freund AM (2010) Staying on and getting back on the wagon: age-related improvement in self-regulation during a low-calorie diet. Psychol Aging 25(4):876–885. https://doi.org/10.1037/a0019935

Impett EA, Gordon AM, Kogan A, Oveis C, Gable SL, Keltner D (2010) Moving toward more perfect unions: daily and long-term consequences of approach and avoidance goals in romantic relationships. J Pers Soc Psychol 99:948. https://doi.org/10.1037/a0020271

Jette AM, Lachman M, Giorgetti MM, Assmann SF, Harris BA, Levenson C et al (1999) Exercise – it's never too late: the strong-for-life program. Am J Public Health 89(1):66–72. https://doi.org/10.2105/ajph.89.1.66

Lachman ME (2006) Perceived control over aging-related declines: adaptive beliefs and behaviors. Curr Dir Psychol Sci 15(6):282–286. https://doi.org/10.1111/j.1467-8721.2006.00453.x

Levy R (1994) Aging-associated cognitive decline. Int Psychogeriatr 6(1):63–68. https://doi.org/10.1017/s1041610294001626

Little BR (1989) Personal projects analysis. Trivial pursuits, magnificient obsessions, and the search for coherence. In: Buss DM, Cantor N (eds) Personality psychology: recent trends and emerging directions. Springer, New York

Locke EA, Latham GP (2006) New directions in goal-setting theory. Curr Dir Psychol Sci 15(5):265–268. https://doi.org/10.1111/j.1467-8721.2006.00449.x

MacLeod AK, Coates E, Hetherton J (2008) Increasing well-being through teaching goal-setting and planning skills: results of a brief intervention. J Happiness Stud 9(2):185–196. https://doi.org/10.1007/s10902-007-9057-2

Mann T, De Ridder DTD, Fujita K (2013) Self-regulation and health behavior: social psychological approaches to goal setting and goal striving. Health Psychol 32:487–498. https://doi.org/10.1037/a0028533

Nurmi JE, Pulliainen H, Salmela-Aro K (1992) Age differences in adults' control beliefs related to life goals and concerns. Psychol Aging 7(2):194–196. https://doi.org/10.1037/0882-7974.7.2.194

Parsons J, Rouse P, Robinson EM, Sheridan N, Connolly MJ (2012) Goal setting as a feature of homecare services for older people: does it make a difference? Age Ageing 41(1):24–29. https://doi.org/10.1093/ageing/afr118

Pearson ES (2012) Goal setting as a health behavior change strategy in overweight and obese adults: a systematic literature review examining intervention components. Patient Educ Couns 87(1):32–42. https://doi.org/10.1016/j.pec.2011.07.018

Penningroth SL, Scott WD (2012) Age-related differences in goals: testing predictions from selection, optimization, and compensation theory and socioemotional selectivity theory. Int J Aging Hum Dev 74(2):87–111. https://doi.org/10.2190/AG.74.2.a

Riediger M, Freund AM (2006) Focusing and restricting: two aspects of motivational selectivity in adulthood. Psychol Aging 21(1):173–185. https://doi.org/10.1037/0882-7974.21.1.173

Riediger M, Freund AM, Baltes PB (2005) Managing life through personal goals: integral facilitation and intensity of goal pursuit in younger and older adulthood. J Gerontol Ser B Psychol Sci Soc Sci 60(2):84–91. https://doi.org/10.1093/geronb/60.2.P84

Saajanaho M, Rantakokko M, Portegijs E, Törmäkangas T, Eronen, J, Tsai LT, . . . Rantanen T (2016) Life resources and personal goals in old age. Eur J Ageing 13(3):195–208. https://doi.org/10.1007/s10433-016-0382-3

Sheldon KM, Kasser T (2001) Getting older, getting better? Personal strivings and psychological maturity across the life span. Dev Psychol 37(4):491–501. https://doi.org/10.1037/0012-1649.37.4.491

West RL, Welch DC, Thorn RM (2001) Effects of goal-setting and feedback on memory performance and beliefs among older and younger adults. Psychol Aging 16(2):240–250. https://doi.org/10.1037/0882-7974.16.2.240

Westerhof G, Miche M, Brothers A, Barrett A, Diehl M, Montepare JM, . . . Wurm S (2014) The influence of subjective aging on health and longevity: a meta-analysis of longitudinal data. Psychol Aging 29:793–802. https://doi.org/10.1037/a0038016

Wurm S, Diehl M, Kornadt A, Westerhof G, Wahl HW (2017) How do views on aging affect health outcomes in adulthood and late life? Explanations for a well-established connection. Dev Rev 46:27–43. https://doi.org/10.1016/j.dr.2017.08.002

Ziegelmann JP, Lippke S, Schwarzer R (2006) Adoption and maintenance of physical activity: planning interventions in young, middle-aged, and older adults. Psychol Health 22(2):145–163. https://doi.org/10.1080/1476832050018891

Goals

▶ Motivation: Theory/Human Model

Good Death

Wallace Chi Ho Chan
Department of Social Work, The Chinese University of Hong Kong, Hong Kong, China

Synonyms

Dignified death; Dying well; Good quality of death and dying; Peaceful death; Successful dying

Definition

Good death refers to a situation in which the dying process and the moment of death are in accordance with a person's preferences and wishes. The definition may vary among different stakeholders, e.g., patients, caregivers, and health-care professionals, in different sociocultural contexts, and may change over time.

Overview

The concept of a good death has gained more discussion as a result of the hospice movement (McNamara et al. 1994). The focus is on the quality of dying and death. The general consensus on the conceptualization of a good death is on minimizing and avoiding pain and suffering in death and dying, as well as respecting the preferences of patients for the dying process (Meier et al. 2016).

Key Research Findings

One of the widely cited articles on good death is written by Steinhauser et al. (2000), in which six themes of a good death are found: (1) pain and symptom management, (2) clear decision-making, (3) preparation for death, (4) completion, (5) contributing to others, and (6) affirmation of the whole person. A more recent review examined 36 articles and found that there are 11 core themes of a good death and/or successful dying: (1) preference for dying process, (2) pain-free status, (3) emotional well-being, (4) family, (5) dignity, (6) life completion, (7) religiosity/spirituality, (8) treatment preferences, (9) quality of life, (10) relationship with health-care professionals, and (11) others (e.g., recognition of culture, physical touch, being with pets, and health-care costs) (Meier et al. 2016).

Despite the fact that what constitutes a good death may vary in different cultural and historical contexts (Walter 2003), it seems that the desire for having a good death without pain and suffering is universal even in other non-Western cultures (Mak 2002). Yet, cultural differences have been observed, such as more emphasis on the family in constituting what a good death is in the Chinese context, e.g., highlighting the importance of family welfare after one's death (Chan et al. 2006; Chan and Epstein 2011–2012).

The literature also indicates the importance of addressing the differences in the views on a good death between patients and family members. For example, Carr (2003) proposed the question "a good death for whom?" and reported that some components which were perceived by patients as constituents of a good death, such as having led a full life, accepting one's impending death, and not being burdensome to family members, may not be necessarily good for bereaved family members – they do not experience a lower level of psychological distress. A recent review also indicated that fewer family members may consider quality of life an important component of a good death than do patients (Meier et al. 2016). From a broader sociological perspective, scholars have also argued that the concurrent ideology and discourse of a good death may limit what a good death can be (e.g., only accepting death is considered a good death) and in turn restrict the choices of dying patients (Hart et al. 1998; Cottrell and Duggleby 2016). The existing conceptualization of a good death, such as the emphasis on decision-making and death preparation, may also be less valid for patients who suffer from illness which may reduce their mental capacity in decision-making, such as those with dementia (Read and MacBride-Stewart 2018). In a previous study, the authors suggested that a good death as defined by palliative care professionals may actually not be fulfilled in reality, as patients may exert their own choices and autonomy and there are so many uncertainties in the dying process (McNamara 2004). Therefore, they proposed the term "good enough death" as a more realistic goal in providing palliative care, in which they can only try their best to meet the patients' preferences.

As for measurement, a review study showed that the Quality of Dying and Death (QODD) Questionnaire was the most widely used and best validated tool for measuring a good death (Hales et al. 2010). The QODD is a 31-item measure developed by Curtis et al. (2002), in which six domains are included for measuring the quality of dying death: (1) symptoms and personal care, (2) preparation for death, (3) moment of death, (4) family, (5) treatment preferences, and (6) whole person concerns. However, a recent study of advanced cancer patients proved a new four-factor structure in this measure: (1) symptoms and functioning, (2) preparation for death, (3) spiritual activities, and (4) acceptance of dying (Mah et al. 2018).

Examples of Application

Understanding what a good death is may have important implications for the provision of end-of-life care. For example, Emanuel and Emanuel (1998) proposed a framework for a good death, in which they suggested that the overall experience of the dying process (outcome) is influenced by three components: (1) fixed characteristics of the patient (e.g., diagnosis, age, and education), (2) modifiable dimensions of the patient's experience (e.g., physical, psychological, social, financial, spiritual, and existential aspects), (3) care system interventions, e.g., family caregiving, advance care planning, hospice, and insurance coverage. This framework may provide a basis for health-care professionals to assess patients' condition, which helps to ensure a better dying experience for patients. It also helps to enhance our awareness that achieving a good death of patients may require joint efforts of multidisciplinary professionals and providing adequate support to family members and caregivers. In fact, different contents of a good death may become a goal of improving end-of-life care. For example, conserving dignity has been considered an important element in ensuring a good death, and it becomes an important goal of palliative care in which specific intervention could be developed (Chochinov 2002; Chochinov et al. 2005). By measuring the quality of dying and death, studies can be conducted to examine the correlates of a good death. For example, a study showed that bereaved family members who perceived the timing of referrals to home-based hospice care as late or too late also rated the quality of dying and death of their deceased family members as poorer (Yamagishi et al. 2015).

Future Directions

The definition of a good death may inevitably change over time in accordance with the changing death and dying experience in this world. Therefore, further efforts are required to examine if the perspective of patients on a good death has been changing. In addition to the typical conceptualization of a good death as mentioned above, voices of minorities and disadvantaged groups should not be ignored. Their views on what a good death means may shed light on care provision of care and policy which may be undermined in the mainstream society. The tensions between the common definition of a good death which is widely adopted in palliative care and the voices of people who desire physician-assisted suicide for enhancing quality of dying and death should be seriously addressed. More dialogue between parties with opposite opinions should be encouraged, to enhance mutual understanding. With reference to changing dying trajectory of different chronic illness, more attention may be given to how the concept of a good death may be changed when people in general may have a longer life expectancy and when impending death may not be easily predicted. A public health approach may also be adopted in order to arouse broader but also more in-depth discussion on what a good death is among the general public in various cultural contexts. The concept of a good death should not be considered a goal in palliative care only, and it may actually be promoted as one of the essential goals in health care. To achieve this, good death could be reconsidered in connection with the end-of-life care and even health-care policy.

Summary

The definitions of a good death are complicate and may be different for different stakeholders and in different cultural contexts. The more common definition of a good death should be considered with caution, to avoid restricting the choices of patients and undermining the needs of patients who may have different preferences for death and dying While achieving a good death may be a universal desire for people, the fact that death may bring tears and grief to patients and family members should be of equal concern. "How good a death can be": this still be the feelings of some bereaved persons who are intensively grieving the loss of their loved ones.

Cross-References

▶ Advance Care Planning
▶ Advance Care Planning: Advance Directives
▶ Advance Care Planning: Medical Orders at the End of Life (MOLST, POLST)
▶ Assisted Dying
▶ Caregiving at the End of Life
▶ Death Anxiety
▶ Death Trajectory
▶ Death with Dignity
▶ End-of-Life Care
▶ End-of-Life Decision-Making in Acute Care Settings
▶ Euthanasia and Senicide
▶ Theory of Harmonious Death

References

Carr D (2003) A "good death" for whom? Quality of spouse's death and psychological distress among older widowed persons. J Health Soc Behav 44:215–232. https://doi.org/10.2307/1519809

Chan WCH, Epstein I (2011) Researching "good death" in a Hong Kong palliative care program: a clinical data-mining study. OMEGA J Death Dying 64:203–222. https://doi.org/10.2190/OM.64.3.b

Chan WCH, Tse HS, Chan THY (2006) What is good death: bridging the gap between research and intervention. In: Chan CLW, Chow AMY (eds) Death, dying and bereavement: a Hong Kong Chinese experience. Hong Kong University Press, Hong Kong, pp 127–135

Chochinov HM (2002) Dignity-conserving care-a new model for palliative care: helping the patient feel valued. JAMA 287:2253–2260. https://doi.org/10.1001/jama.287.17.2253

Chochinov HM, Hack T, Hassard T, Kristjanson LJ, McClement S, Harlos M (2005) Dignity therapy: a novel psychotherapeutic intervention for patients near the end of life. J Clin Oncol 23:5520–5525. https://doi.org/10.1200/JCO.2005.08.391

Cottrell L, Duggleby W (2016) The "good death": an integrative literature review. Palliat Support Care 14:686–712. https://doi.org/10.1017/S1478951515001285

Curtis JR, Patrick DL, Engelberg RA, Norris K, Asp C, Byock I (2002) A measure of the quality of dying and death: initial validation using after-death interviews with family members. J Pain Symptom Manag 24:17–31. https://doi.org/10.1016/S0885-3924(02)00419-0

Emanuel EJ, Emanuel LL (1998) The promise of a good death. Lancet 351:SII21–SII29. https://doi.org/10.1016/S0140-6736(98)90329-4

Hales S, Zimmermann C, Rodin G (2010) The quality of dying and death: a systematic review of measures. Palliat Med 24:127–144. https://doi.org/10.1177/0269216309351783

Hart B, Sainsbury P, Short S (1998) Whose dying? A sociological critique of the 'good death'. Mortality 3:65–77. https://doi.org/10.1080/713685884

Mah K, Hales S, Weerakkody I et al (2018) Measuring the quality of dying and death in advanced cancer: item characteristics and factor structure of the quality of dying and death questionnaire. Palliat Med 33:369–380. https://doi.org/10.1177/0269216318819607

Mak MHJ (2002) Accepting the timing of one's death: an experience of Chinese hospice patients. OMEGA J Death Dying 45:245–260. https://doi.org/10.2190/WVJQ-96WD-5DJJ-F5CU

McNamara B (2004) Good enough death: autonomy and choice in Australian palliative care. Soc Sci Med 58:929–938. https://doi.org/10.1016/j.socscimed.2003.10.042

McNamara B, Waddell C, Colvin M (1994) The institutionalization of the good death. Soc Sci Med 39:1501–1508. https://doi.org/10.1016/0277-9536(94)90002-7

Meier EA, Gallegos JV, Thomas LPM, Depp CA, Irwin SA, Jeste DV (2016) Defining a good death (successful dying): literature review and a call for research and public dialogue. Am J Geriatr Psychiatry 24:261–271. https://doi.org/10.1016/j.jagp.2016.01.135

Read S, MacBride-Stewart S (2018) The 'good death' and reduced capacity: a literature review. Mortality 23:381–395. https://doi.org/10.1080/13576275.2017.1339676

Steinhauser KE, Clipp EC, McNeilly M, Christakis NA, McIntyre LM, Tulsky JA (2000) In search of a good death: observations of patients, families, and providers. Ann Intern Med 132:825–832. https://doi.org/10.7326/0003-4819-132-10-200005160-00011

Walter T (2003) Historical and cultural variants on the good death. BMJ 327:218–220. https://doi.org/10.1136/bmj.327.7408.218

Yamagishi A, Morita T, Kawagoe S et al (2015) Length of home hospice care, family-perceived timing of referrals, perceived quality of care, and quality of death and dying in terminally ill cancer patients who died at home. Support Care Cancer 23:491–499. https://doi.org/10.1007/s00520-014-2397-7

Good Diet

▶ Healthy Diet

Good Dying

▶ Theory of Harmonious Death

Good Quality of Death and Dying

▶ Good Death

Governance of Social Services for Older People

▶ Aging Network

Governmental Social Insurance Programs

▶ Welfare States

Graceful Aging

▶ Representations of Older Women and White Hegemony

Gradual Cell Senescence

Giacinto Libertini
Italian National Health Service, ASL NA2 Nord, Frattamaggiore, Italy
Department of Translational Medical Sciences, Federico II University, Naples, Italy

Synonyms

Gradual senescence; Telomere position effect

Definition

For multicellular eukaryotic species, like humans, gradual cell senescence means progressive alteration of cell functions in relationship with telomere shortening and subtelomere repression. For the yeast, it means progressive alteration of cell functions in relationship with the number of duplications in cells of the mother lineage and subtelomere repression by the accumulation of particular substances on it. It is important to understand the affinities and the differences between the two phenomena.

Overview

In 1990, in the yeast (*Saccharomyces cerevisiae*), the proximity to the telomere of an artificially inserted gene was shown to cause a reversible repression of the gene (Gottschling et al. 1990). This phenomenon, called "telomere position effect" (Gottschling et al. 1990, p. 751; Micheli et al. 2016, p. 325), has also been reported for other species, ours included. A work has shown that "chromosome looping brings the telomere close to genes up to 10 Mb away from the telomere when telomeres are long and that the same loci become separated when telomeres are short" (Robin et al. 2014, p. 2464), and this phenomenon has been suggested as "a potential novel mechanism for how telomere shortening could contribute to aging and disease initiation/progression in human cells long before the induction of a critical DNA damage response" (Robin et al. 2014, p. 2464). However, this mechanism, also defined "telomere position effect over long distances" (Kim and Shay 2018, p. 1), appears too simplistic to explain the several and various regulatory actions that result to be dependent on the subtelomeric DNA, and the following explanation is perhaps more convincing.

Gradual Cell Senescence in Yeast and in Multicellular Eukaryotic Species

There are two main cases:

1. In yeast, a cell reproduces by division into two cells, which are defined as "mother" and "daughter" and are somehow different. While the daughter cell is identical to the parent cell, the cells of the mother lineage manifest some physiological and morphological differences and can reproduce only a limited number of times (about 25–35 duplications (Jazwinski 1993)). In fact, the cells of the mother line, in proportion to the number of previous duplications, show increasing metabolic alterations and an increasing vulnerability to the blocking of replicative abilities (replicative senescence) and to apoptosis (Laun et al. 2001; Lesur and Campbell 2004; Büttner et al. 2006; Fabrizio and Longo 2008). These effects cannot be explained by telomere shortening because in the yeast the telomerase enzyme is always

perfectly active and at each duplication the length of the telomere is restored (D'Mello and Jazwinski 1991; Maringele and Lydall 2004). However, in the cells of the mother lineage (but not in those of the daughter line), at each duplication particular molecules, defined as extrachromosomal ribosomal DNA circles (ERCs), accumulate (Sinclair and Guarente 1997), and this is the probable cause of the phenomenon: "several lines of evidence suggest that accumulation of ERCs is one determinant of life span" (Lesur and Campbell 2004, p. 1297).

2. In multicellular eukaryotic species, like humans, for cell duplications where telomere length is not restored by telomerase enzyme, the telomere is shortened at each replication. Since the telomere is covered by a heterochromatin or nucleoproteinic hood, the progressive shortening of the telomere causes the hood to slide on part of the subtelomere repressing its function. "As the telomere shortens, the hood slides further down the chromosome .... the result is an alteration of transcription from portions of the chromosome immediately adjacent to the telomeric complex, usually causing transcriptional silencing, although the control is doubtless more complex than merely telomere effect through propinquity ... These silenced genes may in turn modulate other, more distant genes (or set of genes). There is some direct evidence for such modulation in the subtelomere ..." (Fossel 2004, p. 50).

This view is confirmed in another paper: "Our results demonstrate that the expression of a subset of subtelomeric genes is dependent on the length of telomeres and that widespread changes in gene expression are induced by telomere shortening long before telomeres become rate-limiting for division or before short telomeres initiate DNA damage signaling. These changes include up-regulation and down-regulation of gene expression levels" (Robin et al. 2014, p. 2471). This means that subtelomere repression modifies gene expression for distant parts of DNA too.

In correlation with telomere shortening, there is a progressive alteration of cell functions and a greater vulnerability to the activation of cell senescence program which involves both replicative senescence and maximal alteration of cell functions. Unlike yeast, however, where there is a greater vulnerability to apoptosis (Laun et al. 2001; Lesur and Campbell 2004; Büttner et al. 2006; Fabrizio and Longo 2008), cell senescence in multicellular eukaryotes determines a greater resistance to apoptosis (He and Sharpless 2017). Telomere position effect and cell senescence, as they alter cell functions, extracellular secretions included with harmful actions on other cells, contribute decisively to the manifestations of aging (Fossel 2004; Libertini 2014; Libertini and Ferrara 2016). To better describe the effects of the phenomenon, the name "gradual cell senescence" (or more succinctly, where there is no possibility of misunderstanding with the aging of the entire organism, "gradual senescence") has been proposed (Libertini 2014, p. 1006, 2015, p. 1536).

These two types of mechanisms underlying gradual cell senescence may seem completely different and without any correlation, but there is a particular type of yeast mutants that allows to frame them in a unitary way.

In yeast, $tlc1\Delta$ mutants have telomerase inactive, and telomeres shorten both in mother and daughter lineage. However, the cells of the daughter cell line, although they have no ERC accumulation, show all the alterations of mother lineage cells with the same number of replications. In particular, the transcriptome, i.e., the overall expression of genes, is similar (Lesur and Campbell 2004). In fact, the repression of the subtelomere is the common key element even if the mechanisms of inhibition are different: (a) for multicellular eukaryotes and yeast $tlc1\Delta$ mutants, telomere shortening, and sliding of the heterochromatin hood over the subtelomere and (b) for nonmutant yeast cells of the mother line, accumulation of ERCs (see Fig. 1 in the entry ▶ "Telomere-Subtelomere-Telomerase System").

The Hypothesis of "r" Sequences

In both cases, it is necessary that the subtelomere has general and essential regulatory functions for the whole cell and that its progressive repression gradually alters cell functions. A possible explanation is that the subtelomere has multiple, equal or similar, sequences ("r") with the same, or analogous, general cell regulating function (Libertini 2017). When none of these sequences is repressed, the cell has the optimal functionality, while the inhibition of a growing number of "r" sequences progressively alters the overall cell functionality.

Moreover, if it is true that the telomere oscillates between two states, one in which it is covered by the heterochromatin hood and is resistant to the passage to cell senescence and the other in which it is temporarily uncovered and vulnerable to cell senescence (Blackburn 2000), these "r" sequences should also regulate the ratio between the durations of the capped and uncapped states (Libertini 2017). Consequently, among the characteristics of gradual cell senescence, it is important to underline the greater vulnerability to cell senescence (see Fig. 2 in the entry ▶ "Telomere-Subtelomere-Telomerase System").

The real existence of these hypothetical "r" sequences appears supported by the structure of the subtelomere: an "unusual structure: patchworks of blocks that are duplicated" (Mefford and Trask 2002, p. 91); "A common feature associated with subtelomeric regions in different eukaryotes is the presence of long arrays of tandemly repeated satellite sequences" (Torres et al. 2011, p. 85).

This "unusual structure", i.e., "long arrays of tandemly repeated satellite sequences," which characterizes telomere sequence without no apparent meaning and utility for the cell, on the contrary would be a pivotal general regulator of all cell functions that is progressively inhibited by the sliding of the heterochromatin hood caused by telomere shortening (or, in yeast, by ERC accumulation).

Gradual Cell Senescence and Evolution

As regards the meaning of gradual cell senescence in terms of evolution, i.e., as a phenomenon determined or not by natural selection, it should be noted: (1) the phenomenon should be evaluated together with cell senescence and the consequent decline in cell turnover capacity in the frame of the mechanisms that underlie aging (Libertini and Ferrara 2016); (2) the presence of regulatory sequences that are essential for the functionality of the cell and are in a vulnerable position that could be avoided indicates that the phenomenon is somehow favored by natural selection (Libertini 2015); (3) in multicellular eukaryotes, the effects of gradual senescence are reversed by telomerase activation (Bodnar et al. 1998). This shows that the phenomenon is not caused by a disorderly accumulation of harmful substances or by other inevitable metabolic effects but dependent on the level of telomerase activity that is under genetic control. As a result, gradual cell senescence must somehow be favored and shaped by natural selection; and (4) In yeast, gradual cell senescence increases the chances of apoptosis, i.e., the death of the unicellular individual. In multicellular organisms, gradual cell senescence increases the resistance to apoptosis. However, in multicellular organisms, the increasing number of altered cells contributes to reduce fitness, i.e., to increase the probability of death of the individual. A common feature of the two cases is that gradual senescence helps to reduce the survival of the individual in proportion to the number of duplications, in yeast, or in proportion to age, in multicellular individuals.

Conclusion

Altogether, the study of gradual cell senescence should not be limited to the uncritical description of its manifestations and should include its explanation in the more general context of other phenomena (such as cell senescence and decline in cell turnover) that gradually compromise the fitness, i.e., the survival capacity of the individual. In this more general assessment, it seems indispensable to conceive this complex of phenomena in one of two entirely alternative interpretative keys: (a) they are the inevitable consequence of metabolic problems which cannot be solved by the evolution; (b) they represent a specific adaptation modelled by particular needs of evolution. The second interpretative key may seem paradoxical or unlikely, but it is the only one that would allow us to interpret the sophisticated nature of the aforementioned phenomena in a logical way.

Cross-References

▶ Aging Mechanisms
▶ Cell Senescence
▶ Telomere-Subtelomere-Telomerase System
▶ Timeline of Aging Research

References

Blackburn EH (2000) Telomere states and cell fates. Nature 408:53–56. https://doi.org/10.1038/35040500

Bodnar AG, Ouellette M, Frolkis M et al (1998) Extension of life-span by introduction of telomerase into normal human cells. Science 279:349–352. https://doi.org/10.1126/science.279.5349.349

Büttner S, Eisenberg T, Herker E et al (2006) Why yeast cells can undergo apoptosis: death in times of peace, love, and war. J Cell Biol 175:521–525. https://doi.org/10.1083/jcb.200608098

D'Mello NP, Jazwinski SM (1991) Telomere length constancy during aging of Saccharomyces cerevisiae. J Bacteriol 173:709–6713. https://doi.org/10.1128/jb.173.21.6709-6713.1991

Fabrizio P, Longo VD (2008) Chronological aging-induced apoptosis in yeast. Biochim Biophys Acta 1783:1280–1285. https://doi.org/10.1016/j.bbamcr.2008.03.017

Fossel MB (2004) Cells, aging and human disease. Oxford University Press, New York

Gottschling DE, Aparicio OM, Billington BL, Zakian VA (1990) Position effect at S. cerevisiae telomeres: reversible repression of Pol II transcription. Cell 63:751–762. https://doi.org/10.1016/0092-8674(90)90141-Z

He S, Sharpless NE (2017) Senescence in health and disease. Cell 169(6):1000–1011. https://doi.org/10.1016/j.cell.2017.05.015

Jazwinski SM (1993) The genetics of aging in the yeast Saccharomyces cerevisiae. Genetica 91:5–51. https://doi.org/10.1007/978-94-017-1671-0_6

Kim W, Shay JW (2018) Long-range telomere regulation of gene expression: telomere looping and telomere position effect over long distances (TPE-OLD). Differentiation 99:1–9. https://doi.org/10.1016/j.diff.2017.11.005

Laun P, Pichova A, Madeo F et al (2001) Aged mother cells of Saccharomyces cerevisiae show markers of oxidative stress and apoptosis. Mol Microbiol 39:1166–1173. https://doi.org/10.1111/j.1365-2958.2001.02317.x

Lesur I, Campbell JL (2004) The transcriptome of prematurely aging yeast cells is similar to that of telomerase-deficient cells. Mol Biol Cell 15:1297–1312. https://doi.org/10.1091/mbc.e03-10-0742

Libertini G (2014) Programmed aging paradigm: how we get old. Biochem Mosc 79(10):1004–1016. https://doi.org/10.1134/S0006297914100034

Libertini G (2015) Phylogeny of aging and related phenoptotic phenomena. Biochem Mosc 80(12):1529–1546. https://doi.org/10.1134/S0006297915120019

Libertini G (2017) The feasibility and necessity of a revolution in geriatric medicine. OBM Geriatrics 1(2). https://doi.org/10.21926/obm.geriat.1702002

Libertini G, Ferrara N (2016) Possible interventions to modify aging. Biochem Mosc 81(12):1413–1428. https://doi.org/10.1134/S0006297916120038

Maringele L, Lydall D (2004) Telomerase- and recombination-independent immortalization of budding yeast. Genes Dev 18:2663–2675. https://doi.org/10.1101/gad.316504

Mefford HC, Trask BJ (2002) The complex structure and dynamic evolution of human subtelomeres. Nat Rev Genet 3(2):91–102. https://doi.org/10.1038/nrg727

Micheli E, Galati A, Cicconi A, Cacchione S (2016) Telomere maintenance in the dynamic nuclear architecture. In: Göndör A (ed) Chromatin regulation and dynamics. Academic – (Elsevier), Cambridge, pp 325–352. https://doi.org/10.1016/B978-0-12-803395-1.00013-7

Robin JD, Ludlow AT, Batten K et al (2014) Telomere position effect: regulation of gene expression with progressive telomere shortening over long distances. Genes Dev 28:2464–2476. https://doi.org/10.1101/gad.251041.114

Sinclair DA, Guarente L (1997) Extrachromosomal rDNA circles – a cause of aging in yeast. Cell 91:1033–1042. https://doi.org/10.1016/S0092-8674(00)80493-6

Torres GA, Gong Z, Iovene M et al (2011) Organization and evolution of subtelomeric satellite repeats in the potato genome. G3 (Bethesda) 1(2):85–92. https://doi.org/10.1534/g3.111.000125

Gradual Senescence

▶ Gradual Cell Senescence

Grandparent-Grandchild Relations

▶ Intergenerational Solidarity

Grandparent-Grandchildren Relationship

▶ Grandparenting

Grandparenting

Montserrat Celdrán and Valentina Cannella
University of Barcelona, Barcelona, Spain

Synonyms

Child-rearing by grandparent; Grandparent-grandchildren relationship; Intergenerational relations; Multigenerational relations

Definition

Being a grandparent involves an intergenerational relationship within a family structure that bonds a person with his/her children's offspring. Grandparenting has been described through four dimensions: behavioral (activities carried out by the grandparent), affective (positive and negative feeling toward grandchildren), symbolic (what this role means for a person), and cognitive (attitudes toward grandparenthood) (Hurme 1991).

Overview

"In the normal course of things, every time a child is born, a grandparent is born and a new three-generational family is formed" (Kornhaber 1985, p. 162). Following on from this traditional scenario, the relevant literature considers the complexity of alternative circumstances in which someone can become and feel like a grandparent, for example, through adoption or being a stepparent. Grandparenthood has usually been portrayed as a positive role and as a new family role in the midst of various social losses that someone has to face while aging, due to retirement, widowhood, or an empty nest among others.

Key Research Findings

The Diversity of Grandparenthood

One of the initial topics of research on grandparenthood was related to the attempt to describe different ways of enacting grandparenthood, based on the types of activities and emotional bonds shared with grandchildren. In this vein Neugarten and Weinstein (1964) provided one of the first classifications, which included the following roles for grandparents: formal (those interested in their grandchildren's well-being and activities but who remain distant from child-rearing and education); fun seeker (those grandparents mostly focused on leisure and recreation activities); distant figure (those who have sporadic contact with their grandchildren); surrogate parent (those grandparents that become their grandchildren's main caregivers due to problems of the legal parents, e.g., health problems); and reservoir of family wisdom (those that provide resources and teach skills to their grandchildren). These different roles could be explained by a set of variables, such as gender, number of grandchildren, and grandparent's age, that would allow us to have a more diverse view of the different kinds of roles and experiences of what it means to be a grandparent (e.g., Reitzes and Mutran 2004; Uhlenberg and Hammill 1998).

Factors Affecting Grandparenthood

- Gender of grandparent. A grandparent's involvement, caregiving tasks, or even affection behavior could be a reflection of gender socialization. This socialization could explain why grandmothers seem to be more involved in their grandchildren's life and are usually identified as the favorite grandparent (MaloneBeach et al. 2018).
- Number of grandchildren. If grandparenthood involvement is conceptualized as a matter of resources, the number of grandchildren could be crucial in terms of time consumption, activities, or even economic involvement with each grandchild (Uhlenberg and Hammill 1998).
- Grandparent's age. Older grandparents are likely to have less energy for being involved in roles such as "fun seeker." However, age is not, strictly speaking, the main cause for reduction in activity. A change in health status and specifically physical or mental dependency could completely change the grandparenthood experience, as is the case for grandparents that have dementia. As the disease progresses, grandchildren may struggle to maintain a relationship with the affected grandparent, especially if they have previously been attached to this grandparent (Celdrán et al. 2011).
- Grandchild's age. Similarly, from a developmental point of view, as grandchildren grow up, the needs that grandparents could satisfy as well as the kind of activities they are involved in change. While the grandchild is a child, help with caring, instrumental help, or leisure activities based on the child's needs and interests are more frequent. However, during adolescence grandparents could act as a mediator during conflicts between adolescents and their parent(s), being a confidant or a source of emotional support for the teenager (Villar et al. 2010).
- Family line. In heterosexual couples, a matrifocal bias that facilitates the relationship between the mother's parents and the newborn has been described. This bias could become an issue in cases of divorce, as the paternal grandparents could be negatively affected, especially in cases of a high-conflict divorce and subsequent tense post-divorce relationships (Albertini and Tosi 2018). Family line and the parents' willingness to facilitate grandparenthood are also mediated by the grandchild's age, as the younger they are, the more dependent they are on their parents for their daily and social activities, which include contact with grandparents.
- Geographic distance. Living closer together facilitates frequent contact and the chance to be involved in grandchildren's lives and, in the long run, could be associated with a more positive and satisfactory relationship between grandparents and grandchildren (Bangerter and Waldron 2014).

Grandparents as Caregivers

Among the different roles a grandparent could have, most research has focused on caring commitments rather than the importance of grandparents in family leisure (Hebblethwaite 2017) or in grandchildren's educational activities (Cantero et al. 2018).

Grandparent involvement in caring can be conceptualized as a continuum (see Fig. 1) (e.g., Hank and Buber 2009; Villar et al. 2012). Some grandparents only provide care sporadically (e.g., when a grandchild is ill and cannot go to school for a couple of days); others, especially when grandchildren are young, provide care more regularly as auxiliary carers, mostly motivated by the wish or need to help balance work and family duties of their adult children (Villar et al. 2012).

Moreover, there are grandparents that have a main role in child-rearing. **Co-parenting** is an appropriate description of those situations in which, although the parents are present, they need assistance with their offspring. This may be the case, for example, of adolescent mothers. In such a three-generational family household, the adult grandparent plays an important role in caring for the baby while also supporting the adolescent in their transitional role into adulthood (Culp et al. 2006).

Finally, **surrogate parenting** describes the situation in which the grandparents become the primary caregiver of their grandchildren because of permanent or prolonged unavailability of one or both parents. Drug addiction, death, mental

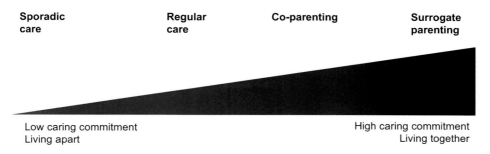

Grandparenting, Fig. 1 Diversity in caregiving by grandparents

problems, and incarceration of one or both parents are the main reasons for becoming the main caregiver of a grandchild. This kind of role has received the majority of attention in research because of the multiple negative consequences that this role could have on the well-being, health, and social conditions of the older person (Hayslip and Kaminski 2005) and, therefore, the greatest need for social support and specific policies (Sumo et al. 2018).

Benefits of Being a Grandparent

Grandparent-grandchild relationships have generally been described as beneficial for both members of the dyad. Both could act as an agent of socialization for each other as they learn new skills or knowledge and this role could have a positive impact on their self-esteem and identity.

Becoming a grandparent has been associated with a broad range of positive consequences, beginning with the fact that grandparenthood could reinforce family ties as a grandchild increases the chances of having more frequent contact with the rest of the family. Being a grandparent is a source of satisfaction and happiness, especially when having the chance to spend time with the grandchild (Bouchard and McNair 2016). It is also seen as a new opportunity to have a positive impact on someone's development or to be more involved in child-rearing for some generations that were unable to be close to their own children because of work duties (Aldous 1985). Having grandchildren is an incentive to improve knowledge and skills in new technologies.

Nevertheless, there are variables that could negatively affect the experience of grandparenthood, as intensive grandchild caregiving could be associated with stress and overload. Also, if grandparenthood occurs at a nonnormative moment (e.g., due to a teenage pregnancy), there could be negative consequences as this could interfere with other adult roles being performed at that time of life.

As for grandchildren, having contact with their grandparents allows them the chance to interact with an adult on different terms than they usually would with their own parents. So, they sometimes enjoy a relationship that is freer of discipline and less rigid. Grandparents could act as an important source of attachment during infancy and adolescence, and their loss is usually the first experience of death that a person has to deal with (Manoogian et al. 2018). Grandparents can serve as role models even in aspects such as the way grandchildren see aging, depending on how the grandparent faces this period of life (Pecchioni and Croghan 2002). Grandchildren could learn new skills, values, and family stories through contact and activities with grandparents. Even during adolescence or young adulthood, when family ties are not as important as those with other social groups such as the peer group, grandparents still have an important role and can help as a counselor to their grandchildren (Villar et al. 2010). With regard to negative influences, there is a lack of studies that explains how grandparenthood could have a negative effect on children's development, although there are some studies that have indicated that a permissive attitude in the form of overfeeding or spoiling could have a negative impact on grandchildren's health and behavior (e.g., Chambers et al. 2017).

Future Directions of Research

Previous studies have concluded that being a grandparent generates diverse roles to play and in normative situations has a positive influence on an older person's well-being and identity. However, this idyllic and win-win relationship could be in part explained by certain methodological limitations, for example, in studies that only asked about the favorite grandchildren or studies that only focused on the grandparents' perspective. In general, grandparents describe a more positive and optimistic view of their relationship with their grandchildren than the other way around (Giarrusso et al. 1995). This could be partly because older members of the dyad may have invested more in the relationship (time, love, patience, money, etc.) than the younger members, so they are more reluctant to see that the relationship is less positive. Therefore, more dyad studies, that is, studies in which both members of the relationship are interviewed, could reveal the inconsistencies and ambiguities that grandparenthood could have.

Aging families will take more various forms in the future, as a result of the diverse family trajectories occurring nowadays during adulthood. This could translate into changes in how grandparenthood is understood and portrayed. For example, more research is needed to understand the challenges faced by LGTBI grandparents or LGBTI adults that want to be parents (and how this decision affects their own parents). Issues such as adoption or stepgrandparenthood will be more frequent, and how older adults adapt to such circumstances should be explored. Finally, grandparents' dependency and the role of grandchildren during this period have been understudied, and in the future grandchildren could have a more important role in families due to the lack of caregivers to take care of the older person. Lastly, although grandparenthood is an important role in family relationships, there have been few attempts to develop programs to enhance or improve this relationship, with the exception of custodial grandparenthood.

Summary

Grandparenthood is usually portrayed as a positive and important role for an older person. It also has been regarded to have a positive influence on the grandchildren's development. This role, however, could be affected by a set of variables not only related with grandparent's characteristics but also with family dynamics, grandchildren's age, or geographical distance.

Cross-References

▶ Intergenerational Exchange and Support
▶ Intergenerational Exchanges
▶ Intergenerational Family Dynamics and Relationships
▶ Intergenerational Family Structures
▶ Intergenerational Housing
▶ Intergenerational Programs
▶ Intergenerational Solidarity
▶ Intergenerational Stake
▶ Intergenerational Stake Hypothesis
▶ Kinship Networks
▶ Multigenerational Families
▶ Parenting and Grandparenting
▶ Sandwich Generation

References

Albertini M, Tosi M (2018) Grandparenting after parental divorce: the association between non-resident parent–child meetings and grandparenting in Italy. Eur J Ageing 15:277–286. https://doi.org/10.1007/s10433-018-0478-z

Aldous J (1985) Parent-adult child relations as affected by the grandparent status. In: Bengtson VL, Robertson JF (eds) Grandparenthood. Berverly Hills, California: Sage, National Institute for the Family, National Jewish Family Center, pp 117–132

Bangerter LR, Waldron VR (2014) Turning points in long distance grandparent–grandchild relationships.

J Aging Stud 29:88–97. https://doi.org/10.1016/j.jaging.2014.01.004

Bouchard G, McNair JL (2016) Dyadic examination of the influence of family relationships on life satisfaction at the empty-nest stage. J Adult Dev 23(3):174–182. https://doi.org/10.1007/s10804-016-9233-x

Cantero N, Pantoja A, Alcaide M (2018) Impacto de la Participación de los Abuelos en una Comunidad de Aprendizaje. Res Ageing Soc Policy 6(2):198–223. https://doi.org/10.17583/rasp.2018.3587

Celdrán M, Triadó C, Villar F (2011) "My grandparent has dementia": how adolescents perceive their relationship with grandparents with a cognitive impairment. J Appl Gerontol 30(3):332–352. https://doi.org/10.1177/0733464810368402

Chambers SA, Rowa-Dewar N, Radley A, Dobbie F (2017) A systematic review of grandparents' influence on grandchildren's cancer risk factors. PLoS One 12(11):e0185420. https://doi.org/10.1371/journal.pone.0185420

Culp AM, Culp RE, Noland D, Anderson JW (2006) Stress, marital satisfaction, and child care provision by mothers of adolescent mothers: considerations to make when providing services. Child Youth Serv Rev 28:673–681. https://doi.org/10.1016/j.childyouth.2005.06.009

Giarrusso R, Stallings M, Bengtson VL (1995) The intergenerational stake hypothesis revisited: parent–child differences in perceptions of relationships 20 years later. In: Bengtson VL, Schaie KW, Burton LM (eds) Adult intergenerational relations: effects of societal change. Springer, New York, pp 227–296

Hank K, Buber I (2009) Grandparents caring for their grandchildren: findings from the 2004 Survey of Health, Ageing, and Retirement in Europe. J Fam Issues 30(1):53–73. https://doi.org/10.1177/0192513X08322627

Hayslip B Jr, Kaminski PL (2005) Grandparents raising their grandchildren: a review of the literature and suggestions for practice. Gerontologist 45(2):262–269. https://doi.org/10.1093/geront/45.2.262

Hebblethwaite S (2017) The (in)visibility of grandparents in family leisure research: a call for expanded conceptualizations of "family". Leis Sci 39(5):415–425. https://doi.org/10.1080/01490400.2017.1333055

Hurme H (1991) Dimensions of the grandparent role in Finland. In: Smith PK (ed) The psychology of grandparenthood: an international perspective. Taylor and Frances/Routledge, Florence, pp 19–31

Kornhaber A (1985) Grandparenthood and the "new social contract". In: Bengtson VL, Robertson JF (eds) Grandparenthood. Bervely Hills, California: Sage, National Institute for the Family, National Jewish Family Center, pp 159–171

MaloneBeach E, Hakoyama M, Shelby A (2018) The good grandparent: perspectives of young adults. Marriage Fam Rev 54(6):582–597. https://doi.org/10.1080/01494929.2017.1414724

Manoogian MM, Vandenbroeke J, Ringering A, Toray T, Cooley E (2018) Emerging adults' experiences of grandparent death. Omega 76(4):351–372. https://doi.org/10.1177/0030222817693140

Neugarten BL, Weinstein KK (1964) The changing American grandparent. J Marriage Fam 26(2):199–204

Pecchioni LL, Croghan JM (2002) Young adults' stereotypes of older adults with their grandparents as the targets. J Commun 52(4):715–730. https://doi.org/10.1111/j.1460-2466.2002.tb02570.x

Reitzes DC, Mutran EJ (2004) Grandparenthood: factors influencing frequency of grandparent-grandchildren contact and grandparent role satisfaction. J Gerontol B Psychol Sci Soc Sci 59(1):S9–S16. https://doi.org/10.1093/geronb/59.1.S9

Sumo J, Wilbur J, Julion W, Buchholz S, Schoeny M (2018) Interventions to improve grandparent caregivers' mental and physical health: an integrative review. West J Nurs Res 40(8):1236–1264. https://doi.org/10.1177/0193945917705376

Uhlenberg P, Hammill BG (1998) Frequency of grandparent contact with grandchild sets: six factors that make a difference. Gerontologist 38(3):276–285

Villar F, Triadó C, Pinazo-Hernandis S, Celdrán M, Solé C (2010) Grandparents and their adolescent grandchildren: generational stake or generational complaint? A study with dyads in Spain. J Intergener Relatsh 8(3):281–297. https://doi.org/10.1080/15350770.2010.498759

Villar F, Celdrán M, Triadó C (2012) Grandmothers offering regular auxiliary care for their grandchildren: an expression of generativity in later life? J Women Aging 24(4):292–312. https://doi.org/10.1080/08952841.2012.708576

Grandparents Caring for Grandchildren

▶ Parenting and Grandparenting

Granny Battering

▶ Financial Exploitation by Family Members

Grant Program

▶ Entitlement Program

Grass Fires

▶ Wildfires and Older People

Gray Consumption

Gary Haq
Stockholm Environment Institute, Department of Environment and Geography, University of York, Heslington, UK

Synonyms

Carbon footprint; Energy use, greenhouse gas emissions; Population aging

Definition

Gray consumption can be defined as the use of goods and services (e.g., housing, food, heating, and personal travel) by older people, which meet basic needs and improve quality of life. These activities require energy and produce greenhouse gas (GHG) emissions. An individual's pattern of consumption changes over the life course, reflecting wealth, age, health, and social needs. Older people's consumption habits therefore contribute to rising GHG emissions and climate change (Estiri and Zagheni 2019; Menz and Welsch 2012).

Overview

Demographic changes such as aging, urbanization, population growth, and household size have implications for consumption patterns and GHG emissions that drive climate change (O′ Neill et al. 2012). There is increasing evidence to suggest that GHG emissions raise as a population grows older (Zagheni 2011; Menz and Welsch 2012). The effects of a changing climate are already impacting natural and human systems with ice caps melting, sea level rise, and more extreme weather events. The global mean temperature is likely to reach 1.5 °C above pre-industrial levels between 2030 and 2052 if it continues to increase at the current rate. Rapid far-reaching and unprecedented changes in all aspects of society are therefore needed to keep below 1.5 °C heating. This will require changes to the lifestyles, consumption preferences, and consumer behavior of older people (IPCC 2018; UN Environment 2019).

Key Research Findings

Population aging and changing consumption patterns are increasing GHG emissions in both developed and developing nations. For example, in China, small and aging households are expected to increase energy consumption and GHG emissions due to changes in time-use and consumption habits (Yu et al. 2019).

The carbon footprint is the total amount of GHG emissions which result from the individual consumption of goods and services. This covers both an individual's direct emissions (e.g., home heating and car use) and indirect emissions arising in the production and supply of goods and services consumed from home and abroad.

An analysis of UK age-related household consumption (Haq et al. 2007) found baby boomers had the largest carbon footprint with car dependency, holidays abroad, and eating out as key carbon-intensive activities, while home energy use was a major contributor to the carbon footprint of older adults (aged 75+) (Hamza and Gilroy 2011). A study of intergenerational emissions found French baby boomers were not only better off than other generations but consumed more and lived in energy-inefficient homes (Chancel 2014).

Age-specific estimates of per capita GHG emissions in the United States for a set of selected carbon-intensive goods (i.e., electricity, natural gas, gasoline, air flights, tobacco products, clothes, food, and cars) also showed that average emissions increase with age, especially home energy use (Estiri and Zagheni 2019). This continues until people reach their late 60s after which per capita emissions decrease, with use of energy-intensive goods decreasing in later life (Zagheni 2011).

Old age can also lead to a reduction in GHG emissions. An increase in the number of older people in European nations including Japan has been associated with a decrease in per capita GHG emissions from the road sector when the share of the aged population reaches over 16 percent (Okada 2012). Early retirement can have a beneficial effect as it means lower energy use and GHG emissions, but if retirement is postponed, emission-reducing effects in later life would be reduced (Wei et al. 2018; O'Neil et al. 2012). Poor health in old age has also been linked to lower home energy use and engagement in all forms of transport. However, when income and other sociodemographic factors are taken into consideration, poor health can result in higher electricity consumption (Buchs et al. 2018).

While many older people are motivated and positive to take action to tackle climate change, they can be frustrated with barriers such as cost, convenience, knowledge, and personal circumstances that prevent them from adopting a low-carbon lifestyle (Haq et al. 2008; also see ▶ "Gray Is the New Green: Opportunities of Population Aging" in this volume). Better public information, incentives, and engagement techniques can enable older people to reduce their climate impact. As the number of older people increases, they are becoming an important resource that can be mobilized to respond to the climate emergency by setting a positive example and taking part in environmental volunteering (Pillemer et al. 2016; Smyer 2017).

Research Challenges

To tackle the climate emergency and meet Sustainable Development Goals (SDG) 12 (responsible consumption and production) and 13 (climate action), additional research is needed to explore how the age-specific consumption patterns might influence GHG emissions (Buchs et al. 2018). In particular, how demographic changes and climate change might amplify the effects of individual consumption patterns such as energy demand because of warmer temperatures (see ▶ "Climate Change, Vulnerability, and Older People" in this volume). Research is also required to examine effective measures to encourage older people to adopt low-carbon lifestyles (Haq et al. 2010) and participate in environmental volunteerism (Pillemer et al. 2016).

Summary

The consumption patterns of an aging population have implications for climate targets and SDGs. Socioeconomic background, values, age, and personal history all influence an individual's lifestyle choices. A better understanding of age-related consumption patterns over the life course is needed to determine what actions older people can take to limit their GHG emissions and contribution to climate change.

Cross-References

▶ Baby Boom/Baby Bust
▶ Climate Change, Vulnerability, and Older People
▶ Climate Resilience and Older People
▶ Gray Is the New Green: Opportunities of Population Aging
▶ Heatwaves and Older People
▶ Older Adults and Environmental Voluntarism

References

Buchs M, Bahaj A, Blunden L, Bourikas L, Falkingham J, James P, Kamanda M, Wu Y (2018) Sick and stuck at home – how poor health increases electricity consumption and reduces opportunities for environmentally friendly travel in the United Kingdom. Energy Res Soc Sci 44:250–259. https://doi.org/10.1016/j.erss.2018.04.041

Chancel L (2014) Are younger generations higher carbon emitters than their elders? Inequalities, generations and CO2 emissions in France and in the USA. Ecol Econ 100:195–207. https://doi.org/10.1016/j.ecolecon.2014.02.009

Estiri H, Zagheni E (2019) Age matters: ageing and household energy demand in the United States. Energy Res Soc Sci 55:62–17. https://doi.org/10.1016/j.erss.2019.05.006

Hamza N, Gilroy R (2011) The challenge to UK energy policy: an ageing population perspective on energy saving measures and consumption. Energy Policy 39:782–789. https://doi.org/10.1016/j.enpol.2010.10.052

Haq G, Minx J, Whitelegg J, Owen A (2007) Greening the greys: climate change and the over 50s. Stockholm Environment Institute, Sweden. https://doi.org/10.13140/RG.2.2.11360.28168

Haq G, Whitelegg J, Kohler M (2008) Growing old in a changing climate: meeting the challenges of an ageing population and climate change. Stockholm Environment Institute, Sweden. https://doi.org/10.13140/RG.2.1.4700.4002

Haq G, Brown D, Hards S (2010) Older people and climate change: the case for better engagement. Stockholm Environment Institute, Sweden. https://doi.org/10.13140/2.1.3935.0728

IPCC (2018) Summary for policymakers. In: global warming of 1.5°C. An IPCC special Report on the impacts of global warming of 1.5°C above pre-industrial levels and related global greenhouse gas emission pathways, in the context of strengthening the global response to the threat of climate change, sustainable development, and efforts to eradicate poverty. Intergovernmental Panel on Climate Change, Geneva

Menz T, Welsch H (2012) Population aging and carbon emissions in OECD countries: accounting for life-cycle and cohort effects. Energy Econ 34(2012):842–849. https://doi.org/10.1016/j.eneco.2011.07.016

Okada A (2012) Is an increased elderly population related to decreased CO_2 emissions from road transportation? Energy Policy 45:286–292. https://doi.org/10.1016/j.enpol.2012.02.033

O'Neil BC, Liddle B, Jiang L, Smith KR, Pachauri S, Dalton M, Fuchs R (2012) Demographic change and carbon dioxide emissions. Lancet 380:157–164. https://doi.org/10.1016/S0140-6736n12)60958-1

Pillemer K, Wells NM, Meador RH, Schultz L, Henderson CR, Tillema Cope M (2016) Engaging Older Adults in Environmental Volunteerism: The retirees in service to the environment program. Gerontologist 57(2):367–375. https://doi.org/10.1093/geront/gnv693

Smyer MA (2017) Greening Grey: climate action for an aging world. Public Policy Aging Rep 1:4–7. https://doi.org/10.1093/ppar/prw028

UN Environment (2019) Global Environment Outlook – GEO-6: Healthy Planet, Healthy People. UN Environment, Nairobi, Kenya. https://www.unenvironment.org/resources/global-environment-outlook-6

Wei T, Zhu Q, Glomsrod S (2018) How will demographic characteristics of the labor force matter for the global economy and carbon dioxide emissions? Ecol Econ 147:197–207. https://doi.org/10.1016/j.ecolecon.2018.01.017

Yu B, Wei Y, Kei G, Matsuoka Y (2019) Future scenarios for energy consumption and carbon emissions due to demographic transitions in Chinese households. Nat Energy 3:109–118. https://doi.org/10.1038/s41560-017-0053-4

Zagheni E (2011) The leverage of demographic dynamics on carbon dioxide emissions: does age structure matter? Demography 48:371–399. https://doi.org/10.1007/s13524-010-0004-1

Gray Economy

▶ Gray Is the New Green: Opportunities of Population Aging

Gray Is the New Green: Opportunities of Population Aging

Karen Sands
Ageless Way Academy and Sands and Associates, LLC, Roxbury, CT, USA

Synonyms

Aging economy; Gray economy; Silver economy

Definition

Gray is the new green (Sands 2018a) is a new idiomatic expression coined by GeroFuturist Karen Sands and describes an emerging economy that offers economic, cultural, organizational, and individual growth as a result of an aging populace. Gray relates to the color of hair of those who are aging, while green is associated with the burgeoning opportunities and development that can come from the aging of business and the business of aging.

The aging of business encompasses both the shift from four to five generations in the US workplace at the same time for the first time ever, as well as the emerging generational divide in the workplace (Bureau of Labor Statistics 2016), creating both challenges and unique opportunities. One welcome opportunity is the new unlimited potential for visionary entrepreneurs in the population aged 40-plus to thrive in their older adult years and reinvent their lives. With a surge in the 40-plus population now, and in the future, there is a wealth of new opportunities for innovation and prosperity across generations. With the average retirement age rising and people living longer, baby boomers are seeking professional and personal fulfillment far beyond the middle age years.

The business of aging refers to the impact of our population aging and increased longevity on every marketplace and in every workplace, thus providing new business development, job growth resulting in greater demand for highly skilled, and more experienced and diverse knowledge workers (Sands 2018a).

Longevity Economy refers to the sum of all economic activity driven by the needs of Americans aged 50 years or older and includes both products and services they purchase directly and the further economic activity this spending generates (Oxford Economics 2013).

Overview

The uncertain, up-and-down stock market, shifting consumer buying preferences, and increasingly diverse workforce of now five generations as of 2020 may intimidate even the most visionary leaders among us. However, trailblazers, no matter their age or stage, will thrive in these discontinuous changing times by applying a foresight approach to every issue – now and into the future. Fresh strategies that respond to these new economic realities of the business of aging and the aging of business will empower budding and established entrepreneurs as well as nurture organizational intrapreneurs. The critical through line is to stay ahead of the curve of change while paying attention to the nuances of emerging futures.

Studies show that the current surge in the population of boomers – the generation of adults born from 1943 to 1967 (Sands 2018b) – is the precursor to a steady rise in the population of older adults that will span several decades, even generations (Sands 2018b). According to a recent study in population changes, by 2030, older adults will outnumber children for the first time in US history (Vespa 2018). By 2035, it is projected that the over 65 population will reach 78.0 million and one in five people will have reached retirement age (Vespa 2018). By 2060, the median age is predicted to increase from 38 to 43 years old. The ratio of working age adults to retirement age adults will decrease from 3.5 to 1 to 2 to 1 (Vespa 2018). The 50-plus age group is projected to grow 45% between 2015 and 2050. Concurrently, the under 50 population will expand by just 13%. As a result, the 50-plus share of the total population will reach 40% (Oxford Economics 2013).

According to research studies, the demographics are now shifting in favor of boomers in the workplace (Arnone et al. 2010). Since the turn of the millennium, there have been increasing calls to prepare for a shortage in workforce knowledge and skill sets (Johnston and Packer 1986). *Workforce 2000,* one of the best treatments on this topic, and its sequel, *Workforce 2020,* were produced by the well-regarded Hudson Institute, a Washington, DC-based policy research center and think tank (Judy and D'Amico 1997). These late 1990s studies foretold the current critical workforce issues of today. The prescriptions for transforming these challenges into opportunities are as relevant today and will be tomorrow, as they were in the mid to late 1990s.

During the late 2000s, the livelihoods of boomers were distinctively threatened by the harmful impact of the Great Recession. Massive layoffs in the US workforce made 40-plus adults more vulnerable to job loss (Sands 2018a). As the recession derailed the overall economy, it triggered a surge of incidences of ageism in the workplace. Older workers who had invested decades in their jobs suddenly found themselves unemployed. Many lost their hard-earned pensions. Just as this population of boomer adults began to move through midlife into their older adulthood, their lifelong financial stability hung in the balance.

The economic recovery of the past decade did not inherently resolve the crisis of ageism in US companies (Sands 2018b). Still reeling out of the recession, now caught in the possible dismantling of democracies around the world, many business leaders today are not yet implementing new strategies to meet the impact of the graying of the globe (Sands 2018b). More people are staying in the workplace either because they love to work or, more recently, because they need to. However, after reaching age 40, many older adults are coming up against a "silver ceiling," whereby their jobs are threatened by the reality that it is cheaper for employers to hire younger employees than to pay more experienced, or "knowledge workers" (Sands 2018b). At the same time, the results of a recent volatility report verified that approximately 90% of current business leaders themselves are over 40 years old (Crist and Kolder Associates 2016).

The Dismantling of the Silver Ceiling

As long as the "silver ceiling" remains intact, executives are putting their own financial security, and that of their organization, at risk. A survey in 2010 revealed that many employers – nearly 40%, in fact – were most worried that the next years will bring a shortage of workers as a result of retirement and the shift in workforce demographics (Arnone et al. 2010). Only 15% of executives in one survey reported that improving the talent pool in their organization was a top priority, even though 75% of those same executives said that "a chronic shortage of talent" was one of the constraints on their company's growth (Scholz 2015). Leaders who do not dismantle this "silver ceiling" will lose their critical base of "knowledge workers" in the older adult population. They also risk alienating the people who keep them in business – 83% of US household wealth overall is held by people over age 50 (Oxford Economics 2013).

Businesses and organizations that recognize the enormous force of new boomer demand in the workplace, the US marketplace, and around the globe will dominate market share and funding. Their ability to monetize will depend on their willingness to serve the needs of older adults. Markets for 40-plus are currently growing at a faster rate than markets for under 40.

The convergence of all these forces is known as the Longevity Economy and is composed of 106 million people age 50 or over who are fueling at least $7.6 trillion in annual economic activity – a figure that is expected to reach well over $13.5 trillion in real terms by 2032 (Oxford Economics 2013). A report from *US News* (2015) reports that boomers of both genders control 70% of all US disposable income – also outpacing other generations in nearly every buying category. Consumers in the 50-plus age group generate 53% of consumer spending, particularly on healthcare, nondurable goods, durable goods, utilities, motor vehicles and parts, financial services, and household goods (Oxford Economics 2013).

Today's older worker population is not limited nor defined by chronological age. People are living longer and healthier lives, resulting in a longevity bonus of greater than 20 years. Boomers and succeeding generations will be working at least 9 years past traditional retirement age, either out of financial necessity or due to a lack of desire to retire (Brandon 2014). According to statistical research, 40-plus adults already dominate the leadership jobs in businesses across the board. The majority of sitting CEOs among 672 Fortune 500 and S&P 500 companies, with 667 known CEO ages, are between the ages 50 and 59, with an average age distribution of 57.2 years old among males and 56.1 years old among females. Over 60% of CFOs are currently between the ages 45 and 55 (Crist and Kolder Associates 2016).

Increase in the Wealth and Power of Women

Among United States encore entrepreneurs, baby boomer women are more prevalent than men. These powerful women – most of them social entrepreneurs – are well-connected and have access to the large amount of capital needed to grow their businesses. Women over 50 control $19 trillion, representing more than three-quarters of US wealth (Massachusetts Mutual Life Insurance Company 2007). Companies across the board and in virtually all industries are seeing a major increase in the involvement of creative, talented, influential women. Between 1997 and 2013, the number of US women-owned businesses grew by 59% vs. 41% for all new businesses in the United States – a rate almost one-and-a-half times the average (American Express Open 2013). The year 2016 was the best year to date to be a woman in business. Woman-owned companies increased by 18% in 2007–2016: the strongest peak was in small businesses with between 50 and 99 employees (Harrington 2016).

Women control the majority of household spending across the world. Globally, women in 2013 controlled 64% of household spending and $29 trillion of consumer spending worldwide. Women globally control about $20 trillion in annual consumer spending along with their $13 trillion in total yearly earnings. In aggregate, women represent a growth market bigger than China and India combined – more than twice as big (Catalyst 2018, 2019).

Data on wealth and spending shared by women in the contemporary world also suggest an increase in wealth shared by women. For example, the purchasing power of women in the United States ranges from $5 trillion to $15 trillion annually (Nielsen 2013); women now control more than 50% of all personal wealth in the United States – and this will only increase (Gorman 2015) – women purchase over 50% of traditional male products, including automobiles, home improvement products, and consumer electronics (Learned and Johnston 2004); approximately 40% of US working women now outearn their husbands (US Bureau of Labor Statistics 2016); the growth rate in the number of women-owned firms between 2007 and 2018 increased the most for these five industries: utilities (151%), other services (126%), construction (94%), accommodations and food services (85%), and administrative, support, and waste management services (70%) (American Express Open 2018); US women who are age 50 years and older control the largest percentage of purchasing power in America. Boomer women account for 93% of the purchase decisions for their households (Pilcher 2013).

Employed boomer women earn 36–59% more per year on average in inflation-adjusted dollars than their pre-boomer counterparts (Harrington 2016). This survey also found that most boomer women are earning their own Social Security benefits and will receive bigger Social Security checks, adjusted for inflation, than women have in the past.

The Reframing of Retirement

However, while women might be enjoying greater financial autonomy through the higher earnings and Social Security benefits, they also face the burden along with their male counterparts of having to finance the elongating retirement years. Add to that, the costs of caring for aging family members that weighs heavily on adults in midlife and beyond (Sands 2018a, b). Sound financial planning is the answer in an ideal world; however, it is not a guarantee in the real world. As the US continues its recovery from the Great Recession, another potentially destructive downturn grips the globe.

Employers can help by enacting change within their companies to ensure that those who want to continue earning money past midlife are not forced into early retirement. These changes can lead to more profit, happier and more productive employees, and improved products and services for the market of the future. For example, businesses that embrace flexible work schedules – including job-sharing, telecommuting, and shorter workweeks – can help turn US workplaces into age-friendly spaces.

Employers and employees alike will benefit from these arrangements, in which those gifted with extra years can use that accumulation of knowledge to benefit their less-experienced colleagues. Increased public funding is available to encourage businesses to start apprentice programs to fill the increasing skills gap. Apprenticeships are registered programs designed to develop skills through a combination of structured on-the-job training and related instruction. Historically, apprenticeships have been mainly focused on skilled trades (Bureau of Labor Statistics 2016). Now, these programs are expanding into areas such as IT, pharmacy technicians, nursing, engineering, and data security (Bureau of Labor Statistics 2016). Business leaders create win-win-win situations for the company, the people they lead, and those they serve by pursuing these opportunities.

Boomers are not leaving senior positions and are seeking ways to stay in their jobs (Arnone et al. 2010). Moreover, the population of younger workers is not yet large enough, or skilled enough, to replace retiring boomers who have developed their knowledge and skills over decades (Arnone et al. 2010). Businesses will find it is more cost-effective to retain and hire mature workers than it is to hire, orient, and train new, younger workers. Consequently, boomers will be filling the need for "knowledge workers" as strategists, consultants, coaches, mentors, and entrepreneurs in the coming decades.

The Multigenerational and Ageless Workplace

Among all generations, diversity, work-life balance, and meaningful work are instrumental to success and happiness. The organizations that will thrive will be multigenerational, utilizing boomers to mentor and collaborate with the upcoming four generations. They will clear the way for employees who wish to work less, but not retire. Adults over 50 may want to keep working, creating, innovating, inspiring, and leading. They have the ability to do so by making adaptations in these pursuits that take into account the realities of their lives – as people at every age need to do (Sands 2018b). However, there is no need or benefit to forcing retirement at a certain age. People are living healthier, more active lives. Someone with the experience and wisdom of their 60 years may functionally be in his or her 40s. The need or desire to work less, or to have more flexible working arrangements, does not necessitate retirement. In fact, people of all ages benefit from and want more work-life balance.

Agelessness in today's world is not a utopian idea. Rather, it constitutes a pragmatic approach to economic growth, global well-being, investment in the future, and innovation. Business author, John Elkington (1998), coined the term, "triple bottom line" in 1994 in reference to a shifting paradigm in corporate conduct. He observed that the paradigm began to shift from one of purely economic values to a combination of economic, social, and environmental values in business models (Mark-Herbert et al. 2010).

Going Gray and Green

This modern standard of social and environmental responsibility has become an integral part of the longevity economy. As our global society is becoming more aware of the effects of climate change, boomers are concurrently becoming more conscientious about saving our environment as well (Sands 2018a, b). To that end, they are donating to nonprofits and buying from companies that are "going green," meaning using practices that are sustainable and ecologically responsible, primarily conserving energy, reducing pollution, and saving money (Davis 2019).

This worldwide "go green" trend is revitalizing the businesses and entrepreneurs that now drive the longevity economy toward prioritizing people, planet, and profit, in that order (Davis 2019). The plethora of twenty-first century "green" businesses, products, and job opportunities has opened doors for the 50-plus population to revive their careers and rejuvenate their lives on the whole. These visionaries who wish to continue working years beyond the traditional retirement age are breaking out of their traditional jobs and starting new businesses, often as first-time entrepreneurs. Others are finding new careers in organizations that provide products and services geared toward their needs. The overall trend in this business of aging has shifted from product-centered to employee- and customer-centered (Sands 2018a). Thus, the force of the longevity economy is moving more toward alignment with the "green" values of the 50-plus generations. Whether they are consumers or new entrepreneurs, the older adult population continues to direct higher revenues into the longevity economy, while simultaneously improving their and our quality of life.

Economic Growth and the Business of Aging

Overall, the longevity economy – also known as the aging or silver economy – portends to benefit all generations. The population boom in the 50-plus age group has created a higher percentage of job openings in fields servicing the needs of the older adult population. Workers of all ages and business leaders have the potential to benefit from the economic growth arising from the business of aging.

Longevity – ever-elongating life spans resulting in large post-50 populations for generations to come – has been shown to have a positive effect on the economy (Murphy and Taylor 2006). Those who serve the 50-plus market are well aware that this demographic group is an unstoppable, powerful force for change. An example of this population's impact on economic growth is the indication of upcoming changes in the employment picture for those who serve the needs of the boomer population.

For instance, the employment outlook is shifting upward in the fields of personal care and home healthcare – jobs for those who assist aging boomers and older adults in their own homes. Job growth will rise by up to an estimated 38% of the current market by the year 2024 (US Bureau of Labor Statistics 2016). Healthcare jobs are expected to see enormous growth from 2016 to 2026. While average job growth will be 7.4%, the following jobs will see an exponential growth: (i) home health workers, 47% increase; (ii) personal aids, 39% increase; (iii) medical assistants, 29% increase; (iv) medical secretaries, 23% increase; (v) registered nurses, 15% increase; (vi) financial managers, 19% increase; and (vii) market research analysts and marketing specialists, 23% increase. The record rise in employment opportunities for people with gerontological training will be the future for employment growth across all generations.

This long-term positive outlook for employment in the aging field promises to elevate career opportunities for women, especially. Women already represent the majority of professionals, employees, and entrepreneurs in the field, along with being the predominant gender of recipients of services and products. Experts report that 60% of women make a career change in midlife. Moreover, 2/3 of these highly educated 40-plus women are driven by the desire to make a positive impact on their communities and the world, whether as an employee or as an entrepreneur (American Express Open 2018). New opportunities in gerontology and in serving the 50-plus market can be the perfect career change for these same women.

Summary

There is limitless potential across the generations as the longevity economy offers opportunities for those who are part of the aging of business and the business of aging. The new economy opens the door to women and men of all ages who are ready to exercise economic power and influence, whether as a senior executive, a sole proprietor, or an innovator of age-friendly technology, healthcare, or urban design. Trailblazers in every phase of the life cycle will be the ones who lead and light the way in this new economy for higher profits, greater sustainability, and robust organizations and communities.

Cross-References

▶ Active Aging and Active Aging Index
▶ Age Discrimination in the Workplace
▶ Age Management and Labor Market Policies
▶ Aging Policy Analysis and Evaluation
▶ Autonomy and Aging
▶ Delaying Retirement
▶ Economics of Aging
▶ Encore Jobs
▶ Financial Gerontology
▶ Gender and Employment in Later Life
▶ Global Agewatch Index and Insights
▶ Human Wealth Span
▶ Late-life Creativity
▶ Longevity Areas and Mass Longevity
▶ Longevity Dividend
▶ Re-employment of (Early) Retirees
▶ Replacement Rate
▶ Self-employment Among Older Adults
▶ Senior Entrepreneurship
▶ Silver Economy
▶ Social Entrepreneurship and Social Innovation in Aging

References

American Express Open (2013) The 2013 state of women-owned business report: a summary of important trends 1997–2013. http://www.womenable.com/content/userfiles/2013_State_of_Women-Owned_Businesses_Report_FINAL.pdf. Accessed 27 Nov 2018

American Express Open (2018) The 2018 state of women-owned business report: a summary of important trends 2007–2016. https://about.americanexpress.com/files/doc_library/file/2018-state-of-women-owned-businesses-report.pdf. Accessed 27 Nov 2018

Arnone J, Leisy W, Conat A et al (2010) The aging of the U.S. workforce: employer challenges and responses. In: Global HR News. https://www.globalbusinessnews.net/the-aging-workforce/. Accessed 23 Nov 2016

Brandon E (2014) The youngest baby boomers turn 50. In: U.S. News and World Report. https://money.usnews.com/money/retirement/articles/2014/06/16/the-youngest-baby-boomers-turn-50. Accessed 10 Dec 2018

Bureau of Labor Statistics (2016) U.S. Department of Labor 2014-2024 occupational outlook handbook. https://www.bls.gov/ooh/. Accessed 30 Nov 2018

Catalyst (2018) Quick take: buying power. https://www.catalyst.org/research/buying-power/. Accessed 9 Dec 2018

Catalyst (2019) Quick take: women in the workforce – United States. https://www.catalyst.org/research/women-in-the-workforce-united-states/. Accessed 10 June 2019

Coffee P (2014) Advertisers don't understand women. In: Adweek. http://www.adweek.com/agencyspy/study-advertisers-don't-understand-women/68869. Accessed 2 Dec 2018

Crist and Kolder Associates (2016) Volatility report 2016. http://www.cristkolder.com/media/1755/volatility-report-americas-leading-companies.pdf. Accessed 30 Nov 2018

Davis A (2019) The definition of "go green." In: Livestrong. https://www.livestrong.com/article/159603-the-definition-of-go-green/. Accessed 2 Dec 2018

Elkington J (1998) Cannibals with forks: the triple bottom line of 21st century business. New Society Publishers, Philadelphia

Gorman R (2015) Women now control more than half of US personal wealth, which will only increase in years to come. In: Business Insider. https://www.businessinsider.com/women-now-control-more-than-half-of-us-personal-wealth-2015-4. Accessed 14 Dec 2018

Greenfield Online for Arnold's Women's Insight Team (2002) What women think. In: Adweek. https://www.adweek.com/brand-marketing/what-women-think-56700/. Accessed 28 Nov 2018

Hannon K (2015) Why 50+ women should take control of their money. In: Next Avenue. https://www.nextavenue.org/why-50-women-should-take-control-their-money/. Accessed 3 Dec 2019

Harrington S (2016) In: Forbes. https://www.forbes.com/sites/samanthaharrington/2016/12/27/was-2016-the-best-year-to-be-a-woman-in-business/709354ff61ca. Accessed 3 Dec 2019

Johnston W, Packer A (1986) Workforce 2000 work and workers in the 21st century. Hudson Institute, Washington, D.C

Judy R, D'Amico C (1997) Workforce 2020 work and workers in the 21st century. Hudson Institute, Washington, DC

Learned A, Johnston L (2004) Don't think pink. Amacom, New York

Mark-Herbert C, Mark-Herbert C, Pakseresht A, Rotter J (2010) A triple bottom line to ensure corporate responsibility. In: Berg P (ed) Timeless Cityland. https://www.researchgate.net/profile/Ashkan_Pakseresht/publication/264883854_A_triple_bottom_line_to_ensure_Corporate_Responsibility/links/543fa10d0cf27832ae8bab8e.pdf. Accessed 12 Dec 2018

Massachusetts Mutual Life Insurance Company (2007) American family business survey: executive summary. https://www.uvm.edu/sites/default/files/2007MassMutualAmericanFamilyBusinessSurvey.pdf. Accessed 27 Nov 2018

Murphy K, Taylor R (2006) The value of health and longevity. J Polit Econ 114(5):871–904. https://doi.org/10.3386/w11405

Nielsen (2013) U.S. women control the purse strings. https://www.nielsen.com/us/en/insights/news/2013/u-s%2D%2Dwomen-control-the-purse-strings.html. Accessed 20 Dec 2018

Oxford Economics (2013) The longevity economy: how people over 50 are driving economic and social value in the U.S. https://www.aarp.org/content/dam/aarp/home-and-family/personal-technology/2013-10/Longevity-Economy-Generating-New-Growth-AARP.pdf. Accessed 29 Nov 2018

Pilcher J (2013) When marketing to women, financial brands fall way short. The financial brand. https://thefinancialbrand.com/35365/marketing-financial-services-banking-to-women/. Accessed 12 Dec 2018

Sands K (2018a) Gray is the new green: rock your revenues in the longevity economy, 2nd edn. Broad Minded, Roxbury

Sands K (2018b) The ageless way: illuminating the story of our age, 2nd edn. Broad Minded, Roxbury

Scholz C (2015) The war for talent. 3PL Americas 7(2):23–25. http://www.iwla.com/assets/1/6/IWLA_2015-2.pdf. Accessed 13 Dec 2018

US News (2015) Baby boomer report. https://www.usnews.com/pubfiles/USNews_Market_Insights_Boomers2015.pdf. Accessed 2 Dec 2018

Vespa J (2018) Older people projected to outnumber children for first time in U.S. History. U.S. Census Bureau Newsroom. https://www.census.gov/newsroom/press-releases/2018/cb18-41-population-projections.html. Accessed 13 Dec 2018

Graying of Hair

Dan Chen[1] and Yuying Tong[2]
[1]The Chinese University of Hong Kong, Shatin, Hong Kong, China
[2]Department of Sociology, The Chinese University of Hong Kong, Sha Tin, Hong Kong

Synonyms

Aging; Population aging

Definition

Graying of hair is an obvious sign of aging process (Tobin and Paus 2001; Kocaman et al. 2012). When age increases, the production of free radicals increases, while the endogenous defense mechanisms decrease (Tobin and Paus 2001; Sabharwal et al. 2014). This imbalance leads to the progressive damage of cellular structure, resulting in the aging phenotype (Trüeb 2006). It is the dysfunction of the melanocyte stem cells that causes the onset of graying (Nishimura et al. 2005). In 1973, grayness of hair was firstly used by Damon and Roen (1973) as a sign of aging. Graying of hair is a physiological process that proved to be positively correlated with chronological age, an indicator of biological age, regardless of gender, ethnicity (Aggarwal et al. 2015; Jo et al. 2012; Kocaman et al. 2012; Lasker and Kaplan 1974; Trüeb 2006; Tobin and Paus 2001), and born hair color (Boas and Michelson 1932; Ortonne et al. 1982; Kocaman et al. 2012).

However, graying of hair is not only a physiological phenomenon but also a social fact, acting as a status symbol of the self and group identity (Synnott 1987). It is a visual feature of later life which often positions old people as outsides (Ward and Holland 2011) under the assumption that people with gray hair are old. At the aggregated level, graying of hair is to describe the population aging trend, the increasing proportion of old people among the population (Bloom et al. 2011).

Overview

Graying of hair is an age-related physiological demonstration. In order to understand hair graying process, studies have explored the intrinsic and external factors for graying of hair (Trüeb 2006) in natural science disciplines (Peters et al. 2011). Premature graying of hair as early onset of graying of hair is often predictive for health status, a risk factor for coronary artery disease (Kocaman et al. 2012). Onset of gray hair depends on age, but there is no significant difference by gender. Incidence of hair graying for men is similar to women (Lapeere et al. 2008; Dawber and Neste 1995). However, it is also reported that early onset does not mean a faster progress, the extent of grayness increases sharply after their 50s regardless of age at onset (Jo et al. 2012).

Graying of hair is also a status symbol. Among social science disciplines, such as sociology and anthropology, researchers tend to focus on examining age discrimination, social meaning of graying of hair, and the significance of hair color, hair aesthetic, and hair management practice (Synott 1987; Ward and Holland 2011; Clarke and Korotchenko 2010). Graying of hair has important social meanings for the collective social consensus and internalized self-perception on individuals' gray hair (Clarke and Korotchenko 2010), while the aggregated level graying of hair means demographic changes and trend of population aging. Studies on graying of hair and population aging have mainly concentrated on the demographic and socioeconomic consequences to emphasize the potential challenges of population aging in restructuring the society (Bloom et al. 2011; Borsch-Supan 2003; Feldstein 1978; Bakshi and Chen 1994; Gavrilov and Heuveline 2003). In youth- and beauty-oriented society, graying of hair is a bad signal for females. Sometimes graying of hair is a symbol of social status, and people with gray hair are perceived to be old, ugly, and unhealthy, resulting in being invisible and being socially isolated (Ward et al. 2008; Ward and Holland 2011; Clarke and Korotchenko 2010).

Theoretical Underpinning

Graying of hair is an indicator of aging, and many studies have been developed to examine whether premature graying of hair would lead to an increased morbidity, mortality, and cause-specific diseases (Netleton and Watson 1998; Clarke and Korotchenko 2010). Graying of hair is often treated as the first physical and public manifestation of mortality (Synnott 1987), and physiological studies often assume graying of hair is a reminder of health decline (Netleton and Watson 1998; Clarke and Korotchenko 2010). Moreover, premature graying of hair is correlated with diseases based on the assumption that premature graying of hair is due to oxidative stress (Tobin and Paus 2001; Trüeb 2006; Sabharwal et al. 2014). Researchers are also interested in determinants and mechanisms underlying hair graying process. Intrinsic factors (e.g., genetic, chronically aging) and extrinsic factors (e.g., environmental, climate, season, infestations, pollutants, toxins, and chemical exposure) are reported to play influential role in graying of hair (Tobin and Paus 2001).

Gray hair could also be a trigger for social discrimination (Bytheway 2005). Individuals internalize ageism and build up their self-identity of gray hair based on social and cultural construction (Clarke and Korotchenko 2010). For the stigmatization of graying of hair, individuals with gray hair have to cope with related stress caused by social disengagement and cultural invisibility (Ward and Holland 2011; Clarke and Korotchenko 2010).

Key Research Findings

Empirical studies on graying of hair often measure rate of graying by the coverage of gray hair: no gray hairs, few gray hairs, moderate gray hairs, and completely gray/white hair (Pomerantz 1962; Glasser 1991; Schnohr et al. 1998). According to Boas and Michelson (1932), there are no significant differences in rate of graying among various social groups in European population, and individuals with higher socioeconomic status enjoy no significant advantages over others.

Physiological significance of early graying of hair is intensively examined. Though graying of hair is assumed to be a first demonstration of mortality (Synnott 1987), association between premature graying of hair and mortality risks is barely significant (Lasker and Kaplan 1974). As reported based on clinic observations, researchers conclude that early graying of hair is not significantly associated with increased morbidity, earlier age at death/life span (Glasser 1991; Schnohr et al. 1998), or specific cause of death.

It is also debatable whether premature graying of hair is a risk marker of health status, independent of chronological age. Some studies favor the positive association between premature graying of hair and risk of coronary artery disease (Aggarwal et al. 2015; Kocaman et al. 2012), while others have not found clear association (Pomerantz 1962). As Pomerantz (1962) maintained, correlation between graying of hair and coronary disease may be contingent on age, since it is only significant for people age over 46. For those aged under 45, there is no significant correlation between grayness and coronary heart disease (Pomerantz 1962).

Another main research focus is to examine the linkage between graying of hair and health behaviors, such as smoking. Studies show that mean percentage of individuals with gray hair is positively associated with risks of cardiovascular disease (Sabharwal et al. 2014), and the association between hair graying and risks of cardiovascular disease could be explained by smoking behavior. Smoking is a well-documented risk factor for coronary artery disease (Trüeb 2006) and positively linked to premature hair graying (Mosley and Gibbs 1996; Zayed et al. 2013). The risk of hair graying among smokers is significantly higher than that among non-smokers among both men and women (Mosley and Gibbs 1996; Jo et al. 2012; Sabharwal et al. 2014).

Studies of graying of hair by social scientists have unfolded social meanings of graying of hair.

It is argued that people with gray hair are discriminated in economic, social, and sexual domains as they are labeled as old, ugly, and unhealthy in youth-, beauty-, and health-oriented culture (Nettleton and Watson 1998; Gilleard and Higgs 2000; Clarke and Kororchenko 2010). In age discrimination literatures, gray hair is associated with a sense of being isolated, and people with gray hair are no longer sexually attractive (Ward and Holland 2011). Respondents report that being old is socially invisible, and "invisibility" is a collective discriminative action toward the old, which conveys the sense of marginalized and discounted (Ward et al. 2008; Ward and Holland 2011). Collective action and consensus on ageism shape personal viewpoint on the physiological change of graying of hair, and people think that "If I look old, I will be treated old" (Ward and Holland 2011). Being treated as old is often unwelcome.

Individual attitudes toward the social constructed culture on graying of hair vary by social status. While white, heterosexual, and middle-class women are more likely to show negative feeling on gray hair (Winterich 2007), marginalized group like lesbian and black women tend to accept their gray hair for their contradictory viewpoint with dominant culture. Women who are averse to gray hair would try to mask gray hair as a coping strategy (Clarke and Korotchenko 2010). Those women coping with stigmatization of graying of hair by dyeing their gray hair and masking their chronological age tend to be more active in social engagement (Synnott 1987; Clarke and Korotchenko 2010). Meanwhile, however, positive implications of graying of hair exist. As graying of hair is age-associated, people with gray hair could also be conceived as experienced. As Ward and Holland (2011) argued, gray hair is a beneficial symbol for professional workers. For example, therapists with gray hair are often linked to "wisdom or extra knowledge."

Future Directions of Research

Epidemiological assumption of graying of hair as a marker of health status is challenged by two methodological issues. One is about confounders. That is, both the incidence of atherosclerotic diseases and graying of hair could be influenced by age and genetic factors. Another methodological pitfall is about reversed causality. It is proved that cardiovascular risk factors, which cause premature atherosclerosis, may also lead to premature and intense hair graying by affecting follicular epithelium and resident stem cells (Misago et al. 2011; Krause and Foitzik 2006). Therefore, future researches on causal analysis are necessary in clarifying if graying of hair could be treated as an independent risk factor from chronological age to predict health status and mortality.

Moreover, social norms on aging and graying of hair tend to marginalize and discount graying of hair. To present negative attitude toward old, age discrimination and burden of aging on socioeconomic development have been suggested in previous studies. As a result, literatures tend to neglect the positive sides of graying of hair. Future studies on how to exploit the value of being old, like being "experienced" at individual level and demonstrating the influential power of "being experienced," are helpful not only to correct age discrimination but also for policymaking. Moreover, future studies should focus more on stress process of aging to unfold the psychological-physiological mechanisms of age discrimination. Such studies will increase knowledge on individuals' strategies in coping with stigmatization.

Summary

Graying of hair is an age-related physiological process, and it occurs regardless of race, gender, socioeconomic conditions, and social context. Graying of hair has its social and epidemiological significances. Socially, graying of hair is marginalized and discounted by ageism culture which constructs graying of hair to be ugly, old, and unhealthy and socially, economically, and sexually irrelevant in youth-oriented society (Nettleton and Watson 1998; Gilleard and Higgs 2000; Clarke and Korotchenko 2010). At the aggregated level, graying of hair reflects a population aging process. Epidemiologically, graying of hair is empirically studied with the assumption that graying of hair is a marker of health risks.

Although association between graying of hair and mortality is barely significant, empirical findings support the positive correlation between graying of hair and coronary artery disease. For the potential methodological challenges including reverse causality and confounding effect, further causal analyses should be conducted to clarify the significance of graying of hair as a marker of health status. Moreover, to overcome limitations of previous literatures for overemphasizing the negative meanings of graying of hair, more studies should examine the positive significances of graying of hair at both individual and population levels to provide evidences for ageism stigmatization management and policies making.

Cross-References

▶ Aging Definition
▶ Gray Is the New Green: Opportunities of Population Aging

References

Aggarwal A, Srivastava S, Agarwal MP, Dwivedi S (2015) Premature graying of hair: an independent risk marker for coronary artery disease in smokers – a retrospective case control study. Ethiopian J Health Sci 25(2):123–128

Bakshi SG, Chen Z (1994) Baby boom, population aging, and capital markets. J Bus 67(2):165–202

Bloom D, Boersch-Supan A, McGee P, Seike A (2011) Population aging: facts, challenges, and responses. PGDA working paper no. 71. http://www.hsph.harvard.edu/pgda/working.htm

Boas F, Michelson N (1932) The greying of hair. Am J Phys Anthropol 17:213–228

Borsch-Supan A (2003) Labor market effects of population aging. Labour 17 (Special Issue) 5–44

Bytheway B (2005) Ageism. In: Johnson ML, Bengston VL, Coleman PG, Kirkwood TBL (eds) Cambridge handbook of age and ageing. Cambridge University Press, Cambridge, pp 338–345

Clarke HL, Korotchenko A (2010) Shades of grey: to dye or not to dye one's hair in later life. Ageing Soc 30(6):1011–1026. https://doi.org/10.1017/S0144686X1000036x

Damon A, Roen JL (1973) Aging in the solomon islands and the United States: tests of Pearl's hypothesis. Human Biology 45:683–693

Dawber R, Neste D (1995) Hair and scalp disorders. Martin Dunitz, London

Feldstein M (1978) Social security. In: Boskin M (ed) The crisis in social security: problems and prospects. Institute for Contemporary Studies, San Francisco

Gavrilov LA, Heuveline P (2003) Aging of population. In: Demeny P, McNicoll G (eds) The encyclopedia of population. Macmillan Reference, New York

Gilleard C, Higgs P (2000) Cultures of ageing: self, citizen and the body. Prentice Hall, Harlow

Glasser M (1991) Is early onset of gray hair a risk factor? Med Hypotheses 36:404–411

Jo SJ, Paik SH, Choi JW, Lee JH, Cho S, Kim KH et al (2012) Hair graying pattern depends on gender, onset age and smoking habits. Acta Derm Venereol 92:160–161

Kocaman SA et al (2012) The degree of premature hair graying as an independent risk marker for coronary artery disease: a predictor of biological age rather than chronological age. Anadolu Kardiyol Derg 12:457–463

Krause K, Foitzik K (2006) Biology of the hair follicle: the basics. Semin Cutan Med Surg 25:2–10

Lapeere H et al (2008) Hypomelanoses and hypermelanoses. In: Wolff K et al (eds) Fitzpatrick's dermatology in general medicine, 7th edn. MaGraw-Hill, New York, p 622–640

Lasker GW, Kaplan B (1974) Graying of the hair and mortality. Soc Biol 21:290–295

Misago N, Toda S, Narisawa Y (2011) CD34 expression in human hair follicles and tricholemmoma: a comprehensive study. J Cutan Pathol 38:609–615

Mosley JG, Gibbs AC (1996) Premature grey hair and hair loss among smokers: a new opportunity for health education? Br Med J 313:1616

Nettleton S, Watson J (1998) The body in everyday life: an introduction. In: Nettleton S, Watson J (eds) The body in everyday life. Routledge, New York, pp 1–24

Nishimura EK, Granter SR, Fisher DE (2005) Mechanisms of hair graying: incomplete melanocyte stem cell maintenance in the niche. Science 307:720–724

Ortonne JP, Thivolet J, Guillet R (1982) Graying of hair with age and sympathectomy. Archives of Dermatology 118:876–877

Peters EMJ, Imfeld D, Graub R (2011) Graying of the human hair follicle. Journal of Cosmetic Science 62:121–125

Pomerantz HZ (1962) The relationship between coronary heart disease and the presence of certain physical characteristics. Can Med Assoc J 86(2):57–60

Sabharwal R et al (2014) Association between use of tobacco and age on graying of hair. Niger J Sur 20(2):83–86

Schnohr P, Nyboe J, Lange P, Jensen G (1998) Longevity and gray hair, baldness, facial wrinkles, and arcus senilis in 13,000 men and women: the Copenhagen City Heart Study. J Gerontol 53A(5):M347–M350

Synnott A (1987) Shame and glory: a sociology of hair. Br J Sociol 38(3):381–483

Tobin DJ, Paus R (2001) Graying: gerontobiology of the hair follicle pigmentary unit. Exp Gerontol 36(1):29–54

Trüeb RM (2006) Pharmacologic interventions in aging hair. Clin Interv Aging 1:121–129

Ward R, Holland C (2011) 'If I look old, I will be treated old': hair and later-life image dilemmas. Ageing Soc 31:288–307. https://doi.org/10.1017/S0144686X10000863

Ward R, Jones R, Hughes J, Humberstone N, Pearson R (2008) Intersections of ageing and sexuality: accounts from older people. In: Ward R, Bytheway B (eds) Researching age and multiple discrimination. Centre for Policy on Ageing, London, pp 45–72

Winterich A (2007) Aging, femininity, and the body: what appearance changes mean to women with age. Gend Issues 24(3):51–69. https://doi.org/10.1007/s12147-007-9045-1

Zayed AA, Shahait AD, Ayoub MN, Yousef AM (2013) Smokers' hair: does smoking cause premature hair graying? Indian Dermatol Online J 4:90–92

Green Volunteerism

▶ Older Adults and Environmental Voluntarism

Grey Entrepreneurship

▶ Senior Entrepreneurship

Grief and Bereavement

▶ Bereavement and Loss

Grief and Loss

▶ Bereavement and Loss

Grief Care

▶ Bereavement Care

Grief Counseling

▶ Bereavement Care

Grief Rumination

▶ Rumination in Bereavement

Grief Support

▶ Bereavement Care

Grossman Model

Audrey Laporte
Institute of Health Policy Management and Evaluation, Dalla Lana School of Public Health, University of Toronto, Toronto, ON, Canada

Definition

The Grossman (1972c) model of investment in health capital is central to the way economists think about and represent the relationship between individual health behaviors, health investments, and health outcomes (Laporte 2015).

Overview

Since the Grossman model explicitly takes a lifetime view, it is well suited to understanding the link between health-related decisions and outcomes and aging not only at the individual but also at the aggregate level. The insights it provides at the aggregate level can be useful for understanding the likely implications of population aging at the level of health systems and overall healthcare expenditure over time.

From the economist point of view, the individual's health evolves over time as a result of things which are unpredictable shocks, some things which are influenced by individual decisions and the individual's context (Grossman 1972c). The most important of these types of factors can be thought of as cumulative so that the health of the older segment of the population is actually a function of the lifetime decisions of that population; the pattern of lifetime shocks experienced by that population; the context faced by a given population, e.g., socioeconomic circumstances; and the way these all interact. The Grossman model is the way economists think about the cumulative pattern of individual health decisions and therefore provides a starting point for how past experience in behavior and context combine to determine the health status of various different age groups in the current population.

The starting point of any economic model of individual behavior is the notion of utility which refers to the individual's subjective assessment of their own well-being (Hashimzade et al. 2017). Clearly good health yields utility in and of itself, and, through its effect on the individual's functionality, it will affect the utility of pretty much everything else they do.

The Grossman model embeds a dynamic long run perspective by being set up as an individual's lifetime utility maximization problem where utility is assumed to be a function of their stock of health as well as non-health-related consumption (Grossman 1972c, p. 225). The model is forward-looking in that the individual is assumed to be aiming to maximize their lifetime utility from the perspective of today and subject to financial constraints as well as the constraint represented by the fact that their health status will evolve over time – they will age and eventually die.

The Grossman model recognizes that individuals will maximize their lifetime utility subject to constraints. The financial or budget constraint in the Grossman model is represented in the form of a lifetime wealth equation (Grossman 1972c, p. 228). The individual earns a certain amount of income in a given period out of which they must pay for food, shelter, consumer goods and health and non-health-related goods and services. Any savings from a given period may be invested (to earn interest or dividends) and carried over to be spent in future periods, say retirement, to allow for increased spending in periods when earned income may be lower.

Health in the Grossman model is characterized as a stock which evolves over time (Grossman 1972a, c, p. 226). The individual is born with an initial level of health which can be augmented by health-related consumption of goods and services and which is subject to depreciation that can be offset to some degree by the health-related investments (Grossman 1972c, p. 226). In each period the individual can either invest in their health through health-related goods and services, e.g., medical care, thereby adding to their stock of health capital and thus to their utility in the present and into the future, or they can spend the money on non-health-related consumption which will increase their utility in the present. Since health-related investments do not enter directly into the individual's utility function, the trade-off the individual faces is as follows: that investment in health-related consumption yields a utility return only to the extent that it increases the individual's stock of health and that additional health stock is bought at the expense of forgone consumption of non-health-related goods in the current period. That is, the individual is assumed not to derive utility from a doctor visit per se but does derive utility from the added health that results from the visit to the doctor; the individual's use of medical care is thus assumed to be a derived demand. The consumption version of the Grossman model only assumes that there is this kind of direct utility return to an increase in the stock of health (Grossman 1972b). The investment version of Grossman's model allows for another return to investments in health – if healthier individuals are able to work more and for longer periods (reduced sick time), then additions to the individual's stock of health will translate into higher earnings (Grossman 1972c, p. 231). These higher earnings will allow for increased purchases of either or both of investments in health- and non-health-related consumption goods – both of which would increase the individual's utility. The Grossman model can also allow for the possibility of interactive effects between the individual's stock of health and consumption of non-health-related consumption. For example, the individual would likely derive more utility from taking a vacation when healthy than when they are less healthy. In essence, the full Grossman model allows for three types of return to the individual from investments in health – the psychic return to utility that comes from feeling healthier, the added income that a healthier individual can earn which will add to their utility because they can afford to buy more goods and services, and the enhanced utility that being healthier confers upon consumption of non-health-related goods and services.

The decision about how much health-related investment to undertake obviously is a function of the weight which individuals place on their own future health status which enters the Grossman model through the discount rate. All else being equal, individuals who discount the future more heavily will be inclined to value non-health-related consumption in the present more highly compared with investments in health which, while undertaken today, yield their returns largely in the form of improved health in future periods of life (Bickel et al. 1999, 2012). Individuals that discount the future heavily are also inclined to save less of their income when young which itself will have implications for their socioeconomic position later in life and by extension their stock of health (Chabris et al. 2008). This would suggest that even relatively high-income individuals, who had not invested much in their health over their life and also had tended not to have accumulated much in the form of financial assets, would be doubly disadvantaged later in life.

Bringing these relationships to bear at the aggregate level can provide insight into the impact of population aging. Aging of the population is comprised of two components – changes in life expectancy and changes in the proportion of the population in older relative to younger age groups. Both of these factors can be looked at through a Grossman lens, allowing us to take an integrated view of the health effects of population aging. In the Grossman model, the individual recognizes that their life is finite. If life is viewed as a trajectory that has a definite start and a definite end, then as the individual ages, the closer they get to the endpoint and by extension, the shorter is the remaining time left over which to recoup the health returns to health-related investments. However, since there are then also fewer remaining periods of life left, at the margin, each additional year at older ages will tend to be highly valued. Thus, the tendency will be for individuals to expend private as well as public healthcare resources to extend and enhance quality of life at the end of life (Becker et al. 2007). While it is observed that health-related expenditure increases toward the end of life (Alemayehu and Warner 2004), conditional on existing medical technology, it is not enough ultimately to prevent the decline in health, although it may affect when an individual dies. As a consequence, changes in life expectancy may themselves change individual patterns of health investment, which would in turn be expected to influence health-related expenditures at the aggregate level. Improved prospects for a longer life at birth are predicted to increase the likelihood that individuals will undertake health investment behavior in earlier periods in their lives (since there is now the prospect of a longer period over which to realize health investment returns) (Oster et al. 2013) which would tend to reinforce increases in life expectancy.

In one sense incorporating this is quite simple. When the individual's expected health trajectory is considered, an aggregate health trajectory can be obtained simply by weighting each segment of the individual trajectory by the proportion of the population that fall within that segment. Arguably, failure to do this simple calculation is the reason that so many countries find themselves scrambling to respond to health policy issues associated with population aging. As one example, health systems based on a pay as you go structure in which the current population funds current healthcare expenditures can run into difficulties if the age distribution changes in ways that result in a smaller younger working age population largely shouldering the public health expenditures that are needed to provide care to larger older segments of the population (Feder et al. 2000; Adams and Vanin 2016). The Grossman model reinforces the notion that it is desirable to think in terms of lifetime trajectories when designing policy – for example, to have individuals pay into financing of services such as long-term care insurance when they are young to finance their own expected expenditures when they are old – so that changes in population distribution do not pose such a challenge for health system financing over time, as some countries have done (Campbell and Ikegami 2000; Evans Cuellar and Wiener 2000).

To this point the focus has been on the demand side and indeed Grossman's original paper referred to the demand for health, but a key part of the Grossman model is the production function for health (Grossman 1972c). Broadly speaking, conditional on the individual's current state of health, they can combine health type labor, e.g., physician services, and capital, e.g., x-rays, to determine their future state of health. The Grossman model emphasizes the role of education in influencing how the individual uses healthcare goods and services (inputs). In the language of economics, the Grossman model hypothesizes that education enables individuals to increase the marginal productivity of inputs into health production (Grossman 1972c, p. 226). Essentially, more educated individuals may, for instance, be more likely to comprehend and to comply with a care plan provided by a physician all else being equal. As the proportion of the educated has increased over time across many countries, the Grossman model raises the possibility that in addition to the expected income returns to education, the nature of care delivery and the ability and preferences of patients to participate in their own care is likely to continue to evolve as well.

Key Research Findings

The Grossman framework while articulated at the individual level is in fact useful for organizing our thoughts on issues that might not immediately appear to be interrelated. For example, it tells us something about why there is an income gradient in health with lower-income individuals generally being less healthy than high-income individuals even if as children they started out as equally healthy. And this in turn lets us think about how the distribution of income in any period feeds into the health status of a population as that population ages. Higher income individuals can invest more in their health in any given period since they have more disposable income once other expenses have been deducted. As noted above, this added investment yields an increased stock of health in the current period but also means that the individual enters the next period of their life with a higher stock of health than they otherwise would have had such that when the depreciation rate is applied in that next period, the individual is left with a higher stock of health. In other words, disparities in the level of health investment in one period will have effects in the current period and will also cumulate over future periods and ultimately lead to differences in quality and quantity of life years (Nordin and Gerdtham 2014; Corman et al. 2018).

The Grossman model highlights the fact that even deprivation that was localized to the earlier periods in an individual's life can result in that individual having a lower stock of health in old age compared to individuals that have the same current socioeconomic status but who did not face the same degree of deprivation in early life. In this way, the model helps us to understand how current observed socioeconomic income gradients across groups in the population are likely to manifest into socioeconomic gradients in health at older ages at the population level (Mackenbach 2006; Chetty et al. 2018). An implication might be that disparities among older individuals which are resistant to current policy changes might have been more sensitive to changes in past behavior and that therefore there should be consideration of the return to policies aimed at reducing income-related health disparities in the young (immediate benefits) also in terms of their potential to alter lifetime health outcomes for these same individuals when they are older (future benefits) (Arpino et al. 2018).

The Grossman model allows us to bring together in an analytical sense, elements which are intuitively plausible but which tend to be thought of in isolation rather than as interacting parts of a whole. For example, as people get older they tend to use more medical care (Alemayehu and Warner 2004). This is one of the key issues with regard to an aging population, not just in the current circumstances but in terms of long-term planning; and indeed, it can be argued that planners in the past have not paid enough attention to this factor. This fits into the overall Grossman framework through an age-related rate of depreciation of the stock of health capital (Grossman 1972c). If the rate of depreciation increases with age, then the individual's stock of health will decline more and more per unit of time which would require increased levels of investment in health to maintain the current stock of health (Grossman 1972c). Given the individual's budget, these added investments in health will come at the expense of consumption of non-health-related consumption from which the individual derives utility in the present. Moreover, the degree to which individuals have health insurance coverage will play into the extent of the opportunity cost the individual faces in terms of foregone consumption of other goods (Paulin and Weber 1995).

The Grossman model, by helping us to think in terms of the production of care over the life span, can be a useful device to frame issues of long-term human resources planning to support care for aging populations. In an accounting sense, the cost of health is the total of what has to be spent on labor and capital and pharmaceuticals. At older ages, the production of health tends to be more labor, or at least provider time intensive and, increasingly these days, more pharmaceutical intensive – even if not necessarily more capital intensive. The whole issue of alternative level of care patients in hospitals can be thought of as using capital intensive facilities to treat patients who do not really need that type of care but for whom there is a lack of a sufficient quantity of appropriate settings of care, e.g., supportive housing and home care. Part of the reason a lot of analysts have claimed the aging population would not have much impact on health expenditures, and part of the reason there has been surprise at its impact is that past research has focused too much on dollars and not enough on the changing mix of inputs on which those dollars have been spent (Weissert et al. 2005; Payne et al. 2007). Changes in the aggregate population age distribution, because it will translate into the particular services being demanded, have an impact not just on cost but on the mix of inputs needed to produce those services. This has been observed in the data on the pressure put on hospital beds by older alternative level of care patients, and it would also, because even ambulatory care of older patients tends to be more time intensive, expect to see it in changes in the demand for different types of healthcare labor inputs. The need to change the mix of inputs could have been better anticipated but was not in most developed countries, such that many jurisdictions are faced with shortages of key health providers such as nurses and personal support workers (see, e.g., Jefferson et al. 2018).

Another key implication is that rather than thinking about how to develop policy for older individuals as though they were simply a subpopulation, the Grossman model encourages thinking in terms of the lifetime trajectory. Older individuals today are individuals who are further along their lifetime trajectory, and many of the causes of gradients in health and socioeconomic outcomes observed among older individuals in any current period can be traced back to the effects of circumstances and behavior that have cumulated over time.

Future Directions of Research

Models that aim to better forecast the demand for various healthcare services could be enhanced by more explicit modeling of dynamic effects at the individual level as implied by the Grossman model. This will necessitate taking full advantage of panel (individual longitudinal) data that allow for estimation of effects of socioeconomic and health circumstances in early life and longer run (later life) outcomes. The original Grossman model emphasized medical care services as an input in the production of health, but it is recognized that informal care provided by family and friends, sometimes referred to as social capital, constitutes a large portion of care provision and thus as an input into health production as well (Laporte et al. 2008; Laporte 2013). Future research should give greater attention to the availability in older age of informal care for today's younger generation; planning models should account for the impact of low birth rates, greater incidence of divorce, and living alone on individual health investment decisions and by extension the required formal supports.

Summary

The Grossman model provides a useful framework within which to conceptualize the impacts of the aging process per se on individual behavior, the impact of an individual's valuation of the present versus the future on their health outcomes at older ages, and the fact of a finite life span on individual choices about health investment behavior. All these elements will interact at the individual level in ways that affect individual outcomes in older age and individual predicted life time public and private health-related spending. The Grossman model highlights that even when analyzing health expenditure patterns at the aggregate level, it is important not to lose sight of the fact that aggregate health expenditures are the summation of the consequences of individual decisions and circumstances. The decisions which individuals make depend on where they are in their lifetime health investment plan. Ultimately, this imparts a fundamentally dynamic element in the relationship between aging and health which is amplified by and interacts with changes in the age distribution of the population over periods of several decades.

Cross-References

▶ Access to Care
▶ Care Needs
▶ Formal and Informal Care
▶ Healthcare Utilization

References

Adams O, Vanin S (2016) Funding long-term care in Canada: issues and options. HealthcarePapers 15:7–19. https://doi.org/10.12927/hcpap.2016.24583

Alemayehu B, Warner KE (2004) The lifetime distribution of healthcare costs. Health Serv Res 39(3):627–642. https://doi.org/10.1111/j.1475-6773.2004.00248.x

Arpino B, Gumà J, Julià A (2018) Early-life conditions and health at older ages: the mediating role of educational attainment, family and employment trajectories. PLoS One 13(4):e0195320. https://doi.org/10.1371/journal.pone.0195320

Becker G, Murphy K, Philipson T (2007, August) The value of life near its end and terminal care. NBER Working Paper No. 13333. https://doi.org/10.3386/w15649

Bickel WK, Odum GJ, Madden J (1999) Impulsivity and cigarette smoking: delay discounting in current, never, and ex-smokers. Psychopharmacology 146(4):447–454. https://doi.org/10.1007/pl00005490

Bickel WK, Jarmolowicz DP, Mueller ET, Koffarnus MN, Gatchalian KM (2012) Excessive discounting of delayed reinforcers as a trans-disease process contributing to addiction and other disease-related vulnerabilities: emerging evidence. Pharmacol Ther 134(3):287–297. https://doi.org/10.1016/j.pharmthera.2012.02.004

Campbell JC, Ikegami N (2000) Long-term care insurance comes to Japan. Health Aff 19(3). https://doi.org/10.1377/hlthaff.19.3.26

Chabris CF, Laibson D, Morris CL, Schuldt JP, Taubinsky D (2008) Individual laboratory-measured discount rates predict field behavior. NBER Working Paper No. 14270, Issued in August. https://doi.org/10.3386/w14270

Chetty R, Friedman J, Hendren N, Jones MR, Porter SR (2018, October) The opportunity atlas: mapping the childhood roots of social mobility. Working Paper. https://doi.org/10.3386/w25147

Corman H, Dhaval D, Reichman NE (2018) Evolution of the infant health production function. South Econ J 85(1):6–47. https://doi.org/10.1002/soej.12279

Evans Cuellar A, Wiener JM (2000) Can social insurance for long-term care work? The experience of Germany. Health Aff 19(3):8–25. https://doi.org/10.1377/hlthaff.19.3.8

Feder J, Komisar HL, Niefeld M (2000) Long-term care in the United States: an overview. Health Aff 19(3):40–56. https://doi.org/10.1377/hlthaff.19.3.40

Grossman M (1972a) A stock approach to the demand for health. In: Grossman M (ed) The demand for health: a theoretical and empirical investigation. NBER, pp 1–10. Volume ISBN: 0-87014-248-8, Volume URL: http://www.nber.org/books/gros72-1

Grossman M (1972b) The pure consumption model. In: Grossman M (ed) The demand for health: a theoretical and empirical investigation. NBER, pp 31–38.

Volume ISBN: 0-87014-248-8, Volume URL: http://www.nber.org/books/gros72-1
Grossman M (1972c) On the concept of health capital and the demand for health. J Polit Econ 80:223–255. https://doi.org/10.1086/259880
Hashimzade N, Myles G, Black J (2017) Utility. In: A dictionary of economics. Oxford University Press. Retrieved 5 Oct 2019, from https://www-oxfordreference-com.myaccess.library.utoronto.ca/view/10.1093/acref/9780198759430.001.0001/acref-9780198759430-e-3278
Jefferson L, Bennett L, Hall P, Cream J, Dale V, Honeyman M, Birks Y, Bloor K, Murray R (2018) Home care in England Views from commissioners and providers (ed: Bottery S). The Kings Fund. https://www.kingsfund.org.uk/sites/default/files/2018-12/Home-care-in-England-report.pdf
Laporte A (2013) Chapter 4: Social capital: an economic perspective. In: Folland S, Rocco L (eds) The economics of social capital and health a conceptual and empirical roadmap. World scientific series. World Scientific. https://doi.org/10.1142/7593
Laporte A (2015) Should the Grossman model of investment in health capital retain its iconic status? Working Papers 140003. Canadian Centre for Health Economics
Laporte A, Nauenberg E, Shen L (2008) Aging, social capital, and health care utilization in Canada. Health Econ Policy Law 3:393–411. https://doi.org/10.1017/S1744133108004568
Mackenbach JP (2006, February) Health inequalities: Europe in profile. Report commissioned by the UK Presidency of the EU. https://www.who.int/social_determinants/media/health_inequalities_europe.pdf
Nordin M, Gerdtham UG (2014) Why a positive link between increasing age and income-related health inequality? Nordic J Health Econ 2(1). https://doi.org/10.5617/njhe.651. https://www.journals.uio.no/index.php/NJHE/article/view/651. Date accessed 29 July 2019
Oster E, Dorsey ER, Shoulson I (2013) Limited life expectancy, human capital and health investments. Am Econ Rev 103(5):1977–2002. https://doi.org/10.1257/aer.103.5.1977
Paulin G, Weber W (1995) The effects of health insurance on consumer spending. Mon Labor Rev 118(3):34–54. Retrieved from http://www.jstor.org/stable/41844408
Payne G, Laporte A, Deber RB, Coyte PC (2007) Counting backward to health care's future: using time-to-death modeling to identify changes in end-of-life morbidity and the impact of aging on health care expenditures. Milbank Q 85(2):213–257. https://doi.org/10.1111/j.1468-0009.2007.00485.x
Weissert WG, Matthews Cready C, Pawelak JE (2005) The past and future of home- and community-based long-term care. Milbank Q 83(4):1–71. https://doi.org/10.1111/j.1468-0009.2005.00434.x

Guardianship

Alexandra Crampton
Department of Social and Cultural Sciences, Marquette University, Milwaukee, WI, USA

Synonyms

Conservatorship; Curatorship

Definition

Guardianship is the legal transfer of authority from an adult to a surrogate decision-maker. The two steps in the legal process are findings of incapacity and guardianship appointment. An adult under guardianship is called a ward (GAO 2016). The surrogate decision-maker is usually called the guardian. However, each state determines legal definitions, capacity assessment, and guardianship appointment criteria (Wood and Quinn 2017). Other terms for guardian are conservator, curator, committee, and tutor (Knapp 2011). Many states distinguish legal authority over finances from care of the person, often using conservatorship for the former and guardianship for the latter. Plenary (or full) guardianship results in complete loss of civil rights and legal decision-making. Limited guardianship restricts a guardian's authority to specific decisions, such as responsibility for finances only (Wright 2010). Less restrictive alternatives to guardianship include powers of attorney and representative payee programs for managing funds from government programs (Wood and Quinn 2017). Older adults are disproportionately represented among wards (Keith and Wacker 1994).

Overview

General Purpose and Challenges
The purpose of guardianship is to ensure that incapacitated adults receive care and protection.

Historically, US guardianship law developed from English law for protecting estates. Guardians had been appointed to prevent "lunatics," adjudicated to have lost reason, and "idiots," adjudicated as having always lacked reason, from dissipating private wealth (Regan 1981). Over time, a legal intervention to protect financial assets has been extended to financial protection of older adults as well overall personal care. Guardianship is a powerful legal tool because full guardianship results in complete loss of rights and responsibilities as an adult. It can be "double-edged" both morally and practically as an intervention to preserve autonomy and provide protection (Crampton 2004). On the one hand, older adults with cognitive incapacity or severe mental illness may need a guardian. The guardian can protect from abuse of older adults. On the other, unnecessary appointments can violate such civil rights as choice in voting, marriage, and residence (Wright 2010). Moreover, the power of a guardian can be abused through negligence and exploitation.

Legal Process for Guardianship Determination

The guardianship appointment process begins with filing a petition in court, typically probate or family court. This can be triggered by a change in the older adult's capacity for decision-making, such as a stroke or decline from a degenerative illness (Crampton 2004). The trigger can also be family conflict over "What do to with Dad" (or Mom), or a change in the caregiver's situation (Wright 2010). Professionals may file as part of an Adult Protective Services case, on behalf of a hospital trying to discharge a patient, or as administrative staff of a long-term care facility seeking a decision-maker for a new resident (Wood and Quinn 2017). Legal professionals may require guardianship of an adult involved in a legal action, such as a divorce or property sale (Wood and Quinn 2017). After case filing, the court typically makes an investigation, such as through an attorney acting as guardian ad litem (GAL) (Wood and Quinn 2017). The GAL may request capacity assessment by a mental health or medical professional. The focus of capacity assessment in most states is on cognitive function in making financial and/or personal care decisions (Falk and Hoffman 2014). Although most cases are effectively uncontested, hearings are conducted as an adversarial process. Due process protections for the proposed ward may include adequate notice, reasonable accommodation to attend the hearing, and legal counsel. Hearings result in two judicial decisions: a finding of incapacity (or not) and a guardian appointment. The preferred court choice of guardian is a family member (GAO 2016). Private professionals and long-term care facilities may also serve as guardian. This is financed by the ward's estate (Wilber and Alkema 2006). Some states support public guardianship programs for "unbefriended" older adults, but these tend to be underfunded and professionals estimate increasing unmet demand (Moye et al. 2017; Teaster et al. 2010).

Alternatives to Plenary Guardianship

Given the severity of legal remedy in guardianship and potential for abuse, elder advocates have promoted limited guardianship and alternatives to guardianship (Wood and Quinn 2017). One goal is to shift from surrogate decision-making to supportive decision-making (Wood 2016). Promoted to support the rights of people with disabilities, this shifts less decision-making power from the older adult to a guardian (Wilber and Alkema 2006; Wood 2016). As a dispute resolution process, mediation can support this option. Mediators facilitate dispute resolution among parties prior to court action or final judicial order. Including older adults in this process can empower them to more directly participate in caregiving decisions (Crampton 2013). Older adults can also be encouraged to plan for incapacity through advance directives and powers of attorney over finances and/or health care (Knapp 2011). Representative payee programs can help with financial protection of benefits and care managers can help with financial and personal care needs (Wood and Quinn 2017). Some courts have developed legal self-help centers and provide legal assistance for older adults who may need help accessing benefits or reporting and filing cases of abuse of older adults (Rothman and Dunlop 2006). There are also specialized older adult protection courts to uphold the rights of vulnerable older adult (Banks et al. 2014).

Key Research Findings

In the 1980s, the first national study of older adults and guardianship triggered calls for reform and need for ongoing evaluation (Associated Press 1987). Research continued investigation identifying who are wards and who are guardians, and how the guardianship system works (Barker and King 2001; Keith and Wacker 1994; Knapp 2011; Reynolds and Carson 1999). Research on capacity assessment has contributed to legal reforms (including removal of advanced age as a criterion) and better assessment tools for capacity assessment (Falk and Hoffman 2014; Moye et al. 2013). Research has also investigated options for less restrictive alternatives (e.g., Wilber 1991) and the demand and use of public guardianship (Teaster et al. 2010). Unfortunately, post-reform studies find ongoing lack of use by courts over accepting a simple medical diagnosis (McSwiggan et al. 2016). Another problem found has been ongoing challenges with guardianship in relation to abuse of older adults (GAO 2016). While guardianship can be used to end abusive relationships, the power of guardianship itself can be abused. Financial fraud is a common concern; guardians are appointed to end exploitation but are also given legal power that allows financial exploitation (Wood and Quinn 2017).

Future Directions of Research

Research has not kept up with advocacy efforts to improve the system to protect rather than abuse older adults (Wilber and Alkema 2006; Wood 2016; Wright 2010). There is ongoing need to track guardianship. There is also ongoing need to identify underuse of limited guardianship and guardianship alternatives. At the same time, population and morbidity trends suggest ongoing and increasing need for intervention to support older adults (Jones and Pastor 2017). Ongoing research is needed in how to effectively and ethically provide support for incapacitated older adults without unnecessary or abusive guardianship practices.

Summary

Guardianship is a legal intervention that disproportionately impacts older adults. Several trends indicate ongoing and growing need for guardianship and guardianship oversight. These include demographic and morbidity trends of an aging population, rising numbers of older adults without nearby informal supports, and ongoing financial pressures on public guardianship programs and court services. Although guardianship is intended as a benevolent intervention to protect vulnerable adults, guardianship can also become a form of marginalization and abuse. Elder advocates have identified alternatives, but research has shown under-utilization of less restrictive interventions.

Cross-References

- ▶ Abuse and Caregiving
- ▶ Adult Protective Services
- ▶ Aging in Place
- ▶ Autonomy and Aging
- ▶ Biosocial Model of Disability
- ▶ Case Management
- ▶ Financial Exploitation by Family Members
- ▶ Formal and Informal Care
- ▶ Intergenerational Family Dynamics and Relationships
- ▶ Laws for Older Adults
- ▶ Older Adults Abuse and Neglect
- ▶ Prevention of Age-Related Cognitive Impairment, Alzheimer's Disease, and Dementia

References

Associated Press (1987) Guardians of the elderly: an ailing system. September

Banks JP, Conger JJ, Cram JJ (2014) Elder protection courts: judicial perspective, holistic approach. Exp Dermatol 24:12–15

Barker J, King M (2001) Taking care of my parents' friends: non-kin guardians and their older female wards. J Elder Abuse Negl 13(1):45–69. https://doi.org/10.1300/J084v13n01_03

Crampton A (2004) The importance of adult guardianship for social work practice. J Gerontol Soc Work 43(2–3):117–129. https://doi.org/10.1300/J083v43n02_08

Crampton A (2013) Elder mediation in theory and practice. J Gerontol Soc Work 56(5):423–437. https://doi.org/10.1080/01634372.2013.777684

Falk E, Hoffman N (2014) The role of capacity assessments in elder abuse investigations and guardianships. Clin Geriatr Med 30:851–868. https://doi.org/10.1016/j.cger.2014.08.009

Government Accounting Office (2016) Elder abuse: the extent of abuse by guardians is unknown, but some measures exist to help protect older adults. Report GA0-17-33. https://www.gao.gov/products/GAO-17-33. Accessed Jan 2019

Jones A, Pastor D (2017) No one wants to help them. Qual Soc Work 16(3):376–393. https://doi.org/10.1177/1473325015620851

Keith P, Wacker R (1994) Older wards and their guardians. Praeger, Westport

Knapp M (2011) Guardianship. In: Maddox G (ed) The encyclopedia of aging, vol A-L. Springer, New York, pp 455–457

McSwiggan S, Meares S, Porter M (2016) Decision-making capacity evaluation in adult guardianship: a systematic review. Int Psychogeriatr 28(3):373–384. https://doi.org/10.1017/S1041610215001490

Moye J, Marson D, Edelstein B (2013) Assessment of capacity in an aging society. Am Psychol 63(3):158–171. https://doi.org/10.1037/a0032159

Moye J, Catlin C, Kwak J, Wood E, Teaster P (2017) Ethical concerns and procedural pathways for patients who are incapacitated and alone. HEC Forum 29(2):171–189. https://doi.org/10.1007/s10730-016-9317-9

Regan J (1981) Protecting the elderly: the new paternalism. Hastings Law J 32(5):1111–1132

Reynolds S, Carson L (1999) Dependent on the kindness of strangers. Aging Ment Health 3(4):301–310

Rothman MB, Dunlop BD (2006) Elder and the courts. J Aging Soc Policy 18(2):31–46. https://doi.org/10.1300/J031v18n02_03

Teaster P, Schmidt W Jr, Wood E, Lawrence S, Mendiondo M (2010) Public guardianship: in the best interests of incapacitated people? Praeger, Santa Barbara

Wilber K (1991) Alternatives to conservatorship: the role of daily money management services. Gerontologist 31(2):150–155. https://doi.org/10.1093/geront/31.2.150

Wilber K, Alkema G (2006) Policies related to compentency and proxy issues. In: Berkman B, D'Ambruoso S (eds) Handbook of social work in health and aging. University of Oxford Press, London. https://doi.org/10.1093/acprof:oso/9780195173727.003.0083

Wood E (2016) Recharging adult guardianship reform: six current paths forward. J Aging Longevity Law Policy 1(1):Article 5. https://digitalcommons.tourolaw.edu/jallp/vol1/iss1/5/. Accessed Jan 2019

Wood E, Quinn M (2017) Guardianship systems. Elder Abuse. https://doi.org/10.1007/978-3-319-47504-2_17

Wright J (2010) Guardianship for your own good. Int J Law Psychiatry 33:350–368. https://doi.org/10.1016/j.ijlp.2010.09.007

GWA

▶ Genome-Wide Association Study

GWAS

▶ Genome-Wide Association Study

Gynecological Cancer

▶ Tumors: Gynecology

Gynecological Tumor

▶ Tumors: Gynecology